D0926954

BOYLESTAD'S

Circuit Analysis

Second Canadian Edition

Robert L. Boylestad

Doug Edgar
Northern Alberta Institute of Technology

John Jenness
Kwantlen University College

Jack Tompkin
British Columbia Institute of Technology

with Practical Business Applications Features by

Carlo Odoardi
Coco Net Inc.

Toronto

Canadian Cataloguing in Publication Data

Boylestad's Circuit Analysis
2nd Canadian ed.
Includes CD-ROM.
Includes index.
ISBN 0-13-085896-X

1. Electric circuits. 2. Electric circuit analysis. I. Boylestad, Robert L. II. Title: Circuit analysis

TK454.B68 2001 621.319′2 C00-930253-0

PHOTO CREDITS
Chapter 1, page 1, The Canadian Press; Chapter 2, page 18, Ron Wood; Chapter 3, page 36, Photo Features/IAMAW; Chapter 4, page 56, Toronto Hydro; Chapter 5, page 74, Bombardier Regional Aircraft, Downsview; Chapter 6, page 108, PhotoDisc, Inc.; Chapter 7, page 143, Bombardier Regional Aircraft, Downsview; Chapter 8, page 179, Bombardier Regional Aircraft, Downsview; Chapter 9, page 238, Tony Stone Images; Chapter 10, page 283, Tony Stone Images; Chapter 11, page 331, Photo courtesy of Telesis, issue 103, page 36, © Nortel/Northern Telecom; Chapter 12, page 341, Bombardier Regional Aircraft, Downsview; Chapter 13, page 377, Ontario Hydro; Chapter 14, page 421, Ontario Hydro; Chapter 15, page 464, Ontario Hydro; Chapter 16, page 520, Ontario Hydro; Chapter 17, page 549, Ontario Hydro Archives, neg. #90.0001-17; Chapter 18, page 590, Canapress/Ryan Remiorz; Chaper 19, page 636, Photo Features/IAMAW; Chapter 20, page 668, John Colletti; Chapter 21, page 708, Courtesy of CN Tower; Chapter 22, page 746, Corbis Digital Stock

ISBN 0-13-085896-X

Vice President, Editorial Director: Michael Young
Executive Acquisitions Editor: David Stover
Marketing Manager: Sophia Fortier
Executive Developmental Editor: Marta Tomins
Production Editor: Matthew Christian
Copy Editor: Tally Morgan
Production Coordinator: Wendy Moran
Page Layout: The PRD Group
Photo Research: Susan Wallace-Cox
Art Director: Mary Opper
Interior and Cover Design: Anthony Leung
Cover Image: Artville LLC, PhotoDisc, Corel Corportion

1 2 3 4 5 05 04 03 02 01

Printed and bound in the United States.

To Else Marie
Alison, Mark, Kelcy, and Morgan
Eric and Rachel
Stacey and Britt

—RLB

Contents

CHAPTER 11: MAGNETIC CIRCUITS 331

CHAPTER 12: INDUCTORS 341

CHAPTER 13: SINUSOIDAL ALTERNATING WAVEFORMS 377

CHAPTER 14: THE BASIC ELEMENTS AND PHASORS 421

CHAPTER 15: SERIES AND PARALLEL ac CIRCUITS 464

CHAPTER 22: TRANSFORMERS 746

APPENDICES

GLOSSARY 805

INDEX 810

CD-ROM

Preface

This second Canadian edition of *Boylestad's Circuit Analysis* builds upon the strengths of the well-received first Canadian edition as well as Robert L. Boylestad's original text, *Introductory Circuit Analysis,* now in its ninth U.S. edition, to strive toward one overarching goal—to provide Canadian students with the clearest, most comprehensive introduction yet available to the fundamentals of electric circuits.

In the more than three decades since its first publication, *Introductory Circuit Analysis* has established itself as the pre-eminent textbook for first courses in electric circuits at colleges and technical institutes across North America and around the world. Hundreds of thousands of students have been introduced to the fundamentals of electricity and electronics by this classic text, with its unrivalled selection of examples, exercises, and problems.

Why then, a generation later, was there a need for a first and now a second Canadian edition of this classic text? And what is so Canadian about circuit analysis anyway?

The answer to that last question is nothing — yet everything. There is nothing uniquely Canadian about the study of electric circuits and their applications in the electricity, electronics, and computer industries. The physical principles underlying the field of electronics are universal. But the students studying electric circuits in Canada do so in a Canadian context. They come to their studies within a unique context that is shaped by their educational, social, and cultural background, as well as by the budgetary and political realities that continue to shape Canadian educational institutions.

Make no mistake about it. Today's students, compared to those who used the first edition of *Introductory Circuit Analysis* thirty years ago, learn differently and face new challenges. Coming of age in an era of television and multimedia, they tend to be more visually-oriented than their predecessors, and perhaps less comfortable with printed text. At the same time, they are conversant with computer technologies and applications that had not even been invented when the first edition of this book saw print.

Today's students face a bewildering array of economic and social pressures in a frantic, fast-paced, globalizing world. Classrooms include increasing numbers of students for whom English is not their first language, as well as many more mature students who have returned to school after years in the workforce.

Educational institutions have changed, too. Colleges and universities must try to do more with less. Class hours have been cut, and class sizes have swelled. Schools are exploring alternative modes of delivering courses, using such resources as the World Wide Web, multimedia materials, and distance learning to make the most of restricted budgets.

Finally, technology is changing. The past thirty years have seen a revolution in the computer industry. What was science fiction when this book first appeared is now technological commonplace. As a result, today's students face new challenges and opportunities in industry, for which their education must prepare them.

A combination of these factors led Prentice Hall to produce the first Canadian edition of *Boylestad's Circuit Analysis.* The book's development was based on feedback from reviewers and focus-group participants from across Canada. The goal was to produce a book that would better fit Canadian curricula; that would be geared to the reading abilities and learning characteristics of today's students; and that would retain the comprehensiveness and authority that are Boylestad trademarks, while at the same time proving more accessible, user-friendly, and oriented toward new multimedia technologies. The enthusiastic response that greeted *Boylestad's Circuit Analysis* suggested to us that, in large measure, we had met that goal — and inspired us to try to do even better the second time around.

Changes to the Second Canadian Edition

As with the first Canadian edition of *Boylestad's Circuit Analysis,* preparation of the second edition was preceded by a series of focus groups and in-depth interviews with instructors and students in electronics, electrical, and telecommunications technology programs nationwide. Reviewers were also asked to provide feedback on the earlier edition. What emerged was a sense that many of the issues that led to the publication of a Canadian edition to begin with had, in fact, been addressed, but that there remained areas that could be improved. Those improvements to the second Canadian edition include:

- **Reinstating the chapter on transformers in the printed book.** Some instructors indicated that this chapter is a crucial component of their course, and that they would prefer to see it included in the printed version of the text rather than on the accompanying CD-ROM.
- **Making every chapter available in printed form.** In the first Canadian edition, the five final chapters of the original text, identified by instructors as topics which, while important, were not covered in detail in an introductory course, were moved in their entirety to the free CD-ROM packaged with every copy of the text, in order to reduce the size and cost of the printed book. While many instructors and students appreciated the convenience of having these chapters on the CD-ROM in multi-platform Adobe Acrobat reader format, others indicated it would be more useful to have them provided in more traditional printed form. So in the second edition we have accommodated both requests. As noted above, the transformers chapter is back in the main text, and the four other concluding chapters — while still available on disk — can also be obtained in printed form as a separately-bound booklet packaged with the text.
- **Enhancing the accompanying CD-ROM.** The CD-ROM itself has been substantially enhanced in this edition. Material found only on the CD-ROM (including computer applications sections for each chapter, as well as the four supplementary chapters) is now referenced in the table of contents of the printed book. The comprehen-

sive set of Weblinks on the CD-ROM has been revised and updated, to provide as complete a guide as possible to electronics technology resources on the Web. New business application discussions tied to each chapter and prepared by Robert Boylestad himself show the practical applications of key theories and concepts. And, most importantly, the CD-ROM now contains **the entire text of the ninth U.S. edition of *Introductory Circuit Analysis* in convenient, searchable Adobe Acrobat Reader format.** That means that, in addition to the ease and accessibility of the printed text, students and instructors now have access to what amounts to an electronics technology encyclopedia on CD, with search functionality allowing for the quick look-up of key terms and concepts.

- **Tying key concepts more closely to the "real world."** In keeping with the first Canadian edition's mandate to make electronics education relevant to real-world success, this new edition includes updated and enhanced *Practical Business Applications* boxes, now more concise and more clearly set off from the chapters in which they appear. As well, new captions tie the subject matter of chapter-opening photographs to the concepts that will be covered in the chapter itself.
- **Enhanced supplement package.** Supplements to this edition include new and improved versions of the *Instructor's Solutions Manual* and *Test Item File* (available in both hard copy and computerized formats), as well as an all-new *Companion Website,* which provides students with self-test questions, links to related sites, and convenient outlines of key concepts.

Other Features of the Second Canadian Edition

Other features of the first Canadian edition which proved useful to instructors and students alike have been retained. They include:

- **Functional use of four-colour photography and artwork.** Today's students are visual learners. Colour photographs, artwork, and other visual elements make the text more useful to students and instructors. We use colour to highlight key concepts, equations, and headings, to divide chapters into digestible sections, and to make diagrams more easily understood. Important rules, laws, and tips are highlighted by use of a second colour. Sections presenting examples, procedures, and derivations now have a colour background that distinguishes them more clearly from the surrounding text.
- **Thorough technical editing and evaluation of reading level.** The Canadian authors and technical editor thoroughly examined the text line-by-line to make sure the reading level is appropriate. At the same time, they have ensured that the technical level and comprehensiveness of the book have *not* been compromised in any way.
- **Coordination with Canadian curricula.** The book has been aligned with the structure and content of circuits courses across the country.
- **Learning Outcomes and Chapter Outlines** provide students with a summary of the major topics they should have mastered by the time they finish each chapter, and an overview-at-a-glance of the major sections in each chapter.

- **End-of-chapter problem sets.** Instructors agree Boylestad's end-of-chapter problems are the "best in the business." This tremendous strength has been retained and enhanced in this edition.
- **Glossary.** Definitions of key terms have been gathered together at the end of the text.

Supplements to the Second Canadian Edition

A comprehensive set of supplements accompanies the text, including:

- **Instructor's Solutions Manual.**
- **Instructor's Problem Supplement.**
- **Test Item File.** Also available in computerized format for Windows.
- ***Experiments in Circuit Analysis*.** A comprehensive lab manual that includes dozens of lab exercises, extra problems, and further exploration of difficult points.
- **Solutions Manual for *Experiments in Circuit Analysis*.**
- **Transparency Masters.** Includes 100 black-line transparency masters for classroom use.
- **Transparencies.** Includes 50 full-colour acetates for classroom use.
- **Multimedia CD-ROM.** Included free in every copy of the text. Includes the entire text of the U.S. ninth edition in Adobe Acrobat Reader format, as well as supplementary chapters, computer applications, Weblinks, and a hotlink to the *Circuit Analysis* Companion Website.
- ***Boylestad's Circuit Analysis* Companion Website.** Provides additional problems, exercises, and self-testing tutorials.

Feedback Welcomed!

We welcome your feedback on *Boylestad's Circuit Analysis, Second Canadian Edition*, whether you're a student, instructor, or interested reader. You can write to us at:

Editorial Department
Higher Education Division
Pearson Education
Canada Inc.
26 Prince Andrew Place,
Don Mills, ON M3C 2T8

Or you can visit our Web site:
www.pearsoned.ca

The Publishers

Acknowledgments

We acknowledge the contributions of the many individuals who helped to make *Boylestad's Circuit Analysis, Second Canadian Edition* a reality.

Reviewers and Focus Group Participants

Focus groups dealing with issues in electronics education as well as in-depth interviews were carried out across the country in preparation for both the first and second Canadian editions. Other instructors prepared written reviews. We would like to acknowledge the invaluable contribution of the following reviewers, focus-group participants, and interviewees:

Philippe Deziel, Algonquin College
John Duffield, Centennial College
J.J. MacQuarrie, Centennial College
Pravin Patel, Durham College
Doug Fuller, Humber College
Shamsh Jiwa, Humber College
Don Matthews, Humber College
Taylor Zomer, Conestoga College
Robert Langlois, Fanshawe College
John Vandersteen, Fanshawe College
Mike Huk, St. Clair College
Marko Jovanovic, St. Clair College
George Peters, St. Clair College
Vince Bennici, Seneca College
Jeremy Clark, Seneca College
John Dandy, Seneca College
John Ebden, Seneca College
David Finlay, Seneca College
Yvette Grimmond, Seneca College
Arjun Rana, Seneca College
Allan Souder, Seneca College
Roddy Turner, Seneca College
Bogdan Pawlowski, Sheridan College
Joyce van de Vegte, Camosun College
George A. Berges, DeVry Institute of Technology (Calgary)
Rafiqui Islam, DeVry Institute of Technology (Calgary)
Lawrence Whitby, DeVry Institute of Technology (Calgary)
Denis Anderson, Southern Alberta Institute of Technology
Eugene Blanchard, Southern Alberta Institute of Technology
Ralph Frese, Southern Alberta Institute of Technology
David Jessop, Southern Alberta Institute of Technology
Dave Lange, Southern Alberta Institute of Technology
Dessie A. Mekuria, Southern Alberta Institute of Technology
Franjo Stvarnik, Southern Alberta Institute of Technology

Marc Anderson, Northern Alberta Institute of Technology
Greg Collins, Northern Alberta Institute of Technology
Doug Edgar, Northern Alberta Institute of Technology
Brian Hadley, Northern Alberta Institute of Technology
Jim Kramps, Northern Alberta Institute of Technology
Ken McDonald, Northern Alberta Institute of Technology
Carl Lindemann, Northern Alberta Institute of Technology
Albert Nadon, Northern Alberta Institute of Technology
Barry Lazzer, Northern Alberta Institute of Technology
Bernie Moisey, Northern Alberta Institute of Technology
Tom Strickland, Saskatchewan Institute of Applied Science and Technology (Palliser Institute)
Kel Ellard, Saskatchewan Institute of Applied Science and Technology (Moose Jaw)
Kris Nanan, Saskatchewan Institute of Applied Science and Technology (Moose Jaw)
Tom Strickland, Saskatchewan Institute of Applied Science and Technology (Moose Jaw)
Gerry Woolsey, Saskatchewan Institute of Applied Science and Technology (Moose Jaw)
Ken Wuschke, Saskatchewan Institute of Applied Science and Technology (Moose Jaw)
M. Jamil Ahmed, British Columbia Institute of Technology
Richard Beketa, British Columbia Institute of Technology
Russ Brendzy, Vancouver Community College & British Columbia Institute of Technology
David Chiu, British Columbia Institute of Technology
Trevor Glave, British Columbia Institute of Technology
Ken Kajiwara, British Columbia Institute of Technology
Jack Tompkin, British Columbia Institute of Technology
John Jenness, Kwantlen University College
Hardev S. Sodhi, Kwantlen University College
Brian Kelley, Cabot College (College of the North Atlantic)
Carlo Odoardi, Coco Net Inc.

1

Introduction

Outline

Learning Outcomes

After completing this chapter, you will be able to

- briefly describe the history and significance of the electrical industries
- use SI units of measurement correctly, including SI prefixes
- perform operations with powers of ten
- understand and correctly apply concepts of significant digits, accuracy and rounding off

After looking at milestones in the history of electricity, we turn to the science of electricity and the units of measure used to study it.

1.1 ELECTRICAL INDUSTRIES

Modern industrialized civilization runs on electricity. From the moment your clock radio wakens you in the morning until you switch off the lights at night, you are surrounded by electrical equipment. Electricity boils the water for your coffee, transports you to work, keeps you in touch with your friends, and provides your entertainment.

There are many industries that can be classified as electrical. They range from power companies that generate and distribute electrical energy to the multitude of manufacturers and users of electrical and electronic equipment. In the electrical power field, they include devices for converting electrical energy to mechanical energy, to heat, and to light. In communications there are telephones, radio, and television, together with their great variety of recording devices. Now, with a computer on every desk, there is another whole cluster of industries involving data manipulation, storage, and transmission.

Beyond the great growth of new industries, electronic control circuits have penetrated into many other fields. Wherever precision and reliability are required, you are likely to find electronic control—from the ignition system of car engines to the almost total operation of plants for refining oil or manufacturing steel. Many specialized electronic components are needed to measure and control a wide range of physical processes.

Every one of these systems and devices contains numerous electrical circuits. The workers who design, manufacture, and service them require a clear understanding of the principles of electrical and electronic circuits. This text is designed to provide you with the foundations for understanding and working with electric circuits. You will find that however much electrical industries grow and diversify, these foundations will continue to supply a solid base for whatever advanced industry you choose to work in.

1.2 A BRIEF HISTORY

Less than two hundred years ago, transportation and communication were limited to the speeds of running horses and sailing ships. Today you can travel in hours, or communicate in seconds, all around the world. Steam engineers and mechanical artisans made the first steps in this transformation. However, that early stage of the industrial revolution could not have advanced to our present stage without electricity. The technologies of steam and iron grew out of the craft traditions of earlier centuries, but electricity was born in the laboratories of scientists—scientists who were curious about the operations of the forces of nature.

Early Electrical Science

After 1700, the scientific works of Isaac Newton provided scientists with techniques for analyzing a variety of motions in terms of the mechanical and gravitational forces that caused them. By the middle of that century, some scientists were applying that analysis to the forces involved in electricity produced by friction and the close contact of dissimilar materials. Benjamin Franklin was able to show that lightning was a form of this **static electricity**. In France, Charles Coulomb invented equipment for measuring tiny forces. He found that electric forces obeyed the same inverse-square law that Newton had demonstrated for gravity. Although a significant amount of mathematical work was done on electric forces, it had very few immediate practical results.

In 1800, Alessandro Volta in Italy found that he could produce a continuous flow of electricity from alternating plates of copper and zinc, separated by cloth soaked in acid. Initially, Volta thought the electricity came from the copper-zinc contact, using his previous experience with static electricity. In the 1830s, Michael Faraday in England showed that the electricity came from chemical activity of the acid on the two metals. In the meantime, Volta's device, the **voltaic cell**, provided scientists with a powerful new tool. A collection of many pairs of copper and zinc plates in jars of dilute acid could produce significant **electric currents**. Applying such currents to molten salts, Humphrey Davy in England isolated the metallic elements sodium and potassium. His work in the early 1800s laid the foundation for the electro-chemical industry—used for metal extraction and electroplating.

In 1820, Hans Christian Oersted in Denmark discovered that wires carrying an electric current were surrounded by a magnetic field. Soon, many laboratories in Europe and America were displaying large electromagnets. Also, André-Marie Ampère in France developed a complex theory of electric currents, and George Simon Ohm in Germany established the relation between current and voltage in electric circuits.

By 1830, electricity had become a field of intense scientific investigation, but was not yet a major force in industry. That began to change when Michael Faraday discovered that the motion of a conductor in a magnetic field generated a current in the conductor. He quickly used this finding to develop electrical generators and basic motors. Many inventors worked to improve generators that were powered by steam engines, and produced respectable amounts of electrical energy. Their earliest practical applications beyond the chemical industry involved powering arc lamps in lighthouses in Britain.

Electrical Communications

The first important electrical industry was the telegraph. Using various electromagnetic devices, Charles Wheatstone in England and Samuel Morse in the United States created systems and codes for communicating along electrical wires. Because these systems did not need much power, they operated on voltaic cells. When the industrial demand increased, a number of workers made various improvements in the electrodes and chemical solutions used in Volta's simple cells.

Telegraph lines were strung along the railroad rights of way. Steam transportation on railroads expanded rapidly starting in the 1830s. During the 1840s, as the railroads spread across Europe and North America, the telegraph lines went along with them. England was linked with Europe in 1851, and with North America in 1866. In 1862, Britain had 24 000 km of telegraph lines, Europe 130 000 km, and America 80 000 km. Undersea cables connected most of the major cities of the world by 1872.

The telephone grew out of the telegraphic industry. Alexander Graham Bell was a Scottish immigrant to Canada working in Boston, MA, teaching deaf children to speak. In his spare time he worked on telegraph inventions. He discovered that wires could carry varying currents that reproduced the sound patterns of speech and music. Following his startling demonstrations in 1876, Bell began installing telephone lines between the major cities of the U.S. eastern seaboard. Because his parents still lived in Brantford, Bell also soon established telephone systems in southern Ontario. After that, telephone lines rapidly grew up within and between cities across the United States and Canada, and soon after in Britain and Europe. Transcontinental telephoning began in 1915.

The quality of telephone equipment improved quickly as industrial research was directed to the many problems of switching and amplification. Much of this work was done at the Bell Telephone Laboratories, which continue today as leaders in research in electronics and communications technology.

Electrical Power Industries

The first major demand for large amounts of electrical energy came from inventions in electrical lighting. Thomas Edison began as an inventor of telegraphic equipment in the United States. About 1880 he began to manufacture the first successful incandescent lamp. His research laboratory also designed the generators driven by steam engines to provide the electricity to light New York City.

At about the same time, European electrical engineers, working to improve generators, realized that electrical motors could have the same basic design as generators. An early major application of motors was in electric street railways. The power of a remote steam engine could be supplied by electrical means on city streets, avoiding smoke and grime in built-up areas. The first commercial street railway was built in Richmond, VA in 1887.

Manufacturers and electrical engineers soon proposed the same technique for factories. Instead of having a steam engine drive all the machinery of a textile mill by rods, pulleys, and belts, each machine could operate from its own electric motor. The great demand for this technology created intense pressure on research and investment in the 1890s. But the fact that all systems up to that time had operated on **direct current** blocked progress. For safety reasons, the machinery had to operate at only one or two hundred volts. But large energy losses in low voltage transmission meant that the distances between generation and use had to be kept short—only a few kilometres.

ac Wins the Battle

The great demand for electrical energy prompted the idea of harnessing water power. Niagara Falls was an early prime target. If the generating installations at Niagara had been confined to direct current, industries would have had to line the banks of the Niagara River. However, electrical engineers decided to install **alternating current** machinery. The recently developed **transformer** permitted generation at relatively low voltages, stepping up to high voltage for transmission over great distances, and stepping down to low voltages for use. This decision was vigorously opposed by Thomas Edison, who had a large investment in direct current systems.

The success of ac was greatly assisted by the invention of a practical ac motor by Nikolai Tesla. Born in Serbia, Tesla moved to the United States to work for Thomas Edison. Soon, he struck out on his own and became an outstanding, if somewhat unstable, inventor of many electrical devices.

The early decades of the twentieth century saw the electrification of most of North America and Europe. Household electrical appliances began to be produced to transform the way we cook and clean, and brush our teeth.

Electronics

Although electronics is strongly associated with radio, radio began before electronics. After starting work on the radio in Italy in the 1890s, Guglielmo Marconi soon moved to Britain, where he demonstrated successful transmissions over a few kilometres. Then, in 1901, from a transmitter in Cornwall, Marconi received the first transatlantic radio signal at his station in St. John's, Newfoundland.

Although Marconi's early equipment used ac machinery to generate gigantic sparks, soon electron tubes were being developed to perform all the radio functions of transmission and reception.

Electron tubes work by having a heated filament gives off electrons into an evacuated space in a glass tube. Other plates and grids within the tube receive and control the electron currents.

After 1910, a great many electron tubes were developed to generate, modulate, detect, and amplify high frequency currents and waves. Until about 1920, radio was used mainly for communication by Morse code with ships at sea. Then, the marriage of radio and telephony produced commercial broadcasting that spread like wildfire across the world.

The most complex electron tubes, developed in the 1930s, are the television camera and picture tubes. Although most other control circuits now use solid state devices, these tubes are still traversed by streams of electrons in evacuated spaces. However, there are now also many solid state imaging and display devices.

Solid-State Electronics

You have seen that electrical technology grew out of scientific research in the early 1800s. As electrical industries grew in the following decades, experimental and theoretical investigations also continued. For example, in the 1860s, the Scot James Clerk Maxwell developed the foundations of electromagnetic theory. It was this work that inspired Marconi and others to develop radio. And it was in scientific laboratories that streams of electrons were investigated, leading to the discovery of electrons and atomic structure, and to x-ray tubes.

In the 1920s, physicists had developed effective theories of atomic structure. In the 1930s, some of them began to investigate more closely the behaviour of electrons in solids. Improved knowledge of the energy levels of electrons in solids inspired workers at the Bell Telephone Laboratories to seek ways to gain fine control of currents in solid-state devices. Their efforts were rewarded in 1947 with the invention of the **transistor** (Fig. 1.1) by William Shockley, John Bardeen, and Walter H. Brattain. Despite massive investments in electron-tube technology, the transition to solid-state devices proceeded rapidly. Before 1970, investment in solid-state devices had surpassed that in electron tubes. Intensive research was needed to produce transistors capable of handling the relatively modest power outputs of stereo amplifiers. That work still continues as solid-state devices gradually replace the electron tubes used for the large power outputs of radio and television transmitters.

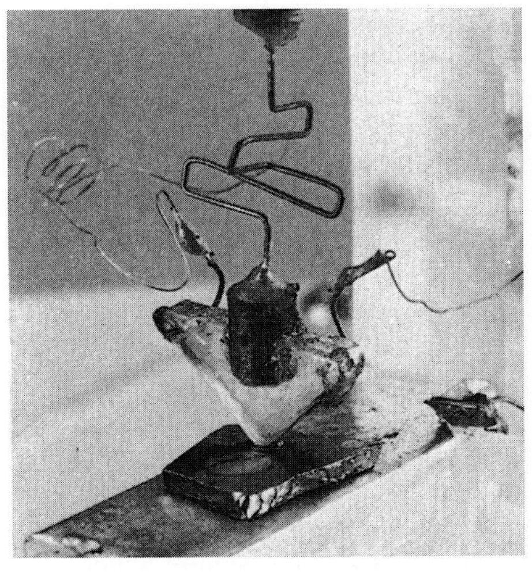

FIG. 1.1
The first transistor. (Courtesy of AT&T, Bell Laboratories.)

The greatest contribution of solid-state devices came from miniaturization. Although even the earliest transistors were much smaller than the tubes they replaced, the drive was on to produce smaller and

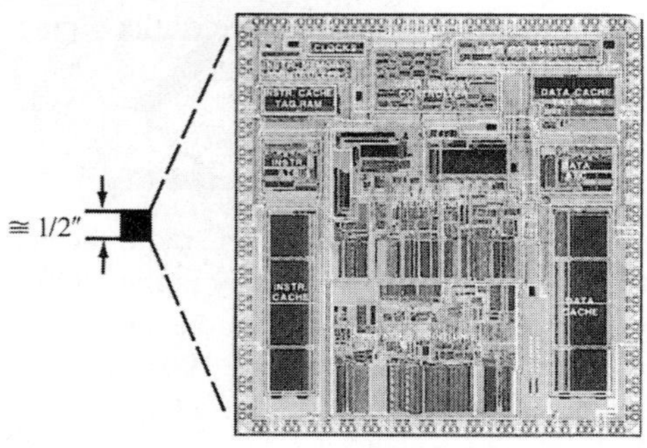

FIG. 1.2
Integrated circuit. (Courtesy of Motorola Semiconductor Products.)

smaller devices. Soon, solid-state devices were developed that combined the actions of many transistors within a single chip. The first **integrated circuit** (IC) was developed in 1958 at Texas Instruments. In 1961 the first commercial integrated circuit was manufactured by the Fairchild Corporation. Since then, integrated circuits have been getting smaller and more powerful (Fig. 1.2).

Integrated circuits and their accompanying printed circuit boards transformed the electronics industry. Besides being involved in all aspects of communication and control devices, ICs formed the basis for the creation of handheld calculators, and soon inspired the take-off of the computer industry.

Computers

Mechanical calculators began to be invented in the 1600s. For a long time they suffered from the limitations of craft skills. In the 1820s, Charles Babbage had designs for complex computing machines that were beyond the abilities of artisans to build. By 1900, electric cash registers and card-sorting machines helped with data manipulation. Under pressures of World War II, great efforts were applied to increasing calculation speeds. In 1946, the first successful electronic system contained 18 000 tubes and weighed thirty tonnes. Today, you can hold in your hand many times the computing power that filled a whole room half a century ago.

Conclusion

Electrical industries are founded on the results of scientific research. As scientists from many countries deepened our knowledge of electrical phenomena, engineers and technologists applied that knowledge to create devices for practical applications. With massive investments by entrepreneurs, whole new industries have been established. Nowadays, many research and development laboratories continue to produce new products that are smaller, faster, and more powerful. Electrical and electronic industries play a central role in modern life.

1.3 SYSTEMS OF UNITS

The design and operation of electric circuits require the measurement and calculation of many different electrical quantities. These quantities are expressed by numbers and **units**. In any measurement, the unit tells you the basic nature of the quantity, and the number tells you how many of those units you have. If you pace across your classroom, you might find its width to be 7 paces. You could convert the unit *pace* by measuring it with a metre stick. If the length of your pace (the unit) is 0.7 m, then the width of the classroom is $7 \times 0.7\ \text{m} = 4.9\ \text{m}$.

Every measurement consists of a number of units.

As in the above example, we express measurements by using **standard units**. The units of electrical measurements belong to the *International System of Units*, abbreviated **SI**. This collection of units is systematic in having a minimum number of independently defined base units, from which all other units are derived. The derivations depend on physical laws, some of which will be introduced in later chapters of this text. The SI based units have developed out of the metric system that was introduced in France in 1792.

Although there are seven SI base units, we only show in Table 1.1 the five that are commonly required for electrical work. Note that each of the units is designated by a symbol, such as "m" for "metre." It is essential that you use the style in which these symbols are expressed—whether lower case or capitalized. The principle is that unit symbols are capitalized only when they are derived from a person's name; for example "A" for "amperes," named after the French physicist André M. Ampère.

TABLE 1.1
SI Base Units

Quantity	Unit	Symbol
Length	metre	m
Mass	kilogram	kg
Time	second	s
Electric current	ampere	A
Temperature	kelvin	K

Each of the SI base units has a very precise physical definition. Length (1 m) and time (1 s) are defined in terms of atomic processes. The mass of 1 kg is that of a specific lump of matter preserved at the International Bureau of Weights and Measures in France. The fundamental definition of 1 A depends on the force between two current-carrying conductors. The different definition we give in Chapter 2 is equivalent to this one. The temperature unit is defined in thermodynamic terms.

All other measurement units are defined by some combination of the base units. The **derived units** needed for electrical analysis are listed in Table 1.2. You will note that they are all named for scientists whose work contributed to their development.

TABLE 1.2
SI Derived Units

Quantity	Unit	Symbol	Definition
Frequency	hertz	Hz	s^{-1}
Force	newton	N	$kg \cdot m/s^2$
Energy	joule	J	$N \cdot m$
Power	watt	W	J/s
Electric charge	coulomb	C	$A \cdot s$
Electric potential difference	volt	V	J/C
Electrical resistance	ohm	Ω	V/A
Electrical conductance	siemens	S	Ω^{-1}
Electrical capacitance	farad	F	C/V
Magnetic flux	weber	Wb	$V \cdot s$
Magnetic flux density	tesla	T	Wb/m^2

You know that there are other systems of units in use. However, with very few exceptions, electrical work is limited to SI units. Two exceptions that you will encounter are the centimetre and the horsepower. They are both simply defined in relation to SI units:

1 cm = 0.01 m, and 1 HP = 746 W.

SI Prefixes

Because there is large variation in the magnitudes of measured quantities, the SI system uses a number of multiplier **prefixes**. The full list ranges in steps of 1000 from 10^{18} to 10^{-18}. Table 1.3 lists the SI prefixes, with their symbols. As with the symbols for units, you should always use the precise symbol shown for the prefixes. Note for example that "m" is the symbol for "milli" (10^{-3}), while "M" is the symbol for "mega" (10^{6}).

TABLE 1.3
SI Prefixes

Name	Symbol	Definition
exa	E	10^{18}
peta	P	10^{15}
tera	T	10^{12}
giga	G	10^{9}
mega	M	10^{6}
kilo	k	10^{3}
milli	m	10^{-3}
micro	μ	10^{-6}
nano	n	10^{-9}
pico	p	10^{-12}
femto	f	10^{-15}
atto	a	10^{-18}

It is important for you to become proficient in using units and prefixes. The next section will review operations with powers of ten and illustrate how to use the SI prefixes.

1.4 POWERS OF TEN

In ordinary decimal notation, large and small numbers are inconvenient to handle because of the need to count place-holding zeros. Power of ten notation helps to avoid errors by counting the zeros for you. For example, the speeds of electrons in a vacuum and in a solid might be found to be

94 300 000 m/s and 0.000 057 m/s respectively

Simply replacing the zeros with powers of ten would result in

943×10^{5} m/s and 0.57×10^{-4} m/s.

However, it is more common to express these values in scientific notation (see page 13) as

9.43×10^{7} m/s and 5.7×10^{-5} m/s.

To make sure that you can handle powers of ten easily, you should review the basic operations given below. The values of m and n can be any real number.

$$\frac{1}{10^{n}} = 10^{-n} \qquad \frac{1}{10^{-n}} = 10^{n} \tag{1.1}$$

Equation (1.1) shows that shifting a power of ten from the denominator to the numerator, or the reverse, requires simply changing the sign of the power.

EXAMPLE 1.1

a. $$\frac{1}{1000} = \frac{1}{10^{+3}} = 10^{-3}$$

b. $$\frac{1}{0.00001} = \frac{1}{10^{-5}} = 10^{+5}$$

The product of powers of ten:

$$(10^n)(10^m) = 10^{(n+m)} \tag{1.2}$$

EXAMPLE 1.2

a. $(1000)(10\,000) = (10^3)(10^4) = 10^{(3+4)} = 10^7$

b. $(0.00001)(100) = (10^{-5})(10^2) = 10^{(-5+2)} = 10^{-3}$

The division of powers of ten:

$$\frac{10^n}{10^m} = 10^{(n-m)} \tag{1.3}$$

EXAMPLE 1.3

a. $$\frac{100\,000}{100} = \frac{10^5}{10^2} = 10^{(5-2)} = 10^3$$

b. $$\frac{1000}{0.0001} = \frac{10^3}{10^{-4}} = 10^{(3-(-4))} = 10^{(3+4)} = 10^7$$

Note the use of parentheses in the second example to ensure that the proper sign is used between operators.

The power of powers of ten:

$$(10^n)^m = 10^{(nm)} \tag{1.4}$$

EXAMPLE 1.4

a. $(100)^4 = (10^2)^4 = 10^{(2)(4)} = 10^8$

b. $(1000)^{-2} = (10^3)^{-2} = 10^{(3)(-2)} = 10^{-6}$

c. $(0.01)^{-3} = (10^{-2})^{-3} = 10^{(-2)(-3)} = 10^6$

Basic Arithmetic Operations

Let us now examine the use of powers of ten to perform some basic arithmetic operations using numbers that are not just powers of ten.

The number 5000 can be written as $5 \times 1000 = 5 \times 10^3$, and the number 0.0004 can be written as $4 \times 0.0001 = 4 \times 10^{-4}$. Of course, 10^5 can also be written as 1×10^5 if it clarifies the operation to be performed.

Addition and Subtraction To perform addition or subtraction using powers of ten, the power of ten *must be the same for each term;* that is,

$$A \times 10^n \pm B \times 10^n = (A \pm B) \times 10^n \tag{1.5}$$

EXAMPLE 1.5

a.
$$\begin{aligned} 6300 + 75\,000 &= (6.3)(1000) + (75)(1000) \\ &= 6.3 \times 10^3 + 75 \times 10^3 \\ &= (6.3 + 75) \times 10^3 \\ &= \mathbf{81.3 \times 10^3} \end{aligned}$$

b.
$$\begin{aligned} 0.000\,96 - 0.0000\,86 &= (96)(0.000\,01) - (8.6)(0.000\,01) \\ &= 96 \times 10^{-5} - 8.6 \times 10^{-5} \\ &= (96 - 8.6) \times 10^{-5} \\ &= \mathbf{87.4 \times 10^{-5}} \end{aligned}$$

Multiplication

In general,

$$(A \times 10^n)(B \times 10^m) = (A)(B) \times 10^{n+m} \tag{1.6}$$

revealing that the *operations with the powers of ten can be separated from the operation with the multipliers.*

EXAMPLE 1.6

a.
$$\begin{aligned} (0.0002)(0.000\,007) &= [(2)(0.0001)][(7)(0.000\,001)] \\ &= (2 \times 10^{-4})(7 \times 10^{-6}) \\ &= (2)(7) \times (10^{-4})(10^{-6}) \\ &= \mathbf{14 \times 10^{-10}} \end{aligned}$$

b.
$$\begin{aligned} (340\,000)(0.000\,61) &= (3.4 \times 10^5)(61 \times 10^{-5}) \\ &= (3.4)(61) \times (10^5)(10^{-5}) \\ &= 207.4 \times 10^0 \\ &= \mathbf{207.4} \end{aligned}$$

Division

In general,

$$\frac{A \times 10^n}{B \times 10^m} = \frac{A}{B} \times 10^{n-m} \tag{1.7}$$

revealing again that the *operations with the powers of ten can be separated from the same operation with the multipliers.*

EXAMPLE 1.7

a. $$\frac{0.000\,47}{0.002} = \frac{47 \times 10^{-5}}{2 \times 10^{-3}} = \left(\frac{47}{2}\right) \times \left(\frac{10^{-5}}{10^{-3}}\right)$$
$$= \mathbf{23.5 \times 10^{-2}}$$

b. $$\frac{690\,000}{0.000\,000\,13} = \frac{69 \times 10^{4}}{13 \times 10^{-8}} = \left(\frac{69}{13}\right) \times \left(\frac{10^{4}}{10^{-8}}\right)$$
$$= \mathbf{5.31 \times 10^{12}}$$

Powers

In general,

$$(A \times 10^n)^m = A^m \times 10^{nm} \tag{1.8}$$

which again permits the separation of the *operation with the powers of ten from the multipliers.*

EXAMPLE 1.8

a. $$(0.000\,03)^3 = (3 \times 10^{-5})^3 = (3)^3 \times (10^{-5})^3$$
$$= \mathbf{27 \times 10^{-15}}$$

b. $$(90\,800\,000)^2 = (9.08 \times 10^7)^2 = (9.08)^2 \times (10^7)^2$$
$$= \mathbf{82.4464 \times 10^{14}}$$

Scientific Notation

Frequently, the most regular way to express measurements in power of ten notation is to choose the power of ten that gives the multiplying number a single digit before the decimal. This is called **scientific notation**. Here are a couple of examples:

The speed of light is $2.997\,924\,58 \times 10^8$ m/s.

The rest mass of an electron is $9.109\,534 \times 10^{-31}$ kg.

The rules for power of ten operations can be used to perform calculations with these numbers. Example 1.9 will show you how it is done.

EXAMPLE 1.9

At speeds up to about one-fifth of the speed of light, the kinetic energy of moving objects is given by the standard formula, $E = \frac{1}{2}mv^2$. Use the values given above to calculate the kinetic energy of an electron travelling at 1/5 the speed of light.

Solution

The speed of the electron is

$$v = \frac{299\,792\,458 \times 10^8}{5} \text{ m/s}$$

continued

The mass of the electron is $m = 9.109\,534 \times 10^{-31}$ kg.

Substitute the values into the formula

$$
\begin{aligned}
E &= \frac{mv^2}{2} \\
&= \frac{9.109\,534 \times 10^{-31}}{2}\left(\frac{299\,792\,458 \times 10^{0}}{5}\right) \text{kg} \cdot \text{m}^2/\text{s}^2 \\
&= 4.554\,767 \times 10^{-31}\left(0.599\,584\,9^2 \times 10^{16}\right) \text{kg} \cdot \text{m}^2/\text{s}^2 \\
&= 1.637\,448 \times 10^{-15}\,\text{J}
\end{aligned}
$$

Here, we have given an intermediate step to show some of the power of ten operations. However, you could perform the whole operation with your scientific calculator in a single sequence of operations. Note that the units for kinetic energy do work out to the definition of *joule* given in Table 1.2. Depending on the number of digits your calculator carries, you may get a slightly different result from the one given here.

SI Prefixes and Engineering Notation

Most often in electrical work, it is convenient to use the SI prefixes to describe very large or very small quantities. You are probably familiar with the units of gigabytes and nanoseconds used for computers. To convert a quantity to this form, you will want to express it with a power of ten that is evenly divisible by 3. In Example 1.9, the final answer had the power of –15, which is divisible by 3. From Table 1.3, the corresponding prefix is *femto*, so the answer could be written as 1.637 448 fJ.

The power of 8 in the speed of light is not evenly divisible by 3. You could express the value either by the power of 6 or of 9 (mega or giga). Rounding the numerical part to 3.00 for convenience, we could write

$$
\begin{aligned}
v = 3.00 \times 10^8\,\text{m/s} &= 300 \times 10^6\,\text{m/s} = 300\,\text{Mm/s} \\
\text{or} \quad &= 0.300 \times 10^9\,\text{m/s} = 0.300\,\text{Gm/s}
\end{aligned}
$$

In the first case, you *reduce* the power by 2, so you have to multiply the numerical part by 100. In the second case, you *increase* the power by 1, so you have to multiply the numerical part by 1/10. Once you have a power of ten divisible by 3, you simply convert it to the appropriate SI prefix. Usual practice is to choose the SI prefix that makes the numerical part fall between 1 and 999. However, the example of 0.300 Gm/s is not unusual.

Engineering notation uses only the powers of ten that are divisible by 3. This makes it easy to convert a quantity in engineering notation to an appropriate SI prefix. For example, the annual energy consumption of a province of 235 billion joules could be expressed variously as:

$$
\begin{aligned}
&235\,000\,000\,000\ \text{J} \\
&235\,000\,000 \times 10^3\ \text{J} = 235\,000\,000\ \text{kJ} \\
&235\,000 \times 10^6\ \text{J} = 235\,000\ \text{MJ} \\
&235 \times 10^9\ \text{J} = 235\ \text{GJ} \\
&0.235 \times 10^{12}\ \text{J} = 0.235\ \text{TJ}
\end{aligned}
$$

Converting Between Powers of Ten

It is often necessary to convert from one power of ten to another. For instance, if a meter measures kilohertz (kHz), it may be necessary to find the corresponding level in megahertz (MHz). If time is measured in milliseconds (ms), it may be necessary to find the corresponding time in microseconds (μs) for a graphical plot. The process is not difficult if you simply keep in mind that an increase or decrease in the power of ten must be associated with the opposite effect on the multiplying factor. The procedure is best described by a few examples.

EXAMPLE 1.10

a. Convert 20 kHz to megahertz.

Solution: In the power-of-ten format:

$$20 \text{ kHz} = 20 \times 10^3 \text{ Hz}$$

The conversion requires that we find the multiplying factor to appear in the space below:

Increase by 3

$$20 \times 10^3 \text{ Hz} \Rightarrow \_\_ \times 10^6 \text{ Hz}$$

Decrease by 3

Since the power of ten will be *increased* by a factor of *three,* the multiplying factor must be *decreased* by moving the decimal point *three* places to the left, as shown below:

$$\underbrace{020.}_{3} = 0.02$$

and $\quad 20 \times 10^3 \text{ Hz} = 0.02 \times 10^6 \text{ Hz} = \mathbf{0.02\ MHz}$

b. Convert 0.01 ms to microseconds.

Solution: In the power-of-ten format:

$$0.01 \text{ ms} = 0.01 \times 10^{-3} \text{ s}$$

Reduce by 3

and $\quad 0.01 \times 10^{-3} \text{ s} = \_\_ \times 10^{-6} \text{ s}$

Increase by 3

Since the power of ten will be *reduced* by a factor of three, the multiplying factor must be *increased* by moving the decimal point three places to the right, as follows:

$$0.\underbrace{010}_{3} = 10$$

and $\quad 0.01 \times 10^{-3} \text{ s} = 10 \times 10^{-6} \text{ s} = \mathbf{10\ \mu s}$

You might think, when comparing -3 to -6, that the power of ten has increased, but keep in mind when making your judgment about increasing or decreasing the magnitude of the multiplier that 10^{-6} is a great deal smaller than 10^{-3}.

continued

c. Convert 0.002 km to millimetres.

Solution:

Reduce by 6

$$0.002 \times 10^{3}\ \text{m} \Rightarrow \_\_\ \times 10^{-3}\ \text{m}$$

Increase by 6

In this example we have to be very careful because the difference between +3 and −3 is a factor of 6, requiring that the multiplying factor be modified as follows:

$$0.002000 = 2000$$

6

and $0.002 \times 10^{3}\ \text{m} = 2000 \times 10^{-3}\ \text{m} = \mathbf{2000\ mm}$

1.5 SIGNIFICANT DIGITS

In making a measurement, you read the digits on a scale or dial. How many digits you read can give you an indication of the accuracy of the measurement. For example, you could measure the length of a room to the nearest metre, 7 m; or the nearest centimetre, 7.35 m; or the nearest millimetre, 7.346 m. Every time you read an additional digit, your measure gets closer to the "true" value of the length. The digits you measure are called **significant digits**. The 7.346 m has 4 significant digits, and is said to have 4-digit accuracy. The value is known to within 1 mm in a total of 7346 mm.

You can see how digits indicate accuracy by considering an airline flight of 7346 km. This is known to within 1 km in 7346 km, the same degree of accuracy as before. However, if you took a flight of 7000 km, you could not tell the degree of accuracy because the zeros may be either measured, or only place-holding. If the value had 3-digit accuracy (to the nearest 10 km), that could be shown using scientific notation: 7.00×10^{3} km. The appearance of zeros after the decimal, indicates that they are significant digits.

Measurements used in this text for examples and problems are often rounded numbers to keep the arithmetic simple. You should consider a value of 100 V as having 3-digit accuracy. For measurements in the laboratory, many electrical meters provide values with 3- or 4-digit accuracy.

In addition, you should not give the results of calculations more accuracy than they deserve. In particular you should realize that

Calculations cannot increase the accuracy of measurements.

Consider a trip of 275 km in 3.7 h. You can calculate your average speed by dividing the distance by the time. The result on a calculator is 74.324 324. Clearly, you cannot know the speed to that degree of accuracy. So, the result should be reported only to 2-digit accuracy, 74 km/h. In general,

The number of significant digits in a result of multiplication or division must be limited to the number of significant digits in the least accurate measurement used in the calculation.

TABLE 1.4

Symbol	Meaning
$\neq$	Not equal to $6.12 \neq 6.13$
$>$	Greater than $4.78 > 4.20$
$\gg$	Much greater than $840 \gg 16$
$<$	Less than $430 < 540$
$\ll$	Much less than $0.002 \ll 46$
$\geq$	Greater than or equal to $x \geq y$ is satisfied for $y = 3$ and $x > 3$ or $x = 3$
$\leq$	Less than or equal to $x \leq y$ is satisfied for $y = 3$ and $x < 3$ or $x = 3$
$\cong$	Approximately equal to $3.14159 \cong 3.14$
Σ	Sum of $\Sigma\,(4 + 6 + 8) = 18$
$\|\;\|$	Absolute magnitude of $\|a\| = 4$, where $a = -4$ or $+4$
$\therefore$	Therefore $x = \sqrt{4} \quad \therefore x = \pm 2$
$\equiv$	By definition Establishes a relationship between two or more quantities

For the addition and subtraction of numbers, you just need to use common sense. If, after an airline trip of 7346 km, you walk an additional 375 m, you would not say the distance you travelled was 7346.375 km. The distance you walked is less than the 1 km uncertainty in the flight distance. The next examples illustrate the proper usage.

For rounding off, drop the unnecessary digits unless the leftmost digit to be dropped is 5 or more. For example, the speed of light to different degrees of accuracy is given in the following list:

$2.997\,924\,58 \times 10^8$ m/s	$2.997\,924\,6 \times 10^8$ m/s
$2.997\,925 \times 10^8$ m/s	$2.997\,9 \times 10^8$ m/s
2.998×10^8 m/s	3.00×10^8 m/s

EXAMPLE 1.11

Perform the indicated operations with the following approximate numbers and round off to the appropriate level of accuracy. The symbol $\cong$ means *approximately equal to.*

a. $532.6 + 4.02 + 0.036 = 536.656 \cong$ **536.7** (as determined by 532.6)
b. $0.04 + 0.003 + 0.0064 = 0.0494 \cong$ **0.05** (as determined by 0.04)
c. $4.632 \times 2.4 = 11.1168 \cong$ **11** (as determined by the two significant digits of 2.4)
d. $3.051 \times 802 = 2446.902 \cong$ **2450** (as determined by the three significant digits of 802)
e. $1402/6.4 = 219.0625 \cong$ **220** (as determined by the two significant digits of 6.4)
f. $0.0046/0.05 = 0.0920 \cong$ **0.09** (as determined by the one significant digit of 0.05)

1.6 SYMBOLS

Throughout the text, various symbols will be used that you may not have seen before. Some are defined in Table 1.4, and others will be defined in the text as they are used.

PROBLEMS

Note: More difficult problems are indicated by an asterisk (*) throughout the text.

SECTION 1.2 A Brief History

1. Write a short essay naming every electrical device you use during the course of a single day. For each, say how it affects your life, or how you would manage without it.
2. Visit your local library and investigate how much help (in print and electronic media) it can give people interested in learning about electrical technologies. Write a report on your findings.
3. Choose an area of particular interest in this field and write a very brief report on the history of the subject.
4. Choose an individual of particular importance in this field and write a very brief review of his or her life and important contributions.

SECTION 1.3 Systems of Units

5. Use the definitions of derived units in Table 1.2 to verify the following relations:
   a. 1 A = 1 C/s
   b. 1 kg·m²/s² = 1 J
   c. 1 W = 1 V·A
   d. 1 A² Ω = 1 W
   e. 1 J = 1 F V²
   f. 1 C²/F = 1 J

SECTION 1.4 Powers of Ten

6. Perform the following operations and express your answer as a power of ten:
   a. 4200 + 6 800 000
   b. $9 \times 10^4 + 3.6 \times 10^3$
   c. $0.5 \times 10^{-3} - 6 \times 10^{-5}$
   d. $1.2 \times 10^3 + 50\ 000 \times 10^{-3} - 0.006 \times 10^5$

rform the following operations and express your wer as a power of ten:
100)(100) **b.** (0.01)(1000)
$)^3)(10^6)$ **d.** (1000)(0.000 01)
$^{-6})(10\ 000\ 000)$ **f.** $(10\ 000)(10^{-8})(10^{35})$

the following operations and express your a power of ten:
(0.0003)
08
(0.000 07)
$0.0002)(7 \times 10^8)$

wing operations and express your of ten:
b. $\dfrac{0.01}{100}$
d. $\dfrac{0.000\ 000\ 1}{100}$
e. $\dfrac{10^{38}}{0.000\ 100}$ **f.** $\dfrac{(100)^{1/2}}{0.01}$

10. Perform the following operations and express your answer as a power of ten:
   a. $\dfrac{2000}{0.000\ 08}$ **b.** $\dfrac{0.004\ 08}{60\ 000}$
   c. $\dfrac{0.000\ 215}{0.000\ 05}$ **d.** $\dfrac{78 \times 10^9}{4 \times 10^{-6}}$

11. Perform the following operations and express your answer as a power of ten:
   a. $(100)^3$ **b.** $(0.0001)^{1/2}$
   c. $(10\ 000)^8$ **d.** $(0.000\ 000\ 10)^9$

12. Perform the following operations and express your answer as a power of ten:
   a. $(2.2 \times 10^3)^3$
   b. $(0.0006 \times 10^2)^4$
   c. $(0.004)(6 \times 10^2)^2$
   d. $((2 \times 10^{-3})(0.8 \times 10^4)(0.003 \times 10^5))^3$

13. Perform the following operations and express your answer in scientific notation:
   a. $(-0.001)^2$ **b.** $\dfrac{(100)(10^{-4})}{10}$
   c. $\dfrac{(0.001)^2(100)}{10\ 000}$ **d.** $\dfrac{(10^2)(10\ 000)}{0.001}$
   e. $\dfrac{(0.0001)^3(100)}{1\ 000\ 000}$ ***f.** $\dfrac{[(100)(0.01)]^{-3}}{[(100)^2][0.001]}$

*14. Perform the following operations and express your answer in engineering notation:
   a. $\dfrac{(300)^2(100)}{10^4}$ **b.** $[(40\ 000)^2][(20)^{-3}]$
   c. $\dfrac{(60\ 000)^2}{(0.02)^2}$ **d.** $\dfrac{(0.000\ 027)^{1/3}}{210\ 000}$
   e. $\dfrac{[(4000)^2][300]}{0.02}$
   f. $[(0.000\ 016)^{1/2}][(100\ 000)^5][0.02]$
   g. $\dfrac{[(0.003)^3][(0.000\ 07)^2][(800)^2]}{[(100)(0.000\ 9)]^{1/2}}$ (a challenge)

15. Fill in the blanks of the following conversions:
   a. $6 \times 10^3 =$ ____ $\times 10^6$
   b. $4 \times 10^{-4} =$ ____ $\times 10^{-6}$
   c. $50 \times 10^5 =$ ____ $\times 10^3 =$ ____ $\times 10^6$ $=$ ____ $\times 10^9$
   d. $30 \times 10^{-8} =$ ____ $\times 10^{-3} =$ ____ $\times 10^{-6}$ $=$ ____ $\times 10^{-9}$

16. Perform the following conversions.
   a. 2000 μs to milliseconds
   b. 0.04 ms to microseconds
   c. 0.06 μF to nanofarads
   d. 8400 ps to microseconds
   e. 0.006 km to millimetres
   f. 260×10^3 mm to kilometres

TABLE 1.4

Symbol	Meaning
$\neq$	Not equal to $6.12 \neq 6.13$
$>$	Greater than $4.78 > 4.20$
$\gg$	Much greater than $840 \gg 16$
$<$	Less than $430 < 540$
$\ll$	Much less than $0.002 \ll 46$
$\geq$	Greater than or equal to $x \geq y$ is satisfied for $y = 3$ and $x > 3$ or $x = 3$
$\leq$	Less than or equal to $x \leq y$ is satisfied for $y = 3$ and $x < 3$ or $x = 3$
$\cong$	Approximately equal to $3.14159 \cong 3.14$
Σ	Sum of $\Sigma\,(4 + 6 + 8) = 18$
$\lvert\ \rvert$	Absolute magnitude of $\lvert a \rvert = 4$, where $a = -4$ or $+4$
$\therefore$	Therefore $x = \sqrt{4} \quad \therefore x = \pm 2$
$\equiv$	By definition Establishes a relationship between two or more quantities

For the addition and subtraction of numbers, you just need to use common sense. If, after an airline trip of 7346 km, you walk an additional 375 m, you would not say the distance you travelled was 7346.375 km. The distance you walked is less than the 1 km uncertainty in the flight distance. The next examples illustrate the proper usage.

For rounding off, drop the unnecessary digits unless the leftmost digit to be dropped is 5 or more. For example, the speed of light to different degrees of accuracy is given in the following list:

$2.997\,924\,58 \times 10^8$ m/s	$2.997\,924\,6 \times 10^8$ m/s
$2.997\,925 \times 10^8$ m/s	$2.997\,9 \times 10^8$ m/s
2.998×10^8 m/s	3.00×10^8 m/s

EXAMPLE 1.11

Perform the indicated operations with the following approximate numbers and round off to the appropriate level of accuracy. The symbol $\cong$ means *approximately equal to.*

a. $532.6 + 4.02 + 0.036 = 536.656 \cong$ **536.7** (as determined by 532.6)
b. $0.04 + 0.003 + 0.0064 = 0.0494 \cong$ **0.05** (as determined by 0.04)
c. $4.632 \times 2.4 = 11.1168 \cong$ **11** (as determined by the two significant digits of 2.4)
d. $3.051 \times 802 = 2446.902 \cong$ **2450** (as determined by the three significant digits of 802)
e. $1402/6.4 = 219.0625 \cong$ **220** (as determined by the two significant digits of 6.4)
f. $0.0046/0.05 = 0.0920 \cong$ **0.09** (as determined by the one significant digit of 0.05)

1.6 SYMBOLS

Throughout the text, various symbols will be used that you may not have seen before. Some are defined in Table 1.4, and others will be defined in the text as they are used.

PROBLEMS

Note: More difficult problems are indicated by an asterisk (*) throughout the text.

SECTION 1.2 A Brief History

1. Write a short essay naming every electrical device you use during the course of a single day. For each, say how it affects your life, or how you would manage without it.
2. Visit your local library and investigate how much help (in print and electronic media) it can give people interested in learning about electrical technologies. Write a report on your findings.
3. Choose an area of particular interest in this field and write a very brief report on the history of the subject.
4. Choose an individual of particular importance in this field and write a very brief review of his or her life and important contributions.

SECTION 1.3 Systems of Units

5. Use the definitions of derived units in Table 1.2 to verify the following relations:
   a. 1 A = 1 C/s
   b. $1\ \text{kg}\cdot\text{m}^2/\text{s}^2 = 1\ \text{J}$
   c. 1 W = 1 V·A
   d. $1\ \text{A}^2\,\Omega = 1\ \text{W}$
   e. $1\ \text{J} = 1\ \text{F V}^2$
   f. $1\ \text{C}^2/\text{F} = 1\ \text{J}$

SECTION 1.4 Powers of Ten

6. Perform the following operations and express your answer as a power of ten:
   a. $4200 + 6\ 800\ 000$
   b. $9 \times 10^4 + 3.6 \times 10^3$
   c. $0.5 \times 10^{-3} - 6 \times 10^{-5}$
   d. $1.2 \times 10^3 + 50\ 000 \times 10^{-3} - 0.006 \times 10^5$
7. Perform the following operations and express your answer as a power of ten:
   a. $(100)(100)$ **b.** $(0.01)(1000)$
   c. $(10^3)(10^6)$ **d.** $(1000)(0.000\ 01)$
   e. $(10^{-6})(10\ 000\ 000)$ **f.** $(10\ 000)(10^{-8})(10^{35})$
8. Perform the following operations and express your answer as a power of ten:
   a. $(50\ 000)(0.0003)$
   b. 2200×0.08
   c. $(0.000\ 082)(0.000\ 07)$
   d. $(30 \times 10^{-4})(0.0002)(7 \times 10^8)$
9. Perform the following operations and express your answer as a power of ten:
   a. $\dfrac{100}{1000}$ **b.** $\dfrac{0.01}{100}$
   c. $\dfrac{10\ 000}{0.000\ 01}$ **d.** $\dfrac{0.000\ 000\ 1}{100}$
   e. $\dfrac{10^{38}}{0.000\ 100}$ **f.** $\dfrac{(100)^{1/2}}{0.01}$
10. Perform the following operations and express your answer as a power of ten:
   a. $\dfrac{2000}{0.000\ 08}$ **b.** $\dfrac{0.004\ 08}{60\ 000}$
   c. $\dfrac{0.000\ 215}{0.000\ 05}$ **d.** $\dfrac{78 \times 10^9}{4 \times 10^{-6}}$
11. Perform the following operations and express your answer as a power of ten:
   a. $(100)^3$ **b.** $(0.0001)^{1/2}$
   c. $(10\ 000)^8$ **d.** $(0.000\ 000\ 10)^9$
12. Perform the following operations and express your answer as a power of ten:
   a. $(2.2 \times 10^3)^3$
   b. $(0.0006 \times 10^2)^4$
   c. $(0.004)(6 \times 10^2)^2$
   d. $((2 \times 10^{-3})(0.8 \times 10^4)(0.003 \times 10^5))^3$
13. Perform the following operations and express your answer in scientific notation:
   a. $(-0.001)^2$ **b.** $\dfrac{(100)(10^{-4})}{10}$
   c. $\dfrac{(0.001)^2(100)}{10\ 000}$ **d.** $\dfrac{(10^2)(10\ 000)}{0.001}$
   e. $\dfrac{(0.0001)^3(100)}{1\ 000\ 000}$ ***f.** $\dfrac{[(100)(0.01)]^{-3}}{[(100)^2][0.001]}$

*14. Perform the following operations and express your answer in engineering notation:
   a. $\dfrac{(300)^2(100)}{10^4}$ **b.** $[(40\ 000)^2][(20)^{-3}]$
   c. $\dfrac{(60\ 000)^2}{(0.02)^2}$ **d.** $\dfrac{(0.000\ 027)^{1/3}}{210\ 000}$
   e. $\dfrac{[(4000)^2][300]}{0.02}$
   f. $[(0.000\ 016)^{1/2}][(100\ 000)^5][0.02]$
   g. $\dfrac{[(0.003)^3][(0.000\ 07)^2][(800)^2]}{[(100)(0.000\ 9)]^{1/2}}$ (a challenge)

15. Fill in the blanks of the following conversions:
   a. $6 \times 10^3 =$ ____ $\times 10^6$
   b. $4 \times 10^{-4} =$ ____ $\times 10^{-6}$
   c. $50 \times 10^5 =$ ____ $\times 10^3 =$ ____ $\times 10^6$ $=$ ____ $\times 10^9$
   d. $30 \times 10^{-8} =$ ____ $\times 10^{-3} =$ ____ $\times 10^{-6}$ $=$ ____ $\times 10^{-9}$
16. Perform the following conversions.
   a. 2000 μs to milliseconds
   b. 0.04 ms to microseconds
   c. 0.06 μF to nanofarads
   d. 8400 ps to microseconds
   e. 0.006 km to millimetres
   f. 260×10^3 mm to kilometres

For Problems 17 and 18, convert the following:

17. **a.** 1.5 min to seconds
b. 0.04 h to seconds
c. 0.05 s to microseconds
d. 0.16 m to millimetres
e. 0.00000012 s to nanoseconds
f. 3 620 000 s to days
g. 1020 mm to metres

18. **a.** 0.1 μF (microfarad) to picofarads
b. 0.467 km to meters
c. 63.9 mm to centimetres
d. 69 cm to kilometres
e. 3.2 h to milliseconds
f. 0.016 mm to μm
g. 60 sq cm (cm^2) to square metres (m^2)

SECTION 1.5 Significant Digits

19. Express the results of the following calculations to the correct number of significant digits:

a. Calculate the distance travelled in 5.27 hours at an average speed of 67 km/h.

b. Calculate the speed of an electron that travels 3.3 mm in 17.58 ns.

c. Calculate the time for an electron to travel 98.7 μm at a speed of 2.998 x 10^7 m/s.

2

Current and Voltage

Outline

Learning Outcomes

After completing this chapter, you will be able to

- describe the structure of atoms
- relate atomic structure to electric charge and force
- define electric charge in relation to electric force
- define potential difference in terms of work on charges
- explain potential difference and its relation to voltage
- explain electric current in relation to charge
- describe a number of dc sources, such as batteries, generators, and solar cells
- explain how atomic structure relates to insulators, conductors, and semiconductors
- describe how ammeters and voltmeters are used to measure circuit quantities

To understand how electricity behaves, it is first necessary to gain an understanding of the basic nature of electricity and the units used to describe it.

2.1 ELECTRICITY

Electricity is a form of energy with a greater versatility than any other form. It can be produced by the transformation of many other forms of energy: chemical energy in batteries, mechanical energy in generators, or light energy in solar cells.

Electrical energy can be stored in batteries or transmitted great distances along transmission lines. When you have it where you want it, you can use electricity to run power tools, illuminate large buildings, operate complex machinery, communicate instantaneously around the world, and perform the many computations inside our computers.

To work in the electrical industry, you need an understanding of a number of concepts, quantities, and relations. In this chapter, we will examine the basic ideas of voltage and current, and their connections with the atomic structure of matter.

2.2 ATOMS AND THEIR STRUCTURE

At the core of every **atom** is a **nucleus** with a positive charge determined by the number of **protons** (symbol p^+) in it. The natural repulsion between protons is offset by the short-range nuclear force provided by neutrons, which have the same mass as protons but no charge. In the lighter atoms, the number of neutrons is approximately the same as the number of protons.

A neutral atom is surrounded by a number of **electrons** (symbol e^-) equal to the number of protons in the nucleus (Fig. 2.1). The mass of an electron is about 1/1800 of the mass of a proton. Each electron has a negative charge equal in magnitude to the charge on each proton. There is an electric force of attraction between a proton and an electron. The electrons are arranged in spherical **shells**, each of which may contain up to a number related to the number, n, of the shell by the formula $2n^2$: 2 in the first shell, 8 in the second, 18 in the third, 32 in the fourth, etc. The shells are filled with electrons from the inside out.

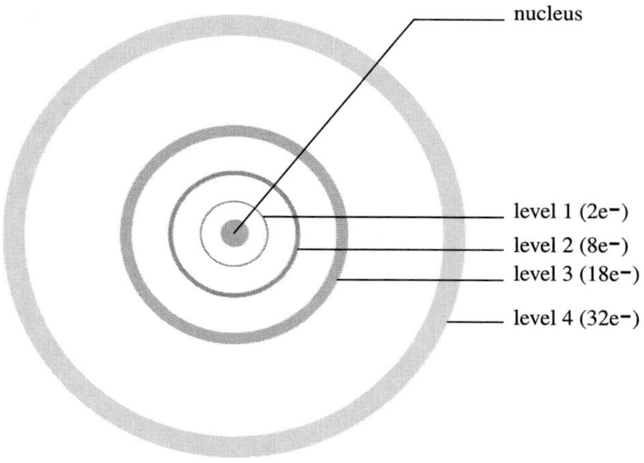

FIG. 2.1
The structure of an atom: a nucleus surrounded by electron shells.

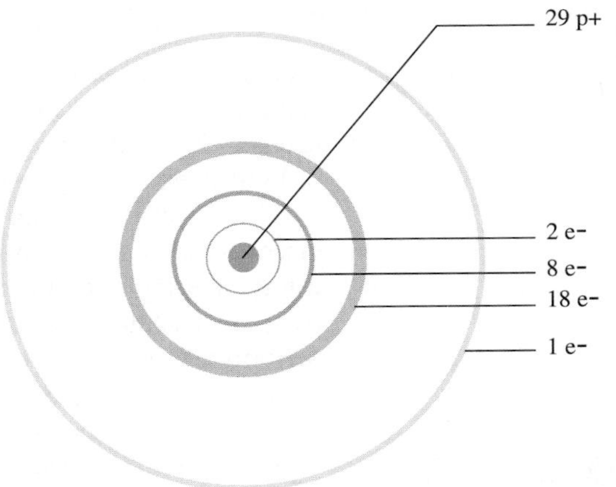

FIG. 2.2
The arrangement of electrons in a copper atom.

With 29 protons in its nucleus, an atom of copper (Fig. 2.2) has a nuclear charge of 29 p^+. Its nucleus has a mass of about 10^{-25} kg. The radius of the copper nucleus is in the range of 10^{-15} m, but the radius of the whole atom is about 10^{-10} m. In the copper atom the first three electron shells are full, making a charge of 28 e^-. The copper atom has only one electron in its fourth shell. Since this electron is so far from the attractive force of the nucleus, it is easily detached to become a "free" electron. When an electron is detached from a neutral atom, the remaining structure is called an **ion**, with a net charge of +1.

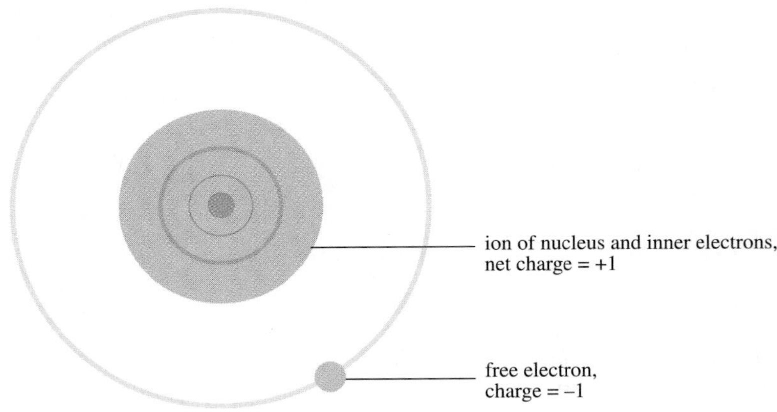

FIG. 2.3
Copper atom (in solid state) ready to conduct electricity.

In solid copper, each nucleus with its 28 inner electrons forms an ion (Fig. 2.3) with a charge of +1; these ions are arranged in an orderly structure. Between the ions, the detached electrons form a "sea" of negative charge that can be made to move along by electric forces.

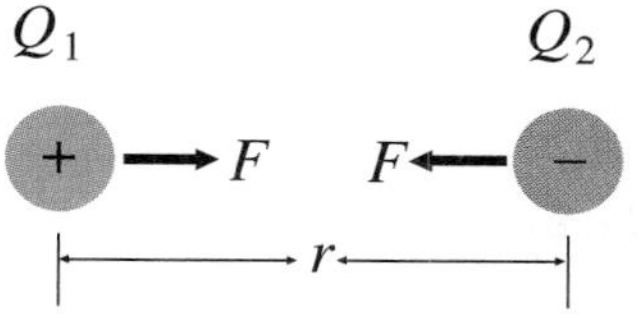

FIG. 2.4
Coulomb's law of electric force between charges.

Electric Force and Charge

Electric forces were first observed in the static electricity that is produced by friction. Even a small proportion of electrons transferred from the surface of one object to that of another results in a force of attraction between the two objects. The quantity of charge involved can be defined in terms of that force. The relation is known as **Coulomb's law**, Eq. (2.1) and Fig. 2.4.

The electric force, F, between two spheres of charges Q_1 and Q_2, separated by a distance r (between centres) is

$$F = \frac{kQ_1Q_2}{r^2} \quad \text{(newtons, N)} \qquad (2.1)$$

The unit of charge is the **coulomb** (C) (see Fig. 2.5). When the two charges are equal and separated by 1 m, and the force between them is 9.0×10^9 N, then the magnitude of each charge is defined to be 1 C. Thus, in Coulomb's law, F is in newtons, Q is in coulombs, r is in metres, and $k = 9.0 \times 10^9\ \text{N}\cdot\text{m}^2/\text{C}^2$. Coulomb's law applies equally to forces of attraction between unlike charges and forces of repulsion between like charges.

If you apply Coulomb's law to the electric force of the nucleus on the outermost electron, you will see that the force is weaker than on any of the inner electrons. In solid copper at room temperature, the agitation of molecular motion is sufficient to allow about 2×10^{29} free electrons per cubic metre.

The charge on an electron (or proton) is considered to be one of the fundamental constants of nature, the **elementary charge**. Its value is $1.602\ 19 \times 10^{-19}$ C. Thus, the number of electrons required to make 1 C is 6.24×10^{18}.

LUMINARIES

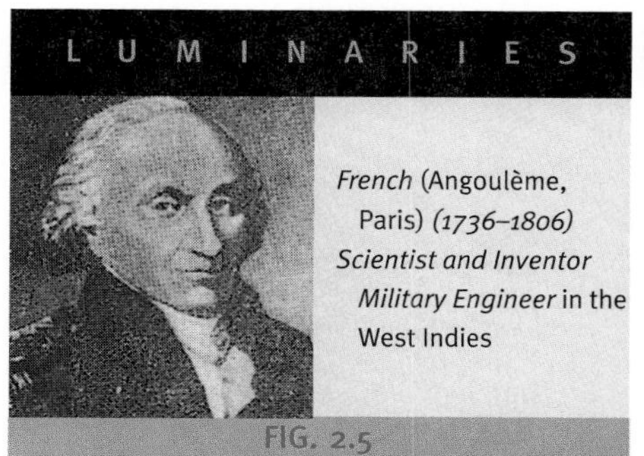

French (Angoulème, Paris) *(1736–1806)*
Scientist and Inventor
Military Engineer in the West Indies

FIG. 2.5
Charles Augustin de Coulomb

Attended the engineering school at Mezieres, the first such school of its kind. Formulated *Coulomb's law*, which defines the force between two electrical charges and is, in fact, one of the principal forces in atomic reactions. Performed extensive research on the friction encountered in machinery and windmills, and the elasticity of metal and silk fibres.

Courtesy of the Smithsonian Institution, Photo No. 52,597

2.3 CURRENT

Consider a short length of copper wire cut with an imaginary perpendicular plane, producing the circular cross-section shown in Fig. 2.6. At room temperature with no external forces applied, there exists within the copper wire the random motion of free electrons created by the thermal energy that the electrons gain from the surrounding medium. When an atom loses its free electron, it acquires a net positive charge and is referred to as a *positive ion.* The free electron is able to move within these positive ions and leave the general area of the parent atom, while the positive ions only oscillate in a mean fixed position. For this reason,

the free electron is the charge carrier in a copper wire or any other solid conductor of electricity.

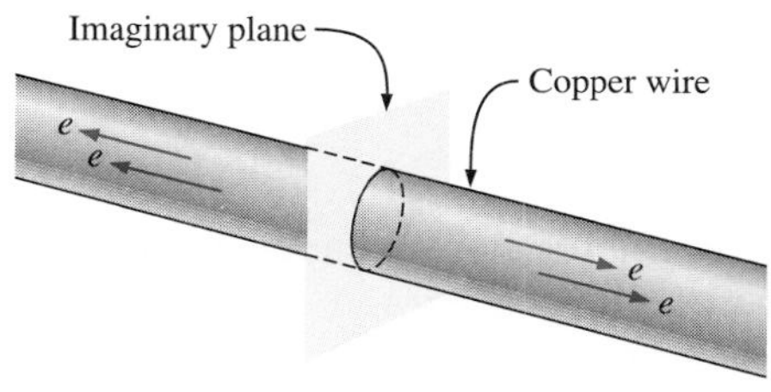

FIG. 2.6
Random motion of electrons in a copper wire with no external "pressure" (voltage) applied.

An array of positive ions and free electrons is shown in Fig. 2.7. Within this array, the free electrons are continually gaining or losing energy by virtue of their changing direction and velocity. Some of the factors responsible for this random motion include (1) the collisions with positive ions and other electrons, (2) the attractive forces of the positive ions, and (3) the force of repulsion that exists between electrons. This random motion of free electrons is such that over a period of time, the number of electrons moving to the right across the circular cross-section of Fig. 2.6 is exactly equal to the number passing over to the left.

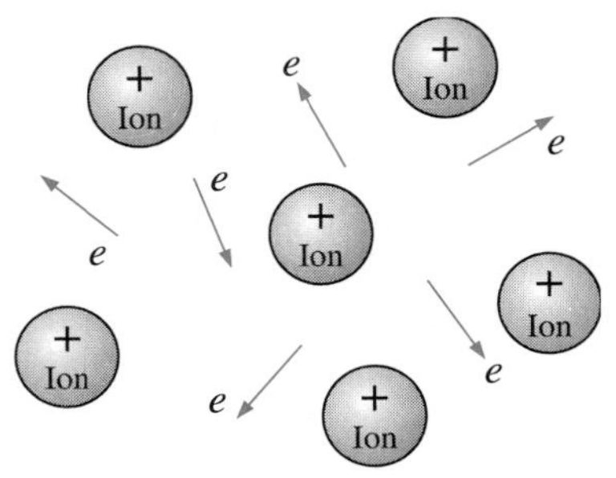

FIG. 2.7
Random motion of free electrons in an atomic structure.

With no external forces applied, the net flow of charge in a conductor in any one direction is zero.

Let us now connect copper wire between two battery terminals and a light bulb, as shown in Fig. 2.8, to create the simplest of electric circuits. The battery places a net positive charge at one terminal and a net negative charge on the other. The instant the final connection is made, the free electrons (of negative charge) will drift toward the positive terminal, while the positive ions left behind in the copper wire will simply oscillate in a mean fixed position. The negative terminal is a "supply" of electrons to be drawn from when the electrons of the copper wire drift toward the positive terminal.

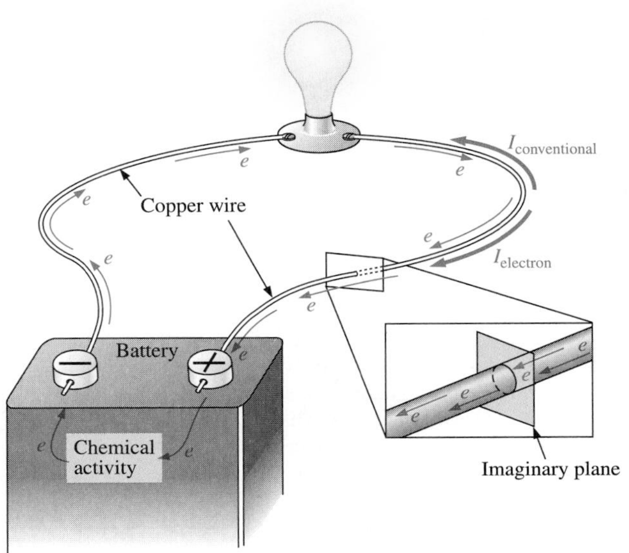

FIG. 2.8
Basic electric circuit.

L U M I N A R I E S

French (Lyon, Paris)
(1775-1836)
Mathematician and Physicist
Professor of Mathematics, École Polytechnique in Paris

FIG. 2.9
André Marie Ampère

On September 18, 1820 introduced a new field of study, *electrodynamics*, devoted to the effect of electricity in motion, including the interaction between currents in adjoining conductors and the interplay of the surrounding magnetic fields. Constructed the first *solenoid* and demonstrated how it could behave like a magnet (the first *electromagnet*). Suggested the name *galvanometer* for an instrument designed to measure current levels.

Courtesy of the Smithsonian Institution, Photo No. 76,524

As shown in Fig. 2.8, the conventional direction for current is chosen to be away from the positive terminal of the battery and toward the negative terminal (that is, opposite to the direction of electron drift).

The rate of flow of charge in a conductor is called the **current**. The unit of current is the **ampere** (A), named in honour of André Marie Ampère (Fig. 2.9). Current is defined by the rate of flow of charge in coulombs per second (C/s).

A current of 1 A is the rate of flow of 1 C/s of charge through any cross-section of a conductor.

From the charge of an electron, you can see that 1 A is the flow of 6.24×10^{18} electrons per second. The symbol I for current was chosen from the French word for current, *intensité*. In equation form, for charge (Q) in colombs, and time (t) in seconds, the current is

$$I = \frac{Q}{t} \qquad \text{(amperes, A)} \tag{2.2}$$

Through algebraic manipulations, the other two quantities can be determined as follows:

$$Q = It \qquad \textit{(coulombs, C)} \tag{2.3}$$

and

$$t = \frac{Q}{I} \qquad \textit{(seconds, s)} \tag{2.4}$$

EXAMPLE 2.1

The charge flowing through the imaginary surface of Fig. 2.8 is 0.16 C every 64 ms. Determine the current in amperes.

Solution: Eq. (2.2):

$$I = \frac{Q}{t} = \frac{0.16\ \text{C}}{64 \times 10^{-3}\ \text{s}} = \frac{160 \times 10^{-3}\ \text{C}}{64 \times 10^{-3}\ \text{s}} = \mathbf{2.50\ A}$$

EXAMPLE 2.2

Determine the time required for 4×10^{16} electrons to pass through the imaginary surface of Fig. 2.8 if the current is 5 mA.

Solution: Determine Q:

$$4 \times 10^{16}\ \cancel{\text{electrons}} \left(\frac{1\ \text{C}}{6.242 \times 10^{18}\ \cancel{\text{electrons}}}\right) = 0.641 \times 10^{-2}\ \text{C}$$

$$= 6.41\ \text{mC}$$

Calculate t [Eq. (2.4)]:

$$t = \frac{Q}{I} = \frac{6.41 \times 10^{-3}\ \text{C}}{5 \times 10^{-3}\ \text{A}} = \mathbf{1.282\ s}$$

A second glance at Fig. 2.8 will reveal that two directions of charge flow have been indicated. One is called *conventional flow,* and the other is called *electron flow.* This text will deal only with conventional flow for a variety of reasons, including the fact that it is the most widely used at educational institutions and in industry, it is employed in the design of all electronic device symbols, and it is the popular choice for all major computer software packages. The flow controversy is a result of an assumption made at the time electricity was discovered that the positive charge was the moving particle in metallic conductors. Be assured that the choice of conventional flow will not create great difficulty and confusion in the chapters to follow. Once the direction of I is established, the issue is dropped and the analysis can continue without confusion.

Safety Considerations

It is important to realize that even small levels of current through the human body can cause serious, dangerous side effects. Experimental results reveal that the human body begins to react to currents of only a few milliamperes. Although most individuals can withstand currents up to perhaps 10 mA for very short periods of time without serious side effects, any current over 10 mA should be considered dangerous. In fact, currents of 50 mA can cause severe shock, and currents of over 100 mA can be fatal. In most cases the skin resistance of the body when dry is high enough to limit the current through the body to relatively safe levels for voltages typically found in the home. However, be aware that when the skin is wet due to perspiration, bathing, etc., or the skin barrier is broken due to an injury, the skin resistance drops dramatically and current levels could rise to dangerous levels for the same voltage shock. In general, therefore, simply remember that *water and electricity don't mix.* Granted, there are safety devices in the home today that are designed specifically for use in wet areas such as the bathroom and kitchen, but accidents can happen. Treat electricity with respect—not fear.

2.4 VOLTAGE

An electric force produces electrical energy when it moves a charge through a distance. This process can be understood by comparison with the action of mechanical forces. Suppose an average force of $F = 50$ N is required to compress a spring by $d = 0.3$ m. The force does an amount of **work**

$$W = F\,d \qquad \text{(joules, J)} \qquad (2.5)$$

of 15 J. That work, or energy, is thereby stored in the spring as **potential energy**. The energy can be transferred to an object placed on the spring when it is released.

Compressing the spring by another distance will create a different amount of potential energy. There is a potential energy difference between the two distances of spring compression.

For electricity, consider the chemical energy found in the materials of a battery. Chemical action in the battery will create a surplus of electrons at the negative terminal, and a deficit of electrons at the positive terminal. These charge accumulations will exert electric forces on electrons in a wire connected between the two terminals (Fig. 2.8). You can say that there is an electric **potential difference** between the two terminals. Just as the spring can transfer energy to an object, the cell can transfer energy to electric charges. The potential difference between the two terminals is defined by the energy transferred per unit charge. In SI units:

A potential difference of 1 V (volt) exists between two points if 1 J of energy is transferred in moving 1 C of charge between the two points.

The volt is named for Alessandro Volta (Fig. 2.10) who developed the voltaic cell in 1800.

Consider a dry cell with a potential difference between its terminals of 1 V. Each coulomb of charge gains 1 J of energy in the cell and transfers 1 J of energy to the lamp. Each coulomb rises through a potential difference of 1 V in the cell (Fig. 2.11), and falls through a potential difference of 1 V in the lamp. The medium of electricity has transformed 1 J of chemical energy in the cell to 1 J of heat and light energy in the lamp. The potential difference is usually called **voltage**.

LUMINARIES

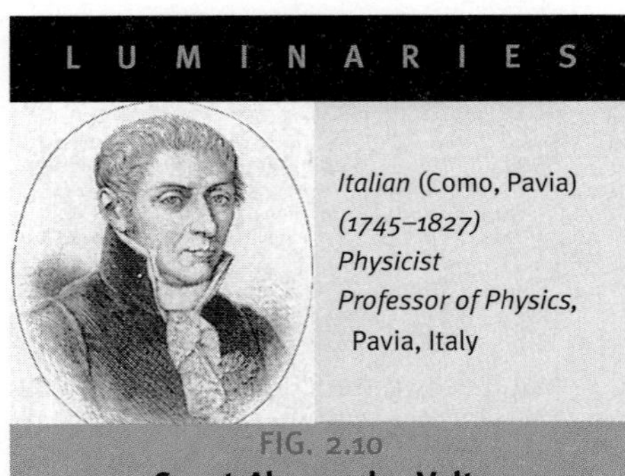

Italian (Como, Pavia)
(1745–1827)
Physicist
Professor of Physics,
Pavia, Italy

FIG. 2.10
Count Alessandro Volta

Began electrical experiments at the age of 18 working with other European investigators. Major contribution was the development of an electrical energy source from chemical action in 1800. For the first time electrical energy was available on a continuous basis and could be used for practical purposes. Developed the first *condenser* known today as the *capacitor.* Was invited to Paris to demonstrate the *voltaic cell* to Napoleon. The International Electrical Congress meeting in Paris in 1881 honoured his efforts by choosing the *volt* as the unit of measure for electromotive force.

Courtesy of the Smithsonian Institution, Photo No. 55,393

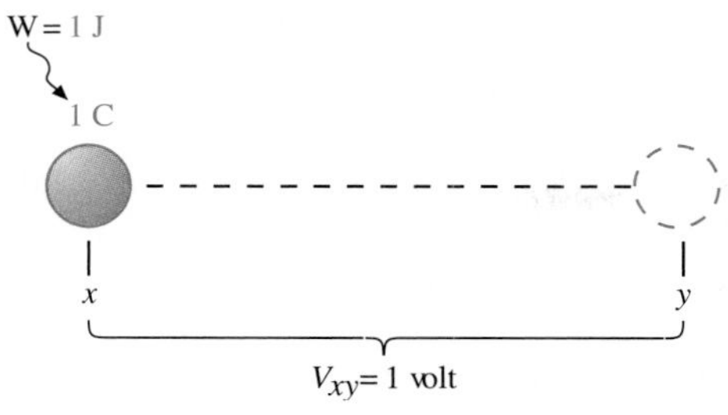

FIG. 2.11
Defining the unit of measurement for voltage.

Note in the above discussion that two points are always involved when talking about voltage or potential difference. In the future, therefore, it is very important to keep in mind that

a potential difference or voltage is always measured between two points in the system. Changing either point may change the potential difference between the two points under investigation.

In general, the potential difference between two points is determined by

$$V = \frac{W}{Q} \qquad \text{(volts, V)} \qquad (2.6)$$

Through algebraic manipulations, we have

$$W = QV \qquad \text{(joules, J)} \qquad (2.7)$$

and

$$Q = \frac{W}{V} \qquad \text{(coulombs, C)} \qquad (2.8)$$

EXAMPLE 2.3

Find the potential difference between two points in an electrical system if 60 J of energy are expended by a charge of 20 C between these two points.

Solution: Eq. (2.6):

$$V = \frac{W}{Q} = \frac{60\text{ J}}{20\text{ C}} = \mathbf{3\ V}$$

EXAMPLE 2.4

Determine the energy expended moving a charge of 50 μC through a potential difference of 6 V.

Solution: Eq. (2.7):

$$W = QV = (50 \times 10^{-6}\text{ C})(6\text{ V}) = 300 \times 10^{-6}\text{ J} = \mathbf{300\ \mu J}$$

To distinguish between sources of voltage (batteries and the like) and losses in potential across dissipative elements, the following notation will be used:

E for voltage sources (volts)
V for voltage drops (volts)

The normal usage for various terms associated with voltage can be seen in the following definitions:

Potential: *The voltage at a point with respect to another point in the electrical system. Typically the reference point is ground, which is at zero potential.*

Potential difference: *The algebraic difference in potential (or voltage) between two points of a network.*

Voltage: *When isolated, like potential, the voltage at a point with respect to some reference such as ground (0 V).*

Voltage difference: *The algebraic difference in voltage (or potential) between two points of the system. A voltage rise is positive and a voltage drop is negative.*

Electromotive force (emf): *The force that establishes the flow of charge (or current) in a system due to the application of a difference in potential. This term is primarily associated with sources of energy.*

The applied potential difference (in volts) of a voltage source in an electric circuit is the "pressure" that causes the flow of charge through the circuit. A mechanical analogy for the applied voltage is the pressure applied to the water in a main. The resulting flow of water through the system is compared to the current through an electric circuit. Without the applied pressure from the spigot, the water will simply sit in the hose, just as the electrons of a copper wire do not have a general direction of drift without an applied voltage.

2.5 FIXED (dc) SUPPLIES

The term **dc** is an abbreviation for **direct current**, found in electrical systems having a *unidirectional* ("one direction") flow of charge.

dc Voltage Sources

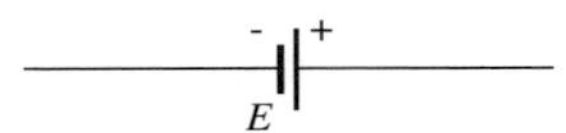

FIG. 2.12
Symbol for a dc voltage source.

The symbol used for all dc voltage supplies in this text appears in Fig. 2.12. The relative lengths of the bars indicate the terminals they represent.

Dc voltage sources can be divided into three broad categories: (1) batteries (chemical action), (2) generators (electromechanical), and (3) power supplies (rectification).

Batteries

General Information Batteries are the most common dc sources. A **battery** consists of a combination of two or more similar **cells.** A cell is the basic source of electrical energy produced by the conversion of chemical or solar energy. All cells can be divided into the **primary** or **secondary** types. The secondary is rechargeable; the primary is not. That is, the chemical reaction of the secondary cell can be reversed to restore its capacity. The two most common rechargeable batteries are the lead-acid unit (used primarily in automobiles) and the nickel-cadmium battery (used in calculators, tools, photoflash units, shavers, and so on). The obvious advantage of the rechargeable unit is the reduced cost associated with not having to continually replace discharged primary cells.

All the cells appearing in this chapter except the solar cell, which absorbs energy from incident light in the form of photons, establish a potential difference at the expense of chemical energy. In addition, each has a positive and a negative electrode and an electrolyte to complete the circuit between electrodes within the battery. The electrolyte is the contact element and the source of ions for conduction between the terminals.

Alkaline and Lithium–Iodine Primary Cells The popular alkaline primary battery uses a powdered zinc anode (+); a potassium (alkali metal) hydroxide electrolyte; and a manganese dioxide, carbon cathode (−) as shown in Fig. 2.13(a). In particular, note in Fig. 2.13(b) that the larger the cylindrical unit, the higher the current capacity. The lantern is designed primarily for long-term use. Figure 2.14 shows two lithium–iodine primary units used in devices where frequent replacement is inconvenient.

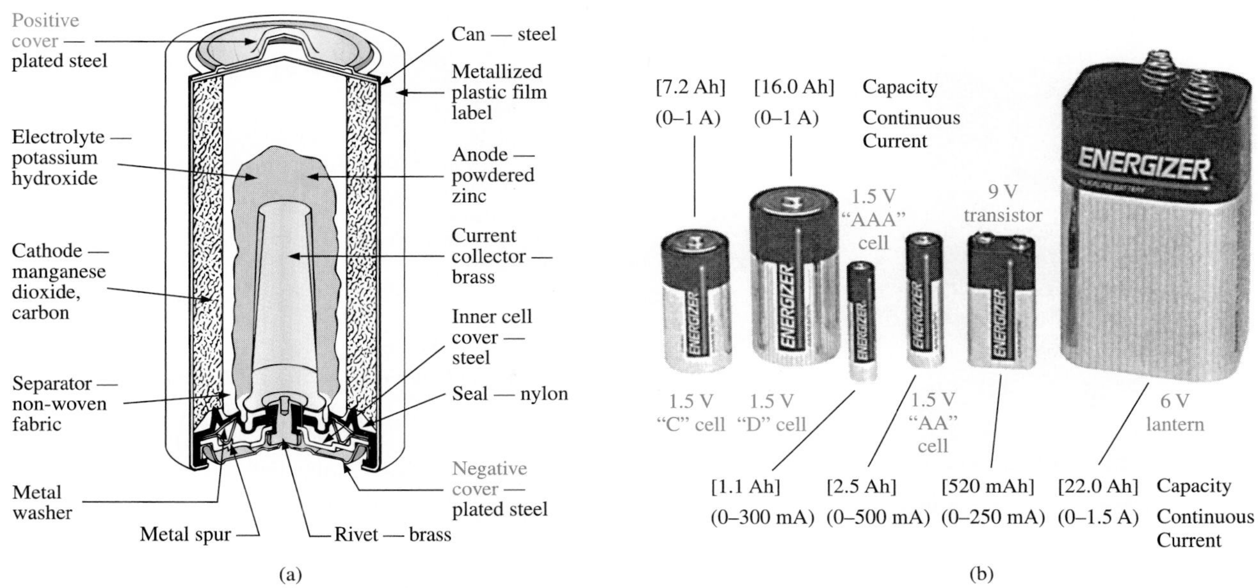

FIG. 2.13
(a) Cutaway of cylindrical Energizer alkaline cell; (b) Eveready Energizer primary cells. (Courtesy of Eveready Battery Company, Inc.)

FIG. 2.14
Lithium–iodine primary cells. (Courtesy of Catalyst Research Corp.)

Nickel–Cadmium Secondary Cell The nickel–cadmium battery is a rechargeable battery that has been receiving enormous interest and development in recent years. For applications such as flashlights, shavers, portable televisions, power drills, and so on, the nickel–cadmium (Ni-Cad) battery of Fig. 2.15 is the secondary battery of choice. Although the current levels are lower, the period of continuous drain is usually longer. A typical nickel–cadmium battery can survive over 1000 charge/discharge cycles over a period of time that can last for years.

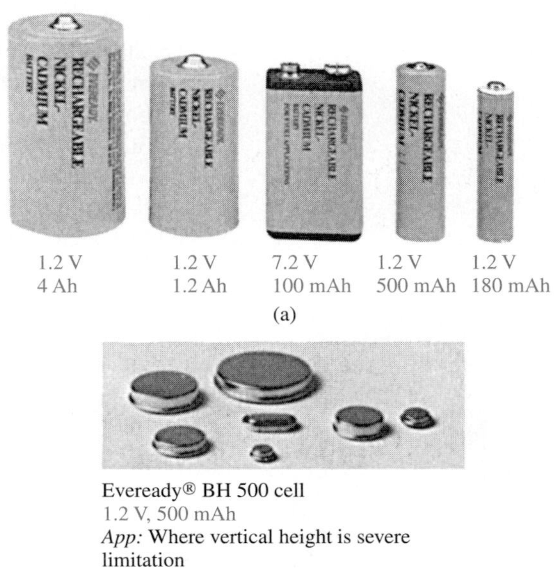

FIG. 2.15
Rechargeable nickel–cadmium batteries. (Courtesy of Eveready Batteries.)

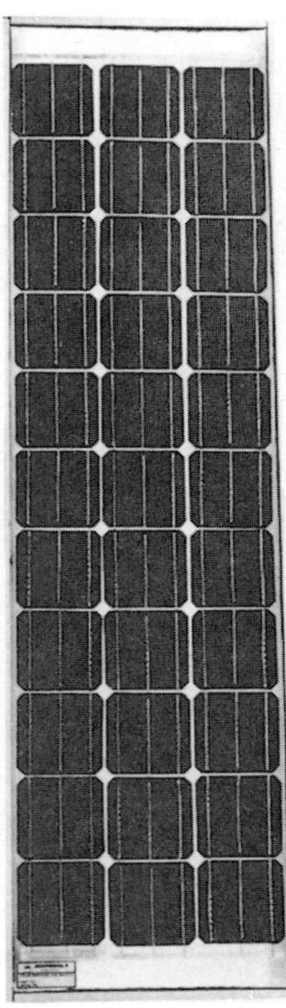

40-W high-density solar module: 100-m × 100-mm square cells are used to provide maximum power in a minimum of space. The 33 series-cell module provides a strong 12-V battery charging current for a wide range of temperatures (−40°C to 60°C).

FIG. 2.16
Solar module. (Courtesy of Motorola Semiconductor Products.)

Solar Cell A high-density, 40-W solar cell appears in Fig. 2.16 with some of its associated data and areas of application. Since the maximum available wattage in an average bright sunlit day is 100 mW/cm^2 and conversion efficiencies are currently between 10% and 14%, the maximum available power per square centimetre from most commercial units is between 10 mW and 14 mW. For a square metre, however, the return would be 100 W to 140 W. A more detailed description of the solar cell will appear in your electronics courses. For now it is important to realize that a fixed illumination of the solar cell will provide a fairly steady dc voltage for driving various loads, from watches to automobiles.

Generators

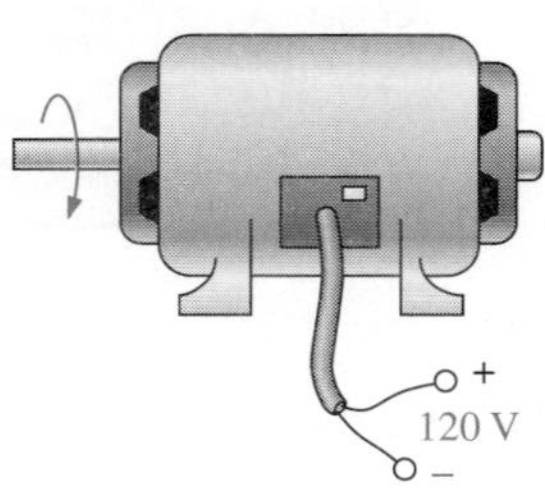

FIG. 2.17
dc generator.

The dc generator is a device that converts mechanical energy to electrical energy (Fig. 2.17). When the shaft of the generator is rotating at the nameplate speed due to the applied torque of some external source of mechanical power, a voltage of rated value will appear across the external terminals. The terminal voltage and power-handling capabilities of the dc generator are typically higher than those of most batteries, and its lifetime is determined only by its construction. Commercially used dc generators are typically of the 120-V or 240-V variety. For the purposes of this text, no distinction will be made between the symbols for a battery and a generator.

Power Supplies

The dc supply encountered most often in the laboratory uses the *rectification* and *filtering* processes to obtain a steady dc voltage. Both processes will be covered in detail in your basic electronics courses. A dc laboratory supply of this type appears in Fig. 2.18.

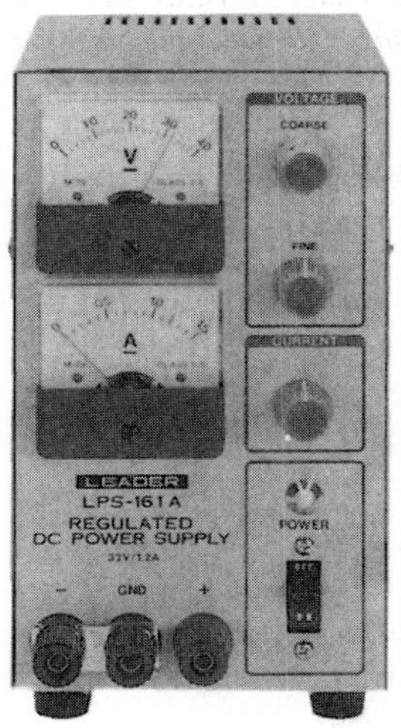

FIG. 2.18
dc laboratory supply. (Courtesy of Leader Instruments Corporation.)

Most dc laboratory supplies have a regulated, adjustable voltage output with three available terminals, as indicated in Figs. 2.18 and 2.19(a). The symbol for ground or zero potential (the reference) is also shown in Fig. 2.19(a). If 10 V above ground potential are required, then the connections are made as shown in Fig. 2.19(b). If 15 V below ground potential are required, then the connections are made as shown in Fig. 2.19(c). If connections are as shown in Fig. 2.19(d), we say we have a "floating" voltage of 5 V since the reference level is not included. The configuration of Fig. 2.19(d) is not often used since it fails to protect the operator by providing a direct low-resistance path to ground and to establish a common ground for the system. In any case, the positive and negative terminals must be part of any circuit configuration.

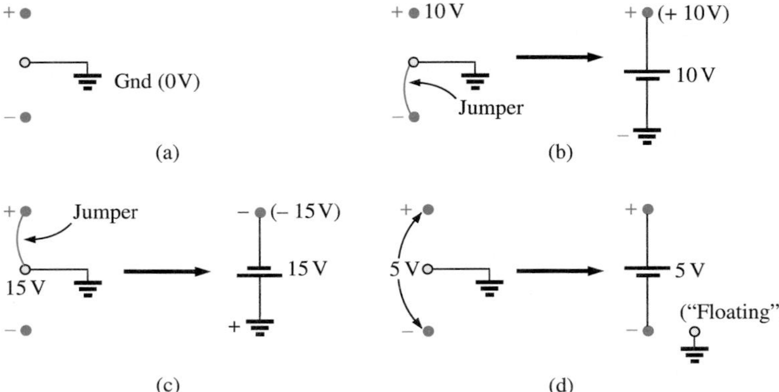

FIG. 2.19
dc laboratory supply: (a) available terminals; (b) positive voltage with respect to (w.r.t.) ground; (c) negative voltage w.r.t. ground; (d) floating supply.

dc Current Sources

The wide variety of types of, and applications for, the dc voltage source have made a rather familiar device, the characteristics of which are understood, at least basically, by the layperson. For example, it is common knowledge that a 12-V car battery has a terminal voltage (at least approximately) of 12 V, even though the current drain by the automobile may vary under different operating conditions. In other words,

a dc voltage source will provide, ideally, a fixed terminal voltage, even though the current demand from the electrical/electronic system may vary,

as shown in Fig. 2.20(a). A dc current source

will supply, ideally, a fixed current to an electrical/electronic system, even though there may be variations in the terminal voltage as determined by the system,

as shown in Fig. 2.20(b).

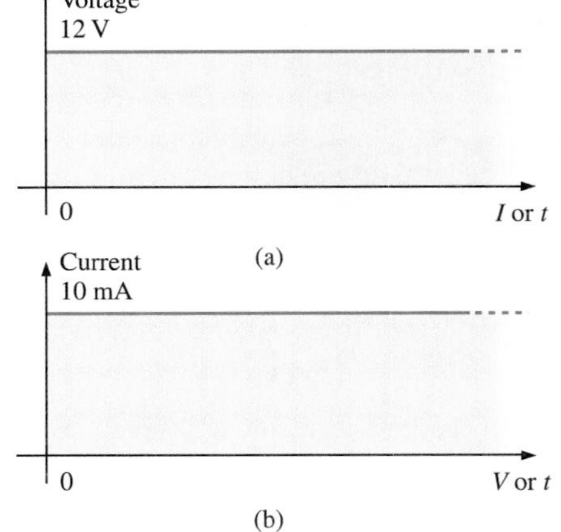

FIG. 2.20
Terminal characteristics: (a) ideal voltage source; (b) ideal current source.

2.6 CONDUCTORS AND INSULATORS

Different wires placed across the same two battery terminals will allow different amounts of charge to flow between the terminals. Many factors, such as the density, mobility, and stability characteristics of a material, account for these variations in charge flow. In general, however, *conductors are those materials that permit a generous flow of electrons with very little external force (voltage) applied.*

In addition,

The atoms of good conductors typically have only one electron in their outermost shell.

Since copper is used most often, it serves as the standard of comparison for the relative conductivity in Table 2.1. Note that aluminum, which has seen some commercial use, has only 61% of the conductivity level of copper, but keep in mind that this must be weighed against the cost and weight factors.

TABLE 2.1
Relative conductivity of various materials.

Metal	Relative Conductivity (%)
Silver	105
Copper	100
Gold	70.5
Aluminum	61
Tungsten	31.2
Nickel	22.1
Iron	14
Constantan	3.52
Nichrome	1.73
Calorite	1.44

Insulators are those materials that have very few free electrons and require a large applied potential (voltage) to establish a measurable current level.

A common use of insulating material is for covering current-carrying wire that would be dangerous if not insulated. Power-line repair people wear rubber gloves and stand on rubber mats as safety measures when working on high-voltage transmission lines. A number of different types of insulators and their applications appear in Fig. 2.21.

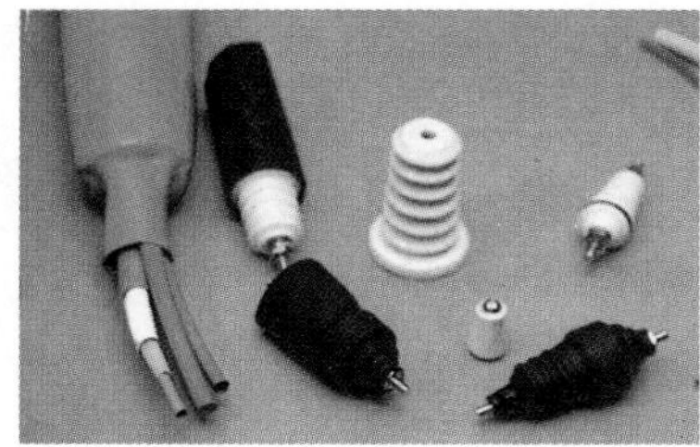

FIG. 2.21
Different types of insulators. (Photo courtesy of Daburn Electronics & Cable Corp.)

It must be pointed out that even the best insulator will break down (permit charge to flow through it) if a sufficiently large potential is applied across it. The breakdown strengths of some common insulators are listed in Table 2.2. According to this table, for insulators with the same geometric shape, it would require 270/30 = 9 times as much potential to pass current through rubber as compared to air and approximately 67 times as much voltage to pass current through mica as through air.

TABLE 2.2
Breakdown strength of some common insulators.

Material	Average Breakdown Strength (kV/cm)
Air	30
Porcelain	70
Oils	140
Bakelite	150
Rubber	270
Paper (paraffin-coated)	500
Teflon	600
Glass	900
Mica	2000

2.7 SEMICONDUCTORS

A **semiconductor** has fewer free electrons than a conductor, but more than an insulator. Materials such as silicon and germanium are semiconductors that are used in diodes and transistors. By varying their physical circumstances, they can control currents in the circuits that use them. Semiconductors are at the heart of the integrated circuits so important in computers and many control devices.

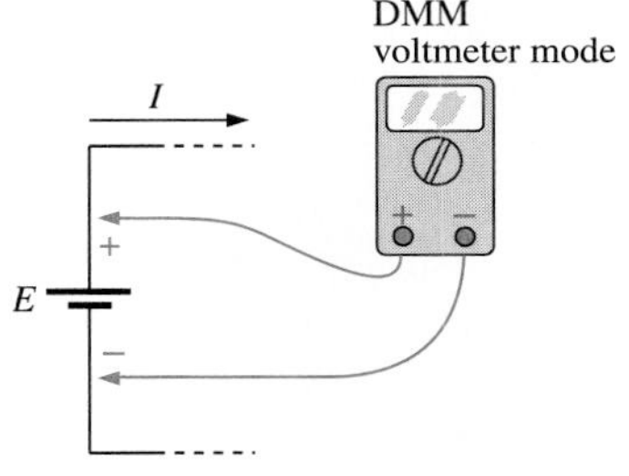

FIG. 2.22
Voltmeter connection for an up-scale (+) reading.

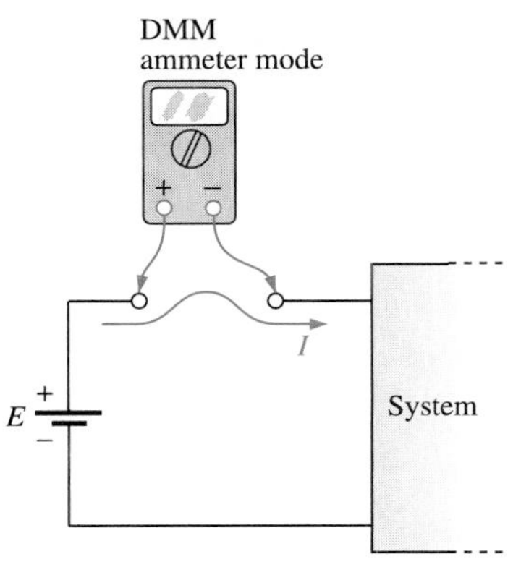

FIG. 2.23
Ammeter connection for an up-scale (+) reading.

2.8 AMMETERS AND VOLTMETERS

It is important to be able to measure the current and voltage levels of an operating electrical system to check its operation, isolate malfunctions, and investigate effects impossible to predict on paper. As the names imply, *ammeters* are used to measure current levels, and *voltmeters,* the potential difference between two points. If the current levels are usually of the order of milliamperes, the instrument will typically be referred to as a milliammeter, and if the current levels are in the microampere range, as a microammeter. Similar statements can be made for voltage levels. Throughout the industry, voltage levels are measured more frequently than current levels primarily because voltmeters do not require that the network connections be disturbed.

The potential difference between two points can be measured by simply connecting the leads of the meter *across the two points,* as indicated in Fig. 2.22. An up-scale reading is obtained by placing the positive lead of the meter to the point of higher potential of the network and the common or negative lead to the point of lower potential. The reverse connection will result in a negative reading or a below-zero indication.

Ammeters are connected as shown in Fig. 2.23. Since ammeters measure the rate of flow of charge, the meter must be placed in the network so that the charge will flow through the meter. The only way this can be accomplished is to open the path in which the current is to be measured and place the meter between the two resulting terminals. For the configuration of Fig. 2.23, the voltage source lead (+) must be disconnected from the system and the ammeter inserted as shown. An up-scale reading will be obtained if the polarities on the terminals of the ammeter are such that the current of the system enters the positive terminal.

The introduction of any meter into an electrical/electronic system raises a concern about whether the meter will affect the behaviour of the system. This question and others will be examined in Chapters 5 and 6 after additional terms and concepts have been introduced. For the moment, let it be said that since voltmeters and ammeters do not have internal sources, they will affect the network when introduced for measurement purposes. They are both designed, however, so that the impact is minimized.

There are instruments designed to measure just current or just voltage levels. However, the most common laboratory meters include the *volt-ohm-milliammeter* (VOM) of Fig. 2.24, and the *digital multimeter* (DMM) of Fig. 2.25. Both instruments will measure voltage and current and a third quantity, resistance, to be introduced in the next chapter. The VOM uses an analog scale, which requires interpreting the position of a pointer on a continuous scale, while the DMM provides a display of numbers with decimal point accuracy determined by the chosen scale. Comments on the characteristics and use of various meters will be made later in the text.

FIG. 2.24
Volt-ohm-milliammeter (VOM) analog meter. (Courtesy of Simpson Electric Co.)

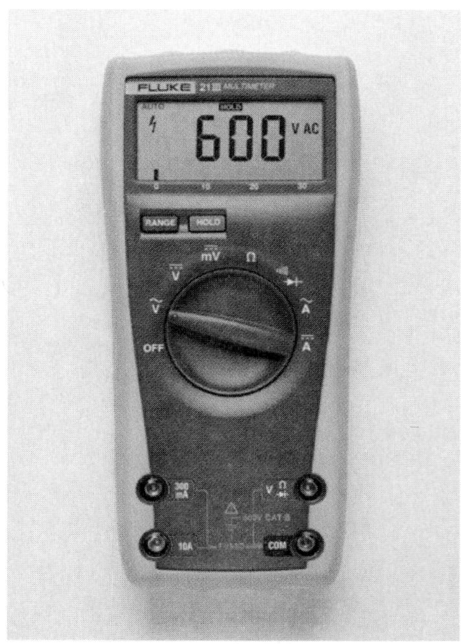

FIG. 2.25
Digital multimeter (DMM). (Courtesy of John Fluke Mfg. Co. Inc.)

Practical Business Applications

Wouldn't you rather be hiking?

Saturday morning, and you are planning a hike on the Bruce Trail. But the grey and stormy skies may ruin your plans. Then as you think about the homework for your course in electricity, you realize that today's rain is part of a real-life example of a system that behaves in a manner similar to all electric circuits.

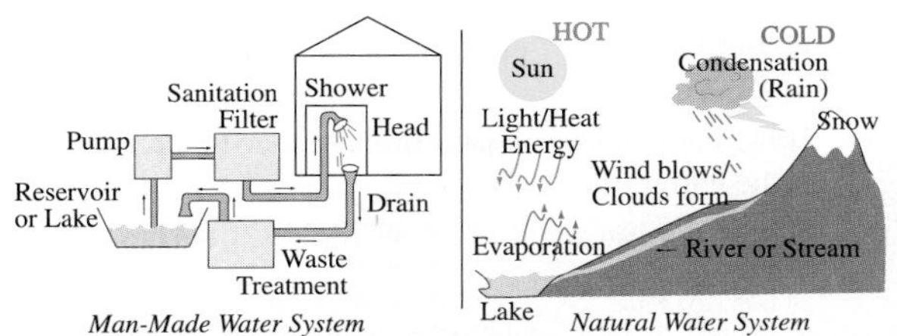

Wouldn't you rather be hiking?

The rain is just one phase of our earth's natural water cycle. This cycle begins with the evaporation of water from lakes, and oceans, up into the clouds. Water in clouds stores energy, often referred to as potential energy. The energy stored in the clouds is comparable to the storage of electrical energy in batteries. When cooled, the water in the clouds will give this energy back to earth in the form of rain. These waters then return to the lakes and oceans, flowing down the streams and rivers.

Oh no! The local news says the rain will continue all day. So, maybe a healthy swim in the local public swimming pool is a better idea. Unable to get your outstanding homework assignments out of your head, you also remember the analogy between electric circuits and circulating water systems.

In a swimming pool, a pump is used to push the water around the system from the pool's inlets, through filtration and heating equipment and back to the pool. Comparing this artificial system to the natural water cycle, you realize that the pump action replaces the evaporative action that lifts water from the lakes and oceans up into the clouds. Pipes are used to contain the water and to move the water through the pool, pump, filter, and heater.

In addition, the pool provides the necessary reservoir of water. Similarly, our domestic water supply illustrates many of the same features, from initial reservoir of water (often our natural lakes and oceans) to filtration equipment. Finally, the piping systems carry the water to and from our homes.

Later, while doing your laps in the pool, your mind turns back to these water-based analogies for electricity. Four basic properties of electrical systems jump into your mind—*current, voltage, resistance,* and *closed circuits.*

Water flowing through rivers and streams, or through pipes in manufactured systems, is often described in terms of the current that is observed. This water current is exactly like electrical current. The current in a water system is

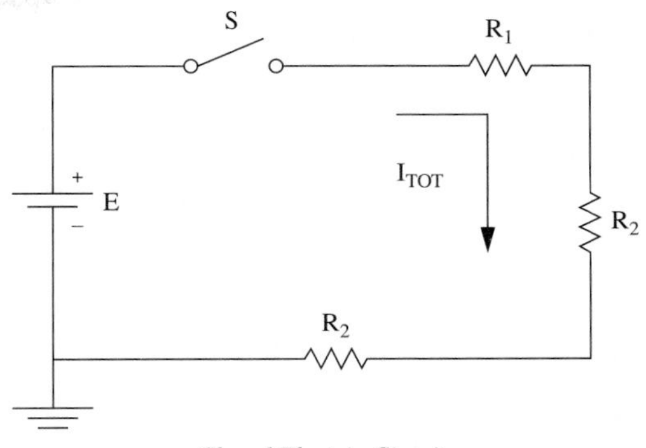

Closed Electric Circuit

defined as the volume of water that flows past any given point over some finite amount of time, i.e., the rate or speed of flow of water or more simply just the flow rate. Similarly, electrical current is the quantity of electric charge that flows past any given point in an electrical circuit in a finite time period.

From your studies, you recall that voltage is also referred to as electromotive force, EMF. This property of electric circuits is comparable to the pressure that is produced by a pump in a manufactured water circuit. Thus, voltage is comparable to water pressure. It represents the degree of push behind each of the moving electrical charges in any circuit.

Water flowing in a stream or river experiences resistance from the frictional effect of the river's bottom and sides. In an artificial water system, the moving water also experiences resistance from the friction produced by the walls of the pipes. Thus in water systems, smaller streams, rivers, or pipes produce more friction relative to larger streams, rivers, or pipes—the net effect is a reduced current or flow rate. This analogy reminds you that electrical resistance can be viewed as a kind of friction opposed to the flow of electrical charge.

The water tap that controls the flow of water in your home is the same as a switch that controls the flow of electrons in an electric circuit. And finally, you realize that all of these systems will break down and become useless unless you have a complete and closed circuit. This water-oriented analogy also demonstrates the need for a complete closed circuit in any type of water system—whether natural or artificial. Without a closed circular flow, the pools, reservoirs, lakes, oceans and even the clouds would, sooner or later, become empty. All related actions would therefore cease.

All electrical circuits have the same requirement. If the flowing electrical charges do not, sooner or later, return to the source from which they came then all electrical current flow will stop.

Wow, are you ever glad you went for that swim! Otherwise, you'd never learn electricity, right?

PROBLEMS

SECTION 2.2 Atoms and Their Structure

1. Calculate the force of repulsion between two adjacent protons in an atomic nucleus, with a distance between their centres of 1.0×10^{-15} m.

2. Find the force of attraction between a proton and an electron separated by a distance equal to the radius of the smallest orbit followed by an electron (5×10^{-11} m) in a hydrogen atom.

3. Plot the force of attraction in newtons between the charges Q_1 and Q_2 in Fig. 2.26 when
a. 1 m **b.** 5 m
c. 8 m **d.** 10m
(Note how quickly the force drops with an increase in *r*.)

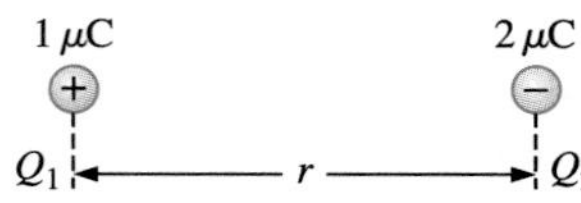

FIG. 2.26
Problem 3.

***4.** Plot the force of repulsion in newtons between Q_1 and Q_2 in Fig. 2.27 when
a. 1 m **b.** 0.5 m
c. 0.25 m **d.** 0.125 m

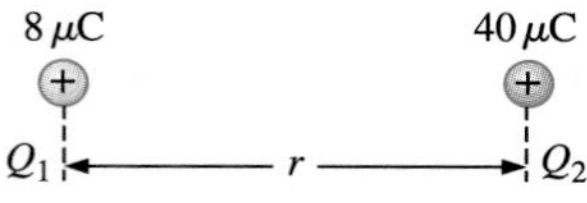

FIG. 2.27
Problem 4.

***5.** Plot the force of attraction (in newtons) versus separation (in metres) for two charges of 2 mC and -4 μC. Set r to 0.5 m and 1 m, followed by 1-m intervals to 10 m. Comment on the shape of the curve. Is it linear or nonlinear? What does it tell you about the force of attraction between charges as they are separated? What does it tell you about any function plotted against a squared term in the denominator?

6. Determine the distance between two charges of 20 μC if the force between the two charges is 3.6×10^4 N.

***7.** Two charged bodies, Q_1 and Q_2, when separated by a distance of 2 m, experience a force of repulsion equal to 1.8 N.
a. What will the force of repulsion be when they are 10 m apart?
b. If the ratio $Q_1/Q_2 = 1/2$, find Q_1 and Q_2 ($r = 10$ m).

SECTION 2.3 Current

8. Find the current in amperes if 650 C of charge pass through a wire in 50 s.

9. If 465 C of charge pass through a wire in 2.5 min, find the current in amperes.

10. If a current of 40 A exists for 1 min, how many coulombs of charge have passed through the wire?

11. How many coulombs of charge pass through a lamp in 2 min if the current is constant at 750 mA?

12. If the current in a conductor is constant at 2 mA, how much time is required for 4600×10^{-6} C to pass through the conductor?

13. If $21.847 \times 10^{+18}$ electrons pass through a wire in 7 s, find the current.

14. How many electrons pass through a conductor in 1 min if the current is 1 A?

15. Will a fuse rated at 1 A "blow" if 86 C pass through it in 1.0 min?

***16.** If $0.784 \times 10^{+18}$ electrons pass through a wire in 643 ms, find the current.

***17.** Which would you prefer?
a. A penny for every electron that passes through a wire in 0.01 μs at a current of 2mA, or
b. A dollar for every electron that passes through a wire in 1.5 ns if the current is 100 μA.

SECTION 2.4 Voltage

18. What is the voltage between two points if 96 mJ of energy are required to move 50×10^{18} electrons between the two points?

19. If the potential difference between two points is 42 V, how much work is required to bring 6 C from one point to the other?

20. Find the charge Q that requires 1205 J of energy to be moved through a potential difference of 24 V.

21. How much charge passes through a battery of 22.5 V if the energy expended is 90 J?

22. If a conductor with a current of 200 mA passing through it converts 40 J of electrical energy into heat in 30 s, what is the potential drop across the conductor?

***23.** Charge is flowing through a conductor at the rate of 420 C/min. If 742 J of electrical energy are converted to heat in 30 s, what is the potential drop across the conductor?

SECTION 2.5 Fixed (dc) Supplies

24. Discuss briefly the difference among the three types of dc voltage supplies (batteries, rectification, and generators).

25. Suggest an application where each of the three types of dc supply could be used to advantage.

26. Compare the characteristics of a dc current source with those of a dc voltage source. How are they similar and how are they different?

SECTION 2.6 Conductors and Insulators

27. Discuss the properties of the atomic structure of gold that make it a good conductor.

28. Name two materials not listed in Table 2.1 that are good conductors of electricity.

29. Explain the terms *insulator* and *breakdown strength.*

30. List three uses of insulators not mentioned in Section 2.6.

SECTION 2.7 Semiconductors

31. What is a semiconductor? How does it compare with a conductor and an insulator?

32. Consult a semiconductor electronics text and note the extensive use of germanium and silicon semiconductor materials. Report the main characteristics of each material.

SECTION 2.8 Ammeters and Voltmeters

33. What are the significant differences in the way ammeters and voltmeters are connected?

34. Explain why a voltmeter must be connected across two points in an electric circuit in order to obtain a correct reading.

35. If an ammeter reads 2.5 A for a period of 4 min, determine the charge that has passed through the meter.

36. Between two points in an electric circuit, a voltmeter reads 12.5 V for a period of 20 s. If the current measured by an ammeter is 10 mA, determine the energy expended and the charge that flowed between the two points.

3

Resistance

Outline

Learning Outcomes

After completing this chapter you will be able to

- explain the nature of resistance
- describe the relationship between resistance and conductance
- determine the resistance of conductors with various cross-sectional areas and with different lengths
- explain the effect of temperature on resistance and solve problems involving temperature change
- explain the difference between fixed resistors and variable resistors
- determine the size and tolerance of commercial resistors using the resistor colour code
- describe the characteristics of various types of resistive devices, such as thermistors, varistors, and photoconductive cells

Resistance is a property of electrical circuits. In this chapter you will learn how temperature affects resistance and how a variety of resistive devices operates.

3.1 INTRODUCTION

The flow of charge through any material encounters an opposing force similar in many respects to mechanical friction. The collisions of electrons with the ions of the material converts electrical energy to heat. The opposition to the current is called **resistance**. The unit of measurement of resistance is the **ohm**, for which the symbol is Ω, the capital Greek letter *omega*. The circuit symbol for resistance appears in Fig. 3.1 with the graphic abbreviation for resistance (R).

FIG. 3.1
Resistance symbol and notation.

The resistance of any material with a uniform cross-sectional area is determined by the following four factors:

1. *Material*
2. *Length*
3. *Cross-sectional area*
4. *Temperature*

Each material, with its unique molecular structure, will react differentially to pressures to establish current flow. Conductors that permit a generous flow of charge with little external pressure will have low resistance, while insulators will have high resistance.

As one might expect, the longer the path the charge must pass through, the higher the resistance, and the larger the area, the lower the resistance. Resistance is thus directly proportional to length and inversely proportional to area.

For most conductors, as the temperature increases, the increased motion of the ions within the molecular structure makes it increasingly difficult for the free electrons to pass through, and the resistance increases.

At a fixed temperature of 20°C (room temperature), the resistance is related to the other three factors by

$$R = \rho\frac{l}{A} \qquad \text{(ohms, } \Omega\text{)} \qquad (3.1)$$

where ρ (Greek letter *rho*) is a characteristic of the material called the **resistivity**, l is the length of the sample, and A is the cross-sectional area of the sample.

The units of measurement for Eq. (3.1) depend on the application. For most applications (e.g., integrated circuits) the units are as defined in Section 3.2.

3.2 RESISTANCE: METRIC UNITS

The design of resistive elements for various areas of application, including circular conductors, thin-film resistors and integrated circuits, uses metric units for the quantities of Eq. (3.1). In SI units, the resistivity would be measured in ohm-metres, the area in square metres, and the length in metres. However, the metre is generally too large a unit of measure for most applications, and so the centimetre is usually used. The resulting dimensions for Eq. (3.1) are therefore

ρ, ohm-centimetres
l, centimetres
A, square centimetres

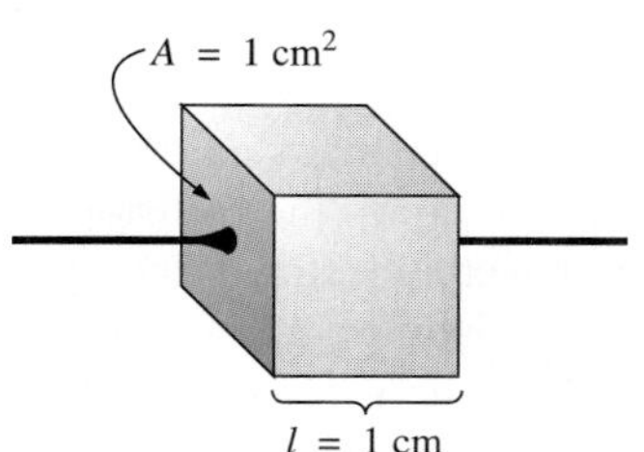

FIG. 3.2
Defining ρ in ohm-centimetres.

TABLE 3.1
Resistivity (r) of various materials in ohm-centimetres.

Material	ρ @ 20°C
Silver	1.645×10^{-6}
Copper	1.723×10^{-6}
Gold	2.443×10^{-6}
Aluminum	2.825×10^{-6}
Tungsten	5.485×10^{-6}
Nickel	7.811×10^{-6}
Iron	12.299×10^{-6}
Tantalum	15.54×10^{-6}
Nichrome	99.72×10^{-6}
Tin oxide	250×10^{-6}
Carbon	3500×10^{-6}

The units for ρ can be derived from

$$\rho = \frac{RA}{l} = \frac{\Omega\cdot\text{cm}^2}{\text{cm}} = \Omega\cdot\text{cm}$$

The resistivity of a material is actually the resistance of a sample such as the one appearing in Fig. 3.2. Table 3.1 provides a list of values of ρ in ohm-centimetres. Note that the area is expressed in square centimetres. For circular wires, the area is related to diameter d by $A = \pi d^2/4$.

EXAMPLE 3.1

What is the resistance of a 100 m length of #22 AWG copper conductor?

Solution:

$$\rho = 1.723\times 10^{-6}\ \Omega\cdot\text{cm}$$

From Table 3.2, the cross-sectional area is 0.0033 cm
From Equation 3.1:

$$R = \rho\frac{l}{A}$$

$$= \frac{(1.723\times 10^{-6}\Omega\cdot\text{cm})\ (10\,000\text{ cm})}{0.0033\text{cm}^2}$$

$$= 5.2\ \Omega$$

EXAMPLE 3.2

An undetermined number of metres have been used from a carton of #14 AWG wire. If the measured resistance is 1.6Ω, determine the length of wire remaining in the carton.

Solution:

$$\rho = 1.723\times 10^{-6}\ \Omega\cdot\text{cm}$$

From Table 3.2, the cross-sectional area is 0.0211 cm^2
From Equation 3.1:

$$R = \rho\frac{l}{A}$$

$$l = \frac{RA}{\rho}$$

$$= \frac{(1.6\ \Omega)\ (0.0211\text{ cm}^2)}{1.723\times 10^{-6}\ \Omega\cdot\text{cm}}$$

$$= 19\,594\text{ cm}$$

$$= 196\text{ m}$$

The resistivity in IC design is typically in ohm-centimetre units, although tables often provide ρ in ohm-metres or microhm-centimetres. The conversion factor between ohm-centimetres and ohm-metres is the following:

$$1.723 \times 10^{-6}\ \Omega\cdot\cancel{\text{cm}}\left[\frac{1\text{ m}}{100\ \cancel{\text{cm}}}\right] = 1.723 \times 10^{-8}\ \Omega\cdot\text{m}$$

or the value in ohm-metres is 1/100 the value in ohm-centimetres, and

$$\rho\ (\Omega\cdot\text{m}) = \left(\frac{1}{100}\right) \times (\text{value in }\Omega\cdot\text{cm})$$

Similarly:

$$\rho\ (\mu\Omega\cdot\text{cm}) = (10^6) \times (\text{value in }\Omega\cdot\text{cm})$$

For comparison purposes, typical values of ρ in ohm-centimetres for conductors, semiconductors, and insulators are provided in Table 3.2.

TABLE 3.2
Comparing levels of ρ in Ω cm.

Conductor	Semiconductor	Insulator
Copper 1.723×10^{-6}	Ge 50 Si 200×10^3 GaAs 70×10^6	In general: 10^{15}

In particular, note the magnitude of difference between conductors and insulators (10^{21})—a huge difference. Resistivities of semiconductors cover a wide range. However, they all differ by a factor of a million or more from both conductors and insulators.

3.3 WIRE TABLES

The wire table was designed primarily to standardize the size of wire produced by manufacturers throughout North America. As a result, the manufacturer has a larger market and the consumer knows that standard wire sizes will always be available. The table was designed to assist the user in every way possible; in Canada, it usually includes data such as the cross-sectional area in cm^2, diameter in mm, ohms per 305 m at 20°C, and weight per 305 m.

The American Wire Gauge (AWG) sizes are given in Table 3.3 for solid round copper wire.

TABLE 3.3
American Wire Gauge (AWG) Sizes.

AWG	Diameter (mm)	Area (cm^2)	Ohms/305m (1000 Ft) @20°C
4/0	1.175	1.084	0.49
3/0	1.046	0.859	0.618
2/0	0.931	0.681	0.78
1/0	0.829	0.5398	0.0983
2	0.658	0.3400	0.1563
4	0.522	0.2140	0.2485
6	0.414	0.1346	0.3951
8	0.328	0.0845	0.6282
10	0.260	0.0531	0.9989
12	0.206	0.0333	1.588
14	0.164	0.0211	2.525
16	0.130	0.0132	4.016
18	0.103	0.0083	6.385
20	0.0816	0.0052	10.15
22	0.0647	0.0033	16.14
24	0.0513	0.0021	25.67
26	0.0407	0.00130	40.81
28	0.0323	0.00082	64.9
30	0.0256	0.00051	103.2
32	0.0203	0.00032	164.1
34	0.0161	0.00020	260.9
36	0.0128	0.00013	414.8
38	0.0101	0.00008	659.6
40	0.00802	0.00005	1049

The chosen sizes have an interesting relationship: For every drop in 3 gauge numbers, the area is doubled; and for every drop in 10 gauge numbers, the area increases by a factor of 10.

Examining Eq. (3.1), we note also that *doubling the area cuts the resistance in half, and increasing the area by a factor of 10 decreases the resistance to 1/10 the original, everything else kept constant.*

3.4 TEMPERATURE EFFECTS

Temperature has a significant effect on the resistance of conductors, semiconductors, and insulators.

Conductors

In conductors there is a large number of free electrons, and any introduction of thermal energy will have little impact on the total number of free carriers. In fact the thermal energy will only increase the intensity of the random motion of the ions within the material and make it increasingly difficult for a general drift of electrons in any one direction to be established. The result is that

for good conductors, an increase in temperature will result in an increase in the resistance level. As a result, conductors have a positive temperature coefficient.

The plot of Fig. 3.3(a) has a positive temperature coefficient.

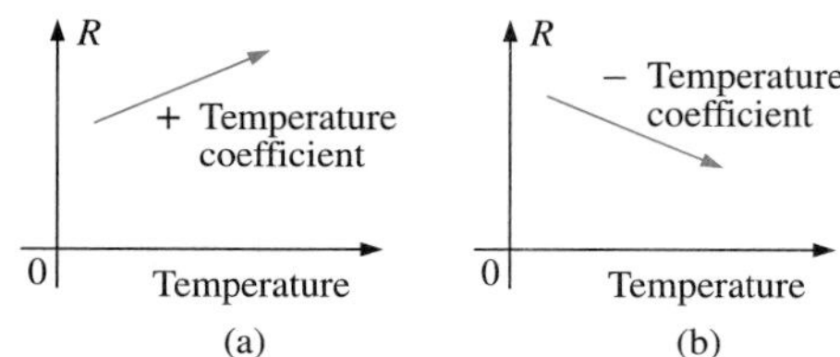

FIG. 3.3
(a) Positive temperature coefficient—conductors; (b) negative temperature coefficient—semiconductors.

Semiconductors

In semiconductors an increase in temperature will give a measure of thermal energy to the system that will result in an increase in the number of free carriers in the material for conduction. The result is that

for semiconductor materials, an increase in temperature will result in a decrease in the resistance level. As a result, semiconductors have negative temperature coefficients.

The thermistor and photoconductive cell of Sections 3.9 and 3.10 of this chapter are excellent examples of semiconductor devices with negative temperature coefficients. The plot of Fig. 3.3(b) has a negative temperature coefficient.

Insulators

As with semiconductors, an increase in temperature will result in a decrease in the resistance of an insulator. The result is a negative temperature coefficient.

Inferred Absolute Temperature

Figure 3.4 reveals that for copper (and most other metallic conductors), the resistance increases almost linearly (in a straight-line relationship) with an increase in temperature. Since temperature can have such a strong effect on the resistance of a conductor, it is important that we have some method of determining the resistance at any temperature within operating limits. An equation for this purpose can be obtained by approximating the curve of Fig. 3.4 by the straight dashed line that intersects the temperature scale at −234.5°C. Although the actual curve extends to *absolute zero* (−273.15°C, or 0 K), the straight-line approximation is quite accurate for the normal operating temperature range. At two different temperatures, t_1 and t_2, the resistance of copper is R_1 and R_2, as indicated on the curve. Using a property of similar triangles, we may develop a mathematical relationship between these values of resistances at different temperatures. Let x equal the distance from −234.5°C to t_1 and y the distance from −234.5°C to t_2, as shown in Fig. 3.4. From similar triangles,

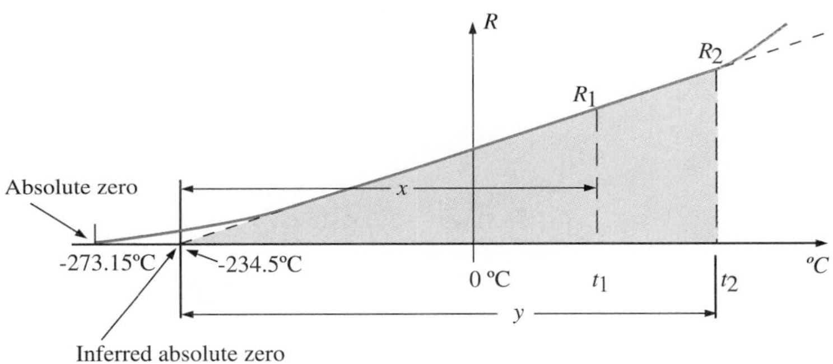

FIG. 3.4
Effect of temperature on the resistance of copper.

$$\frac{x}{R_1} = \frac{y}{R_2}$$

or

$$\frac{234.5 + t_1}{R_1} = \frac{234.5 + t_2}{R_2} \tag{3.2}$$

The temperature of −234.5°C is called the *inferred absolute temperature* of copper. For different conducting materials, the intersection of the straight-line approximation will occur at different temperatures. A few typical values are listed in Table 3.4.

The minus sign does not appear with the inferred absolute temperature on either side of Eq. (3.2) because x and y are the *distances* from −234.5°C to t_1 and t_2, respectively, and therefore are simply magnitudes. For t_1 and t_2 less than zero, x and y are less than −234.5°C and the distances are the differences between the inferred absolute temperature and the temperature of interest.

Equation (3.2) can easily be adapted to any material by inserting the proper inferred absolute temperature. It may therefore be written as follows:

$$\frac{|T| + t_1}{R_1} = \frac{|T| + t_2}{R_2} \tag{3.3}$$

TABLE 3.4
Inferred absolute temperatures.

Material	°C
Silver	−243
Copper	−234.5
Gold	−274
Aluminum	−236
Tungsten	−204
Nickel	−147
Iron	−162
Nichrome	−2 250
Constantan	−125 000

where $|T|$ indicates that the inferred absolute temperature of the material involved is inserted as a positive value in the equation. In general, therefore, associate the sign only with t_1 and t_2.

EXAMPLE 3.3

If the resistance of a copper wire is 50 Ω at 20°C, what is its resistance at 100°C (boiling point of water)?

Solution: Eq. (3.2):

$$\frac{234.5°\text{C} + 20°\text{C}}{50\ \Omega} = \frac{234.5°\text{C} + 100°\text{C}}{R_2}$$

$$R_2 = \frac{(50\ \Omega)(334.5°\text{C})}{254.5°\text{C}} = \mathbf{65.72\ \Omega}$$

EXAMPLE 3.4

If the resistance of a copper wire at freezing (0°C) is 30 Ω, what is its resistance at −40°C?

Solution: Eq. (3.2):

$$\frac{234.5°\text{C} + 0}{30\ \Omega} = \frac{234.5°\text{C} - 40°\text{C}}{R_2}$$

$$R_2 = \frac{(30\ \Omega)(194.5°\text{C})}{234.5°\text{C}} = \mathbf{24.88\ \Omega}$$

EXAMPLE 3.5

If the resistance of an aluminum wire at room temperature (20°C) is 100 mΩ (measured by a milliohmmeter), at what temperature will its resistance increase to 120 mΩ?

Solution: Eq. (3.3):

$$\frac{236°\text{C} + 20°\text{C}}{100\ \text{m}\Omega} = \frac{236°\text{C} + t_2}{120\ \text{m}\Omega}$$

and

$$t_2 = 120\ \text{m}\Omega\left(\frac{256°\text{C}}{100\ \text{m}\Omega}\right) - 236°\text{C}$$

$$t_2 = \mathbf{71.2°C}$$

Temperature Coefficient of Resistance

There is a second popular equation for calculating the resistance of a conductor at different temperatures. Defining

$$\alpha_{20} = \frac{1}{|T| + 20°\text{C}} \qquad (\Omega/°\text{C}/\Omega) \qquad (3.4)$$

as the **temperature coefficient of resistance** at a temperature of 20°C, and R_{20} as the resistance of the sample at 20°C, the resistance R at a temperature t is determined by

$$R = R_{20}[1 + \alpha_{20}(t - 20°\text{C})] \qquad (3.5)$$

TABLE 3.5

Temperature coefficient of resistance for various conductors at 20°C.

Material	Temperature Coefficient (α_{20})
Silver	0.0038
Copper	0.00393
Gold	0.0034
Aluminum	0.00391
Tungsten	0.005
Nickel	0.006
Iron	0.0055
Constantan	0.000008
Nichrome	0.00044

The values of α_{20} for different materials have been evaluated, and a few are listed in Table 3.5.

Equation (3.5) can be written in the following form:

$$\alpha_{20} = \frac{\left(\dfrac{R - R_{20}}{t - 20°\text{C}}\right)}{R_{20}} = \frac{\dfrac{\Delta R}{\Delta T}}{R_{20}}$$

from which the units of Ω/°C/Ω for α_{20} are defined.

Since $\Delta R/\Delta T$ is the slope of the curve of Fig. 3.4, we can conclude that

the higher the temperature coefficient of resistance for a material, the more sensitive the resistance level to changes in temperature.

Looking at Table 3.5, we find that copper is more sensitive to temperature variations than is silver, gold, or aluminum, although the differences are quite small. The slope defined by α_{20} for constantan is so small the curve is almost horizontal.

Since R_{20} of Eq. (3.5) is the resistance of the conductor at 20°C and $t - 20°\text{C}$ is the change in temperature from 20°C, Eq. (3.5) can be written in the following form:

$$R = \rho\frac{l}{A}[1 + \alpha_{20}\,\Delta T] \qquad (3.6)$$

providing an equation for resistance in terms of all the controlling parameters.

PPM/°C

For resistors, as for conductors, resistance changes with a change in temperature. The specification is normally provided in parts per million per degree Celsius (PPM/°C). This gives an immediate indication of the sensitivity level of the resistor to temperature. For resistors, a 5000-PPM level is considered high, whereas 20 PPM is quite low. A 1000-PPM/°C characteristic reveals that a 1° change in temperature will result in a change in resistance equal to 1000 PPM, or 1000/1 000 000 = 1/1000 of its nameplate value—not a significant change for most applications. However, a 10° change would result in a change equal to 1/100 (1%) of its nameplate value, which is becoming significant. The concern, therefore, lies not only with the PPM level but with the range of expected temperature variation.

In equation form, the change in resistance is given by

$$\Delta R = \frac{R_{\text{nominal}}}{10^6}(\text{PPM})(\Delta T) \qquad (3.7)$$

where R_{nominal} is the nameplate value of the resistor at room temperature and ΔT is the change in temperature from the reference level of 20°C.

3.5 TYPES OF RESISTORS

Fixed Resistors

Resistors are made in many forms, but all belong in either of two groups: fixed or variable. The most common of the low-wattage, fixed-

EXAMPLE 3.6

For a 1-kΩ carbon composition resistor with a PPM of 2500, determine the resistance at 60°C.

Solution:

$$\Delta R = \frac{1000\ \Omega}{10^6}(2500)(60°\text{C} - 20°\text{C}) = 100\ \Omega$$

and

$$R = R_{\text{nominal}} + \Delta R = 1000\ \Omega + 100\ \Omega = \mathbf{1100\ \Omega}$$

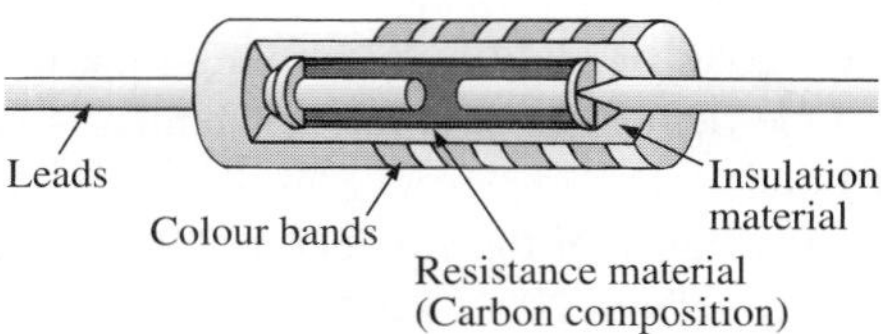

FIG. 3.5
Fixed composition resistor.

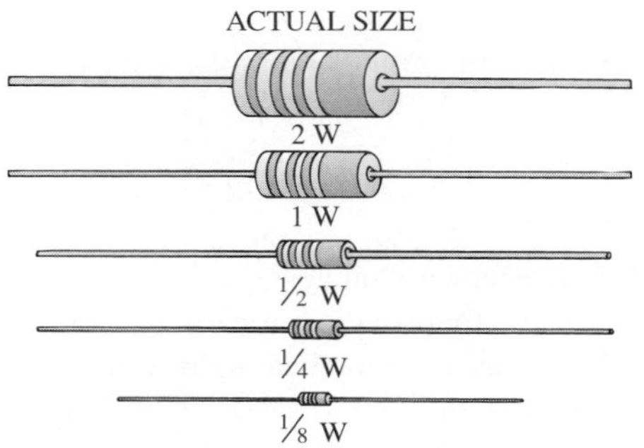

FIG. 3.6
Fixed composition resistors of different wattage ratings.

type resistors is the molded carbon composition resistor. The basic construction is shown in Fig. 3.5.

Resistors rated for higher powers need to be larger to withstand greater heat dissipation. The relative sizes of the molded composition resistors for different power ratings (wattage) are shown in Fig. 3.6. Resistors of this type are readily available in values ranging from 2.7 Ω to 22 MΩ.

The temperature-versus-resistance curve for a 10 000-Ω and 0.5-MΩ composition-type resistor is shown in Fig. 3.7. Note the small percent resistance change in the normal temperature operating range. Several other types of fixed resistors using high resistance wire or metal films are shown in Fig. 3.8.

The miniaturization of parts—used quite extensively in computers—requires that resistances of different values be placed in very small packages. Some examples appear in Fig. 3.9.

For use with printed circuit boards, fixed resistor networks in a variety of configurations are available in miniature packages, such as those shown in Fig. 3.10. The figure includes a photograph of three different casings and the internal resistor configuration for the single in-line structure to the right.

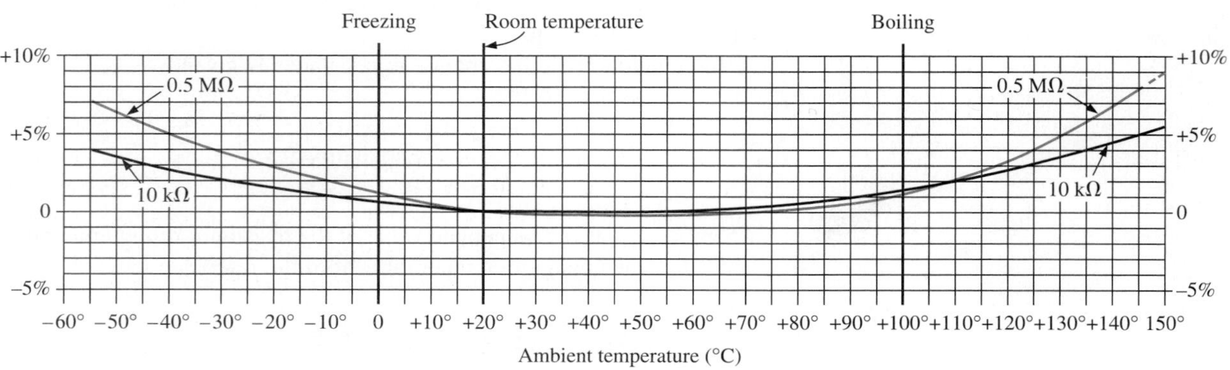

FIG. 3.7
Curves showing percentage temporary resistance changes from +20°C values. (Courtesy of Allen-Bradley Co.)

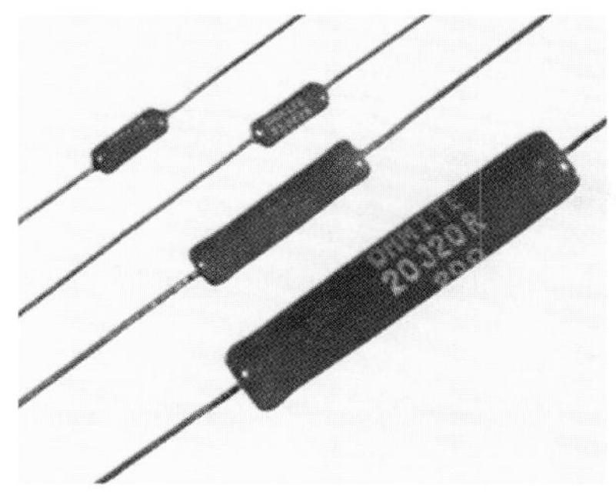

(a) Vitreous conformal wire resistor

(b) Power wire-wound stand-up cemented leaded fixed resistors

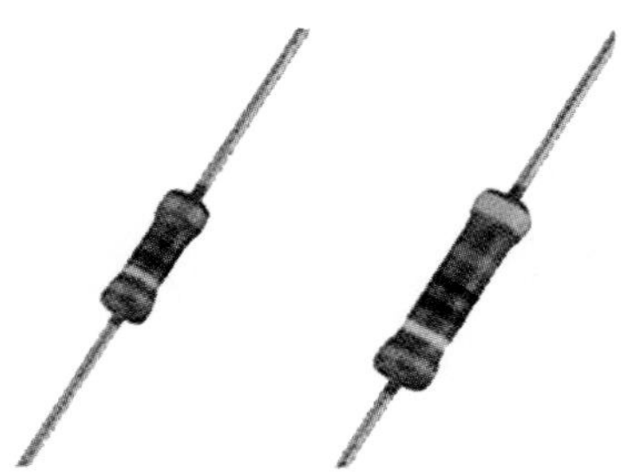

(c) High precision and ultra high precision metal film leaded fixed resistors

FIG. 3.8
Resistors. [Part (a) courtesy of Ohmite Manufacturing Co. Part (b) and (c) courtesy of Philips Components Inc.]

(a) Surface mount power resistors ideal for printed circuit boards

(b) Surface mount resistors

(c) Thick-film chip resistors for design flexiblity with hybrid circuitry. Pretinned, gold and silver electrodes available. Operating temperature range $-55°$ to $+150°C$

FIG. 3.9
Miniature fixed resistors. [Part (a) courtesy of Ohmite Manufacturing Co. Parts (b) courtesy of Philips Components Inc. (c) courtesy of Vishay Dale Electronics, Inc.]

FIG. 3.10
Thick-film resistor networks. (Courtesy of Vishay Dale Electronics, Inc.)

Variable Resistors

Variable resistors, as the name suggests, have a terminal resistance that can be varied by turning a dial, knob, or screw. They can have two or three terminals, but most have three terminals. If the two- or three-terminal device is used as a variable resistor, it is usually referred to as a **rheostat**. If the three-terminal device is used for controlling potential levels, it is then commonly called a **potentiometer**. Even though a three-terminal device can be used as a rheostat or potentiometer (depending on how it is connected), it is typically called a *potentiometer* in trade magazines or when requested for a particular application.

The symbol for a three-terminal potentiometer appears in Fig. 3.11(a). When used as a variable resistor (or rheostat), it can be hooked up in one of two ways, as shown in Fig. 3.11(b) and (c). In Fig. 3.11(b), points *a* and *b* are hooked up to the circuit, and the remaining terminal is left hanging. The resistance introduced is determined by

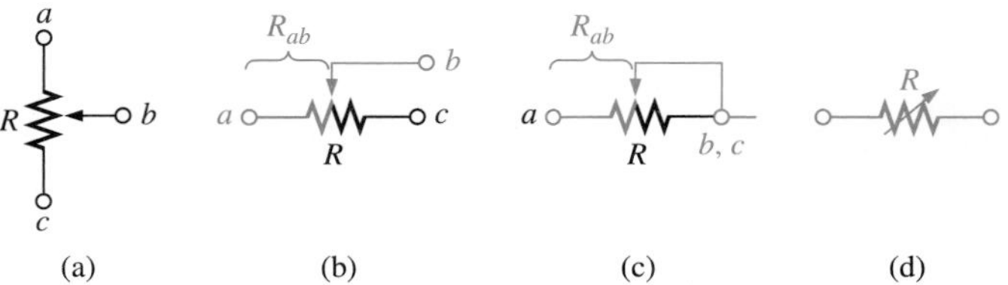

(a) (b) (c) (d)

FIG. 3.11
Potentiometer: (a) symbol; (b) and (c) rheostat connections; (d) rheostat symbol.

that portion of the resistive element between points *a* and *b*. In Fig. 3.11(c), the resistance is again between points *a* and *b*, but now the remaining resistance is "shorted-out" (effect removed) by the connection from *b* to *c*. The universally accepted symbol for a rheostat appears in Fig. 3.11(d).

Most potentiometers have three terminals in the relative positions shown in Fig. 3.12. The knob, dial, or screw in the centre of the housing controls the motion of a contact that can move along the resistive element connected between the outer two terminals. The contact is connected to the centre terminal, establishing a resistance from movable contact to each outer terminal.

The resistance between the outside terminals a and c of Fig. 3.13(a) (and Fig. 3.12) is always fixed at the full rated value of the potentiometer, regardless of the position of the wiper arm b.

In other words, the resistance between terminals *a* and *c* of Fig. 3.13(a) for a 1-MΩ potentiometer will always be 1 MΩ, no matter how we turn the control element and move the contact. In Fig. 3.13(a) the centre contact is not part of the network configuration.

The resistance between the wiper arm and either outside terminal can be varied from a minimum of 0 Ω to a maximum value equal to the full rated value of the potentiometer.

In Fig. 3.13(b) the wiper arm has been placed 1/4 of the way down from point *a* to point *c*. The resulting resistance between points *a* and *b* will therefore be 1/4 of the total, or 250 kΩ (for a 1-MΩ potentiometer), and the resistance between *b* and *c* will be 3/4 of the total, or 750 kΩ.

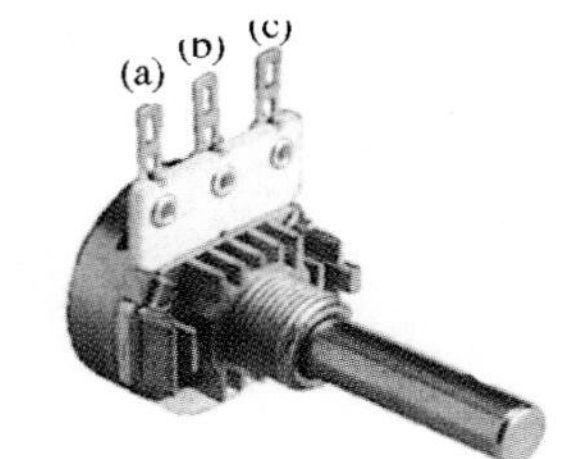

(a) Cermet control potentiometer

(b) Cermet single turn trimming potentiometer

(c) Cermet multiturn trimming potentiometer

FIG. 3.12
Potentiometer. (Courtesy of Phillips Components Inc.)

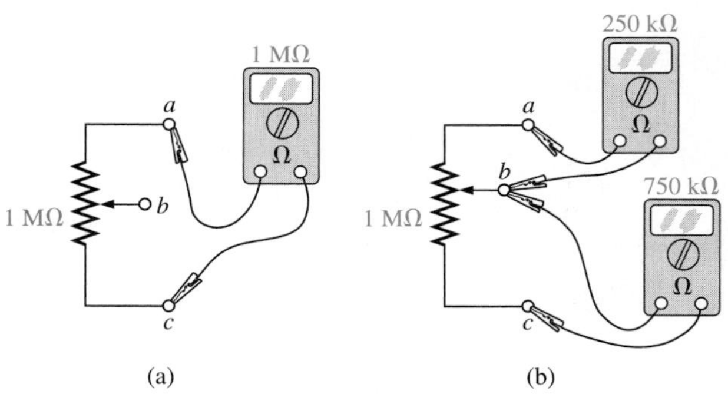

(a) (b)

FIG. 3.13
Terminal resistance of a potentiometer: (a) between outside terminals; (b) among all three terminals.

The sum of the resistances between the wiper arm and each outside terminal will equal the full rated resistance of the potentiometer.

This was demonstrated by Fig. 3.13(b), where 250 kΩ + 750 kΩ = 1 MΩ. Specifically:

$$R_{ac} = R_{ab} + R_{bc} \tag{3.8}$$

Therefore, as the resistance from the wiper arm to one outside contact increases, the resistance between the wiper arm and the other outside terminal must decrease accordingly. For example, if R_{ab} of a 1-kΩ potentiometer is 200 Ω, then the resistance R_{bc} must be 800 Ω. If R_{ab} is further decreased to 50 Ω, then R_{bc} must increase to 950 Ω, and so on.

The molded carbon composition potentiometer is typically applied in networks with smaller power demands, and it ranges in size from 20 Ω to 22 MΩ (maximum values). Other commercially available potentiometers appear in Fig. 3.14.

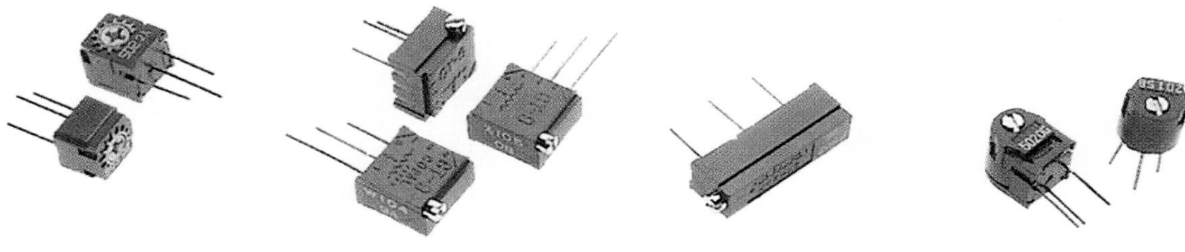

FIG. 3.14
Trimming potentiometers. (Courtesy of Phillips Components Inc.)

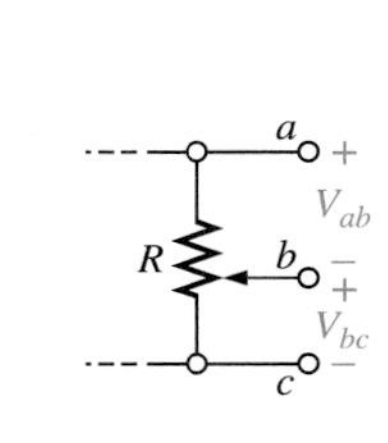

FIG. 3.15
Potentiometer control of voltage levels.

When the device is used as a potentiometer, the connections are as shown in Fig. 3.15. It can be used to control the level of V_{ab}, V_{bc}, or both, depending on the application. Additional discussion of the potentiometer in a loaded situation can be found in the chapters that follow.

3.6 COLOUR CODING AND STANDARD RESISTOR VALUES

Many resistors, fixed or variable, are large enough to have their resistance in ohms printed on the casing. Some, however, are too small to have numbers printed on them, so a system of colour coding is used. For the fixed molded composition resistor, four or five colour bands are printed on one end of the outer casing, as shown in Fig. 3.16. Each colour has the numerical value indicated in Table 3.6. The colour bands are always read from the end that has the band closest to it, as shown in Fig. 3.16. The first and second bands represent the first and second digits, respectively. The third band determines the power-of-10 multiplier for the first two digits (actually, the number of zeros that follow the second digit) or a multiplying factor if gold or silver. The fourth band is the manufacturer's tolerance, which is an indication of the precision by which the resistor was made. If the fourth band is omitted, the tolerance is assumed to be ±20%. The fifth band is a reliability factor, which gives the percentage of failure per 1000 hours of use. For instance, a 1% failure rate would reveal that one out of every 100 (or 10 out of every 1000) will fail to fall within the tolerance range after 1000 hours of use.

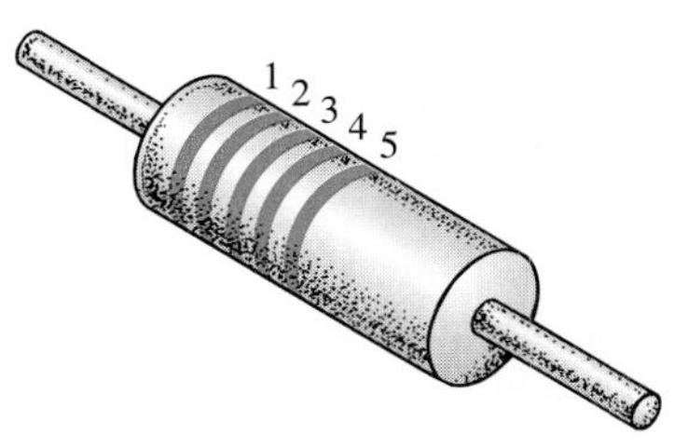

FIG. 3.16
Colour coding—fixed molded composition resistor.

TABLE 3.6

Resistor colour coding.

Bands 1–3	Band 3	Band 4	Band 5
0 Black	0.1 Gold (multiplying factors)	5% Gold	1% Brown
1 Brown	0.01 Silver (multiplying factors)	10% Silver	0.1% Red
2 Red		20% No band	0.01% Orange
3 Orange			0.001% Yellow
4 Yellow			
5 Green			
6 Blue			
7 Violet			
8 Grey			
9 White			

EXAMPLE 3.7

Find the range in which a resistor having the following colour bands (Fig. 3.17) must exist to satisfy the manufacturer's tolerance:

a.

1st band	2nd band	3rd band	4th band	5th band
Grey	Red	Black	Gold	Brown
8	2	0	±5%	1%

82 Ω ± 5% (1% reliability)

Since 5% of 82 = 4.10, the resistor should be within the range 82 Ω ± 4.10 Ω, or *between 77.90 and 86.10 Ω.*

b.

1st band	2nd band	3rd band	4th band	5th band
Orange	White	Gold	Silver	No colour
3	9	0.1	±10%	

3.9 Ω ± 10% = 3.9 ± 0.39 Ω

The resistor should lie somewhere *between 3.51 and 4.29 Ω.*

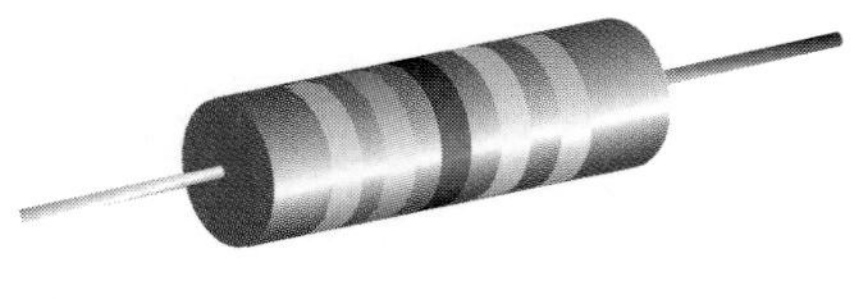

a)

b)

FIG. 3.17
Resistors for Example 3.7.

One might expect that resistors would be available for a full range of values such as 10 Ω, 20 Ω, 30 Ω, 40 Ω, 50 Ω, and so on. However, this is not the case—some typical commercial values are 27 Ω, 56 Ω, and 68 Ω. This may seem odd, but there is a reason for the chosen values. It can be demonstrated by examining the list of standard values of commercially available resistors in Table 3.7. The values in boldface blue are available with 5%, 10%, and 20% tolerances, making them the most common of the commercial variety. The values in boldface black are typically available with 5% and 10% tolerances, and those in normal print are available only in the 5% variety. If we separate the values available into tolerance levels, we have Table 3.8, which clearly reveals how few are available up to 100 Ω with 20% tolerances.

An examination of the impact of the tolerance level will now help explain the choice of numbers for the commercial values. Take the sequence 47 Ω–68 Ω–100 Ω, which are all available with 20% tolerances. In Fig. 3.18(a), the tolerance band for each has been determined and plotted on a single axis. Take note that, with this tolerance (which is all the manufacturer will guarantee), the full range of resistor values is available from 37.6 Ω to 120 Ω. In other words, the manufacturer is guaranteeing the full range, using the tolerances to fill in the gaps. Dropping to the 10% level introduces the 56-Ω and 82-Ω resistors to fill in the gaps, as shown in Fig. 3.18(b). Dropping to the 5% level would require additional resistor values to fill in the gaps. In total, therefore,

TABLE 3.7

Standard values of commercially available resistors.

Ohms (Ω)					Kilohms (kΩ)		Megohms (MΩ)	
0.10	1.0	10	100	1000	10	100	1.0	10.0
0.11	1.1	11	110	1100	11	110	1.1	11.0
0.12	**1.2**	**12**	**120**	**1200**	**12**	**120**	**1.2**	**12.0**
0.13	1.3	13	130	1300	13	130	1.3	13.0
0.15	1.5	15	150	1500	15	150	1.5	15.0
0.16	1.6	16	160	1600	16	160	1.6	16.0
0.18	**1.8**	**18**	**180**	**1800**	**18**	**180**	**1.8**	**18.0**
0.20	2.0	20	200	2000	20	200	2.0	20.0
0.22	2.2	22	220	2200	22	220	2.2	22.0
0.24	2.4	24	240	2400	24	240	2.4	
0.27	**2.7**	**27**	**270**	**2700**	**27**	**270**	**2.7**	
0.30	3.0	30	300	3000	30	300	3.0	
0.33	3.3	33	330	3300	33	330	3.3	
0.36	3.6	36	360	3600	36	360	3.6	
0.39	**3.9**	**39**	**390**	**3900**	**39**	**390**	**3.9**	
0.43	4.3	43	430	4300	43	430	4.3	
0.47	4.7	47	470	4700	47	470	4.7	
0.51	5.1	51	510	5100	51	510	5.1	
0.56	**5.6**	**56**	**560**	**5600**	**56**	**560**	**5.6**	
0.62	6.2	62	620	6200	62	620	6.2	
0.68	6.8	68	680	6800	68	680	6.8	
0.75	7.5	75	750	7500	75	750	7.5	
0.82	**8.2**	**82**	**820**	**8200**	**82**	**820**	**8.2**	
0.91	9.1	91	910	9100	91	910	9.1	

TABLE 3.8

Standard values and their tolerances.

±5%	±10%	±20%
10	**10**	10
11		
12	**12**	
13		
15	**15**	15
16		
18	**18**	
20		
22	**22**	22
24		
27	**27**	
30		
33	**33**	33
36		
39	**39**	
43		
47	**47**	47
51		
56	**56**	
62		
68	**68**	68
75		
82	**82**	
91		

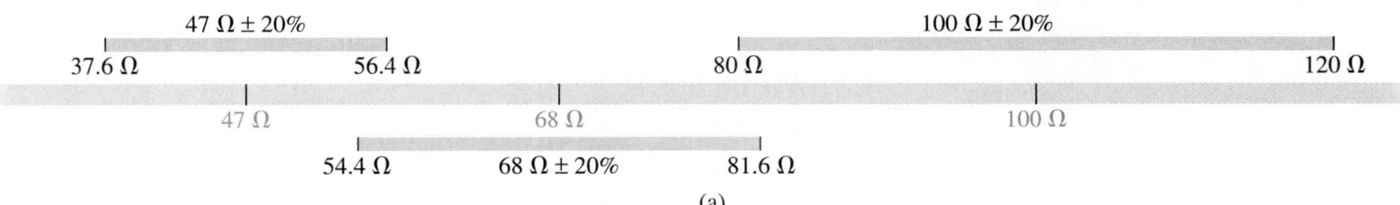

(a)

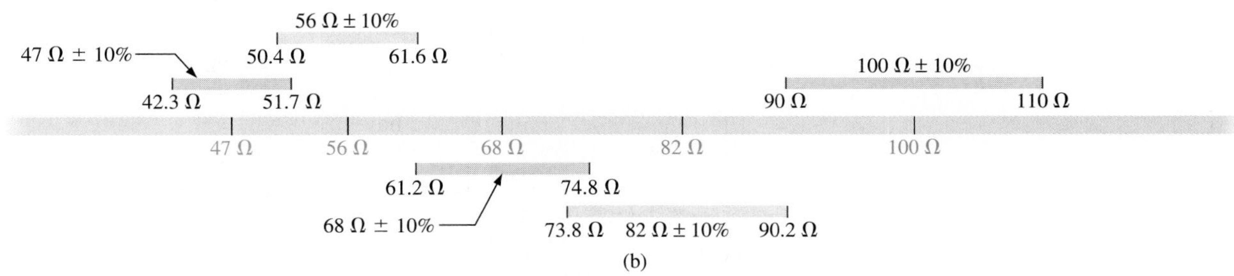

(b)

FIG. 3.18

Guaranteeing the full range of resistor values for the given tolerance: (a) 20%; (b) 10%.

the resistor values were chosen to ensure that the full range was covered, as determined by the tolerances used. Of course, if a specific value is desired but is not one of the standard values, combinations of standard values will often result in a total resistance very close to the desired level. If this approach is still not satisfactory, a potentiometer can be set to the exact value and then inserted in the network.

Throughout the text you will find that many of the resistor values are not standard values. This was done to reduce the mathematical complexity, which might cloud the procedure being introduced. In the problem sections, however, standard values are frequently used to help you to become familiar with the commercial values available.

LUMINARIES

German (Lenthe, Berlin)
(1816-1892)
Electrical Engineer
Telegraph Manufacturer:
Siemens & Halske AG

FIG. 3.19
Werner von Siemens

Developed an *electroplating process* during a brief stay in prison for acting as a second in a duel between fellow officers of the Prussian army. Inspired by the electric telegraph invented by Sir Charles Wheatstone in 1837, he improved on the design and proceeded to lay cable with the help of his brother Carl across the Mediterranean and from Europe to India. His inventions included the first *self-excited generator*, which depended on the *residual* magnetism of its electromagnet rather than an inefficient permanent magnet. In 1888 he was raised to the rank of nobility with the addition of *von* to his name. The current firm of Siemens AG has manufacturing outlets in 35 countries with sales offices in 125 countries.

Bettman Archives Photo Number 336.19

3.7 CONDUCTANCE

Sometimes, instead of resistance it is useful to know how well a material will *conduct* current. This property is known as **conductance**, and is defined as the reciprocal of resistance. Conductance has the symbol G, and is measured in **siemens** (S) (note Fig. 3.19). In equation form, conductance is

$$G = \frac{1}{R} \qquad \text{(siemens, S)} \qquad (3.9)$$

A resistance of 1 MΩ is equivalent to a conductance of 10^{-6} S, and a resistance of 10 Ω is equivalent to a conductance of 10^{-1} S. The larger the conductance, therefore, the less the resistance and the greater the conductivity.

In equation form, the conductance is determined by

$$G = \frac{A}{\rho l} \qquad \text{(S)} \qquad (3.10)$$

indicating that increasing the area or decreasing either the length or the resistivity will increase the conductance.

EXAMPLE 3.8

What is the relative increase or decrease in conductivity of a conductor if the area is reduced by 30% and the length is increased by 40%? The resistivity is fixed.

Solution: Eq. (3.10):

$$G_i = \frac{A_i}{\rho_i l_i}$$

with the subscript i for the initial value. Using the subscript n for new value:

$$G_n = \frac{A_n}{\rho_n l_n} = \frac{0.70A_i}{\rho_i(1.4l_i)} = \frac{0.70}{1.4}\frac{A_i}{\rho_i l_i} = \frac{0.70}{1.4\,G_i}$$

and $\quad G_n = 0.5G_i$

3.8 OHMMETERS

The **ohmmeter** is an instrument used to measure resistance and to perform other useful tasks:

1. *Measure the resistance of individual or combined elements*
2. *Detect open-circuit (high-resistance) and short-circuit (low-resistance) situations*
3. *Check continuity of network connections and identify wires of a multilead cable*

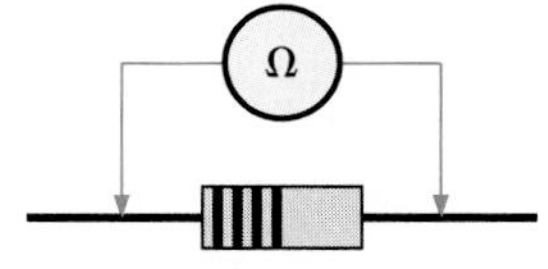

FIG. 3.20
Measuring the resistance of a single element.

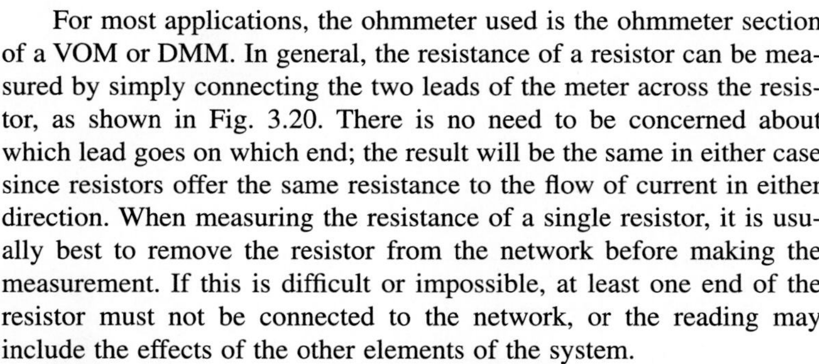

For most applications, the ohmmeter used is the ohmmeter section of a VOM or DMM. In general, the resistance of a resistor can be measured by simply connecting the two leads of the meter across the resistor, as shown in Fig. 3.20. There is no need to be concerned about which lead goes on which end; the result will be the same in either case since resistors offer the same resistance to the flow of current in either direction. When measuring the resistance of a single resistor, it is usually best to remove the resistor from the network before making the measurement. If this is difficult or impossible, at least one end of the resistor must not be connected to the network, or the reading may include the effects of the other elements of the system.

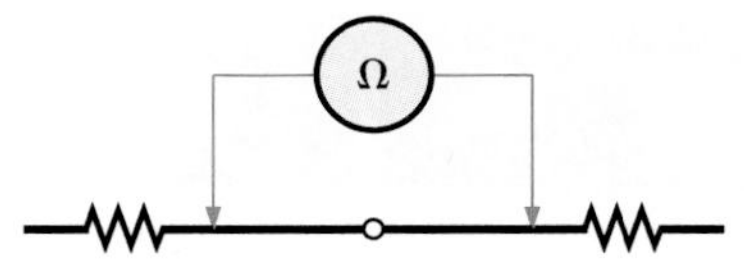

FIG. 3.21
Checking the continuity of a connection.

If the two leads of the meter are touching in the ohmmeter mode, the resulting resistance is zero. A connection can be checked as shown in Fig. 3.21 by simply hooking up the meter to both sides of the connection. If the resistance is zero, the connection is secure. If it is other than zero, it could be a weak connection, and, if it is infinite, there is no connection at all.

If one wire of a harness is known, a second can be found as shown in Fig. 3.22. Simply connect the end of the known lead to the end of any other lead. When the ohmmeter indicates zero ohms (or very low resistance), the second lead has been identified. The above procedure can also be used to determine the first known lead by simply connecting the meter to any wire at one end and then touching all the leads at the other end until a zero-ohm indication is obtained.

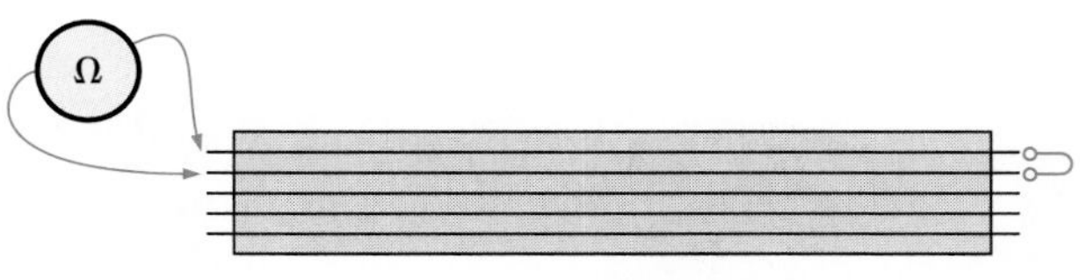

FIG. 3.22
Identifying the leads of a multilead cable.

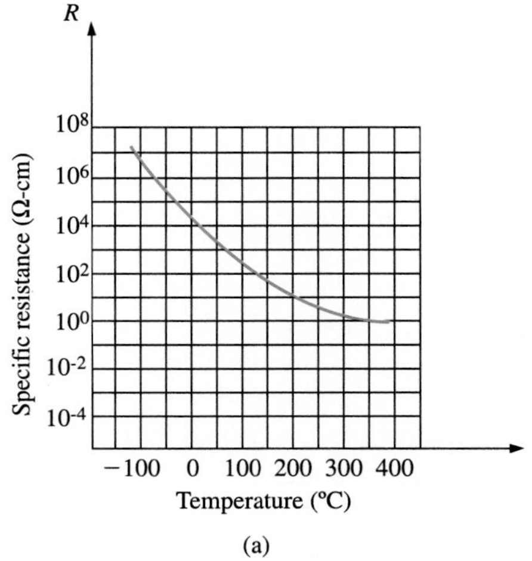

(a)

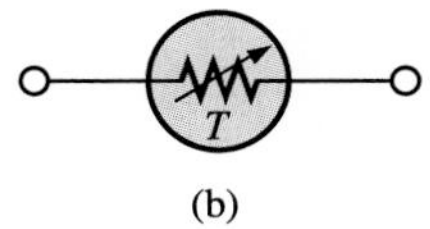

(b)

FIG. 3.23
Thermistor: (a) characteristics; (b) symbol.

3.9 THERMISTORS

The **thermistor** is a two-terminal semiconductor device whose resistance, as the name suggests, is temperature sensitive. A representative characteristic appears in Fig. 3.23 with the graphic symbol for the device. Note the nonlinearity of the curve and the drop in specific resistance from about 5000 Ω-cm to 100 Ω-cm for an increase in temperature from 20°C to 100°C. The decrease in resistance with an increase in temperature indicates a negative temperature coefficient.

The temperature of the device can be changed internally or externally. An increase in current through the device will raise its temperature, causing a drop in its terminal resistance. Any externally applied heat source will result in an increase in its body temperature and a drop in resistance. This type of action (internal or external) lends itself well to control mechanisms. Many different types of thermistors are shown in Fig. 3.24. Materials used to make thermistors include oxides of cobalt, nickel, strontium, and manganese.

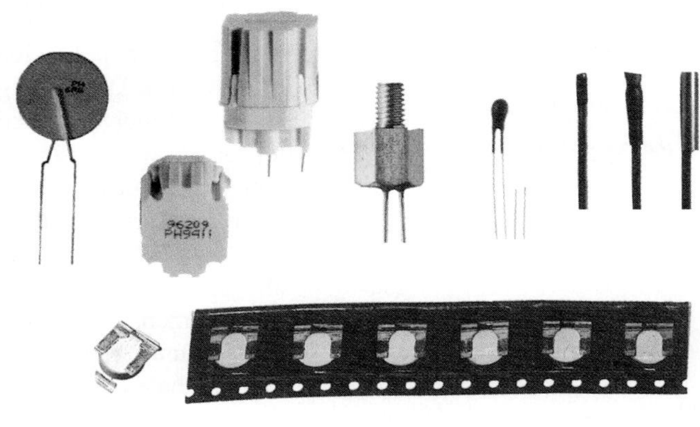

FIG. 3.24
NTC (negative temperature coefficient) and PTC (positive temperature coefficient) thermistors. (Courtesy of Philips Components Inc.)

R
100 kΩ
10 kΩ
1 kΩ
0.1 kΩ
Cell resistance
0.1 1.0 10 100 1000
Illumination (foot-candles)
(a)

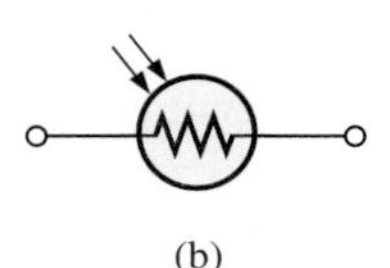

(b)

FIG. 3.25
Photoconductive cell: (a) characteristics; (b) symbol.

Note the use of a log scale (to be discussed in Chapter 21) in Fig. 3.23 for the vertical axis. The log scale makes it possible to show a wider range of specific resistance levels than a linear scale such as the horizontal axis. Note that it extends from 0.0001 V·cm to 100 000 000 V·cm over a very short interval. The log scale is used for both the vertical and the horizontal axis of Fig. 3.25, which appears in the next section.

3.10 PHOTOCONDUCTIVE CELL

The **photoconductive cell** is a two-terminal semiconductor device with a terminal resistance that is determined by the intensity of the incident light on its exposed surface. As the applied illumination increases in intensity, the energy state of the surface electrons and atoms increases. The result is that the number of "free carriers" increases and the resistance drops. A typical set of characteristics and the photoconductive cell's graphic symbol appear in Fig. 3.25. Note the negative illumination coefficient. Several cadmium sulfide photoconductive cells appear in Fig. 3.26.

FIG. 3.26
Street lighting photocontrol that uses a photoconductive cell (visible in the window of the casings). (Courtesy of Precision.)

3.11 VARISTORS

Varistors are voltage-dependent, nonlinear resistors used to suppress high-voltage transients. In other words, their characteristics limit the voltage that can appear across the terminals of a sensitive device or system. A typical set of characteristics appears in Fig. 3.27(a), along with a linear resistance characteristic for comparison purposes. Note that at a particular "firing voltage," the current rises rapidly but the voltage is limited to a level just above this firing potential. In other words, the magnitude of the voltage that can appear across this device cannot be greater than the level defined by its characteristics. Through proper design techniques this device can therefore limit the voltage appearing across sensitive regions of a network. The current is simply limited by the network to which it is connected. A photograph of a number of commercial units appears in Fig. 3.27.

FIG. 3.27
Zinc-oxide varistors for overvoltage protection. (Courtesy of Philips Components, Inc.)

PROBLEMS

SECTION 3.2 Resistance: Metric Units

*1. **a.** What is the resistance of a copper bus-bar with the dimensions shown in Fig. 3.28 ($T = 20°C$)?
b. Repeat (a) for aluminum and compare the results.
c. Without working out the numerical solution, determine whether the resistance of the bar (aluminum or copper) will increase or decrease with an increase in length.
d. Repeat (c) for an increase in cross-sectional area.

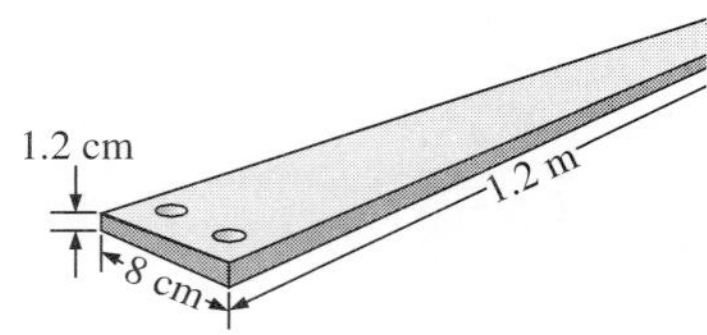

FIG. 3.28
Problem 1.

2. Using metric units, determine the length of a copper wire that has a resistance of 0.2 Ω and a diameter of 0.25 cm.

3. If the sheet resistance of a tin oxide sample is 100 Ω, what is the thickness of the oxide layer?

4. Determine the width of a carbon resistor having a sheet resistance of 150 Ω if the length is 0.125 cm and the resistance is 500 Ω.

SECTION 3.3 Wire Tables

5. **a.** Using Table 3.3, find the resistance of 450 m of #11 and #14 AWG wires.
b. Compare the resistances of the two wires.
c. Compare the areas of the two wires.

6. **a.** Using Table 3.3, find the resistance of 1800 m of #8 and #18 AWG wires.
b. Compare the resistances of the two wires.
c. Compare the areas of the two wires.

SECTION 3.4 Temperature Effects

7. The resistance of a copper wire is 2 Ω at 10°C. What is its resistance at 60°C?

8. The resistance of an aluminum bus-bar is 0.02 Ω at 0°C. What is its resistance at 100°C?

9. The resistance of a copper wire is 4 Ω at 21°C. What is its resistance at 0°C?

10. The resistance of a copper wire is 0.76 Ω at 30°C. What is its resistance at −40°C?

11. If the resistance of a silver wire is 0.04 Ω at −30°C, what is its resistance at 0°C?

*12. **a.** The resistance of a copper wire is 0.002 Ω at room temperature (20°C). What is its resistance at 0°C (freezing) and 100°C (boiling)?
b. For (a), determine the change in resistance for each 5° change in temperature between room temperature and 100°C.

13. **a.** The resistance of a copper wire is 0.92 Ω at 4°C. At what temperature (°C) will it be 1.06 Ω?
b. At what temperature will it be 0.15 Ω?

*14. **a.** If the resistance of a 300-m length of copper wire is 10 Ω at room temperature (20°C), what will its resistance be at 50 K (kelvin units) using Eq. (3.3)?
b. Repeat part (a) for a temperature of 38.65 K. Comment on the results obtained by reviewing the curve of Fig. 3.4.

15. A 100 W 120 V lamp has a resistance of 12Ω at 22°C. During normal operation, the resistance is 144 Ω. Find the filament temperature of the operating lamp.

16. **a.** Verify the value of α_{20} for copper in Table 3.7 by substituting the inferred absolute temperature into Eq. (3.4).
b. Using Eq. (3.5), find the temperature at which the resistance of a copper conductor will increase to 1 Ω from a level of 0.8 Ω at 20°C.

17. Using Eq. (3.5), find the resistance of a copper wire at 16°C if its resistance at 20°C is 0.4 Ω.

*18. Determine the resistance of a 300-m coil of 2.00 mm diameter copper wire sitting in the desert at a temperature of 45°C.

19. A 22-Ω wire-wound resistor is rated at +200 PPM for a temperature range of −10°C to +75°C. Determine its resistance at 65°C.

20. Determine the PPM rating of the 10-kΩ carbon composition resistor of Fig. 3.7 using the resistance level determined at 90°C.

SECTION 3.5 Types of Resistors

21. **a.** What is the approximate increase in size from a 1-W to a 2-W carbon resistor?
b. What is the approximate increase in size from a 1/2-W to a 2-W carbon resistor?

22. If the 10-kΩ resistor of Fig. 3.7 is exactly 10 kΩ at room temperature, what is its approximate resistance at −30°C and 100°C (boiling)?

23. Repeat Problem 22 at a temperature of 0° and 75°.

24. If the resistance between the outside terminals of a linear potentiometer is 10 kΩ, what is its resistance between the wiper (movable) arm and an outside terminal if the resistance between the wiper arm and the other outside terminal is 3.5 kΩ?

25. If the wiper arm of a linear potentiometer is one-quarter the way around the contact surface, what is the resistance between the wiper arm and each terminal if the total resistance is 25 kΩ?

*26. Show the connections required to establish 4 kΩ between the wiper arm and one outside terminal of a 10-kΩ potentiometer while having only zero ohms between the other outside terminal and the wiper arm.

27. A portion of a 1000 foot roll of #14 AWG wire has been used. What is the length of the remaining wire if the measured resistance is 1.68 Ω?

SECTION 3.6 Colour Coding and Standard Resistor Values

28. Find the range in which a resistor having the following colour bands must exist to satisfy the manufacturer's tolerance:

	1st band	2nd band	3rd band	4th band
a.	green	blue	orange	gold
b.	red	red	brown	silver
c.	brown	black	black	—

29. Find the colour code for the following 10% resistors:

a. 22 Ω
b. 4700 Ω
c. 56 kΩ
d. 1 MΩ

30. Is there an overlap in coverage between 20% resistors? That is, determine the tolerance range for a 10-Ω 20% resistor and a 15-Ω 20% resistor, and note whether their tolerance ranges overlap.

31. Repeat Problem 30 for 10% resistors of the same value.

SECTION 3.7 Conductance

32. Find the conductance of each of the following resistances:

a. 0.086 Ω **b.** 4 kΩ
c. 2.2 MΩ

Compare the three results.

33. Find the conductance of 300 m of 1.00 mm diameter wire made of

a. copper
b. aluminum
c. iron

***34.** The conductance of a wire is 100 S. If the area of the wire is increased by 2/3 and the length is reduced by the same amount, find the new conductance of the wire if the temperature remains fixed.

SECTION 3.8 Ohmmeters

35. How would you check the status of a fuse with an ohmmeter?

36. How would you determine the on and off states of a switch using an ohmmeter?

37. How would you use an ohmmeter to check the status of a light bulb?

38. Why should an ohmmeter never be used in an energized circuit?

SECTION 3.9 Thermistors

***39. a.** Find the specific resistance of the thermistor having the characteristics of Fig. 3.23 at −50°C, 50°C, and 100°C. Note that it is a log scale. If necessary, consult a reference with an expanded log scale.

b. Does the thermistor have a positive or negative temperature coefficient?

c. Is the coefficient a fixed value for the range −100°C to 400°C? Why?

d. What is the approximate rate of change of ρ with temperature at 100°C?

SECTION 3.10 Photoconductive Cell

***40. a.** Using the characteristics of Fig. 3.25, determine the resistance of the photoconductive cell at 10 and 100 foot-candle illumination. As in Problem 39, note that it is a log scale.

b. Does the cell have a positive or negative illumination coefficient?

c. Is the coefficient a fixed value for the range 0.1 to 1000 foot-candles? Why?

d. What is the approximate rate of change of ρ with illumination at 10 foot-candles?

SECTION 3.11 Varistors

41. a. Referring to Fig. 3.27(a), find the terminal voltage of the device at 0.5, 1, 3, and 5 mA.

b. What is the total change in voltage for the indicated range of current levels?

c. Compare the ratio of maximum to minimum current levels above to the corresponding ratio of voltage levels.

4

Ohm's Law, Power, and Energy

Outline

Learning Outcomes

After completing this chapter you will be able to

- explain Ohm's Law
- solve Ohm's Law problems involving current, voltage, and resistance in simple circuits
- describe the behaviour of a circuit graphically by plotting Ohm's Law
- explain the difference between power and energy
- solve problems involving power and energy
- define efficiency and solve related problems

Ohm's Law describes the impact of resistance on circuit current and voltage. Power and energy measurements allow us to quantify, in electrical terms, how energy is converted from one form to another.

4.1 OHM'S LAW

LUMINARIES

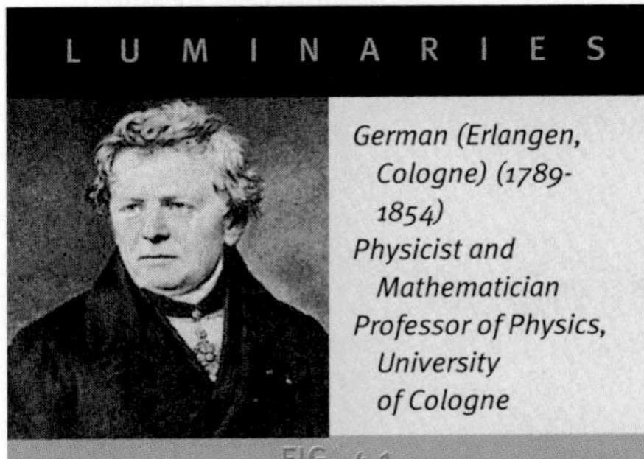

German (Erlangen, Cologne) (1789-1854)
Physicist and Mathematician
Professor of Physics, University of Cologne

FIG. 4.1
Georg Simon Ohm

In 1827, developed one of the most important laws of electric circuits: *Ohm's law*. When first introduced, the supporting documentation was considered lacking and foolish. This caused him to lose his teaching position and search for a living doing odd jobs and some tutoring. It took 22 years for his work to be recognized as a major contribution to the field. He was then awarded a chair at the University of Munich and received the Copley Medal of the Royal Society in 1841. His research also extended into the areas of molecular physics, acoustics, and telegraphic communication.

Courtesy of the Smithsonian Institution, Photo No. 51,145

Consider the following relationship:

$$\text{Effect} = \frac{\text{cause}}{\text{opposition}} \tag{4.1}$$

Every conversion of energy from one form to another can be related to this equation. In electric circuits, the *effect* we are trying to establish is the flow of charge, or *current*. The *potential difference*, or voltage, between two points is the *cause* ("pressure"), and the opposition is the *resistance* encountered.

Substituting these electrical terms into Eq. (4.1) results in

$$\text{Current} = \frac{\text{potential difference}}{\text{resistance}}$$

and

$$I = \frac{E}{R} \qquad \text{(amperes, A)} \tag{4.2}$$

The relation of Eq. (4.2) is known as **Ohm's law** in honour of Georg Simon Ohm (Fig. 4.1). The law clearly reveals that for a fixed resistance, the greater the voltage (or pressure) across a resistor, the more the current, and the more the resistance for the same voltage, the less the current. In other words, the current is proportional to the applied voltage and inversely proportional to the resistance.

By simple mathematical manipulation of Eq. (4.2), the voltage and resistance can be found in terms of the other two quantities:

$$E = IR \qquad \text{(volts, V)} \tag{4.3}$$

$$R = \frac{E}{I} \qquad \text{(ohms, } \Omega\text{)} \tag{4.4}$$

The three quantities of Eqs. (4.2) through (4.4) are defined by the simple circuit of Fig. 4.2. The current I of Eq. (4.2) results from applying a dc supply of E volts across a network having a resistance R. Equation (4.3) determines the voltage E required to establish a current I through a network with a total resistance R, and Eq. (4.4) provides the resistance of a network that results in a current I due to an impressed voltage E.

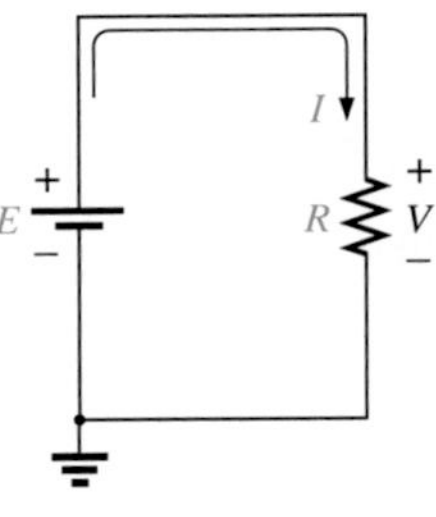

FIG. 4.2
Basic circuit.

Note in Fig. 4.2 that within the voltage source, current passes from the negative to the positive terminal of the battery. This will always be the case for single-source circuits. The effect of more than one source in the network will be examined in the chapters to follow. The symbol for the voltage of the battery (a source of electrical energy) is the uppercase letter E, and the potential drop across the resistor is given the symbol V. The polarity of the voltage drop across the resistor is defined by the applied source because the two terminals of the battery are connected directly across the resistive element.

EXAMPLE 4.1

Determine the current resulting from the application of a 9-V battery across a network with a resistance of 2.2 Ω.

Solution: Eq. (4.2):

$$I = \frac{E}{R} = \frac{9\text{ V}}{2.2\ \Omega} = \mathbf{4.09\ A}$$

EXAMPLE 4.2

Calculate the resistance of a 60-W bulb if a current of 500 mA results from an applied voltage of 120 V.

Solution: Eq. (4.4):

$$R = \frac{E}{I} = \frac{120\text{ V}}{500 \times 10^{-3}\text{ A}} = \mathbf{240\ \Omega}$$

FIG. 4.3
Defining polarities.

For an isolated resistive element, the polarity of the voltage drop is shown in Fig. 4.3(a) for the indicated current direction. A reversal in current will reverse the polarity, as shown in Fig. 4.3(b). In general, the current is from a high (+) to a low (−) potential. Polarities as established by current direction will become increasingly important in the analysis to follow.

EXAMPLE 4.3

Calculate the current through the 2-kΩ resistor of Fig. 4.4 if the voltage drop across it is 16 V.

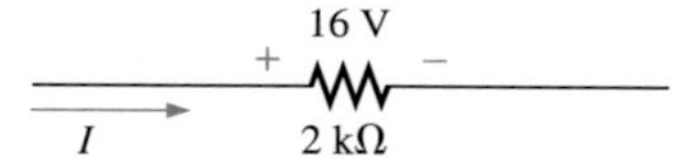

FIG. 4.4
Example 4.3.

Solution:

$$I = \frac{V}{R} = \frac{16\text{ V}}{2 \times 10^{3}\ \Omega} = \mathbf{8\ mA}$$

EXAMPLE 4.4

Calculate the voltage that must be applied across the soldering iron of Fig. 4.5 to establish a current of 1.5 A through the iron if its internal resistance is 80 Ω.

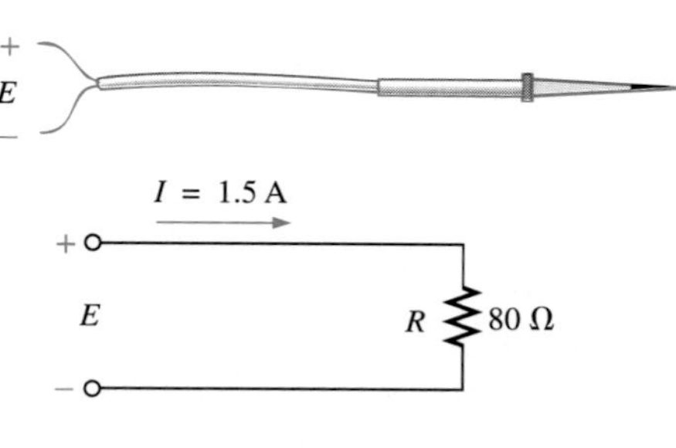

FIG. 4.5
Example 4.4.

Solution:

$$E = IR = (1.5\text{ A})(80\ \Omega) = \mathbf{120\ V}$$

4.2 PLOTTING OHM'S LAW

Graphs, characteristics, plots, and the like play an important role in every technical field as a way of displaying the broad picture of the behaviour or response of a system. It is therefore very important to develop the skills necessary both to read data and to plot them in such a way that they can be interpreted easily.

For most sets of characteristics for electronic devices, the current is represented by the vertical axis, and the voltage by the horizontal axis, as shown in Fig. 4.6. Note that the vertical axis is in amperes and the horizontal axis is in volts. The linear (straight-line) graph reveals that the resistance is not changing with current or voltage level; rather, it is a fixed quantity throughout. The current direction and the voltage polarity shown at the right of Fig. 4.6 are the defined direction and polarity for the provided plot.

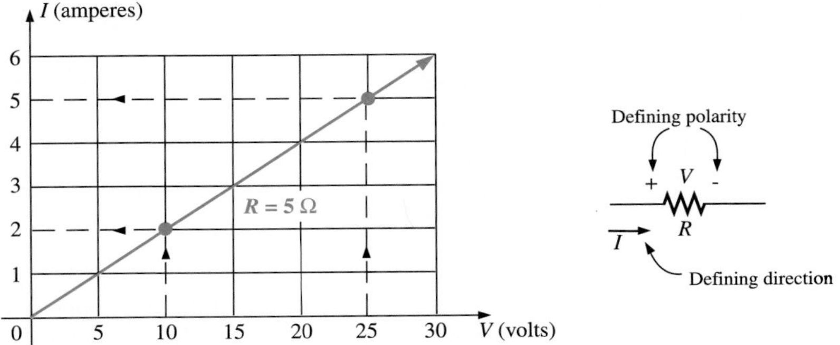

FIG. 4.6
Plotting Ohm's law.

Once a graph such as Fig. 4.6 is developed, the current or voltage at any level can be found from the other quantity by simply using the resulting plot. For instance, at $V = 25$ V, if a vertical line is drawn on Fig. 4.6 to the curve as shown, the resulting current can be found by drawing a horizontal line over to the current axis, where a result of 5 A is obtained. Similarly, at $V = 10$ V, a vertical line to the plot and a horizontal line to the current axis will result in a current of 2 A, as determined by Ohm's law.

If the resistance of a plot is unknown, it can be determined at any point on the plot since a straight line indicates a fixed resistance. At any point on the plot, find the resulting current and voltage and simply substitute into the following equation:

$$R_{dc} = \frac{V}{I} \tag{4.5}$$

To test Eq. (4.5) consider a point on the plot where $V = 20$ V and $I = 4$ A. The resulting resistance is $R_{dc} = V/I = 20\text{ V}/4\text{ A} = 5\ \Omega$. For comparison purposes, a 1-Ω and 10-Ω resistor were plotted on the graph of Fig. 4.7. Note that the less the resistance, the steeper the slope (closer to the vertical axis) of the curve.

If we write Ohm's law in the following way and relate it to the basic straight-line equation

$$\begin{array}{ccccccc} I & = & \frac{1}{R} & \cdot & E & + & 0 \\ \downarrow & & \downarrow & & \downarrow & & \downarrow \\ y & = & m & \cdot & x & + & b \end{array}$$

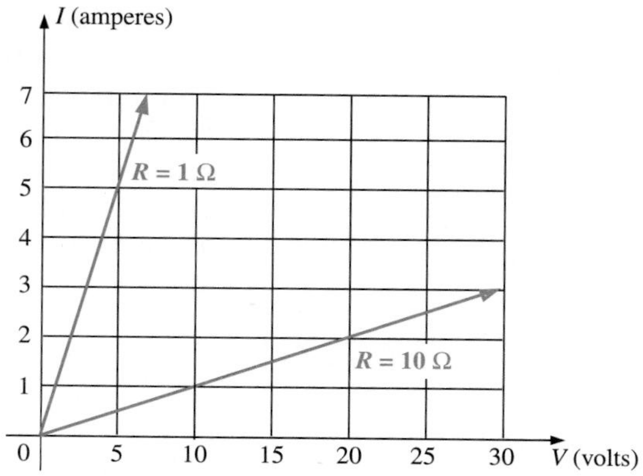

FIG. 4.7
Demonstrating on an I-V plot that the less the resistance, the steeper the slope.

we find that the slope is equal to 1 divided by the resistance value, as indicated by the following:

$$m = \text{slope} = \frac{\Delta y}{\Delta x} = \frac{\Delta I}{\Delta V} = \frac{1}{R} \tag{4.6}$$

where Δ signifies a small, finite change in the variable.

Equation (4.6) clearly reveals that the greater the resistance, the less the slope. If written in the following form, Eq. (4.6) can be used to determine the resistance from the linear curve:

$$R = \frac{\Delta V}{\Delta I} \quad \text{(ohms)} \tag{4.7}$$

The equation states that by choosing a particular ΔV (or ΔI), the corresponding ΔI (or ΔV, respectively) can be obtained from the graph, as shown in Fig. 4.8, and the resistance can be determined. If the plot is a straight line, Eq. (4.7) will provide the same result no matter where the equation is applied. However, if the plot curves at all, the resistance will change.

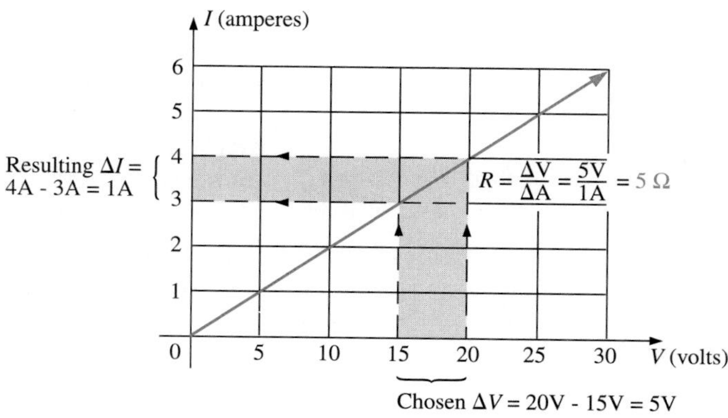

FIG. 4.8
Applying Eq. (4.6).

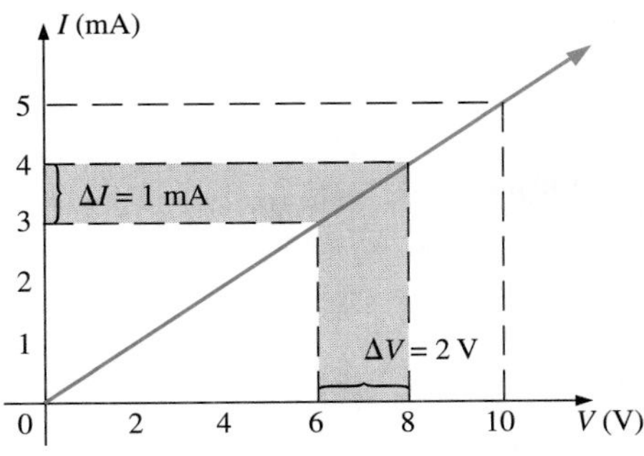

FIG. 4.9
Example 4.5.

EXAMPLE 4.5

Determine the resistance associated with the curve of Fig. 4.9 using Eqs. (4.5) and (4.7), and compare results.

Solution: At $V = 6$ V, $I = 3$ mA, and

$$R_{dc} = \frac{V}{I} = \frac{6\text{ V}}{3\text{ mA}} = \mathbf{2\ k\Omega}$$

For the interval between 6 V and 8 V,

$$R = \frac{\Delta V}{\Delta I} = \frac{2\text{ V}}{1\text{ mA}} = \mathbf{2\ k\Omega}$$

The results are equivalent.

Before leaving the subject, let us first look at the characteristics of a very important semiconductor device called the *diode,* which has a nonlinear *I-V* plot. A typical set of characteristics appears in Fig. 4.10. Without any mathematical calculations, the closeness of the characteristic to the voltage axis for negative values of applied voltage indicates that this is the low conductance (high resistance, switch opened) region. Note that this region extends to approximately 0.7 V positive. However, for values of applied voltage greater than 0.7 V, the vertical rise in the characteristics indicates a high conductivity (low resistance, switch closed) region. Application of Ohm's law will now verify the above conclusions.

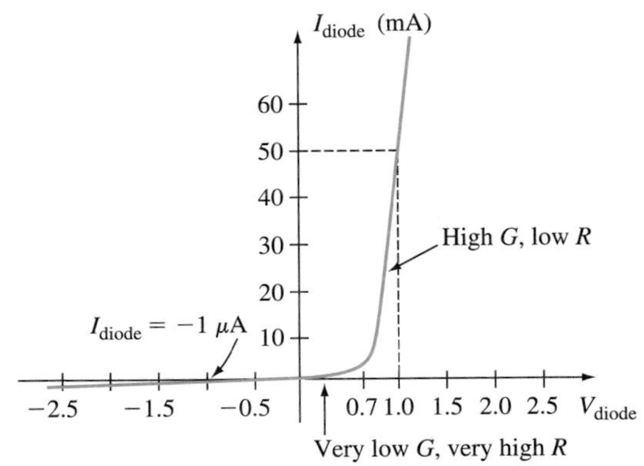

FIG. 4.10
Semiconductor diode characteristics.

At $V = +1$ V,

$$R_{\text{diode}} = \frac{V}{I} = \frac{1\text{ V}}{50\text{ mA}} = \frac{1\text{ V}}{50 \times 10^{-3}\text{ A}}$$
$$= 20\ \Omega$$

(a relatively low value for most applications)

At $V = -1$ V,

$$R_{\text{diode}} = \frac{V}{I} = \frac{1\text{ V}}{1\ \mu\text{A}}$$
$$= 1\ \text{M}\Omega$$

(which is often represented by an open-circuit equivalent)

4.3 POWER

Power is the measure of the *rate* at which energy is transferred, or converted from one form to another. For instance, a large motor has more power than a small motor because it can convert more electrical energy into mechanical energy in the same period of time. Since converted energy is measured in *joules* (J) and time in seconds (s), power is measured in joules/second (J/s). The electrical unit of measurement for power is the **watt** (W), defined by

$$1 \text{ watt (W)} = 1 \text{ joule/second (J/s)} \qquad (4.8)$$

In equation form, power is determined by

$$P = \frac{W}{t} \qquad \text{(watts, W, or joules/second, J/s)} \qquad (4.9)$$

with the energy W measured in joules and the time t in seconds.

Throughout the text, the symbol for energy (W) can be distinguished from that for the watt (W) by the fact that one is in italics while the other is in roman type. In fact, all variables in the dc section appear in italics while the units appear in roman.

The unit of measurement, the watt, is derived from the surname of James Watt (Fig. 4.11), who was instrumental in establishing the standards for power measurements. He introduced *horsepower* (hp) as a measure of the average power of a strong dray horse over a full working day. It is approximately 50% more than can be expected from the average horse. The horsepower and watt are related in the following manner:

$$1 \text{ horsepower} \cong 746 \text{ watts}$$

LUMINARIES

Scottish (Greenock, Birmingham) (1736-1819)
Instrument maker and inventor
Elected fellow of the Royal Society of London in 1785

FIG. 4.11
James Watt

In 1757, at the age of 21, used his innovative talents to design mathematical instruments such as the *quadrant, compass*, and various *scales*. In 1765, introduced the use of a separate *condenser* to increase the efficiency of steam engines. In the years to follow he received a number of important patents on improved engine design, including a rotary motion for the steam engine (versus the reciprocating action) and a double-action engine, in which the piston pulled as well as pushed in its cyclic motion. Introduced the term *horsepower* as the average power of a strong dray (small cart) horse over a full working day.

Courtesy of the Smithsonian Institution, Photo No. 30,391

The power dissipated by an electrical device or system can be found in terms of the current and voltage by first substituting Eq. (2.7) into Eq. (4.9):

$$P = \frac{W}{t} = \frac{QV}{t} = V\frac{Q}{t}$$

But

$$I = \frac{Q}{t}$$

so that

$$P = VI \qquad \text{(watts)} \qquad (4.10)$$

By direct substitution of Ohm's law, the equation for power can be obtained in two other forms:

$$P = VI = V\left(\frac{V}{R}\right)$$

and

$$P = \frac{V^2}{R} \qquad \text{(watts)} \qquad (4.11)$$

or

$$P = VI = (IR)I$$

and

$$P = I^2R \qquad \text{(watts)} \qquad (4.12)$$

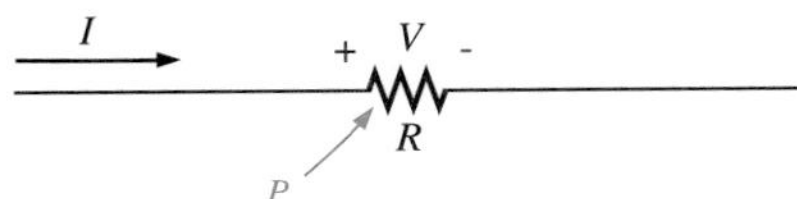

FIG. 4.12
Defining the power to a resistive element.

The result is that the power absorbed by the resistor of Fig. 4.12 can be found directly depending on the information available. In other words, if the current and resistance are known, it pays to use Eq. (4.12) directly, and if V and I are known, Eq. (4.10) is appropriate. It saves having to apply Ohm's law before determining the power.

Power can be delivered or absorbed as defined by the polarity of the voltage and the direction of the current. For all dc voltage sources, power is being *delivered* by the source if the current has the direction appearing in Fig. 4.13(a). Note that the current has the same direction as established by the source in a single-source network. If the current direction and polarity are as shown in Fig. 4.13(b) due to a multisource network, the battery is absorbing power much like when a battery is being charged.

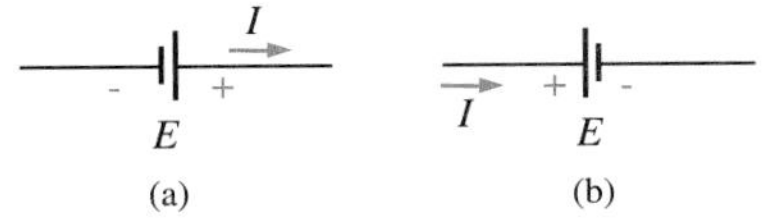

FIG. 4.13
Battery power: (a) supplied; (b) absorbed.

For resistive elements, all the power is dissipated in the form of heat because the voltage polarity is defined by the current direction (and vice versa), and current will always enter the terminal of higher potential corresponding with the absorbing state of Fig. 4.13(b). A reversal of the current direction in Fig. 4.12 will also reverse the polarity of the voltage across the resistor and match the conditions of Fig. 4.13(b).

The magnitude of the power delivered or absorbed by a battery is given by

$$P = EI \quad \text{(watts)} \tag{4.13}$$

where E is the battery terminal voltage and I is the current through the source.

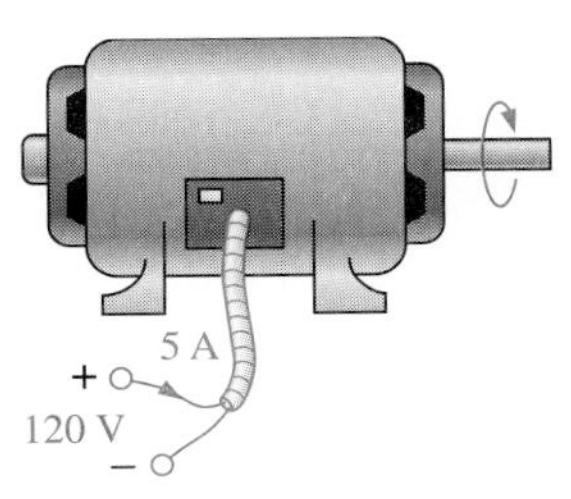

FIG. 4.14
Example 4.6.

EXAMPLE 4.6

Find the power delivered to the dc motor of Fig. 4.14.

Solution:

$$P = VI = (120 \text{ V})(5 \text{ A}) = 600 \text{ W} = \mathbf{0.6\ kW}$$

EXAMPLE 4.7

What is the power dissipated by a 5-Ω resistor if the current is 4 A?

Solution:

$$P = I^2R = (4 \text{ A})^2(5\ \Omega) = \mathbf{80\ W}$$

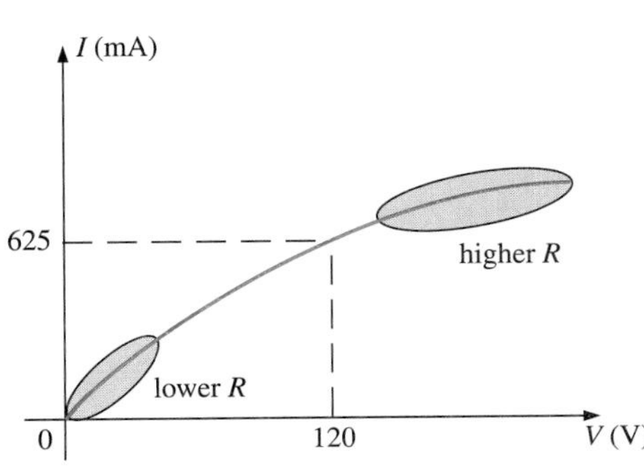

FIG. 4.15
The nonlinear I-V characteristics of a 75-W light bulb.

EXAMPLE 4.8

The I-V characteristics of a light bulb are provided in Fig. 4.15. Note the nonlinearity of the curve, indicating a wide range in resistance of the bulb with applied voltage as defined by the discussion of Section 4.2. If the rated voltage is 120 V, find the wattage rating of the bulb. Also calculate the resistance of the bulb under rated conditions.

Solution: At 120 V,

$$I = 0.625 \text{ A}$$

and

$$P = VI = (120 \text{ V})(0.625 \text{ A}) = \mathbf{75\ W}$$

At 120 V,

$$R = \frac{V}{I} = \frac{120 \text{ V}}{0.625 \text{ A}} = \mathbf{192\ \Omega}$$

Sometimes the power is given and the current or voltage must be determined. Through algebraic manipulations, an equation for each variable is derived as follows:

$$P = I^2R \Rightarrow I^2 = \frac{P}{R}$$

and

$$I = \sqrt{\frac{P}{R}} \qquad \text{(amperes)} \qquad (4.14)$$

$$P = \frac{V^2}{R} \Rightarrow V^2 = PR$$

and

$$V = \sqrt{PR} \qquad \text{(volts)} \qquad (4.15)$$

EXAMPLE 4.9

Determine the current through a 5-kΩ resistor when the power dissipated by the element is 20 mW.

Solution: Eq. (4.14):

$$I = \sqrt{\frac{P}{R}} = \sqrt{\frac{20 \times 10^{-3}\text{ W}}{5 \times 10^{3}\ \Omega}} = \sqrt{4 \times 10^{-6}} = 2 \times 10^{-3}\text{ A}$$
$$= \mathbf{2\ mA}$$

4.4 WATTMETERS

A **wattmeter** measures the power delivered by a source or to a dissipative element. One such instrument appears in Fig. 4.16. Since power is a function of both the current and the voltage levels, a wattmeter must measure both of them. As a result, the wattmeter has four terminals, two for the current coils (CC) and two for the potential coils (PC).

If the current coils and potential coils of the wattmeter are connected as shown in Fig. 4.17, there will be an up-scale reading on the wattmeter. A reversal of either coil will result in a below-zero indication.

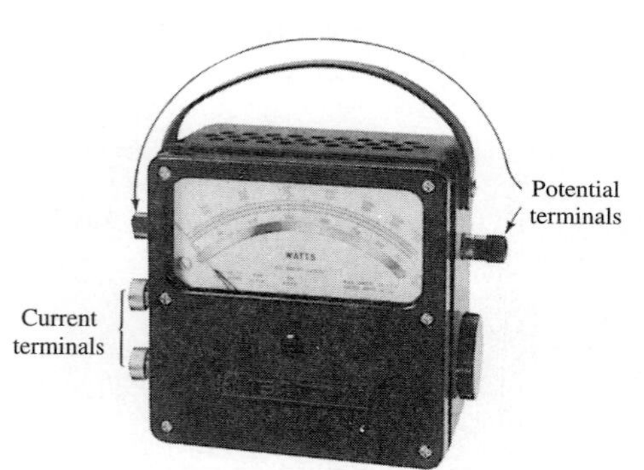

FIG. 4.16
Wattmeter. (Courtesy of Electrical Instrument Service, Inc.)

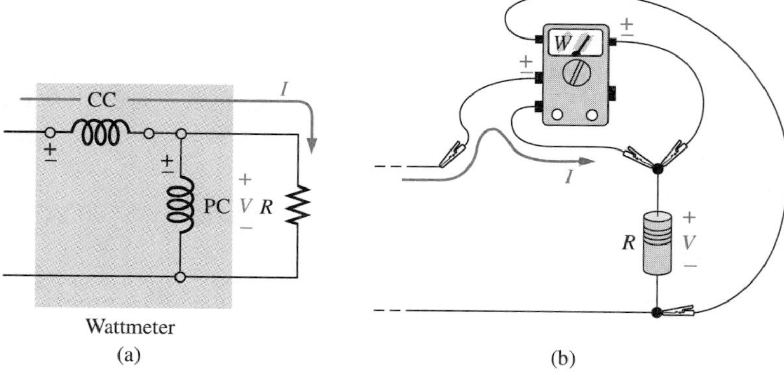

FIG. 4.17
Wattmeter connections.

4.5 EFFICIENCY

A flowchart for the energy levels associated with any system that converts energy from one form to another is provided in Fig. 4.18. Note that the output energy level must always be less than the applied energy due to losses and storage within the system. The best one can hope for is that W_o and W_i are relatively close in magnitude.

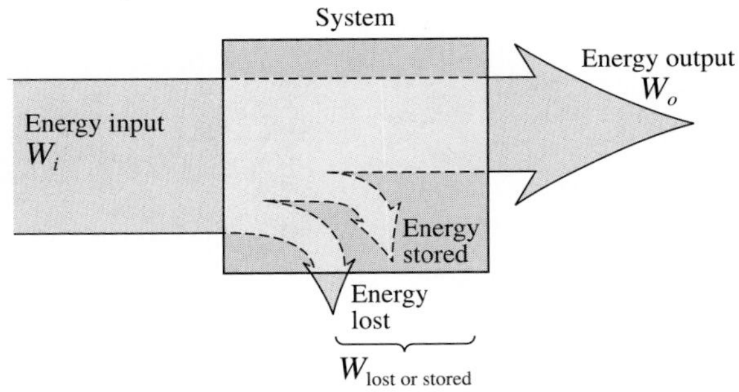

FIG. 4.18
Energy flow through a system.

Conservation of energy requires that

Energy input = energy output + energy lost or stored in the system

Dividing both sides of the relationship by t gives

$$\frac{W_{\text{in}}}{t} = \frac{W_{\text{out}}}{t} + \frac{W_{\text{lost or stored by the system}}}{t}$$

Since $P = W/t$, we have the following:

$$P_i = P_o + P_{\text{lost or stored}} \quad \text{(W)} \qquad (4.16)$$

The efficiency (η) of the system is then determined by the following equation:

$$\text{Efficiency} = \frac{\text{power output}}{\text{power input}}$$

and

$$\eta = \frac{P_o}{P_i} \quad \text{(decimal number)} \qquad (4.17)$$

where η (lowercase Greek letter eta) is a decimal number. Expressed as a percentage,

$$\eta\% = \frac{P_o}{P_i} \times 100\% \quad \text{(percent)} \qquad (4.18)$$

In terms of the input and output energy, the efficiency in percent is given by

$$\eta\% = \frac{W_o}{W_i} \times 100\% \quad \text{(percent)} \qquad (4.19)$$

The maximum possible efficiency is 100%, which occurs when $P_o = P_i$, or when the power lost or stored in the system is zero. Obviously, the greater the internal losses of the system in generating the necessary output power or energy, the lower the net efficiency.

EXAMPLE 4.10

A 2-hp motor operates at an efficiency of 75%. What is the power input in watts? If the applied voltage is 220 V, what is the input current?

Solution:

$$\eta\% = \frac{P_o}{P_i} \times 100\%$$

$$0.75 = \frac{(2\text{ hp})(746\text{ W/hp})}{P_i}$$

and

$$P_i = \frac{1492\text{ W}}{0.75} = \mathbf{1989.33\text{ W}}$$

$$P_i = EI \text{ or } I = \frac{P_i}{E} = \frac{1989.33\text{ W}}{220\text{ V}} = \mathbf{9.04\text{ A}}$$

EXAMPLE 4.11

What is the output in horsepower of a motor with an efficiency of 80% and an input current of 8 A at 120 V?

Solution:

$$\eta\% = \frac{P_o}{P_i} \times 100\%$$

$$0.80 = \frac{P_o}{(120\text{ V})(8\text{ A})}$$

and

$$P_o = (0.80)(120\text{ V})(8\text{ A}) = 768\text{ W}$$

with

$$768\text{ W}\left(\frac{1\text{ hp}}{746\text{ W}}\right) = \mathbf{1.029\text{ hp}}$$

EXAMPLE 4.12

If $\eta = 0.85$, determine the output energy level if the applied energy is 50 J.

Solution:

$$\eta = \frac{W_o}{W_i} \Rightarrow W_o = \eta W_i$$

$$= (0.85)(50\text{ J})$$

$$= \mathbf{42.5\text{ J}}$$

The very basic components of a generating (voltage) system are shown in Fig. 4.19. The source of mechanical power is a structure such as a paddlewheel that is turned by water rushing over the dam. The gear train will then ensure that the rotating member of the generator is turning at rated speed. The output voltage must then be fed through a trans-

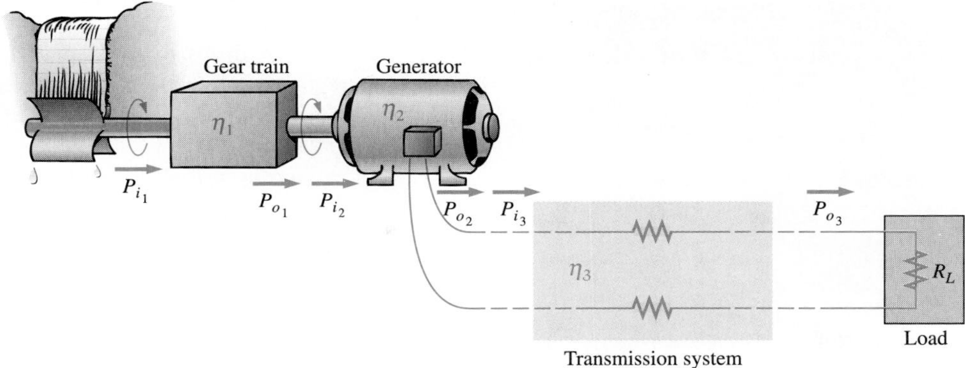

FIG. 4.19
Basic components of a generating system.

mission system to the load. For each component of the system, an input and output power have been indicated. In this system, η_1 represents the efficiency of the gear train; η_2 describes how well mechanical energy is converted to electrical energy by the generator; and η_3 represents how efficiently energy is delivered by the transmission system to the load, R_L. The power loss in the transmission system is referred to as **line loss**.

The efficiency of each system is given by

$$\eta_1 = \frac{P_{o_1}}{P_{i_1}} \qquad \eta_2 = \frac{P_{o_2}}{P_{i_2}} \qquad \eta_3 = \frac{P_{o_3}}{P_{i_3}}$$

If we form the product of these three efficiencies,

$$\eta_1 \cdot \eta_2 \cdot \eta_3 = \frac{P_{o_1}}{P_{i_1}} \cdot \frac{P_{o_2}}{P_{i_2}} \cdot \frac{P_{o_3}}{P_{i_3}}$$

and substitute the fact that $P_{i_2} = P_{o_1}$ and $P_{i_3} = P_{o_2}$, we find that the quantities indicated above will cancel, resulting in P_{o_3}/P_{i_1}, which is a measure of the efficiency of the entire system. In general, for the representative cascaded system of Fig. 4.20,

$$\eta_{\text{total}} = \eta_1 \cdot \eta_2 \cdot \eta_3 \cdots \eta_n \tag{4.20}$$

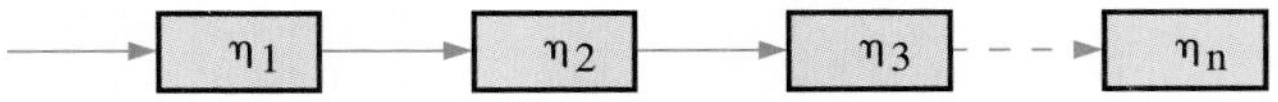

$\rightarrow$ η_1 $\rightarrow$ η_2 $\rightarrow$ η_3 - - $\rightarrow$ η_n

FIG. 4.20
Cascaded system.

EXAMPLE 4.13

Find the overall efficiency of the system of Fig. 4.19 if $\eta_1 = 90\%$, $\eta_2 = 85\%$, and $\eta_3 = 95\%$.

Solution:

$$\eta_T = \eta_1 \cdot \eta_2 \cdot \eta_3 = (0.90)(0.85)(0.95) = 0.727, \text{ or } \mathbf{72.7\%}$$

LUMINARIES

British (Salford, Manchester)
(1818-1889)
Physicist
Honorary doctorates from the Universities of Dublin and Oxford

FIG. 4.21
James Prescott Joule

Contributed to the important fundamental *law of conservation of energy* by establishing that various forms of energy (electrical, mechanical, or heat) are in the same family and can be exchanged from one form to another. In 1841 introduced *Joule's law*, which stated that the heat developed by electric current in a wire is proportional to the product of the current squared and the resistance of the wire (I^2R). He further determined that the heat emitted was equivalent to the power absorbed and therefore heat is a form of energy.

Bettman Archive Photo Number 076800P

EXAMPLE 4.14

If the efficiency η_1 drops to 40%, find the new overall efficiency and compare the result with that obtained in Example 4.13.

Solution:

$$\eta_T = \eta_1 \cdot \eta_2 \cdot \eta_3 = (0.40)(0.85)(0.95) = 0.323, \text{ or } \mathbf{32.3\%}$$

Certainly 32.3% is noticeably less than 72.7%. The total efficiency of a cascaded system is therefore determined primarily by the lowest efficiency (weakest link) and is less than (or equal to if the remaining efficiencies are 100%) the least efficient link of the system.

4.6 ENERGY

For power, which is the rate of doing work, to produce an energy conversion of any form, it must be *used over a period of time.* For example, a motor may have the horsepower to run a heavy load, but unless the motor is *used* over a period of time, there will be no energy conversion. In addition, the longer the motor is used to drive the load, the greater will be the energy expended.

The energy lost or gained by any system is therefore determined by

$$W = Pt \qquad \text{(wattseconds, Ws, or joules)} \qquad (4.21)$$

Since power is measured in watts (or joules per second) and time in seconds, the unit of energy is the *wattsecond* or *joule* (note Fig. 4.21) as indicated above. The wattsecond, however, is too small a quantity for most practical purposes, so the *watthour* (Wh) and *kilowatthour* (kWh) were defined, as follows:

$$\text{Energy (Wh)} = \text{power (W)} \times \text{time (h)} \qquad (4.22)$$

$$\text{Energy (kWh)} = \frac{\text{power (W)} \times \text{time (h)}}{1000} \qquad (4.23)$$

Note that the energy in kilowatthours is simply the energy in watthours divided by 1000. To develop some sense for the kilowatthour energy level, consider that 1 kWh is the energy dissipated by a 100-W bulb in 10 h.

While the kilowatthour is commonly used to measure electrical energy, it is not an SI unit. Since the SI unit for energy, the *joule*, is too small for practical applications, the *megajoule* (MJ) is frequently used for billing purposes in Canada. The kilowatthour can be expressed in megajoules as follows:

$$\begin{aligned} 1 \text{ kWh} &= 1000 \text{ (J/s) } (3600 \text{ s}) \\ &= 3\,600\,000 \text{ J} \\ &= 3.6 \text{ MJ} \end{aligned}$$

The *kilowatthour meter* is an instrument for measuring the energy supplied to the residential or commercial user of electricity. It is normally connected directly to the lines at a point just before entering the power distribution panel of the building. A typical set of dials is shown in Fig. 4.22(a) with a photograph of an analog kilowatthour meter. As indicated, each power of 10 below a dial is in kilowatthours. The quicker the aluminum disc rotates, the greater the energy demand. The dials are connected through a set of gears to the rotation of this disc. A solid-state digital meter with an extended range of capabilities is shown in Fig. 4.22(b).

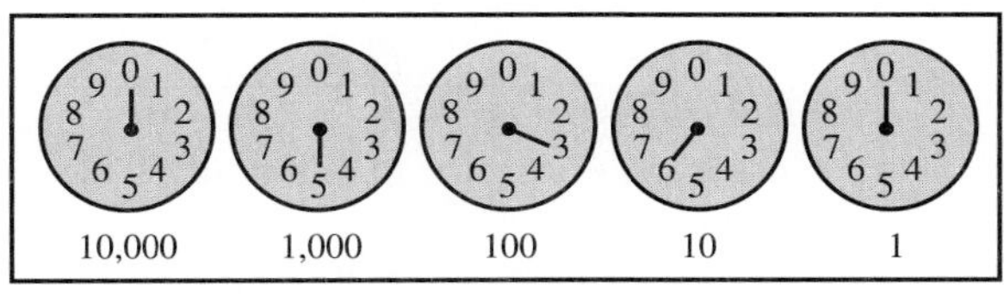

(a)

(b)

FIG. 4.22
Kilowatthour meters: (a) analog; (b) digital. (Courtesy of ABB Electric Metering Systems.)

EXAMPLE 4.15

For the dial positions of Fig. 4.22(a), calculate the electricity bill if the previous reading was 4650 kWh and the average cost is 9¢ per kilowatthour.

Solution:

$$5360 \text{ kWh} - 4650 \text{ kWh} = 710 \text{ kWh used}$$

$$710 \cancel{\text{kWh}}\left(\frac{9¢}{\cancel{\text{kWh}}}\right) = \mathbf{\$63.90}$$

EXAMPLE 4.16

How much energy (in kilowatthours) is required to light a 60-W bulb continuously for 1 year (365 days)?

Solution:

$$W = \frac{Pt}{1000} = \frac{(60 \text{ W})(24 \text{ h/day})(365 \text{ days})}{1000} = \frac{525\,600 \text{ Wh}}{1000}$$

$$= \mathbf{525.60\ kWh}$$

EXAMPLE 4.17

How long can a 205-W television set be on before using more than 4 kWh of energy?

Solution:

$$W = \frac{Pt}{1000} \Rightarrow t \text{ (hours)} = \frac{(W)(1000)}{P}$$

$$= \frac{(4 \text{ kWh})(1000)}{205 \text{ W}} = \mathbf{19.51\ h}$$

EXAMPLE 4.18

What is the cost of using a 5-hp motor for 2 h if the rate is 9¢ per kilowatthour?

Solution:

$$W\text{ (kilowatthours)} = \frac{Pt}{1000} = \frac{(5\text{ hp} \times 746\text{ W/hp})(2\text{ h})}{1000} = 7.46\text{ kWh}$$

$$\text{Cost} = (7.46\text{ kWh})(9¢/\text{kWh}) = \mathbf{67.14¢}$$

EXAMPLE 4.19

What is the total cost of using all of the following at 9¢ per kilowatthour?

A 1200-W toaster for 30 min
Six 50-W bulbs for 4 h
A 400-W washing machine for 45 min
A 4800-W electric clothes dryer for 20 min

Solution:

$$W = \frac{(1200\text{ W})(\tfrac{1}{2}\text{ h}) + (6)(50\text{ W})(4\text{ h}) + (400\text{ W})(\tfrac{3}{4}\text{ h}) + (4800\text{ W})(\tfrac{1}{3}\text{ h})}{1000}$$

$$= \frac{600\text{ Wh} + 1200\text{ Wh} + 300\text{ Wh} + 1600\text{ Wh}}{1000} = \frac{3700\text{ Wh}}{1000}$$

$$W = 3.7\text{ kWh}$$

$$\text{Cost} = (3.7\text{ kWh})(9¢/\text{kWh}) = \mathbf{33.3¢}$$

Table 4.1 lists some common household items with their typical wattage ratings. You might find it interesting to calculate the cost of operating some of these appliances over a period of time.

TABLE 4.1

Typical wattage ratings of some common household appliances.

Appliance	Wattage Rating	Appliance	Wattage Rating
Air conditioner	860	Microwave oven	800
Blow dryer	1 300	Phonograph	75
Cassette player/ recorder	5	Projector	1 200
Clock	2	Radio	70
Clothes dryer (electric)	4 800	Range (self-cleaning)	12 200
Coffee maker	900	Refrigerator (automatic defrost)	1 800
Dishwasher	1 200	Shaver	15
Fan:		Stereo equipment	110
Portable	90	Sun lamp	280
Window	200	Toaster	1 200
Heater	1 322	Trash compactor	400
Heating equipment:		TV (colour)	250
Furnace fan	320	Videocassette recorder	110
Oil-burner motor	230	Washing machine	400
Iron, dry or steam	1 100	Water heater	2 500

Courtesy of General Electric Co.

Practical Business Applications

Anyone for baked potato?

In today's fast-paced world, could any of us survive without modern conveniences such as the microwave oven? Who hasn't used this appliance? It only takes about 5 min. to cook a potato. What a great time saver!

So, how much power and energy are we consuming with a microwave? And how does this compare to the amount of nutritional energy our bodies receive by digesting a potato?

From your studies of electricity you know that energy, W, is just the product of power, P, and time, t.

$$W = Pt$$

Thus, the microwave energy, W_1, required to cook a potato for a typical 700 W (P_1) microwave oven is given by:

$$\begin{aligned} W_1 &= P_1 t_1 \\ &= (700\text{ W})(5\text{ min.})(60\text{ s/min.}) \\ &= 210\text{ kJ} \end{aligned}$$

But there's more to a microwave oven than just microwaves. It consists of a dc power supply, a magnetron (the microwave energy source), control circuits, and the oven enclosure. The dc power supply converts the 120 V ac household electricity from the wall outlet in your kitchen to approximately 4 kV dc.

This voltage then supplies the needed dc power to the magnetron, which generates the microwave energy at a frequency of 2.45 GHz (i.e., the frequency that will shake the water molecule's hydrogen–oxygen bonds so that water boils).

These magnetrons are typically only 55% efficient ($\eta_1 = 0.55$) since a significant amount of the input power is dissipated as self-heat. Thus, much more dc power (P_2), and hence energy (W_2), must go into the magnetron than we get out of it as microwave energy. This is given by the following:

$$\begin{aligned} P_2 &= P_1/\eta_1 = 700\text{ W}/0.55 = 1273\text{ W} \\ W_2 &= W_1/\eta_1 = 210\text{ kJ}/0.55 = 382\text{ kJ} \end{aligned}$$

Similarly, the dc power supply has an efficiency factor (η_2) which needs to be taken into account —it's approximately 95%. Thus, the ac power, P_3, and energy, W_3, we need from our kitchen outlet are as follows:

$$\begin{aligned} P_3 &= P_2/\eta_2 = 1273\text{ W}/0.95 = 1340\text{ W} \\ W_3 &= W_2/\eta_2 = 382\text{ kJ}/0.95 = 402\text{ kJ} \end{aligned}$$

By comparison, the nutritional value we get from digesting one medium-sized 150-g potato is 460 kJ. So, about 87% as much energy as we receive nutritionally is needed just to cook this potato. And that doesn't even consider all the energy needed for growing, cultivating, harvesting, and shipping the potato to our kitchen. We live in a very energy-dependent world.

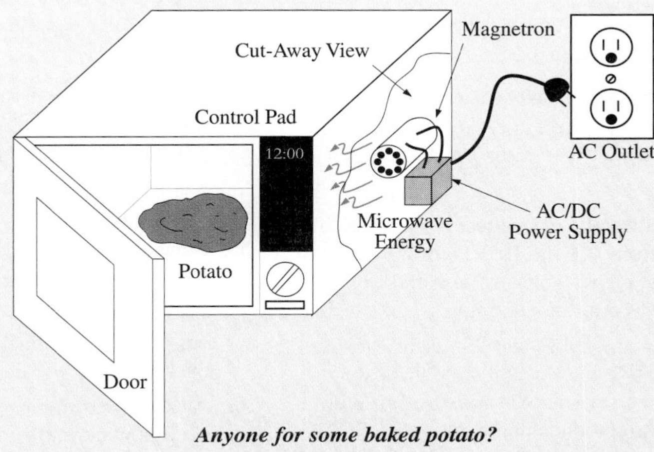

Anyone for some baked potato?

PROBLEMS

SECTION 4.1 Ohm's Law

1. What is the potential drop across a 6-Ω resistor if the current through it is 2.5 A?
2. What is the current through a 72-Ω resistor if the voltage drop across it is 12 V?
3. How much resistance is required to limit the current to 1.5 mA if the potential drop across the resistor is 6 V?
4. At starting, what is the current drain on a 12-V car battery if the resistance of the starting motor is 0.056 Ω?
5. If the current through a 0.02-MΩ resistor is 3.6 μA, what is the voltage drop across the resistor?
6. If a voltmeter has an internal resistance of 15 kΩ, find the current through the meter when it reads 62 V.
7. If a refrigerator draws 2.2 A at 120 V, what is its resistance?
8. If a clock has an internal resistance of 7.5 kΩ, find the current through the clock if it is plugged into a 120-V outlet.
9. A washing machine is rated at 4.2 A at 240 V. What is its internal resistance?
10. If a soldering iron draws 0.76 A at 120 V, what is its resistance?
11. The input current to a transistor is 20 μA. If the applied (input) voltage is 24 mV, determine the input resistance of the transistor.
12. When rated voltage is applied to a 120V, 100W lamp, the initial current is 12A. After 0.5 seconds, the current is 0.83A. Determine the change in resistance.
13. The internal resistance of a dc generator is 0.5 Ω. Determine the loss in terminal voltage across this internal resistance if the current is 15 A.

*14. **a.** If an electric heater draws 9.5 A when connected to a 120-V supply, what is the internal resistance of the heater?
b. Using the basic relationships of Chapter 2, how much energy is converted in 1 h?

SECTION 4.2 Plotting Ohm's Law

15. Plot the linear curves of a 100-Ω, a 10-Ω and 1.0-Ω resistor on the graph of Fig. 4.6. If necessary, reproduce the graph.
16. Sketch the characteristics of a device that has an internal resistance of 20 Ω from 0 V to 10 V and an internal resistance of 2 Ω for higher voltages. Use the axes of Fig. 4.6. If necessary, reproduce the graph.
17. Plot the linear curves of a 2-kΩ and 20-kΩ resistor on the graph of Fig. 4.6. Use a horizontal scale that extends from 0 V to 20 V and a vertical axis scaled off in milliamperes. If necessary, reproduce the graph.
18. What is the change in voltage across a 2-kΩ resistor established by a change in current of 400 mA through the resistor?

*19. **a.** Using the axis of Fig. 4.10, sketch the characteristics of a device that has an internal resistance of 500 Ω from 0 V to 1 V and 50 Ω between 1 and 2 V. Its resistance then changes to −20 Ω for higher voltages. The result is a set of characteristics very similar to those of an electronic device called a tunnel diode.
b. Using the above characteristics, determine the resulting current at voltages of 0.7 V, 1.5 V, and 2.5 V.

20. A 48V DC source has an internal resistance of 0.8Ω. If the maximum load is 10A, plot the change in load voltage as the load current changes from 0 to 10A.

SECTION 4.3 Power

21. If 420 J of energy are absorbed by a resistor in 7 min, what is the power to the resistor?
22. The power to a device is 40 J/s. How long will it take to deliver 640 J?
23. **a.** How many joules of energy does a 2-W nightlight dissipate in 8 h?
b. How many kilowatthours does it dissipate?
24. A resistor of 10 Ω has charge flowing through it at the rate of 300 C/min. How much power is dissipated?
25. How long must a steady current of 2 A exist in a resistor that has 3 V across it to dissipate 12 J of energy?
26. What is the power delivered by a 6-V battery if the charge flows at the rate of 48 C/min?
27. The current through a 4-Ω resistor is 7 mA. What is the power delivered to the resistor?
28. A 12-V battery has a 0.2 Ω internal resistance. When supplying a 1.5A load, determine the power dissipated by the internal resistance.
29. The voltage drop across a 3-Ω resistor is 9 mV. What is the power input to the resistor?
30. If the power input to a 4-Ω resistor is 64 W, what is the current through the resistor?
31. A 1/2-W resistor has a resistance of 1000 Ω. What is the maximum current that it can safely handle?
32. A 2.2-kΩ resistor in a stereo system dissipates 42 mW of power. What is the voltage across the resistor?
33. A dc battery can deliver 45 mA at 1.5 V. What is the power rating?
34. What are the "hot" resistance level and current rating of a 120-V, 100-W bulb?
35. A 150 W, 120 V resistive load is connected across a 100 V supply. Determine the power dissipated by the load at the lower voltage.
36. What are the internal resistance and voltage ratings of a 450-W automatic washer that draws 3.75 A?
37. A calculator with an internal 3-V battery draws 0.4 mW when fully functional. What is the current demand from the supply?
38. A 20-kΩ resistor has a rating of 100 W. What are the maximum current and the maximum voltage that can be applied to the resistor?

*39. **a.** Plot power versus current for a 100-Ω resistor. Use a power scale from 0 to 1 W and a current scale from 0 to 100 mA with divisions of 0.1 W and 10 mA, respectively.

b. Is the curve linear or nonlinear?
c. Using the resulting plot, determine the current at a power level of 500 mW.

*40. A small portable black-and-white television draws 0.455 A at 9 V.
a. What is the power rating of the television?
b. What is the internal resistance of the television?
c. What is the energy converted in 6 h of typical battery life?

*41. a. If a home is supplied with a 120-V, 100-A service, find the maximum power capability.
b. Can the homeowner safely operate the following loads at the same time?
A 5-hp motor
A 3000-W clothes dryer
A 2400-W electric range
A 1000-W steam iron

SECTION 4.5 Efficiency

42. What is the efficiency of a motor that has an output of 0.5 hp with an input of 450 W?

43. The motor of a power saw is rated 68.5% efficient. If 1.8 hp are required to cut a particular piece of lumber, what is the current drawn from a 120-V supply?

44. A 2 horsepower motor dissipates 100 W of energy as heat; a further 50W supplies the motor's internal losses. Calculate the efficiency of the motor.

45. What is the efficiency of a dryer motor that delivers 1 hp when the input current and voltage are 4 A and 220 V, respectively?

46. A stereo system draws 2.4 A at 120 V. If the audio output power is 50 W,
a. How much power is lost in the form of heat in the system?
b. What is the efficiency of the system?

47. If an electric motor having an efficiency of 87% and operating off a 220-V line delivers 3.6 hp, what input current does the motor draw?

48. A motor is rated to deliver 2 hp.
a. If it runs on 110 V and is 90% efficient, how many watts does it draw from the power line?
b. What is the input current?
c. What is the input current if the motor is only 70% efficient?

49. An electric motor used in an elevator system has an efficiency of 90%. If the input voltage is 220 V, what is the input current when the motor is delivering 15 hp?

50. A 2-hp motor drives a sanding belt. If the efficiency of the motor is 87% and that of the sanding belt 75% due to slippage, what is the overall efficiency of the system?

51. In testing, a 5 horsepower motor is found to be 85 percent efficient. Further testing shows that 475W was lost as heat. How much energy is stored within the motor?

52. If two systems in cascade each have an efficiency of 80% and the input energy is 60 J, what is the output energy?

53. The overall efficiency of two systems in cascade is 72%. If the efficiency of one is 0.9, what is the efficiency in percent of the other?

*54. If the total input and output power of two systems in cascade are 400 W and 128 W, respectively, what is the efficiency of each system if one has twice the efficiency of the other?

55. a. What is the total efficiency of three systems in cascade with efficiencies of 98%, 87%, and 21%?
b. If the system with the least efficiency (21%) were removed and replaced by one with an efficiency of 90%, what would be the percentage increase in total efficiency?
c. What would be the effect on the input current?

56. a. Perform the following conversions:
1 Wh to joules
1 kWh to joules
b. Based on the results of part (a), discuss when it is more appropriate to use one unit versus the other.

SECTION 4.6 Energy

57. A 10-Ω resistor is connected across a 15-V battery.
a. How many joules of energy will it dissipate in 1 min?
b. If the resistor is left connected for 2 min instead of 1 min, will the energy used increase? Will the power dissipation level increase?

58. How much energy in kilowatthours is required to keep a 230-W oil-burner motor running 12 h a week for 5 months (using $4\frac{1}{3}$ weeks = 1 month)?

59. How long can a 1500-W heater be on before using more than 36 MJ of energy?

60. How much does it cost to use a 30-W radio for 3 h at 9¢ per kilowatthour?

61. Over a 14-day period, a 10kW load consumed 2250 MJ of energy. How many hours did the load operate?

62. a. In 10 h an electrical system converts 500 kWh of electrical energy into heat. What is the power level of the system?
b. If the applied voltage is 208 V, what is the current drawn from the supply?
c. If the efficiency of the system is 82%, how much energy is lost or stored in 10 h?

63. a. At 9¢ per kilowatthour, how long can one play a 250-W colour television for $1?
b. For $1, how long can one use a 4.8-kW dryer?
c. Compare the results of parts (a) and (b) and comment on the effect of the wattage level on the relative cost of using an appliance.

64. What is the total cost of using the following at 9¢ per kilowatthour?
860-W air conditioner for 24 h
4800-W clothes dryer for 30 min
400-W washing machine for 1 h
1200-W dishwasher for 45 min

*65. What is the total cost of using the following at 2.5¢ per megajoule?
110-W stereo set for 4 h
1200-W projector for 20 min
60-W tape recorder for 1.5 h
150-W colour television set for 3 h 45 min

5

Series Circuits

Outline

Learning Outcomes

After completing this chapter you will be able to

- describe the characteristics of a series circuit
- explain the relationship between total resistance of a circuit and the individual resistances in a series circuit
- apply Ohm's law to series circuits
- determine the power consumed by the components in a series circuit
- explain Kirchhoff's voltage law and apply the law to series circuits
- determine the voltage across a resistive component using the voltage divider rule
- describe several types of circuit notation and their application in series circuits
- determine the effect of internal resistance upon the terminal voltage of a source
- explain the term voltage regulation and its relationship to internal resistance
- describe how meters are connected in order to determine circuit quantities

In a series circuit, the current is constant through all components. In this chapter, you will learn how voltage is distributed across the components that make up a series circuit, and how this relates to Ohm's law.

5.1 INTRODUCTION

Two types of current are readily available to the consumer today, **dc** and **ac**. In **direct current** (dc), the flow of current does not change in magnitude (or direction) with time. In **sinusoidal alternating current** (ac), the flow of current continually changes in magnitude (and direction) with time. The next few chapters introduce circuit analysis purely from a dc approach. Aspects that are different for ac will be mentioned where appropriate.

The battery of Fig. 5.1 has the ability to cause charge to flow through the simple circuit. The positive terminal attracts the electrons through the wire at the same rate that electrons are supplied by the negative terminal. As long as the battery is connected in the circuit and maintains its terminal characteristics, the current (dc) through the circuit will not change in magnitude or direction.

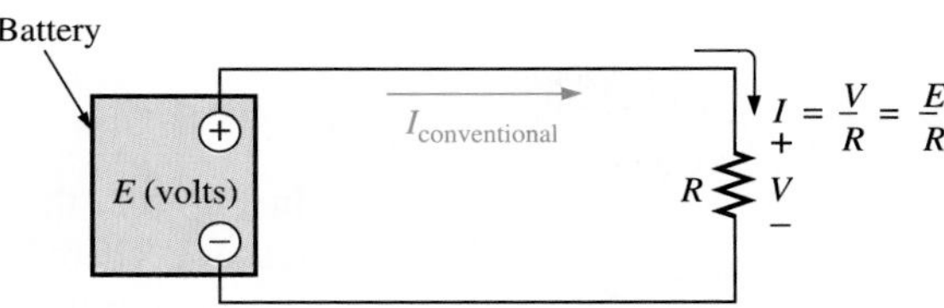

FIG. 5.1
Introducing the basic components of an electric circuit.

If we consider the wire to be an ideal conductor (having no resistance) the potential difference V across the resistor will equal the applied voltage of the battery: V (volts) $= E$ (volts).

The current is limited only by the resistor R. The higher the resistance, the less the current, as determined by Ohm's law.

The circuit of Fig. 5.1 is the simplest possible configuration. This chapter and the following chapters will add elements to the system in a very specific manner to introduce the concepts that will form a major part of the foundation needed to analyze the most complex system. Be aware that the laws, rules, and so on, introduced in Chapters 5 and 6 will be used throughout your studies of electrical, electronic, or computer systems. They will not be dropped for a more advanced set as you progress to more sophisticated material. It is therefore critical that you understand the concepts thoroughly and that you are able to apply the various procedures and methods with confidence.

5.2 SERIES CIRCUITS

A **circuit** consists of any number of elements joined at terminal points, providing at least one closed path through which charge can flow. The circuit of Fig. 5.2 has three elements joined at three terminal points (*a, b,* and *c*) to provide a closed path for the current I.

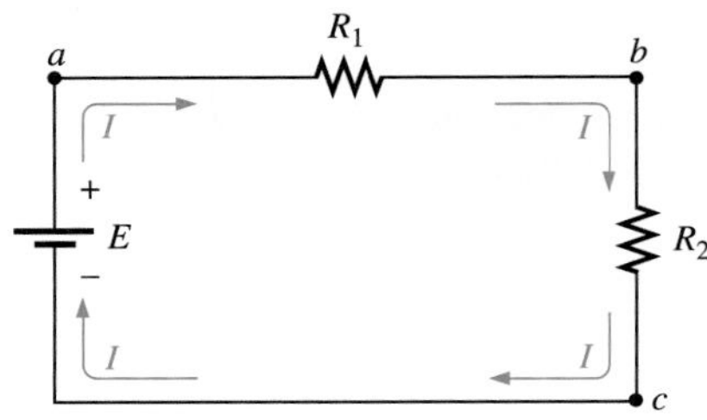

FIG. 5.2
Series circuit.

Two elements are in series if

1. *They have only one terminal in common (i.e., one lead of one is connected to only one lead of the other).*
2. *The common point between the two elements is not connected to another current-carrying element.*

In Fig. 5.2, the resistors R_1 and R_2 are in series because they have *only* point *b* in common. The other ends of the resistors are connected elsewhere in the circuit. For the same reason, the battery *E* and resistor R_1 are in series (terminal *a* in common) and the resistor R_2 and the battery *E* are in series (terminal *c* in common). Since all the elements are in series, the network is called a **series circuit**.

The current is the same through series elements.

Since there is only one path in Fig. 5.2, the current must flow through each resistor. Therefore, the current *I* through each resistor is the same as that through the battery. The fact that the current is the same through series elements is often used to determine whether two elements are in series.

A **branch** of a circuit is any portion of the circuit that has one or more elements in series. In Fig. 5.2, the resistor R_1 forms one branch of the circuit, the resistor R_2 another, and the battery *E* a third.

The total resistance of a series circuit is the sum of the separate resistances.

In Fig. 5.2, the current flowing in the circuit must pass through each resistor. Therefore, the total resistance encountered by the current must be equal to the sum of R_1 and R_2.

To find the total resistance of *N* resistors in series, the following equation is applied:

$$R_T = R_1 + R_2 + R_3 + \cdots + R_N \qquad \text{(ohms, } \Omega) \qquad (5.1)$$

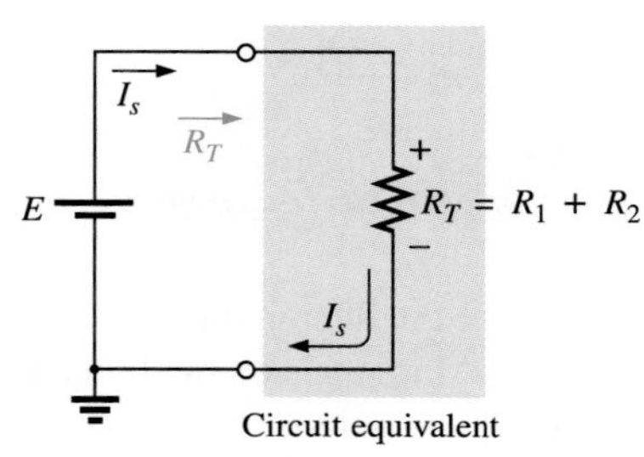

FIG. 5.3
Replacing the series resistors R_1 and R_2 of Fig. 5.2 with the total resistance.

Once the total resistance is known the circuit of Fig. 5.2 can be redrawn as shown in Fig. 5.3, clearly revealing that the only resistance the source "sees" is the total resistance. It is totally unaware of how the elements are connected to establish R_T. Once R_T is known, the current drawn from the source can be determined using Ohm's law, as follows:

$$I_s = \frac{E}{R_T} \qquad \text{(amperes, A)} \qquad (5.2)$$

Since *E* is fixed, the magnitude of the source current will be totally dependent on the magnitude of R_T. A larger R_T will result in a relatively small value of I_s, while lesser values of R_T will result in increased current levels.

The fact that the current is the same through each element of Fig. 5.2 permits a direct calculation of the voltage across each resistor using Ohm's law; that is,

$$V_1 = IR_1,\ V_2 = IR_2,\ V_3 = IR_3,\ \cdots,\ V_N = IR_N \qquad \text{(volts, V)} \qquad (5.3)$$

The power dissipated by each resistor can then be determined using any one of three equations as listed below for R_1.

$$P_1 = V_1 I_1 = I_1^2 R_1 = \frac{V_1^2}{R_1} \qquad \text{(watts, W)} \qquad (5.4)$$

The power delivered by the source is

$$P_T = EI \qquad \text{(watts, W)} \qquad (5.5)$$

Power Relations in a Series Circuit

The relation between the total power delivered by the source and the individual powers dissipated in the elements of a series circuit can be easily derived from the foregoing equations.

The same current I flows through every element of the circuit. The total resistance R_T is given by

$$R_T = R_1 + R_2 + R_3 + \dots + R_N$$

The total power delivered by the source is given by

$$P_T = 1^2R$$

since

$$R_T = R_1 + R_2 + \dots + R_N$$

Then

$$P_T = I^2(R_1 + R_2 + \dots + R_N)$$

$$P_T = I^2R_1 + I^2R_2 + \dots + I^2R_N$$

and

$$I = I_1 = I_2 = \dots = I_N$$

then

$$P_T = I_1^2R_1 + I_2^2R_2 + \dots + I_N^2R_N$$

where

$$P_x = I_x^2R_x$$

therefore

$$P_T = \mathrm{P}_1 + \mathrm{P}_2 + \dots + P_N \tag{5.6}$$

The total power delivered to a resistive network is equal to the total power dissipated by the resistive elements.

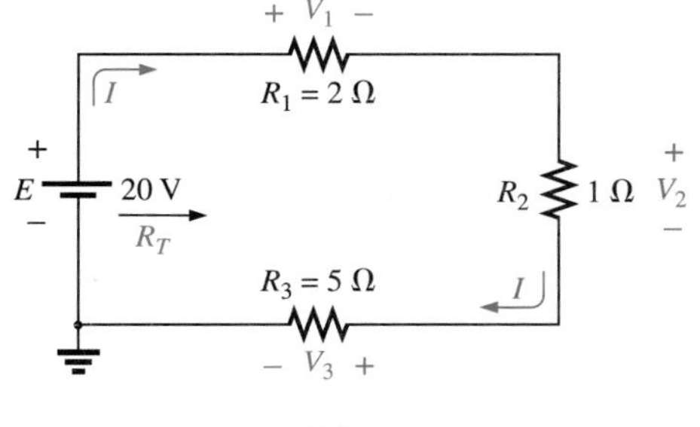

FIG. 5.4
Example 5.1.

EXAMPLE 5.1

a. Find the total resistance for the series circuit of Fig. 5.4.
b. Calculate the source current I_s.
c. Determine the voltages V_1, V_2, and V_3.
d. Calculate the power dissipated by R_1, R_2, and R_3.
e. Determine the power delivered by the source and compare it to the sum of the power levels of part (d).

Solution:

a. $R_T = R_1 + R_2 + R_3 = 2\ \Omega + 1\ \Omega + 5\ \Omega = \mathbf{8\ \Omega}$

b. $I_s = \dfrac{E}{R_T} = \dfrac{20\ \text{V}}{8\ \Omega} = \mathbf{2.5\ A}$

continued

c. $V_1 = IR_1 = (2.5\text{ A})(2\ \Omega) = \mathbf{5\ V}$
$V_2 = IR_2 = (2.5\text{ A})(1\ \Omega) = \mathbf{2.5\ V}$
$V_3 = IR_3 = (2.5\text{ A})(5\ \Omega) = \mathbf{12.5\ V}$

d. $P_1 = V_1I_1 = (5\text{ V})(2.5\text{ A}) = \mathbf{12.5\ W}$
$P_2 = I_2^2R_2 = (2.5\text{ A})^2(1\ \Omega) = \mathbf{6.25\ W}$
$P_3 = V_3^2/R_3 = (12.5\text{ V})^2/5\ \Omega = \mathbf{31.25\ W}$

e. $P_{del} = EI = (20\text{ V})(2.5\text{ A}) = \mathbf{50\ W}$
$P_{del} = P_1 + P_2 + P_3$
$50\text{ W} = 12.5\text{ W} + 6.25\text{ W} + 31.25\text{ W}$
$50\text{ W} = 50\text{ W}$ (checks)

To find the total resistance of N resistors of the same value in series, simply multiply the value of *one* of the resistors by the number in series; that is,

$$R_T = NR \tag{5.7}$$

EXAMPLE 5.2

Determine R_T, I, and V_2 for the circuit of Fig. 5.5.

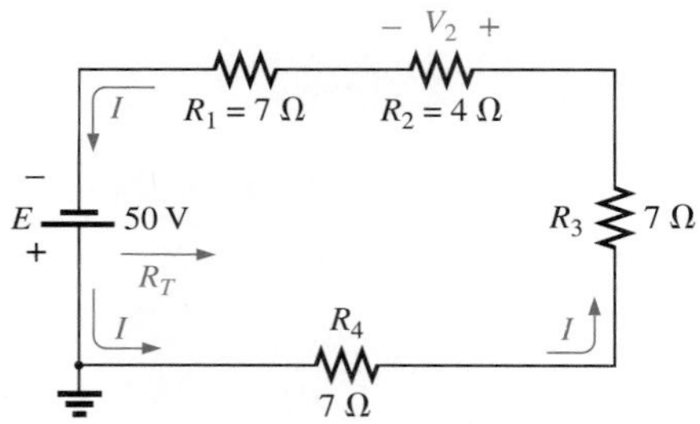

Solution: Note the current direction as established by the battery and the polarity of the voltage drops across R_2 as determined by the current direction. Since $R_1 = R_3 = R_4$,

$$R_T = NR_1 + R_2 = (3)(7\ \Omega) + 4\ \Omega = 21\ \Omega + 4\ \Omega = \mathbf{25\ \Omega}$$

$$I = \frac{E}{R_T} = \frac{50\text{ V}}{25\ \Omega} = \mathbf{2\ A}$$

$$V_2 = IR_2 = (2\text{ A})(4\ \Omega) = \mathbf{8\ V}$$

FIG. 5.5
Example 5.2.

Examples 5.1 and 5.2 are straightforward substitution-type problems that are relatively easy to solve with some practice. Example 5.3, however, is a different type of problem that requires a firm grasp of the fundamental equations and an ability to identify which equation to use first. The best preparation for this type of exercise is simply to work through as many problems of this kind as possible.

EXAMPLE 5.3

Given R_T and I, calculate R_1 and E for the circuit of Fig. 5.6.

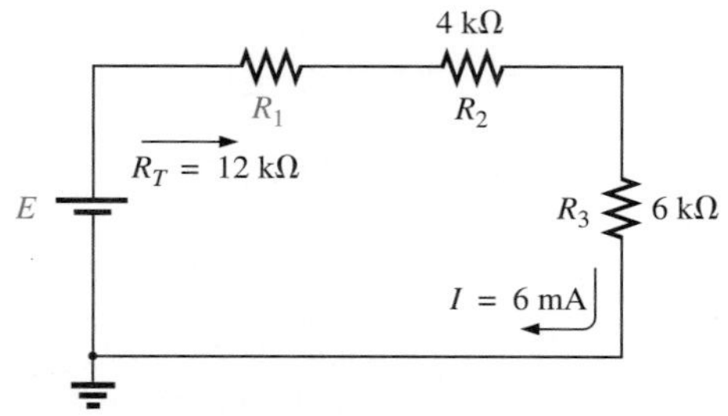

Solution:

$$R_T = R_1 + R_2 + R_3$$
$$12\text{ k}\Omega = R_1 + 4\text{ k}\Omega + 6\text{ k}\Omega$$
$$R_1 = 12\text{ k}\Omega - 10\text{ k}\Omega = \mathbf{2\ k\Omega}$$
$$E = IR_T = (6 \times 10^{-3}\text{ A})(12 \times 10^3\ \Omega) = \mathbf{72\ V}$$

FIG. 5.6
Example 5.3.

5.3 VOLTAGE SOURCES IN SERIES

Voltage sources can be connected in series, as shown in Fig. 5.7, to increase or decrease the total voltage applied to a system. The net voltage is determined simply by summing the sources with the same polarity and subtracting the total of the sources with the opposite polarity. The net polarity is the polarity of the larger sum.

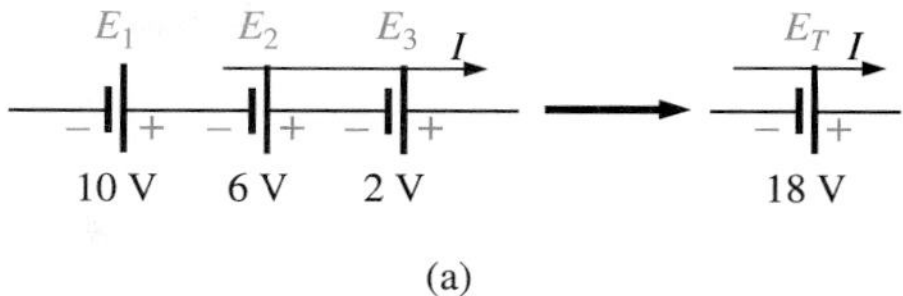

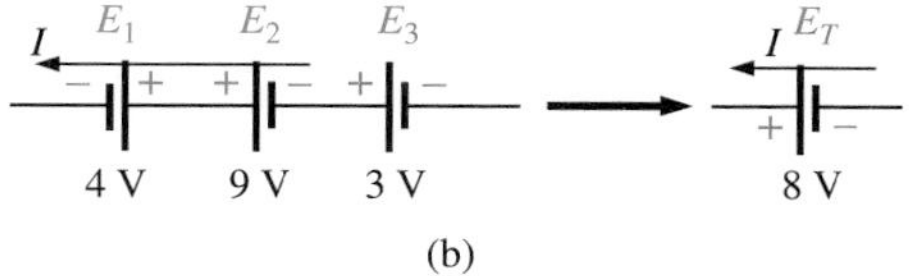

FIG. 5.7
Reducing series dc voltage sources to a single source.

Contributed to a number of areas in the physics domain, but is best known for his work in the electrical area where he defined the relationships between the currents and voltages of a network in 1847. Did extensive research with German chemist Robert Bunsen (who developed the *Bunsen burner*), resulting in the discovery of the important elements *cesium* and *rubidium*.

Courtesy of the Smithsonian Institution, Photo No. 58,283

In Fig. 5.7(a), for example, the sources are all "pressuring" current to the right, so the net voltage is

$$E_T = E_1 + E_2 + E_3 = 10\text{ V} + 6\text{ V} + 2\text{ V} = 18\text{ V}$$

as shown in the figure. In Fig. 5.7(b), however, the greater "pressure" is to the left, with a net voltage of

$$E_T = E_2 + E_3 - E_1 = 9\text{ V} + 3\text{ V} - 4\text{ V} = 8\text{ V}$$

and the polarity shown in the figure.

5.4 KIRCHHOFF'S VOLTAGE LAW

Note Fig. 5.8.

Kirchhoff's voltage law (KVL) states that the algebraic sum of the potential rises and drops around a closed loop (or path) is zero.

A **closed loop** is any continuous path that leaves a point in one direction and returns to that same point from another direction without leaving the circuit. In Fig. 5.9, by following the current, we can trace a continuous path that leaves point *a* through R_1 and returns through E without leaving the circuit. Therefore, *abcda* is a closed loop. For us to be able to apply Kirchhoff's voltage law, the summation of potential rises and drops must be made in one direction around the closed loop.

For consistency, the clockwise direction will be used throughout the text for all applications of Kirchhoff's voltage law. However, you will get the same result if you choose the counterclockwise direction and apply the law correctly.

A plus sign is used for a potential rise ($-$ to $+$) and a minus sign for a potential drop ($+$ to $-$). If we follow the current in Fig. 5.9 from point *a,* we first see a potential drop V_1 ($+$ to $-$) across R_1 and then another potential drop V_2 across R_2. Continuing through the voltage source, we have a potential rise E ($-$ to $+$) before returning to point *a*. Using symbols, where Σ represents summation, $\circlearrowright$ the closed loop, and V the potential drops and rises, we have

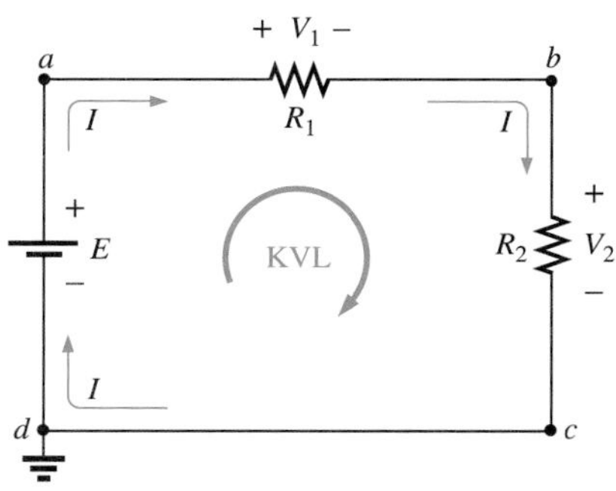

FIG. 5.9
Applying Kirchhoff's voltage law to a series dc circuit.

$$\Sigma_{\circlearrowright} V = 0 \tag{5.8}$$

which gives the following for the circuit of Fig. 5.9 (clockwise direction, following the current I and starting at point d):

$$+E - V_1 - V_2 = 0$$

or

$$E = V_1 + V_2$$

revealing that

the applied voltage of a series circuit equals the sum of the voltage drops across the series elements.

Kirchhoff's voltage law can also be stated in the following form:

$$E_1 + E_2 + \ldots + E_N = V_1 + V_2 + \ldots + V_N$$

or

$$\Sigma_{\circlearrowright} V_{\text{rises}} = \Sigma_{\circlearrowright} V_{\text{drops}} \tag{5.9}$$

which states that the sum of the rises around a closed loop must equal the sum of the drops in potential. The text will emphasize the use of Eq. (5.8), however.

If the loop is taken in the counterclockwise direction starting at point a, the following would result:

$$\Sigma_{\circlearrowright} V = 0$$

$$-E + V_2 + V_1 = 0$$

or, as before,

$$E = V_1 + V_2$$

The application of Kirchhoff's voltage law does not need to follow a path that includes current-carrying elements.

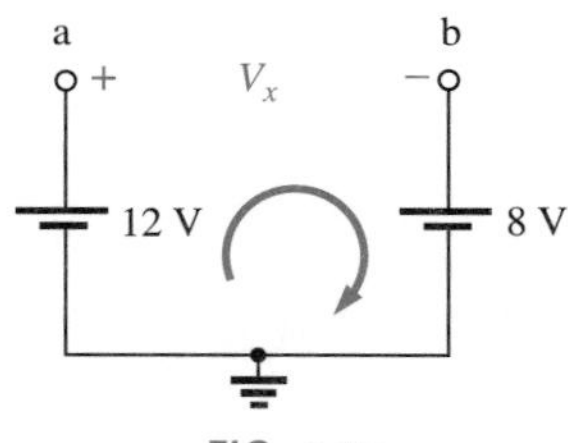

FIG. 5.10
Demonstration that a voltage can exist between two points not connected by a current-carrying conductor.

For example, in Fig. 5.10 there is a difference in potential between points a and b, even though the two points are not connected by a current-carrying element. Applying Kirchhoff's voltage law around the closed loop will result in a difference in potential of 4 V between the two points. That is, using the clockwise direction:

$$+12\text{ V} - V_x - 8\text{ V} = 0$$

and

$$V_x = \mathbf{4\text{ V}}$$

EXAMPLE 5.4

Determine the unknown voltages for the networks of Fig. 5.11.

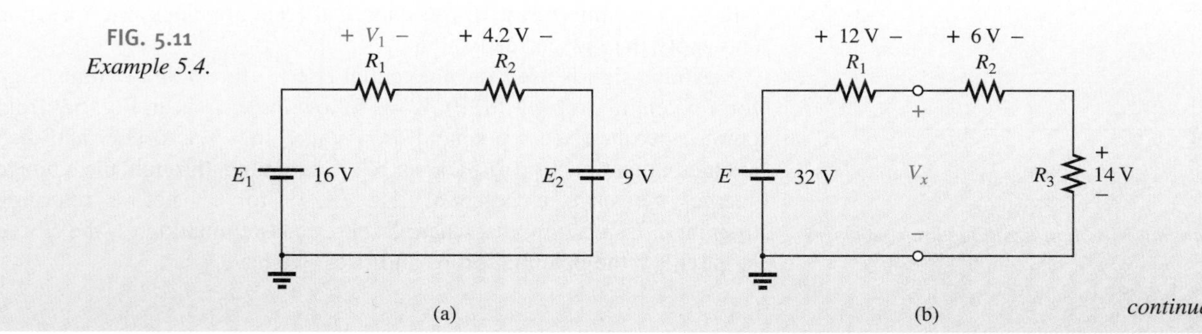

FIG. 5.11
Example 5.4.

continued

Solution: When applying Kirchhoff's voltage law, be sure to concentrate on the polarities of the voltage rise or drop rather than the type of element. In other words, do not treat a voltage drop across a resistive element differently from a voltage drop across a source. If the polarity shows that a drop has occurred, that is the important fact when applying the law. In Fig. 5.11(a), for instance, if we choose the clockwise direction, we will find that there is a drop across the resistors R_1 and R_2 and a drop across the source E_2. All will therefore have a minus sign when Kirchhoff's voltage law is applied.

Applying Kirchhoff's voltage law to the circuit of Fig. 5.11(a) in the clockwise direction will result in

$$+E_1 - V_1 - V_2 - E_2 = 0$$

and

$$V_1 = E_1 - V_2 - E_2 = 16\text{ V} - 4.2\text{ V} - 9\text{ V}$$
$$= \mathbf{2.8\ V}$$

The result clearly indicates that there was no need to know the values of the resistors or the current to determine the unknown voltage. The other voltage levels carried enough information to determine the unknown.

In Fig. 5.11(b) the unknown voltage is not across a current-carrying element. However, as indicated in the paragraphs above, Kirchhoff's voltage law is not limited to current-carrying elements. In this case there are two possible paths for finding the unknown. Using the clockwise path, including the voltage source E, will result in

$$+E - V_1 - V_x = 0$$

and

$$V_x = E - V_1 = 32\text{ V} - 12\text{ V}$$
$$= \mathbf{20\ V}$$

Using the clockwise direction for the other loop involving R_2 and R_3 will result in

$$+V_x - V_2 - V_3 = 0$$

and

$$V_x = V_2 + V_3 = 6\text{ V} + 14\text{ V}$$
$$= \mathbf{20\ V}$$

matching the result above.

EXAMPLE 5.5

Find V_1 and V_2 for the network of Fig. 5.12.

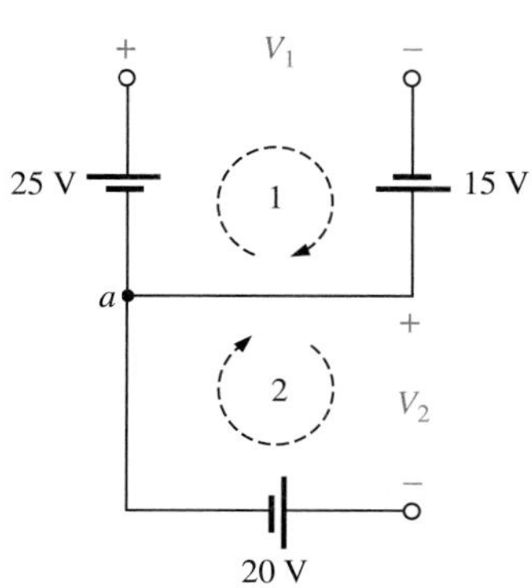

FIG. 5.12
Example 5.5.

Solution: For path 1, starting at point a in a clockwise direction:

$$+25\text{ V} - V_1 + 15\text{ V} = 0$$

and

$$V_1 = \mathbf{40\ V}$$

For path 2, starting at point a in a clockwise direction:

$$-V_2 - 20\text{ V} = 0$$

and

$$V_2 = \mathbf{-20\ V}$$

The minus sign simply indicates that the actual polarities of the potential difference are opposite the assumed polarity indicated in Fig. 5.12.

The next example will emphasize that the polarities of the voltage rise or drop are the important parameters when applying Kirchhoff's voltage law, not the type of element involved.

EXAMPLE 5.6

Using Kirchhoff's voltage law, determine the unknown voltages for the networks of Fig. 5.13.

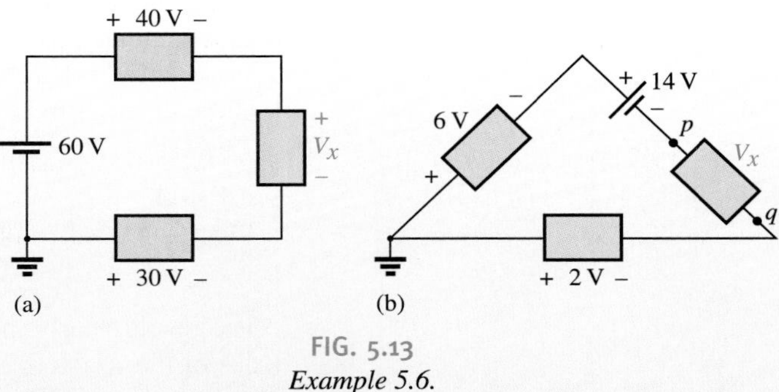

FIG. 5.13
Example 5.6.

Solutions: Note in each circuit that there are various polarities across the unknown elements, since they can contain any mixture of components. Applying Kirchhoff's voltage law to the network of Fig. 5.13(a) in the clockwise direction will result in

$$60\text{ V} - 40\text{ V} - V_x + 30\text{ V} = 0$$

and
$$V_x = 60\text{ V} + 30\text{ V} - 40\text{ V} = 90\text{ V} - 40\text{ V}$$
$$= \mathbf{50\text{ V}}$$

In Fig. 5.13(b) the polarity of the unknown voltage is not provided. In such cases, make an assumption about the polarity and apply Kirchhoff's voltage law as before. If the result has a plus sign, the assumed polarity was correct. If it has a minus sign, the magnitude is correct but the assumed polarity has to be reversed. In this case if we assume *p* to be positive and *q* to be negative, applying Kirchhoff's voltage law in the clockwise direction will result in

$$-6\text{ V} - 14\text{ V} - V_x + 2\text{ V} = 0$$

and
$$V_x = -20\text{ V} + 2\text{ V}$$
$$= \mathbf{-18\text{ V}}$$

Since the result is negative, we know that *p* should be negative and *q* should be positive, but the magnitude of 18 V is correct.

EXAMPLE 5.7

For the circuit of Fig. 5.14:

a. Find R_T.
b. Find I.
c. Find V_1 and V_2.
d. Find the power to the 4-Ω and 6-Ω resistors.
e. Find the power delivered by the battery, and compare it to that dissipated by the 4-Ω and 6-Ω resistors combined.
f. Verify Kirchhoff's voltage law (clockwise direction).

continued

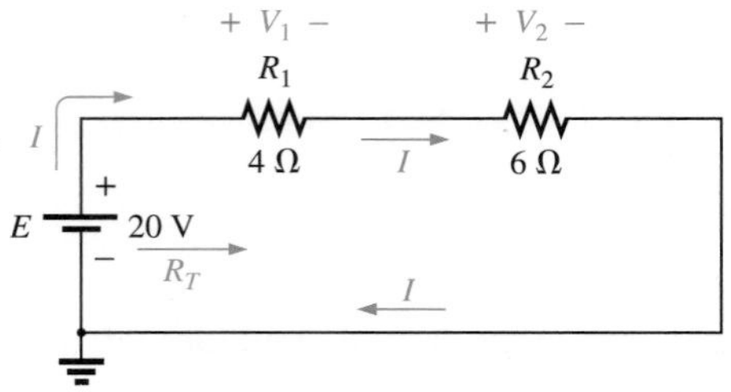

FIG. 5.14
Example 5.7.

Solutions:

a. $R_T = R_1 + R_2 = 4\ \Omega + 6\ \Omega = \mathbf{10\ \Omega}$

b. $I = \dfrac{E}{R_T} = \dfrac{20\text{ V}}{10\ \Omega} = \mathbf{2\text{ A}}$

c. $V_1 = IR_1 = (2\text{ A})(4\ \Omega) = \mathbf{8\text{ V}}$
$V_2 = IR_2 = (2\text{ A})(6\ \Omega) = \mathbf{12\text{ V}}$

d. $P_{4\Omega} = \dfrac{V_1^2}{R_1} = \dfrac{(8\text{ V})^2}{4} = \dfrac{64}{4} = \mathbf{16\text{ W}}$

$P_{6\Omega} = I^2R_2 = (2\text{ A})^2(6\ \Omega) = (4)(6) = \mathbf{24\text{ W}}$

e. $P_E = EI = (20\text{ V})(2\text{ A}) = \mathbf{40\text{ W}}$
$P_E = P_{4\Omega} + P_{6\Omega}$
$40\text{ W} = 16\text{ W} + 24\text{ W}$
$40\text{ W} = 40\text{ W}$ (checks)

f. $\Sigma_{\circlearrowright} V = +E - V_1 - V_2 = 0$
$E = V_1 + V_2$
$20\text{ V} = 8\text{ V} + 12\text{ V}$
$20\text{ V} = 20\text{ V}$ (checks)

EXAMPLE 5.8

For the circuit of Fig. 5.15:

a. Determine V_2 using Kirchhoff's voltage law.
b. Determine I.
c. Find R_1 and R_3.

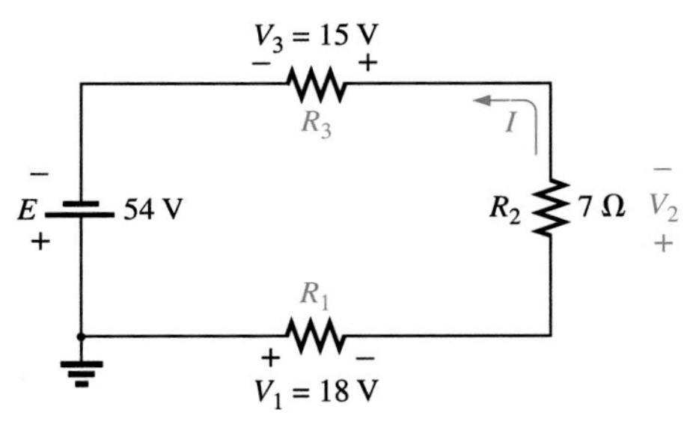

FIG. 5.15
Example 5.8.

Solutions:

a. Kirchhoff's voltage law (clockwise direction):

$$-E + V_3 + V_2 + V_1 = 0$$

or
$$E = V_1 + V_2 + V_3$$

and $V_2 = E - V_1 - V_3 = 54\text{ V} - 18\text{ V} - 15\text{ V} = \mathbf{21\text{ V}}$

b. $I = \dfrac{V_2}{R_2} = \dfrac{21\text{ V}}{7\ \Omega} = \mathbf{3\text{ A}}$

c. $R_1 = \dfrac{V_1}{I} = \dfrac{18\text{ V}}{3\text{ A}} = \mathbf{6\ \Omega}$

$R_3 = \dfrac{V_3}{I} = \dfrac{15\text{ V}}{3\text{ A}} = \mathbf{5\ \Omega}$

5.5 INTERCHANGING SERIES ELEMENTS

The elements of a series circuit can be interchanged without affecting the total resistance, current, or power to each element. For instance, the network of Fig. 5.16 can be redrawn as shown in Fig. 5.17 without affecting I or V_2. The total resistance R_T is 35 Ω in both cases, and I = 70 V/35 Ω = 2 A. The voltage $V_2 = IR_2 = (2\text{ A})(5\ \Omega) = 10\text{ V}$ for both configurations.

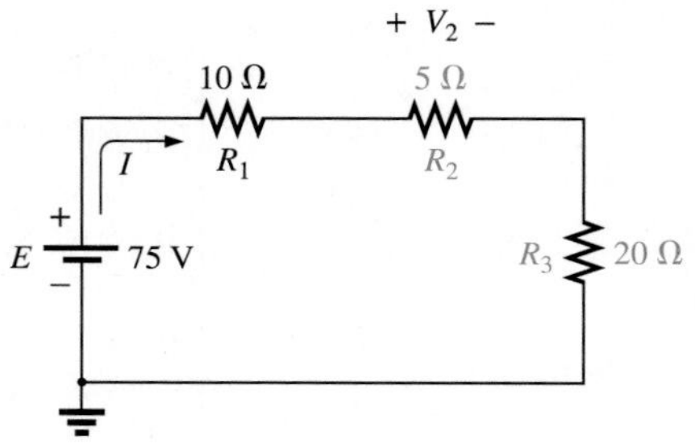

FIG. 5.16
Series dc circuit with elements to be interchanged.

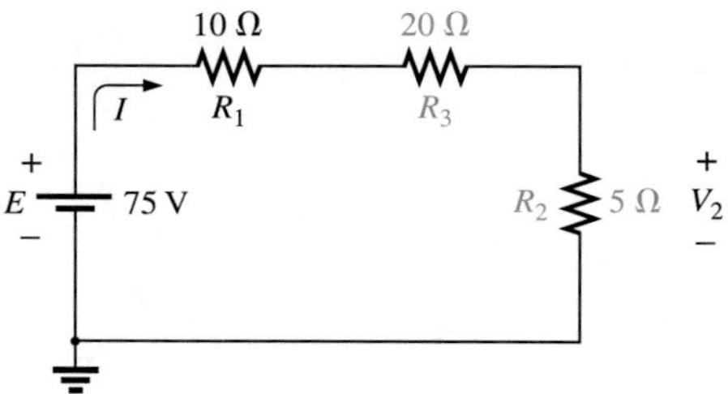

FIG. 5.17
Circuit of Fig. 5.16 with R_2 and R_3 interchanged.

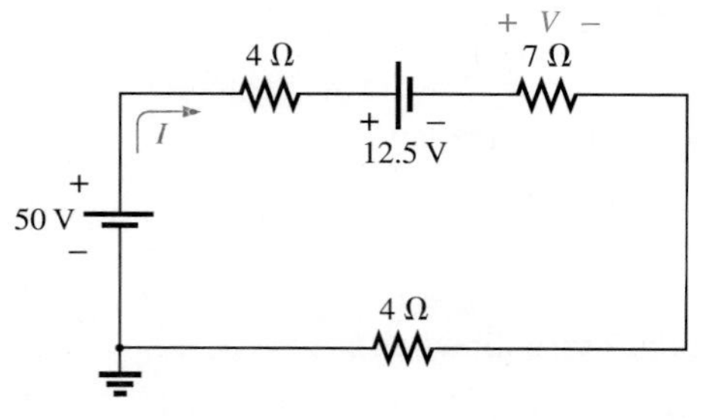

FIG. 5.18
Example 5.9.

EXAMPLE 5.9

Determine I and the voltage across the 7-Ω resistor for the network of Fig. 5.18.

Solution: The network is redrawn in Fig. 5.19.

$$R_T = (2)(4\ \Omega) + 7\ \Omega = 15\ \Omega$$

$$I = \frac{E}{R_T} = \frac{37.5\ \text{V}}{15\ \Omega} = \mathbf{2.5\ A}$$

$$V_{7\Omega} = IR = (2.5\ \text{A})(7\ \Omega) = \mathbf{17.5\ V}$$

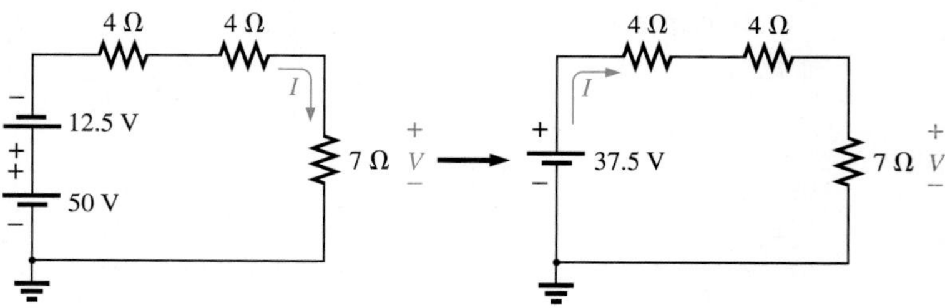

FIG. 5.19
Redrawing the circuit of Fig. 5.18.

5.6 VOLTAGE DIVIDER RULE

In a series circuit,

the voltage across the resistive elements will divide in proportion to the magnitude of the resistances.

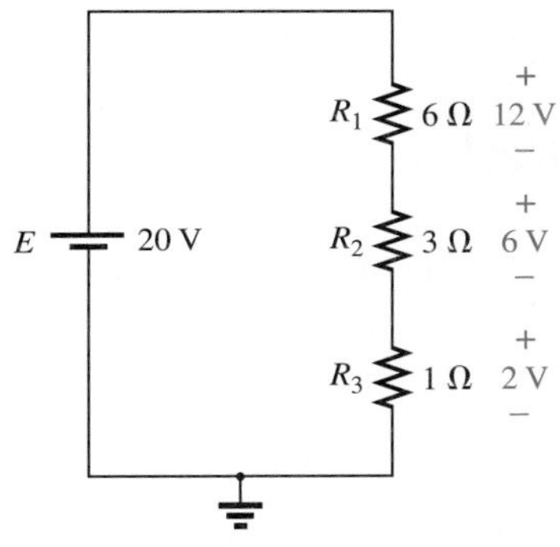

FIG. 5.20
Revealing how the voltage will divide across series resistive elements.

For example, the voltages across the resistive elements of Fig. 5.20 are provided. The largest resistor of 6 Ω captures most of the applied voltage, while the smallest resistor R_3 has the least. Note in addition that, since the resistance of R_1 is 6 times that of R_3, the voltage across R_1 is 6 times that of R_3. The fact that the resistance level of R_2 is 3 times that of R_1 results in three times the voltage across R_2. Finally, since R_1 is twice R_2, the voltage across R_1 is twice that of R_2. In general, therefore, the voltage across series resistors will have the same ratio as their resistances.

It is particularly interesting to note that, if the resistances of all the resistors of Fig. 5.20 are multiplied by the same factor, as shown in Fig. 5.21, the voltage levels will all remain the same. In other words, even though the resistances were increased by a factor of 1 million, the voltage ratios remain the same. Clearly, therefore, it is the ratio of resistor values that counts when it comes to voltage division and not the relative magnitude of all the resistors. The current level of the network will be severely affected by the change in resistance from Fig. 5.20 to Fig. 5.21, but the voltage levels will remain the same.

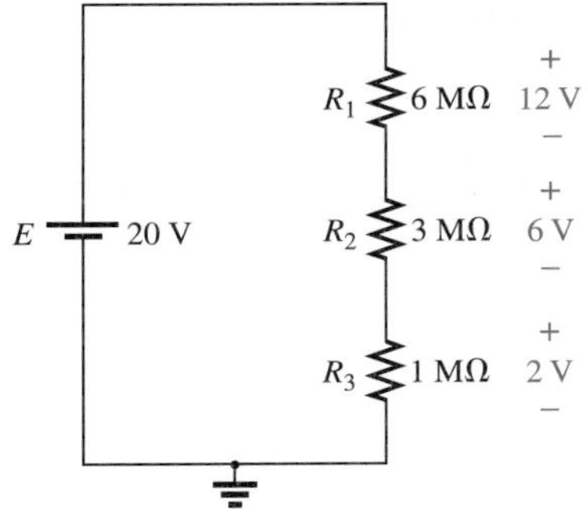

FIG. 5.21
The ratio of the resistive values determines the voltage division of a series dc circuit.

Based on the above, a first glance at the series network of Fig. 5.22 should suggest that the major part of the applied voltage will appear across the 1-MΩ resistor and very little across the 100-Ω resistor. In fact 1 MΩ = (1000)1 kΩ = (10 000)100 Ω, revealing that $V_1 = 1000V_2 = 10\,000V_3$.

Solving for the current and then the three voltage levels will result in

$$I = \frac{E}{R_T} = \frac{100\text{ V}}{1\,001\,100\ \Omega} \cong 99.89\ \mu\text{A}$$

and

$$V_1 = IR_1 = (99.89\ \mu\text{A})(1\text{ M}\Omega) = \mathbf{99.89\ V}$$
$$V_2 = IR_2 = (99.89\ \mu\text{A})(1\text{ k}\Omega) = \mathbf{99.89\ mV} = 0.099\,89\text{ V}$$
$$V_3 = IR_3 = (99.89\ \mu\text{A})(100\ \Omega) = \mathbf{9.989\ mV} = 0.009\,989\text{ V}$$

clearly supporting the above conclusions. For the future, therefore, use this approach to estimate the share of the input voltage across series elements as a check against the actual calculations or to obtain an estimate easily.

In the above discussion the current was determined before the voltages of the network were determined. There is, however, a method referred to as the *voltage divider rule* that lets you determine the voltage levels without first finding the current.

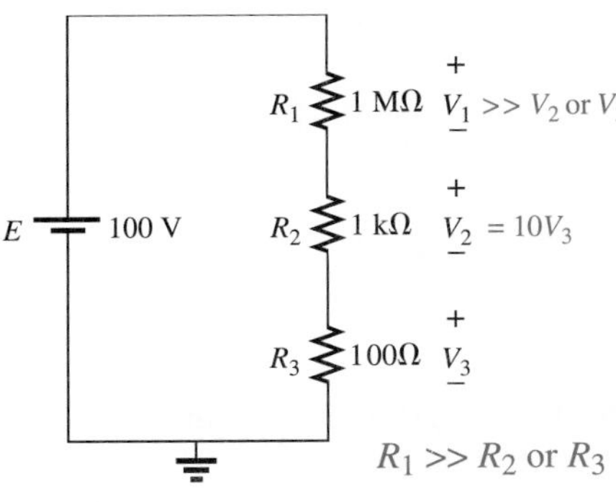

FIG. 5.22
The largest of the series resistive elements will capture the major share of the applied voltage.

The rule can be derived by analyzing the network of Fig. 5.23.

$$R_T = R_1 + R_2$$

and

$$I = \frac{E}{R_T}$$

Applying Ohm's law:

$$V_1 = IR_1 = \left(\frac{E}{R_T}\right)R_1 = \frac{R_1E}{R_T}$$

with

$$V_2 = IR_2 = \left(\frac{E}{R_T}\right)R_2 = \frac{R_2E}{R_T}$$

continued

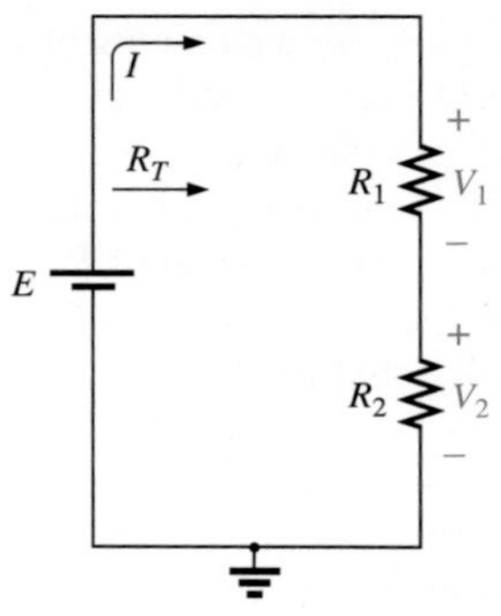

FIG. 5.23
Developing the voltage divider rule.

Note that the format for V_1 and V_2 is

$$V_x = \frac{R_x E}{R_T} \quad \text{(voltage divider rule)} \quad (5.10)$$

where V_x is the voltage across R_x, E is the impressed voltage across the series elements, and R_T is the total resistance of the series circuit.

In words, the **voltage divider rule** states that

the voltage across a resistor in a series circuit is equal to the value of that resistor times the total impressed voltage across the series elements divided by the total resistance of the series elements.

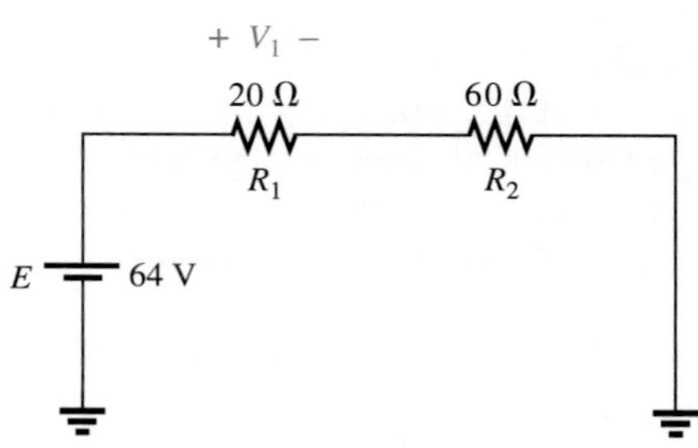

FIG. 5.24
Example 5.10.

EXAMPLE 5.10

Determine the voltage V_1 for the network of Fig. 5.24.

Solution: Eq. (5.10):

$$V_1 = \frac{R_1 E}{R_T} = \frac{R_1 E}{R_1 + R_2} = \frac{(20\ \Omega)(64\ \text{V})}{20\ \Omega + 60\ \Omega} = \frac{1280\ \text{V}}{80} = \mathbf{16\ V}$$

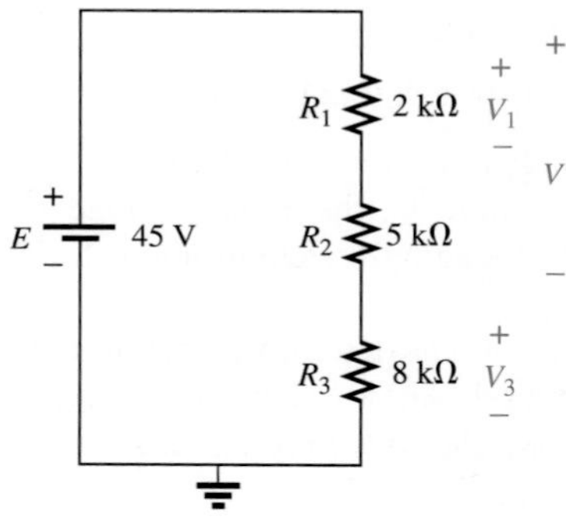

FIG. 5.25
Example 5.11.

EXAMPLE 5.11

Using the voltage divider rule, determine the voltages V_1 and V_3 for the series circuit of Fig. 5.25.

Solution:

$$V_1 = \frac{R_1 E}{R_T} = \frac{(2\ \text{k}\Omega)(45\ \text{V})}{2\ \text{k}\Omega + 5\ \text{k}\Omega + 8\ \text{k}\Omega} = \frac{(2\ \text{k}\Omega)(45\ \text{V})}{15\ \text{k}\Omega}$$

$$= \frac{(2 \times 10^3\ \Omega)(45\ \text{V})}{15 \times 10^3\ \Omega} = \frac{90\ \text{V}}{15} = \mathbf{6\ V}$$

$$V_3 = \frac{R_3 E}{R_T} = \frac{(8\ \text{k}\Omega)(45\ \text{V})}{15\ \text{k}\Omega} = \frac{(8 \times 10^3\ \Omega)(45\ \text{V})}{15 \times 10^3\ \Omega}$$

$$= \frac{360\ \text{V}}{15} = \mathbf{24\ V}$$

The rule can be extended to the voltage across two or more series elements by expanding the resistance in the numerator of Eq. (5.10) to include the total resistance of the series elements that the voltage is to be found across (R'); that is,

$$V' = \frac{R'E}{R_T} \quad \text{(volts)} \quad (5.11)$$

EXAMPLE 5.12

Determine the voltage V' in Fig. 5.25 across resistors R_1 and R_2.

Solution:

$$V' = \frac{R'E}{R_T} = \frac{(2\ \text{k}\Omega + 5\ \text{k}\Omega)(45\ \text{V})}{15\ \text{k}\Omega} = \frac{(7\ \text{k}\Omega)(45\ \text{V})}{15\ \text{k}\Omega} = \mathbf{21\ V}$$

There is also no need for the voltage E in the equation to be the source voltage of the network. For example, if V is the total voltage across a number of series elements such as those shown in Fig. 5.26, then

$$V_{2\Omega} = \frac{(2\ \Omega)(27\ \text{V})}{4\ \Omega + 2\ \Omega + 3\ \Omega} = \frac{54\ \text{V}}{9} = \mathbf{6\ V}$$

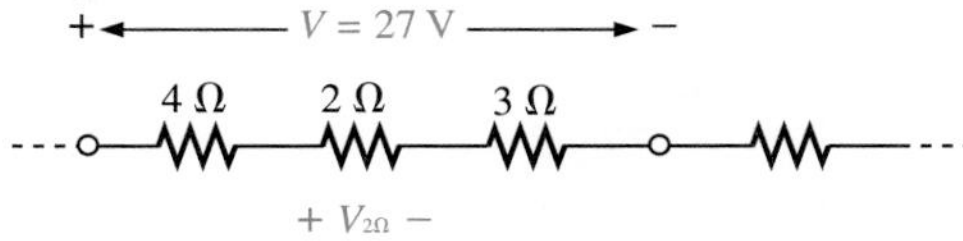

FIG. 5.26
The total voltage across series elements need not be an independent voltage source.

EXAMPLE 5.13

Design the voltage divider of Fig. 5.27 such that

$V_{R_1} = 4V_{R_2}$

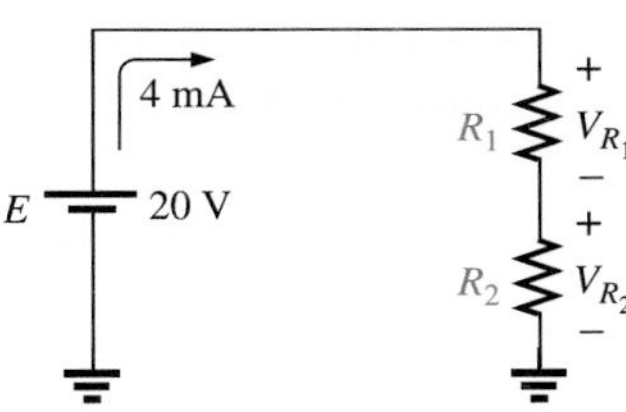

FIG. 5.27
Example 5.13.

Solution: The total resistance is defined by

$$R_T = \frac{E}{I} = \frac{20\ \text{V}}{4\ \text{mA}} = 5\ \text{k}\Omega$$

Since $V_{R_1} = 4V_{R_2}$,

$$R_1 = 4R_2$$

Thus
$$R_T = R_1 + R_2 = 4R_2 + R_2 = 5R_2$$

and
$$5R_2 = 5\ \text{k}\Omega$$
$$R_2 = \mathbf{1\ k\Omega}$$

and
$$R_1 = 4R_2 = \mathbf{4\ k\Omega}$$

5.7 NOTATION

Notation will play an increasingly important role in the analysis to follow. It is important, therefore, that we begin to examine the notation used throughout the industry.

Voltage Sources and Ground

FIG. 5.28
Ground potential.

Except for a few special cases, electrical and electronic systems are grounded for reference and safety purposes. The symbol for the ground connection appears in Fig. 5.28 with its defined potential level—zero volts. If we redrew Fig. 5.2 with a grounded supply, it might look like Fig. 5.29(a) or (b) or (c). In every case, the negative terminal of the battery and the bottom of the resistor R_2 are at ground potential. Although Fig. 5.29(c) shows no connection between the two grounds, it is recognized that such a connection exists for the continuous flow of charge. If $E = 12$ V, then point a is 12 V positive with respect to ground potential, and 12 V exist across the series combination of resistors R_1 and R_2. If a voltmeter placed from point b to ground reads 4 V, then the voltage across R_2 is 4 V, with the higher potential at point b.

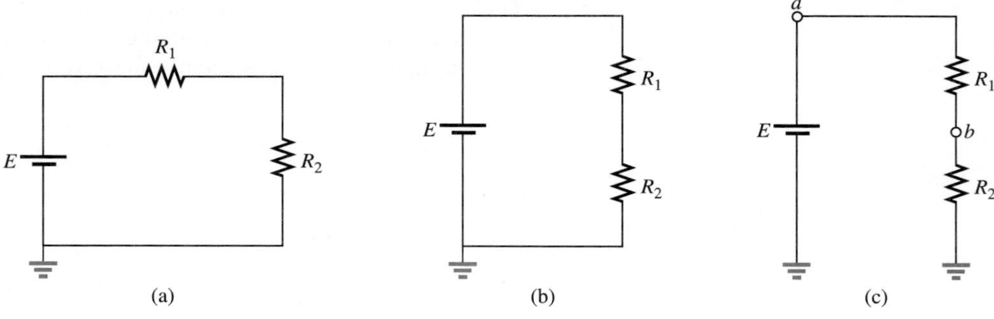

FIG. 5.29
Three ways to sketch the same series dc circuit.

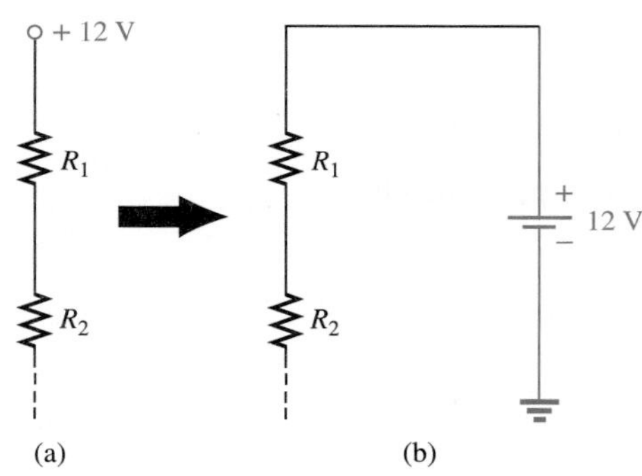

FIG. 5.30
Replacing the special notation for a dc voltage source with the standard symbol.

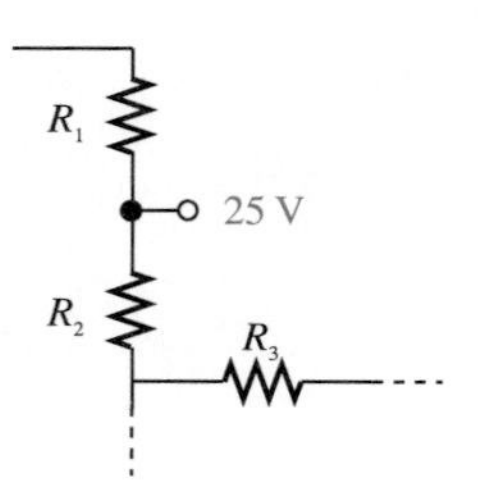

FIG. 5.32
The expected voltage level at a particular point in a network if the system is functioning properly.

On large schematics where space is limited and clarity is important, voltage sources may be indicated as shown in Figs. 5.30(a) and 5.31(a) rather than as illustrated in Figs. 5.30(b) and 5.31(b). In addition, potential levels may be indicated as in Fig. 5.32, to permit a rapid check of the potential levels at various points in a network with respect to ground to ensure that the system is operating properly.

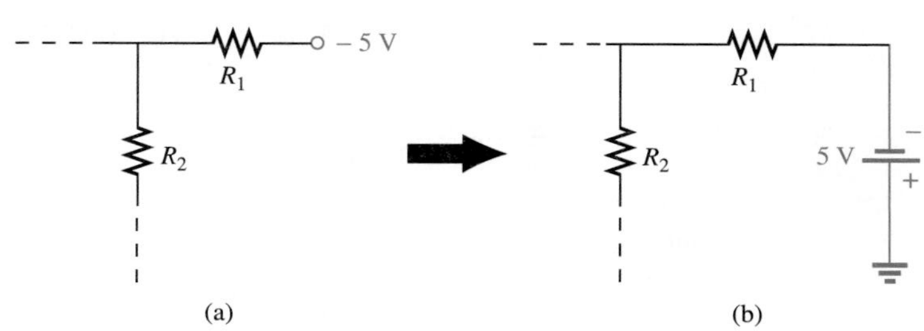

FIG. 5.31
Replacing the notation for a negative dc supply with the standard notation.

Double-Subscript Notation

The fact that voltage exists *between* two points has resulted in a double-subscript notation that defines the first subscript as the higher potential. In Fig. 5.33(a), the two points that define the voltage across the resistor R are indicated by a and b. Since a is the first subscript for V_{ab}, point a must have a higher potential than point b if V_{ab} is to have a positive value. If, in fact, point b is at a higher potential than point a, V_{ab} will have a negative value, as indicated in Fig. 5.33(b).

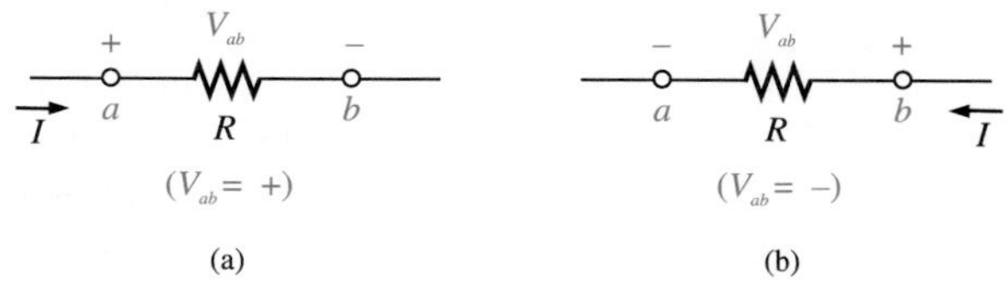

FIG. 5.33
Defining the sign for double-subscript notation.

In summary:

The double-subscript notation V_{ab} specifies point ***a*** as the higher potential. If this is not the case, a negative sign must be associated with the magnitude of V_{ab}.

In other words,

the voltage V_{ab} is the voltage at point ***a*** with respect to point ***b***.

Single-Subscript Notation

If point b of the notation V_{ab} is specified as ground potential (zero volts), then a single-subscript notation can be used to provide the voltage at a point with respect to ground.

In Fig. 5.34, V_a is the voltage from point a to ground. In this case it is obviously 10 V since it is right across the source voltage E. The voltage V_b is the voltage from point b to ground. Because it is directly across the 4-Ω resistor, $V_b = 4$ V.

In summary:

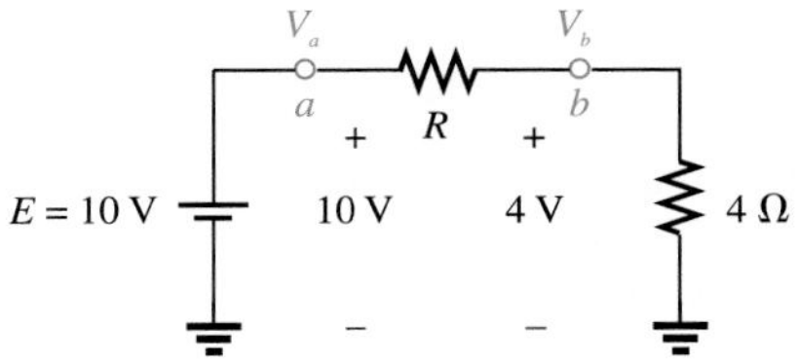

FIG. 5.34
Defining the use of single-subscript notation for voltage leveis.

The single-subscript notation V_a specifies the voltage at point **a** with respect to ground (zero volts). If the voltage is less than zero volts, a negative sign must be associated with the magnitude of V_a.

General Comments

We can now establish a very useful relationship that will have extensive applications in the analysis of electronic circuits. For the above notational standards, the following relationship exists:

$$V_{ab} = V_a - V_b \tag{5.12}$$

In other words, if the voltage at points a and b is known with respect to ground, then the voltage V_{ab} can be determined using the above equation. In Fig. 5.34, for example,

$$\begin{aligned} V_{ab} &= V_a - V_b = 10\text{ V} - 4\text{ V} \\ &= 6\text{ V} \end{aligned}$$

EXAMPLE 5.14

Find the voltage V_{ab} for the conditions of Fig. 5.35.

FIG. 5.35
Example 5.14.

Solution: Applying Eq. (5.12):

$$\begin{aligned} V_{ab} &= V_a - V_b = 16\text{ V} - 20\text{ V} \\ &= \mathbf{-4\text{ V}} \end{aligned}$$

Note the negative sign to reflect the fact that point b is at a higher potential than point a.

EXAMPLE 5.15

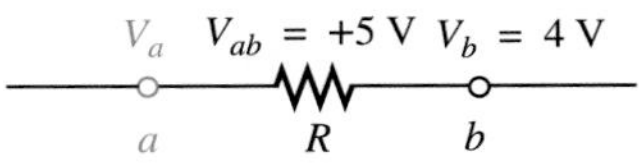

Find the voltage V_a for the configuration of Fig. 5.36.

FIG. 5.36
Example 5.15.

Solution: Applying Eq. (5.12):

$$V_{ab} = V_a - V_b$$

and

$$\begin{aligned} V_a &= V_{ab} + V_b = 5\text{ V} + 4\text{ V} \\ &= \mathbf{9\text{ V}} \end{aligned}$$

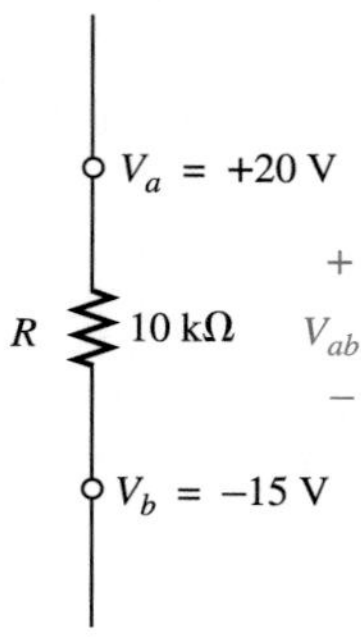

FIG. 5.37
Example 5.16.

EXAMPLE 5.16

Find the voltage V_{ab} for the configuration of Fig. 5.37.

Solution: Applying Eq. (5.12):

$$V_{ab} = V_a - V_b = 20\text{ V} - (-15\text{ V}) = 20\text{ V} + 15\text{ V}$$
$$= \mathbf{35\ V}$$

Note in Example 5.16 that you must be very careful with the signs when applying the equation. The voltage is dropping from a high level of +20 V to a negative voltage of −15 V. As shown in Fig. 5.38, this represents a drop in voltage of 35 V. In some ways it's like going from a positive chequing balance of \$20 to owing \$15; the total expenditure is \$35.

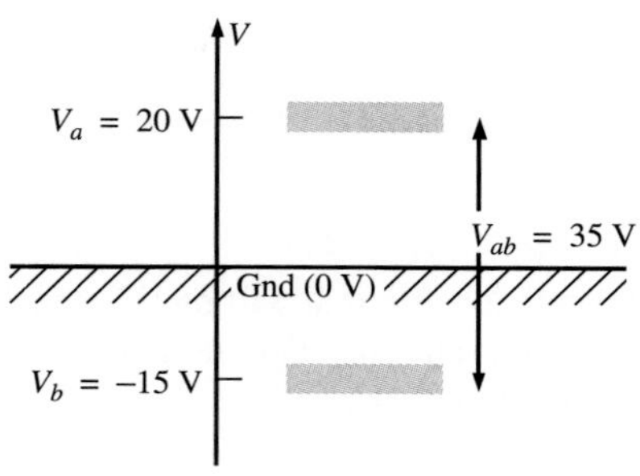

FIG. 5.38
The impact of positive and negative voltages on the total voltage drop.

EXAMPLE 5.17

Find the voltages V_b, V_c, and V_{ac} for the network of Fig. 5.39.

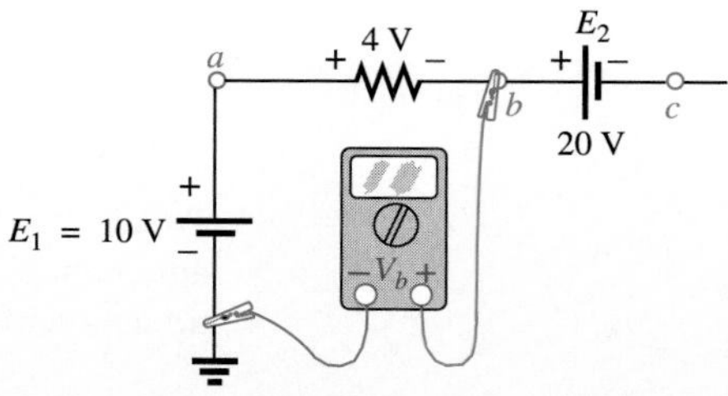

FIG. 5.39
Example 5.17.

Solution: Starting at ground potential (zero volts), we proceed through a rise of 10 V to reach point *a* and then pass through a drop in potential of 4 V to point *b*. The result is that the meter will read

$$V_b = +10\text{ V} - 4\text{ V} = \mathbf{6\ V}$$

as clearly demonstrated by Fig. 5.40.

If we then proceed to point *c*, there is an additional drop of 20 V, resulting in

$$V_c = V_b - 20\text{ V} = 6\text{ V} - 20\text{ V} = \mathbf{-14\ V}$$

as shown in Fig. 5.41.

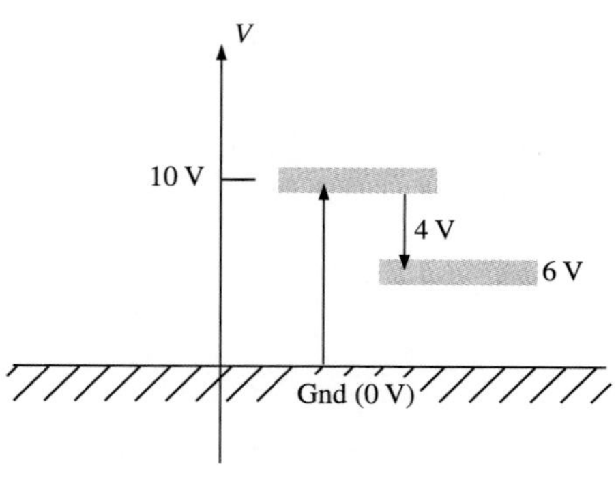

FIG. 5.40
Determining V_b using the defined voltage levels.

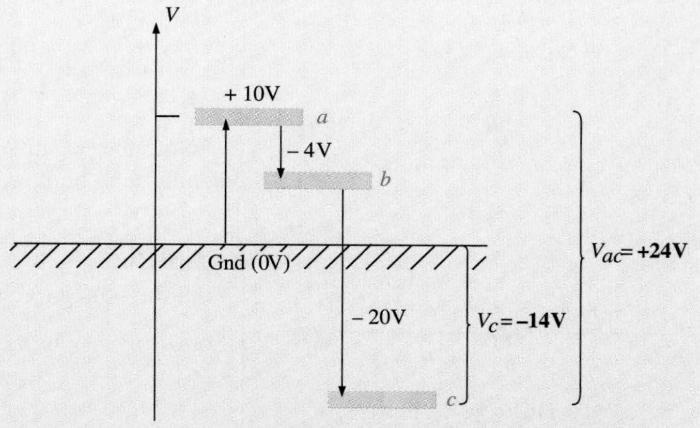

FIG. 5.41
Review of the potential levels for the circuit of Fig. 5.39.

continued

The voltage V_{ac} can be obtained using Eq. (5.12) or by simply referring to Fig. 5.41:

$$V_{ac} = V_a - V_c = 10\text{ V} - (-14\text{ V})$$
$$= \mathbf{24\text{ V}}$$

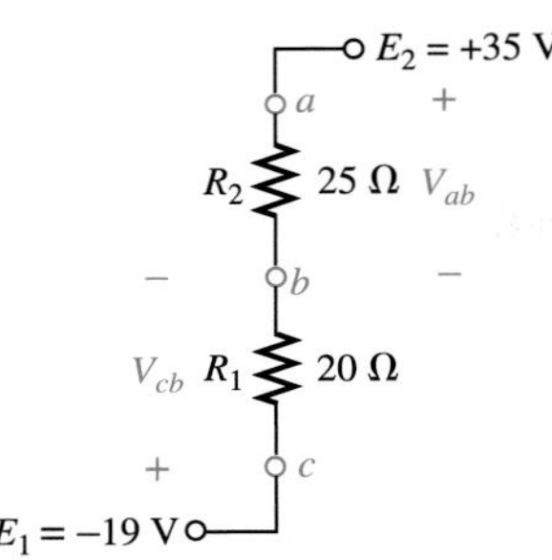

FIG. 5.42
Example 5.18.

EXAMPLE 5.18

Determine V_{ab}, V_{cb}, and V_c for the network of Fig. 5.42.

Solution: There are two ways to approach this problem. The first is to sketch the diagram of Fig. 5.43 and note that there is a 54-V drop across the series resistors R_1 and R_2. The current can then be determined using Ohm's law and the voltage levels as follows:

$$I = \frac{54\text{ V}}{45\ \Omega} = 1.2\text{ A}$$

$$V_{ab} = IR_2 = (1.2\text{ A})(25\ \Omega) = \mathbf{30\text{ V}}$$
$$V_{cb} = -IR_1 = -(1.2\text{ A})(20\ \Omega) = \mathbf{-24\text{ V}}$$
$$V_c = E_1 = \mathbf{-19\text{ V}}$$

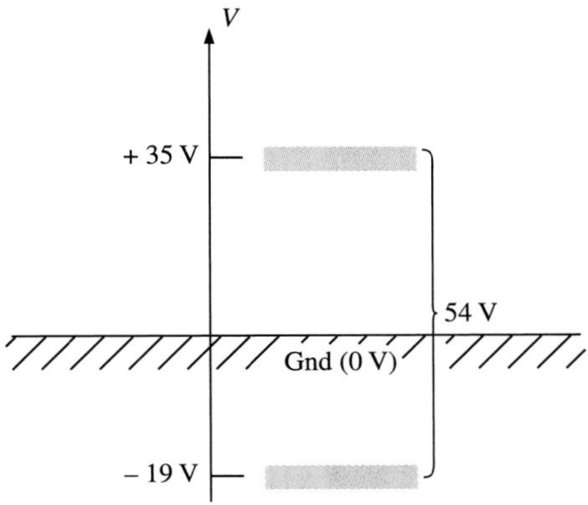

FIG. 5.43
Determining the total voltage drop across the resistive elements of Fig. 5.42.

The other approach is to redraw the network as shown in Fig. 5.44 to clearly establish the aiding effect of E_1 and E_2 and solve the resulting series circuit.

$$I = \frac{E_1 + E_2}{R_T} = \frac{19\text{ V} + 35\text{ V}}{45\ \Omega} = \frac{54\text{ V}}{45\ \Omega} = 1.2\text{ A}$$

and $\qquad V_{ab} = \mathbf{30\text{ V}}, \quad V_{cb} = \mathbf{-24\text{ V}}, \quad V_c = \mathbf{-19\text{ V}}$

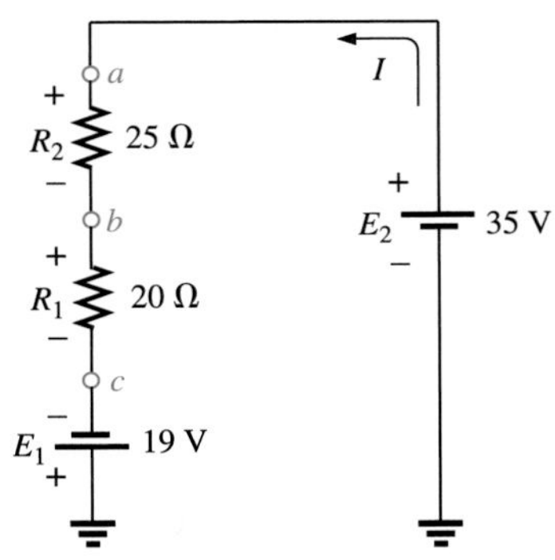

FIG. 5.44
Redrawing the circuit of Fig. 5.42 using standard dc voltage supply symbols.

EXAMPLE 5.19

Using the voltage divider rule, determine the voltages V_1 and V_2 of Fig. 5.45.

Solution: Redrawing the network with the standard battery symbol will result in the network of Fig. 5.46. Applying the voltage divider rule,

$$V_1 = \frac{R_1E}{R_1 + R_2} = \frac{(4\ \Omega)(24\text{ V})}{4\ \Omega + 2\ \Omega} = \mathbf{16\text{ V}}$$

$$V_2 = \frac{R_2E}{R_1 + R_2} = \frac{(2\ \Omega)(24\text{ V})}{4\ \Omega + 2\ \Omega} = \mathbf{8\text{ V}}$$

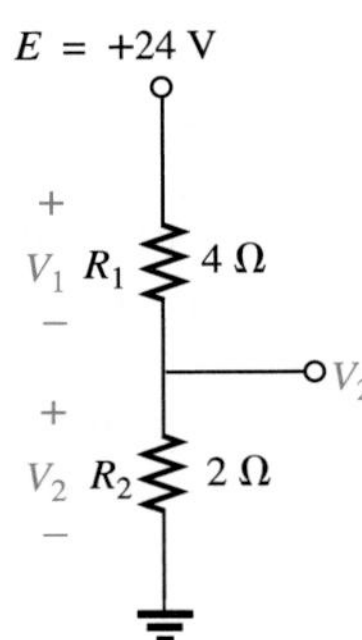

FIG. 5.45
Example 5.19.

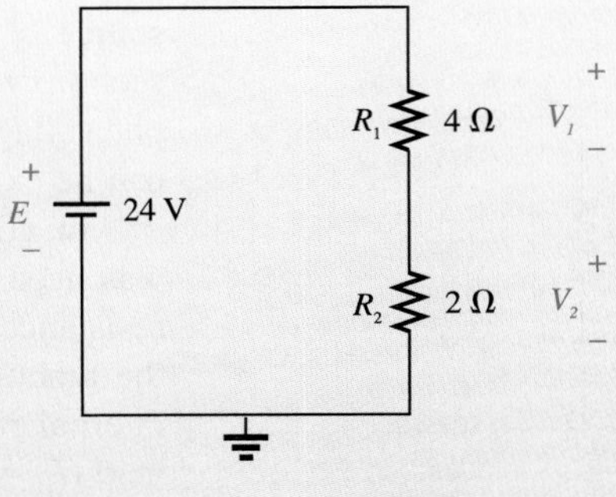

FIG. 5.46
Circuit of Fig. 5.45 redrawn.

EXAMPLE 5.20

For the network of Fig. 5.47:

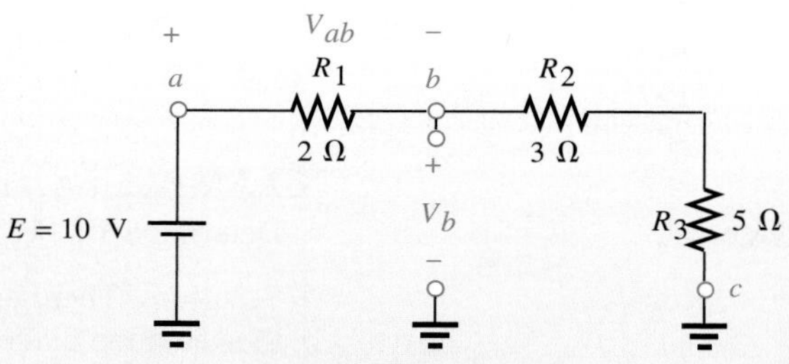

FIG. 5.47

a. Calculate V_{ab}.
b. Determine V_b.
c. Calculate V_c.

Solution:

a. Voltage divider rule:

$$V_{ab} = \frac{R_1 E}{R_T} = \frac{(2\ \Omega)(10\ \text{V})}{2\ \Omega + 3\ \Omega + 5\ \Omega} = \mathbf{+2\ V}$$

b. Voltage divider rule:

$$V_b = V_{R_2} + V_{R_3} = \frac{(R_2 + R_3)E}{R_T} = \frac{(3\ \Omega + 5\ \Omega)(10\ \text{V})}{10\ \Omega} = \mathbf{8\ V}$$

or $\quad V_b = V_a - V_{ab} = E - V_{ab} = 10\ \text{V} - 2\ \text{V} = \mathbf{8\ V}$

c. V_c = ground potential = **0 V**

5.8 INTERNAL RESISTANCE OF VOLTAGE SOURCES

Every source of voltage, whether a generator, battery, or laboratory supply, will have some internal resistance. The equivalent circuit of any source of voltage will therefore appear as shown in Fig. 5.48. In this section, we will examine the effect of the internal resistance on the output voltage so that any unexpected changes in terminal characteristics can be explained.

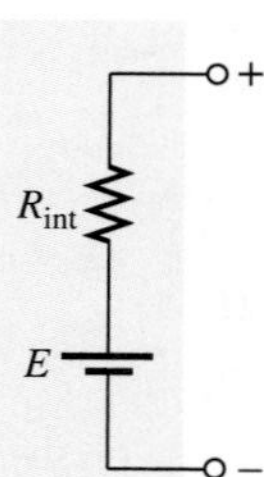

FIG. 5.48
Equivalent circuit for a dc source.

In all the circuit analyses to this point, the ideal voltage source was used [see Fig. 5.49(a)]. The ideal voltage source has no internal resistance and an output voltage of E volts with no load or full load. In the practical case [Fig. 5.49(b)], where we consider the effects of the internal resistance, the output voltage will be E volts only when no-load ($I_L = 0$) conditions exist. When a load is connected [Fig. 5.49(c)], the output voltage of the voltage source will decrease due to the voltage drop across the internal resistance.

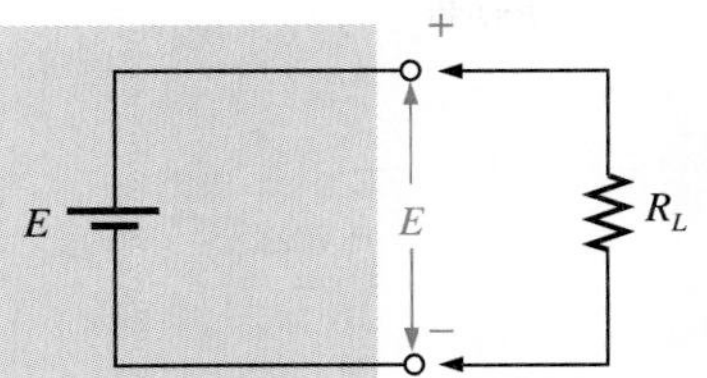

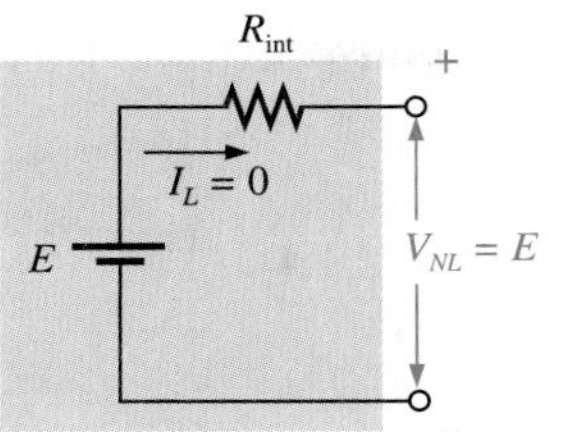

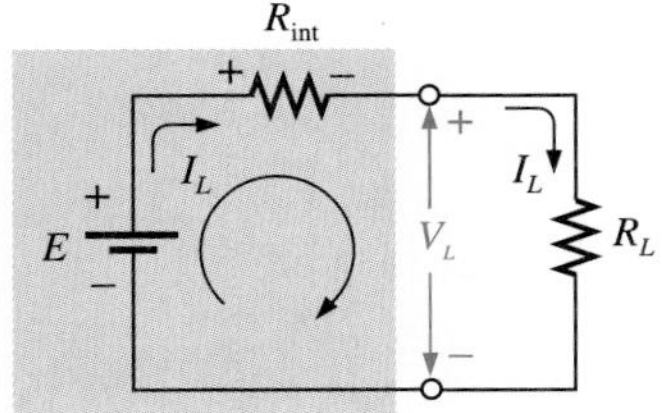

FIG. 5.49
Voltage source: (a) ideal, $R_{int} = 0\ \Omega$; (b) determining V_{NL}; (c) determining R_{int}.

By applying Kirchhoff's voltage law around the indicated loop of Fig. 5.49(c), we obtain

$$E - I_L R_{\text{int}} - V_L = 0$$

or, since

$$E = V_{NL}$$

we have

$$V_{NL} - I_L R_{\text{int}} - V_L = 0$$

and

$$V_L = V_{NL} - I_L R_{\text{int}} \tag{5.13}$$

If the value of R_{int} is not available, it can be found by first solving for R_{int} in the equation just derived for V_L; that is,

$$R_{\text{int}} = \frac{V_{NL} - V_L}{I_L} = \frac{V_{NL}}{I_L} - \frac{I_L R_L}{I_L}$$

and

$$R_{\text{int}} = \frac{V_{NL}}{I_L} - R_L \tag{5.14}$$

Fig. 5.50 shows a plot of the output voltage versus current for a dc source with the circuit representation of Fig. 5.48. Note that any increase in load demand, starting at any level, causes an additional drop in terminal voltage due to the increasing loss in potential across the internal resistance. At maximum current, denoted by I_{FL}, the voltage across the internal resistance is $V_{\text{int}} = I_{FL}R_{\text{int}} = (10\text{ A})(2\ \Omega) = 20$ V. The terminal voltage has dropped to 100 V. This is a significant difference when you would ideally expect a 120-V generator to provide the full 120 V if you stay below the listed full-load current. Eventually, if the load current were permitted to increase without limit, the voltage across the internal resistance would equal the supply voltage, and the terminal voltage would be zero. The larger the internal resistance, the steeper the slope of the characteristic of Fig. 5.50. In fact, for any chosen interval of voltage or current, the magnitude of the internal resistance is given by

$$R_{\text{int}} = \frac{\Delta V_L}{\Delta I_L} \tag{5.15}$$

For the chosen interval of 5–7 A ($\Delta I_L = 2$ A) on Fig. 5.50, ΔV_L is 4 V, and $R_{\text{int}} = \Delta V_L/\Delta I_L = 4\text{ V}/2\text{ A} = 2\ \Omega$.

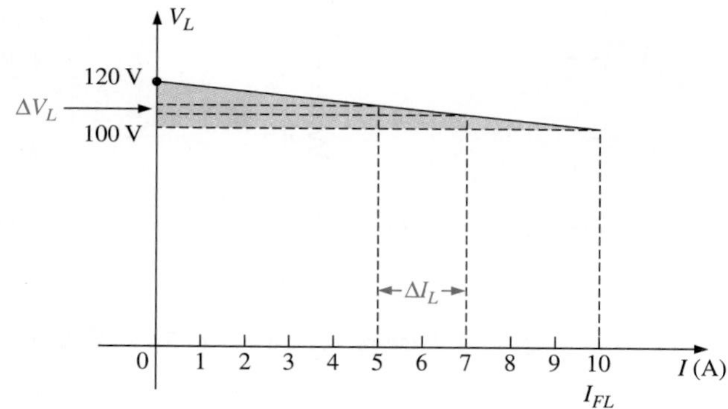

FIG. 5.50
V_L versus I_L for a dc generator with $R_{int} = 2\ \Omega$.

The loss in output voltage results directly in a loss in power delivered to the load. Multiplying both sides of Eq. (5.13) by the current I_L in the circuit, we obtain

$$\underbrace{I_L V_L}_{\text{Power to load}} = \underbrace{I_L V_{NL}}_{\text{Power output by battery}} - \underbrace{I_L^2 R_{\text{int}}}_{\text{Power loss in the form of heat}} \tag{5.16}$$

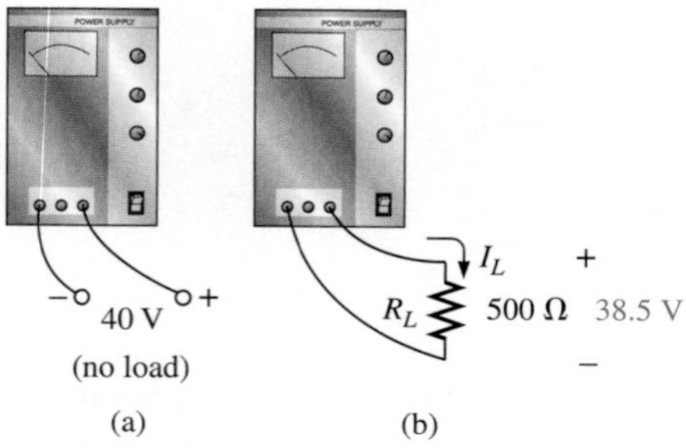

FIG. 5.51
Example 5.21.

EXAMPLE 5.21

Before a load is applied, the terminal voltage of the power supply of Fig. 5.51(a) is set to 40 V. When a load of 500 Ω is attached, as shown in Fig. 5.51(b), the terminal voltage drops to 38.5 V. What happened to the remainder of the no-load voltage, and what is the internal resistance of the source?

Solution: The difference of 40 V − 38.5 V = 1.5 V now appears across the internal resistance of the source. The load current is 38.5 V/0.5 kΩ = 77 mA. Applying Eq. (5.14),

$$R_{\text{int}} = \frac{V_{NL}}{I_L} - R_L = \frac{40\text{ V}}{77\text{ mA}} - 0.5\text{ k}\Omega$$
$$= 519.48\ \Omega - 500\ \Omega = \mathbf{19.48\ \Omega}$$

FIG. 5.52
Example 5.22.

EXAMPLE 5.22

The battery of Fig. 5.52 has an internal resistance of 2 Ω. Find the voltage V_L and the power lost to the internal resistance if the applied load is a 13-Ω resistor.

Solution:

$$I_L = \frac{30\text{ V}}{2\ \Omega + 13\ \Omega} = \frac{30\text{ V}}{15\ \Omega} = 2\text{ A}$$
$$V_L = V_{NL} - I_L R_{\text{int}} = 30\text{ V} - (2\text{ A})(2\ \Omega) = \mathbf{26\ V}$$
$$P_{\text{lost}} = I_L^2 R_{\text{int}} = (2\text{ A})^2(2\ \Omega) = (4)(2) = \mathbf{8\ W}$$

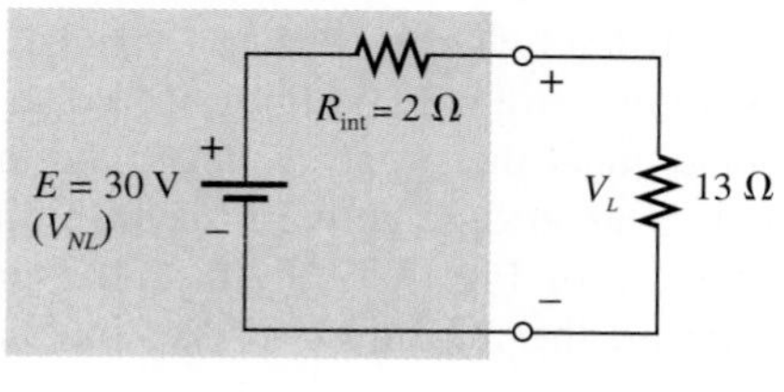

Procedures for measuring R_{int} will be described in Section 5.10.

5.9 VOLTAGE REGULATION

For any supply, ideal conditions dictate that for the range of load demand (I_L), the terminal voltage remains fixed in magnitude. In other words, if a supply is set for 12 V, it should maintain this terminal voltage, even though the current demand on the supply may vary. A measure of how close a supply will come to ideal conditions is given by the voltage regulation characteristic. By definition, the **voltage regulation** of a supply between the limits of full-load and no-load conditions (Fig. 5.53) is given by the following:

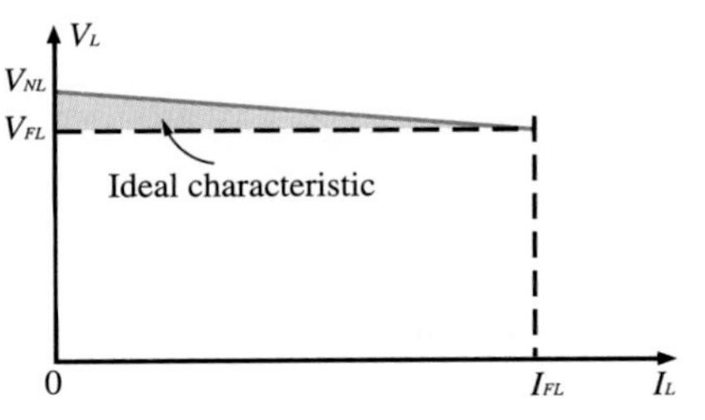

FIG. 5.53
Defining voltage regulation.

$$\text{Voltage regulation } (VR)\% = \frac{V_{NL} - V_{FL}}{V_{FL}} \times 100\% \qquad (5.17)$$

For ideal conditions, $V_{FL} = V_{NL}$ and $VR\% = 0$. Therefore, *the smaller the voltage regulation, the less the variation in terminal voltage with change in load.*

Voltage regulation is also given by

$$VR\% = \frac{V_{NL} - V_{FL}}{V_{FL}} \times 100\%$$

$$V_{\text{int}} = V_{NL} - V_{FL}$$

Therefore

$$VR\% = \frac{V_{\text{int}}}{V_{FL}} \times 100\%$$

and

$$VR\% = \frac{\cancel{I_T} R_{\text{int}}}{\cancel{I_T} R_L} \times 100\%$$

Thus

$$VR\% = \frac{R_{\text{int}}}{R_L} \times 100\% \qquad (5.18)$$

In other words, the smaller the internal resistance for the same load, the smaller the regulation and the more ideal the output.

EXAMPLE 5.23

Calculate the voltage regulation of a supply having the characteristic of Fig. 5.50.

Solution:

$$VR\% = \frac{V_{NL} - V_{FL}}{V_{FL}} \times 100\% = \frac{120\text{ V} - 100\text{ V}}{100\text{ V}} \times 100\%$$

$$= \frac{20}{100} \times 100\% = \mathbf{20\%}$$

EXAMPLE 5.24

Determine the voltage regulation of the supply of Fig. 5.51.

Solution:

$$VR\% = \frac{R_{int}}{R_L} \times 100\% = \frac{19.48\ \Omega}{500\ \Omega} \times 100\% \cong \mathbf{3.9\%}$$

5.10 MEASUREMENT TECHNIQUES

Ammeters are inserted in the branch in which the current is to be measured. Such a condition specifies that

ammeters are placed in series with the branch in which the current is to be measured

as shown in Fig. 5.54.

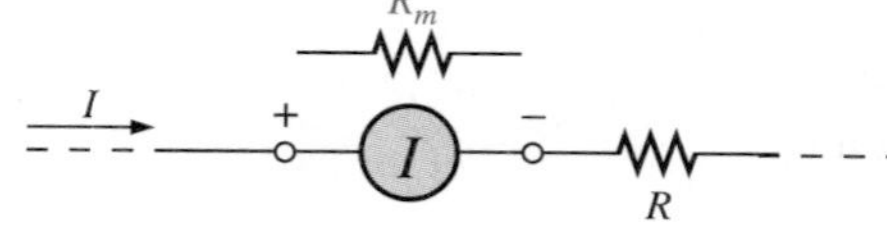

FIG. 5.54
Series connection of an ammeter.

If the ammeter is to have minimal impact on the behaviour of the network, its resistance should be very small (ideally, zero ohms) compared to the other series elements of the branch such as the resistor R of Fig. 5.54. If the meter resistance approaches or exceeds 10% of R, it would have a significant impact on the current level it is measuring. Note also that the resistances of the separate current scales of the same meter are usually not the same. In fact, the meter resistance normally increases with decreasing current levels. However, for the majority of situations you can simply assume that the internal ammeter resistance is small enough compared to the other circuit elements to be ignored.

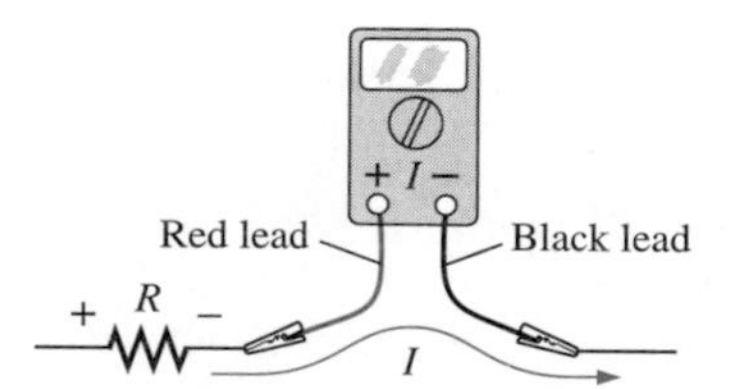

FIG. 5.55
Connecting an ammeter for an up-scale (positive) reading.

For an up-scale (analog meter) or positive (digital meter) reading, an ammeter must be connected with current entering the positive terminal of the meter and leaving the negative terminal, as shown in Fig. 5.55. Since most meters employ a red lead for the positive terminal and a black lead for the negative, simply ensure that current enters the red lead and leaves the black one.

Voltmeters are always connected across the element for which the voltage is to be determined.

You will obtain an up-scale or positive reading on a voltmeter if the positive terminal (red lead) is connected to the point of higher potential and the negative terminal (black lead) is connected to the lower potential, as shown in Fig. 5.56.

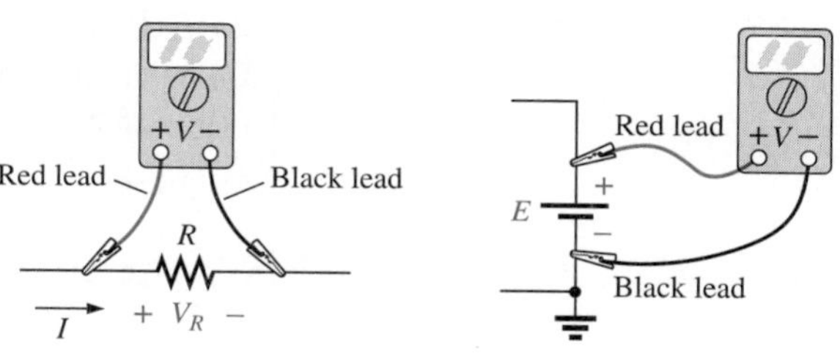

FIG. 5.56
Hooking up a voltmeter to obtain an up-scale (positive) reading.

For the double-subscript notation, always hook up the red lead to the first subscript and the black lead to the second; that is, to measure the voltage V_{ab} in Fig. 5.57, the red lead is connected to point a and the black lead is connected to point b. For single-subscript notation, hook up the red lead to the point of interest and the black lead to ground, as shown in Fig. 5.57 for V_a and V_b.

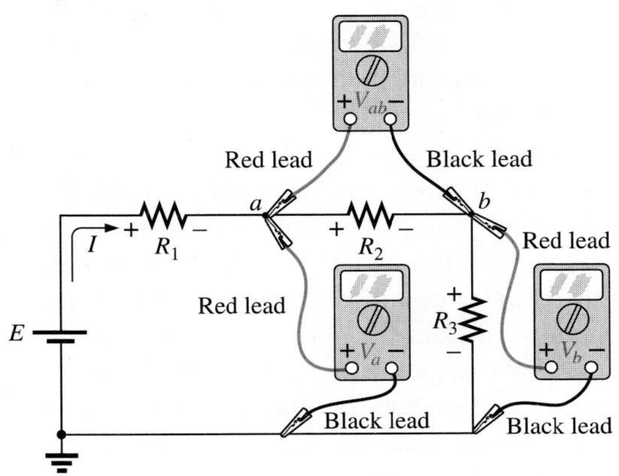

FIG. 5.57
Measuring voltages with double- and single-subscript notation.

The internal resistance of a supply cannot be measured with an ohmmeter due to the voltage present. However, the no-load voltage can be measured by simply hooking up the voltmeter as shown in Fig. 5.58(a). Do not be concerned about the apparent path for current that the meter seems to provide by completing the circuit. The internal resistance of the meter is usually sufficiently high to ensure that the resulting current is so small that it can be ignored. (Voltmeter loading effects will be discussed in detail in Section 6.9.) An ammeter could then be placed directly across the supply, as shown in Fig. 5.58(b), to measure the short-circuit current I_{SC} and R_{int} as determined by Ohm's law: $R_{int} = E_{NL}/I_{sc}$. However, since the internal resistance of the supply may be very low, performing the measurement could result in high current levels that could damage the meter and supply, and possibly cause dangerous side effects. The setup of Fig. 5.58(b) is dangerous and therefore *not* recommended. It would better to apply a resistive load that will result in a supply current of about half the maximum rated value and measure the terminal voltage. Then use Eq. (5.14).

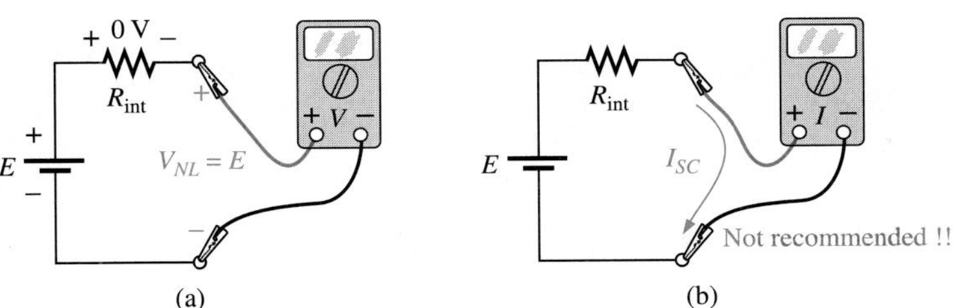

FIG. 5.58
(a) Measuring the no-load voltage E; (b) measuring the short-circuit current.

Practical Business Applications

Why doesn't this thing work?

Okay, so you went to the store and bought a fresh new pack of D-size batteries for your flashlight. You open it up, pop them in, screw on the cover and turn it on. **Nothing**! You hit it and shake it, but, it still won't light. So, why doesn't this thing work? Where do you start to try fixing it?

Well, for one thing, you should check out the battery configuration diagram. All flashlight manufacturers should have one stamped or stuck on the inside of the unit. Notice that the batteries have to be put in with the little nipples pointing in the same direction. This little nipple indicates the positive (+) pole, while the other side is the negative (−) pole of the battery.

So, why must they both be pointing in the same direction? Because we want the battery voltages to be series-aiding rather than series-opposing in the flashlight circuit. Most batteries have a voltage rating of about 1.5 V. If you put two of them in series-aiding, you'll get: $+\ 1.5\text{ V} + 1.5\text{ V} = 3\text{ V}$. If the circuit path is complete and the light bulb in the flashlight is rated at 3 V and isn't burned out, these two series-aiding batteries will supply enough potential to make the flashlight work.

According to Kirchhoff's Voltage Law, if we accidentally put the batteries into the flashlight series-opposing, then what you'll get is: $+\ 1.5\text{ V} - 1.5\text{ V} = 0\text{ V}$. And that's why your flashlight isn't working. How can it? One battery's potential, + 1.5 V, is being applied to the circuit so that it exactly cancels the effect of the other one. Assuming the light bulb isn't polarity sensitive, all you have to do is reverse either one of the two batteries and the flashlight will work.

Will reversing either one of the two batteries always work? Yes and no. Notice we said "assuming the light bulb isn't polarity sensitive." That's because if the device you're trying to energize has required positive and negative terminals, even if you place both batteries in the same way, they may not work. For that matter, you may even physically destroy the device. For this reason, the manufacturer will recommend that you insert the batteries in a particular way.

Loads that are polarity sensitive are dc in nature. Some dc loads that can't tolerate reverse series-aiding potentials are certain toys, portable stereos, notebook computers, calculators, cellular phones, etc.

Suppose your portable stereo took six C-size batteries. Say that they're all new and you put them all into the unit the way the manufacturer shows in the manual except for one. Will the unit work? The answer is maybe but probably not. This is because the voltage that the unit receives is $+\ 1.5\text{ V} + 1.5\text{V} + 1.5\text{ V} + 1.5\text{ V} + 1.5\text{ V} - -1.5\text{ V} = +\ 6\text{ V}$, not + 9 V, as should be the case. In other words, the unit is operating at only two thirds its normal capacity. Therefore, it probably won't work properly.

If you reverse two batteries in the unit, the total voltage is only + 3 V and if you reverse three batteries, the total voltage is **zero**. That is, it's as if **no batteries were in the unit at all**. Worse still, if you put in more than three of the six batteries backwards, the total voltage will be **negative**. Now you run the risk of destroying the unit because it probably wasn't designed to handle reverse polarity applied to it. Not only that, but if it blows up, you or someone near the unit can get seriously hurt if it starts to burn or parts go flying across the room.

So, in summary, it's absolutely crucial that you know how to put the batteries into portable devices. Not knowing the proper polarity of the device can be costly in time, money and safety!

PROBLEMS

SECTION 5.2 Series Circuits

1. Find the total resistance and current I for each circuit of Fig. 5.59.

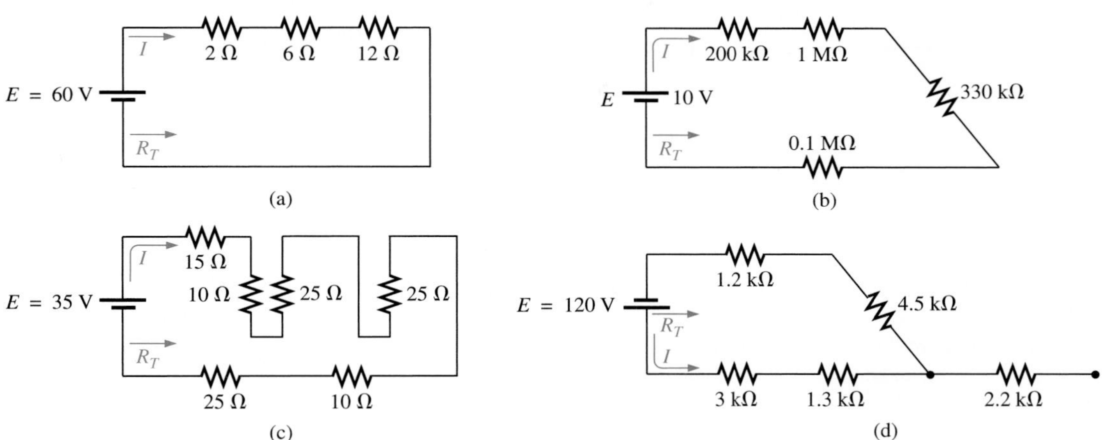

FIG. 5.59
Problem 1.

2. For the circuits of Fig. 5.60, the total resistance is specified. Find the unknown resistances and the current I for each circuit.

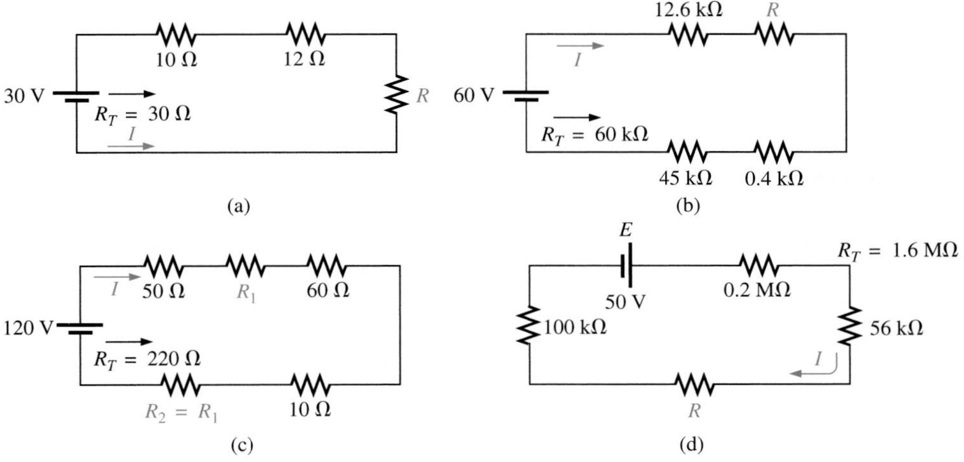

FIG. 5.60
Problem 2.

3. Find the applied voltage E necessary to develop the current specified in each network of Fig. 5.61.

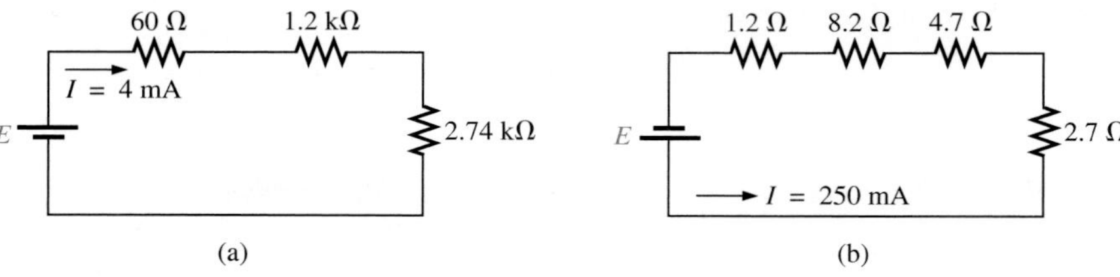

FIG. 5.61
Problem 3.

*4. For each network of Fig. 5.62, determine R_{TOTAL}, the source voltage E, the unknown resistance, and the voltage across each element.

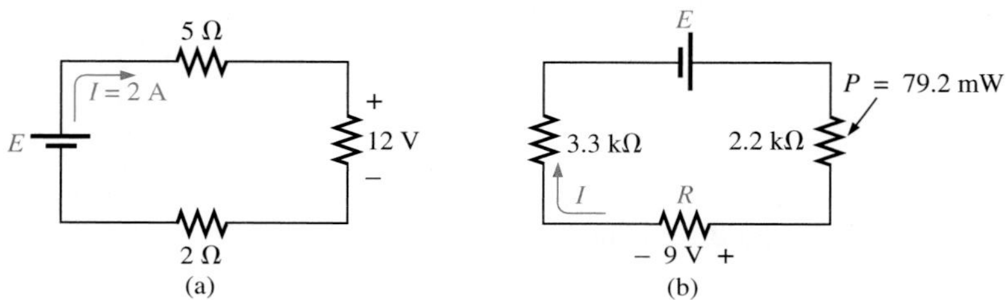

FIG. 5.62
Problem 4.

SECTION 5.3 Voltage Sources in Series

5. For the network in Fig. 5.63, determine the current (I) and the source voltage (E).

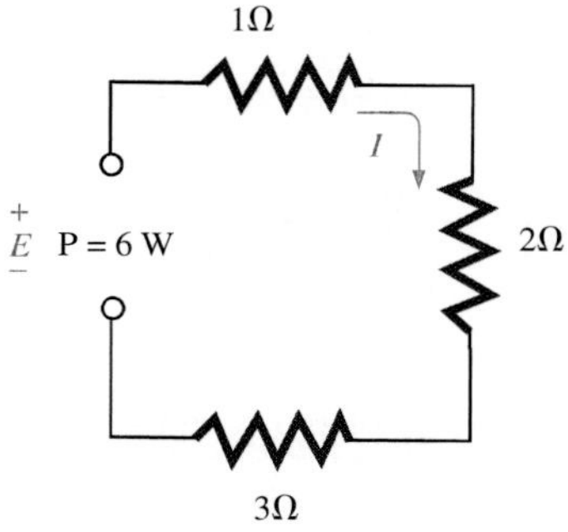

FIG. 5.63
Problem 5.

***6.** Find the unknown voltage source and resistor for the networks of Fig. 5.64. Also indicate the direction of the resulting current.

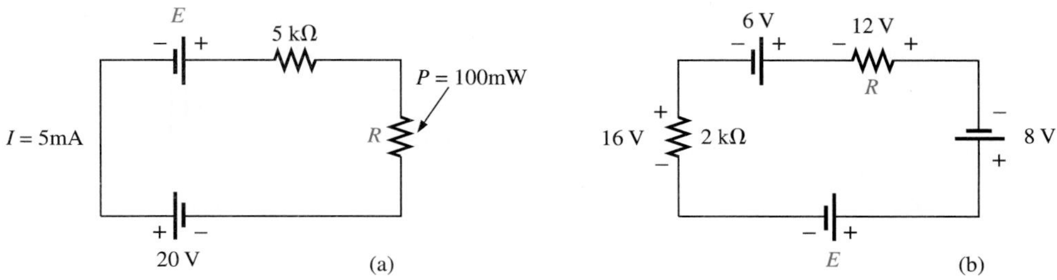

FIG. 5.64
Problem 6.

SECTION 5.4 Kirchhoff's Voltage Law

7. Find V_{ab} with polarity for the circuits of Fig. 5.65. Each box can contain a load or a power supply, or a combination of both.

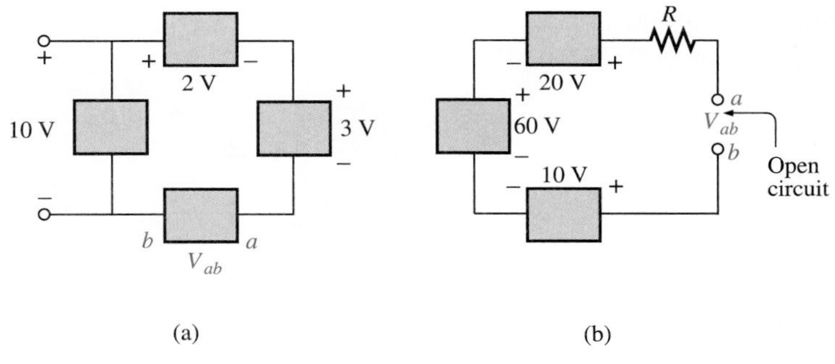

FIG. 5.65
Problem 7.

8. Although the networks of Fig. 5.66 are not simple series circuits, determine the unknown voltages using Kirchhoff's voltage law.

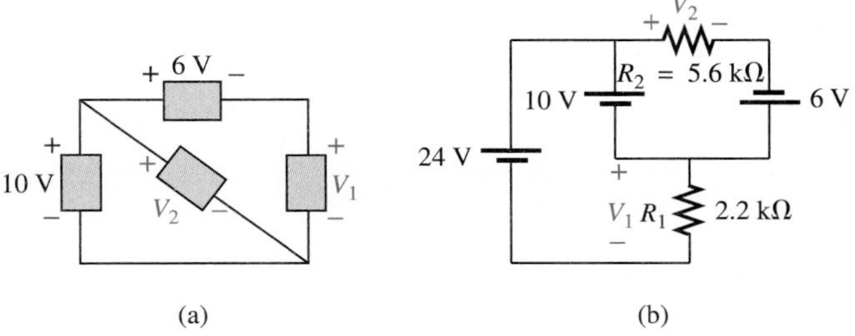

FIG. 5.66
Problem 8.

9. For the network in Fig. 5.67, find V_1 and V_2.

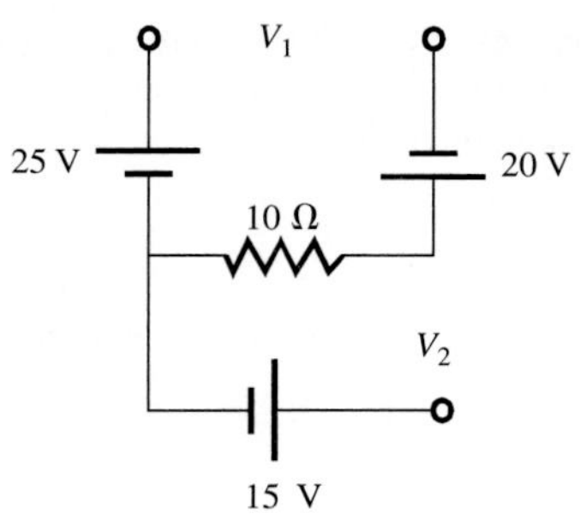

FIG. 5.67
Problem 9.

10. For the circuit of Fig. 5.68:
    a. Find the total resistance, current, and unknown voltage drops.
    b. Verify Kirchhoff's voltage law around the closed loop.
    c. Find the power dissipated by each resistor, and note whether the power delivered is equal to the power dissipated.
    d. If the resistors are available with wattage ratings of 1/2, 1, and 2 W, what minimum wattage rating can be used for each resistor in this circuit?

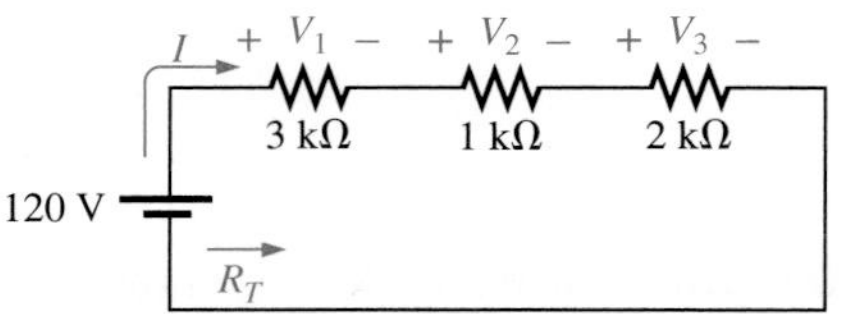

FIG. 5.68
Problem 10.

11. Repeat Problem 10 for the circuit of Fig. 5.69.

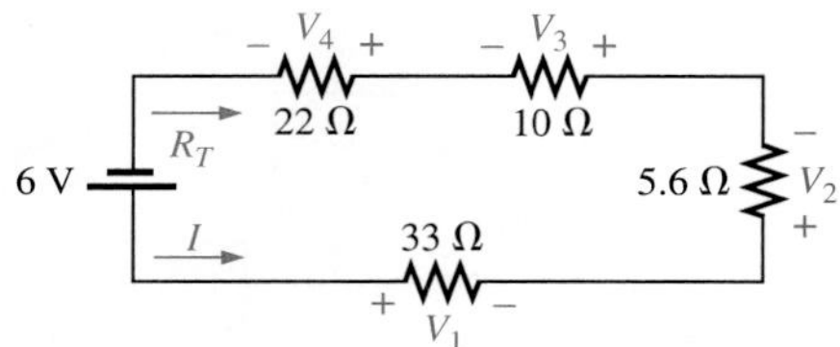

FIG. 5.69
Problem 11.

*12. Find the unknown quantities in the circuits of Fig. 5.70 using the information provided.

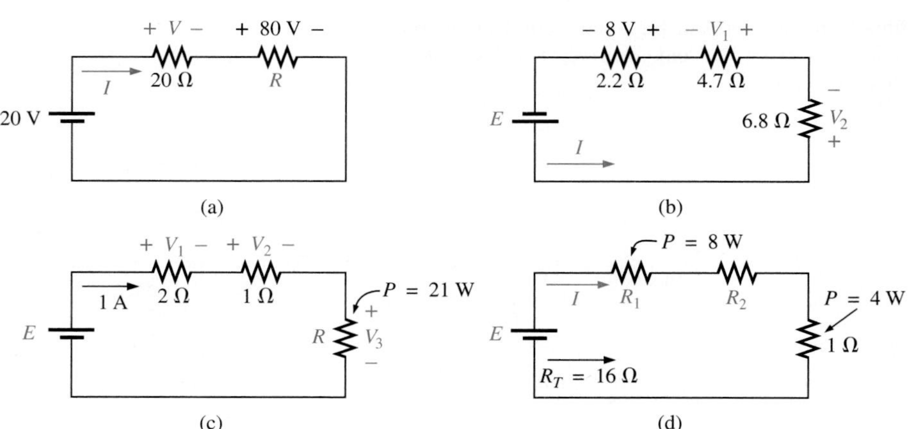

FIG. 5.70
Problem 12.

13. Eight holiday lights are connected in series as shown in Fig. 5.71.
   a. If the set is connected to a 120-V source, what is the current through the bulbs if each bulb has an internal resistance of $28\frac{1}{8}\ \Omega$?
   b. Determine the power delivered to each bulb.
   c. Calculate the voltage drop across each bulb.
   d. If one bulb burns out (that is, the filament opens), what is the effect on the remaining bulbs?

FIG. 5.71
Problem 13.

*14. For the conditions specified in Fig. 5.72, determine the unknown resistance.

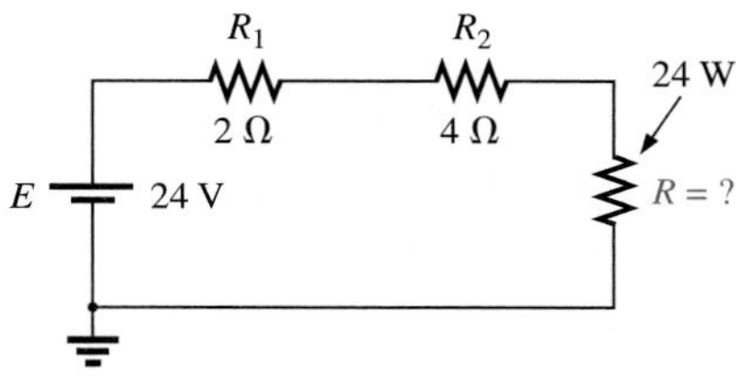

FIG. 5.72
Problem 14.

SECTION 5.6 Voltage Divider Rule

15. Using the voltage divider rule, find V_{ab} (with polarity) for the circuits of Fig. 5.73.

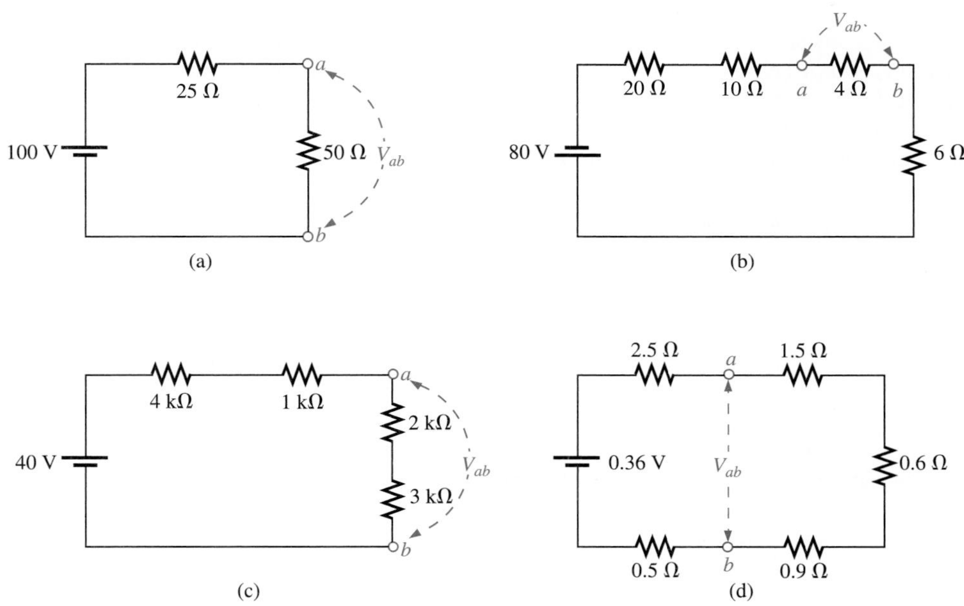

FIG. 5.73
Problem 15.

16. Find the unknown resistance using the voltage divider rule and the information provided for the circuits of Fig. 5.74.

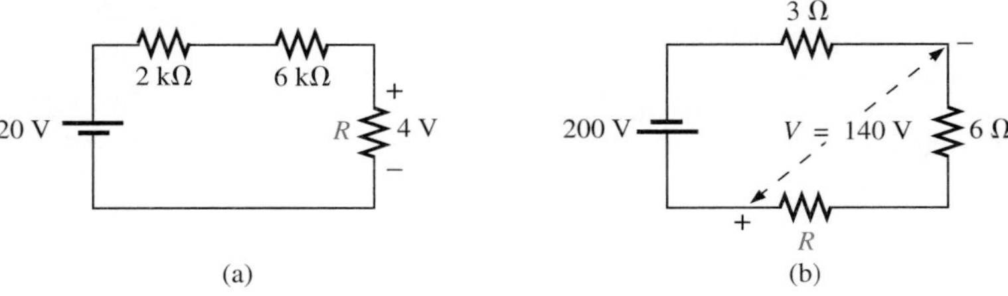

FIG. 5.74
Problem 16.

17. Referring to Fig. 5.75:
    a. Determine V_2 by simply noting that $R_2 = 3R_1$.
    b. Calculate V_3.
    c. Noting the magnitude of V_3 as compared to V_2 or V_1, determine R_3 by inspection.
    d. Calculate the source current I.
    e. Calculate the resistance R_3 using Ohm's law and compare it to the result of part (c).

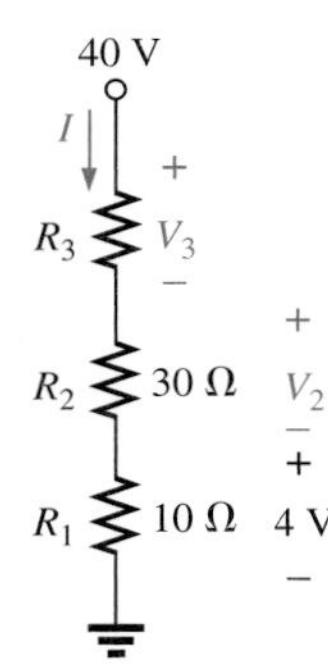

FIG. 5.75
Problem 17.

18. If one of the bulbs in Fig. 5.71 should burn out, how would you determine which lamp was faulty?

19. a. Design a voltage divider circuit that will permit the use of an 8-V, 50-mA bulb in an automobile with a 12-V electrical system.
    b. What is the minimum wattage rating of the chosen resistor if (1/4)-W, (1/2)-W and 1-W resistors are available?

20. Determine the values of R_1, R_2, R_3, and R_4 for the voltage divider of Fig. 5.76 if the source current is 16 mA.

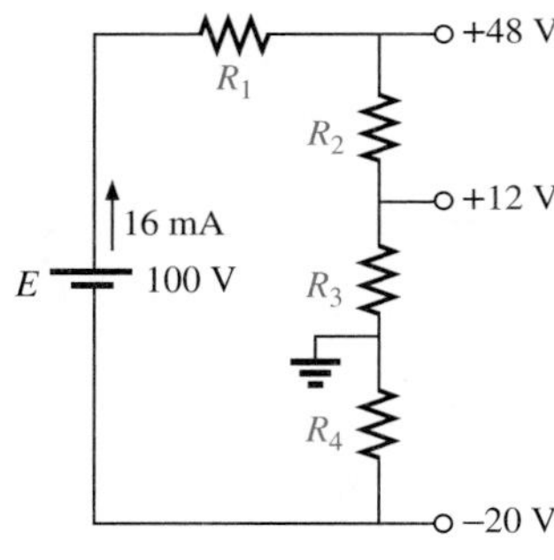

FIG. 5.76
Problem 20.

21. Design the voltage divider of Fig. 5.77 such that $V_{R_1} = (1/4)V_{R_2}$ if $I = 100$ mA.

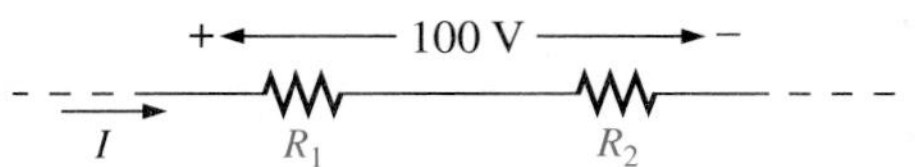

FIG. 5.77
Problem 21.

22. Find the voltage across each resistor of Fig. 5.78 if $R_1 = 2R_3$ and $R_2 = 7R_3$.

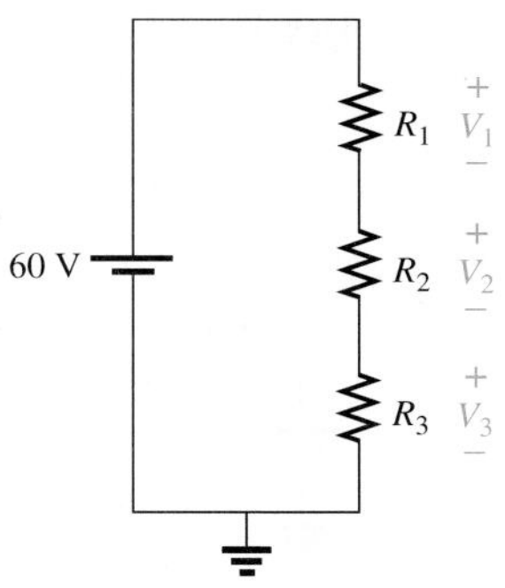

FIG. 5.78
Problem 22.

23. **a.** Design the circuit of Fig. 5.79 such that $V_{R_2} = 3V_{R_1}$ and $V_{R_3} = 4V_{R_2}$.
b. If the current I is reduced to 10 μA, what are the new values of R_1, R_2, and R_3? How do they compare to the results of part (a)?

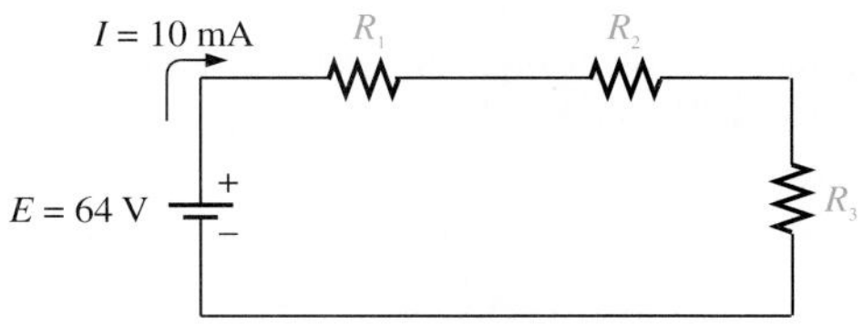

FIG. 5.79
Problem 23.

SECTION 5.7 Notation

24. Determine the voltages V_a, V_b, and V_{ab} for the networks of Fig. 5.80.

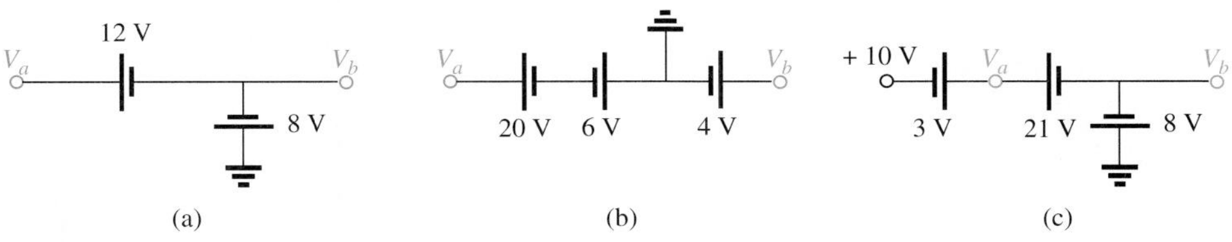

FIG. 5.80
Problem 24.

25. Determine the current I (with direction) and the voltage V (with polarity) for the networks of Fig. 5.81.

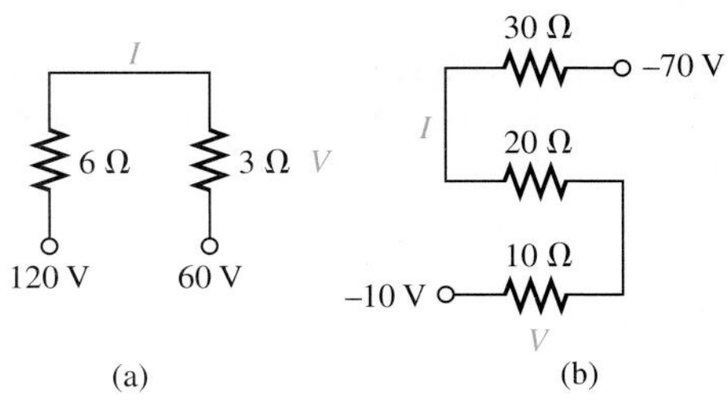

FIG. 5.81
Problem 25.

26. Determine the voltages V_a and V_1 for the networks of Fig. 5.82.

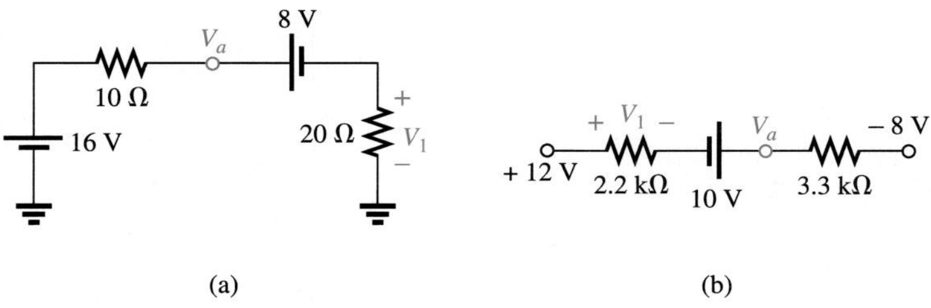

FIG. 5.82
Problem 26.

***27.** For the network of Fig. 5.83, determine the voltages:

a. V_a, V_b, V_c, V_d, V_e

b. V_{ab}, V_{dc}, V_{cb}

c. V_{ac}, V_{db}

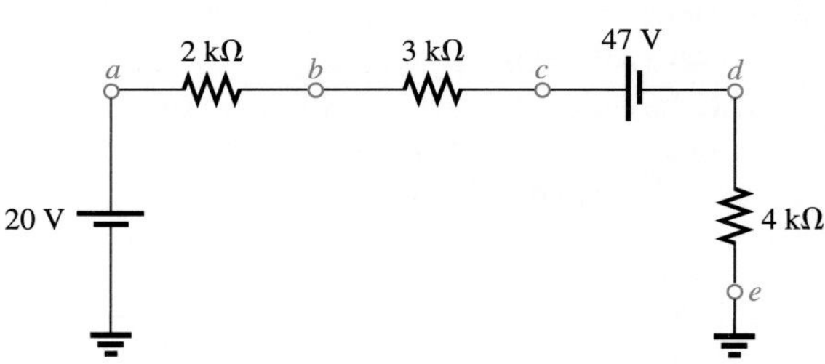

FIG. 5.83
Problem 27.

***28.** For the network of Fig. 5.84, determine the voltages:

a. V_a, V_b, V_c, V_d

b. V_{ab}, V_{cb}, V_{cd}

c. V_{ad}, V_{ca}

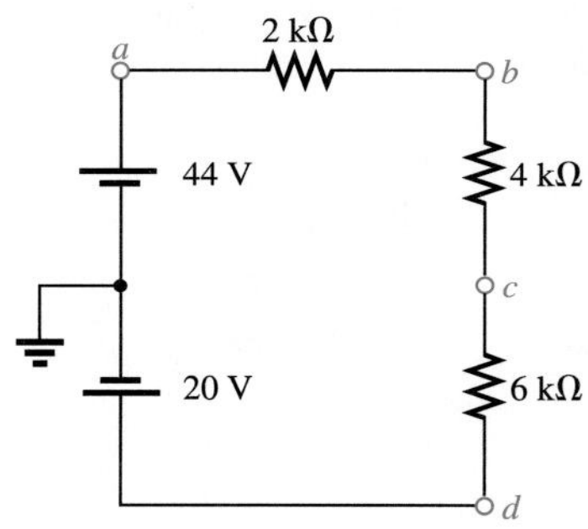

FIG. 5.84
Problem 28.

***29.** For the integrated circuit of Fig. 5.85, determine V_0, V_4, V_7, V_{10}, V_{23}, V_{30}, V_{67}, V_{56}, and I (magnitude and direction).

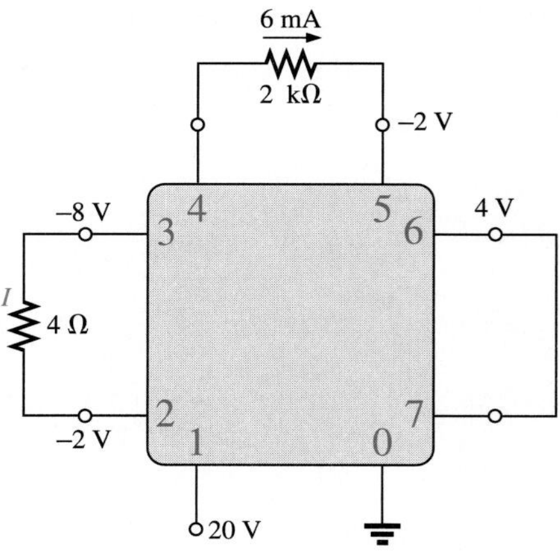

FIG. 5.85
Problem 29.

*30. For the integrated circuit of Fig. 5.86, determine V_0, V_{03}, V_2, V_{23}, V_{12}, and I_i.

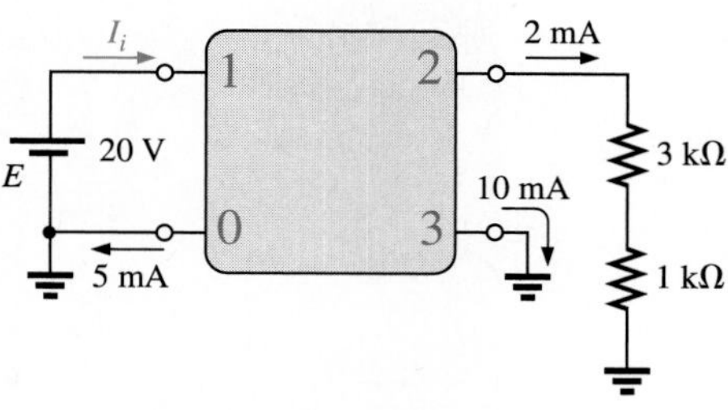

FIG. 5.86
Problem 30.

SECTION 5.8 Internal Resistance of Voltage Sources

31. Find the internal resistance of a battery that has a no-load output voltage of 60 V and supplies a current of 2 A to a load of 28 Ω.

32. Determine the voltage regulation for the battery in Problem 31, if the load resistance is dropped to 14 Ω.

33. Find the internal resistance of a battery that has a no-load output voltage of 6 V and supplies a current of 10 mA to a load of 1/2 kΩ.

SECTION 5.9 Voltage Regulation

34. Plot the voltage regulation for the battery of Problem 31.

35. Calculate the voltage regulation for the supply of Fig. 5.87.

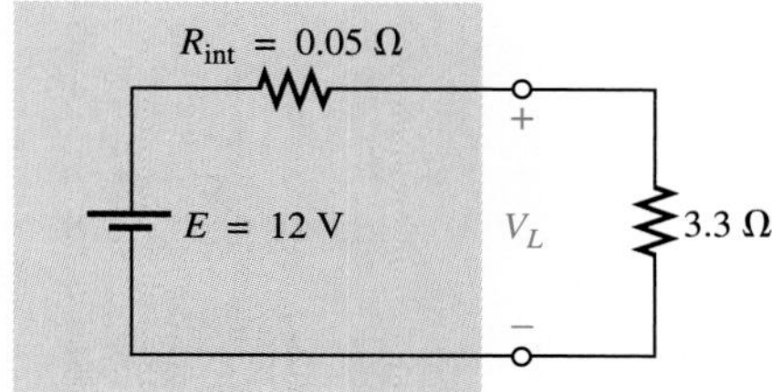

FIG. 5.87
Problem 32.

6

Parallel Circuits

Outline

Learning Outcomes

After completing this chapter you will be able to

- describe the characteristics of a parallel circuit
- explain the relationship between total circuit resistance and the individual resistances in a parallel circuit
- explain Kirchhoff's current law and apply the law to parallel circuits
- determine the branch circuit currents using the current divider rule
- explain the effect of series and parallel connected sources
- determine the power consumed by the branches of a parallel circuit and their relationship to total power of a circuit
- explain the terms open circuit and short circuit and their effect upon a parallel circuit
- determine the loading effect of a voltmeter on a circuit

The voltage across the components in a parallel circuit will remain constant. In this chapter you will learn how current flows in a parallel circuit and how this relates to Ohm's law.

6.1 INTRODUCTION

Two network configurations form the framework for some of the most complex network structures. It will be very important to understand both of these as as more complex methods and networks are examined. The *series* connection was discussed in detail in the last chapter. We will now examine the **parallel** connection and all the methods and laws associated with this important configuration.

6.2 PARALLEL ELEMENTS

Two elements, branches, or networks are in parallel if they have two points in common.

In Fig. 6.1, for example, elements 1 and 2 have terminals *a* and *b* in common; they are therefore in parallel.

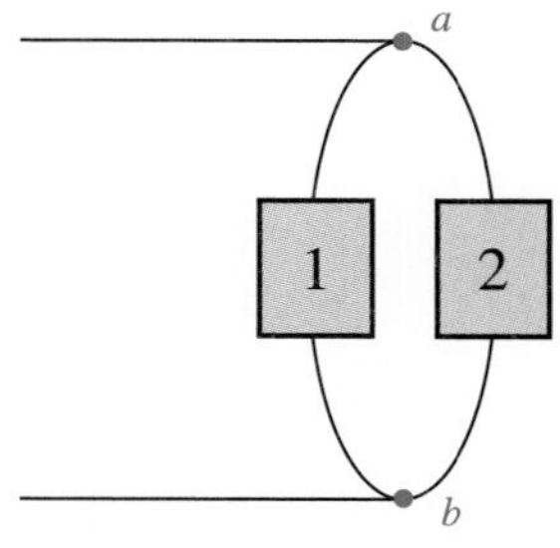

FIG. 6.1
Parallel elements.

In Fig. 6.2, all the elements are in parallel because they satisfy the above criterion. Three configurations are provided to show how the parallel networks can be drawn. Do not let the squaring of the connection at the top and bottom of Fig. 6.2(a) and (b) confuse the fact that all the elements are connected to one terminal point at the top and bottom, as shown in Fig. 6.2(c).

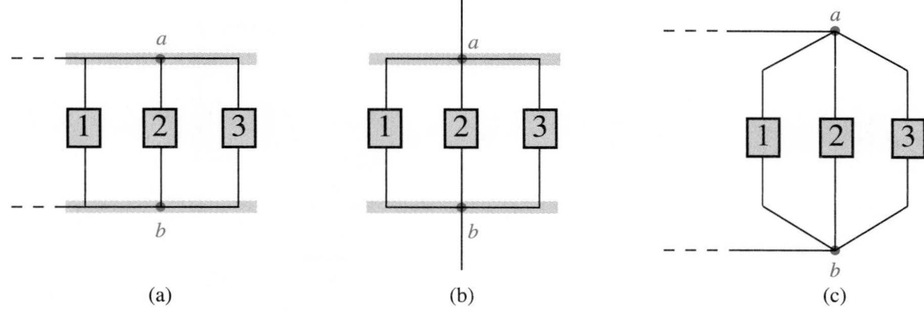

FIG. 6.2
Different ways in which three parallel elements may appear.

In Figs. 6.1 and 6.2, the numbered boxes are used as a general symbol representing either single resistive elements, batteries, or complex network configurations.

Common examples of parallel elements include the rungs of a ladder, the tying of more than one rope between two points to increase the strength of the connection, and the use of pipes between two points to split the water between the two points at a ratio determined by the area of the pipes.

6.3 TOTAL CONDUCTANCE AND RESISTANCE

When resistors are in series, the total resistance is the sum of the resistor values. However, in a parallel circuit, each additional branch adds another conducting path. To evaluate the net effect, we can use the values of *conductance*. As described in Section 3.7, conductance is the reciprocal of resistance, $G = 1/R$.

Therefore, in a parallel network (Fig. 6.3), the total conductance is the sum of the individual conductances:

$$G_T = G_1 + G_2 + G_3 + \cdots + G_N \tag{6.1}$$

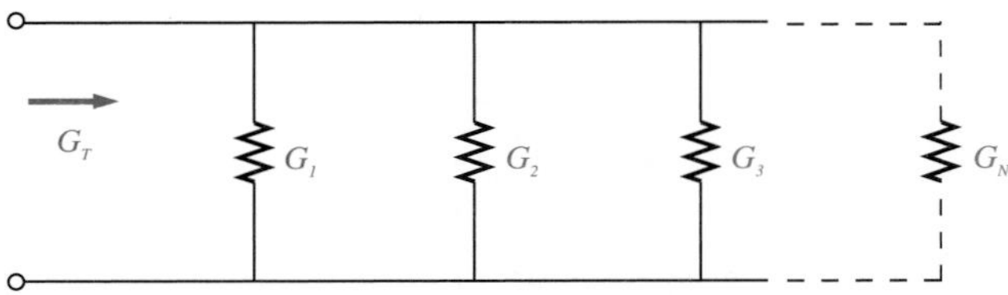

FIG. 6.3
Determining the total conductance of parallel conductances.

Since increasing levels of conductance will establish higher current levels, the more terms appearing in Eq. (6.1), the higher the input current level. In other words, as the number of resistors in parallel increases, the input current level will increase for the same applied voltage—the opposite effect of increasing the number of resistors in series.

Substituting resistor values for the network of Fig. 6.3 will result in the network of Fig. 6.4. Since $G = 1/R$, the total resistance for the network can be determined by direct substitution into Eq. (6.1):

$$\frac{1}{R_T} = \frac{1}{R_1} + \frac{1}{R_2} + \frac{1}{R_3} + \cdots + \frac{1}{R_N} \tag{6.2}$$

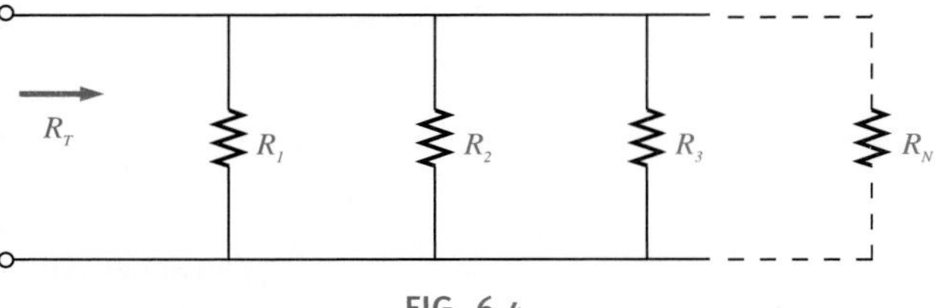

FIG. 6.4
Determining the total resistance of parallel resistors.

Note that Eq. (6.2) provides the value of $1/R_T$ (equal to G_T). To calculate R_T, *both* sides of the equation must be inverted:

$$\frac{1}{\frac{1}{R_T}} = \frac{1}{\frac{1}{R_1} + \frac{1}{R_2} + \cdots + \frac{1}{R_N}}$$

$$R_T = \frac{1}{\frac{1}{R_1} + \frac{1}{R_2} + \cdots + \frac{1}{R_N}}$$

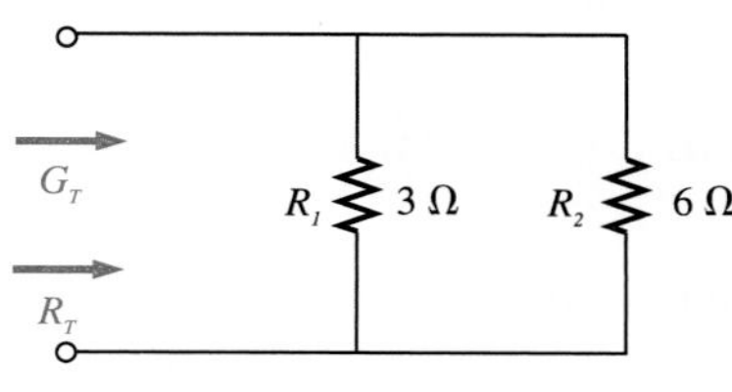

FIG. 6.5
Example 6.1.

EXAMPLE 6.1

Determine the total conductance and resistance for the parallel network of Fig. 6.5.

Solution:

$$G_T = G_1 + G_2 = \frac{1}{3\ \Omega} + \frac{1}{6\ \Omega} = 0.333\ \text{S} + 0.167\ \text{S} = \mathbf{0.5\ S}$$

and
$$R_T = \frac{1}{G_T} = \frac{1}{0.5\ \text{S}} = \mathbf{2\ \Omega}$$

EXAMPLE 6.2

Determine the effect on the total conductance and resistance of the network of Fig. 6.5 if another resistor of 10 Ω were added in parallel with the other elements.

Solution:

$$G_T = 0.5\ \text{S} + \frac{1}{10\ \Omega} = 0.5\ \text{S} + 0.1\ \text{S} = \mathbf{0.6\ S}$$

$$R_T = \frac{1}{G_T} = \frac{1}{0.6\ \text{S}} \cong \mathbf{1.667\ \Omega}$$

Note, as mentioned above, that adding additional terms increases the conductance level and decreases the resistance level.

EXAMPLE 6.3

Determine the total resistance for the network of Fig. 6.6.

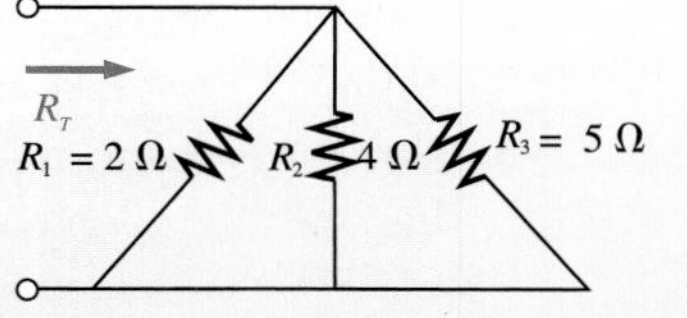

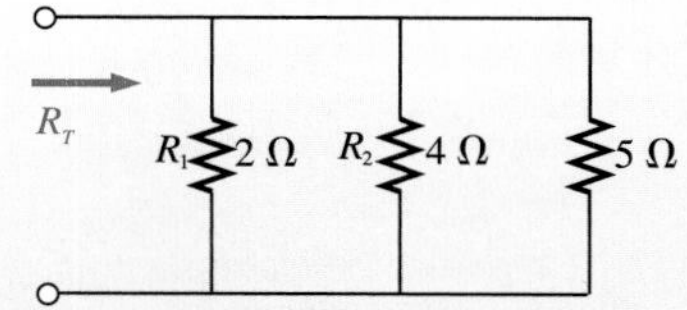

FIG. 6.6
Example 6.3.

Solution:

$$\frac{1}{R_T} = \frac{1}{R_1} + \frac{1}{R_2} + \frac{1}{R_3}$$

continued

$$= \frac{1}{2\ \Omega} + \frac{1}{4\ \Omega} + \frac{1}{5\ \Omega} = 0.5\ \text{S} + 0.25\ \text{S} + 0.2\ \text{S}$$

$$= 0.95\ \text{S}$$

and $$R_T = \frac{1}{0.95\ \text{S}} = \mathbf{1.053\ \Omega}$$

The above examples demonstrate an interesting and useful characteristic of parallel resistors:

The total resistance of parallel resistors is always less than the value of the smallest resistor.

In addition, the wider the spread in numerical value between two parallel resistors, the closer the total resistance will be to the smaller resistor. For instance, the total resistance of 3 Ω in parallel with 6 Ω is 2 Ω, as demonstrated in Example 6.1. However, the total resistance of 3 Ω in parallel with 60 Ω is 2.85 Ω, which is much closer to the value of the smaller resistor.

For *equal* resistors in parallel, the equation becomes significantly easier to apply. For N equal resistors in parallel, Eq. (6.2) becomes

$$\frac{1}{R_T} = \underbrace{\frac{1}{R} + \frac{1}{R} + \frac{1}{R} + \cdots + \frac{1}{R}}_{N}$$

$$= N\left(\frac{1}{R}\right)$$

and $$R_T = \frac{R}{N} \tag{6.3}$$

In other words, the total resistance of N parallel resistors of equal value is the resistance of *one* resistor divided by the number (N) of parallel elements.

For conductance levels, we have

$$G_T = NG \tag{6.4}$$

EXAMPLE 6.4

a. Find the total resistance of the network of Fig. 6.7.
b. Calculate the total resistance for the network of Fig. 6.9.

Solution:

a. Fig. 6.7 is redrawn in Fig. 6.8:

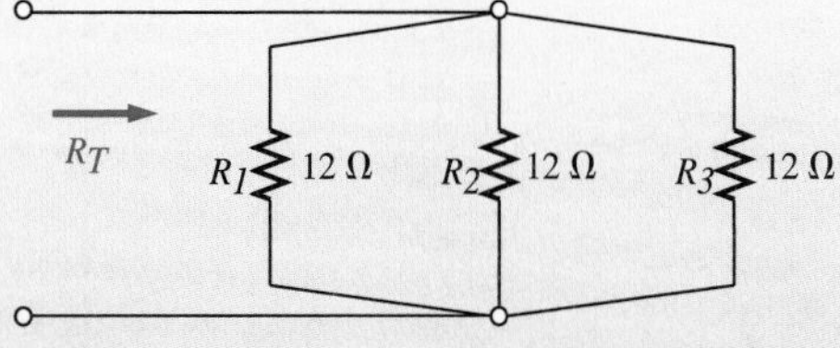

FIG. 6.7
Example 6.4(a): three parallel resistors of equal value.

continued

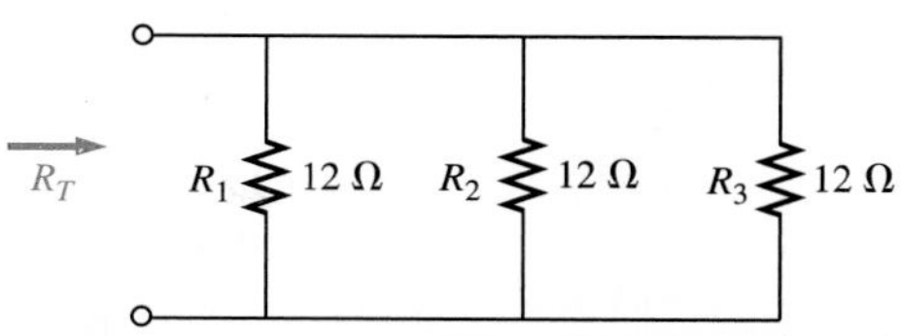

FIG. 6.8
Redrawing the network of Fig. 6.7.

$$R_T = \frac{R}{N} = \frac{12\ \Omega}{3} = \mathbf{4\ \Omega}$$

b. Fig. 6.9 is redrawn in Fig. 6.10:

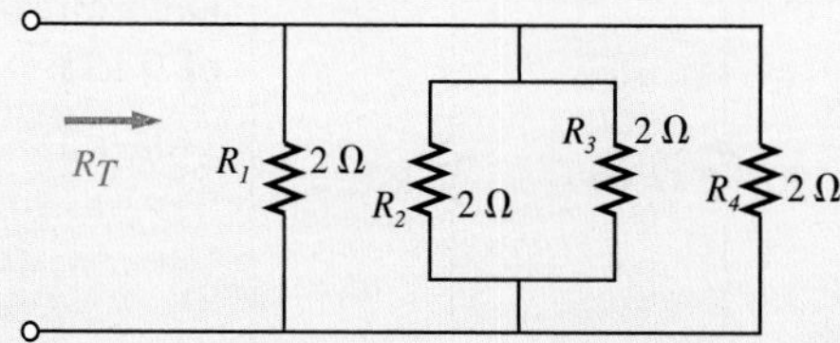

FIG. 6.9
Example 6.4(b): four parallel resistors of equal value.

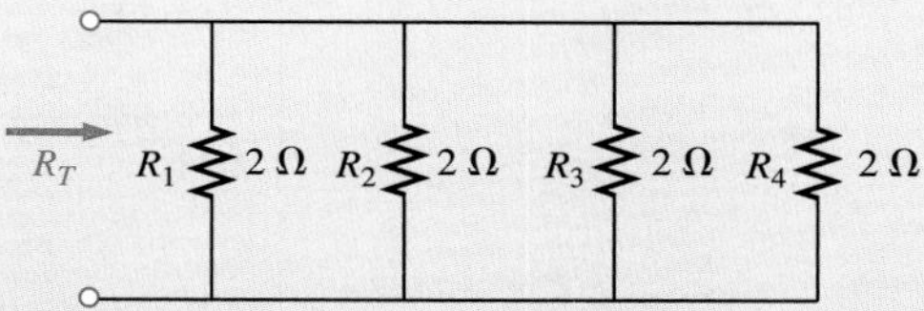

FIG. 6.10
Redrawing the network of Fig. 6.9.

$$R_T = \frac{R}{N} = \frac{2\ \Omega}{4} = \mathbf{0.5\ \Omega}$$

In almost all situations, only two or three parallel resistive elements need to be combined. With this in mind, the following equation was developed to reduce the complication of the inverse relationship when determining R_T.

For two parallel resistors, we write

$$\frac{1}{R_T} = \frac{1}{R_1} + \frac{1}{R_2}$$

Multiplying the top and bottom of each term of the right side of the equation by the other resistor will result in

$$\frac{1}{R_T} = \left(\frac{R_2}{R_2}\right)\frac{1}{R_1} + \left(\frac{R_1}{R_1}\right)\frac{1}{R_2} = \frac{R_2}{R_1R_2} + \frac{R_1}{R_1R_2}$$

$$= \frac{R_2 + R_1}{R_1 R_2}$$

and $$R_T = \frac{R_1 R_2}{R_1 + R_2} \tag{6.5}$$

In words,

the total resistance of two parallel resistors is the product of the two divided by their sum.

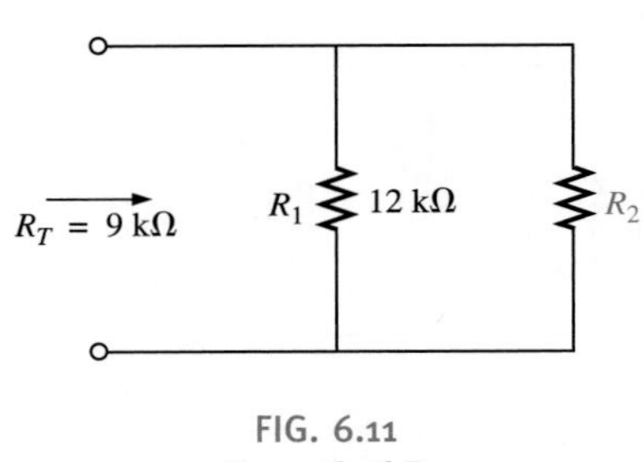

FIG. 6.11
Example 6.5.

EXAMPLE 6.5

Determine the value of R_2 in Fig. 6.11 to establish a total resistance of 9 kΩ.

Solution:

$$R_T = \frac{R_1 R_2}{R_1 + R_2}$$

$$R_T(R_1 + R_2) = R_1 R_2$$

$$R_T R_1 + R_T R_2 = R_1 R_2$$

$$R_T R_1 = R_1 R_2 - R_T R_2$$

$$R_T R_1 = (R_1 - R_T) R_2$$

and $$R_2 = \frac{R_T R_1}{R_1 - R_T} \tag{6.6}$$

Substituting values:

$$R_2 = \frac{(9\ \text{k}\Omega)(12\ \text{k}\Omega)}{12\ \text{k}\Omega - 9\ \text{k}\Omega}$$

$$= \frac{108\ \text{k}\Omega}{3} = \mathbf{36\ k\Omega}$$

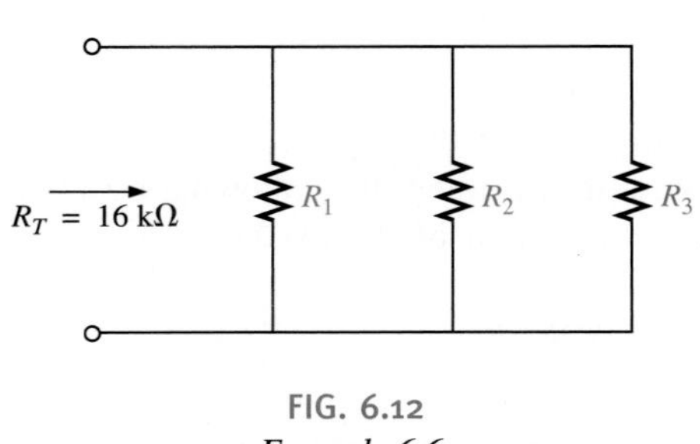

FIG. 6.12
Example 6.6.

EXAMPLE 6.6

Determine the values of R_1, R_2, and R_3 in Fig. 6.12 if $R_2 = 2R_1$ and $R_3 = 2R_2$ and the total resistance is 16 kΩ.

Solution:

$$\frac{1}{R_T} = \frac{1}{R_1} + \frac{1}{R_2} + \frac{1}{R_3}$$

$$\frac{1}{16\ \text{k}\Omega} = \frac{1}{R_1} + \frac{1}{2R_1} + \frac{1}{4R_1}$$

since $$R_3 = 2R_2 = 2(2R_1) = 4R_1$$

and $$\frac{1}{16\ \text{k}\Omega} = \frac{1}{R_1} + \frac{1}{2}\left(\frac{1}{R_1}\right) + \frac{1}{4}\left(\frac{1}{R_1}\right)$$

continued

$$\frac{1}{16\text{ k}\Omega} = 1.75\left(\frac{1}{R_1}\right)$$

with $$R_1 = 1.75(16\text{ k}\Omega) = \mathbf{28\text{ k}\Omega}$$

Recall for series circuits that the total resistance will always increase as additional elements are added in series.

For parallel resistors, the total resistance will always decrease as additional elements are added in parallel.

The next example demonstrates this unique characteristic of parallel resistors. Note the use of the || symbol. 5 Ω || 7 Ω means a 5-Ω resistor in parallel with a 7-Ω resistor.

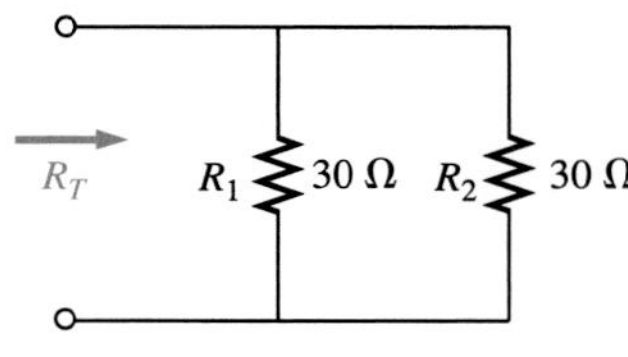

FIG. 6.13
Example 6.7: two equal, parallel resistors.

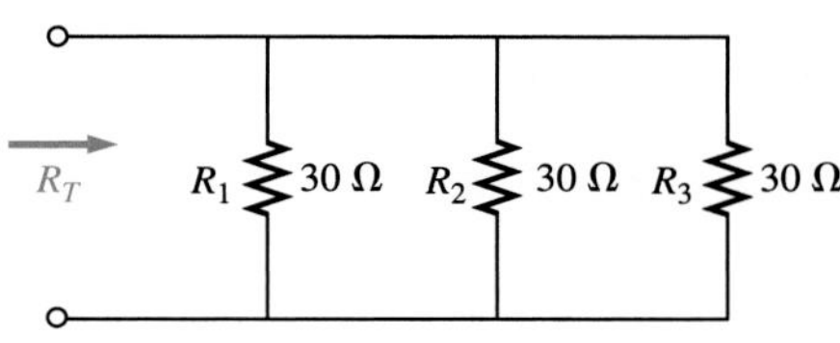

FIG. 6.14
Adding a third parallel resistor of equal value to the network of Fig. 6.13.

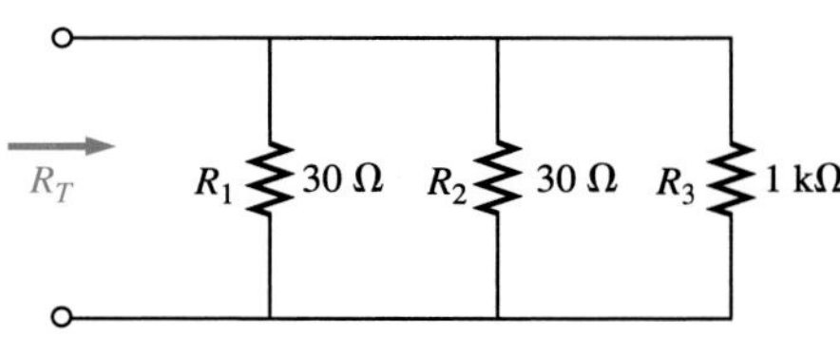

FIG. 6.15
Adding a much larger parallel resistor to the network of Fig. 6.13.

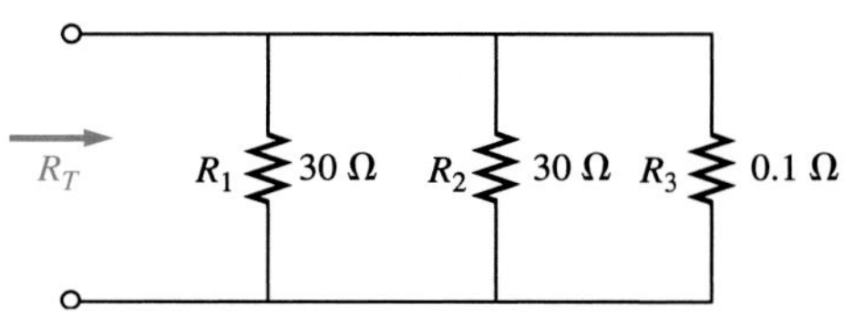

FIG. 6.16
Adding a much smaller parallel resistor to the network of Fig. 6.13.

EXAMPLE 6.7

a. Determine the total resistance of the network of Fig. 6.13.
b. What is the effect on the total resistance of the network of Fig. 6.13 if an additional resistor of the same value is added, as shown in Fig. 6.14?
c. What is the effect on the total resistance of the network of Fig. 6.13 if a very large resistance is added in parallel, as shown in Fig. 6.15?
d. What is the effect on the total resistance of the network of Fig. 6.13 if a very small resistance is added in parallel, as shown in Fig. 6.16?

Solution:

a. $$R_T = 30\ \Omega \parallel 30\ \Omega = \frac{30\ \Omega}{2} = \mathbf{15\ \Omega}$$

b. $$R_T = 30\ \Omega \parallel 30\ \Omega \parallel 30\ \Omega = \frac{30\ \Omega}{3} = \mathbf{10\ \Omega} < 15\ \Omega$$

R_T decreased

c. $$R_T = 30\ \Omega \parallel 30\ \Omega \parallel 1\text{ k}\Omega = 15\ \Omega \parallel 1\text{ k}\Omega$$

$$= \frac{(15\ \Omega)(1000\ \Omega)}{15\ \Omega + 1000\ \Omega} = \mathbf{14.778\ \Omega} \lesssim 15\ \Omega$$

Small decrease in R_T

d. $$R_T = 30\ \Omega \parallel 30\ \Omega \parallel 0.1\ \Omega = 15\ \Omega \parallel 0.1\ \Omega$$

$$= \frac{(15\ \Omega)(0.1\ \Omega)}{15\ \Omega + 0.1\ \Omega} = \mathbf{0.099\ \Omega} \ll 15\ \Omega$$

Significant decrease in R_T

In each case the total resistance of the network decreased with the increase of an additional parallel resistive element, no matter how large or small. Note also that the total resistance is also smaller than the smallest parallel element.

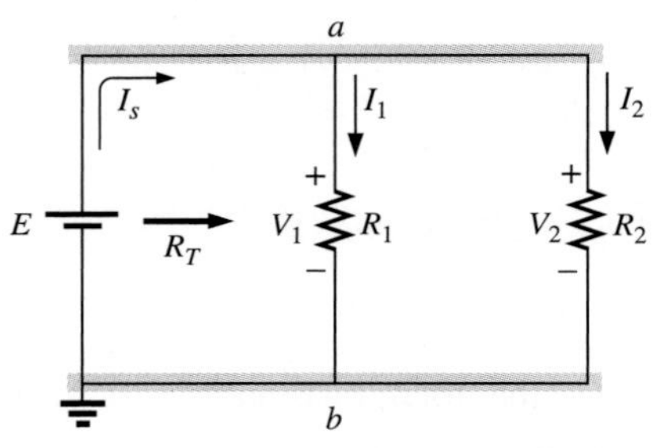

FIG. 6.17
Parallel network.

6.4 PARALLEL NETWORKS

The network of Fig. 6.17 is the simplest of parallel networks. All the elements have terminals a and b in common. The total resistance is determined by $R_T = R_1R_2/(R_1 + R_2)$, and the source current by $I_s = E/R_T$. Throughout the text, the subscript s will be used to denote a property of the source. Since the terminals of the battery are connected directly across the resistors R_1 and R_2,

the voltage across parallel elements is the same.

Using this fact will result in

$$V_1 = V_2 = E$$

and
$$I_1 = \frac{V_1}{R_1} = \frac{E}{R_1}$$

with
$$I_2 = \frac{V_2}{R_2} = \frac{E}{R_2}$$

If we take the equation for the total resistance and multiply both sides by the applied voltage, we obtain

$$E\left(\frac{1}{R_T}\right) = E\left(\frac{1}{R_1} + \frac{1}{R_2}\right)$$

and
$$\frac{E}{R_T} = \frac{E}{R_1} + \frac{E}{R_2}$$

Substituting the Ohm's law relationships appearing above, we find that the source current

$$I_s = I_1 + I_2$$

permitting the following conclusion:

For single-source parallel networks, the source current (I_s) is equal to the sum of the individual branch currents.

The power dissipated by the resistors and delivered by the source can be determined from

$$P_1 = V_1I_1 = I_1^2R_1 = \frac{V_1^2}{R_1}$$

$$P_2 = V_2I_2 = I_2^2R_2 = \frac{V_2^2}{R_2}$$

$$P_s = EI_s = I_s^2R_T = \frac{E^2}{R_T}$$

The total power dissipated in a parallel network can be determined, noting that the voltage is the same across each element:

Since
$$P_T = VI_T$$

and
$$I_T = I_1 + I_2 + I_3 + \cdot \cdot \cdot + I_N$$

continued

Then $$P_T = V(I_1 + I_2 + I_3 + \cdots + I_N)$$
$$= VI_1 + VI_2 + VI_3 + \cdots + VI_N$$
$$= P_1 + P_2 + P_3 + \cdots + P_N$$

The total power dissipated in a parallel network is the sum of the powers dissipated in the individual elements.

EXAMPLE 6.8

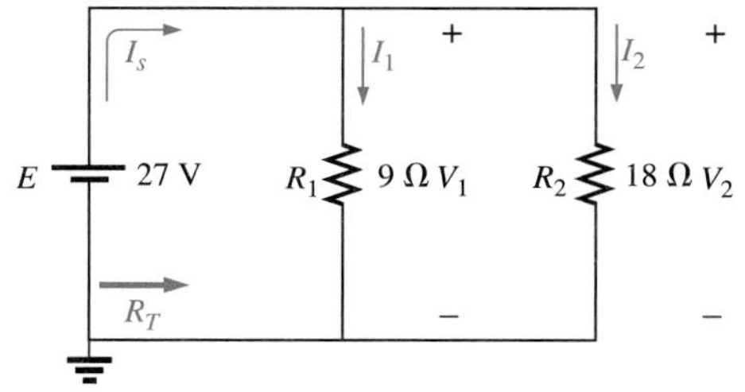

For the parallel network of Fig. 6.18:

a. Calculate R_T.
b. Determine I_s.
c. Calculate I_1 and I_2 and demonstrate that $I_s = I_1 + I_2$.
d. Determine the power to each resistive load.
e. Determine the power delivered by the source and compare it to the total power dissipated by the resistive elements.

FIG. 6.18
Example 6.8.

Solution:

a. $R_T = \dfrac{R_1R_2}{R_1 + R_2} = \dfrac{(9\ \Omega)(18\ \Omega)}{9\ \Omega + 18\ \Omega} = \dfrac{162\ \Omega}{27} = \mathbf{6\ \Omega}$

b. $I_s = \dfrac{E}{R_T} = \dfrac{27\ \text{V}}{6\ \Omega} = \mathbf{4.5\ A}$

c. $I_1 = \dfrac{V_1}{R_1} = \dfrac{E}{R_1} = \dfrac{27\ \text{V}}{9\ \Omega} = \mathbf{3\ A}$

$I_2 = \dfrac{V_2}{R_2} = \dfrac{E}{R_2} = \dfrac{27\ \text{V}}{18\ \Omega} = \mathbf{1.5\ A}$

$I_s = I_1 + I_2$
$4.5\ \text{A} = 3\ \text{A} + 1.5\ \text{A}$
$\mathbf{4.5\ A} = \mathbf{4.5\ A}$ (checks)

d. $P_1 = V_1I_1 = EI_1 = (27\ \text{V})(3\ \text{A}) = \mathbf{81\ W}$
$P_2 = V_2I_2 = EI_2 = (27\ \text{V})(1.5\ \text{A}) = \mathbf{40.5\ W}$
e. $P_s = EI_s = (27\ \text{V})(4.5\ \text{A}) = \mathbf{121.5\ W}$
$P_s = P_1 + P_2 = 81\ \text{W} + 40.5\ \text{W} = \mathbf{121.5\ W}$

EXAMPLE 6.9

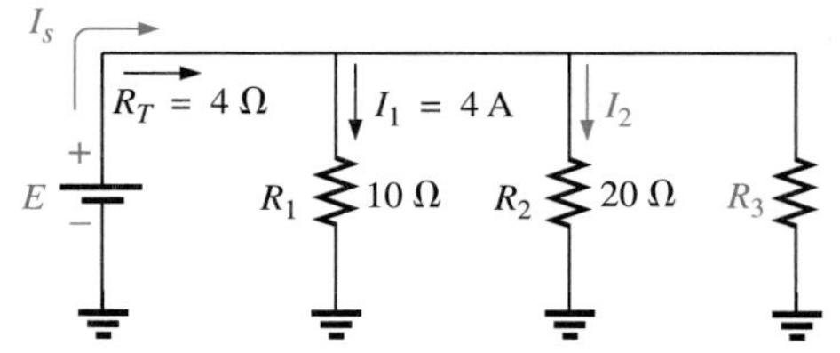

Given the information provided in Fig. 6.19:
a. Determine R_3.
b. Calculate E.
c. Find I_s.
d. Find I_2.
e. Determine P_2.

FIG. 6.19
Example 6.9.

Solution:

a. $$\frac{1}{R_T} = \frac{1}{R_1} + \frac{1}{R_2} + \frac{1}{R_3}$$

continued

$$\frac{1}{4\ \Omega} = \frac{1}{10\ \Omega} + \frac{1}{20\ \Omega} + \frac{1}{R_3}$$

$$0.25\ \text{S} = 0.1\ \text{S} + 0.05\ \text{S} + \frac{1}{R_3}$$

$$0.25\ \text{S} = 0.15\ \text{S} + \frac{1}{R_3}$$

$$\frac{1}{R_3} = 0.1\ \text{S}$$

$$R_3 = \frac{1}{0.1\ \text{S}} = \mathbf{10\ \Omega}$$

b. $E = V_1 = I_1R_1 = (4\ \text{A})(10\ \Omega) = \mathbf{40\ V}$

c. $I_s = \dfrac{E}{R_T} = \dfrac{40\ \text{V}}{4\ \Omega} = \mathbf{10\ A}$

d. $I_2 = \dfrac{V_2}{R_2} = \dfrac{E}{R_2} = \dfrac{40\ \text{V}}{20\ \Omega} = \mathbf{2\ A}$

e. $P_2 = I_2^2R_2 = (2\ \text{A})^2(20\ \Omega) = \mathbf{80\ W}$

6.5 KIRCHHOFF'S CURRENT LAW

Kirchhoff's voltage law provides an important relationship among voltage levels around any closed loop of a network. We now consider **Kirchhoff's current law**, which provides an equally important relationship among current levels at any junction.

Kirchhoff's current law (KCL) states that the algebraic sum of the currents entering and leaving an area, system, or junction is zero.

In other words,

the sum of the currents entering an area, system, or junction must equal the sum of the currents leaving the area, system, or junction.

In equation form, for entering currents, I_{Ej}, and leaving currents, I_{Lj}, ($j = 1, 2, \cdots, N$):

$$I_{E1} + I_{E2} + \cdots + I_{EN} = I_{L1} + I_{L2} + \cdots + I_{LN}$$

or

$$\Sigma I_{\text{entering}} = \Sigma I_{\text{leaving}} \tag{6.7}$$

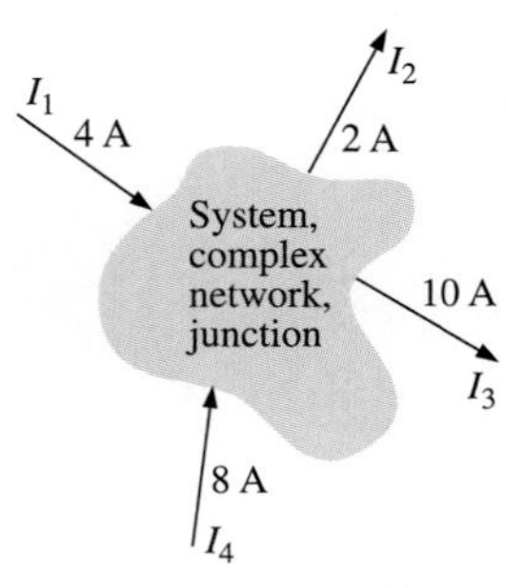

FIG. 6.20
Introducing Kirchhoff's current law.

In Fig. 6.20, for instance, the shaded area can enclose an entire system, a complex network, or simply a junction of two or more paths. In each case the current entering must equal that leaving, as witnessed by the fact that

$$I_1 + I_4 = I_2 + I_3$$

$$4\ \text{A} + 8\ \text{A} = 2\ \text{A} + 10\ \text{A}$$

$$12\ \text{A} = 12\ \text{A}$$

The most common application of the law will be at the junction of two or more paths of current flow, as shown in Fig. 6.21. For some students it is difficult initially to determine whether a current is entering or leaving a junction. One approach that may help is to picture yourself as standing on the junction and treating the path currents as arrows. If the arrow appears to be heading toward you, as is the case for I_1 in Fig. 6.21, then it is entering the junction. If you see the tail of the arrow (from the junction) as it travels down its path away from you, it is leaving the junction, as is the case for I_2 and I_3 in Fig. 6.21.

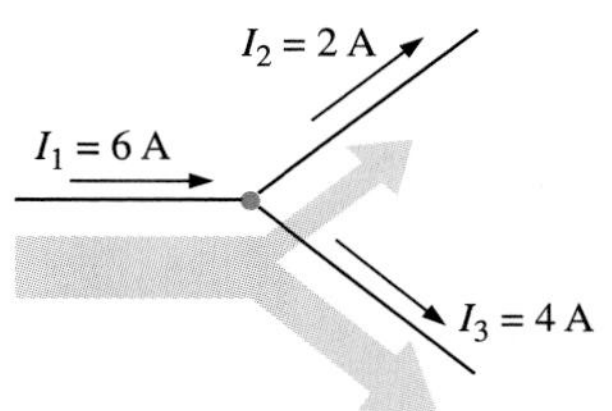

FIG. 6.21
Demonstrating Kirchhoff's current law.

Applying Kirchhoff's current law to the junction of Fig. 6.21:

$$\Sigma I_{\text{entering}} = \Sigma I_{\text{leaving}}$$
$$6\text{ A} = 2\text{ A} + 4\text{ A}$$
$$6\text{ A} = 6\text{ A} \quad \text{(checks)}$$

In the next two examples, unknown currents can be determined by applying Kirchhoff's current law. Simply remember to place all current levels entering a junction to the left of the equals sign and the sum of all currents leaving a junction to the right of the equals sign. The water-in-the-pipe analogy helps to support and clarify the preceding law. Quite obviously, the sum total of the water entering a junction must equal the total of the water leaving the exit pipes.

EXAMPLE 6.10

Determine the currents I_3 and I_4 of Fig. 6.22 using Kirchhoff's current law.

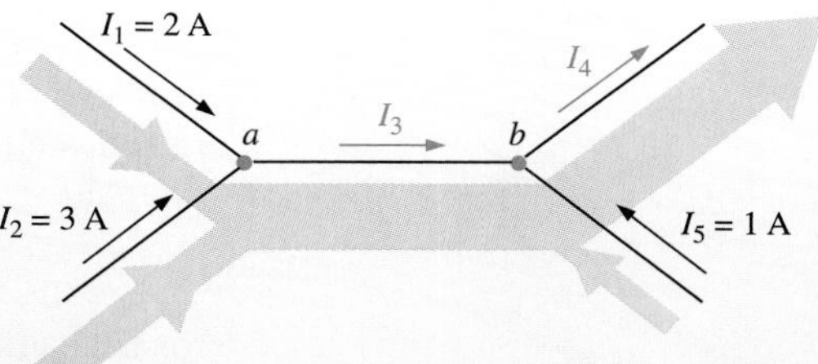

FIG. 6.22
Example 6.10.

Solution: We must first work with junction a since the only unknown is I_3. At junction b there are two unknowns, and both cannot be determined from one application of the law.

At a:

$$\Sigma I_{\text{entering}} = \Sigma I_{\text{leaving}}$$
$$I_1 + I_2 = I_3$$
$$2\text{ A} + 3\text{ A} = I_3$$
$$I_3 = \mathbf{5\text{ A}}$$

At b:

$$\Sigma I_{\text{entering}} = \Sigma I_{\text{leaving}}$$
$$I_3 + I_5 = I_4$$
$$5\text{ A} + 1\text{ A} = I_4$$
$$I_4 = \mathbf{6\text{ A}}$$

EXAMPLE 6.11

Determine I_1, I_3, I_4, and I_5 for the network of Fig. 6.23.

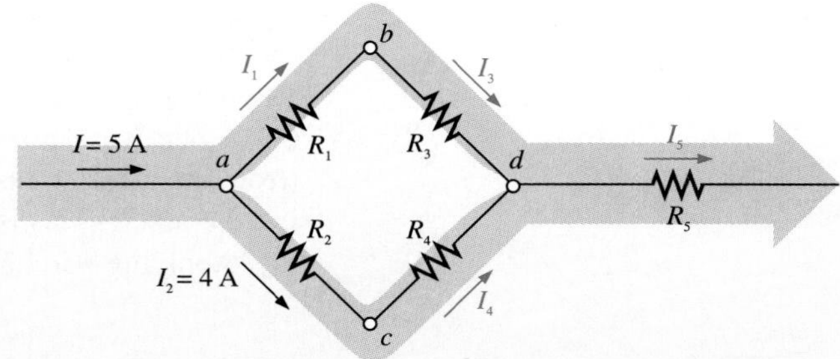

FIG. 6.23
Example 6.11.

Solution: At *a:*

$$\Sigma I_{\text{entering}} = \Sigma I_{\text{leaving}}$$
$$I = I_1 + I_2$$
$$5\text{ A} = I_1 + 4\text{ A}$$

Subtracting 4 A from both sides gives

$$5\text{ A} - 4\text{ A} = I_1 + 4\cancel{\text{A}} - 4\cancel{\text{A}}$$
$$I_1 = 5\text{ A} - 4\text{ A} = \mathbf{1\text{ A}}$$

At *b:*

$$\Sigma I_{\text{entering}} = \Sigma I_{\text{leaving}}$$
$$I_1 = I_3 = \mathbf{1\text{ A}}$$

as it should, since R_1 and R_3 are in series and the current is the same in series elements.

At *c:*

$$I_2 = I_4 = \mathbf{4\text{ A}}$$

for the same reasons given for junction *b.*

At *d:*

$$\Sigma I_{\text{entering}} = \Sigma I_{\text{leaving}}$$
$$I_3 + I_4 = I_5$$
$$1\text{ A} + 4\text{ A} = I_5$$
$$I_5 = \mathbf{5\text{ A}}$$

If we enclose the entire network, we find that the current entering is $I = 5$ A; the net current leaving from the far right is $I_5 = 5$ A. The two must be equal, since the net current entering any system must equal that leaving.

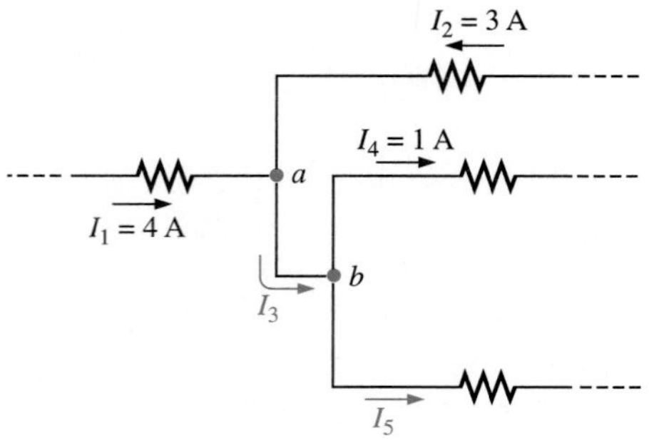

FIG. 6.24
Example 6.12.

EXAMPLE 6.12

Determine the currents I_3 and I_5 of Fig. 6.24 through applications of Kirchhoff's current law.

Solution: Note that since node *b* has two unknown quantities and node *a* has only one, we must first apply Kirchhoff's current law to node *a*. The result can then be applied to node *b*. For node *a,*

continued

$$I_1 + I_2 = I_3$$
$$4\text{ A} + 3\text{ A} = I_3$$

and $$I_3 = \mathbf{7\text{ A}}$$

For node b,

$$I_3 = I_4 + I_5$$
$$7\text{ A} = 1\text{ A} + I_5$$

and $$I_5 = 7\text{ A} - 1\text{ A} = \mathbf{6\text{ A}}$$

EXAMPLE 6.13

Find the magnitude and direction of the currents I_3, I_4, I_6, and I_7 for the network of Fig. 6.25. Even though the elements are not in series or parallel, Kirchhoff's current law can be applied to determine all the unknown currents.

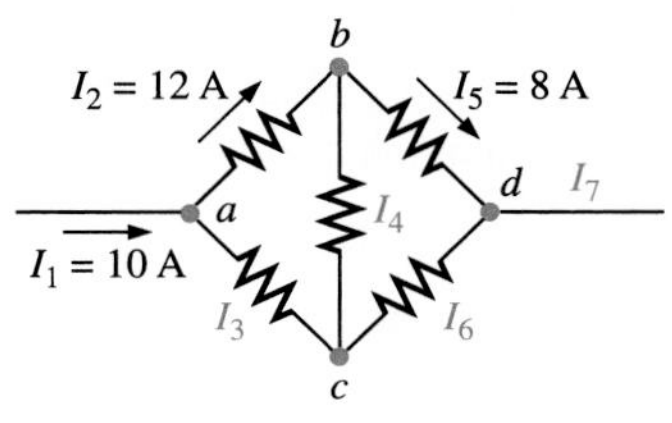

FIG. 6.25
Example 6.13.

Solution: Considering the overall system, we know that the current entering must equal that leaving. Therefore,

$$I_7 = I_1 = \mathbf{10\text{ A}}$$

Since 10 A are entering node a and 12 A are leaving, I_3 must be supplying current to the node. Applying Kirchhoff's current law at node a,

$$I_1 + I_3 = I_2$$
$$10\text{ A} + I_3 = 12\text{ A}$$

and $$I_3 = 12\text{ A} - 10\text{ A} = \mathbf{2\text{ A}}$$

At node b, since 12 A are entering and 8 A are leaving, I_4 must be leaving. Therefore,

$$I_2 = I_4 + I_5$$
$$12\text{ A} = I_4 + 8\text{ A}$$

and $$I_4 = 12\text{ A} - 8\text{ A} = \mathbf{4\text{ A}}$$

At node c, I_3 is leaving at 2 A and I_4 is entering at 4 A, requiring that I_6 be leaving. Applying Kirchhoff's current law at node c,

$$I_4 = I_3 + I_6$$
$$4\text{ A} = 2\text{ A} + I_6$$

and $$I_6 = 4\text{ A} - 2\text{ A} = \mathbf{2\text{ A}}$$

As a check at node d,

$$I_5 + I_6 = I_7$$
$$8\text{ A} + 2\text{ A} = 10\text{ A}$$
$$10\text{ A} = 10\text{ A} \quad \text{(checks)}$$

Look back at Example 6.8 to see that Kirchhoff's current law applied, with the current entering the top node being equal to the sum of the two currents leaving: 3 A + 1.5 A = 4.5 A.

Also, in Example 6.9 you could use Kirchhoff's current law to calculate that $I_3 = 4$ A, making $I_1 + I_2 + I_3 = I_s$ at the top node in Fig. 6.19.

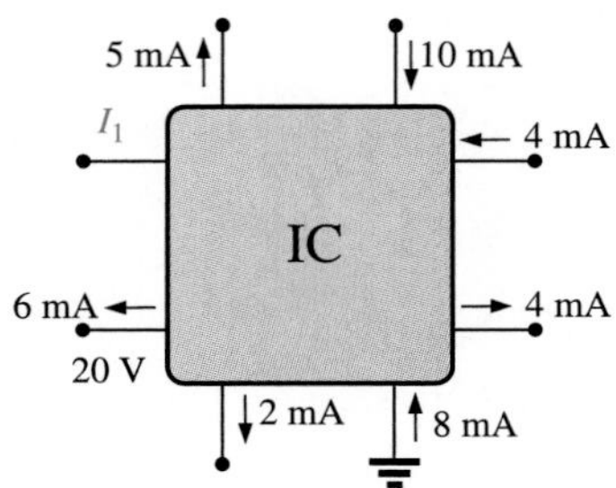

FIG. 6.26
Integrated circuit.

Applying of Kirchhoff's current law is not limited to networks where all the internal connections are known or visible. For instance, all the currents of the integrated circuit of Fig. 6.26 are known except I_1. By treating the system as a single node, we can apply Kirchhoff's current law using the following table to ensure an accurate listing of all known quantities:

I_i	I_o
10 mA	5 mA
4 mA	4 mA
8 mA	2 mA
22 mA	6 mA
	17 mA

Noting the total input current versus that leaving clearly reveals that I_1 is a current of 22 mA − 17 mA = 5 mA leaving the system.

6.6 CURRENT DIVIDER RULE

As the name suggests, the **current divider rule** (CDR) will determine how the current entering a set of parallel branches will split between the elements.

For two parallel elements of equal value, the current will divide equally.

For parallel elements with different values, the smaller the resistance, the greater the share of input current.

For parallel elements of different values, the current will split with a ratio equal to the inverse of their resistor values.

For example, if one of two parallel resistors is twice the other, then the current through the larger resistor will be half the other.

In Fig. 6.27, since I_1 is 1 mA and R_1 is six times that of R_3, the current through R_3 must be 6 mA (without making any other calculations including the total current or the actual resistance levels). For R_2 the current must be 2 mA since R_1 is twice R_2. The total current must then be the sum of I_1, I_2, and I_3, or 9 mA. In total, therefore, knowing only the current through R_1, we were able to find all the other currents of the configuration without knowing anything more about the network.

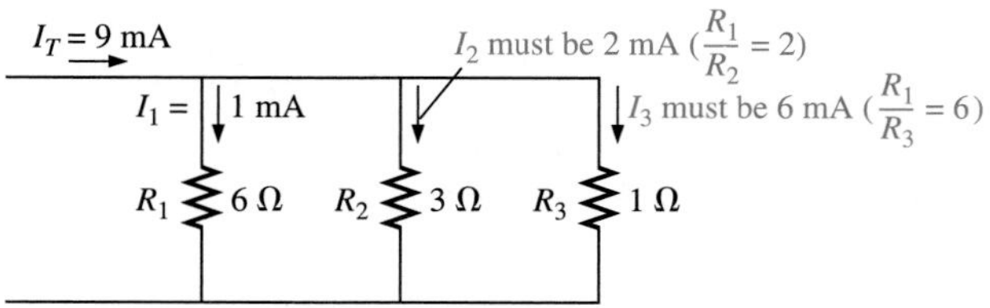

FIG. 6.27
Demonstrating how current will divide between unequal resistors.

For networks in which only the resistor values are given along with the input current, the *current divider rule* should be applied to determine the various branch currents. It can be derived using the network of Fig. 6.28. The input current I equals V/R_T, where R_T is the total resistance of the

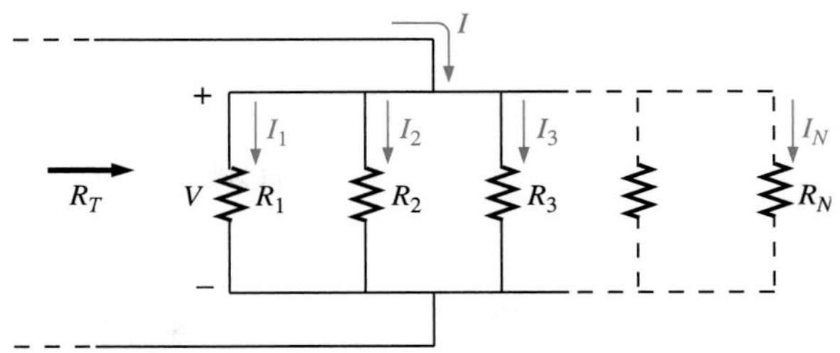

FIG. 6.28
Deriving the current divider rule.

parallel branches. Substituting $V = I_x R_x$ into the above equation, where I_x refers to the current through a parallel branch of resistance R_x, we have

$$I = \frac{V}{R_T} = \frac{I_x R_x}{R_T}$$

and

$$I_x = \frac{R_T}{R_x} I \tag{6.8}$$

which is the general form for the current divider rule. In words, the current through any parallel branch is equal to the product of the *total* resistance of the parallel branches and the input current divided by the resistance of the branch through which the current is to be determined.

For the current I_1,

$$I_1 = \frac{R_T}{R_1} I$$

and for I_2,

$$I_2 = \frac{R_T}{R_2} I$$

and so on.

For the particular case of *two parallel resistors,* as shown in Fig. 6.29,

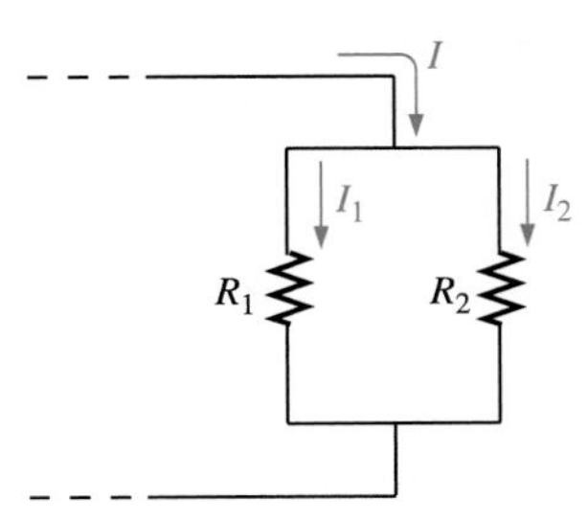

FIG. 6.29
Developing an equation for current division between two parallel resistors.

$$R_T = \frac{R_1 R_2}{R_1 + R_2}$$

and

$$I_1 = \frac{R_T}{R_1} I = \frac{\dfrac{R_1 R_2}{R_1 + R_2}}{R_1} I$$

Note difference in subscripts.

and

$$I_1 = \frac{R_2 I}{R_1 + R_2} \tag{6.9}$$

Similarly for I_2,

$$I_2 = \frac{R_1 I}{R_1 + R_2} \tag{6.10}$$

In words, for two parallel branches, the current through either branch is equal to the product of the *other* parallel resistor and the input current divided by the *sum* (not total parallel resistance) of the two parallel resistances.

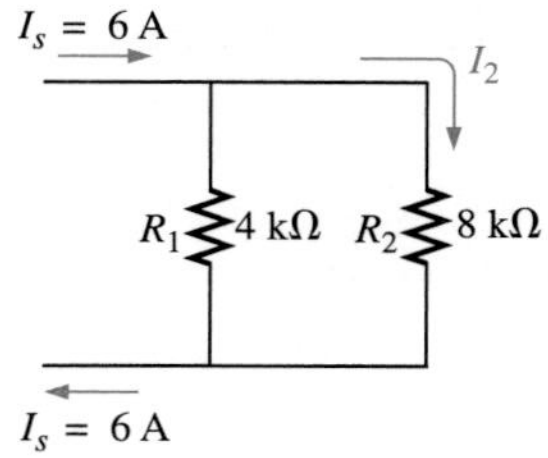

FIG. 6.30
Example 6.14.

EXAMPLE 6.14

Determine the current I_2 for the network of Fig. 6.30 using the current divider rule.

Solution:

$$I_2 = \frac{R_1 I_s}{R_1 + R_2} = \frac{(4\text{ k}\Omega)(6\text{ A})}{4\text{ k}\Omega + 8\text{ k}\Omega} = \frac{4}{12}(6\text{ A}) = \frac{1}{3}(6\text{ A})$$

$$= \mathbf{2\text{ A}}$$

EXAMPLE 6.15

Find the current I_1 for the network of Fig. 6.31.

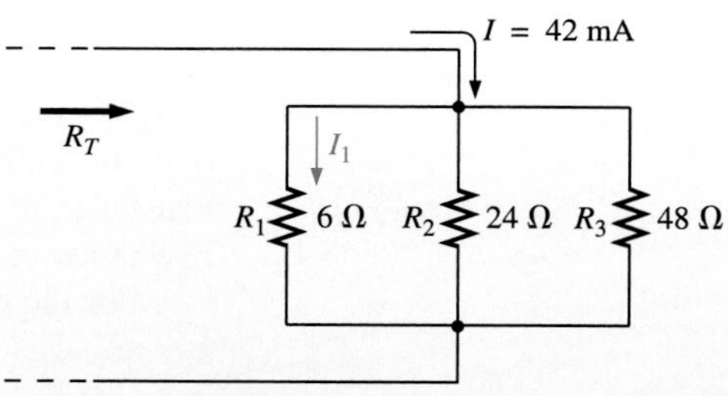

FIG. 6.31
Example 6.15.

Solution: There are two options for solving this problem. The first is to use Eq. (6.8) as follows:

$$\frac{1}{R_T} = \frac{1}{6\,\Omega} + \frac{1}{24\,\Omega} + \frac{1}{48\,\Omega} = 0.1667\text{ S} + 0.0417\text{ S} + 0.0208\text{ S}$$

$$= 0.2292\text{ S}$$

and
$$R_T = \frac{1}{0.2292\text{ S}} = 4.363\,\Omega$$

with
$$I_1 = \frac{R_T}{R_1} I = \frac{4.363\,\Omega}{6\,\Omega}(42\text{ mA}) = \mathbf{30.54\text{ mA}}$$

or apply Eq. (6.9) once after combining R_2 and R_3 as follows:

$$24\,\Omega \parallel 48\,\Omega = \frac{(24\,\Omega)(48\,\Omega)}{24\,\Omega + 48\,\Omega} = 16\,\Omega$$

and
$$I_1 = \frac{16\,\Omega(42\text{ mA})}{16\,\Omega + 6\,\Omega} = \mathbf{30.54\text{ mA}}$$

Both generated the same answer, leaving you with a choice for future calculations involving more than two parallel resistors.

EXAMPLE 6.16

Determine the magnitude of the currents I_1, I_2, and I_3 for the network of Fig. 6.32.

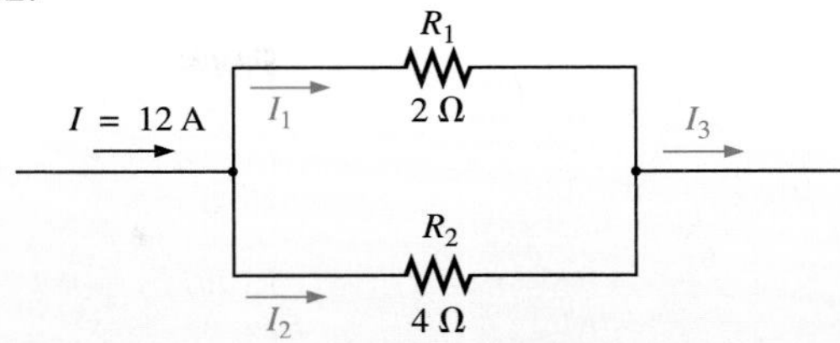

FIG. 6.32
Example 6.16.

Solution: By Eq. (6.9), the current divider rule,

$$I_1 = \frac{R_2 I}{R_1 + R_2} = \frac{(4\ \Omega)(12\ \text{A})}{2\ \Omega + 4\ \Omega} = \mathbf{8\ A}$$

Applying Kirchhoff's current law,

$$I = I_1 + I_2$$

and $$I_2 = I - I_1 = 12\ \text{A} - 8\ \text{A} = \mathbf{4\ A}$$

or, using the current divider rule again,

$$I_2 = \frac{R_1 I}{R_1 + R_2} = \frac{(2\ \Omega)(12\ \text{A})}{2\ \Omega + 4\ \Omega} = \mathbf{4\ A}$$

The total current entering the parallel branches must equal that leaving. Therefore,

$$I_3 = I = \mathbf{12\ A}$$

or $$I_3 = I_1 + I_2 = 8\ \text{A} + 4\ \text{A} = \mathbf{12\ A}$$

EXAMPLE 6.17

Determine the resistance R_1 to effect the division of current in Fig. 6.33.

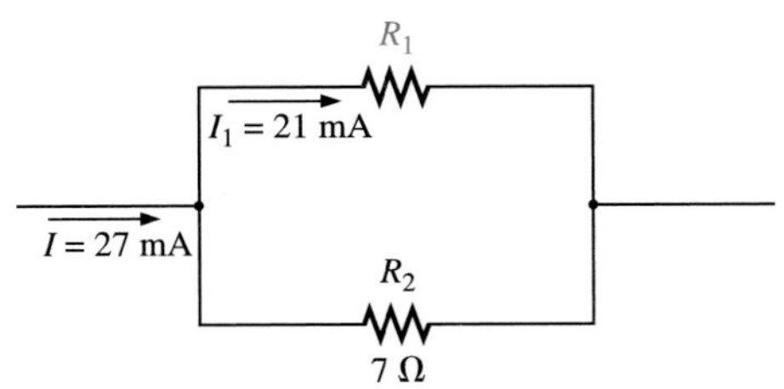

FIG. 6.33
Example 6.17.

Solution: Applying the current divider rule,

$$I_1 = \frac{R_2 I}{R_1 + R_2}$$

and
$$(R_1 + R_2)I_1 = R_2 I$$
$$R_1 I_1 + R_2 I_1 = R_2 I$$
$$R_1 I_1 = R_2 I - R_2 I_1$$
$$R_1 = \frac{R_2(I - I_1)}{I_1}$$

Substituting values:

$$R_1 = \frac{7\ \Omega(27\ \text{mA} - 21\ \text{mA})}{21\ \text{mA}}$$
$$= 7\ \Omega\left(\frac{6}{21}\right) = \frac{42\ \Omega}{21} = \mathbf{2\ \Omega}$$

continued

An alternative approach is

$$I_2 = I - I_1 \quad \text{(Kirchhoff's current law)}$$
$$= 27\text{ mA} - 21\text{ mA} = 6\text{ mA}$$
$$V_2 = I_2R_2 = (6\text{ mA})(7\ \Omega) = 42\text{ mV}$$
$$V_1 = I_1R_1 = V_2 = 42\text{ mV}$$

and
$$R_1 = \frac{V_1}{I_1} = \frac{42\text{ mV}}{21\text{ mA}} = \mathbf{2\ \Omega}$$

From the examples just described, note the following:

Current seeks the path of least resistance.

That is,

1. More current passes through the smaller of two parallel resistors.
2. The current entering any number of parallel resistors divides into these resistors as the inverse ratio of their ohmic values. This relationship is depicted in Fig. 6.34.

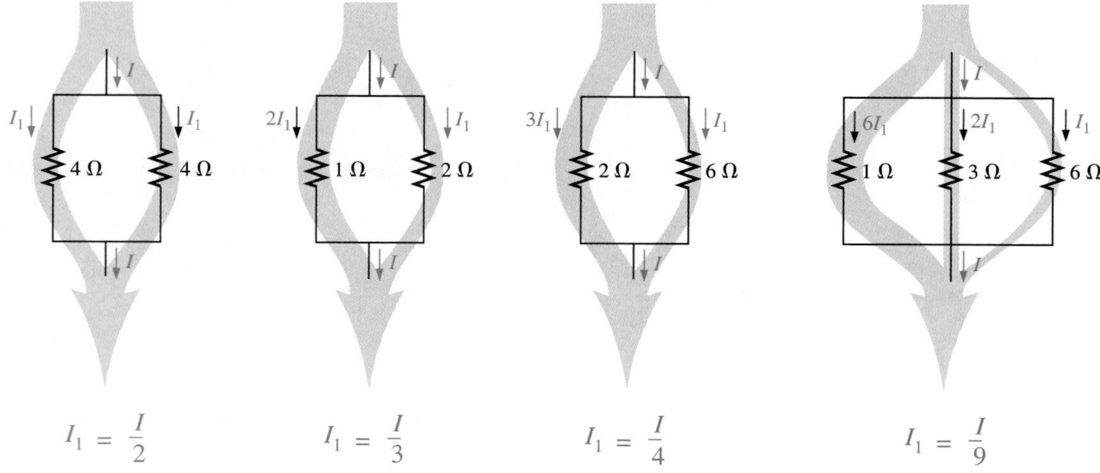

FIG. 6.34
Current division through parallel branches.

6.7 VOLTAGE SOURCES IN PARALLEL

Voltage sources are placed in parallel as shown in Fig. 6.35 only if they have the same voltage rating. The main reason for placing two or more batteries in parallel of the same terminal voltage would be to increase

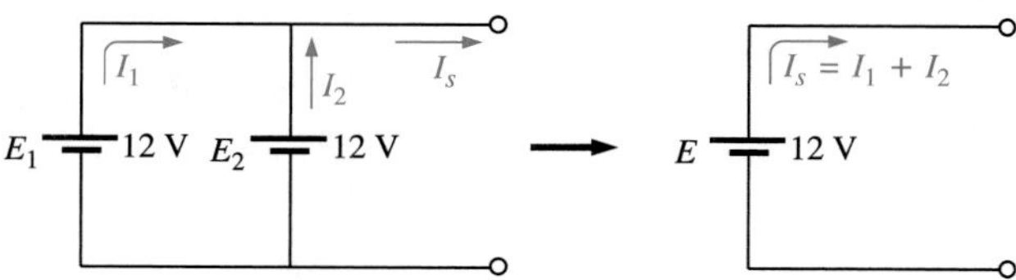

FIG. 6.35
Parallel voltage sources.

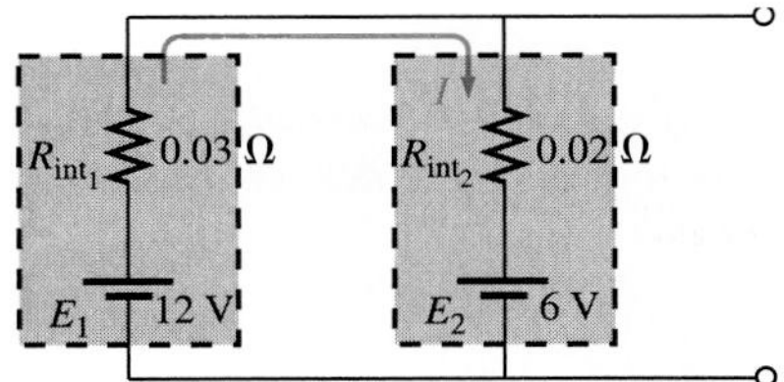

FIG. 6.36
Parallel batteries of different terminal voltages.

the current rating (and, therefore, the power rating) of the source. As shown in Fig. 6.35, the current rating of the combination is determined by $I_s = I_1 + I_2$ at the same terminal voltage. The resulting power rating is twice that available with one supply.

If two batteries of different terminal voltages are placed in parallel, both would be left ineffective or damaged because the terminal voltage of the larger battery would try to drop rapidly to that of the lower supply. Consider two lead-acid car batteries of different terminal voltage placed in parallel, as shown in Fig. 6.36.

The relatively small internal resistances of the batteries are the only current-limiting elements of the resulting series circuit. The current

$$I = \frac{E_1 - E_2}{R_{int_1} + R_{int_2}} = \frac{12\text{ V} - 6\text{ V}}{0.03\ \Omega + 0.02\ \Omega} = \frac{6\text{ V}}{0.05\ \Omega} = \mathbf{120\ A}$$

far exceeds the continuous drain rating of the larger supply, resulting in a rapid discharge of E_1 and a destructive impact on the smaller supply.

6.8 OPEN AND SHORT CIRCUITS

Open circuits and short circuits can often cause more confusion and difficulty in the analysis of a system than standard series or parallel configurations. This will become more obvious in the chapters to follow when we apply some of the methods and theorems.

An open circuit is simply two isolated terminals not connected by an element of any kind, as shown in Fig. 6.37(a). Since a path for conduction does not exist, the current associated with an open circuit must always be zero. The voltage across the open circuit, however, can be any value, as determined by the system it is connected to. In summary, therefore,

an open circuit can have a potential difference (voltage) across its terminals, but the current is always zero amperes.

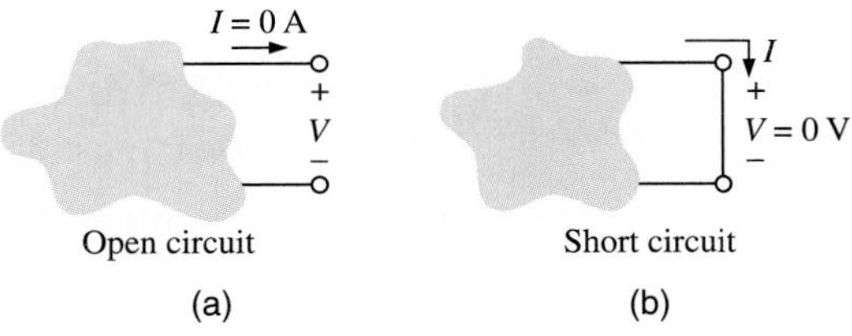

FIG. 6.37
Two special network configurations.

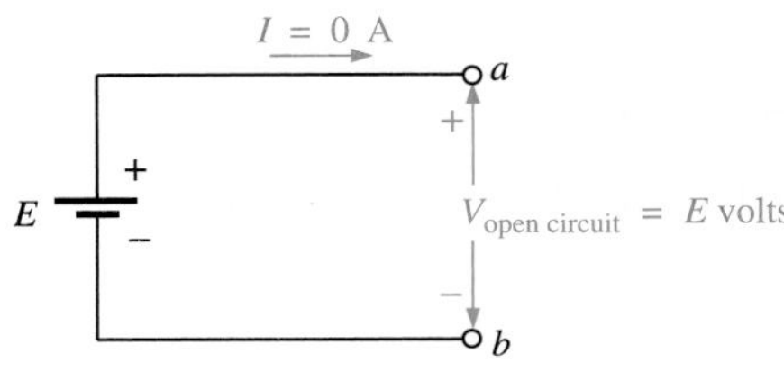

FIG. 6.38
Demonstrating the characteristics of an open circuit.

In Fig. 6.38, an open circuit exists between terminals a and b. As shown in the figure, the voltage across the open-circuit terminals is the supply voltage, but the current is zero due to the absence of a complete circuit.

A short circuit is a very low resistance, direct connection between two terminals of a network, as shown in Fig. 6.37(b). The current through the short circuit can be any value, as determined by the system it is connected to, but the voltage across the short circuit will always be zero volts because the resistance of the short circuit is assumed to be essentially zero ohms and $V = IR = I(0\ \Omega) = 0$ V.

In summary, therefore,

a short circuit can carry a current of a level determined by the external circuit, but the potential difference (voltage) across its terminals is always zero volts.

In Fig. 6.39(a), the current through the 2-Ω resistor is 5 A. If a short circuit should develop across the 2-Ω resistor, the total resistance of the parallel combination of the 2-Ω resistor and the short (of essentially zero ohms) is $2\ \Omega \parallel 0\ \Omega = \dfrac{(2\ \Omega)(0\ \Omega)}{2\ \Omega + 0\ \Omega} = 0\ \Omega$, and the current will rise to very high levels, as determined by Ohm's law:

$$I = \frac{E}{R} = \frac{10\text{ V}}{0\ \Omega} \rightarrow \infty\text{ A}$$

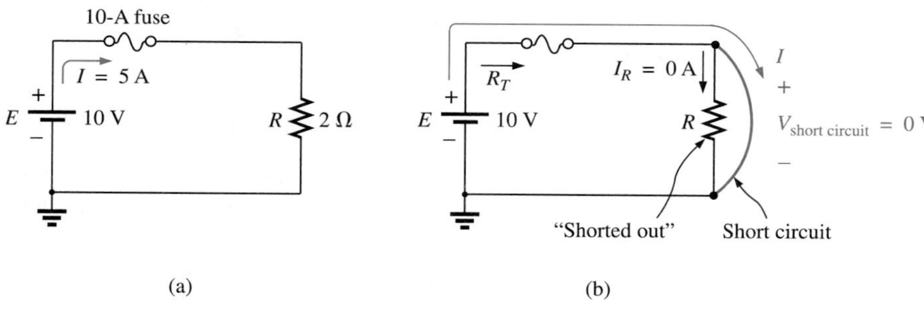

FIG. 6.39
Demonstrating the effect of a short circuit on current levels.

The effect of the 2-Ω resistor has effectively been "shorted out" by the low-resistance connection. The maximum current is now limited only by the circuit breaker or fuse in series with the source.

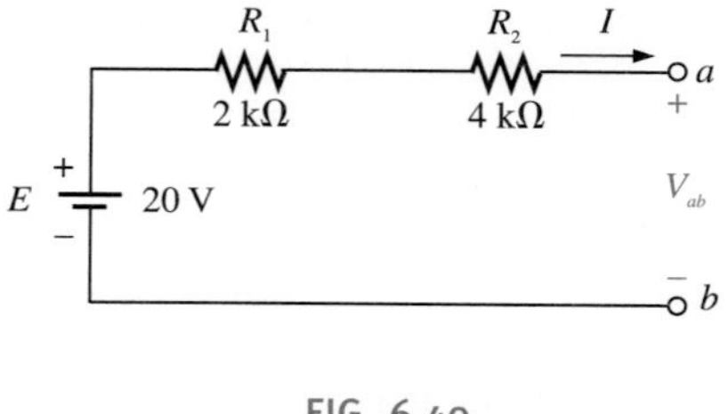

FIG. 6.40
Example 6.18.

EXAMPLE 6.18

Determine the voltage V_{ab} for the network of Fig. 6.40.

Solution: The open circuit requires that I be zero amperes. The voltage drop across both resistors is therefore zero volts since $V = IR = (0)R = 0$ V. Applying Kirchhoff's voltage law around the closed loop,

$$V_{ab} = E = \mathbf{20\ V}$$

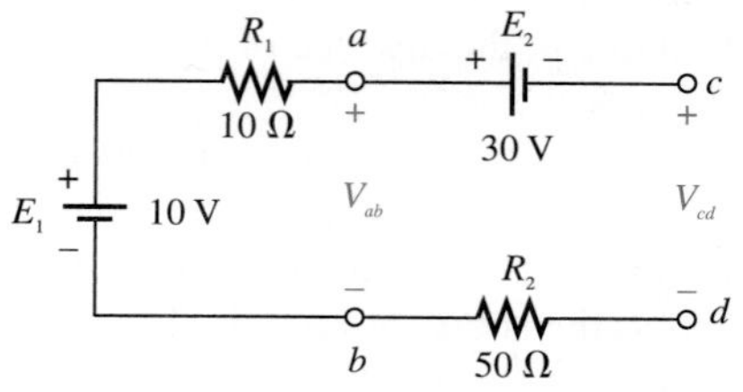

FIG. 6.41
Example 6.19.

EXAMPLE 6.19

Determine the voltages V_{ab} and V_{cd} for the network of Fig. 6.41.

Solution: The current through the system is zero amperes due to the open circuit, resulting in a 0-V drop across each resistor. Both resistors can therefore be replaced by short circuits, as shown in Fig. 6.42. The voltage V_{ab} is then directly across the 10-V battery, and

$$V_{ab} = E_1 = \mathbf{10\ V}$$

continued

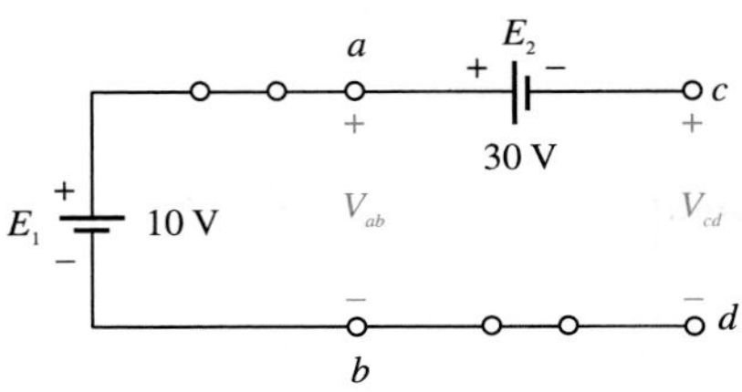

FIG. 6.42
Circuit of Fig. 6.45 redrawn.

The voltage V_{cd} requires an application of Kirchhoff's voltage law:

$$+E_1 - E_2 - V_{cd} = 0$$

or

$$V_{cd} = E_1 - E_2 = 10\text{ V} - 30\text{ V} = \mathbf{-20\text{ V}}$$

The negative sign in the solution simply indicates that the actual voltage V_{cd} has the opposite polarity of that appearing in Fig. 6.41.

EXAMPLE 6.20

Determine the unknown voltage and current for the network of Fig. 6.43.

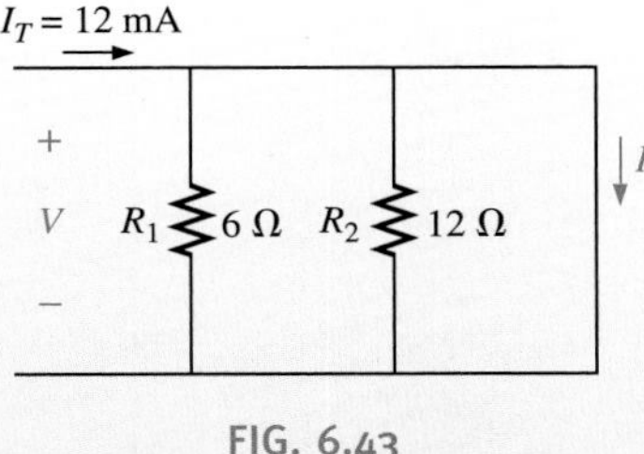

FIG. 6.43
Example 6.20.

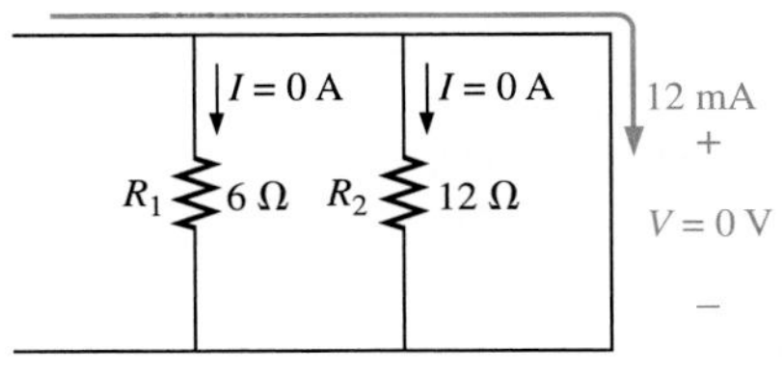

FIG. 6.44
Solution to Example 6.20.

Solution: For the network of Fig. 6.43, the current I_T will take the path of least resistance, and, since the short-circuit condition at the end of the network is the least resistance path, all the current will pass through the short circuit. This conclusion can be verified using Eq. (6.8). The voltage across the network is the same as that across the short circuit and will be zero volts, as shown in Fig. 6.44.

EXAMPLE 6.21

Calculate the current I and the voltage V for the network of Fig. 6.45.

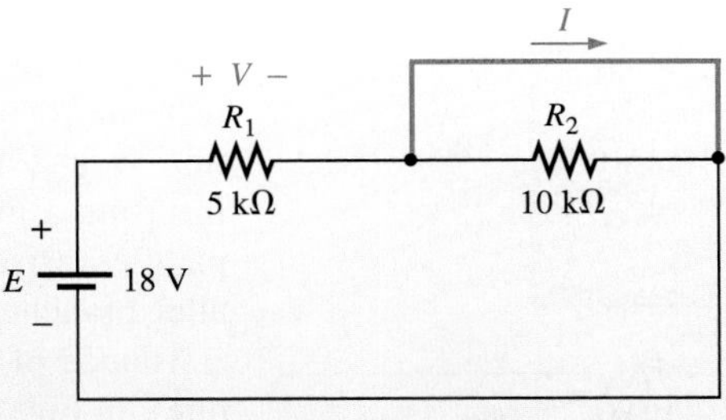

FIG. 6.45
Example 6.21.

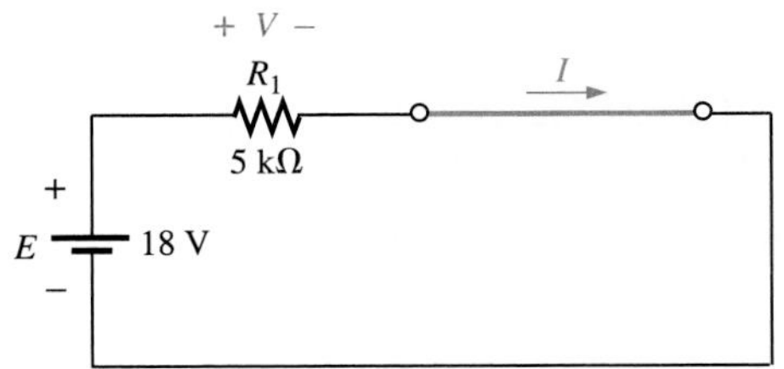

FIG. 6.46
Network of Fig. 6.45 redrawn.

Solution: The 10-kΩ resistor has been effectively shorted out by the jumper, resulting in the equivalent network of Fig. 6.46. Using Ohm's law,

$$I = \frac{E}{R_1} = \frac{18\text{ V}}{5\text{ k}\Omega} = \mathbf{3.6\text{ mA}}$$

and

$$V = E = \mathbf{18\text{ V}}$$

EXAMPLE 6.22

Determine V and I for the network of Fig. 6.47 if the resistor R_2 is shorted out.

Solution: The redrawn network appears in Fig. 6.48. The current through the 3-Ω resistor is zero due to the open circuit, causing all the current I to pass through the jumper. Since $V_{3\Omega} = IR = (0)R = 0$ V, the voltage V is directly across the short, and

$$V = \mathbf{0\ V}$$

with

$$I = \frac{E}{R_1} = \frac{6\text{ V}}{2\ \Omega} = \mathbf{3\ A}$$

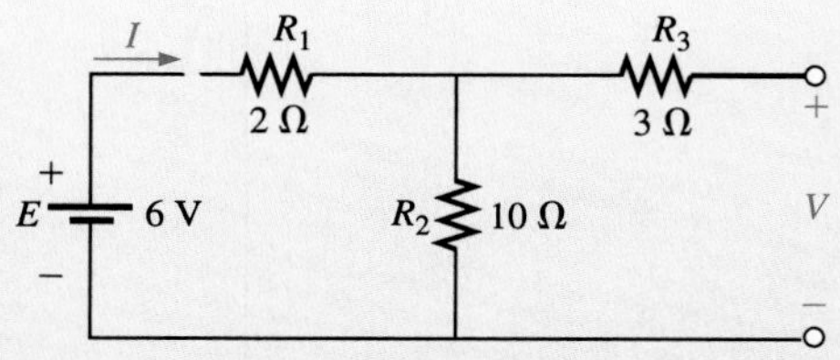

FIG. 6.47
Example 6.22.

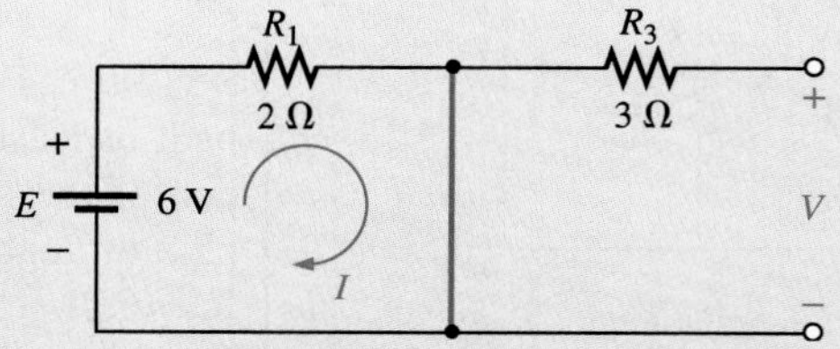

FIG. 6.48
Network of Fig. 6.47 with R_2 replaced by a jumper.

6.9 VOLTMETERS: LOADING EFFECT

You have seen that voltmeters are always placed across an element to measure the potential difference. Now you can realize that this connection is just like placing the voltmeter in parallel with the element. Inserting a meter in parallel with a resistor results in a combination of parallel resistors as shown in Fig. 6.49. Since the resistance of two parallel branches is always less than the smaller parallel resistance, the resistance of the voltmeter should be as large as possible (ideally infinite). In Fig. 6.49, a DMM with an internal resistance of 11 MΩ is measuring the voltage across a 10-kΩ resistor. The total resistance of the combination is

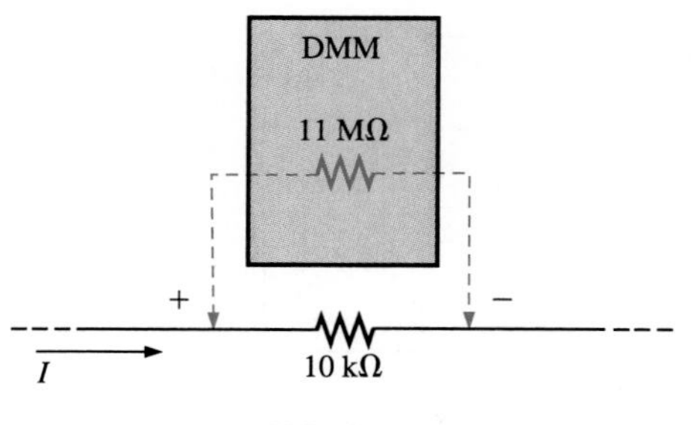

FIG. 6.49
Voltmeter loading.

$$R_T = 10\text{ k}\Omega \parallel 11\text{ M}\Omega = \frac{(10^4\ \Omega)(11 \times 10^6\ \Omega)}{10^4\ \Omega + (11 \times 10^6\ \Omega)} = 9.99\text{ k}\Omega$$

and we find that the network is essentially undisturbed. However, if we use a VOM with an internal resistance of 50 kΩ on the 2.5-V scale, the parallel resistance is

$$R_T = 10\text{ k}\Omega \parallel 50\text{ k}\Omega = \frac{(10^4\ \Omega)(50 \times 10^3\ \Omega)}{10^4\ \Omega + (50 \times 10^3\ \Omega)} = 8.33\text{ k}\Omega$$

and the behaviour of the network will be altered somewhat since the 10-kΩ resistor will now appear to be 8.33 kΩ to the rest of the network.

Loading a network by inserting meters is not to be taken lightly, especially in research efforts where accuracy is essential. It is good practice always to check the meter resistance level against the resistive elements of the network before making measurements. A factor of 10 between resistance levels will usually provide fairly accurate meter readings for a wide range of applications.

Most DMMs have internal resistance levels in excess of 10 MΩ on all voltage scales, while the internal resistance of VOMs is sensitive to the chosen scale. To determine the resistance of each scale setting of a VOM in the voltmeter mode, simply multiply the maximum voltage of the scale setting by the ohm/volt (Ω/V) rating of the meter, normally found at the bottom of the face of the meter.

For a typical ohm/volt rating of 20 000, the 2.5-V scale would have an internal resistance of

$$(2.5\text{ V})(20\,000\ \Omega/\text{V}) = 50\text{ k}\Omega$$

whereas for the 100-V scale, it would be

$$(100\text{ V})(20\,000\ \Omega/\text{V}) = 2\text{ M}\Omega$$

and for the 250-V scale,

$$(250\text{ V})(20\,000\ \Omega/\text{V}) = 5\text{ M}\Omega$$

EXAMPLE 6.23

For the relatively simple network of Fig. 6.50:

a. *What is the open-circuit voltage V_{ab}?*

b. What will a DMM indicate if it has an internal resistance of 11 MΩ? Compare your answer to the results of part (a).

c. Repeat part (b) for a VOM with an Ω/V rating of 20 000 on the 100-V scale.

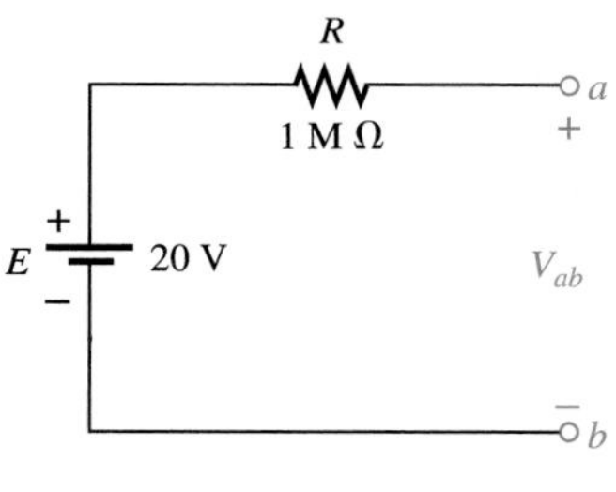

FIG. 6.50
Example 6.23.

Solution:

a. $V_{ab} = \mathbf{20\ V}$.

b. The meter will complete the circuit as shown in Fig. 6.51; using the voltage divider rule,

$$V_{ab} = \frac{11\text{ M}\Omega(20\text{ V})}{11\text{ M}\Omega + 1\text{ M}\Omega} = \mathbf{18.33\ V}$$

c. For the VOM, the internal resistance of the meter is

$$R_m = 100\text{ V}(20{,}000\ \Omega/\text{V}) = 2\text{ M}\Omega$$

and
$$V_{ab} = \frac{2\text{ M}\Omega(20\text{ V})}{2\text{ M}\Omega + 1\text{ M}\Omega} = \mathbf{13.33\ V}$$

revealing the need to consider carefully the internal resistance of the meter in some instances.

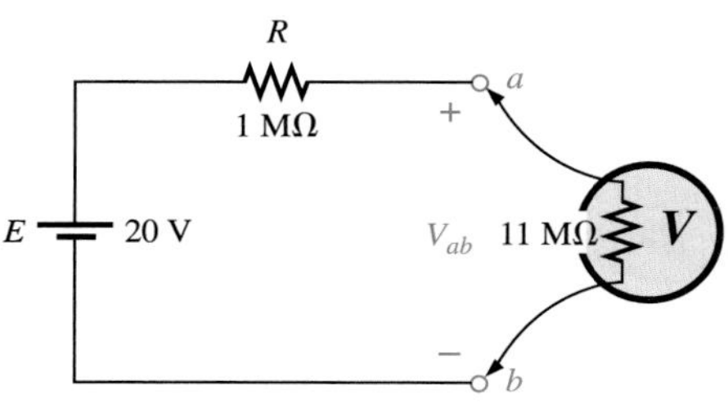

FIG. 6.51
Applying a DMM to the circuit of Fig. 6.50.

PROBLEMS

SECTION 6.2 Parallel Elements

1. For each configuration of Fig. 6.52, determine which elements are in series or parallel.

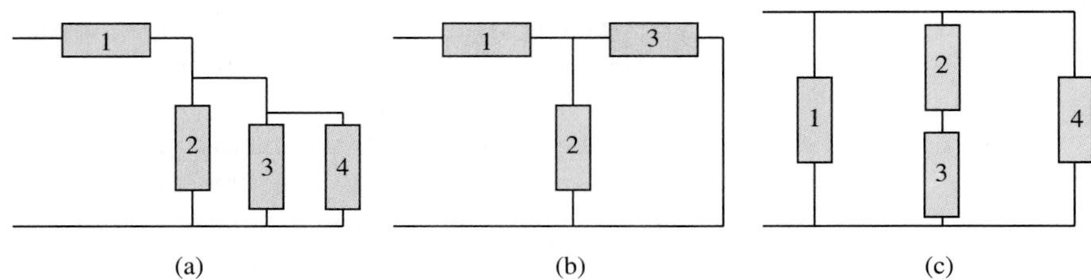

FIG. 6.52
Problem 1.

2. For the network of Fig. 6.53:

a. Which elements are in parallel?
b. Which elements are in series?
c. Which branches are in parallel?

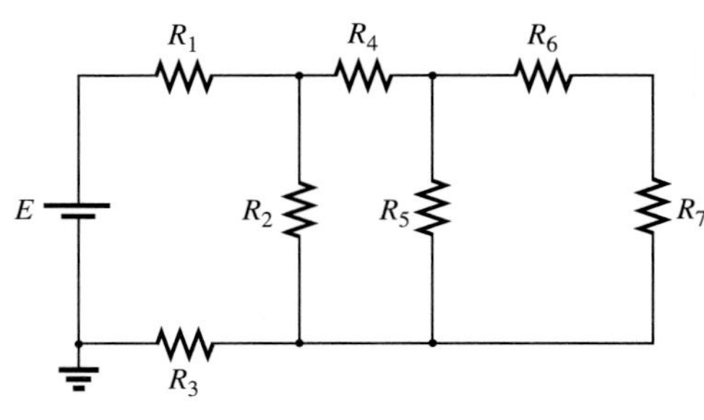

FIG. 6.53
Problem 2.

SECTION 6.3 Total Conductance and Resistance

3. Find the total conductance and resistance for the networks of Fig. 6.54.

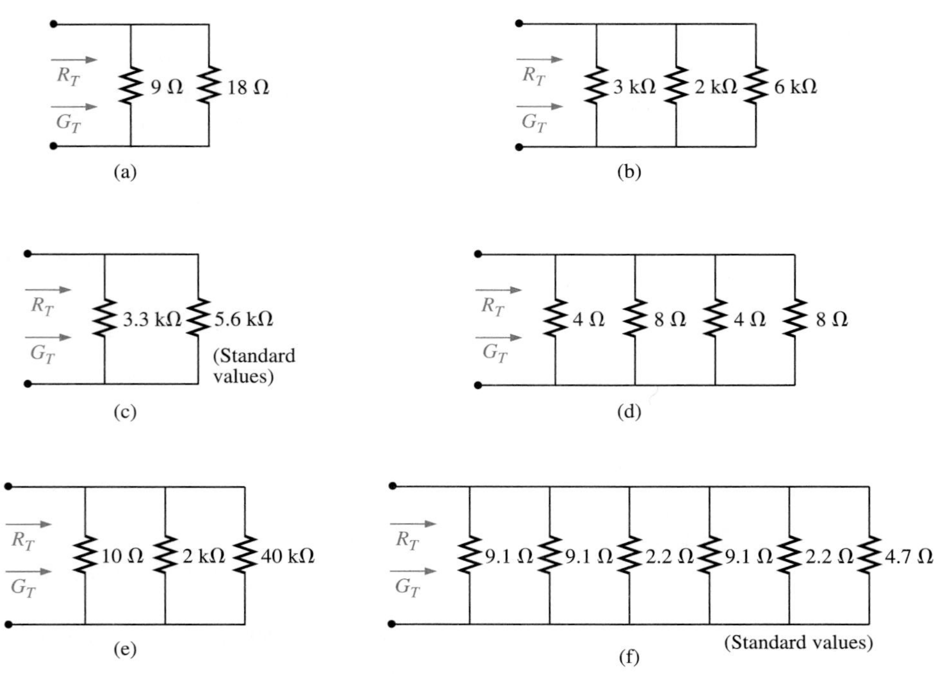

FIG. 6.54
Problem 3.

4. The total conductance of the networks of Fig. 6.55 is specified. Find the value in ohms of the unknown resistances.

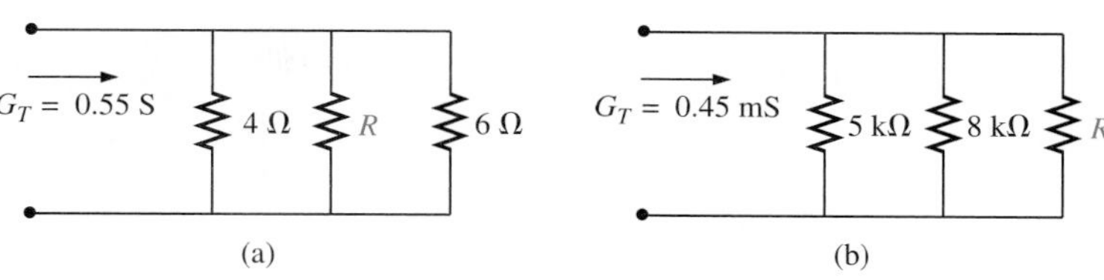

FIG. 6.55
Problem 4.

5. The total resistance of the circuits of Fig. 6.56 is specified. Find the value in ohms of the unknown resistances.

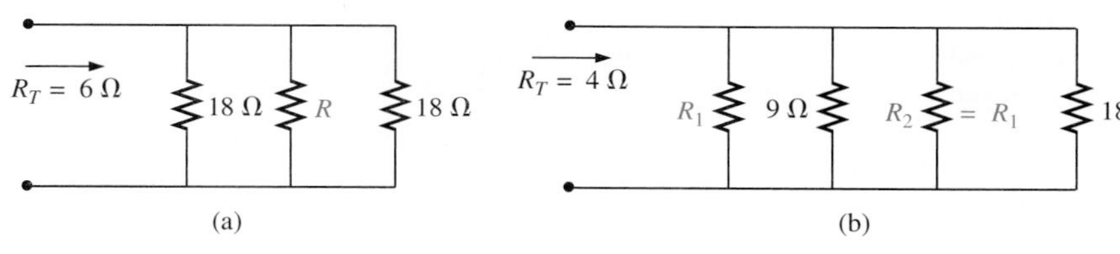

FIG. 6.56
Problem 5.

*6. Determine the unknown resistors of Fig. 6.57 given the fact that $R_2 = 5R_1$ and $R_3 = (1/2)R_1$.

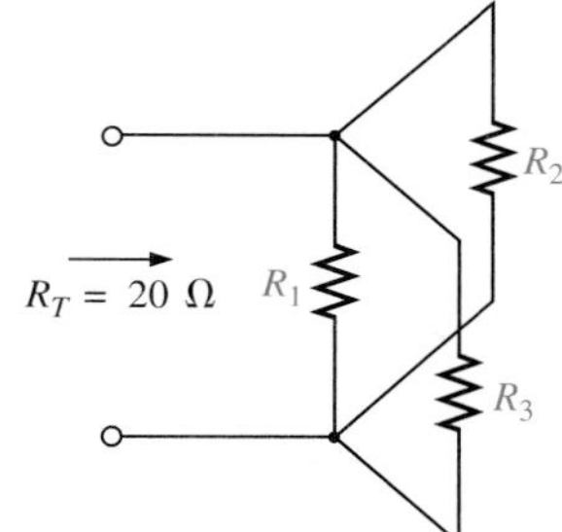

FIG. 6.57
Problem 6.

*7. Determine R_1 for the network of Fig. 6.58.

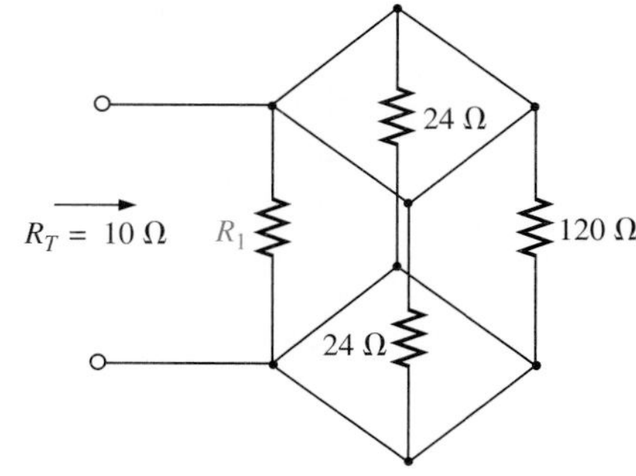

FIG. 6.58
Problem 7.

8. For the network in Fig. 6.59, determine I_1, I_2, R_2, R_3, and I_T.

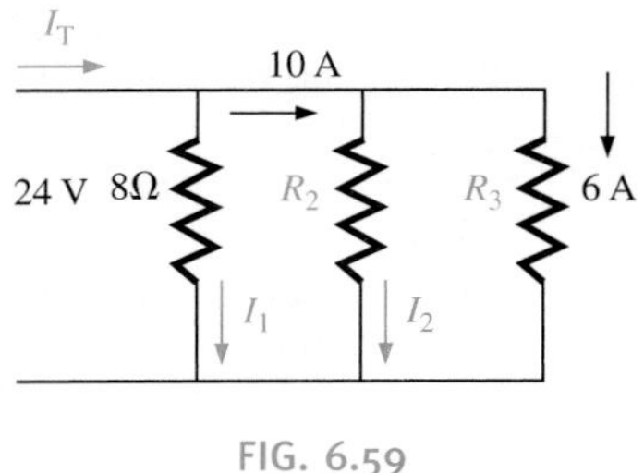

FIG. 6.59
Problem 8.

SECTION 6.4 Parallel Networks

9. For the network of Fig. 6.60:
   a. Find the total conductance and resistance.
   b. Determine I_s and the current through each parallel branch.
   c. Verify that the source current equals the sum of the parallel branch currents.
   d. Find the power dissipated by each resistor, and note whether the power delivered is equal to the power dissipated.
   e. If the resistors are available with wattage ratings of 1/2, 1, 2, and 50 W, what is the minimum wattage rating for each resistor?

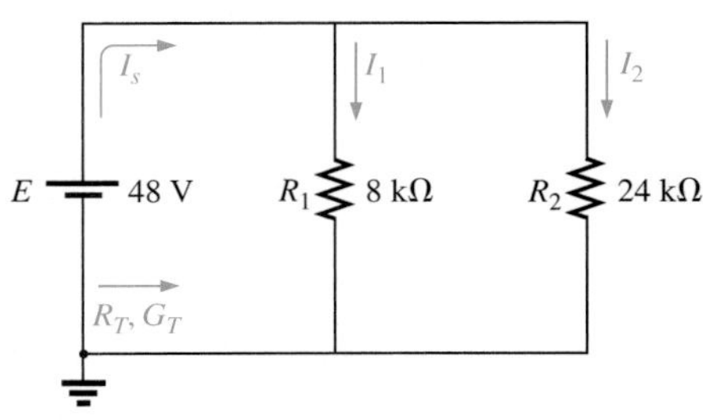

FIG. 6.60
Problem 9.

10. Repeat Problem 9 for the network of Fig. 6.61.

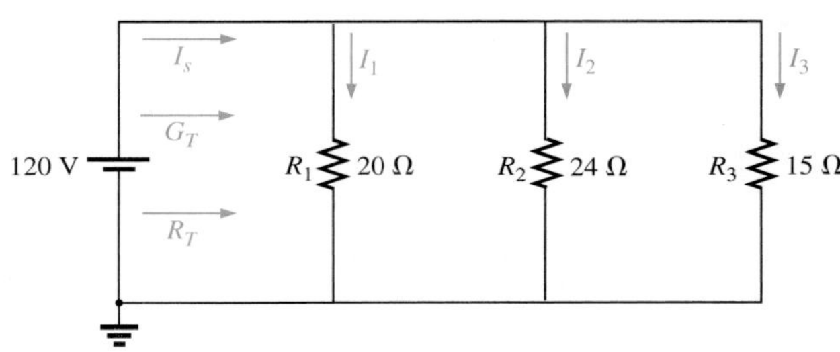

FIG. 6.61
Problem 10.

11. Repeat Problem 9 for the network of Fig. 6.62 constructed of standard resistor values.

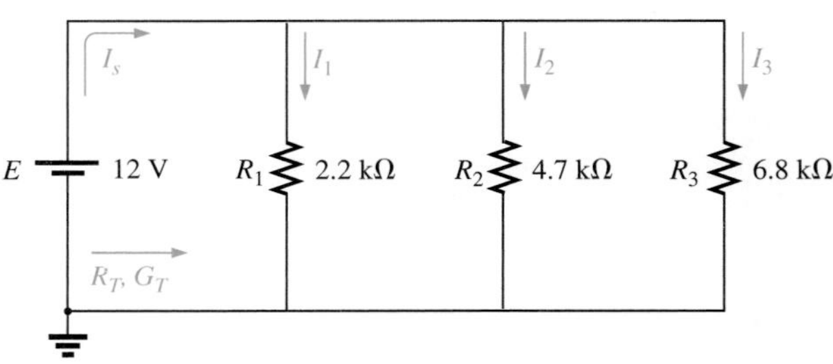

FIG. 6.62
Problem 11.

12. There are eight holiday lights connected in parallel as shown in Fig. 6.63.
    a. If the set is connected to a 120-V source, what is the current through each bulb if each bulb has an internal resistance of 1.8 kΩ?
    b. Determine the total resistance of the network.
    c. Find the power delivered to each bulb.
    d. If one bulb burns out (that is, the filament opens), what is the effect on the remaining bulbs?
    e. Compare the parallel arrangement of Fig. 6.63 to the series arrangement of Fig. 5.71. What are the relative advantages and disadvantages of the parallel system as compared to the series arrangement?

FIG. 6.63
Problem 12.

13. A portion of a residential service to a home is depicted in Fig. 6.64.
    a. Determine the current through each parallel branch of the network.
    b. Calculate the current drawn from the 120-V source. Will the 20-A circuit breaker trip?
    c. What is the total resistance of the network?
    d. Determine the power supplied by the 120-V source. How does it compare to the total power of the load?

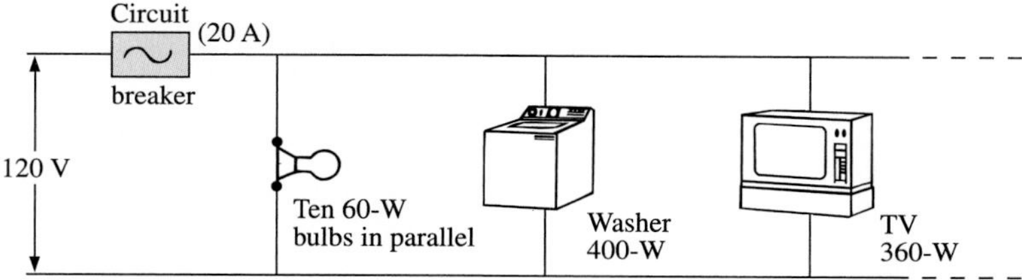

FIG. 6.64
Problem 13.

14. Determine the currents I_1 and I_s for the networks of Fig. 6.65.

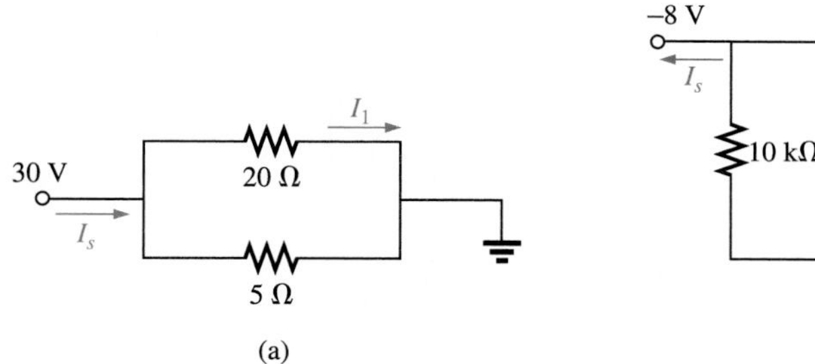

FIG. 6.65
Problem 14.

15. Using the information provided, determine the resistance R_1 for the network of Fig. 6.66.

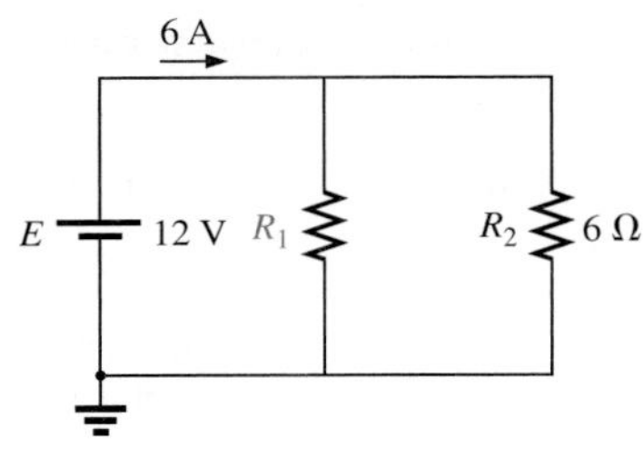

FIG. 6.66
Problem 15.

***16.** Determine the power delivered by the dc battery in Fig. 6.67.

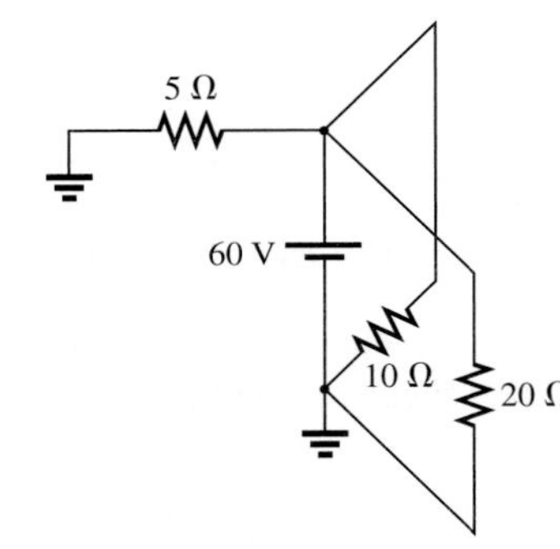

FIG. 6.67
Problem 16.

***17.** For the network of Fig. 6.68:

a. Find the current I_1.
b. Calculate the power dissipated by the 4-Ω resistor.
c. Find the current I_2.

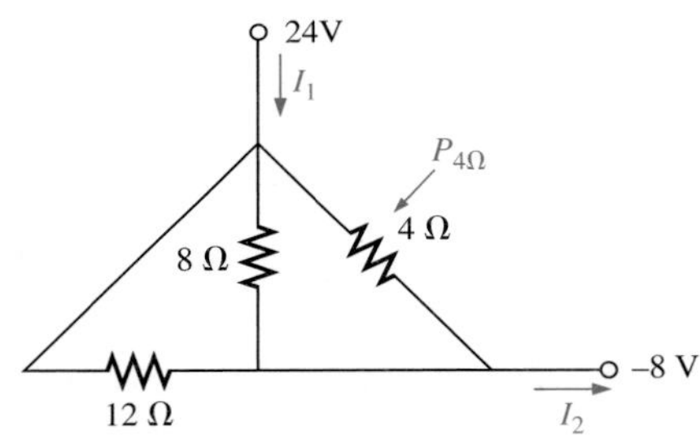

FIG. 6.68
Problem 17.

***18.** For the network of Fig. 6.69:

a. Find the current *I*.
b. Determine the voltage *V*.
c. Calculate the source current I_s.

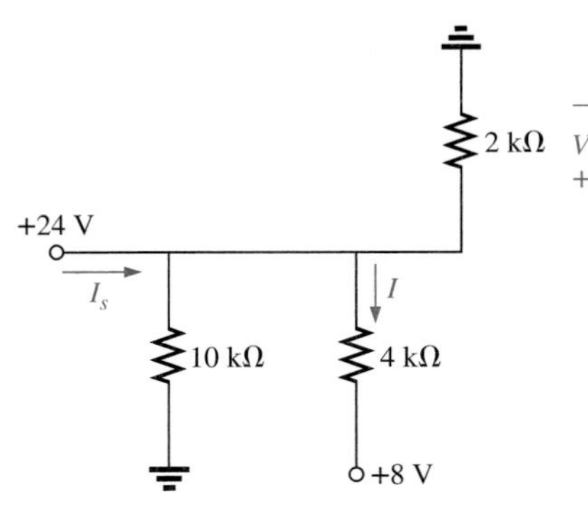

FIG. 6.69
Problem 18.

SECTION 6.5 Kirchhoff's Current Law

19. Find all unknown currents and their directions in the circuits of Fig. 6.70.

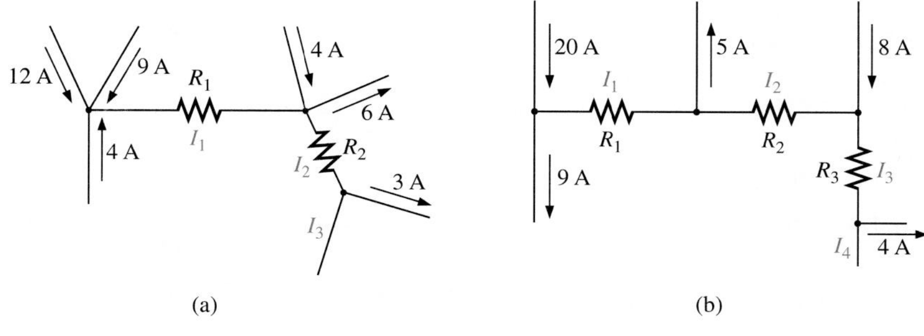

FIG. 6.70
Problem 19.

***20.** Using Kirchhoff's current law, determine the unknown currents for the networks of Fig. 6.71.

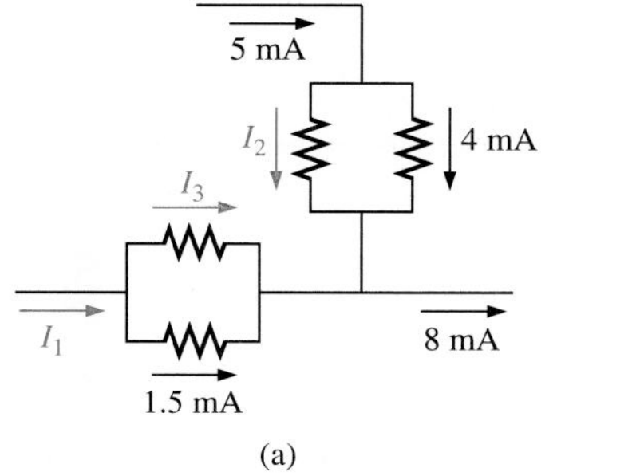

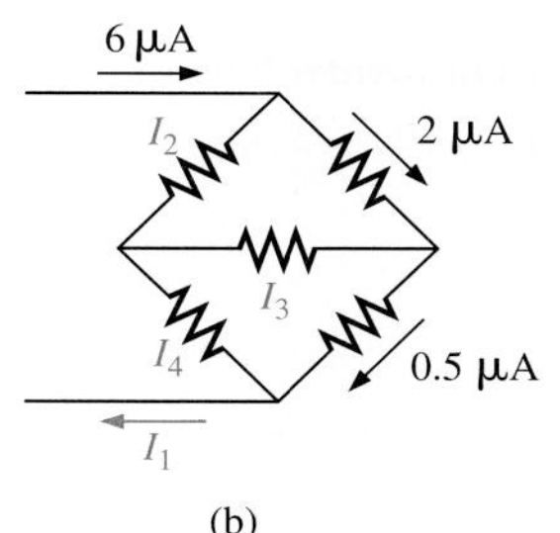

FIG. 6.71
Problem 20.

21. Using the information provided in Fig. 6.72, find the branch resistors R_1 and R_3, the total resistance R_T, and the voltage source E.

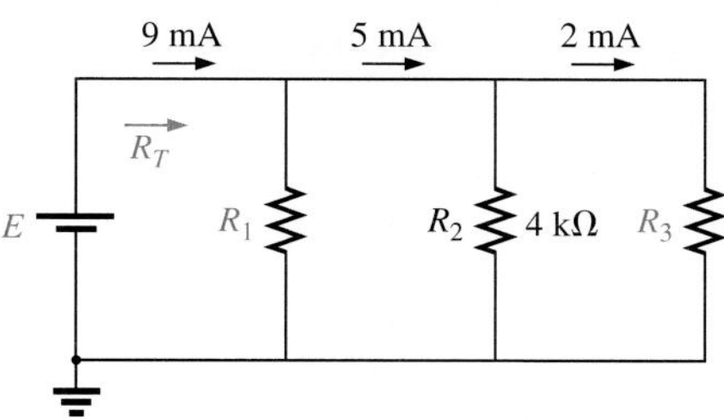

FIG. 6.72
Problem 21.

*22. Find the unknown quantities for the circuits of Fig. 6.73 using the information provided.

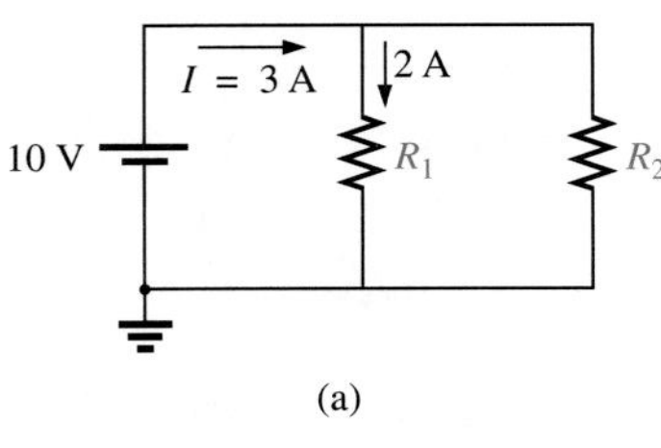

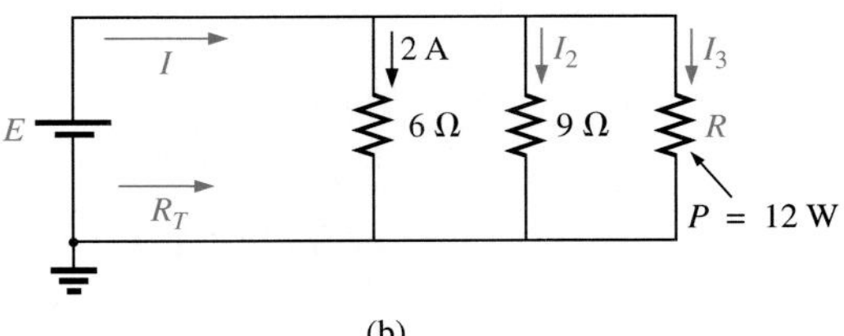

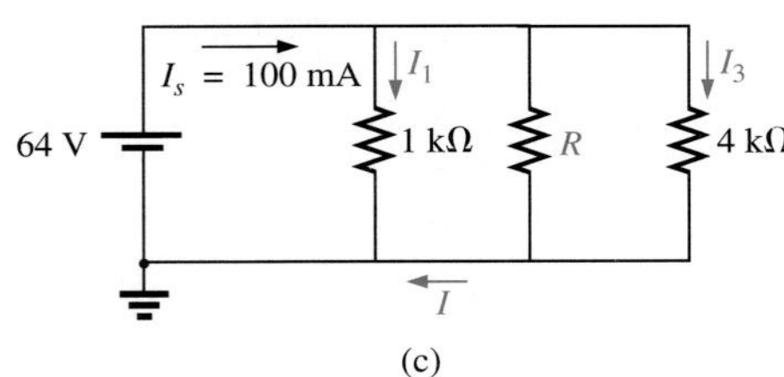

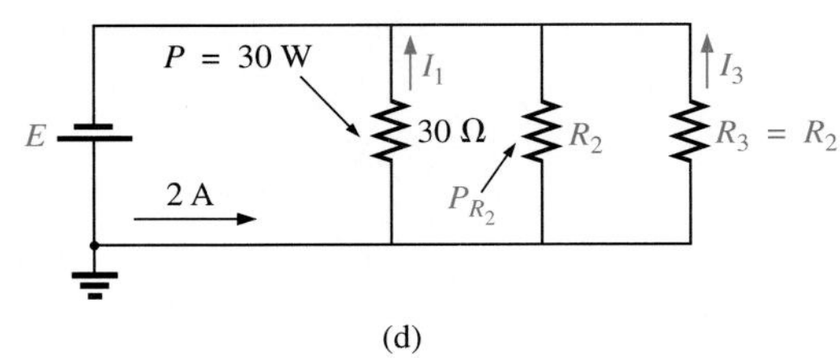

FIG. 6.73
Problem 22.

SECTION 6.6 Current Divider Rule

23. For the circuit in Fig. 6.74 determine I_1 and I_2.

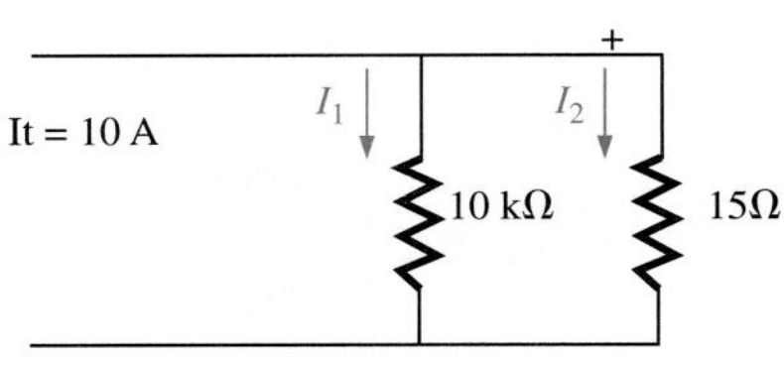

FIG. 6.74
Problem 23.

24. Using the information provided in Fig. 6.75, determine the current through each branch using simply the ratio of parallel resistor values. Then determine the total current I_T.

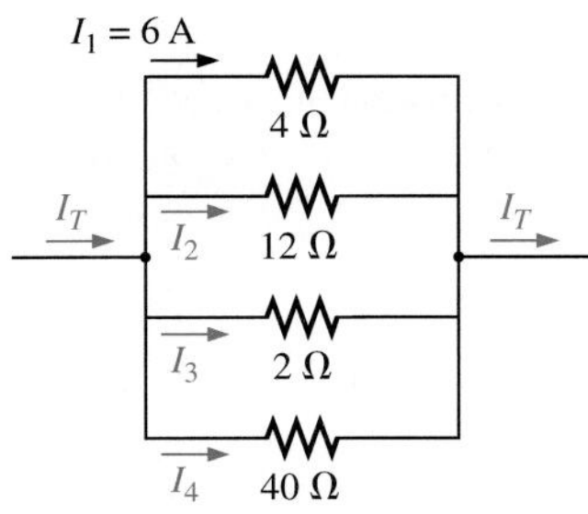

FIG. 6.75
Problem 24.

25. Using the current divider rule, find the unknown currents for the networks of Fig. 6.76.

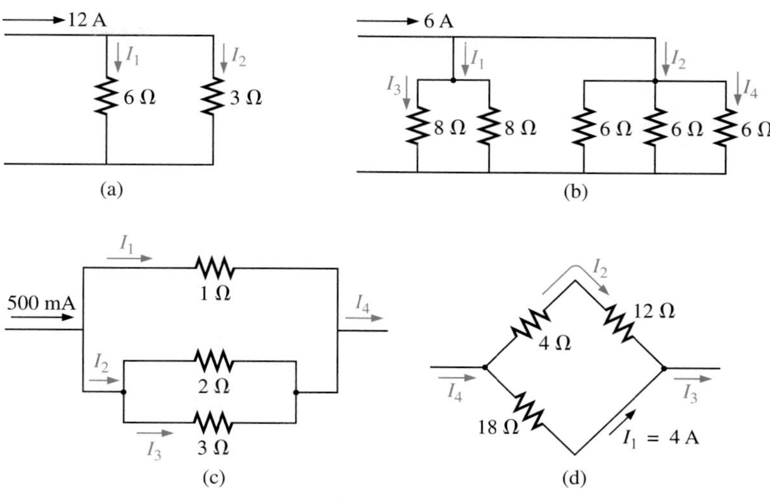

FIG. 6.76
Problem 25.

***26.** Parts (a), (b), and (c) of this problem should be done by inspection—that is, mentally. The intent is to obtain a solution without a lengthy series of calculations. For the network of Fig. 6.77:

a. What is the approximate value of I_1 considering the magnitude of the parallel elements?

b. What are the ratios I_1/I_2 and I_3/I_4?

c. What are the ratios I_2/I_3 and I_1/I_4?

d. Calculate the current I_1 and compare it to the result of part (a).

e. Determine the current I_4 and calculate the ratio I_1/I_4. How does the ratio compare to the result of part (c)?

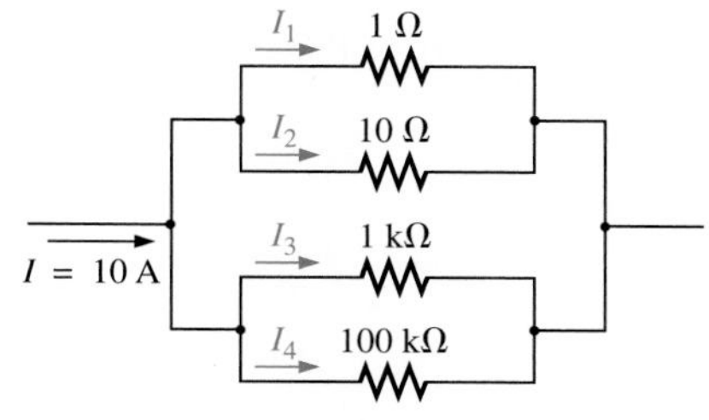

FIG. 6.77
Problem 26.

27. Find the unknown quantities using the information provided for the networks of Fig. 6.78.

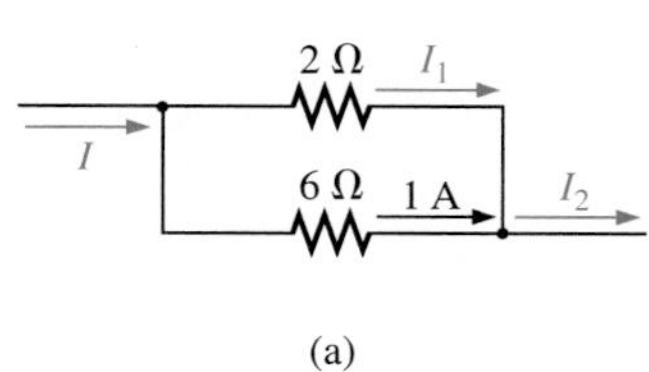

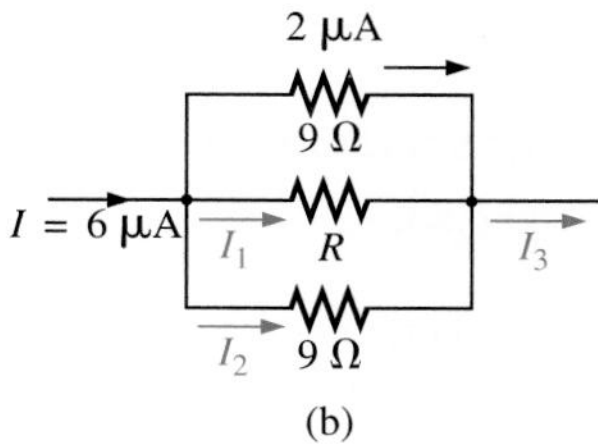

FIG. 6.78
Problem 27.

***28.** Calculate the resistor R for the network of Fig. 6.79 that will ensure the current $I_1 = 3I_2$.

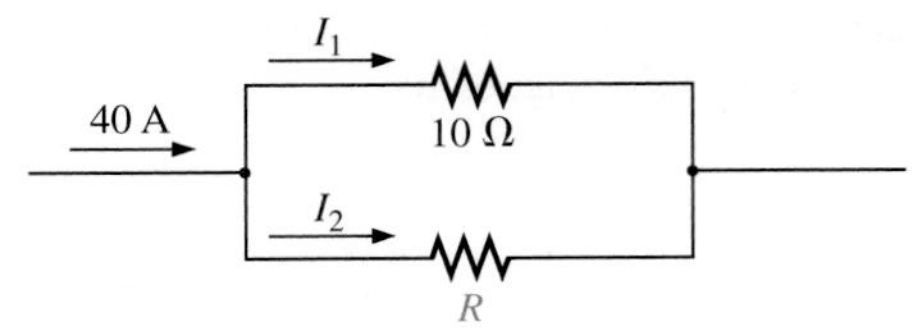

FIG. 6.79
Problem 28.

***29.** Design the network of Fig. 6.80 such that $I_2 = 4I_1$ and $I_3 = 3I_2$.

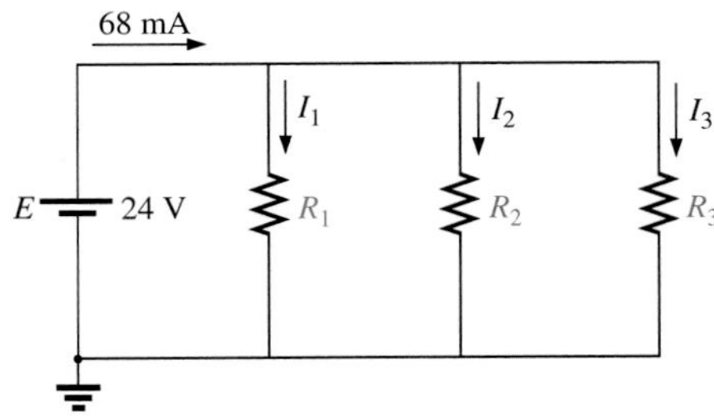

FIG. 6.80
Problem 29.

SECTION 6.7 Voltage Sources in Parallel

30. Assuming identical supplies, determine the currents I_1 and I_2 for the network of Fig. 6.81.

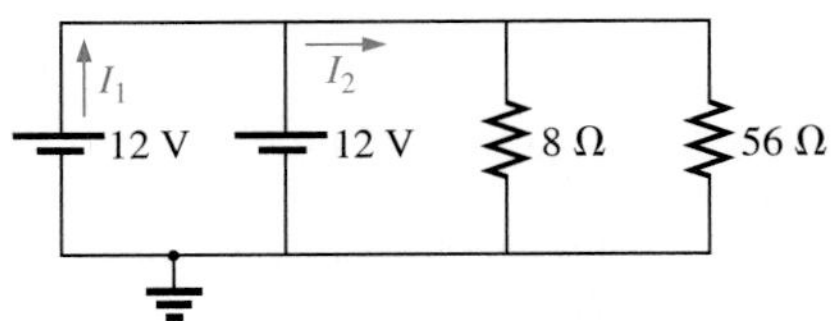

FIG. 6.81
Problem 30.

31. Assuming identical supplies, determine the current I and resistance R for the parallel network of Fig. 6.82.

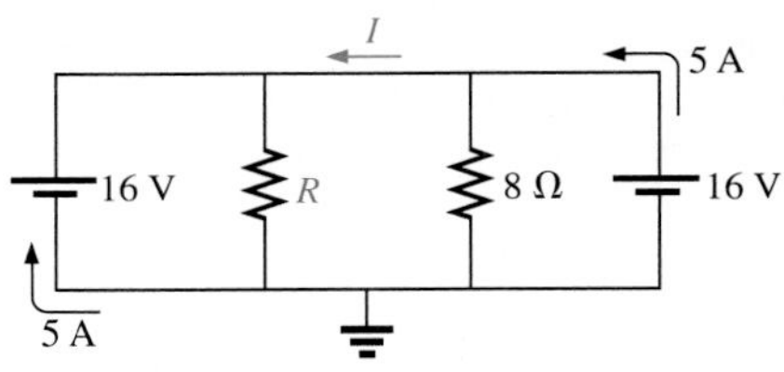

FIG. 6.82
Problem 31.

SECTION 6.8 Open and Short Circuits

32. For the network of Fig. 6.83:

a. Determine I_s and V_L.

b. Determine I_s if R_L is shorted out.

c. Determine V_L if R_L is replaced by an open circuit.

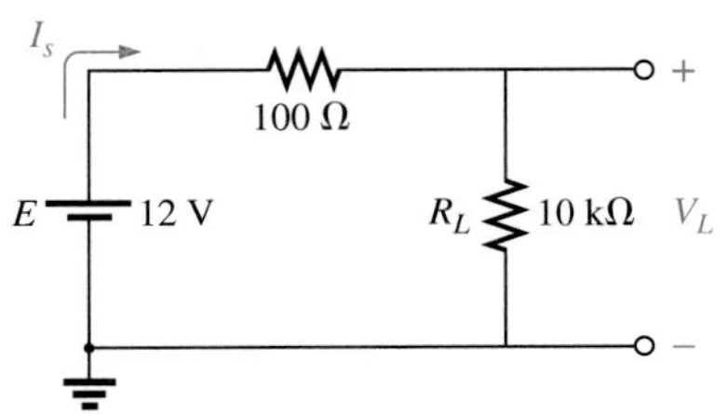

FIG. 6.83
Problem 32.

33. For the network of Fig. 6.84:
   a. Determine the open-circuit voltage V_L.
   b. If the 2.2-kΩ resistor is short circuited, what is the new value of V_L?
   c. Determine V_L if the 4.7-kΩ resistor is replaced by an open circuit.

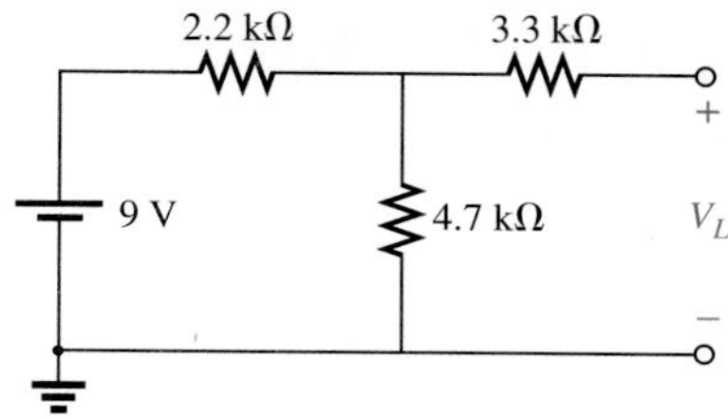

FIG. 6.84
Problem 33.

***34.** For the network of Fig. 6.85, determine
   a. The short-circuit currents I_1 and I_2.
   b. The voltages V_1 and V_2.
   c. The source current I_s.

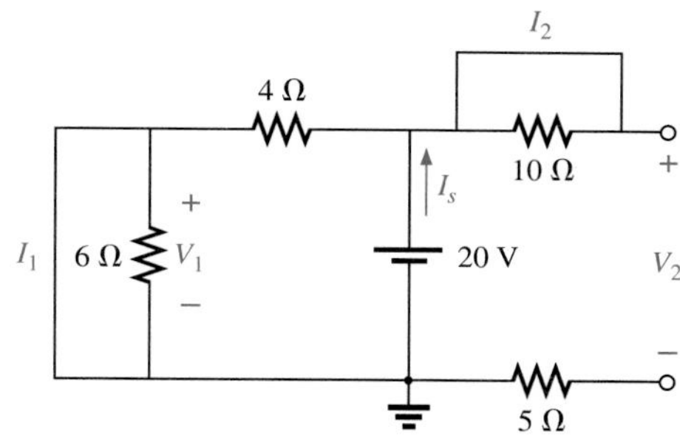

FIG. 6.85
Problem 34.

SECTION 6.9 Voltmeters: Loading Effect

35. For the network of Fig. 6.86:
   a. Determine the voltage V_2.
   b. Determine the reading of a DMM having an internal resistance of 11 MΩ when used to measure V_2.
   c. Repeat part (b) with a VOM having an ohm/volt rating of 20 000 using the 10-V scale. Compare the results of parts (b) and (c). Explain any difference.
   d. Repeat part (c) with $R_1 = 100$ kΩ and $R_2 = 200$ kΩ.
   e. Based on the above, can you make any general conclusions about the use of a voltmeter?

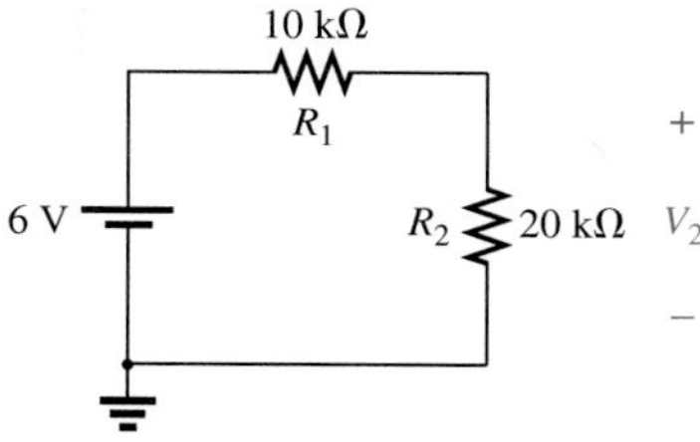

FIG. 6.86
Problem 35.

36. Determine the current flowing through the voltmeter in Fig. 6.87 if the voltmeter has a sensitivity of 5 kΩ/volt using the 15 V range.

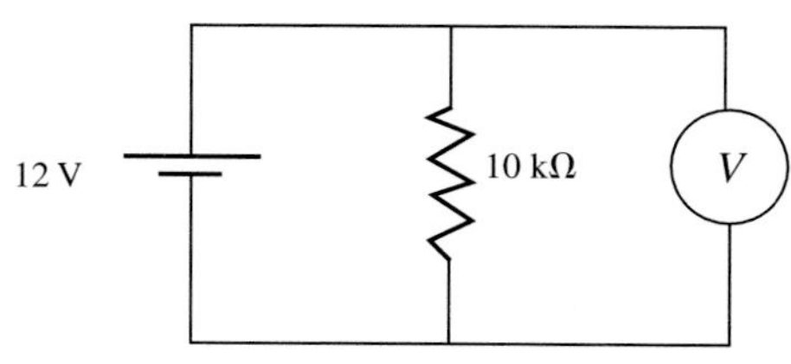

FIG. 6.87
Problem 36.

37. For the network in Fig. 6.88, determine I_4 and I_5.

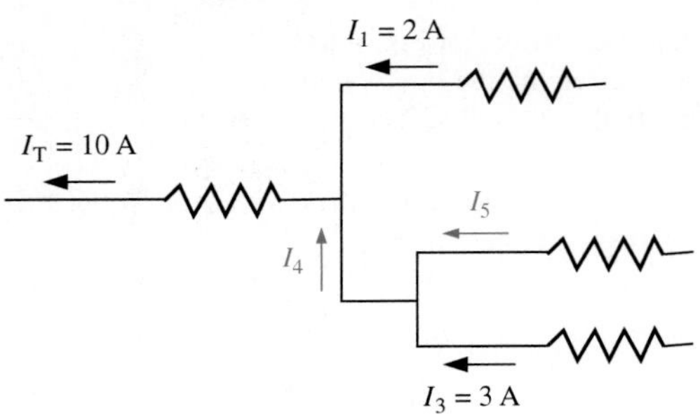

FIG. 6.88
Problem 37.

7 Series-Parallel Networks

Outline

Learning Outcomes

After completing this chapter you will be able to

- describe the characteristics of series-parallel circuits
- describe several methods of solving series-parallel problems
- solve network problems using these methods
- solve ladder network problems
- explain the purpose of a voltage-divider supply
- solve problems involving loaded and unloaded voltage dividers
- explain the operation of moving-coil or d'Arsonval meter movements
- describe the basic operation of an electrodynamometer
- design parallel resistance ammeter and Ayrton shunts
- design voltage multipliers to extend the measuring range of a voltmeter
- determine the loading effect of a voltmeter on a circuit
- explain the operation of an ohmmeter with a d'Arsonval movement
- explain the significance of grounding as applied to metering

Series-parallel networks are very common complex circuits. Using the theory developed in chapters 5 and 6, you will be able to analyze and describe the operation of these circuits.

7.1 SERIES-PARALLEL NETWORKS

A firm understanding of the basic principles associated with series and parallel circuits is a sufficient background to approach most series-parallel networks (a network being a combination of any number of series and parallel elements) with *one* source of voltage. Multisource networks are considered in detail in Chapters 8 and 9. In general,

in a series-parallel network a portion of the circuit is connected in series, and other portions are connected in parallel.

You will become very good at analyzing series-parallel networks only through exposure, practice, and experience. In time the path to the desired unknown becomes more obvious as you remember similar configurations and the frustration of choosing the wrong approach. There are a few steps that can be helpful in getting started on the first few exercises. With experience you will discover the value of these steps.

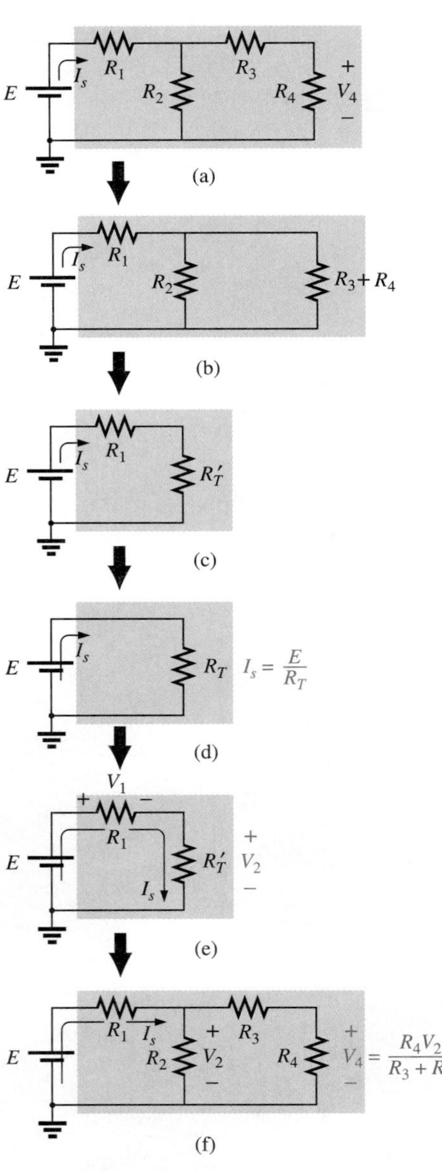

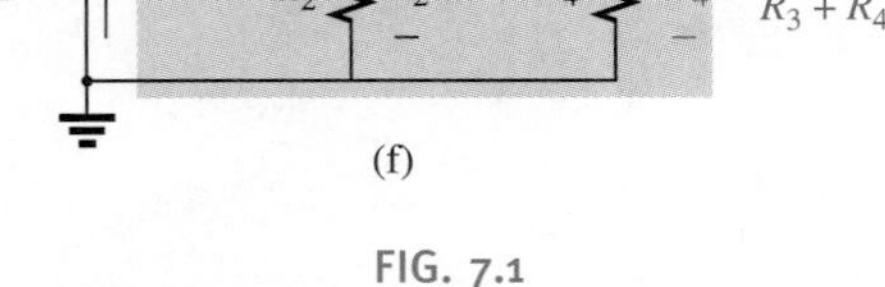

FIG. 7.1
Introducing the reduce and return approach.

General Approach

1. Take a moment to study the problem "in total" and make a brief mental sketch of the overall approach you plan to use. The result may be time- and energy-saving shortcuts.
2. Next examine each region of the network independently before tying them together in series-parallel combinations. This will usually simplify the network and may reveal a direct approach toward obtaining one or more desired unknowns. It also eliminates many of the errors caused by not having a systematic approach.
3. Redraw the network as often as possible with the reduced branches and undisturbed unknown quantities to keep it clear and provide the reduced networks for the trip back to unknown quantities from the source.
4. When you have a solution, check that it is reasonable by considering the magnitudes of the energy source and the elements in the network. If it does not seem reasonable, either solve the circuit using another approach or check over your work very carefully.

Reduce and Return Approach

In many single-source series-parallel networks, you can reduce the network in order to work back to the source. Then, determine the source current and return to the network to find the desired unknown. In Fig. 7.1(a), suppose you want to find the voltage V_4 across resistor R_4. Without a single series or parallel path to R_4 from the source, you cannot use the methods of the last two chapters.

You can begin to reduce the circuit by combining the series elements R_3 and R_4 to form the single element $R_3 + R_4$ in parallel with R_2 as in Fig. 7.1(b). Then, combine those parallel elements to make R'_T in series with R_1 as in Fig. 7.1(c). Next, combine those two series elements to form the simple configuration of Fig. 7.1(d). Now, you can find the source current using Ohm's law, and then work back through the network.

In Fig. 7.1(e), find the voltage drop V_2 across R'_T. Then redraw the original network as in Fig. 7.1(f). Since V_2 is known, you can find V_4 using the voltage divider rule. The problem is solved.

Note that the networks drawn during the reduction phase are often also used in the return phase, as can be seen from the similarities of Figs. 7.1(a) and (f), and of Figs. 7.1(c) and (e). You will find that you can analyze many series-parallel network problems in this way—reduce the network to get back to I_s, and then work back to return to the desired unknown. Some problems may be simpler than this one. However, this reduce and return approach is suitable for certain problems that appear frequently.

Block Diagram Approach

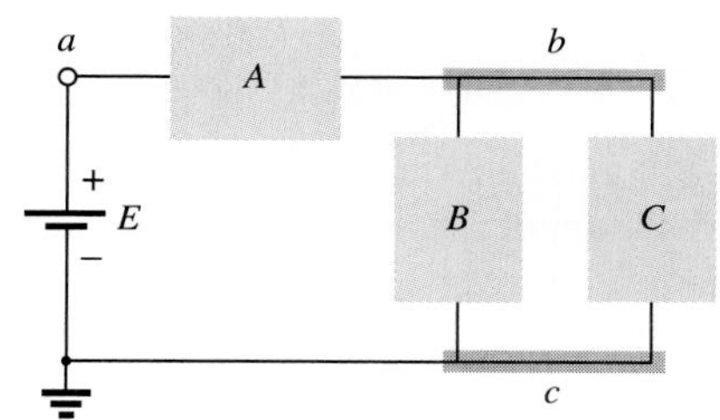

FIG. 7.2
Introducing the block diagram approach.

The block-diagram approach will be used throughout to stress that *combinations* of elements (not just single resistors) can be connected in series and parallel. It will show that many networks that seem different have the same basic structure, and can be analyzed in a similar way.

At first, you may be concerned about identifying the best way to combine elements to arrive at the solution. Soon, however, you will develop routines that will increase your confidence in your ability to analyze.

In Fig. 7.2, blocks B and C are in parallel (points b and c in common), and the voltage source E is in series with block A (point a in common). The parallel combination of B and C is also in series with A and the voltage source E due to the common points b and c, respectively.

To make the analysis as clear as possible, we use the following notation to distinguish series and parallel combinations. For resistances in series, the combined value will be shown with a subscript having a comma between the separate identifiers:

$$R_{1,2} = R_1 + R_2$$

For resistances in parallel, the combined value will be shown with a subscript having the parallel symbol between the separate identifiers:

$$R_{1\|2} = R_1 \| R_2 = \frac{R_1 R_2}{R_1 + R_2}$$

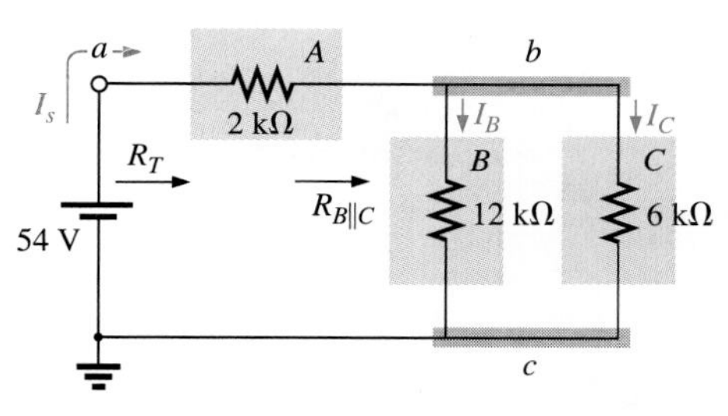

FIG. 7.3
Example 7.1.

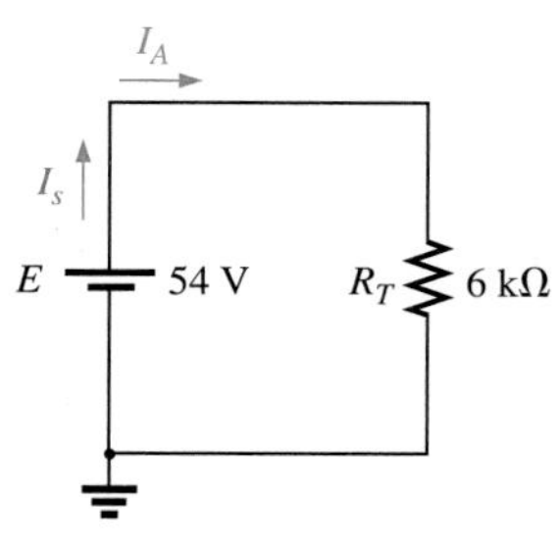

FIG. 7.4
Reduced equivalent of Fig. 7.3.

EXAMPLE 7.1

If each block of Fig. 7.2 were a single resistive element, you might end up with the network of Fig. 7.3.

The parallel combination of R_B and R_C results in

$$R_{B\|C} = R_B \| R_C = \frac{(12\text{ k}\Omega)(6\text{ k}\Omega)}{12\text{ k}\Omega + 6\text{ k}\Omega} = 4\text{ k}\Omega$$

The equivalent resistance $R_{B\|C}$ is then in series with R_A and the total resistance "seen" by the source is

$$R_T = R_A + R_{B\|C} = 2\text{ k}\Omega + 4\text{ k}\Omega = \mathbf{6\text{ k}\Omega}$$

The result is an equivalent network, as shown in Fig. 7.4, that allows you to determine the source current I_s.

$$I_s = \frac{E}{R_T} = \frac{54\text{ V}}{6\text{ k}\Omega} = \mathbf{9\text{ mA}}$$

continued

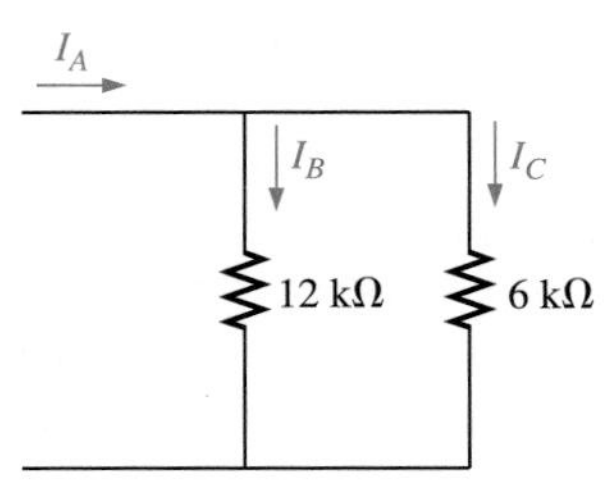

FIG. 7.5
Determining I_B and I_C for the network of Fig. 7.3.

and, since the source and R_A are in series,

$$I_A = I_s = 9\text{ mA}$$

We can then use the equivalent network of Fig. 7.5 to determine I_B and I_C using the current divider rule:

$$I_B = \frac{6\text{ k}\Omega(I_s)}{6\text{ k}\Omega + 12\text{ k}\Omega} = \frac{6}{18}I_s = \frac{1}{3}(9\text{ mA}) = \mathbf{3\text{ mA}}$$

$$I_C = \frac{12\text{ k}\Omega(I_s)}{12\text{ k}\Omega + 6\text{ k}\Omega} = \frac{12}{18}I_s = \frac{2}{3}(9\text{ mA}) = \mathbf{6\text{ mA}}$$

or, applying Kirchhoff's current law,

$$I_C = I_s - I_B = 9\text{ mA} - 3\text{ mA} = \mathbf{6\text{ mA}}$$

Note in this solution that we worked back to the source to obtain the source current or total current supplied by the source. We then determined the remaining unknowns by working back through the network to find the other unknowns.

EXAMPLE 7.2

It is also possible that the blocks *A*, *B*, and *C* of Fig. 7.2 contain the elements and configurations of Fig. 7.6. Working with each region:

$$A\text{: } R_A = 4\ \Omega$$

$$B\text{: } R_B = R_2 \parallel R_3 = R_{2\parallel3} = \frac{R}{N} = \frac{4\ \Omega}{2} = 2\ \Omega$$

$$C\text{: } R_C = R_4 + R_5 = R_{4,5} = 0.5\ \Omega + 1.5\ \Omega = 2\ \Omega$$

Blocks *B* and *C* are still in parallel and

$$R_{B\parallel C} = \frac{R}{N} = \frac{2\ \Omega}{2} = 1\ \Omega$$

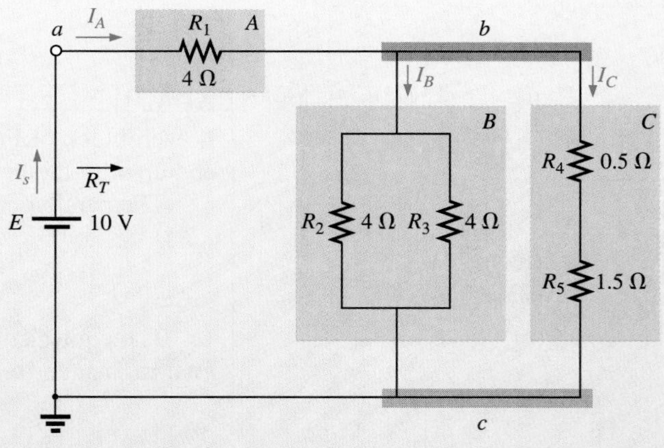

FIG. 7.6
Example 7.2.

with

$$R_T = R_A + R_{B\parallel C} = 4\ \Omega + 1\ \Omega = \mathbf{5\ \Omega}$$

(Note how similar this equation is to the one obtained for Example 7.1.)

continued

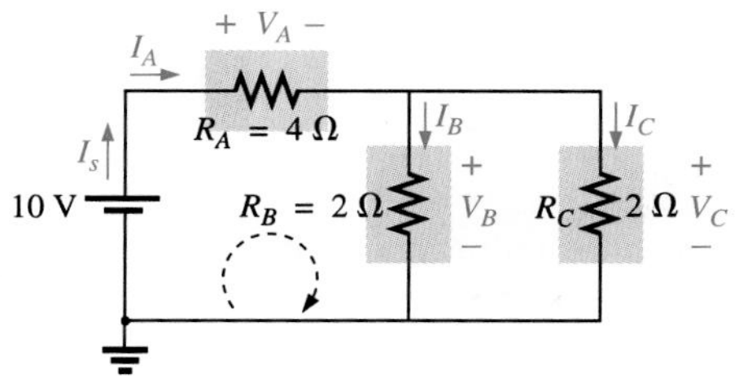

FIG. 7.7
Reduced equivalent of Fig. 7.6.

and
$$I_s = \frac{E}{R_T} = \frac{10\ \text{V}}{5\ \Omega} = \mathbf{2\ A}$$

We can find the currents I_A, I_B, and I_C using the reduction of the network of Fig. 7.6 (recall Step 3) as found in Fig. 7.7. Note that I_A, I_B, and I_C are the same in Figs. 7.6 and 7.7 and therefore also appear in Fig. 7.7. In other words, the currents I_A, I_B, and I_C of Fig. 7.7 will have the same magnitude as the same currents of Fig. 7.6.

$$I_A = I_s = \mathbf{2\ A}$$

and
$$I_B = I_C = \frac{I_A}{2} = \frac{I_s}{2} = \frac{2\ \text{A}}{2} = \mathbf{1\ A}$$

Returning to the network of Fig. 7.6, we have

$$I_{R_2} = I_{R_3} = \frac{I_B}{2} = \mathbf{0.5\ A}$$

The voltages V_A, V_B, and V_C from either figure are

$$V_A = I_A R_A = (2\ \text{A})(4\ \Omega) = \mathbf{8\ V}$$
$$V_B = I_B R_B = (1\ \text{A})(2\ \Omega) = \mathbf{2\ V}$$
$$V_C = V_B = \mathbf{2\ V}$$

Applying Kirchhoff's voltage law for the loop indicated in Fig. 7.7, we obtain

$$\Sigma_{\circlearrowright} V = E - V_A - V_B = 0$$
$$E = V_A + V_B = 8\ \text{V} + 2\ \text{V}$$
or
$$\underline{10\ \text{V} = 10\ \text{V} \quad \text{(checks)}}$$

EXAMPLE 7.3

Another possible variation of Fig. 7.2 appears in Fig. 7.8.

$$R_A = R_{1\|2} = \frac{(9\ \Omega)(6\ \Omega)}{9\ \Omega + 6\ \Omega} = \frac{54\ \Omega}{15} = 3.6\ \Omega$$

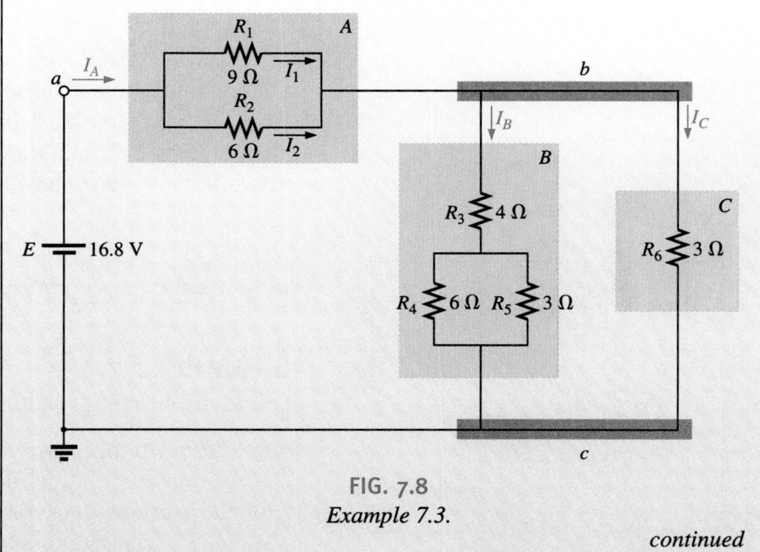

FIG. 7.8
Example 7.3.

continued

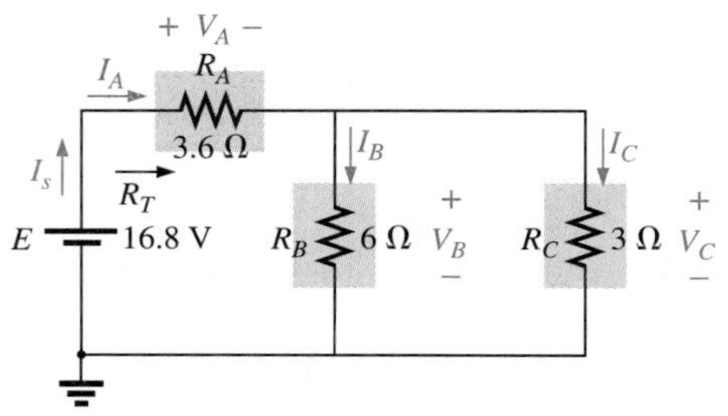

FIG. 7.9
Reduced equivalent of Fig. 7.8.

$$R_B = R_3 + R_{4\|5} = 4\,\Omega + \frac{(6\,\Omega)(3\,\Omega)}{6\,\Omega + 3\,\Omega} = 4\,\Omega + 2\,\Omega = 6\,\Omega$$

$$R_C = 3\,\Omega$$

The network of Fig. 7.8 can then be redrawn in reduced form, as shown in Fig. 7.9. Note the similarities between this circuit and those of Fig. 7.3 and 7.7.

$$R_T = R_A + R_{B\|C} = 3.6\,\Omega + \frac{(6\,\Omega)(3\,\Omega)}{6\,\Omega + 3\,\Omega}$$

$$= 3.6\,\Omega + 2\,\Omega = \mathbf{5.6\,\Omega}$$

$$I_s = \frac{E}{R_T} = \frac{16.8\text{ V}}{5.6\,\Omega} = \mathbf{3\text{ A}}$$

$$I_A = I_s = \mathbf{3\text{ A}}$$

Applying the current divider rule gives us

$$I_B = \frac{R_C I_A}{R_C + R_B} = \frac{(3\,\Omega)(3\text{ A})}{3\,\Omega + 6\,\Omega} = \frac{9\text{ A}}{9} = \mathbf{1\text{ A}}$$

By Kirchhoff's current law,

$$I_C = I_A - I_B = 3\text{ A} - 1\text{ A} = \mathbf{2\text{ A}}$$

By Ohm's law,

$$V_A = I_A R_A = (3\text{ A})(3.6\,\Omega) = \mathbf{10.8\text{ V}}$$

$$V_B = I_B R_B = V_C = I_C R_C = (2\text{ A})(3\,\Omega) = \mathbf{6\text{ V}}$$

Returning to the original network (Fig. 7.8) and applying the current divider rule,

$$I_1 = \frac{R_2 I_A}{R_2 + R_1} = \frac{(6\,\Omega)(3\text{ A})}{6\,\Omega + 9\,\Omega} = \frac{18\text{ A}}{15} = \mathbf{1.2\text{ A}}$$

By Kirchhoff's current law,

$$I_2 = I_A - I_1 = 3\text{ A} - 1.2\text{ A} = \mathbf{1.8\text{ A}}$$

There are many ways to arrange the blocks of Fig. 7.2. In fact, there is no limit to the number of series-parallel configurations within a given network. In reverse, the block diagram approach can be used effectively to simplify a system by identifying the major series and parallel components of the network. This approach will be demonstrated in the next few examples.

7.2 DESCRIPTIVE EXAMPLES

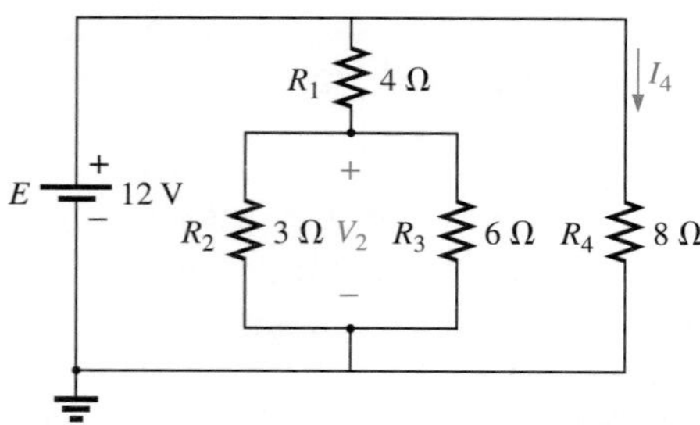

FIG. 7.10
Example 7.4.

EXAMPLE 7.4

Find the current I_4 and the voltage V_2 for the network of Fig. 7.10.

Solution: In this case, you have to find particular unknowns instead of a complete solution. It would, therefore, be a waste of time to find all the currents and voltages of the network. The method you use should concentrate on obtaining only the unknowns requested. With

continued

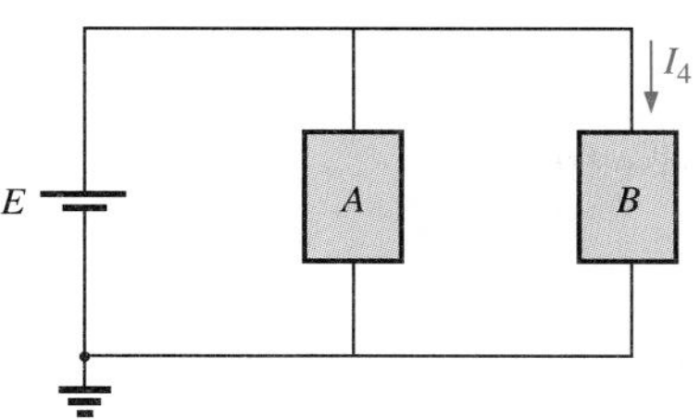

FIG. 7.11
Block diagram of Fig. 7.10.

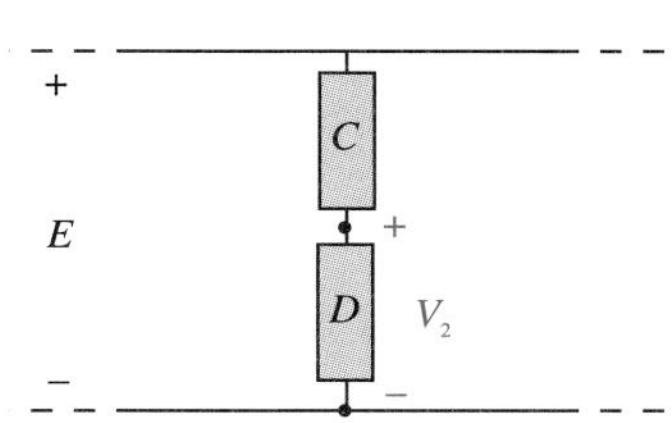

FIG. 7.12
Alternative block diagram for the first parallel branch of Fig. 7.10.

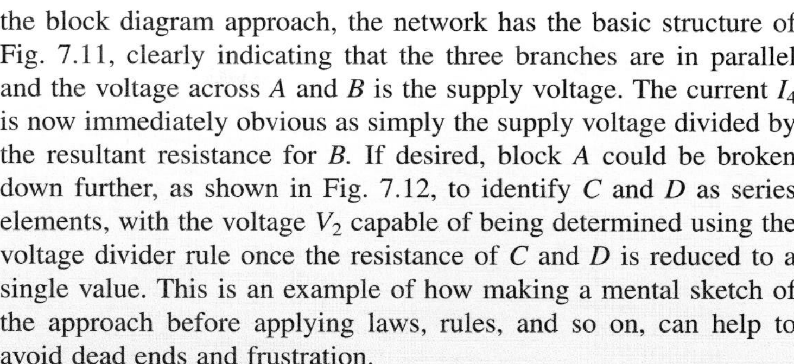

the block diagram approach, the network has the basic structure of Fig. 7.11, clearly indicating that the three branches are in parallel and the voltage across A and B is the supply voltage. The current I_4 is now immediately obvious as simply the supply voltage divided by the resultant resistance for B. If desired, block A could be broken down further, as shown in Fig. 7.12, to identify C and D as series elements, with the voltage V_2 capable of being determined using the voltage divider rule once the resistance of C and D is reduced to a single value. This is an example of how making a mental sketch of the approach before applying laws, rules, and so on, can help to avoid dead ends and frustration.

Applying Ohm's law,

$$I_4 = \frac{E}{R_B} = \frac{E}{R_4} = \frac{12\text{ V}}{8\ \Omega} = \mathbf{1.5\ A}$$

Combining the resistors R_2 and R_3 of Fig. 7.10 will result in

$$R_D = R_2 \parallel R_3 = 3\ \Omega \parallel 6\ \Omega = \frac{(3\ \Omega)(6\ \Omega)}{3\ \Omega + 6\ \Omega} = \frac{18\ \Omega}{9} = 2\ \Omega$$

and, applying the voltage divider rule,

$$V_2 = \frac{R_D E}{R_D + R_C} = \frac{(2\ \Omega)(12\text{ V})}{2\ \Omega + 4\ \Omega} = \frac{24\text{ V}}{6} = \mathbf{4\ V}$$

EXAMPLE 7.5

Find the indicated currents and voltages for the network of Fig. 7.13.

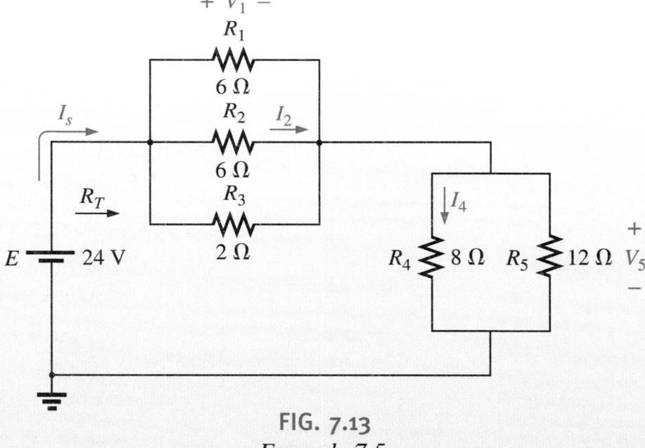

FIG. 7.13
Example 7.5.

Solution: Again, only specific unknowns are needed. When the network is redrawn, you should note particularly which unknowns are preserved and which will have to be determined using the original configuration. The block diagram of the network may appear as shown in Fig. 7.14, clearly revealing that A and B are in series. Note in this form the number of unknowns that have been preserved. The voltage V_1 will be the same across the three parallel branches of Fig. 7.13, and V_5 will be the same across R_4 and R_5. The unknown currents I_2 and I_4 are lost since they represent the currents through only one of the parallel branches. However, once V_1 and V_5 are known, the required currents can be found using Ohm's law.

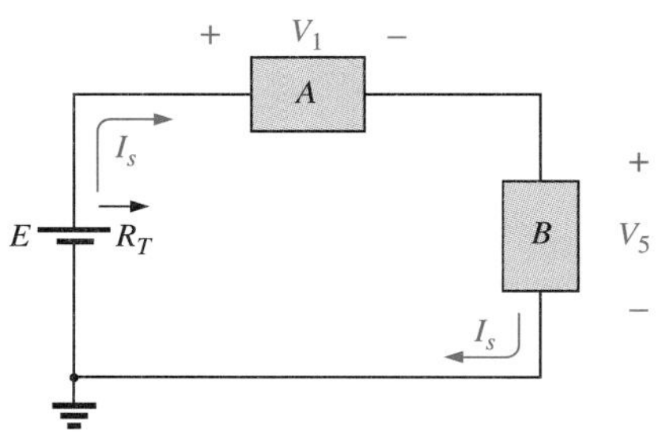

FIG. 7.14
Block diagram for Fig. 7.13.

continued

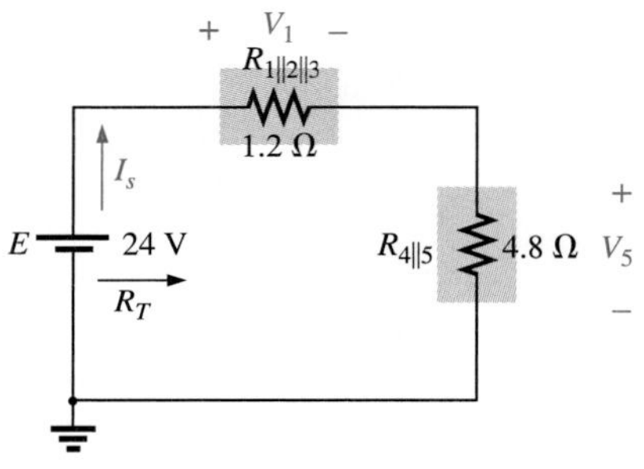

FIG. 7.15
Reduced form of Fig. 7.13.

$$R_{1\|2} = \frac{R}{N} = \frac{6\ \Omega}{2} = 3\ \Omega$$

$$R_A = R_{1\|2\|3} = \frac{(3\ \Omega)(2\ \Omega)}{3\ \Omega + 2\ \Omega} = \frac{6\ \Omega}{5} = 1.2\ \Omega$$

$$R_B = R_{4\|5} = \frac{(8\ \Omega)(12\ \Omega)}{8\ \Omega + 12\ \Omega} = \frac{96\ \Omega}{20} = 4.8\ \Omega$$

The reduced form of Fig. 7.13 will then appear as shown in Fig. 7.15, and

$$R_T = R_{1\|2\|3} + R_{4\|5} = 1.2\ \Omega + 4.8\ \Omega = \mathbf{6\ \Omega}$$

$$I_s = \frac{E}{R_T} = \frac{24\ \text{V}}{6\ \Omega} = \mathbf{4\ A}$$

with

$$V_1 = I_s R_{1\|2\|3} = (4\ \text{A})(1.2\ \Omega) = \mathbf{4.8\ V}$$

$$V_5 = I_s R_{4\|5} = (4\ \text{A})(4.8\ \Omega) = \mathbf{19.2\ V}$$

Applying Ohm's law,

$$I_4 = \frac{V_5}{R_4} = \frac{19.2\ \text{V}}{8\ \Omega} = \mathbf{2.4\ A}$$

$$I_2 = \frac{V_2}{R_2} = \frac{V_1}{R_2} = \frac{4.8\ \text{V}}{6\ \Omega} = \mathbf{0.8\ A}$$

The next example shows that unknown voltages do not have to be across elements. They can exist between any two points in a network. In addition, the analysis to follow clearly reveals the importance of redrawing the network in a more familiar form.

EXAMPLE 7.6

a. Find the voltages V_1, V_3, and V_{ab} for the network of Fig. 7.16.
b. Calculate the source current I_s.

Solution: This is one of those situations where it might be best to redraw the network before beginning the analysis. Since combining both sources will not affect the unknowns, the network is redrawn as shown in Fig. 7.17. This establishes a parallel network with the total source voltage across each parallel branch. The net source voltage is the difference between the two with the polarity of the larger.

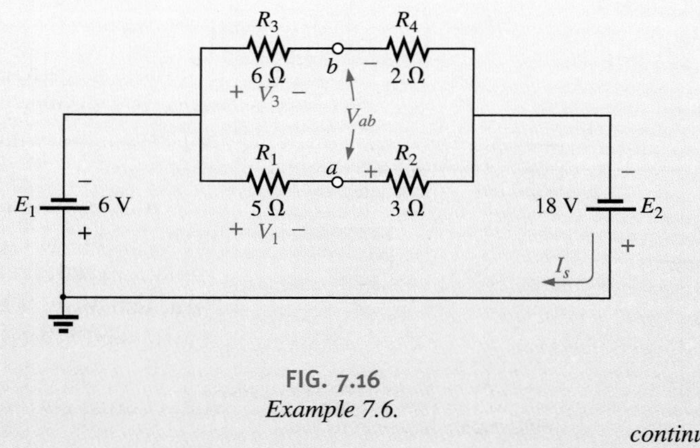

FIG. 7.16
Example 7.6.

continued

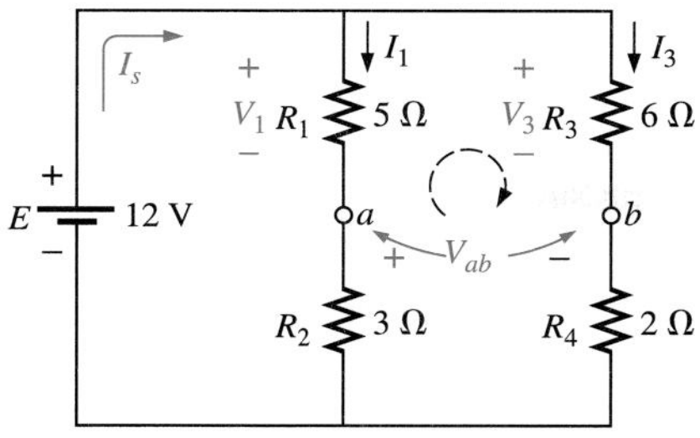

FIG. 7.17
Network of Fig. 7.16 redrawn.

a. Note that because this is similar to Fig. 7.12, you can use the voltage divider rule to determine V_1 and V_3:

$$V_1 = \frac{R_1E}{R_1 + R_2} = \frac{(5\ \Omega)(12\ \text{V})}{5\ \Omega + 3\ \Omega} = \frac{60\ \text{V}}{8} = \mathbf{7.5\ V}$$

$$V_3 = \frac{R_3E}{R_3 + R_4} = \frac{(6\ \Omega)(12\ \text{V})}{6\ \Omega + 2\ \Omega} = \frac{72\ \text{V}}{8} = \mathbf{9\ V}$$

The open-circuit voltage V_{ab} is determined by applying Kirchhoff's voltage law around the indicated loop of Fig. 7.17 in the clockwise direction starting at terminal *a*.

$$+V_1 - V_3 + V_{ab} = 0$$

and $$V_{ab} = V_3 - V_1 = 9\ \text{V} - 7.5\ \text{V} = \mathbf{1.5\ V}$$

b. By Ohm's law,

$$I_1 = \frac{V_1}{R_1} = \frac{7.5\ \text{V}}{5\ \Omega} = 1.5\ \text{A}$$

$$I_3 = \frac{V_3}{R_3} = \frac{9\ \text{V}}{6\ \Omega} = 1.5\ \text{A}$$

Applying Kirchhoff's current law,

$$I_s = I_1 + I_3 = 1.5\ \text{A} + 1.5\ \text{A} = \mathbf{3\ A}$$

EXAMPLE 7.7

For the network of Fig. 7.18, determine the voltages V_1 and V_2 and the current *I*.

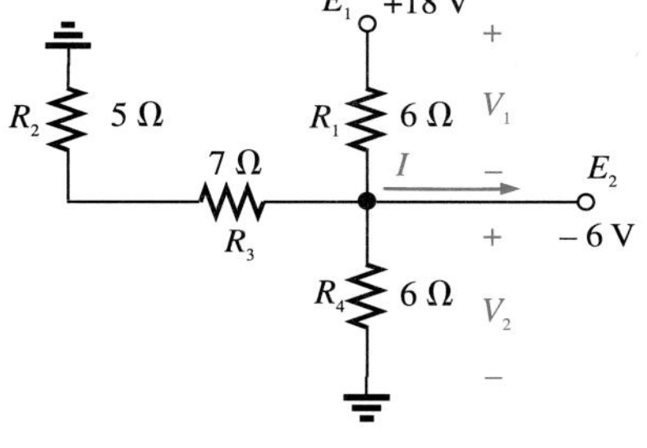

FIG. 7.18
Example 7.7.

Solution: It would be difficult to analyze the network in the form of Fig. 7.18 with the symbolic notation for the sources and the reference or ground connection in the upper left-hand corner of the diagram. However, when the network is redrawn as shown in Fig. 7.19, the unknowns and the relationship between branches become much clearer. Note the common connection of the grounds and the replacing of the terminal notation by actual supplies.

It is now obvious that

$$V_2 = -E_1 = \mathbf{-6\ V}$$

The minus sign simply indicates that the chosen polarity for V_2 in Fig. 7.18 is opposite to the actual voltage. Applying Kirchhoff's voltage law to the loop indicated, we obtain

$$-E_1 + V_1 - E_2 = 0$$

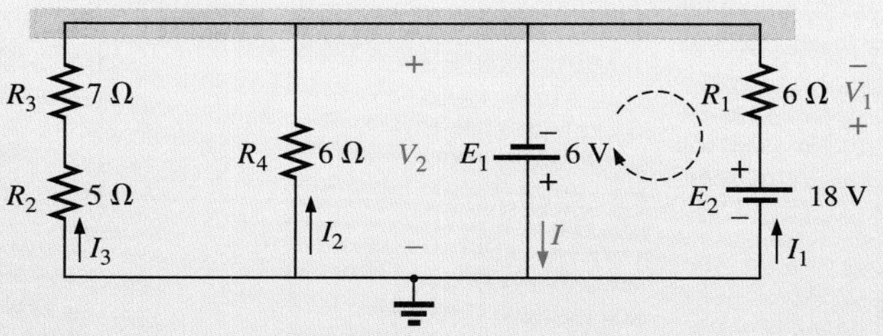

FIG. 7.19
Network of Fig. 7.18 redrawn.

continued

and $V_1 = E_2 + E_1 = 18\text{ V} + 6\text{ V} = \mathbf{24\text{ V}}$

Applying Kirchhoff's current law to node a gives us

$$\begin{aligned} I &= I_1 + I_2 + I_3 \\ &= \frac{V_1}{R_1} + \frac{E_1}{R_4} + \frac{E_1}{R_2 + R_3} \\ &= \frac{24\text{ V}}{6\ \Omega} + \frac{6\text{ V}}{6\ \Omega} + \frac{6\text{ V}}{12\ \Omega} \\ &= 4\text{ A} + 1\text{ A} + 0.5\text{ A} \\ I &= \mathbf{5.5\text{ A}} \end{aligned}$$

EXAMPLE 7.8

Calculate the indicated currents and voltage of Fig. 7.20.

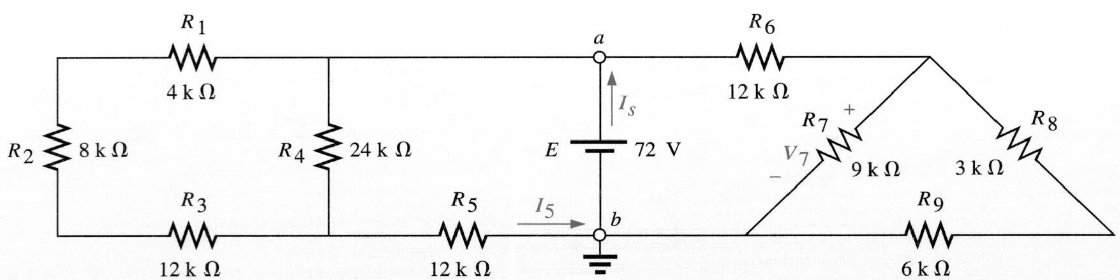

FIG. 7.20
Example 7.8.

Solution: Redrawing the network after combining series elements gives us Fig. 7.21, and

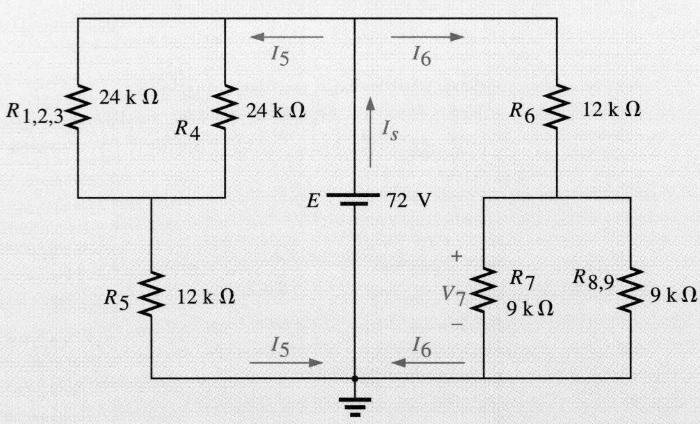

FIG. 7.21
Network of Fig. 7.20 redrawn.

continued

$$I_5 = \frac{E}{R_{(1,2,3)\|4} + R_5} = \frac{72 \text{ V}}{12 \text{ k}\Omega + 12 \text{ k}\Omega} = \frac{72 \text{ V}}{24 \text{ k}\Omega} = \mathbf{3 \text{ mA}}$$

with

$$V_7 = \frac{R_{7\|(8,9)}E}{R_{7\|(8,9)} + R_6} = \frac{(4.5 \text{ k}\Omega)(72 \text{ V})}{4.5 \text{ k}\Omega + 12 \text{ k}\Omega} = \frac{324 \text{ V}}{16.5} = \mathbf{19.6 \text{ V}}$$

$$I_6 = \frac{V_7}{R_{7\|(8,9)}} = \frac{19.6 \text{ V}}{4.5 \text{ k}\Omega} = \mathbf{4.35 \text{ mA}}$$

and $I_s = I_5 + I_6 = 3 \text{ mA} + 4.35 \text{ mA} = \mathbf{7.35 \text{ mA}}$

Since the potential difference between points *a* and *b* of Fig. 7.20 is fixed at *E* volts, the circuit to the right or left is unaffected if the network is reconstructed as shown in Fig. 7.22.

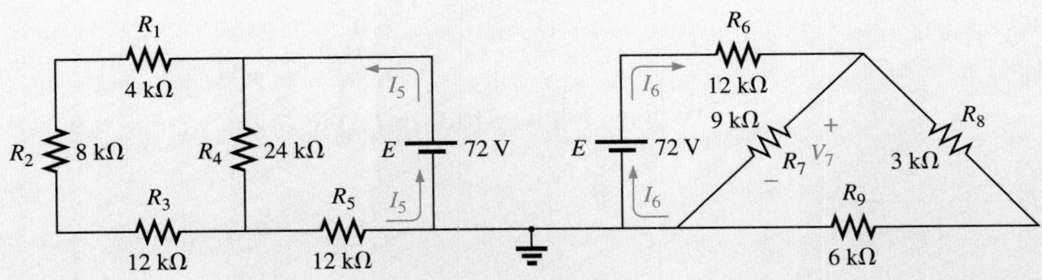

FIG. 7.22
An alternative approach to Example 7.8.

We can find each quantity required, except I_s, by analyzing each circuit independently. To find I_s, we must find the source current for each circuit and add it as in the above solution; that is, $I_s = I_5 + I_6$.

7.3 LADDER NETWORKS

A three-section **ladder network** appears in Fig. 7.23. The reason for the name is quite obvious from the repetitive structure. There are basically two approaches used to solve this type of network.

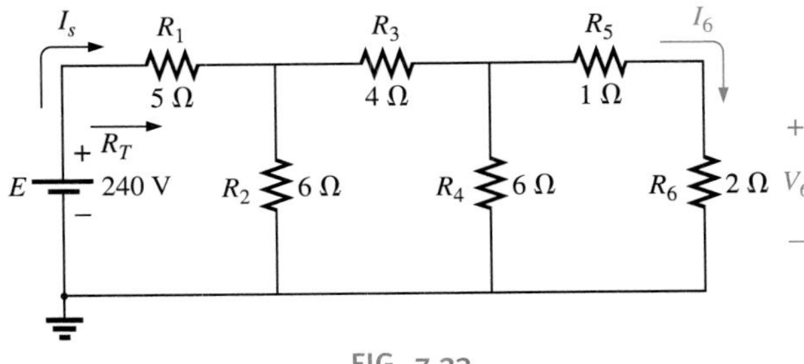

FIG. 7.23
Ladder network.

Method 1

Calculate the total resistance and resulting source current, and then work back through the ladder until the desired current or voltage is obtained. We will now use this method to determine V_6 in Fig. 7.23.

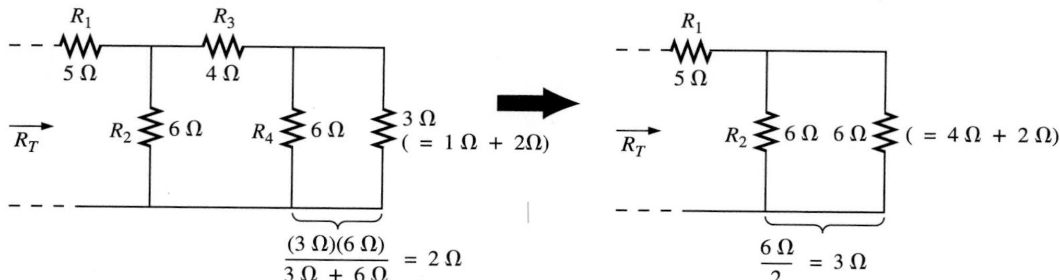

FIG. 7.24
Working back to the source to determine R_T for the network of Fig. 7.23.

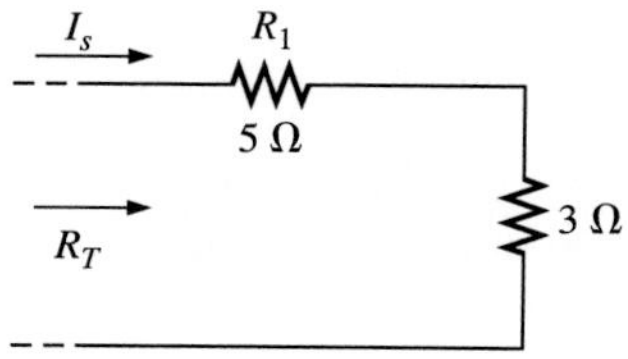

FIG. 7.25
Calculating R_T and I_s.

Combining parallel and series elements as shown in Fig. 7.24 will result in the reduced network of Fig. 7.25, and

$$R_T = 5\ \Omega + 3\ \Omega = 8\ \Omega$$

$$I_s = \frac{E}{R_T} = \frac{240\ \text{V}}{8\ \Omega} = 30\ \text{A}$$

Working our way back to I_6 (Fig. 7.26), we find that

$$I_1 = I_s$$

and

$$I_3 = \frac{I_s}{2} = \frac{30\ \text{A}}{2} = 15\ \text{A}$$

and, finally (Fig. 7.27),

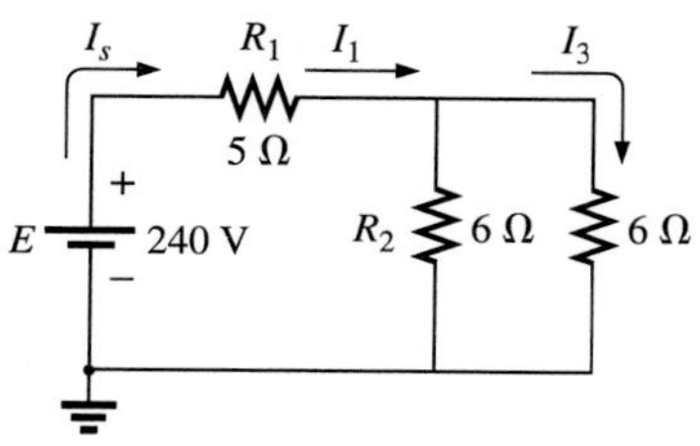

FIG. 7.26
Working back toward I_6.

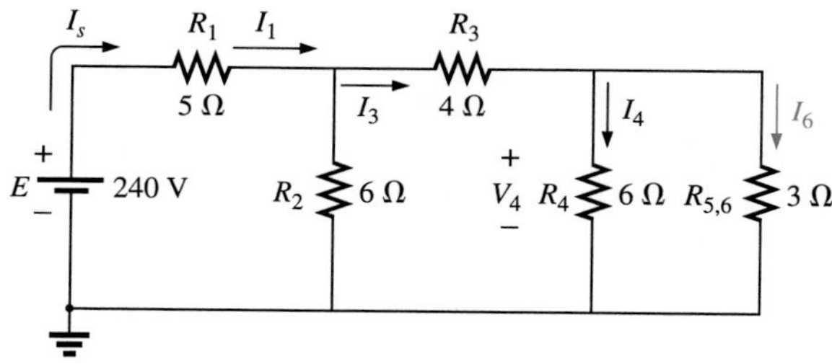

FIG. 7.27
Calculating I_6.

$$I_6 = \frac{(6\ \Omega)I_3}{6\ \Omega + 3\ \Omega} = \frac{6}{9}(15\ \text{A}) = 10\ \text{A}$$

and

$$V_6 = I_6 R_6 = (10\ \text{A})(2\ \Omega) = \mathbf{20\ V}$$

Method 2

Assign a letter symbol to the last branch current and work back through the network to the source, maintaining this assigned current or other current of interest. You can then find the desired current directly. The best way to describe this method is to analyze the same network considered above in Fig. 7.23, as redrawn in Fig. 7.28.

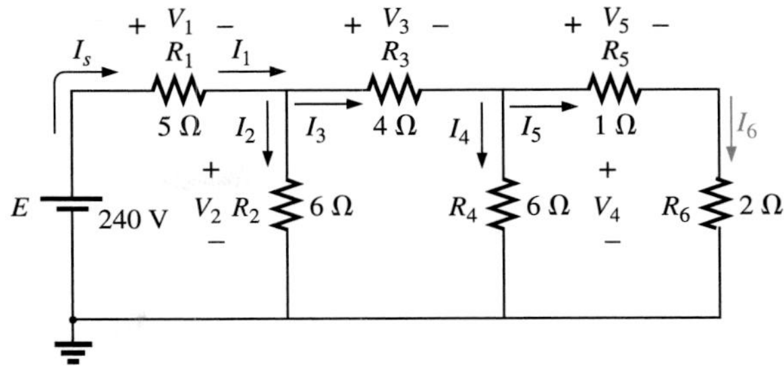

FIG. 7.28
An alternative approach for ladder networks.

The assigned notation for the current through the final branch is I_6:

$$I_6 = \frac{V_4}{R_5 + R_6} = \frac{V_4}{1\ \Omega + 2\ \Omega} = \frac{V_4}{3\ \Omega}$$

or $$V_4 = (3\ \Omega)I_6$$

so that $$I_4 = \frac{V_4}{R_4} = \frac{(3\ \Omega)I_6}{6\ \Omega} = 0.5I_6$$

and $$I_3 = I_4 + I_6 = 0.5I_6 + I_6 = 1.5I_6$$

$$V_3 = I_3R_3 = (1.5I_6)(4\ \Omega) = (6\ \Omega)I_6$$

Also, $$V_2 = V_3 + V_4 = (6\ \Omega)I_6 + (3\ \Omega)I_6 = (9\ \Omega)I_6$$

so that $$I_2 = \frac{V_2}{R_2} = \frac{(9\ \Omega)I_6}{6\ \Omega} = 1.5I_6$$

and $$I_s = I_2 + I_3 = 1.5I_6 + 1.5I_6 = 3I_6$$

with $$V_1 = I_1R_1 = I_sR_1 = (5\ \Omega)I_s$$

so that $$E = V_1 + V_2 = (5\ \Omega)I_s + (9\ \Omega)I_6$$
$$= (5\ \Omega)(3I_6) + (9\ \Omega)I_6 = (24\ \Omega)I_6$$

and $$I_6 = \frac{E}{24\ \Omega} = \frac{240\ \text{V}}{24\ \Omega} = 10\ \text{A}$$

with $$V_6 = I_6R_6 = (10\ \text{A})(2\ \Omega) = \mathbf{20\ V}$$

as we obtained using method 1.

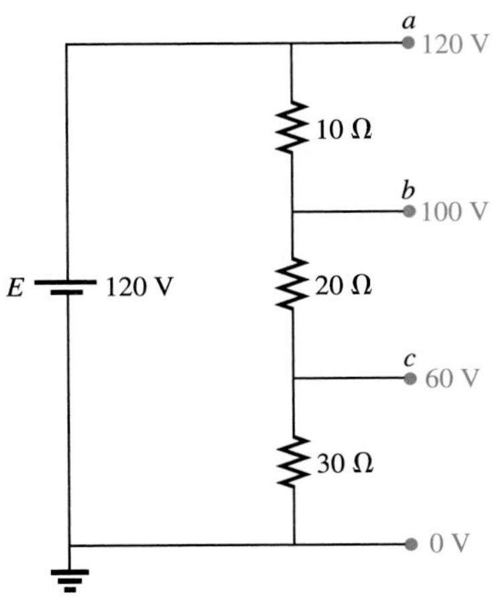

FIG. 7.29
Voltage divider supply.

7.4 VOLTAGE DIVIDER SUPPLY (UNLOADED AND LOADED)

The term **loaded** refers to the application of an element, network, or system to a supply that will draw current from the supply. As pointed out in Section 5.8, the application of a load can affect the terminal voltage of the supply.

Through a voltage divider network such as the one in Fig. 7.29, a number of terminal voltages can be made available from a single supply. The voltage levels shown (with respect to ground) are determined by a direct application of the voltage divider rule. Figure 7.29 reflects a no-load situation due to the absence of any current-drawing elements connected between terminals *a, b,* or *c* and ground.

The larger the resistance level of the applied loads compared to the resistance level of the voltage divider network, the closer the resulting terminal voltage to the no-load levels. In other words, the lower the current demand from a supply, the closer the terminal characteristics to the no-load levels.

To demonstrate the above statement, we will consider the network of Fig. 7.29 with resistive loads that are the average value of the resistive elements of the voltage divider network, as shown in Fig. 7.30.

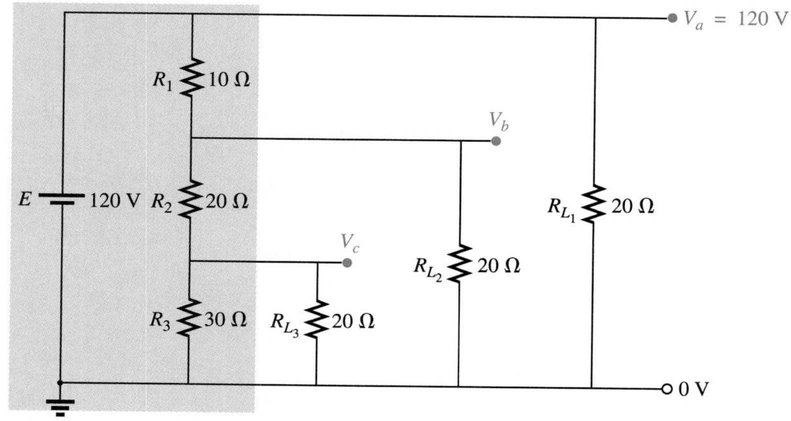

FIG. 7.30
Voltage divider supply with loads equal to the average value of the resistive elements that make up the supply.

The voltage V_a is unaffected by the load R_{L_1} since the load is in parallel with the supply voltage E. The result is $V_a = 120$ V, which is the same as the no-load level. To determine V_b, we must first note that R_3 and R_{L_3} are in parallel and $R'_3 = R_3 \parallel R_{L_3} = 30\ \Omega \parallel 20\ \Omega = 12\ \Omega$. The parallel combination $R'_2 = (R_2 + R'_3) \parallel R_{L_2} = (20\ \Omega + 12\ \Omega) \parallel 20\ \Omega = 32\ \Omega \parallel 20\ \Omega = 12.31\ \Omega$. Applying the voltage divider rule gives

$$V_b = \frac{(12.31\ \Omega)(120\ \text{V})}{12.31\ \Omega + 10\ \Omega} = 66.21\ \text{V}$$

versus 100 V under no-load conditions.

The voltage V_c is

$$V_c = \frac{(12\ \Omega)(66.21\ \text{V})}{12\ \Omega + 20\ \Omega} = 24.83\ \text{V}$$

versus 60 V under no-load conditions.

Thus, load resistors close in value to the resistor used in the voltage divider network significantly decrease some of the terminal voltages.

If the load resistors are changed to the 1-kΩ level, the terminal voltages will all be relatively close to the no-load values. The analysis is similar to the above, with the following results:

$$V_a = 120\ \text{V}, \quad V_b = 98.88\ \text{V}, \quad V_c = 58.63\ \text{V}$$

If we compare current drains established by the applied loads, we find for the network of Fig. 7.30 that

$$I_{L_2} = \frac{V_{L_2}}{R_{L_2}} = \frac{66.21\text{ V}}{20\ \Omega} = 3.31\text{ A}$$

and for the 1-kΩ level,

$$I_{L_2} = \frac{98.88\text{ V}}{1\text{ k}\Omega} = 98.88\text{ mA} \lesssim 0.1\text{ A}$$

As noted above in the highlighted statement, the more the current drain, the greater the change in terminal voltage with the application of the load. This is certainly shown by the fact that I_{L_2} is about 33.5 times larger with the 20-Ω loads.

The next example is a design exercise. The voltage and current rating of each load is provided, along with the terminal ratings of the supply. You must find the required voltage divider resistors.

EXAMPLE 7.9

Determine R_1, R_2, and R_3 for the voltage divider supply of Fig. 7.31. Can 2-W resistors be used in the design?

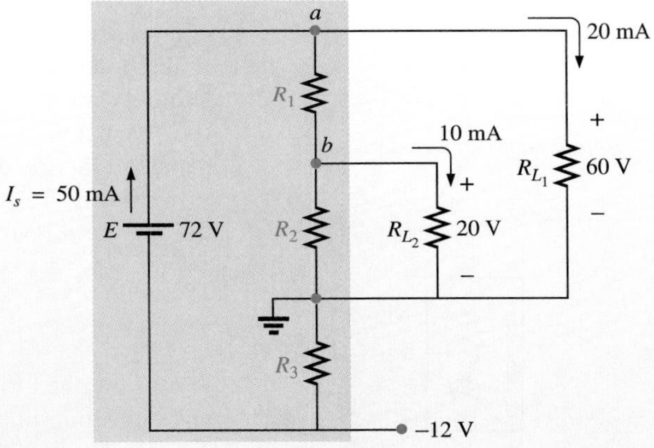

FIG. 7.31
Example 7.9.

Solution: R_3:

$$R_3 = \frac{V_{R_3}}{I_{R_3}} = \frac{V_{R_3}}{I_s} = \frac{12\text{ V}}{50\text{ mA}} = \mathbf{240\ \Omega}$$

$$P_{R_3} = (I_{R_3})^2 R_3 = (50\text{ mA})^2\, 240\ \Omega = 0.6\text{ W} < 2\text{ W}$$

R_1: Applying Kirchhoff's current law to node a:

$$I_s - I_{R_1} - I_{L_1} = 0$$

and $$I_{R_1} = I_s - I_{L_1} = 50\text{ mA} - 20\text{ mA}$$
$$= 30\text{ mA}$$

$$R_1 = \frac{V_{R_1}}{I_{R_1}} = \frac{V_{L_1} - V_{L_2}}{I_{R_1}} = \frac{60\text{ V} - 20\text{ V}}{30\text{ mA}} = \frac{40\text{ V}}{30\text{ mA}}$$
$$= \mathbf{1.33\text{ k}\Omega}$$

$$P_{R_1} = (I_{R_1})^2 R_1 = (30\text{ mA})^2\, 1.33\text{ k}\Omega = 1.197\text{ W} < 2\text{ W}$$

continued

R_2: Applying Kirchhoff's current law at node *b:*

$$I_{R_1} - I_{R_2} - I_{L_2} = 0$$

and $$I_{R_2} = I_{R_1} - I_{L_2} = 30 \text{ mA} - 10 \text{ mA} = 20 \text{ mA}$$

$$R_2 = \frac{V_{R_2}}{I_{R_2}} = \frac{20 \text{ V}}{20 \text{ mA}} = \mathbf{1\ k\Omega}$$

$$P_{R_2} = (I_{R_2})^2 R_2 = (20 \text{ mA})^2\, 1 \text{ k}\Omega = 0.4 \text{ W} < 2 \text{ W}$$

Since P_{R_1}, P_{R_2}, and P_{R_3} are less than 2 W, 2-W resistors can be used for the design.

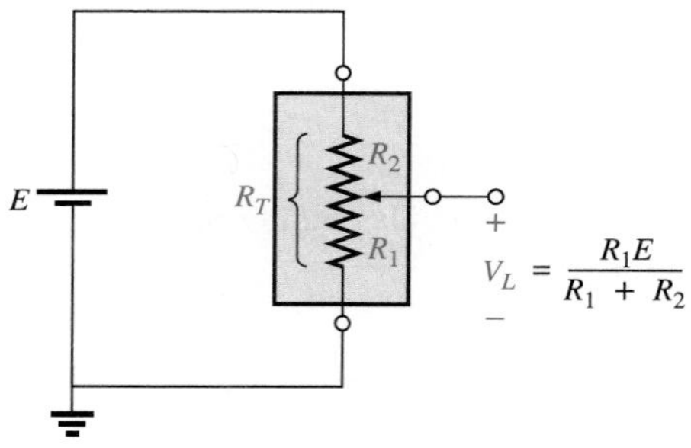

FIG. 7.32
Unloaded potentiometer.

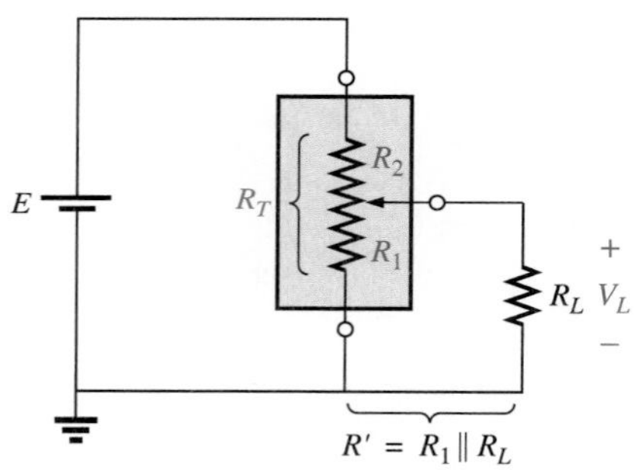

FIG. 7.33
Loaded potentiometer.

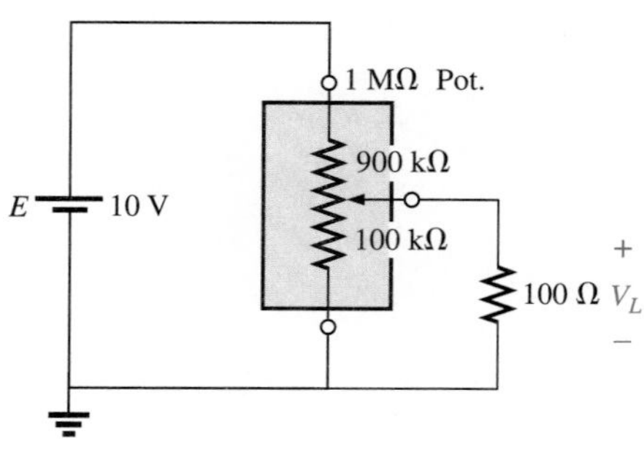

FIG. 7.34
$R_T > R_L$.

7.5 POTENTIOMETER LOADING

For the unloaded potentiometer of Fig. 7.32, the output voltage is determined by the voltage divider rule, with R_T in the figure representing the total resistance of the potentiometer. Too often it is assumed that the voltage across a load connected to the wiper arm is determined solely by the potentiometer and the effect of the load can be ignored. This is definitely not the case, as is demonstrated in the next few paragraphs.

When a load is applied as shown in Fig. 7.33, the output voltage V_L is now a function of the magnitude of the load applied since R_1 is not as shown in Fig. 7.32 but is instead the parallel combination of R_1 and R_L.

The output voltage is now:

$$V_L = \frac{R'E}{R' + R_2} \quad \text{with } R' = R_1 \| R_L \tag{7.1}$$

If you need to have good control of the output voltage V_L through the controlling dial, knob, screw, or whatever, choose a load or potentiometer that satisfies the following relationship:

$$R_L \geq R_T \tag{7.2}$$

For example, if we ignore Eq. (7.2) and choose a 1-MΩ potentiometer with a 100-Ω load and set the wiper arm to 1/10 the total resistance, as shown in Fig. 7.34, then

$$R' = 100 \text{ k}\Omega \,\|\, 100\ \Omega = 99.9\ \Omega$$

and $$V_L = \frac{99.9\ \Omega(10 \text{ V})}{99.9\ \Omega + 900 \text{ k}\Omega} \cong 0.001 \text{ V} = 1 \text{ mV}$$

which is extremely small compared to the expected level of 1 V.

In fact, if we move the wiper arm to the midpoint,

$$R' = 500 \text{ k}\Omega \,\|\, 100\ \Omega = 99.98\ \Omega$$

and $$V_L = \frac{(99.98\ \Omega)(10 \text{ V})}{99.98\ \Omega + 500 \text{ k}\Omega} \cong 0.002 \text{ V} = 2 \text{ mV}$$

which is negligible compared to the expected level of 5 V. Even at $R_1 = 900$ kΩ, V_L is simply 0.01 V, or 1/1000 of the available voltage.

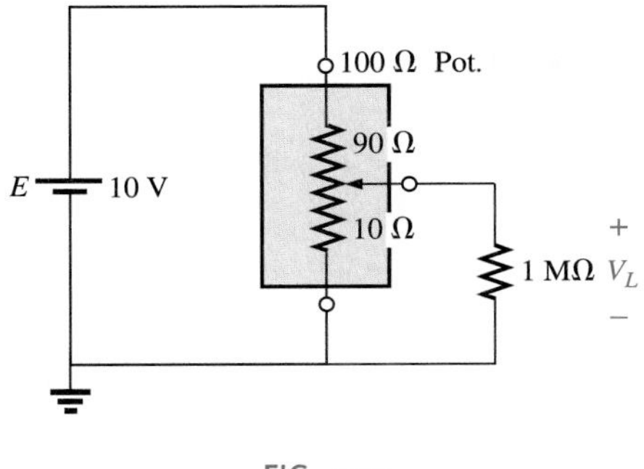

FIG. 7.35
$R_L > R_T$.

Using the reverse situation of $R_T = 100\ \Omega$ and $R_L = 1\ \text{M}\Omega$ and the wiper arm at the 1/10 position, as in Fig. 7.35, we find

$$R' = 10\ \Omega \parallel 1\ \text{M}\Omega \cong 10\ \Omega$$

and
$$V_L = \frac{10\ \Omega(10\ \text{V})}{10\ \Omega + 90\ \Omega} = 1\ \text{V}$$

as desired.

For the lower limit (worst-case design) of $R_L = R_T = 100\ \Omega$, as defined by Eq. (7.2) and the halfway position of Fig. 7.33,

$$R' = 50\ \Omega \parallel 100\ \Omega = 33.33\ \Omega$$

and
$$V_L = \frac{33.33\ \Omega(10\ \text{V})}{33.33\ \Omega + 50\ \Omega} \cong 4\ \text{V}$$

It may not be the ideal level of 5 V, but at least 40% of the voltage E has been achieved at the halfway position rather than the 0.02% obtained with $R_L = 100\ \Omega$ and $R_T = 1\ \text{M}\Omega$.

In general, therefore, try to establish a situation for potentiometer control in which Eq. (7.2) is satisfied to the highest degree possible.

Someone might suggest that we make R_T as small as possible to bring the percent result as close to the ideal as possible. Keep in mind, however, that the potentiometer has a power rating, and for networks such as Fig. 7.35, $P_{max} \cong E^2/R_T = (10\ \text{V})^2/100\ \Omega = 1\ \text{W}$. If R_T is reduced to 10 Ω, $P_{max} = (10\ \text{V})^2/10\ \Omega = 10\ \text{W}$, which would require a *much larger* unit.

EXAMPLE 7.10

Find the voltages V_1 and V_2 for the loaded potentiometer of Fig. 7.36.

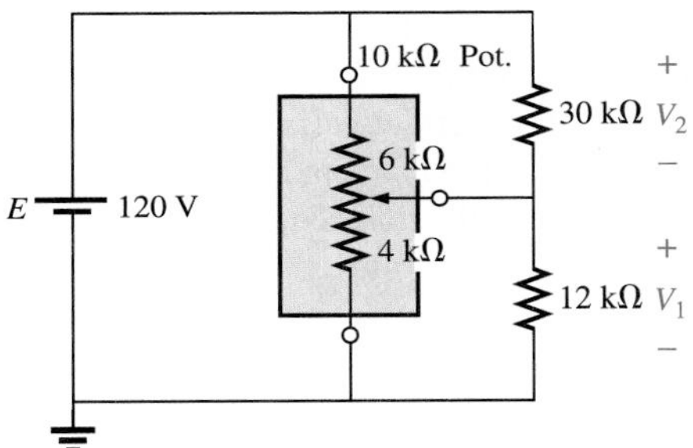

FIG. 7.36
Example 7.10.

Solution:

Ideal (no load): $$V_1 = \frac{4\ \text{k}\Omega(120\ \text{V})}{10\ \text{k}\Omega} = 48\ \text{V}$$

$$V_2 = \frac{6\ \text{k}\Omega(120\ \text{V})}{10\ \text{k}\Omega} = 72\ \text{V}$$

Loaded: $$R' = 4\ \text{k}\Omega \parallel 12\ \text{k}\Omega = 3\ \text{k}\Omega$$

$$R'' = 6\ \text{k}\Omega \parallel 30\ \text{k}\Omega = 5\ \text{k}\Omega$$

$$V_1 = \frac{3\ \text{k}\Omega(120\ \text{V})}{8\ \text{k}\Omega} = \mathbf{45\ V}$$

$$V_2 = \frac{5\ \text{k}\Omega(120\ \text{V})}{8\ \text{k}\Omega} = \mathbf{75\ V}$$

The ideal and loaded voltage levels are so close that the design can be considered a good one for the applied loads. A slight variation in the position of the wiper arm will establish the ideal voltage levels across the two loads.

7.6 AMMETER, VOLTMETER, AND OHMMETER DESIGN

These basic electrical measuring instruments are often combined into a single meter. Two types of instrument are widely used: analog meters such as the VOM and electronic meters such as the DMM. Since the DMM is a complex electronic device, we will only consider the analog meter.

The Moving-Coil Movement

The most sensitive, accurate, and common analog meter type is the **D'Arsonval** or **moving coil movement**. Its construction is shown in Fig. 7.37. The movement operates by the interaction of the magnetic field of the moving coil with that of the permanent magnet.

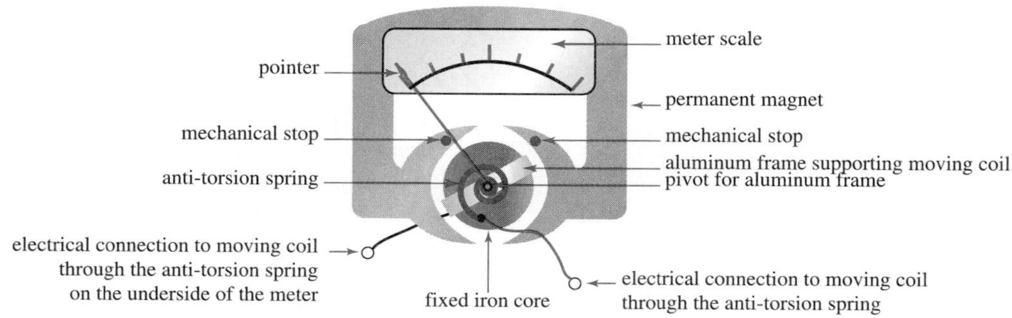

FIG. 7.37
Physical construction of a D'Arsonval meter movement.

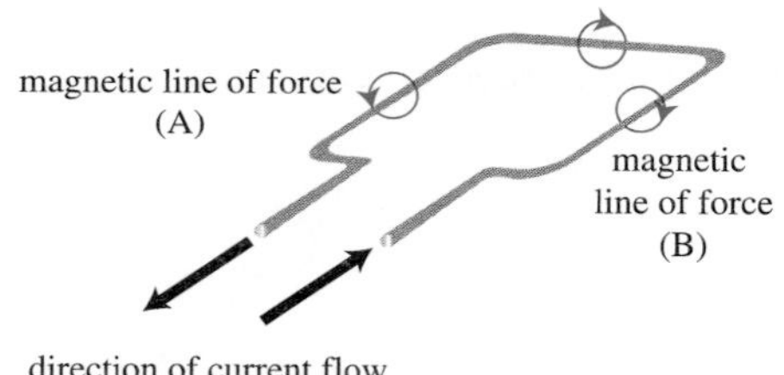

FIG. 7.38
Direction of the magnetic lines of force surrounding a simple current-carrying loop.

Any conductor carrying a current is surrounded by a cylindrical **magnetic field**. The strength of the field is proportional to the current (see Section 11.2). The magnetic field is represented by **lines of force**. Several of these are shown in Fig. 7.38 for a looped current-carrying conductor. In the meter movement, the pointer is attached to a pivoted coil composed of a number of such loops. The force between the magnetic fields of the permanent magnet and the coil causes the physical movement of the pointer.

All the magnetic lines of force behave in the same way. The following behaviours can explain the basic operation of the moving-coil movement:

- magnetic lines of force always follow a path from north to south
- magnetic lines of force always follow the shortest path from north to south
- like magnetic fields repel each other, while unlike magnetic fields attract each other
- magnetic lines of force cannot cross.

FIG. 7.39
(a) The interacting fields of the permanent magnet and the current-carrying loop exert a force indicated by the red arrows. (b) The rotation of the loop reduces the distortion in the magnetic field.

In the d'Arsonval movement, the field of the permanent magnet is distorted by the presence of the field generated by the moving coil [Fig. 7.39(a)] since the fields may not cross. Being distorted, the lines of force of the permanent magnet do not follow the shortest path between its poles. The distortion is such that the permanent magnet's lines of force will pass above the right-most conductor and below the left-most conductor. Since the moving coil is free to pivot, the path length of those lines of force will be shortened by displacing the moving coil clockwise on its pivots [Fig. 7.39(b)]. This motion of the coil will be seen in the rotation of the pointer.

Since the magnetic field of the moving coil is proportional to the current, the extent of pointer movement is proportional to the amount of current in the moving coil. Thus, a large current results in a large pointer deflection, while a small current will result in a small deflection. There is, however, a limiting factor—since the moving coil is made of very fine wire, it is not able to carry large currents.

Physically, the pointer rests against the left stop when no current flows in the moving coil. A reverse current flow would cause the pointer to move anti-clockwise against the stop. That could damage the mechanical structure of the meter movement. For that reason, dc meters are polarized.

The Electrodynamometer

This is a variation on the d'Arsonval movement. In the **electrodynamometer**, the permanent magnet is replaced by an electromagnet which is fixed in place. The moving coil remains unchanged. The field strength of the electromagnet is determined by the current flowing in its winding. Thus, the amount of pointer movement is again determined by the interaction of the magnetic fields generated by the two coils.

Although the electrodynamometer functions in the same way as a moving-coil movement, it usually has four terminals: two for the moving coil and two for the fixed coil. This type of movement is used in **wattmeters**. In a typical wattmeter, the moving coil senses the voltage across a circuit component, and the fixed coil senses the current flowing through the circuit. The moving coil functions as a voltmeter, while the fixed coil functions as an ammeter. Since the physical movement of the pointer is determined by the interaction of the two magnetic fields, both coils must be connected with the same polarity. For this reason, one end of each coil is identified with a white dot or an asterisk (*), or a plus/minus symbol (±).

The Ammeter

The specifications of a typical d'Arsonval movement is 1 mA, 50 Ω. The 1 mA is the *current sensitivity* of the movement, which is the current required for a full-scale deflection. It will be denoted by the symbol I_{CS}. The 50 Ω represents the internal resistance (R_m) of the movement.

The maximum current that the d'Arsonval movement can read independently is equal to the current sensitivity of the movement. However, higher currents can be measured if additional circuitry is introduced. This additional circuitry, as shown in Fig. 7.40, results in the basic construction of an ammeter.

The resistance R_{shunt} is chosen for the ammeter of Fig. 7.40 to allow 1 mA to flow through the movement when a maximum current of 1 A enters the ammeter. If less than 1 A flows through the ammeter, the movement will have less than 1 mA flowing through it and will indicate less than full-scale deflection.

Since the voltage across parallel elements must be the same, the potential drop across *a-b* in Fig. 7.40 must equal that across *c-d*; that is,

$$(1\ \text{mA})(50\ \Omega) = R_{\text{shunt}} I_s$$

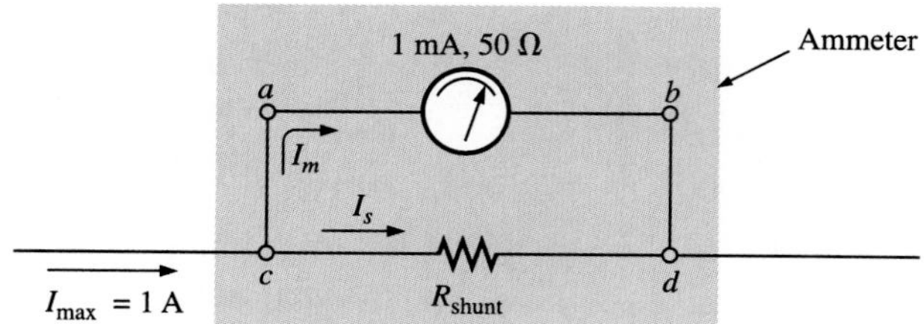

FIG. 7.40
Basic ammeter.

Also, I_s must equal 1 A − 1 mA = 999 mA if the current is to be limited to 1 mA through the movement (Kirchhoff's current law). Therefore,

$$(1\text{ mA})(50\ \Omega) = R_{shunt}(999\text{ mA})$$

$$R_{shunt} = \frac{(1\text{ mA})(50\ \Omega)}{999\text{ mA}}$$

$$\cong 0.05\ \Omega$$

In general,

$$R_{shunt} = \frac{R_m I_{CS}}{I_{max} - I_{CS}} \tag{7.3}$$

One method of constructing a multirange ammeter is shown in Fig. 7.41. Here the rotary switch determines the R_{shunt} to be used for the maximum current indicated on the face of the meter. Most meters use the same scale for various values of maximum current. If you read 375 on the 0–5 mA scale with the switch on the 5 setting, the current is 3.75 mA; on the 50 setting, the current is 37.5 mA; and so on.

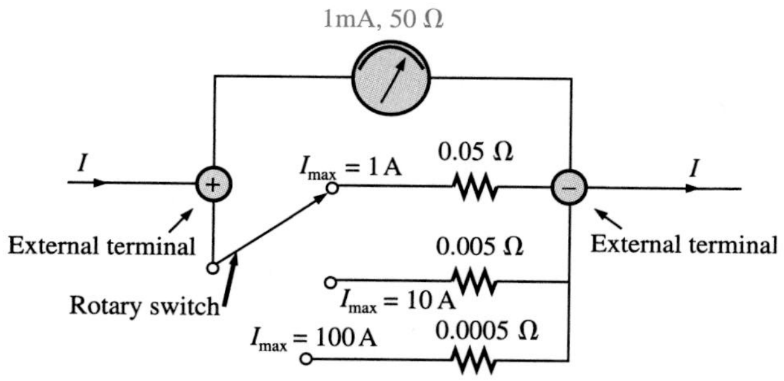

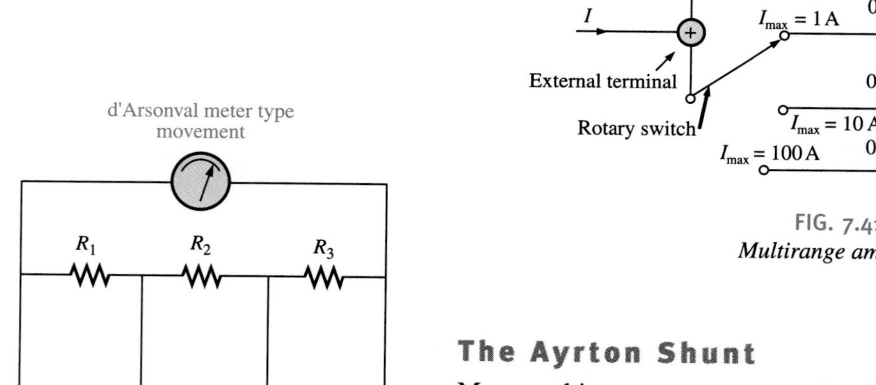

FIG. 7.41
Multirange ammeter.

FIG. 7.42
The Ayrton-shunt ammeter.

The Ayrton Shunt

Many multirange ammeters use the **Ayrton shunt**. As shown in Fig. 7.42, this shunt uses three series resistors paralleled with the meter movement. Thus, the meter movement cannot be connected in a circuit without its shunt. This gives the meter movement some protection from excessive currents; however, this protection is provided at the expense of a higher resistance in the meter branch.

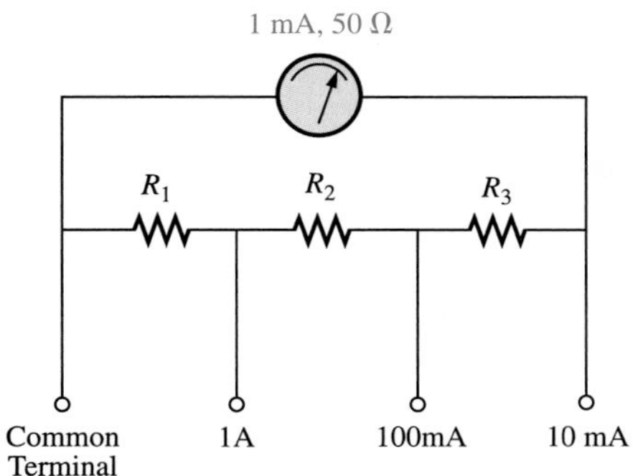

FIG. 7.43
Example 7.11.

EXAMPLE 7.11

The d'Arsonval movement of Fig. 7.43 has a resistance of 50 Ω and a current sensitivity of 1 mA. We need to select values for the three individual Ayrton-shunt resistors in order to establish current scales of 10 mA, 100 mA, and 1 A.

Solution:

In the 10 mA range, the shunt is formed by $R_1 + R_2 + R_3$. Of the total current of 10 mA, only 1 mA can go through the meter movement. The shunt must conduct the other 9/10. Therefore the shunt must have 1/9 of the resistance of the meter movement:

$$R_1 + R_2 + R_3 = \frac{50\ \Omega}{9} = 5.56\ \Omega \qquad (1)$$

For the 100 mA range, the shunt ($R_1 + R_2$) must carry 99/100 of the total 100 mA current. The other branch has a resistance of 50 Ω + R_3. Therefore:

$$R_1 + R_2 = \frac{50\ \Omega + R_3}{99} \qquad (2)$$

On the 1 A range (1000 mA), the shunt (R_1) must carry 999/1000 of the total current. Therefore:

$$R_1 = \frac{50\ \Omega + R_2 + R_3}{999} \qquad (3)$$

Now, use simultaneous equations to solve for R_3:

$$\begin{aligned} R_1 + R_2 + R_3 &= 5.56\ \Omega & (1)\\ 99R_1 + 99R_2 - R_3 &= 50\ \Omega & (2) \end{aligned}$$

Multiplying (1) by 99 gives us:

$$\begin{aligned} 99R_1 + 99R_2 + 99R_3 &= 550.4\ \Omega & (1)\\ 99R_1 + 99R_2 - R_3 &= 50\ \Omega & (2) \end{aligned}$$

Subtracting (2) from (1):

$$\begin{aligned} 100R_3 &= 500.4\ \Omega\\ R_3 &= 5.004\ \Omega \end{aligned}$$

To solve for R_2:

$$\begin{aligned} R_1 + R_2 + R_3 &= 5.56\ \Omega & (1)\\ 999R_1 - R_2 - R_3 &= 50\ \Omega & (3) \end{aligned}$$

Multiplying (1) by 999 gives us:

$$\begin{aligned} 999R_1 + 999R_2 + 999R_3 &= 5554\ \Omega & (1)\\ 999R_1 - R_2 - R_3 &= 50\ \Omega & (3) \end{aligned}$$

Subtracting (3) from (1):

$$\begin{aligned} 1000R_2 + 1000R_3 &= 5504\ \Omega\\ 1000R_2 + (1000 \times 5.004\ \Omega) &= 5504\ \Omega\\ 1000R_2 &= 500\ \Omega\\ R_2 &= 0.500\ \Omega \end{aligned}$$

continued

Substituting R_2 and R_3 in (1) will give R_1:

$$R_1 + R_2 + R_3 = 5.56\ \Omega \qquad (1)$$
$$R_1 + 0.500\ \Omega + 5.004\ \Omega = 5.56\ \Omega$$
$$R_1 = 5.56\ \Omega - 0.500\ \Omega - 5.004\ \Omega$$
$$R_1 = 0.056\ \Omega$$

For this meter, the Ayrton-shunt ammeter will be able to measure 10 mA, 100 mA, and 1A, when $R_1 = 0.056\ \Omega$, $R_2 = 0.500\ \Omega$, and $R_3 = 5.004\ \Omega$.

Both forms of shunt described above show that ammeter shunts have quite small resistances. For this reason, ammeters can be connected in series with a load without having a significant effect on the current in the circuit. For the same reason, connecting an ammeter in parallel with a load would result in excessive current that would probably damage the meter.

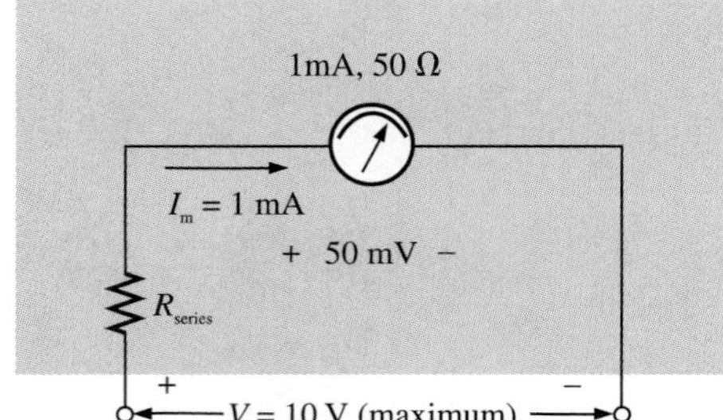

FIG. 7.44
Basic voltmeter.

The Voltmeter

A variation in the additional circuitry will permit the use of the d'Arsonval movement in the design of a voltmeter. The 1-mA, 50-Ω movement can also be rated as a 50-mV (1 mA × 50 Ω), 50-Ω movement, indicating that the maximum voltage (V_m) that the movement can measure independently is 50 mV. The basic construction of the voltmeter is shown in Fig. 7.44.

The R_{series} is adjusted to limit the current through the movement to 1 mA when the maximum voltage is applied across the voltmeter. A lesser voltage would simply reduce the current in the circuit and thereby the deflection of the movement.

Applying Kirchhoff's voltage law around the closed loop of Fig. 7.45, we obtain

$$[10\text{ V} - (1\text{ mA})(R_{series})] - 50\text{ mV} = 0$$

or

$$R_{series} = \frac{10\text{ V} - (50\text{ mV})}{1\text{ mA}} = 9950\ \Omega$$

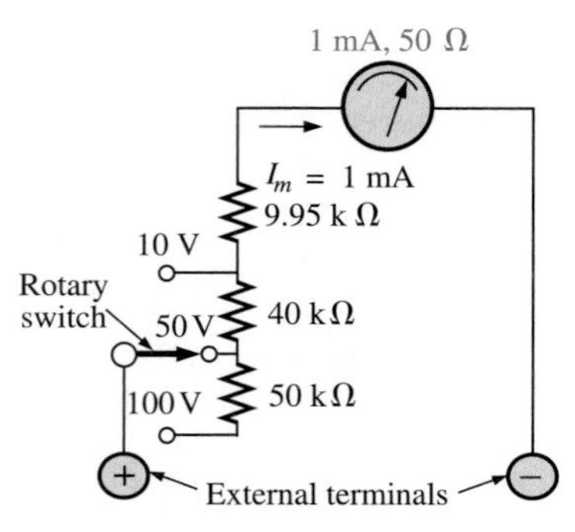

FIG. 7.45
Multirange voltmeter.

In general,

$$R_{series} = \frac{V_{max} - V_m}{I_{CS}} \qquad (7.4)$$

One method of constructing a multirange voltmeter is shown in Fig. 7.45. If the rotary switch is at 10 V, $R_{series} = 9.950\text{ k}\Omega$; at 50 V, $R_{series} = 40\text{ k}\Omega + 9.950\text{ k}\Omega = 49.950\text{ k}\Omega$; and at 100 V, $R_{series} = 50\text{ k}\Omega + 40\text{ k}\Omega + 9.950\text{ k}\Omega = 99.950\text{ k}\Omega$.

The Ohmmeter

In general, ohmmeters are designed to measure resistance in the low, mid-, or high range. The most common is the **series ohmmeter**, designed to read resistance levels in the midrange. It uses the series configuration of Fig. 7.46. The design is quite different from the ammeter or voltmeter design because it will show a full-scale deflection for zero ohms and no deflection for infinite resistance.

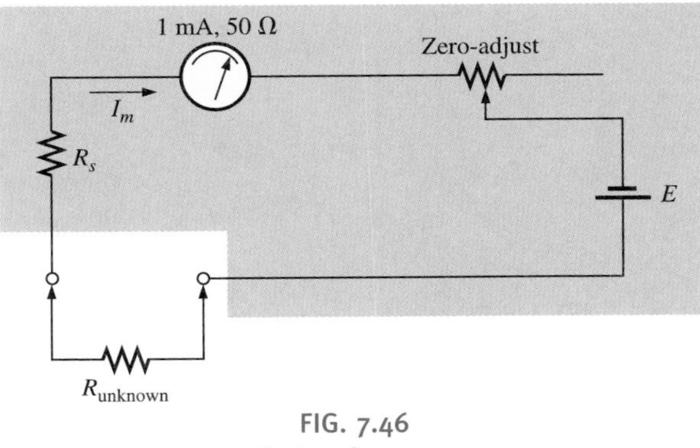

FIG. 7.46
Series ohmmeter.

To determine the series resistance R_s, the external terminals are shorted (a direct connection of zero ohms between the two) to simulate zero ohms, and the zero-adjust is set to half its maximum value. The resistance R_s is then adjusted to allow a current equal to the current sensitivity of the movement (1 mA) to flow in the circuit. The zero-adjust is set to half its value so that any variation in the components of the meter that may produce a current more or less than the current sensitivity can be compensated for. The current I_m is

$$I_m \text{ (full scale)} = I_{CS} = \frac{E}{R_s + R_m + \dfrac{\text{zero-adjust}}{2}} \tag{7.5}$$

and

$$R_s = \frac{E}{I_{CS}} - R_m - \frac{\text{zero-adjust}}{2} \tag{7.6}$$

If an unknown resistance is then placed between the external terminals, the current will be reduced, causing a deflection less than full scale. If the terminals are left open, simulating infinite resistance, the pointer will not deflect since the current through the circuit is zero.

An instrument designed to read very low values of resistance appears in Fig. 7.47. Because of its low-range capability, the network design must be much more sophisticated than described above. It uses electronic components that eliminate the inaccuracies introduced by lead and contact resistances. It is similar to the above system in the sense that it is completely portable and does require a dc battery to establish measurement conditions. Special leads are used to limit any introduced resistance levels. The maximum scale setting can be set as low as 3.52 mΩ.

FIG. 7.47
Milliohmmeter. (Courtesy of Keithley Instruments, Inc.)

The megohmmeter is an instrument for measuring very high resistance values. It is mainly used to test the insulation found in power transmission systems, electrical machinery, transformers, and so on. To measure the high-resistance values, a high dc voltage is established by a hand-driven generator. If the shaft is rotated above some set value, the output of the generator will be fixed at one selectable voltage, typically 250, 500, or 1000 V. A photograph of the commercially available tester is shown in Fig. 7.48. For this instrument, the range is zero to 5000 MΩ.

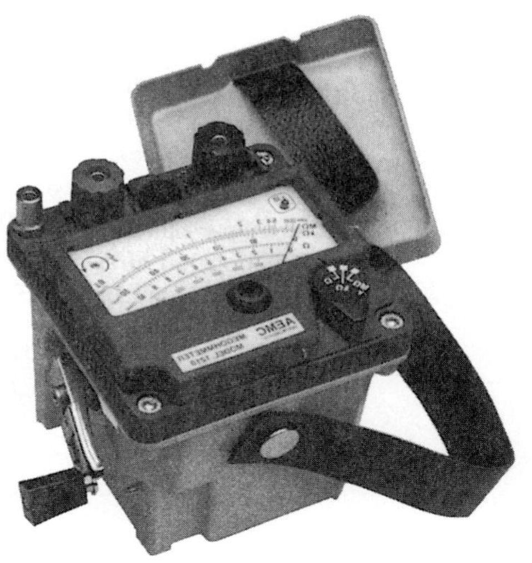

FIG. 7.48
Megohmmeter. (Courtesy of AEMC Corp.)

7.7 GROUNDING

The impact of the ground connection and how it can provide a measure of safety to a design are usually treated lightly in most introductory electrical or electronics texts. However, these are very important topics. Ground potential is 0 V at every point in a network that has a ground symbol. Since they are all at the same potential, they can all be connected together, but for purposes of clarity most are left isolated on a large schematic. On a schematic the voltage levels provided are always with respect to ground. A system can therefore be checked quite rapidly by simply connecting the black lead of the voltmeter to the ground connection and placing the red lead at the various points where the typical operating voltage is provided. A close match normally means that that portion of the system is operating properly.

Practical Business Applications

Vroom! Vroom! This is cool!

Remember when the coolest thing to do on a Saturday afternoon was to call your friends over to play with your racing-car set? It was neat putting it together in all sorts of different layouts and then grabbing the hand controls to give it a try. True, it was neat, but how did those hand controls vary the speed of the racecars on the track?

Well, first of all, realize that the racecars probably had permanent magnet dc motors in them. So, they needed a dc power supply in order to work right. You probably had a big square unit with two cords on it. One cord plugged into the wall and the other into the racetrack. This was the ac to dc converter. It was made up of a small ac step-down transformer and a dc bridge rectifier. This was enough to provide the racecars with energy.

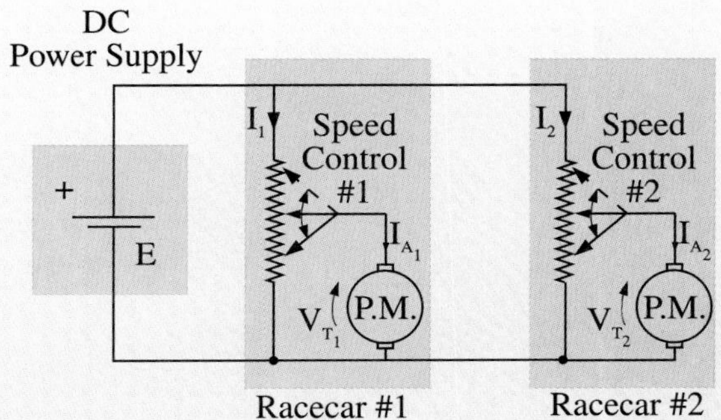

Now, a separate voltage divider network for each car determined the speed of the individual racecars. If you recall, you had a spring-loaded trigger on the hand control. That trigger was attached to the wiper of a rheostat or potentiometer (*pot*) so that you could control the amount of voltage applied to the racecar.

For instance, when you first turned the racecar set on, the hand controls would be in a position where the pot's wiper was at the bottom. Here, the pot transfers no voltage to the racecar and since the dc motor inside the racecar needs voltage to work, it's stopped. A good thing too! Could you imagine if the pot's wiper were at the top? Full voltage would be applied to the racecar at startup and you'd never get a chance to play with it because the racecar would just zoom off.

So, this is what happens: as you squeeze the trigger of the hand control, you move the wiper on the pot upward. When you release the trigger, the wiper moves back down the pot. By doing this repeatedly, you pick off a higher and lower voltage from the dc power supply, respectively, and apply it to the racecar motor. The higher the voltage to the racecar motor, the faster the racecar's speed. The lower the voltage to the racecar motor, the slower its speed.

Your friend's hand control works like yours except it's wired in parallel. This means that your friend can do the same thing you can and that's why you can race. That is, each of you can independently control your own racecar's speed.

By the way, did you notice the dc power supply connectors were polarized? That's because if you connected the dc power supply with reverse polarity then the dc motors inside the racecars would spin the opposite direction to normal. In other words, the racecars would go backwards!

Neat, eh? Okay, on your mark, get set, go!

PROBLEMS

SECTION 7.2 Descriptive Examples

1. Determine the total resistance for the networks in Fig. 7.49.

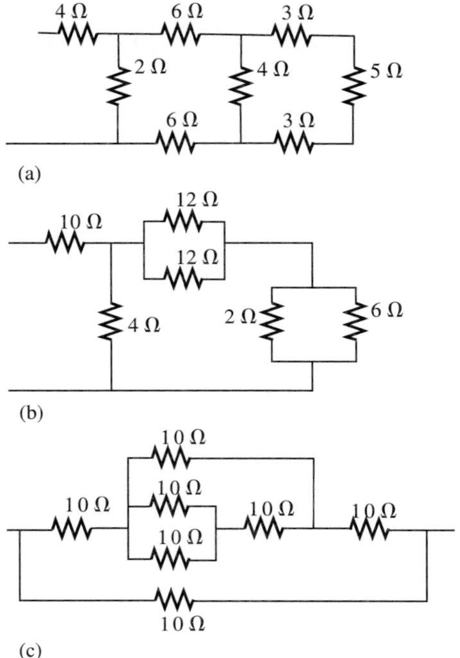

FIG. 7.49
Problem 1.

2. Which elements of the following networks (Fig. 7.50) are in series or parallel? In other words, which elements of the following networks have the same current (series) or voltage (parallel)? Restrict your decision to single elements and do not include combined elements.

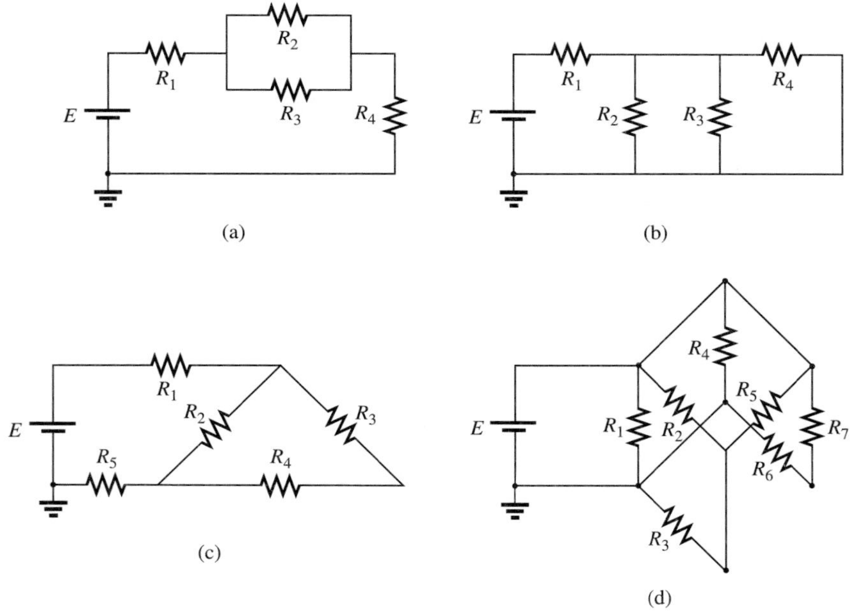

FIG. 7.50
Problem 2.

3. Determine the total resistance for the networks in Fig. 7.51.

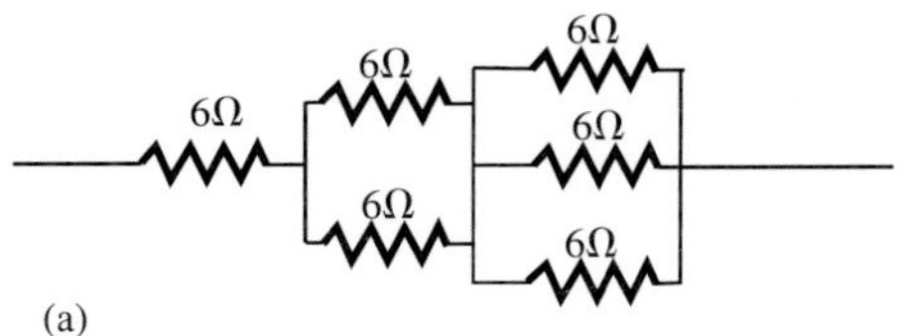

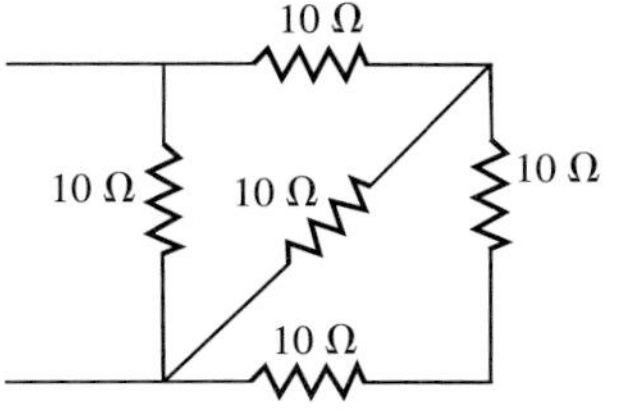

FIG. 7.51
Problem 3.

4. Determine R_T for the networks of Fig. 7.52.

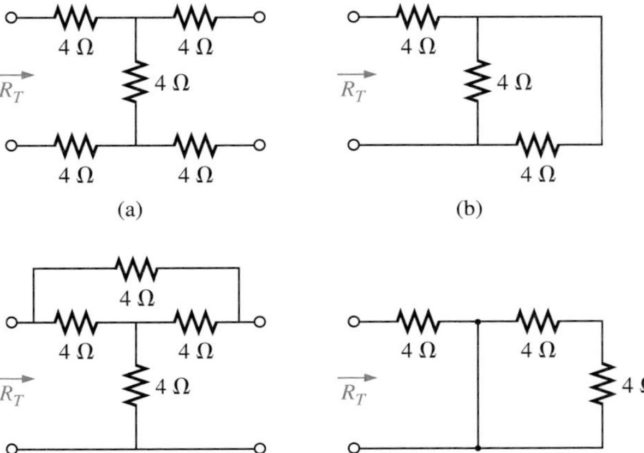

FIG. 7.52
Problem 4.

5. For the network of Fig. 7.53:
   a. Does $I = I_3 = I_6$? Explain.
   b. If $I = 5$ A and $I_1 = 2$ A, find I_2.
   c. Does $I_1 + I_2 = I_4 + I_5$? Explain.
   d. If $V_1 = 6$ V and $E = 10$ V, find V_2.
   e. If $R_1 = 3\ \Omega$, $R_2 = 2\ \Omega$, $R_3 = 4\ \Omega$, and $R_4 = 1\ \Omega$, what is R_T?
   f. If the resistors have the values given in part (e) and $E = 10$ V, what is the value of I in amperes?
   g. Using values given in parts (e) and (f), find the power delivered by the battery E and dissipated by the resistors R_1 and R_2.

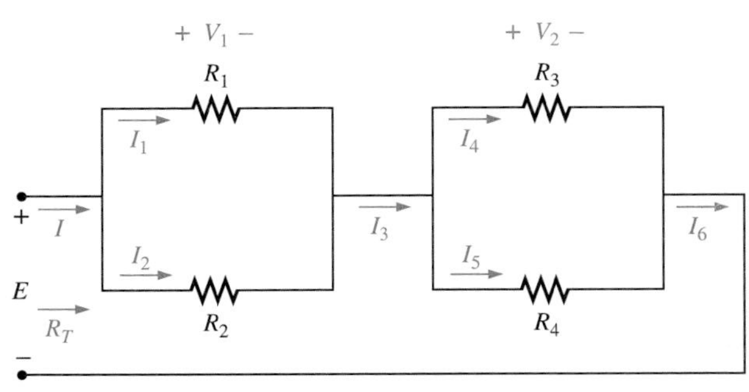

FIG. 7.53
Problem 5.

6. For the network of Fig. 7.54:
  a. Calculate R_T.
  b. Determine I and I_1.
  c. Find V_3.

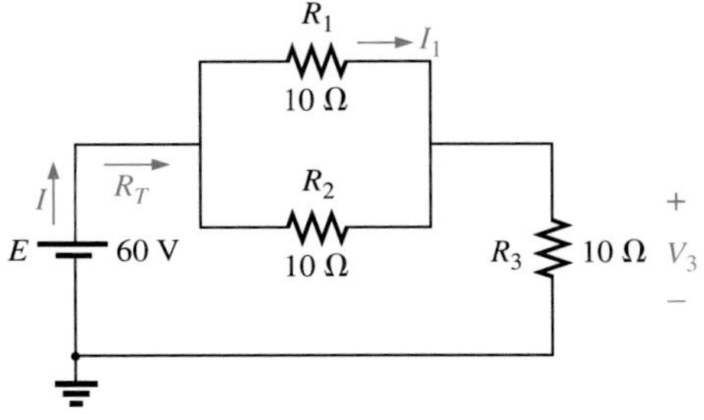

FIG. 7.54
Problem 6.

7. For the network of Fig. 7.55:
  a. Determine R_T.
  b. Find I_s, I_1, and I_2.
  c. Calculate V_a.

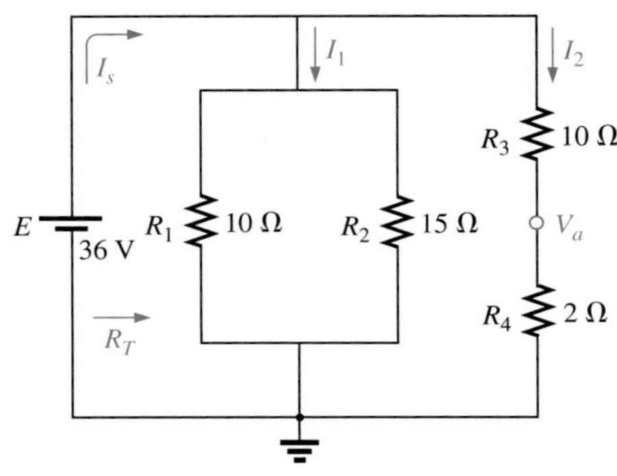

FIG. 7.55
Problem 7.

8. For the network in Fig. 7.56:
  a. Determine the total resistance.
  b. Calculate V_{ab}.
  c. Determine I.

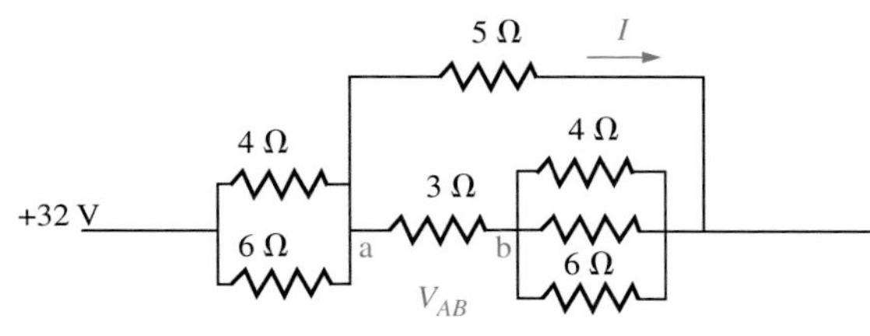

FIG. 7.56
Problem 8.

9. Determine the currents I_1 and I_2 for the network of Fig. 7.57.

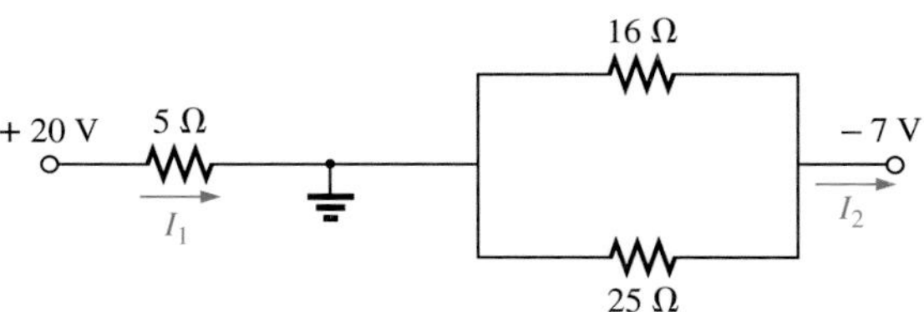

FIG. 7.57
Problem 9.

10. a. Find the magnitude and direction of the currents I, I_1, I_2, and I_3 for the network of Fig. 7.58.
b. Indicate their direction on Fig. 7.58.

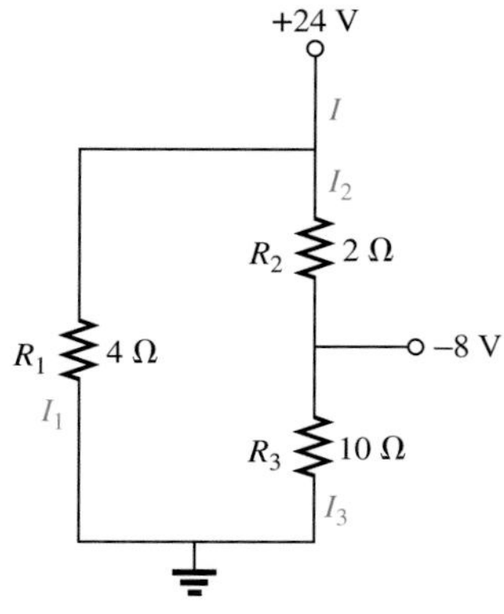

FIG. 7.58
Problem 10.

***11.** For the network of Fig. 7.59:
a. Determine the currents I_s, I_1, I_3, and I_4.
b. Calculate V_a and V_{bc}.

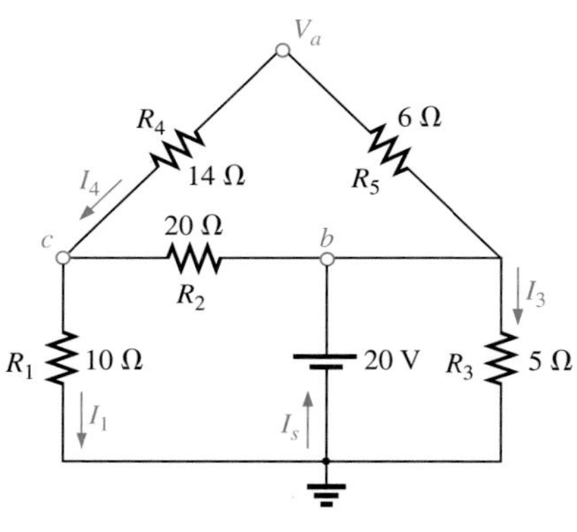

FIG. 7.59
Problem 11.

12. For the network of Fig. 7.60:
a. Determine the current I_1.
b. Calculate the currents I_2 and I_3.
c. Determine the voltage levels V_a and V_b.

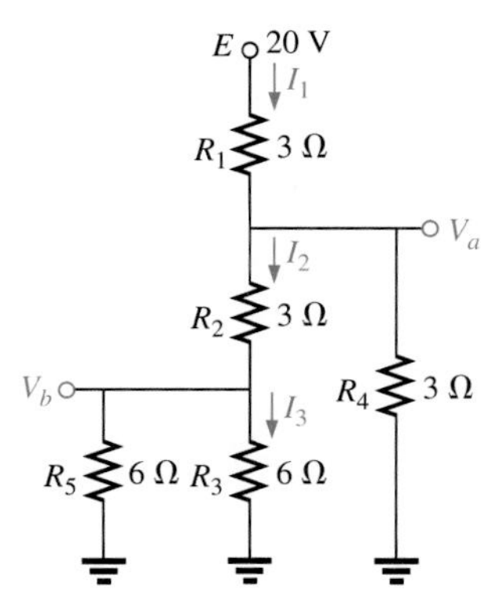

FIG. 7.60
Problem 12.

13. For the network of Fig. 7.61:
  a. Find the currents I and I_6.
  b. Find the voltages V_1 and V_5.
  c. Find the power delivered to the 6-kΩ resistor.

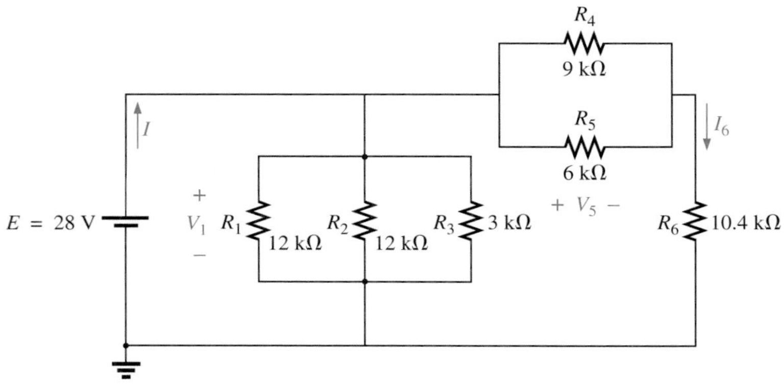

FIG. 7.61
Problem 13.

***14.** For the series-parallel network of Fig. 7.62:
  a. Find the current I.
  b. Find the currents I_3 and I_9.
  c. Find the current I_8.
  d. Find the voltage V_{ab}.

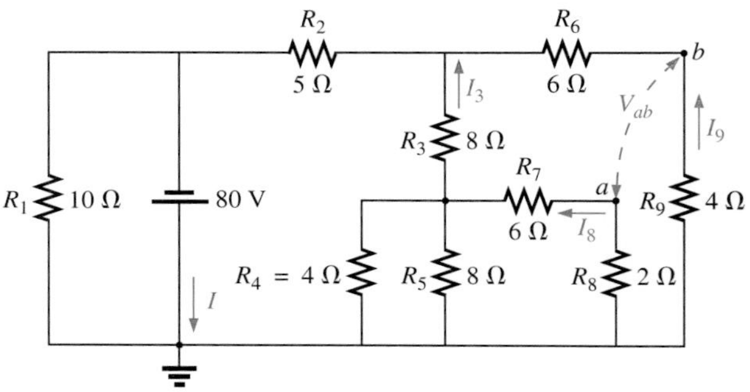

FIG. 7.62
Problem 14.

15. For the network in Fig. 7.63, find:
  a. V_{AB}.
  b. I and its direction.

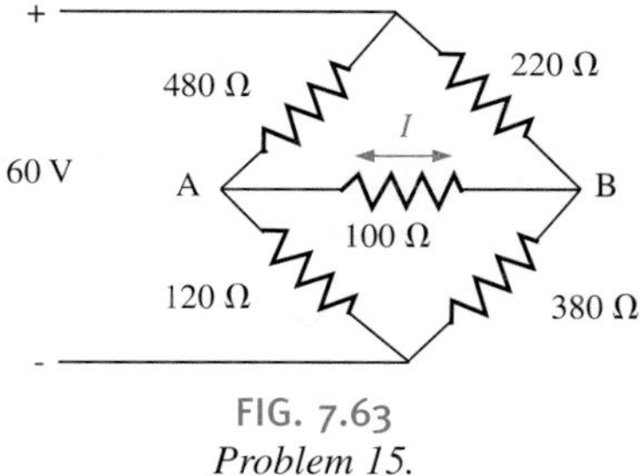

FIG. 7.63
Problem 15.

***16.** For the network of Fig. 7.64:
  a. Determine R_T.
  b. Calculate V_a.
  c. Find V_1.
  d. Calculate V_2.
  e. Determine I (with direction).

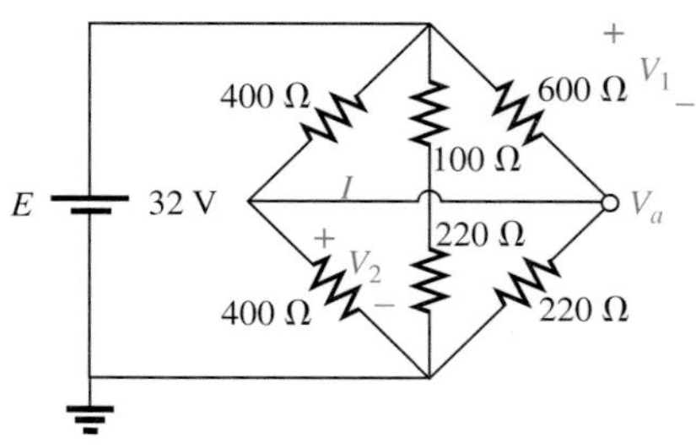

FIG. 7.64
Problem 16.

17. For the network of Fig. 7.65:
   a. Determine the current I.
   b. Find V.

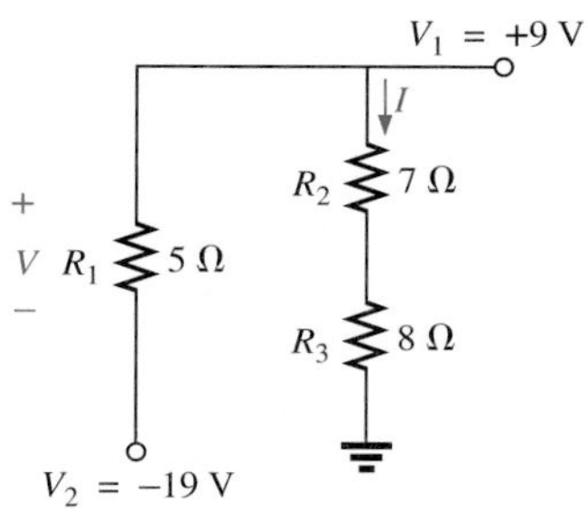

FIG. 7.65
Problem 17.

*18. Determine the current I and the voltages V_a, V_b, and V_{ab} for the network of Fig. 7.66.

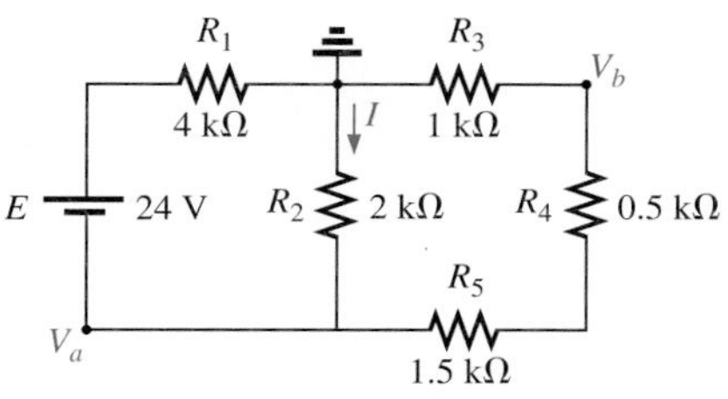

FIG. 7.66
Problem 18.

19. For the configuration of Fig. 7.67:
   a. Find the currents I_2, I_6, and I_8.
   b. Find the voltages V_4 and V_8.

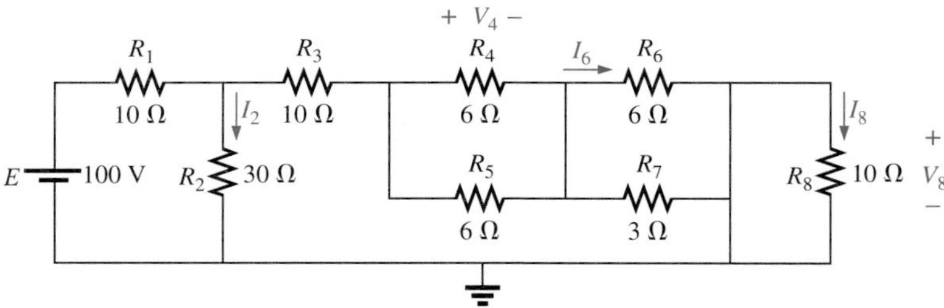

FIG. 7.67
Problem 19.

20. Determine the voltage V and the current I for the network of Fig. 7.68.

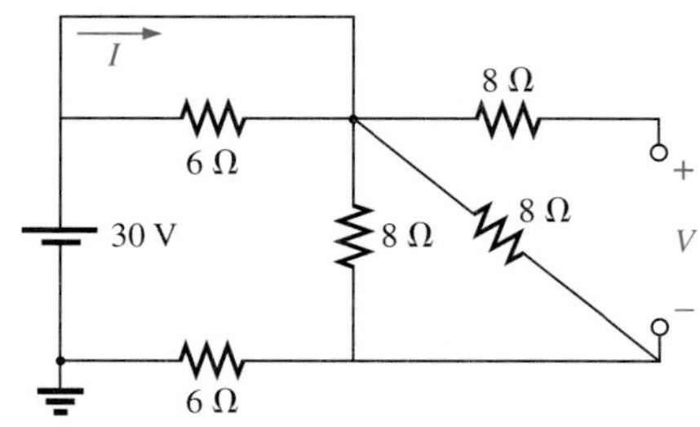

FIG. 7.68
Problem 20.

***21.** For the network of Fig. 7.69:

a. Determine R_T by combining resistive elements.

b. Find V_1 and V_4.

c. Calculate I_3 (with direction).

d. Determine I_s by finding the current through each element and then applying Kirchhoff's current law. Then calculate R_T from $R_T = E/I_s$ and compare with the solution of part (a).

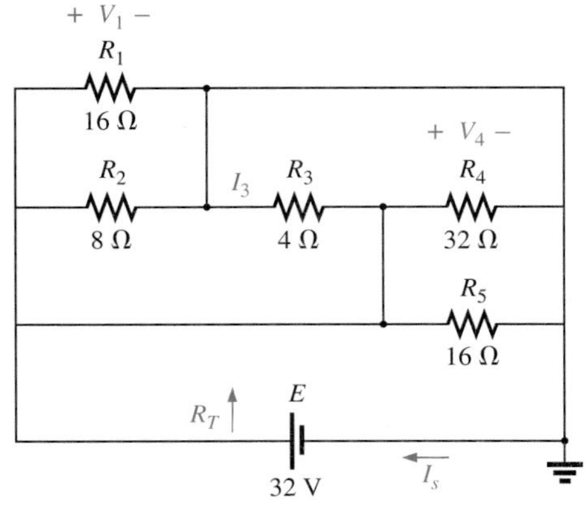

FIG. 7.69
Problem 21.

22. For the network of Fig. 7.70:

a. Determine the voltage V_{ab}. (*Hint:* Just use Kirchhoff's voltage law.)

b. Calculate the current I.

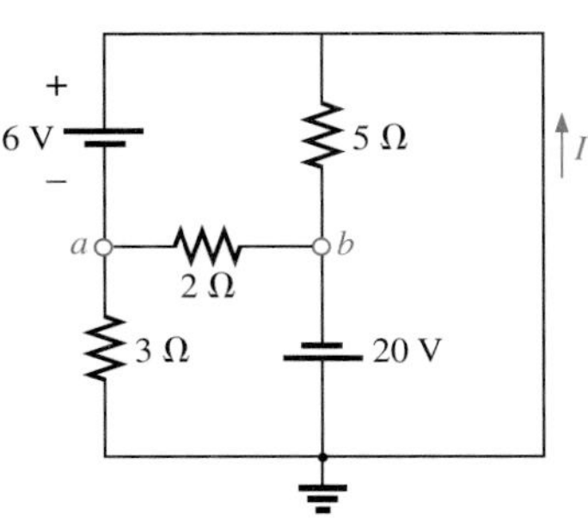

FIG. 7.70
Problem 22.

***23.** For the network of Fig. 7.71:

a. Determine the current I.

b. Calculate the open-circuit voltage V.

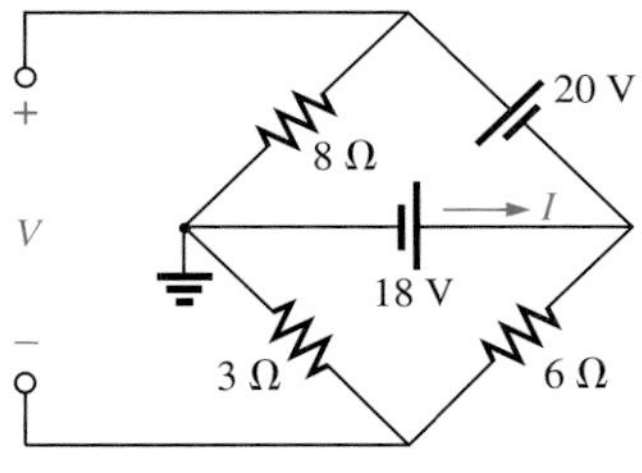

FIG. 7.71
Problem 23.

***24.** For the network of Fig. 7.72, find the resistance R_3 if the current through it is 2 A.

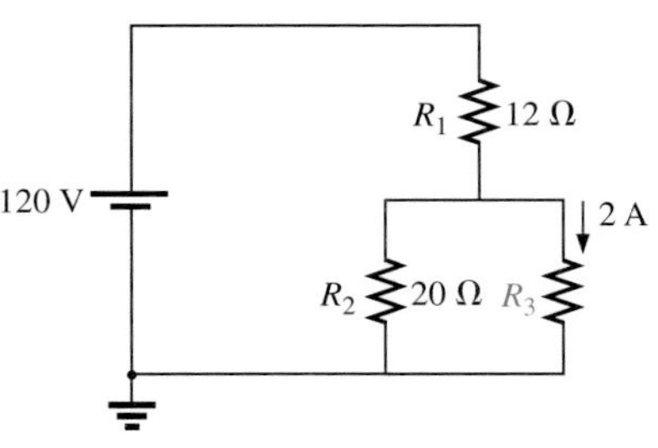

FIG. 7.72
Problem 24.

***25.** If all the resistors of the cube (Fig. 7.73) are 10 Ω, what is the total resistance? (*Hint:* Make some basic assumptions about current division through the cube.)

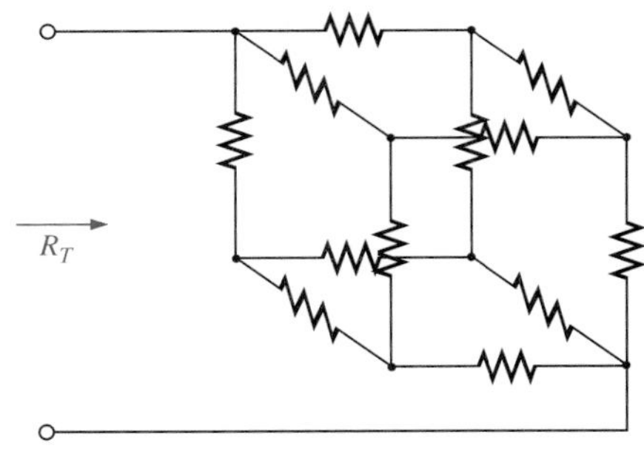

FIG. 7.73
Problem 25.

***26.** Given the voltmeter reading $V = 27$ V in Fig. 7.74:

a. Is the network operating properly?

b. If not, what could be the cause of the incorrect reading?

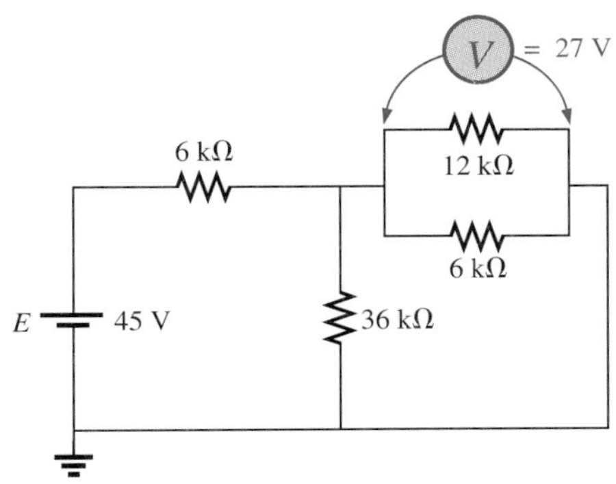

FIG. 7.74
Problem 26.

SECTION 7.3 Ladder Networks

27. For the ladder network of Fig. 7.75:

a. Find the current I.

b. Find the current I_7.

c. Determine the voltages V_3, V_5, and V_7.

d. Calculate the power delivered to R_7 and compare it to the power delivered by the 240-V supply.

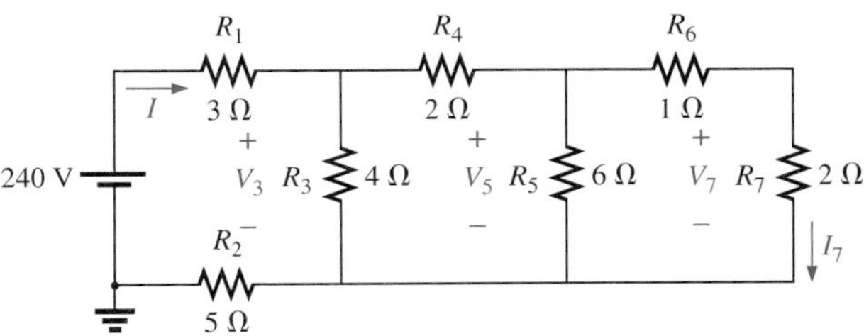

FIG. 7.75
Problem 27.

28. For the ladder network of Fig. 7.76:

a. Determine R_T.

b. Calculate I.

c. Find I_8.

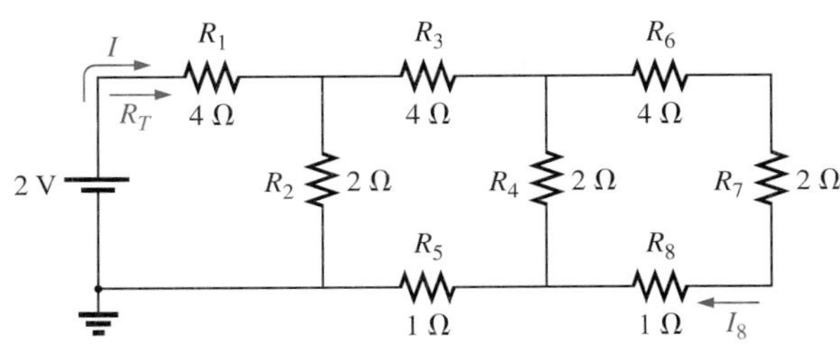

FIG. 7.76
Problem 28.

***29.** Determine the power delivered to the 10-Ω load of Fig. 7.77.

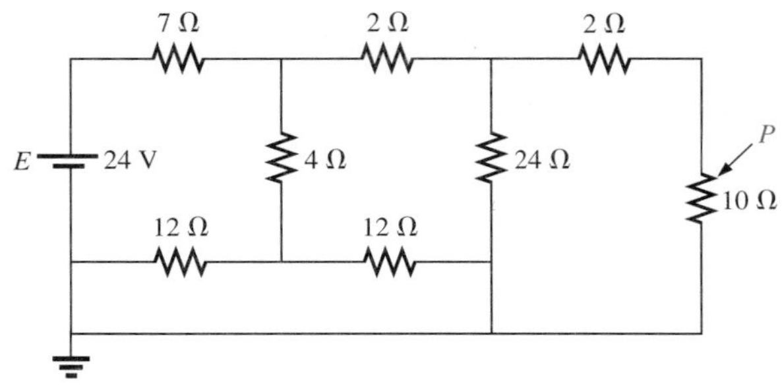

FIG. 7.77
Problem 29.

***30.** For the multiple ladder configuration of Fig. 7.78:

a. Determine I.
b. Calculate I_4.
c. Find I_6.
d. Find I_{10}.

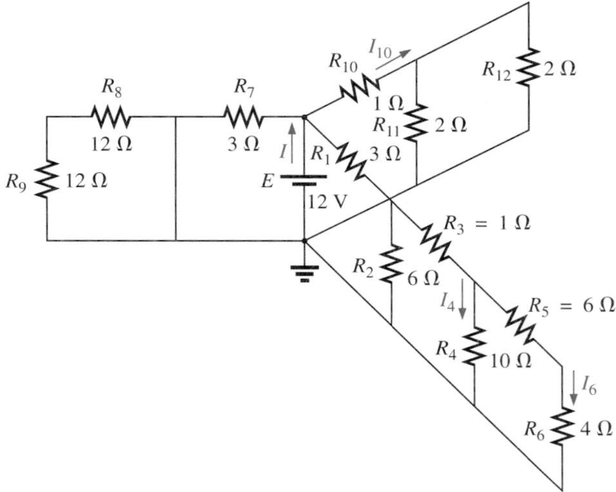

FIG. 7.78
Problem 30.

31. Determine the resistance for V_1, V_2 and V_3 for Fig. 7.79.

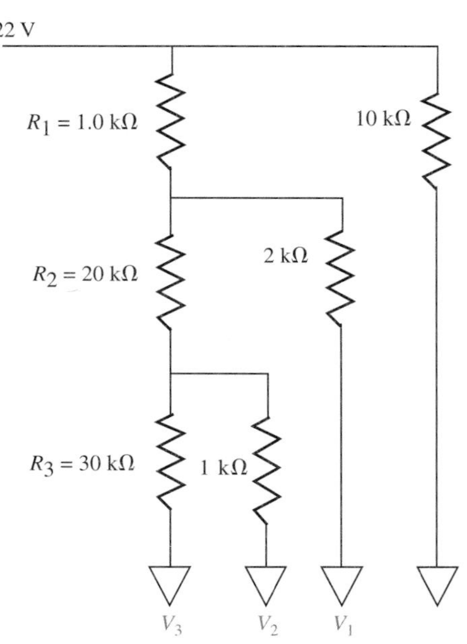

FIG. 7.79
Problem 31.

SECTION 7.4 Voltage Divider Supply (Unloaded and Loaded)

32. Given the voltage divider supply of Fig. 7.80:
a. Determine the supply voltage E.
b. Find the load resistors R_{L_2} ad R_{L_3}.
c. Determine the voltage divider resistors R_1, R_2, and R_3.

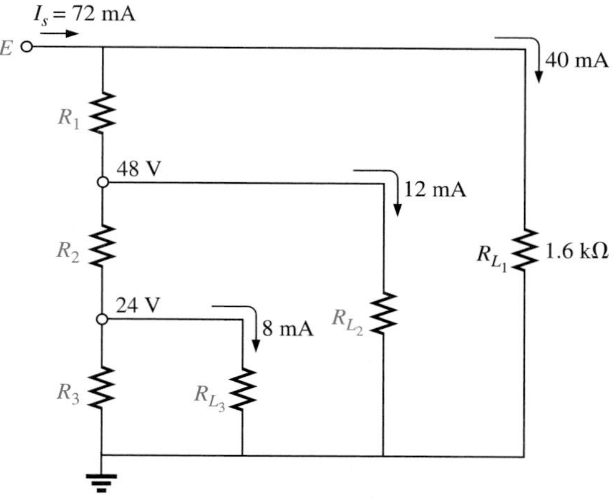

FIG. 7.80
Problem 32.

***33.** Determine the voltage divider supply resistors for the configuration of Fig. 7.81. Also determine the required wattage rating for each resistor and compare their levels.

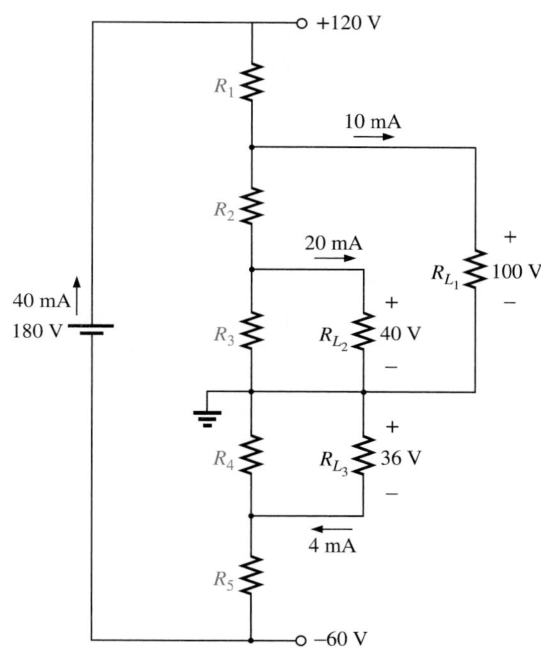

FIG. 7.81
Problem 33.

SECTION 7.5 Potentiometer Loading

***34.** For the system of Fig. 7.82:
a. At first exposure, does the design appear to be a good one?
b. In the absence of the 10-kΩ load, what are the values of R_1 and R_2 to establish 3 V across R_2?
c. Determine the values of R_1 and R_2 when the load is applied and compare to the results of part (b).

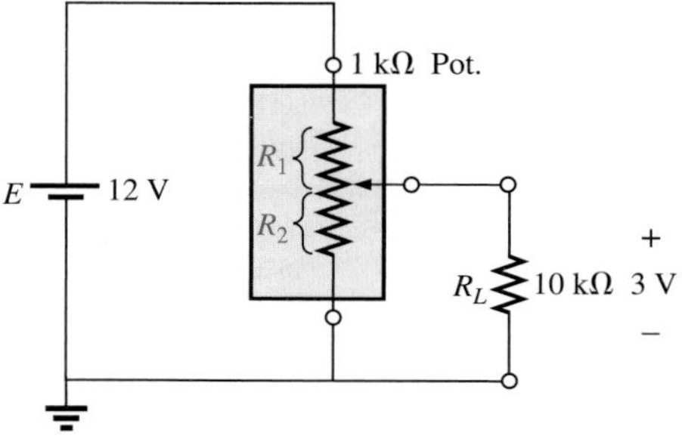

FIG. 7.82
Problem 34.

*35. For the potentiometer of Fig. 7.83:
   a. What are the voltages V_{ab} and V_{bc} with no load applied?
   b. What are the voltages V_{ab} and V_{bc} with the indicated loads applied?
   c. What is the power dissipated by the potentiometer under the loaded conditions of Fig. 7.83?
   d. What is the power dissipated by the potentiometer with no loads applied? Compare to the results of part (c).

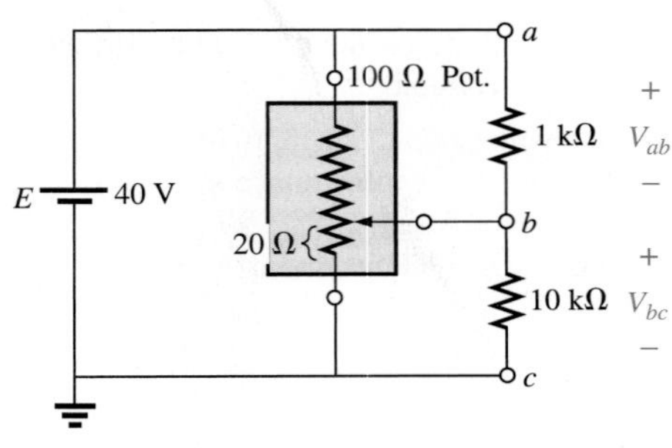

FIG. 7.83
Problem 35.

SECTION 7.6 Ammeter, Voltmeter, and Ohmmeter Design

36. A d'Arsonval movement is rated 1 mA, 100 Ω.
   a. What is the current sensitivity?
   b. Design a 20-A ammeter using the above movement. Show the circuit and component values.

37. Using a 50-μA, 1000-Ω d'Arsonval movement, design a multirange milliammeter having scales of 25, 50, and 100 mA. Show the circuit and component values.

38. Design an ammeter with an Ayrton shunt using a 1 mA, 100 Ω d'Arsonval meter movement. The ammeter is to have three ranges: 1A, 5 A, and 10 A.

39. Design an ammeter with an Ayrton shunt using a 2 mA, 30 Ω d'Arsonval meter movement, for ranges of 10 mA, 30 mA, and 100 mA.

40. A d'Arsonval movement is rated 50 μA, 1000 Ω.
   a. Design a 15-V dc voltmeter. Show the circuit and component values.
   b. What is the ohm/volt rating of the voltmeter?

41. Using a 1-mA, 100-Ω d'Arsonval movement, design a multirange voltmeter having scales of 5, 50, and 500 V. Show the circuit and component values.

42. A digital meter has an internal resistance of 10 MΩ on its 0.5-V range. If you had to build a voltmeter with a d'Arsonval movement, what current sensitivity would you need if the meter were to have the same internal resistance on the same voltage scale?

***43.** **a.** Design a series ohmmeter using a 100-μA, 1000-Ω movement; a zero-adjust with a maximum value of 2 kΩ; a battery of 3 V; and a series resistor whose value is to be determined.
   b. Find the resistance required for full-scale, 3/4-scale, 1/2-scale, and 1/4-scale deflection.
   c. Using the results of part (b), draw the scale to be used with the ohmmeter.

44. Describe the basic construction and operation of the megohmmeter.

8 Methods of Analysis and Selected Topics (dc)

Outline

Learning Outcomes

After completing this chapter you will be able to

- explain the nature of constant current sources
- convert constant voltage sources to constant current sources
- explain the effect of connecting current sources in series and parallel
- solve network problems involving multiple sources using branch-current analysis
- solve network problems involving multiple sources using mesh analysis
- solve network problems involving multiple sources using nodal analysis
- solve bridge network problems using mesh equations and nodal analysis
- solve network problems which are series or parallel connections using Δ-Y and Y-Δ conversions

Using the analytic techniques in this chapter, you will be able to simplify and describe the operation of very complex electrical systems.

8.1 INTRODUCTION

The circuits described in the previous chapters had only one source or two or more sources in series or parallel present. The step-by-step procedure outlined in those chapters cannot be applied if the sources are not in series or parallel. There will be an *interaction of sources* that will not permit the reduction technique used in Chapter 7 to find quantities such as the total resistance and source current.

Methods of analysis have been developed that allow us to approach, in a systematic manner, a network with any number of sources in any arrangement. Fortunately, these methods can also be applied to networks with only one source. The methods to be discussed in detail in this chapter include **branch-current analysis, mesh analysis,** and **nodal analysis**. Each can be applied to the same network. The best method cannot be defined by a set of rules. It can be determined only by acquiring a firm understanding of the relative advantages of each method. All the methods can be applied to *linear bilateral* networks. The term **linear** indicates that the characteristics of the network elements (such as the resistors) are independent of the voltage across or current through them. The second term, **bilateral**, refers to the fact that there is no change in the behaviour or characteristics of an element if the current through or voltage across the element is reversed. Of the three methods listed above, the branch-current method is the only one not restricted to bilateral devices. Before discussing the methods in detail, we will consider the current source and conversions between voltage and current sources. At the end of the chapter we will consider bridge networks and Δ-Y and Y-Δ conversions. Chapter 9 will present the important theorems of network analysis that can also be used to solve networks with more than one source.

8.2 CURRENT SOURCES

The concept of the current source was introduced in Section 2.5 with the photograph of a commercially available unit. We must now investigate its characteristics in greater detail so that we can properly determine its effect on the networks to be examined in this chapter.

The current source is often referred to as the **dual** of the voltage source. A battery supplies a *fixed* voltage and the source current can vary, but the current source supplies a *fixed* current to the branch in which it is located, while its terminal voltage may vary as determined by the network to which it is applied. Note from the above that **duality** simply means an interchange of current and voltage to distinguish the characteristics of one source from the other.

The interest in the current source is due mainly to semiconductor devices such as the transistor. In basic electronics courses, you will find that the transistor is a current-controlled device. In the physical model (equivalent circuit) of a transistor used in the analysis of transistor networks, there is a current source as indicated in Fig. 8.1. The symbol for a current source appears in Fig. 8.1. The direction of the arrow within the circle indicates the direction current is being supplied.

For further comparison, the terminal characteristics of an **ideal dc** voltage and current source are presented in Fig. 8.2. The term *ideal* refers to perfect sources, or no internal losses sensitive to the demand from the applied load. Note that for the voltage source, the terminal voltage is fixed at E volts independent of the direction of the current I. The direction and magnitude of I will be determined by the network to which the supply is connected.

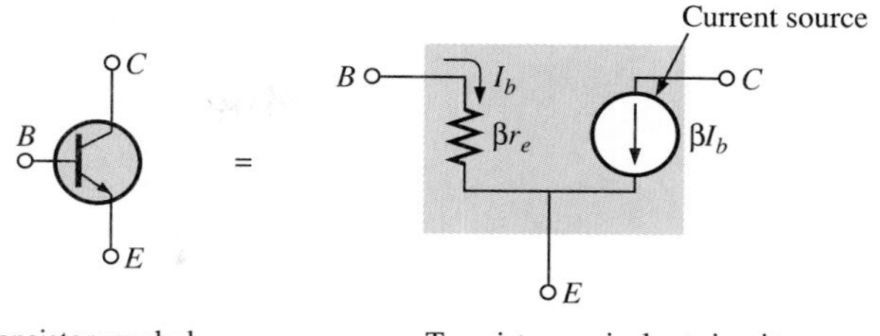

FIG. 8.1
Current source within the transistor equivalent circuit.

The characteristics of the ideal source, shown in Fig. 8.2(b), show that the magnitude of the supply current is independent of the polarity of the voltage across the source. The polarity and magnitude of the source voltage V_s will be determined by the network to which the source is connected.

For all one-voltage-source networks the current will have the direction indicated to the right of the battery in Fig. 8.2(a). For all single-current-source networks, it will have the polarity indicated to the right of the current source in Fig. 8.2(b).

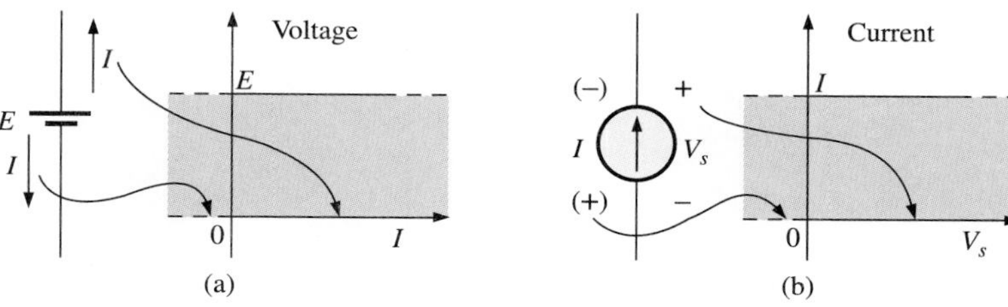

FIG. 8.2
Comparing the characteristics of an ideal voltage and current source.

In review:

A current source determines the current in the branch in which it is located

and

the magnitude and polarity of the voltage across a current source are a function of the network to which it is applied.

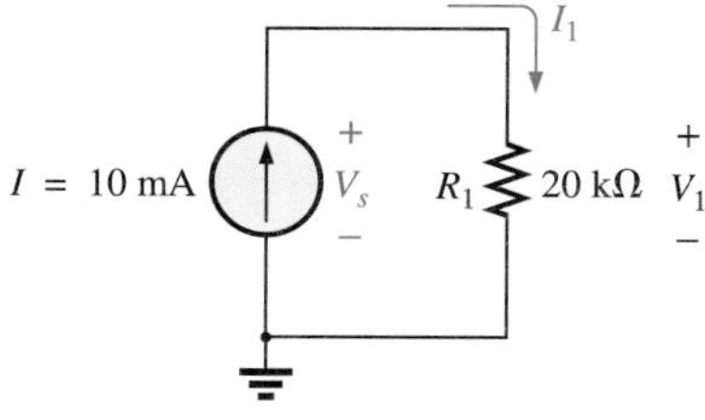

FIG. 8.3
Example 8.1.

EXAMPLE 8.1

Find the source voltage V_s and the current I_1 for the circuit of Fig. 8.3.

Solution:

$$I_1 = I = \mathbf{10\ mA}$$

$$V_s = V_1 = I_1R_1 = (10\ \text{mA})(20\ \text{k}\Omega) = \mathbf{200\ V}$$

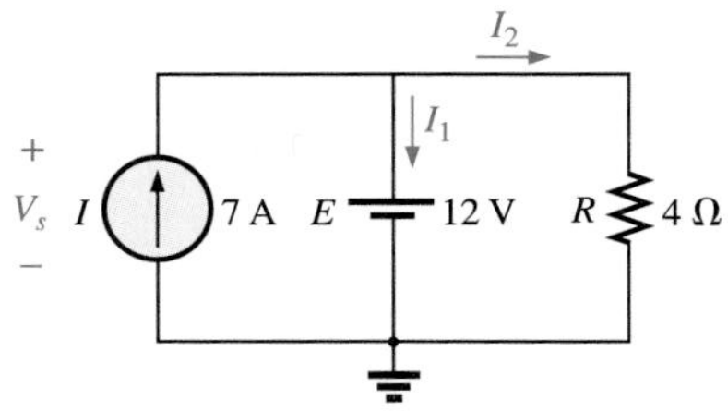

FIG. 8.4
Example 8.2.

EXAMPLE 8.2

Find the voltage V_s and the currents I_1 and I_2 for the network of Fig. 8.4.

Solution:

$$V_s = E = \mathbf{12\ V}$$

$$I_2 = \frac{V_R}{R} = \frac{E}{R} = \frac{12\ \text{V}}{4\ \Omega} = \mathbf{3\ A}$$

Applying Kirchhoff's current law:

$$I = I_1 + I_2$$

and
$$I_1 = I - I_2 = 7\ \text{A} - 3\ \text{A} = \mathbf{4\ A}$$

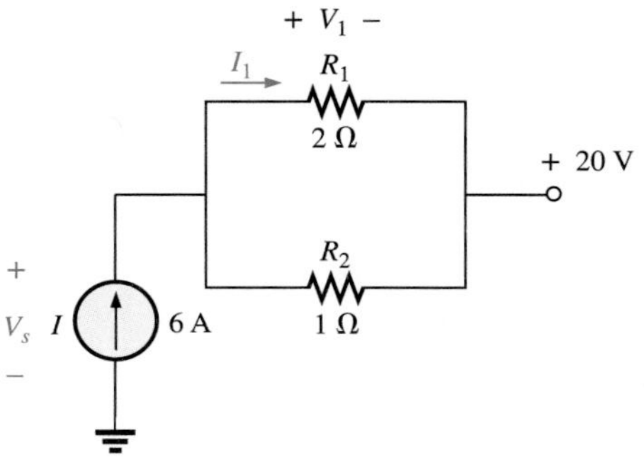

FIG. 8.5
Example 8.3.

EXAMPLE 8.3

Determine the current I_1 and the voltage V_s for the network of Fig. 8.5.

Solution: Using the current divider rule:

$$I_1 = \frac{R_2 I}{R_2 + R_1} = \frac{(1\ \Omega)(6\ \text{A})}{1\ \Omega + 2\ \Omega} = \mathbf{2\ A}$$

The voltage V_1 is

$$V_1 = I_1 R_1 = (2\ \text{A})(2\ \Omega) = 4\ \text{V}$$

and, applying Kirchhoff's voltage law,

$$+V_s - V_1 - 20\ \text{V} = 0$$

and
$$V_s = V_1 + 20\ \text{V} = 4\ \text{V} + 20\ \text{V} = \mathbf{24\ V}$$

Note the polarity of V_s as determined by the multisource network.

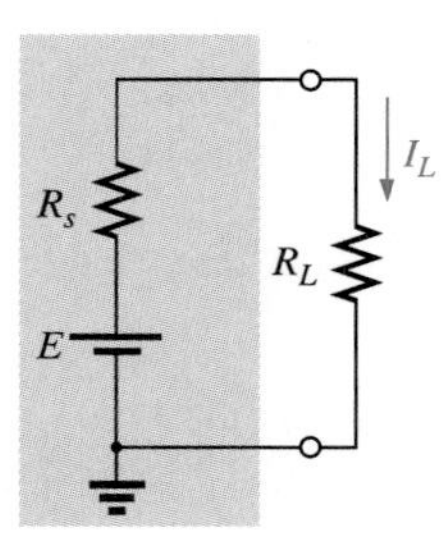

FIG. 8.6
Practical voltage source.

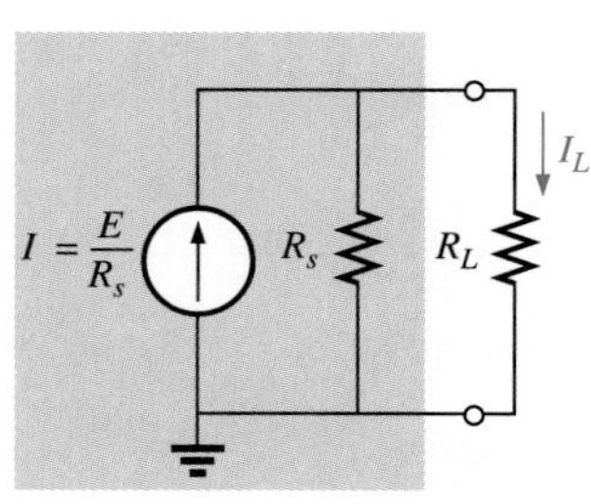

FIG. 8.7
Practical current source.

8.3 SOURCE CONVERSIONS

The current source described in the previous section is called an **ideal source** because there is no internal resistance. In reality, all sources—voltage or current—have some internal resistance in the relative positions shown in Figs. 8.6 and 8.7. For the voltage source, if $R_s = 0\ \Omega$ or is so small compared to any series resistor that it can be ignored, then we have an "ideal" voltage source. For the current source, if $R_s = \infty\ \Omega$ or is large enough compared to other parallel elements that it can be ignored, then we have an "ideal" current source.

If the internal resistance is included with either source, then that source can be converted to the other type using the procedure to be described in this section. Since it is often a good idea to do this, we'll focus this entire section on making sure the steps are understood. It is important to realize, however, as we proceed through this section that

source conversions are equivalent only at their external terminals.

The internal characteristics of each are quite different.

We want the equivalence to ensure that the applied load of Figs. 8.6 and 8.7 will receive the same current, voltage, and power from each source and in effect not know, or care, which source is present.

In Fig. 8.6 if we solve for the load current I_L, we get

$$I_L = \frac{E}{R_s + R_L} \tag{8.1}$$

If we multiply this by a factor of 1, which we can choose to be R_s/R_s, we get

$$I_L = \frac{(1)E}{R_s + R_L} = \frac{(R_s/R_s)E}{R_s + R_L} = \frac{R_s(E/R_s)}{R_s + R_L} = \frac{R_s I}{R_s + R_L} \tag{8.2}$$

If we define $I = E/R_s$, Eq. (8.2) is the same as that obtained by applying the current divider rule to the network of Fig. 8.7. The result is an equivalence between the networks of Fig. 8.6 and Fig. 8.7 that simply requires that $I = E/R_s$ and the series resistor R_s of Fig. 8.6 be placed in parallel, as in Fig. 8.7. The validity of this is demonstrated in the first example of this section.

For clarity, the equivalent sources, *as far as terminals* a *and* b *are concerned,* are repeated in Fig. 8.8 with the equations for converting in either direction. Note, as we just discussed, that the resistor R_s is the same in each source; only its position changes. The current of the current source or the voltage of the voltage source is determined using Ohm's law and the parameters of the other configuration. We showed in detail in Chapter 6 that every source of voltage has some internal series resistance. *For the current source, some internal parallel resistance will always exist in the practical world.* However, in many cases, it is an excellent approximation to drop the internal resistance of a source due to the magnitude of the elements of the network to which it is applied. For this reason, in the analyses to follow, voltage sources may appear without a series resistor, and current sources may appear without a parallel resistance. Realize, however, that to perform a conversion from one type of source to another, a voltage source must have a resistor in series with it, and a current source must have a resistor in parallel.

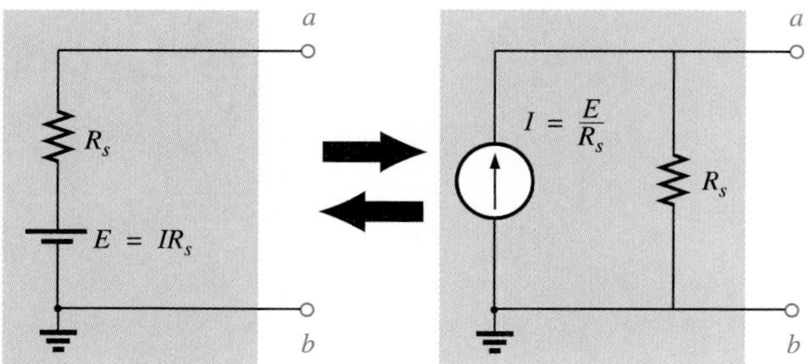

FIG. 8.8
Source conversion.

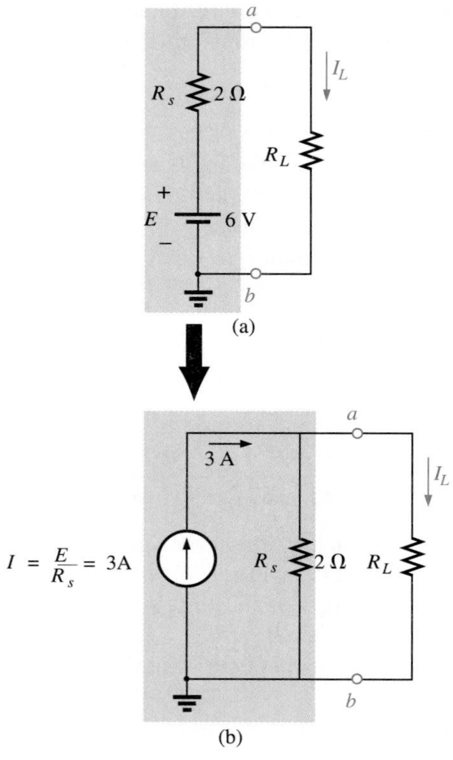

FIG. 8.9
Example 8.4.

EXAMPLE 8.4

a. Convert the voltage source of Fig. 8.9(a) to a current source and calculate the current through the 4-Ω (R_L) load for each source.
b. Replace the 4-Ω load with a 1-kΩ load and calculate the current I_L for the voltage source.
c. Repeat the calculation of part (b) assuming the voltage source is ideal ($R_s = 0\ \Omega$) because R_L is so much larger than R_s. Is this one of those situations where assuming the source is ideal is an appropriate approximation?

Solution:

a. See Fig. 8.9.

$$\text{Fig. 8.9(a): } I_L = \frac{E}{R_s + R_L} = \frac{6\text{ V}}{2\ \Omega + 4\ \Omega} = \mathbf{1\ A}$$

$$\text{Fig. 8.9(b): } I_L = \frac{R_s I}{R_s + R_L} = \frac{(2\ \Omega)(3\text{ A})}{2\ \Omega + 4\ \Omega} = \mathbf{1\ A}$$

b. $I_L = \dfrac{E}{R_s + R_L} = \dfrac{6\text{ V}}{2\ \Omega + 1\text{ k}\Omega} \cong \mathbf{5.99\ mA}$

c. $I_L = \dfrac{E}{R_L} = \dfrac{6\text{ V}}{1\text{ k}\Omega} = \mathbf{6\ mA} \cong 5.99\text{ mA}$

Yes, $R_L \gg R_s$ (voltage source).

EXAMPLE 8.5

a. Convert the current source of Fig. 8.10(a) to a voltage source and find the load current for each source.
b. Replace the 6-kΩ load with a 10-Ω load and calculate the current I_L for the current source.
c. Repeat the calculation of part (b) assuming the current source is ideal ($R_s = \infty\ \Omega$) because R_L is so much smaller than R_s. Is this one of those situations where assuming the source is ideal is an appropriate approximation?

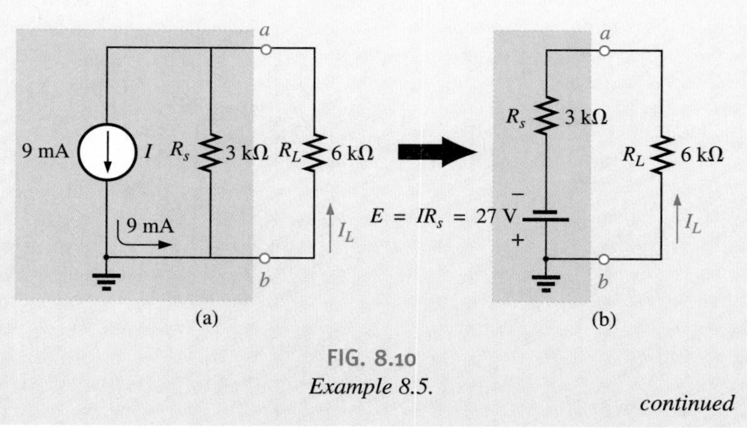

FIG. 8.10
Example 8.5.

continued

Solution:

a. See Fig. 8.10.

Fig. 8.10(a): $I_L = \dfrac{R_s I}{R_s + R_L} = \dfrac{(3\text{ k}\Omega)(9\text{ mA})}{3\text{ k}\Omega + 6\text{ k}\Omega} = \mathbf{3\text{ mA}}$

Fig. 8.10(b): $I_L = \dfrac{E}{R_s + R_L} = \dfrac{27\text{ V}}{3\text{ k}\Omega + 6\text{ k}\Omega} = \dfrac{27\text{ V}}{9\text{ k}\Omega} = \mathbf{3\text{ mA}}$

b. $I_L = \dfrac{R_s I}{R_s + R_L} = \dfrac{(3\text{ k}\Omega)(9\text{ mA})}{3\text{ k}\Omega + 10\ \Omega} = \mathbf{8.97\text{ mA}}$

c. $I_L = I = \mathbf{9\text{ mA}} \cong 8.97\text{ mA}$

Yes, $R_s \gg R_L$ (current source).

8.4 CURRENT SOURCES IN PARALLEL

If two or more current sources are in parallel, they may all be replaced by one current source having the magnitude and direction of the resultant. This can be found by summing the currents in one direction and subtracting the sum of the currents in the opposite direction. The new parallel resistance is determined by methods described in the discussion of parallel resistors in Chapter 5. Consider the following examples.

EXAMPLE 8.6

Reduce the parallel current sources of Figs. 8.11 and 8.12 to a single current source.

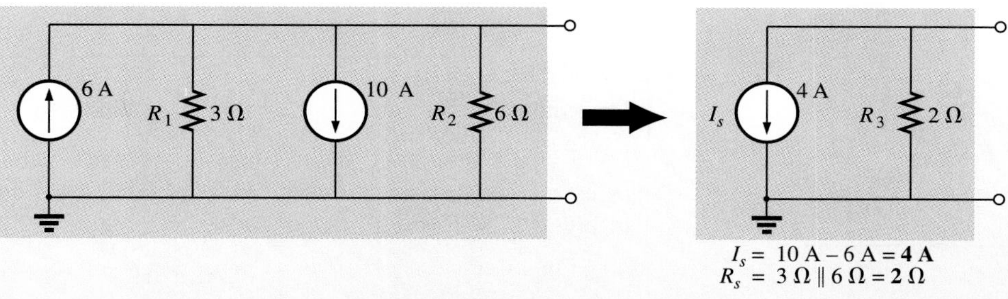

FIG. 8.11
Example 8.6.

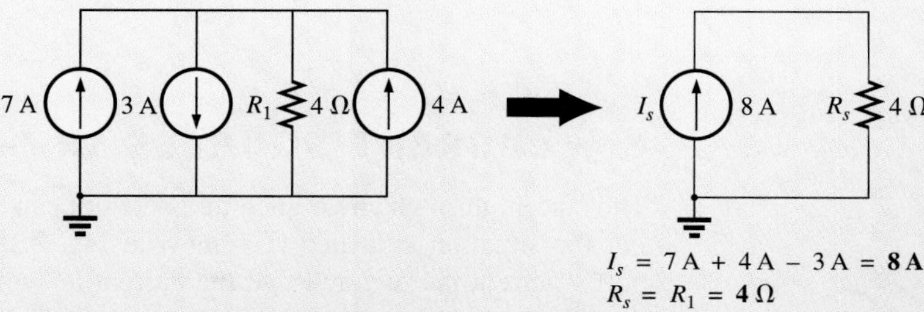

FIG. 8.12
Example 8.6.

Solution: Note the solution in each figure.

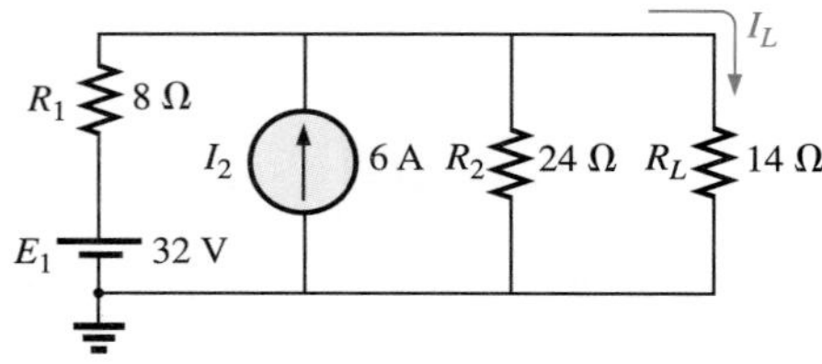

FIG. 8.13
Example 8.7.

EXAMPLE 8.7

Reduce the network of Fig. 8.13 to a single current source and calculate the current through R_L.

Solution: In this example, the voltage source will first be converted to a current source as shown in Fig. 8.14. Combining current sources,

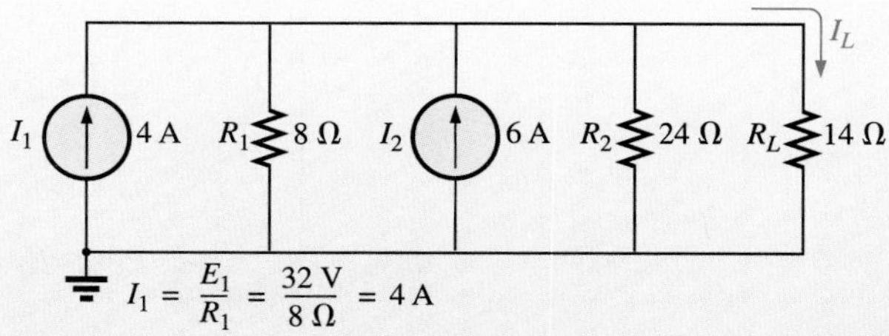

FIG. 8.14
Network of Fig. 8.13 following the conversion of the voltage source to a current source.

$$I_s = I_1 + I_2 = 4\text{ A} + 6\text{ A} = \mathbf{10\text{ A}}$$

and

$$R_s = R_1 \parallel R_2 = 8\text{ }\Omega \parallel 24\text{ }\Omega = \mathbf{6\text{ }\Omega}$$

Applying the current divider rule to the resulting network of Fig. 8.15,

$$I_L = \frac{R_s I_s}{R_s + R_L} = \frac{(6\text{ }\Omega)(10\text{ A})}{6\text{ }\Omega + 14\text{ }\Omega} = \frac{60\text{ A}}{20} = \mathbf{3\text{ A}}$$

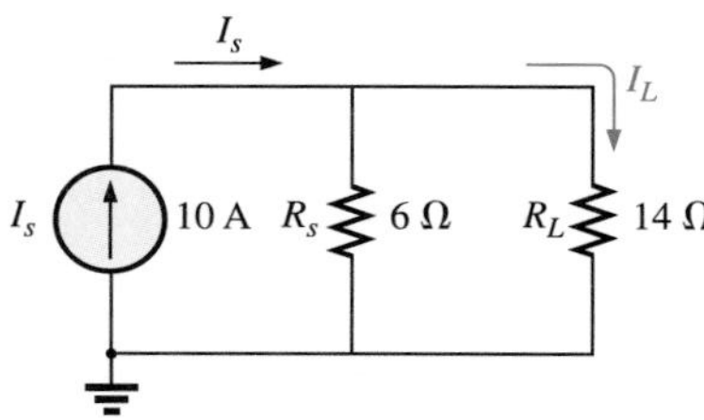

FIG. 8.15
Network of Fig. 8.14 reduced to its simplest form.

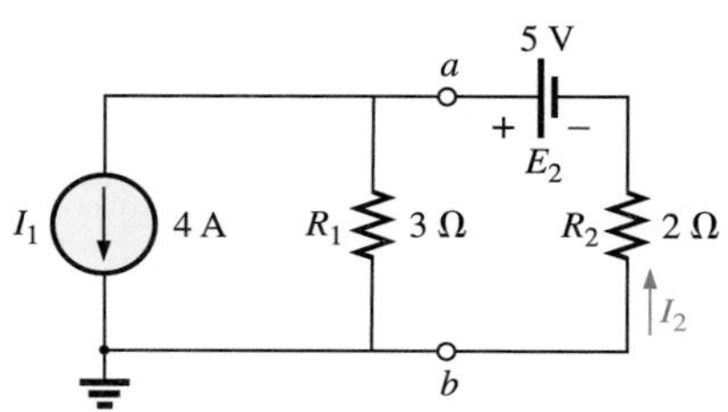

FIG. 8.16
Example 8.8.

EXAMPLE 8.8

Determine the current I_2 in the network of Fig. 8.16.

Solution: Although it seems that the network cannot be solved using methods we have introduced, one source conversion as shown in Fig. 8.17 will result in a simple series circuit:

$$E_s = I_1 R_1 = (4\text{ A})(3\text{ }\Omega) = 12\text{ V}$$

and

$$R_s = R_1 = 3\text{ }\Omega$$

and

$$I_2 = \frac{E_s + E_2}{R_s + R_2} = \frac{12\text{ V} + 5\text{ V}}{3\text{ }\Omega + 2\text{ }\Omega} = \frac{17\text{ V}}{5\text{ }\Omega} = \mathbf{3.4\text{ A}}$$

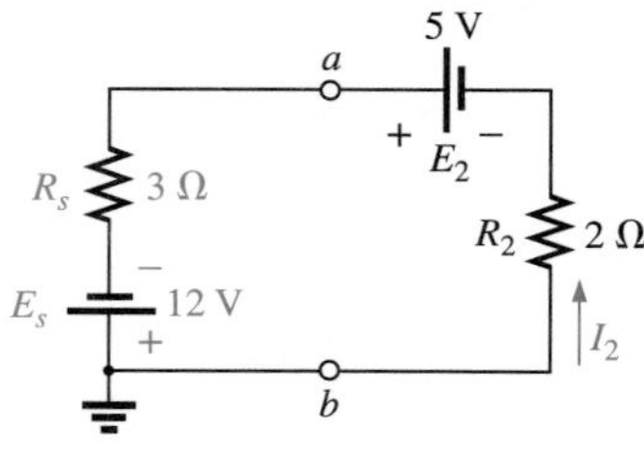

FIG. 8.17
Network of Fig. 8.16 following the conversion of the current source to a voltage source.

8.5 CURRENT SOURCES IN SERIES

The current through any branch of a network can be only single-valued. For the situation indicated at point *a* in Fig. 8.18, if we apply Kirchhoff's current law we find that the current leaving that point is greater than that entering—an impossible situation. Therefore,

current sources of different current ratings are not connected in series,

just as voltage sources of different voltage ratings are not connected in parallel.

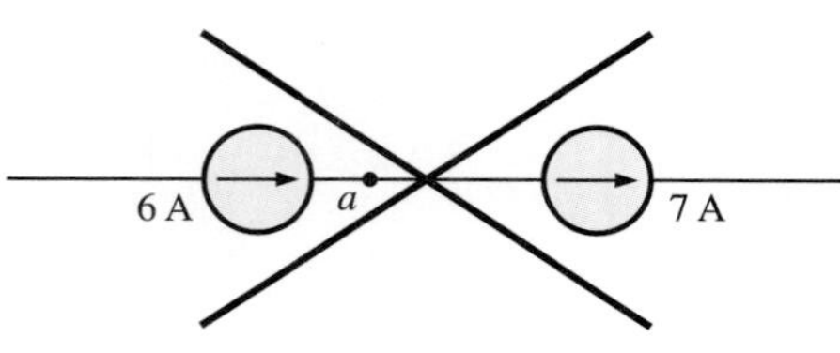

FIG. 8.18
Invalid situation.

8.6 BRANCH-CURRENT ANALYSIS

We will now consider the first in a series of methods for solving networks with two or more sources. Once you understand this method, there is no linear dc network for which you cannot find a solution. Keep in mind that networks with two isolated voltage sources cannot be solved using the approach of Chapter 7. For additional evidence of this fact, try solving for the unknown elements of Example 8.9 using the methods introduced in Chapter 7. The network of Fig. 8.21 can be solved using the source conversions described in the last section, but the method to be described in this section has applications far beyond the configuration of this network. The most direct way to introduce a method of this type is to list the series of steps required to apply it. There are four steps, as indicated below. Before continuing, understand that this method will produce the current through each branch of the network, the **branch current**. Once this is known, all other quantities, such as voltage or power, can be determined.

1. Assign a distinct current of arbitrary direction to each branch of the network.
2. Indicate the polarities for each resistor as determined by the assumed current direction.
3. Apply Kirchhoff's voltage law around each closed, independent loop of the network.

The best way to determine how many times you will have to apply Kirchhoff's voltage law is to determine the number of "windows" in the network. The network of Example 8.9 has a definite similarity to the two-window configuration of Fig. 8.19(a). Therefore, you need to apply Kirchhoff's voltage law twice. For networks with three windows, as shown in Fig. 8.19(b), you would need to apply Kirchhoff's voltage law three times, and so on.

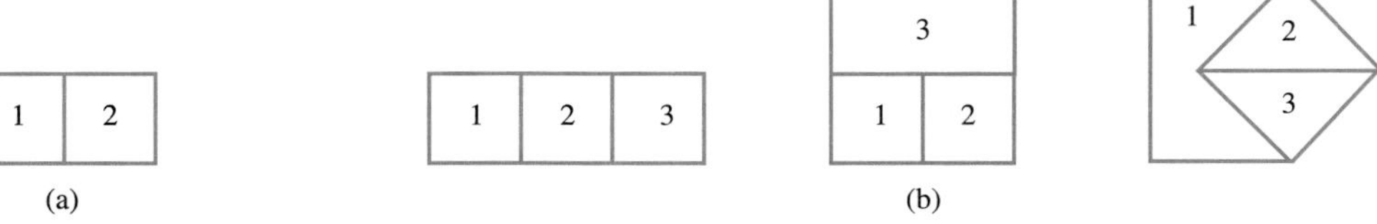

(a)

(b)

FIG. 8.19
Determining the number of independent closed loops.

4. Apply Kirchhoff's current law at the minimum number of nodes that will include all the branch currents of the network.

The minimum number is one less than the number of independent nodes of the network. For the purposes of this analysis, a node is a junction of two or more branches, where a branch is any combination of series elements. Figure 8.20 defines the number of applications of Kirchhoff's current law for each configuration of Fig. 8.19.

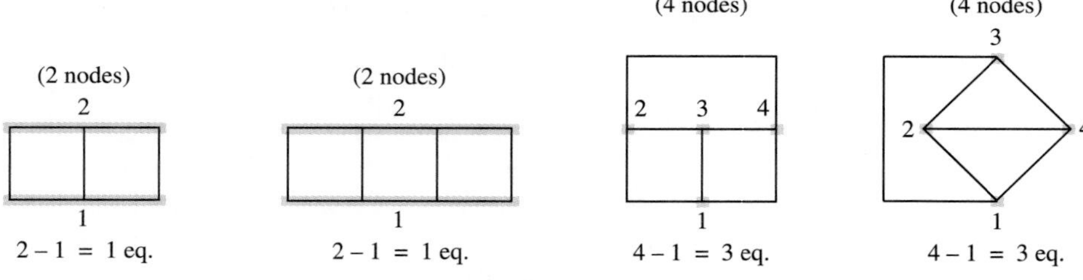

FIG. 8.20
Determining the number of applications of Kirchhoff's current law required.

5. Solve the resulting simultaneous linear equations for assumed branch currents.

It is assumed that you understand how to use determinants to solve for the currents I_1, I_2, and I_3 and that it is a part of your mathematical background. If not, a detailed explanation of the procedure is provided in Appendix A. Calculators and computer software packages such as MathCad can find the solutions quickly and accurately.

EXAMPLE 8.9

Apply the branch-current method to the network of Fig. 8.21.

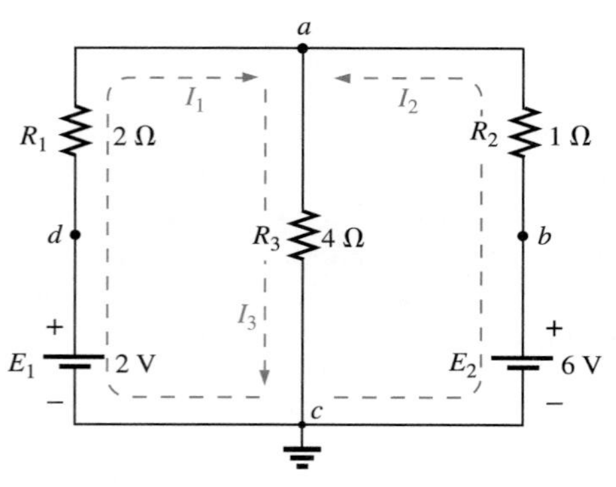

FIG. 8.21
Example 8.9.

Solution:

Step 1: Since there are three distinct branches (*cda, cba, ca*), three currents of arbitrary directions (I_1, I_2, I_3) are chosen, as indicated in Fig. 8.21. The current directions for I_1 and I_2 were chosen to match the "pressure" applied by sources E_1 and E_2, respectively. Since both I_1 and I_2 enter node *a*, I_3 is leaving.

Step 2: Polarities for each resistor are drawn to agree with assumed current directions, as indicated in Fig. 8.22.

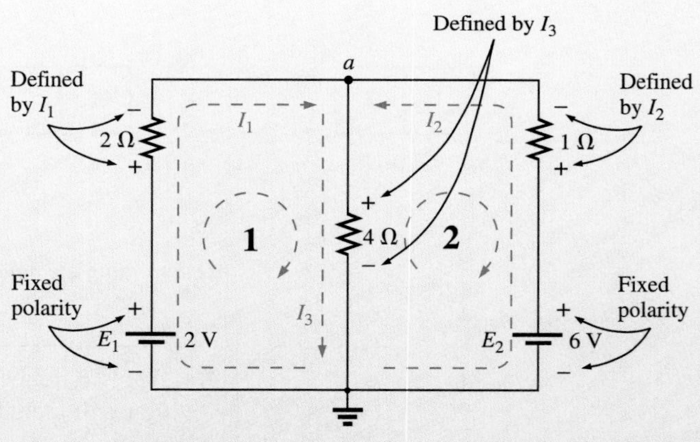

FIG. 8.22
Inserting the polarities across the resistive elements as defined by the chosen branch currents.

Step 3: Kirchhoff's voltage law is applied around each closed loop (1 and 2) in the clockwise direction:

continued

Calculators

The TI-85 calculator is only one of numerous calculators that can convert from one form to another and perform lengthy calculations with complex numbers in a concise, neat form. All the details of using a specific calculator will not be included here because each has its own format and sequence of steps. However, the basic operations with the TI-85 will be included primarily to demonstrate the ease with which the conversions can be made and the format for more complex operations.

For the TI-85 calculator, one must first call up the 2nd function CPLX from the keyboard, which results in a menu at the bottom of the display including conj, real, imag, abs, and angle. By choosing the key MORE, ▶ Rec and ▶ Pol will appear as options (for the conversion process). To convert from one form to another, simply enter the current form in brackets with a comma between components for the rectangular form and an angle symbol for the polar form. Follow this form with the operation to be performed and press the ENTER key—the result will appear on the screen in the desired format.

The example below is for demonstration purposes only. It is not expected that all the readers will have a TI-85 calculator. The sole purpose of the example is to demonstrate the power of today's calculators.

Using the TI-85 calculator, perform the following conversions:

a. $3 - j4$ to polar form.
b. $0.006 \angle 20.6°$ to rectangular form.

Solutions:

a. The TI-85 display for part (a) is the following:

```
(3, −4) ▶ Pol (ENTER)
(5.000E0∠−53.130E0)
```

b. The TI-85 display for part (b) is the following:

```
(0.006∠20.6) ▶ Rec (ENTER)
(5.616E−3, 2.111E−3)
```

and

$$\text{loop 1: } \Sigma_{\circlearrowright} V = +E_1 - V_{R_1} - V_{R_3} = 0$$

(Rise in potential: $+E_1$; Drop in potential: V_{R_1}, V_{R_3})

$$\text{loop 2: } \Sigma_{\circlearrowright} V = +V_{R_3} + V_{R_2} - E_2 = 0$$

(Rise in potential: V_{R_3}, V_{R_2}; Drop in potential: E_2)

$$\text{loop 1: } \Sigma_{\circlearrowright} V = \underbrace{+2\text{ V}}_{\text{Battery potential}} - \underbrace{(2\,\Omega)I_1}_{\text{Voltage drop across 2-}\Omega\text{ resistor}} - \underbrace{(4\,\Omega)I_3}_{\text{Voltage drop across 4-}\Omega\text{ resistor}} = 0$$

$$\text{loop 2: } \Sigma_{\circlearrowright} V = (4\,\Omega)I_3 + (1\,\Omega)I_2 - 6\text{ V} = 0$$

Step 4: Applying Kirchhoff's current law at node *a* (in a two-node network, the law is applied at only one node),

$$I_1 + I_2 = I_3$$

Step 5: There are three equations and three unknowns (units removed for clarity):

$$\begin{aligned} 2 - 2I_1 - 4I_3 &= 0 \\ 4I_3 + 1I_2 - 6 &= 0 \\ I_1 + I_2 &= I_3 \end{aligned} \qquad \text{Rewritten:} \qquad \begin{aligned} 2I_1 + 0 + 4I_3 &= 2 \\ 0 + I_2 + 4I_3 &= 6 \\ I_1 + I_2 - I_3 &= 0 \end{aligned}$$

Using third-order determinants (Appendix A), we have

$$I_1 = \frac{\begin{vmatrix} 2 & 0 & 4 \\ 6 & 1 & 4 \\ 0 & 1 & -1 \end{vmatrix}}{D = \begin{vmatrix} 2 & 0 & 4 \\ 0 & 1 & 4 \\ 1 & 1 & -1 \end{vmatrix}} = -1\text{ A}$$

A negative sign in front of a branch current indicates only that the actual current is in the direction opposite to that assumed.

$$I_2 = \frac{\begin{vmatrix} 2 & 2 & 4 \\ 0 & 6 & 4 \\ 1 & 0 & -1 \end{vmatrix}}{D} = 2\text{ A}$$

$$I_3 = \frac{\begin{vmatrix} 2 & 0 & 2 \\ 0 & 1 & 6 \\ 1 & 1 & 0 \end{vmatrix}}{D} = 1\text{ A}$$

Instead of using third-order determinants, we could reduce the three equations to two by substituting the third equation in the first and second equations:

$$\left.\begin{aligned} 2 - 2I_1 - 4\overbrace{(I_1 + I_2)}^{I_3} &= 0 \\ 4\overbrace{(I_1 + I_2)}^{I_3} + I_2 - 6 &= 0 \end{aligned}\right\} \quad \begin{aligned} 2 - 2I_1 - 4I_1 - 4I_2 &= 0 \\ 4I_1 + 4I_2 + I_2 - 6 &= 0 \end{aligned}$$

continued

or

$$\begin{aligned} -6I_1 - 4I_2 &= -2 \\ +4I_1 + 5I_2 &= +6 \end{aligned}$$

Multiplying through by -1 in the top equation yields

$$\begin{aligned} 6I_1 + 4I_2 &= +2 \\ 4I_1 + 5I_2 &= +6 \end{aligned}$$

and using determinants,

Using the TI-85 calculator:

det[[2,4][6,5]]/	−1
det[[6,4][4,5]] (ENTER)	−1

CALC. 8.1

$$I_1 = \frac{\begin{vmatrix} 2 & 4 \\ 6 & 5 \end{vmatrix}}{\begin{vmatrix} 6 & 4 \\ 4 & 5 \end{vmatrix}} = \frac{10 - 24}{30 - 16} = \frac{-14}{14} = \mathbf{-1\,A}$$

$$I_2 = \frac{\begin{vmatrix} 6 & 2 \\ 4 & 6 \end{vmatrix}}{14} = \frac{36 - 8}{14} = \frac{28}{14} = \mathbf{2\,A}$$

$$I_3 = I_1 + I_2 = -1 + 2 = \mathbf{1\,A}$$

It is now important that you understand the impact of the results you obtained. The currents I_1, I_2, and I_3 are the actual currents in the branches in which they were defined. A negative sign in the solution simply reveals that the actual current has the opposite direction than initially defined—the magnitude is correct. Once the actual current directions and their magnitudes are inserted in the original network, the various voltages and power levels can be determined. For this example, the actual current directions and their magnitudes have been entered on the original network in Fig. 8.23. Note that the current through the series elements R_1 and E_1 is 1 A; the current through R_3, 1 A; and the current through the series elements R_2 and E_2, 2 A. Due to the minus sign in the solution, the direction of I_1 is opposite to that shown in Fig. 8.21. You can now use Ohm's law to find the voltage across any resistor and use the appropriate power equation to find the power delivered by either source or to any one of the three resistors.

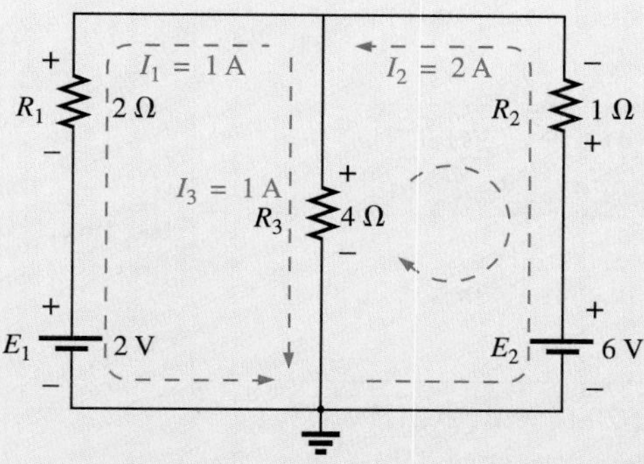

FIG. 8.23
Reviewing the results of the analysis of the network of Fig. 8.21.

continued

Applying Kirchhoff's voltage law around the loop indicated in Fig. 8.23,

$$\Sigma_{\circlearrowright} V = +(4\ \Omega)I_3 + (1\ \Omega)I_2 - 6\ \text{V} = 0$$

or
$$(4\ \Omega)I_3 + (1\ \Omega)I_2 = 6\ \text{V}$$

and
$$(4\ \Omega)(1\ \text{A}) + (1\ \Omega)(2\ \text{A}) = 6\ \text{V}$$
$$4\ \text{V} + 2\ \text{V} = 6\ \text{V}$$
$$6\ \text{V} = 6\ \text{V} \quad \text{(checks)}$$

EXAMPLE 8.10

Apply branch-current analysis to the network of Fig. 8.24.

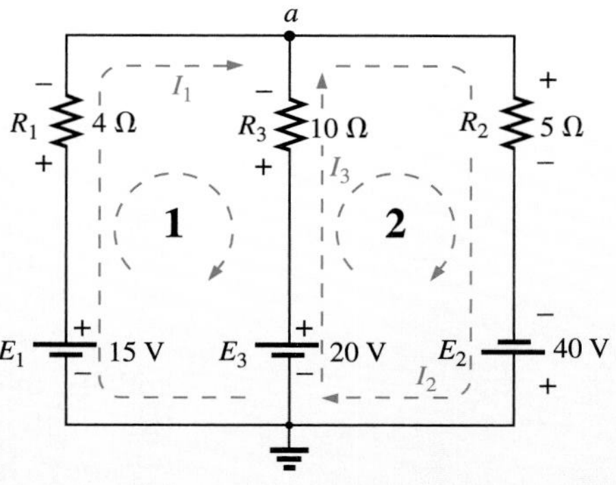

FIG. 8.24
Example 8.10.

Solution: Again, the current directions were chosen to match the "pressure" of each battery. The polarities are then added and Kirchhoff's voltage law is applied around each closed loop in the clockwise direction. The result is as follows:

$$\text{loop 1:}\ +15\ \text{V} - (4\ \Omega)I_1 + (10\ \Omega)I_3 - 20\ \text{V} = 0$$
$$\text{loop 2:}\ +20\ \text{V} - (10\ \Omega)I_3 - (5\ \Omega)I_2 + 40\ \text{V} = 0$$

Applying Kirchhoff's current law at node *a*,

$$I_1 + I_3 = I_2$$

Substituting the third equation into the other two gives (with units removed for clarity):

$$\left.\begin{aligned} 15 - 4I_1 + 10I_3 - 20 &= 0 \\ 20 - 10I_3 - 5(I_1 + I_3) + 40 &= 0 \end{aligned}\right\} \text{Substituting for } I_2 \text{ (since it occurs only once in the two equations)}$$

or
$$-4I_1 + 10I_3 = 5$$
$$-5I_1 - 15I_3 = -60$$

Multiplying the lower equation by -1, we have

$$-4I_1 + 10I_3 = 5$$
$$5I_1 + 15I_3 = 60$$

continued

$$I_1 = \frac{\begin{vmatrix} 5 & 10 \\ 60 & 15 \end{vmatrix}}{\begin{vmatrix} -4 & 10 \\ 5 & 15 \end{vmatrix}} = \frac{75 - 600}{-60 - 50} = \frac{-525}{-110} = \mathbf{4.773\ A}$$

$$I_3 = \frac{\begin{vmatrix} -4 & 5 \\ 5 & 60 \end{vmatrix}}{-110} = \frac{-240 - 25}{-110} = \frac{-265}{-110} = \mathbf{2.409\ A}$$

$$I_2 = I_1 + I_3 = 4.773 + 2.409 = \mathbf{7.182\ A}$$

revealing that the assumed directions were the actual directions, with I_2 equal to the sum of I_1 and I_3.

8.7 MESH ANALYSIS (GENERAL APPROACH)

The second method of analysis to be described is called **mesh analysis.** The term **mesh** comes from the similarities in appearance between the closed loops of a network and a wire mesh fence. Although this approach is more advanced than the branch-current method, it uses many of the ideas just developed. Of the two methods, mesh analysis is the one more frequently applied today. Branch-current analysis is introduced as a stepping stone to mesh analysis because branch currents are initially more "real" than the loop currents used in mesh analysis. Essentially, the mesh-analysis approach simply eliminates the need to substitute the results of Kirchhoff's current law into the equations derived from Kirchhoff's voltage law. It is now done in the initial writing of the equations. The systematic approach outlined below should be followed when applying this method.

1. Assign a distinct current in the clockwise direction to each independent, closed loop of the network. It is not absolutely necessary to choose the clockwise direction for each loop current. In fact, any direction can be chosen for each loop current with no loss in accuracy, as long as the remaining steps are followed properly. However, by choosing the clockwise direction as a standard, we can develop a shorthand method (Section 8.8) for writing the required equations that will save time and possibly prevent some common errors.

The best way to accomplish this first step is to place a loop current *within* each "window" of the network, as shown in the previous section, to ensure that they are all independent. There is a variety of other loop currents that can be assigned. In each case, however, be sure that the information carried by any one loop equation is not included in a combination of the other network equations. This is the point of the term: *independent.* No matter how you choose your loop currents, the number of loop currents required is always equal to the number of windows of a planar (no-crossovers) network. Sometimes a network may appear to be nonplanar. However, redrawing the network may reveal that it is, in fact, planar. This may be the case in one or two problems at the end of the chapter.

Before continuing to the next step, let us ensure that the concept of a loop current is clear. For the network of Fig. 8.25, the loop current I_1 is

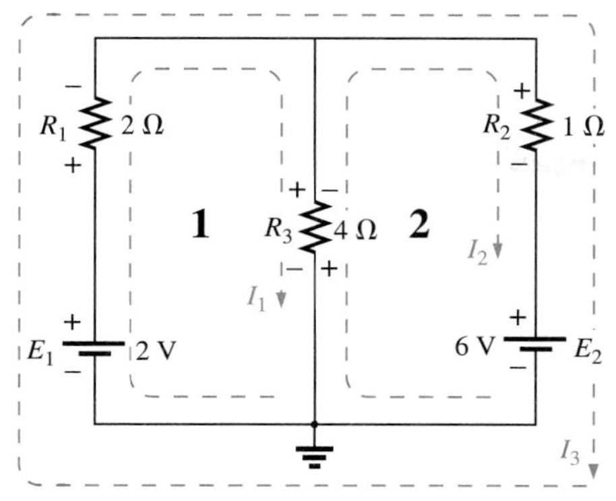

FIG. 8.25
Defining the mesh currents for a "two-window" network.

the branch current of the branch containing the 2-Ω resistor and 2-V battery. The current through the 4-Ω resistor is not I_1, however, since there is also a loop current I_2 through it. Since they have opposite directions, $I_{4\Omega}$ equals the difference between the two, $I_1 - I_2$ or $I_2 - I_1$, depending on which you choose to be the defining direction. In other words,

a loop current is a branch current only when it is the only loop current assigned to that branch.

2. Indicate the polarities within each loop for each resistor as determined by the assumed direction of loop current for that loop. Note that the polarities must be placed within each loop. This means, as shown in Fig. 8.25, that the 4-V resistor must have two sets of polarities across it.
3. Apply Kirchhoff's voltage law around each closed loop in the clockwise direction. Again, the clockwise direction was chosen for uniformity and to prepare us for the method to be introduced in the next section.
   a. If a resistor has two or more assumed currents through it, the total current through the resistor is the assumed current of the loop in which Kirchhoff's voltage law is being applied, plus the assumed currents of the other loops passing through in the same direction, minus the assumed currents in the opposite direction.
   b. The polarity of a voltage source is unaffected by the direction of the assigned loop currents.
4. Solve the resulting simultaneous linear equations for the assumed loop currents.

EXAMPLE 8.11

Consider the same basic network as in Example 8.9 of the preceding section, now shown in Fig. 8.25.

Solution:

Step 1: Assign loop currents (I_1 and I_2) in the clockwise direction in the windows of the network. A third loop (I_3) could have been included around the entire network, but the information carried by this loop is already included in the other two.

Step 2: Draw polarities within each window to agree with assumed current directions. Note that for this case, the polarities across the 4-Ω resistor are the opposite for each loop current.

Step 3: Apply Kirchhoff's voltage law around each loop in the clockwise direction. Keep in mind as you do this step, that the law is concerned only with the magnitude and polarity of the voltages around the closed loop and not with whether a voltage rise or drop is due to a battery or resistive element. The voltage across each resistor is determined by $V = IR$, and for a resistor with more than one current through it, the current is the loop current of the loop being examined plus or minus the other loop currents as determined by their directions. If Kirchhoff's voltage law is always applied clockwise, the other loop currents will always be subtracted from the loop current of the loop being analyzed.

continued

loop 1: $+E_1 - V_1 - V_3 = 0$ (clockwise starting at point *a*)

$$+2\text{ V} - (2\ \Omega)I_1 - \underbrace{(4\ \Omega)\overbrace{(I_1 - I_2)}}= 0$$

Voltage drop across 4-Ω resistor; Total current through 4-Ω resistor; Subtracted since I_2 is opposite in direction to I_1.

loop 2: $-V_3 - V_2 - E_2 = 0$ (clockwise starting at point *b*)

$$-(4\ \Omega)(I_2 - I_1) - (1\ \Omega)I_2 - 6\text{ V} = 0$$

Step 4: Rewrite equations as follows (without units for clarity):

$$\text{loop 1: } +2 - 2I_1 - 4I_1 + 4I_2 = 0$$
$$\text{loop 2: } -4I_2 + 4I_1 - 1I_2 - 6 = 0$$

and

$$\text{loop 1: } +2 - 6I_1 + 4I_2 = 0$$
$$\text{loop 2: } -5I_2 + 4I_1 - 6 = 0$$

or

$$\text{loop 1: } -6I_1 + 4I_2 = -2$$
$$\text{loop 2: } +4I_1 - 5I_2 = +6$$

Applying determinants will result in

$$I_1 = \mathbf{-1\ A} \text{ and } I_2 = \mathbf{-2\ A}$$

The minus signs indicate that the currents have a direction opposite to that indicated by the assumed loop current.

The actual current through the 2-V source and 2-Ω resistor is therefore 1 A in the other direction, and the current through the 6-V source and 1-Ω resistor is 2 A in the opposite direction indicated on the circuit. The current through the 4-Ω resistor is determined by the following equation from the original network:

$$\text{loop 1: } I_{4\Omega} = I_1 - I_2 = -1\text{ A} - (-2\text{ A}) = -1\text{ A} + 2\text{ A}$$
$$= \mathbf{1\ A} \text{ (in the direction of } I_1)$$

The outer loop (I_3) and *one* inner loop (either I_1 or I_2) would also have produced the correct results. This approach, however, will often lead to errors since the loop equations may be more difficult to write. The best method of picking these loop currents is to use the window approach.

EXAMPLE 8.12

Find the current through each branch of the network of Fig. 8.26.

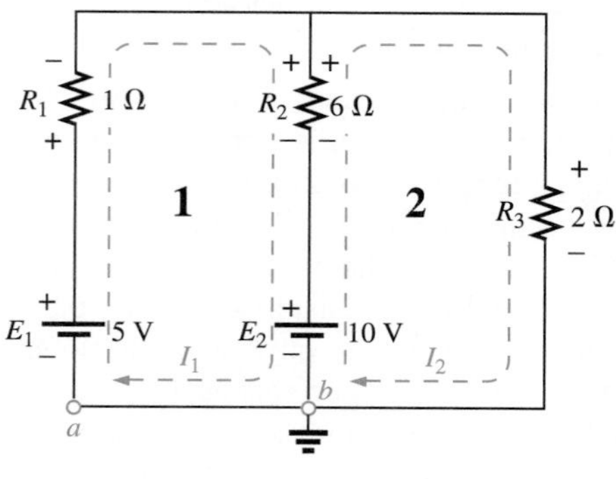

FIG. 8.26
Example 8.12.

Solution:

Steps 1 and 2 are shown in the circuit. Note that the polarities of the 6-Ω resistor are different for each loop current.

Step 3: Apply Kirchhoff's voltage law around each closed loop in the clockwise direction:

loop 1: $+E_1 - V_1 - V_2 - E_2 = 0$ (clockwise starting at point *a*)

$$+5\text{ V} - (1\ \Omega)I_1 - (6\ \Omega)(I_1 - I_2) - 10\text{ V} = 0$$

I_2 flows through the 6-Ω resistor in the direction opposite to I_1.

continued

loop 2: $E_2 - V_2 - V_3 = 0$ (clockwise starting at point b)

$$+10\text{ V} - (6\ \Omega)(I_2 - I_1) - (2\ \Omega)I_2 = 0$$

Rewrite the equations as

$$\left.\begin{aligned}5 - I_1 - 6I_1 + 6I_2 - 10 &= 0\\ 10 - 6I_2 + 6I_1 - 2I_2 &= 0\end{aligned}\right\}\begin{aligned}-7I_1 + 6I_2 &= 5\\ +6I_1 - 8I_2 &= -10\end{aligned}$$

$$I_1 = \frac{\begin{vmatrix}5 & 6\\ -10 & -8\end{vmatrix}}{\begin{vmatrix}-7 & 6\\ 6 & -8\end{vmatrix}} = \frac{-40 + 60}{56 - 36} = \frac{20}{20} = \mathbf{1\ A}$$

$$I_2 = \frac{\begin{vmatrix}-7 & 5\\ 6 & -10\end{vmatrix}}{20} = \frac{70 - 30}{20} = \frac{40}{20} = \mathbf{2\ A}$$

Since I_1 and I_2 are positive and flow in opposite directions through the 6-Ω resistor and 10-V source, the total current in this branch is equal to the difference of the two currents in the direction of the larger:

$$I_2 > I_1 \quad (2\text{ A} > 1\text{ A})$$

Therefore, $I_{R_2} = I_2 - I_1 = 2\text{ A} - 1\text{ A} = \mathbf{1\ A}$ in the direction of I_2.

It is sometimes impractical to draw all the branches of a circuit at right angles to one another. The next example shows how a portion of a network may appear due to various constraints. The method of analysis does not change with this change in configuration.

EXAMPLE 8.13

Find the branch currents of the network of Fig. 8.27.

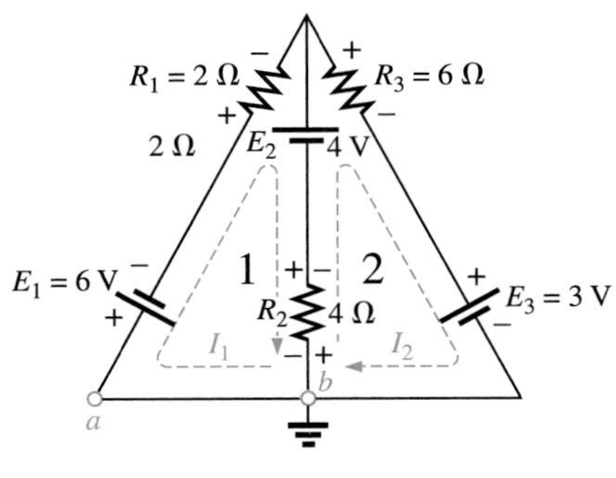

FIG. 8.27
Example 8.13.

Solution:

Steps 1 and 2 are shown in the circuit.

Step 3: Apply Kirchhoff's voltage law around each closed loop:

loop 1: $-E_1 - I_1R_1 - E_2 - V_2 = 0$ (clockwise from point a)

$$-6\text{ V} - (2\ \Omega)I_1 - 4\text{ V} - (4\ \Omega)(I_1 - I_2) = 0$$

loop 2: $-V_2 + E_2 - V_3 - E_3 = 0$ (clockwise from point b)

$$-(4\ \Omega)(I_2 - I_1) + 4\text{ V} - (6\ \Omega)(I_2) - 3\text{ V} = 0$$

which are rewritten as

$$\left.\begin{aligned}-10 - 4I_1 - 2I_1 + 4I_2 &= 0\\ +1 + 4I_1 - 4I_2 - 6I_2 &= 0\end{aligned}\right\}\begin{aligned}-6I_1 + 4I_2 &= +10\\ +4I_1 - 10I_2 &= -1\end{aligned}$$

or, by multiplying the top equation by -1, we obtain

$$\begin{aligned}6I_1 - 4I_2 &= -10\\ 4I_1 - 10I_2 &= -1\end{aligned}$$

continued

and
$$I_1 = \frac{\begin{vmatrix} -10 & -4 \\ -1 & -10 \end{vmatrix}}{\begin{vmatrix} 6 & -4 \\ 4 & -10 \end{vmatrix}} = \frac{100 - 4}{-60 + 16} = \frac{96}{-44} = \mathbf{-2.182\ A}$$

$$I_2 = \frac{\begin{vmatrix} 6 & -10 \\ 4 & -1 \end{vmatrix}}{-44} = \frac{-6 + 40}{-44} = \frac{34}{-44} = \mathbf{-0.773\ A}$$

Using the TI-85 calculator:

det[[−10,−4][−1,−10]]/

det[[6,−4][4,−10]] (ENTER) −2.182

CALC. 8.2

The current in the 4-Ω resistor and 4-V source for loop 1 is

$$\begin{aligned} I_1 - I_2 &= -2.182\ \text{A} - (-0.773\ \text{A}) \\ &= -2.182\ \text{A} + 0.773\ \text{A} \\ &= \mathbf{-1.409\ A} \end{aligned}$$

revealing that it is 1.409 A in a direction opposite (due to the minus sign) to I_1 in loop 1.

Supermesh Currents

Sometimes there will be current sources in the network to which you need to apply mesh analysis. In such cases you can convert the current source to a voltage source (if a parallel resistor is present) and proceed as before or use a *supermesh current* and proceed as follows.

Start as before and assign a mesh current to each independent loop, including the current sources, as if they were resistors or voltage sources. Then mentally remove the current sources (replace with open-circuit equivalents) and apply Kirchhoff's voltage law to all the remaining independent paths of the network using the mesh currents just defined. Any resulting open window, including two or more mesh currents, is said to be the path of a **supermesh current**. Then relate the chosen mesh currents of the network to the independent current sources of the network and solve for the mesh currents. The next example will clarify the definition of a supermesh current and the procedure.

EXAMPLE 8.14

Using mesh analysis, determine the currents of the network of Fig. 8.28.

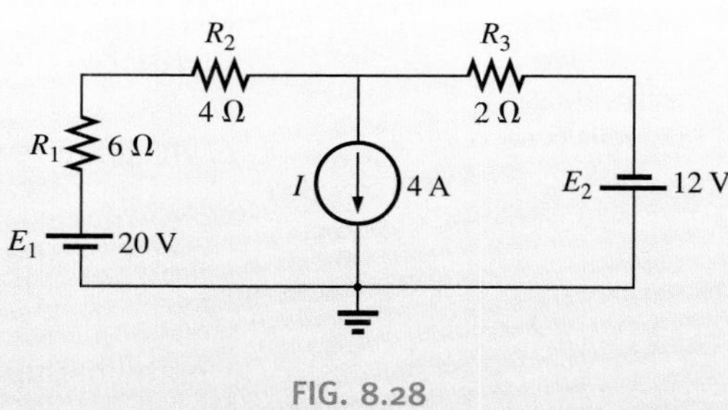

FIG. 8.28
Example 8.14.

continued

Solution: First, define the mesh currents for the network, as shown in Fig. 8.29. Then mentally remove the current source, as shown in Fig. 8.30, and apply Kirchhoff's voltage law to the resulting network. The single path now including the effects of two mesh currents is referred to as the path of a **supermesh** current.

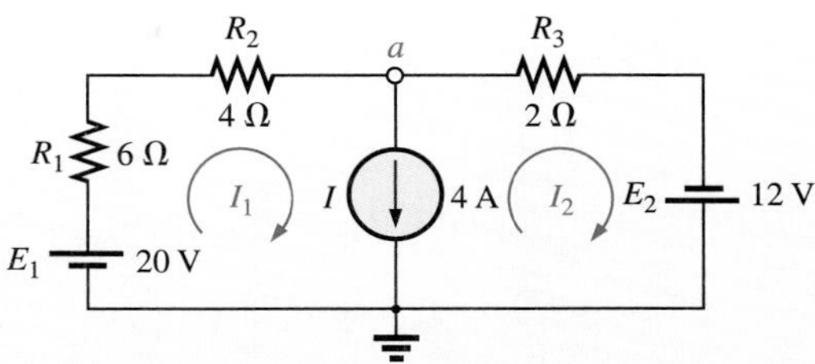

FIG. 8.29
Defining the mesh currents for the network of Fig. 8.28.

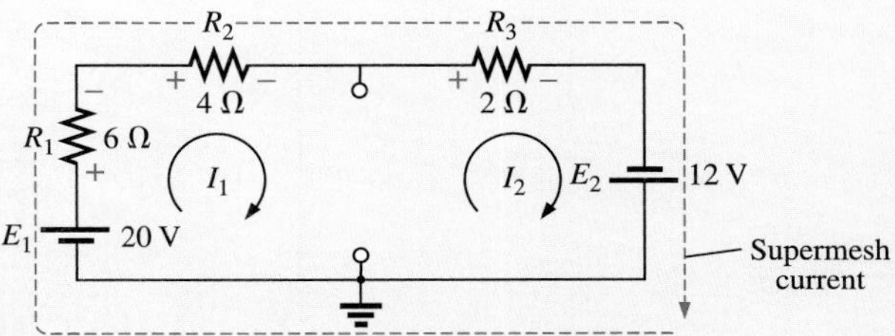

FIG. 8.30
Defining the supermesh current.

Applying Kirchhoff's law:

$$20\text{ V} - I_1(6\ \Omega) - I_1(4\ \Omega) - I_2(2\ \Omega) + 12\text{ V} = 0$$

or

$$10I_1 + 2I_2 = 32$$

Use node a to relate the mesh currents and the current source using Kirchhoff's current law:

$$I_1 = I + I_2$$

The result is two equations and two unknowns:

$$\begin{aligned} 10I_1 + 2I_2 &= 32 \\ I_1 - I_2 &= 4 \end{aligned}$$

Applying determinants:

$$I_1 = \frac{\begin{vmatrix} 32 & 2 \\ 4 & -1 \end{vmatrix}}{\begin{vmatrix} 10 & 2 \\ 1 & -1 \end{vmatrix}} = \frac{(32)(-1) - (2)(4)}{(10)(-1) - (2)(1)} = \frac{40}{12} = \mathbf{3.33\ A}$$

and

$$I_2 = I_1 - I = 3.33\text{ A} - 4\text{ A} = \mathbf{-0.67\ A}$$

EXAMPLE 8.15

Using mesh analysis, determine the currents for the network of Fig. 8.31.

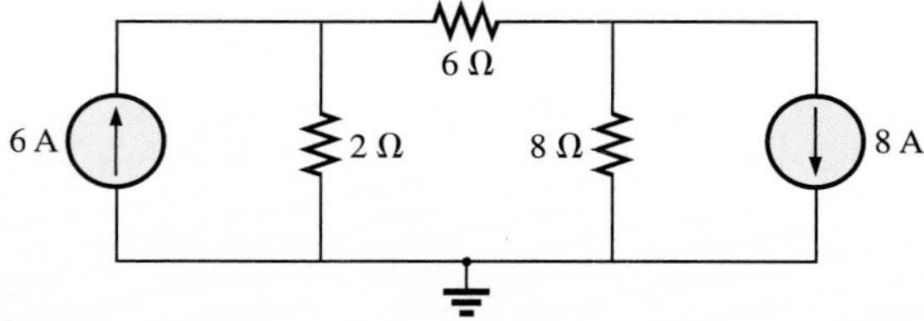

FIG. 8.31
Example 8.15.

Solution: Define the mesh currents in Fig. 8.32. Remove the current sources and define the single supermesh path in Fig. 8.33.

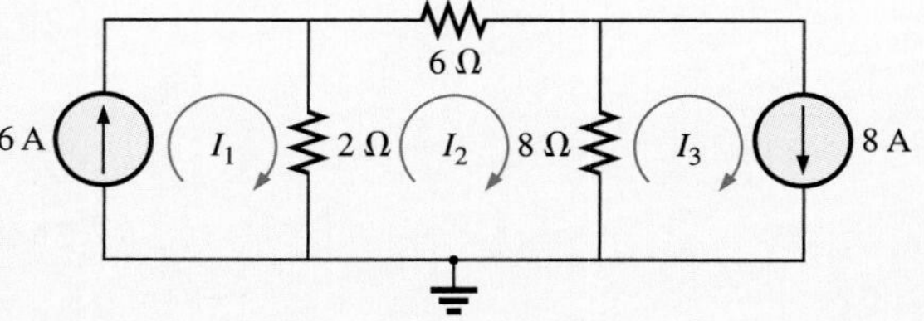

FIG. 8.32
Defining the mesh currents for the network of Fig. 8.31.

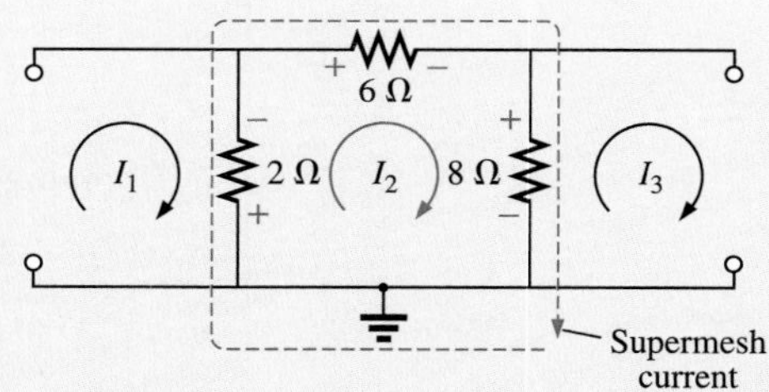

FIG. 8.33
Defining the supermesh current for the network of Fig. 8.31.

Applying Kirchhoff's voltage law around the supermesh path:

$$-V_{2\Omega} - V_{6\Omega} - V_{8\Omega} = 0$$
$$-(I_2 - I_1)2\ \Omega - I_2(6\ \Omega) - (I_2 - I_3)8\ \Omega = 0$$
$$-2I_2 + 2I_1 - 6I_2 - 8I_2 + 8I_3 = 0$$
$$2I_1 - 16I_2 + 8I_3 = 0$$

Introducing the relationship between the mesh currents and the current sources:

$$I_1 = 6\text{ A}$$
$$I_3 = 8\text{ A}$$

results in the following solutions:

$$2I_1 - 16I_2 + 8I_3 = 0$$
$$2(6\text{ A}) - 16I_2 + 8(8\text{ A}) = 0$$

continued

and $$I_2 = \frac{76\text{ A}}{16} = \mathbf{4.75\text{ A}}$$

Then $$I_{2\Omega}\downarrow = I_1 - I_2 = 6\text{ A} - 4.75\text{ A} = \mathbf{1.25\text{ A}}$$

and $$I_{8\Omega}\uparrow = I_3 - I_2 = 8\text{ A} - 4.75\text{ A} = \mathbf{3.25\text{ A}}$$

8.8 MESH ANALYSIS (FORMAT APPROACH)

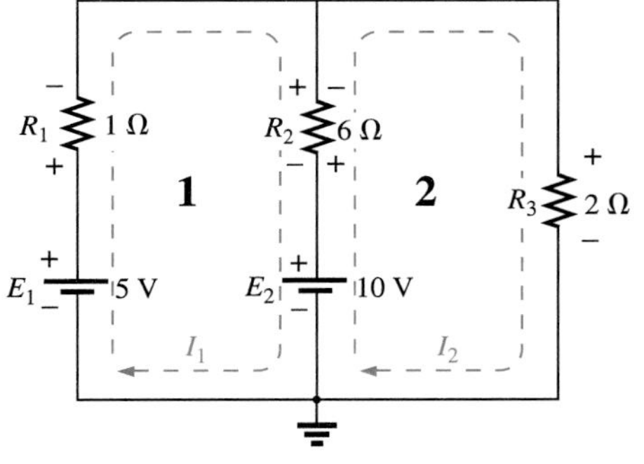

FIG. 8.34
Network of Fig. 8.26 redrawn with assigned loop currents.

Now that we have established the basis for the mesh-analysis approach, we will examine a technique for writing the mesh equations more rapidly and usually with fewer errors. To help introduce the procedure, we have redrawn the network of Example 8.12 (Fig. 8.26) in Fig. 8.34 with the assigned loop currents. (Note that each loop current has a clockwise direction.)

The equations obtained are

$$\begin{aligned} -7I_1 + 6I_2 &= 5 \\ 6I_1 - 8I_2 &= -10 \end{aligned}$$

which can also be written as

$$\begin{aligned} 7I_1 - 6I_2 &= -5 \\ 8I_2 - 6I_1 &= 10 \end{aligned}$$

and expanded as

Col. 1	Col. 2	Col. 3
$(1 + 6)I_1 -$	$6I_2$	$= (5 - 10)$
$(2 + 6)I_2 -$	$6I_1$	$= 10$

Note in the above equations that column 1 is made up of a loop current that is times the sum of the resistors through which that loop current passes. Column 2 is the product of the resistors common to another loop current times that other loop current. Note that in each equation, this column is subtracted from column 1. Column 3 is the *algebraic* sum of the voltage sources through which the loop current of interest passes. A source is assigned a positive sign if the loop current passes from the negative to the positive terminal, and a negative value is assigned if the polarities are reversed. The comments above are correct only for a standard direction of loop current in each window, in this case, the clockwise direction.

The above statements can be extended to develop the following **format approach** to mesh analysis:

1. Assign a loop current to each independent, closed loop (as in the previous section) in a clockwise direction.
2. The number of required equations is equal to the number of chosen independent, closed loops. Column 1 of each equation is formed by summing the resistance values of those resistors through which the loop current of interest passes and multiplying the result by that loop current.
3. We must now consider the mutual terms, which, as noted in the examples above, are always subtracted from the first column. A mutual term is simply any resistive element having an additional loop current passing through it. It is possible to have more than one

continued

mutual term if the loop current of interest has an element in common with more than one other loop current. This will be demonstrated in an example to follow. Each term is the product of the mutual resistor and the other loop current passing through the same element.

4. The column to the right of the equals sign is the algebraic sum of the voltage sources through which the loop current of interest passes. Positive signs are assigned to those sources of voltage having a polarity such that the loop current passes from the negative to the positive terminal. A negative sign is assigned to those potentials for which the reverse is true.
5. Solve the resulting simultaneous equations for the desired loop currents.

Before considering a few examples, be aware that since the column to the right of the equals sign is the algebraic sum of the voltages sources in that loop,

the format approach can be applied only to networks in which all current sources have been converted to their equivalent voltage source.

EXAMPLE 8.16

Write the mesh equations for the network of Fig. 8.35 and find the current through the 7-Ω resistor.

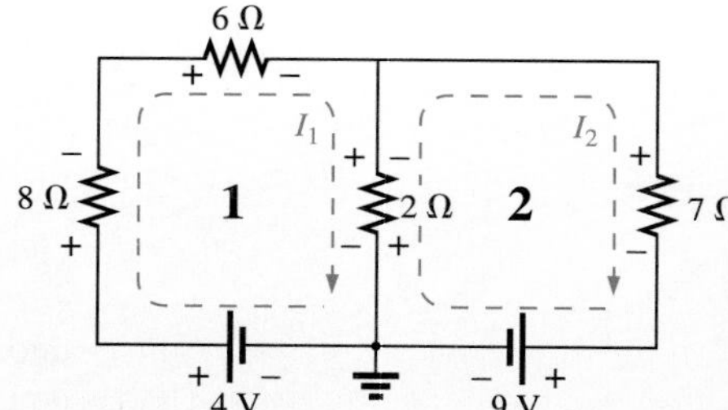

FIG. 8.35
Example 8.16.

Solution:

Step 1: As indicated in Fig. 8.35, each assigned loop current has a clockwise direction.

Steps 2 to 4:

$$\begin{aligned} I_1&: (8\ \Omega + 6\ \Omega + 2\ \Omega)I_1 - (2\ \Omega)I_2 = 4\ \text{V} \\ I_2&: \qquad (7\ \Omega + 2\ \Omega)I_2 - (2\ \Omega)I_1 = -9\ \text{V} \end{aligned}$$

and

$$\begin{aligned} 16I_1 - 2I_2 &= 4 \\ 9I_2 - 2I_1 &= -9 \end{aligned}$$

which, for determinants, are

$$\begin{aligned} 16I_1 - 2I_2 &= 4 \\ -2I_1 + 9I_2 &= -9 \end{aligned}$$

and

$$I_2 = I_{7\Omega} = \frac{\begin{vmatrix} 16 & 4 \\ -2 & -9 \end{vmatrix}}{\begin{vmatrix} 16 & -2 \\ -2 & 9 \end{vmatrix}} = \frac{-144 + 8}{144 - 4} = \frac{-136}{140}$$

$$= \mathbf{-0.971\ A}$$

EXAMPLE 8.17

Write the mesh equations for the network of Fig. 8.36.

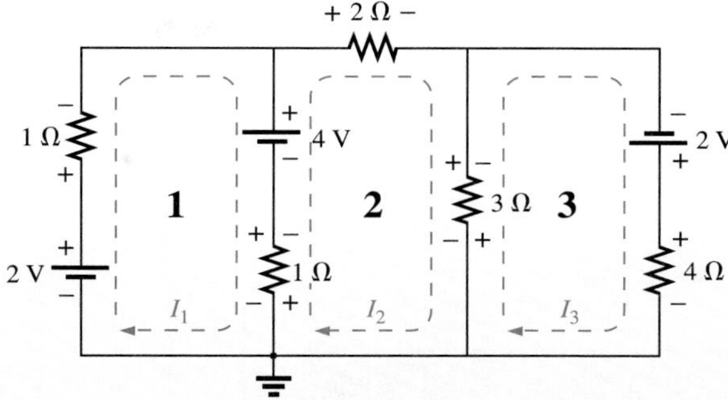

FIG. 8.36
Example 8.17.

Solution:

Step 1: Assign each window to a loop current in the clockwise direction:

I_1 does not pass through an element mutual with I_3.

$$
\begin{aligned}
I_1&: \quad (1\ \Omega + 1\ \Omega)I_1 - (1\ \Omega)I_2 + 0 = 2\ \text{V} - 4\ \text{V}\\
I_2&: \quad (1\ \Omega + 2\ \Omega + 3\ \Omega)I_2 - (1\ \Omega)I_1 - (3\ \Omega)I_3 = 4\ \text{V}\\
I_3&: \quad (3\ \Omega + 4\ \Omega)I_3 - (3\ \Omega)I_2 + 0 = 2\ \text{V}
\end{aligned}
$$

I_3 does not pass through an element mutual with I_1.

Summing terms gives

$$
\begin{aligned}
2I_1 - I_2 + 0 &= -2\\
6I_2 - I_1 - 3I_3 &= 4\\
7I_3 - 3I_2 + 0 &= 2
\end{aligned}
$$

which are rewritten for determinants as

$$
\begin{aligned}
2I_1 - I_2 + 0 &= -2\\
-I_1 + 6I_2 - 3I_3 &= 4\\
0 - 3I_2 + 7I_3 &= 2
\end{aligned}
$$

(diagonals labeled c, b, a)

Note that the coefficients of the *a* and *b* diagonals are equal. This *symmetry* about the *c* axis will always be true for equations written using the format approach. It is a check on whether the equations were obtained correctly.

We will now consider a network with only one source of voltage to point out that mesh analysis can be useful in other than multisource networks.

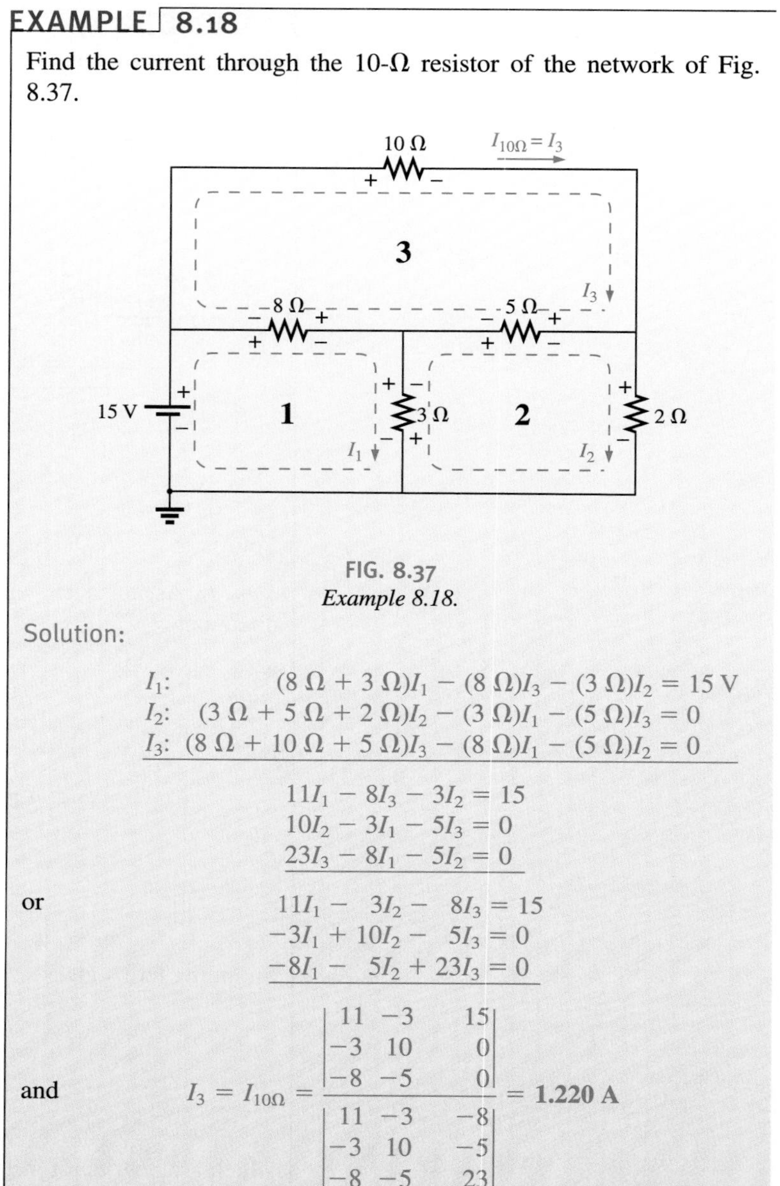

EXAMPLE 8.18

Find the current through the 10-Ω resistor of the network of Fig. 8.37.

FIG. 8.37
Example 8.18.

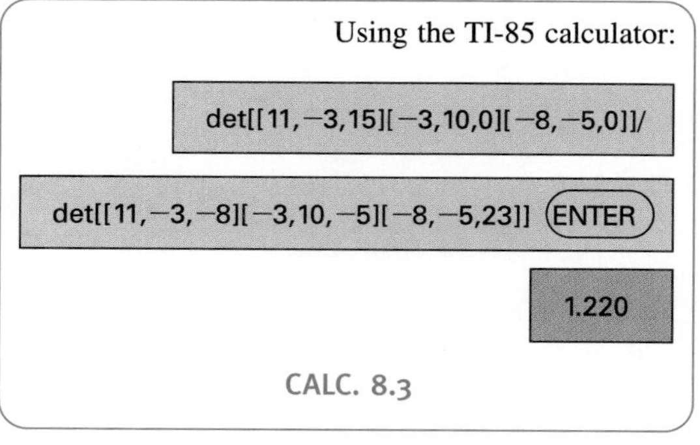

Using the TI-85 calculator:

```
det[[11,−3,15][−3,10,0][−8,−5,0]]/
det[[11,−3,−8][−3,10,−5][−8,−5,23]] ENTER
1.220
```

CALC. 8.3

Solution:

$$
\begin{aligned}
I_1&: \quad (8\ \Omega + 3\ \Omega)I_1 - (8\ \Omega)I_3 - (3\ \Omega)I_2 = 15\ \text{V} \\
I_2&: \quad (3\ \Omega + 5\ \Omega + 2\ \Omega)I_2 - (3\ \Omega)I_1 - (5\ \Omega)I_3 = 0 \\
I_3&: \quad (8\ \Omega + 10\ \Omega + 5\ \Omega)I_3 - (8\ \Omega)I_1 - (5\ \Omega)I_2 = 0
\end{aligned}
$$

$$
\begin{aligned}
11I_1 - 8I_3 - 3I_2 &= 15 \\
10I_2 - 3I_1 - 5I_3 &= 0 \\
23I_3 - 8I_1 - 5I_2 &= 0
\end{aligned}
$$

or

$$
\begin{aligned}
11I_1 - 3I_2 - 8I_3 &= 15 \\
-3I_1 + 10I_2 - 5I_3 &= 0 \\
-8I_1 - 5I_2 + 23I_3 &= 0
\end{aligned}
$$

and

$$
I_3 = I_{10\Omega} = \frac{\begin{vmatrix} 11 & -3 & 15 \\ -3 & 10 & 0 \\ -8 & -5 & 0 \end{vmatrix}}{\begin{vmatrix} 11 & -3 & -8 \\ -3 & 10 & -5 \\ -8 & -5 & 23 \end{vmatrix}} = \mathbf{1.220\ A}
$$

8.9 NODAL ANALYSIS (GENERAL APPROACH)

Recall from the development of loop analysis that the general network equations were obtained by applying Kirchhoff's voltage law around each closed loop. We will now use Kirchhoff's current law to develop a method called **nodal analysis**.

A **node** is defined as a junction of two or more branches. If we now define one node of any network as a reference (that is, a point of zero potential or ground), the remaining nodes of the network will all have a fixed potential relative to this reference. For a network of N nodes, therefore, there will exist $(N - 1)$ nodes with a fixed potential relative to the assigned reference node. We can write equations relating these nodal voltages by applying Kirchhoff's current law at each of the $(N - 1)$ nodes. To obtain the complete solution of a network, we then evaluate these nodal voltages in the same manner in which we found loop currents in loop analysis.

The nodal analysis method is applied as follows:

1. Determine the number of nodes within the network.
2. Pick a reference node and label each remaining node with a subscripted value of voltage: V_1, V_2, and so on.
3. Apply Kirchhoff's current law at each node except the reference. Assume all unknown currents leave the node for each application of Kirchhoff's current law. In other words, for each node, don't be influenced by the direction an unknown current for another node may have had. Each node is to be treated as a separate entity, independent of the application of Kirchhoff's current law to the other nodes.
4. Solve the resulting equations for the nodal voltages.

A few examples will clarify the procedure defined by step 3. It will initially take some practice writing the equations for Kirchhoff's current law correctly, but in time you will see the advantage of assuming all the currents leave a node rather than identifying a specific direction for each branch. (The same type of advantage is associated with assuming that all the mesh currents are clockwise when applying mesh analysis.)

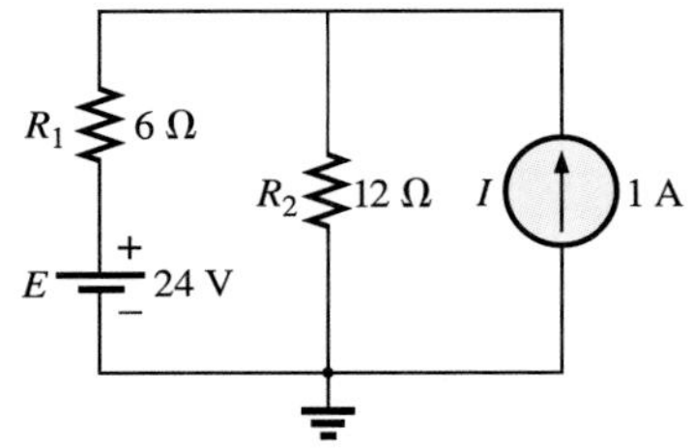

FIG. 8.38
Example 8.19.

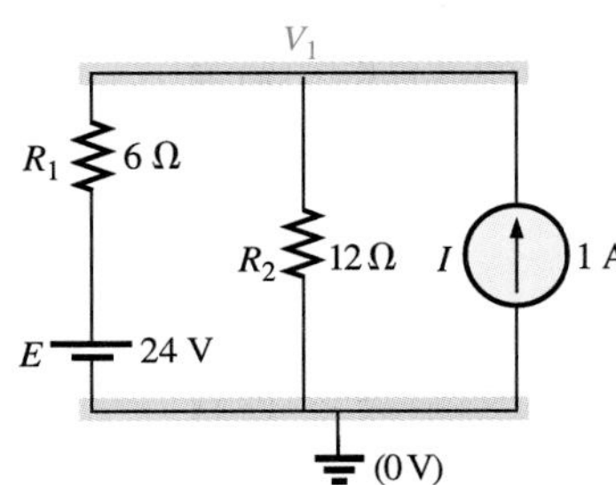

FIG. 8.39
Network of Fig. 8.38 with assigned nodes.

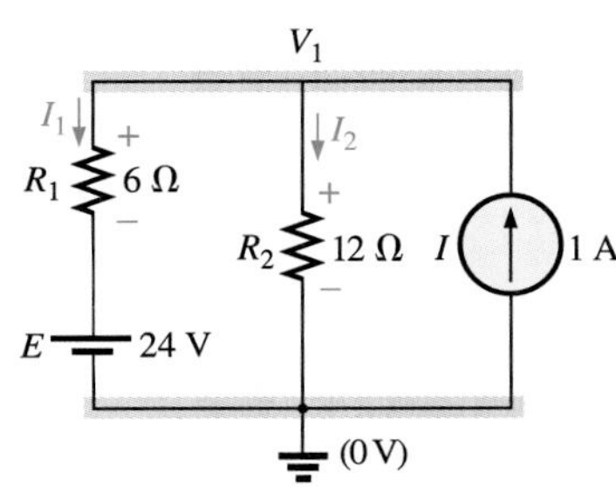

FIG. 8.40
Applying Kirchhoff's current law to the node V_1.

EXAMPLE 8.19

Apply nodal analysis to the network of Fig. 8.38.

Solution:

Steps 1 and 2: The network has two nodes, as shown in Fig. 8.39. Define the lower node as the reference node at ground potential (zero volts) and the other V_1, the voltage from node 1 to ground.

Step 3: Define I_1 and I_2 as leaving the node in Fig. 8.40, and apply Kirchhoff's current law as follows:

$$I = I_1 + I_2$$

Relate the current I_2 to the nodal voltage V_1 by Ohm's law:

$$I_2 = \frac{V_{R_2}}{R_2} = \frac{V_1}{R_2}$$

Determine the current I_1 by Ohm's law as follows:

$$I_1 = \frac{V_{R_1}}{R_1}$$

with

$$V_{R_1} = V_1 - E$$

Substituting into the Kirchhoff's current law equation:

$$I = \frac{V_1 - E}{R_1} + \frac{V_1}{R_2}$$

and rearranging, we have:

$$I = \frac{V_1}{R_1} - \frac{E}{R_1} + \frac{V_1}{R_2} = V_1\left(\frac{1}{R_1} + \frac{1}{R_2}\right) - \frac{E}{R_1}$$

or

$$V_1\left(\frac{1}{R_1} + \frac{1}{R_2}\right) = \frac{E}{R_1} + I$$

continued

Substituting numerical values,

$$V_1\left(\frac{1}{6\ \Omega} + \frac{1}{12\ \Omega}\right) = \frac{24\ \text{V}}{6\ \Omega} + 1\ \text{A} = 4\ \text{A} + 1\ \text{A}$$

$$V_1\left(\frac{1}{4\ \Omega}\right) = 5\ \text{A}$$

$$V_1 = \mathbf{20\ V}$$

The currents I_1 and I_2 can then be determined using the preceding equations:

$$I_1 = \frac{V_1 - E}{R_1} = \frac{20\ \text{V} - 24\ \text{V}}{6\ \Omega} = \frac{-4\ \text{V}}{6\ \Omega}$$
$$= \mathbf{-0.667\ A}$$

The minus sign indicates simply that the current I_1 has a direction opposite to that appearing in Fig. 8.40.

$$I_2 = \frac{V_1}{R_2} = \frac{20\ \text{V}}{12\ \Omega} = \mathbf{1.667\ A}$$

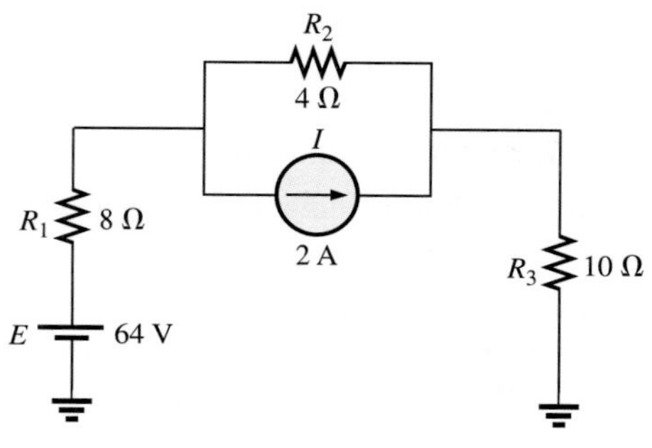

FIG. 8.41
Example 8.20.

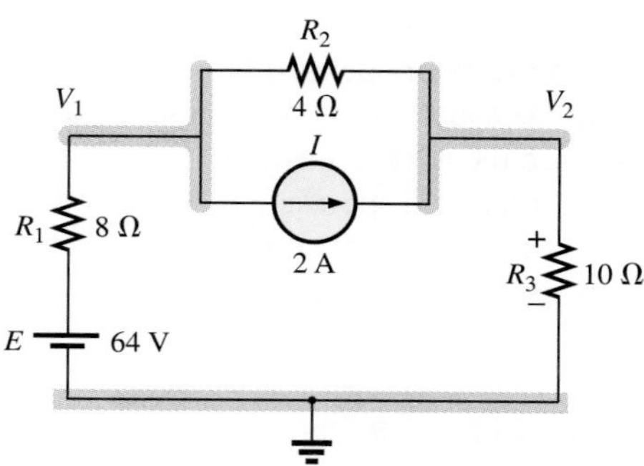

FIG. 8.42
Defining the nodes for the network of Fig. 8.41.

EXAMPLE 8.20

Apply nodal analysis to the network of Fig. 8.41.

Solution:

Steps 1 and 2: The network has three nodes, as defined in Fig. 8.42, with the bottom node again defined as the reference node (at ground potential, or zero volts) and the other nodes V_1 and V_2.

Step 3: For node V_1 define the currents as shown in Fig. 8.43, and apply Kirchhoff's current law:

$$0 = I_1 + I_2 + I$$

with
$$I_1 = \frac{V_1 - E}{R_1}$$

and
$$I_2 = \frac{V_{R_2}}{R_2} = \frac{V_1 - V_2}{R_2}$$

so that
$$\frac{V_1 - E}{R_1} + \frac{V_1 - V_2}{R_2} + I = 0$$

or
$$\frac{V_1}{R_1} - \frac{E}{R_1} + \frac{V_1}{R_2} - \frac{V_2}{R_2} + I = 0$$

and
$$V_1\left(\frac{1}{R_1} + \frac{1}{R_2}\right) - V_2\left(\frac{1}{R_2}\right) = -I + \frac{E}{R_1}$$

Substituting values:

$$V_1\left(\frac{1}{8\ \Omega} + \frac{1}{4\ \Omega}\right) - V_2\left(\frac{1}{4\ \Omega}\right) = -2\ \text{A} + \frac{64\ \text{V}}{8\ \Omega} = 6\ \text{A}$$

For node V_2 define the current as shown in Fig. 8.44, and apply Kirchhoff's current law:

$$I = I_2 + I_3$$

continued

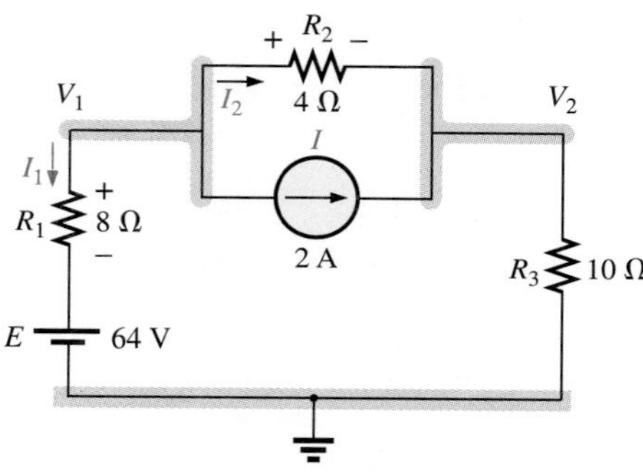

FIG. 8.43
Applying Kirchhoff's current law to node V_1.

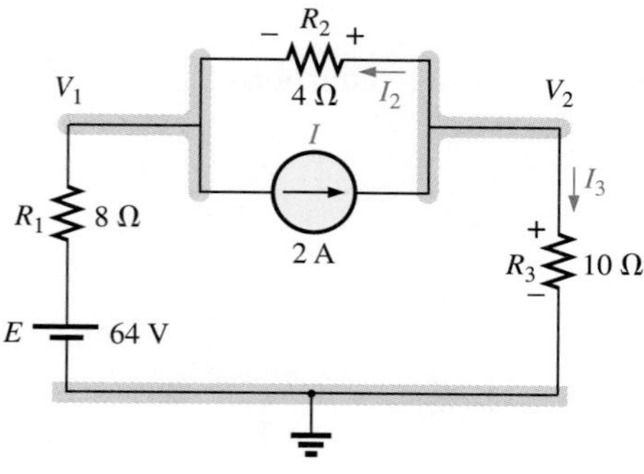

FIG. 8.44
Applying Kirchhoff's current law to node V_2.

with
$$I = \frac{V_2 - V_1}{R_2} + \frac{V_2}{R_3}$$

or
$$I = \frac{V_2}{R_2} - \frac{V_1}{R_2} + \frac{V_2}{R_3}$$

and
$$V_2\left(\frac{1}{R_2} + \frac{1}{R_3}\right) - V_1\left(\frac{1}{R_2}\right) = I$$

Substituting values:

$$V_2\left(\frac{1}{4\ \Omega} + \frac{1}{10\ \Omega}\right) - V_1\left(\frac{1}{4\ \Omega}\right) = 2\text{ A}$$

Step 4: The result is two equations and two unknowns:

$$V_1\left(\frac{1}{8\ \Omega} + \frac{1}{4\ \Omega}\right) - V_2\left(\frac{1}{4\ \Omega}\right) = 6\text{ A}$$
$$-V_1\left(\frac{1}{4\ \Omega}\right) + V_2\left(\frac{1}{4\ \Omega} + \frac{1}{10\ \Omega}\right) = 2\text{ A}$$

which become

$$0.375V_1 - 0.25V_2 = 6$$
$$-0.25V_1 + 0.35V_2 = 2$$

Using determinants,

$$V_1 = \mathbf{37.818\ V}$$
$$V_2 = \mathbf{32.727\ V}$$

Since E is greater than V_1, the current I_1 flows from ground to V_1 and is equal to

$$I_{R_1} = \frac{E - V_1}{R_1} = \frac{64\text{ V} - 37.818\text{ V}}{8\ \Omega} = \mathbf{3.273\ A}$$

The positive value for V_2 results in a current I_{R_3} from node V_2 to ground equal to

$$I_{R_3} = \frac{V_{R_3}}{R_3} = \frac{V_2}{R_3} = \frac{32.727\text{ V}}{10\ \Omega} = \mathbf{3.273\ A}$$

Since V_1 is greater than V_2, the current I_{R_2} flows from V_1 to V_2 and is equal to

$$I_{R_2} = \frac{V_1 - V_2}{R_2} = \frac{37.818\text{ V} - 32.727\text{ V}}{4\ \Omega} = \mathbf{1.273\ A}$$

EXAMPLE 8.21

Determine the nodal voltages for the network of Fig. 8.45.

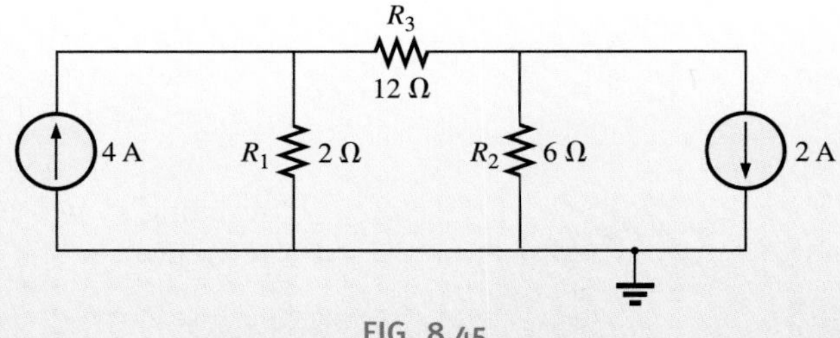

FIG. 8.45
Example 8.21.

continued

Solution:

Steps 1 and 2: As indicated in Fig. 8.46.

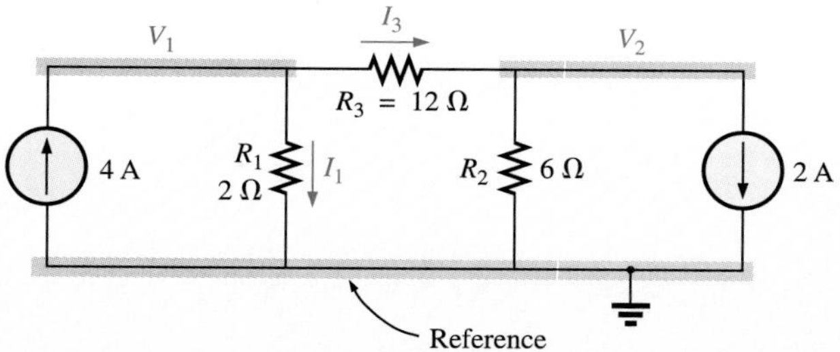

FIG. 8.46
Defining the nodes and applying Kirchhoff's current law to the node V_1.

Step 3: Included in Fig. 8.46 for the node V_1. Applying Kirchhoff's current law:

$$4\text{ A} = I_1 + I_3$$

and
$$4\text{ A} = \frac{V_1}{R_1} + \frac{V_1 - V_2}{R_3} = \frac{V_1}{2\ \Omega} + \frac{V_1 - V_2}{12\ \Omega}$$

Expanding and rearranging:

$$V_1\left(\frac{1}{2\ \Omega} + \frac{1}{12\ \Omega}\right) - V_2\left(\frac{1}{12\ \Omega}\right) = 4\text{ A}$$

For node V_2 define the currents as in Fig. 8.47.

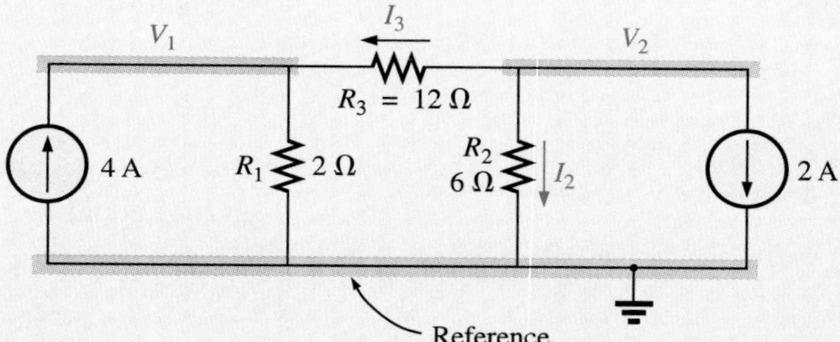

FIG. 8.47
Applying Kirchhoff's current law to the node V_2.

Applying Kirchhoff's current law:

$$0 = I_3 + I_2 + 2\text{ A}$$

and
$$\frac{V_2 - V_1}{R_3} + \frac{V_2}{R_2} + 2\text{ A} = 0 \quad \rightarrow \quad \frac{V_2 - V_1}{12\ \Omega} + \frac{V_2}{6\ \Omega} + 2\text{ A} = 0$$

Expanding and rearranging:

$$V_2\left(\frac{1}{12\ \Omega} + \frac{1}{6\ \Omega}\right) - V_1\left(\frac{1}{12\ \Omega}\right) = -2\text{ A}$$

resulting in two equations and two unknowns (numbered for later reference):

continued

$$V_1\left(\frac{1}{2\ \Omega} + \frac{1}{12\ \Omega}\right) - V_2\left(\frac{1}{12\ \Omega}\right) = +4\ \text{A}$$

$$V_2\left(\frac{1}{12\ \Omega} + \frac{1}{6\ \Omega}\right) - V_1\left(\frac{1}{12\ \Omega}\right) = -2\ \text{A} \qquad (8.3)$$

producing

$$\frac{7}{12}V_1 - \frac{1}{12}V_2 = +4 \qquad 7V_1 - V_2 = 48$$

$$-\frac{1}{12}V_1 + \frac{3}{12}V_2 = -2 \qquad -1V_1 + 3V_2 = -24$$

and

$$V_1 = \frac{\begin{vmatrix} -48 & -1 \\ -24 & -3 \end{vmatrix}}{\begin{vmatrix} -7 & -1 \\ -1 & 3 \end{vmatrix}} = \frac{120}{20} = \mathbf{+6\ V}$$

$$V_2 = \frac{\begin{vmatrix} -7 & 48 \\ -1 & -24 \end{vmatrix}}{20} = \frac{-120}{20} = \mathbf{-6\ V}$$

Since V_1 is greater than V_2, the current through R_3 passes from V_1 to V_2. Its value is:

$$I_{R_3} = \frac{V_1 - V_2}{R_3} = \frac{6\ \text{V} - (-6\ \text{V})}{12\ \Omega} = \frac{12\ \text{V}}{12\ \Omega} = \mathbf{1\ A}$$

The fact that V_1 is positive results in a current I_{R_1} from V_1 to ground equal to:

$$I_{R_1} = \frac{V_{R_1}}{R_1} = \frac{V_1}{R_1} = \frac{6\ \text{V}}{2\ \Omega} = \mathbf{3\ A}$$

Finally, since V_2 is negative, the current I_{R_2} flows from ground to V_2 and is equal to:

$$I_{R_2} = \frac{V_{R_2}}{R_2} = \frac{V_2}{R_2} = \frac{6\ \text{V}}{6\ \Omega} = \mathbf{1\ A}$$

Supernode

There will sometimes be independent voltage sources in the network to which nodal analysis is to be applied. In such cases you can convert the voltage source to a current source (if a series resistor is present) and proceed as before, or introduce the concept of a *supernode* and proceed as follows.

Start as before and assign a nodal voltage to each independent node of the network, including each independent voltage source as if it were a resistor or current source. Then mentally replace the independent voltage sources with short-circuit equivalents and apply Kirchhoff's current law to the defined nodes of the network. Any node including the effect of elements tied only to *other* nodes is referred to as a **supernode** (since it has an additional number of terms). Finally, relate the defined nodes to the independent voltage sources of the network and solve for the nodal voltages. The next example will clarify the definition of supernode.

EXAMPLE 8.22

Determine the nodal voltages V_1 and V_2 of Fig. 8.48 using the concept of a *supernode.*

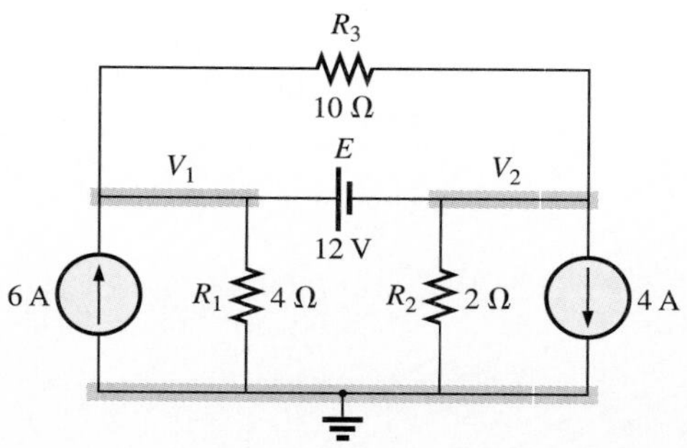

FIG. 8.48
Example 8.22.

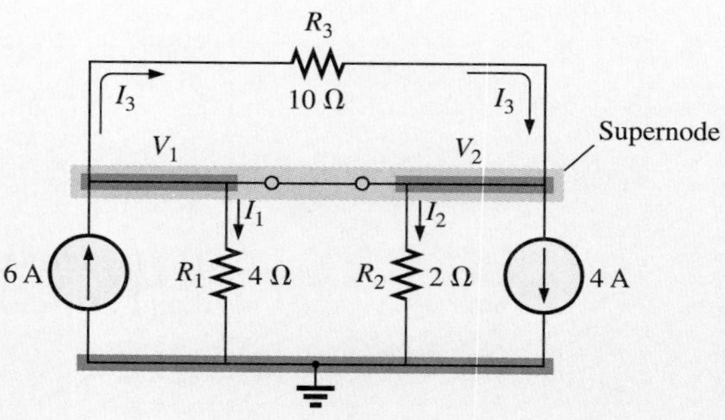

FIG. 8.49
Defining the supernode for the network of Fig. 8.48.

Solution: Replacing the independent voltage source of 12 V with a short-circuit equivalent will result in the network of Fig. 8.49. Even though mentally applying a short-circuit equivalent is discussed above, it would be wise in the early stage of development to redraw the network as shown in Fig. 8.49. The result is a single supernode for which you must apply Kirchhoff's current law. Be sure to leave the other defined nodes in place and use them to define the currents from that region of the network. In particular, note that the current I_3 will leave the supernode at V_1 and then enter the same supernode at V_2. It must therefore appear twice when applying Kirchhoff's current law, as shown below:

$$\Sigma I_i = \Sigma I_o$$

$$6\text{ A} + I_3 = I_1 + I_2 + 4\text{ A} + I_3$$

or

$$I_1 + I_2 = 6\text{ A} - 4\text{ A} = 2\text{ A}$$

Then,

$$\frac{V_1}{R_1} + \frac{V_2}{R_2} = 2\text{ A}$$

continued

and $$\frac{V_1}{4\ \Omega} + \frac{V_2}{2\ \Omega} = 2\text{ A}$$

Relating the defined nodal voltages to the independent voltage source:

$$V_1 - V_2 = E = 12\text{ V}$$

which results in two equations and two unknowns:

$$\begin{aligned} 0.25V_1 + 0.5V_2 &= 2 \\ V_1 - V_2 &= 12 \end{aligned}$$

Substituting:

$$V_1 = V_2 + 12$$
$$0.25(V_2 + 12) + 0.5V_2 = 2$$

and $$0.75V_2 = 2 - 3 = -1$$

so that $$V_2 = \frac{-1}{0.75} = \mathbf{-1.333\ V}$$

and $$V_1 = V_2 + 12\text{ V} = -1.333\text{ V} + 12\text{ V} = \mathbf{+10.667\ V}$$

The current of the network can then be determined as follows:

$$I_1\downarrow = \frac{V}{R_1} = \frac{10.667\text{ V}}{4\ \Omega} = \mathbf{2.667\ A}$$

$$I_2\uparrow = \frac{V_2}{R_2} = \frac{1.333\text{ V}}{2\ \Omega} = \mathbf{0.667\ A}$$

$$I_3 = \frac{V_1 - V_2}{10\ \Omega} = \frac{10.667\text{ V} - (-1.333\text{ V})}{10\ \Omega} = \frac{12\text{ V}}{10\ \Omega} = \mathbf{1.2\ A}$$

Carefully examining the network at the beginning of the analysis would have revealed that the voltage across the resistor R_3 must be 12 V and I_3 must be equal to 1.2 A.

8.10 NODAL ANALYSIS (FORMAT APPROACH)

If you closely examine Eq. (8.3) in Example 8.21 you will see that the subscripted voltage at the node in which Kirchhoff's current law is applied is multiplied by the sum of the conductances attached to that node. Note also that the other nodal voltages within the same equation are multiplied by the negative of the conductance between the two nodes. The current sources are represented to the right of the equals sign with a positive sign if they supply current to the node and with a negative sign if they draw current from the node.

We can expand these conclusions to include networks with any number of nodes. This will allow us to write nodal equations rapidly and in a form that is convenient for the use of determinants. A major requirement, however, is that

all voltage sources must first be converted to current sources before the procedure is applied.

Note the parallelism between the following four steps of application and those required for mesh analysis in Section 8.8:

1. Choose a reference node and assign a subscripted voltage label to the (N − 1) remaining nodes of the network.
2. The number of equations required for a complete solution is equal to the number of subscripted voltages (N − 1). Column 1 of each equation is formed by summing the conductances tied to the node of interest and multiplying the result by that subscripted nodal voltage.
3. We must now consider the mutual terms that, as noted in the preceding example, are always subtracted from the first column. It is possible to have more than one mutual term if the nodal voltage of interest has an element in common with more than one other nodal voltage. We will demonstrate this in an example to follow. Each mutual term is the product of the mutual conductance and the other nodal voltage tied to that conductance.
4. The column to the right of the equals sign is the algebraic sum of the current sources tied to the node of interest. A current source is assigned a positive sign if it supplies current to a node and a negative sign if it draws current from the node.
5. Solve the resulting simultaneous equations for the desired voltages.

Let us now consider a few examples.

EXAMPLE 8.23

Write the nodal equations for the network of Fig. 8.50.

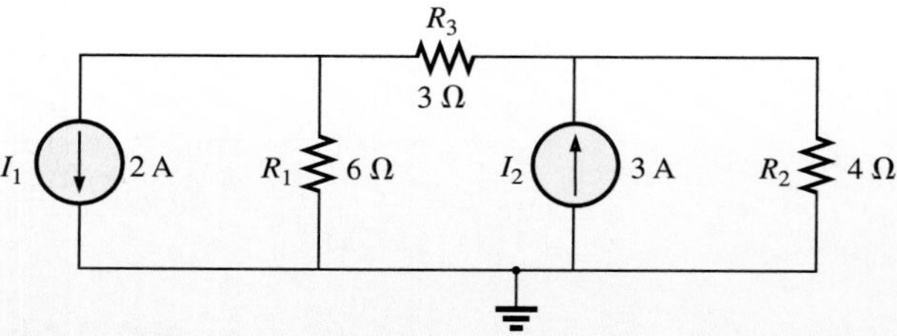

FIG. 8.50
Example 8.23.

Solution:

Step 1: Redraw the figure with assigned subscripted voltages, as in Fig. 8.51.

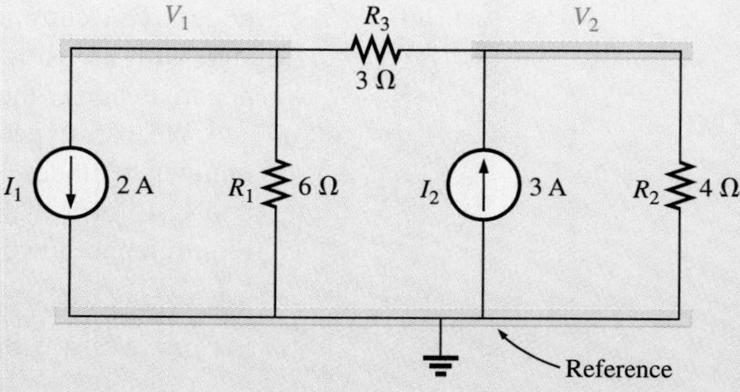

FIG. 8.51
Defining the nodes for the network of Fig. 8.50.

continued

Steps 2 to 4:

Drawing current from node 1

$$V_1: \underbrace{\left(\frac{1}{6\,\Omega} + \frac{1}{3\,\Omega}\right)}_{\text{Sum of conductances connected to node 1}} V_1 - \underbrace{\left(\frac{1}{3\,\Omega}\right)}_{\text{Mutual conductance}} V_2 = -2\text{ A}$$

Supplying current to node 2

$$V_2: \underbrace{\left(\frac{1}{4\,\Omega} + \frac{1}{3\,\Omega}\right)}_{\text{Sum of conductances connected to node 2}} V_2 - \underbrace{\left(\frac{1}{3\,\Omega}\right)}_{\text{Mutual conductance}} V_1 = +3\text{ A}$$

and

$$\frac{1}{2}V_1 - \frac{1}{3}V_2 = -2$$
$$-\frac{1}{3}V_1 + \frac{7}{12}V_2 = 3$$

EXAMPLE 8.24

Find the voltage across the 3-Ω resistor of Fig. 8.52 by nodal analysis.

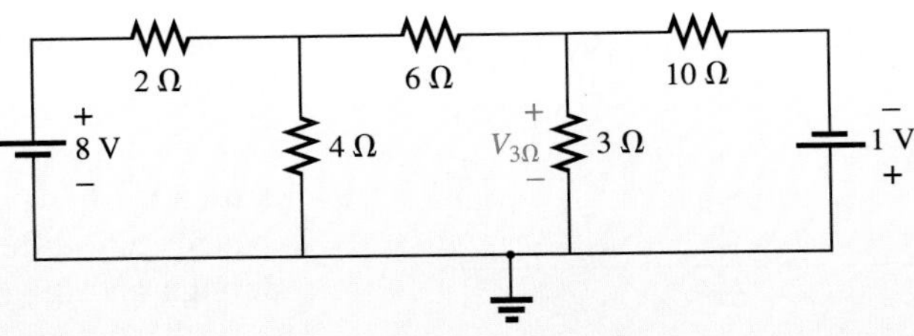

FIG. 8.52
Example 8.24.

Solution: Converting sources and choosing nodes (Fig. 8.53), we have

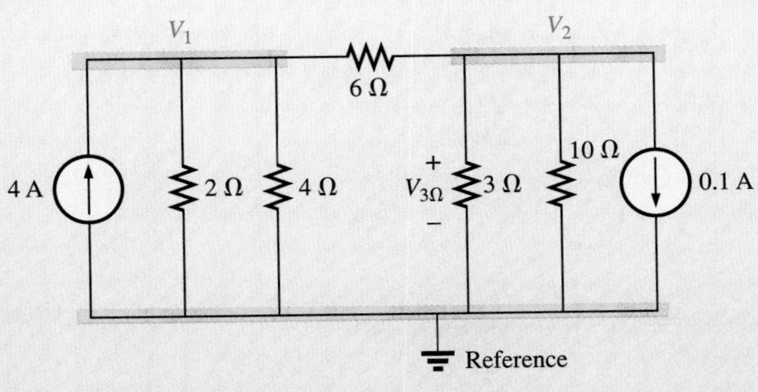

FIG. 8.53
Defining the nodes for the network of Fig. 8.52.

continued

$$\left(\frac{1}{2\,\Omega}+\frac{1}{4\,\Omega}+\frac{1}{6\,\Omega}\right)V_1-\left(\frac{1}{6\,\Omega}\right)V_2=+4\text{ A}$$

$$\left(\frac{1}{10\,\Omega}+\frac{1}{3\,\Omega}+\frac{1}{6\,\Omega}\right)V_2-\left(\frac{1}{6\,\Omega}\right)V_1=-0.1\text{ A}$$

$$\frac{11}{12}V_1-\frac{1}{6}V_2=4$$

$$-\frac{1}{6}V_1+\frac{3}{5}V_2=-0.1$$

resulting in

$$11V_1-2V_2=+48$$
$$-5V_1+18V_2=-3$$

and

$$V_2=V_{3\Omega}=\frac{\begin{vmatrix}11 & 48\\ -5 & -3\end{vmatrix}}{\begin{vmatrix}11 & -2\\ -5 & 18\end{vmatrix}}=\frac{-33+240}{198-10}=\frac{207}{188}=\mathbf{1.101\ V}$$

Like mesh analysis, nodal analysis can also be a very useful technique for solving networks with only one source.

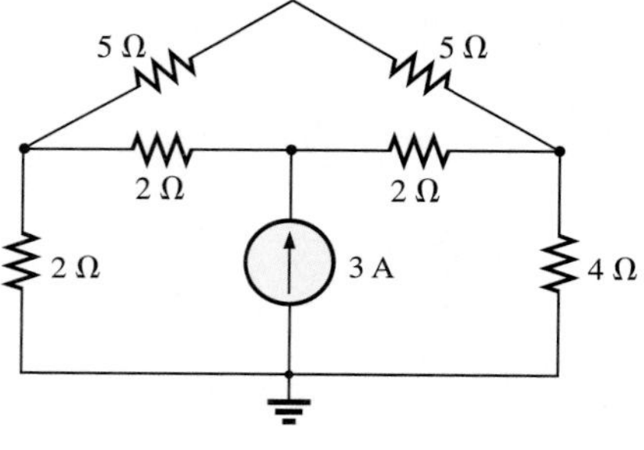

FIG. 8.54
Example 8.25.

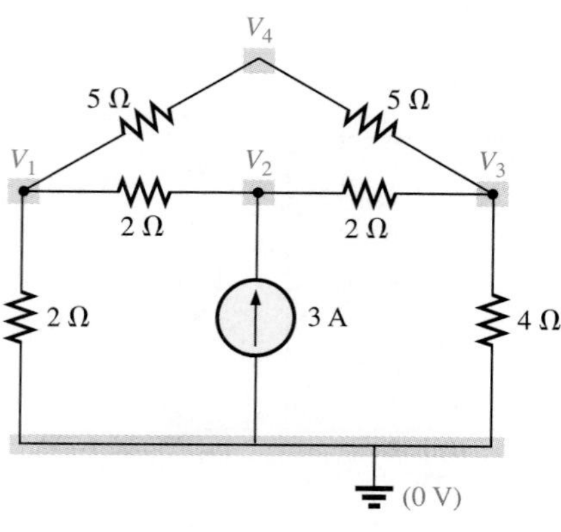

FIG. 8.55
Defining the nodes for the network of Fig. 8.54.

EXAMPLE 8.25

Using nodal analysis, determine the potential across the 4-Ω resistor in Fig. 8.54.

Solution: The reference and four subscripted voltage levels were chosen as shown in Fig. 8.55. A moment of reflection should reveal that for any difference in potential between V_1 and V_3, the current through and the potential drop across each 5-Ω resistor will be the same. Therefore, V_4 is simply a midvoltage level between V_1 and V_3 and is known if V_1 and V_3 are available. We will therefore not include it in a nodal voltage and will redraw the network as shown in Fig. 8.56. Understand, however, that V_4 can be included if desired, although four nodal voltages will result rather than the three to be obtained in the solution of this problem.

$$V_1\text{:}\quad\left(\frac{1}{2\,\Omega}+\frac{1}{2\,\Omega}+\frac{1}{10\,\Omega}\right)V_1-\left(\frac{1}{2\,\Omega}\right)V_2-\left(\frac{1}{10\,\Omega}\right)V_3=0$$

$$V_2\text{:}\quad\left(\frac{1}{2\,\Omega}+\frac{1}{2\,\Omega}\right)V_2-\left(\frac{1}{2\,\Omega}\right)V_1-\left(\frac{1}{2\,\Omega}\right)V_3=3\text{ A}$$

$$V_3\text{:}\quad\left(\frac{1}{10\,\Omega}+\frac{1}{2\,\Omega}+\frac{1}{4\,\Omega}\right)V_3-\left(\frac{1}{2\,\Omega}\right)V_2-\left(\frac{1}{10\,\Omega}\right)V_1=0$$

which are rewritten as

$$1.1V_1-0.5V_2-0.1V_3=0$$
$$V_2-0.5V_1-0.5V_3=3$$
$$0.85V_3-0.5V_2-0.1V_1=0$$

continued

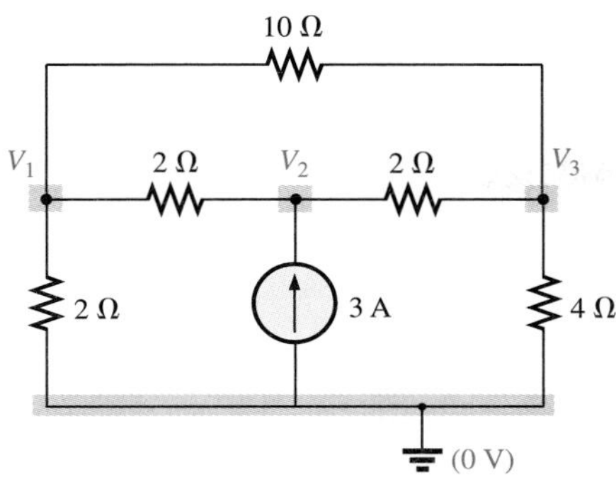

FIG. 8.56
Reducing the number of nodes for the network of Fig. 8.54 by combining the two 5-Ω resistors.

For determinants,

$$
\begin{aligned}
1.1V_1 - 0.5V_2 - 0.1V_3 &= 0 \\
-0.5V_1 + 1V_2 - 0.5V_3 &= 3 \\
-0.1V_1 - 0.5V_2 + 0.85V_3 &= 0
\end{aligned}
$$

Before continuing, note the symmetry about the major diagonal in the equation above. Recall a similar result for mesh analysis. Examples 8.23 and 8.24 also show this symmetry in the resulting equations. Keep this thought in mind as a check on future applications of nodal analysis.

$$
V_3 = V_{4\Omega} = \frac{\begin{vmatrix} 1.1 & -0.5 & 0 \\ -0.5 & +1 & 3 \\ -0.1 & -0.5 & 0 \end{vmatrix}}{\begin{vmatrix} 1.1 & -0.5 & -0.1 \\ -0.5 & +1 & -0.5 \\ -0.1 & -0.5 & +0.85 \end{vmatrix}} = \mathbf{4.645\ V}
$$

The next example has only one source applied to a ladder network.

EXAMPLE 8.26

Write the nodal equations and find the voltage across the 2-Ω resistor for the network of Fig. 8.57.

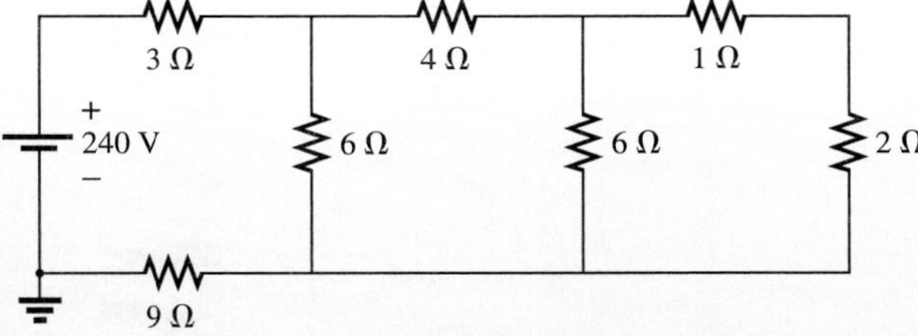

FIG. 8.57
Example 8.26.

Solution: The nodal voltages are chosen as shown in Fig. 8.58.

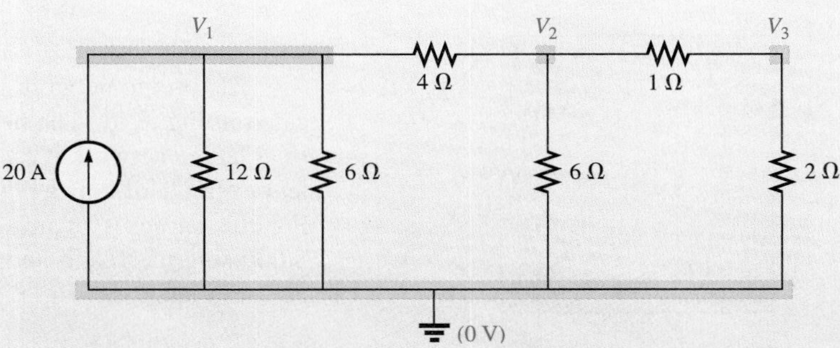

FIG. 8.58
Converting the voltage source to a current source and defining the nodes for the network of Fig. 8.57.

continued

$$V_1: \left(\frac{1}{12\,\Omega} + \frac{1}{6\,\Omega} + \frac{1}{4\,\Omega}\right)V_1 - \left(\frac{1}{4\,\Omega}\right)V_2 + \quad 0 \quad = 20\text{ V}$$

$$V_2: \quad \left(\frac{1}{4\,\Omega} + \frac{1}{6\,\Omega} + \frac{1}{1\,\Omega}\right)V_2 - \left(\frac{1}{4\,\Omega}\right)V_1 - \left(\frac{1}{1\,\Omega}\right)V_3 = 0$$

$$V_3: \quad \left(\frac{1}{1\,\Omega} + \frac{1}{2\,\Omega}\right)V_3 - \left(\frac{1}{1\,\Omega}\right)V_2 + \quad 0 \quad = 0$$

and

$$\begin{aligned} 0.5V_1 \quad -0.25V_2 \quad\quad +0 &= 20 \\ -0.25V_1 \quad +\frac{17}{12}V_2 \quad -1V_3 &= 0 \\ 0 \quad -1V_2 \quad +1.5V_3 &= 0 \end{aligned}$$

Note the symmetry about the major axis. Applying determinants reveals that

$$V_3 = V_{2\Omega} = \mathbf{10.667\ V}$$

8.11 BRIDGE NETWORKS

This section introduces the **bridge network**, a configuration that has many applications. In the chapters to follow, it will be used in both dc and ac meters. In electronics courses it will be encountered early in the discussion of rectifying circuits used in converting a varying signal to one of a steady nature (such as dc). There is a number of other areas of application that require some knowledge of ac networks, which will be discussed later.

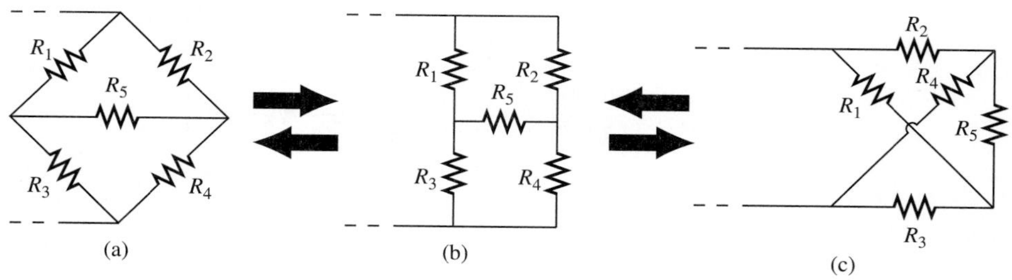

FIG. 8.59
Various formats for a bridge network.

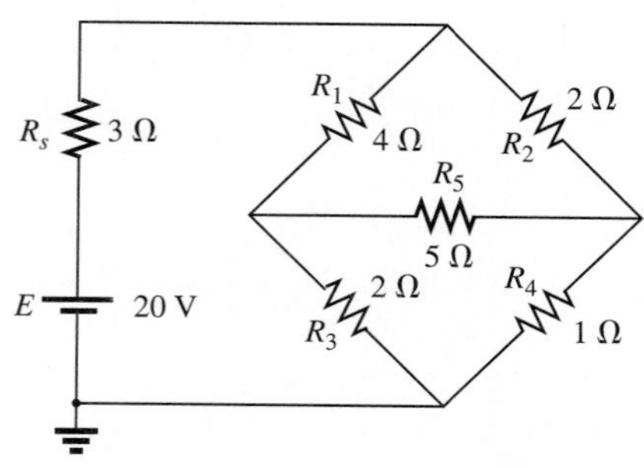

FIG. 8.60
Standard bridge configuration.

The bridge network may appear in one of the three forms as indicated in Fig. 8.59. The network of Fig. 8.59(c) is also called a symmetrical lattice network if $R_2 = R_3$ and $R_1 = R_4$. Figure 8.59(c) is an excellent example of how a planar network can be made to appear nonplanar. For the purposes of investigation, let us examine the network of Fig. 8.60 using mesh and nodal analysis.

Mesh analysis (Fig. 8.61) gives

$$\begin{aligned} (3\,\Omega + 4\,\Omega + 2\,\Omega)I_1 - (4\,\Omega)I_2 - (2\,\Omega)I_3 &= 20\text{ V} \\ (4\,\Omega + 5\,\Omega + 2\,\Omega)I_2 - (4\,\Omega)I_1 - (5\,\Omega)I_3 &= 0 \\ (2\,\Omega + 5\,\Omega + 1\,\Omega)I_3 - (2\,\Omega)I_1 - (5\,\Omega)I_2 &= 0 \end{aligned}$$

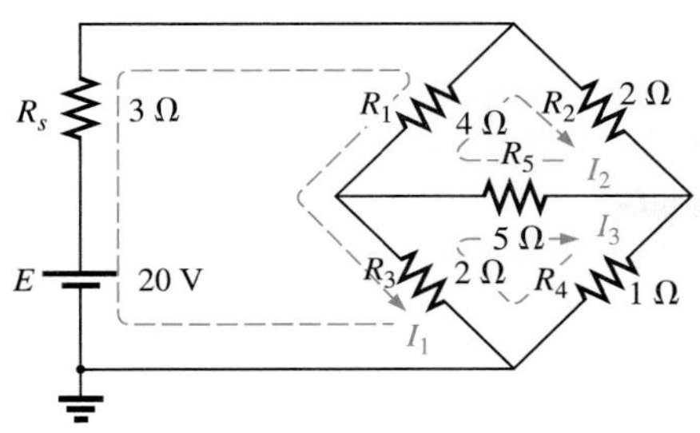

FIG. 8.61
Assigning the mesh currents to the network of Fig. 8.60.

and

$$\begin{aligned} 9I_1 - 4I_2 - 2I_3 &= 20 \\ -4I_1 + 11I_2 - 5I_3 &= 0 \\ -2I_1 - 5I_2 + 8I_3 &= 0 \end{aligned}$$

with the result that

$$I_1 = \mathbf{4\ A}$$
$$I_2 = \mathbf{2.667\ A}$$
$$I_3 = \mathbf{2.667\ A}$$

The net current through the 5-Ω resistor is

$$I_{5\Omega} = I_2 - I_3 = \mathbf{2.667\ A - 2.667\ A = 0\ A}$$

Nodal analysis (Fig. 8.62) gives

$$\left(\frac{1}{3\ \Omega} + \frac{1}{4\ \Omega} + \frac{1}{2\ \Omega}\right)V_1 - \left(\frac{1}{4\ \Omega}\right)V_2 - \left(\frac{1}{2\ \Omega}\right)V_3 = \frac{20}{3}\ \text{A}$$
$$\left(\frac{1}{4\ \Omega} + \frac{1}{2\ \Omega} + \frac{1}{5\ \Omega}\right)V_2 - \left(\frac{1}{4\ \Omega}\right)V_1 - \left(\frac{1}{5\ \Omega}\right)V_3 = 0$$
$$\left(\frac{1}{5\ \Omega} + \frac{1}{2\ \Omega} + \frac{1}{1\ \Omega}\right)V_3 - \left(\frac{1}{2\ \Omega}\right)V_1 - \left(\frac{1}{5\ \Omega}\right)V_2 = 0$$

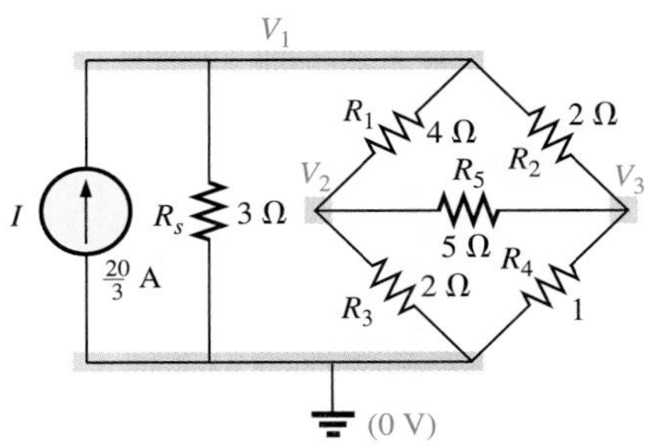

FIG. 8.62
Defining the nodal voltages for the network of Fig. 8.60.

and

$$\left(\frac{1}{3\ \Omega} + \frac{1}{4\ \Omega} + \frac{1}{2\ \Omega}\right)V_1 - \left(\frac{1}{4\ \Omega}\right)V_2 - \left(\frac{1}{2\ \Omega}\right)V_3 = \frac{20}{3}\ \text{A}$$
$$-\left(\frac{1}{4\ \Omega}\right)V_1 + \left(\frac{1}{4\ \Omega} + \frac{1}{2\ \Omega} + \frac{1}{5\ \Omega}\right)V_2 - \left(\frac{1}{5\ \Omega}\right)V_3 = 0$$
$$-\left(\frac{1}{2\ \Omega}\right)V_1 - \left(\frac{1}{5\ \Omega}\right)V_2 + \left(\frac{1}{5\ \Omega} + \frac{1}{2\ \Omega} + \frac{1}{1\ \Omega}\right)V_3 = 0$$

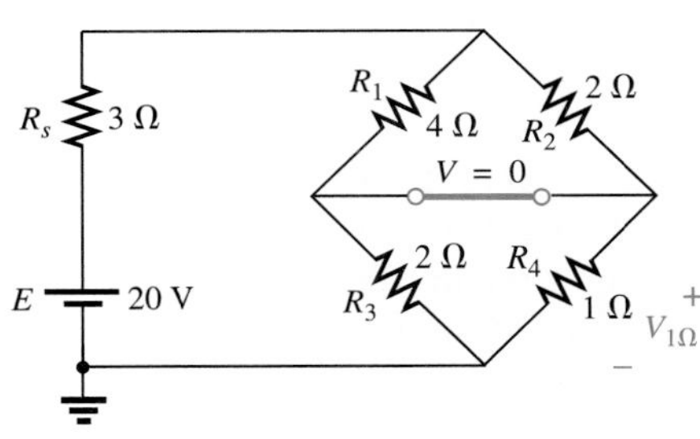

FIG. 8.63
Substituting the short-circuit equivalent for the balance arm of a balanced bridge.

Note the symmetry of the solution.

Application of determinants yields

$$V_1 = \frac{10.5}{1.312} = \mathbf{8\ V}$$

Similarly, $V_2 = \mathbf{2.667\ V}$ and $V_3 = \mathbf{2.667\ V}$

and the voltage across the 5-Ω resistor is

$$V_{5\Omega} = V_2 - V_3 = \mathbf{2.667\ V - 2.667\ V = 0\ V}$$

Since $V_{5\Omega} = 0$ V, we can insert a short in place of the bridge arm without affecting the network behaviour. (Certainly $V = IR = I\cdot(0) = 0$ V.) In Fig. 8.63, a short circuit has replaced the resistor R_5, and the voltage across R_4 is to be determined. The network is redrawn in Fig. 8.64, and

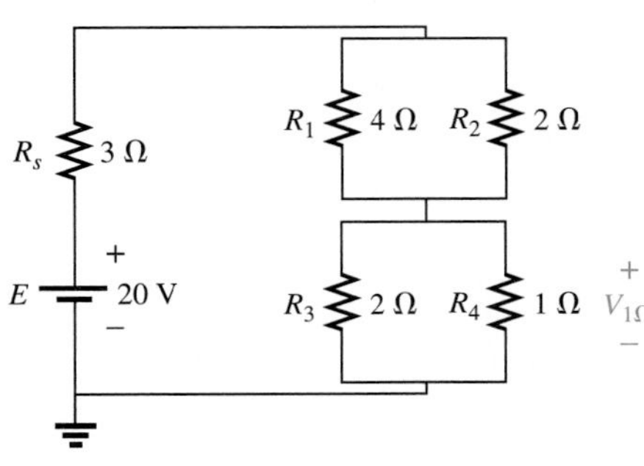

FIG. 8.64
Redrawing the network of Fig. 8.63.

$$V_{1\Omega} = \frac{(2\ \Omega \parallel 1\ \Omega)20\ \text{V}}{(2\ \Omega \parallel 1\ \Omega) + (4\ \Omega \parallel 2\ \Omega) + 3\ \Omega} \qquad \text{(voltage divider rule)}$$

$$= \frac{\frac{2}{3}(20\ \text{V})}{\frac{2}{3} + \frac{8}{6} + 3} = \frac{\frac{2}{3}(20\ \text{V})}{\frac{2}{3} + \frac{4}{3} + \frac{9}{3}}$$

$$= \frac{2(20\ \text{V})}{2 + 4 + 9} = \frac{40\ \text{V}}{15} = \mathbf{2.667\ V}$$

as obtained earlier.

We found through mesh analysis that $I_{5\Omega} = 0$ A, which has as its equivalent an open circuit as shown in Fig. 8.65(a). (Certainly $I = V/R = 0/(\infty\ \Omega) = 0$ A.) The voltage across the resistor R_4 will again be determined and compared with the result above.

The network is redrawn after combining series elements, as shown in Fig. 8.65(b), and

$$V_{3\Omega} = \frac{(6\ \Omega \parallel 3\ \Omega)(20\text{ V})}{6\ \Omega \parallel 3\ \Omega + 3\ \Omega} = \frac{2\ \Omega(20\text{ V})}{2\ \Omega + 3\ \Omega} = 8\text{ V}$$

and

$$V_{1\Omega} = \frac{1\ \Omega(8\text{ V})}{1\ \Omega + 2\ \Omega} = \frac{8\text{ V}}{3} = \mathbf{2.667\ V}$$

as above.

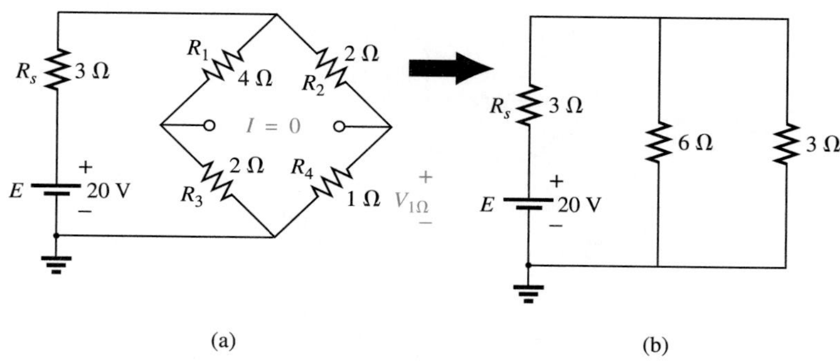

FIG. 8.65
Substituting the open-circuit equivalent for the balance arm of a balanced bridge.

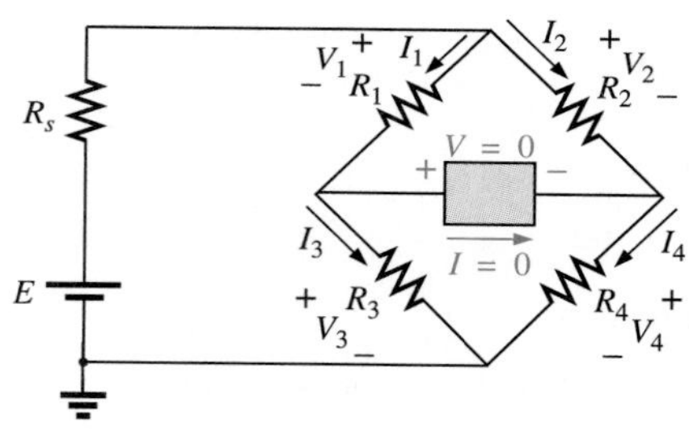

FIG. 8.66
Establishing the balance criteria for a bridge network.

The condition $V_{5\Omega} = 0$ V or $I_{5\Omega} = 0$ A exists only for a particular relationship between the resistors of the network. Let us now derive this relationship using the network of Fig. 8.66, in which it is indicated that $I = 0$ A and $V = 0$ V. Note that resistor R_s of the network of Fig. 8.65 will not appear in the following analysis.

The bridge network is said to be *balanced* when the condition of $I = 0$ A or $V = 0$ V exists.

If $V = 0$ V (short circuit between *a* and *b*), then

$$V_1 = V_2$$

and

$$I_1R_1 = I_2R_2$$

or

$$I_1 = \frac{I_2R_2}{R_1}$$

In addition, when $V = 0$ V,

$$V_3 = V_4$$

and

$$I_3R_3 = I_4R_4$$

If we set $I = 0$ A, then $I_3 = I_1$ and $I_4 = I_2$, with the result that the above equation becomes

$$I_1R_3 = I_2R_4$$

Substituting for I_1 from above gives

$$\left(\frac{I_2 R_2}{R_1}\right) R_3 = I_2 R_4$$

or, rearranging, we have

$$\frac{R_1}{R_3} = \frac{R_2}{R_4} \tag{8.4}$$

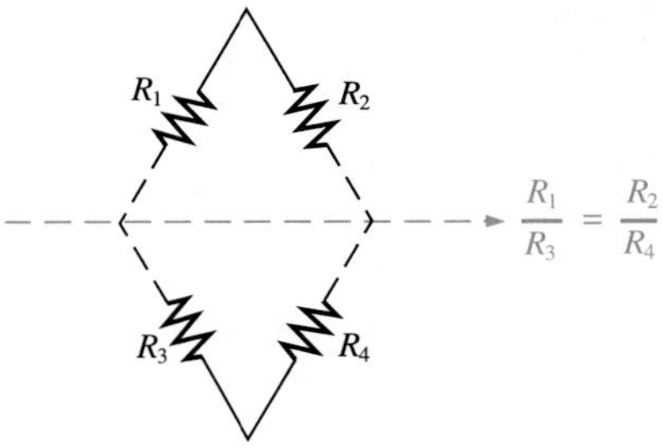

FIG. 8.67
A visual approach to remembering the balance condition.

This conclusion states that if the ratio of R_1 to R_3 is equal to that of R_2 to R_4, the bridge will be balanced, and $I = 0$ A or $V = 0$ V. A method of memorizing this form is indicated in Fig. 8.67.

For the example above, $R_1 = 4\ \Omega$, $R_2 = 2\ \Omega$, $R_3 = 2\ \Omega$, $R_4 = 1\ \Omega$, and

$$\frac{R_1}{R_3} = \frac{R_2}{R_4} \qquad \frac{4\ \Omega}{2\ \Omega} = \frac{2\ \Omega}{1\ \Omega} = 2$$

The emphasis in this section has been on the balanced situation. Understand that if the ratio is not satisfied, there will be a potential drop across the balance arm and a current through it. The methods just described (mesh and nodal analysis) will yield any and all potentials or currents desired, just as they did for the balanced situation.

8.12 Y-Δ (T-π) AND Δ-Y (π-T) CONVERSIONS

You will often encounter circuit configurations in which the resistors do not appear to be in series or parallel. Under these conditions, it may be necessary to convert the circuit from one form to another to solve for any unknown quantities if mesh or nodal analysis is not applied. Two circuit configurations that often account for these difficulties are the *wye* (Y) and *delta* (Δ), depicted in Fig. 8.68(a). They are also referred to as the *tee* (T) and *pi* (π), respectively, as indicated in Fig. 8.68(b). Note that the pi is actually an inverted delta.

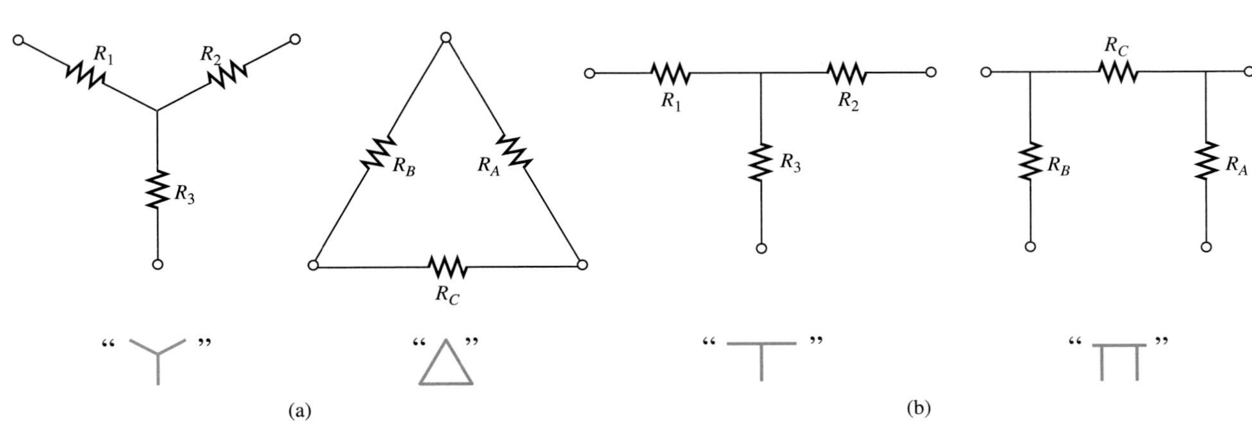

FIG. 8.68
The Y (T) and Δ (π) configurations.

The purpose of this section is to develop the equations for converting from Δ to Y, or vice versa. This type of conversion will normally lead to a network that can be solved using techniques such as those

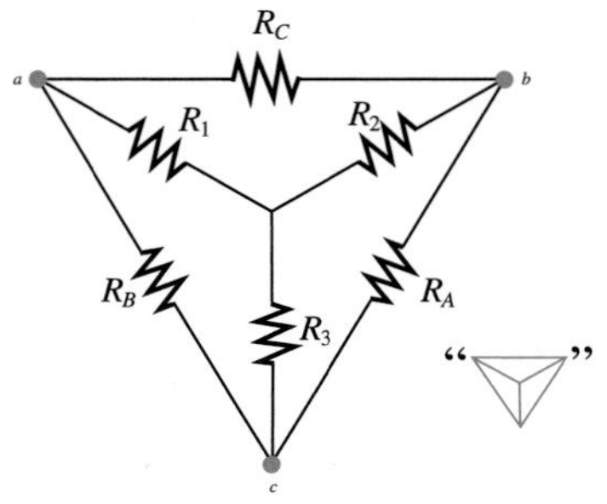

FIG. 8.69
Introducing the concept of Δ-Y or Y-Δ conversions.

described in Chapter 7. In other words, in Fig. 8.69, with terminals *a*, *b*, and *c* held fast, if the wye (Y) configuration were desired *instead of* the inverted delta (Δ) configuration, all that would be necessary is a direct application of the equations to be derived. The phrase *instead of* is emphasized to ensure that you understand that only one of these configurations is to appear at one time between the indicated terminals.

It is our purpose (referring to Fig. 8.69) to find some expression for R_1, R_2, and R_3 in terms of R_A, R_B, and R_C, and vice versa, that will ensure that the resistance between any two terminals of the Y configuration will be the same with the Δ configuration inserted in place of the Y configuration (and vice versa). If the two circuits are to be equivalent, the total resistance between any two terminals must be the same. Consider terminals *a*-*c* in the Δ-Y configurations of Fig. 8.70.

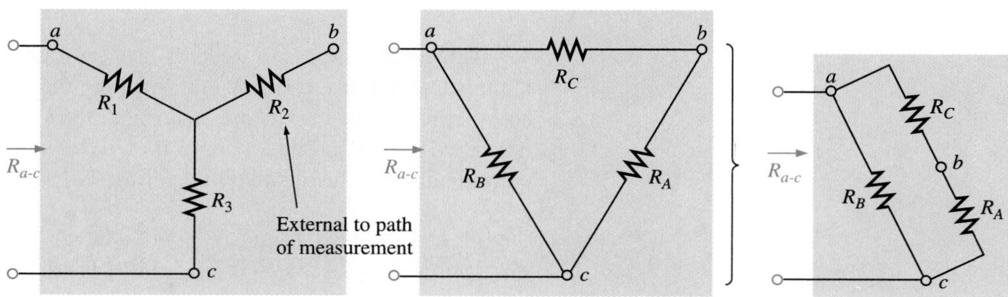

FIG. 8.70
Finding the resistance $R_{a\text{-}c}$ for the Y and Δ configurations.

Let us first assume that we want to convert the Δ (R_A, R_B, R_C) to the Y (R_1, R_2, R_3). This requires that we have a relationship for R_1, R_2, and R_3 in terms of R_A, R_B, and R_C. If the resistance is to be the same between terminals *a*-*c* for both the Δ and the Y, the following must be true:

$$R_{a\text{-}c}\,(\text{Y}) = R_{a\text{-}c}\,(\Delta)$$

so that

$$R_{a\text{-}c} = R_1 + R_3 = \frac{R_B(R_A + R_C)}{R_B + (R_A + R_C)} \qquad (8.5a)$$

Using the same approach for *a*-*b* and *b*-*c*, we obtain the following relationships:

$$R_{a\text{-}b} = R_1 + R_2 = \frac{R_C(R_A + R_B)}{R_C + (R_A + R_B)} \qquad (8.5b)$$

and

$$R_{b\text{-}c} = R_2 + R_3 = \frac{R_A(R_B + R_C)}{R_A + (R_B + R_C)} \qquad (8.5c)$$

Subtracting Eq. (8.5a) from Eq. (8.5b), we have

$$(R_1 + R_2) - (R_1 + R_3) = \left(\frac{R_C R_B + R_C R_A}{R_A + R_B + R_C}\right) - \left(\frac{R_B R_A + R_B R_C}{R_A + R_B + R_C}\right)$$

so that

$$R_2 - R_3 = \frac{R_A R_C - R_B R_A}{R_A + R_B + R_C} \qquad (8.5d)$$

continued

Subtracting Eq. (8.5d) from Eq. (8.5c) gives

$$(R_2 + R_3) - (R_2 - R_3) = \left(\frac{R_A R_B + R_A R_C}{R_A + R_B + R_C}\right) - \left(\frac{R_A R_C - R_B R_A}{R_A + R_B + R_C}\right)$$

so that

$$2R_3 = \frac{2R_B R_A}{R_A + R_B + R_C}$$

resulting in the following expression for R_3 in terms of R_A, R_B, and R_C:

$$R_3 = \frac{R_A R_B}{R_A + R_B + R_C} \tag{8.6a}$$

Following the same procedure for R_1 and R_2, we have

$$R_1 = \frac{R_B R_C}{R_A + R_B + R_C} \tag{8.6b}$$

and

$$R_2 = \frac{R_A R_C}{R_A + R_B + R_C} \tag{8.6c}$$

Note that each resistor of the Y is equal to the product of the resistors in the two closest branches of the D divided by the sum of the resistors in the D.

To obtain the relationships necessary to convert from a Y to a Δ, first divide Eq. (8.6a) by Eq. (8.6b):

$$\frac{R_3}{R_1} = \frac{(R_A R_B)/(R_A + R_B + R_C)}{(R_B R_C)/(R_A + R_B + R_C)} = \frac{R_A}{R_C}$$

or

$$R_A = \frac{R_C R_3}{R_1}$$

Then divide Eq. (8.6a) by Eq. (8.6c):

$$\frac{R_3}{R_2} = \frac{(R_A R_B)/(R_A + R_B + R_C)}{(R_A R_C)/(R_A + R_B + R_C)} = \frac{R_B}{R_C}$$

or

$$R_B = \frac{R_3 R_C}{R_2}$$

Substituting for R_A and R_B in Eq. (8.6c) gives

$$R_2 = \frac{(R_C R_3/R_1)R_C}{(R_3 R_C/R_2) + (R_C R_3/R_1) + R_C}$$

$$= \frac{(R_3/R_1)R_C}{(R_3/R_2) + (R_3/R_1) + 1}$$

Placing these over a common denominator, we obtain

$$R_2 = \frac{(R_3 R_C/R_1)}{(R_1 R_2 + R_1 R_3 + R_2 R_3)/(R_1 R_2)}$$

$$= \frac{R_2 R_3 R_C}{R_1 R_2 + R_1 R_3 + R_2 R_3}$$

and

$$R_C = \frac{R_1 R_2 + R_1 R_3 + R_2 R_3}{R_3} \tag{8.7a}$$

continued

We follow the same procedure for R_B and R_A:

$$R_A = \frac{R_1R_2 + R_1R_3 + R_2R_3}{R_1} \tag{8.7b}$$

and

$$R_B = \frac{R_1R_2 + R_1R_3 + R_2R_3}{R_2} \tag{8.7c}$$

Note that the value of each resistor of the Δ is equal to the sum of the possible product combinations of the resistances of the Y divided by the resistance of the Y farthest from the resistor to be determined.

Let us consider what would occur if all the values of a Δ or Y were the same. If $R_A = R_B = R_C$, Eq. (8.6a) would become (using R_A only) the following:

$$R_3 = \frac{R_AR_B}{R_A + R_B + R_C} = \frac{R_AR_A}{R_A + R_A + R_A} = \frac{R_A^2}{3R_A} = \frac{R_A}{3}$$

and, following the same procedure,

$$R_1 = \frac{R_A}{3} \qquad R_2 = \frac{R_A}{3}$$

In general, therefore,

$$R_Y = \frac{R_\Delta}{3} \tag{8.8a}$$

or

$$R_\Delta = 3R_Y \tag{8.8b}$$

which indicates that

for a Y of three equal resistors, the value of each resistor of the Δ is equal to three times the value of any resistor of the Y.

If only two elements of a Y or a Δ are the same, the corresponding Δ or Y of each will also have two equal elements. Converting the equations will be left as an exercise for the reader.

The Y and the Δ will often appear as shown in Fig. 8.71. They are then referred to as a *tee* (T) and *pi* (π) network. The equations used to convert from one form to the other are exactly the same as those developed for the Y and Δ transformation.

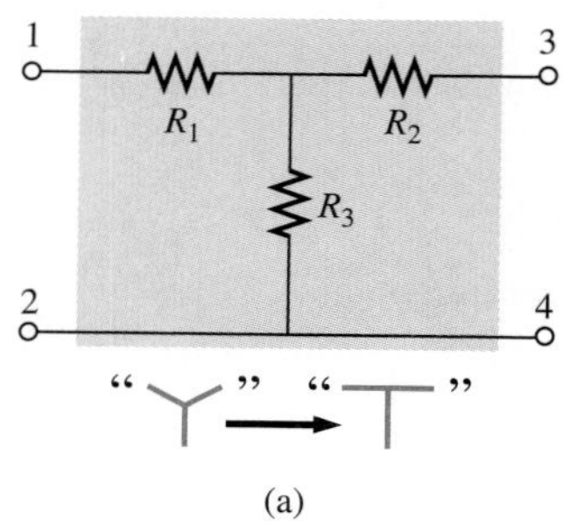

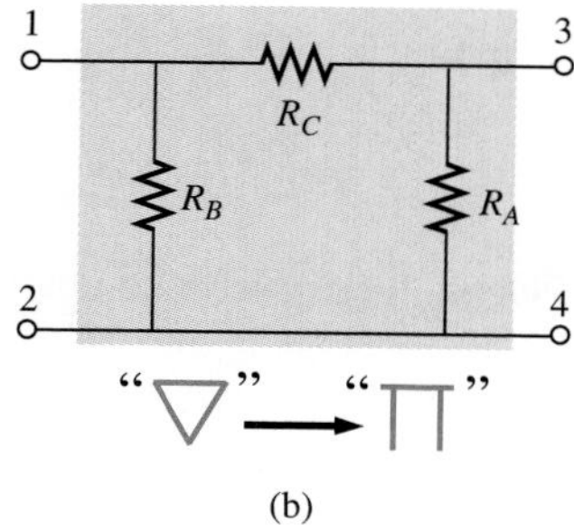

FIG. 8.71
The relationship between the Y and T configurations and the Δ and π configurations.

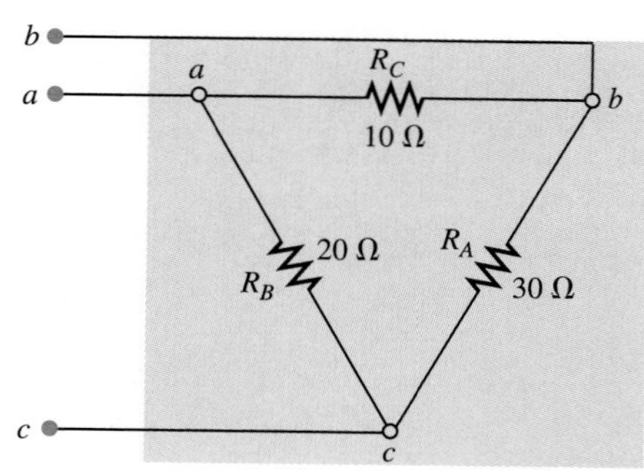

FIG. 8.72
Example 8.27.

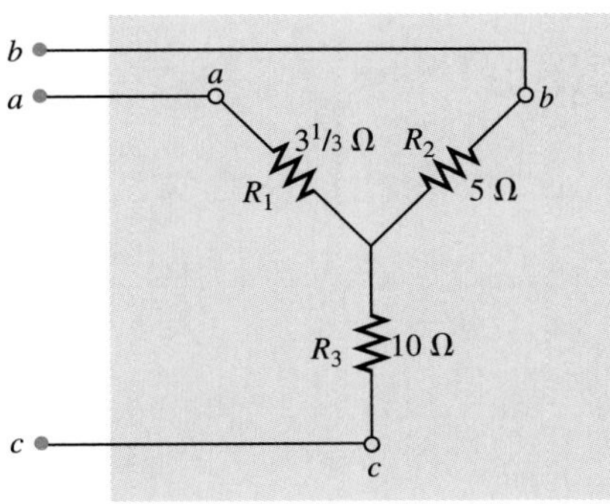

FIG. 8.73
The Y equivalent for the Δ of Fig. 8.72.

EXAMPLE 8.27

Convert the Δ of Fig. 8.72 to a Y.

Solution:

$$R_1 = \frac{R_B R_C}{R_A + R_B + R_C} = \frac{(20\ \Omega)(10\ \Omega)}{30\ \Omega + 20\ \Omega + 10\ \Omega} = \frac{200\ \Omega}{60} = \mathbf{3\tfrac{1}{3}\ \Omega}$$

$$R_2 = \frac{R_A R_C}{R_A + R_B + R_C} = \frac{(30\ \Omega)(10\ \Omega)}{60\ \Omega} = \frac{300\ \Omega}{60} = \mathbf{5\ \Omega}$$

$$R_3 = \frac{R_A R_B}{R_A + R_B + R_C} = \frac{(20\ \Omega)(30\ \Omega)}{60\ \Omega} = \frac{600\ \Omega}{60} = \mathbf{10\ \Omega}$$

The equivalent network is shown in Fig. 8.73.

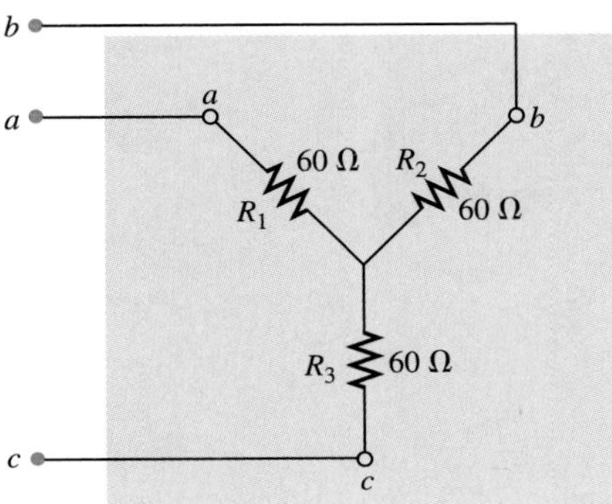

FIG. 8.74
Example 8.28.

EXAMPLE 8.28

Convert the Y of Fig. 8.74 to a Δ.

Solution:

$$R_A = \frac{R_1R_2 + R_1R_3 + R_2R_3}{R_1}$$

$$= \frac{(60\ \Omega)(60\ \Omega) + (60\ \Omega)(60\ \Omega) + (60\ \Omega)(60\ \Omega)}{60\ \Omega}$$

$$= \frac{3600\ \Omega + 3600\ \Omega + 3600\ \Omega}{60} = \frac{10{,}800\ \Omega}{60}$$

$$R_A = \mathbf{180\ \Omega}$$

However, the three resistors for the Y are equal, which permits the use of Eq. (8.8) and giving

$$R_\Delta = 3R_Y = 3(60\ \Omega) = 180\ \Omega$$

and

$$R_B = R_C = \mathbf{180\ \Omega}$$

The equivalent network is shown in Fig. 8.75.

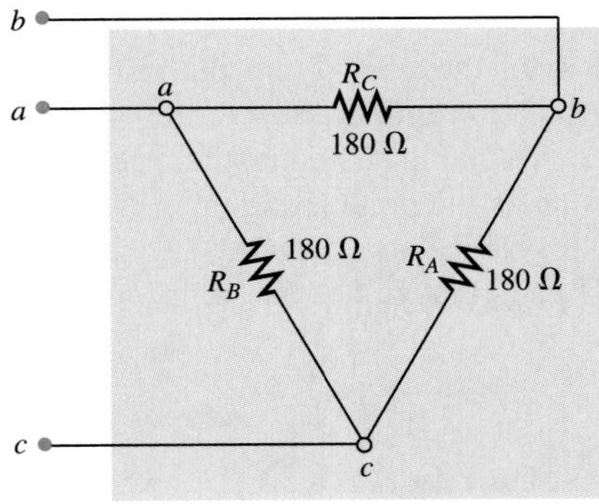

FIG. 8.75
The Δ equivalent for the Y of Fig. 8.74.

EXAMPLE 8.29

Find the total resistance of the network of Fig. 8.76, where $R_A = 3\ \Omega$, $R_B = 3\ \Omega$, and $R_C = 6\ \Omega$.

Solution:

Two resistors of the Δ were equal; therefore, two resistors of the Y will be equal.

$$R_1 = \frac{R_B R_C}{R_A + R_B + R_C} = \frac{(3\ \Omega)(6\ \Omega)}{3\ \Omega + 3\ \Omega + 6\ \Omega} = \frac{18\ \Omega}{12} = \mathbf{1.5\ \Omega}$$

$$R_2 = \frac{R_A R_C}{R_A + R_B + R_C} = \frac{(3\ \Omega)(6\ \Omega)}{12\ \Omega} = \frac{18\ \Omega}{12} = \mathbf{1.5\ \Omega}$$

continued

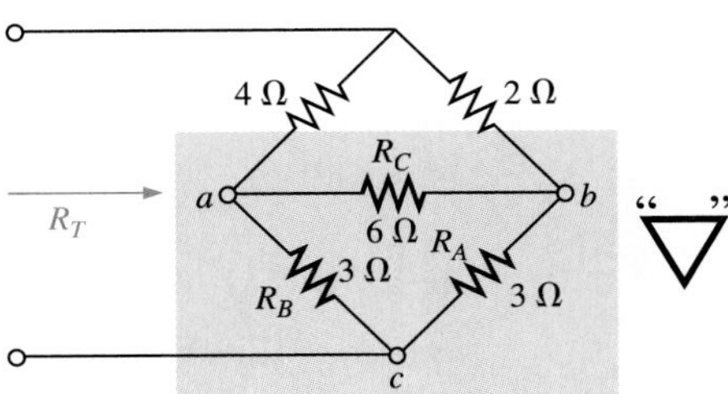

FIG. 8.76
Example 8.29.

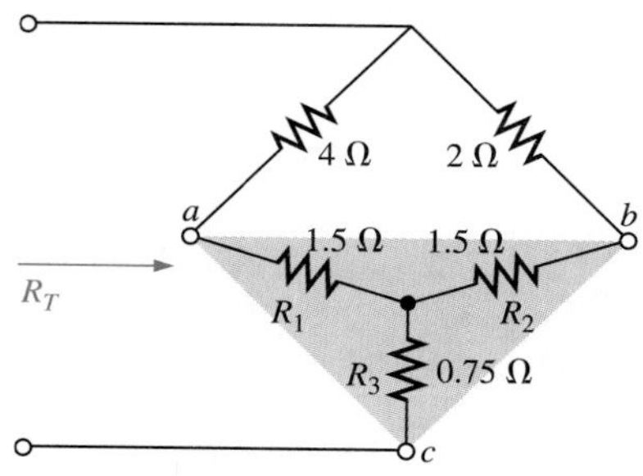

FIG. 8.77
Substituting the Y equivalent for the bottom Δ of Fig. 8.76.

$$R_3 = \frac{R_A R_B}{R_A + R_B + R_C} = \frac{(3\ \Omega)(3\ \Omega)}{12\ \Omega} = \frac{9\ \Omega}{12} = \mathbf{0.75\ \Omega}$$

Replacing the Δ by the Y, as shown in Fig. 8.77, results in

$$R_T = 0.75\ \Omega + \frac{(4\ \Omega + 1.5\ \Omega)(2\ \Omega + 1.5\ \Omega)}{(4\ \Omega + 1.5\ \Omega) + (2\ \Omega + 1.5\ \Omega)}$$

$$= 0.75\ \Omega + \frac{(5.5\ \Omega)(3.5\ \Omega)}{5.5\ \Omega + 3.5\ \Omega}$$

$$= 0.75\ \Omega + 2.139\ \Omega$$

$$R_T = \mathbf{2.889\ \Omega}$$

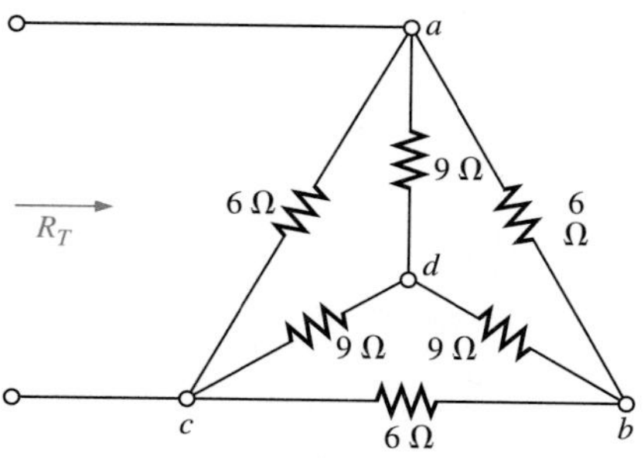

FIG. 8.78
Example 8.30.

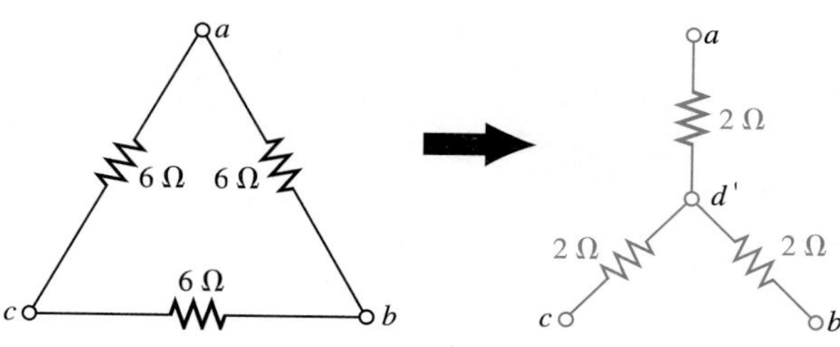

FIG. 8.79
Converting the Δ configuration of Fig. 8.78 to a Y configuration.

EXAMPLE 8.30

Find the total resistance of the network of Fig. 8.78.

Solution: Since all the resistors of the Δ or Y are the same, Eqs. (8.8a) and (8.8b) can be used to convert either form to the other.

a. *Converting the Δ to a Y. Note:* When this is done, the resulting d' of the new Y will be the same as the point d shown in the original figure, only because both systems are "balanced." That is, the resistance in each branch of each system has the same value:

$$R_Y = \frac{R_\Delta}{3} = \frac{6\ \Omega}{3} = 2\ \Omega \quad \text{(Fig. 8.79)}$$

The network then appears as shown in Fig. 8.80.

$$R_T = 2\left[\frac{(2\ \Omega)(9\ \Omega)}{2\ \Omega + 9\ \Omega}\right] = \mathbf{3.2727\ \Omega}$$

b. *Converting the Y to a Δ.*

$$R_\Delta = 3R_Y = (3)(9\ \Omega) = 27\ \Omega \quad \text{(Fig. 8.81)}$$

$$R'_T = \frac{(6\ \Omega)(27\ \Omega)}{6\ \Omega + 27\ \Omega} = \frac{162\ \Omega}{33} = 4.9091\ \Omega$$

$$R_T = \frac{R'_T(R'_T + R'_T)}{R'_T + (R'_T + R'_T)} = \frac{R'_T 2R'_T}{3R'_T} = \frac{2R'_T}{3}$$

$$= \frac{2(4.9091\ \Omega)}{3} = \mathbf{3.2727\ \Omega}$$

which checks with the previous solution.

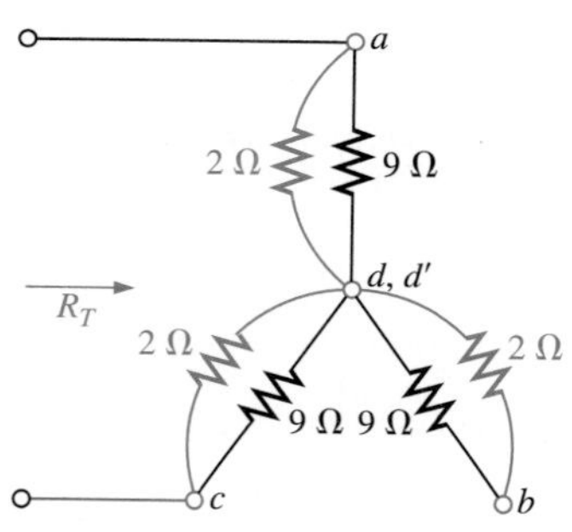

FIG. 8.80
Substituting the Y configuration for the converted Δ into the network of Fig. 8.78.

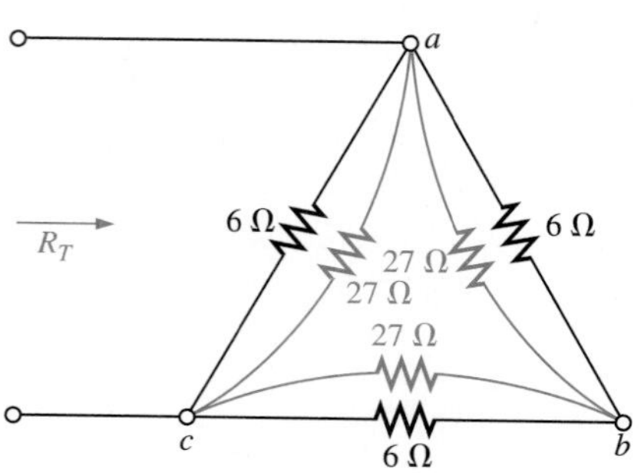

FIG. 8.81
Substituting the converted Y configuration into the network of Fig. 8.78.

Practical Business Applications

Oh no! I didn't, did I?

With three shopping days to Christmas, it's 9:30 PM in a mall parking lot and it's snowing—a blizzard. You walk to your parked car and notice that your headlights are glowing very dimly. "Oh no! I didn't, did I?" You frantically reach for your keys, open the door, throw in your shopping and test the ignition.

Too late! You should have known better. Your battery's almost flat dead. So you think "Oh well, I guess it's fix-it time. Who can I find to give me a boost?" You look around the parking lot and find a kindly older fellow who is enough in the spirit to help you in the miserable weather. "What a nice guy!" you think to yourself. But then you realize that he doesn't know too much about jump-starting cars.

"No problem! I've got a college diploma in Electrical Engineering Technology. What can possibly go wrong? After all, I studied stuff ten times harder than this, right?" you think to yourself. So, you grab your flashlight from your glove compartment (with fresh batteries installed in it the correct way) and set out to exhibit your skill.

By this time, the Good Samaritan has parked his Mercedes Benz in front of your car and has popped his hood open. You pop your hood too, quickly grab your brand new booster cables and meet him up front. It's time to play ball.

Now, you are standing in snowy –10°C weather; it's dark and you're testing the patience of a kind-hearted person with a really expensive car just a few short days before the biggest holiday of the year. He turns to you and asks "Are you sure you learned how to do this in college?" You almost choke on your words as you turn to him and say "Absolutely! Watch this."

What you do in the next few minutes will result either in glory or disaster. So, let's go over the details now, before it's too late.

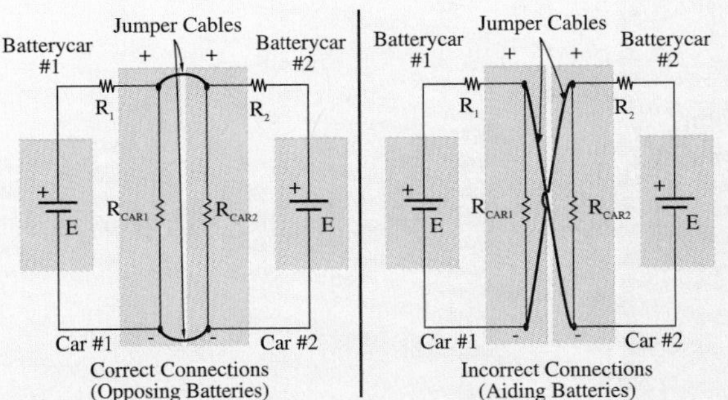

Jump-starting a car is essentially putting two dc batteries in parallel. You use the black jumper cable to connect both cars' negative (–) polarity together and the red jumper cable to connect both cars' positive (+) polarity together. Sounds simple. But wait.

Let's work it through. The simplified circuit for modelling a car battery is an ideal dc voltage source (12 V) in series with a small internal resistance (a few milliohms). Normally, this power supply has a much larger load on it than the internal resistance. Most of the voltage is applied to the load, the car's electrical system.

In our story, two such equivalent battery circuits are placed in parallel. It should be easy to see that both voltage sources are available for powering the starter motor in your car. If one of these two batteries is effectively dead (the one in your car), then the total available power comes from the good battery (the one in the Mercedes).

Now, when you disconnect the booster cables, you're effectively removing one of the two parallel sources from the circuit. But won't the car stop running since the remaining car battery is dead? No, because the car is designed for the alternator to carry the load after ignition. The alternator then starts to charge the flat car battery while the engine is running.

You're probably saying "Too easy! It's a piece of cake. Who needs a college diploma to do this?" Right? Wrong! You see, something terrible can happen if you reverse the polarity of the cables. Look at the outer circuit if you parallel the sources incorrectly. Can you see the problem?

The problem is that you would form a low resistance path with two series-aiding sources circulating a current around the normal load. (Done correctly, the two dc batteries will form a series-opposing circuit around the normal load.) The circulating currents can be quite significant, and since the car's electrical system wasn't designed to handle these currents, failure may result. Most important, people nearby may suffer severe personal injury.

So take it seriously when it's time to jump-start a car. Don't be nervous, just make sure you understand both the right way **and** the wrong way. After all, can you imagine how angry the Mercedes owner would be if a college graduate blew up his car?

PROBLEMS

SECTION 8.2 Current Sources

1. Find the voltage and current I_1 for the circuit in Fig. 8.82.

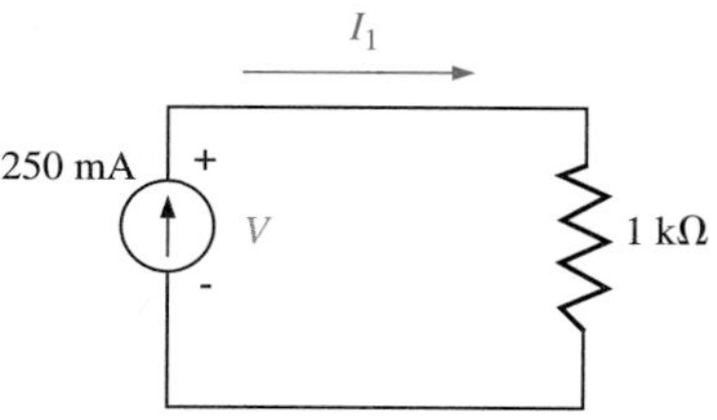

FIG. 8.82
Problem 1.

2. Find the voltage V_{ab} (with polarity) across the ideal current source of Fig. 8.83.

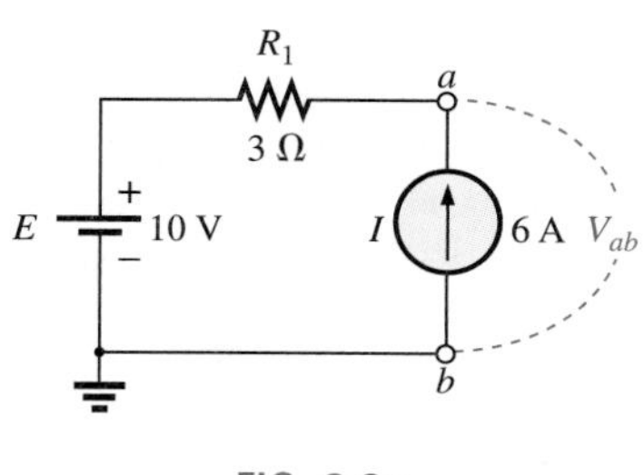

FIG. 8.83
Problem 2.

3. **a.** Determine V for the current source of Fig. 8.84(a) with an internal resistance of 10 kΩ.
   b. The source of part (a) is approximated by an ideal current source in Fig. 8.84(b) since the source resistance is much larger than the applied load. Determine the resulting voltage V for Fig. 8.84(b) and compare to that obtained in part (a). Is the use of the ideal current source a good approximation?

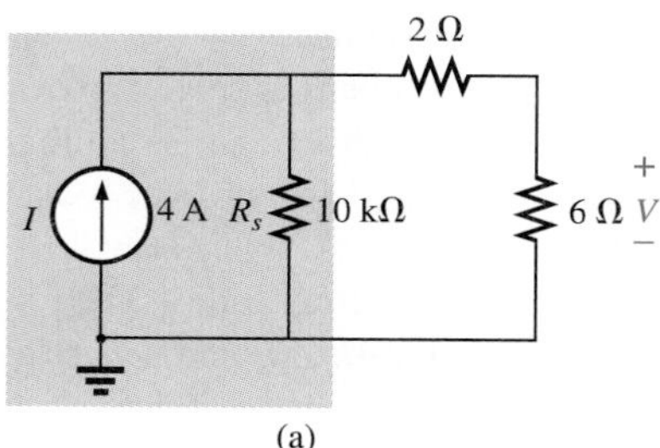

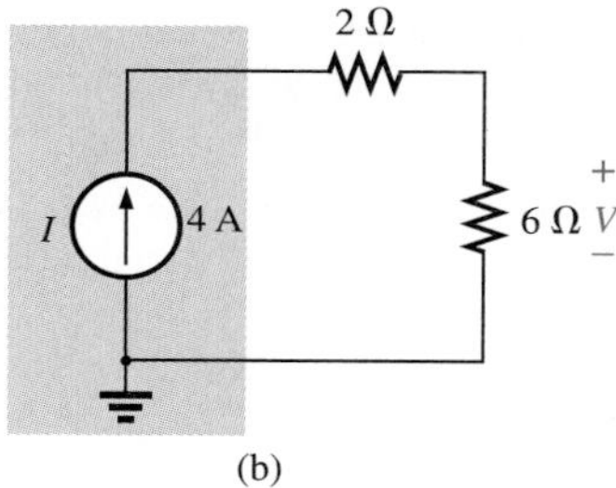

FIG. 8.84
Problem 3.

4. For the network of Fig. 8.85:
   a. Find the currents I_1 and I_s.
   b. Find the voltages V_s and V_3.

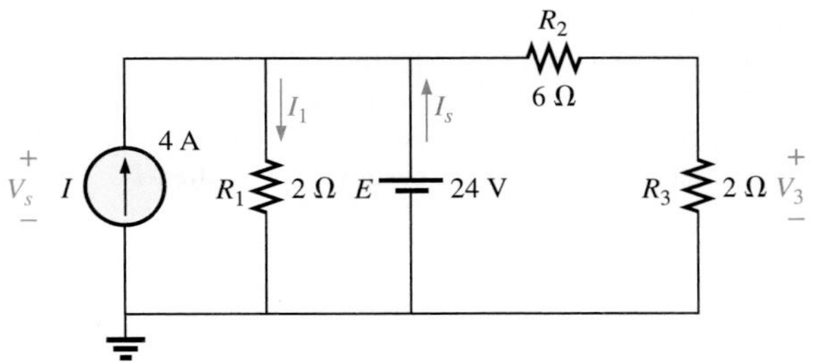

FIG. 8.85
Problem 4.

5. Find the voltage V_3 and the current I_2 for the network of Fig. 8.86.

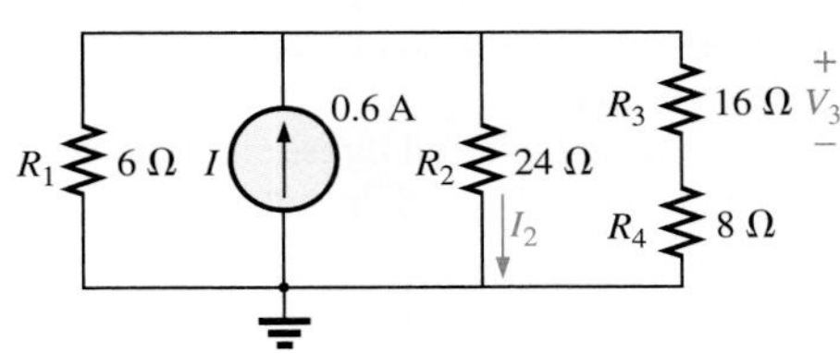

FIG. 8.86
Problem 5.

SECTION 8.3 Source Conversions

6. Convert the voltage sources of Fig. 8.87 to current sources.

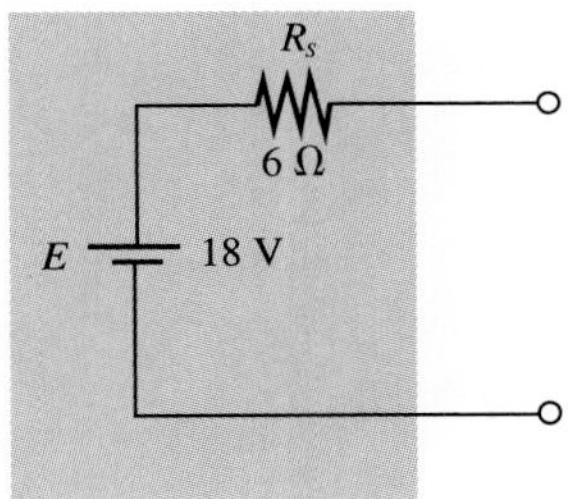

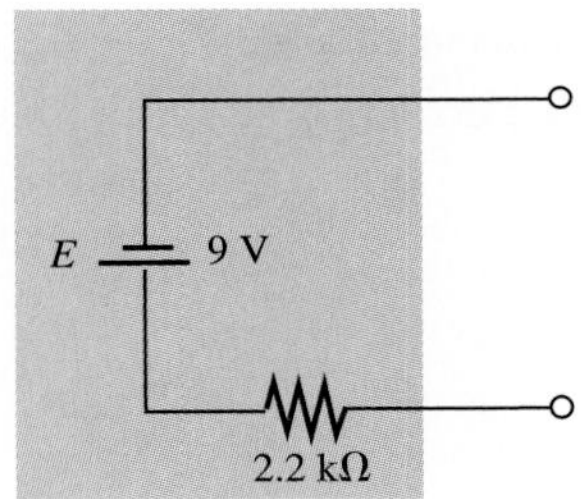

FIG. 8.87
Problem 6.

7. Convert the current sources of Fig. 8.88 to voltage sources.

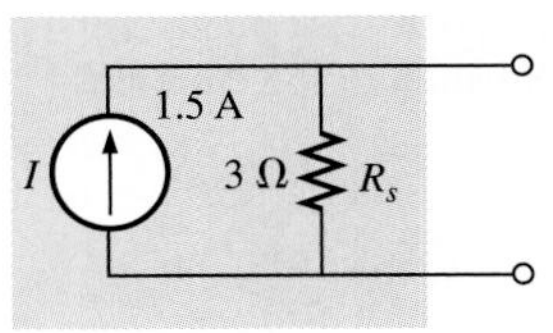

(a)

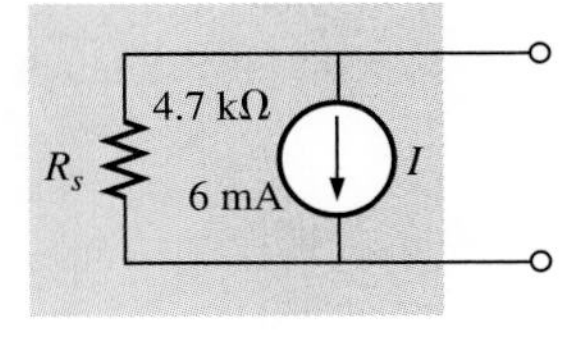

(b)

FIG. 8.88
Problem 7.

8. For the network of Fig. 8.89:

a. Find the current through the 4-Ω resistor.

b. Convert the current source and 2-Ω resistor to a voltage source, and again solve for the current in the 4-Ω resistor. Compare the results.

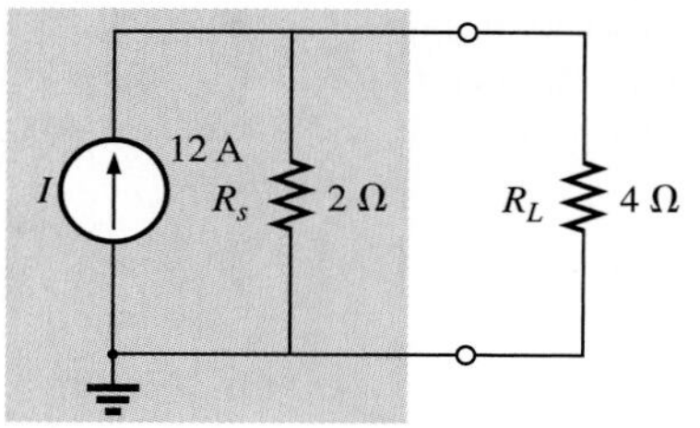

FIG. 8.89
Problem 8.

9. For the configuration of Fig. 8.90:

a. Convert the current source and 6.8-Ω resistor to a voltage source.

b. Find the magnitude and direction of the current I_1.

c. Find the voltage V_{ab} and the polarity of points a and b.

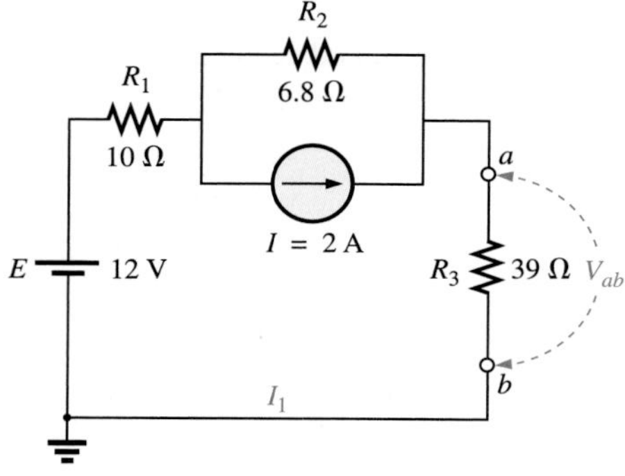

FIG. 8.90
Problem 9.

SECTION 8.4 Current Sources in Parallel

10. Find the voltage V_2 and the current I_1 for the network of Fig. 8.91.

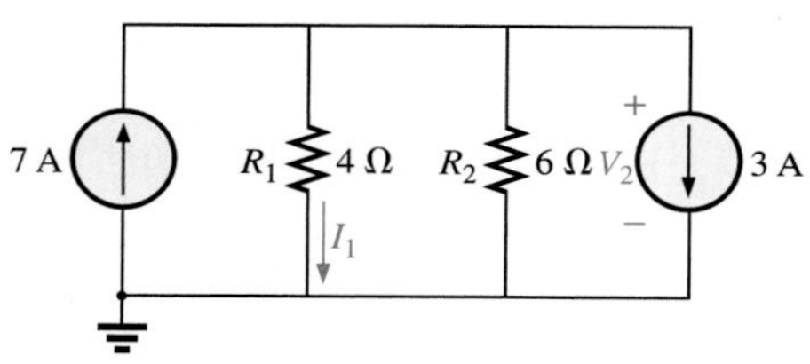

FIG. 8.91
Problem 10.

11. a. Convert the voltage sources of Fig. 8.92 to current sources.

b. Find the voltage V_{ab} and the polarity of points a and b.

c. Find the magnitude and direction of the current I.

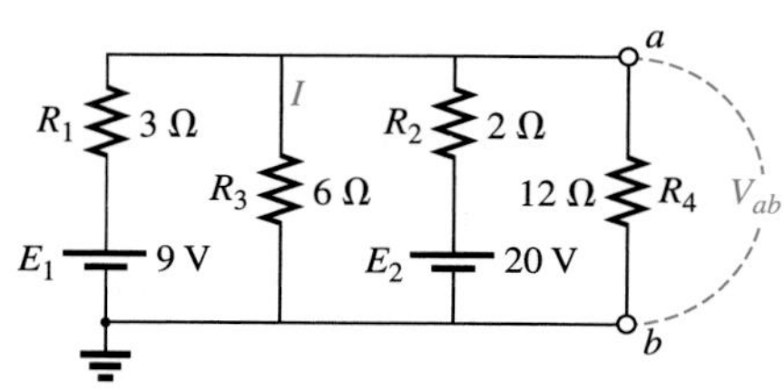

FIG. 8.92
Problem 11.

12. For the network of Fig. 8.93:
a. Convert the voltage source to a current source.
b. Reduce the network to a single current source and determine the voltage V_1.
c. Using the results of part (b), determine V_2.
d. Calculate the current I_2.

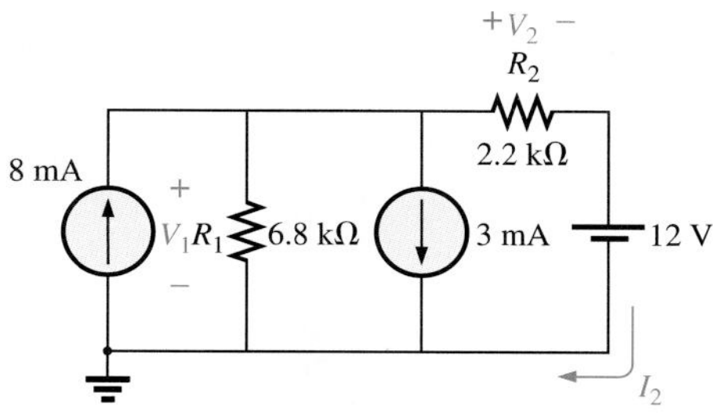

FIG. 8.93
Problem 12.

SECTION 8.6 Branch-Current Analysis

13. Using branch-current analysis, find the magnitude and direction of the current through each resistor for the networks of Fig. 8.94.

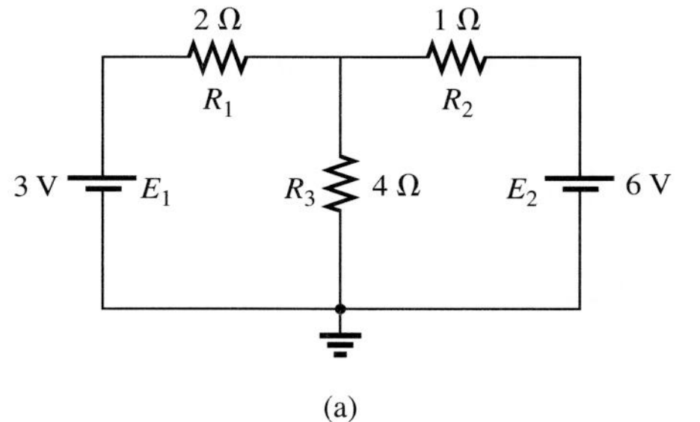

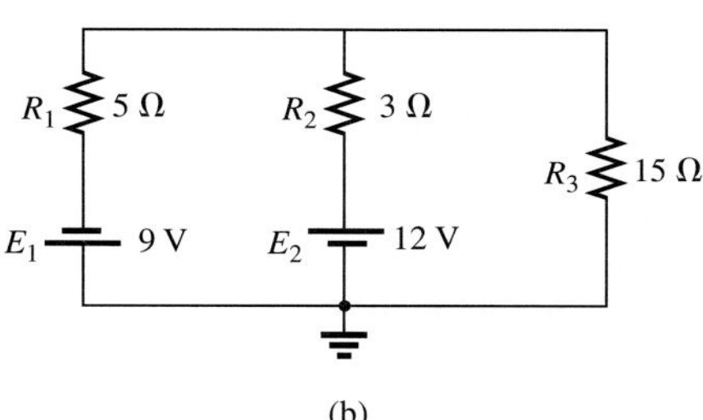

FIG. 8.94
Problems 13, 19, and 27.

***14.** Using branch-current analysis, find the current through each resistor for the networks of Fig. 8.95. The resistors are all standard values.

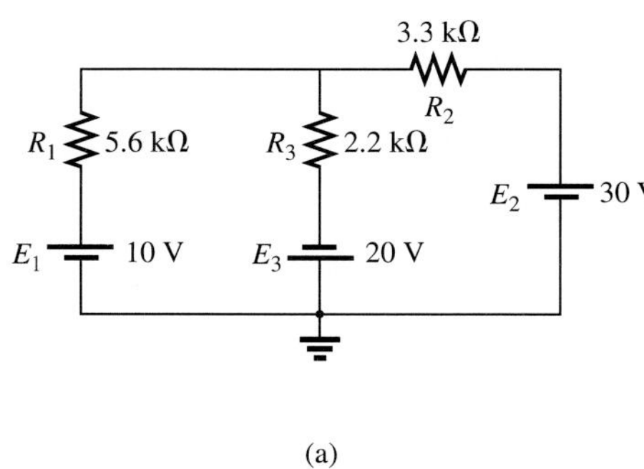

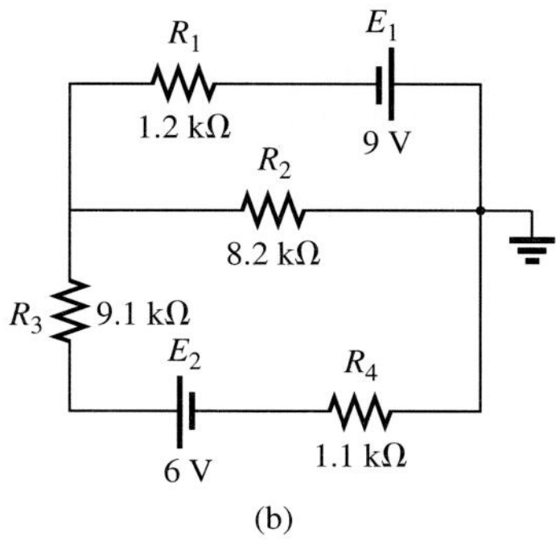

FIG. 8.95
Problems 14, 20, and 28.

*15. For the networks of Fig. 8.96, determine the current I_2 using branch-current analysis, and then find the voltage V_{ab}.

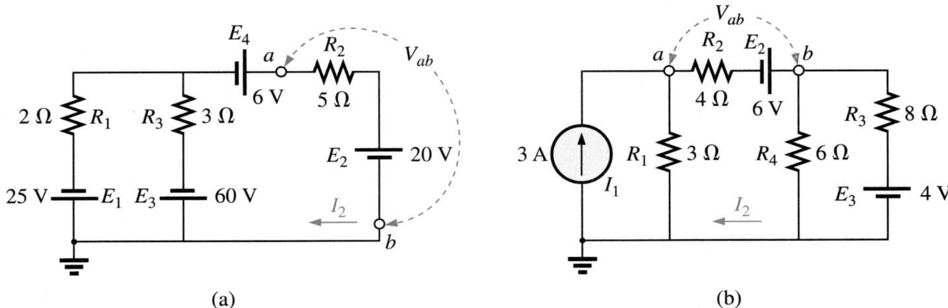

FIG. 8.96
Problems 14, 21, and 19.

*16. For the network of Fig. 8.97:
  a. Write the equations necessary to solve for the branch currents.
  b. By substitution of Kirchhoff's current law, reduce the set to three equations.
  c. Rewrite the equations in a format that can be solved using third-order determinants.
  d. Solve for the branch current through the resistor R_3.

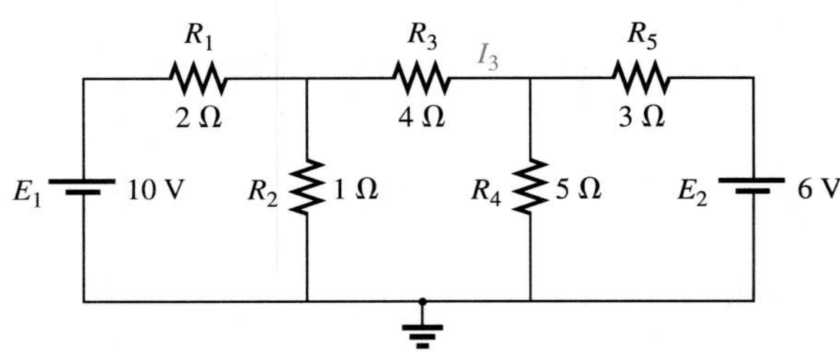

FIG. 8.97
Problems 16 and 30.

*17. For the transistor configuration of Fig. 8.98:
  a. Solve for the currents I_B, I_C, and I_E using the fact that $V_{BE} = 0.7$ V and $V_{CE} = 8$ V.
  b. Find the voltages V_B, V_C, and V_E with respect to ground.
  c. What is the ratio of output current I_C to input current I_B? [*Note:* In transistor analysis this ratio is referred to as the dc beta of the transistor (β_{dc}).]

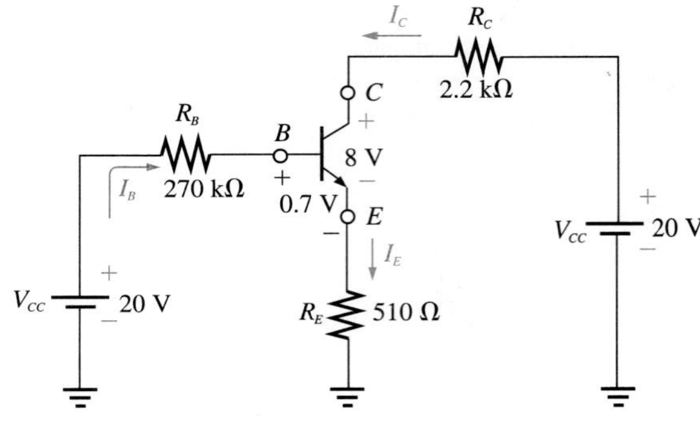

FIG. 8.98
Problem 17.

18. For the network in Fig. 8.99, find the current through each of the branches.

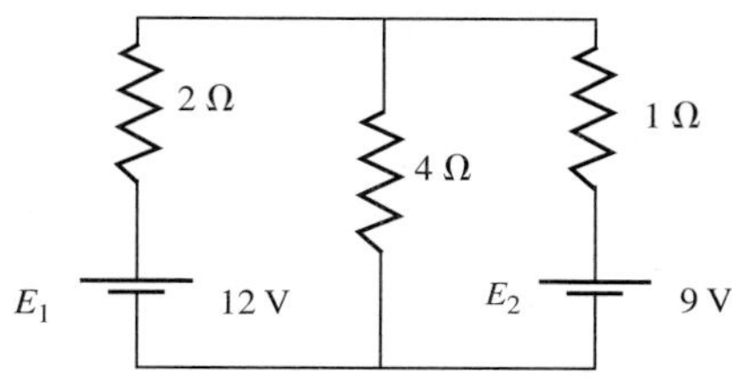

FIG. 8.99
Problem 18.

SECTION 8.7 Mesh Analysis (General Approach)

19. Find the current through each resistor for the networks of Fig. 8.94.
20. Find the current through each resistor for the networks of Fig. 8.95.
21. Find the mesh currents and the voltage V_{ab} for each network of Fig. 8.96. Use clockwise mesh currents.
22. **a.** Find the current I_3 for the network of Fig. 8.97 using mesh analysis.
    b. Based on the results of part (a), how would you compare the application of mesh analysis to the branch-current method?

*23. Using mesh analysis, determine the current through the 5-Ω resistor for each network of Fig. 8.100. Then determine the voltage V_a.

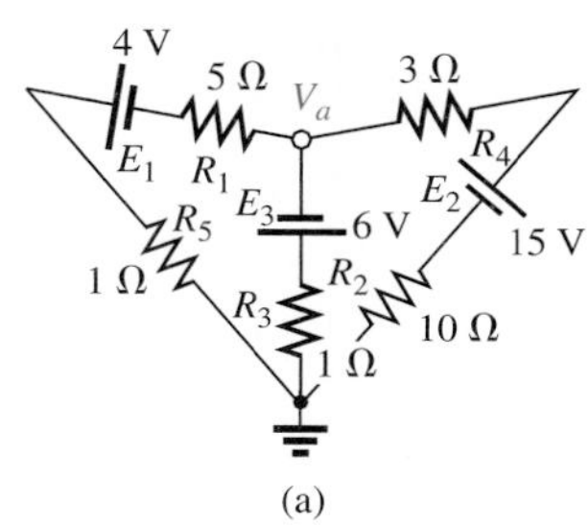

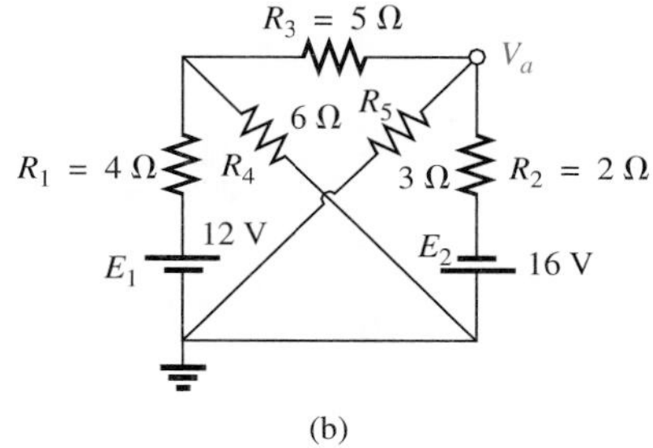

FIG. 8.100
Problems 23 and 31.

*24. Write the mesh equations for each of the networks of Fig. 8.101 and, using determinants, solve for the loop currents in each network. Use clockwise mesh currents.

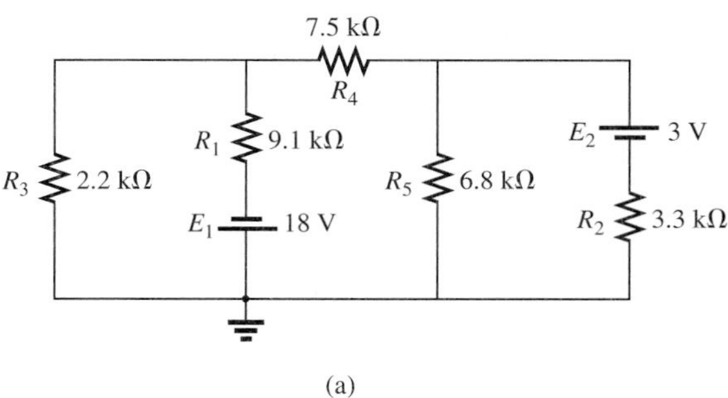

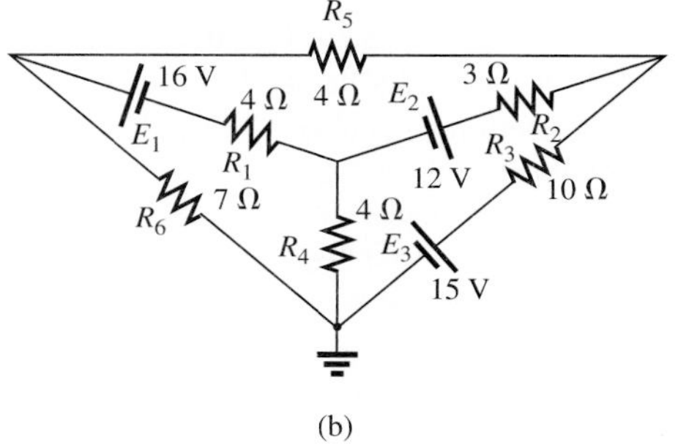

FIG. 8.101
Problems 24, 32, and 37.

***25.** Write the mesh equations for each of the networks of Fig. 8.102 and, using determinants, solve for the loop currents in each network.

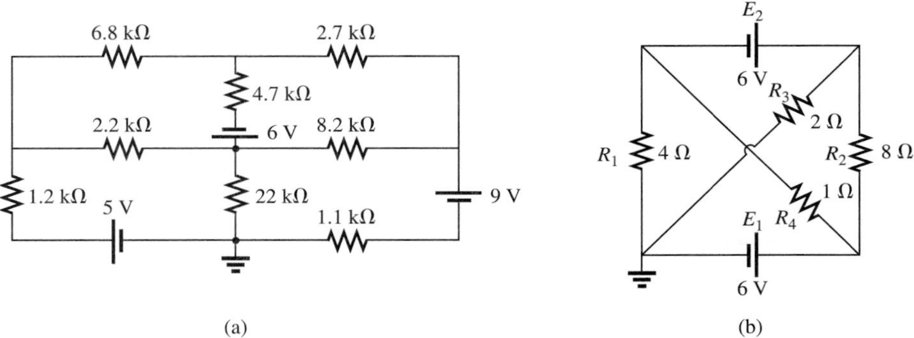

FIG. 8.102
Problems 25 and 33.

***26.** Using the supermesh approach, find the current through each element of the networks of Fig. 8.103.

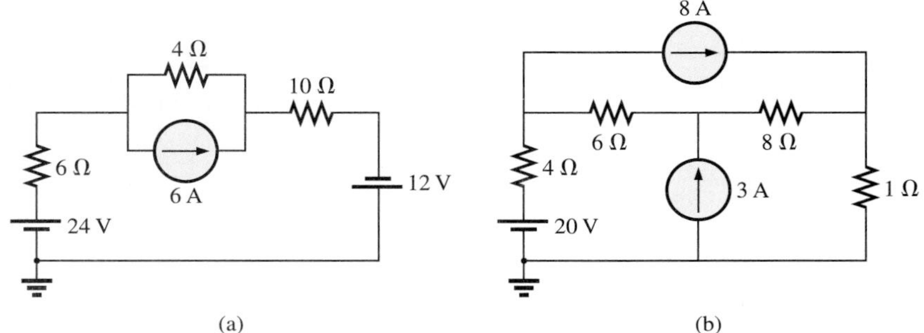

FIG. 8.103
Problem 26.

SECTION 8.8 Mesh Analysis (Format Approach)

27. Using the format approach, write the mesh equations for the networks of Fig. 8.94. Is symmetry present? Using determinants, solve for the mesh currents.

28. a. Using the format approach, write the mesh equations for the networks of Fig. 8.95.

b. Using determinants, solve for the mesh currents.

c. Determine the magnitude and direction of the current through each resistor.

29. a. Using the format approach, write the mesh equations for the networks of Fig. 8.96.

b. Using determinants, solve for the mesh currents.

c. Determine the magnitude and direction of the current through each resistor.

30. Determine the current I_3 for the network of Fig. 8.97 using mesh analysis and compare to the solution of Problem 15.

31. Using mesh analysis, determine $I_{5\Omega}$ and V_a for the network of Fig. 8.100(b).

32. Using mesh analysis, determine the mesh currents for the networks of Fig. 8.101.

33. Using mesh analysis, determine the mesh currents for the networks of Fig. 8.102.

SECTION 8.9 Nodal Analysis (General Approach)

34. Apply nodal analysis to the network in Fig. 8.104.

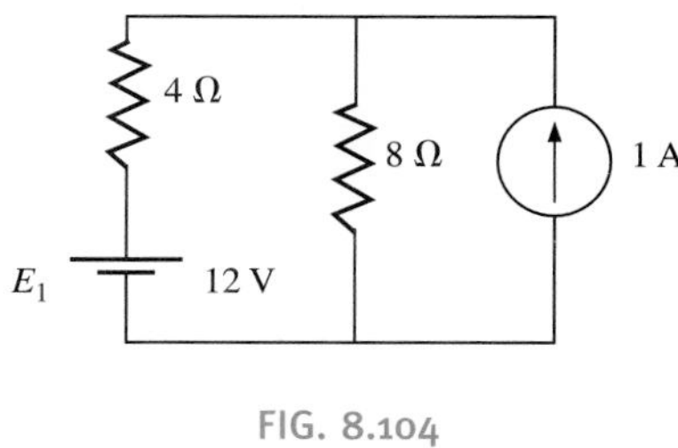

FIG. 8.104
Problem 34.

35. Write the nodal equations for the networks of Fig. 8.105 and, using determinants, solve for the nodal voltages. Is symmetry present?

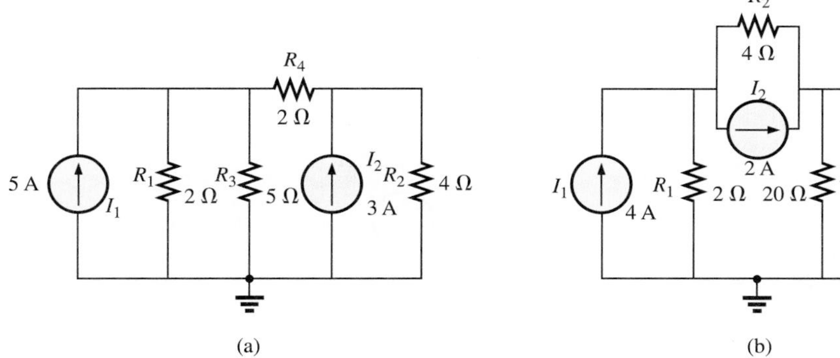

FIG. 8.105
Problems 35 and 41.

36. a. Write the nodal equations for the networks of Fig. 8.106.

b. Using determinants, solve for the nodal voltages.

c. Determine the magnitude and polarity of the voltage across each resistor.

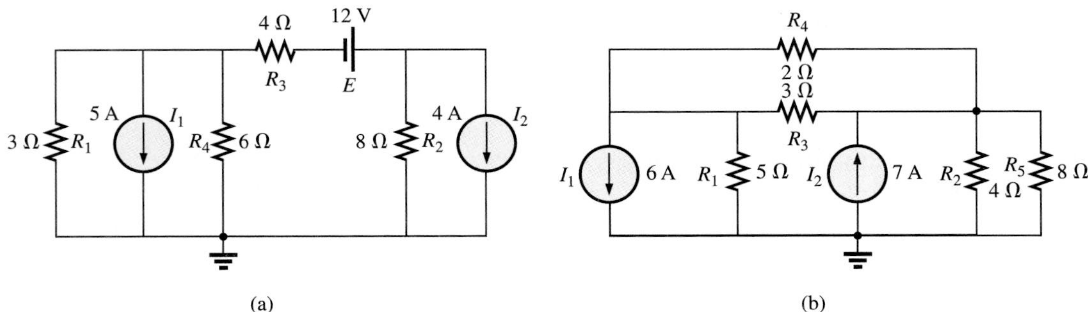

FIG. 8.106
Problems 36 and 42.

37. a. Write the nodal equations for the networks of Fig. 8.101.

b. Using determinants, solve for the nodal voltages.

c. Determine the magnitude and polarity of the voltage across each resistor.

***38.** For the networks of Fig. 8.107, write the nodal equations and solve for the nodal voltages.

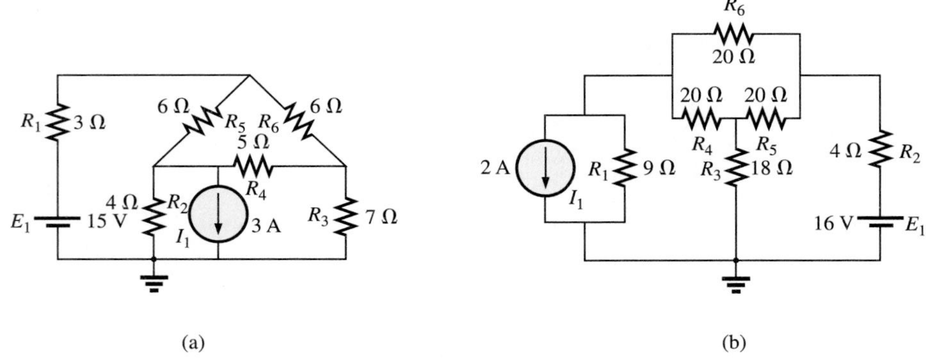

FIG. 8.107
Problems 38 and 43.

39. a. Determine the nodal voltages for the networks of Fig. 8.108.
b. Find the voltage across each current source.

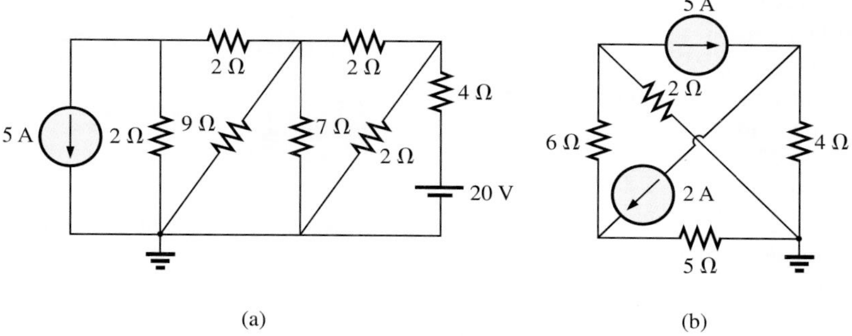

FIG. 8.108
Problems 39 and 44.

***40.** Using the supernode approach, determine the nodal voltages for the networks of Fig. 8.109.

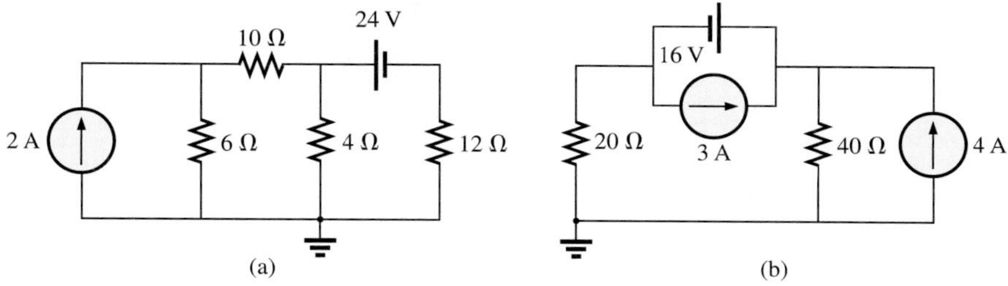

FIG. 8.109
Problem 40.

SECTION 8.10 Nodal Analysis (Format Approach)

41. Using the format approach, write the nodal equations for the networks of Fig. 8.105. Is symmetry present? Using determinants, solve for the nodal voltages.

42. a. Write the nodal equations for the networks of Fig. 8.106.
b. Solve for the nodal voltages.
c. Find the magnitude and polarity of the voltage across each resistor.

43. a. Write the nodal equations for the networks of Fig. 8.107.
b. Solve for the nodal voltages.
c. Find the magnitude and polarity of the voltage across each resistor.

44. Determine the nodal voltages for the networks of Fig. 8.108. Then determine the voltage across each current source.

SECTION 8.11 Bridge Networks

45. For the bridge network of Fig. 8.110:
  a. Write the mesh equations using the format approach.
  b. Determine the current through R_5.
  c. Is the bridge balanced?
  d. Is Eq. (8.4) satisfied?

46. For the network of Fig. 8.110:
  a. Write the nodal equations using the format approach.
  b. Determine the voltage across R_5.
  c. Is the bridge balanced?
  d. Is Eq. (8.4) satisfied?

47. For the bridge of Fig. 8.111:
  a. Write the mesh equations using the format approach.
  b. Determine the current through R_5.
  c. Is the bridge balanced?
  d. Is Eq. (8.4) satisfied?

48. For the bridge network of Fig. 8.111:
  a. Write the nodal equations using the format approach.
  b. Determine the current across R_5.
  c. Is the bridge balanced?
  d. Is Eq. (8.4) satisfied?

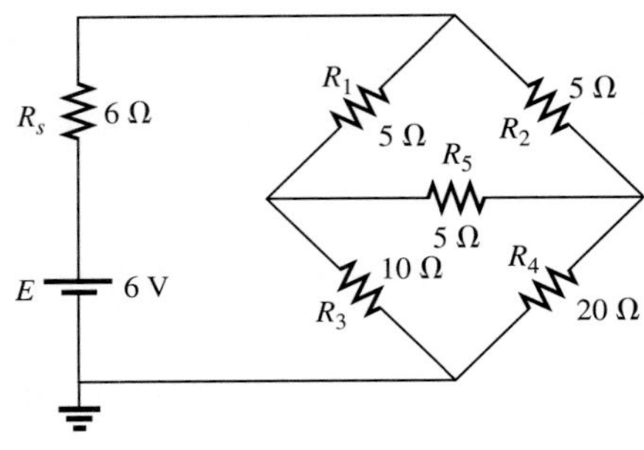

FIG. 8.110
Problems 45 and 46.

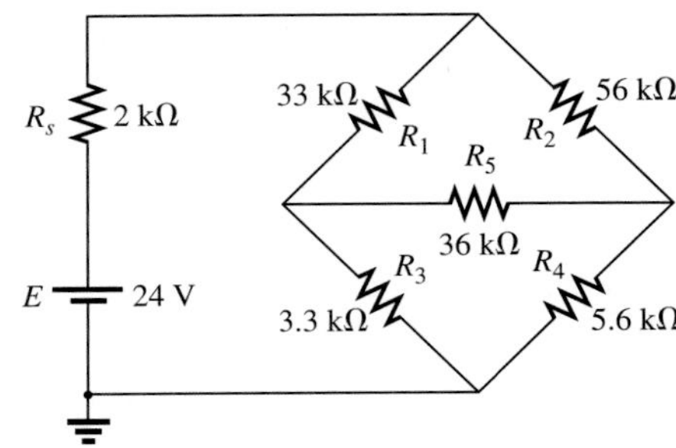

FIG. 8.111
Problems 47 and 48.

49. Write the nodal equations for the bridge configuration of Fig. 8.112. Use the format approach.

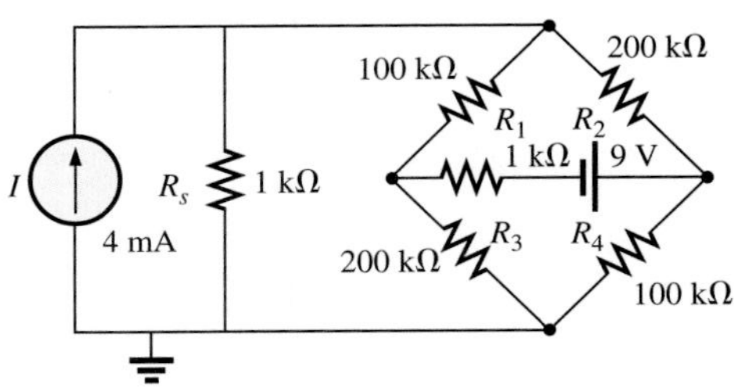

FIG. 8.112
Problem 49.

***50.** Determine the current through the source resistor R_s of each network of Fig. 8.113 using either mesh or nodal analysis. Discuss why you chose one method over the other.

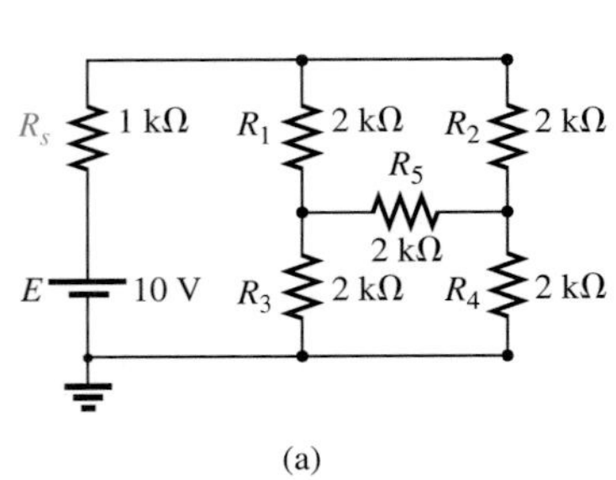

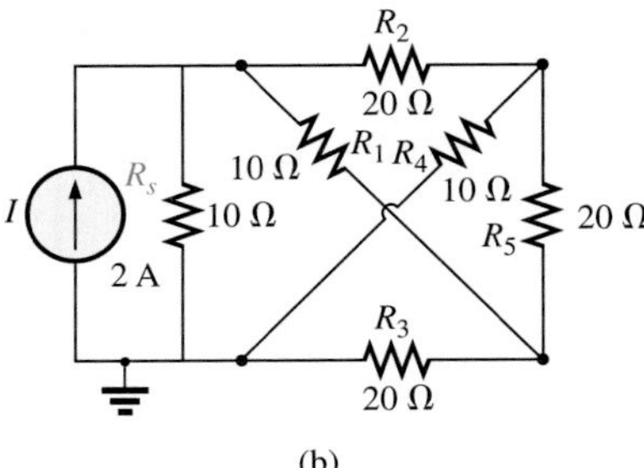

FIG. 8.113
Problem 50.

SECTION 8.12 Y-Δ (T-π) and Δ-Y (π-T) Conversions

51. Convert the network in Fig. 8.114 to the equivalent delta network.

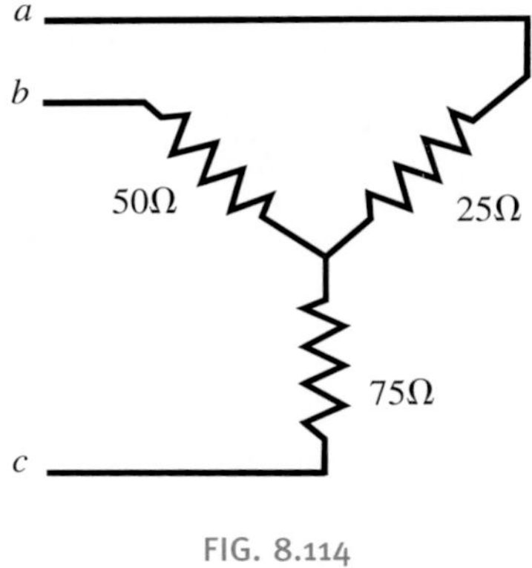

FIG. 8.114
Problem 51.

52. Convert the delta network in Fig. 8.115 to the equivalent wye network.

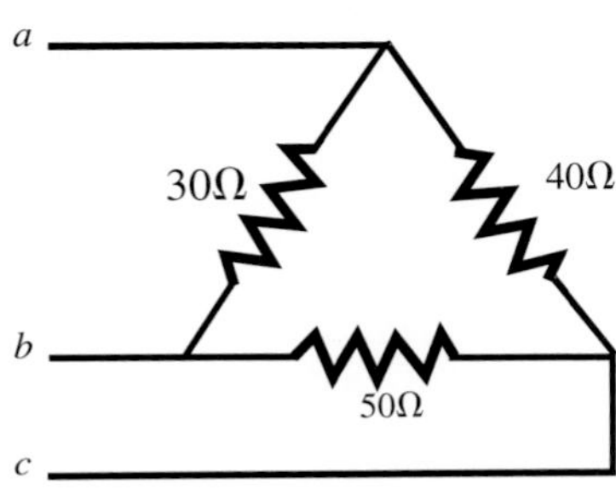

FIG. 8.115
Problem 52.

53. Using a Δ-Y or Y-Δ conversion, find the current I in each of the networks of Fig. 8.116.

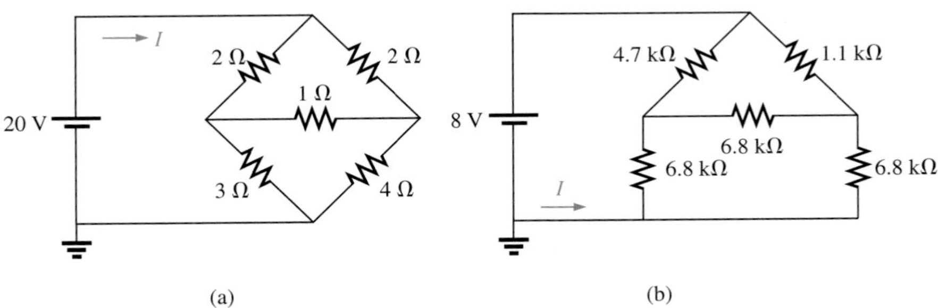

FIG. 8.116
Problem 53.

*54. Repeat Problem 53 for the networks of Fig. 8.117.

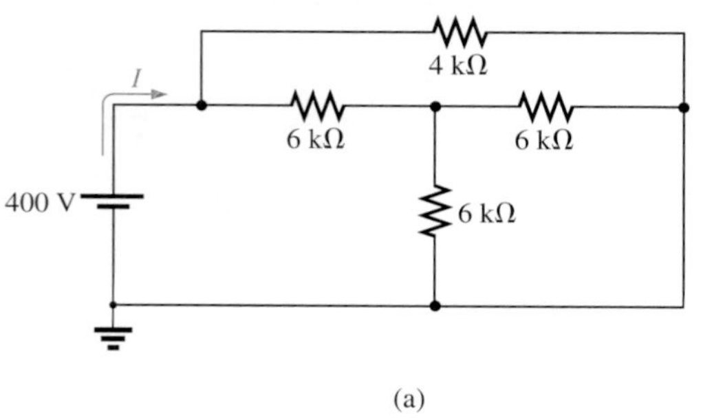

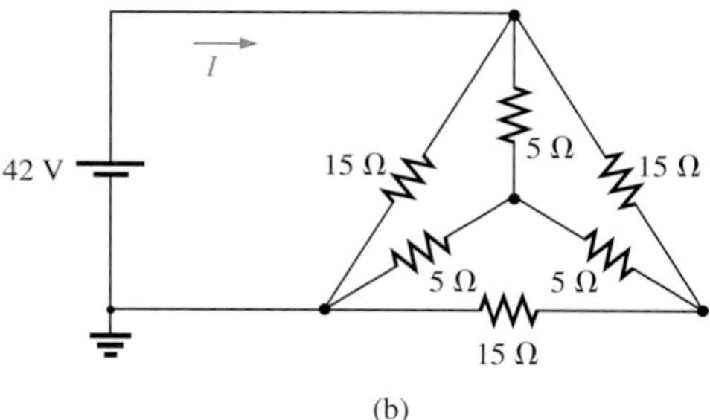

FIG. 8.117
Problem 54.

*55. Determine the current I for the network of Fig. 8.118.

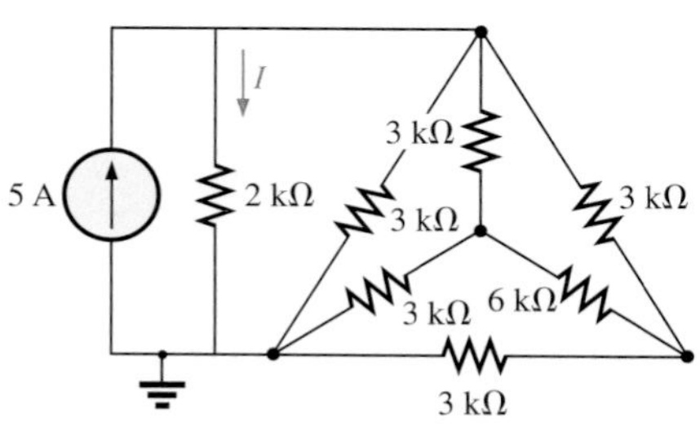

FIG. 8.118
Problem 55.

*56. a. Replace the T configuration of Fig. 8.119 (composed of 6-kΩ resistors) with a π configuration.
b. Solve for the source current I_{s_1}.

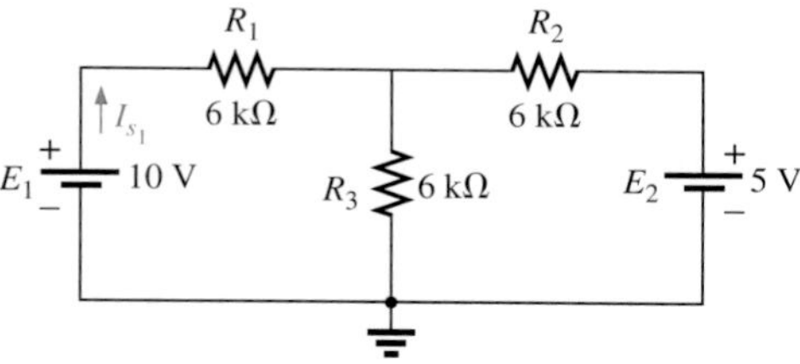

FIG. 8.119
Problem 56.

*57. a. Replace the π configuration of Fig. 8.120 (composed of 3-kΩ resistors) with a T configuration.
b. Solve for the source current I_s.

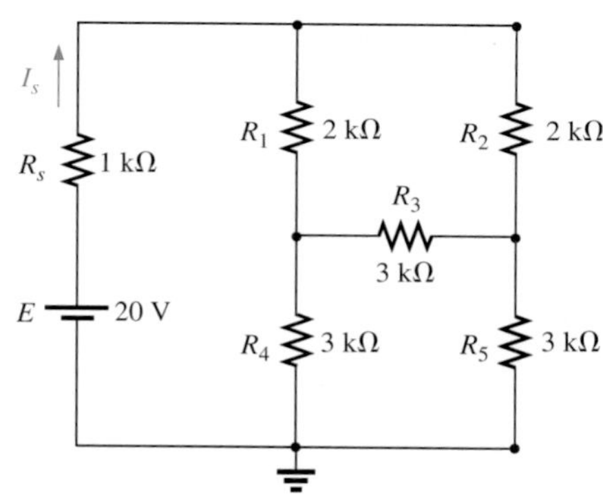

FIG. 8.120
Problem 57.

***58.** Using Y-Δ or Δ-Y conversions, determine the total resistance of the network of Fig. 8.121.

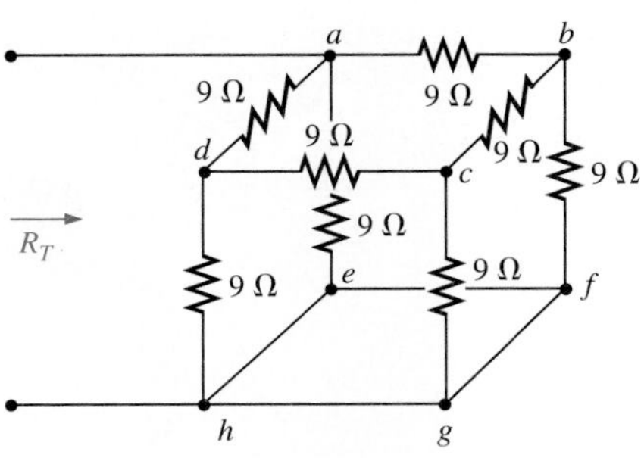

FIG. 8.121
Problem 58.

9 Network Theorems

Outline

9.1 Introduction
9.2 Superposition Theorem
9.3 Thévenin's Theorem
9.4 Norton's Theorem
9.5 Maximum Power Transfer Theorem
9.6 Millman's Theorem
9.7 Substitution Theorem
9.8 Reciprocity Theorem

Learning Outcomes

After completing this chapter you will be able to

- solve network problems using the superposition theorem
- explain Thévenin's theorem and apply the theorem to network problems
- state Norton's theorem and apply the theorem to network problems
- explain the maximum power transfer theorem and its relationship to circuit efficiency
- solve network problems using Millman's theorem
- solve network problems using the substitution theorem
- solve network problems using the reciprocity theorem

The network theorems described in this chapter will provide you with a number of techniques which can simplify complex electrical circuits.

9.1 INTRODUCTION

This chapter will introduce the important fundamental theorems of network analysis. Included are the **superposition, Thévenin's, Norton's, maximum power transfer, Millman's**, **substitution,** and **reciprocity** theorems. We will consider a number of areas of application for each. It is important that you thoroughly understand each theorem because we will apply a number of them repeatedly in the material to follow.

9.2 SUPERPOSITION THEOREM

The superposition theorem, like the methods of the last chapter, can be used to find the solution to networks with two or more sources that are not in series or parallel. The most obvious advantage of this method is that it does not require the use of a mathematical technique such as determinants to find the required voltages or currents. Instead, each source is treated independently, and the algebraic sum is found to determine a particular unknown quantity of the network.

The **superposition theorem** states the following:

The current through, or voltage across, an element in a linear bilateral network is equal to the algebraic sum of the currents or voltages produced independently by each source.

When applying the theorem it is possible to consider the effects of two sources at the same time and reduce the number of networks that have to be analyzed, but, in general,

$$\text{Number of networks to be analyzed} = \text{Number of independent sources} \tag{9.1}$$

To consider the effects of each source independently requires that sources be removed and replaced without affecting the final result. To remove a voltage source when applying this theorem, the difference in potential between the terminals of the voltage source must be set to zero (short circuited); removing a current source requires that its terminals be opened (open circuit). Any internal resistance or conductance associated with the displaced sources is not eliminated but must still be considered.

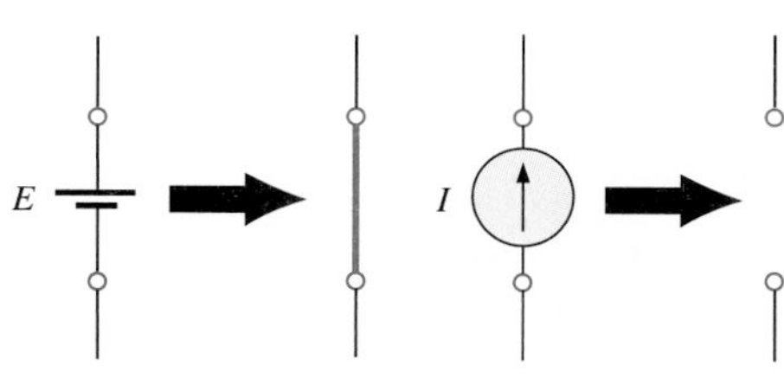

FIG. 9.1
Removing the effects of ideal sources.

Figure 9.1 reviews the substitutions required when removing an ideal source, and Fig. 9.2 reviews the substitutions with practical sources that have an internal resistance.

The total current through any portion of the network is equal to the algebraic sum of the currents produced independently by each source. That is, for a two-source network,

if the current produced by one source is in one direction, while that produced by the other is in the opposite direction through the same resistor, the resulting current is the difference of the two and has the direction of the larger.

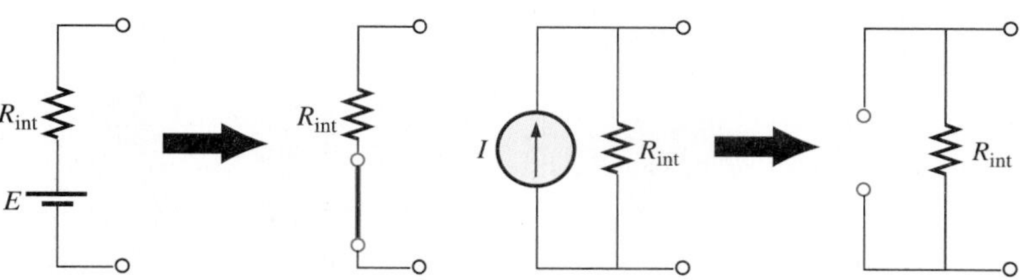

FIG. 9.2
Removing the effects of practical sources.

If the individual currents are in the same direction, the resulting current is the sum of two in the direction of either current.

This rule holds true for the voltage across a portion of a network as determined by polarities, and it can be extended to networks with any number of sources.

The superposition principle cannot be applied to power effects since the power loss in a resistor varies as the square (nonlinear) of the current or voltage. For instance, the current through the resistor R of Fig. 9.3(a) is I_1 due to one source of a two-source network. The current through the same resistor due to the other source is I_2 as shown in Fig. 9.3(b). Applying the superposition theorem, the total current through the resistor due to both sources is I_T, as shown in Fig. 9.3(c) with

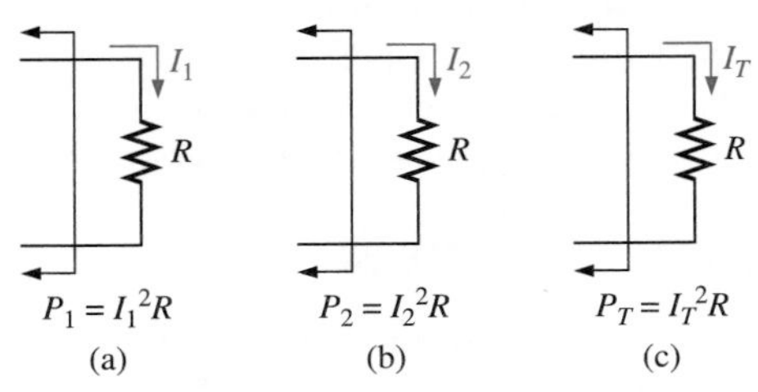

FIG. 9.3
Demonstration of the fact that superposition is not applicable to power effects.

$$I_T = I_1 + I_2$$

The power delivered to the resistor in Fig. 9.3(a) is

$$P_1 = I_1^2R$$

while the power delivered to the same resistor in Fig. 9.3(b) is

$$P_2 = I_2^2R$$

If we assume that the total power delivered in Fig. 9.3(c) can be obtained by simply adding the power delivered due to each source, we find that

$$P_T = P_1 + P_2 = I_1^2R + I_2^2R = I_T^2R$$

or

$$I_T^2 = I_1^2 + I_2^2$$

This final relationship between current levels is *incorrect,* however, as we can show by taking the total current determined by the superposition theorem and squaring it as follows:

$$I_T^2 = (I_1 + I_2)^2 = I_1^2 + I_2^2 + 2I_1I_2$$

which is certainly different from the expression obtained when we added power levels.

In general, therefore,

the total power delivered to a resistive element must be determined using the total current through or the total voltage across the element and cannot be determined by a simple sum of the power levels established by each source.

EXAMPLE 9.1

Determine I_1 for the network of Fig. 9.4.

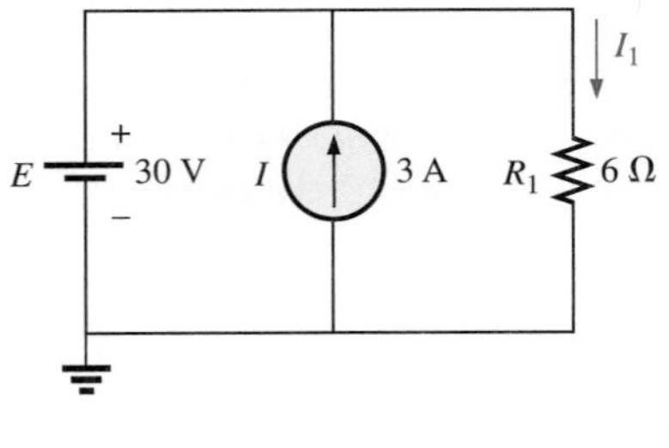

FIG. 9.4
Example 9.1.

Solution: Setting $E = 0$ V for the network of Fig. 9.4 results in the network of Fig. 9.5(a), where a short-circuit equivalent has replaced the 30-V source.

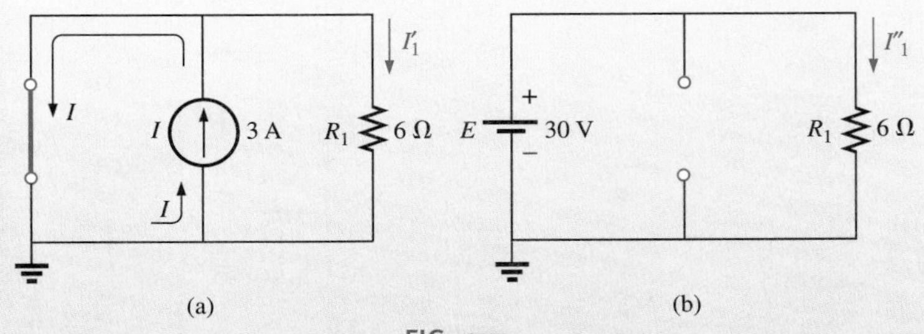

FIG. 9.5
(a) The contribution of I to I_1; (b) the contribution of E to I_1.

continued

As shown in Fig. 9.5(a), the source current will choose the short-circuit path, and $I'_1 = 0$ A. If we applied the current divider rule,

$$I'_1 = \frac{R_{sc}I}{R_{sc} + R_1} = \frac{(0\ \Omega)I}{0\ \Omega + 6\ \Omega} = 0\ \text{A}$$

Setting I to zero amperes will result in the network of Fig. 9.5(b), with the current source replaced by an open circuit. Applying Ohm's law,

$$I''_1 = \frac{E}{R_1} = \frac{30\ \text{V}}{6\ \Omega} = 5\ \text{A}$$

Since I'_1 and I''_1 have the same defined direction in Figs. 9.5(a) and (b), the current I_1 is the sum of the two, and

$$I_1 = I'_1 + I''_1 = 0\ \text{A} + 5\ \text{A} = \mathbf{5\ A}$$

Note in this case that the current source has no effect on the current through the 6-Ω resistor. The voltage across the resistor must be fixed at 30 V because they are parallel elements.

EXAMPLE 9.2

Using superposition, determine the current through the 4-Ω resistor of Fig. 9.6. Note that this is a two-source network of the type considered in Chapter 8.

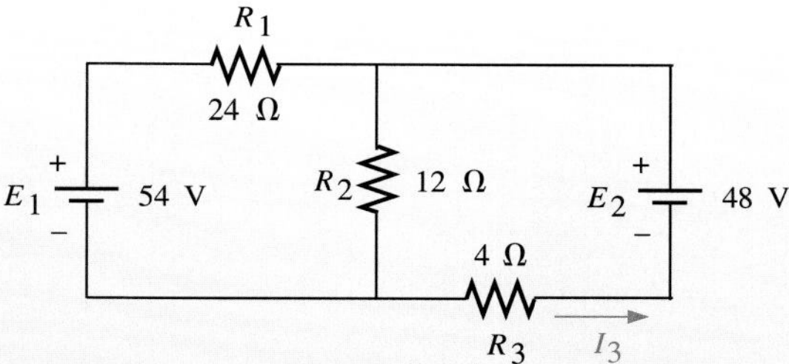

FIG. 9.6
Example 9.2.

Solution: *Considering the effects of a* 54-V *source* (Fig. 9.7):

$$R_T = R_1 + R_2 \parallel R_3 = 24\ \Omega + 12\ \Omega \parallel 4\ \Omega = 24\ \Omega + 3\ \Omega = 27\ \Omega$$

$$I = \frac{E_1}{R_T} = \frac{54\ \text{V}}{27\ \Omega} = 2\ \text{A}$$

Using the current divider rule,

$$I'_3 = \frac{R_2 I}{R_2 + R_3} = \frac{(12\ \Omega)(2\ \text{A})}{12\ \Omega + 4\ \Omega} = \frac{24\ \text{A}}{16} = 1.5\ \text{A}$$

continued

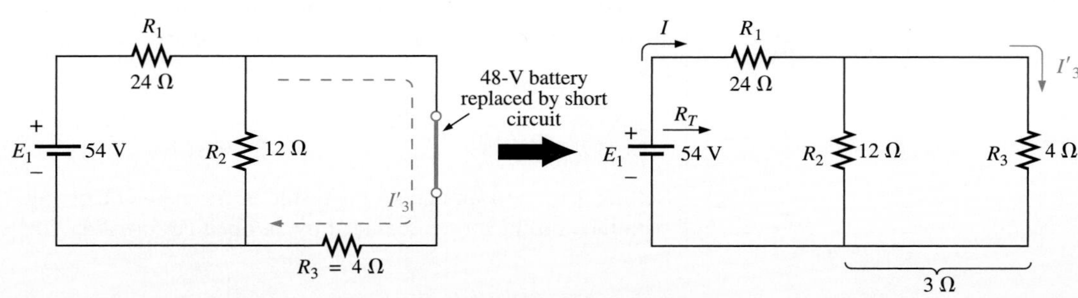

FIG. 9.7
The effect of E_1 on the current I_3.

Considering the effects of the 48-V *source* (Fig. 9.8):

$$R_T = R_3 + R_1 \parallel R_2 = 4\ \Omega + 24\ \Omega \parallel 12\ \Omega = 4\ \Omega + 8\ \Omega = 12\ \Omega$$

$$I''_3 = \frac{E_2}{R_T} = \frac{48\ \text{V}}{12\ \Omega} = 4\ \text{A}$$

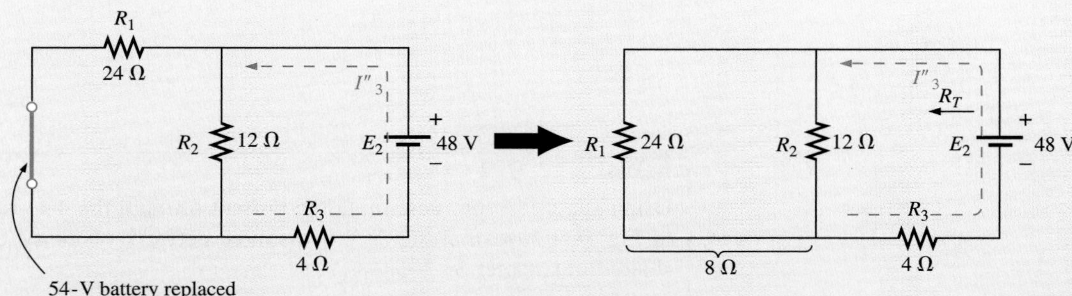

FIG. 9.8
The effect of E_2 on the current I_3.

The total current through the 4-Ω resistor (Fig. 9.9) is

$$I_3 = I''_3 - I'_3 = 4\ \text{A} - 1.5\ \text{A} = \mathbf{2.5\ A} \quad \text{(direction of } I''_3\text{)}$$

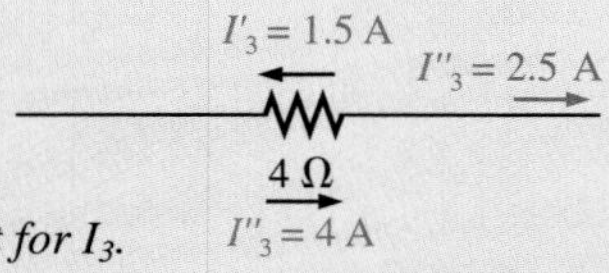

$I'_3 = 1.5$ A
$I''_3 = 2.5$ A
4 Ω
$I''_3 = 4$ A

FIG. 9.9
The resultant current for I_3.

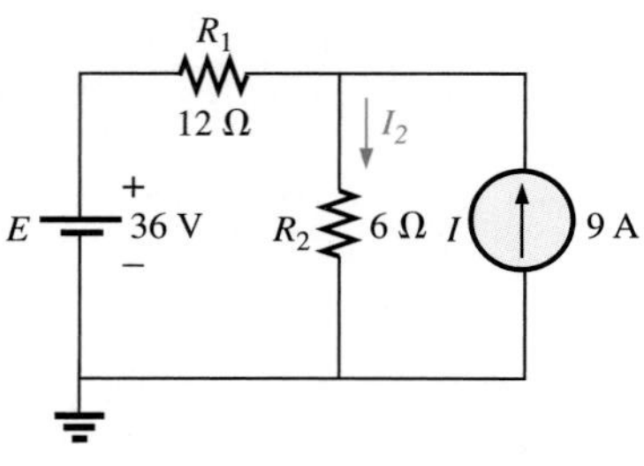

FIG. 9.10
Example 9.3.

EXAMPLE 9.3

a. Using superposition, find the current through the 6-Ω resistor of the network of Fig. 9.10.
b. Demonstrate that superposition is not applicable to power levels.

Solution: a. *Considering the effect of the* 36-V *source* (Fig. 9.11):

$$I'_2 = \frac{E}{R_T} = \frac{E}{R_1 + R_2} = \frac{36\ \text{V}}{12\ \Omega + 6\ \Omega} = 2\ \text{A}$$

continued

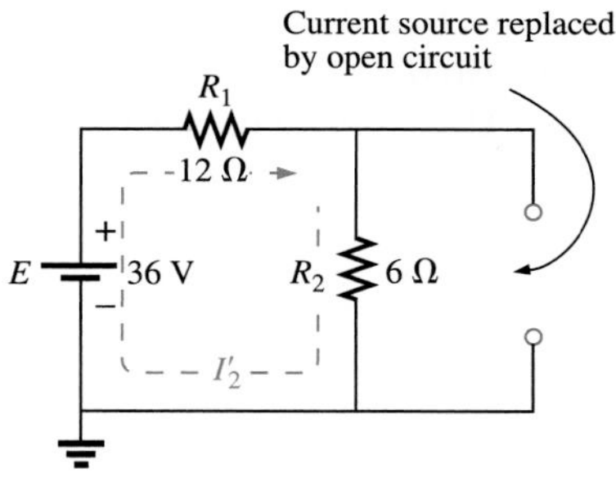

FIG. 9.11
The contribution of E to I_2.

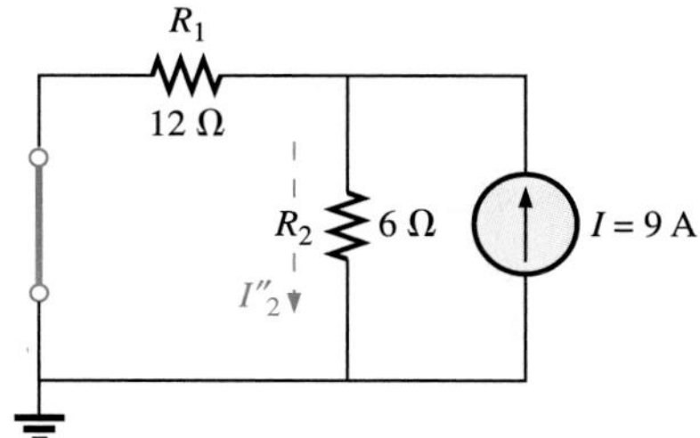

FIG. 9.12
The contribution of I to I_2.

Considering the effect of the 9-A *source* (Fig. 9.12): Applying the current divider rule,

$$I''_2 = \frac{R_1 I}{R_1 + R_2} = \frac{(12\ \Omega)(9\ \text{A})}{12\ \Omega + 6\ \Omega} = \frac{108\ \text{A}}{18} = 6\ \text{A}$$

The total current through the 6-Ω resistor (Fig. 9.13) is

$$I_2 = I'_2 + I''_2 = 2\ \text{A} + 6\ \text{A} = \mathbf{8\ A}$$

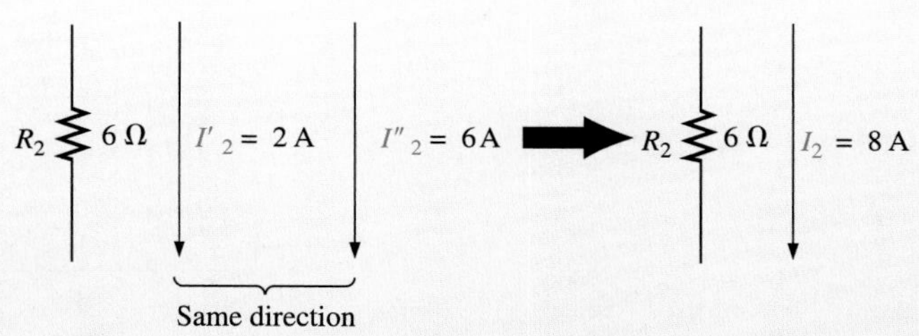

FIG. 9.13
The resultant current for I_2.

b. The power to the 6-Ω resistor is

$$\text{Power} = I^2R = (8\ \text{A})^2(6\ \Omega) = \mathbf{384\ W}$$

The calculated power to the 6-Ω resistor due to each source, *misusing* the principle of superposition, is

$$P_1 = (I'_2)^2R = (2\ \text{A})^2(6\ \Omega) = 24\ \text{W}$$
$$P_2 = (I''_2)^2R = (6\ \text{A})^2(6\ \Omega) = 216\ \text{W}$$
$$P_1 + P_2 = 240\ \text{W} \neq 384\ \text{W}$$

This results because 2 A + 6 A = 8 A, but

$$(2\ \text{A})^2 + (6\ \text{A})^2 \neq (8\ \text{A})^2$$

As mentioned previously, the superposition principle is not applicable to power effects since power is proportional to the square of the current or voltage (I^2R or V^2/R).

Figure 9.14 is a plot of the power delivered to the 6-Ω resistor versus current.

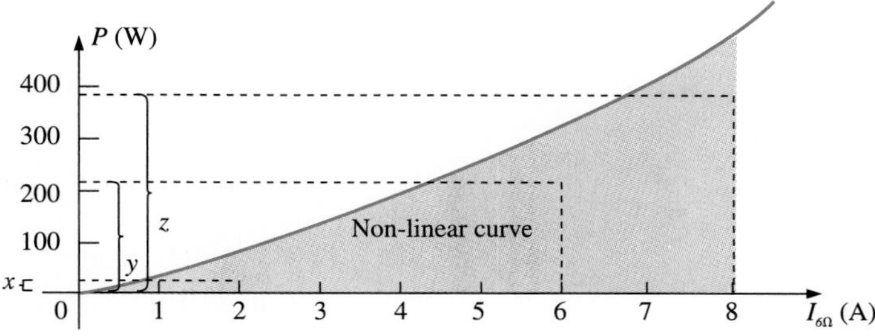

FIG. 9.14
Plotting the power delivered to the 6-Ω resistor versus current through the resistor.

Since, $x + y \neq z$, or 24 W + 216 W ≠ 384 W, superposition does not hold. However, for a linear relationship, such as that between the voltage and current of the fixed-type 6-Ω resistor, superposition can be applied, as shown by the graph of Fig. 9.15, where $a + b = c$, or 2 A + 6 A = 8 A.

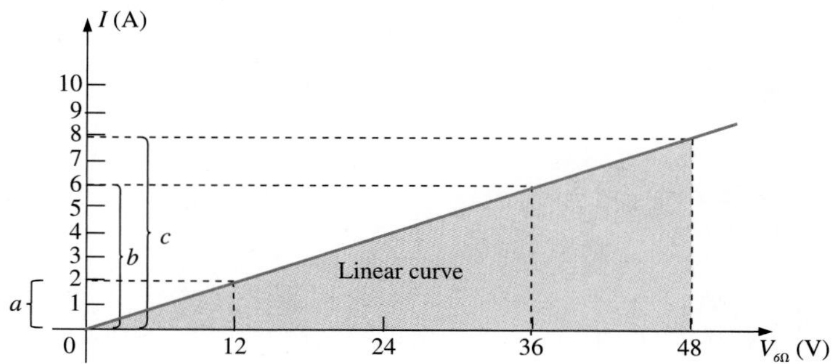

FIG. 9.15
Plotting I versus V for the 6-Ω resistor.

EXAMPLE 9.4

Using the principle of superposition, find the current I_2 through the 12-kΩ resistor of Fig. 9.16.

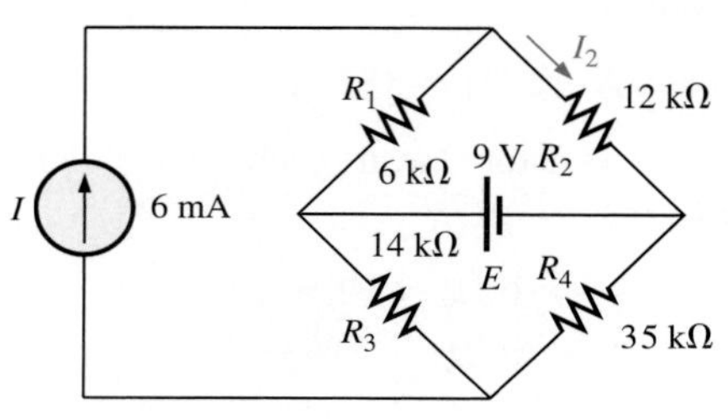

FIG. 9.16
Example 9.4.

Solution: *Considering the effect of the* 6-mA *current source* (Fig. 9.17):

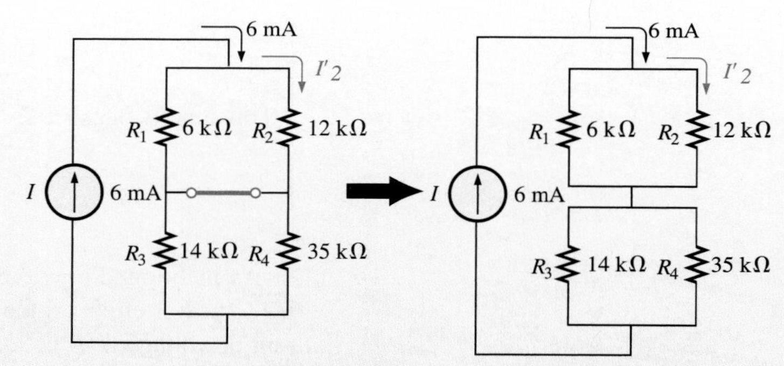

FIG. 9.17
The effect of the current source I on the current I_2.

Current divider rule:

$$I'_2 = \frac{R_1 I}{R_1 + R_2} = \frac{(6\text{ k}\Omega)(6\text{ mA})}{6\text{ k}\Omega + 12\text{ k}\Omega} = 2\text{ mA}$$

Considering the effect of the 9-V *voltage source* (Fig. 9.18):

$$I''_2 = \frac{E}{R_1 + R_2} = \frac{9\text{ V}}{6\text{ k}\Omega + 12\text{ k}\Omega} = 0.5\text{ mA}$$

continued

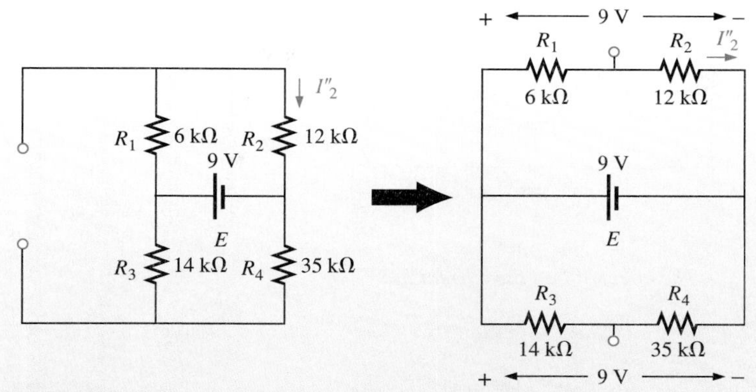

FIG. 9.18
The effect of the voltage source E on the current I_2.

Since I'_2 and I''_2 have the same direction through R_2, the desired current is the sum of the two:

$$\begin{aligned} I_2 &= I'_2 + I''_2 \\ &= 2 \text{ mA} + 0.5 \text{ mA} \\ &= \mathbf{2.5\ mA} \end{aligned}$$

EXAMPLE 9.5

Find the current through the 2-Ω resistor of the network of Fig. 9.19. The presence of three sources will result in three different networks to be analyzed.

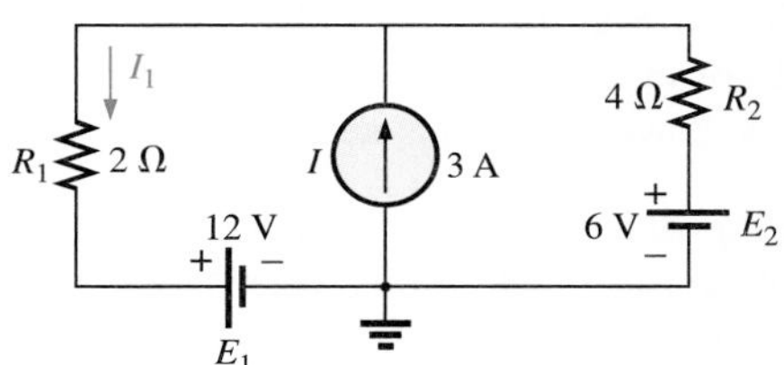

FIG. 9.19
Example 9.5.

Solution: *Considering the effect of the* 12-V *source* (Fig. 9.20):

$$I'_1 = \frac{E_1}{R_1 + R_2} = \frac{12 \text{ V}}{2 \ \Omega + 4 \ \Omega} = \frac{12 \text{ V}}{6 \ \Omega} = 2 \text{ A}$$

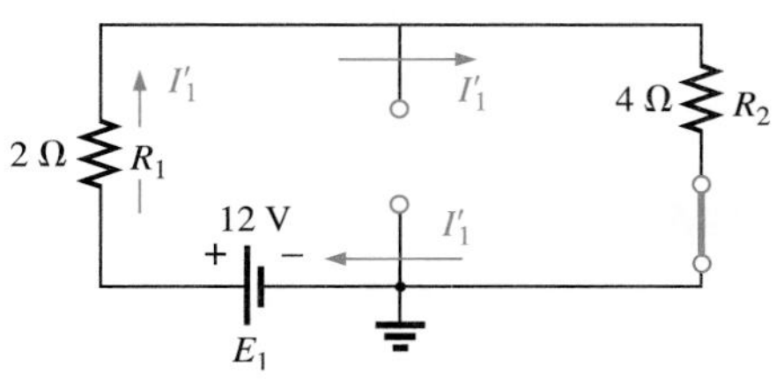

FIG. 9.20
The effect of E_1 on the current I.

Considering the effect of the 6-V *source* (Fig. 9.21):

$$I''_1 = \frac{E_2}{R_1 + R_2} = \frac{6 \text{ V}}{2 \ \Omega + 4 \ \Omega} = \frac{6 \text{ V}}{6 \ \Omega} = 1 \text{ A}$$

Considering the effect of the 3-A *source* (Fig. 9.22):
Applying the current divider rule,

$$I'''_1 = \frac{R_2 I}{R_1 + R_2} = \frac{(4 \ \Omega)(3 \text{ A})}{2 \ \Omega + 4 \ \Omega} = \frac{12 \text{ A}}{6} = 2 \text{ A}$$

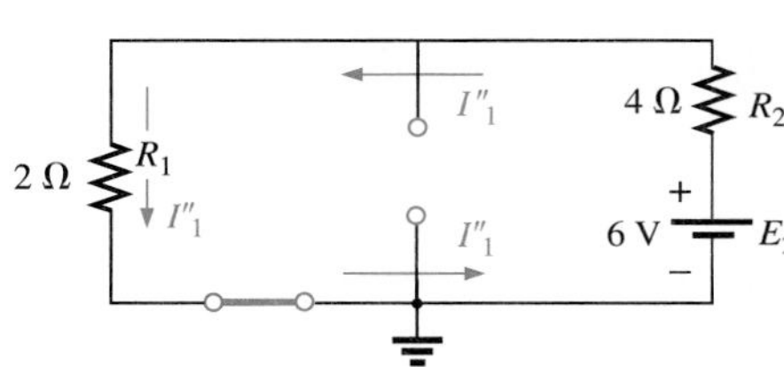

FIG. 9.21
The effect of E_2 on the current I_1.

The total current through the 2-Ω resistor appears in Fig. 9.23, and

Same direction as I_1 in Fig. 9.19 — Opposite direction to I_1 in Fig. 9.19

$$\begin{aligned} I_1 &= I''_1 + I'''_1 - I'_1 \\ &= 1 \text{ A} + 2 \text{ A} - 2 \text{ A} = \mathbf{1\ A} \end{aligned}$$

continued

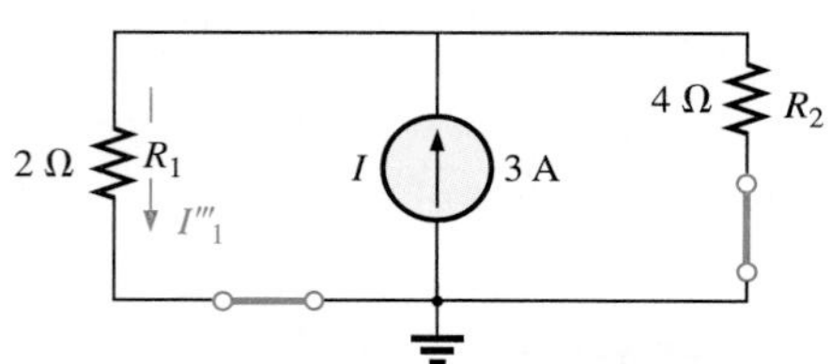

FIG. 9.22
The effect of I on the current I_1.

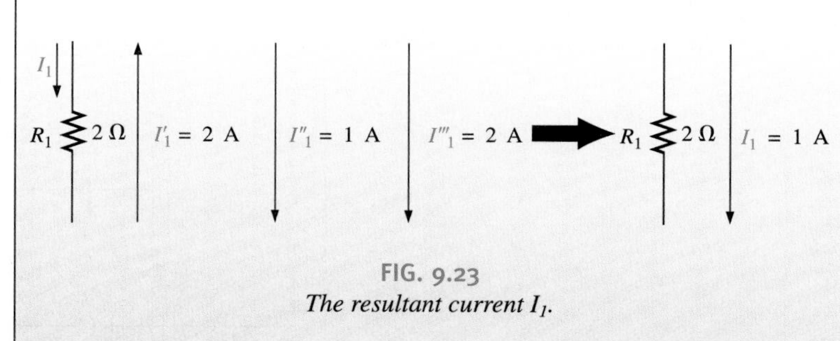

FIG. 9.23
The resultant current I_1.

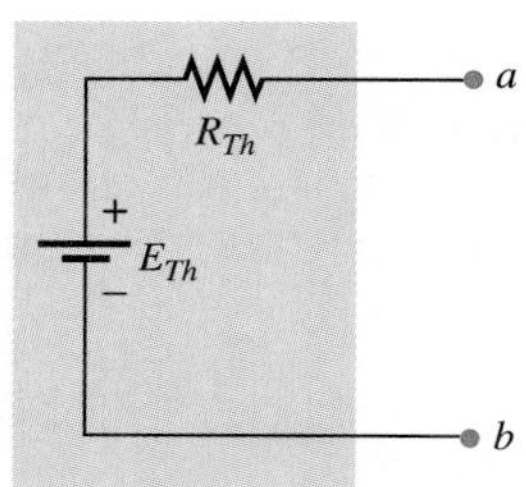

FIG. 9.24
Thévenin equivalent circuit.

9.3 THÉVENIN'S THEOREM

Thévenin's theorem states the following:

Any two-terminal linear bilateral dc network can be replaced by an equivalent circuit consisting of a voltage source and a series resistor, as shown in Fig. 9.24.

In Fig. 9.25(a), for example, the network inside the container has only two terminals available to the outside world, labeled *a* and *b*. It is possible using Thévenin's theorem to replace everything in the container with one source and one resistor, as shown in Fig. 9.25(b), and keep the same terminal characteristics at terminals *a* and *b*. That is, any load connected to terminals *a* and *b* will not know whether it is hooked up to the network of Fig. 9.25(a) or Fig. 9.25(b). The load will receive the same current, voltage, and power from either configuration of Fig. 9.25. In the discussion to follow, however, always keep in mind that

the Thévenin equivalent circuit provides an equivalence at the terminals only—the internal construction and characteristics of the original network and the Thévenin equivalent are usually quite different.

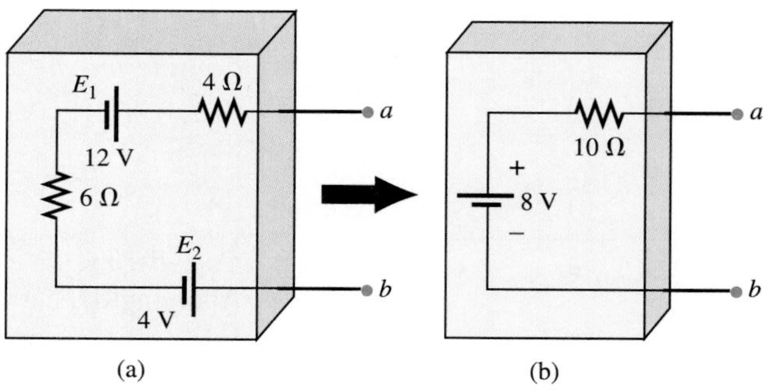

FIG. 9.25
The effect of applying Thévenin's theorem.

For the network of Fig. 9.25(a), you can find the Thévenin equivalent circuit quite directly by simply combining the series batteries and resistors. Note that the network of Fig. 9.25(b) looks exactly like the Thévenin configuration of Fig. 9.24. The method described on the following pages will allow us to extend the procedure just applied to more complex configurations and still end up with the relatively simple network of Fig. 9.24.

LUMINARIES

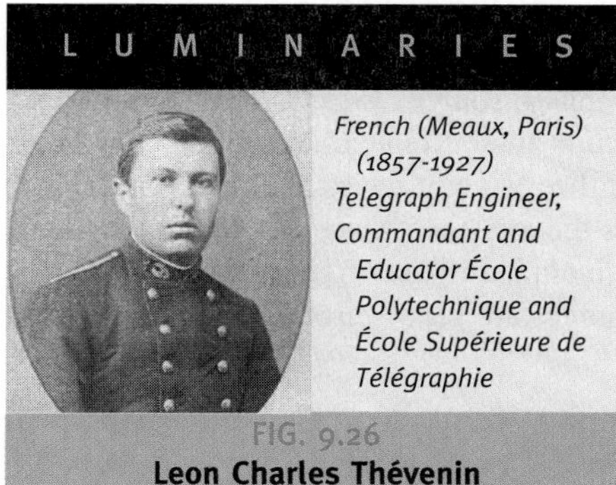

French (Meaux, Paris) (1857-1927) Telegraph Engineer, Commandant and Educator École Polytechnique and École Supérieure de Télégraphie

FIG. 9.26
Leon Charles Thévenin

Although he was active in the study and design of telegraphic systems (including underground transmission), cylindrical condensers (capacitors), and electromagnetism, he is best known for a theorem first presented in the *French Journal of Physics–Theory and Applications* in 1883. It appeared under the heading of "Sur un nouveau théorème d'électricité dynamique" (On a new theorem of dynamic electricity) and was originally referred to as the equivalent generator theorem. There is some evidence that a similar theorem was introduced by Hermann von Helmholtz in 1853. However, Professor Helmholtz applied the theorem to animal physiology and not to communication or generator systems, and therefore has not received the credit in this field that he might deserve. In the early 1920s AT&T did some pioneering work using the equivalent circuit and may have started the reference to the theorem as simply Thévenin's theorem. In fact, Edward L. Norton, an engineer at AT&T at the time, introduced a current source equivalent of the Thévenin equivalent currently referred to as the Norton equivalent circuit. As an aside, Commandant Thévenin was an avid skier and in fact was commissioner of an international ski competition in Chamonix, France in 1912.

Courtesy of the the Bibliothèque École Polytechnique, Paris, France

In most cases, there will be other elements connected to the right of terminals a and b in Fig. 9.25. To apply the theorem, however, the network to be reduced to the Thévenin equivalent form must be isolated as shown in Fig. 9.25 and the two "holding" terminals identified. Once the proper Thévenin equivalent circuit has been determined, the voltage, current, or resistance readings between the two "holding" terminals will be the same whether the original or Thévenin equivalent circuit is connected to the left of terminals a and b in Fig. 9.25. Any load connected to the right of terminals a and b of Fig. 9.25 will receive the same voltage or current with either network.

This theorem achieves two important objectives. First, as was true for all the methods previously described, it allows us to find any particular voltage or current in a linear network with one, two, or any other number of sources. Second, we can concentrate on a specific portion of a network by replacing the remaining network with an equivalent circuit. In Fig. 9.27, for example, by finding the Thévenin equivalent circuit for the network in the shaded area, we can quickly calculate the change in current through or voltage across the variable resistor R_L for the various values that it may assume. This is demonstrated in Example 9.6.

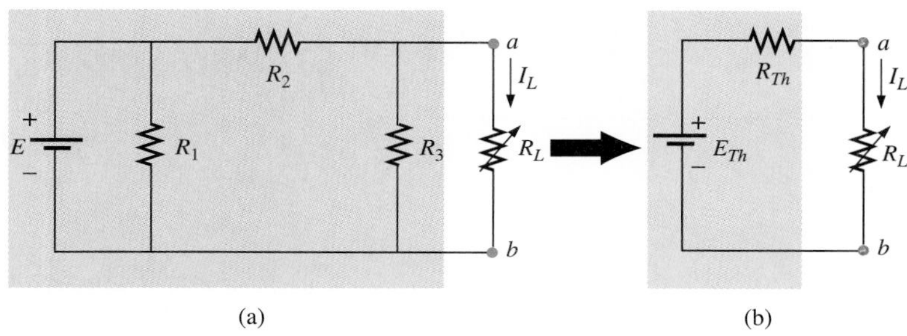

FIG. 9.27
Substituting the Thévenin equivalent circuit for a complex network.

Before we examine the steps involved in applying this theorem, we want to be sure that the implications of the Thévenin equivalent circuit are clear. In Fig. 9.27, the entire network, except R_L, is to be replaced by a single series resistor and battery as shown in Fig. 9.24. The values of these two elements of the Thévenin equivalent circuit must be chosen to ensure that the resistor R_L will react to the network of Fig. 9.27(a) in the same way it will react to the network of Fig. 9.27(b). In other words, the current through or voltage across R_L must be the same for either network for any value of R_L.

The following sequence of steps will lead to the proper value of R_{Th} and E_{Th}.

Preliminary:

1. Remove the portion of the network across which the Thévenin equivalent circuit is to be found. In Fig. 9.27(a), this requires that you temporarily remove the load resistor R_L from the network.
2. Mark the terminals of the remaining two-terminal network. (You will see the importance of this step as we progress through some complex networks.)

continued

R_{Th}:

3. Calculate R_{Th} by first setting all sources to zero (voltage sources are replaced by short circuits and current sources by open circuits) and then finding the resultant resistance between the two marked terminals. (If the internal resistance of the voltage and/or current sources is included in the original network, it must remain when the sources are set to zero.)

E_{Th}:

4. Calculate E_{Th} by first returning all sources to their original position and finding the open-circuit voltage between the marked terminals. (This step is the one that will lead to the most confusion and errors. In all cases, keep in mind that it is the open-circuit potential between the two terminals marked in step 2.)

Conclusion:

5. Draw the Thévenin equivalent circuit with the portion of the circuit previously removed replaced between the terminals of the equivalent circuit. This step is indicated by the placement of the resistor R_L between the terminals of the Thévenin equivalent circuit as shown in Fig. 9.27(b).

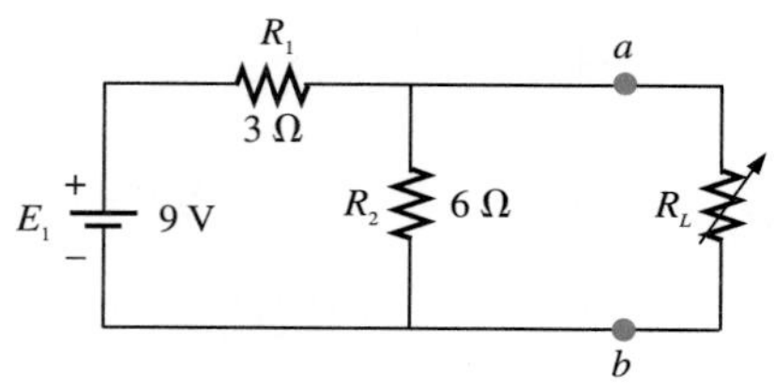

FIG. 9.28
Example 9.6.

EXAMPLE 9.6

Find the Thévenin equivalent circuit for the network of Fig. 9.28. Then find the current through R_L for values of 2 Ω, 10 Ω, and 100 Ω.

Solution:

Steps 1 and 2 produce the network of Fig. 9.29. Note that the load resistor R_L has been removed and the two "holding" terminals have been defined as *a* and *b*.

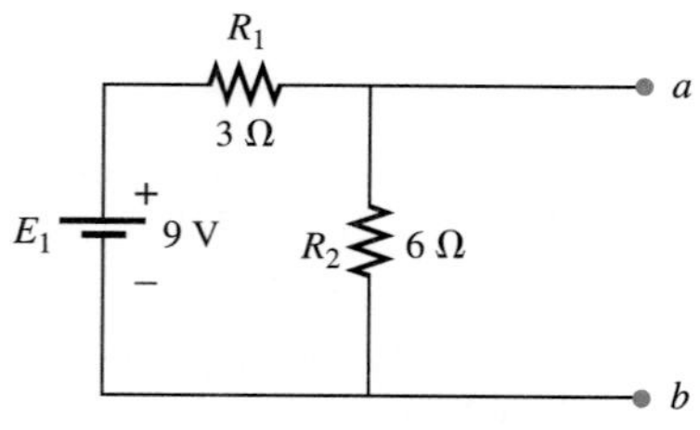

FIG. 9.29
Identifying the terminals of particular importance when applying Thévenin's theorem.

Step 3: Replacing the voltage source E_1 with a short-circuit equivalent gives us the network of Fig. 9.30(a), where

$$R_{Th} = R_1 \parallel R_2 = \frac{(3\ \Omega)(6\ \Omega)}{3\ \Omega + 6\ \Omega} = \mathbf{2\ \Omega}$$

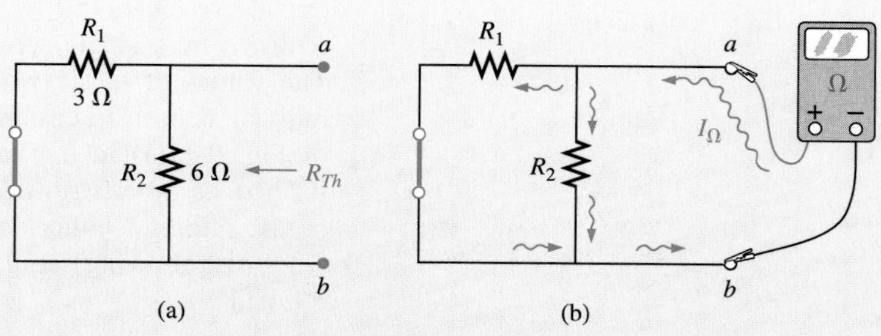

FIG. 9.30
Determining R_{Th} for the network of Fig. 9.29.

You will now begin to see why the two marked terminals are so important. They are the two terminals across which the Thévenin resistance is measured. It is no longer the total resistance as seen by the source, as determined in the majority of problems of Chapter 7. When determining R_{Th}, if you don't know whether the resistive

continued

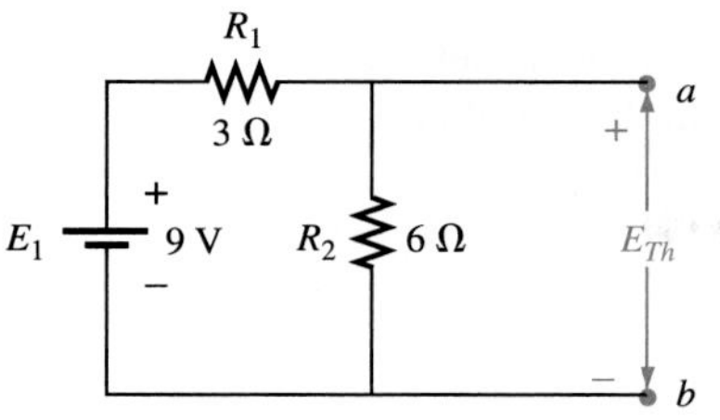

FIG. 9.31
Determining E_{Th} for the network of Fig. 9.27.

elements are in series or parallel, recall that the ohmmeter sends out a trickle current into a resistive combination and senses the level of the resulting voltage to establish the measured resistance level. In Fig. 9.30(b), the trickle current of the ohmmeter approaches the network through terminal *a,* and when it reaches the junction of R_1 and R_2, it splits as shown. The fact that the trickle current splits and then recombines at the lower node shows that the resistors are in parallel as far as the ohmmeter reading is concerned. The path of the sensing current of the ohmmeter shows how the resistors are connected to the two terminals of interest and how the Thévenin resistance should be determined. Keep this in mind as you work through the examples in this section.

Step 4: Replace the voltage source (Fig. 9.31). For this case, the open-circuit voltage E_{Th} is the same as the voltage drop across the 6-Ω resistor. Applying the voltage divider rule,

$$E_{Th} = \frac{R_2E_1}{R_2 + R_1} = \frac{(6\ \Omega)(9\ \text{V})}{6\ \Omega + 3\ \Omega} = \frac{54\ \text{V}}{9} = \mathbf{6\ V}$$

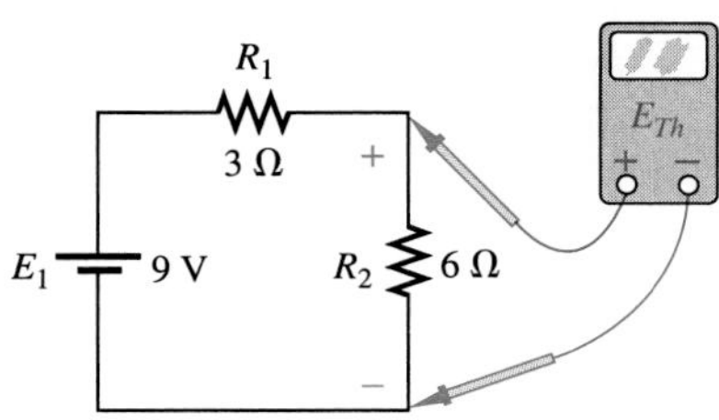

FIG. 9.32
Measuring E_{Th} for the network of Fig. 9.29.

It is important to recognize that E_{Th} is the open-circuit potential between points *a* and *b*. Remember that an open circuit can have any voltage across it, but the current must be zero. In fact, the current through any element in series with the open circuit must be zero also. The use of a voltmeter to measure E_{Th} appears in Fig. 9.32. Note that it is placed directly across the resistor R_2 since E_{Th} and V_{R_2} are in parallel.

Step 5 (Fig. 9.33):

$$I_L = \frac{E_{Th}}{R_{Th} + R_L}$$

$$R_L = 2\ \Omega: \quad I_L = \frac{6\ \text{V}}{2\ \Omega + 2\ \Omega} = \mathbf{1.5\ A}$$

$$R_L = 10\ \Omega: \quad I_L = \frac{6\ \text{V}}{2\ \Omega + 10\ \Omega} = \mathbf{0.5\ A}$$

$$R_L = 100\ \Omega: \quad I_L = \frac{6\ \text{V}}{2\ \Omega + 100\ \Omega} = \mathbf{0.059\ A}$$

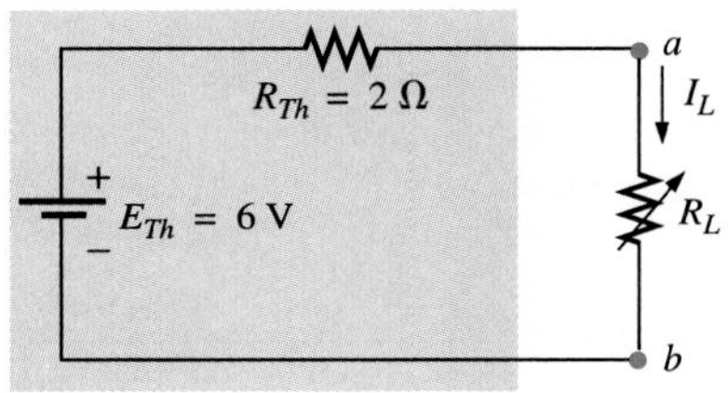

FIG. 9.33
Substituting the Thévenin equivalent circuit for the network external to R_L in Fig. 9.28.

Without Thévenin's theorem, you would have to reexamine the entire network of Fig. 9.28 for each change in R_L to find the new value of I_L.

EXAMPLE 9.7

Find the Thévenin equivalent circuit for the network in the shaded area of the network of Fig. 9.34.

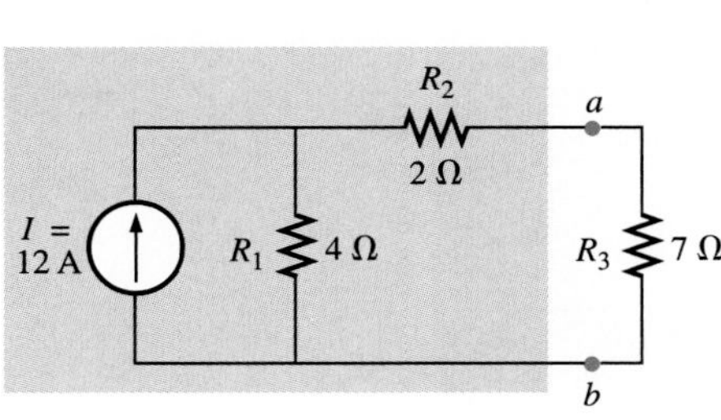

FIG. 9.34
Example 9.7.

Solution:

Steps 1 and 2 are shown in Fig. 9.35.

Step 3 is shown in Fig. 9.36. The current source has been replaced with an open-circuit equivalent and the resistance determined between terminals *a* and *b.*

continued

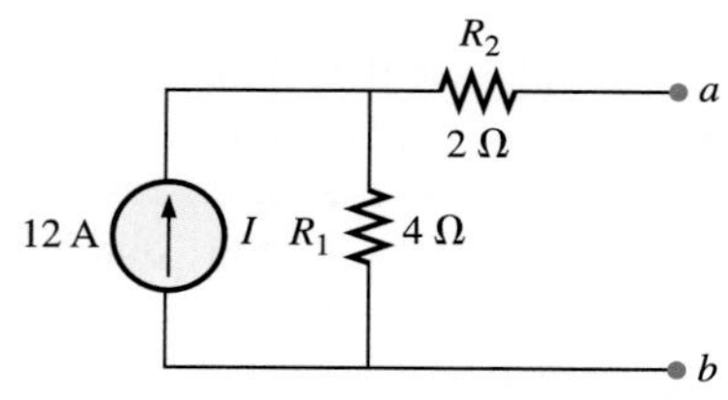

FIG. 9.35
Establishing the terminals of particular interest for the network of Fig. 9.34.

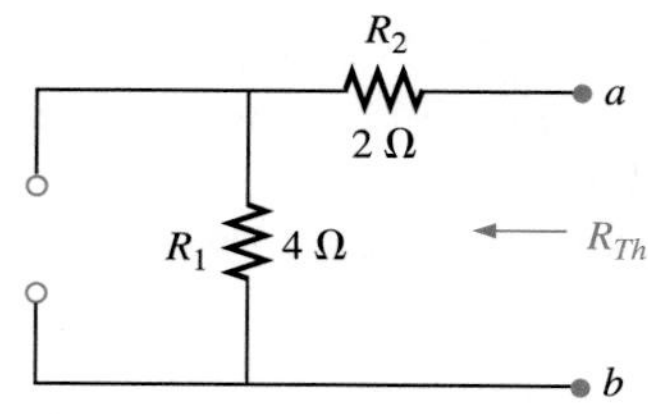

FIG. 9.36
Determining R_{Th} for the network of Fig. 9.35.

In this case an ohmmeter connected between terminals a and b would send out a sensing current that would flow directly through R_1 and R_2 (at the same level). The result is that R_1 and R_2 are in series and the Thévenin resistance is the sum of the two.

$$R_{Th} = R_1 + R_2 = 4\ \Omega + 2\ \Omega = \mathbf{6\ \Omega}$$

Step 4 (Fig. 9.37): In this case, since there is an open circuit between the two marked terminals, the current is zero between these terminals and through the 2-Ω resistor. The voltage drop across R_2 is, therefore,

$$V_2 = I_2R_2 = (0)R_2 = 0\ \text{V}$$

and
$$E_{Th} = V_1 = I_1R_1 = IR_1 = (12\ \text{A})(4\ \Omega) = \mathbf{48\ V}$$

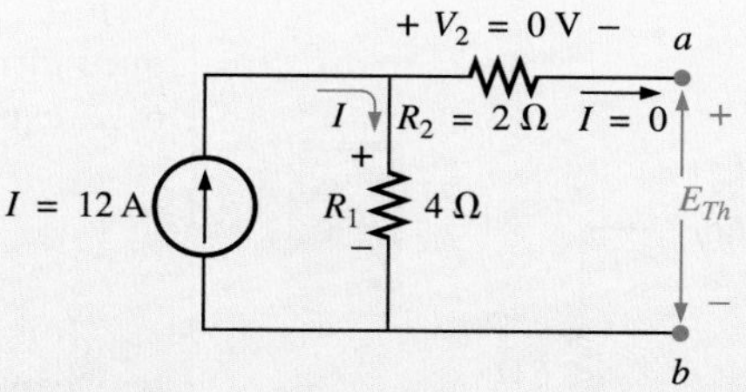

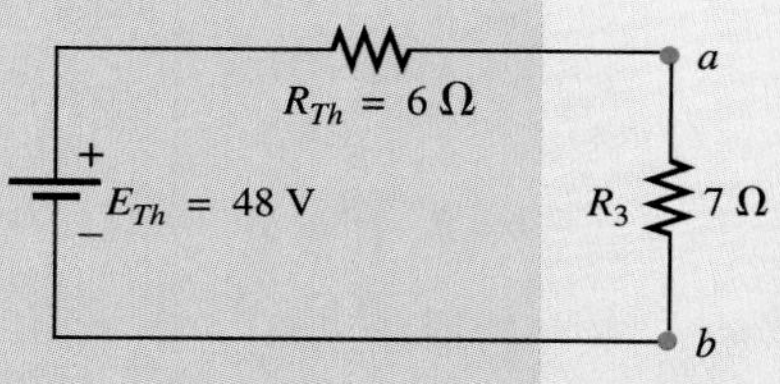

FIG. 9.37
Determining E_{Th} for the network of Fig. 9.35.

Step 5 is shown in Fig. 9.38.

FIG. 9.38
Substituting the Thévenin equivalent circuit in the network external to the resistor R_3 of Fig. 9.34.

EXAMPLE 9.8

Find the Thévenin equivalent circuit for the network in the shaded area of the network of Fig. 9.39. Note in this example that the section of the network to be preserved does not need to be at the "end" of the configuration.

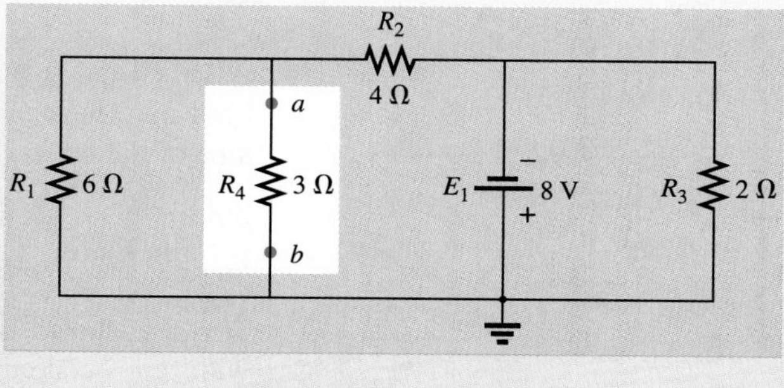

FIG. 9.39
Example 9.8.

continued

Solution:

Steps 1 and 2: See Fig. 9.40.

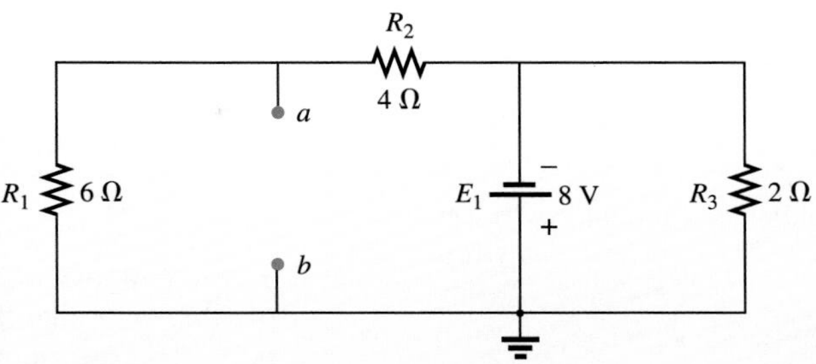

FIG. 9.40
Identifying the terminals of particular interest for the network of Fig. 9.39.

Step 3: See Fig. 9.41. Steps 1 and 2 are relatively easy to apply, but now we must be careful to "hold" onto the terminals a and b as the Thévenin resistance and voltage are determined. In Fig. 9.41, all the remaining elements turn out to be in parallel, and the network can be redrawn as shown.

$$R_{Th} = R_1 \parallel R_2 = \frac{(6\ \Omega)(4\ \Omega)}{6\ \Omega + 4\ \Omega} = \frac{24\ \Omega}{10} = \mathbf{2.4\ \Omega}$$

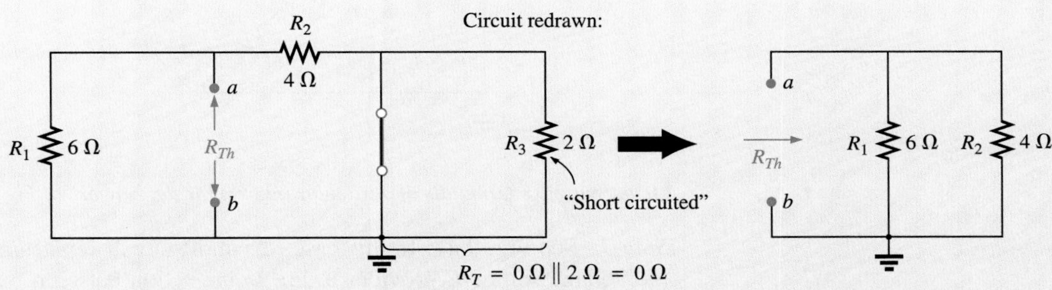

FIG. 9.41
Determining R_{Th} for the network of Fig. 9.40.

Step 4: See Fig. 9.42. In this case, the network can be redrawn as shown in Fig. 9.43, and since the voltage is the same across parallel

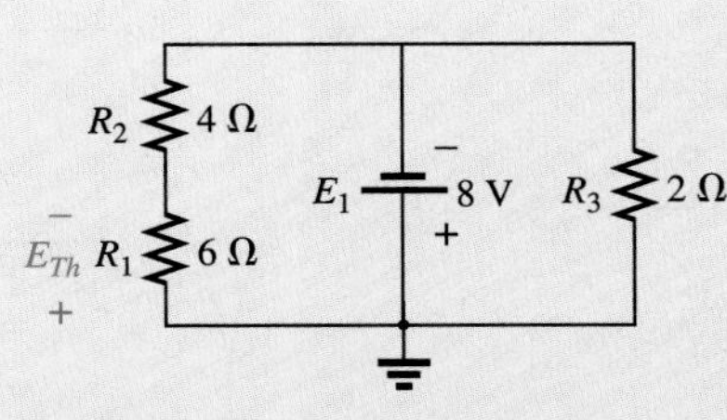

FIG. 9.42
Network of Fig. 9.43 redrawn.

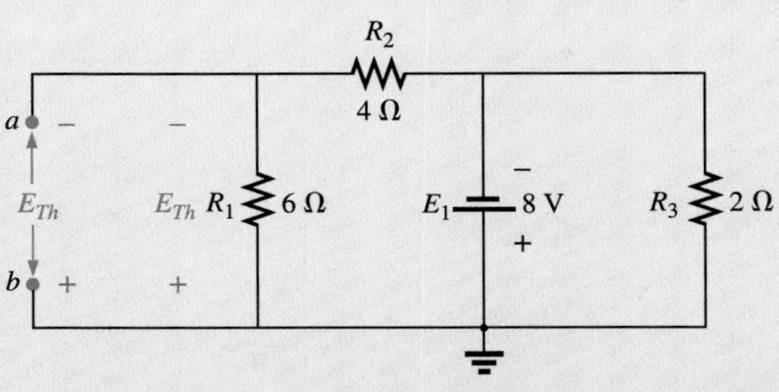

FIG. 9.43
Determining E_{Th} for the network of Fig. 9.40.

continued

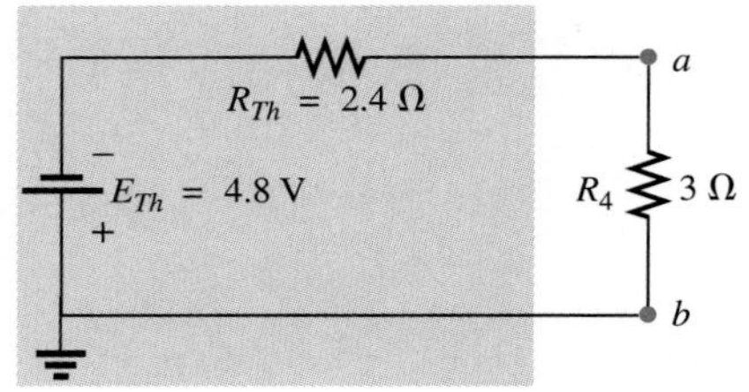

FIG. 9.44
Substituting the Thévenin equivalent circuit for the network external to the resistor R_4 of Fig. 9.39.

elements, the voltage across the series resistors R_1 and R_2 is E_1, or 8 V. Applying the voltage divider rule,

$$E_{Th} = \frac{R_1E_1}{R_1 + R_2} = \frac{(6\ \Omega)(8\ \text{V})}{6\ \Omega + 4\ \Omega} = \frac{48\ \text{V}}{10} = \mathbf{4.8\ V}$$

Step 5: See Fig. 9.44.

The importance of marking the terminals should be obvious from Example 9.8. Note that the Thévenin voltage does not need to have the same polarity as the equivalent circuit originally introduced.

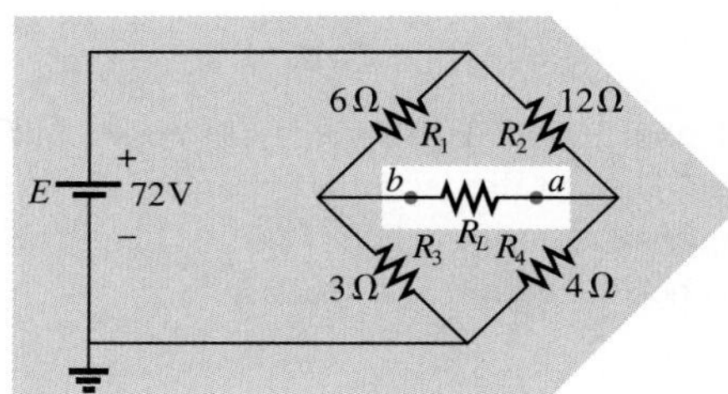

FIG. 9.45
Example 9.9.

EXAMPLE 9.9

Find the Thévenin equivalent circuit for the network in the shaded area of the bridge network of Fig. 9.45.

Solution:

Steps 1 and 2 are shown in Fig. 9.46.

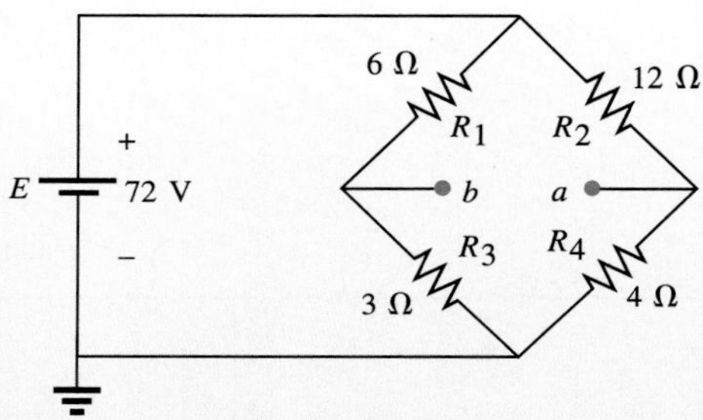

FIG. 9.46
Identifying the terminals of particular interest for the network of Fig. 9.45.

Step 3: See Fig. 9.47. In this case, the short-circuit replacement of the voltage source E provides a direct connection between c and c' of Fig. 9.47(a), This allows a "folding" of the network around the horizontal line of a-b to produce the configuration of Fig. 9.47(b).

$$\begin{aligned} R_{Th} = R_{a\text{-}b} &= R_1 \parallel R_3 + R_2 \parallel R_4 \\ &= 6\ \Omega \parallel 3\ \Omega + 4\ \Omega \parallel 12\ \Omega \\ &= 2\ \Omega + 3\ \Omega = \mathbf{5\ \Omega} \end{aligned}$$

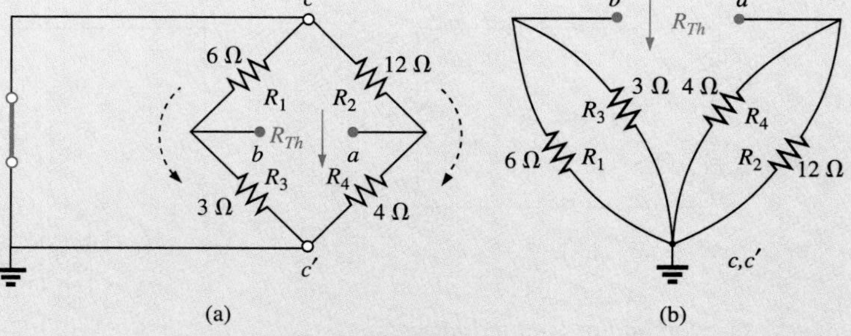

FIG. 9.47
Solving for R_{Th} for the network of Fig. 9.46.

continued

Step 4: The circuit is redrawn in Fig. 9.48. The absence of a direct connection between a and b results in a network with three parallel branches. The voltages V_1 and V_2 can therefore be determined using the voltage divider rule:

$$V_1 = \frac{R_1E}{R_1 + R_3} = \frac{(6\ \Omega)(72\ \text{V})}{6\ \Omega + 3\ \Omega} = \frac{432\ \text{V}}{9} = 48\ \text{V}$$

$$V_2 = \frac{R_2E}{R_2 + R_4} = \frac{(12\ \Omega)(72\ \text{V})}{12\ \Omega + 4\ \Omega} = \frac{864\ \text{V}}{16} = 54\ \text{V}$$

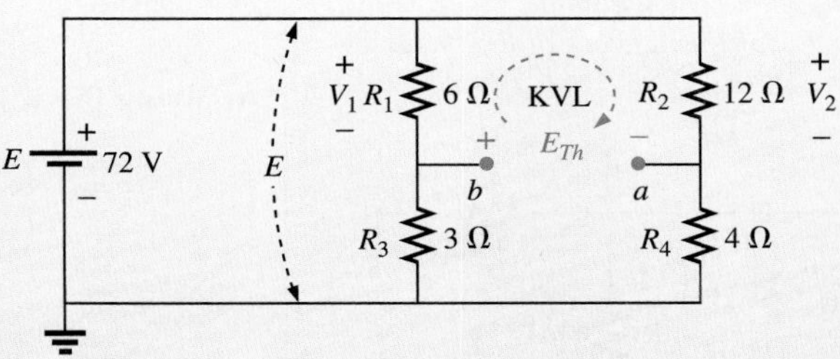

FIG. 9.48
Determining E_{Th} for the network of Fig. 9.46.

Assuming the polarity shown for E_{Th} and applying Kirchhoff's voltage law to the top loop in the clockwise direction will result in

$$\Sigma_C V = +E_{Th} + V_1 - V_2 = 0$$

and $$E_{Th} = V_2 - V_1 = 54\ \text{V} - 48\ \text{V} = \mathbf{6\ V}$$

Step 5 is shown in Fig. 9.49.

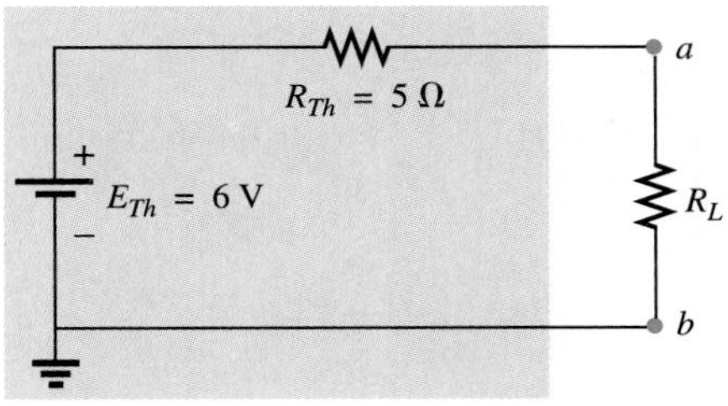

FIG. 9.49
Substituting the Thévenin equivalent circuit for the network external to the resistor R_L of Fig. 9.45.

Thévenin's theorem is not restricted to a single passive element, as shown in the preceding examples. It can be applied across sources, whole branches, portions of networks, or any circuit configuration, as shown in the following example. It is also possible that one of the methods previously described, such as mesh analysis or superposition, may have to be used to find the Thévenin equivalent circuit.

EXAMPLE 9.10

(Two sources) Find the Thévenin circuit for the network within the shaded area of Fig. 9.50.

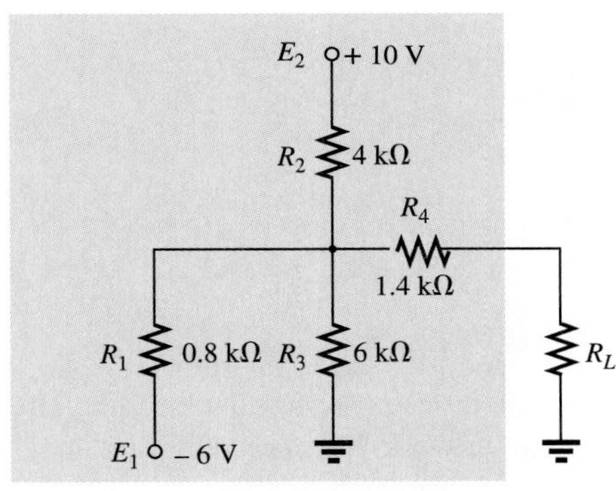

FIG. 9.50
Example 9.10.

Solution: The network is redrawn and steps 1 and 2 are applied as shown in Fig. 9.51.

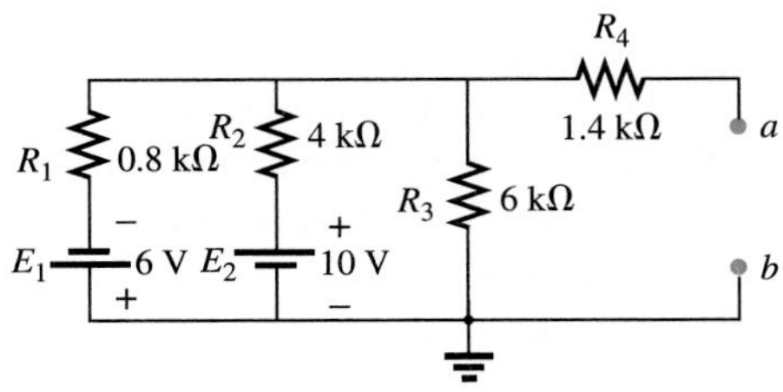

FIG. 9.51
Identifying the terminals of particular interest for the network of Fig. 9.50.

Step 3: See Fig. 9.52.

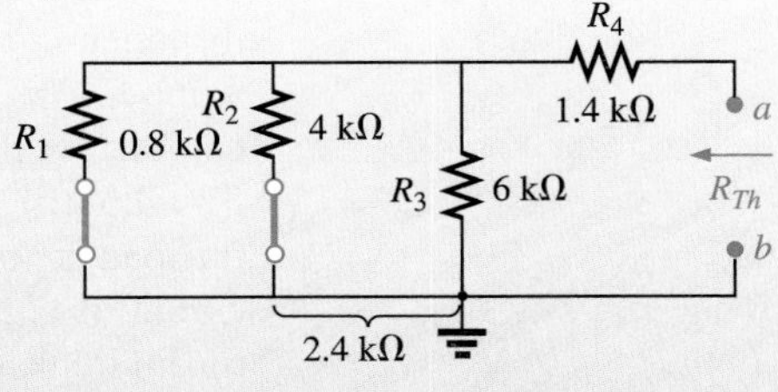

FIG. 9.52
Determining R_{Th} for the network of Fig. 9.51.

continued

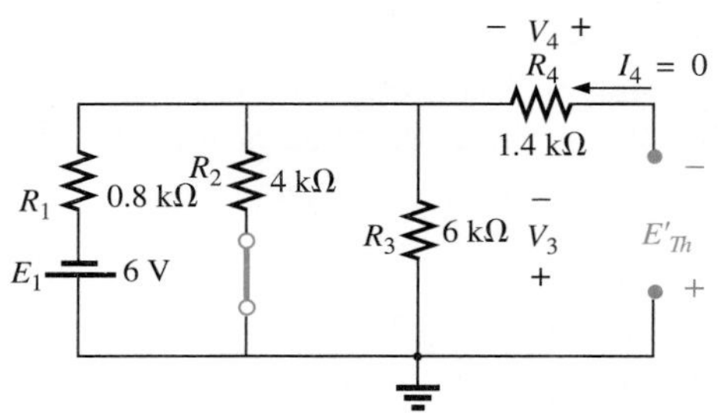

FIG. 9.53
Determining the contribution to E_{Th} from the source E_1 for the network of Fig. 9.51.

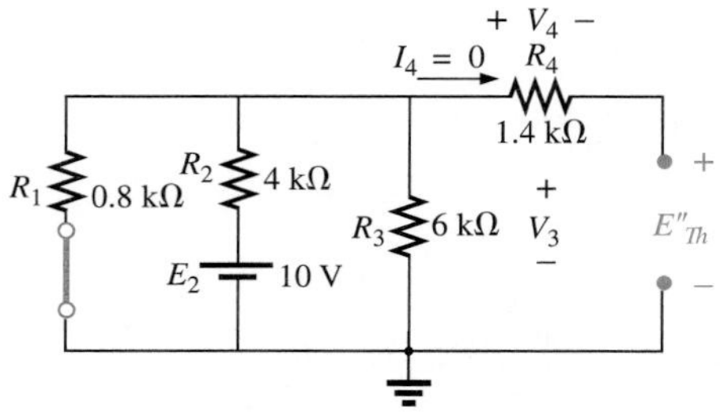

FIG. 9.54
Determining the contribution to E_{Th} from the source E_2 for the network of Fig. 9.51.

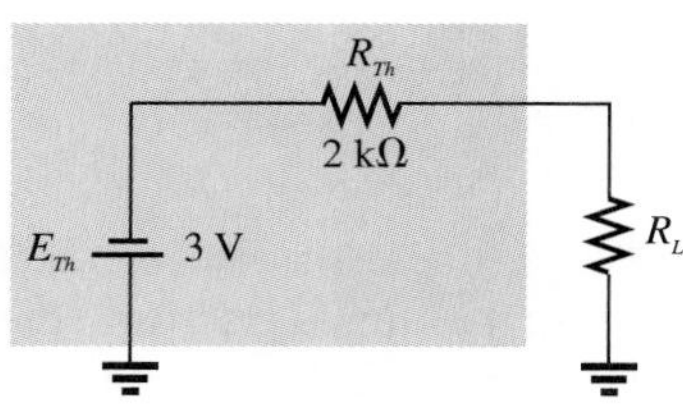

FIG. 9.55
Substituting the Thévenin equivalent circuit for the network external to the resistor R_L of Fig. 9.50.

$$\begin{aligned} R_{Th} &= R_4 + R_1 \parallel R_2 \parallel R_3 \\ &= 1.4\text{ k}\Omega + 0.8\text{ k}\Omega \parallel 4\text{ k}\Omega \parallel 6\text{ k}\Omega \\ &= 1.4\text{ k}\Omega + 0.8\text{ k}\Omega \parallel 2.4\text{ k}\Omega \\ &= 1.4\text{ k}\Omega + 0.6\text{ k}\Omega \\ &= \mathbf{2\text{ k}\Omega} \end{aligned}$$

Step 4: Applying superposition, we will consider the effects of the voltage source E_1 first. Note Fig. 9.53. The open circuit requires that $V_4 = I_4R_4 = (0)R_4 = 0$ V, and

$$E'_{Th} = V_3$$
$$R'_T = R_2 \parallel R_3 = 4\text{ k}\Omega \parallel 6\text{ k}\Omega = 2.4\text{ k}\Omega$$

Applying the voltage divider rule,

$$V_3 = \frac{R'_T E_1}{R'_T + R_1} = \frac{(2.4\text{ k}\Omega)(6\text{ V})}{2.4\text{ k}\Omega + 0.8\text{ k}\Omega} = \frac{14.4\text{ V}}{3.2} = 4.5\text{ V}$$

and $$E'_{Th} = V_3 = 4.5\text{ V}$$

For the source E_2, the network of Fig. 9.54 will result. Again, $V_4 = I_4R_4 = (0)R_4 = 0$ V, and

$$E''_{Th} = V_3$$
$$R'_T = R_1 \parallel R_3 = 0.8\text{ k}\Omega \parallel 6\text{ k}\Omega = 0.706\text{ k}\Omega$$

and $$V_3 = \frac{R'_T E_2}{R'_T + R_2} = \frac{(0.706\text{ k}\Omega)(10\text{ V})}{0.706\text{ k}\Omega + 4\text{ k}\Omega} = \frac{7.06\text{ V}}{4.706} = 1.5\text{ V}$$

and $$E''_{Th} = V_3 = 1.5\text{ V}$$

Since E'_{Th} and E''_{Th} have opposite polarities,

$$\begin{aligned} E_{Th} &= E'_{Th} - E''_{Th} \\ &= 4.5\text{ V} - 1.5\text{ V} \\ &= \mathbf{3\text{ V}} \quad (\text{polarity of } E'_{Th}) \end{aligned}$$

Step 5: See Fig. 9.55.

Experimental Procedures

There are two popular experimental procedures for determining the parameters of a Thévenin equivalent network. The procedure for measuring the Thévenin voltage is the same for each, but the approach for determining the Thévenin resistance is quite different for each.

Direct measurement of E_{Th} and R_{Th} For any physical network, the value of E_{Th} can be determined experimentally by measuring the open-circuit voltage across the load terminals, as shown in Fig. 9.56; $E_{Th} = V_{oc} = V_{ab}$. The value of R_{Th} can then be determined by completing the network with a variable R_L such as the potentiometer of Fig. 9.57(b). R_L can then be varied until the voltage appearing across the load is one-half the open-circuit value, or $V_L = E_{Th}/2$. For the series circuit of Fig. 9.57(a), when the load voltage is reduced to one-half the open-circuit level, the voltage across R_{Th} and R_L must be the same. If we read the value of R_L [as shown in Fig. 9.57(c)] that resulted in the preceding calculations, we will also have the value of R_{Th}, since $R_L = R_{Th}$ if V_L equals the voltage across R_{Th}.

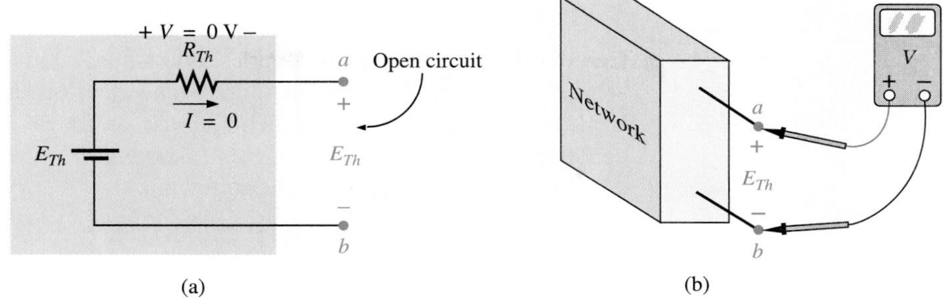

FIG. 9.56
Determining E_{Th} experimentally.

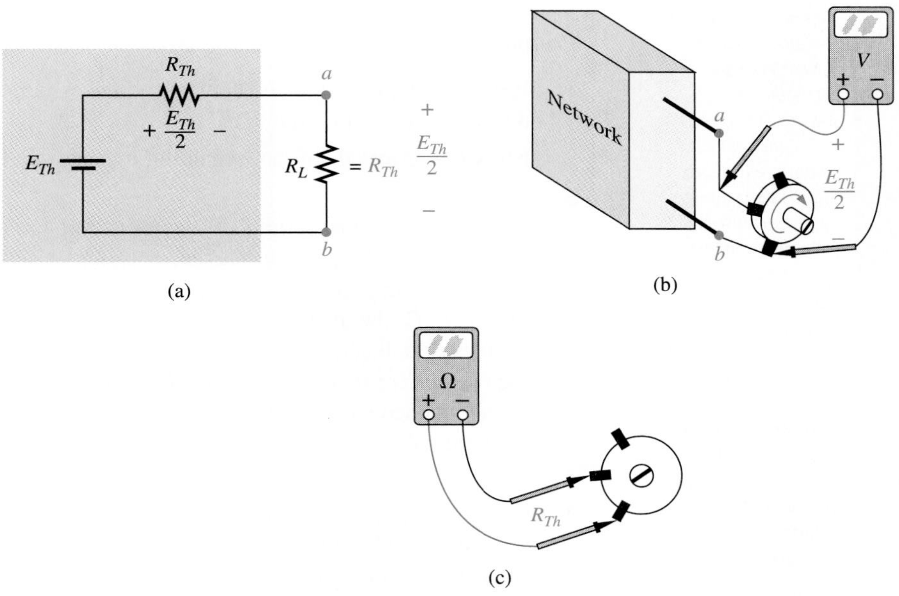

FIG. 9.57
Determining R_{Th} experimentally.

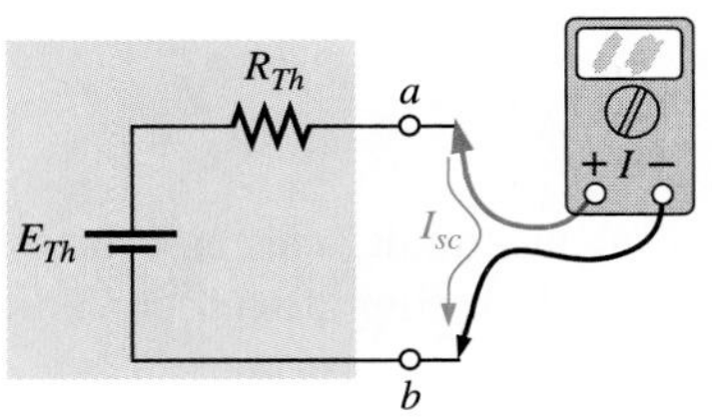

FIG. 9.58
Measuring I_{sc}.

Measuring V_{oc} and I_{sc} The Thévenin voltage is again determined by measuring the open-circuit voltage across the terminals of interest; that is, $E_{Th} = V_{oc}$. To determine R_{Th}, a short-circuit condition is established across the terminals of interest, as shown in Fig. 9.58, and the current through the short circuit is measured with an ammeter. Using Ohm's law, we find that the short-circuit current is determined by

$$I_{sc} = \frac{E_{Th}}{R_{Th}}$$

and the Thévenin resistance by

$$R_{Th} = \frac{E_{Th}}{I_{sc}}$$

However, $E_{Th} = V_{oc}$ resulting in the following equation for R_{Th}:

$$R_{Th} = \frac{V_{oc}}{I_{sc}} \tag{9.2}$$

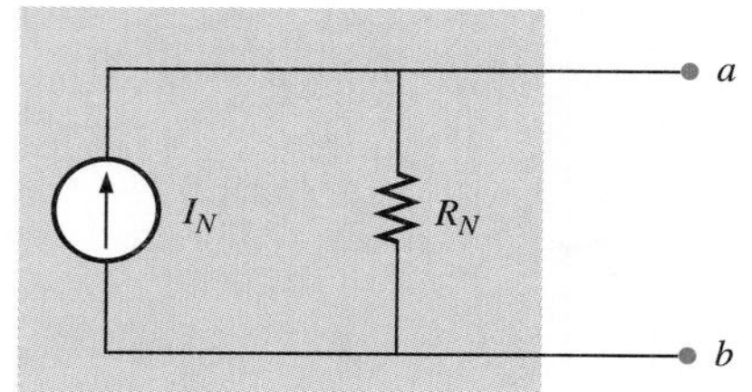

FIG. 9.59
Norton equivalent circuit.

9.4 NORTON'S THEOREM

Every voltage source with a series internal resistance has a current source equivalent. The current source equivalent of the Thévenin network (which satisfies the above conditions), as shown in Fig. 9.59, can be determined by Norton's (Fig. 9.60) theorem.

Norton's theorem states the following:

Any two-terminal linear bilateral dc network can be replaced by an equivalent circuit consisting of a current source and a parallel resistor, as shown in Fig. 9.60.

The discussion of Thévenin's theorem with respect to the equivalent circuit can also be applied to the Norton equivalent circuit. The steps leading to the proper values of I_N and R_N are now listed.

Preliminary:

1. Remove the portion of the network across which the Norton equivalent circuit is found.

2. Mark the terminals of the remaining two-terminal network.

R_N:

3. Calculate R_N by first setting all sources to zero (voltage sources are replaced with short circuits and current sources with open circuits) and then finding the resultant resistance between the two marked terminals. (If the internal resistance of the voltage and/or current sources is included in the original network, it must remain when the sources are set to zero.) Since $R_N = R_{Th}$, the procedure and value obtained using the approach described for Thévenin's theorem will determine the proper value of R_N.

I_N:

4. Calculate I_N by first returning all sources to their original position and then finding the short-circuit current between the marked terminals. It is the same current that would be measured by an ammeter placed between the marked terminals.

Conclusion:

5. Draw the Norton equivalent circuit with the portion of the circuit previously removed replaced between the terminals of the equivalent circuit.

LUMINARIES

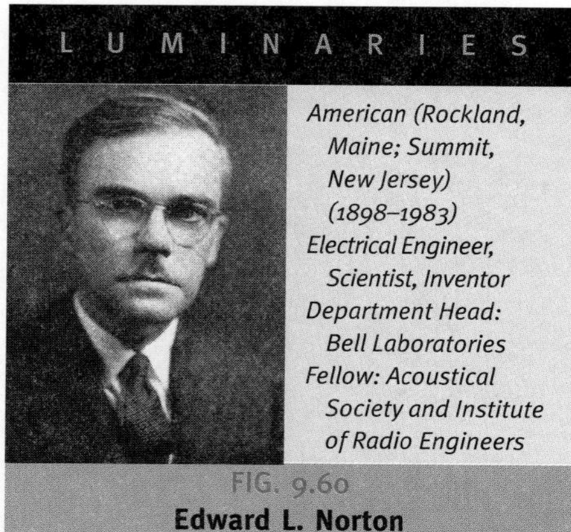

American (Rockland, Maine; Summit, New Jersey) (1898–1983)
Electrical Engineer, Scientist, Inventor
Department Head: Bell Laboratories
Fellow: Acoustical Society and Institute of Radio Engineers

FIG. 9.60
Edward L. Norton

Although he was primarily interested in communications circuit theory and the transmission of data at high speeds over telephone lines, Edward L. Norton is best remembered for development of the dual of Thévenin's equivalent circuit, currently referred to as Norton's equivalent circuit. In fact, Norton and his associates at AT&T in the early 1920s are recognized as some of the first people to perform pioneering work applying Thévenin's equivalent circuit. They were also the first to call this concept Thévenin's theorem. In 1926 Norton proposed the equivalent circuit using a current source and parallel resistor to assist in the design of recording instrumentation that was primarily current driven. He began his telephone career in 1922 with the Western Electric Company's Engineering Department, which later became Bell Laboratories. His areas of active research included network theory, acoustical systems, electromagnetic apparatus, and data transmission. A graduate of MIT and Columbia University, he held nineteen patents on his work.

The Norton and Thévenin equivalent circuits can also be found from each other by using the source transformation discussed earlier in this chapter and reproduced in Fig. 9.61.

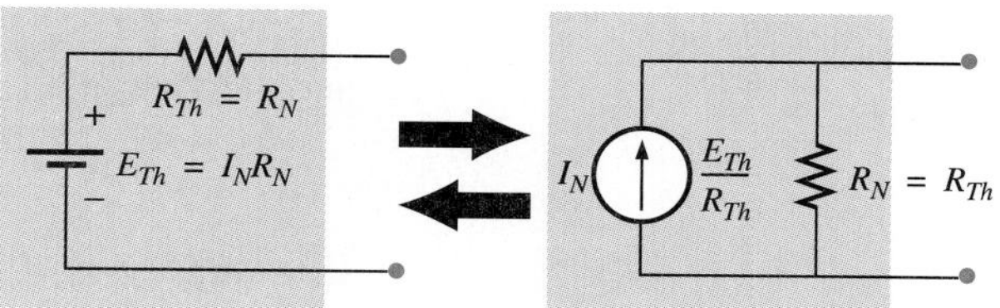

FIG. 9.61
Converting between Thévenin and Norton equivalent circuits.

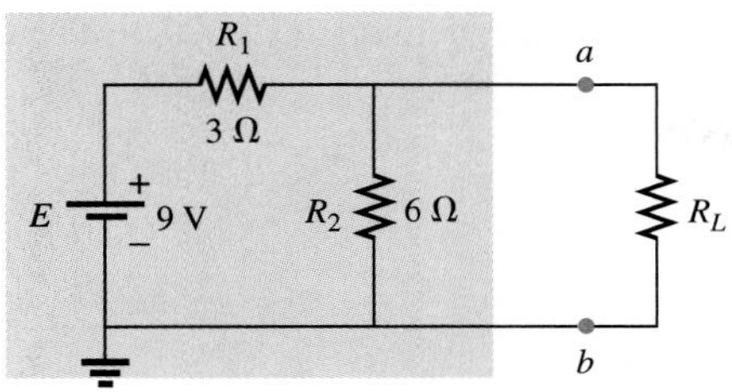

FIG. 9.62
Example 9.11.

EXAMPLE 9.11

Find the Norton equivalent circuit for the network in the shaded area of Fig. 9.62.

Solution:

Steps 1 and 2 are shown in Fig. 9.63.

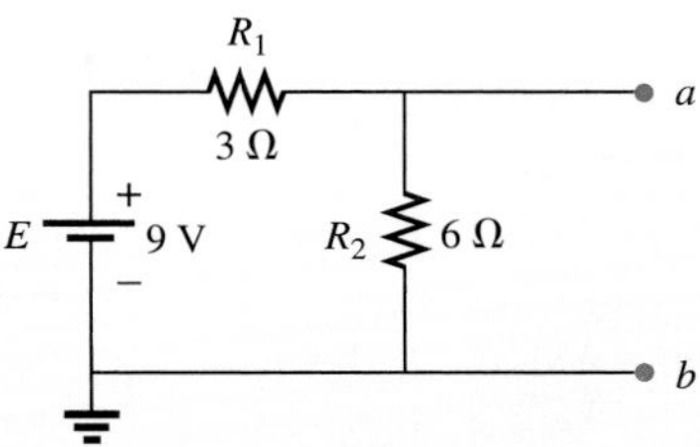

FIG. 9.63
Identifying the terminals of interest for the network of Fig. 9.62.

Step 3 is shown in Fig. 9.64, and

$$R_N = R_1 \parallel R_2 = 3\ \Omega \parallel 6\ \Omega = \frac{(3\ \Omega)(6\ \Omega)}{3\ \Omega + 6\ \Omega} = \frac{18\ \Omega}{9} = \mathbf{2\ \Omega}$$

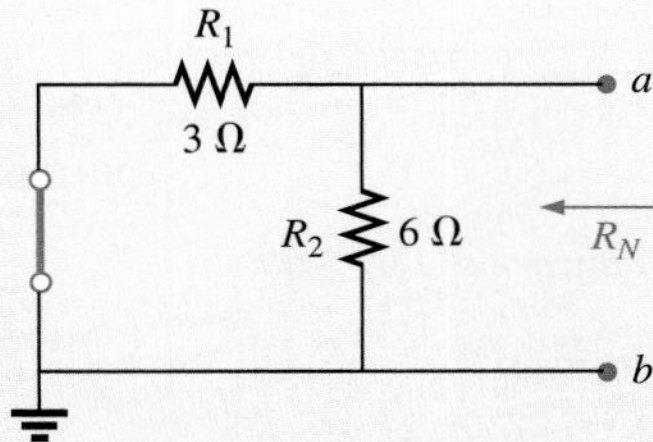

FIG. 9.64
Determining R_N for the network of Fig. 9.63.

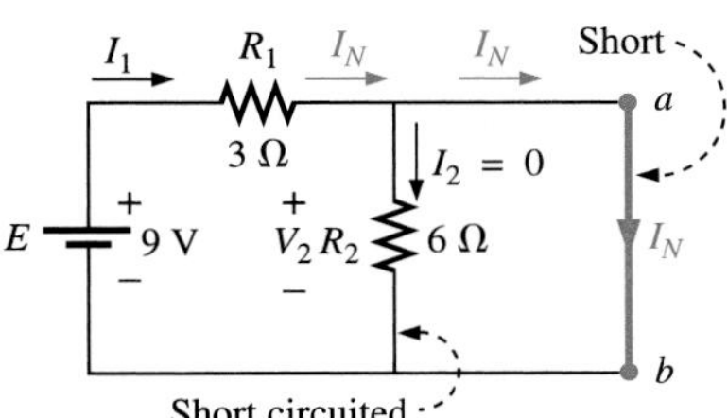

FIG. 9.65
Determining I_N for the network of Fig. 9.63.

Step 4 is shown in Fig. 9.65, clearly indicating that the short-circuit connection between terminals *a* and *b* is in parallel with R_2 and eliminates its effect. I_N is therefore the same as through R_1, and the full battery voltage appears across R_1, since

$$V_2 = I_2R_2 = (0)6\ \Omega = 0\ \text{V}$$

Therefore,

$$I_N = \frac{E}{R_1} = \frac{9\ \text{V}}{3\ \Omega} = \mathbf{3\ A}$$

Step 5: See Fig. 9.66. This circuit is the same as the first one considered in the development of Thévenin's theorem. A simple conversion indicates that the Thévenin circuits are, in fact, the same (Fig. 9.67).

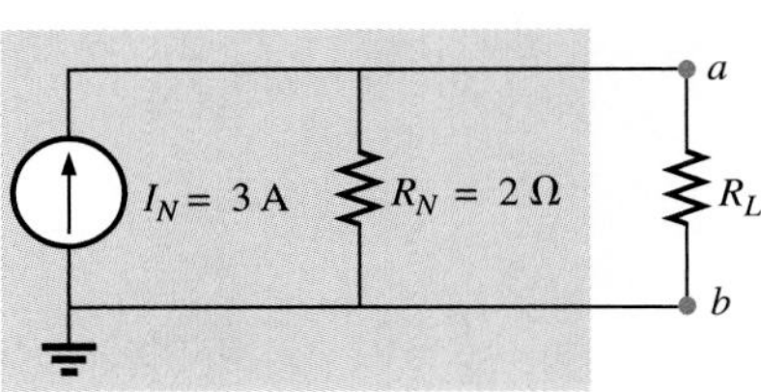

FIG. 9.66
Substituting the Norton equivalent circuit for the network external to the resistor R_L of Fig. 9.62.

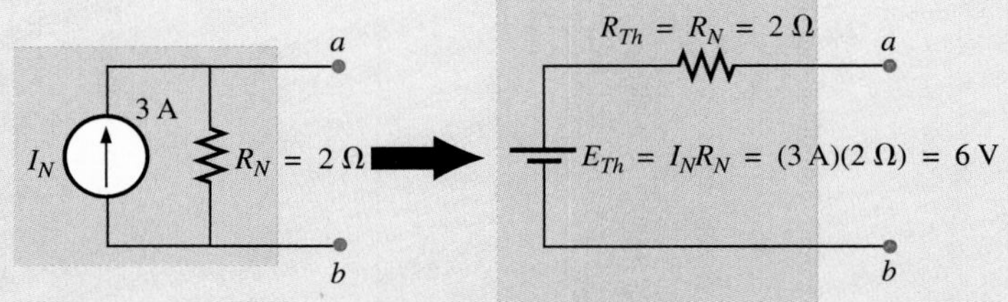

FIG. 9.67
Converting the Norton equivalent circuit of Fig. 9.66 to a Thévenin equivalent circuit.

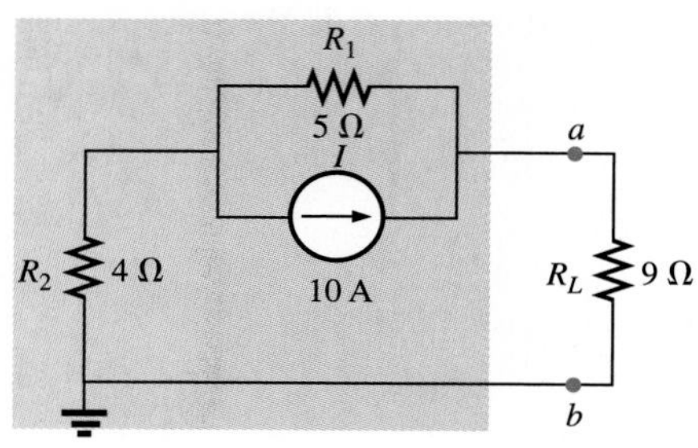

FIG. 9.68
Example 9.12.

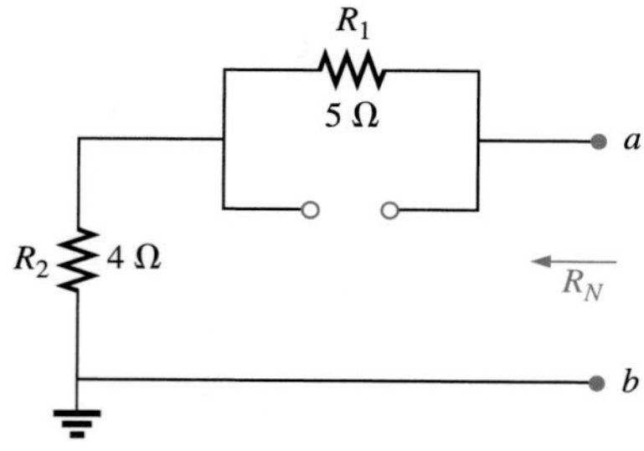

FIG. 9.70
Determining R_N for the network of Fig. 9.69.

EXAMPLE 9.12

Find the Norton equivalent circuit for the network external to the 9-Ω resistor in Fig. 9.68.

Solution:

Steps 1 and 2: See Fig. 9.69.

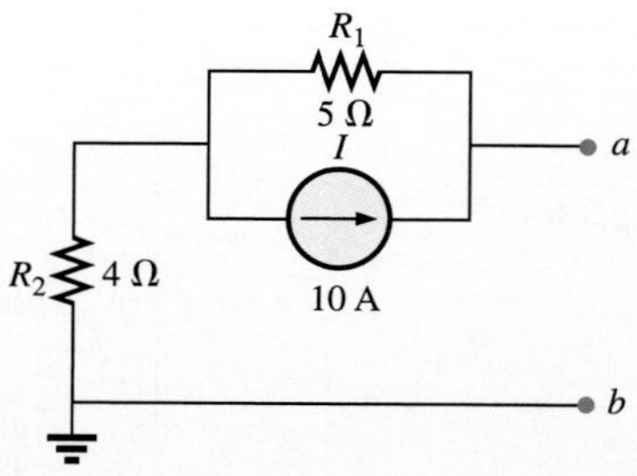

FIG. 9.69
Identifying the terminals of interest for the network of Fig. 9.68.

Step 3: See Fig. 9.70, and

$$R_N = R_1 + R_2 = 5\ \Omega + 4\ \Omega = \mathbf{9\ \Omega}$$

Step 4: As shown in Fig. 9.71, the Norton current is the same as the current through the 4-Ω resistor. Applying the current divider rule,

$$I_N = \frac{R_1 I}{R_1 + R_2} = \frac{(5\ \Omega)(10\ \text{A})}{5\ \Omega + 4\ \Omega} = \frac{50\ \text{A}}{9} = \mathbf{5.556\ A}$$

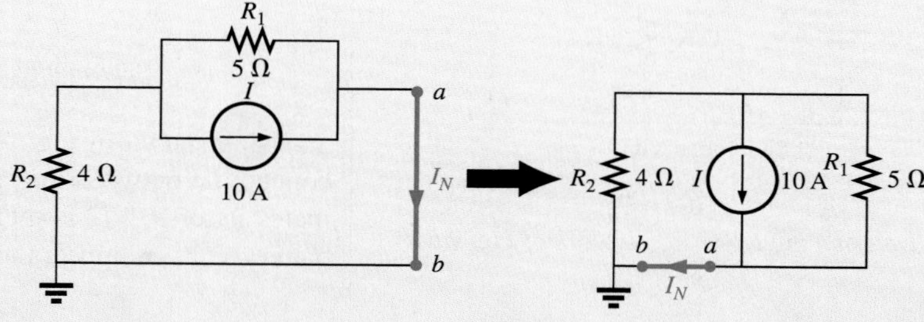

FIG. 9.71
Determining I_N for the network of Fig. 9.69.

Step 5: See Fig. 9.72.

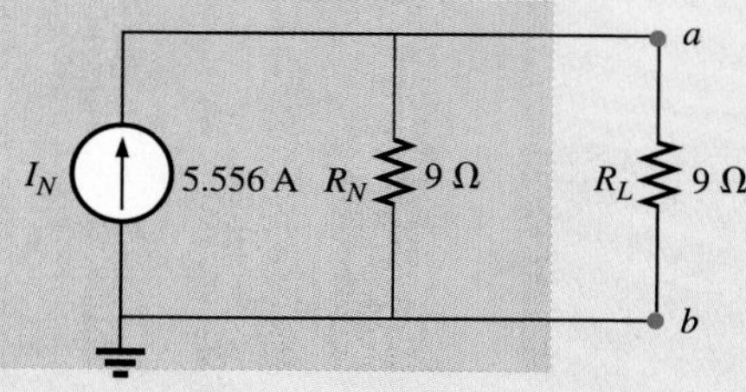

FIG. 9.72
Substituting the Norton equivalent circuit for the network external to the resistor R_L of Fig. 9.68.

EXAMPLE 9.13

(*Two sources*) Find the Norton equivalent circuit for the portion of the network to the left of *a-b* in Fig. 9.73.

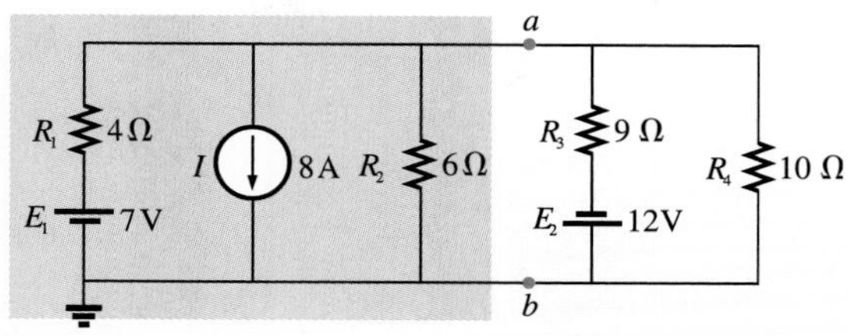

FIG. 9.73
Example 9.13.

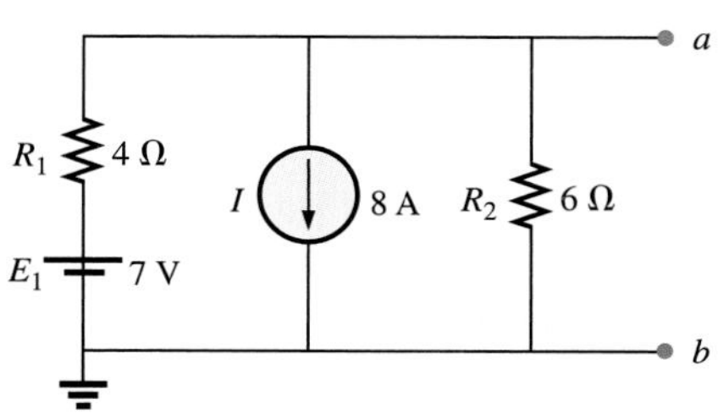

FIG. 9.74
Identifying the terminals of interest for the network of Fig. 9.73.

Solution:

Steps 1 and 2: See Fig. 9.74.

Step 3 is shown in Fig. 9.75, and

$$R_N = R_1 \parallel R_2 = 4\ \Omega \parallel 6\ \Omega = \frac{(4\ \Omega)(6\ \Omega)}{4\ \Omega + 6\ \Omega} = \frac{24\ \Omega}{10} = \mathbf{2.4\ \Omega}$$

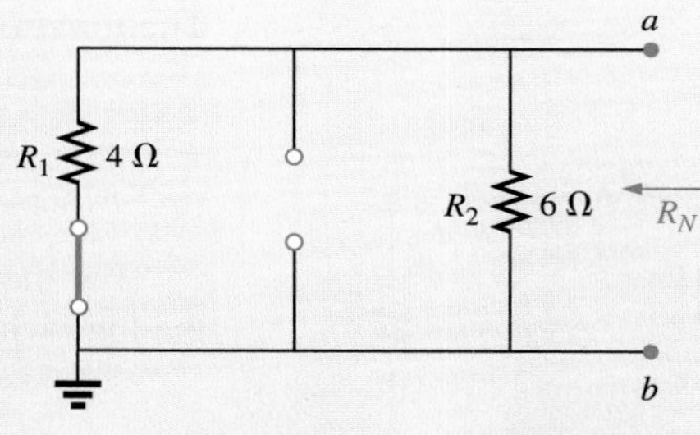

FIG. 9.75
Determining R_N for the network of Fig. 9.74.

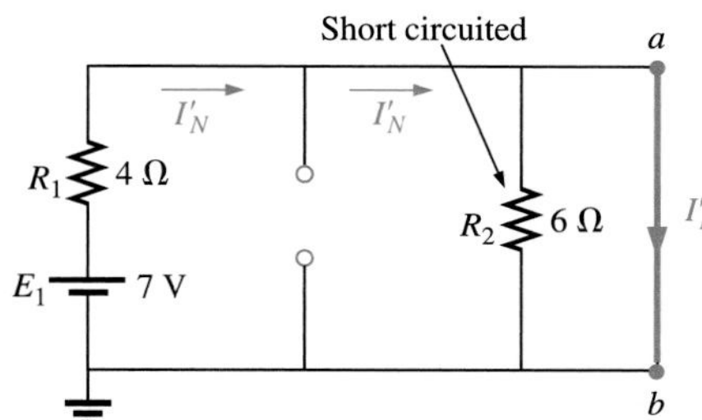

FIG. 9.76
Determining the contribution to I_N from the voltage source E_1.

Step 4: (*Using superposition*) For the 7-V battery (Fig. 9.76),

$$I'_N = \frac{E_1}{R_1} = \frac{7\ \text{V}}{4\ \Omega} = 1.75\ \text{A}$$

For the 8-A source (Fig. 9.77), we find that both R_1 and R_2 have been "short circuited" by the direct connection between *a* and *b,* and

$$I''_N = I = 8\ \text{A}$$

The result is

$$I_N = I''_N - I'_N = 8\ \text{A} - 1.75\ \text{A} = \mathbf{6.25\ A}$$

Step 5: See Fig. 9.78.

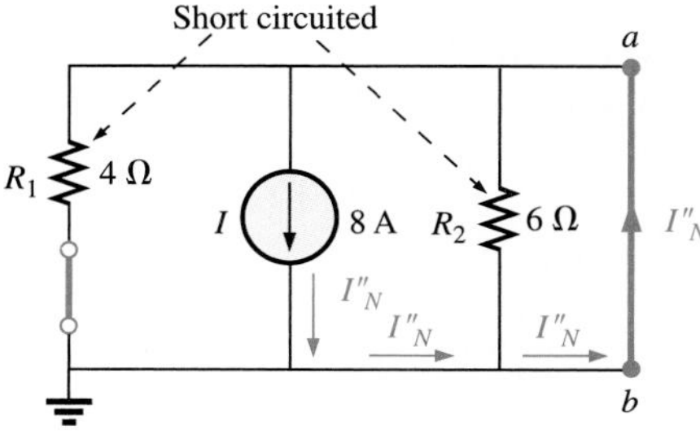

FIG. 9.77
Determining the contribution to I_N from the current source I.

continued

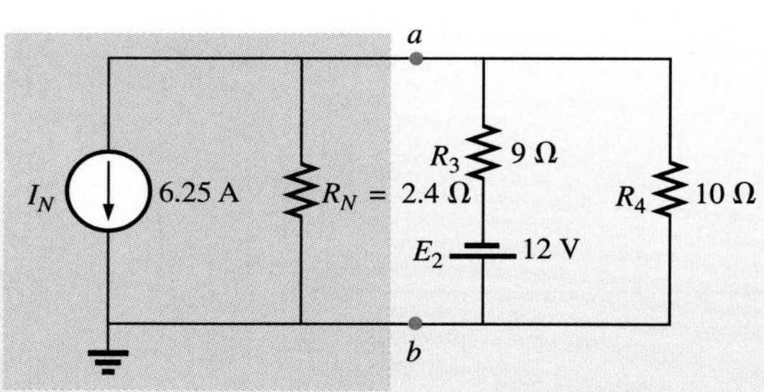

FIG. 9.78
Substituting the Norton equivalent circuit for the network to the left of terminals a-b in Fig. 9.73.

Experimental Procedure

The Norton current is measured in the same way as the short-circuit current for the Thévenin network. Since the Norton and Thévenin resistances are the same, the same procedures can be used for the Thévenin network.

9.5 MAXIMUM POWER TRANSFER THEOREM

The **maximum power transfer theorem** states the following:

A load will receive maximum power from a linear bilateral dc network when its total resistive value is exactly equal to the Thévenin resistance of the network as "seen" by the load.

For the network of Fig. 9.79, maximum power will be delivered to the load when

$$R_L = R_{Th} \tag{9.3}$$

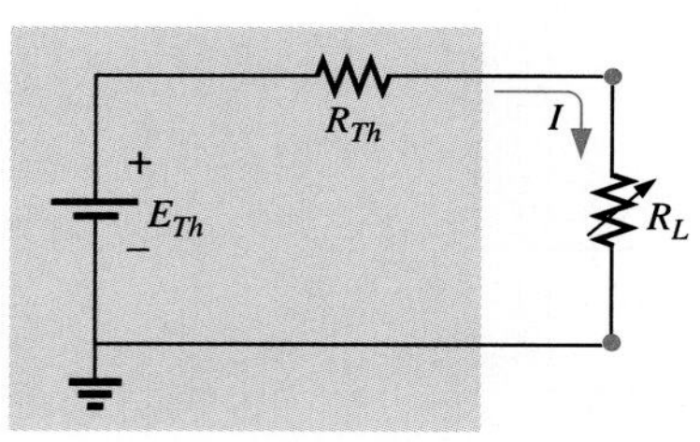

FIG. 9.79
Defining the conditions for maximum power to a load using the Thévenin equivalent circuit.

We know that a Thévenin equivalent circuit can be found across any element or group of elements in a linear bilateral dc network. Therefore, if we consider the case of the Thévenin equivalent circuit in terms of the maximum power transfer theorem, we are considering the *total* effects of any network across a resistor R_L, such as in Fig. 9.79.

For the Norton equivalent circuit of Fig. 9.80, maximum power will be delivered to the load when

$$R_L = R_N \tag{9.4}$$

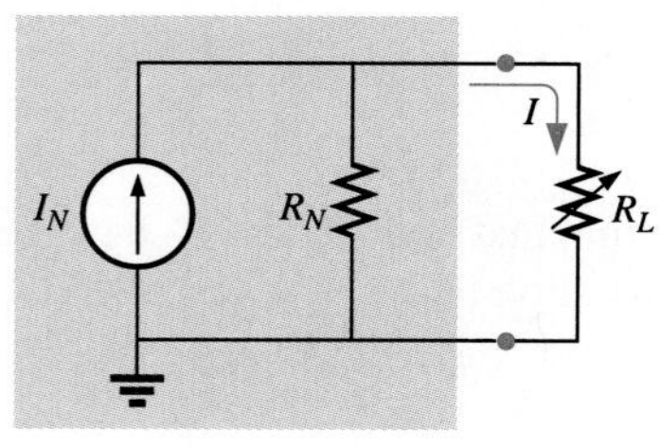

FIG. 9.80
Defining the conditions for maximum power to a load using the Norton equivalent circuit.

This result [Eq. (9.4)] will be very useful in analyzing transistor networks, where the most frequently applied transistor circuit model uses a current source rather than a voltage source.

For the network of Fig. 9.79,

$$I = \frac{E_{Th}}{R_{Th} + R_L}$$

and

$$P_L = I^2R_L = \left(\frac{E_{Th}}{R_{Th} + R_L}\right)^2 R_L$$

so that

$$P_L = \frac{E_{Th}^2 R_L}{(R_{Th} + R_L)^2}$$

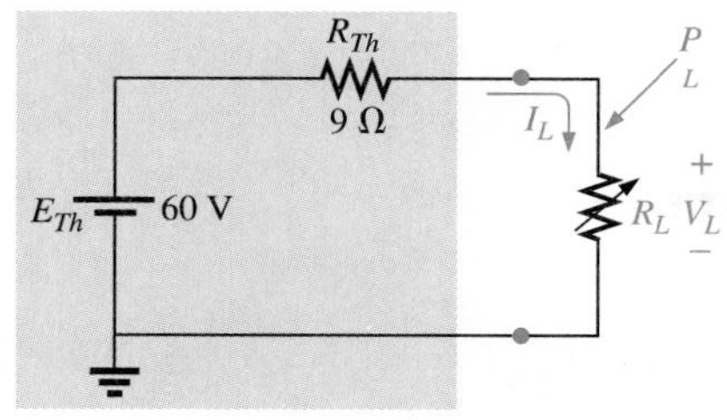

FIG. 9.81
Thévenin equivalent network to be used to validate the maximum power transfer theorem.

We will now consider an example where $E_{Th} = 60$ V and $R_{Th} = 9\ \Omega$, as shown in Fig. 9.81.

The power to the load is determined by:

$$P_L = \frac{E_{Th}^2 R_L}{(R_{Th} + R_L)^2} = \frac{3600 R_L}{(9\ \Omega + R_L)^2}$$

with

$$I_L = \frac{E_{Th}}{R_{Th} + R_L} = \frac{60\ \text{V}}{9\ \Omega + R_L}$$

and

$$V_L = \frac{R_L(60\ \text{V})}{R_{Th} + R_L} = \frac{R_L(60\ \text{V})}{9\ \Omega + R_L}$$

A tabulation of P_L for a range of values of R_L is shown in Table 9.1. A plot of P_L versus R_L using the data of Table 9.1 will result in the plot of Fig. 9.82 for the range $R_L = 0.1\ \Omega$ to $30\ \Omega$.

TABLE 9.1

R_L (Ω)	P_L (W)		I_L (A)		V_L (V)	
0.1	4.35		6.59		0.66	
0.2	8.51		6.52		1.30	
0.5	19.94		6.32		3.16	
1	36.00		6.00		6.00	
2	59.50		5.46		10.91	
3	75.00		5.00		15.00	
4	85.21		4.62		18.46	
5	91.84	Increase	4.29	Decrease	21.43	Increase
6	96.00		4.00		24.00	
7	98.44		3.75		26.25	
8	99.65 ↓		3.53 ↓		28.23 ↓	
9(R_{Th})	100.00	(Maximum)	3.33	($I_{max}/2$)	30.00	($E_{Th}/2$)
10	99.72		3.16		31.58	
11	99.00		3.00		33.00	
12	97.96		2.86		34.29	
13	96.69		2.73		35.46	
14	95.27		2.61		36.52	
15	93.75		2.50		37.50	
16	92.16		2.40		38.40	
17	90.53	Decrease	2.31	Decrease	39.23	Increase
18	88.89		2.22		40.00	
19	87.24		2.14		40.71	
20	85.61		2.07		41.38	
25	77.86		1.77		44.12	
30	71.00		1.54		46.15	
40	59.98		1.22		48.98	
100	30.30		0.55		55.05	
500	6.95		0.12		58.94	
1000	3.54 ↓		0.06 ↓		59.47 ↓	

Note, in particular, that P_L is, in fact, a maximum when $R_L = R_{Th} = 9\ \Omega$. The power curve increases more rapidly toward its maximum value than it decreases after the maximum point, clearly revealing that a small change in load resistance for levels of R_L below R_{Th} will have a more dramatic effect on the power delivered than similar changes in R_L above the R_{Th} level.

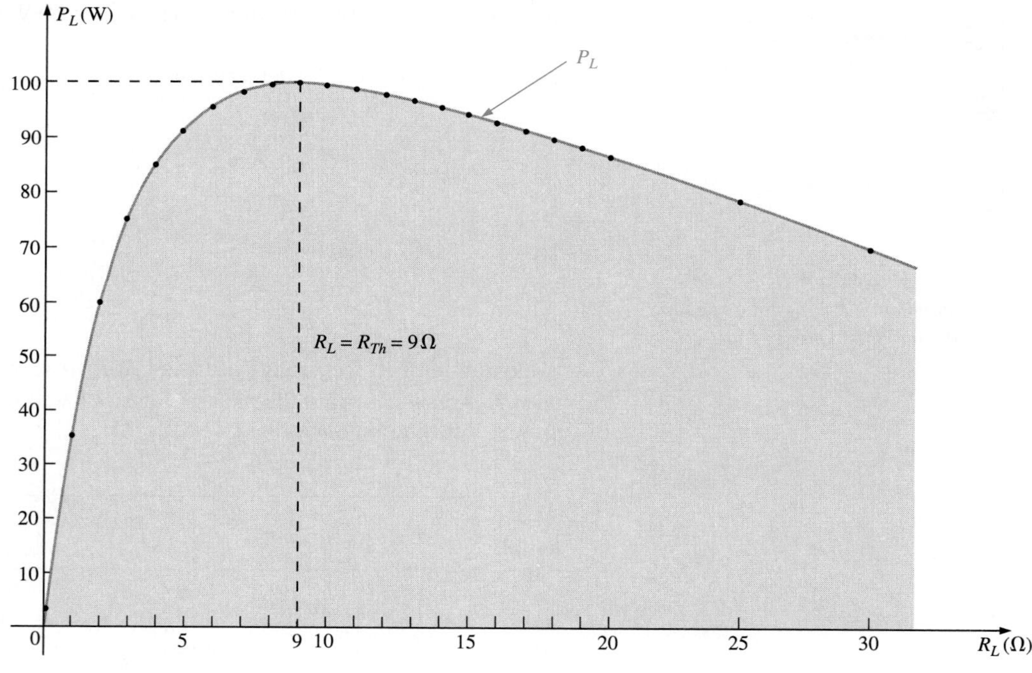

FIG. 9.82
P_L versus R_L for the network of Fig. 9.81.

If we plot V_L and I_L versus the same resistance scale (Fig. 9.83), we find that both change nonlinearly, with the terminal voltage increasing with an increase in load resistance as the current decreases. Note again that the most dramatic changes in V_L and I_L occur for levels of R_L less than R_{Th}. As pointed out on the plot, when $R_L = R_{Th}$, $V_L = E_{Th}/2$ and $I_L = I_{max}/2$, with $I_{max} = E_{Th}/R_{Th}$.

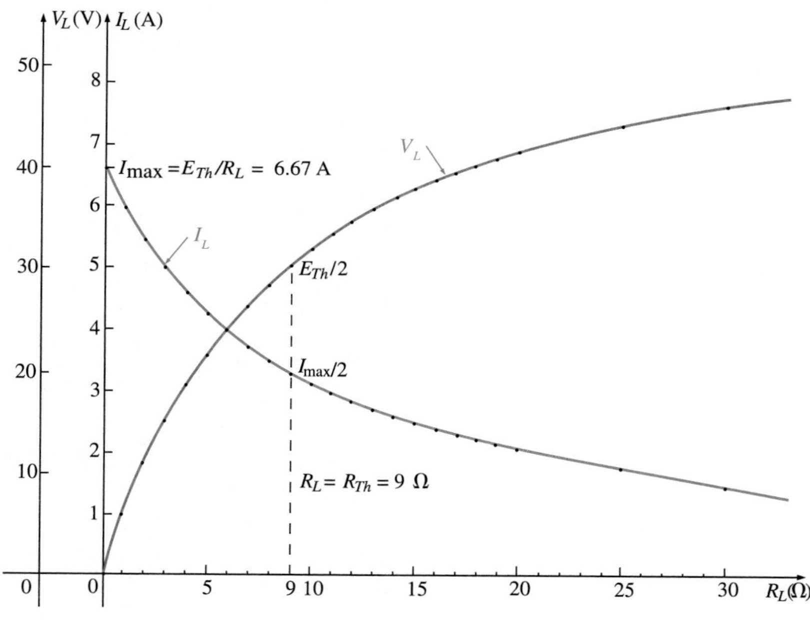

FIG. 9.83
V_L and I_L versus R_L for the network of Fig. 9.81.

The dc operating efficiency of a system is defined by the ratio of the power delivered to the load to the power supplied by the source; that is,

$$\eta\% = \frac{P_L}{P_s} \times 100\% \tag{9.5}$$

For the situation defined by Fig. 9.79,

$$\eta\% = \frac{P_L}{P_s} \times 100\% = \frac{I_L^2 R_L}{I_L^2 R_T} \times 100\%$$

and

$$\eta\% = \frac{R_L}{R_{Th} + R_L} \times 100\%$$

For R_L that is small compared to R_{Th}, $R_{Th} \gg R_L$ and $R_{Th} + R_L \cong R_{Th}$, with

$$\eta\% \cong \frac{R_L}{R_{Th}} \times 100\% = \underbrace{\left(\frac{1}{R_{Th}}\right)}_{\text{Constant}} R_L \times 100\% = kR_L \times 100\%$$

The resulting percentage efficiency, therefore, will be relatively low (since k is small) and will increase almost linearly as R_L increases.

For situations where the load resistance R_L is much larger than R_{Th}, $R_L \gg R_{Th}$ and $R_{Th} + R_L \cong R_L$.

$$\eta\% = \frac{R_L}{R_L} \times 100\% = 100\%$$

The efficiency therefore increases linearly and dramatically for small levels of R_L and then begins to level off as it approaches the 100% level for very large values of R_L, as shown in Fig. 9.84. Keep in mind, however, that the efficiency criterion is sensitive only to the ratio of P_L to P_s and not to their actual levels. At efficiency levels approaching 100%, the power delivered to the load may be so small that it has little practical value. Note the low level of power to the load in Table 9.1 when $R_L = 1000\ \Omega$ even though the efficiency level will be

$$\eta\% = \frac{R_L}{R_{Th} + R_L} \times 100\% = \frac{1000}{1009} \times 100\% = 99.11\%$$

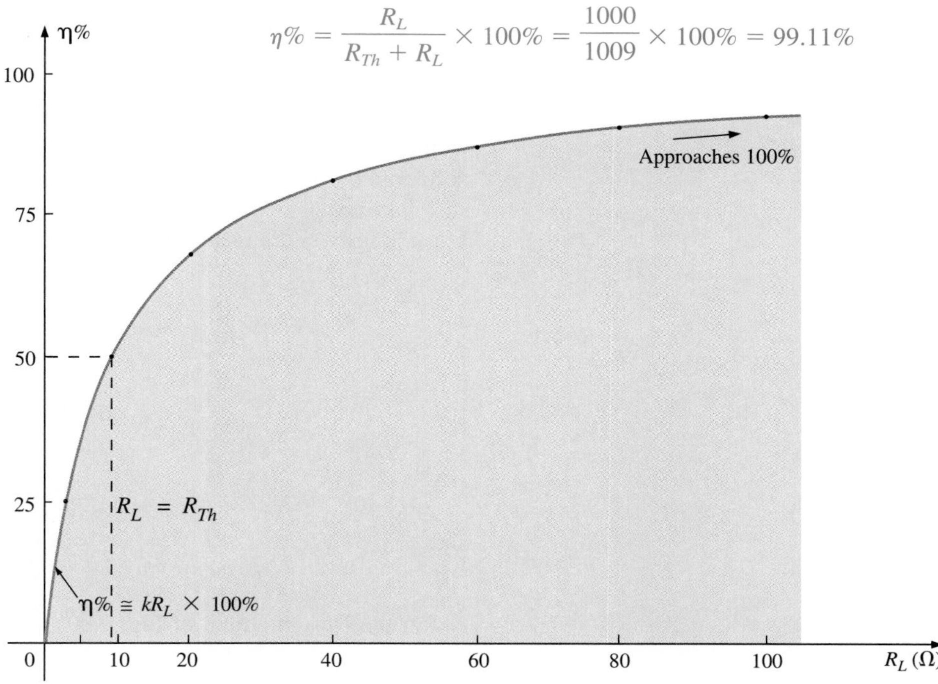

FIG. 9.84
Efficiency of operation versus increasing values of R_L.

When $R_L = R_{Th}$,

$$\eta\% = \frac{R_L}{R_{Th} + R_L} \times 100\% = \frac{R_L}{2R_L} \times 100\% = \mathbf{50\%}$$

Under maximum power transfer conditions, therefore, P_L is a maximum, but the dc efficiency is only 50%; that is, only half the power delivered by the source is getting to the load.

A relatively low efficiency of 50% can be tolerated in situations where power levels are relatively low, such as in a wide variety of electronic systems. However, when large power levels are involved, such as at generating stations, efficiencies of 50% would not be acceptable. In fact, a great deal of expense and research is dedicated to raising power-generating and transmission efficiencies a few percentage points. Raising an efficiency level of a 10-GW power plant from 94% to 95% (a 1% increase) can save 0.1 GW, or 100 MW, of power—an enormous saving!

Consider a change in load levels from 9 Ω to 20 Ω. In Fig. 9.82, the power level has dropped from 100 W to 85.61 W (a 14.4% drop), but the efficiency has increased to 69% (a 38% increase); see Fig. 9.84. For each application, we must find a balance point where the efficiency is high enough without reducing the power to the load to insignificant levels.

Figure 9.85 is a semilog plot of P_L and the power delivered by the source $P_s = E_{Th}I_L$ versus R_L for $E_{Th} = 60$ V and $R_{Th} = 9$ Ω. A semilog graph uses one log scale and one linear scale, as indicated by the prefix *semi,* meaning *half.* Log scales are discussed in detail in Chapter 21. For the moment, note the wide range of R_L permitted using the log scale as compared to Figs. 9.82 through 9.84.

It is now clear that the P_L curve has only one maximum (at $R_L = R_{Th}$), whereas P_s decreases for every increase in R_L. Note that for low levels of R_L, only a small portion of the power delivered by the source makes it to the load. Even when $R_L = R_{Th}$, the source is generating twice the power absorbed by the load. For values of R_L greater than R_{Th}, the two curves approach each other until they are essentially the same at high levels of R_L. For the range $R_L = R_{Th} = 9$ Ω to $R_L = 100$ Ω, P_L and P_s are relatively close in magnitude. This would be an appropriate range of operation, since most of the power delivered by the source is getting to the load and the power levels are still significant.

The power delivered to R_L under maximum power conditions ($R_L = R_{Th}$) is

$$I = \frac{E_{Th}}{R_{Th} + R_L} = \frac{E_{Th}}{2R_{Th}}$$

$$P_L = I^2R_L = \left(\frac{E_{Th}}{2R_{Th}}\right)^2 R_{Th} = \frac{E_{Th}^2 R_{Th}}{4R_{Th}^2}$$

and

$$P_{L_{\max}} = \frac{E_{Th}^2}{4R_{Th}} \quad \text{(watts, W)} \tag{9.6}$$

For the Norton circuit of Fig. 9.80,

$$P_{L_{\max}} = \frac{I_N^2 R_N}{4} \quad \text{(W)} \tag{9.7}$$

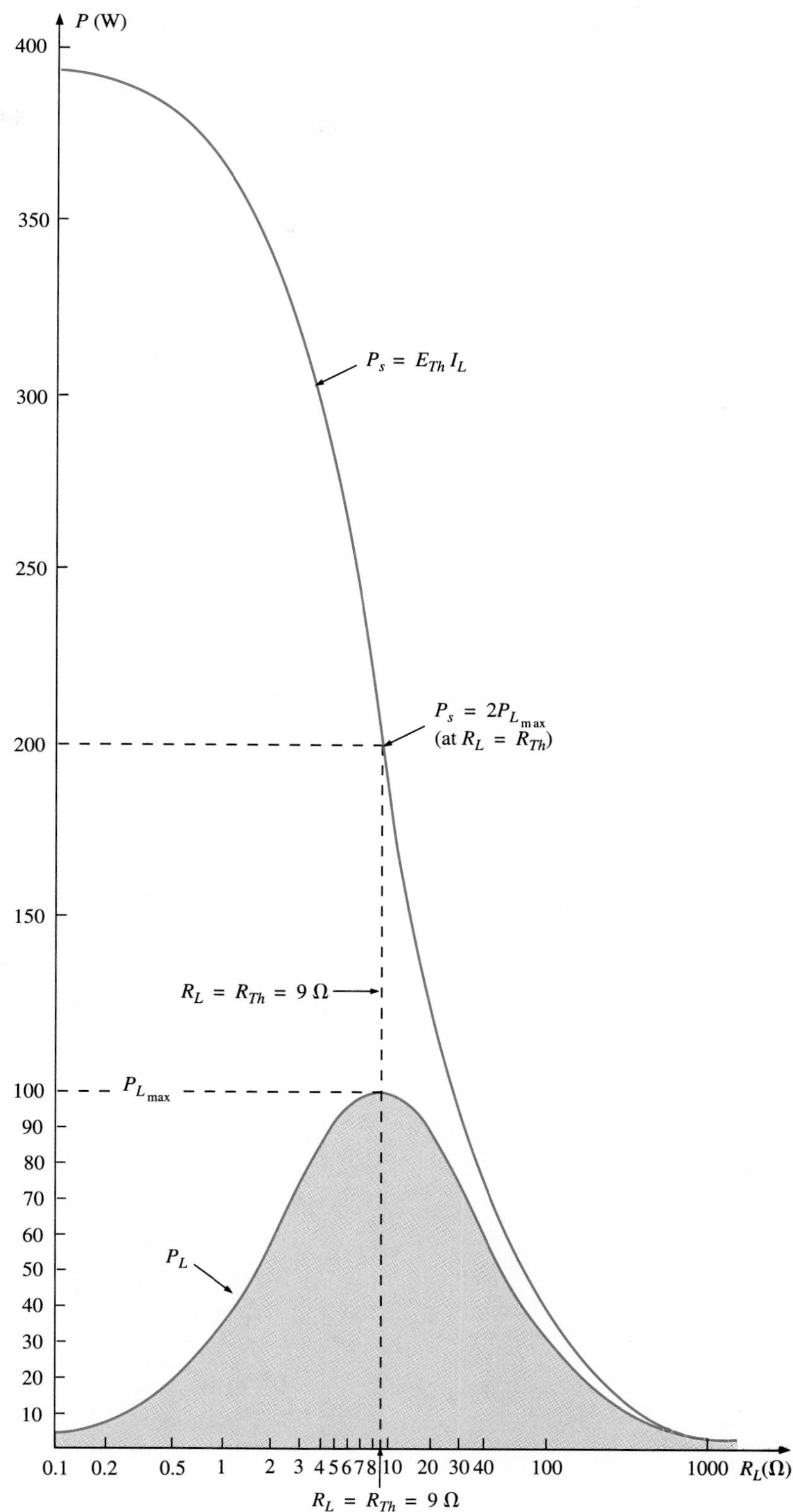

FIG. 9.85
P_s and P_L versus R_L for the network of Fig. 9.81.

EXAMPLE 9.14

A dc generator, battery, and laboratory supply are connected to a resistive load R_L in Fig. 9.86(a), (b), and (c), respectively.

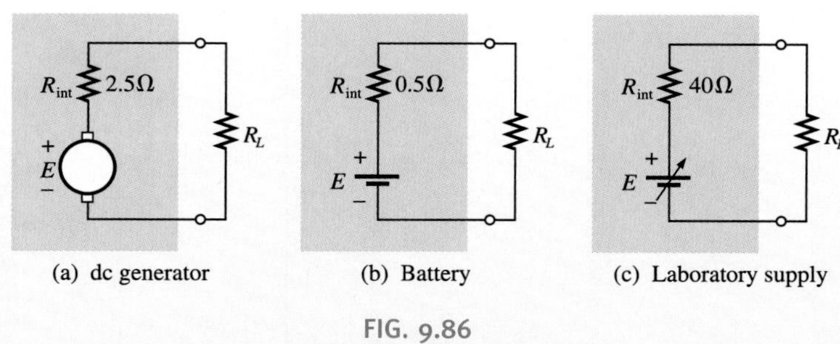

FIG. 9.86
Example 9.14.

a. For each, determine the value of R_L for maximum power transfer to R_L.
b. Determine R_L for 75% efficiency.

Solution:

a. For the dc generator,

$$R_L = R_{Th} = R_{int} = \mathbf{2.5\ \Omega}$$

For the battery,

$$R_L = R_{Th} = R_{int} = \mathbf{0.5\ \Omega}$$

For the laboratory supply,

$$R_L = R_{Th} = R_{int} = \mathbf{40\ \Omega}$$

b. For the dc generator,

$$\eta = \frac{P_o}{P_s} \quad (\eta \text{ in decimal form})$$

$$\eta = \frac{R_L}{R_{Th} + R_L}$$

$$\eta(R_{Th} + R_L) = R_L$$

$$\eta R_{Th} + \eta R_L = R_L$$

$$R_L(1 - \eta) = \eta R_{Th}$$

and

$$R_L = \frac{\eta R_{Th}}{1 - \eta} \tag{9.8}$$

$$R_L = \frac{0.75(2.5\ \Omega)}{1 - 0.75} = \mathbf{7.5\ \Omega}$$

For the battery,

$$R_L = \frac{0.75(0.5\ \Omega)}{1 - 0.75} = \mathbf{1.5\ \Omega}$$

continued

For the laboratory supply,

$$R_L = \frac{0.75(40\ \Omega)}{1 - 0.75} = \mathbf{120\ \Omega}$$

The results of this example show that the following modified form of the maximum power transfer theorem is valid:

For loads connected directly to a dc voltage supply, maximum power will be delivered to the load when the load resistance is equal to the internal resistance of the source;

that is, when $$R_L = R_{int} \tag{9.9}$$

EXAMPLE 9.15

Analyzing a transistor network resulted in the reduced configuration of Fig. 9.87. Determine the R_L necessary to transfer maximum power to R_L, and calculate the power of R_L under these conditions.

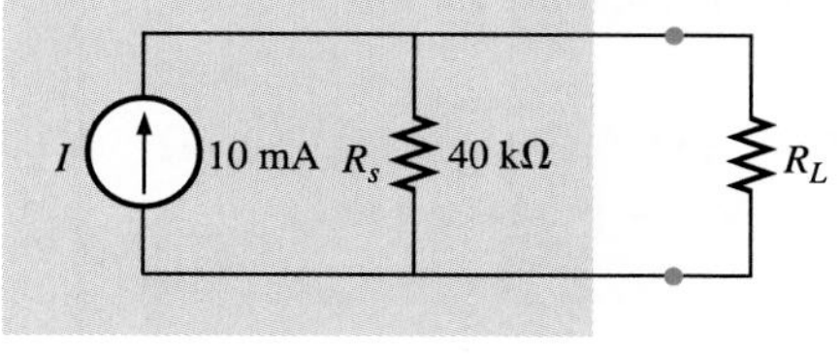

FIG. 9.87
Example 9.15.

Solution: Eq. (9.4):

$$R_L = R_s = \mathbf{40\ k\Omega}$$

Eq. (9.7):

$$P_{L_{max}} = \frac{I_N^2 R_N}{4} = \frac{(10\ \text{mA})^2(40\ \text{k}\Omega)}{4} = \mathbf{1\ W}$$

EXAMPLE 9.16

For the network of Fig. 9.88, determine the value of R for maximum power to R, and calculate the power delivered under these conditions.

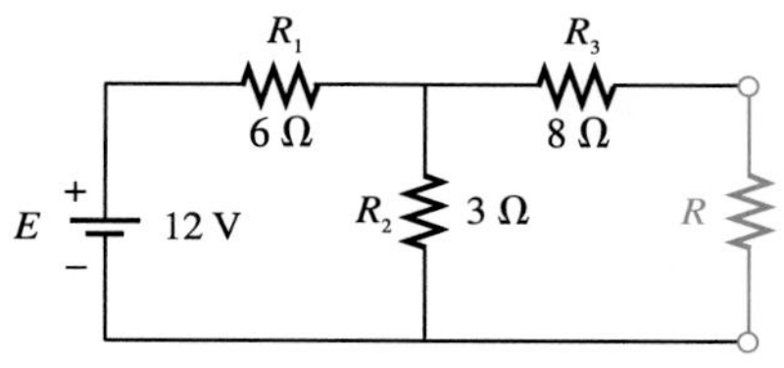

FIG. 9.88
Example 9.16.

Solution: See Fig. 9.89;

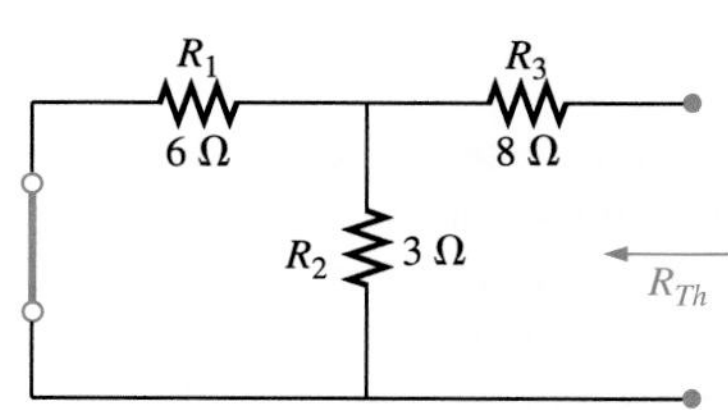

FIG. 9.89
Determining R_{Th} for the network external to the resistor R of Fig. 9.88.

$$R_{Th} = R_3 + R_1 \| R_2 = 8\ \Omega + \frac{(6\ \Omega)(3\ \Omega)}{6\ \Omega + 3\ \Omega} = 8\ \Omega + 2\ \Omega$$

and $$R = R_{Th} = \mathbf{10\ \Omega}$$

See Fig. 9.90;

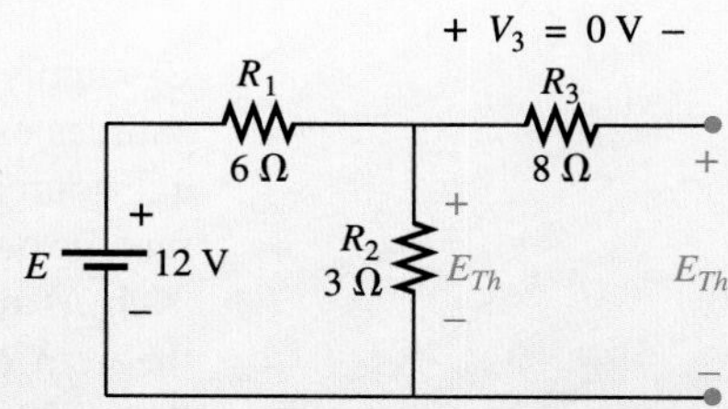

FIG. 9.90
Determining E_{Th} for the network external to the resistor R of Fig. 9.88.

$$E_{Th} = \frac{R_2E}{R_2 + R_1} = \frac{(3\ \Omega)(12\ \text{V})}{3\ \Omega + 6\ \Omega} = \frac{36\ \text{V}}{9} = \mathbf{4\ V}$$

and, by Eq. (9.6),

$$P_{L_{max}} = \frac{E_{Th}^2}{4R_{Th}} = \frac{(4\ \text{V})^2}{4(10\ \Omega)} = \mathbf{0.4\ W}$$

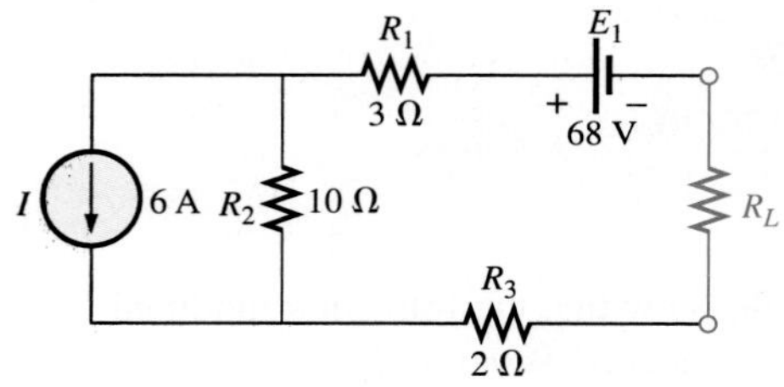

FIG. 9.91
Example 9.17.

EXAMPLE 9.17

Find the value of R_L in Fig. 9.91 for maximum power to R_L, and determine the maximum power.

Solution: See Fig. 9.92:

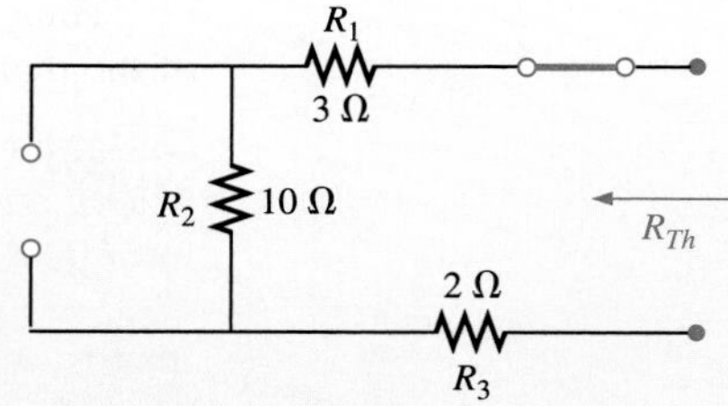

FIG. 9.92
Determining R_{Th} for the network external to the resistor R_L of Fig. 9.91.

$$R_{Th} = R_1 + R_2 + R_3 = 3\ \Omega + 10\ \Omega + 2\ \Omega = 15\ \Omega$$

and
$$R_L = R_{Th} = \mathbf{15\ \Omega}$$

Note Fig. 9.93, where

$$V_1 = V_3 = 0\ \text{V}$$

and
$$V_2 = I_2R_2 = IR_2 = (6\ \text{A})(10\ \Omega) = 60\ \text{V}$$

Applying Kirchhoff's voltage law,

$$\Sigma_{\circlearrowleft} V = -V_2 - E_1 + E_{Th} = 0$$

and
$$E_{Th} = V_2 + E_1 = 60\ \text{V} + 68\ \text{V} = 128\ \text{V}$$

Thus,
$$P_{L_{\max}} = \frac{E_{Th}^2}{4R_{Th}} = \frac{(128\ \text{V})^2}{4(15\ \Omega)} = \mathbf{273\ W}$$

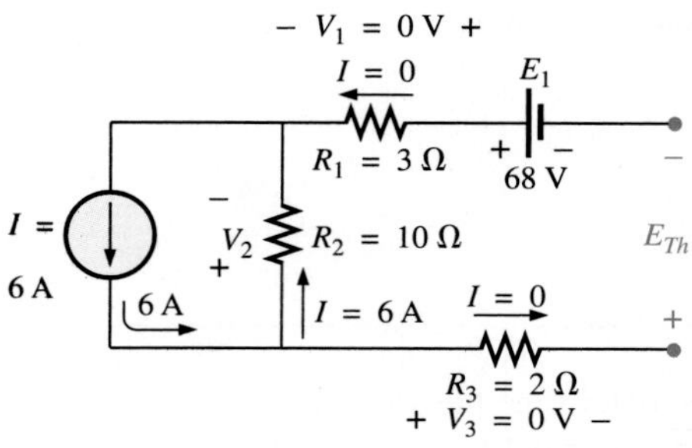

FIG. 9.93
Determining E_{Th} for the network external to the resistor R_L of Fig. 9.91.

9.6 MILLMAN'S THEOREM

By applying **Millman's theorem**, any number of parallel voltage sources can be reduced to one. In Fig. 9.94, for example, the three voltage sources can be reduced to one. This would allow us to find the current through or voltage across R_L without having to apply a method such as mesh analysis, nodal analysis, superposition, and so on. The theorem can best be described by applying it to the network of Fig. 9.94. There are basically three steps included in applying it.

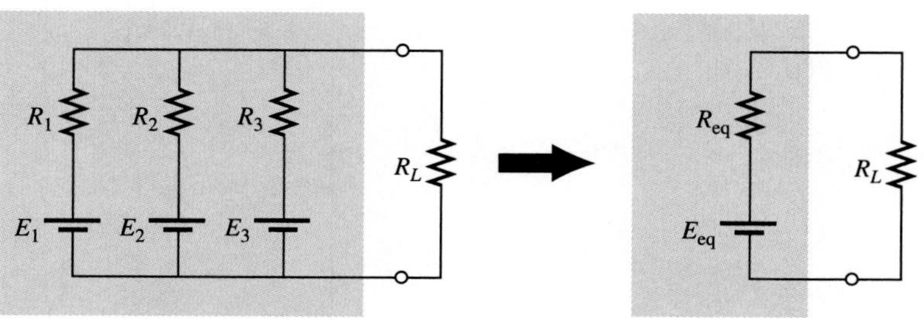

FIG. 9.94
Demonstrating the effect of applying Millman's theorem.

Step 1: Convert all voltage sources to current sources as outlined in Section 8.3. This is done in Fig. 9.95 for the network of Fig. 9.94.

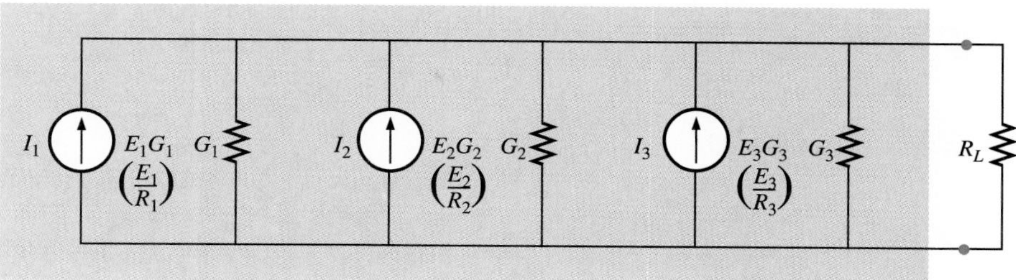

FIG. 9.95
Converting all the sources of Fig. 9.94 to current sources.

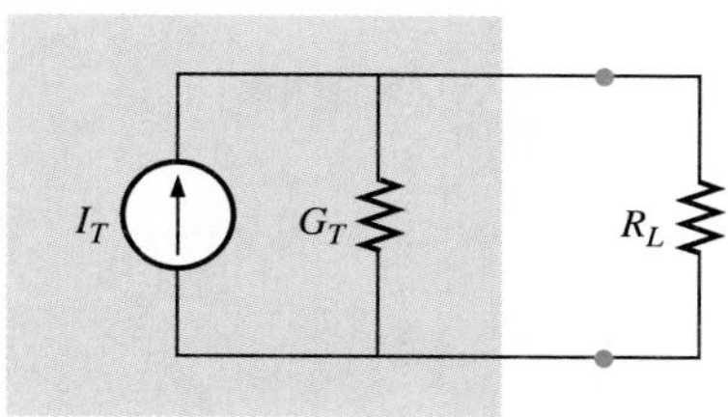

FIG. 9.96
Reducing all the current sources of Fig. 9.95 to a single current source.

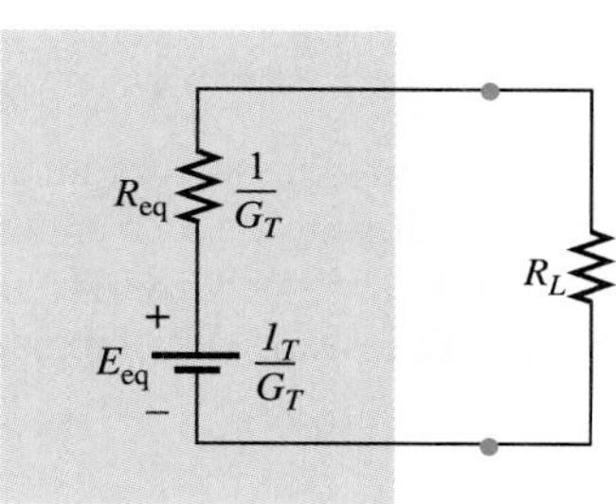

FIG. 9.97
Converting the current source of Fig. 9.96 to a current source.

Step 2: Combine parallel current sources as described in Section 8.4. The resulting network is shown in Fig. 9.96, where

$$I_T = I_1 + I_2 + I_3 \qquad G_T = G_1 + G_2 + G_3$$

Step 3: Convert the resulting current source to a voltage source, and the desired single-source network is obtained, as shown in Fig. 9.97.

In general, Millman's theorem states that for any number of parallel voltage sources,

$$E_{eq} = \frac{I_T}{G_T} = \frac{\pm I_1 \pm I_2 \pm I_3 \pm \cdots \pm I_N}{G_1 + G_2 + G_3 + \cdots + G_N}$$

or

$$E_{eq} = \frac{\pm E_1G_1 \pm E_2G_2 \pm E_3G_3 \pm \cdots \pm E_NG_N}{G_1 + G_2 + G_3 + \cdots + G_N} \tag{9.10}$$

The plus and minus signs appear in Eq. (9.10) to include those cases where the sources may not be supplying energy in the same direction. (Note Example 9.18.)

The equivalent resistance is

$$R_{eq} = \frac{1}{G_T} = \frac{1}{G_1 + G_2 + G_3 + \cdots + G_N} \tag{9.11}$$

In terms of the resistance values,

$$E_{eq} = \frac{\pm \dfrac{E_1}{R_1} \pm \dfrac{E_2}{R_2} \pm \dfrac{E_3}{R_3} \pm \cdots \pm \dfrac{E_N}{R_N}}{\dfrac{1}{R_1} + \dfrac{1}{R_2} + \dfrac{1}{R_3} + \cdots + \dfrac{1}{R_N}} \tag{9.12}$$

and

$$R_{eq} = \frac{1}{\dfrac{1}{R_1} + \dfrac{1}{R_2} + \dfrac{1}{R_3} + \cdots + \dfrac{1}{R_N}} \tag{9.13}$$

The small number of direct steps suggests that you should apply each step rather than memorizing and using Eqs. (9.10) through (9.13).

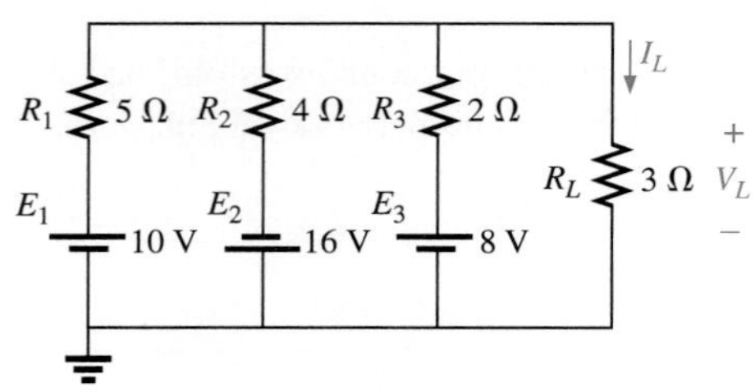

FIG. 9.98
Example 9.18.

EXAMPLE 9.18

Using Millman's theorem, find the current through and voltage across the resistor R_L of Fig. 9.98.

Solution: By Eq. (9.12),

$$E_{eq} = \frac{+\dfrac{E_1}{R_1} - \dfrac{E_2}{R_2} + \dfrac{E_3}{R_3}}{\dfrac{1}{R_1} + \dfrac{1}{R_2} + \dfrac{1}{R_3}}$$

The minus sign is used for E_2/R_2 because that supply has the opposite polarity of the other two. The chosen reference direction is therefore that of E_1 and E_3. The total conductance is unaffected by the direction, and

$$E_{eq} = \frac{+\dfrac{10\text{ V}}{5\ \Omega} - \dfrac{16\text{ V}}{4\ \Omega} + \dfrac{8\text{ V}}{2\ \Omega}}{\dfrac{1}{5\ \Omega} + \dfrac{1}{4\ \Omega} + \dfrac{1}{2\ \Omega}} = \frac{2\text{ A} - 4\text{ A} + 4\text{ A}}{0.2\text{ S} + 0.25\text{ S} + 0.5\text{ S}}$$

$$= \frac{2\text{ A}}{0.95\text{ S}} = \mathbf{2.105\ V}$$

with
$$R_{eq} = \frac{1}{\dfrac{1}{5\ \Omega} + \dfrac{1}{4\ \Omega} + \dfrac{1}{2\ \Omega}} = \frac{1}{0.95\text{ S}} = \mathbf{1.053\ \Omega}$$

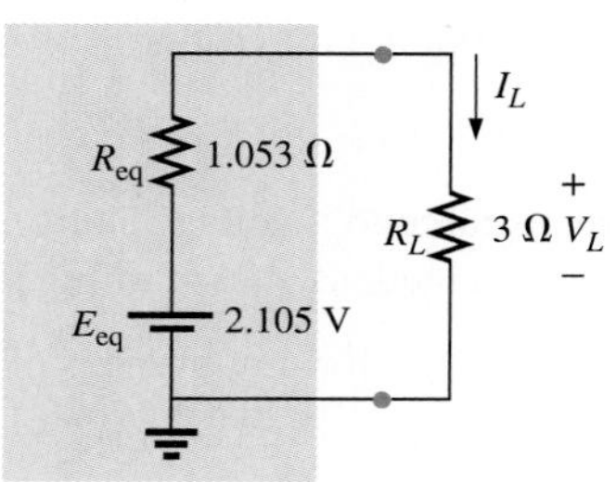

FIG. 9.99
The result of applying Millman's theorem to the network of Fig. 9.98.

The resulting source is shown in Fig. 9.99, and

$$I_L = \frac{2.105\text{ V}}{1.053\ \Omega + 3\ \Omega} = \frac{2.105\text{ V}}{4.053\ \Omega} = \mathbf{0.519\ A}$$

with
$$V_L = I_L R_L = (0.519\text{ A})(3\ \Omega) = \mathbf{1.557\ V}$$

EXAMPLE 9.19

Let us now consider the type of problem we saw in the introduction to mesh and nodal analysis in Chapter 8. Mesh analysis was applied to the network of Fig. 9.100 (Example 8.12). We will now use Millman's theorem to find the current through the 2-Ω resistor and compare the results.

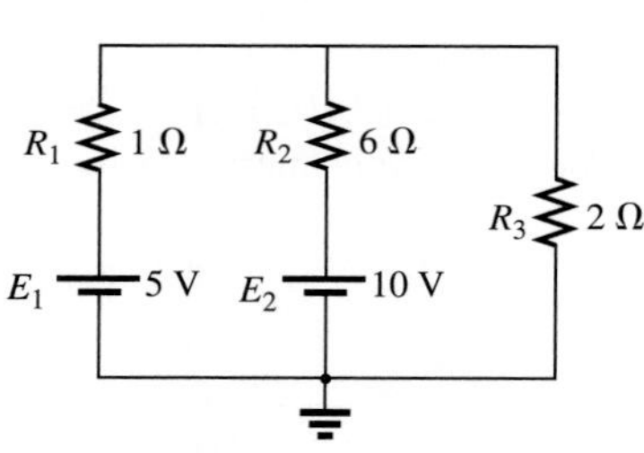

FIG. 9.100
Example 9.19.

Solution:

a. We will first apply each step and, in the (b) solution, we will apply Eq. (9.12). Converting sources gives us the results shown in Fig. 9.101.

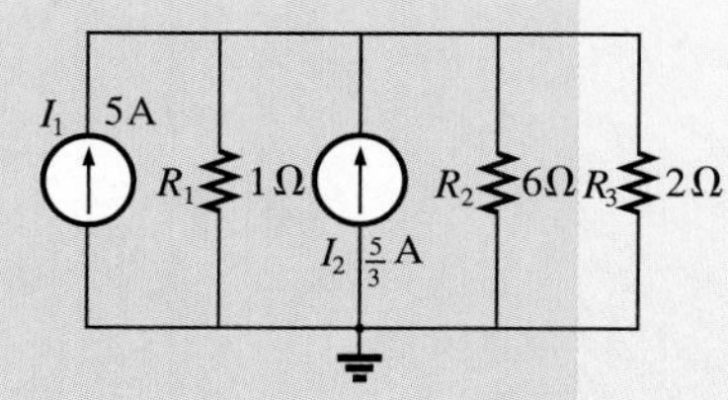

FIG. 9.101
Converting the sources of Fig. 9.100 to current sources.

continued

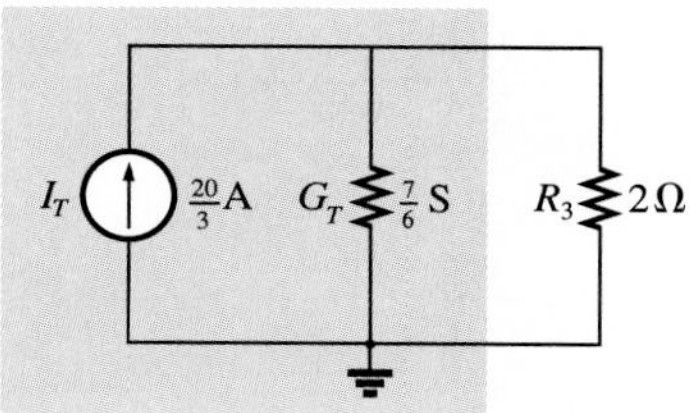

FIG. 9.102
Reducing the current sources of Fig. 9.101 to a single source.

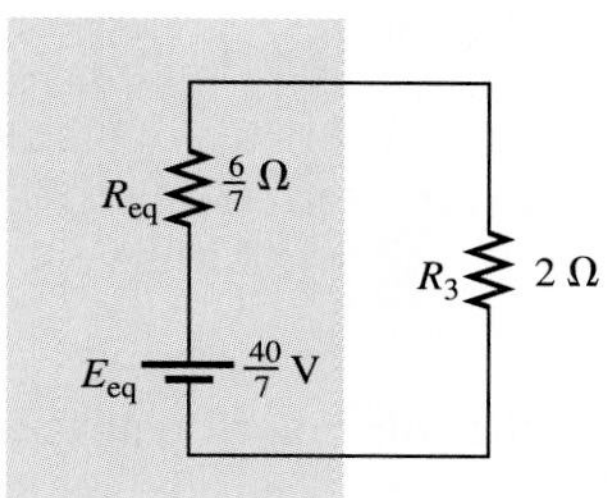

FIG. 9.103
Converting the current source of Fig. 9.102 to a voltage source.

Combining sources and parallel conductance branches (see Fig. 9.102) results in

$$I_T = I_1 + I_2 = 5\text{ A} + \frac{5}{3}\text{ A} = \frac{15}{3}\text{ A} + \frac{5}{3}\text{ A} = \frac{20}{3}\text{ A}$$

$$G_T = G_1 + G_2 = 1\text{ S} + \frac{1}{6}\text{ S} = \frac{6}{6}\text{ S} + \frac{1}{6}\text{ S} = \frac{7}{6}\text{ S}$$

Converting the current source to a voltage source (Fig. 9.103), we get

$$E_{eq} = \frac{I_T}{G_T} = \frac{\frac{20}{3}\text{ A}}{\frac{7}{6}\text{ S}} = \frac{(6)(20)}{(3)(7)}\text{ V} = \mathbf{\frac{40}{7}\text{ V}}$$

and

$$R_{eq} = \frac{1}{G_T} = \frac{1}{\frac{7}{6}\text{ S}} = \mathbf{\frac{6}{7}\ \Omega}$$

so that

$$I_{2\Omega} = \frac{E_{eq}}{R_{eq} + R_3} = \frac{\frac{40}{7}\text{ V}}{\frac{6}{7}\ \Omega + 2\ \Omega} = \frac{\frac{40}{7}\text{ V}}{\frac{6}{7}\ \Omega + \frac{14}{7}\ \Omega} = \frac{40\text{ V}}{20\ \Omega} = \mathbf{2\text{ A}}$$

which agrees with the result we got in Example 8.18.

b. We will now simply apply the proper equation, Eq. (9.12):

$$E_{eq} = \frac{+\frac{5\text{ V}}{1\ \Omega} + \frac{10\text{ V}}{6\ \Omega}}{\frac{1}{1\ \Omega} + \frac{1}{6\ \Omega}} = \frac{\frac{30\text{ V}}{6\ \Omega} + \frac{10\text{ V}}{6\ \Omega}}{\frac{6}{6\ \Omega} + \frac{1}{6\ \Omega}} = \mathbf{\frac{40}{7}\text{ V}}$$

and

$$R_{eq} = \frac{1}{\frac{1}{1\ \Omega} + \frac{1}{6\ \Omega}} = \frac{1}{\frac{6}{6\ \Omega} + \frac{1}{6\ \Omega}} = \frac{1}{\frac{7}{6}\text{ S}} = \mathbf{\frac{6}{7}\ \Omega}$$

which are the same values obtained above.

The dual effect of Millman's theorem (Fig. 9.94) appears in Fig. 9.104. It can be shown that I_{eq} and R_{eq}, as in Fig. 9.104, are given by

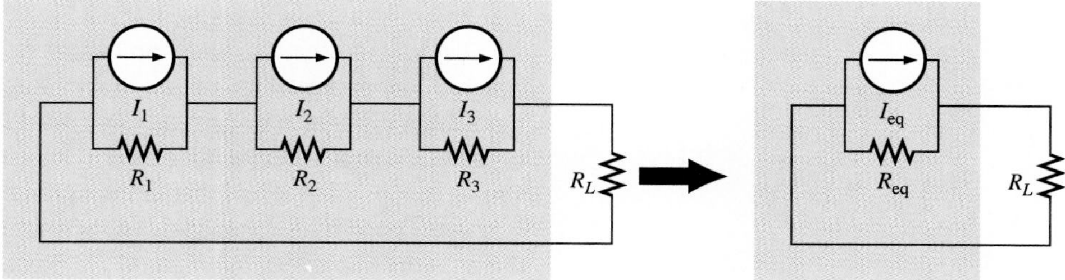

FIG. 9.104
The dual effect of Millman's theorem.

$$I_{eq} = \frac{\pm I_1R_1 \pm I_2R_2 \pm I_3R_3}{R_1 + R_2 + R_3} \tag{9.14}$$

and

$$R_{eq} = R_1 + R_2 + R_3 \tag{9.15}$$

The derivation will appear as a problem at the end of the chapter.

9.7 SUBSTITUTION THEOREM

The **substitution theorem** states the following:

If the voltage across and current through any branch of a dc bilateral network are known, this branch can be replaced by any combination of elements that will maintain the same voltage across and current through the chosen branch.

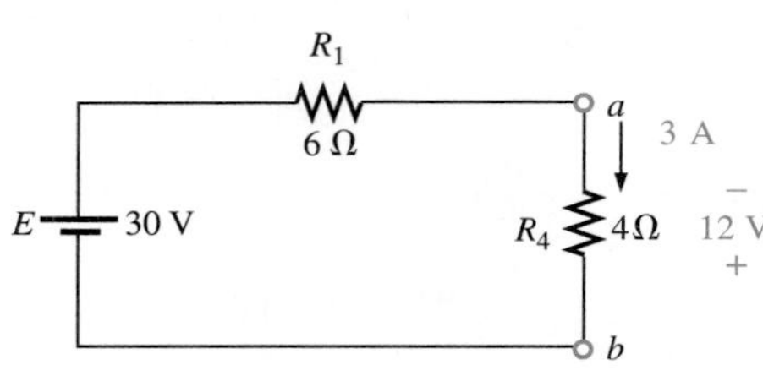

FIG. 9.105
Demonstrating the effect of the substitution theorem.

More simply, the theorem states that for branch equivalence, the terminal voltage and current must be the same. Consider the circuit of Fig. 9.105, in which the voltage across and current through the branch *a-b* are determined. Through the use of the substitution theorem, a number of equivalent *a-b* branches are shown in Fig. 9.106.

Note that for each equivalent, the terminal voltage and current are the same. Also consider that the response of the remainder of the circuit of Fig. 9.105 is unchanged by substituting any one of the equivalent branches. As demonstrated by the single-source equivalents of Fig. 9.106,

a known potential difference and current in a network can be replaced by an ideal voltage source and current source, respectively.

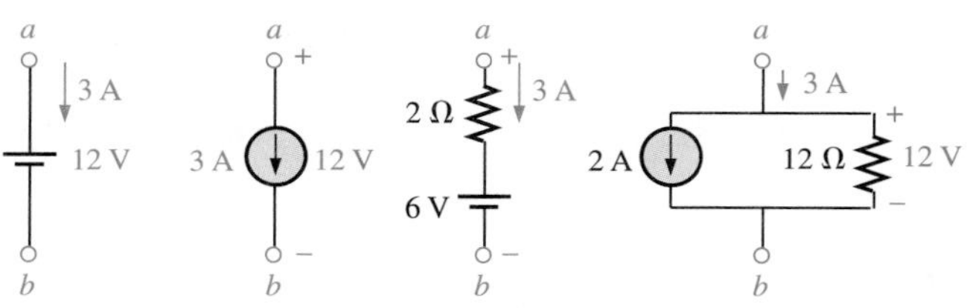

FIG. 9.106
Equivalent branches for the branch a-b of Fig. 9.105.

Understand that this theorem cannot be used to *solve* networks with two or more sources that are not in series or parallel. For it to be applied, a potential difference or current value must be known or found using one of the techniques discussed earlier. One application of the theorem is shown in Fig. 9.107. Note that in the figure the known potential difference *V* was replaced by a voltage source, permitting us to isolate the portion of the network including R_3, R_4, and R_5. Recall that this was basically the approach used in the analysis of the ladder network as we worked our way back toward the terminal resistance R_5.

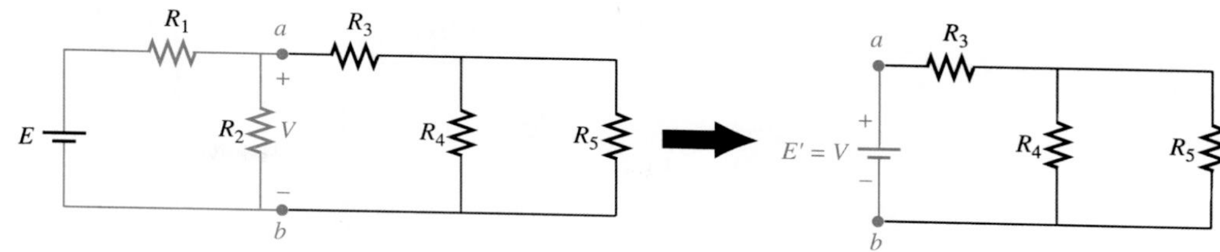

FIG. 9.107
Demonstrating the effect of knowing a voltage at some point in a complex network.

The current source equivalence of the above is shown in Fig. 9.108, where a known current is replaced by an ideal current source, permitting R_4 and R_5 to be isolated.

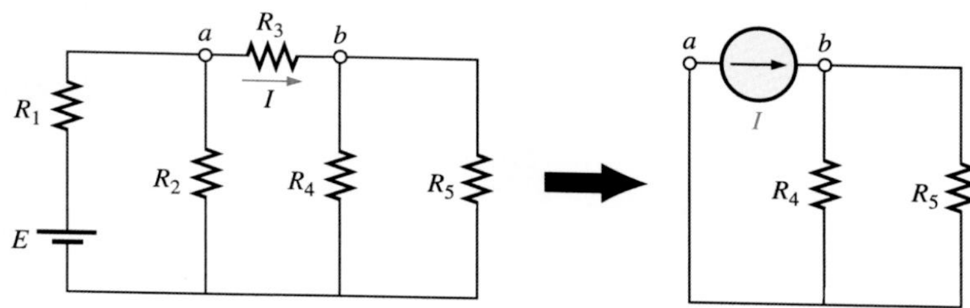

FIG. 9.108
Demonstrating the effect of knowing a current at some point in a complex network.

You will also recall from the discussion of bridge networks that $V = 0$ and $I = 0$ were replaced by a short circuit and an open circuit, respectively. This substitution is a very specific application of the substitution theorem.

9.8 RECIPROCITY THEOREM

The **reciprocity theorem** applies only to single-source networks. It is, therefore, not a theorem used in analyzing the multisource networks we have already described. The theorem states the following:

The current I in any branch of a network, due to a single voltage source E anywhere else in the network, will equal the current through the branch in which the source was originally located if the source is placed in the branch in which the current I was originally measured.

In other words, the location of the voltage source and the resulting current may be interchanged without a change in current. The theorem requires that the polarity of the voltage source have the same correspondence with the direction of the branch current in each position.

In the representative network of Fig. 9.109(a), the current I due to the voltage source E was determined. If the position of each is interchanged as shown in Fig. 9.109(b), the current I will be the same value as indicated. To demonstrate the truth of this statement and the theorem, consider the network of Fig. 9.110, in which values for the elements of Fig. 9.109(a) have been assigned.

The total resistance is

$$\begin{aligned} R_T &= R_1 + R_2 \parallel (R_3 + R_4) = 12\ \Omega + 6\ \Omega \parallel (2\ \Omega + 4\ \Omega) \\ &= 12\ \Omega + 6\ \Omega \parallel 6\ \Omega = 12\ \Omega + 3\ \Omega = 15\ \Omega \end{aligned}$$

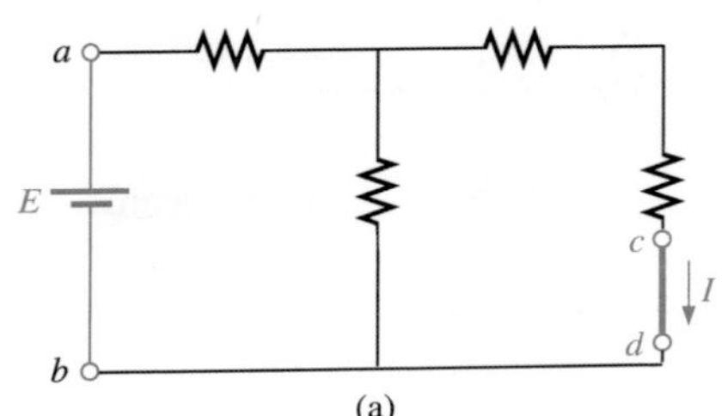

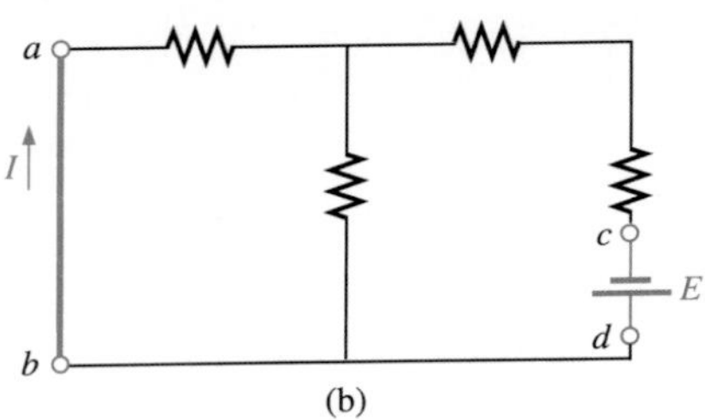

FIG. 9.109
Demonstrating the impact of the reciprocity theorem.

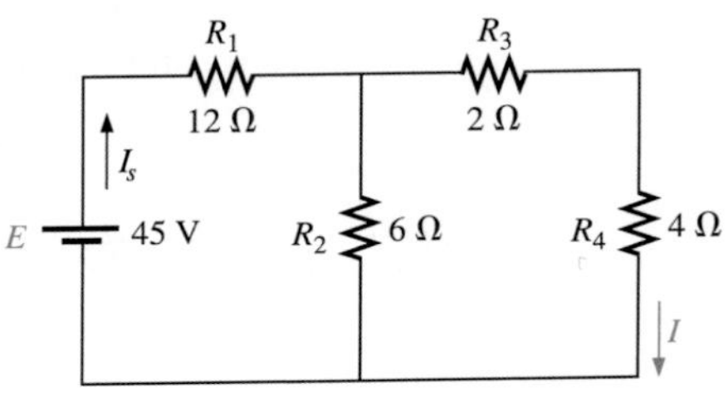

FIG. 9.110
Finding the current I due to a source E.

and
$$I_s = \frac{E}{R_T} = \frac{45\text{ V}}{15\ \Omega} = 3\text{ A}$$

with
$$I = \frac{3\text{ A}}{2} = \mathbf{1.5\ A}$$

For the network of Fig. 9.111, which corresponds to that of Fig. 9.109(b), we find

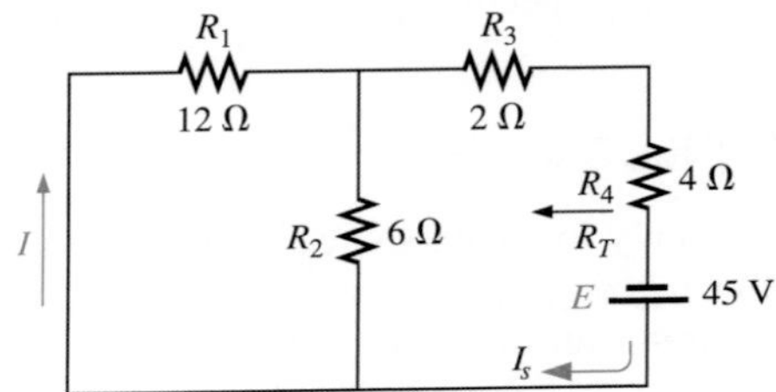

FIG. 9.111
Interchanging the location of E and I of Fig. 9.110 to demonstrate the reciprocity theorem.

$$R_T = R_4 + R_3 + R_1 \parallel R_2$$
$$= 4\ \Omega + 2\ \Omega + 12\ \Omega \parallel 6\ \Omega = 10\ \Omega$$

and
$$I_s = \frac{E}{R_T} = \frac{45\text{ V}}{10\ \Omega} = 4.5\text{ A}$$

so that
$$I = \frac{(6\ \Omega)(4.5\text{ A})}{12\ \Omega + 6\ \Omega} = \frac{4.5\text{ A}}{3} = \mathbf{1.5\ A}$$

which agrees with the above.

The best way to show the uniqueness and power of such a theorem is to consider a complex, single-source network such as the one shown in Fig. 9.112.

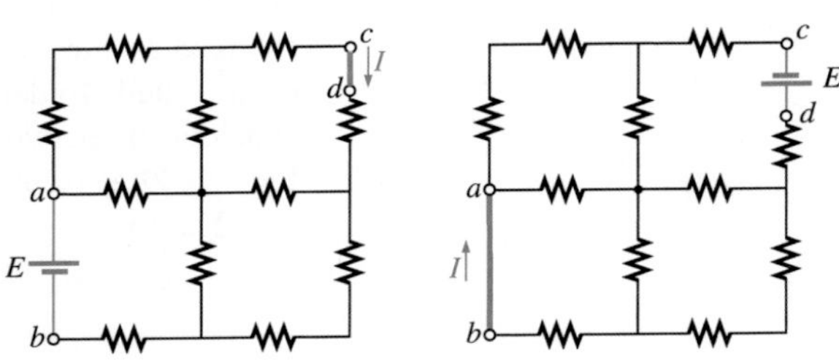

FIG. 9.112
Demonstrating the power and uniqueness of the reciprocity theorem.

PROBLEMS

SECTION 9.2 Superposition Theorem

1. a. Using superposition, find the current through each resistor of the network of Fig. 9.113.
   b. Find the power delivered to R_1 for each source.
   c. Find the power delivered to R_1 using the total current through R_1.
   d. Does superposition apply to power effects? Explain.

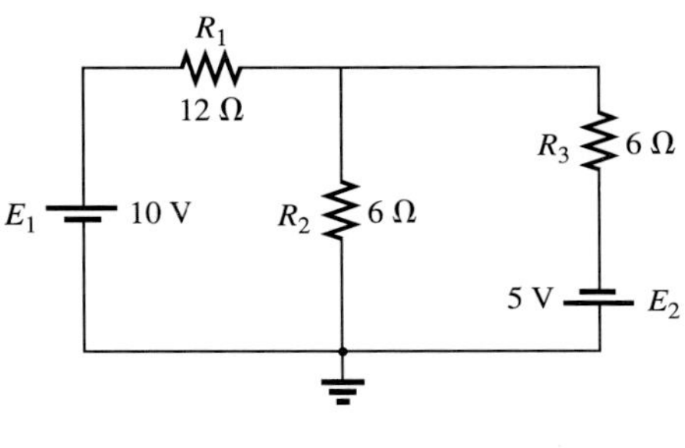

FIG. 9.113
Problem 1.

2. Using superposition, find the current I through the 10-Ω resistor for each of the networks of Fig. 9.114.

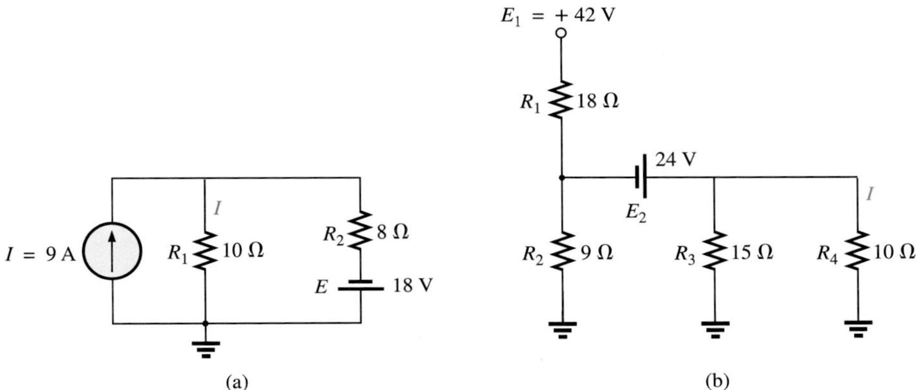

FIG. 9.114
Problem 2.

*3. Using superposition, find the current through R_1 for each network of Fig. 9.115.

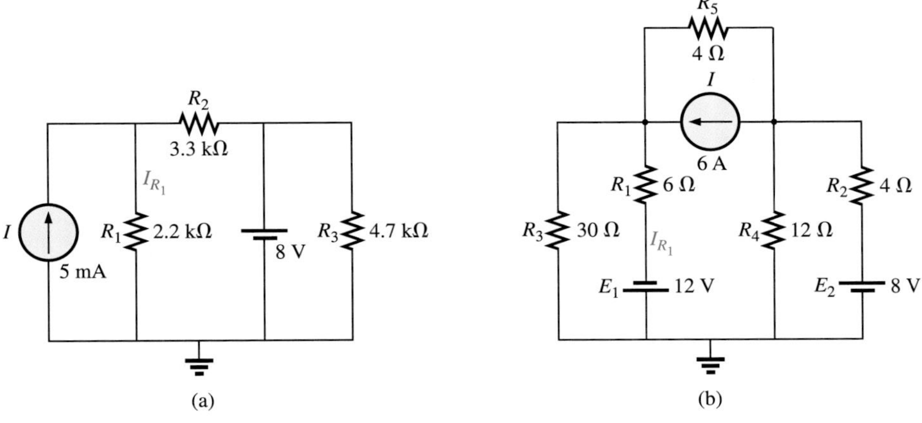

FIG. 9.115
Problem 3.

4. Using superposition, find the voltage V_2 for the network of Fig. 9.116.

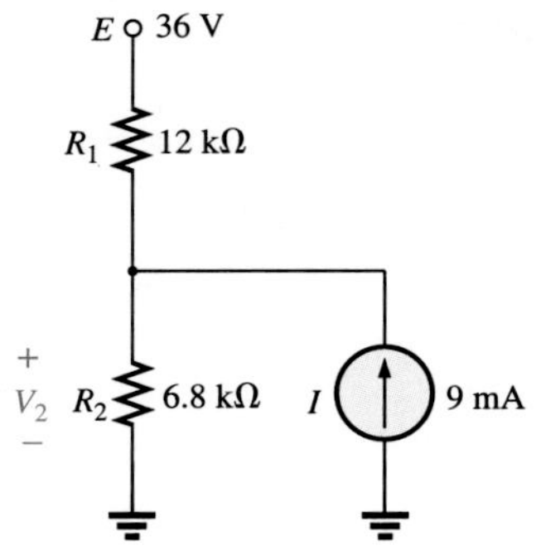

FIG. 9.116
Problem 4.

SECTION 9.3 Thévenin's Theorem

5. a. Find the Thévenin equivalent circuit for the network external to the resistor R of Fig. 9.117.

b. Find the current through R when R is 2, 30, and 100 Ω.

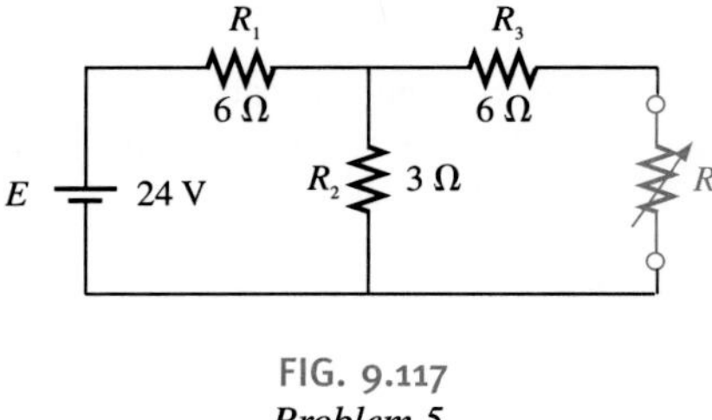

FIG. 9.117
Problem 5.

6. a. Find the Thévenin equivalent circuit for the network external to the resistor R in each of the networks of Fig. 9.118.

b. Find the power delivered to R when R is 2 Ω and 100 Ω.

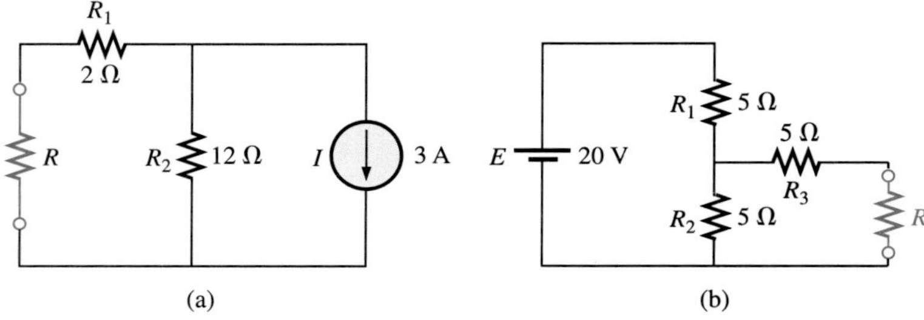

FIG. 9.118
Problems 6, 13, and 19.

7. Find the Thévenin equivalent circuit for the network external to the resistor R in each of the networks of Fig. 9.119.

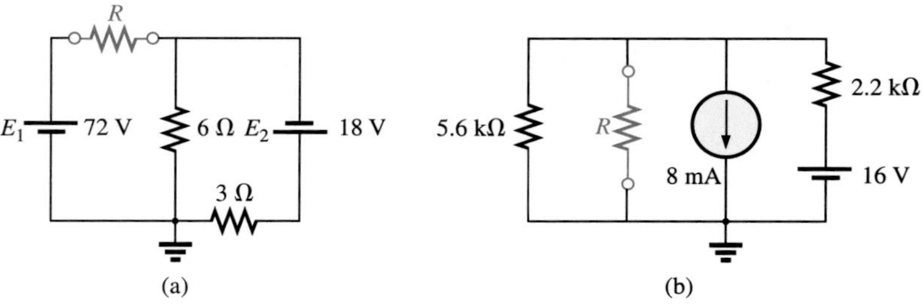

FIG. 9.119
Problems 7, 14, and 20.

***8.** Find the Thévenin equivalent circuit for the network external to the resistor R in each of the networks of Fig. 9.120.

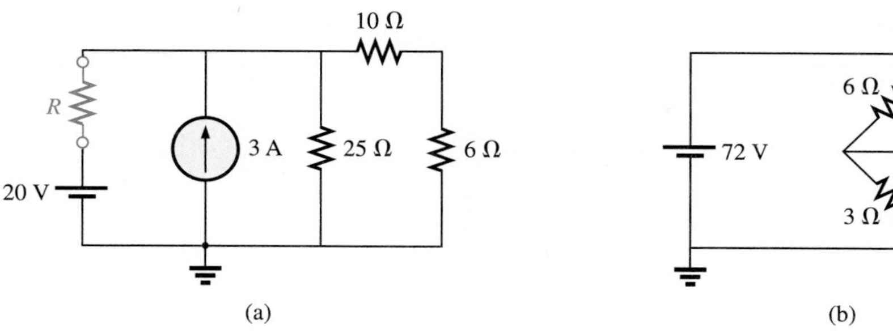

FIG. 9.120
Problems 8, 15, and 21.

***9.** Find the Thévenin equivalent circuit for the portions of the networks of Fig. 9.121 external to points a and b.

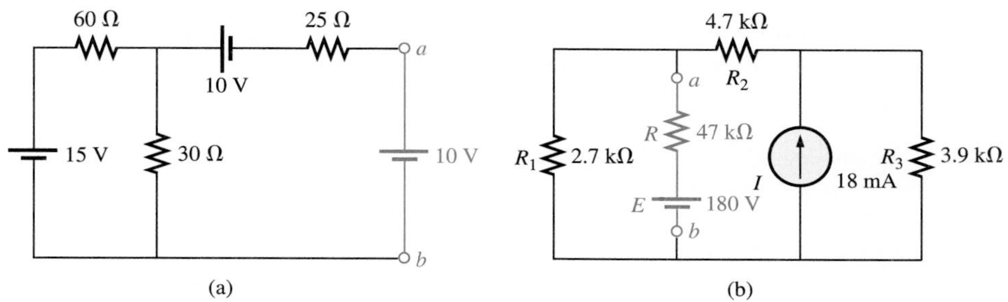

FIG. 9.121
Problems 9 and 16.

***10.** Determine the Thévenin equivalent circuit for the network external to the resistor R in both networks of Fig. 9.122.

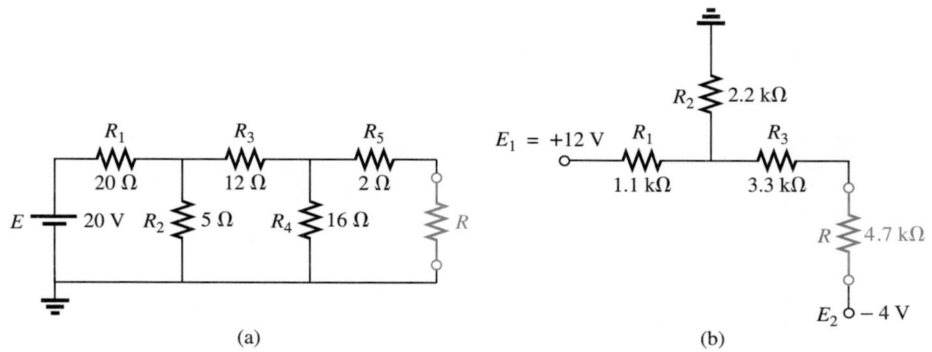

FIG. 9.122
Problems 10 and 17.

***11.** For the network of Fig. 9.123, find the Thévenin equivalent circuit for the network external to the load resistor R_L.

***12.** For the transistor network of Fig. 9.124,

a. Find the Thévenin equivalent circuit for that portion of the network to the left of the base (B) terminal.

b. Using the fact that $I_C = I_E$ and $V_{CE} = 8$ V, determine the magnitude of I_E.

c. Using the results of parts (a) and (b), calculate the base current I_B if $V_{BE} = 0.7$ V.

d. What is the voltage V_C?

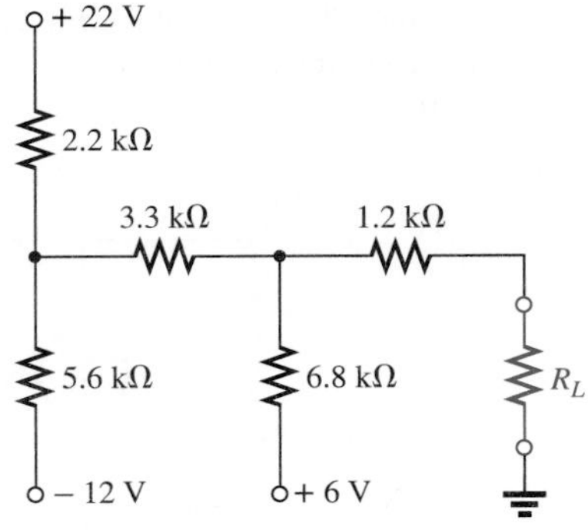

FIG. 9.123
Problem 11.

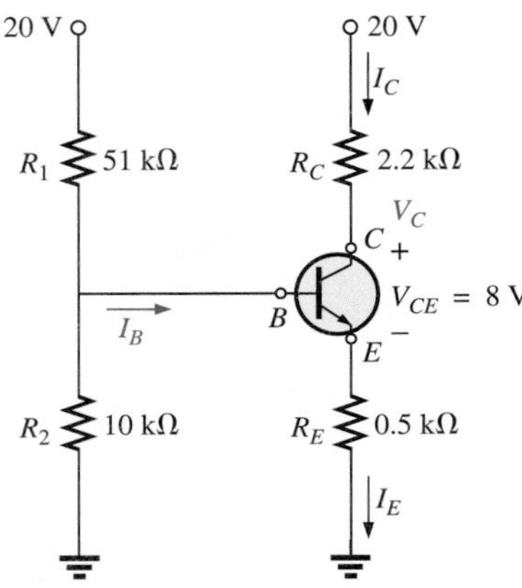

FIG. 9.124
Problem 12.

SECTION 9.4 Norton's Theorem

13. Find the Norton equivalent circuit for the network external to the resistor R in each network of Fig. 9.118.

14. a. Find the Norton equivalent circuit for the network external to the resistor R for each network of Fig. 9.119.

b. Convert to the Thévenin equivalent circuit and compare your solution for E_{Th} and R_{Th} to that appearing in the solutions for Problem 7.

15. Find the Norton equivalent circuit for the network external to the resistor R for each network of Fig. 9.120.

16. a. Find the Norton equivalent circuit for the network external to the resistor R for each network of Fig. 9.121.

b. Convert to the Thévenin equivalent circuit and compare your solution for E_{Th} and R_{Th} to that appearing in the solutions for Problem 9.

17. Find the Norton equivalent circuit for the network external to the resistor R for each network of Fig. 9.122.

18. Find the Norton equivalent circuit for the portions of the networks of Fig. 9.125 external to branch *a-b*.

SECTION 9.5 Maximum Power Transfer Theorem

19. a. For each network of Fig. 9.118, find the value of R for maximum power to R.

b. Determine the maximum power to R for each network.

20. a. For each network of Fig. 9.119, find the value of R for maximum power to R.

b. Determine the maximum power to R for each network.

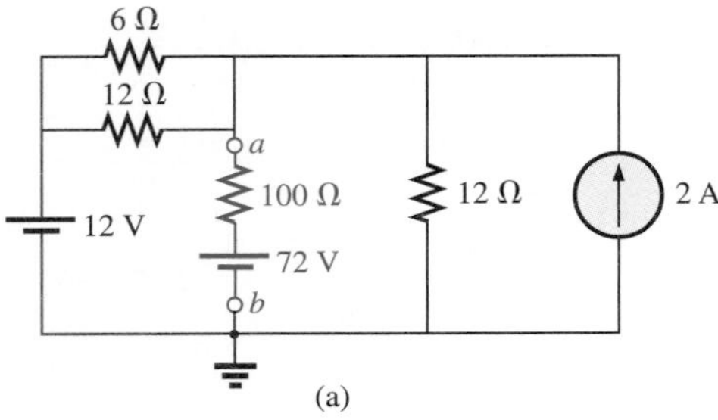

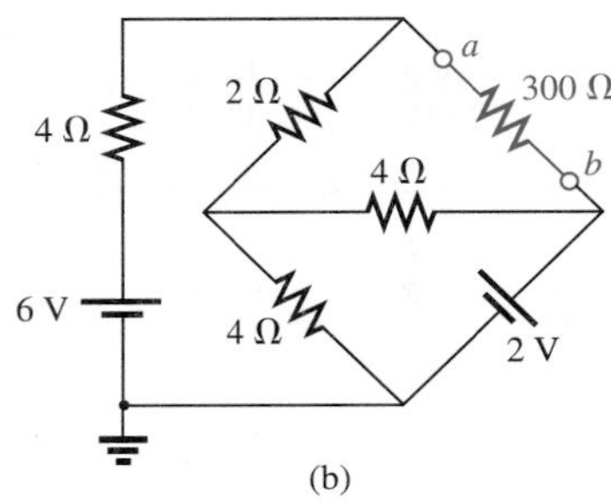

FIG. 9.125
Problem 18.

21. For each network of Fig. 9.120, find the value of R for maximum power to R and determine the maximum power to R for each network.

22. a. For the network of Fig. 9.126, determine the value of R for maximum power to R.

b. Determine the maximum power to R.

c. Plot a curve of power to R versus R for R equal to $\frac{1}{4}$, $\frac{1}{2}$, $\frac{3}{4}$, 1, $1\frac{1}{4}$, $1\frac{1}{2}$, $1\frac{3}{4}$, and 2 times the value obtained in part (a).

***23.** Find the resistance R_1 of Fig. 9.127 such that the resistor R_4 will receive maximum power. Think!

***24. a.** For the network of Fig. 9.128, determine the value R_2 for maximum power to R_4.

b. Is there a general statement that can be made about situations such as those presented here and in Problem 23?

***25.** For the network of Fig. 9.129, determine the level of R that will ensure maximum power to the 100-Ω resistor.

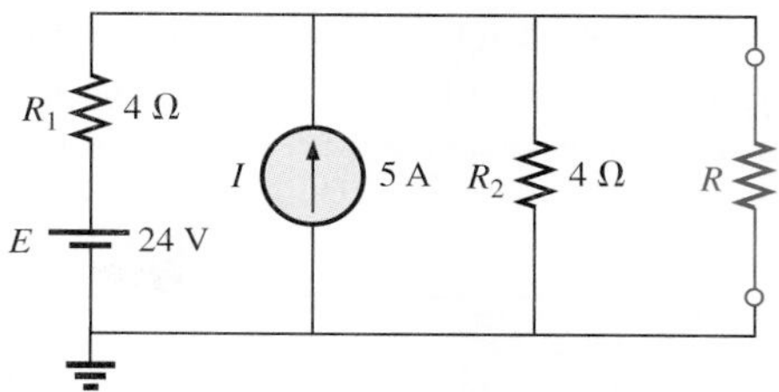

FIG. 9.126
Problem 22.

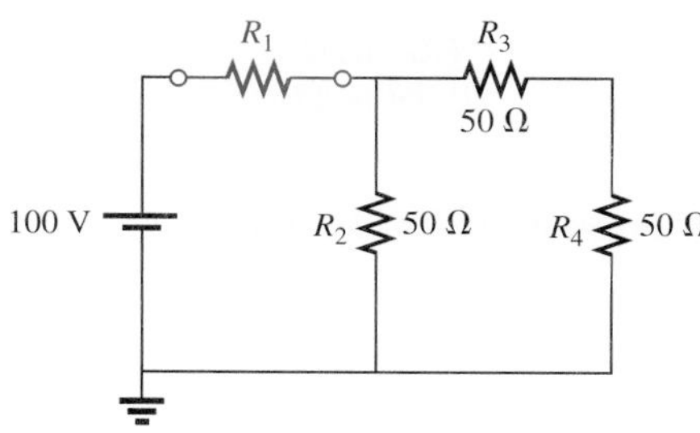

FIG. 9.127
Problem 23.

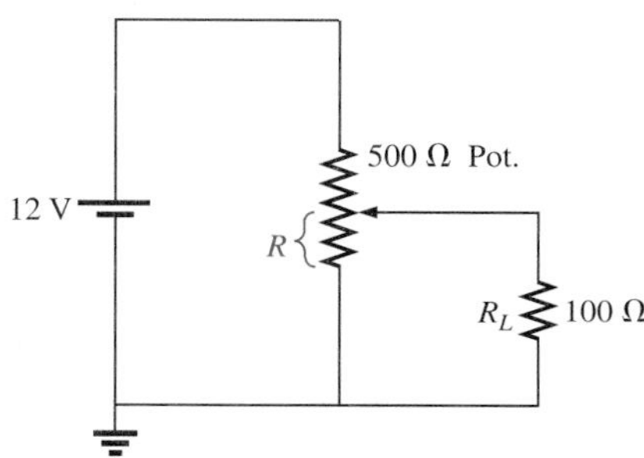

FIG. 9.129
Problem 25.

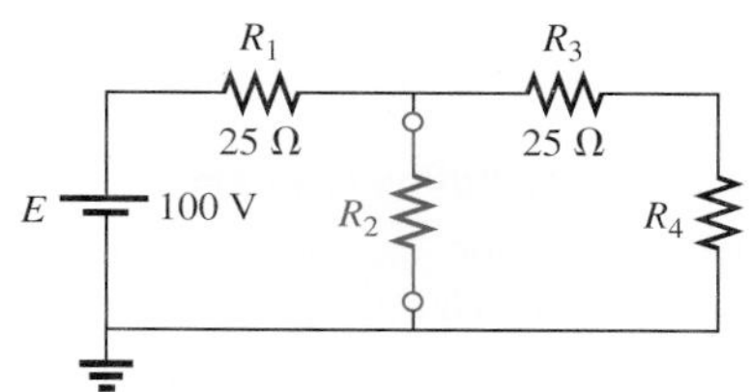

FIG. 9.128
Problem 24.

SECTION 9.6 Millman's Theorem

26. Using Millman's theorem, find the current through and voltage across the resistor R_L of Fig. 9.130.

27. Repeat Problem 26 for the network of Fig. 9.131.

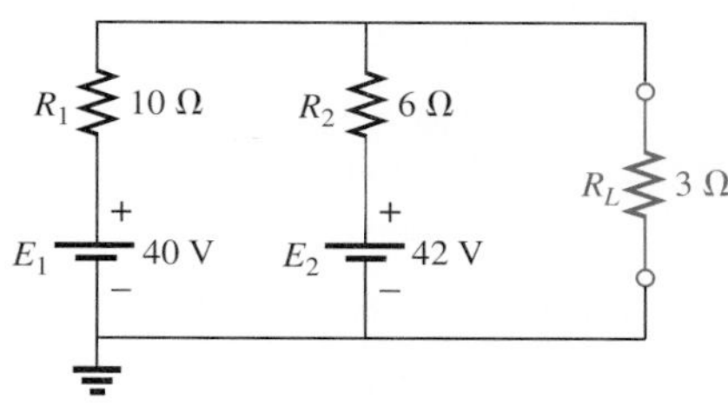

FIG. 9.130
Problem 26.

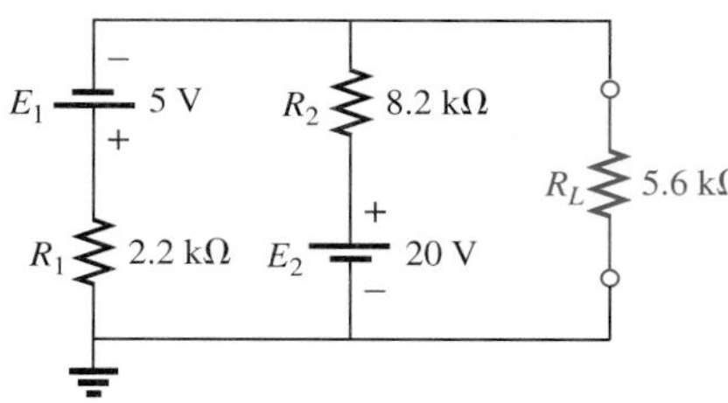

FIG. 9.131
Problem 27.

28. Repeat Problem 26 for the network of Fig. 9.132.

29. Using the dual of Millman's theorem, find the current through and voltage across the resistor R_L of Fig. 9.133.

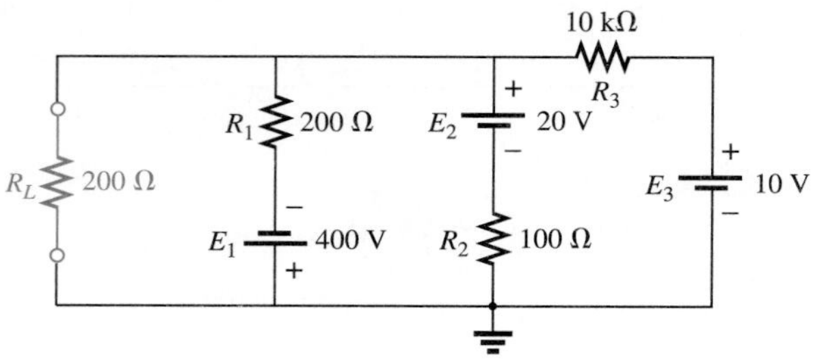

FIG. 9.132
Problem 28.

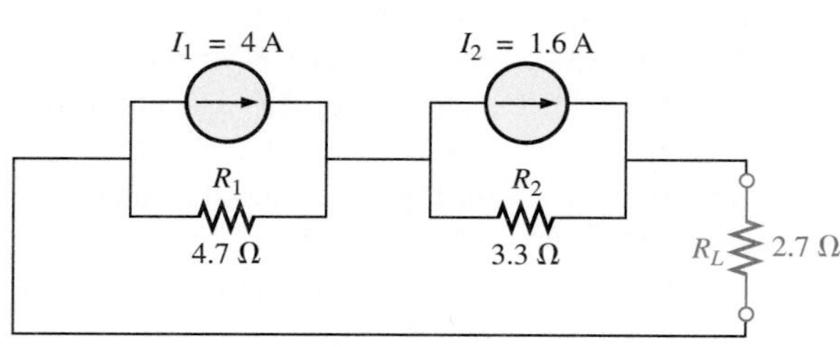

FIG. 9.133
Problem 29.

***30.** Repeat Problem 29 for the network of Fig. 9.134.

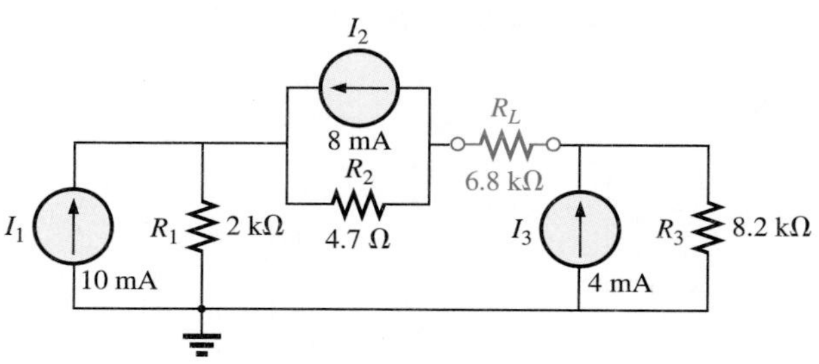

FIG. 9.134
Problem 30.

SECTION 9.7 Substitution Theorem

31. Using the substitution theorem, draw three equivalent branches for the branch *a-b* of the network of Fig. 9.135.

32. Repeat Problem 31 for the network of Fig. 9.136.

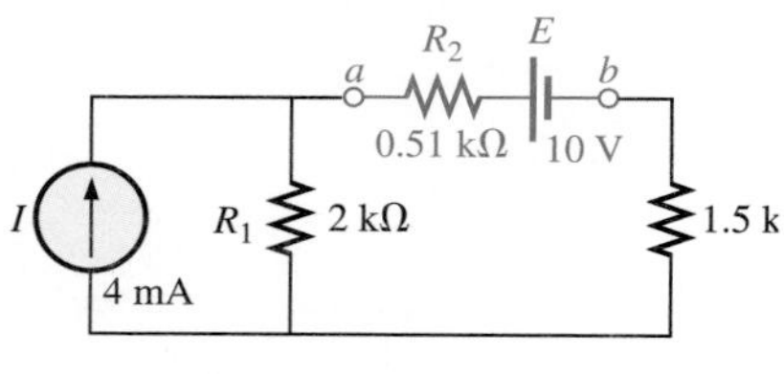

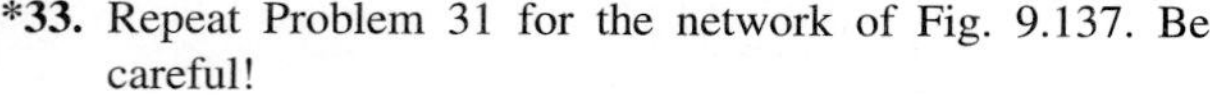

FIG. 9.136
Problem 32.

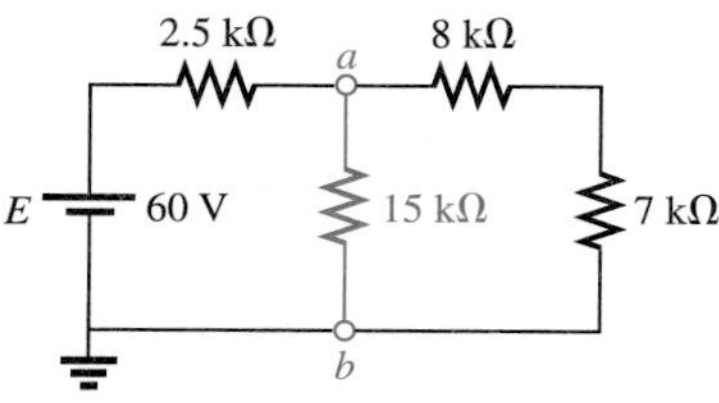

FIG. 9.135
Problem 31.

***33.** Repeat Problem 31 for the network of Fig. 9.137. Be careful!

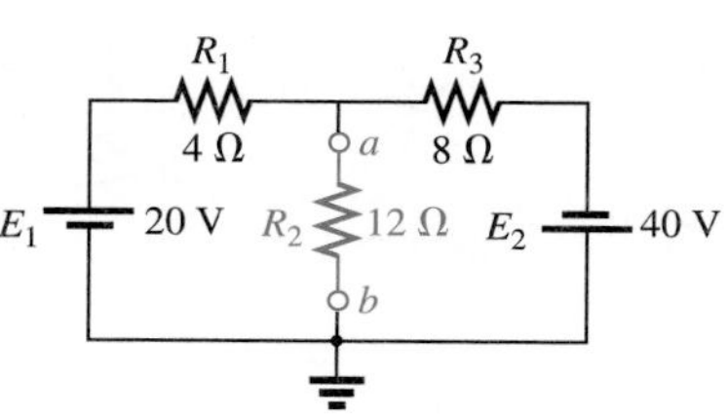

FIG. 9.137
Problem 33.

SECTION 9.8 Reciprocity Theorem

34. **a.** For the network of Fig. 9.138(a), determine the current *I*.
b. Repeat part (a) for the network of Fig. 9.138(b).
c. Is the reciprocity theorem satisfied?

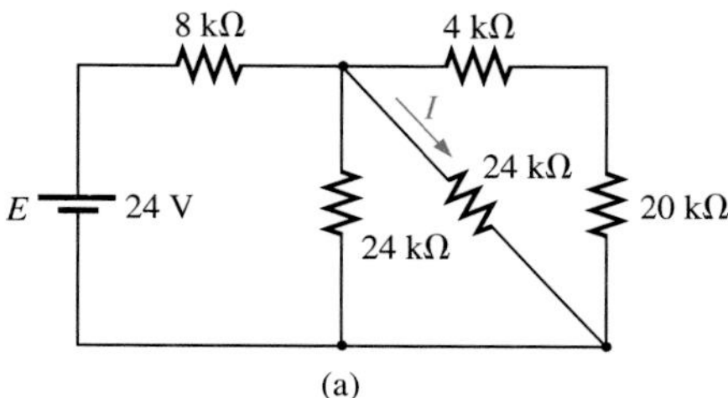

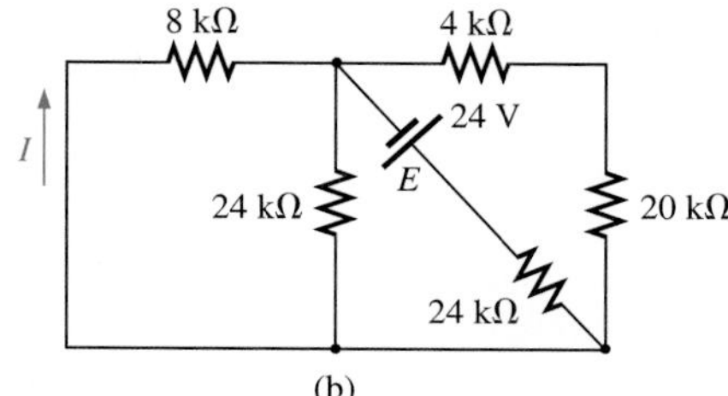

FIG. 9.138
Problem 34.

35. Repeat Problem 34 for the networks of Fig. 9.139.

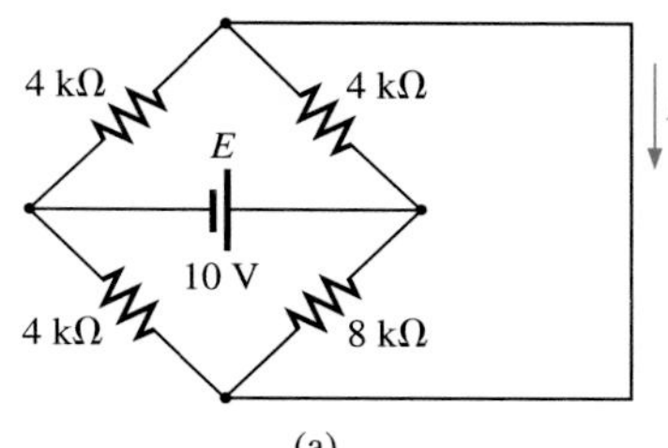

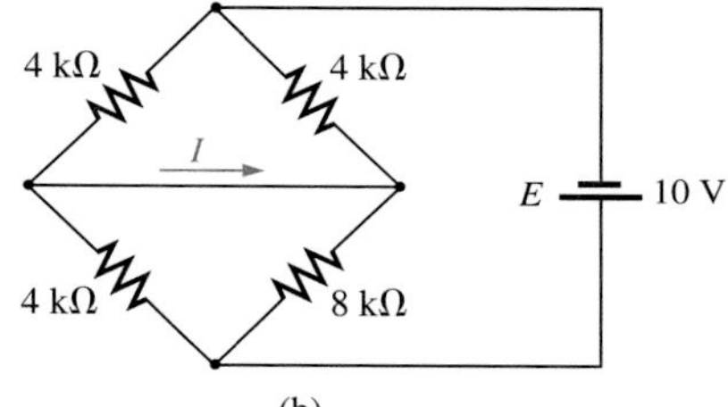

FIG. 9.139
Problem 35.

36. **a.** Determine the voltage *V* for the network of Fig. 9.140(a).
b. Repeat part (a) for the network of Fig. 9.140(b).
c. Is the dual of the reciprocity theorem satisfied?

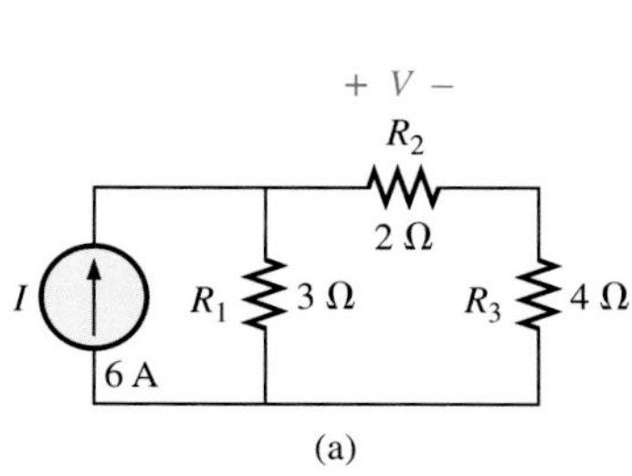

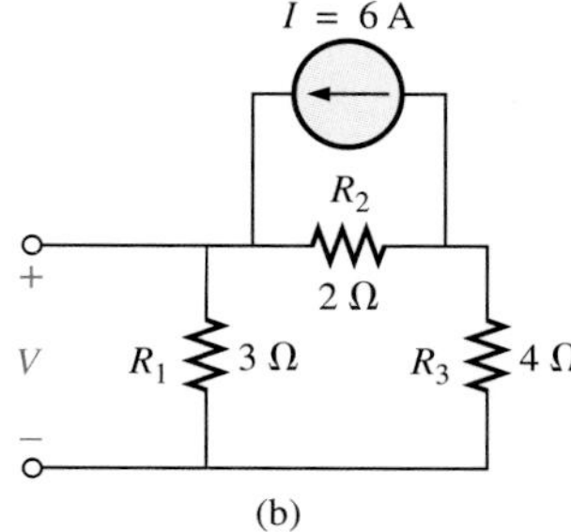

FIG. 9.140
Problem 36.

37. Determine I_1 for the network in Fig. 9.141.

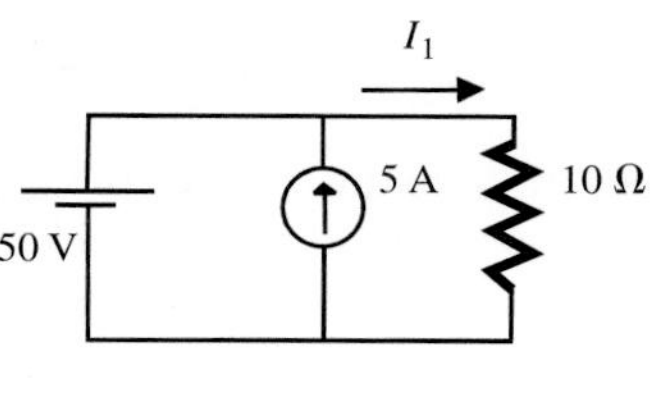

FIG. 9.141
Problem 37.

38. Find the Norton equivalent circuit for the network in Fig. 9.141.

39. Find the equivalent Thévenin circuit for the network in Fig. 9.141.

40. A continuity tester had a terminal voltage of 120V and a short-circuit current of 20 mA. Determine the Thévenin resistance.

41. For the network in Fig. 9.142.

a. Find the value of R which will have the maximum power transfer.

b. Determine the maximum power to R.

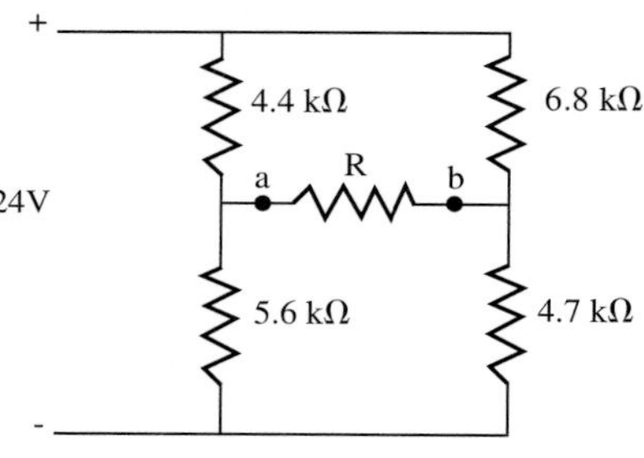

FIG. 9.142
Problem 41.

10

Capacitors

Outline

Learning Outcomes

After completing this chapter you will be able to

- explain the nature of capacitance using electric field theory
- determine the breakdown voltage for different capacitors
- describe the construction of a number of different types of capacitor
- explain circuit operation as a capacitor is charging
- determine the instantaneous voltage across a charging capacitor
- explain circuit operation as a capacitor is discharging
- determine the instantaneous voltage across a discharging capacitor
- explain the relationship between capacitor voltage and current
- explain the effect of connecting capacitors in series and parallel
- determine the energy stored by a capacitor

Capacitance is a property found in electrical circuits. In this chapter you will learn how capacitance behaves when connected across a dc source.

10.1 INTRODUCTION

Thus far, the resistor is the only **passive device** we have introduced. Now we consider two more—the **capacitor** and the **inductor**. They both are quite different from resistors in their construction, operation, and purpose.

Unlike the resistor, both elements display their total characteristics only when a change in voltage or current is made in the circuit where they exist. In addition, if we consider the *ideal* situation, they do not dissipate energy like the resistor but store it in a form that can be returned to the circuit whenever required by the circuit design.

Proper treatment of each device requires that we devote this entire chapter to the capacitor and Chapter 12 to the inductor. Since electromagnetic effects are a major consideration in the design of inductors, this topic will be covered in Chapter 11.

10.2 THE ELECTRIC FIELD

Recall from Chapter 2 that a force of attraction or repulsion exists between two charged bodies. Now consider the **electric field** that exists in the region around any charged body. This electric field is represented by **electric flux lines**, which are drawn to indicate the strength of the electric field at any point around the charged body; that is, the denser the lines of flux, the stronger the electric field. In Fig. 10.1, the electric field strength is stronger at position *a* than at position *b* because the flux lines are denser at *a* than at *b*. The symbol for **electric flux** is the Greek letter ψ (psi). The **flux per unit area** *(flux density)* is represented by the capital letter *D* and is determined by

$$D = \frac{\psi}{A} \quad \text{(flux/unit area)} \tag{10.1}$$

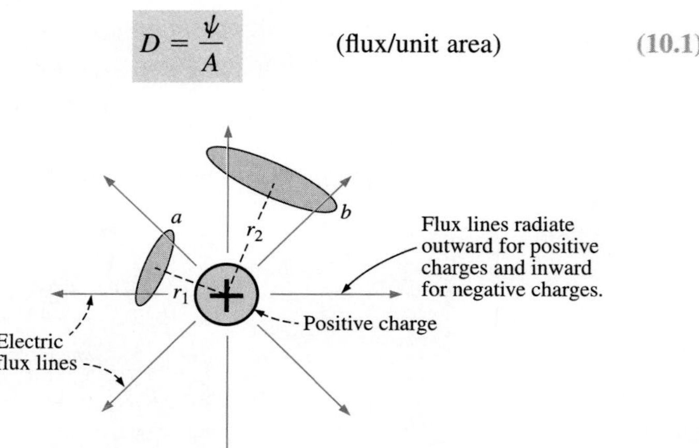

FIG. 10.1
Flux distribution from an isolated positive charge.

The larger the charge *Q* in coulombs, the greater the number of flux lines per unit area, independent of the surrounding medium. Twice the charge will produce twice the flux per unit area. The two can therefore be equated:

$$\psi \equiv Q \quad \text{(coulombs, C)} \tag{10.2}$$

By definition, the **electric field strength** ($\mathscr{E}$) at a point is the force acting on a unit positive charge at that point; that is,

$$\mathscr{E} = \frac{F}{Q} \quad \text{(newtons/coulomb, N/C)} \tag{10.3}$$

The force exerted on a unit positive charge ($Q_2 = 1$ C), by a charge Q_1, r metres away, as determined by Coulomb's law, is

$$F = \frac{kQ_1Q_2}{r^2} = \frac{kQ_1(1)}{r^2} = \frac{kQ_1}{r^2} \quad (k = 9 \times 10^9 \text{ N·m}^2/\text{C}^2)$$

Substituting this force F into Eq. (10.3) gives us

$$\mathscr{E} = \frac{F}{Q_2} = \frac{kQ_1/r^2}{1}$$

$$\mathscr{E} = \frac{kQ_1}{r^2} \quad \text{(N/C)} \tag{10.4}$$

We can therefore conclude that

the electric field strength at any point distance r from a point charge of Q coulombs is directly proportional to the magnitude of the charge and inversely proportional to the distance squared from the charge.

The squared term in the denominator will result in a rapid decrease in the strength of the electric field with distance from the point charge. In Fig. 10.1, substituting distances r_1 and r_2 into Eq. (10.4) will verify our previous conclusion that the electric field strength is greater at a than at b.

Electric flux lines always extend from a positively charged to a negatively charged body, always extend or terminate perpendicular to the charged surfaces, and never intersect.

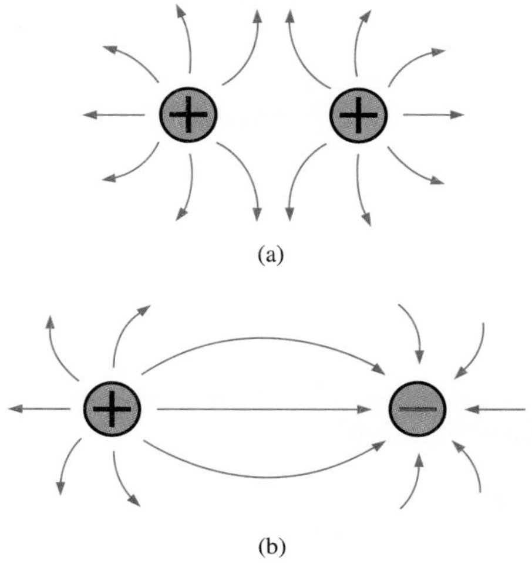

(a)

(b)

FIG. 10.2
Electric flux distribution: (a) like charges; (b) opposite charges.

For two charges of similar and opposite polarities, the flux distribution would appear as shown in Fig. 10.2.

The attraction and repulsion between charges can now be explained in terms of the electric field and its flux lines. In Fig. 10.2(a), the flux lines are not interlocked but tend to act as a buffer, preventing attraction and causing repulsion. Since the electric field strength is stronger (flux lines denser) for each charge the closer we are to the charge, the more we try to bring the two charges together, the stronger will be the force of repulsion between them. In Fig. 10.2(b), the flux lines extending from the positive charge are terminated at the negative charge. A basic law of physics states that electric flux lines always tend to be as short as possible. The two charges will therefore be drawn to each other. Again, the closer the two charges, the stronger the attraction between the two charges due to the increased field strengths.

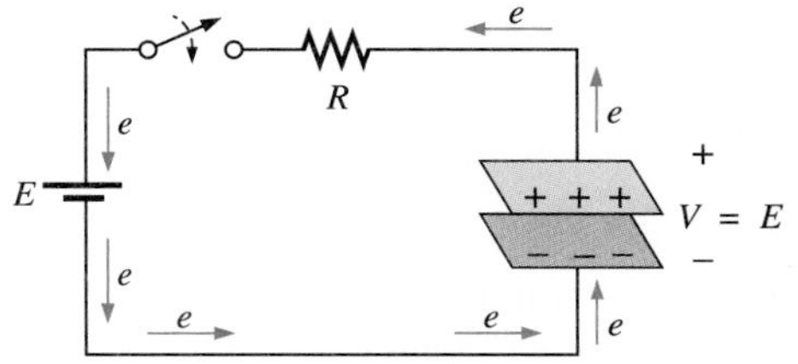

FIG. 10.3
Fundamental charging network.

10.3 CAPACITANCE

Up to this point we have considered only isolated positive and negative spherical charges, but the analysis can be extended to charged surfaces of any shape and size. In Fig. 10.3, for example, two parallel plates of a conducting material separated by an air gap have been connected

LUMINARIES

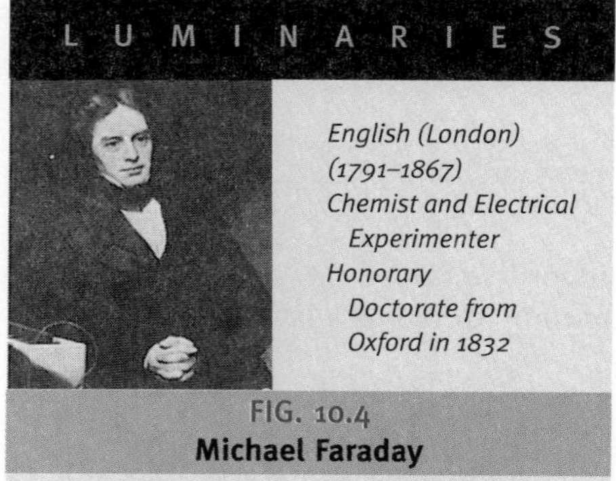

FIG. 10.4
Michael Faraday

An experimenter with no formal education, he began his research career at the Royal Institution in London as a laboratory assistant. Intrigued by the interaction between electrical and magnetic effects, he discovered *electromagnetic induction*, demonstrating that electrical effects can be generated from a magnetic field (the birth of the generator as we know it today). He also discovered *self-induced currents* and introduced the concept of *lines and fields of magnetic force*. He received over a hundred academic and scientific honours, and became a Fellow of the Royal Society in 1824 at the young age of 32.

Courtesy of the Smithsonian Institution, Photo No. 51,147

through a switch and a resistor to a battery. If the parallel plates are initially uncharged and the switch is left open, no net positive or negative charge will exist on either plate. The instant the switch is closed, however, electrons are drawn from the upper plate through the resistor to the positive terminal of the battery. There will be a surge of current at first, limited in magnitude by the resistance present. The level of flow will then decline, as will be demonstrated in the sections to follow. This action creates a net positive charge on the top plate. Electrons are being repelled by the negative terminal through the lower conductor to the bottom plate at the same rate they are being drawn to the positive terminal. This transfer of electrons continues until the potential difference across the parallel plates is exactly equal to the battery voltage. The final result is a net positive charge on the top plate and a negative charge on the bottom plate, very similar in many respects to the two isolated charges of Fig. 10.2(b).

This element, constructed simply of two parallel conducting plates separated by an insulating material (in this case, air), is called a **capacitor. Capacitance** is a measure of a capacitor's ability to store charge on its plates.

A capacitor has a capacitance of 1 farad if 1 C (coulomb) of charge is deposited on the plates by a potential difference of 1 V (volt) across the plates.

The farad is named after Michael Faraday (Fig. 10.4), a nineteenth-century English chemist and physicist. The farad, however, is generally too large a measure of capacitance for most practical applications, so the microfarad (10^{-6} F) or picofarad (10^{-12} F) is more commonly used. Expressed as an equation, the capacitance is determined by

$$C = \frac{Q}{V} \qquad \text{(farads, F)} \tag{10.5}$$

Different capacitors for the same voltage across their plates will acquire greater or lesser amounts of charge on their plates. In other words, the capacitors have a greater or lesser capacitance, respectively.

A cross-sectional view of the parallel plates is shown with the distribution of electric flux lines in Fig. 10.5(a). The number of flux lines per unit area (D) between the two plates is quite uniform. At the edges, the flux lines extend outside the common surface area of the plates, producing an effect known as *fringing*. This effect, which reduces the capacitance somewhat, can be neglected for most practical applications. For the analysis to follow, we will assume that all the flux lines leaving the positive plate will pass directly to the negative plate within the common surface area of the plates [Fig. 10.5(b)].

If a potential difference of V volts is applied across the two plates separated by a distance of d, the electric field strength between the plates is determined by

$$\mathscr{E} = \frac{V}{d} \qquad \text{(volts/metre, V/m)} \tag{10.6}$$

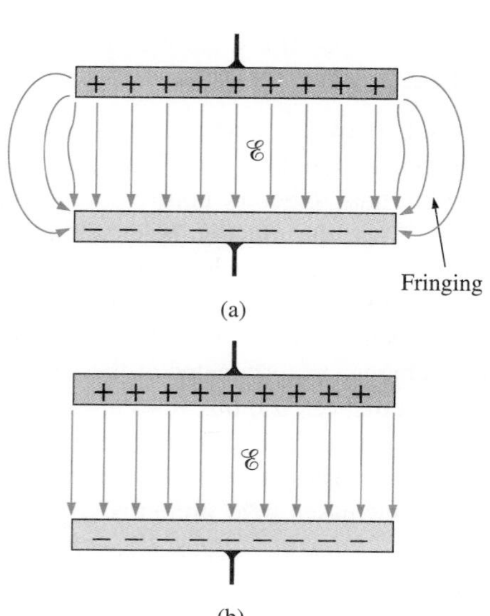

FIG. 10.5
Electric flux distribution between the plates of a capacitor: (a) including fringing; (b) ideal.

The uniformity of the flux distribution in Fig. 10.5(b) also indicates that the electric field strength is the same at any point between the two plates.

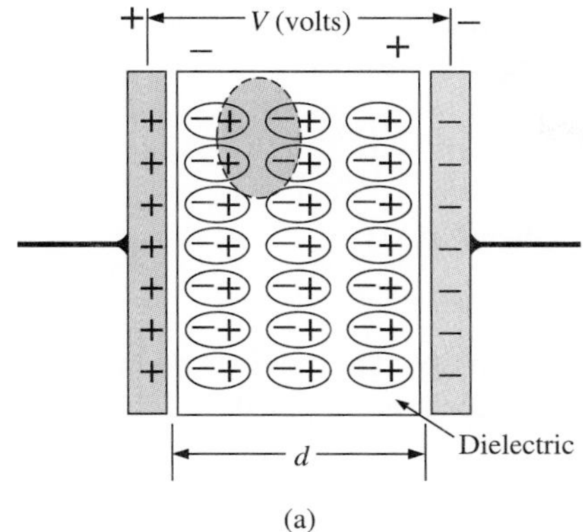

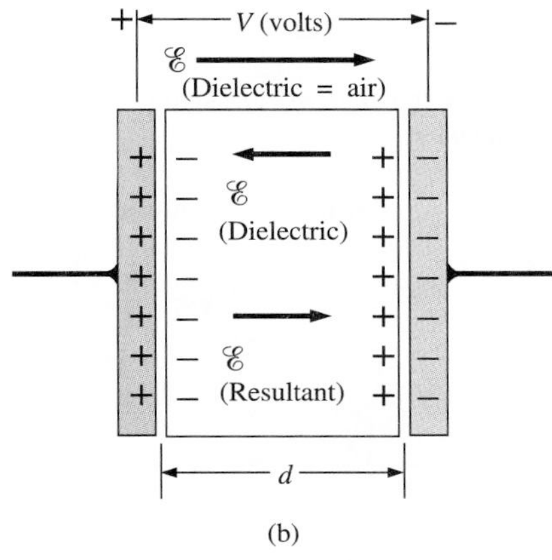

FIG. 10.6
Effect of a dielectric on the field distribution between the plates of a capacitor: (a) alignment of dipoles in the dielectric; (b) electric field components between the plates of a capacitor with a dielectric present.

Many values of capacitance can be obtained for the same set of parallel plates by adding certain insulating materials between the plates. In Fig. 10.6(a), an insulating material has been placed between a set of parallel plates having a potential difference of V volts across them.

Since the material is an insulator, the electrons within the insulator are unable to leave the parent atom and travel to the positive plate. The positive components (protons) and negative components (electrons) of each atom do shift, however [as shown in Fig. 10.6(a)], to form **dipoles**.

When the dipoles align themselves as shown in Fig. 10.6(a), the material is **polarized**. If we closely examine this polarized material, we see that the positive and negative components of adjoining dipoles are neutralizing the effects of each other [note the dashed area in Fig. 10.6(a)]. The layer of positive charge on one surface and the negative charge on the other are not neutralized, however, resulting in the establishment of an electric field within the insulator [$\mathscr{E}_{\text{dielectric}}$, Fig. 10.6(b)]. The net electric field between the plates ($\mathscr{E}_{\text{resultant}} = \mathscr{E}_{\text{air}} - \mathscr{E}_{\text{dielectric}}$) would therefore be reduced due to the insertion of the dielectric.

The purpose of the dielectric, therefore, is to create an electric field to oppose the electric field set up by free charges on the parallel plates. For this reason, the insulating material is referred to as a **dielectric**: **di** for *opposing* and **electric** for *electric field.*

In either case—with or without the dielectric—if the potential across the plates is kept constant and the distance between the plates is fixed, the net electric field within the plates must remain the same, as determined by the equation $\mathscr{E} = V/d$. We just saw, however, that the net electric field between the plates would decrease with insertion of the dielectric for a fixed amount of free charge on the plates. To compensate and keep the net electric field equal to the value determined by V and d, more charge must be deposited on the plates. [Look ahead to Eq. (10.11).] This additional charge for the same potential across the plates increases the capacitance, as determined by the following equation:

$$C\uparrow = \frac{Q\uparrow}{V}$$

For different dielectric materials between the same two parallel plates, different amounts of charge will be deposited on the plates. But $\psi \equiv Q$, so the dielectric is also determining the number of flux lines between the two plates and consequently the flux density ($D = \psi/A$), since A is fixed.

The ratio of the flux density to the electric field intensity in the dielectric is called the **permittivity** (ϵ) of the dielectric:

$$\epsilon = \frac{D}{\mathscr{E}} \quad \text{(farads/metre, F/m)} \qquad (10.7)$$

Permittivity is a measure of how easily the dielectric will allow the establishment of flux lines within the dielectric. The greater its value, the greater the amount of charge deposited on the plates and, consequently, the greater the flux density for a fixed area.

For a vacuum, the value of ϵ (denoted by ϵ_o) is 8.85×10^{-12} F/m. The ratio of the permittivity of any dielectric to that of a vacuum is called the **relative permittivity**, ϵ_r. It simply compares the permittivity of the dielectric to that of air. In equation form,

$$\epsilon_r = \frac{\epsilon}{\epsilon_o} \qquad (10.8)$$

The value of ϵ for any material, therefore, is

$$\epsilon = \epsilon_r \epsilon_o$$

Note that ϵ_r is a dimensionless quantity. The relative permittivity, or **dielectric constant**, as it is often called, is provided in Table 10.1 for various dielectric materials.

TABLE 10.1
Relative permittivity (dielectric constant) of various dielectrics.

Dielectric	**ϵ_r (Average Values)**
Vacuum	1.0
Air	1.0006
Teflon	2.0
Paper, paraffined	2.5
Rubber	3.0
Transformer oil	4.0
Mica	5.0
Porcelain	6.0
Bakelite	7.0
Glass	7.5
Distilled water	80.0
Barium-strontium titanite (ceramic)	7500.0

Substituting for D and $\mathscr{E}$ in Eq. (10.7), we have

$$\epsilon = \frac{D}{\mathscr{E}} = \frac{\psi/A}{V/d} = \frac{Q/A}{V/d} = \frac{Qd}{VA}$$

But

$$C = \frac{Q}{V}$$

and, therefore,

$$\epsilon = \frac{Cd}{A}$$

and

$$C = \epsilon\frac{A}{d} \quad \text{(F)} \qquad (10.9)$$

or

$$C = \epsilon_o\epsilon_r\frac{A}{d} = 8.85 \times 10^{-12}\epsilon_r\frac{A}{d} \quad \text{(F)} \qquad (10.10)$$

where A is the area in square metres of the plates, d is the distance in metres between the plates, and ϵ_r is the relative permittivity. The

continued

capacitance, therefore, will be greater if the area of the plates is increased, or the distance between the plates is decreased, or the dielectric is changed so that ϵ_r is increased.

Solving for the distance d in Eq. (10.9), we have

$$D = \frac{\epsilon A}{C}$$

and substituting into Eq. (10.6) gives us

$$\mathscr{E} = \frac{V}{d} = \frac{V}{\epsilon A/C} = \frac{CV}{\epsilon A}$$

But $Q = CV$, and therefore

$$\mathscr{E} = \frac{Q}{\epsilon A} \quad \text{(V/m)} \qquad (10.11)$$

which gives the electric field intensity between the plates in terms of the permittivity ϵ, the charge Q, and the surface area A of the plates. The ratio

$$\frac{C}{C_o} = \frac{\epsilon A/d}{\epsilon_o A/d} = \frac{\epsilon}{\epsilon_o} = \epsilon_r$$

or

$$C = \epsilon_r C_o \qquad (10.12)$$

In words, Eq. (10.12) states that

for the same set of parallel plates, the capacitance using a dielectric (of relative permittivity ϵ_r) is ϵ_r times that obtained for a vacuum (or air, approximately) between the plates.

This relationship between ϵ_r and the capacitances provides an excellent experimental method for finding the value of ϵ_r for various dielectrics.

EXAMPLE 10.1

Determine the capacitance of each capacitor on the right side of Fig. 10.7.

Solutions:

a. $C = 3(5\ \mu\text{F}) = \mathbf{15\ \mu F}$

b. $C = \frac{1}{2}(0.1\ \mu\text{F}) = \mathbf{0.05\ \mu F}$

c. $C = 2.5(20\ \mu\text{F}) = \mathbf{50\ \mu F}$

d. $C = (5)\frac{4}{(1/8)}(1000\ \text{pF}) = (160)(1000\ \text{pF}) = \mathbf{0.16\ \mu F}$

continued

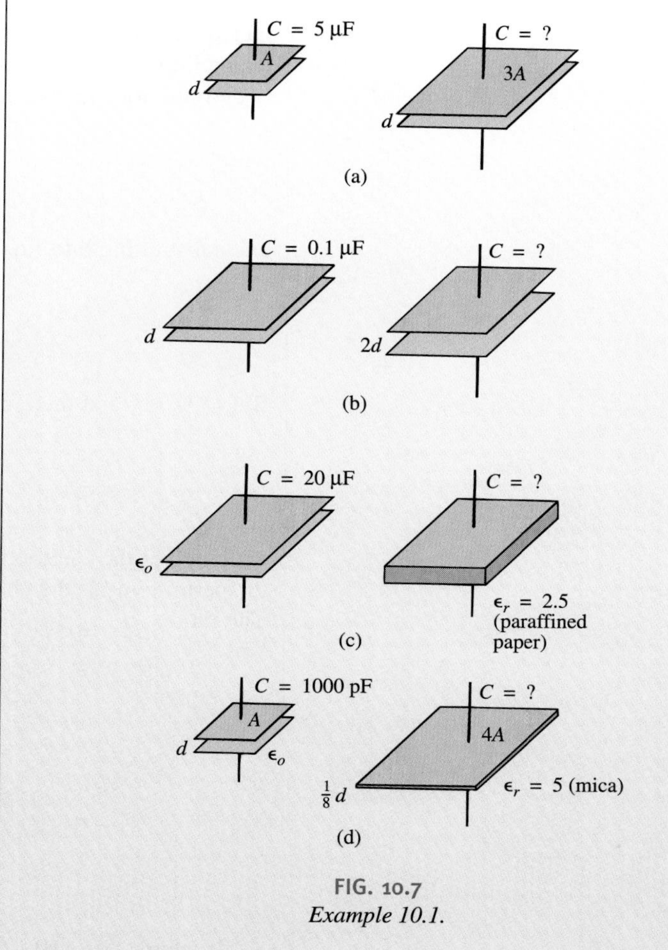

FIG. 10.7
Example 10.1.

EXAMPLE 10.2

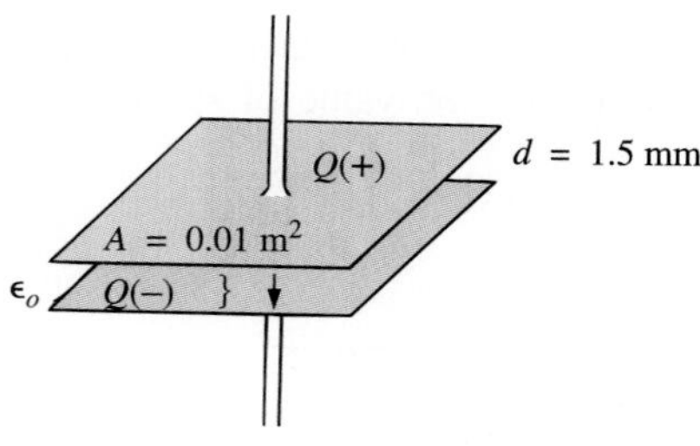

FIG. 10.8
Example 10.2.

For the capacitor of Fig. 10.8:

a. Determine the capacitance.
b. Determine the electric field strength between the plates if 450 V are applied across the plates.
c. Find the resulting charge on each plate.

Solutions:

a. $C_o = \dfrac{\epsilon_o A}{d} = \dfrac{(8.85 \times 10^{-12}\text{ F/m})(0.01\text{ m}^2)}{1.5 \times 10^{-3}\text{ m}} = 59.0 \times 10^{-12}\text{ F}$

$= \mathbf{59\ pF}$

b. $\mathscr{E} = \dfrac{V}{d} = \dfrac{450\text{ V}}{1.5 \times 10^{-3}\text{ m}}$

$\cong \mathbf{300 \times 10^3\ V/m}$

c. $C = \dfrac{Q}{V}$

or $Q = CV = (59.0 \times 10^{-12}\text{ F})(450\text{ V})$

$= 26.550 \times 10^{-9}\text{ C}$

$= \mathbf{26.55\ nC}$

EXAMPLE 10.3

A sheet of mica 1.5 mm thick having the same area as the plates is inserted between the plates of Example 10.2.

a. Find the electric field strength between the plates.
b. Find the charge on each plate.
c. Find the capacitance.

Solutions:

a. $\mathscr{E}$ is fixed by

$$\mathscr{E} = \frac{V}{d} = \frac{450 \text{ V}}{1.5 \times 10^{3} \text{ m}}$$
$$\cong \mathbf{300 \times 10^3 \ V/m}$$

b. $\mathscr{E} = \dfrac{Q}{\epsilon A}$ or

$$Q = \epsilon \mathscr{E} A = \epsilon_r \epsilon_o \mathscr{E} A$$
$$= (5)(8.85 \times 10^{-12} \text{ F/m})(300 \times 10^3 \text{ V/m})(0.01 \text{ m}^2)$$
$$= 132.75 \times 10^{-9} \text{ C} = \mathbf{132.75 \ nC}$$

(five times the amount for air between the plates)

c. $C = \epsilon_r C_o$

$$= (5)(59 \times 10^{-12} \text{ F}) = \mathbf{295 \ pF}$$

10.4 DIELECTRIC STRENGTH

For every dielectric there is a potential that, if applied across the dielectric, will break the bonds within the dielectric and cause current to flow. The voltage required per unit length (electric field intensity) to establish conduction in a dielectric is an indication of its **dielectric strength** and is called the **breakdown voltage**. When breakdown occurs, the capacitor has characteristics very similar to those of a conductor. A typical example of breakdown is lightning, which occurs when the potential between the clouds and the earth is so high that charge can pass from one to the other through the atmosphere, which acts as the dielectric.

The average dielectric strengths for various dielectrics are shown in volts/micrometre in Table 10.2. The relative permittivity appears in parentheses to emphasize the importance of considering both factors in the design of capacitors. Take particular note of barium-strontium titanite and mica.

TABLE 10.2

Dielectric strength of some dielectric materials.

Dielectric	Dielectric Strength (Average Value), in V/μm	(ϵ_r)
Air	3	(1.0006)
Barium-strontium titanite (ceramic)	3	(7500)
Porcelain	8	(6.0)
Transformer oil	16	(4.0)
Bakelite	16	(7.0)
Rubber	30	(3.0)
Paper, paraffined	50	(2.5)
Teflon	60	(2.0)
Glass	120	(7.5)
Mica	200	(5.0)

EXAMPLE 10.4

Find the maximum voltage that can be applied across a 0.2-μF capacitor having a plate area of 0.3 m^2. The dielectric is porcelain. Assume a linear relationship between the dielectric strength and the thickness of the dielectric.

Solution:

$$C = 8.85 \times 10^{-12}\epsilon_r \frac{A}{d}$$

or $$d = \frac{8.85\epsilon_r A}{10^{12}C} = \frac{(8.85)(6)(0.3 \text{ m}^2)}{(10^{12})(0.2 \times 10^{-6}\text{ F})} = 7.965 \times 10^{-5}\text{ m}$$

$$\cong \mathbf{79.65\ \mu m}$$

$$\text{Dielectric strength} = 8\text{ V}/\mu\text{m}$$

Therefore, $$\left(\frac{8\text{ V}}{\mu\text{m}}\right)(79.65\mu\text{m}) = \mathbf{637\ V}$$

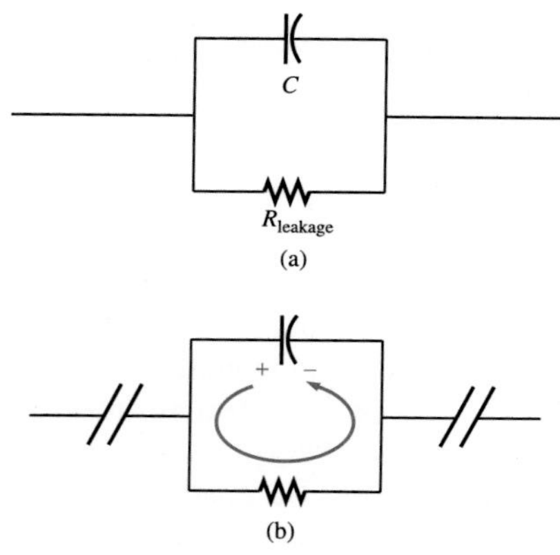

FIG. 10.9
Demonstrating the effect of the leakage current.

10.5 LEAKAGE CURRENT

Up to this point, we have assumed that the flow of electrons will occur in a dielectric only when the breakdown voltage is reached. This is the ideal case. In reality, there are free electrons in every dielectric because of impurities in the dielectric and forces within the material itself.

When a voltage is applied across the plates of a capacitor, a **leakage current** of free electrons passes from one plate to the other. The current is usually so small, however, that it can be neglected for most practical applications. This effect is shown by a resistor in parallel with the capacitor, as in Fig. 10.9(a), whose value is typically more than 100 MΩ. However, some capacitors, such as the electrolytic type, have high leakage currents. When charged and then disconnected from the charging circuit, these capacitors lose their charge in a matter of seconds because of the leakage current from one plate to the other [Fig. 10.9(b)].

10.6 TYPES OF CAPACITORS

Like resistors, all capacitors can be included under two general headings: fixed or variable. The symbol for a fixed capacitor is ⊣⊢ and for a variable capacitor, ⊣⊬. The curved line represents the plate that is usually connected to the point of lower potential. The following descriptions of varieties of capacitors are summarized in Fig. 10.20, at the end of this section.

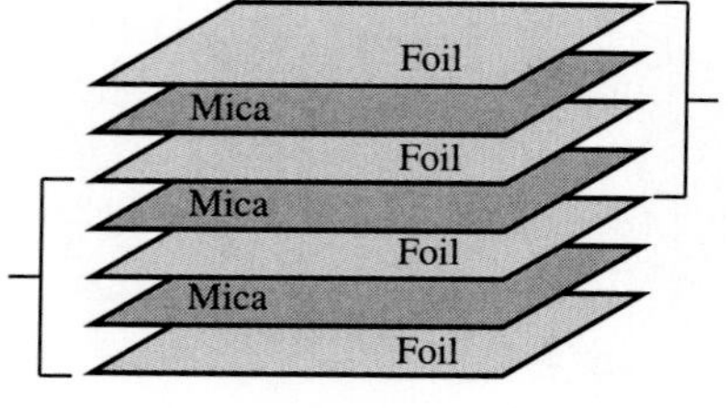

FIG. 10.10
Basic structure of a stacked mica capacitor.

Fixed Capacitors

Many types of fixed capacitors are available today. Some of the most common are the mica, ceramic, electrolytic, tantalum, and polyester-film capacitors. The typical **mica capacitor** consists basically of mica sheets separated by sheets of metal foil. The plates are connected to two electrodes, as shown in Fig. 10.10. The total area is the area of one sheet times the number of dielectric sheets. The entire system is encased in a plastic insulating material. The mica capacitor exhibits excellent charac-

FIG. 10.11
Mica capacitors. (Courtesy of Custom Electronics Inc.)

teristics under stress of temperature variations and high voltage applications (its dielectric strength is 200 V/μm). Its leakage current is also very small (R_{leakage} about 1000 MΩ).

Mica capacitors are typically between a few picofarads and 0.2 μF, with voltages of 100 V or more.

A second type of mica capacitor appears in Fig. 10.11. Note in particular the cylindrical unit in the bottom left-hand corner of the figure. The ability to "roll" the mica to form the cylindrical shape is due to a process of removing the soluble contaminants in natural mica, leaving a paperlike structure due to the cohesive forces in natural mica. It is commonly referred to as *reconstituted mica,* although this does not mean "recycled" or "second-hand" mica. For some of the units in the photograph, different levels of capacitance are available between different sets of terminals.

The **ceramic capacitor** is made in many shapes and sizes, some of which are shown in Fig. 10.12. The basic construction, however, is about the same for each, as shown in Fig. 10.13. A ceramic base is coated on two sides with a metal, such as copper or silver, to act as the two plates. The leads are then attached through electrodes to the plates. An insulating coating of ceramic or plastic is then applied over the plates and dielectric. Ceramic capacitors also have a very low leakage current (R_{leakage} about 1000 MΩ) and can be used in both dc and ac networks. They can be found in values ranging from a few picofarads to perhaps 2 μF, with very high working voltages such as 5000 V or more.

In recent years there has been increasing interest in monolithic (single-structure) chip capacitors such as those appearing in Fig. 10.14(a) due to their application on hybrid circuitry [networks using both discrete and integrated circuit (IC) components]. There has also been increasing use of microstrip (strip-line) circuitry such as the one in Fig. 10.14(b). Note the small chips in this cutaway section. The L and H of Fig. 10.14(a) indicate the level of capacitance. For example, the letter H in black letters represents 16 units of capacitance (in picofarads), or 16 pF. If blue ink is used, a multiplier of 100 is applied, resulting in 1600 pF. Although the size is similar, the type of ceramic material controls the capacitance level.

The **electrolytic capacitor** is used most commonly in situations where capacitances of the order of one to several thousand microfarads are required. They are designed mainly for use in networks where only dc voltages will be applied across the capacitor because they have good

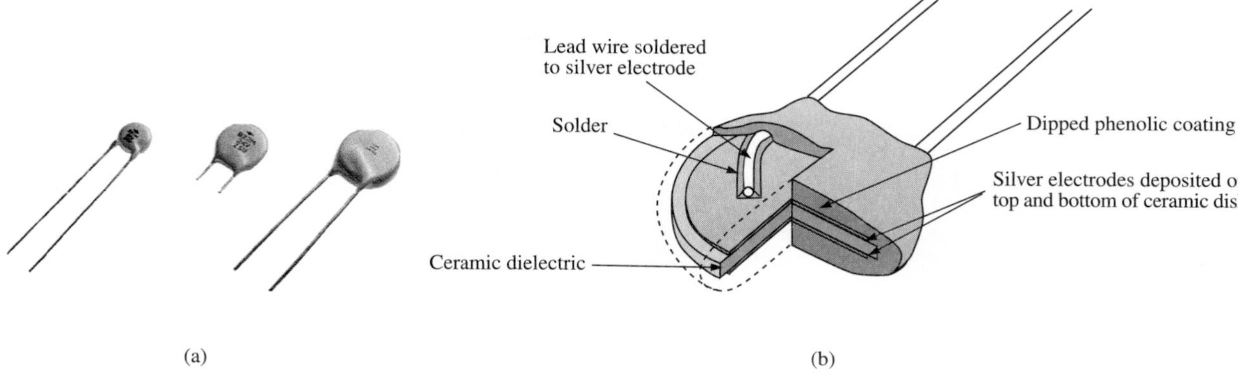

FIG. 10.12
Ceramic disc capacitors: (a) photograph (Courtesy of Philips Components, Inc.); (b) construction.

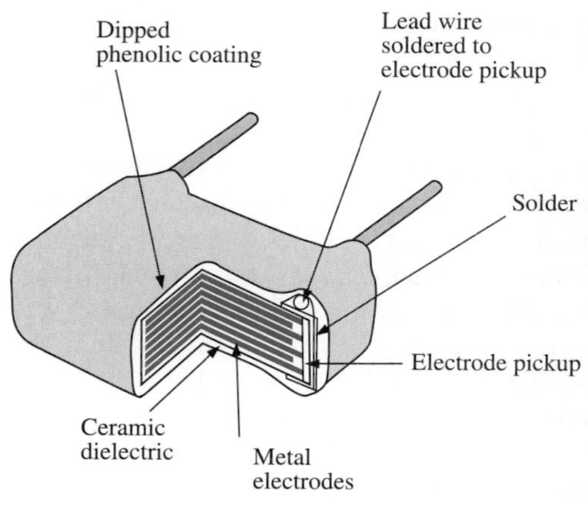

FIG. 10.13
Multilayer, radial-lead ceramic capacitor.

(a)

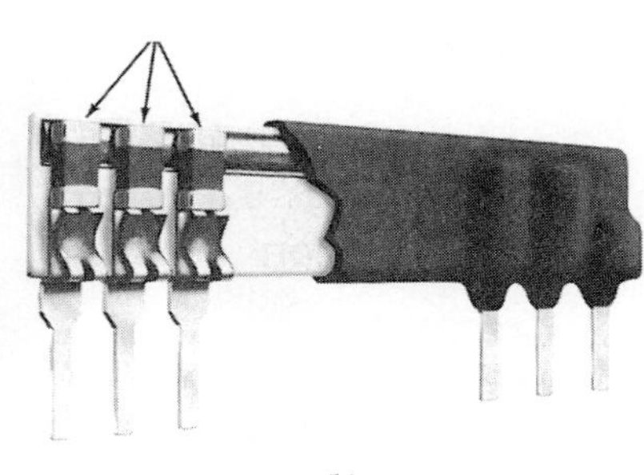

(b)

FIG. 10.14
Monolithic chip capacitors. (Courtesy of Vitramon, Inc.)

insulating characteristics (high leakage current) between the plates in one direction but take on the characteristics of a conductor in the other direction. There are electrolytic capacitors available that can be used in ac circuits (for starting motors) and in cases where the polarity of the dc voltage will reverse across the capacitor for short periods of time.

The basic construction of the electrolytic capacitor consists of a roll of aluminum foil coated on one side with an aluminum oxide, the aluminum being the positive plate and the oxide the dielectric. A layer of paper or gauze saturated with an electrolyte is placed over the aluminum oxide on the positive plate. Another layer of aluminum without the oxide coating is then placed over this layer to assume the role of the negative plate. In most cases the negative plate is connected directly to the aluminum container, which then serves as the negative terminal for external connections. Because of the size of the roll of aluminum foil, the overall area of this capacitor is large; and due to the use of an oxide as the dielectric, the distance between the plates is extremely small. The negative terminal of the electrolytic capacitor is usually the one with no visible identification on the casing. The positive is usually indicated by such designs as +, △, □, and so on. Due to the polarity requirement, the symbol for an electrolytic capacitor will normally appear as ⊤⊥+.

Associated with each electrolytic capacitor are the dc working voltage and the surge voltage. The **working voltage** is the voltage that can be applied across the capacitor for long periods of time without breakdown. The **surge voltage** is the maximum dc voltage that can be applied for a short period of time. Electrolytic capacitors are characterized as having low breakdown voltages and high leakage currents (R_{leakage} about 1 MΩ). Various types of electrolytic capacitors are shown in Fig. 10.15. They can be found in values extending from a few microfarads to several thousand microfarads and working voltages as high as 500 V. However, increased levels of voltage are normally associated with lower values of available capacitance.

There are fundamentally two types of **tantalum capacitors**: the **solid** and the **wet-slug**. In each case, tantalum powder of high purity is

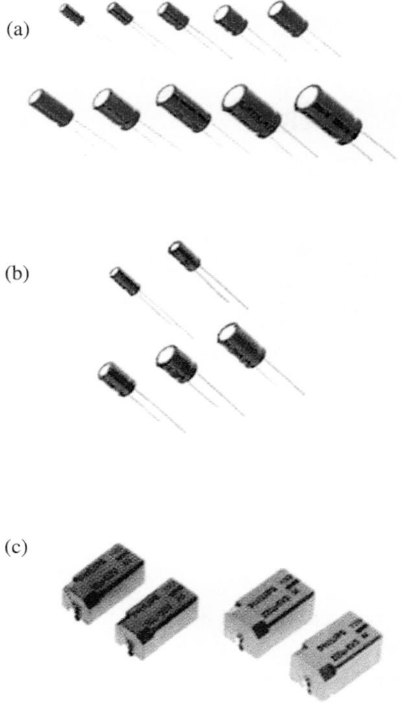

FIG. 10.15
Electrolytic capacitors: (a) radial standard miniature; (b) radial standard high voltage; (c) chip long life. (Courtesy of Philips Components, Inc.)

pressed into a rectangular or cylindrical shape, as shown in Fig. 10.16. The anode (+) connection is then simply pressed into the resulting structures, as shown in the figure. The resulting unit is then sintered (baked) in a vacuum at very high temperatures to establish a very porous material. The result is a structure with a very large surface area in a limited volume. Through immersion in an acid solution, a very thin manganese dioxide (MnO_2) coating is established on the large, porous surface area. An electrolyte is then added to establish contact between the surface area and the cathode, producing a solid tantalum capacitor. If an appropriate "wet" acid is introduced, it is called a *wet-slug* tantalum capacitor.

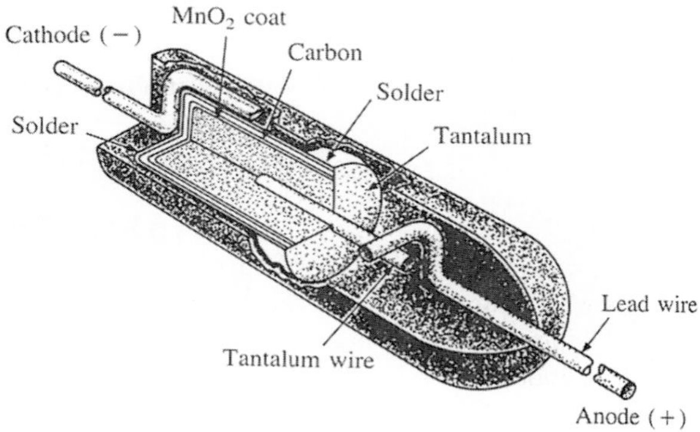

FIG. 10.16
Tantalum capacitor. (Courtesy of Union Carbide Corp.)

The last type of fixed capacitor to be introduced is the **polyester-film capacitor**, the basic construction of which is shown in Fig. 10.17. It consists simply of two metal foils separated by a strip of polyester material such as Mylar®. The outside layer of polyester is applied to act as an insulating jacket. Each metal foil is connected to a lead that extends either axially or radially from the capacitor. The rolled construction results in a large surface area, and the use of the plastic dielectric results in a very thin layer between the conducting surfaces.

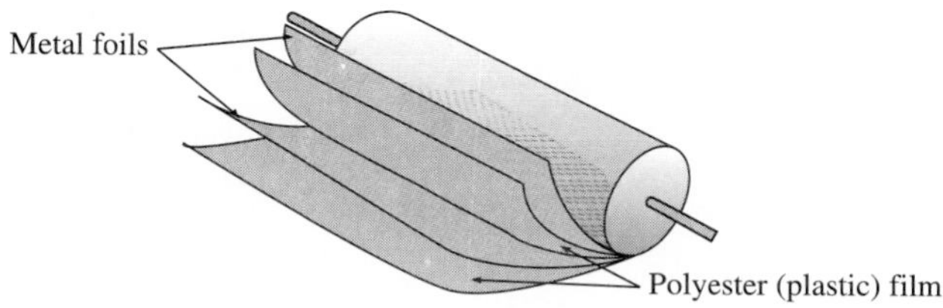

FIG. 10.17
Polyester-film capacitor.

Data such as capacitance and working voltage are printed on the outer wrapping if the polyester capacitor is large enough. This capacitor can be used for both dc and ac networks. Its leakage resistance is of the order of 100 MΩ. An axial lead and radial lead polyester-film capacitor appear in Fig. 10.18. The axial lead variety is available with capacitance levels of 0.1 μF to 18 μF, with working voltages extending to 630 V. The radial lead variety has a capacitance range of 0.01 μF to 10 μF, with working voltages extending to 1000 V.

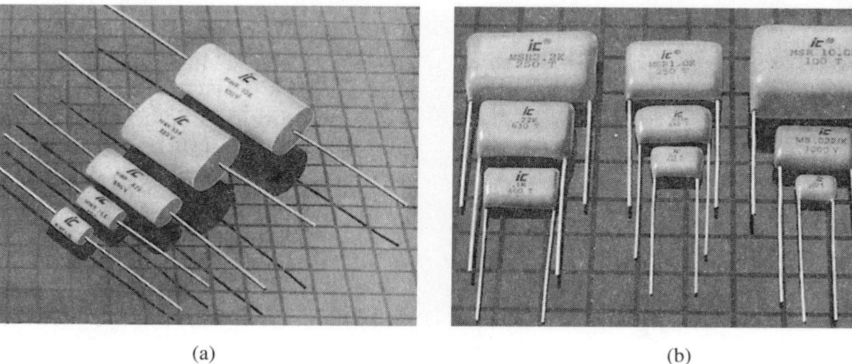

(a) (b)

FIG. 10.18
Polyester-film capacitors: (a) axial lead; (b) radial lead. (Courtesy of Illinois Capacitor, Inc.)

Variable Capacitors

The most common of the **variable-type capacitors** is shown in Fig. 10.19. The dielectric for each capacitor is air. The capacitance in Fig. 10.19(a) is changed by turning the shaft at one end to vary the common area of the movable and fixed plates. The greater the common area, the larger the capacitance, as determined by Eq. (10.10). The capacitance of the **trimmer capacitor** in Fig. 10.19(b) is changed by turning the screw, which will vary the distance between the plates and thereby the capacitance.

(a) (b)

FIG. 10.19
Variable air capacitors. (Part (a) courtesy of James Millen Manufacturing Co.; part (b) courtesy of Johnson Manufacturing Co.)

10.7 TRANSIENTS IN CAPACITIVE NETWORKS: CHARGING PHASE

Section 10.3 described how a capacitor acquires its charge. Let us now extend this discussion to include the potentials and current developed within the network of Fig. 10.21 (on the next page) following the closing of the switch (to position 1).

You will recall that the instant the switch is closed, electrons are drawn from the top plate and deposited on the bottom plate by the battery, resulting in a net positive charge on the top plate, and a negative charge on the bottom plate. The transfer of electrons is very rapid at

Type: Miniature Axial Electrolytic
Typical Values: 0.1 μF to 15,000 μF
Typical Voltage Range: 5 V to 450 V
Capacitor Tolerance: ±20%
Description: Polarized, used in DC power supplies, bypass filters, DC blocking.

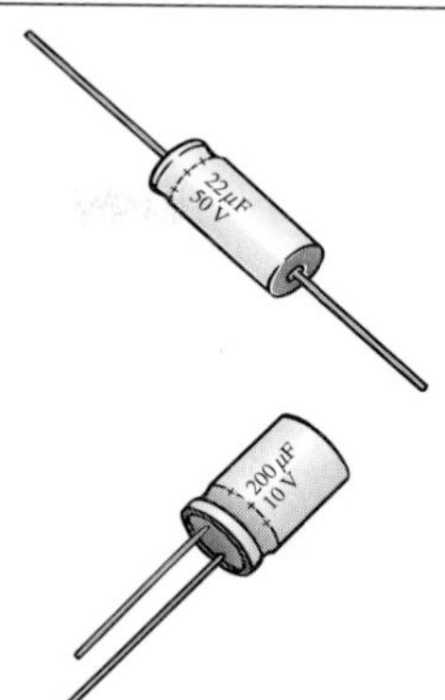

Type: Miniature Radial Electrolyte
Typical Values: 0.1 μF to 15,000 μF
Typical Voltage Range: 5 V to 450 V
Capacitor Tolerance: ±20%
Description: Polarized, used in DC power supplies, bypass filters, DC blocking.

Type: Ceramic Disc
Typical Values: 10 pF to 0.047 μF
Typical Voltage Range: 100 V to 6 kV
Capacitor Tolerance: ±5%, ±10%
Description: Non-polarized, NPO type, stable for a wide range of temperatures. Used in oscillators, noise filters, circuit coupling, tank circuits.

Type: Dipped Tantalum
Typical Values: 0.047 μF to 470 μF
Typical Voltage Range: 6.3 V to 50 V
Capacitor Tolerance: ±10%, ±20%
Description: Polarized, low leakage current, used in power supplies, high frequency noise filters, bypass filter.

Type: Surface Mount Type (SMT)
Typical Values: 10 pF to 10 μF
Typical Voltage Range: 6.3 V to 16 V
Capacitor Tolerance: ±10%
Description: Polarized and non-polarized, used in all types of circuits, requires a minimum amount of PC board real estate.

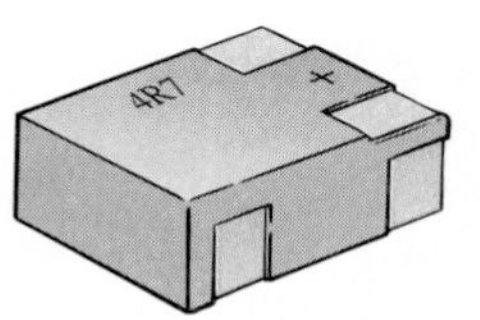

Type: Silver Mica
Typical Value: 10 pF to 0.001 μF
Typical Voltage Range: 50 V to 500 V
Capacitor Tolerance: ±5%
Description: Non-polarized, used in oscillators, in circuits that require a stable component over a range of temperatures and voltages.

Type: Mylar Paper
Typical Value: 0.001 μF to 0.68 μF
Typical Voltage Range: 50 V to 600 V
Capacitor Tolerance: ±22%
Description: Non-polarized, used in all types of circuits, moisture resistant.

Type: AC/DC Motor Run
Typical Value: 0.25 μF to 1200 μF
Typical Voltage Range: 240 V to 660 V
Capacitor Tolerance: ±10%
Description: Non-polarized, used in motor run-start, high intensity lighting supplies, AC noise filtering.

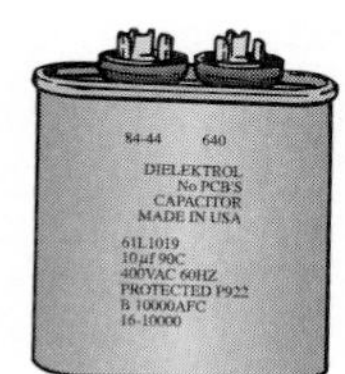

Type: Trimmer Variable
Typical Value: 1.5 pF to 600 pF
Typical Voltage Range: 5 V to 100 V
Capacitor Tolerance: ±10%
Description: Non-polarized, used in oscillators, tuning circuits, AC filters.

Type: Tuning variable
Typical Value: 10 pF to 600 pF
Typical Voltage Range: 5 V to 100 V
Capacitor Tolerance: ±10%
Description: Non-polarized, used in oscillators, radio tuning circuit.

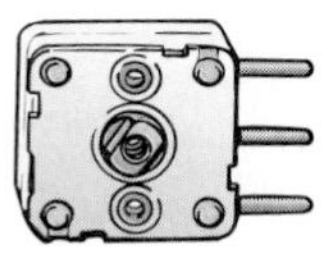

FIG. 10.20
Summary of capacitive elements.

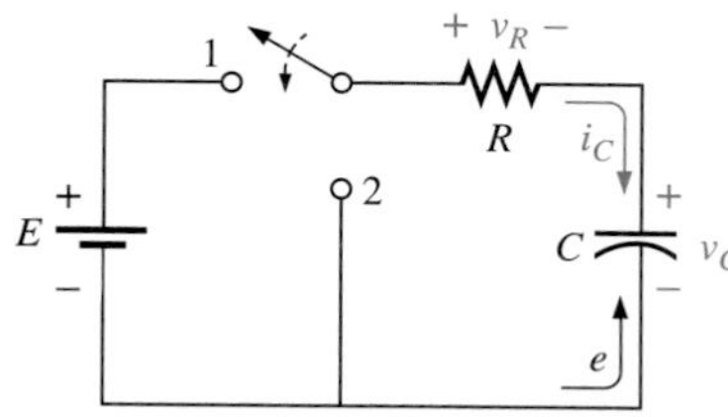

FIG. 10.21
Basic charging network.

first, slowing down as the potential across the capacitor approaches the applied voltage of the battery. When the voltage across the capacitor equals the battery voltage, the transfer of electrons will stop and the plates will have a net charge determined by $Q = CV_C = CE$.

Plots of the changing current and voltage appear in Figs. 10.22 and 10.23, respectively. When the switch is closed at $t = 0$ s, the current jumps to a value limited only by the resistance of the network and then decays to zero as the plates are charged. Note the rapid decay in current level, revealing that the amount of charge deposited on the plates per unit time is rapidly decaying also. Since the voltage across the plates is directly related to the charge on the plates by $v_C = q/C$, the rapid rate with which charge is initially deposited on the plates will result in a rapid increase in v_C. Obviously, as the rate of flow of charge (I) decreases, the rate of change in voltage will follow. Eventually, the flow of charge will stop, the current I will be zero, and the voltage will stop changing in magnitude—the *charging phase* has passed. At this point

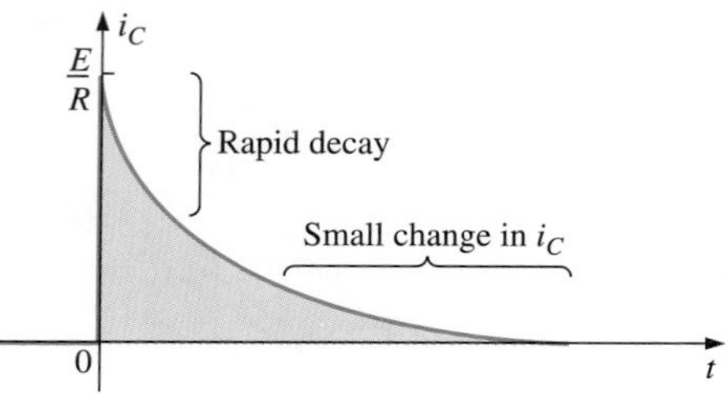

FIG. 10.22
i_C during the charging phase.

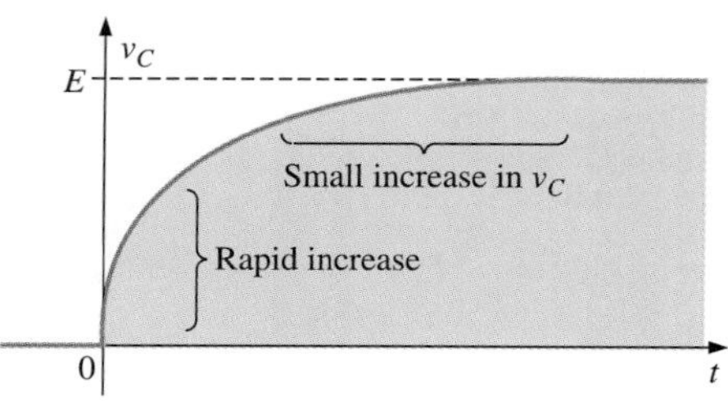

FIG. 10.23
v_C during the charging phase.

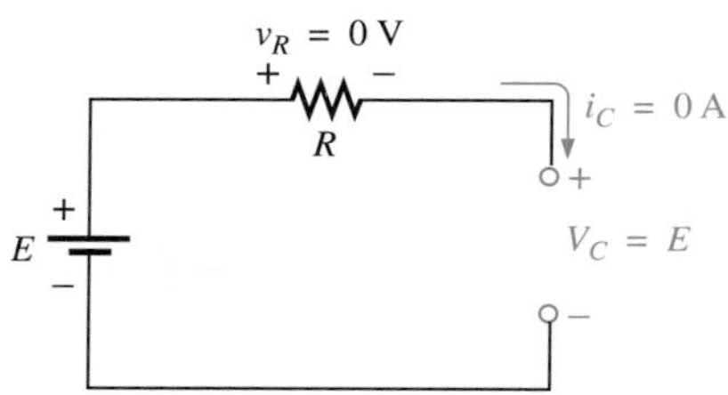

FIG. 10.24
Open-circuit equivalent for a capacitor following the charging phase.

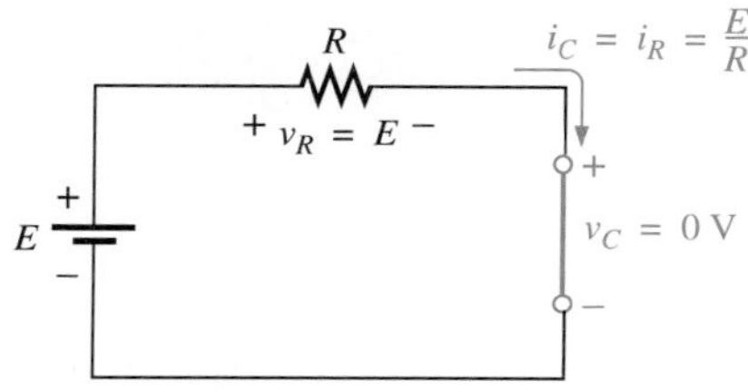

FIG. 10.25
Short-circuit equivalent for a capacitor (switch closed, $t = 0$).

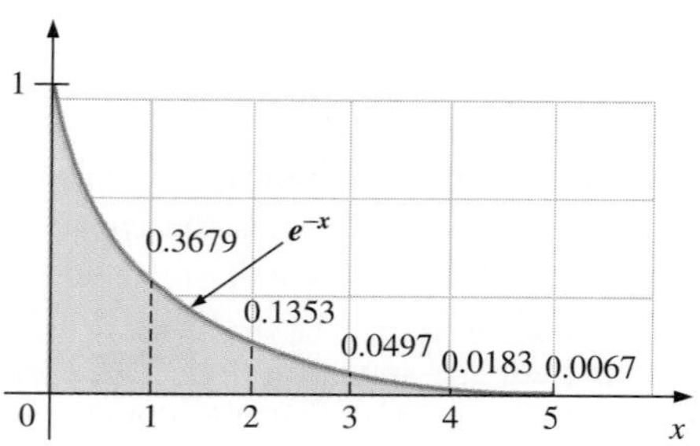

FIG. 10.26
The e^{-x} function ($x \geq 0$).

the capacitor takes on the characteristics of an open circuit: a voltage drop across the plates without a flow of charge "between" the plates. As demonstrated in Fig. 10.24, the voltage across the capacitor is the source voltage since $i = i_C = i_R = 0$ A and $v_R = i_R R = (0)R = 0$ V. For all future analysis:

A capacitor can be replaced by an open-circuit equivalent once the charging phase in a dc network has passed.

Looking back at the instant the switch is closed, we can also assume that a capacitor behaves like a short circuit the moment the switch is closed in a dc charging network, as shown in Fig. 10.25. The current $i = i_C = i_R = E/R$, and the voltage $v_C = E - v_R = E - i_R R = E - (E/R)R = E - E = 0$ V at $t = 0$ s.

Through the use of calculus, the following mathematical equation for the charging current i_C can be obtained:

$$i_C = \frac{E}{R}e^{-t/RC} \tag{10.13}$$

The factor $e^{-t/RC}$ is an exponential function of the form e^{-x}, where $x = -t/RC$ and $e = 2.71828 \ldots$. A plot of e^{-x} for $x \geq 0$ appears in Fig. 10.26. Exponentials are mathematical functions that all students of electrical, electronic, or computer systems must become very familiar with. They will appear throughout the analysis to follow in this course, and in later courses.

Our interest in the function e^{-x} is limited to values of x greater than zero, as noted by the curve of Fig. 10.22. All modern-day scientific calculators have the function e^x. To obtain e^{-x}, the sign of x must be changed using the sign key before the exponential function is keyed in. The magnitude of e^{-x} has been listed in Table 10.3 for a range of values of x. Note the rapidly decreasing magnitude of e^{-x} with increasing value of x.

TABLE 10.3
Selected values of e^{2x}.

$x = 0$	$e^{-x} = e^{-0} = \frac{1}{e^0} = \frac{1}{1} = 1$
$x = 1$	$e^{-1} = \frac{1}{e} = \frac{1}{2.71828\ldots} = 0.3679$
$x = 2$	$e^{-2} = \frac{1}{e^2} = 0.1353$
$x = 5$	$e^{-5} = \frac{1}{e^5} = 0.00674$
$x = 10$	$e^{-10} = \frac{1}{e^{10}} = 0.0000454$
$x = 100$	$e^{-100} = \frac{1}{e^{100}} = 3.72 \times 10^{-44}$

The factor RC in Eq. (10.13) is called the **time constant** of the system and has the units of time as follows:

$$RC = \left(\frac{V}{I}\right)\left(\frac{Q}{V}\right) = \left(\frac{\not V}{\not Q/t}\right)\left(\frac{\not Q}{\not V}\right) = t$$

Its symbol is the Greek letter τ (tau), and its unit of measure is the second. Thus,

$$\tau = RC \qquad \text{(seconds, s)} \qquad (10.14)$$

TABLE 10.4
i_C versus ϕ (charging phase).

t	Magnitude	
0	100%	
1τ	36.8%	
2τ	13.5%	
3τ	5.0%	
4τ	1.8%	
5τ	**0.67%** ←	Less than 1% of maximum
6τ	0.24%	

TABLE 10.5
Change in i_C between time constants.

$(0 \rightarrow 1)\tau$	63.2%	
$(1 \rightarrow 2)\tau$	23.3%	
$(2 \rightarrow 3)\tau$	8.6%	
$(3 \rightarrow 4)\tau$	3.0%	
$(4 \rightarrow 5)\tau$	**1.2%**	
$(5 \rightarrow 6)\tau$	0.4%	← Less than 1%

If we substitute $\tau = RC$ into the exponential function $e^{-t/RC}$, we obtain $e^{-t/\tau}$. In one time constant, $e^{-t/\tau} = e^{-\tau/\tau} = e^{-1} = 0.3679$, or the function equals 36.79% of its maximum value of 1. At $t = 2\tau$, $e^{-t/\tau} = e^{-2\tau/\tau} = e^{-2} = 0.1353$, and the function has decayed to only 13.53% of its maximum value.

The magnitude of $e^{-t/\tau}$ and the percentage change between time constants are shown in Tables 10.4 and 10.5, respectively. Note that the current has dropped 63.2% (100% − 36.8%) in the first time constant but only 0.4% between the fifth and sixth time constants. The rate of change of i_C is therefore quite sensitive to the time constant determined by the network parameters R and C. For this reason, the universal time constant chart of Fig. 10.27 is provided to allow you to estimate more accurately the value of the function e^{-x} for specific time intervals related to the time constant. The term *universal* is used because the axes are not scaled to specific values.

Returning to Eq. (10.13), we find that the multiplying factor E/R is the maximum value the current i_C can attain, as shown in Fig. 10.22. Substituting $t = 0$ s into Eq. (10.13) gives

$$i_C = \frac{E}{R}e^{-t/RC} = \frac{E}{R}e^{-0} = \frac{E}{R}$$

verifying our earlier conclusion.

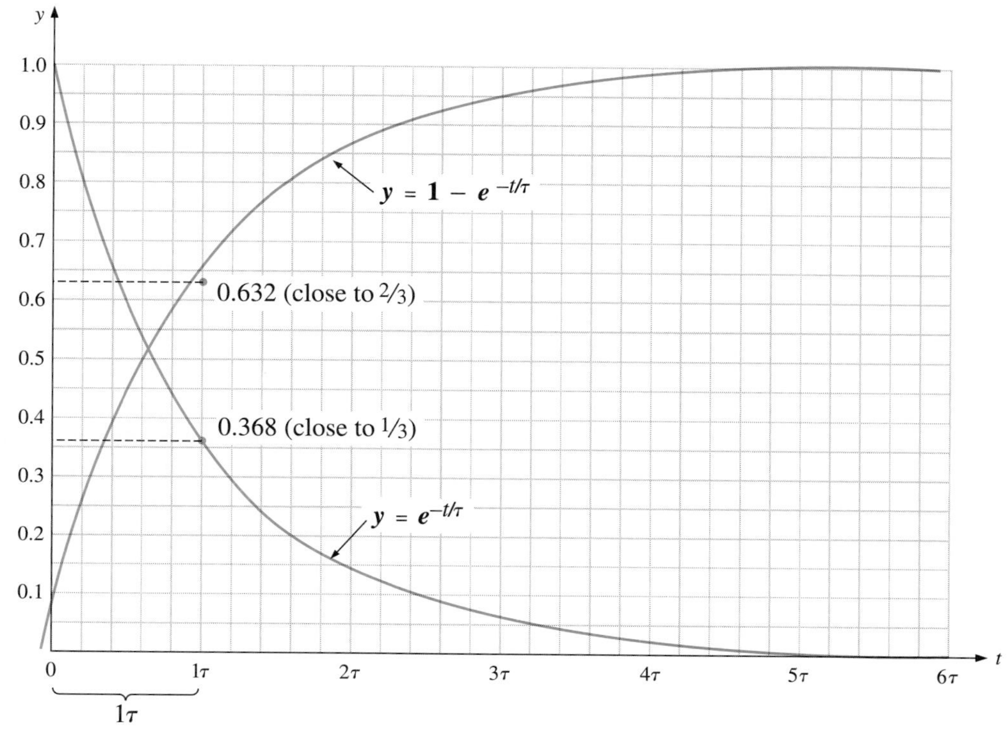

FIG. 10.27
Universal time constant chart.

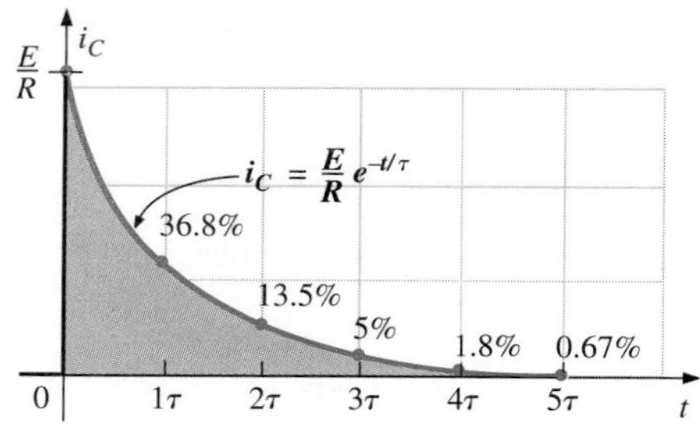

FIG. 10.28
i_C versus t during the charging phase.

For increasing values of *t*, the magnitude of $e^{-t/\tau}$, and therefore the value of i_C, will decrease, as shown in Fig. 10.28. Since the magnitude of i_C is less than 1% of its maximum after five time constants, we will assume for future analysis that:

The current iC of a capacitive network is essentially zero after five time constants of the charging phase have passed in a dc network.

Since C is usually found in microfarads or picofarads, the time constant $\tau = RC$ will never be greater than a few seconds unless R is very large.

Let us now turn our attention to the charging voltage across the capacitor. Through further mathematical analysis, the following equation for the voltage across the capacitor can be determined:

$$v_C = E(1 - e^{-t/RC}) \qquad (10.15)$$

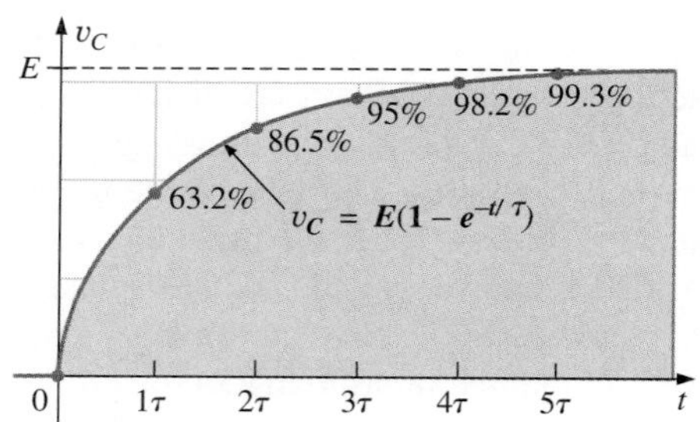

FIG. 10.29
v_C versus t during the charging phase.

Note the presence of the same factor $e^{-t/RC}$ and the function $(1 - e^{-t/RC})$ appearing in Fig. 10.27. Since $e^{-t/\tau}$ is a decaying function, the factor $(1 - e^{-t/\tau})$ will grow toward a maximum value of 1 with time, as shown in Fig. 10.27. In addition, since E is the multiplying factor, we can conclude that, for all practical purposes, the voltage v_C is E volts after five time constants of the charging phase. A plot of v_C versus t is provided in Fig. 10.29.

If we keep R constant and reduce C, the product RC will decrease, and the rise time of five time constants will decrease. The change in transient behaviour of the voltage v_C is plotted in Fig. 10.30 for various values of C. The product RC will always have some numerical value, even though it may be very small in some cases. For this reason:

The voltage across a capacitor cannot change instantaneously.

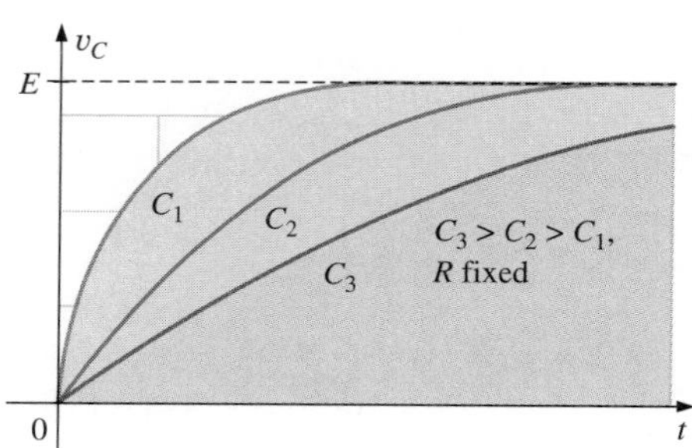

FIG. 10.30
Effect of C on the charging phase.

In fact, the capacitance of a network is also a measure of how much it will oppose a change in voltage across the network. The larger the capacitance, the larger the time constant and the longer it takes to charge up to its final value (curve of C_3 in Fig. 10.30). A lesser capacitance would permit the voltage to build up more quickly since the time constant is less (curve of C_1 in Fig. 10.30).

The rate at which charge is deposited on the plates during the charging phase can be found by substituting the following for v_C in Eq. (10.15):

$$v_C = \frac{q}{C}$$

and

$$q = Cv_C = CE(1 - e^{-t/\tau}) \quad \text{charging} \qquad (10.16)$$

indicating that the charging rate is very high during the first few time constants and less than 1% after five time constants.

The voltage across the resistor is determined by Ohm's law:

$$v_R = i_R R = Ri_C = R\frac{E}{R}e^{-t/\tau}$$

or

$$v_R = Ee^{-t/\tau} \qquad (10.17)$$

A plot of v_R appears in Fig. 10.31.

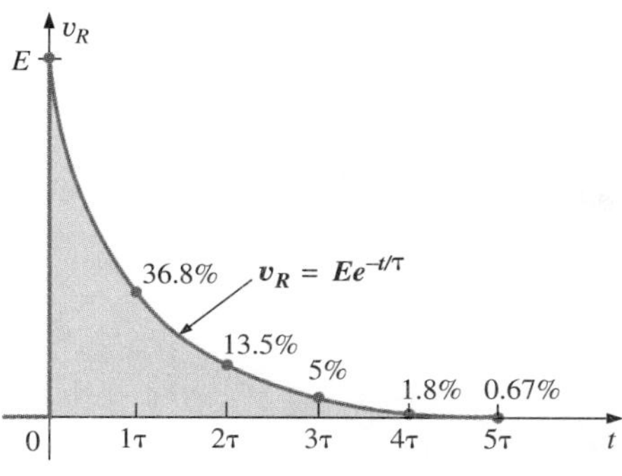

FIG. 10.31
v_R versus t during the charging phase.

Applying Kirchhoff's voltage law to the circuit of Fig. 10.21 will result in

$$v_C = E - v_R$$

Substituting Eq. (10.17):

$$v_C = E - Ee^{-t/\tau}$$

Factoring gives $v_C = E(1 - e^{-t/\tau})$, as obtained earlier.

EXAMPLE 10.5

a. Find the mathematical expressions for the transient behaviour of v_C, i_C, and v_R for the circuit of Fig. 10.32 when the switch is moved to position 1. Plot the curves of v_C, i_C, and v_R.
b. How much time must pass before it can be assumed, for all practical purposes, that $i_C \cong 0$ A and $v_C \cong E$ volts?

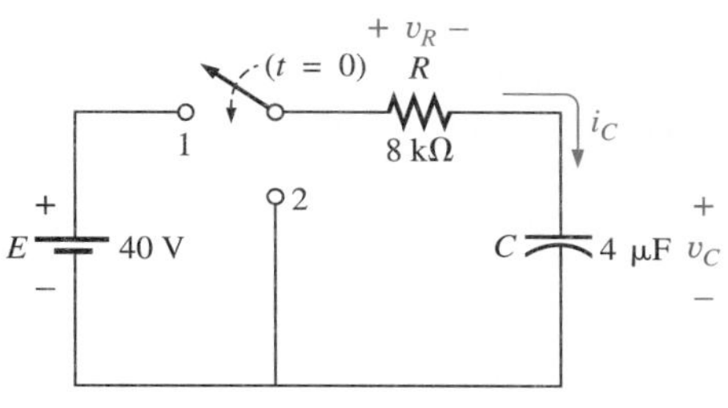

FIG. 10.32
Example 10.5.

Solutions:

a. $\tau = RC = (8 \times 10^3\ \Omega)(4 \times 10^{-6}\ \text{F}) = 32 \times 10^{-3}\ \text{s} = \mathbf{32\ ms}$

By Eq. (10.15),

$$v_C = E(1 - e^{-t/\tau}) = \mathbf{40(1 - e^{-t/(32\times10^{-3})})}$$

By Eq. (10.13),

$$i_C = \frac{E}{R}e^{-t/\tau} = \frac{40\ \text{V}}{8\ \text{k}\Omega}e^{-t/(32\times10^{-3})}$$
$$= \mathbf{(5 \times 10^{-3})e^{-t/(32\times10^{-3})}}$$

By Eq. (10.17),

$$v_R = Ee^{-t/\tau} = \mathbf{40e^{-t/(32\times10^{-3})}}$$

The curves appear in Fig. 10.33.

b. $5\tau = 5(32\ \text{ms}) = \mathbf{160\ ms}$

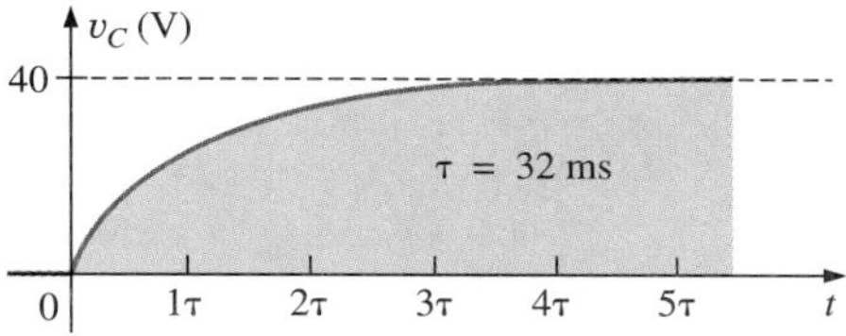

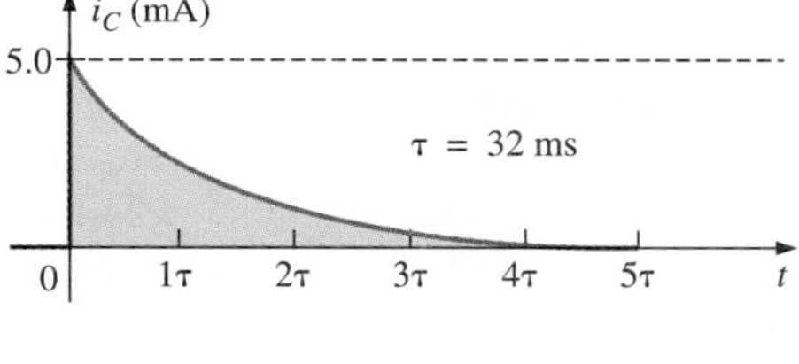

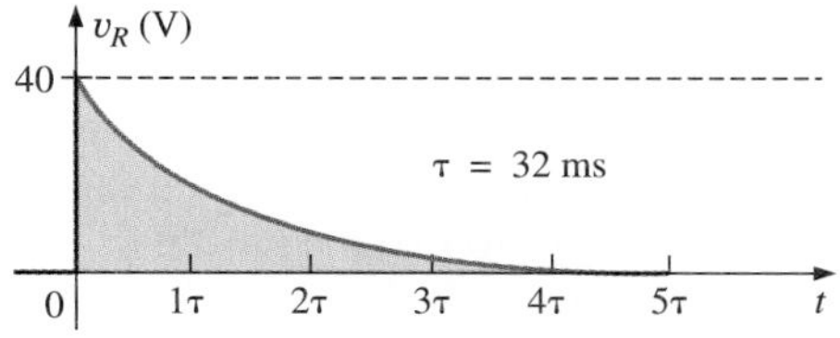

FIG. 10.33
Waveforms for the network of Fig. 10.32.

Once the voltage across the capacitor has reached the input voltage *E*, the capacitor is fully charged and will remain in this state if no further changes are made in the circuit.

If the switch of Fig. 10.21 is opened, as shown in Fig. 10.34(a), the capacitor will retain its charge for a period of time determined by its leakage current. For capacitors such as the mica and ceramic, the leakage current ($i_{\text{leakage}} = v_C/R_{\text{leakage}}$) is very small, enabling the capacitor to retain its charge, and hence the potential difference across its plates, for a long time. For electrolytic capacitors, which have very high leakage currents, the capacitor will discharge more rapidly, as shown in Fig. 10.34(b). In any event, to ensure that they are completely discharged, capacitors should be shorted by a lead or a screwdriver before they are handled.

10.8 DISCHARGE PHASE

The network of Fig. 10.21 is designed to both charge and discharge the capacitor. When the switch is placed in position 1, the capacitor will charge toward the supply voltage, as described in the last section. At any point in the charging process, if the switch is moved to position 2, the capacitor will begin to discharge at a rate sensitive to the same time

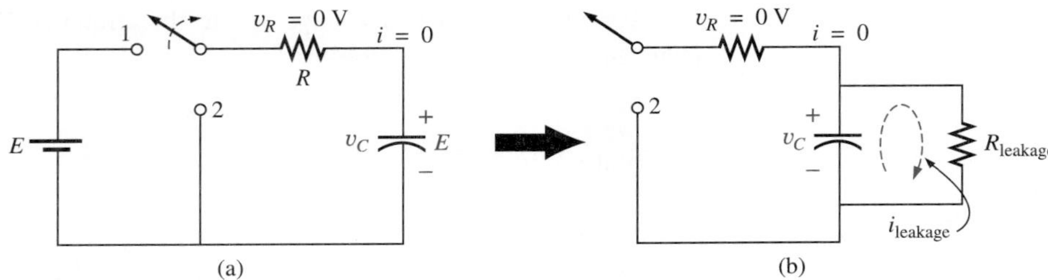

FIG. 10.34
Effect of the leakage current on the steady-state behaviour of a capacitor.

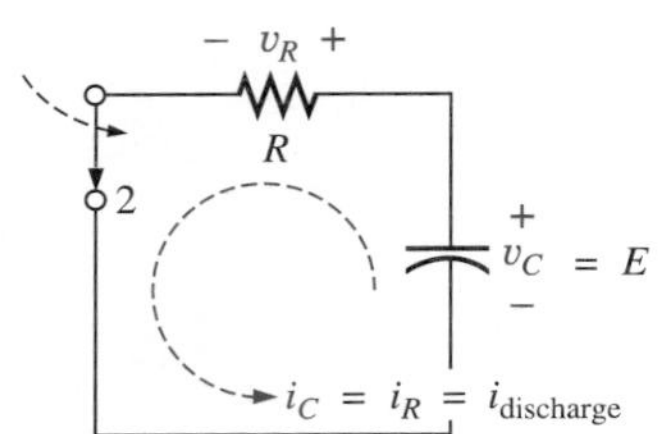

FIG. 10.35
Demonstrating the discharge behaviour of a capacitive network.

constant $\tau = RC$. The established voltage across the capacitor will create a flow of charge in the closed path that will eventually discharge the capacitor completely. The capacitor functions like a battery with a decreasing terminal voltage. Note in particular that the current i_C has reversed direction, changing the polarity of the voltage across R.

If the capacitor had charged to the full battery voltage as indicated in Fig. 10.35, the equation for the decaying voltage across the capacitor would be the following:

$$v_C = Ee^{-t/RC} \quad \textit{discharging} \tag{10.18}$$

which uses the function e^{-x} and the same time constant used above. The resulting curve will have the same shape as the curve for i_C and v_R in the last section. During the discharge phase, the current i_C will also decrease with time, as defined by the following equation:

$$i_C = \frac{E}{R}e^{-t/RC} \quad \textit{discharging} \tag{10.19}$$

The voltage $v_R = v_C$, and

$$v_R = Ee^{-t/RC} \quad \textit{discharging} \tag{10.20}$$

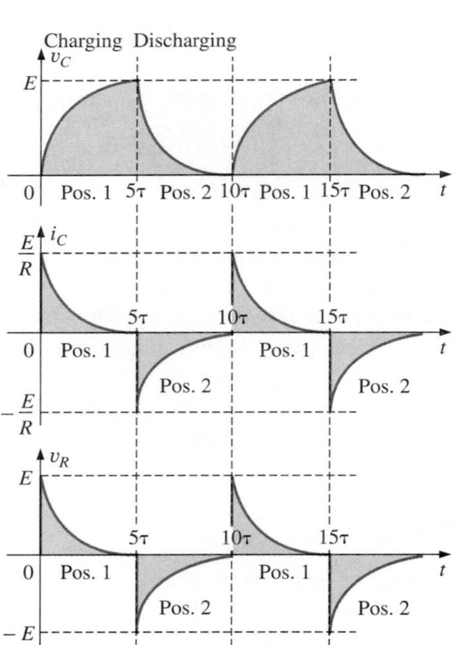

FIG. 10.36
The charging and discharging cycles for the network of Fig. 10.21.

The complete discharge will occur, for all practical purposes, in five time constants. If the switch is moved between terminals 1 and 2 every five time constants, the waveshapes of Fig. 10.36 will result for v_C, i_C, and v_R. For each curve, the current direction and voltage polarities were defined by Fig. 10.21. Since the polarity of v_C is the same for both the charging and discharging phases, the entire curve lies above the axis. The current i_C reverses direction during the charging and discharging phases, producing a negative pulse for both the current and the voltage v_R. Note that the voltage v_C never changes magnitude instantaneously but that the current i_C has the ability to change instantaneously, as demonstrated by its vertical rises and drops to maximum values.

EXAMPLE 10.6

After v_C in Example 10.5 has reached its final value of 40 V, the switch is thrown into position 2, as shown in Fig. 10.37. Find the mathematical expressions for the transient behaviour of v_C, i_C, and v_R after the closing of the switch. Plot the curves for v_C, i_C, and v_R using the defined directions and polarities of Fig. 10.32. Assume $t = 0$ when the switch is moved to position 2.

continued

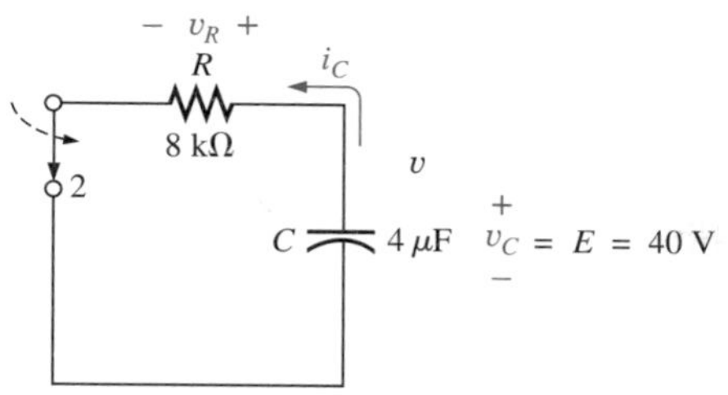

FIG. 10.37
Example 10.6.

Solution:

$$\tau = 32 \text{ ms}$$

By Eq. (10.18),

$$v_C = Ee^{-t/\tau} = \mathbf{40}e^{-t/(32\times10^{-3})}$$

By Eq. (10.19),

$$i_C = -\frac{E}{R}e^{-t/\tau} = -(5 \times 10^{-3})e^{-t/(32\times10^{-3})}$$

By Eq. (10.20),

$$v_R = -Ee^{-t/\tau} = -\mathbf{40}e^{-t/(32\times10^{-3})}$$

The curves appear in Fig. 10.38.

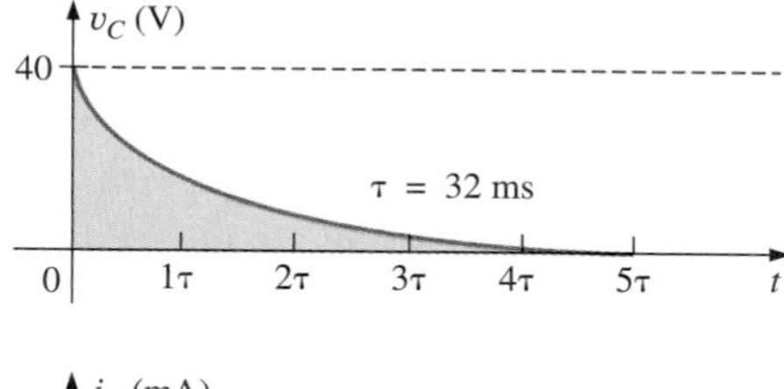

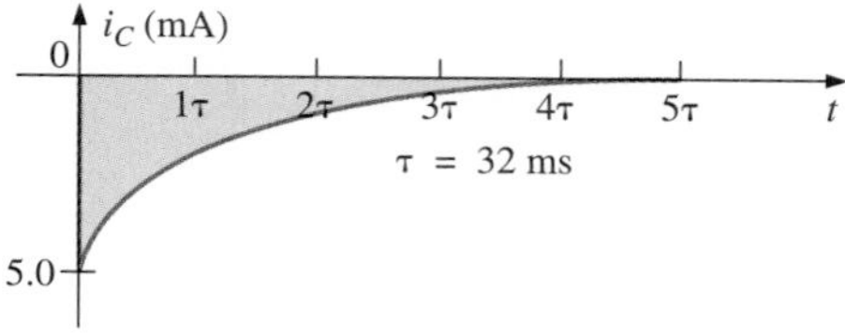

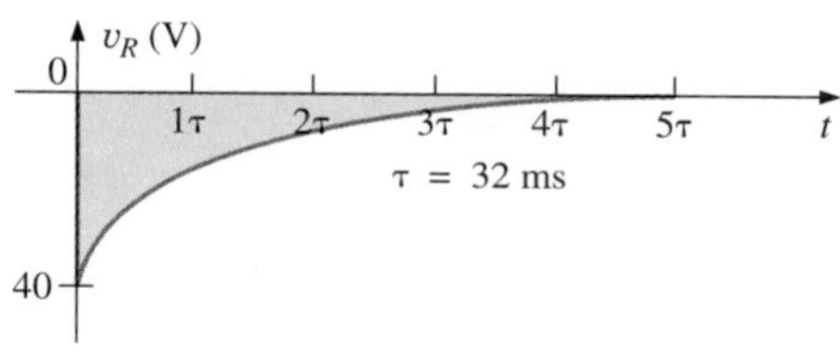

FIG. 10.38
The waveforms for the network of Fig. 10.37.

The preceding discussion and examples apply to situations in which the capacitor charges to the battery voltage. If the charging phase is disrupted before reaching the supply voltage, the capacitive voltage will be less, and the equation for the discharging voltage v_C will take on the form

$$v_C = V_i e^{-t/RC} \tag{10.21}$$

where V_i is the starting or initial voltage for the discharge phase. The equation for the decaying current is also modified by simply substituting V_i for E; that is,

$$i_C = \frac{V_i}{R}e^{-t/\tau} = I_i e^{-t/\tau} \tag{10.22}$$

Use of the above equations will be demonstrated in Examples 10.7 and 10.8.

EXAMPLE 10.7

a. Find the mathematical expression for the transient behaviour of the voltage across the capacitor of Fig. 10.39 if the switch is thrown into position 1 at $t = 0$ s.
b. Repeat part (a) for i_C.
c. Find the mathematical expressions for the response of v_C and i_C if the switch is thrown into position 2 at 30 ms (assuming the leakage resistance of the capacitor is infinite ohms).
d. Find the mathematical expressions for the voltage v_C and current i_C if the switch is thrown into position 3 at $t = 48$ ms.
e. Plot the waveforms obtained in parts (a) through (d) on the same time axis for the voltage v_C and the current i_C using the defined polarity and current direction of Fig. 10.39.

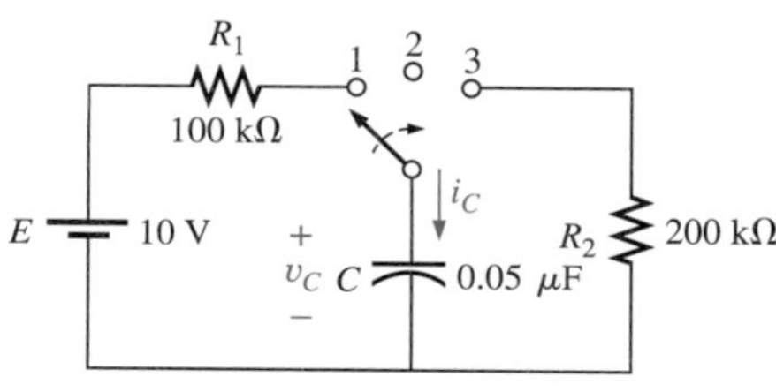

FIG. 10.39
Example 10.7.

Solutions:

a. Charging phase:

$$v_C = E(1 - e^{-t/\tau})$$

$$\tau = R_1C = (100 \times 10^3\ \Omega)(0.05 \times 10^{-6}\text{ F}) = 5 \times 10^{-3}\text{ s}$$

$$= 5 \text{ ms}$$

$$v_C = \mathbf{10(1 - e^{-t/(5\times10^{-3})})}$$

continued

b. $i_C = \frac{E}{R_1} e^{-t/\tau}$

$$= \frac{10 \text{ V}}{100 \times 10^3 \ \Omega} e^{-t/(5 \times 10^{-3})}$$

$$i_C = \mathbf{(0.1 \times 10^{-3}) e^{-t/(5 \times 10^{-3})}}$$

c. Storage phase:

$$v_C = E = \mathbf{10 \ V}$$
$$i_C = \mathbf{0 \ A}$$

d. Discharge phase (starting at 48 ms with $t = 0$ s for the following equations):

$$v_C = Ee^{-t/\tau'}$$
$$\tau' = R_2C = (200 \times 10^3 \ \Omega)(0.05 \times 10^{-6} \text{ F}) = 10 \times 10^{-3} \text{ s}$$
$$= 10 \text{ ms}$$
$$v_C = \mathbf{10e^{-t/(10 \times 10^{-3})}}$$
$$i_C = -\frac{E}{R_2} e^{-t/\tau'}$$
$$= -\frac{10 \text{ V}}{200 \times 10^3 \ \Omega} e^{-t/(10 \times 10^{-3})}$$
$$i_C = \mathbf{-(0.05 \times 10^{-3}) e^{-t/(10 \times 10^{-3})}}$$

e. See Fig. 10.40.

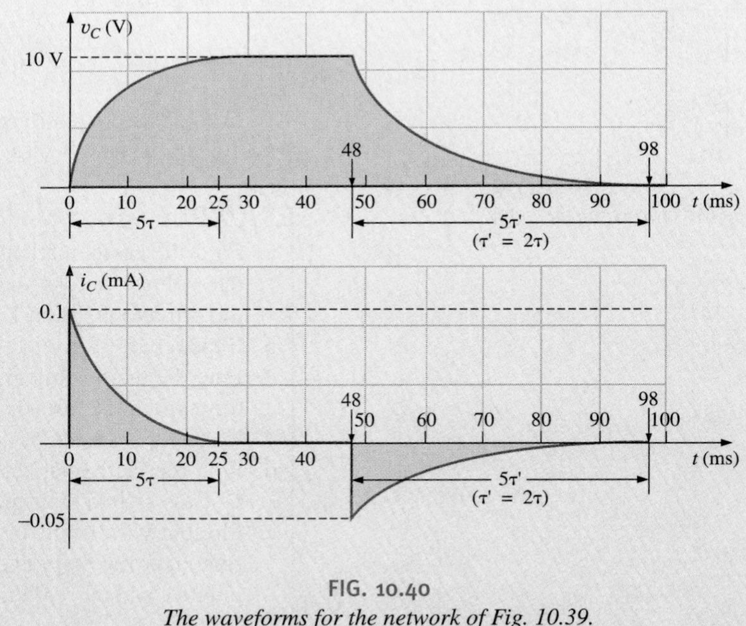

FIG. 10.40
The waveforms for the network of Fig. 10.39.

EXAMPLE 10.8

a. Find the mathematical expression for the transient behaviour of the voltage across the capacitor of Fig. 10.41 if the switch is thrown into position 1 at $t = 0$ s.

continued

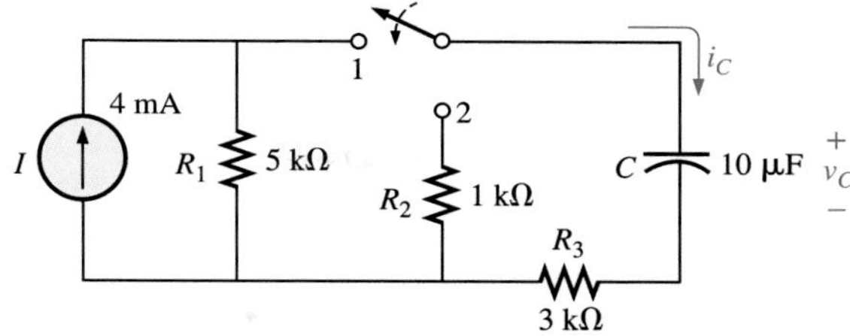

FIG. 10.41
Example 10.8.

b. Repeat part (a) for i_C.
c. Find the mathematical expression for the response of v_C and i_C if the switch is thrown into position 2 at $t = 1\tau$ of the charging phase.
d. Plot the waveforms obtained in parts (a) through (c) on the same time axis for the voltage v_C and the current i_C using the defined polarity and current direction of Fig. 10.41.

Solutions:

a. *Charging phase:* Converting the current source to a voltage source will result in the network of Fig. 10.42.

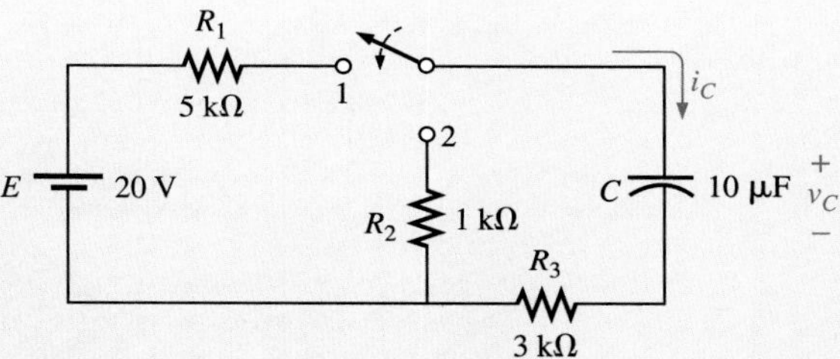

FIG. 10.42
The charging phase for the network of Fig. 10.41.

$$v_C = E(1 - e^{-t/\tau_1})$$

$$\tau_1 = (R_1 + R_3)C = (5\text{ k}\Omega + 3\text{ k}\Omega)(10 \times 10^{-6}\text{ F}) = 80\text{ ms}$$

$$\mathbf{v_C = 20(1 - e^{-t/(80\times 10^{-3})})}$$

b. $$i_C = \frac{E}{R_1 + R_3}e^{-t/\tau_1} = \frac{20\text{ V}}{8\text{ k}\Omega}e^{-t/(80\times 10^{-3})}$$

$$\mathbf{i_C = (2.5 \times 10^{-3})e^{-t/(80\times 10^{-3})}}$$

c. At $t = 1\tau_1$, $v_C = 0.632E = 0.632(20\text{ V}) = 12.64\text{ V}$, resulting in the network of Fig. 10.43. Then $v_C = V_i e^{-t/\tau_2}$

with $\tau_2 = (R_2 + R_3)C = (1\text{ k}\Omega + 3\text{ k}\Omega)(10 \times 10^{-6}\text{ F}) = 40\text{ ms}$

and $\mathbf{v_C = 12.64e^{-t/(40\times 10^{-3})}}$

continued

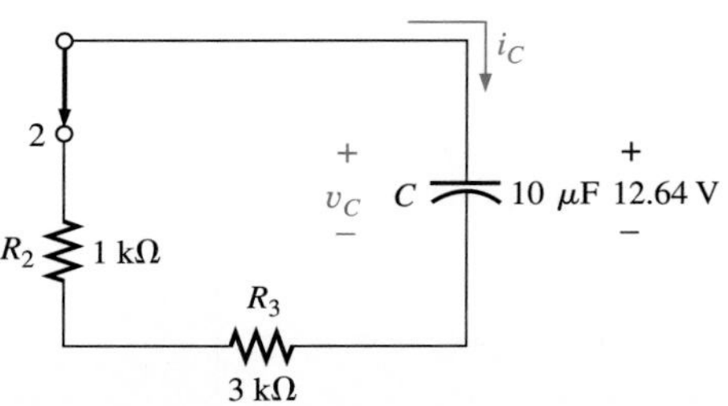

FIG. 10.43
Network of Fig. 10.42 when the switch is moved to position 2 at $t = 1\tau_1$.

At $t = 1\tau_1$, i_C drops to (0.368)(2.5 mA) = 0.92 mA. Then it switches to

$$i_C = -I_i e^{-t/\tau_2}$$

$$= -\frac{V_i}{R_2 + R_3} e^{-t/\tau_2} = -\frac{12.64\ \text{V}}{1\ \text{k}\Omega + 3\ \text{k}\Omega} e^{-t/(40\times 10^{-3})}$$

$$i_C = \mathbf{-3.16 \times 10^{-3} e^{-t/(40\times 10^{-3})}}$$

d. See Fig. 10.44.

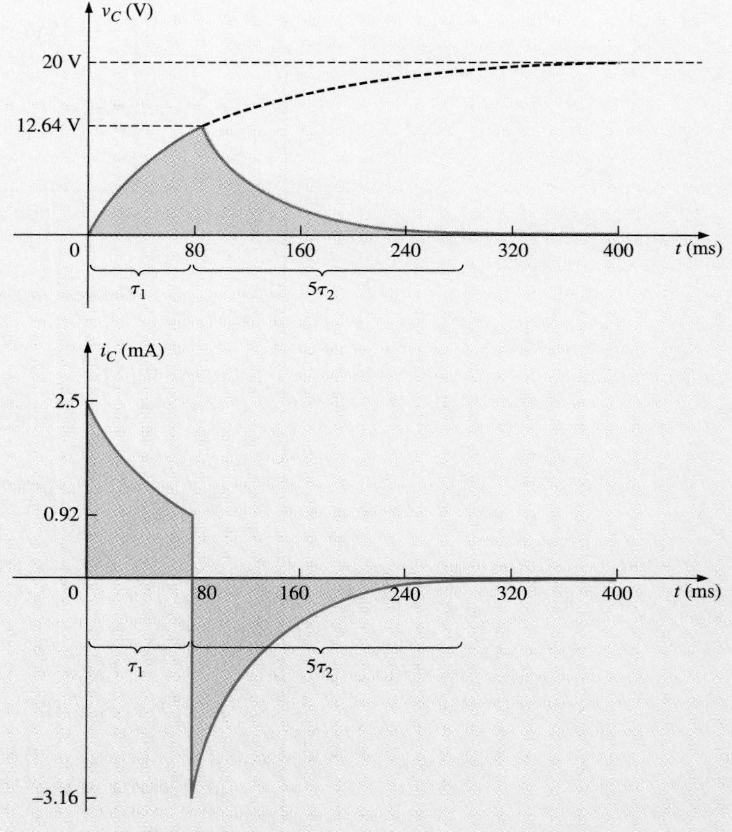

FIG. 10.44
The waveforms for the network of Fig. 10.41.

10.9 INITIAL VALUES

In all the examples examined in the previous sections, the capacitor was uncharged before the switch was thrown. We will now examine the effect of a charge, and therefore a voltage ($V = Q/C$), on the plates at the instant the switching action takes place. The voltage across the capacitor at this instant is called the **initial value**, as shown for the general waveform of Fig. 10.45. Once the switch is thrown, the transient phase will begin until a leveling off occurs after 5 time constants. This region of relatively fixed value that follows the transient response is called the **steady-state region**, and the resulting value is called the **steady-state** or **final value**. The steady-state value is found by simply substituting the open-circuit equivalent for the capacitor and finding the voltage across the plates. Using the transient equation developed in Section 10.7,

$$v_c = V(1 - e^{-t/\tau})$$

an equation for the voltage v_C can be written for the entire time interval of Fig. 10.45; that is,

$$v_c = V_i + (V_f - V_i)(1 - e^{-t/\tau})$$

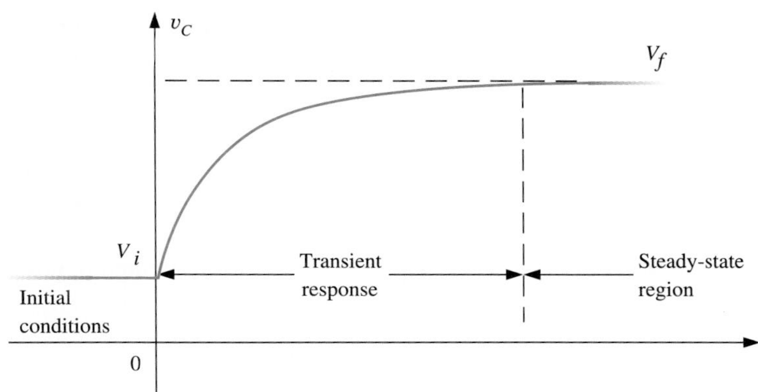

FIG. 10.45
Defining the regions associated with a transient response.

However, by multiplying through and rearranging terms:

$$\begin{aligned} v_C &= V_i + V_f - V_f e^{-t/\tau} - V_i + V_i e^{-t/\tau} \\ &= V_f - V_f e^{-t/\tau} + V_i e^{-t/\tau} \end{aligned}$$

we find

$$v_C = V_f + (V_i - V_f)e^{-t/\tau} \qquad (10.23)$$

If you are required to draw the waveform for the voltage v_C from the initial value to the final value, start by drawing a line at the initial and steady-state levels and then add the transient response (sensitive to the time constant) between the two levels. The example to follow will clarify the procedure.

EXAMPLE 10.9

The capacitor of Fig. 10.46 has an initial voltage of 4 volts.

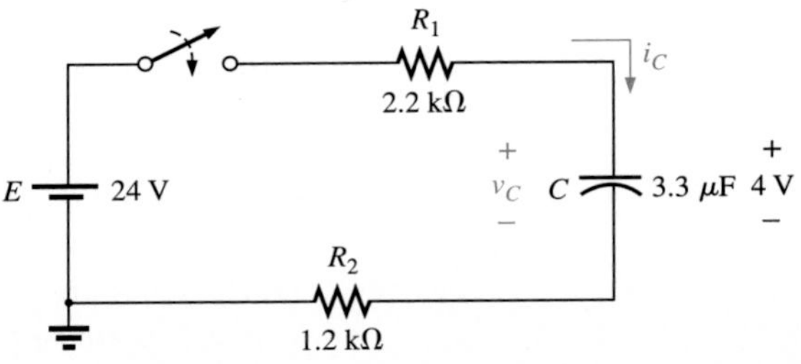

FIG. 10.46
Example 10.9.

a. Find the mathematical expression for the voltage across the capacitor once the switch is closed.
b. Find the mathematical expression for the current during the transient period.
c. Sketch the waveform for each from initial value to final value.

Solution:

a. Substituting the open-circuit equivalent for the capacitor will result in a final or steady-state voltage v_C of 24 V.

The time constant is determined by

$$\tau = (R_1 + R_2)C$$
$$= (2.2\text{ k}\Omega + 1.2\text{ k}\Omega)(3.3\ \mu\text{F})$$
$$= 11.22\text{ ms}$$

with $$5\tau = 56.1\text{ ms}$$

Applying Eq. (10.23):

$$v_C = V_f + (V_i - V_f)e^{-t/\tau}$$
$$= 24\text{ V} + (4\text{ V} - 24\text{ V})e^{-t/11.22\text{ms}}$$

and $$v_C = \mathbf{24\ V - 20\ V}e^{-t/11.22\text{ms}}$$

b. Since the voltage across the capacitor is constant at 4 V before the closing of the switch, the current (whose level is sensitive only to changes in voltage across the capacitor) must have an initial value of 0 mA. At the instant the switch is closed, the voltage across the capacitor cannot change instantaneously, so the voltage across the resistive elements at this instant is the applied voltage less the initial voltage across the capacitor. The resulting peak current is

$$I_m = \frac{E - V_C}{R_1 + R_2} = \frac{24\text{ V} - 4\text{ V}}{2.2\text{ k}\Omega + 1.2\text{ k}\Omega} = \frac{20\text{ V}}{3.4\text{ k}\Omega} = 5.88\text{ mA}$$

The current will then decay (with the same time constant as the voltage v_C) to zero because the capacitor is approaching its open-circuit equivalence.

The equation for i_C is therefore:

$$i_C = \mathbf{5.88\ mA}e^{-t/11.22\text{ms}}$$

c. See Fig. 10.47.

continued

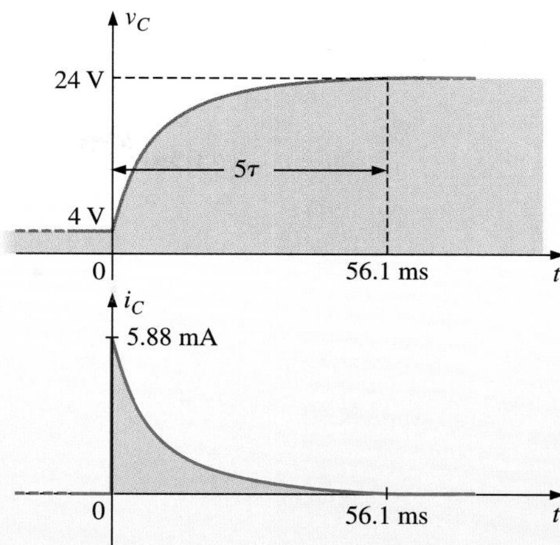

FIG. 10.47
v_C and i_C for the network of Fig. 10.46.

The initial and final values of the voltage were drawn first and then the transient response was included between these levels. For the current, the waveform begins and ends at zero, with the peak value having a sign sensitive to the defined direction of i_C in Fig. 10.46.

Let us now test the validity of the equation for v_C by substituting $t = 0$ s to reflect the instant the switch is closed.

$$e^{-t/\tau} = e^{-0} = 1$$

and $$v_C = 24\text{ V} - 20\text{ V}e^{-t/\tau} = 24\text{ V} - 20\text{ V} = 4\text{ V}$$

When $$t > 5\tau,\ e^{-t/\tau} \cong 0$$

and $$v_C = 24\text{ V} - 20\text{ V}e^{-t/\tau} = 24\text{ V} - 0\text{ V} = 24\text{ V}$$

10.10 INSTANTANEOUS VALUES

You will sometimes have to determine the voltage or current at a particular instant of time that is not an integral multiple of τ, as in the previous sections. For example, if

$$v_C = 20(1 - e^{-t/(2\times 10^{-3})})$$

the voltage v_C may be required at $t = 5$ ms, which does not correspond to an integral multiple of τ. Figure 10.27 reveals that $(1 - e^{-t/\tau})$ is approximately 0.93 at $t = 5\text{ ms} = 2.5\tau$, resulting in $v_C = 20(0.93) = 18.6$ V. You can get additional accuracy simply by substituting $t = 5$ ms into the equation and solving for v_C using a calculator or table to determine $e^{-2.5}$. Thus,

$$\begin{aligned} v_C &= 20(1 - e^{-5\text{ms}/2\text{ms}}) \\ &= 20(1 - e^{-2.5}) \\ &= 20(1 - 0.082) \\ &= 20(0.918) \\ &= \mathbf{18.36\ V} \end{aligned}$$

The results are close, but accuracy beyond the tenths' place is suspect using Fig. 10.27. The above procedure can also be applied to any other equation introduced in this chapter for currents or other voltages.

You will also sometimes have to find the time to reach a particular voltage or current. The procedure is complicated somewhat by the use of natural logs ($\log_e$, or ln), but today's calculators are equipped to handle the operation with ease. There are two forms that require some development. First, consider the following sequence:

$$v_C = E(1 - e^{-t/\tau})$$

$$\frac{v_C}{E} = 1 - e^{-t/\tau}$$

$$1 - \frac{v_C}{E} = e^{-t/\tau}$$

$$\log_e\left(1 - \frac{v_C}{E}\right) = \log_e e^{-t/\tau}$$

$$\log_e\left(1 - \frac{v_C}{E}\right) = -\frac{t}{\tau}$$

and

$$t = -\tau \log_e\left(1 - \frac{v_C}{E}\right)$$

but

$$-\log_e\frac{x}{y} = +\log_e\frac{y}{x}$$

Therefore,

$$t = \tau \log_e\left(\frac{E}{E - v_C}\right) \tag{10.24}$$

The second form is as follows:

$$v_C = Ee^{-t/\tau}$$

$$\frac{v_C}{E} = e^{-t/\tau}$$

$$\log_e\frac{v_C}{E} = \log_e e^{-t/\tau}$$

$$\log_e\frac{v_C}{E} = -\frac{t}{\tau}$$

and

$$t = -\tau \log_e\frac{v_C}{E}$$

or

$$t = \tau \log_e\frac{E}{v_C} \tag{10.25}$$

For $i_C = (E/R)e^{-t/\tau}$,

$$t = \tau \log_e\frac{E}{i_C R} \tag{10.26}$$

For example, suppose

$$v_C = 20(1 - e^{-t/(2\times10^{-3})})$$

and the time to reach 10 V is required. Substituting into Eq. (10.24), we have

$$t = (2\text{ ms})\log_e\left(\frac{20\text{ V}}{20\text{ V} - 10\text{ V}}\right)$$

$$= (2\ \text{ms})\log_e 2$$
$$= (2\ \text{ms})(0.693)$$
$$= \mathbf{1.386\ ms}$$ ← [ln] key on calculator

Using Fig. 10.27, we find at $(1 - e^{-t/\tau}) = v_C/E = 0.5$ that $t \cong 0.7\tau = 0.7(2\ \text{ms}) = 1.4$ ms, which is relatively close to the above.

10.11 $\tau = R_{Th}C$

Sometimes the network will not have the simple series form of Fig. 10.21. In these cases you will first have to find the Thévenin equivalent circuit for the network external to the capacitive element. E_{Th} will then be the source voltage E of Eqs. (10.15) through (10.20) and R_{Th} will be the resistance R. The time constant is then $\tau = R_{Th}C$.

EXAMPLE 10.10

For the network of Fig. 10.48:

a. Find the mathematical expression for the transient behaviour of the voltage v_C and the current i_C following the closing of the switch (position 1 at $t = 0$ s).

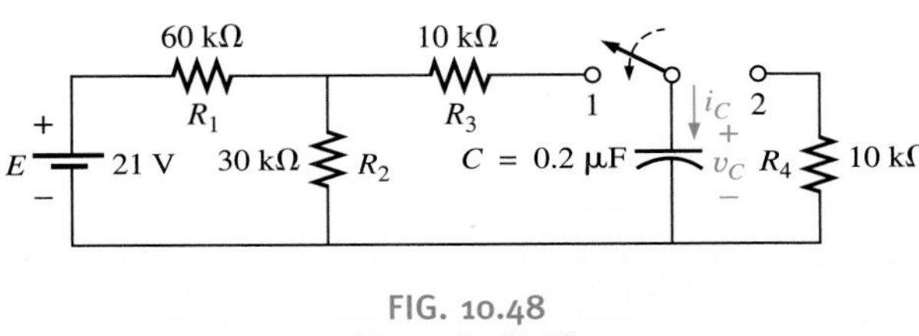

FIG. 10.48
Example 10.10.

b. Find the mathematical expression for the voltage v_C and current i_C as a function of time if the switch is thrown into position 2 at $t = 9$ ms.
c. Draw the resulting waveforms of parts (a) and (b) on the same time axis.

Solutions:

a. Applying Thévenin's theorem to the 0.2-μF capacitor, we obtain Fig. 10.49:

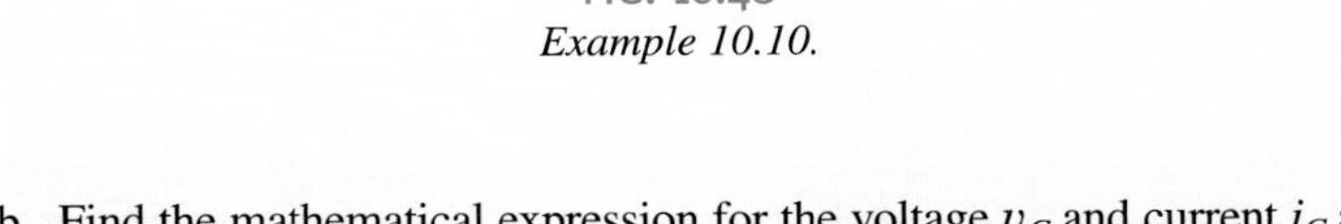

$$R_{Th} = R_1 \parallel R_2 + R_3 = \frac{(60\ \text{k}\Omega)(30\ \text{k}\Omega)}{90\ \text{k}\Omega} + 10\ \text{k}\Omega$$
$$= 20\ \text{k}\Omega + 10\ \text{k}\Omega$$
$$R_{Th} = 30\ \text{k}\Omega$$
$$E_{Th} = \frac{R_2E}{R_2 + R_1} = \frac{(30\ \text{k}\Omega)(21\ \text{V})}{30\ \text{k}\Omega + 60\ \text{k}\Omega} = \frac{1}{3}(21\ \text{V}) = 7\ \text{V}$$

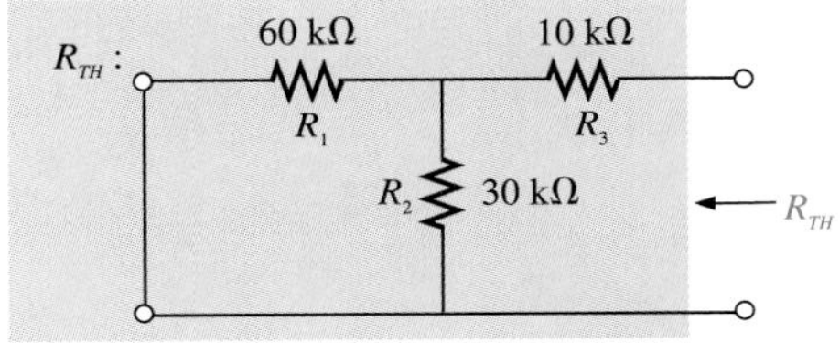

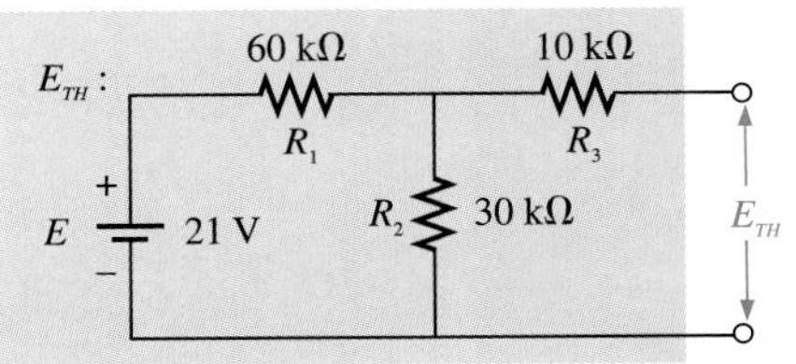

FIG. 10.49
Applying Thévenin's theorem to the network of Fig. 10.48.

The resultant Thévenin equivalent circuit with the capacitor replaced is shown in Fig. 10.50:

$$v_C = E_{Th}(1 - e^{-t/\tau})$$
$$\tau = R_{Th}C = (30\ \text{k}\Omega)(0.2\ \mu\text{F})$$
$$= (30 \times 10^3\ \Omega)(0.2 \times 10^{-6}\ \text{F}) = 6 \times 10^{-3}\ \text{s}$$

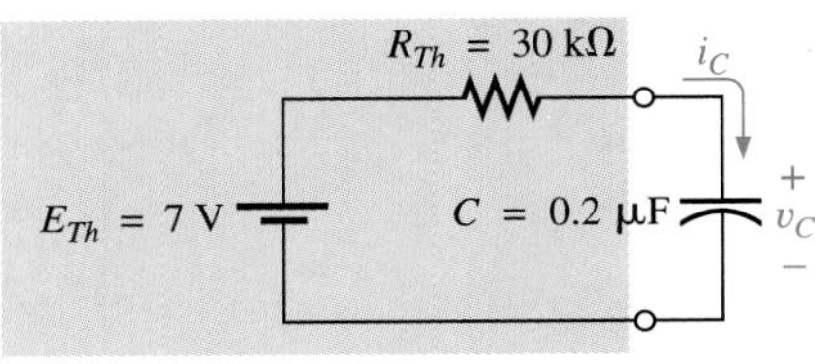

FIG. 10.50
Substituting the Thévenin equivalent for the network of Fig. 10.48.

continued

$$\tau = 6 \text{ ms}$$
$$\upsilon_C = \mathbf{7(1 - e^{-t/(6\times10^{-3})})}$$

and
$$i_C = \frac{E_{Th}}{R}e^{-t/RC}$$
$$= \frac{7\text{ V}}{30\text{ k}\Omega}e^{-t/(6\times10^{-3})}$$
$$i_C = \mathbf{(0.233\times10^{-3})e^{-t/(6\times10^{-3})}}$$

b. At t = 9 ms,

$$\upsilon_C = E_{Th}(1 - e^{-t/\tau}) = 7(1 - e^{-(9\times10^{-3})/(6\times10^{-3})})$$
$$= 7(1 - e^{-1.5}) = 7(1 - 0.223)$$
$$\upsilon_C = 7(0.777) = 5.44\text{ V}$$
$$i_C = \frac{E_{Th}}{R}e^{-t/\tau} = (0.233\times10^{-3})e^{-1.5}$$
$$= (0.233\times10^{-3})(0.223)$$
$$i_C = 0.052\times10^{-3} = 0.052\text{ mA}$$

By Eq. (10.21),
$$\upsilon_C = V_i e^{-t/\tau'}$$

with
$$\tau' = R_4C = (10\times10^3\ \Omega)(0.2\times10^{-6}\text{ F}) = 2\times10^{-3}\text{ s}$$
$$= 2\text{ ms}$$

and
$$\upsilon_C = \mathbf{5.44e^{-t/(2\times10^{-3})}}$$

By Eq. (10.22),
$$I_i = \frac{5.44\text{ V}}{10\text{ k}\Omega} = 0.054\text{ mA}$$

and
$$i_C = I_i e^{-t/\tau} = \mathbf{-(0.54\times10^{-3})e^{-t/(2\times10^{-3})}}$$

c. See Fig. 10.51.

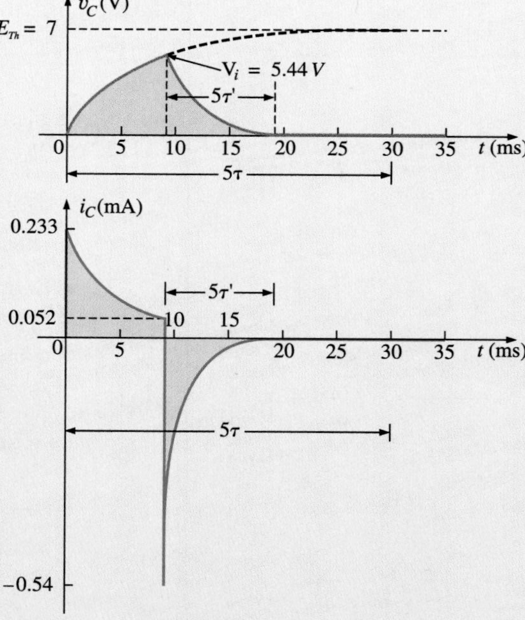

FIG. 10.51
The resulting waveforms for the network of Fig. 10.48.

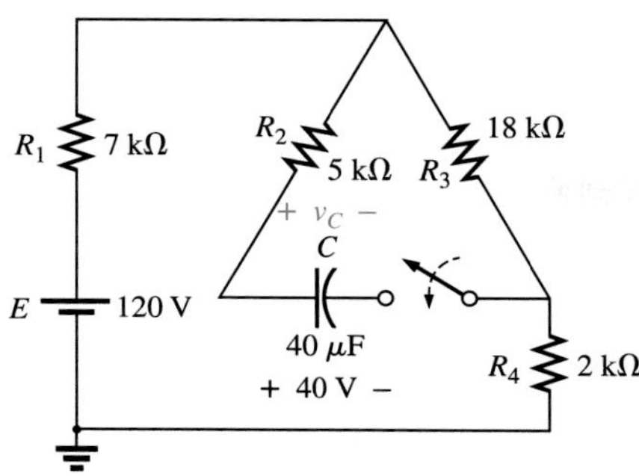

FIG. 10.52
Example 10.11.

EXAMPLE 10.11

The capacitor of Fig. 10.52 is initially charged to 40 V. Find the mathematical expression for v_C after the closing of the switch.

Solution: The network is redrawn in Fig. 10.53:

E_{Th}:

$$E_{Th} = \frac{R_3E}{R_3 + R_1 + R_4} = \frac{18\ \text{k}\Omega(120\ \text{V})}{18\ \text{k}\Omega + 7\ \text{k}\Omega + 2\ \text{k}\Omega}$$
$$= 80\ \text{V}$$

R_{Th}:

$$R_{Th} = 5\ \text{k}\Omega + 18\ \text{k}\Omega \parallel (7\ \text{k}\Omega + 2\ \text{k}\Omega)$$
$$= 5\ \text{k}\Omega + 6\ \text{k}\Omega$$
$$= 11\ \text{k}\Omega$$

Therefore, $$V_i = 40\ \text{V},\ V_f = 80\ \text{V}$$

and $$\tau = R_{Th}C = (11\ \text{k}\Omega)(40\ \mu\text{F}) = 0.44\ \text{s}$$

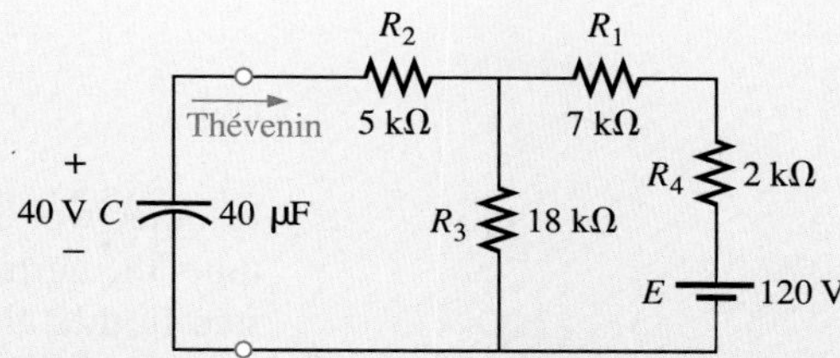

FIG. 10.53
Network of Fig. 10.52 redrawn.

Eq. (10.23): $$v_C = V_f + (V_i - V_f)e^{-t/\tau}$$
$$= 80\ \text{V} + (40\ \text{V} - 80\ \text{V})e^{-t/0.44\text{s}}$$

and $$v_C = \mathbf{80\ V - 40\ V}e^{-t/0.44\text{s}}$$

EXAMPLE 10.12

For the network of Fig. 10.54, find the mathematical expression for the voltage v_C after the closing of the switch (at $t = 0$).

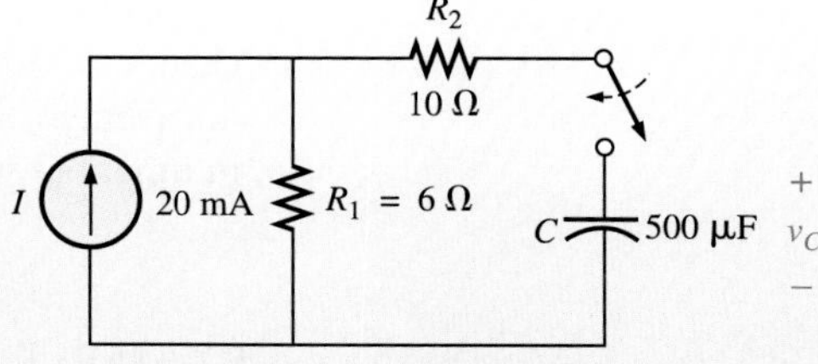

FIG. 10.54
Example 10.12.

Solution:

$$R_{Th} = R_1 + R_2 = 6\ \Omega + 10\ \Omega = 16\ \Omega$$
$$E_{Th} = V_1 + V_2 = IR_1 + 0$$
$$= (20 \times 10^{-3}\ \text{A})(6\ \Omega) = 120 \times 10^{-3}\ \text{V} = 0.12\ \text{V}$$

continued

and $\tau = R_{Th}C = (16\ \Omega)(500 \times 10^{-6}\ \text{F}) = 8\ \text{ms}$

so that $v_C = \mathbf{0.12(1 - e^{-t/(8\times10^{-3})})}$

10.12 THE CURRENT i_C

The current i_C associated with a capacitance C is related to the voltage across the capacitor by

$$i_C = C\frac{dv_C}{dt} \tag{10.27}$$

where dv_C/dt is a measure of the change in v_C in a vanishingly small period of time. The function dv_C/dt is called the **derivative** of the voltage v_C with respect to time t.

If the voltage fails to change at a particular instant, then

$$dv_C = 0$$

and

$$i_C = C\frac{dv_C}{dt} = 0$$

In other words, if the voltage across a capacitor fails to change with time, the current i_C associated with the capacitor is zero. To take this a step further, the equation also states that the more rapid the change in voltage across the capacitor, the greater the resulting current.

To make Eq. (10.27) clearer, let us calculate the average current associated with a capacitor for various voltages impressed across the capacitor. The average current is defined by the equation

$$i_{C\text{av}} = C\frac{\Delta v_C}{\Delta t} \tag{10.28}$$

where Δ indicates a finite (measurable) change in charge, voltage, or time. The instantaneous current can be derived from Eq. (10.28) by letting Δt become vanishingly small; that is,

$$i_{C\text{inst}} = \lim_{\Delta t \to 0} C\frac{\Delta v_C}{\Delta t} = C\frac{dv_C}{dt}$$

In the following example, the change in voltage Δv_C will be considered for each slope of the voltage waveform. If the voltage increases with time, the average current is the change in voltage divided by the change in time, with a positive sign. If the voltage decreases with time, the average current is again the change in voltage divided by the change in time, but with a negative sign.

EXAMPLE 10.13

Find the waveform for the average current if the voltage across a 2-μF capacitor is as shown in Fig. 10.55.

continued

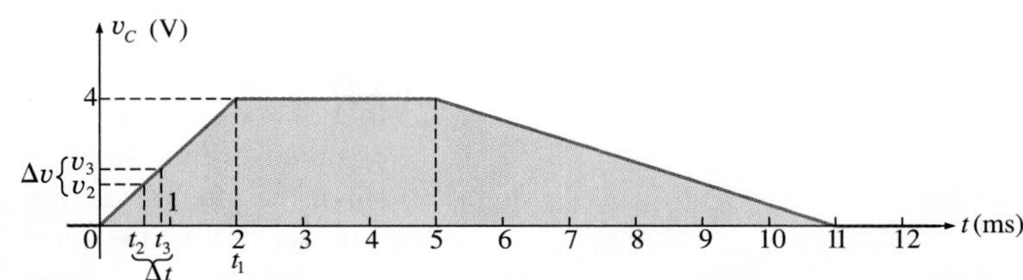

FIG. 10.55
Example 10.13.

Solution:

a. From 0 to 2 ms, the voltage increases linearly from 0 to 4 V, the change in voltage $\Delta v = 4\text{ V} - 0 = 4\text{ V}$ (with a positive sign since the voltage increases with time). The change in time $\Delta t = 2\text{ ms} - 0 = 2\text{ ms}$, and

$$i_{C_{av}} = C\frac{\Delta v_C}{\Delta t} = (2 \times 10^{-6}\text{ F})\left(\frac{4\text{ V}}{2 \times 10^{-3}\text{ s}}\right)$$
$$= 4 \times 10^{-3}\text{ A} = 4\text{ mA}$$

b. From 2 to 5 ms, the voltage remains constant at 4 V; the change in voltage $\Delta v = 0$. The change in time $\Delta t = 3$ ms, and

$$i_{C_{av}} = C\frac{\Delta v_C}{\Delta t} = C\frac{0}{\Delta t} = 0$$

c. From 5 to 11 ms, the voltage decreases from 4 to 0 V. The change in voltage Δv is, therefore, $4\text{ V} - 0 = 4\text{ V}$ (with a negative sign since the voltage is decreasing with time). The change in time $\Delta t = 11\text{ ms} - 5\text{ ms} = 6\text{ ms}$, and

$$i_{C_{av}} = C\frac{\Delta v_C}{\Delta t} = -(2 \times 10^{-6}\text{ F})\left(\frac{4\text{ V}}{6 \times 10^{-3}\text{ s}}\right)$$
$$= -1.33 \times 10^{-3}\text{ A} = -1.33\text{ mA}$$

d. From 11 ms on, the voltage remains constant at 0 and $\Delta v = 0$, so $i_{C_{av}} = 0$. The waveform for the average current for the impressed voltage is as shown in Fig. 10.56.

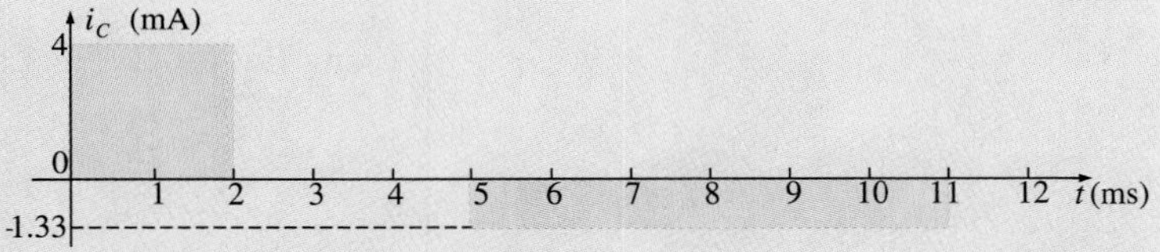

FIG. 10.56
The resulting current i_C for the applied voltage of Fig. 10.55.

Note in Example 10.13 that, in general, the steeper the slope, the greater the current, and when the voltage fails to change, the current is zero. In addition, the average value is the same as the instantaneous value at any point along the slope over which the average value was found. For example, if the interval Δt is reduced from $0 \rightarrow t_1$ to $t_2 - t_3$, as noted in Fig. 10.55, $\Delta v/\Delta t$ is still the same. In fact, no matter how small the interval Δt, the slope will be the same, and therefore the current $i_{C_{av}}$ will be the same. If we consider the limit as $\Delta t \rightarrow 0$, the slope will still remain the same, and therefore $i_{C_{av}} = i_{C_{inst}}$ at any instant of time between 0 and t_1. The same can be said about any portion of the voltage waveform that has a constant slope.

An important point to understand from this discussion is that it is not the magnitude of the voltage across a capacitor that determines the current but rather how quickly the voltage *changes* across the capacitor. An applied steady dc voltage of 10 000 V would (ideally) not create any flow of charge (current), but a change in voltage of 1 V in a very brief period of time could create a significant current.

The method described above is only for waveforms with straight-line (linear) segments. For nonlinear (curved) waveforms, a method of **calculus** (differentiation) must be used.

10.13 CAPACITORS IN SERIES AND PARALLEL

Capacitors, like resistors, can be placed in series and in parallel. Increasing levels of capacitance can be obtained by placing capacitors in parallel, while decreasing levels can be obtained by placing capacitors in series.

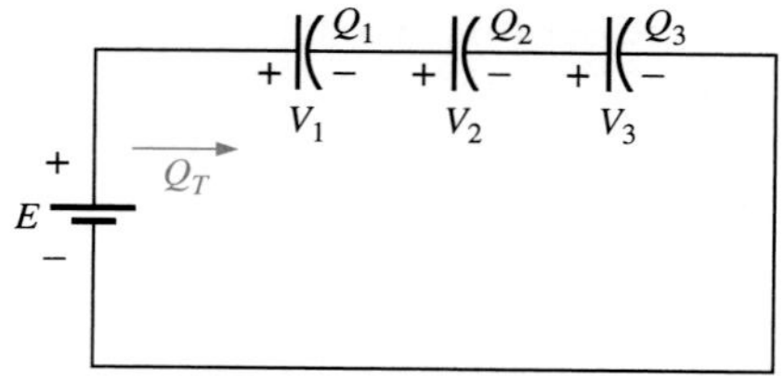

FIG. 10.57
Series capacitors.

For capacitors in series, the charge is the same on each capacitor (Fig. 10.57):

$$Q_T = Q_1 = Q_2 = Q_3 \qquad (10.29)$$

Applying Kirchhoff's voltage law around the closed loop gives

$$E = V_1 + V_2 + V_3$$

However,

$$V = \frac{Q}{C}$$

so that

$$\frac{Q_T}{C_T} = \frac{Q_1}{C_1} + \frac{Q_2}{C_2} + \frac{Q_3}{C_3}$$

Using Eq. (10.29) and dividing both sides by Q gives

$$\frac{1}{C_T} = \frac{1}{C_1} + \frac{1}{C_2} + \frac{1}{C_3} \qquad (10.30)$$

which is similar to the way we found the total resistance of a parallel resistive circuit. The total capacitance of two capacitors in series is

$$C_T = \frac{C_1 C_2}{C_1 + C_2} \qquad (10.31)$$

continued

The voltage across each capacitor of Fig. 10.57 can be found by first recognizing that

$$Q_T = Q_1$$

or

$$C_T E = C_1 V_1$$

Solving for V_1:

$$V_1 = \frac{C_T E}{C_1}$$

and substituting for C_T:

$$V_1 = \frac{\frac{1}{C_1}(E)}{\frac{1}{C_1} + \frac{1}{C_2} + \frac{1}{C_3}} \qquad (10.32)$$

A similar equation will result for each capacitor of the network.

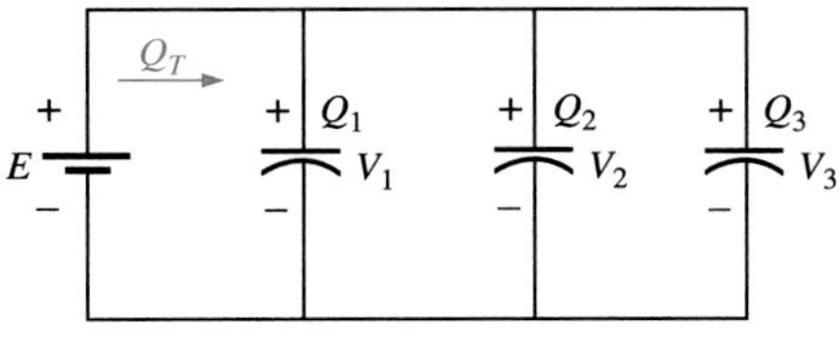

FIG. 10.58
Parallel capacitors.

For capacitors in parallel, as shown in Fig. 10.58, the voltage is the same across each capacitor, and the total charge is the sum of that on each capacitor:

$$Q_T = Q_1 + Q_2 + Q_3 \qquad (10.33)$$

However,

$$Q = CV$$

Therefore,

$$C_T E = C_1 V_1 + C_2 V_2 + C_3 V_3$$

but

$$E = V_1 = V_2 = V_3$$

Thus,

$$C_T = C_1 + C_2 + C_3 \qquad (10.34)$$

which is similar to the way that the total resistance of a series circuit is found.

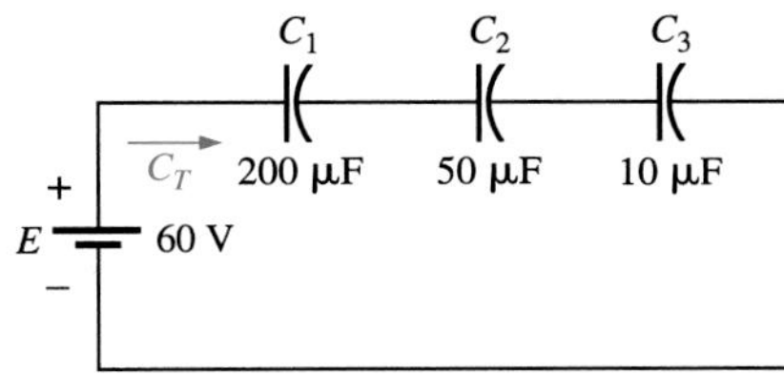

FIG. 10.59
Example 10.14.

EXAMPLE 10.14

For the circuit of Fig. 10.59:

a. Find the total capacitance.
b. Determine the charge on each plate.
c. Find the voltage across each capacitor.

Solutions:

a. $\frac{1}{C_T} = \frac{1}{C_1} + \frac{1}{C_2} + \frac{1}{C_3}$

$$= \frac{1}{200 \times 10^{-6}\,\text{F}} + \frac{1}{50 \times 10^{-6}\,\text{F}} + \frac{1}{10 \times 10^{-6}\,\text{F}}$$

$$= 0.005 \times 10^{6} + 0.02 \times 10^{6} + 0.1 \times 10^{6}$$

$$= 0.125 \times 10^{6}$$

continued

and $$C_T = \frac{1}{0.125 \times 10^6} = \mathbf{8\ \mu F}$$

b. $Q_T = Q_1 = Q_2 = Q_3$

$$Q_T = C_T E = (8 \times 10^{-6}\ \text{F})(60\ \text{V}) = \mathbf{480\ \mu C}$$

c. $$V_1 = \frac{Q_1}{C_1} = \frac{480 \times 10^{-6}\ \text{C}}{200 \times 10^{-6}\ \text{F}} = \mathbf{2.4\ V}$$

$$V_2 = \frac{Q_2}{C_2} = \frac{480 \times 10^{-6}\ \text{C}}{50 \times 10^{-6}\ \text{F}} = \mathbf{9.6\ V}$$

$$V_3 = \frac{Q_3}{C_3} = \frac{480 \times 10^{-6}\ \text{C}}{10 \times 10^{-6}\ \text{F}} = \mathbf{48.0\ V}$$

and $$E = V_1 + V_2 + V_3 = 2.4\ \text{V} + 9.6\ \text{V} + 48\ \text{V} = \mathbf{60\ V} \quad \text{(checks)}$$

EXAMPLE 10.15

For the network of Fig. 10.60:

a. Find the total capacitance.
b. Determine the charge on each plate.
c. Find the total charge.

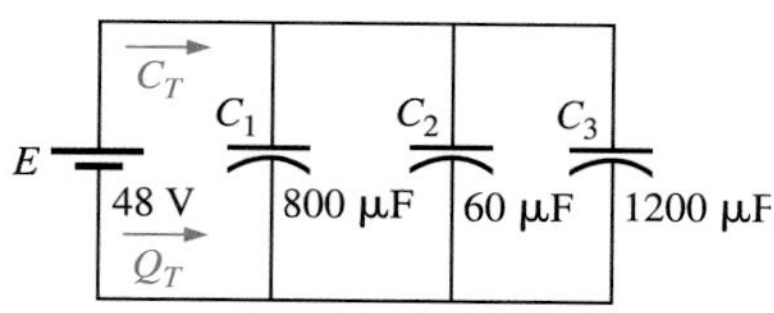

FIG. 10.60
Example 10.15.

Solutions:

a. $$C_T = C_1 + C_2 + C_3 = 800\ \mu\text{F} + 60\ \mu\text{F} + 1200\ \mu\text{F} = \mathbf{2060\ \mu F}$$

b. $$Q_1 = C_1 E = (800 \times 10^{-6}\ \text{F})(48\ \text{V}) = \mathbf{38.4\ mC}$$
$$Q_2 = C_2 E = (60 \times 10^{-6}\ \text{F})(48\ \text{V}) = \mathbf{2.88\ mC}$$
$$Q_3 = C_3 E = (1200 \times 10^{-6}\ \text{F})(48\ \text{V}) = \mathbf{57.6\ mC}$$

c. $$Q_T = Q_1 + Q_2 + Q_3 = 38.4\ \text{mC} + 2.88\ \text{mC} + 57.6\ \text{mC} = \mathbf{98.88\ mC}$$

EXAMPLE 10.16

Find the voltage across and charge on each capacitor for the network of Fig. 10.61.

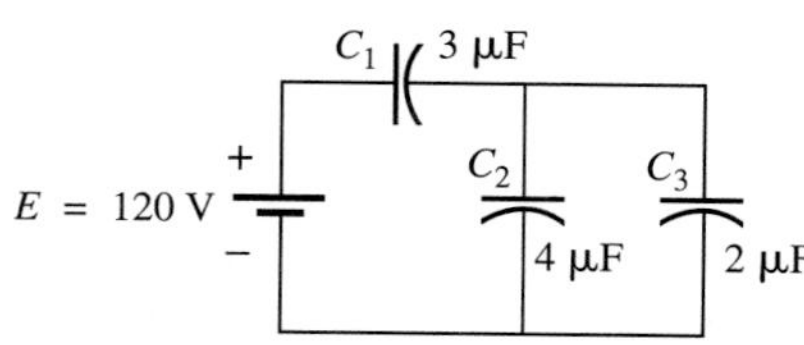

FIG. 10.61
Example 10.16.

Solutions:

$$C'_T = C_2 + C_3 = 4\ \mu\text{F} + 2\ \mu\text{F} = 6\ \mu\text{F}$$
$$C_T = \frac{C_1 C'_T}{C_1 + C'_T} = \frac{(3\ \mu\text{F})(6\ \mu\text{F})}{3\ \mu\text{F} + 6\ \mu\text{F}} = 2\ \mu\text{F}$$
$$Q_T = C_T E = (2 \times 10^{-6}\ \text{F})(120\ \text{V}) = \mathbf{240\ \mu C}$$

An equivalent circuit (Fig. 10.62) has

$$Q_T = Q_1 = Q'_T$$

continued

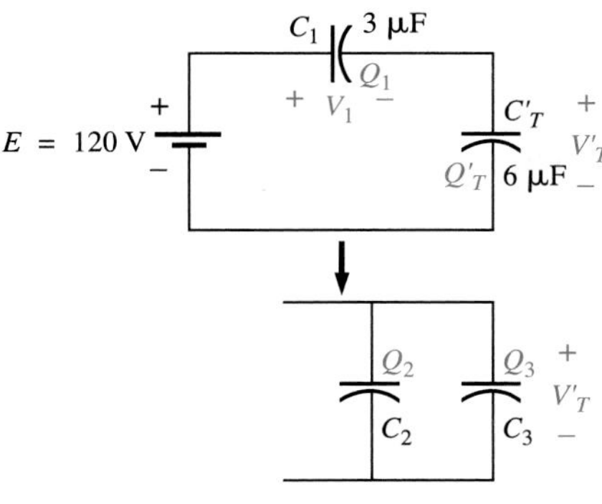

FIG. 10.62
Reduced equivalent for the network of Fig. 10.61.

and, therefore,
$$Q_1 = \mathbf{240\ \mu C}$$

and
$$V_1 = \frac{Q_1}{C_1} = \frac{240 \times 10^{-6}\ \text{C}}{3 \times 10^{-6}\ \text{F}} = \mathbf{80\ V}$$
$$Q'_T = 240\ \mu\text{C}$$

and, therefore,
$$V'_T = \frac{Q'_T}{C'_T} = \frac{240 \times 10^{-6}\ \text{C}}{6 \times 10^{-6}\ \text{F}} = \mathbf{40\ V}$$

and
$$Q_2 = C_2V'_T = (4 \times 10^{-6}\ \text{F})(40\ \text{V}) = \mathbf{160\ \mu C}$$
$$Q_3 = C_3V'_T = (2 \times 10^{-6}\ \text{F})(40\ \text{V}) = \mathbf{80\ \mu C}$$

EXAMPLE 10.17

Find the voltage across and charge on capacitor C_1 of Fig. 10.63 after it has charged up to its final value.

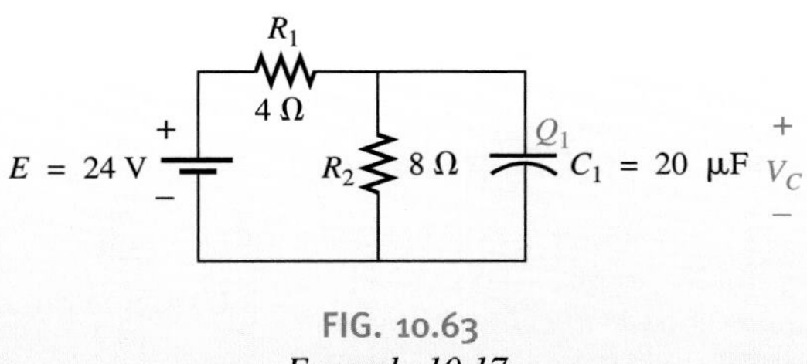

FIG. 10.63
Example 10.17.

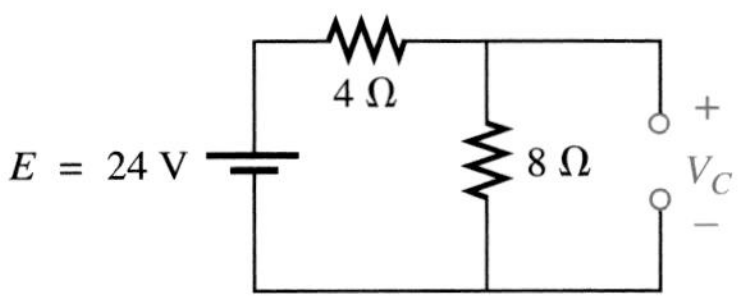

FIG. 10.64
Determining the final (steady-state) value for v_C.

Solution: As previously discussed, the capacitor is effectively an open circuit for dc after charging up to its final value (Fig. 10.64). Therefore,

$$V_C = \frac{(8\ \Omega)(24\ \text{V})}{4\ \Omega + 8\ \Omega} = \mathbf{16\ V}$$
$$Q_1 = C_1V_C = (20 \times 10^{-6}\ \text{F})(16\ \text{V}) = \mathbf{320\ \mu C}$$

EXAMPLE 10.18

Find the voltage across and charge on each capacitor of the network of Fig. 10.65 after each has charged up to its final value.

Solution:

$$V_{C_2} = \frac{(7\ \Omega)(72\ \text{V})}{7\ \Omega + 2\ \Omega} = \mathbf{56\ V}$$
$$V_{C_1} = \frac{(2\ \Omega)(72\ \text{V})}{2\ \Omega + 7\ \Omega} = \mathbf{16\ V}$$
$$Q_1 = C_1V_{C_1} = (2 \times 10^{-6}\ \text{F})(16\ \text{V}) = \mathbf{32\ \mu C}$$
$$Q_2 = C_2V_{C_2} = (3 \times 10^{-6}\ \text{F})(56\ \text{V}) = \mathbf{168\ \mu C}$$

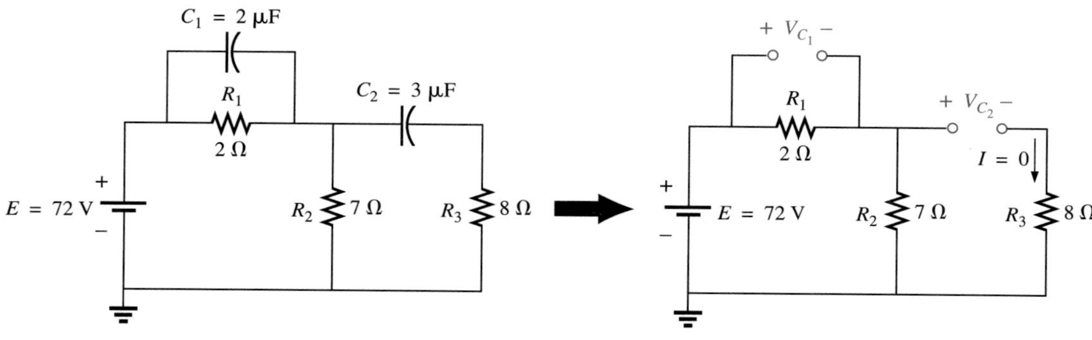

FIG. 10.65
Example 10.18.

10.14 ENERGY STORED BY A CAPACITOR

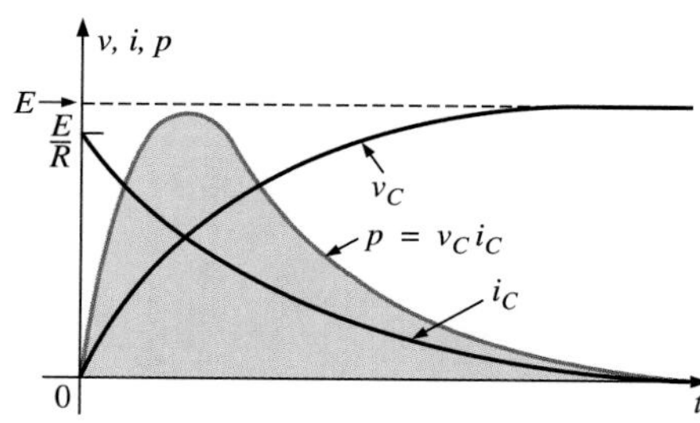

FIG. 10.66
Plotting the power to a capacitive element during the transient phase.

The ideal capacitor does not dissipate any of the energy supplied to it. It stores the energy in the form of an electric field between the conducting surfaces. A plot of the voltage, current, and power to a capacitor during the charging phase is shown in Fig. 10.66. The power curve can be obtained by finding the product of the voltage and current at selected instants of time and connecting the points obtained. The energy stored is represented by the shaded area under the power curve.

Using calculus, we can determine the area under the curve:

$$W_C = \frac{1}{2}CE^2$$

In general,

$$W_C = \frac{1}{2}CV^2 \qquad \text{(J)} \qquad (10.35)$$

where V is the steady-state voltage across the capacitor. In terms of Q and C,

$$W_C = \frac{1}{2}C\left(\frac{Q}{C}\right)^2$$

or

$$W_C = \frac{Q^2}{2C} \qquad \text{(J)} \qquad (10.36)$$

EXAMPLE 10.19

For the network of Fig. 10.65, determine the energy stored by each capacitor.

Solution:

For C_1, $W_C = \frac{1}{2}CV^2$

$$= \frac{1}{2}(2 \times 10^{-6}\ \text{F})(16\ \text{V})^2 = (1 \times 10^{-6})(256)$$

$$= \mathbf{256\ \mu J}$$

continued

For C_2, $W_C = \frac{1}{2}CV^2$

$$= \frac{1}{2}(3 \times 10^{-6}\ \text{F})(56\ \text{V})^2 = (1.5 \times 10^{-6})(3136)$$

$$= \mathbf{4704\ \mu J}$$

Due to the squared term, note the difference in energy stored because of a higher voltage.

PROBLEMS

SECTION 10.2 The Electric Field

1. Find the electric field strength at a point 2 m from a charge of 4 μC.
2. The electric field strength is 36 newtons/coulomb (N/C) at a point *r* metres from a charge of 0.064 μC. Find the distance *r*.
3. Determine the charge at 3 metres if the field strength is 50 N/C.

SECTION 10.3 Capacitance

4. Find the capacitance of a parallel plate capacitor if 1400 μC of charge are deposited on its plates when 20 V are applied across the plates.
5. How much charge is deposited on the plates of a 0.05-μF capacitor if 45 V are applied across the capacitor?
6. Find the electric field strength between the plates of a parallel plate capacitor if 100 mV are applied across the plates and the plates are 2 mm apart.
7. Determine the plate area for a 1-farad parallel plate capacitor in which the plates are separated by 10mm. The dielectric is air.
8. Repeat Problem 6 if the plates are separated by 100 μm.
9. A 4-μF parallel plate capacitor has 160 μC of charge on its plates. If the plates are 5 mm apart, find the electric field strength between the plates.
10. Find the capacitance of a parallel plate capacitor if the area of each plate is 0.075 m^2 and the distance between the plates is 1.77 mm. The dielectric is air.
11. Repeat Problem 10 if the dielectric is paraffin-coated paper.

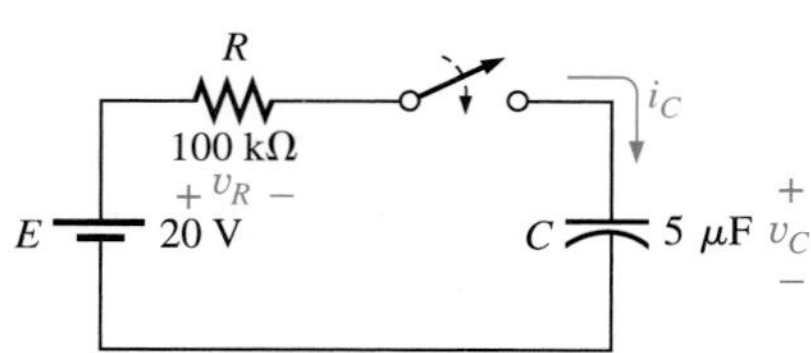

FIG. 10.67
Problems 19 and 20.

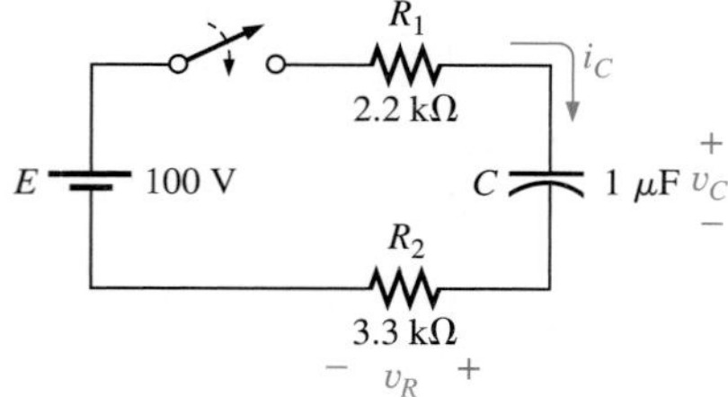

FIG. 10.68
Problem 21.

12. Find the distance in micrometres between the plates of a 2-μF capacitor if the area of each plate is 0.09 m^2 and the dielectric is transformer oil.
13. The capacitance of a capacitor with a dielectric of air is 1200 pF. When a dielectric is inserted between the plates, the capacitance increases to 0.006 μF. What material is the dielectric made of?
14. The plates of a parallel plate air capacitor are 0.2 mm apart and have an area of 0.08 m^2, and 200 V are applied across the plates.
    - **a.** Determine the capacitance.
    - **b.** Find the electric field intensity between the plates.
    - **c.** Find the charge on each plate if the dielectric is air.
15. A sheet of Bakelite 0.2 mm thick having an area of 0.08 m^2 is inserted between the plates of Problem 14.
    - **a.** Find the electric field strength between the plates.
    - **b.** Determine the charge on each plate.
    - **c.** Determine the capacitance.

SECTION 10.4 Dielectric Strength

16. Find the maximum voltage ratings of the capacitors of Problems 14 and 15 assuming a linear relationship between the breakdown voltage and the thickness of the dielectric.
17. Find the maximum voltage that can be applied across a parallel plate capacitor of 0.006 μF. The area of one plate is 0.02 m^2 and the dielectric is mica. Assume a linear relationship between the dielectric strength and the thickness of the dielectric.
18. Find the distance in millimetres between the plates of a parallel plate capacitor if the maximum voltage that can be applied across the capacitor is 1000 V. The dielectric is Teflon. Assume a linear relationship between the breakdown strength and the thickness of the dielectric.

SECTION 10.7 Transients in Capacitive Networks: Charging Phase

19. For the circuit of Fig. 10.67:
    - **a.** Determine the time constant of the circuit.
    - **b.** Write the mathematical equation for the voltage v_C following the closing of the switch.
    - **c.** Determine the voltage v_C after one, three, and five time constants.
    - **d.** Write the equations for the current i_C and the voltage v_R.
    - **e.** Sketch the waveforms for v_C and i_C.
20. Repeat Problem 19 for $R = 1\ \text{M}\Omega$ and compare the results.
21. For the circuit of Fig. 10.68:
    - **a.** Determine the time constant of the circuit.
    - **b.** Write the mathematical equation for the voltage v_C following the closing of the switch.
    - **c.** Determine v_C after one, three, and five time constants.
    - **d.** Write the equations for the current i_C and the voltage v_R.
    - **e.** Sketch the waveforms for v_C and i_C.

22. Repeat Problem 21 for a source voltage of 48V.

23. For the circuit of Fig. 10.69:
   a. Determine the time constant of the circuit.
   b. Write the mathematical equation for the voltage v_C following the closing of the switch.
   c. Write the mathematical expression for the current i_C following the closing of the switch.
   d. Sketch the waveforms of v_C and i_C.

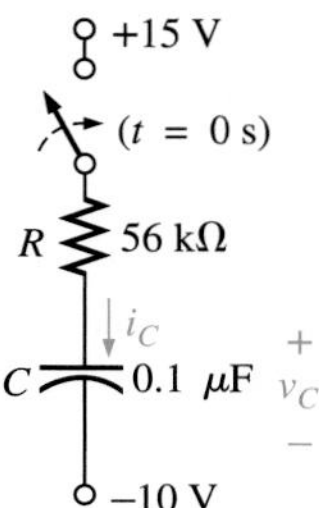

FIG. 10.69
Problem 23.

SECTION 10.8 Discharge Phase

24. For the circuit of Fig. 10.70:
   a. Determine the time constant of the circuit when the switch is thrown into position 1.
   b. Find the mathematical expression for the voltage across the capacitor after the switch is thrown into position 1.
   c. Determine the mathematical expression for the current following the closing of the switch (position 1).
   d. Determine the voltage v_C and the current i_C if the switch is thrown into position 2 at $t = 100$ ms.
   e. Determine the mathematical expressions for the voltage v_C and the current i_C if the switch is thrown into position 3 at $t = 200$ ms.
   f. Plot the waveforms of v_C and i_C for a period of time extending from $t = 0$ to $t = 300$ ms.

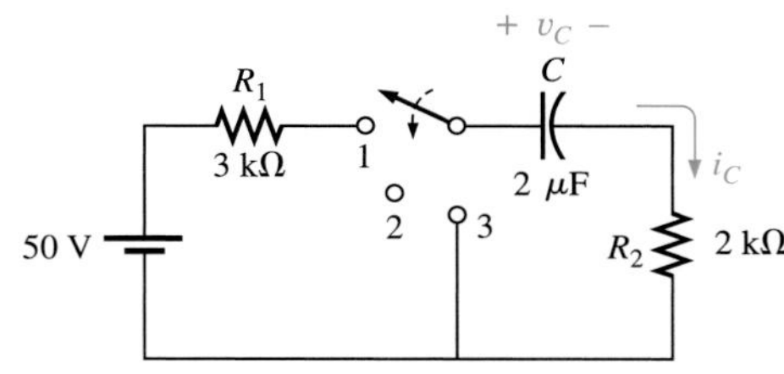

FIG. 10.70
Problems 24 and 25.

25. Repeat Problem 24 for a capacitance of 20 μF.

26. Repeat Problem 24 for a source voltage of 18V and a capacitor of 5 μF.

*27. For the network of Fig. 10.71:
   a. Find the mathematical expression for the voltage across the capacitor after the switch is thrown into position 1.
   b. Repeat part (a) for the current i_C.
   c. Find the mathematical expressions for the voltage v_C and current i_C if the switch is thrown into position 2 at a time equal to five time constants of the charging circuit.
   d. Plot the waveforms of v_C and i_C for a period of time extending from $t = 0$ to $t = 30$ μs.

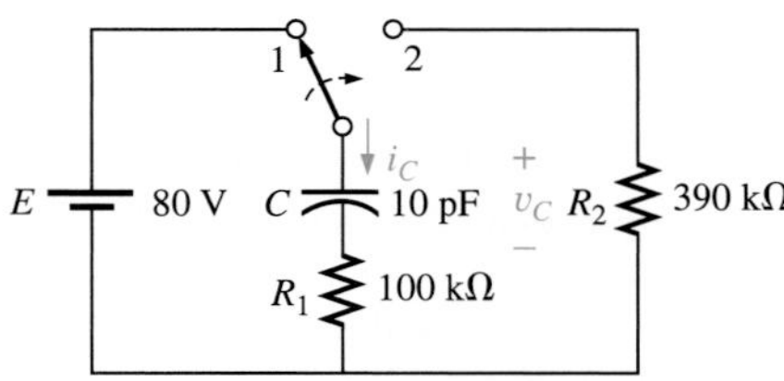

FIG. 10.71
Problem 27.

28. The capacitor of Fig. 10.72 is initially charged to 40 V before the switch is closed. Write the expressions for the voltages v_C and v_R and the current i_C for the decay phase.

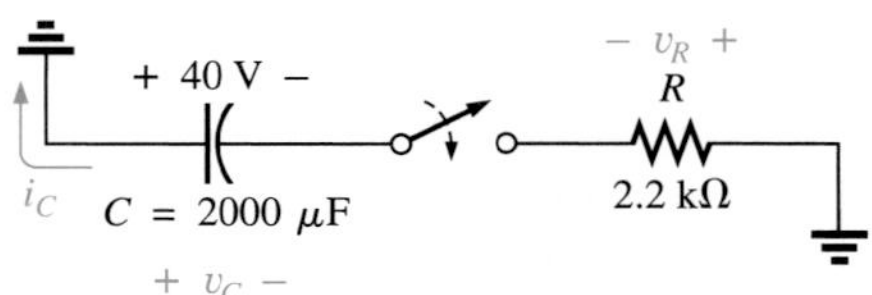

FIG. 10.72
Problem 28.

29. The 1000-μF capacitor of Fig. 10.73 is charged to 6 V. To discharge the capacitor before further use, a wire with a resistance of 0.002 Ω is placed across the capacitor.

a. How long will it take to discharge the capacitor?

b. What is the peak value of the current?

c. Based on the answer to part (b), would you expect a spark when contact is made with both ends of the capacitor?

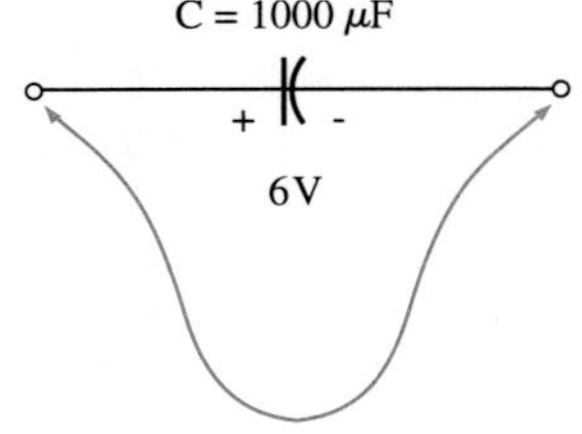

FIG. 10.73
Problems 29 and 33.

SECTION 10.9 Initial Values

30. The capacitor in Fig. 10.74 is initially charged to 3 V with the polarity shown.

a. Find the mathematical expressions for the voltage v_C and the current i_C when the switch is closed.

b. Sketch the waveforms for v_C and i_C.

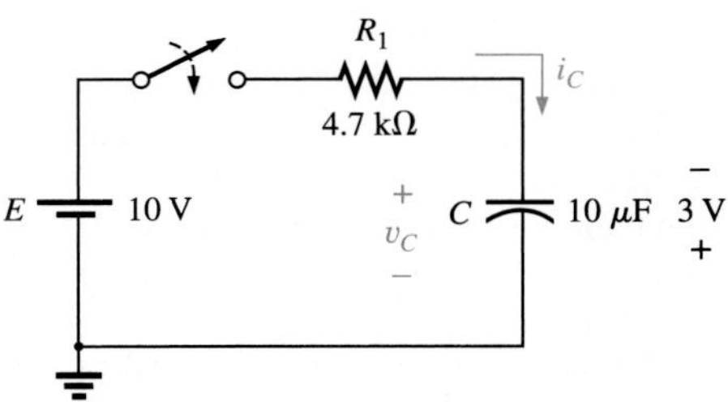

FIG. 10.74
Problem 30.

***31.** The capacitor of Fig. 10.75 is initially charged to 12 V with the polarity shown.

a. Find the mathematical expressions for the voltage v_C and the current i_C when the switch is closed.

b. Sketch the waveforms for v_C and i_C.

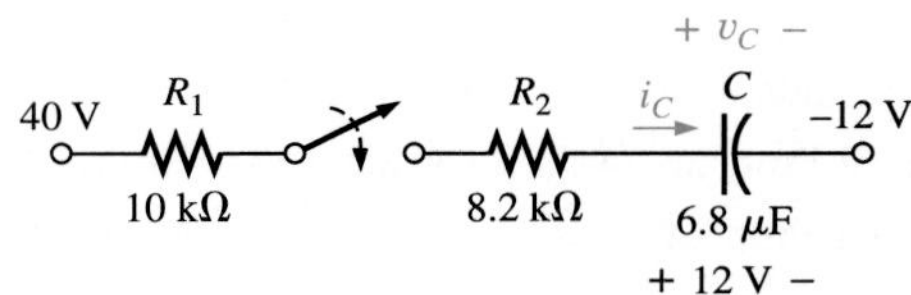

FIG. 10.75
Problem 31.

SECTION 10.10 Instantaneous Values

32. Given the expression $v_C = 8(1 - e^{-t/(20\times10^{-6})})$:

a. Determine v_C after five time constants.

b. Determine v_C after 10 time constants.

c. Determine v_C at $t = 5\ \mu$s.

33. For the situation of Problem 29, determine when the discharge current is one-half its maximum value if contact is made at $t = 0$ s.

34. For the network of Fig. 10.76, V_L must be 8 V before the system is activated. If the switch is closed at $t = 0$ s, how long will it take for the system to be activated?

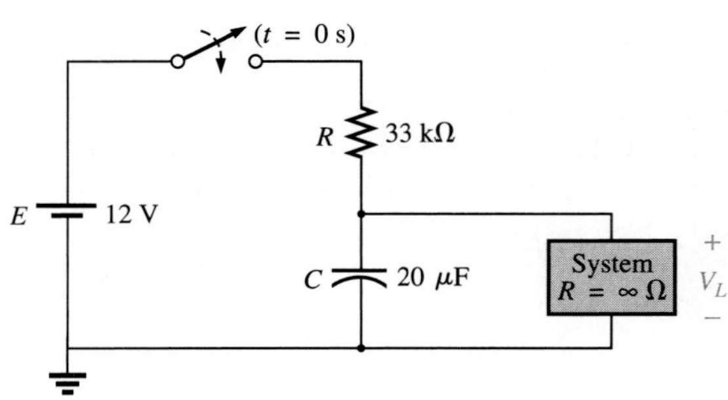

FIG. 10.76
Problem 34.

***35.** Design the network of Fig. 10.77 so that the system will turn on 10 s after the switch is closed.

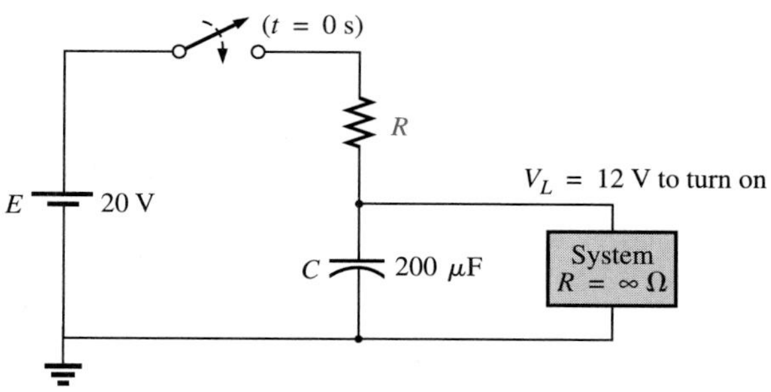

FIG. 10.77
Problem 35.

36. For the circuit of Fig. 10.78:

a. Find the time required for v_C to reach 60 V following the closing of the switch.

b. Calculate the current i_C at the instant $v_C = 60$ V.

c. Determine the power delivered by the source at the instant $t = 2\tau$.

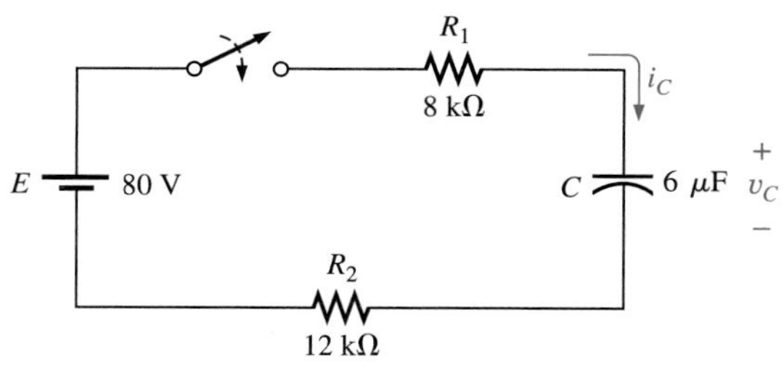

FIG. 10.78
Problem 36.

***37.** For the network of Fig. 10.79:

a. Calculate v_C, i_C, and v_{R_1} at 0.5 s and 1 s after the switch makes contact with position 1.

b. The network sits in position 1 10 min before the switch is moved to position 2. How long after making contact with position 2 will it take for the current i_C to drop to 8 μA? How much *longer* will it take for v_C to drop to 10 V?

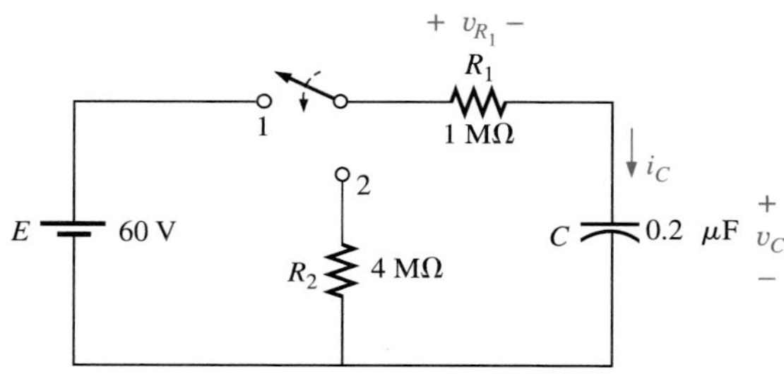

FIG. 10.79
Problem 37.

38. For the system of Fig. 10.80, using a DMM with a 10-MΩ internal resistance in the voltmeter mode:

a. Determine the voltmeter reading 1 time constant after the switch is closed.

b. Find the current i_C 2 time constants after the switch is closed.

c. Calculate the time that must pass after the closing of the switch for the voltage v_C to be 50 V.

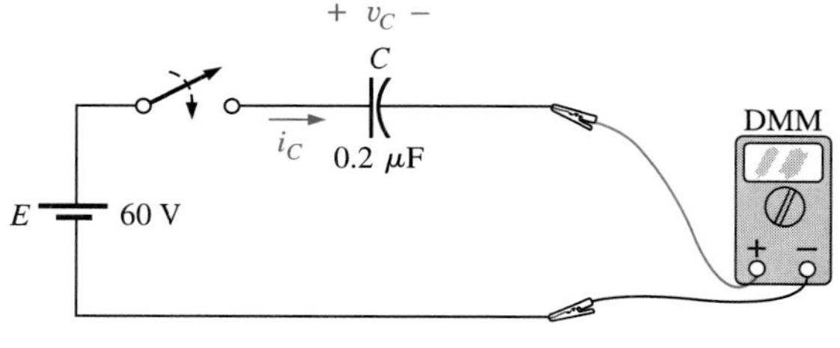

FIG. 10.80
Problem 38.

SECTION 10.11 $\tau = R_{Th}C$

39. For the system of Fig. 10.81, using a DMM with a 10-MΩ internal resistance in the voltmeter mode:

a. Determine the voltmeter reading 4 time constants after the switch is closed.

b. Find the time that must pass before i_C drops to 3 μA.

c. Find the time that must pass after the closing of the switch for the voltage across the meter to reach 10 V.

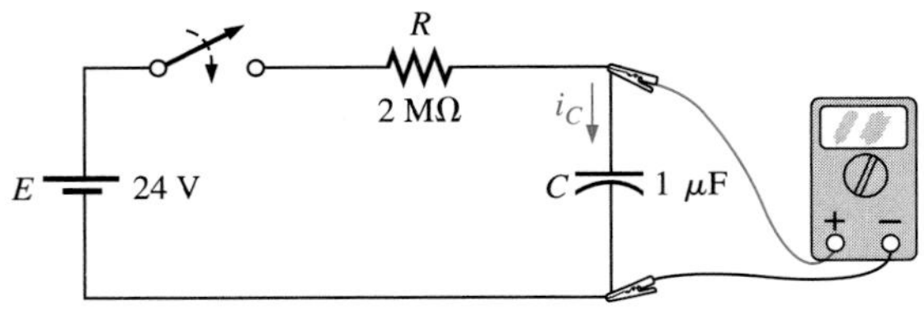

FIG. 10.81
Problem 39.

40. For the circuit of Fig. 10.82:

a. Find the mathematical expressions for the transient behaviour of the voltage v_C and the current i_C following the closing of the switch.

b. Sketch the waveforms of v_C and i_C.

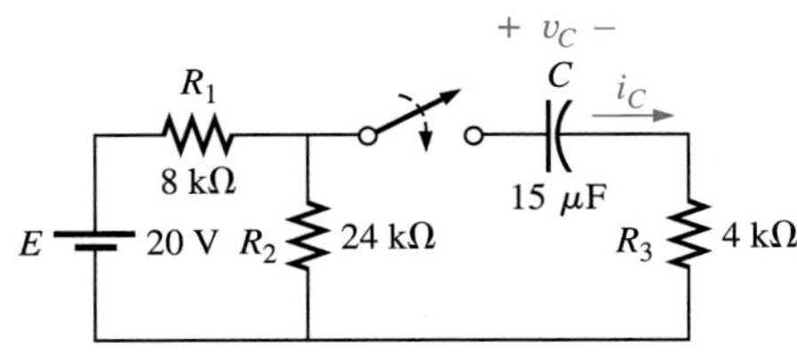

FIG. 10.82
Problem 40.

***41.** Repeat Problem 40 for the circuit of Fig. 10.83.

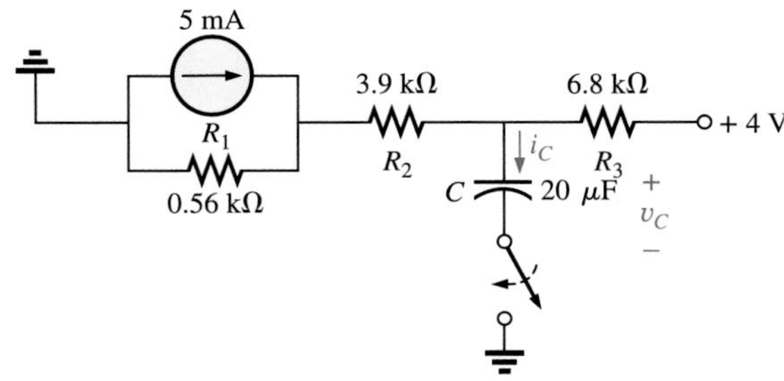

FIG. 10.83
Problem 41.

42. The capacitor of Fig. 10.84 is initially charged to 4 V with the polarity shown.

a. Write the mathematical expressions for the voltage v_C and the current i_C when the switch is closed.

b. Sketch the waveforms of v_C and i_C.

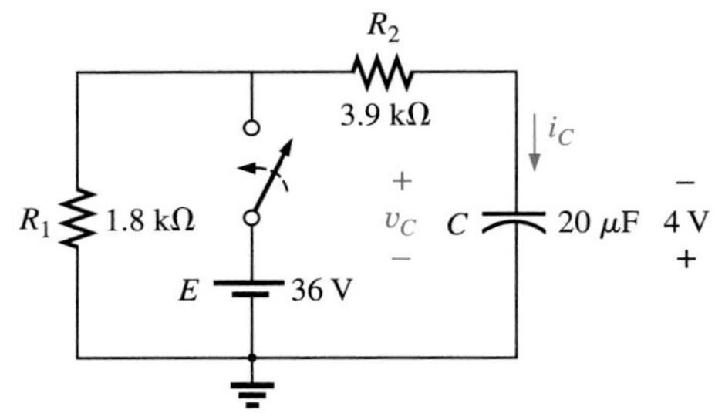

FIG. 10.84
Problem 42.

43. The capacitor of Fig. 10.85 is initially charged to 2 V with the polarity shown.

a. Write the mathematical expressions for the voltage v_C and the current i_C when the switch is closed.

b. Sketch the waveforms of v_C and i_C.

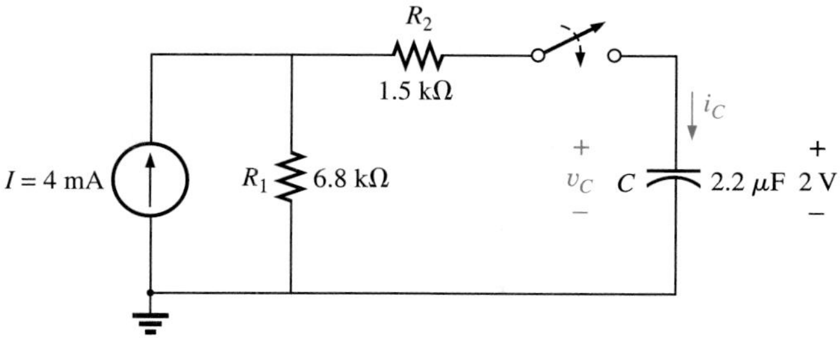

FIG. 10.85
Problem 43.

***44.** The capacitor of Fig. 10.86 is initially charged to 3 V with the polarity shown.

a. Write the mathematical expressions for the voltage v_C and the current i_C when the switch is closed.

b. Sketch the waveforms of v_C and i_C.

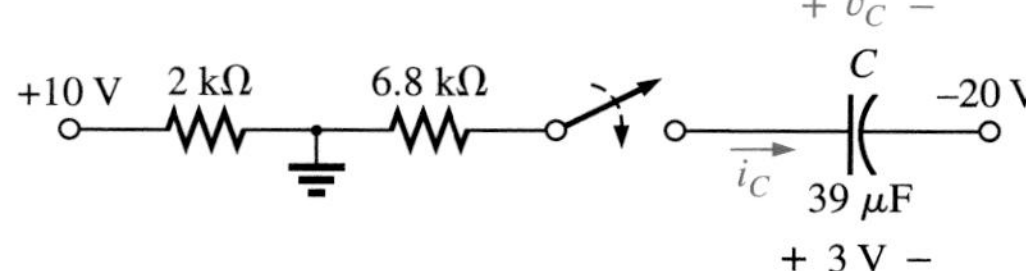

FIG. 10.86
Problem 44.

SECTION 10.12 The Current i_C

45. Draw the waveform for the average current if the voltage across a 5 μF capacitor is as shown in Fig. 10.87.

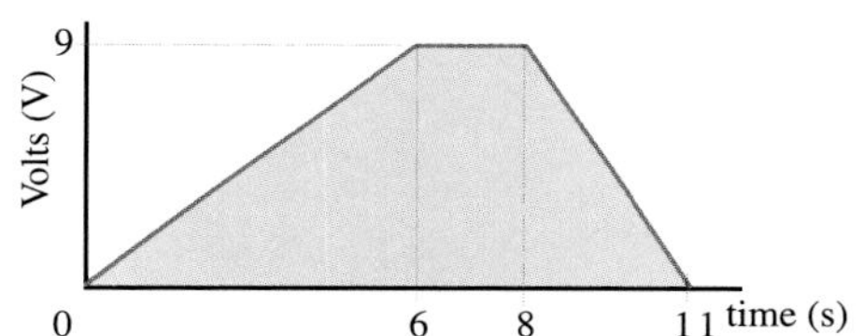

FIG. 10.87
Problem 45.

46. Find the waveform for the average current if the voltage across a 0.06-μF capacitor is as shown in Fig. 10.88.

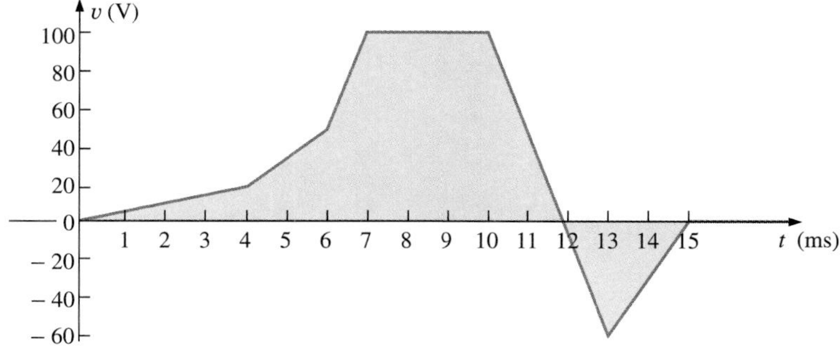

FIG. 10.88
Problem 46.

47. Repeat Problem 46 for the waveform of Fig. 10.89.

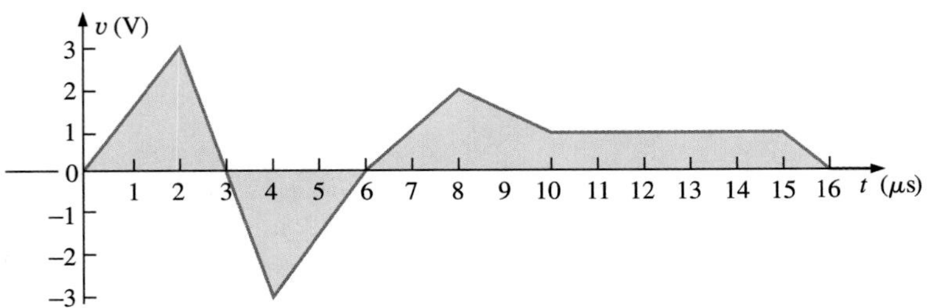

FIG. 10.89
Problem 47.

***48.** Given the following waveform (Fig. 10.90) for the current of a 20-μF capacitor, sketch the waveform of the voltage v_C across the capacitor if $v_C = 0$ V at $t = 0$ s.

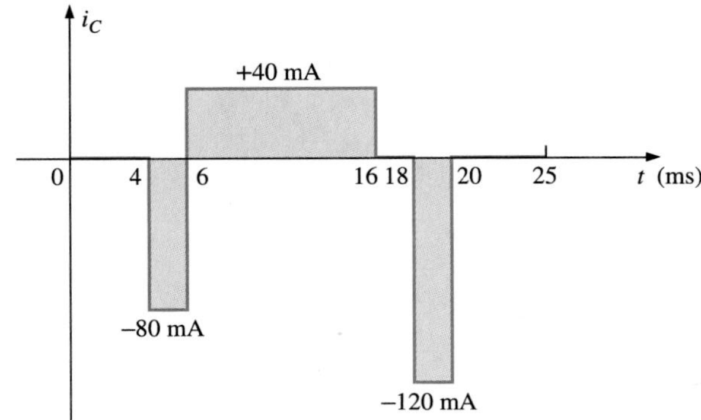

FIG. 10.90
Problem 48.

SECTION 10.13 Capacitors in Series and Parallel

49. Determine the total capacitance for the networks of Fig. 10.91.

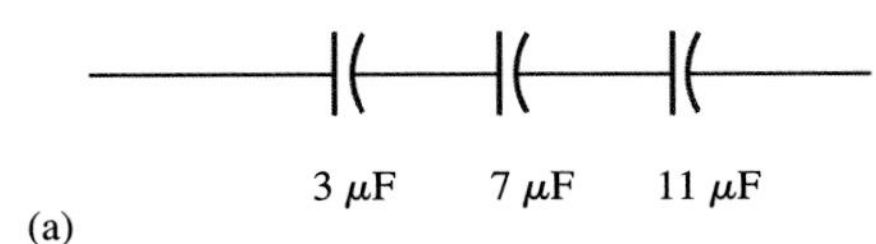

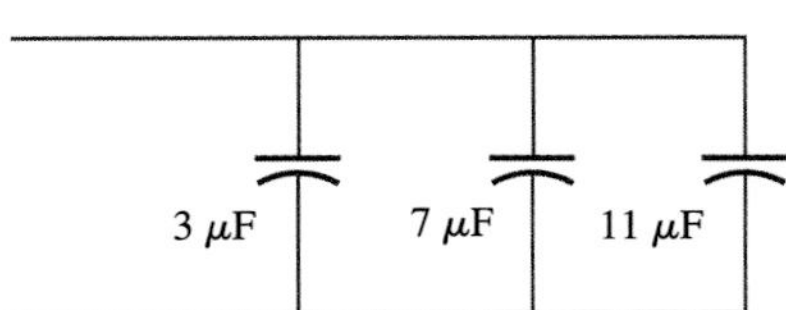

(b)

FIG. 10.91
Problem 49.

50. Find the total capacitance C_T between points a and b of the circuits of Fig. 10.92.

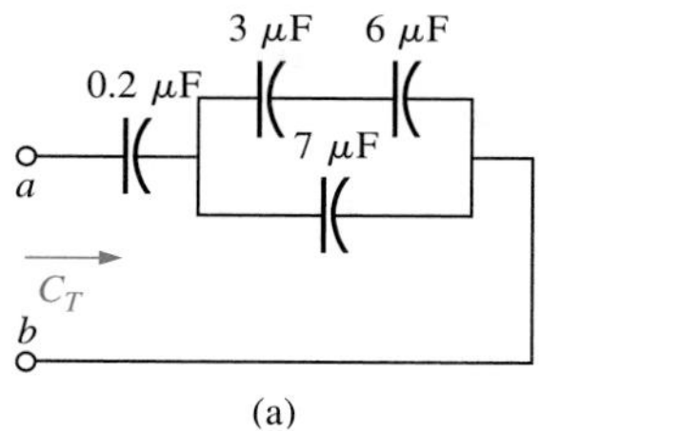

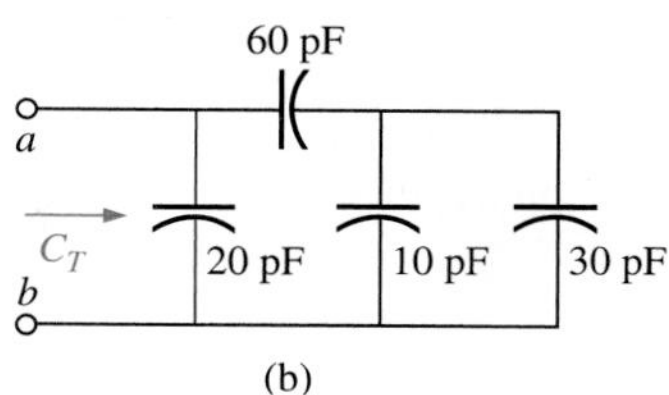

FIG. 10.92
Problem 50.

51. Find the voltage across and charge on each capacitor for the circuits of Fig. 10.93.

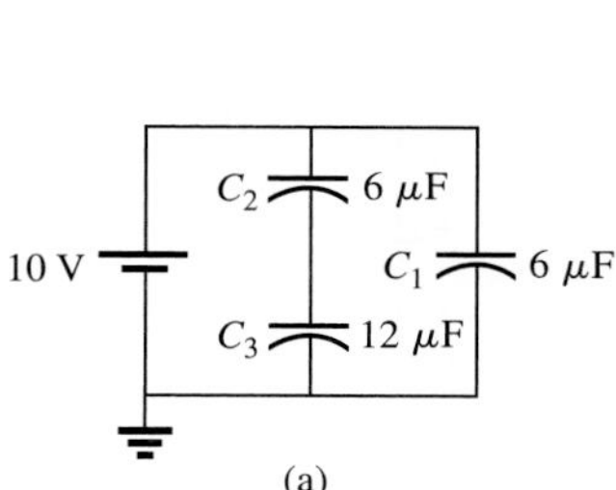

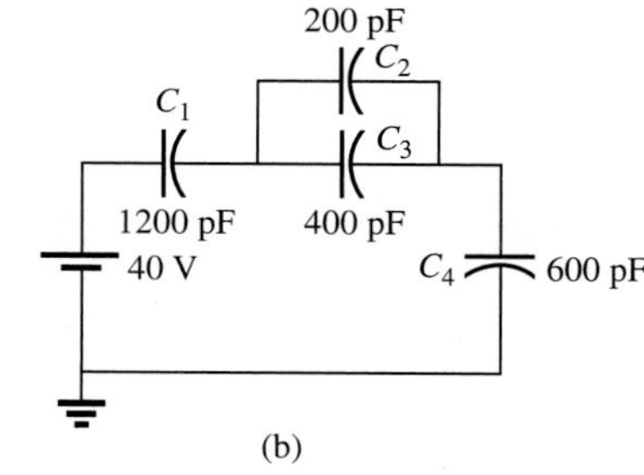

FIG. 10.93
Problem 51.

***52.** For each configuration of Fig. 10.94, determine the voltage across each capacitor and the charge on each capacitor.

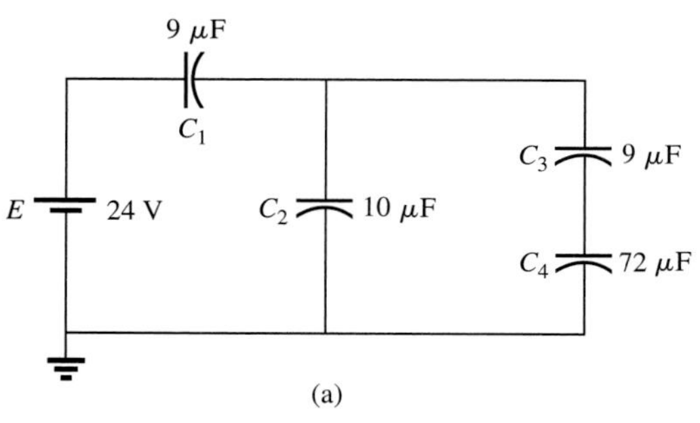

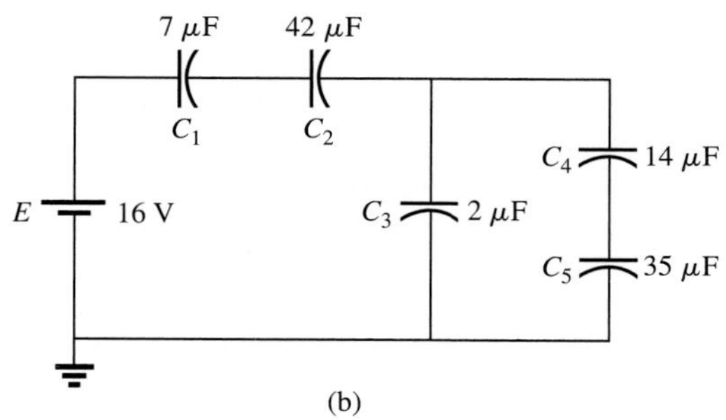

FIG. 10.94
Problem 52.

*53. For the network of Fig. 10.95, determine the following 100 ms after the switch is closed.

a. V_{ab}

b. V_{ac}

c. V_{cb}

d. V_{da}

e. If the switch is moved to position 2 one hour later, find the time required for v_{R_2} to drop to 20 V.

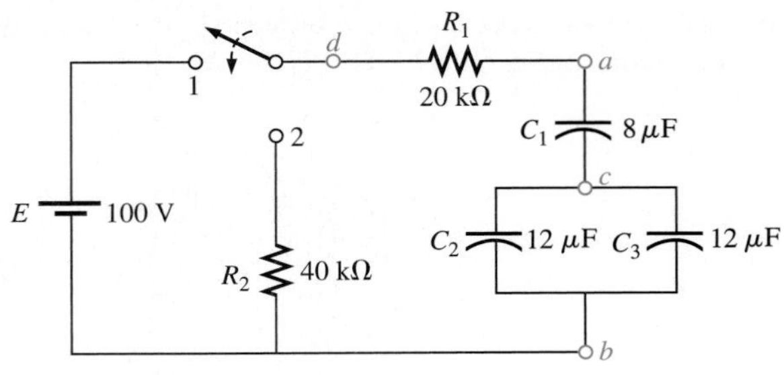

FIG. 10.95
Problem 53.

54. For the circuits of Fig. 10.96, find the voltage across and charge on each capacitor after each capacitor has charged to its final value.

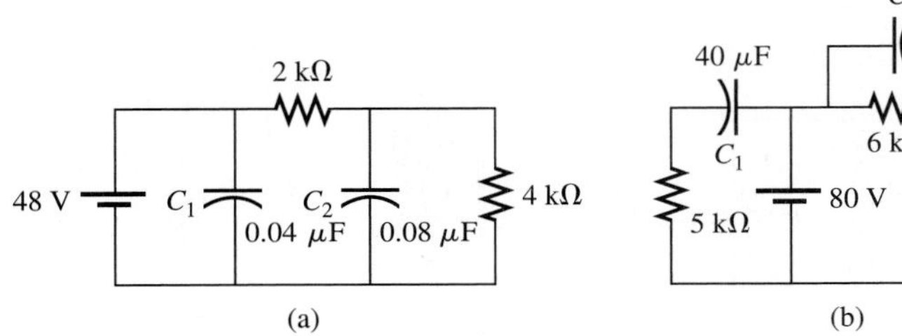

FIG. 10.96
Problem 54.

SECTION 10.14 Energy Stored by a Capacitor

55. Find the energy stored by a 120-pF capacitor with 12 V across its plates.

56. If the energy stored by a 5-μF capacitor is 1400 J, find the charge Q on each plate of the capacitor.

*57. An electronic flashgun has a 1000-μF capacitor that is charged to 100 V.

a. How much energy is stored by the capacitor?

b. What is the charge on the capacitor?

c. When the photographer takes a picture, the flash fires for 1/2000 s. What is the average current through the flashtube?

d. Find the power delivered to the flashtube.

e. After a picture is taken, the capacitor has to be recharged by a power supply that delivers a maximum current of 10 mA. How long will it take to charge the capacitor?

58. For the network of Fig. 10.97:

a. Determine the energy stored by each capacitor under steady-state conditions.

b. Repeat part (a) if the capacitors are in series.

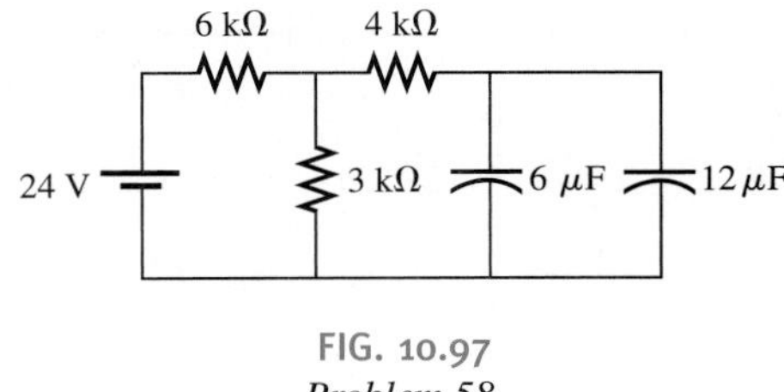

FIG. 10.97
Problem 58.

11 Magnetic Circuits

Outline

Learning Outcomes

After completing this chapter you will be able to

- describe the distribution of flux in a permanent magnet
- predict the results when the fields of two magnets interact
- describe and predict the effect of placing soft iron or other magnetic materials in a magnetic field
- describe and predict the direction of the magnetic field caused by passing current through a wire and a coil
- describe and calculate the relationship between flux, flux density, and the dimensions of a magnetic circuit
- explain and describe the factors that affect the strength of a magnetic circuit or electromagnet
- analyze a simple magnetic circuit using the same principle as Ohm's law

This chapter describes magnetic circuits and the factors that determine the amount of magnetic flux.

11.1 INTRODUCTION

Magnetism plays an important part in almost every electrical device used today in industry, research, or the home. Generators, motors, transformers, circuit breakers, televisions, computers, tape recorders, and telephones all use magnetic effects to perform a variety of important tasks.

The compass, used by Chinese sailors as early as the second century A.D., relies on a **permanent magnet** for indicating direction. The permanent magnet is made of a material, such as steel or iron, that will remain magnetized for long periods of time without the need for an external source of energy.

In 1820, the Danish physicist Hans Christian Oersted discovered that the needle of a compass would deflect if brought near a current-carrying conductor. For the first time it was shown that electricity and magnetism were related, and in the same year the French physicist André-Marie Ampère performed experiments in this area and developed what is now known as **Ampère's circuital law**. In later years, men such as Michael Faraday, Karl Friedrich Gauss, Wilhelm Eduard Weber, and James Clerk Maxwell continued to experiment in this area and developed many of the basic concepts of **electromagnetism**—magnetic effects created by the flow of charge, or current.

The analyses of electric circuits and magnetic circuits are very similar. This will be demonstrated later in this chapter when we compare the basic equations and methods used to solve magnetic circuits with those used for electric circuits.

11.2 MAGNETIC FIELDS

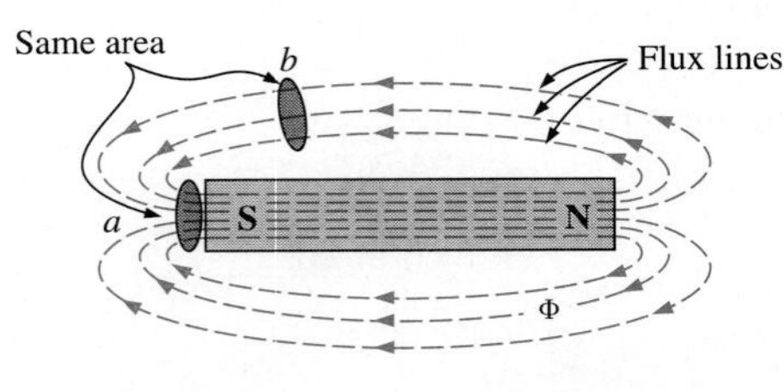

FIG. 11.1
Flux distribution for a permanent magnet.

In the region surrounding a permanent magnet there is a magnetic field, which can be represented by magnetic flux lines similar to electric flux lines. Magnetic flux lines, however, do not have origins or terminating points like electric flux lines but exist in continuous loops, as shown in Fig. 11.1. The symbol for magnetic flux is the Greek letter Φ (phi).

The magnetic flux lines form loops as they radiate from the north pole to the south pole, returning to the north pole through the metallic bar. The flux lines have characteristics that account for the behaviour of magnets and magnetic materials that are near one another. First, they try to enclose as small an area as possible. This means the loops try to be as short as possible. Second, the lines never cross one another. Third, the lines try to put as much space between parallel lines as possible.

Figure 11.2 shows the flux pattern when the unlike poles of two permanent magnets are brought together. As the loops of flux try to make themselves as short as possible, the magnets will be dragged towards one another if possible, leading to the characteristic that "unlike poles attract." Figure 11.3 shows the flux pattern when the like

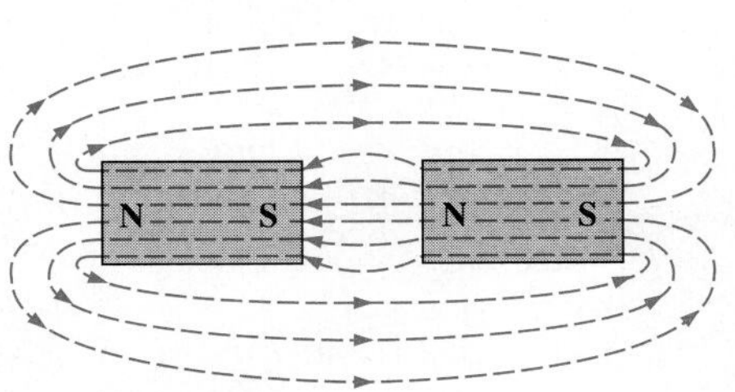

FIG. 11.2
Flux distribution for two adjacent, opposite poles.

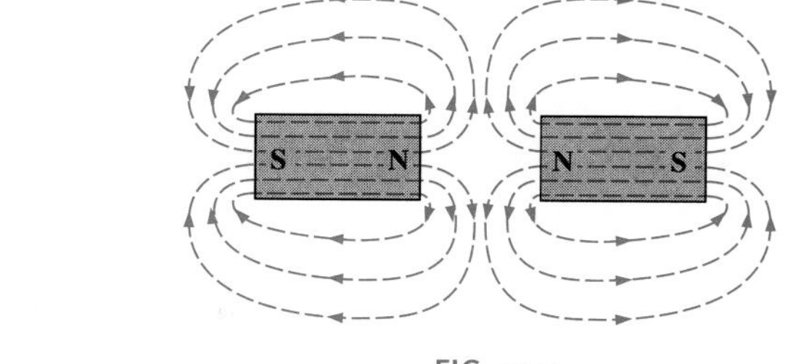

FIG. 11.3
Flux distribution for two adjacent, like poles.

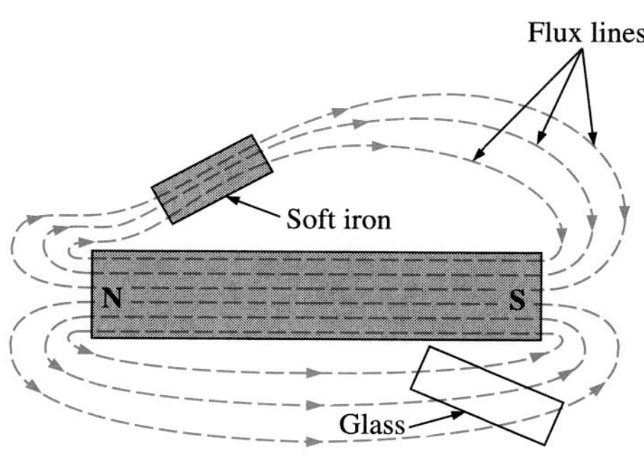

FIG. 11.4
Effect of a ferromagnetic sample on the flux distribution of a permanent magnet.

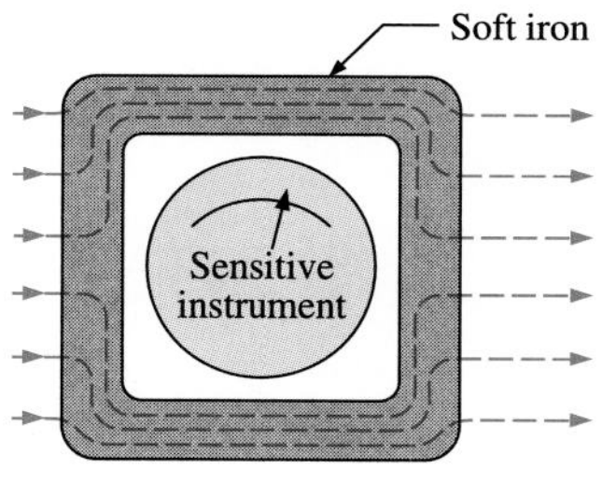

FIG. 11.5
Effect of a magnetic shield on the flux distribution.

poles are brought together. The lines radiating from the two north poles cannot cross one another, and the lines will try to put as much space between one another as possible. Therefore, the magnets will tend to push apart, leading to the characteristic that "like poles repel."

If a nonmagnetic material, such as glass or copper, is placed in the flux paths surrounding a permanent magnet, there will be an almost unnoticeable change in the flux distribution (Fig. 11.4). However, if a magnetic material, such as soft iron, is placed in the flux path, the flux lines will pass through the soft iron rather than the surrounding air because flux lines pass with greater ease through magnetic materials than through air. This principle is put to use in shielding sensitive electrical elements and instruments that can be affected by stray magnetic fields (Fig. 11.5). Materials that have this property are said to be **ferromagnetic** materials.

As we said in the introduction, there is a magnetic field around every wire that carries an electric current. The direction of the magnetic flux lines can be found by simply placing the thumb of the right hand in the direction of conventional current flow and noting the direction of the fingers. (This is called the **right-hand rule.**) In Fig. 11.6(a), the current causes a flux represented by the flux line. In Fig. 11.6(b), the current is doubled, causing a second line of flux to form which pushes the first line of flux outward and concentric with the first line. This means that the number of lines of flux is proportional to the current.

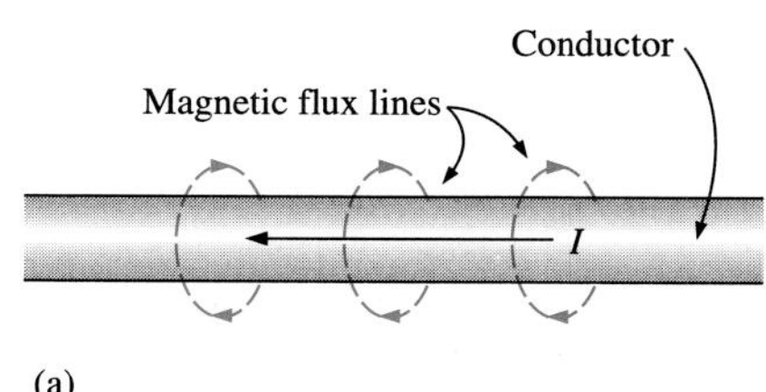

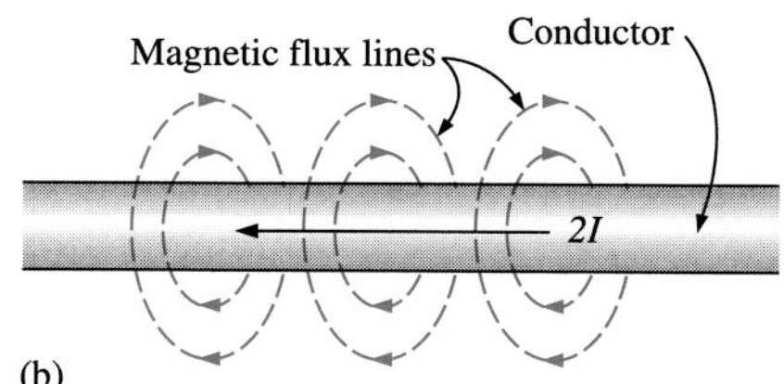

FIG. 11.6
Magnetic flux lines around a conductor; (a) with current I, (b) with current 2I.

If the conductor is wound in a single-turn coil (Fig. 11.7), there will be a resultant flux in a common direction through the centre of the coil. If a second turn is added to the coil (Fig. 11.8), and the current is unchanged, each turn causes the same amount of resultant flux. Therefore, the overall resultant flux in the centre is doubled. This means that the number of lines of flux in a coil is proportional to the number of turns and the amount of current. Thus, a measure of the magnetic flux can be given the unit **ampere-turn** (symbol At).

FIG. 11.7
Flux distribution of a single-turn coil.

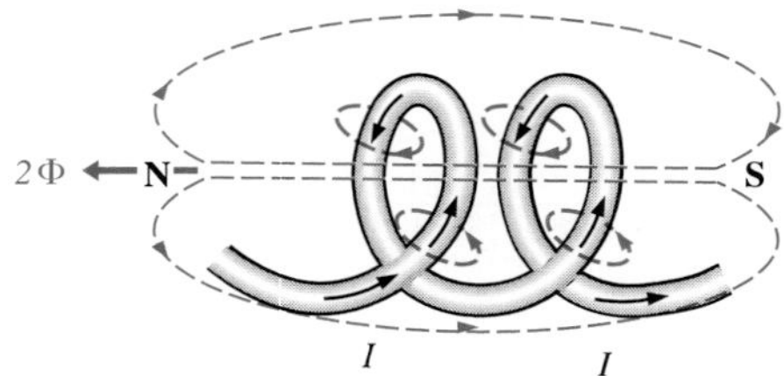

FIG. 11.8
Flux distribution of a two-turn coil.

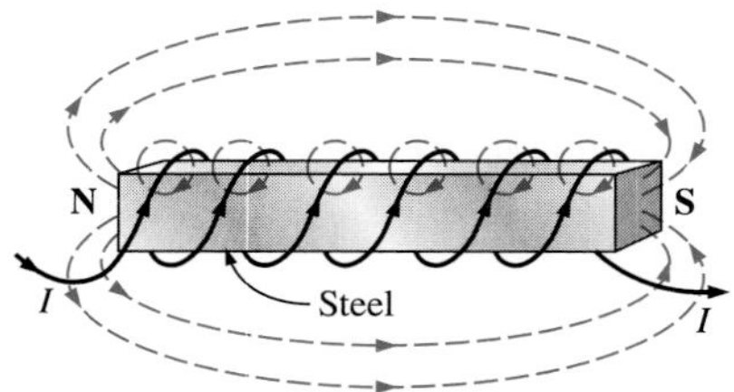

FIG. 11.9
Electromagnet.

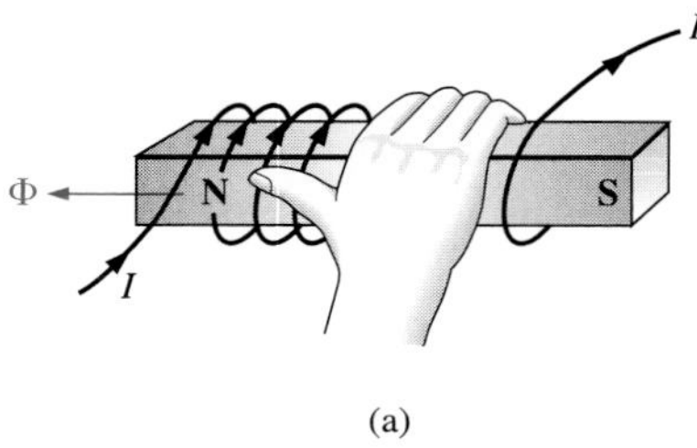

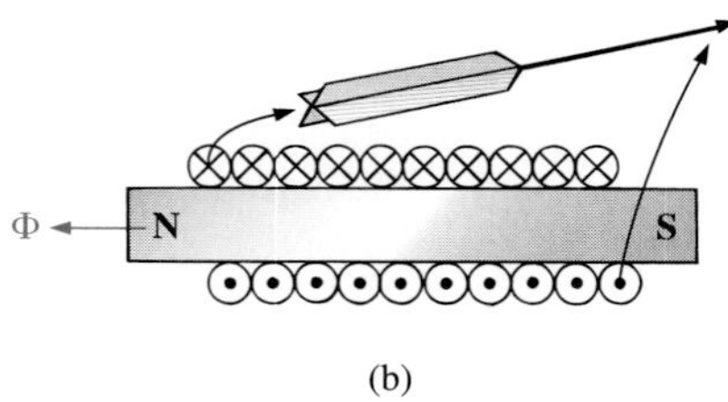

FIG. 11.10
Determining the direction of flux for an electromagnet: (a) method, (b) notation.

The flux distribution of the coil is quite similar to that of the permanent magnet. The flux lines leaving the coil from the left and entering to the right simulate a north and south pole, respectively. The field strength of the coil can be effectively increased by placing certain materials, such as iron, steel, or cobalt, within the coil to increase the flux density within the coil. The addition of one of these core materials results in an electromagnet (Fig. 11.9) that has all of the properties of a permanent magnet, but whose field strength can be varied by changing either the current or the number of turns. The direction of the flux lines can be determined by letting the fingers of the right hand point in the direction of current in the coil. The thumb will then point in the direction of the north pole of the induced magnetic flux, as demonstrated in Fig. 11.10(a). A cross section of the same electromagnet is included in Fig. 11.10(b) to introduce the convention for directions perpendicular to the page. The cross and dot refer to the tail and head of the arrow, respectively.

Other areas of application for electromagnetic effects are shown in Fig. 11.11. The flux path for each is indicated in each figure.

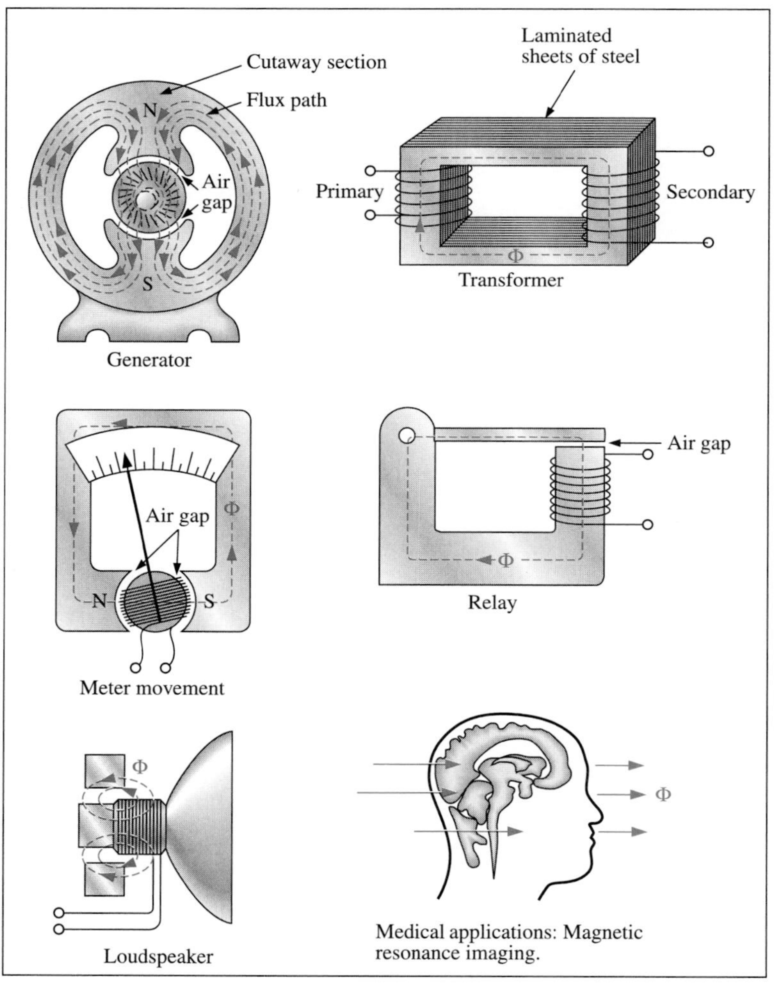

FIG. 11.11
Some areas of application of magnetic effects.

LUMINARIES

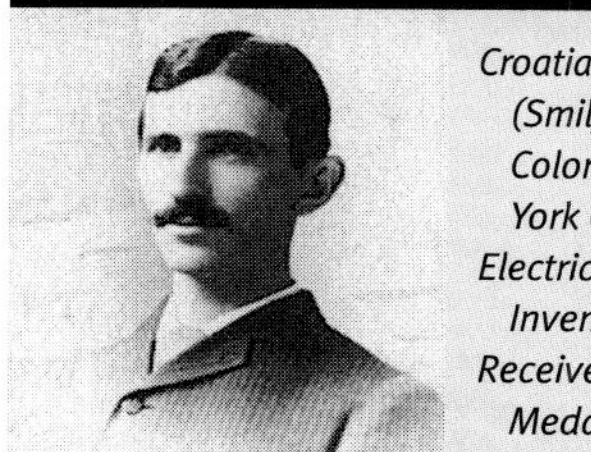

Croatian-American (Smiljan, Paris, Colorado Springs, New York City) (1856–1943) Electrical Engineer and Inventor Received the Edison Medal in 1917

FIG. 11.12
Nikola Tesla

Nikola Tesla is often regarded as one of the most innovative and inventive people in the history of the sciences. He was the first to introduce the *alternating-current machine,* removing the need for commutator bars of dc machines. After emigrating to the United States in 1884, he sold a number of his patents on *ac machines*, *transformers*, and *induction coils* (including the *Tesla coil,* as we know it today) to the Westinghouse Electric Company. Some say that his most important discovery was made at his laboratory in Colorado Springs, where he discovered *terrestrial stationary waves* in 1900. The range of his discoveries and inventions is too large to list here but extends from lighting systems to *polyphase power systems* to a *wireless world broadcasting system.*

Courtesy of the Smithsonian Institution, Photo No. 52,223

11.3 FLUX DENSITY

The unit of **flux** (symbol Φ) in SI units is the *weber* (symbol Wb), named for Wilhelm Eduard Weber. The number of flux lines per unit area (A in m^2) is called the **flux density** (symbol B), and is measured in *teslas* (symbol T), named for Nikola Tesla (Fig. 11.12). Note that 1 T = 1 Wb/m^2 as shown in the following equation that defines flux density:

$$B = \frac{\Phi}{A} \qquad \text{(teslas, T)} \qquad (11.1)$$

where Φ is the number of flux lines passing through the area A (Fig. 11.13). The flux density at position a in Fig. 11.1 is twice that at b because twice as many flux lines are passing through the same area.

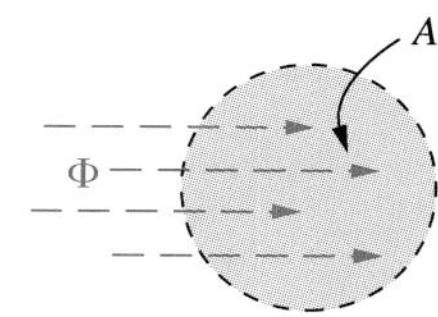

FIG. 11.13
Defining the flux density B.

By definition,

$$1\ \text{T} = 1\ \text{Wb/m}^2$$

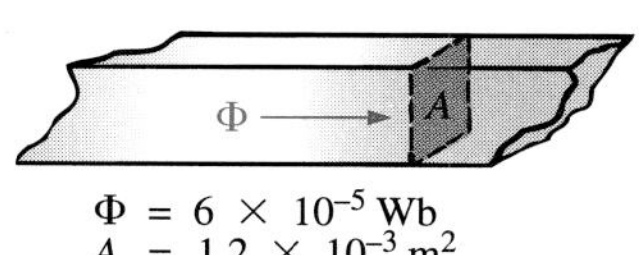

$\Phi = 6 \times 10^{-5}$ Wb
$A = 1.2 \times 10^{-3}\ m^2$

FIG. 11.14
Example 11.1.

EXAMPLE 11.1

For the core of Fig. 11.14, determine the flux density B in teslas.

Solution:

$$B = \frac{\Phi}{A} = \frac{6 \times 10^{-5}\ \text{Wb}}{1.2 \times 10^{-3}\ \text{m}^2} = \mathbf{5 \times 10^{-2}\ T}$$

EXAMPLE 11.2

In Fig. 11.14, if the flux density is 1.2 T and the area is 0.25 in.^{2}, determine the flux through the core.

Solution: By Eq. (11.1),

$$\Phi = BA$$

However, converting 0.25 in.^{2} to metric units,

$$A = 0.25\ \text{in.}^2 \left(\frac{1\ \text{m}}{39.37\ \text{in.}}\right)\left(\frac{1\ \text{m}}{39.37\ \text{in.}}\right) = 1.613 \times 10^{-4}\ \text{m}^2$$

and

$$\Phi = (1.2\ \text{T})(1.613 \times 10^{-4}\ \text{m}^2) = \mathbf{1.936 \times 10^{-4}\ Wb}$$

FIG. 11.15
Digital display gaussmeter. (Courtesy of LDJ Electronics, Inc.)

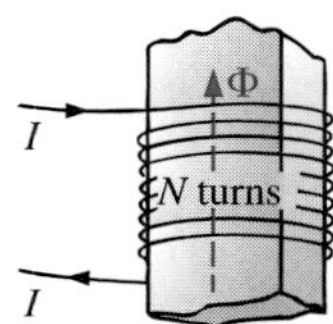

FIG. 11.16
Defining the components of a magnetomotive force.

Magnetic flux density can be measured with an instrument called a **gaussmeter** (Fig. 11.15). Some gaussmeters are calibrated in the non-SI unit, the *gauss*. To convert to SI, you need to know that 1 T = 10^4 gauss. Hence, the value of the reading on the face of the meter in Fig. 11.15 is

$$1.964 \text{ gauss}\left(\frac{1 \text{ T}}{10^4 \text{ gauss}}\right) = 1.964 \times 10^{-4} \text{ T}$$

11.4 OHM'S LAW FOR MAGNETIC CIRCUITS

Recall the equation

$$\text{Effect} = \frac{\text{cause}}{\text{opposition}}$$

used in Chapter 4 to introduce Ohm's law for electric circuits. For magnetic circuits, the effect desired is the flux Φ. We saw in Section 11.2 that the number of flux lines is proportional to the number of ampere-turns which we call the **magnetomotive force** (mmf) $\mathcal{F}$ (Fig. 11.16). This is similar to current in an electric circuit being proportional to the applied voltage or *electromotive force* (emf) *E*. In the electric circuit, it is the resistance (*R*) that determines the amount of current that will flow for a given applied voltage. In the magnetic circuit it is the reluctance ($\mathcal{R}$) that determines the amount of flux that can be set up for a given number of ampere-turns.

This leads us to the Ohm's law equation of a magnetic circuit

$$\Phi = \frac{\mathcal{F}}{\mathcal{R}} \tag{11.2}$$

where

$$\mathcal{F} = NI \qquad \text{(ampere-turns, At)} \tag{11.3}$$

The equation clearly shows that an increase in the number of turns or the current through the wire will result in an increased "pressure" on the system to establish flux lines through the core.

Similarly, we can define the magnetic effects of materials by comparison with the material effects on currents. Recall that the resistance of a material to the flow of charge (current) is determined by the equation

$$R = \frac{\rho l}{A} \qquad \text{(ohms, } \Omega\text{)}$$

The equivalent magnetic term is **reluctance**. The reluctance of a material to the setting up of magnetic flux lines in the material is determined by the equation

$$\mathcal{R} = \frac{l}{\mu A} \qquad \text{(ampere-turns/weber, At/Wb)} \tag{11.4}$$

where l is the length of the magnetic path, A is its cross-sectional area, and μ is the *permeability* of the magnetic path. The unit *ampere-turns/weber* is derived from equation 11.2. By substituting the units into Eq. (11.4) and solving for μ, you can show that the unit for permeability is *weber per ampere-turn-metre*, Wb/(At.m) or Wb/At.m.

Note that both resistance and reluctance are inversely proportional to the area. So, an increase in the area of an element will decrease resistance and reluctance, and increase the desired effects, current and flux, respectively. For an increase in length the opposite is true, and the desired effect in each is reduced. The type of material affects the *resistance* by the property known as **resistivity** (ρ). Copper has a lower resistivity than aluminum. Similarly, the property of the material to contain the flux in the magnetic circuit or core is called its **permeability** μ.

The **permeability** (μ) is a measure of the ease with which magnetic flux lines can be established in a material. It is similar in many respects to conductivity in electric circuits. The permeability of free space (vacuum) is ρ_0, where

$$\mu_o = 4\pi \times 10^{-7} \text{ Wb/At.m}$$

Practically speaking, the permeability of all nonmagnetic materials, such as copper, aluminum, wood, glass, and air is the same as that of free space. The permeability of magnetic materials such as iron and steel is hundreds—or even thousands—of times larger than that of air or free space. For this reason, iron and steel have much lower reluctance than air. Therefore, the flux being set up by the coil in Fig. 11.17 stays in the cast-steel core and does not return by the shorter path in the air in the center of the core. In other words, we can say that magnetic flux lines will always take the path of least reluctance. The relative permeability, μ_r, is the ratio of the permeability of a material to that of free space. We will show elsewhere (on the CD-ROM) that the permeability of magnetic core material is very nonlinear and varies as the amount of flux caused by the coil.

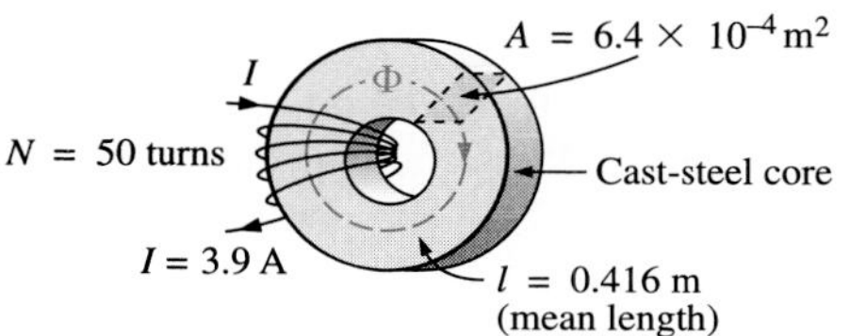

FIG. 11.17
Example 11.3.

EXAMPLE 11.3

For the magnetic circuit of Fig. 11.17, the coil has 50 turns and the current is 3.9A. The flux is 3 $\times$ 10^{-4}Wb and the dimensions are as shown.

a) Calculate the reluctance and value of μ and μ_r for the core.

b) *Estimate* the value of the flux if the iron core is removed.

Solution:

a) $\mathcal{R} = \mathcal{F}/\Phi = 3.9\text{A} \times 50\text{t}/3 \times 10^{-4} \text{ Wb} = 65 \times 10^4 \text{ At/Wb}$

$\mu = 1/\mathcal{R}\text{A} = 0.416\text{m}/(65 \times 10^4 \text{ At/Wb} \times 6.4 \times 10^{-4} \text{ m}^2)$

$= 0.001 \text{ Wb/At.m}$

Note: This value of μ of the iron is $0.001/4\pi \times 10^{-7} = 796$ times larger than that of air.

Therefore, μ_r is 796.

b) If the iron core were removed, the permeability would become

$\mu_o = 4\pi \times 10^{-7} \text{ Wb/At.m}$

But both the area and the length of the flux path would also change, and change significantly. However, if we assume the area and length remain unchanged—a poor assumption at best—the reluctance will become 796 times larger. Therefore, the flux will become that much smaller.

Notes: Example 11.3 gives the impression it would be easy to find the flux for a different value of current and/or turns. However, the value of μ varies with the flux, the quantity being determined in the first place. This means other methods must be used to compensate for the reluctance being dependent on flux. A more complete treatment is given on the CD-ROM.

Also, you must remember that while electric and magnetic circuits are very similar, the flux is not a flow variable such as charge in an electric circuit.

11.5 SATURATION AND HYSTERESIS

Two practical considerations when using coils with ferromagnetic cores are saturation and hysteresis. The magnetization curve (Fig. 11.18) shows a core with no residual flux. As we increase the current from zero and plot the flux vs. the magnetomotive force, we can see that the graph rises linearly for low values of current. Increasing the current by a certain amount increases the number of lines of flux by the same amount (*a* to *b* and *b* to *c*). However, a point is reached where increasing the current by the same amount as before does not result in a corresponding increase in the lines of flux (*c* to *d*). Eventually, the graph flattens out (*e* to *f*). When this happens, the coil or magnetic circuit is said to be **saturated**. This means that no matter how much more current is forced through the coil, the iron cannot set up any more flux. In fact some lines will set up outside the iron in the air. Therefore, for many applications, it is common to use coils in the linear region.

Another practical consideration when using coils is the effect known as **hysteresis**. In the same Fig. 11.18, we see that when the current is decreased, the graph follows a different path from *f* to *g* as the current decreases to zero. As can be seen, the magnetic circuit has retained a residual flux as indicated by point *g* on the graph. This means that before any current is passed through the coil the next time, the core will already have some flux. This is caused by the particles making up the iron core not returning to the original position they had at point *a* on the graph.

If the coil is subjected to an alternating current, then the hysteresis curve takes on the full loop shown in Fig. 11.18 as both current and flux alternately reverse directions (polarity).

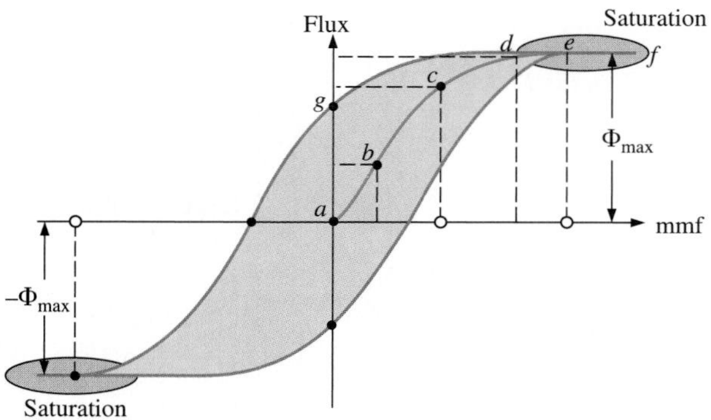

FIG. 11.18
Magnetization curve showing hysteresis and saturation.

Practical Business Applications

Hey, ding, dong! How does it do that?

Ding, dong goes the old familiar chime as you announce your visit at your friend's home. As you wait for someone to answer the door, you begin to think about how electricity makes the doorbell ring.

"It's all very simple," you convince yourself. It starts with the normally open switch at the front door. You know from your electrical studies that switches are used to complete a closed circuit. This permits the current to flow from a power supply to the doorbell unit that is usually mounted high on a wall in the front hall of our homes. But, what's going on inside this doorbell unit? Just as the door opens, you decide to figure it out when you get back home.

Later that night, back at your own home, you decide to sneak a peek under the cover of your doorbell unit. You're not afraid of the potential inside because you know that there's a step-down (bell) transformer that takes the 120V ac and reduces it to manageable voltage levels. You notice that it looks something like this:

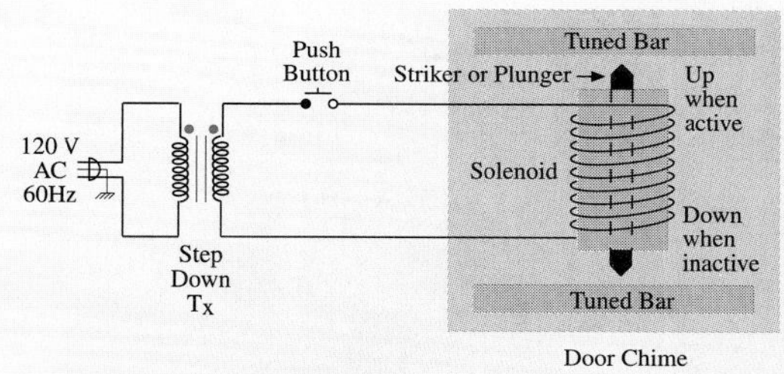

Door Chime

With the cover removed, you see a vertically oriented solenoid with two metal bars, one below the solenoid and the other above it. The solenoid is made up of copper wire in the form of an electrical coil on a hollow metal cylinder with a movable, but solid, rubber-tipped metal rod called a striker or plunger.

Using your electrical knowledge, you realize that when the current flows through an electromagnet, a flux field is produced. This flux field will, in turn, produce a force on the metal rod if it is ferromagnetic (can be magnetized). This force will cause the iron rod to move vertically until it strikes the upper metal bar like a hammer. Also, because the movable iron rod is mounted on springs, it should return to its normal position when it is de-energized.

The metal bars above and below the solenoid have been designed to give specific tones when struck by the iron rod. Every manufacturer designs their doorbell a little differently because they want you to buy their doorbell rather than their competitors'. It's really quite something when you think about how such a simple concept as a doorbell can lead to so many neat products that we consumers can buy, and that manufacturers can profit from.

So, now that you've thought it through and armed yourself with knowledge, why not test your theory? It's as simple as watching the doorbell in action. So, you leave the cover off, and get someone to push the front door switch while you observe the response. Unless you've got really long arms.

Sure enough, when the switch is pressed, the coil is energized and the iron rod is propelled upward within the coil, striking the upper bar, creating that familiar "Ding." Then, as the switch is released, the rod falls back through the solenoid due to inertia, striking the lower horizontal bar and thus completing the final "Dong" of the door chime.

But what about the side door? When you push that button, only one "Ding" is heard. Hmmm. Let's see now. Ah yes, that's it. There's a second solenoid that drives another iron rod. It's designed so it can only strike one of the metal bars but can't reach (or is prevented from striking) the other one.

"Finally, it all makes sense," you say. "They can't call me a Dingdong anymore!"

PROBLEMS

SECTION 11.3 Flux Density

1. Calculate the area through which a magnetic flux of 5×10^{-4} Wb has a flux density of 8×10^{-4} T.
2. Calculate the flux in a core of square cross-section 1.5 cm on a side with a flux density of 2×10^{-2} T.
3. For the electromagnet of Fig. 11.19:
   a. Find the flux density in the core.
   b. Sketch the magnetic flux lines and indicate their direction.
   c. Indicate the north and south poles of the magnet.

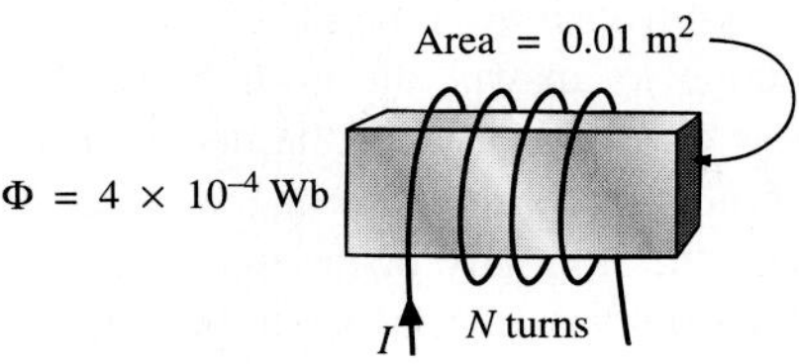

FIG. 11.19
Problem 3.

SECTION 11.4 Ohm's Law for Magnetic Circuits

4. Which section of Fig. 11.20 [(a), (b), or (c)] has the largest reluctance to the setting up of flux lines through its longest dimension?
5. Find the reluctance of a magnetic circuit if a magnetic flux $\Phi = 4.2 \times 10^{-4}$ Wb is established by an impressed mmf of 400 At.

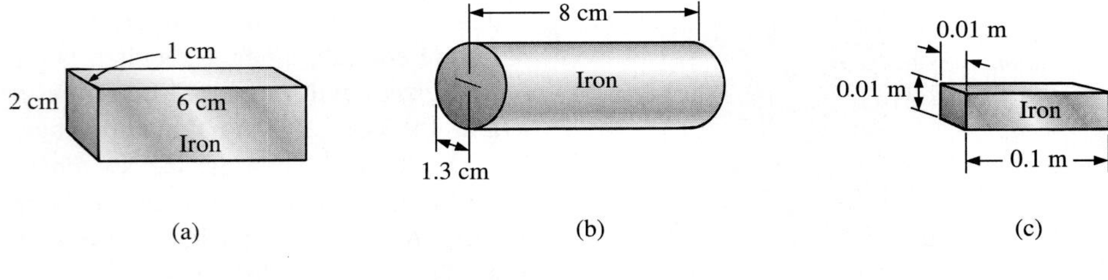

FIG. 11.20
Problem 4.

6. Repeat Problem 5 for $\Phi = 72 \times 10^{-4}$ Wb and an impressed mmf of 120 At.

*7. The force carried by the plunger of the door chime of Fig. 11.21 is determined by

$$f = \frac{1}{2}NI\frac{d\phi}{dx} \quad \text{(newtons)}$$

where $d\phi/dx$ is the rate of change of flux linking the coil as the core is drawn into the coil. The greatest rate of change of flux will occur when the core is ¼ to ¾ the way through. In this region, if Φ changes from 0.5×10^{-4} Wb to 8×10^{-4} Wb, what is the force carried by the plunger?

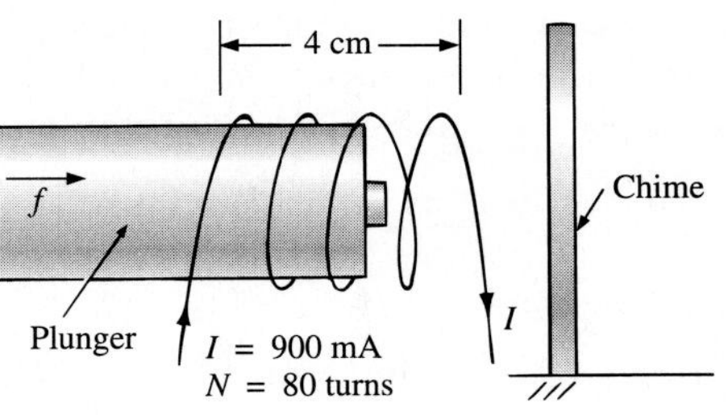

FIG. 11.21
Door chime for Problem 7.

12 Inductors

Outline

Learning Outcomes

After completing this chapter you will be able to

- discuss the important aspects of the third basic element of electric circuits: *the inductor*
- show how the speed and the direction of motion of a conductor through a magnetic field affects the resulting induced current
- show how the magnitude of magnetic field affects the induced *emf* in a conductor
- show the effects of different core materials on inductance
- identify inductors by type, typical values, and typical applications
- identify and verify *R-L* transient profiles for storage and decay cycles
- show the relationship between inductors in series and an equivalent inductance
- show the relationship between inductors in parallel and an equivalent inductance
- show that for most practical purposes, inductors of *R-L* and *R-L-C* circuits with dc inputs can be replaced by a short circuit
- show that the energy stored in an inductor is related to the current and voltage applied

Inductors are the third basic electrical component. Together with resistors and capacitors, they can be built into circuits used in the navigation and communication equipment of all modern aircraft.

12.1 INTRODUCTION

We have examined the resistor and the capacitor in detail. In this chapter we consider a third element, the **inductor**, which has a number of response characteristics similar in many respects to those of the capacitor. Sections of this chapter are in the same order as those of Chapter 10 (Capacitance) to emphasize the similarity between capacitors and inductors.

12.2 ELECTROMAGNETIC INDUCTION—FARADAY'S AND LENZ'S LAWS

If a conductor is moved through a magnetic field so that it cuts magnetic lines of flux, a voltage will be induced across the conductor, as shown in Fig. 12.1. The induced voltage across a conductor can be increased by:

a) increasing the speed with which the conductor passes through a field, that is, increasing the number of flux lines cut per unit time, or

b) increasing the strength of the magnetic field (for a constant traversing speed), or

c) both (a) and (b).

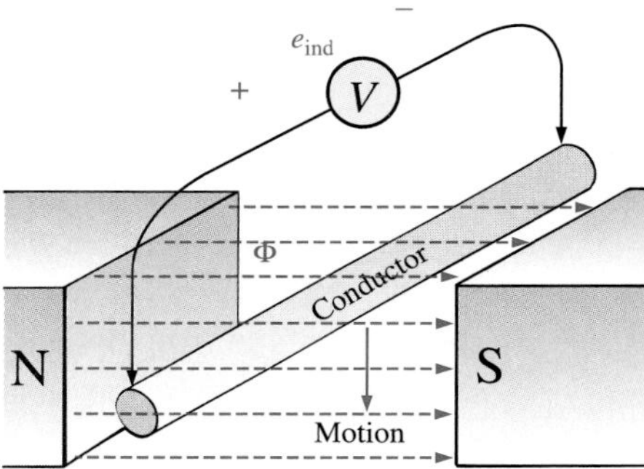

FIG. 12.1
Generating an induced voltage by moving a conductor through a magnetic field.

If the conductor is held fixed and the magnetic field is moved so that its flux lines cut the conductor, the same effect will be produced. This is a practical explanation underlying Faraday's Law.

Faraday's Law

Faraday's law states that:

The voltage induced in an electric circuit is proportional to the rate of change of the flux of magnetic induction linking the turns of the coil within a circuit.

If a coil of N turns is placed in the region of a changing flux, as in Fig. 12.2, a voltage will be induced across the coil as determined by **Faraday's law**:

$$e = N\frac{d\phi}{dt} \quad \text{(volts, V)} \tag{12.1}$$

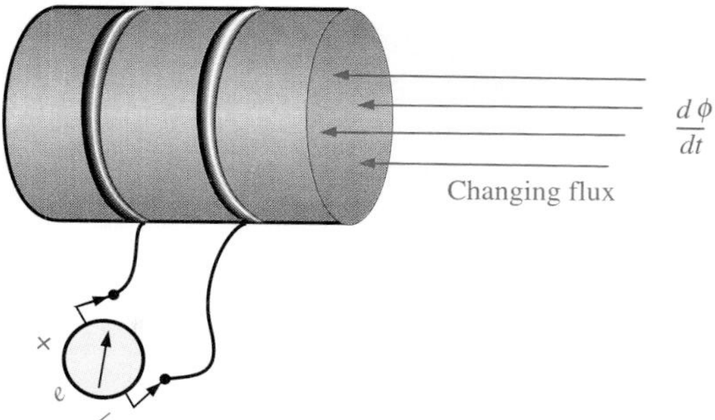

FIG. 12.2
Demonstrating Faraday's law.

where N represents the number of turns of the coil and $d\phi/dt$ is the instantaneous change in flux (in webers) linking the coil.

The term **linking** refers to the flux within the turns of wire. The term **changing** simply indicates that either the strength of the field linking the coil changes in magnitude or the coil is moved through the field in such a way that the number of flux lines through the coil changes with time.

If the flux linking the coil ceases to change, such as when the coil simply sits still in a magnetic field of fixed strength, $d\phi/dt = 0$, and the induced voltage $e = N(d\phi/dt) = N(0) = 0$.

Lenz's Law

In Section 11.2 it was shown that the magnetic flux linking a coil of N turns with a current I has the distribution of Fig. 12.3.

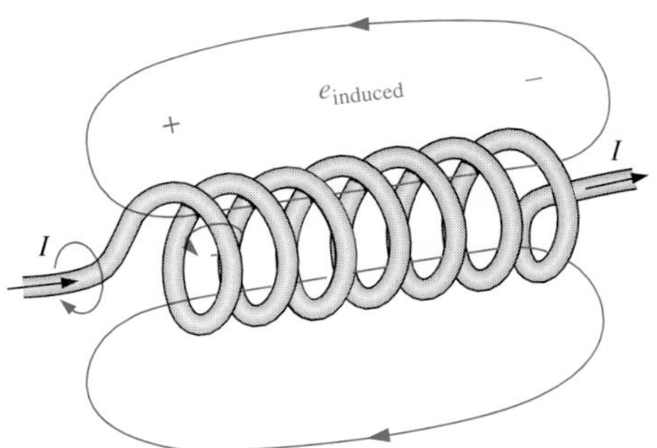

FIG. 12.3
Demonstrating the effect of Lenz's law.

If the current increases in magnitude, the flux linking the coil also increases. It was also shown above that a changing flux linking a coil induces a voltage across the coil. Therefore, for this coil an induced voltage (e_{ind}) is developed *across* the coil due to the change in current *through* the coil. The polarity of this induced voltage tends to establish a current in the coil that produces a flux that will oppose any change in the original flux.

However, the resulting induced voltage (e_{ind}) will tend to establish a current that will oppose the increasing change in current through the coil the instant the current begins to increase in magnitude. For this

LUMINARIES

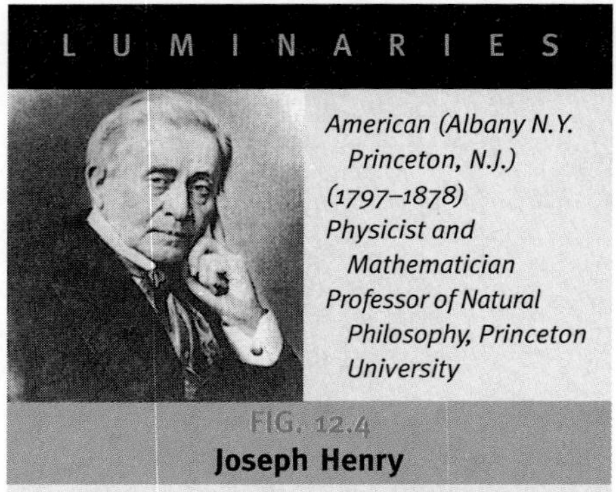

FIG. 12.4
Joseph Henry

As student and teacher at the Albany Academy, Henry performed extensive research in electromagnetism. He improved the design of *electromagnets* by insulating the wire to permit a tighter wrap on the core. An early design could lift almost two tonnes. In 1832 he discovered *self-induction*. He soon constructed an effective *electric telegraph system*. In 1845 he was appointed the first Secretary of the Smithsonian Institution. Note that his title of Professor of Natural Philosophy was common for science educators in the early 1800s. Even today, a doctoral degree in science and engineering is often called Doctor of Philosophy.

Courtesy of the Smithsonian Institution, Photo No. 59,054

reason, there will be an opposing effect trying to limit the change. In other words, the induced effect (e_{ind}) is the result of the increasing (or decreasing) current through the coil. This opposing effect is "choking" the change in current through the coil. Hence, the term *choke* is often applied to the inductor or coil.

The reaction above is true when the current is increasing or decreasing through the coil. This effect is an example of a general principle known as **Lenz's law**, which states that:

an induced effect is always such as to oppose the cause that produced it.

Recall in Chapter 10 that we found that the voltage across a capacitor cannot change instantaneously. Similarly, we will find that the current through a coil cannot change instantaneously. A period of time determined by the coil and the resistance of the circuit is required before the inductor discontinues its opposition to a momentary change in current. This is referred to as the **time constant** of the system.

Self-Inductance

The ability of a coil to oppose any change in current is a measure of the **self-inductance** L of the coil. For brevity, the prefix *self* is usually dropped. Inductance is measured in henries (H), after the American physicist Joseph Henry (Fig. 12.4).

Inductors are coils of various dimensions designed to introduce specified amounts of inductance into a circuit. The inductance of a coil varies directly with the magnetic properties of the coil. Ferromagnetic materials, therefore, are frequently used to increase the inductance by increasing the flux linking the coil.

The following equation gives us a close approximation for the inductance of the coils of Fig. 12.5:

$$L = \frac{N^2\mu A}{l} \quad \text{(henries, H)} \qquad (12.2)$$

where N represents the number of turns; μ, the permeability of the core (as introduced in Section 11.4; μ is not a constant but depends on the strength of the magnetomotive force); A, the area of the core in square metres; and l, the mean length of the core in metres.

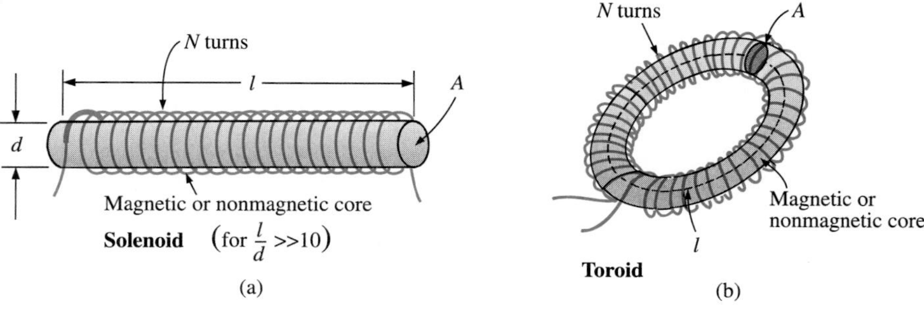

FIG. 12.5
Inductor configurations for which Eq. (12.2) is appropriate.

Substituting $\mu = \mu_r\mu_o$ into Eq. (12.2) gives

$$L = \frac{N^2\mu_r\mu_o A}{l} = \mu_r\frac{N^2\mu_o A}{l}$$

and

$$L = \mu_r L_o \tag{12.3}$$

where L_o is the inductance of the coil with an air core. In other words, the inductance of a coil with a ferromagnetic core is the relative permeability of the core times the inductance achieved with an air core.

Equations for the inductance of coils different from those shown above can be found in reference handbooks. Most of the equations are more complex than those just described.

EXAMPLE 12.1

Find the inductance of the air-core coil of Fig. 12.6.

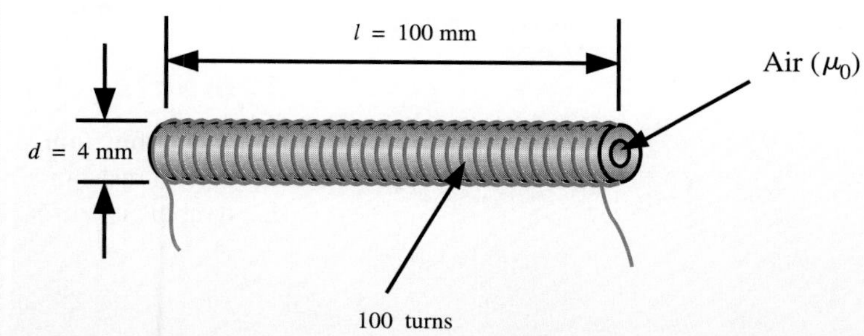

FIG. 12.6
Example 12.1.

Solution:

$$\mu = \mu_r\mu_o = (1)(\mu_o) = \mu_o$$

$$A = \frac{\pi d^2}{4} = \frac{(\pi)(4 \times 10^{-3}\text{ m})^2}{4} = 12.57 \times 10^{-6}\text{ m}^2$$

$$L_o = \frac{N^2\mu_o A}{l} = \frac{(100\text{ t})^2(4\pi \times 10^{-7}\text{ Wb/A·m})(12.57 \times 10^{-6}\text{ m}^2)}{0.1\text{ m}}$$

$$= \mathbf{1.58\ \mu H}$$

EXAMPLE 12.2

Repeat Example 12.1, but with an iron core and conditions such that $\mu_r = 2000$.

Solution: By Eq. (12.3),

$$L = \mu_r L_o = (2000)(1.58 \times 10^{-6}\text{ H}) = \mathbf{3.16\ mH}$$

12.3 TYPES OF INDUCTORS

Practical Equivalent

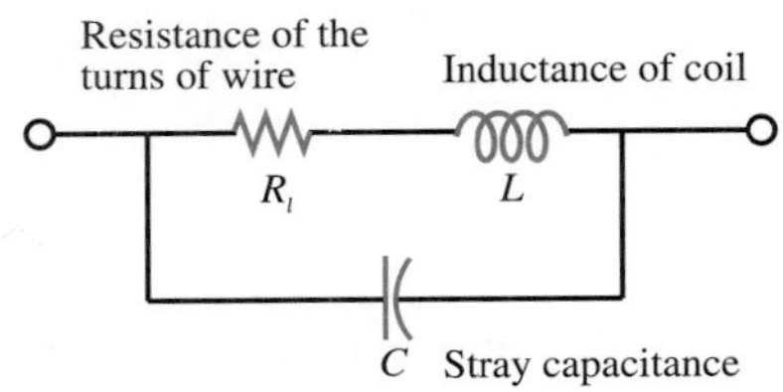

FIG. 12.7
Complete equivalent model for an inductor.

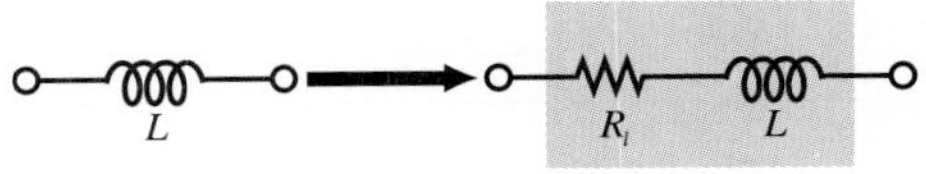

FIG. 12.8
Practical equivalent model for an inductor.

Inductors, like capacitors, are not ideal. Associated with every inductor are a resistance equal to the resistance of the turns and a stray capacitance due to the capacitance between the turns of the coil. To include these effects, the equivalent circuit for the inductor is as shown in Fig. 12.7. However, for most applications considered in this text, the stray capacitance appearing in Fig. 12.7 can be ignored, resulting in the equivalent model of Fig. 12.8. The resistance R_l can play an important role in the analysis of networks with inductive elements.

For most applications, we have been able to treat the capacitor as an ideal element and maintain a high degree of accuracy. For the inductor, however, R_l must often be included in the analysis and can have a significant effect on the response of a system (see Chapter 20, Resonance). The level of R_l can extend from a few ohms to a few hundred ohms. Keep in mind that the longer or thinner the wire used in the construction of the inductor, the greater will be the dc resistance as determined by $R = \rho l/A$. Our initial analysis will treat the inductor as an ideal element. Once a general feeling for the response of the element is established, we will include the effects of R_l.

Symbols

The main function of the inductor, however, is to introduce inductance—not resistance or capacitance—into the network. For this reason, the symbols used for inductance are as shown in Fig. 12.9.

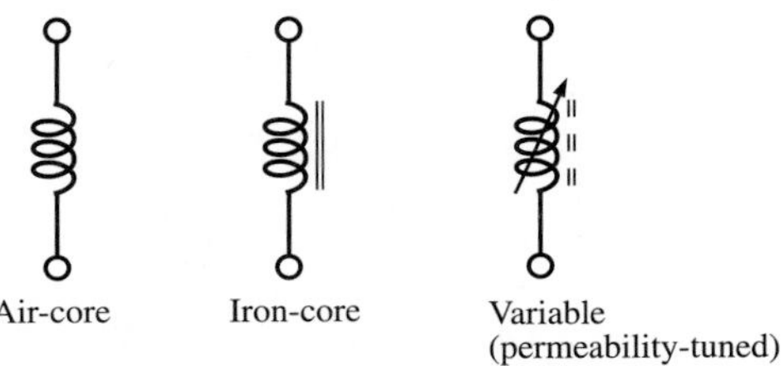

FIG. 12.9
Inductor symbols.

Appearance

All inductors, like capacitors, can be listed under two general headings: *fixed* and *variable*. The fixed air-core and iron-core inductors were described in the last section. The permeability-tuned variable coil has a ferromagnetic shaft that can be moved within the coil to vary the flux linkages of the coil and thereby its inductance. A few types of inductors are shown in Fig. 12.10. Some typical applications of various inductors are given in Fig. 12.11.

12.4 INDUCED VOLTAGE

The inductance of a coil is also a measure of the change in flux linking a coil due to a change in current through the coil; that is,

$$L = N\frac{d\phi}{di} \qquad \text{(henries, H)} \qquad (12.4)$$

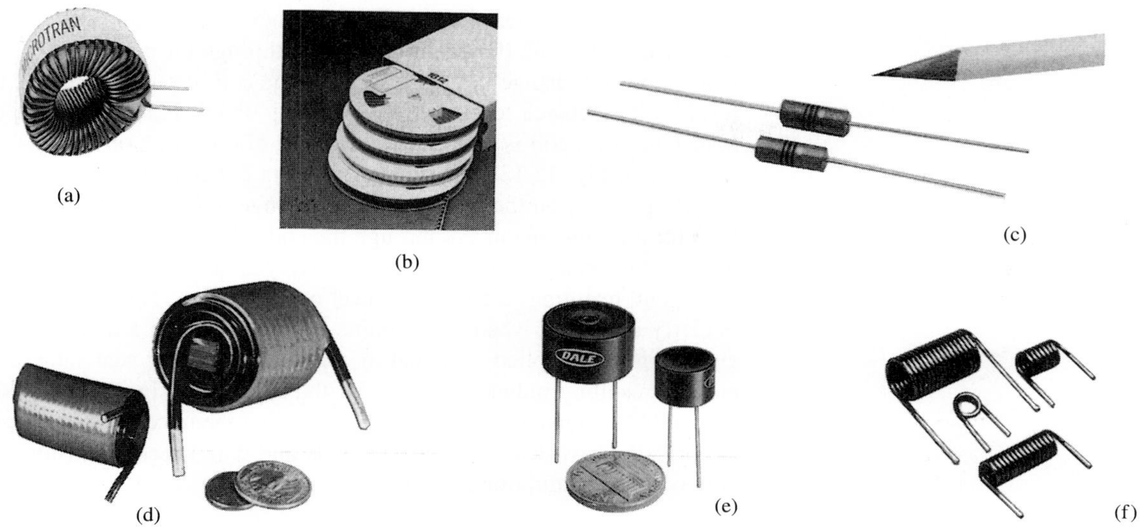

FIG. 12.10
Various types of inductors: (a) toroidal power inductor (1.4 μH to 5.6 mH) (courtesy of Microtan Co., Inc.); (b) surface mount inductors on reels (0.1 μH through 1000 μH on 500-piece reels in 46 values) (courtesy of Bell Industries); (c) molded inductors (0.1 μH to 10 μH); (d) high current filter inductors (24 μH at 60 A to 500 μH at 15 A); (e) toroid filter inductors (40 μH to 5 H); (f) air-core inductors (1 to 32 turns) for high-frequency applications. (Parts (c) through (f) courtesy of Dale Electronics, Inc.)

Type: Open Core Coil
Typical Values: 3 mH to 40 mH
Description: Used in low-pass filter circuits. Found in speaker crossover networks.

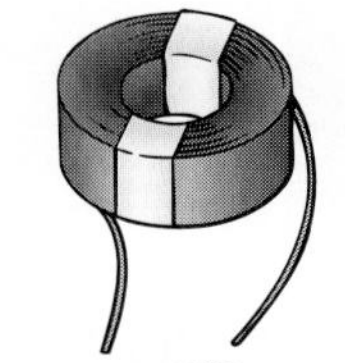

Type: Toroid Coil
Typical Values: 1 mH to 30 mH
Description: Used as a choke in AC power lines circuits to filter transient and reduce EMI interference. This coil is found in many electronic appliances.

Type: Hash Choke Coil
Typical Values: 3 μH to 1 mH
Description: Used in AC supply lines that deliver high currents.

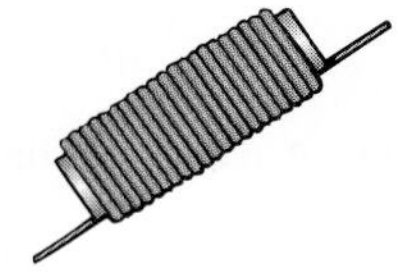

Type: Delay Line Coil
Typical Values: 10 μH to 50 μH
Description: Used in color televisions to correct for timing differences between the colour signal and black and white signal.

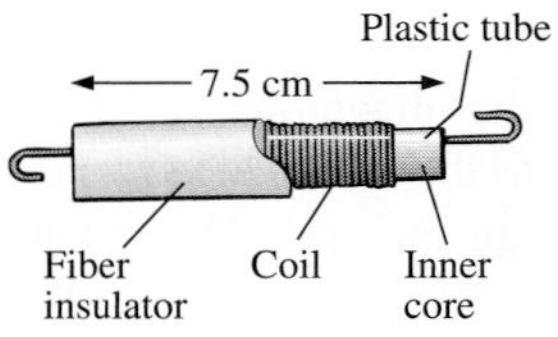

Plastic tube
7.5 cm
Fiber insulator
Coil
Inner core

Type: Common Mode Choke Coil
Typical Values: 0.6 mH to 50 mH
Description: Used in AC line filters, switching power supplies, battery charges and other electronic equipment.

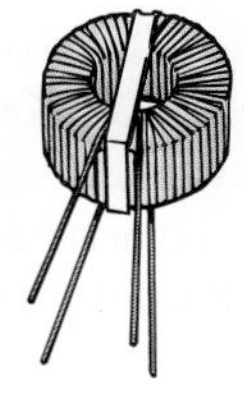

Type: RF Chokes
Typical Values: 10 μH to 50 μH
Description: Used in radio, television, and communication circuits. Found in AM, FM, and UHF circuits.

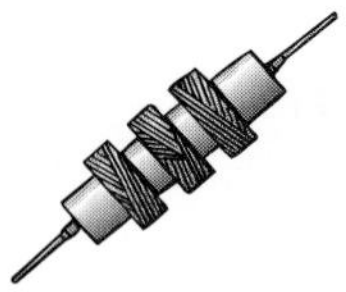

Type: Molded Coils
Typical Values: 0.1 μH to 100 μH
Description: Used in a wide variety of circuit such as oscillators, filters, pass-band filters, and others.

Type: Surface Mounted Inductors
Typical Values: 0.01 μH to 100 μH
Description: Found in many electronic circuits that require miniature components on multilayered PCB.

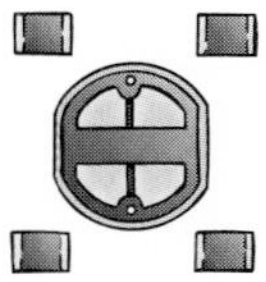

Type: Adjustable RF Coil
Typical Values: 1 μH to 100 μH
Description: Variable inductor used in oscillators and various RF circuits such as CB transceivers, televisions, and radios.

FIG. 12.11
Typical areas of application for inductive elements.

where N is the number of turns, ϕ is the flux in webers, and i is the current through the coil. If a change in current through the coil fails to result in a significant change in the flux linking the coil through its centre, the resulting inductance level will be relatively small. For this reason the inductance of a coil is sensitive to the point of operation on the hysteresis curve of Fig. 12.12 (described in detail in Section 11.5).

If operating on the steep slope, the change in flux will be relatively high for a change in current through the coil. If operating near or in saturation, the change in flux will be relatively small for the same change in current, resulting in a reduced level of inductance. This effect is particularly important when we examine ac circuits, since a dc level associated with the applied ac signal may put the coil at or near saturation, and the resulting inductance level for the applied ac signal will be significantly less than expected. You will find that the maximum dc current is normally provided in supply manuals and data sheets to ensure that you avoid the saturation region.

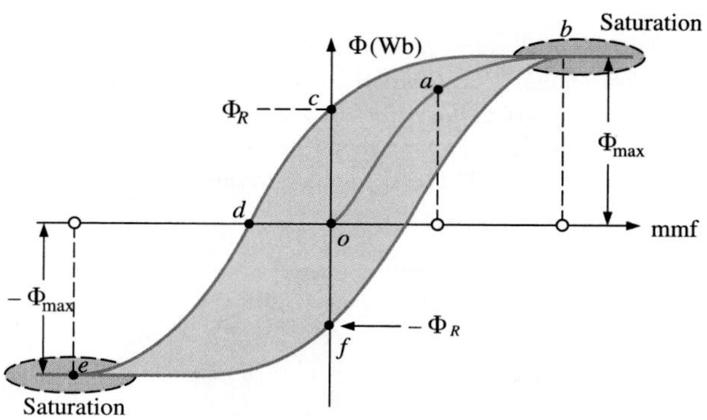

FIG. 12.12
Hysteresis curve.

Equation (12.4) also reveals that the larger the inductance of a coil (with N fixed), the larger will be the instantaneous change in flux linking the coil due to an instantaneous change in current through the coil.

If we write Eq. (12.1) as

$$e_L = N\frac{d\phi}{dt} = \left(N\frac{d\phi}{di}\right)\left(\frac{di}{dt}\right)$$

and substitute Eq. (12.4), we then have

$$e_L = L\frac{di}{dt} \qquad \text{(volts, V)} \qquad (12.5)$$

revealing that the magnitude of the voltage across an inductor is directly related to the inductance L and the instantaneous rate of change of current through the coil. The greater the *rate* of change of current through the coil, the greater will be the induced voltage, which agrees with Lenz's law.

When induced effects are used in generating voltages such as those available from dc or ac generators, the symbol e is appropriate for the induced voltage. In network analysis, since the voltage across an inductor will always have a polarity such as to oppose the source that produced it, we use v_L in the following analysis:

$$v_L = L\frac{di}{dt} \qquad \text{(volts,V)} \qquad (12.6)$$

If the current through the coil fails to change at a particular instant, the induced voltage across the coil will be zero. For dc applications, after the transient effect has passed, $di/dt = 0$, and the induced voltage is

$$v_L = L\frac{di}{dt} = L(0) = 0 \text{ V}$$

Recall that the equation for the current of a capacitor is the following:

$$i_C = C\frac{dv_C}{dt}$$

Note the similarity between this equation and Eq. (12.6). If we apply the duality $v \leftrightarrows i$ (that is, interchange the two) and $L \leftrightarrows C$ for capacitance and inductance, each equation can be derived from the other.

The average voltage across the coil is defined by the equation

$$v_{L_{av}} = L\frac{\Delta i}{\Delta t} \qquad \text{(volts, V)} \qquad (12.7)$$

where Δ signifies finite change (a measurable change). Compare this to $i_C = C(\Delta v/\Delta t)$, and the meaning of Δ and application of this equation should be clarified from Chapter 10. An example follows.

EXAMPLE 12.3

Find the waveform for the average voltage across the coil if the current through a 4-mH coil is as shown in Fig. 12.13.

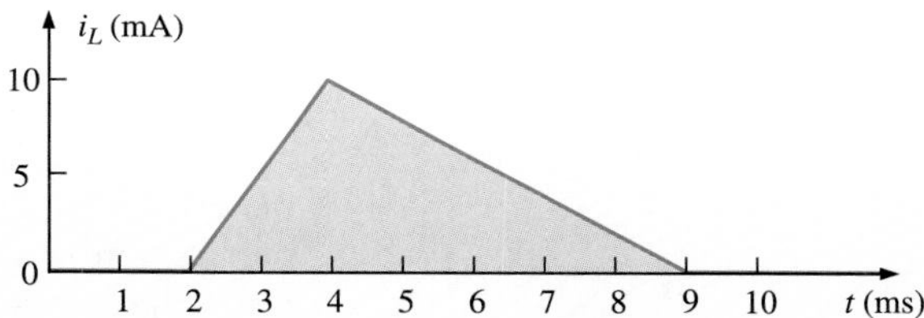

FIG. 12.13
Example 12.3.

Solution:

a. 0 to 2 ms: Since there is no change in current through the coil, there is no voltage induced across the coil; that is,

$$v_L = L\frac{\Delta i}{\Delta t} = L\frac{0}{\Delta t} = \mathbf{0}$$

b. 2 ms to 4 ms:

$$v_L = L\frac{\Delta i}{\Delta t} = (4 \times 10^{-3}\text{ H})\left(\frac{10 \times 10^{-3}\text{ A}}{2 \times 10^{-3}\text{ s}}\right) = 20 \times 10^{-3}\text{ V}$$
$$= \mathbf{20\ mV}$$

c. 4 ms to 9 ms:

$$v_L = L\frac{\Delta i}{\Delta t} = (-4 \times 10^{-3}\text{ H})\left(\frac{10 \times 10^{-3}\text{ A}}{5 \times 10^{-3}\text{ s}}\right) = -8 \times 10^{-3}\text{ V}$$
$$= \mathbf{-8\ mV}$$

continued

d. 9 ms to ∞:

$$v_L = L\frac{\Delta i}{\Delta t} = L\frac{0}{\Delta t} = \mathbf{0}$$

The waveform for the average voltage across the coil is shown in Fig. 12.14.

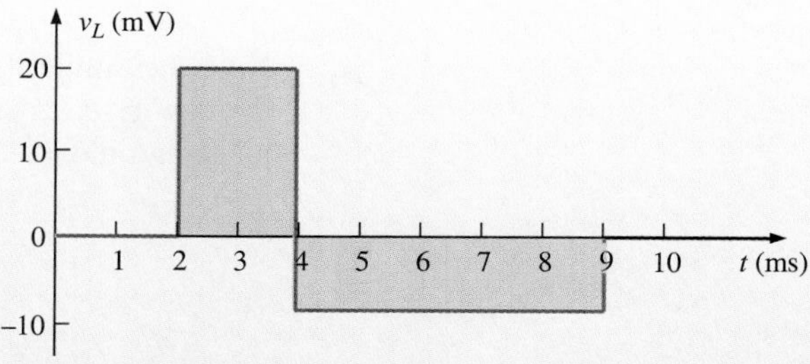

FIG. 12.14
Voltage across a 4-mH coil due to the current of Fig. 12.13.

Note from the curve that the voltage across the coil is not determined solely by the magnitude of the change in current through the coil (Δi), but by the rate of change of current through the coil (Δi/Δt).

A similar statement was made for the current of a capacitor due to a change in voltage across the capacitor.

If you carefully examine Fig. 12.14 you will see that the area under the positive pulse from 2 ms to 4 ms equals the area under the negative pulse from 4 ms to 9 ms. In Section 12.9, we will find that the area under the curves represents the energy stored or released by the inductor. From 2 ms to 4 ms, the inductor is storing energy, whereas from 4 ms to 9 ms, the inductor is releasing the energy stored. For the full period zero to 10 ms, energy has simply been stored and released; there has been no dissipation as experienced for resistive elements. Over a full cycle, both the ideal capacitor and inductor do not consume energy but simply store and release it in their respective forms.

12.5 *R-L* TRANSIENTS

Storage Cycle

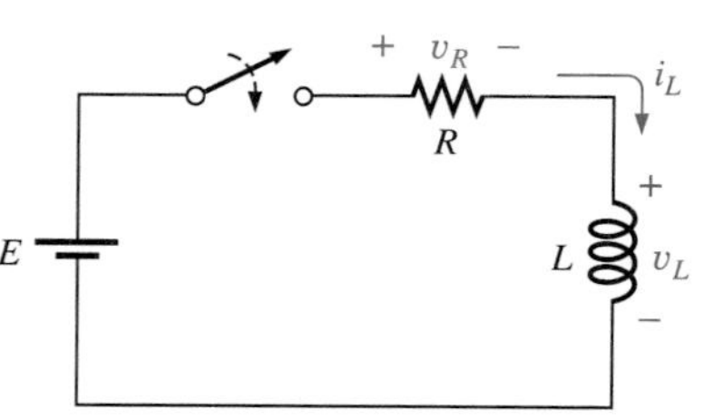

FIG. 12.15
Basic R-L transient network.

The changing voltages and currents that result during the storing of energy in the form of a magnetic field by an inductor in a dc circuit can best be described using the circuit of Fig. 12.15. At the instant the switch is closed, the inductance of the coil will prevent an instantaneous change in current through the coil. The potential drop across the coil, v_L, will equal the impressed voltage E as determined by Kirchhoff's voltage law since $v_R = iR = (0)R = 0$ V. The current i_L will then build up from zero, establishing a voltage drop across the resistor and a corresponding drop in v_L. The current will continue to increase until the voltage across the inductor drops to zero volts and the full impressed voltage appears across the resistor. Initially, the current i_L increases quite rapidly, followed by a continually decreasing rate until it reaches its maximum value of E/R.

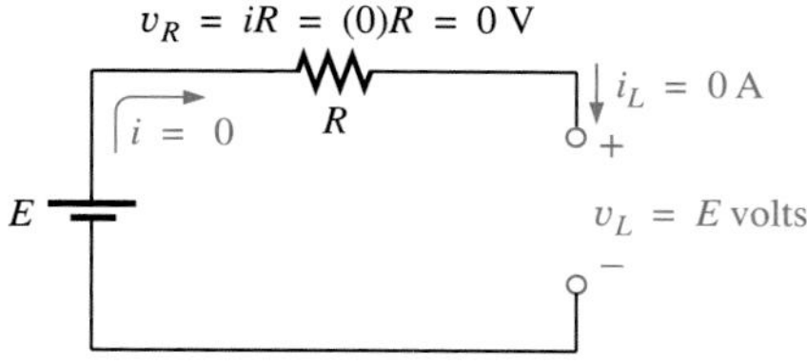

FIG. 12.16
Circuit of Fig. 12.15 the instant the switch is closed.

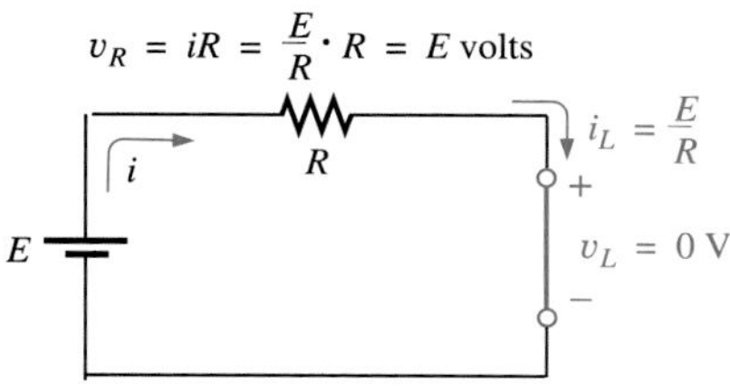

FIG. 12.17
Circuit of Fig. 12.15 under steady-state conditions.

You will recall from the discussion of capacitors that a capacitor has a short-circuit equivalent when the switch is first closed and an open-circuit equivalent when steady-state conditions are established. The inductor assumes the opposite equivalents for each stage. The instant the switch of Fig. 12.15 is closed, the equivalent network will appear as shown in Fig. 12.16. Note the connection with the earlier comments about the levels of voltage and current. The inductor meets all the requirements for an open-circuit equivalent, $v_L = E$ volts, $i_L = 0$ A.

When steady-state conditions have been established and the storage phase is complete, the "equivalent" network will appear as shown in Fig. 12.17. The network clearly reveals that:

An ideal inductor ($R_l = 0$ V) assumes a short-circuit equivalent in a dc network once steady-state conditions have been established.

Fortunately, the mathematical equations for the voltages and current for the storage phase are similar to those for the *R-C* network. The experience you gained with these equations in Chapter 10 will help you understand the analysis of *R-L* networks.

The equation for the current i_L during the storage phase is the following:

$$i_L = I_m(1 - e^{-t/\tau}) = \frac{E}{R}(1 - e^{-t/(L/R)}) \qquad (12.8)$$

Note the factor $(1 - e^{-t/\tau})$, which also appeared for the voltage v_C of a capacitor during the charging phase. Also note that the *e* in the expression $(1 - e^{-t/\tau})$ is the base of the natural logarithms. It is *not* the *e* referring to an induced voltage.

A plot of Eq. (12.8) is given in Fig. 12.18, indicating that the maximum steady-state value of i_L is E/R, and that the rate of change in current decreases as time passes. The horizontal axis is scaled in time constants, with τ for inductive circuits defined by the following:

$$\tau = \frac{L}{R} \quad \text{(seconds, s)} \qquad (12.9)$$

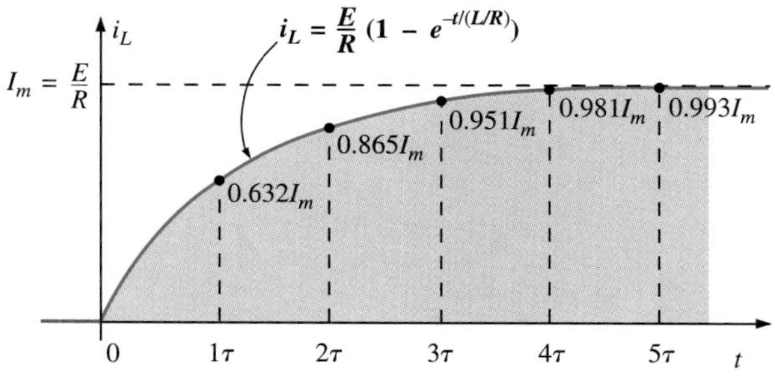

FIG. 12.18
Plotting the waveform for i_L during the storage cycle.

The fact that τ has the units of time can be verified by taking the equation for the induced voltage

$$v_L = L\frac{di}{dt}$$

and solving for L: $$L = \frac{v_L}{di/dt}$$

which leads to the ratio

$$\tau = \frac{L}{R} = \frac{\dfrac{v_L}{di/dt}}{R} = \frac{v_L}{\dfrac{di}{dt}R} \rightarrow \frac{V}{\dfrac{IR}{t}} = \frac{\cancel{V}}{\dfrac{\cancel{V}}{t}} = t \text{ (s)}$$

Our experience with the factor $(1 - e^{-t/\tau})$ verifies the level of 63.2% after one time constant, 86.5% after two time constants, and so on. For convenience, Fig. 10.27 is repeated as Fig. 12.19 to evaluate the functions $(1 - e^{-t/\tau})$ and $e^{-t/\tau}$ at various values of τ.

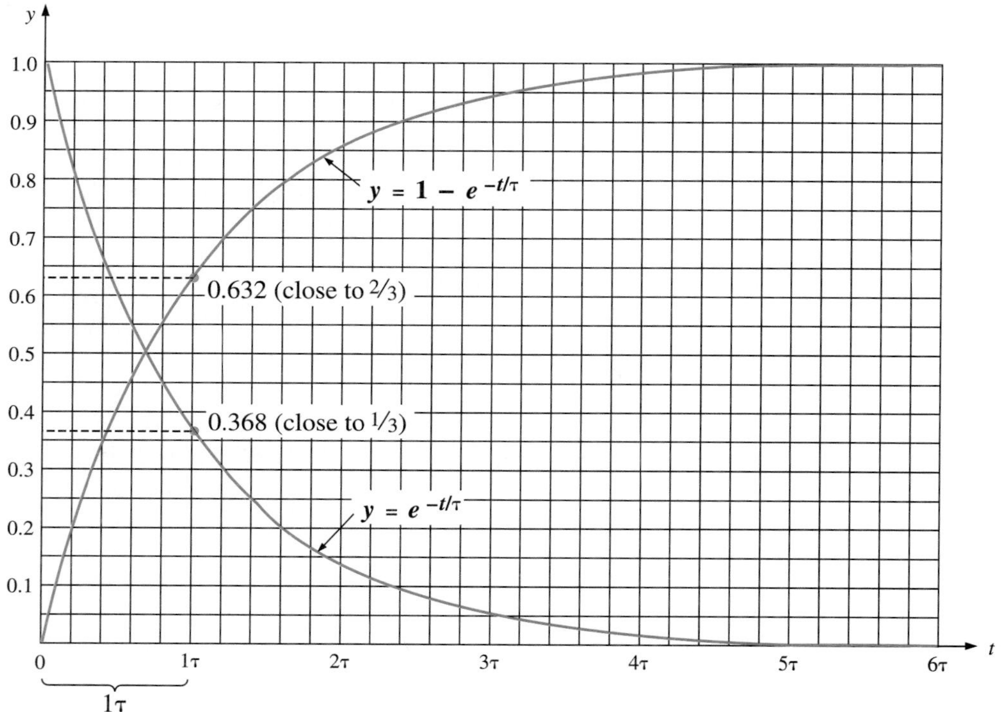

FIG. 12.19
Plotting the functions $y = 1 - e^{-t/\tau}$ and $y = e^{-t/\tau}$.

If we keep R constant and increase L, the ratio L/R increases and the rise time increases. The change in transient behaviour for the current i_L is plotted in Fig. 12.20 for various values of L. Note again the similarity between these curves and those obtained for the R-C network in Fig. 10.30.

For most practical applications, we will assume that:

The storage phase has passed and steady-state conditions have been established once a period of time equal to five time constants has occurred.

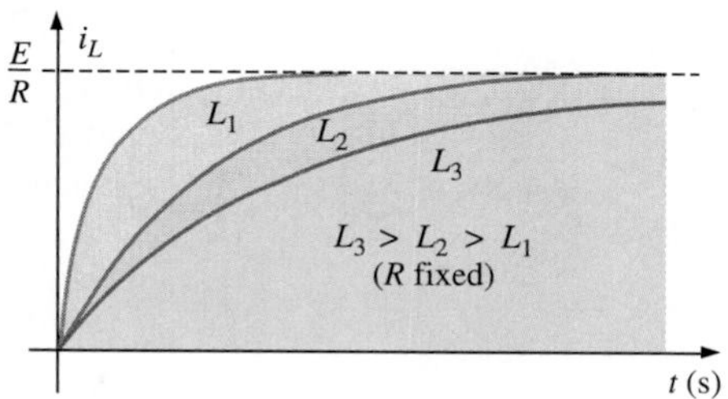

FIG. 12.20
Effect of L on the shape of the i_L storage waveform.

In addition, since L/R will always have some numerical value, even though it may be very small, the period 5τ will always be greater than zero, confirming the fact that

the current cannot change instantaneously in an inductive network.

The larger the inductance, the more the circuit will oppose a rapid buildup in current level.

Figures 12.16 and 12.17 show that the voltage across the coil jumps to E volts when the switch is closed and decays to zero volts with time. The decay occurs in an exponential manner, and v_L during the storage phase can be described mathematically by the following equation:

$$v_L = Ee^{-t/\tau} \tag{12.10}$$

A plot of v_L appears in Fig. 12.21 with the time axis again divided into equal increments of τ. The voltage v_L will decrease to zero volts at the same rate the current presses toward its maximum value.

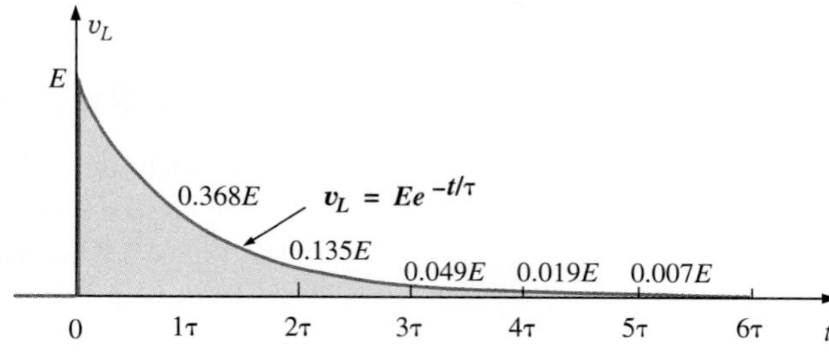

FIG. 12.21
Plotting the voltage v_R versus time for the network of Fig. 12.15.

In five time constants, $i_L = E/R$, $v_L = 0\ \Omega$, and the inductor can be replaced by its short-circuit equivalent.

Since $$v_R = i_R R = i_L R$$

then $$v_R = \left[\frac{E}{R}(1 - e^{-t/\tau})\right]R$$

and $$v_R = E(1 - e^{-t/\tau}) \tag{12.11}$$

and the curve for v_R will have the same shape as obtained for i_L.

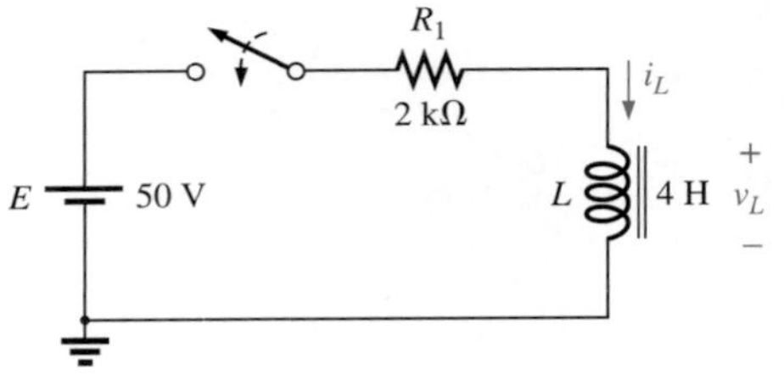

FIG. 12.22
Example 12.4.

EXAMPLE 12.4

Find the mathematical expressions for the transient behaviour of i_L and v_L for the circuit of Fig. 12.22 after the closing of the switch. Sketch the resulting curves.

Solution:

$$\tau = \frac{L}{R_1} = \frac{4\ \text{H}}{2\ \text{k}\Omega} = 2\ \text{ms}$$

a. By Eq (12.8),

$$I_m = \frac{E}{R_1} = \frac{50}{2\ \text{k}\Omega} = 25 \times 10^{-3}\ \text{A} = 25\ \text{mA}$$

and
$$\mathbf{i_L = (25 \times 10^{-3})(1 - e^{-t/(2\times 10^{-3})})}$$

b. By Eq. (12.10),

$$\mathbf{v_L = 50e^{-t/(2\times 10^{-3})}}$$

Both waveforms appear in Fig. 12.23.

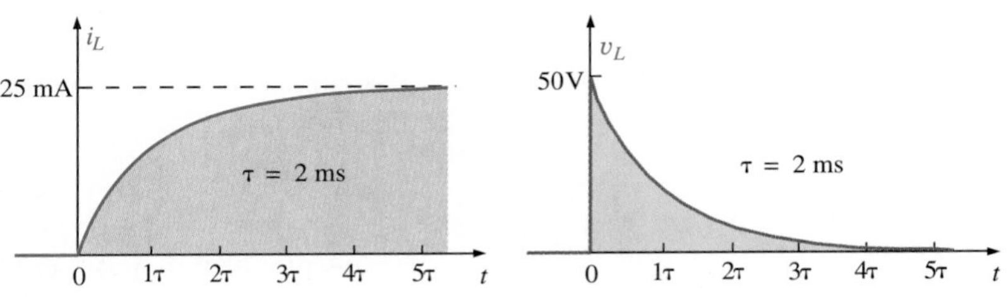

FIG. 12.23
i_L and v_L for the network of Fig. 12.22.

Decay Phase

In the analysis of *R-C* circuits, we found that the capacitor could hold its charge and store energy in the form of an electric field for a period of time determined by the leakage factors. In *R-L* circuits, the energy is stored in the form of a magnetic field established by the current through the coil. Unlike the capacitor, however, an isolated inductor cannot continue to store energy since the absence of a closed path would cause the current to drop to zero, releasing the energy that was stored in the form of a magnetic field.

If the switch of Fig. 12.22 were opened quickly, a spark would probably occur across the contacts due to the rapid change in current from a maximum of E/R to zero amperes. The change in current di/dt of the equation $v_L = L(di/dt)$ would establish a high voltage v_L across the coil that would discharge across the points of the switch. This is the same mechanism as applied in the ignition system of a car to ignite the fuel in the cylinder. Some 25 000 V are generated by the rapid decrease in ignition coil current that occurs when the switch in the system is opened. (In older systems, the points in the distributor served as the switch.) This inductive reaction is significant when you consider that the only independent source in a car is a 12-V battery.

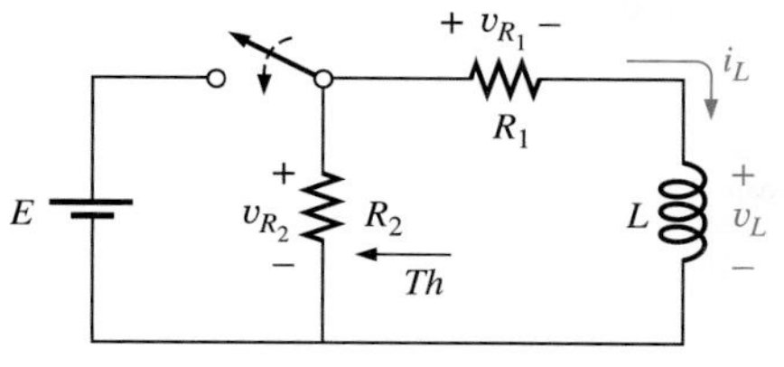

FIG. 12.24
Initiating the storage phase for the inductor L by closing the switch.

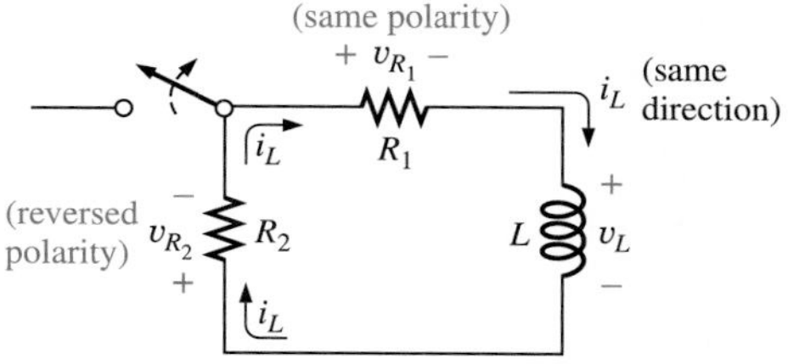

FIG. 12.25
Network of Fig. 12.24 the instant the switch is opened.

If opening the switch to move it to another position will cause such a rapid discharge in stored energy, how can the decay phase of an *R-L* circuit be analyzed in the same way as the *R-C* circuit? The solution is to use a network like the one in Fig. 12.24. When the switch is closed, the voltage across the resistor R_2 is E volts and the *R-L* branch will respond as described above, with the same waveforms and levels. A Thévenin network of E in parallel with R_2 becomes only the source, since R_2 is shorted out by the short-circuit replacement of the voltage source E when the Thévenin resistance is determined.

After the storage phase has passed and steady-state conditions are established, the switch can be opened without the sparking effect or rapid discharge due to the resistor R_2, which provides a complete path for the current i_L. For clarity the discharge path is isolated in Fig. 12.25. The voltage v_L across the inductor will reverse polarity and have a magnitude determined by

$$v_L = v_{R_1} + v_{R_2} \tag{12.12}$$

Recall that the voltage across an inductor can change instantaneously but the current cannot. The result is that the current i_L must maintain the same direction and magnitude as shown in Fig. 12.25. Therefore, the instant after the switch is opened, i_L is still $I_m = E/R_1$, and

$$\begin{aligned} v_L &= v_{R_1} + v_{R_2} = i_1R_1 + i_2R_2 \\ &= i_L(R_1 + R_2) = \frac{E}{R_1}(R_1 + R_2) = \left(\frac{R_1}{R_1} + \frac{R_2}{R_1}\right)E \end{aligned}$$

and

$$v_L = \left(1 + \frac{R_2}{R_1}\right)E \tag{12.13}$$

which is bigger than E volts by the ratio R_2/R_1. When the switch is opened, the voltage across the inductor will reverse polarity and drop instantaneously from E to $-[1 + (R_2/R_1)]E$ volts. The minus sign is a result of v_L having a polarity opposite to the defined polarity of Fig. 12.25.

As an inductor releases its stored energy, the voltage across the coil will decay to zero in the following way:

$$v_L = V_i e^{-t/\tau'} \tag{12.14}$$

with

$$V_i = \left(1 + \frac{R_2}{R_1}\right)E$$

and

$$\tau' = \frac{L}{R_T} = \frac{L}{R_1 + R_2}$$

The current will decay from a maximum of $I_m = E/R_1$ to zero, in the following way:

$$i_L = I_m e^{-t/\tau'} \tag{12.15}$$

with

$$I_m = \frac{E}{R_1} \quad \text{and} \quad \tau' = \frac{L}{R_1 + R_2}$$

The mathematical expression for the voltage across either resistor can then be determined using Ohm's law:

$$\begin{aligned} v_{R_1} &= i_{R_1}R_1 = i_L R_1 \\ &= I_m e^{-t/\tau'} R_1 \\ &= \frac{E}{R_1} R_1 e^{-t/\tau'} \end{aligned}$$

and

$$v_{R_1} = Ee^{-t/\tau'} \qquad (12.16)$$

The voltage v_{R_1} has the same polarity as during the storage phase since the current i_L has the same direction. The voltage v_{R_2} is expressed as follows:

$$\begin{aligned} v_{R_2} &= i_{R_2}R_2 = i_L R_2 \\ &= I_m e^{-t/\tau'} R_2 \\ &= \frac{E}{R_1} R_2 e^{-t/\tau'} \end{aligned}$$

and

$$v_{R_2} = \frac{R_2}{R_1} Ee^{-t/\tau'} \qquad (12.17)$$

with the polarity indicated in Fig. 12.25.

EXAMPLE 12.5

The network of Fig. 12.26 is modified from that of Fig. 12.22 by adding the resistor, $R_2 = 3\text{ k}\,\Omega$. For this new network:

a. Find the mathematical expressions for i_L, v_L, v_{R_1}, and v_{R_2} after the storage phase has been completed and the switch is opened.
b. Sketch the waveforms for each voltage and current for both phases covered by this example and Example 12.4 if five time constants pass between phases. Use the defined polarities of Fig. 12.24.

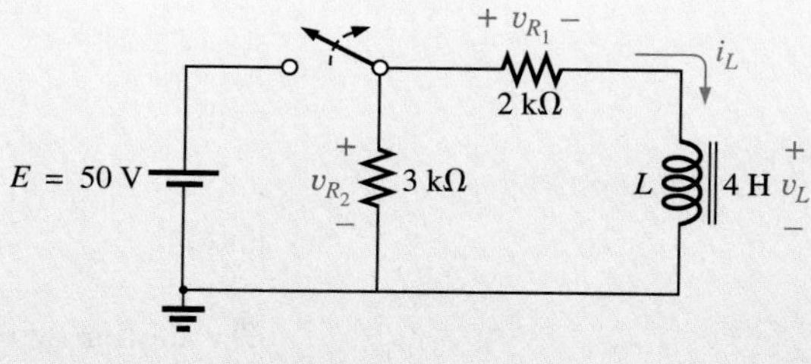

FIG. 12.26
Example 12.5.

continued

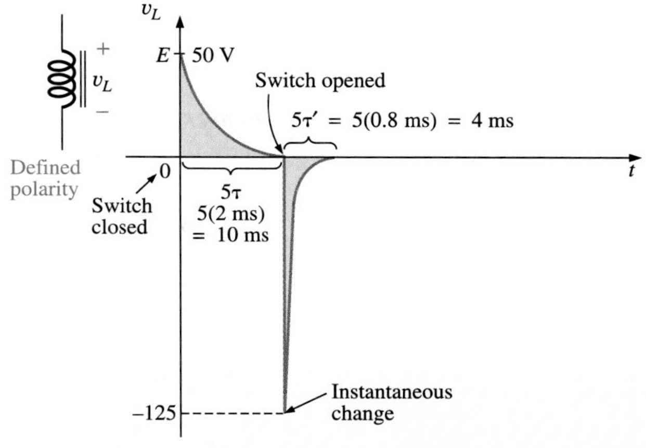

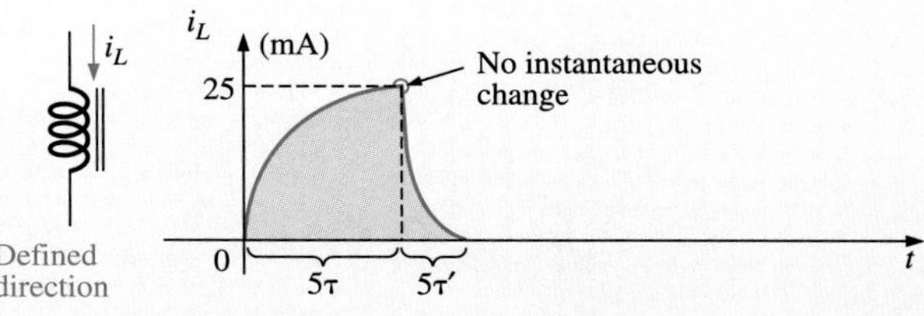

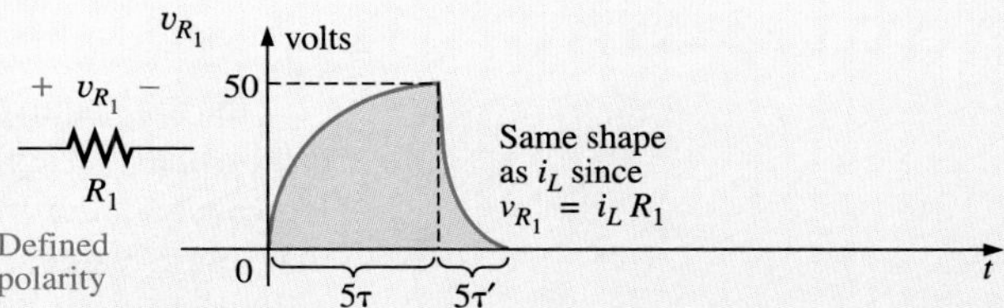

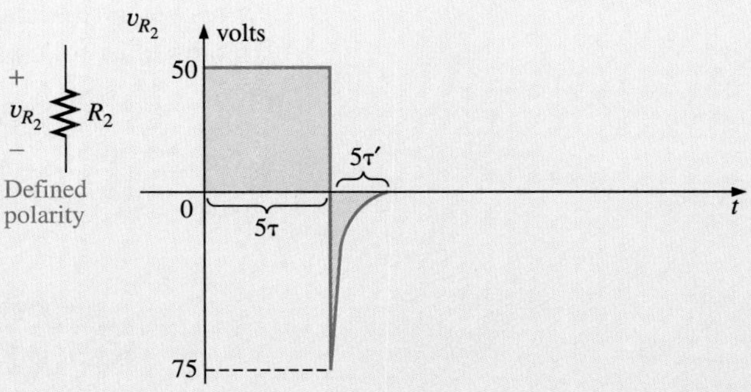

FIG. 12.27
The various voltages and the current for the network of Fig. 12.26.

continued

Solution:

a. $\tau' = \dfrac{L}{R_1 + R_2} = \dfrac{4\text{ H}}{2\text{ k}\Omega + 3\text{ k}\Omega} = \dfrac{4\text{ H}}{5 \times 10^3\ \Omega} = 0.8 \times 10^{-3}\text{ s}$

$= 0.8\text{ ms}$

By Eq. (12.14),

$$V_i = \left(1 + \frac{R_2}{R_1}\right)E = \left(1 + \frac{3\text{ k}\Omega}{2\text{ k}\Omega}\right)(50\text{ V}) = 125\text{ V}$$

and $$v_L = -V_i e^{-t/\tau'} = \mathbf{-125}\boldsymbol{e}^{-t/(0.8\times 10^{-3})}$$

By Eq. (12.15),

$$I_m = \frac{E}{R_1} = \frac{50\text{ V}}{2\text{ k}\Omega} = 25\text{ mA}$$

and $$i_L = I_m e^{-t/\tau'} = \mathbf{(25 \times 10^{-3})}\boldsymbol{e}^{-t/(0.8\times 10^{-3})}$$

By Eq. (12.16),

$$v_{R_1} = Ee^{-t/\tau'} = \mathbf{50}\boldsymbol{e}^{-t/(0.8\times 10^{-3})}$$

By Eq. (12.17),

$$v_{R_2} = -\frac{R_2}{R_1}Ee^{-t/\tau'} = -\frac{3\text{ k}\Omega}{2\text{ k}\Omega}(50\text{ V})e^{-t/\tau'} = \mathbf{-75}\boldsymbol{e}^{-t/(0.8\times 10^{-3})}$$

b. See Fig. 12.27 on previous page.

In this analysis, we assumed that steady-state conditions were established during the charging phase and $I_m = E/R_1$, with $v_L = 0$ V. However, if the switch of Fig. 12.25 is opened before i_L reaches its maximum value, the equation for the decaying current of Fig. 12.25 must change to

$$i_L = I_i e^{-t/\tau'} \tag{12.18}$$

where I_i is the starting or initial current. Equation (12.14) would be modified as follows:

$$v_L = V_i e^{-t/\tau'} \tag{12.19}$$

with $$V_i = I_i(R_1 + R_2)$$

12.6 CALCULATING INSTANTANEOUS VALUES AND TIME CONSTANTS

Initial Values

This section will parallel Section 10.9 on the effect of **initial values** on the transient phase. Since the current through a coil cannot change instantaneously, the current through a coil will begin the **transient phase** at the *initial value* established by the network (note Fig. 12.28) before the switch was closed. It will then pass through the transient phase until it reaches the **steady-state** (or *final*) level after about five time constants. The steady-state level of the inductor current can be found by simply substituting its short-circuit equivalent (or R_l for the practical equivalent) and finding the resulting current through the element.

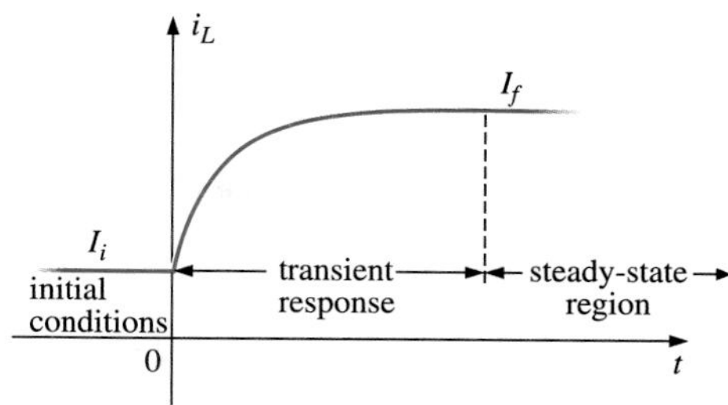

FIG. 12.28
Defining the three phases of a transient waveform.

Using the transient equation developed in the previous section, we can write an equation for the current i_L for the entire time interval of Fig. 12.28; that is,

$$i_L = I_i + (I_f - I_i)(1 - e^{-t/\tau})$$

with $(I_f - I_i)$ representing the total change during the transient phase. However, by multiplying through and rearranging terms:

$$\begin{aligned} i_L &= I_i + I_f - I_f e^{-t/\tau} - I_i + I_i e^{-t/\tau} \\ &= I_f - I_f e^{-t/\tau} + I_i e^{-t/\tau} \end{aligned}$$

we find

$$i_L = I_f + (I_i - I_f)e^{-t/\tau} \qquad (12.20)$$

If you have to draw the waveform for the current i_L from initial value to final value, start by drawing a line at the initial value and steady-state levels, and then add the transient response (sensitive to the time constant) between the two levels. The following example will help you understand the procedure.

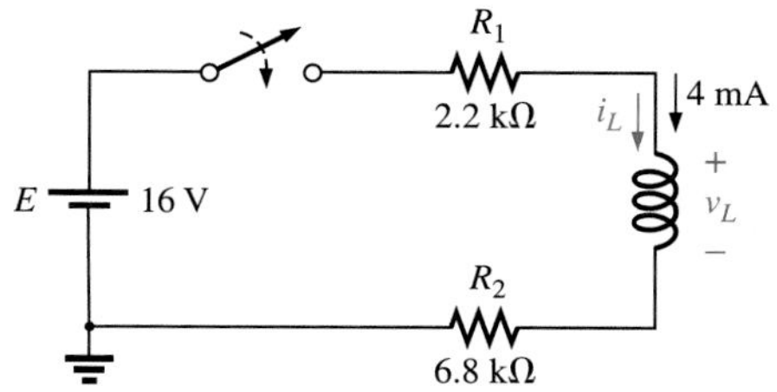

FIG. 12.29
Example 12.6.

EXAMPLE 12.6

The inductor of Fig. 12.29 has an initial current level of 4 mA in the direction shown. (We will present specific methods to establish the initial current in the sections and problems to follow.) Determine the transient current and voltage waveforms when the switch is closed. Follow these three steps:

1. **Find the mathematical expression for the current through the coil once the switch is closed.**
2. **Find the mathematical expression for the voltage across the coil during the same transient period.**
3. **Sketch the waveform for each from initial value to final value.**

Solution:

1. Substituting the short-circuit equivalent for the inductor will result in a final or steady-state current determined by Ohm's law:

$$I_f = \frac{E}{R_1 + R_2} = \frac{16\text{ V}}{2.2\text{ k}\Omega + 6.8\text{ k}\Omega} = \frac{16\text{ V}}{9\text{ k}\Omega} = 1.78\text{ mA}$$

The time constant is determined by:

$$\tau = \frac{L}{R_T} = \frac{100\text{ mH}}{2.2\text{ k}\Omega + 6.8\text{ k}\Omega} = \frac{100\text{ mH}}{9\text{ k}\Omega} = 11.11\ \mu\text{s}$$

continued

Applying Eq. (12.20):

$$i_L = I_f + (I_i - I_f)e^{-t/\tau}$$
$$= 1.78\text{ mA} + (4\text{ mA} - 1.78\text{ mA})e^{-t/11.11\ \mu s}$$
$$= \mathbf{1.78\ mA + 2.22\ mA\ } e^{-t/11.11\ \mu s}$$

2. Since the current through the inductor is constant at 4 mA before closing the switch, the voltage (whose level is sensitive only to changes in current through the coil) must have an initial value of 0 V. At the instant the switch is closed, the current through the coil cannot change instantaneously, so the current through the resistive elements will be 4 mA. You can then find the resulting peak voltage at $t = 0$ s using Kirchhoff's voltage law (Section 5.4) as follows:

$$V_m = E - V_{R_1} - V_{R_2}$$
$$= 16\text{ V} - (4\text{ mA})(2.2\text{ k}\Omega) - (4\text{ mA})(6.8\text{ k}\Omega)$$
$$= 16\text{ V} - 8.8\text{ V} - 27.2\text{ V} = 16\text{ V} - 36\text{ V}$$
$$= -20\text{ V}$$

Note the minus sign to indicate that the polarity of the voltage v_L is opposite to the defined polarity of Fig. 12.29.

The voltage will then decay (with the same time constant as the current i_L) to zero because the inductor is approaching its short-circuit equivalence.

The equation for v_L is therefore:

$$v_L = \mathbf{-20}e^{-t/11.11\ \mu s}$$

3. See Fig. 12.30. The initial and final values of the current were drawn first and then the transient response was included between these levels. For the voltage, the waveform begins and ends at zero, and the peak value has a sign sensitive to the defined polarity of v_L in Fig. 12.29.

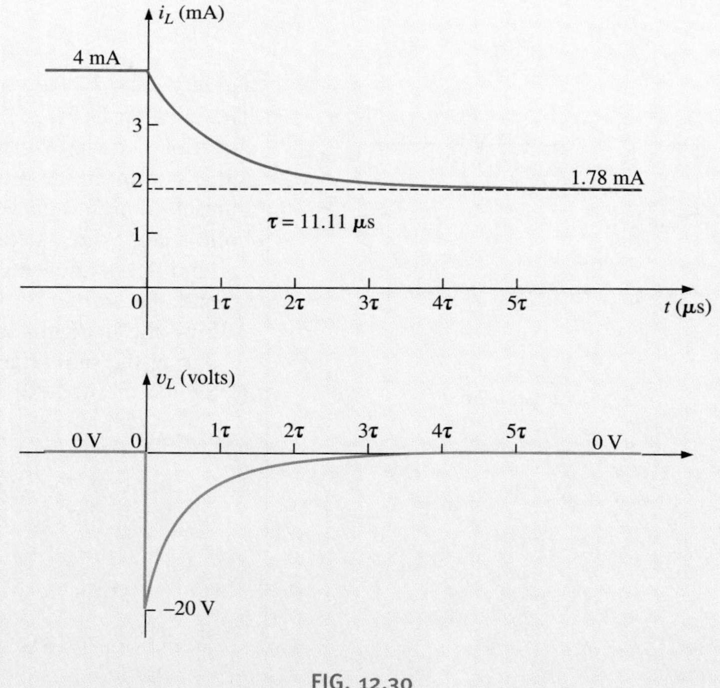

FIG. 12.30
i_L and v_L for the network of Fig. 12.29.

continued

Let us now test the validity of the equation for i_L by substituting $t = 0$ s to reflect the instant the switch is closed.

$$e^{-t/\tau} = e^{-0} = 1$$

and $i_L = 1.78\text{ mA} + 2.22\text{ mA}\, e^{-t/\tau} = 1.78\text{ mA} + 2.22\text{ mA}$
$= 4\text{ mA}$

when $\quad t > 5\tau,\ e^{-t/\tau} \cong 0$

and $\quad i_L = 1.78\text{ mA} + 2.22\text{ mA}\, e^{-t/\tau} = 1.78\text{ mA}$

Instantaneous Values

The development presented in Section 10.10 for capacitive networks can also be applied to *R-L* networks to determine instantaneous voltages, currents, and times. The instantaneous values of any voltage or current can be determined by simply inserting *t* into the equation and using a calculator or table to determine the magnitude of the exponential term.

The similarity between the equations $v_C = E(1 - e^{-t/\tau})$ and $i_L = I_m(1 - e^{-t/\tau})$ results in a derivation of the following for *t* that is identical to the one used to obtain Eq. (10.24).

$$t = \tau \log_e \left(\frac{I_m}{I_m - i_L}\right) \tag{12.21}$$

For the other form, the equation $v_C = Ee^{-t/\tau}$ is a close match with $v_L = Ee^{-t/\tau}$, permitting a derivation similar to the one used for Eq. (10.25):

$$t = \tau \log_e \frac{E}{v_L} \tag{12.22}$$

The similarities between the above and the equations in Chapter 10 should make the equation for *t* fairly easy to obtain.

Time Constant: $\tau = L/R_{Th}$

To analyze an inductive network that is not a simple series circuit, you must first reduce it to a Thévenin equivalent circuit. The Thévenin procedure was introduced in Section 9.3, and was applied to a capacitive network in Section 10.11. Once the Thévenin equivalent circuit is found, you can continue the analysis as described above. The time constant will be $\tau = L/R_{Th}$ as shown in the next example.

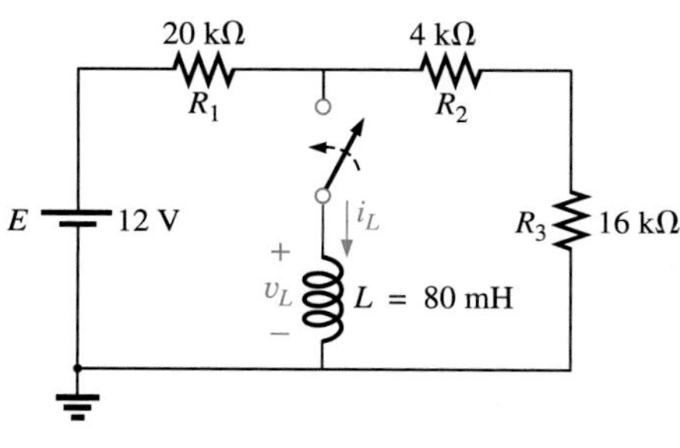

FIG. 12.31
Example 12.7.

EXAMPLE 12.7

For the network of Fig. 12.31:

a. Find the mathematical expression for the transient behaviour of the current i_L and the voltage v_L after the closing of the switch ($I_i = 0$ mA).
b. Draw the resulting waveform for each.

continued

Solution:

a. Applying Thévenin's theorem to the 80-mH inductor (Fig. 12.32) gives

$$R_{Th} = \frac{R}{N} = \frac{20\text{ k}\Omega}{2} = 10\text{ k}\Omega$$

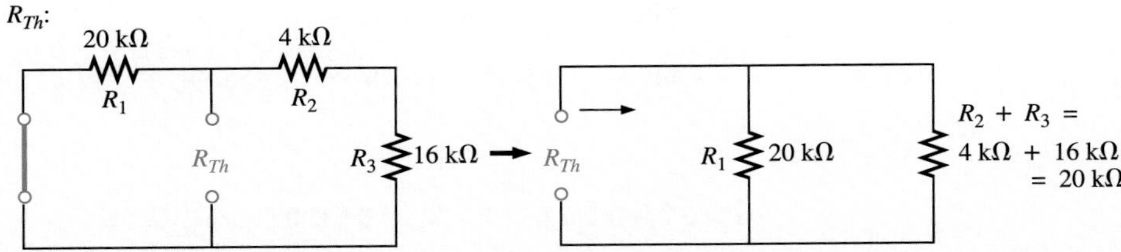

FIG. 12.32
Determining R_{Th} for the network of Fig. 12.31.

Applying the voltage divider rule (Fig. 12.33),

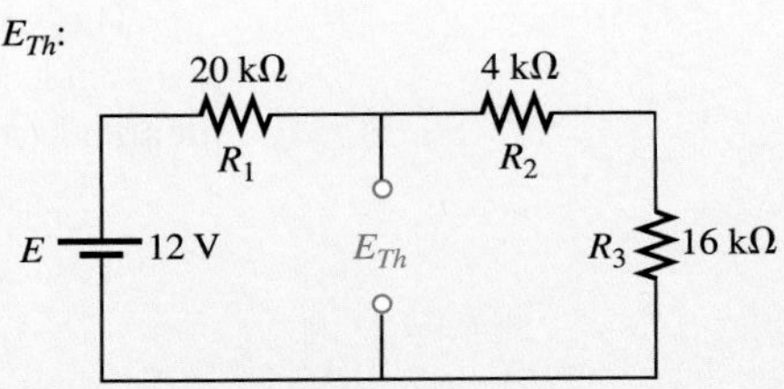

FIG. 12.33
Determining E_{Th} for the network of Fig. 12.31.

$$E_{Th} = \frac{(R_2 + R_3)E}{R_1 + R_2 + R_3}$$

$$= \frac{(4\text{ k}\Omega + 16\text{ k}\Omega)(12\text{ V})}{20\text{ k}\Omega + 4\text{ k}\Omega + 16\text{ k}\Omega} = \frac{(20\text{ k}\Omega)(12\text{ V})}{40\text{ k}\Omega} = 6\text{ V}$$

The Thévenin equivalent circuit is shown in Fig. 12.34. Using Eq. (12.8),

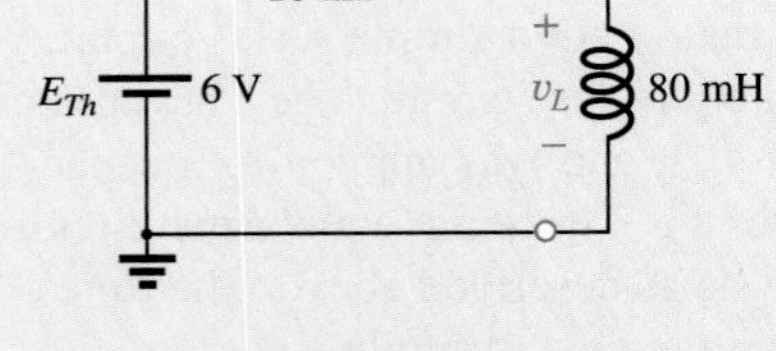

FIG. 12.34
The resulting Thévenin equivalent circuit for the network of Fig. 12.31.

$$i_L = \frac{E_{Th}}{R}(1 - e^{-t/\tau})$$

$$\tau = \frac{L}{R_{Th}} = \frac{80 \times 10^{-3}\text{ H}}{10 \times 10^{3}\ \Omega} = 8 \times 10^{-6}\text{ s}$$

$$I_m = \frac{E_{Th}}{R_{Th}} = \frac{6\text{ V}}{10 \times 10^{3}\ \Omega} = 0.6 \times 10^{-3}\text{ A}$$

and $\mathbf{i_L = (0.6 \times 10^{-3})(1 - e^{-t/(8\times10^{-6})})}$

Using Eq. (12.10),

$$v_L = E_{Th}e^{-t/\tau}$$

so that $\mathbf{v_L = 6e^{-t/(8\times10^{-6})}}$

b. See Fig. 12.35.

continued

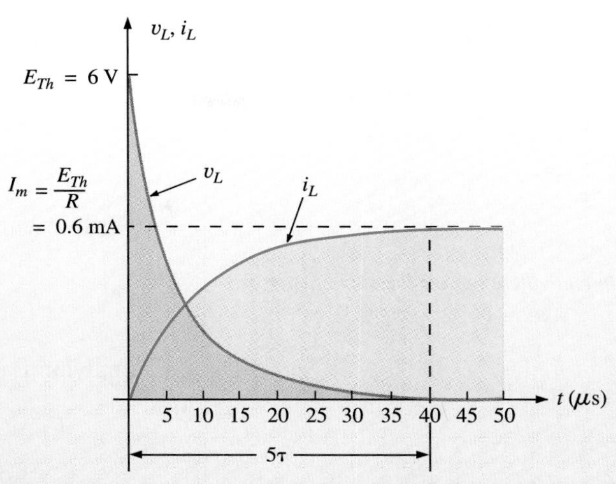

FIG. 12.35
The resulting waveforms for i_L and v_L for the network of Fig. 12.31.

EXAMPLE 12.8

The switch S_1 of Fig. 12.36 has been closed for a long time. At t = 0 s, S_1 is opened at the same instant S_2 is closed to avoid an interruption in current through the coil.

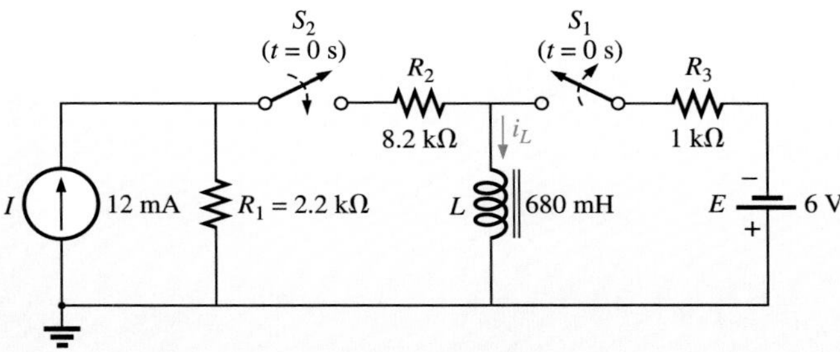

FIG. 12.36
Example 12.8.

a. Find the initial current through the coil. Pay particular attention to its direction.
b. Find the mathematical expression for the current i_L following the closing of the switch S_2.
c. Sketch the waveform for i_L.

Solution:

a. Using Ohm's law the initial current through the coil is determined by:

$$I_i = -\frac{E}{R_3} = -\frac{6\text{ V}}{1\text{ k}\Omega} = -6\text{ mA}$$

b. Applying Thévenin's theorem:

$$R_{Th} = R_1 + R_2 = 2.2\text{ k}\Omega + 8.2\text{ k}\Omega = 10.4\text{ k}\Omega$$
$$E_{Th} = IR_1 = (12\text{ mA})(2.2\text{ k}\Omega) = 26.4\text{ V}$$

continued

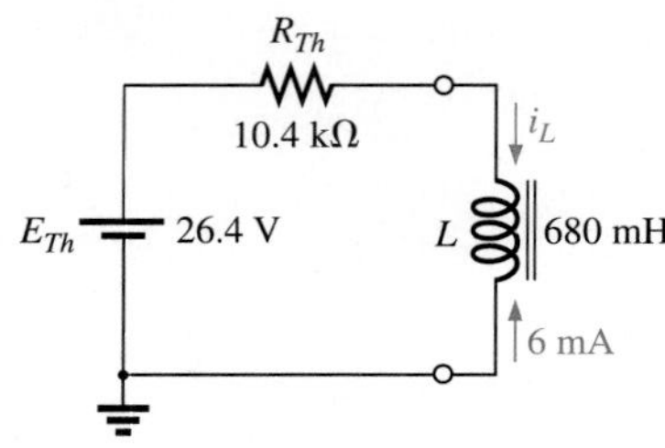

FIG. 12.37
Thévenin equivalent circuit for the network of Fig. 12.36 for $t \geq 0$ s.

The Thévenin equivalent network appears in Fig. 12.37.

The steady-state current can then be determined by substituting the short-circuit equivalent for the inductor:

$$I_f = \frac{E}{R_{Th}} = \frac{26.4\text{ V}}{10.4\text{ k}\Omega} = 2.54\text{ mA}$$

The time constant:

$$\tau = \frac{L}{R_{Th}} = \frac{680\text{ mH}}{10.4\text{ k}\Omega} = 65.39\ \mu\text{s}$$

Applying Eq. (12.20):

$$\begin{aligned} i_L &= I_f + (I_i - I_f)e^{-t/\tau} \\ &= 2.54\text{ mA} + (-6\text{ mA} - 2.54\text{ mA})e^{-t/65.39\ \mu\text{s}} \\ &= \mathbf{2.54\ mA - 8.54\ mA\ } \boldsymbol{e^{-t/(65.39\ \mu s)}} \end{aligned}$$

c. Note Fig. 12.38.

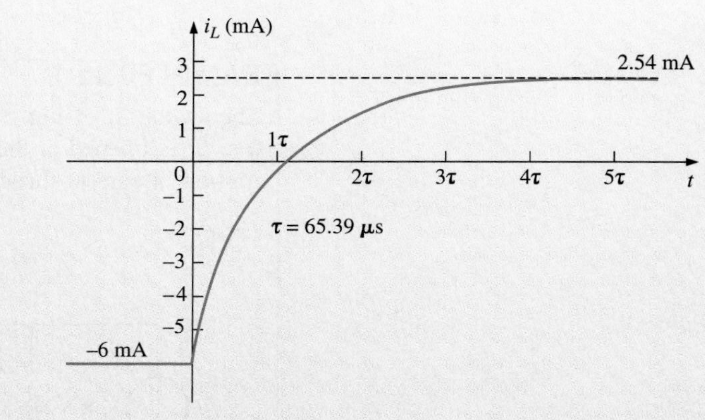

FIG. 12.38
The current i_L for the network of Fig. 12.37.

12.7 INDUCTORS IN SERIES AND PARALLEL

Inductors, like resistors and capacitors, can be placed in series or parallel. Increasing levels of inductance can be obtained by placing inductors in series, while decreasing levels can be obtained by placing inductors in parallel.

For inductors in series, the total inductance is found in the same manner as the total resistance of resistors in series (Fig. 12.39):

$$L_T = L_1 + L_2 + L_3 + \cdots + L_N \tag{12.23}$$

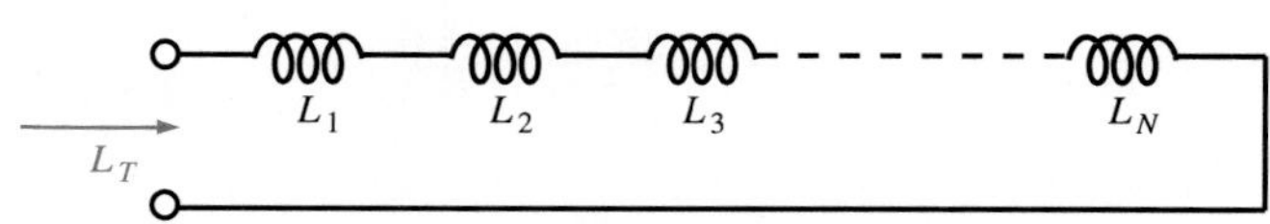

FIG. 12.39
Inductors in series.

For inductors in parallel, the total inductance is found the same way as the total resistance of resistors in parallel (Fig. 12.40):

$$\frac{1}{L_T} = \frac{1}{L_1} + \frac{1}{L_2} + \frac{1}{L_3} + \cdots + \frac{1}{L_N} \tag{12.24}$$

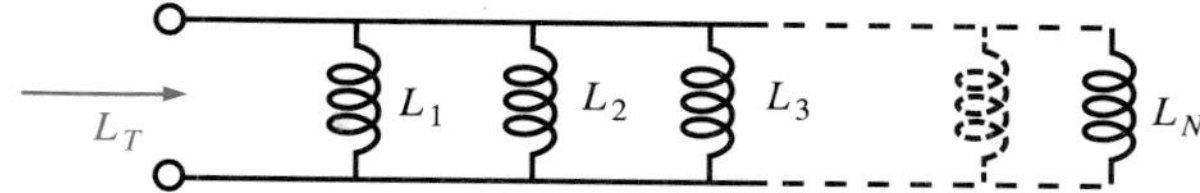

FIG. 12.40
Inductors in parallel.

For two inductors in parallel, Eq. (12.24) reduces to

$$L_T = \frac{L_1 L_2}{L_1 + L_2} \tag{12.25}$$

and for three parallel inductors the equation simplifies to

$$L_T = \frac{L_1 L_2 L_3}{L_1 L_2 + L_1 L_3 + L_2 L_3} \tag{12.26}$$

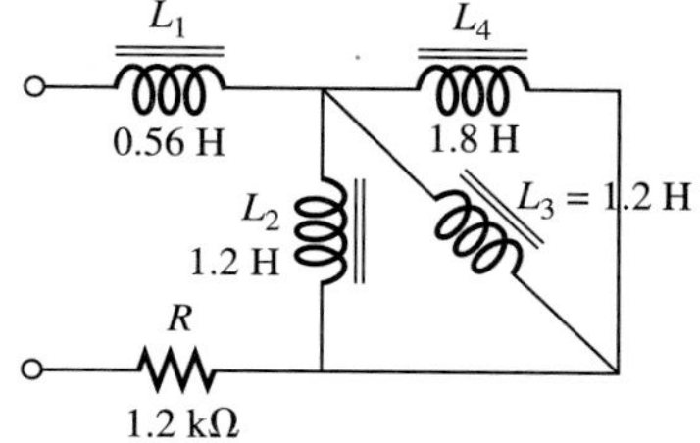

FIG. 12.41
Example 12.9.

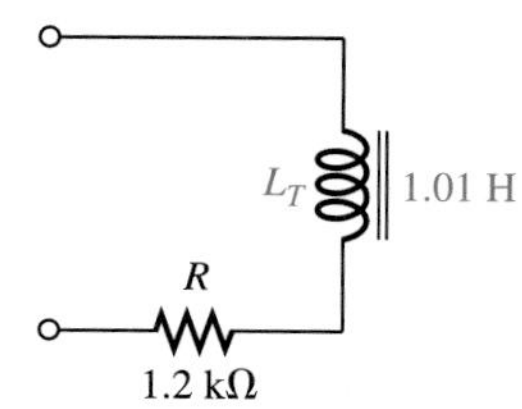

FIG. 12.42
Terminal equivalent of the network of Fig. 12.41.

EXAMPLE 12.9

Reduce the network of Fig. 12.41 to its simplest form.

Solution: The inductors L_2 and L_3 are equal in value and they are in parallel, resulting in an equivalent parallel value of

$$L'_T = \frac{L}{N} = \frac{1.2\text{ H}}{2} = 0.6\text{ H}$$

The resulting 0.6 H is then in parallel with the 1.8-H inductor, and

$$L''_T = \frac{(L'_T)(L_4)}{L'_T + L_4} = \frac{(0.6\text{ H})(1.8\text{ H})}{0.6\text{ H} + 1.8\text{ H}}$$
$$= 0.45\text{ H}$$

The inductor L_1 is then in series with the equivalent parallel value, and

$$L_T = L_1 + L''_T = 0.56\text{ H} + 0.45\text{ H}$$
$$= 1.01\text{ H}$$

The reduced equivalent network appears in Fig. 12.42.

An alternate solution method starts off with the simplified equation [Eq. (12.26)] because $L_2 L_3$ and L_4 are all in parallel.

$$L'_T = \frac{(L_2)\,(L_3)\,(L_4)}{L_2L_3 + L_2L_4 + L_3L_4} = 0.45\text{ H}$$

Once again the inductor L_1 is then in series with the equivalent parallel value L'_T and the final equivalent inductance is 0.45 H, which is the same as before.

12.8 *R-L* AND *R-L-C* CIRCUITS WITH dc INPUTS

We found in Section 12.5 that, for all practical purposes, an inductor can be replaced by a short circuit in a dc circuit after a period of time greater than five time constants has passed. If in the following circuits we assume that all of the currents and voltages have reached their final values, the current through each inductor can be found by replacing each inductor with a short circuit. For the circuit of Fig. 12.43, for example,

$$I_1 = \frac{E}{R_1} = \frac{10\text{ V}}{2\ \Omega} = \mathbf{5\text{ A}}$$

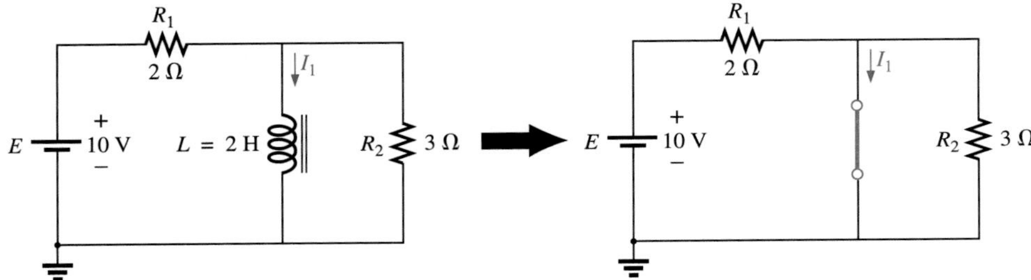

FIG. 12.43
Substituting the short-circuit equivalent for the inductor for $t > 5\tau$.

For the circuit of Fig. 12.44,

$$I = \frac{E}{R_2 \parallel R_3} = \frac{21\text{ V}}{2\ \Omega} = \mathbf{10.5\ A}$$

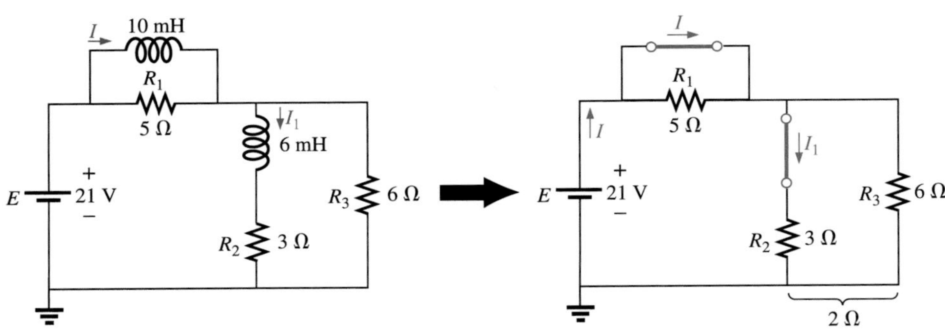

FIG. 12.44
Establishing the equivalent network for $t > 5\tau$.

Applying the current divider rule,

$$I_1 = \frac{R_3 I}{R_3 + R_2} = \frac{(6\ \Omega)(10.5\text{ A})}{6\ \Omega + 3\ \Omega} = \frac{63\text{ A}}{9} = \mathbf{7\ A}$$

In the following examples we will assume that the voltage across the capacitors and the current through the inductors have reached their final values. Under these conditions, the inductors can be replaced with short circuits, and the capacitors can be replaced with open circuits.

EXAMPLE 12.10

Find the current I_L and the voltage V_C for the network of Fig. 12.45.

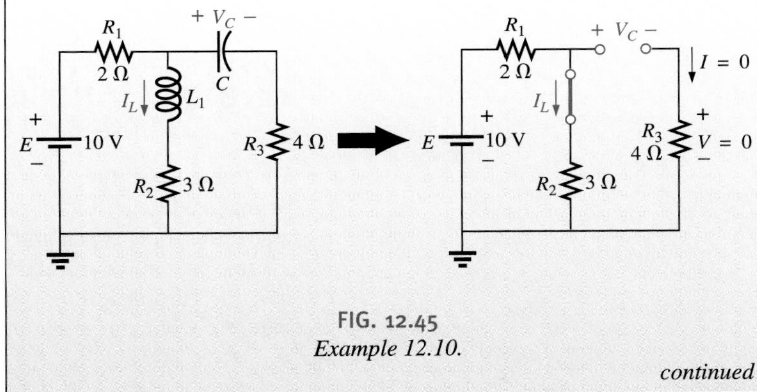

FIG. 12.45
Example 12.10.

continued

Solution:

$$I_L = \frac{E}{R_1 + R_2} = \frac{10\text{ V}}{5\ \Omega} = \mathbf{2\ A}$$

$$V_C = \frac{R_2 E}{R_2 + R_1} = \frac{(3\ \Omega)(10\text{ V})}{3\ \Omega + 2\ \Omega} = \mathbf{6\ V}$$

EXAMPLE 12.11

Find the currents I_1 and I_2 and the voltages V_1 and V_2 for the network of Fig. 12.46.

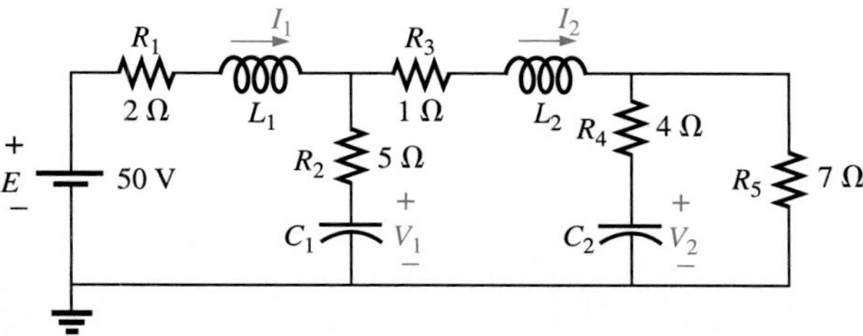

FIG. 12.46
Example 12.11.

Solution: Note Fig. 12.47:

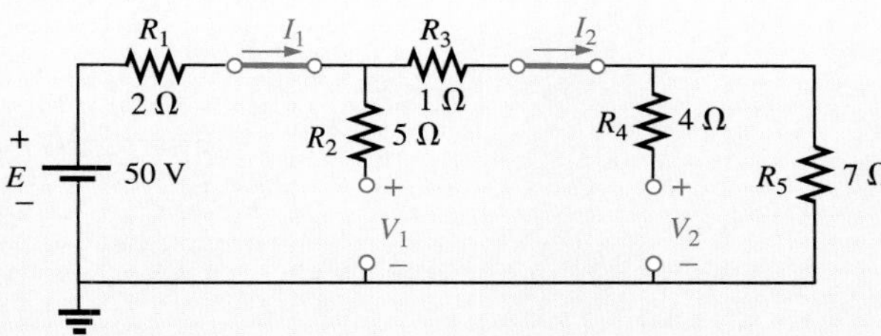

FIG. 12.47
Substituting the short-circuit equivalents for the inductors and open-circuit equivalents for the capacitor for $t > 5\tau$.

$$I_1 = I_2$$

$$I_1 = \frac{E}{R_1 + R_3 + R_5} = \frac{50\text{ V}}{2\ \Omega + 1\ \Omega + 7\ \Omega} = \frac{50\text{ V}}{10\ \Omega} = \mathbf{5\ A}$$

$$V_2 = I_2 R_5 = (5\text{ A})(7\ \Omega) = \mathbf{35\ V}$$

Applying the voltage divider rule,

$$V_1 = \frac{(R_3 + R_5)E}{R_1 + R_3 + R_5} = \frac{(1\ \Omega + 7\ \Omega)(50\text{ V})}{2\ \Omega + 1\ \Omega + 7\ \Omega} = \frac{(8\ \Omega)(50\text{ V})}{10\ \Omega} = \mathbf{40\ V}$$

12.9 ENERGY STORED BY AN INDUCTOR

The ideal inductor, like the ideal capacitor, does not dissipate the electrical energy supplied to it. It stores the energy in the form of a magnetic field. A plot of the voltage, current, and power to an inductor is shown in Fig. 12.48 during the buildup of the magnetic field surrounding the inductor. The energy stored is represented by the shaded area under the power curve. Using calculus, we can show that the evaluation of the area under the curve yields

$$W_{stored} = \frac{1}{2}LI_m^2 \qquad \text{(joules, J)} \tag{12.27}$$

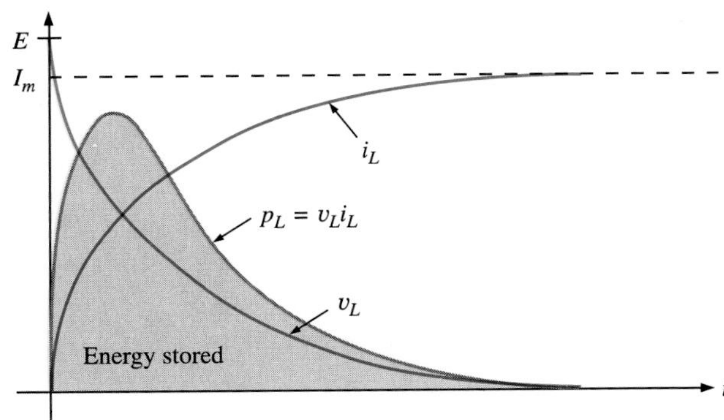

FIG. 12.48
The power curve for an inductive element under transient conditions.

EXAMPLE 12.12

Find the energy stored by the inductor in the circuit of Fig. 12.49 when the current through it has reached its final value.

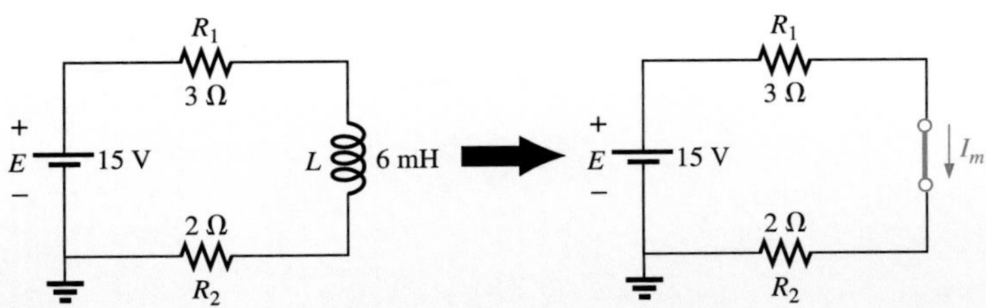

FIG. 12.49
Example 12.12.

Solution:

$$I_m = \frac{E}{R_1 + R_2} = \frac{15\text{ V}}{3\ \Omega + 2\ \Omega} = \frac{15\text{ V}}{5\ \Omega} = 3\text{ A}$$

$$W_{stored} = \frac{1}{2}LI_m^2 = \frac{1}{2}(6 \times 10^{-3}\text{ H})(3\text{ A})^2 = \frac{54}{2} \times 10^{-3}\text{ J}$$

$$= \mathbf{27\ mJ}$$

PROBLEMS

SECTION 12.2 Electromagnetic Induction

1. If the flux linking a coil changes at a rate of 0.085 Wb/s, what is the induced voltage across a coil of:
   a. 50 turns?
   b. 100 turns?
   c. 200 turns?

2. Determine the rate of change of flux linking a coil if 20 V are induced across a coil of:
   a. 40 turns
   b. 125 turns
   c. 350 turns

3. How many turns does a coil have if 420 mV are induced across the coil by a changing flux of:
   a. 0.003 Wb/s
   b. 0.06 Wb/s
   c. 0.17 Wb/s

4. Find the inductance L in henries of the inductor of Fig. 12.50.

5. Repeat Problem 4 with $l = 10$ cm. and $d = 6$ mm.

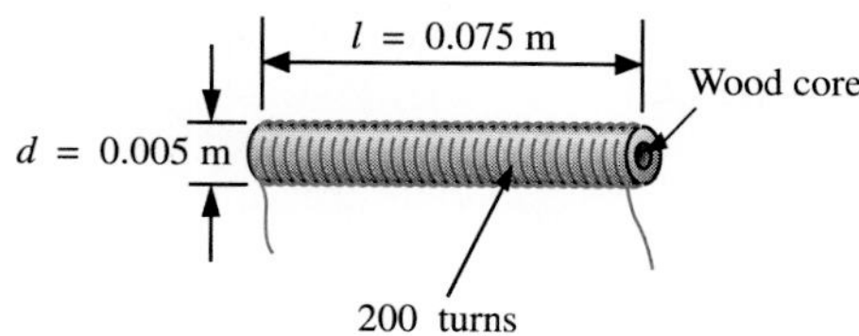

FIG. 12.50
Problems 4 and 5.

6. a. Find the inductance L in henries of the inductor of Fig. 12.51.
   b. Repeat part (a) if a ferromagnetic core is added having a μ_r of 2000.

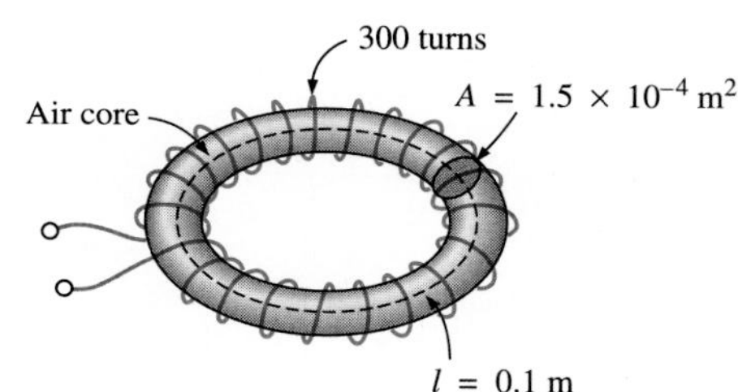

FIG. 12.51
Problem 6.

SECTION 12.4 Induced Voltage

7. Find the voltage induced across a coil of 5 H if the rate of change of current through the coil is
   a. 0.5 A/s
   b. 60 mA/s
   c. 0.04 A/ms

8. Find the induced voltage across a 50-mH inductor if the current through the coil changes at a rate of 0.1 mA/μs.

9. Find the waveform for the voltage induced across a 200-mH coil if the current through the coil is as shown in Fig. 12.52.

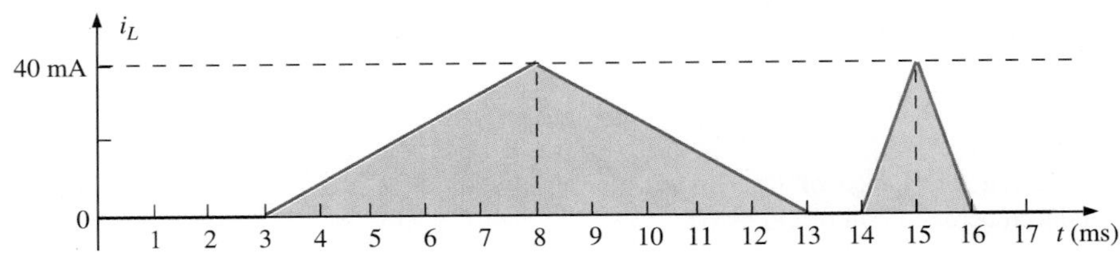

FIG. 12.52
Problem 9.

10. Sketch the waveform for the voltage induced across a 0.2-H coil if the current through the coil is as shown in Fig. 12.53.

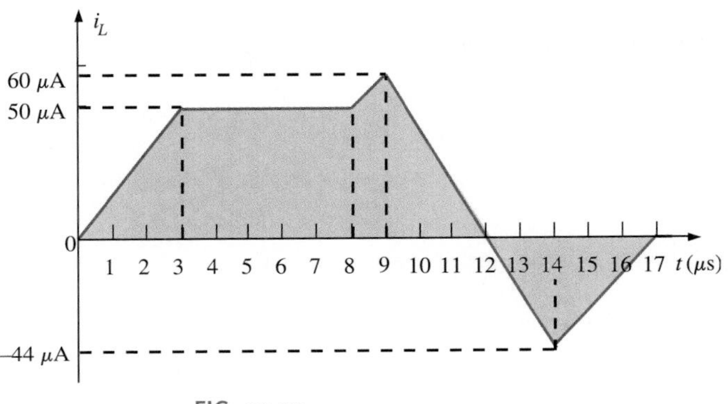

FIG. 12.53
Problem 10.

***11.** Find the waveform for the current of a 10-mH coil if the voltage across the coil follows the pattern of Fig. 12.54. The current i_L is 4 mA at $t = 0$ s.

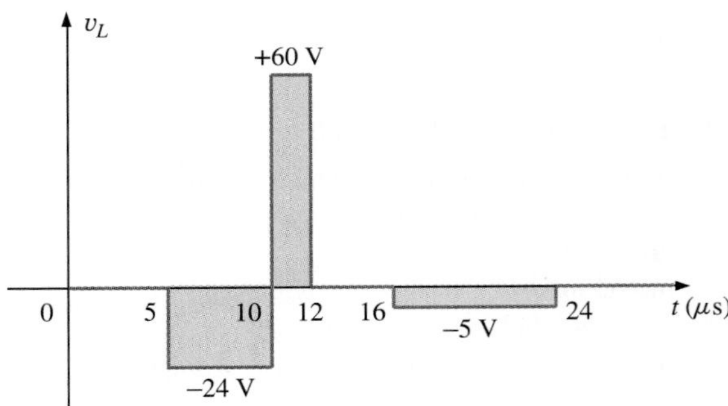

FIG. 12.54
Problem 11.

SECTION 12.5 *R-L* Transients: Storage and Decay Cycles

12. For the circuit of Fig. 12.55:

a. Determine the time constant.

b. Write the mathematical expression for the current i_L after the switch is closed.

c. Repeat part (b) for v_L and v_R.

d. Determine i_L and v_L at one, three, and five time constants.

e. Sketch the waveforms of i_L, v_L, and v_R.

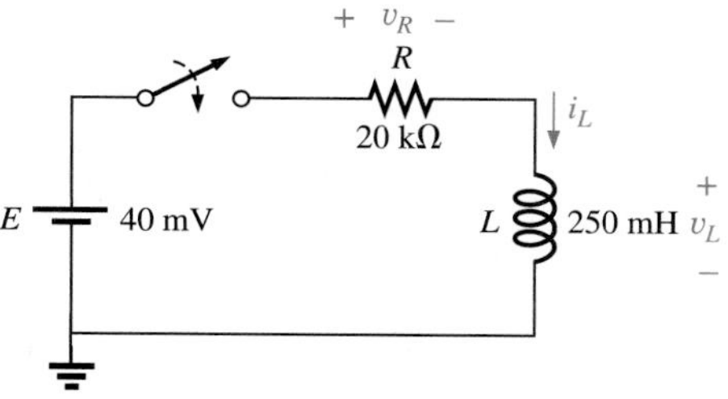

FIG. 12.55
Problem 12.

13. For the circuit of Fig. 12.56:

a. Determine τ.

b. Write the mathematical expression for the current i_L after the switch is closed at $t = 0$ s.

c. Write the mathematical expressions for v_L and v_R after the switch is closed at $t = 0$ s.

d. Determine i_L and v_L at $t = 1\tau$, 3τ, and 5τ.

e. Sketch the waveforms of i_L, v_L, and v_R for the storage phase.

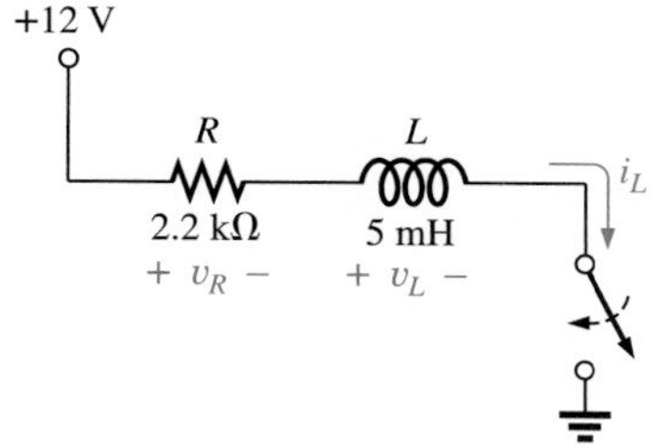

FIG. 12.56
Problem 13.

14. For the network of Fig. 12.57:

a. Determine the mathematical expressions for the current i_L and the voltage v_L when the switch is closed.

b. Repeat part (a) if the switch is opened after a period of five time constants has passed.

c. Sketch the waveforms of parts (a) and (b) on the same axis.

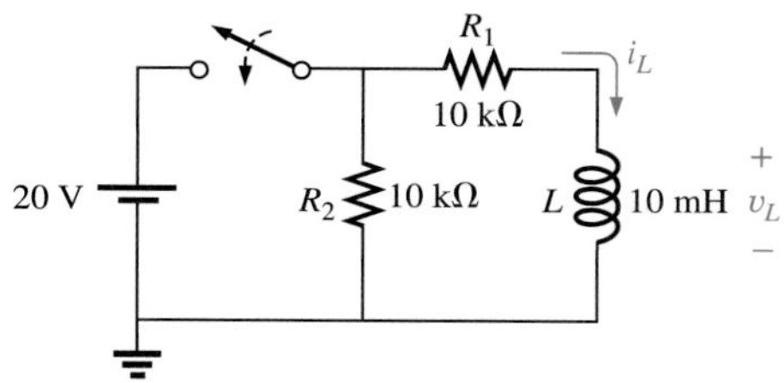

FIG. 12.57
Problem 14.

***15.** For the network of Fig. 12.58:

a. Write the mathematical expression for the current i_L and the voltage v_L following the closing of the switch.

b. Determine the mathematical expressions for i_L and v_L if the switch is opened after a period of five time constants has passed.

c. Sketch the waveforms of i_L and v_L for the time periods defined by parts (a) and (b).

d. Sketch the waveform for the voltage across R_2 for the same period of time covered by i_L and v_L. Take careful note of the defined polarities and directions of Fig. 12.58.

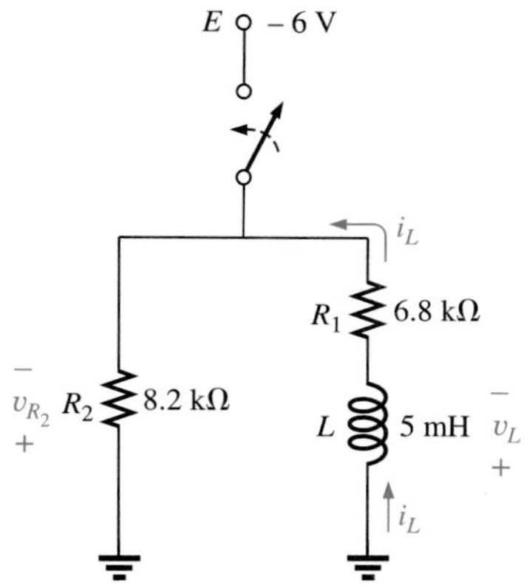

FIG. 12.58
Problem 15.

***16.** For the network of Fig. 12.59:

a. Determine the mathematical expressions for the current i_L and the voltage v_L following the closing of the switch.

b. Repeat part (a) if the switch is opened at $t = 1\ \mu s$.

c. Sketch the waveforms of parts (a) and (b) on the same axis.

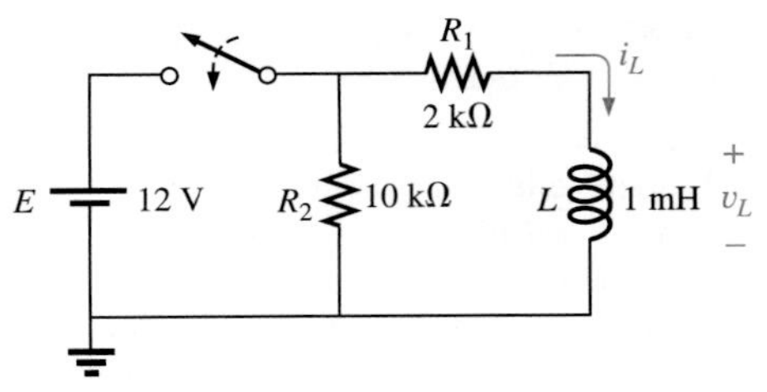

FIG. 12.59
Problem 16.

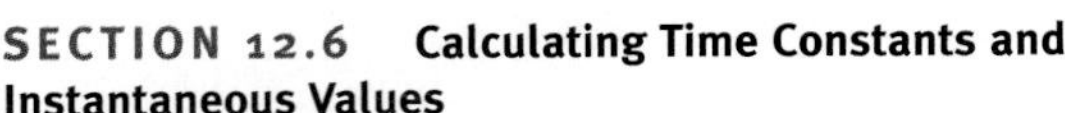

SECTION 12.6 Calculating Time Constants and Instantaneous Values

17. For the network of Fig. 12.60:

a. Write the mathematical expressions for the current i_L and the voltage v_L following the closing of the switch. Note the magnitude and direction of the initial current.

b. Sketch the waveform of i_L and v_L for the entire period from initial value to steady-state level.

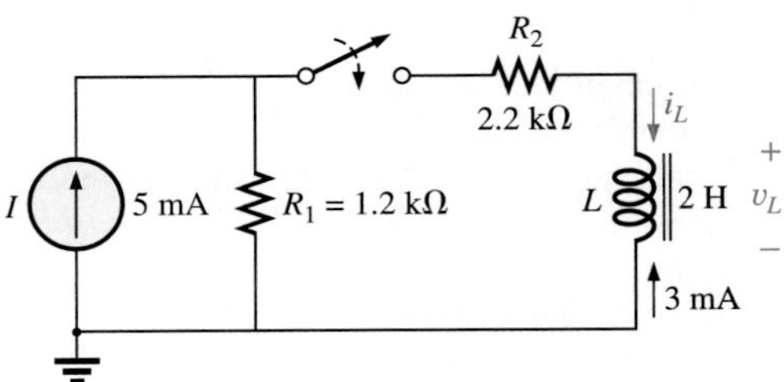

FIG. 12.60
Problem 17.

18. For the network of Fig. 12.61:

a. Write the mathematical expressions for the current i_L and the voltage v_L following the closing of the switch. Note the magnitude and direction of the initial current.

b. Sketch the waveform of i_L and v_L for the entire period from initial value to steady-state level.

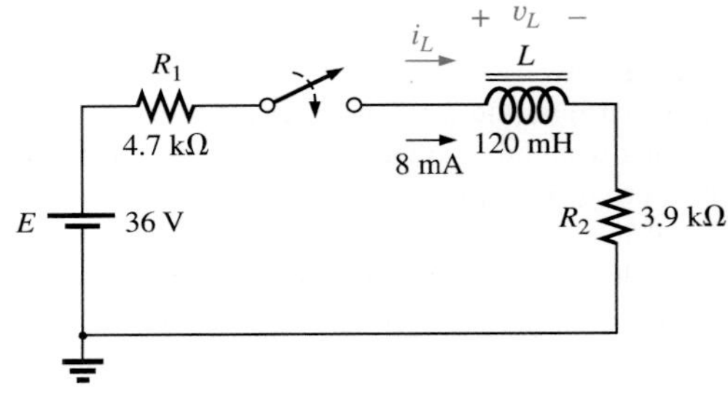

FIG. 12.61
Problem 18.

***19.** For the network of Fig. 12.62:

a. Write the mathematical expressions for the current i_L and the voltage v_L following the closing of the switch. Note the magnitude and direction of the initial current.

b. Sketch the waveform of i_L and v_L for the entire period from initial value to steady-state level.

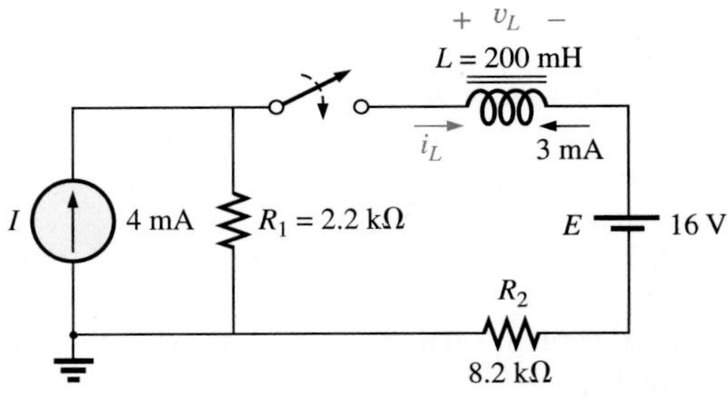

FIG. 12.62
Problem 19.

20. Referring to the solution to Example 12.4, determine the time when the current i_L reaches a level of 10 mA. Then determine the time when the voltage drops to a level of 10 V.

21. Referring to the solution to Example 12.6, determine the time when the current i_L drops to 2 mA.

22. **a.** Determine the mathematical expressions for i_L and v_L following the closing of the switch in Fig. 12.63.
b. Determine i_L and v_L at $t = 100$ ns.

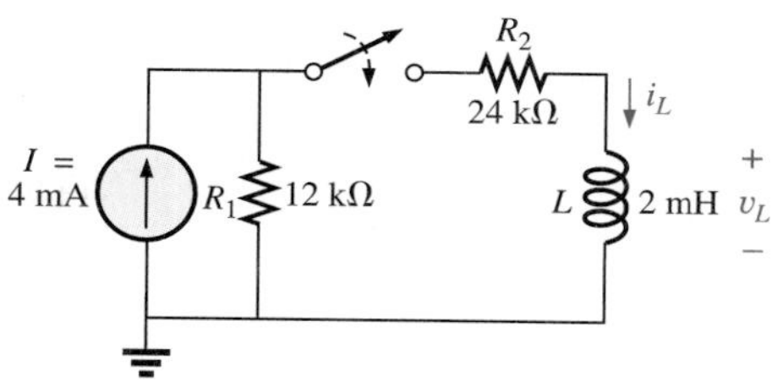

FIG. 12.63
Problem 22.

*23. **a.** Determine the mathematical expressions for i_L and v_L following the closing of the switch in Fig. 12.64.
b. Calculate i_L and v_L at $t = 10\ \mu s$.
c. Write the mathematical expressions for the current i_L and the voltage v_L if the switch is opened at $t = 10\ \mu s$.
d. Sketch the waveforms of i_L and v_L for parts (a) and (c).

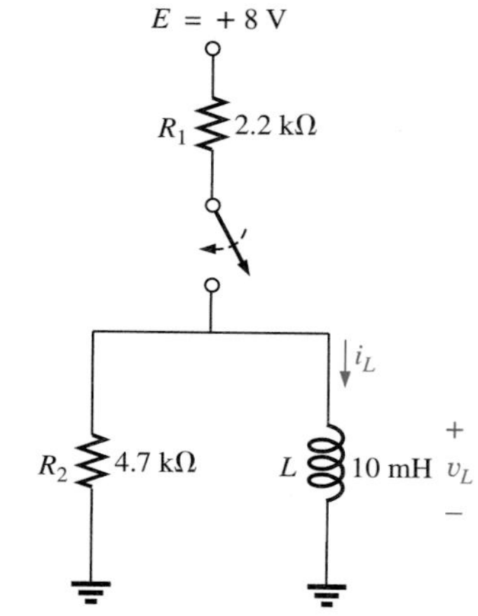

FIG. 12.64
Problem 23.

*24. **a.** Determine the mathematical expressions for i_L and v_L following the closing of the switch in Fig. 12.65.
b. Determine i_L and v_L after two time constants of the storage phase.
c. Write the mathematical expressions for the current i_L and the voltage v_L if the switch is opened at the instant defined by part (b).
d. Sketch the waveforms of i_L and v_L for parts (a) and (c).

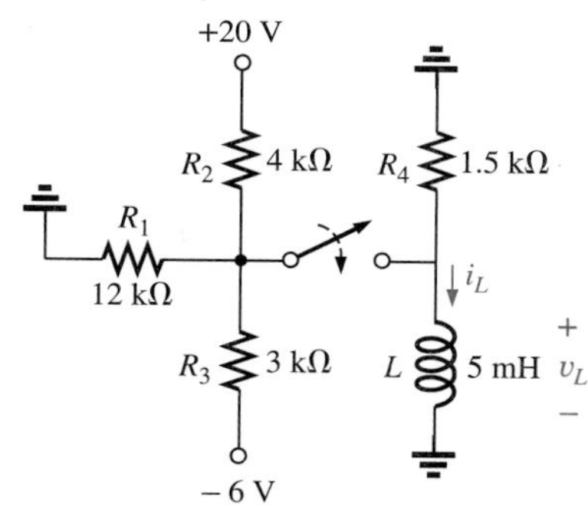

FIG. 12.65
Problem 24.

*25. For the network of Fig. 12.66, the switch is closed at $t = 0$ s.
a. Find v_L at $t = 1$ ms.
b. Determine v_L at $t = 25$ ms.
c. Calculate v_{R_1} at $t = 3\tau$.
d. Find the time required for the current i_L to reach 100 mA.

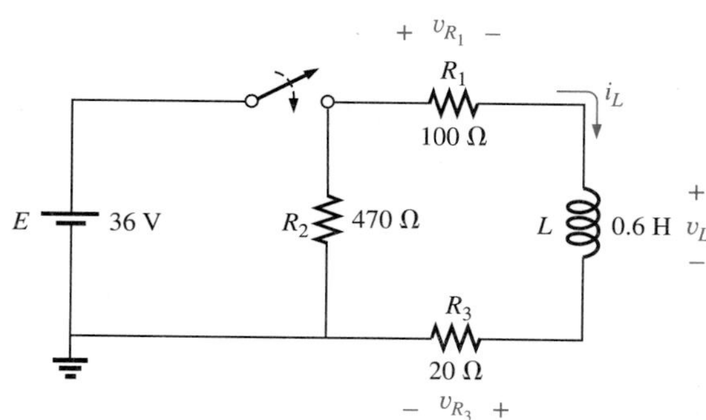

FIG. 12.66
Problem 25.

*26. The switch for the network of Fig. 12.67 has been closed for about 1 h. It is then opened at the time defined as $t = 0$ s.
  a. Determine the time required for the current i_R to drop to 2.5 mA.
  b. Find the voltage v_L at $t = 1$ ms.
  c. Calculate v_{R_3} at $t = 5\tau$.

27. The network of Fig. 12.67 uses a DMM with an internal resistance of 10 MΩ in the voltmeter mode. The switch is closed at $t = 0$ s.
  a. Find the voltage across the coil the instant after the switch is closed.
  b. What is the final value of the current i_L?
  c. How much time must pass before i_L reaches 10 μA?
  d. What is the voltmeter reading at $t = 12$ μs?

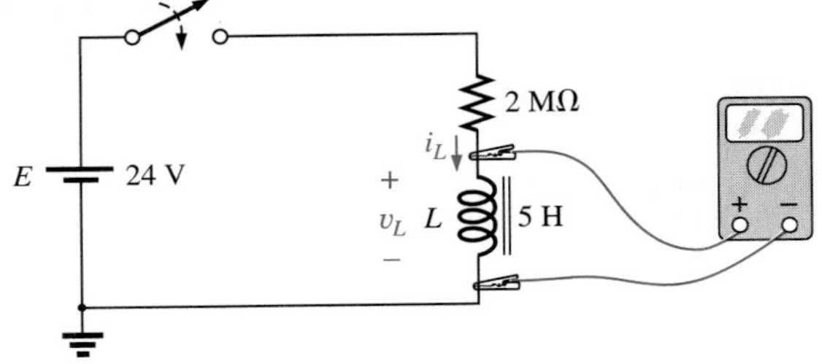

FIG. 12.67
Problems 26 and 27.

*28. The switch in Fig. 12.68 has been open for a long time. It is then closed at $t = 0$ s.
  a. Write the mathematical expression for the current i_L and the voltage v_L after the switch is closed.
  b. Sketch the waveform of i_L and v_L from the initial value to the steady-state level.

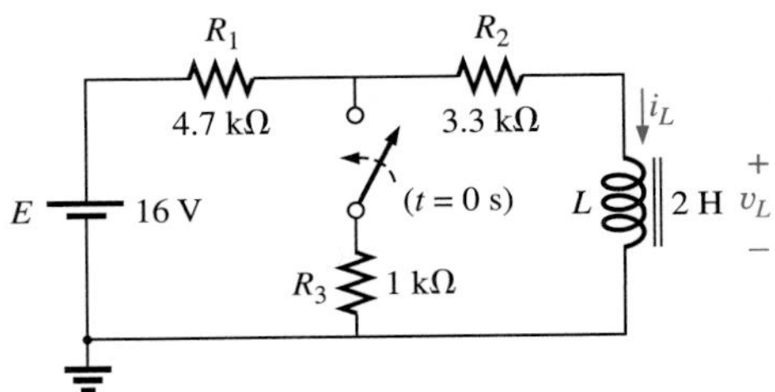

FIG. 12.68
Problem 28.

*29. The switch of Fig. 12.69 has been closed for a long time. It is then opened at $t = 0$ s.
  a. Write the mathematical expression for the current i_L and the voltage v_L after the switch is opened.
  b. Sketch the waveform of i_L and v_L from initial value to the steady-state level.

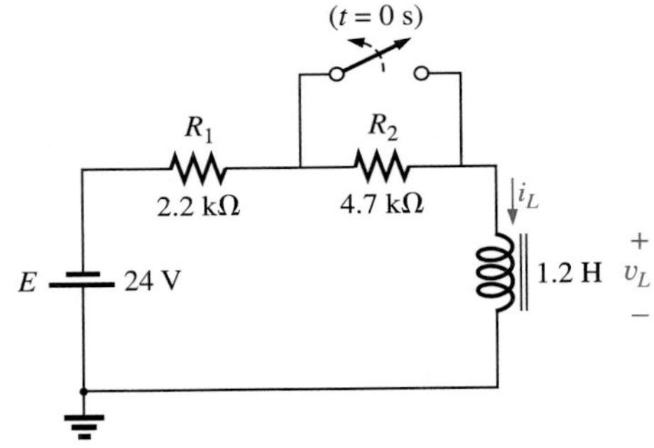

FIG. 12.69
Problem 29.

*30. The switch of Fig. 12.70 has been open for a long time. It is then closed at $t = 0$ s.
  a. Write the mathematical expression for the current i_L and the voltage v_L after the switch is closed.
  b. Sketch the waveform of i_L and v_L from initial value to the steady-state level.

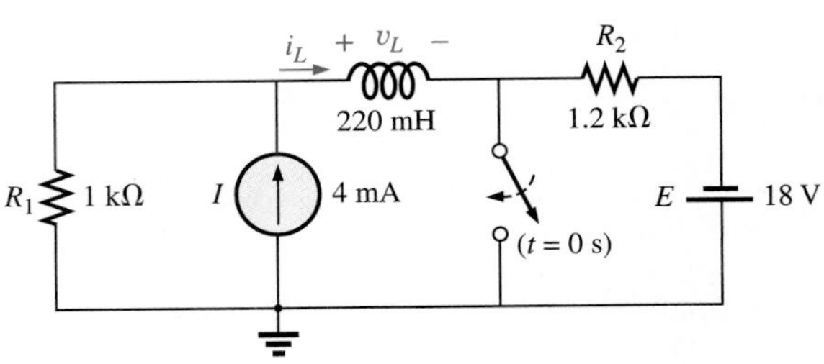

FIG. 12.70
Problem 30.

SECTION 12.7 Inductors in Series and Parallel

31. Find the total inductance of the circuits of Fig. 12.71.

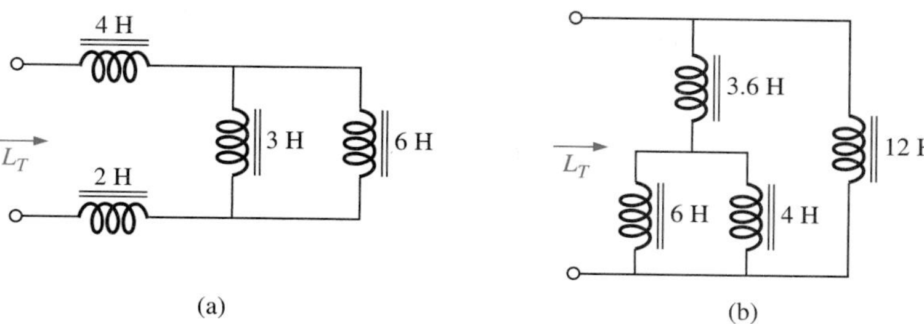

FIG. 12.71
Problem 31.

32. Reduce the networks of Fig. 12.72 to the fewest elements.

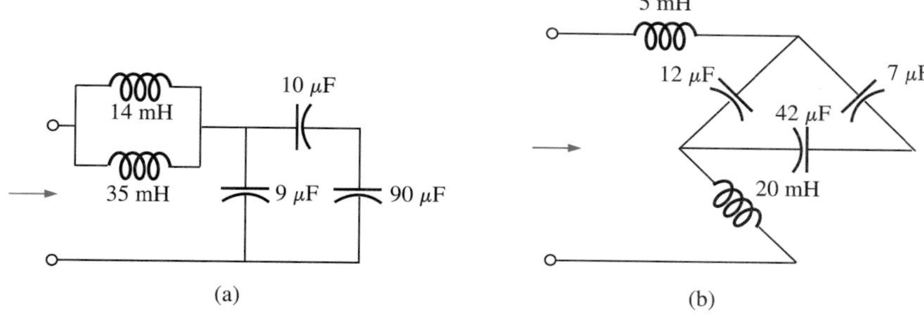

FIG. 12.72
Problem 32.

33. Reduce the network of Fig. 12.73 to the fewest components.

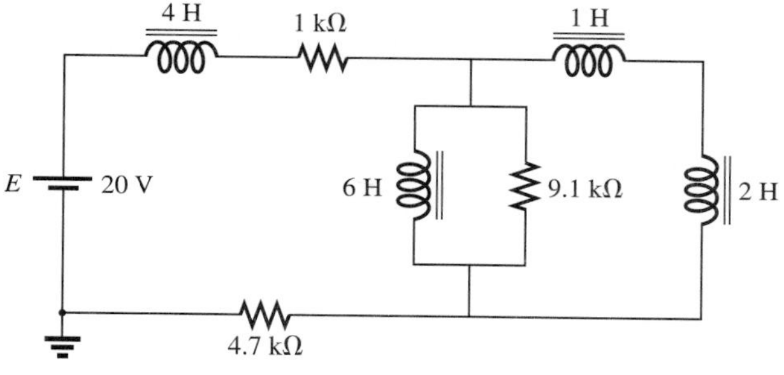

FIG. 12.73
Problem 33.

***34.** For the network of Fig. 12.74:

a. Find the mathematical expressions for the voltage v_L and the current i_L following the closing of the switch.

b. Sketch the waveforms of v_L and i_L obtained in part (a).

c. Determine the mathematical expression for the voltage v_{L_3} following the closing of the switch, and sketch the waveform.

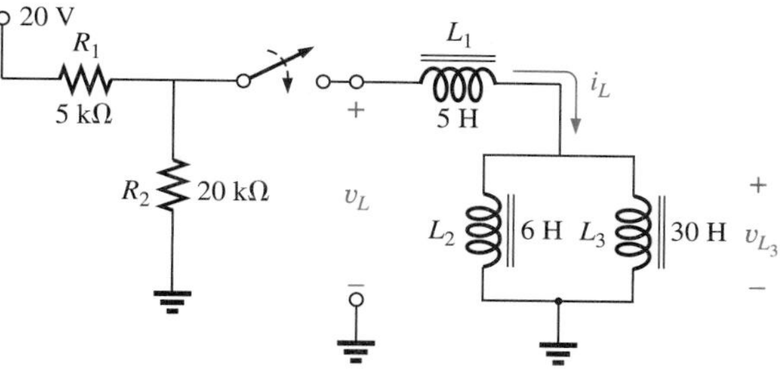

FIG. 12.74
Problem 34.

SECTION 12.8 *R-L* and *R-L-C* Circuits with dc Inputs

For Problems 35 through 37, assume that the voltage across each capacitor and the current through each inductor have reached their final values.

35. a. Find the voltage V_1, V_2 and the current I_1 for the circuit of Fig. 12.75

b. Repeat part (a) but replace the voltage source with a 12 V source.

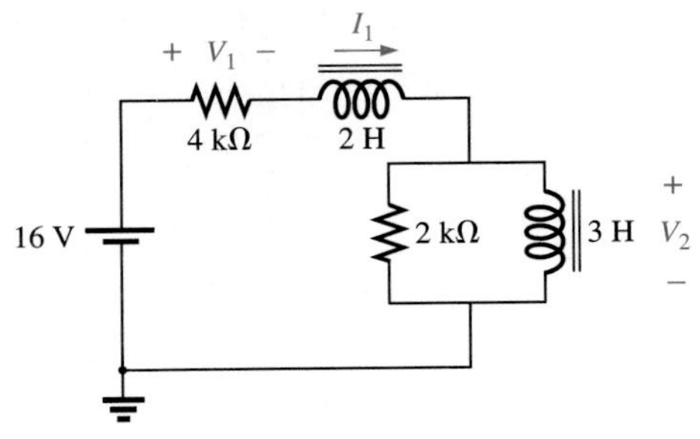

FIG. 12.75
Problems 35 and 38.

36. a. Find the current I_1 and the voltage V_1 for the circuit of Fig. 12.76

b. Repeat part (a) but replace the voltage source with a 24 V source.

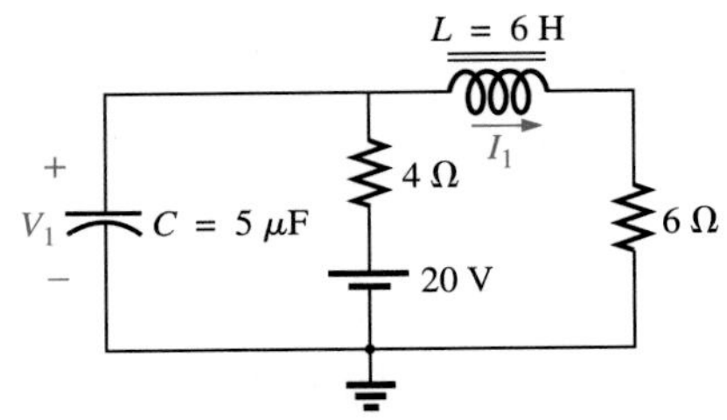

FIG. 12.76
Problems 36 and 39.

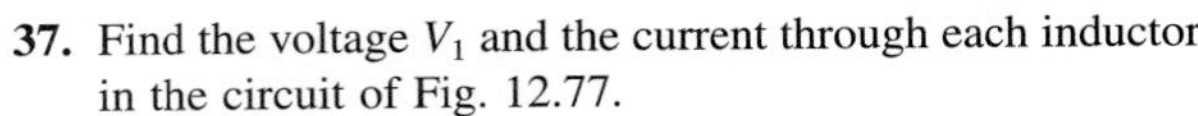
37. Find the voltage V_1 and the current through each inductor in the circuit of Fig. 12.77.

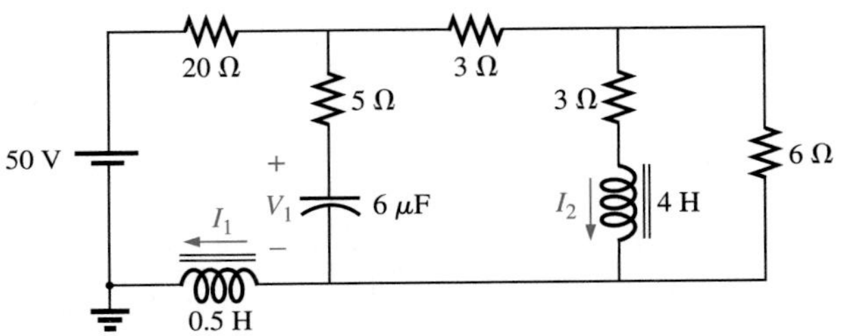

FIG. 12.77
Problems 37 and 40.

SECTION 12.9 Energy Stored by an Inductor

38. a. Find the energy stored in each inductor of Problem 35(a).

b. Find the energy stored in each inductor of Problem 35(b).

39. a. Find the energy stored in the capacitor and inductor of Problem 36(a).

b. Find the energy stored in the capacitor and inductor of Problem 36(b).

40. Find the energy stored in each inductor of Problem 37.

13

Sinusoidal Alternating Waveforms

Outline

Learning Outcomes

After completing this chapter you will be able to

- describe basic components of ac voltage and current
- describe the effects of changing the frequency, period, and amplitude of an ac sine wave
- convert between angular and radian measure for determining voltage and current values
- analyze the effect of phase on the relationship between two sine waves of the same frequency
- describe the phase relationships between two waveforms
- determine that there is more than one correct representation of a sinusoidal function
- examine and describe **average value** as it relates to an ac voltage
- identify the characteristics of an ac waveform on an oscilloscope
- examine and describe **effective value** of an ac voltage
- identify how meters and instruments measure ac voltage and current values

An abundance of hydro power is used to generate ac power which is delivered to our homes. This power, consisting of ac voltage and current, is used in all aspects of our life from heating or cooling to playing computer games.

13.1 INTRODUCTION

Our analysis so far has been limited to dc networks, where currents or voltages are fixed in magnitude except for transient effects. We will now turn our attention to the analysis of networks in which the magnitude of the source varies in a set manner. Of particular interest is the time-varying voltage that is commercially available in large quantities and is commonly called the **ac voltage**. (The letters *ac* are an abbreviation for *alternating current.*)

Terminology

The term **alternating** indicates only that the wave amplitude changes continuously back and forth between two prescribed levels in a set time sequence. To describe the *shape* of the waveform, a term such as *sinusoidal, square wave*, or *triangular* must also be included. For example, each shape in Fig.13.1 is an alternating waveform available from commercial suppliers. However, the pattern of particular interest is the **sinusoidal** ac voltage of Fig.13.1, since this is the most common type of signal. The abbreviated phrases *ac voltage* and *ac current* are usually used when referring to the ac sinusoidal waveforms. For the other two patterns of Fig.13.1, the ac abbreviation is often dropped resulting in the terms **square-wave** or **triangular** waveforms.

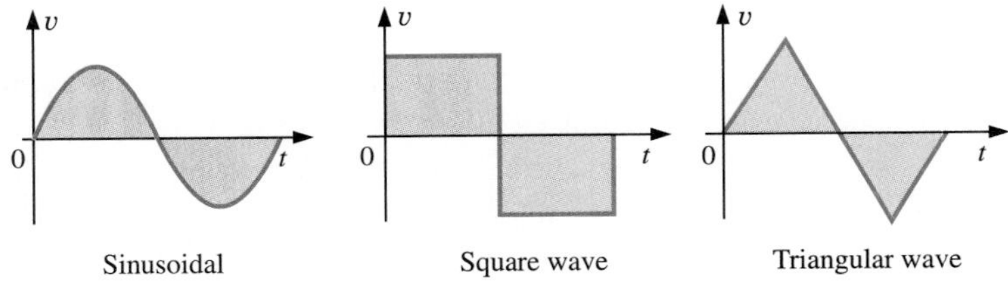

FIG. 13.1
Alternating waveforms.

One of the important reasons for concentrating on the sinusoidal ac voltage is that it is the voltage generated by electrical utilities throughout the world. Other reasons include its application throughout electrical, electronic, communications, and industrial systems. In addition, following chapters will reveal that the waveform itself has a number of characteristics that result in a unique response when it is applied to basic electrical elements.

The many theorems and procedures introduced for dc networks can also be applied to sinusoidal ac systems. This requires introducing the mathematical concepts associated with phasor notation in Chapter 14. After that, most of the concepts introduced in the dc chapters can be applied to ac networks with very little difficulty.

Pulse waveforms are being used by an increasing number of computer systems instead of sinusoidal ac waveforms. They will be discussed in the supplementary CD-ROM to show the response of some fundamental configurations to pulse waveforms.

13.2 SINUSOIDAL ac VOLTAGE CHARACTERISTICS AND DEFINITIONS

Generation

Sinusoidal ac voltages are available from a variety of sources (Fig. 13.2). The most common source is the typical home outlet, which provides an ac voltage that originates at a power plant, fueled by water

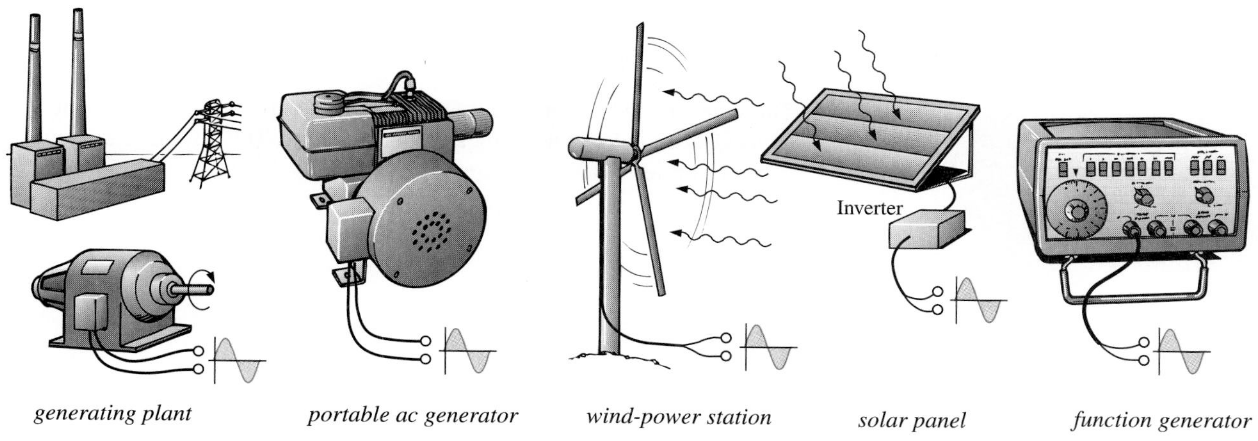

FIG. 13.2
Various sources of ac power.

power, oil, gas, or nuclear fission. In each case an **ac generator** (also called **alternator**) is the primary component in the energy-conversion process. The power to the shaft developed by one of the energy sources listed will turn a **rotor** (constructed of alternating magnetic poles) inside a set of windings housed in the **stator** (the stationary part of the dynamo) and induce a voltage across the windings of the stator, as defined by Faraday's law (Section 12.2),

$$e = N \frac{d\phi}{dt}$$

Through proper design of the generator, a sinusoidal ac voltage is developed that can be transformed to higher levels for distribution through the power lines to the consumer. For isolated locations where power lines have not been installed, portable ac generators that run on gasoline are available. As in the larger power plants, however, an ac generator is a basic part of the design.

In an effort to conserve our natural resources, wind power and solar energy are receiving increasing interest from parts of the world that have the level and duration of such energy sources to make the conversion process viable. The turning propellers of the windpower station are connected directly to the shaft of an ac generator to provide the ac voltage described above. Through light energy absorbed in the form of **photons**, solar cells can generate dc voltages. Through an electronic package called an **inverter**, the dc voltage can be converted to a sinusoidal one. Boats, recreational vehicles (RVs), etc., often use the inversion process in isolated areas.

Sinusoidal ac voltages with characteristics that can be controlled by the user are available from **function generators**. By setting the various switches and controlling the position of the knobs on the face of the instrument, sinusoidal voltages of different peak values and different repetition rates can be made available. The function generator plays an important role in the theorems, methods of analysis, and topics to be introduced in the chapters that follow.

Definitions

The sinusoidal waveform of Fig. 13.3 with its additional notation will now be used as a model in defining a few basic terms. These terms can, however, be applied to any alternating waveform.

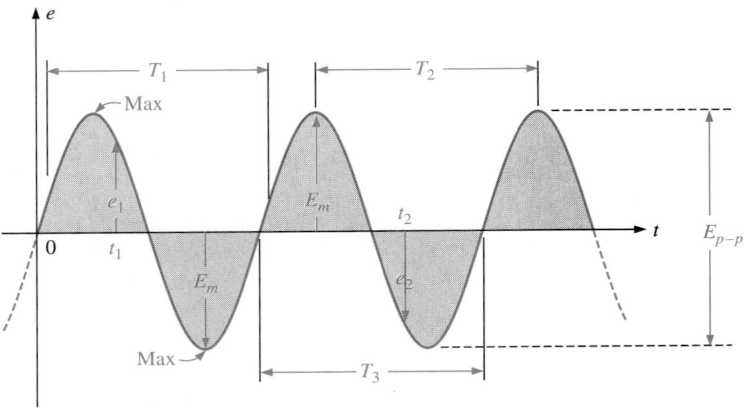

FIG. 13.3
Important parameters for a sinusoidal voltage.

It is important to remember as you proceed through the various definitions that the vertical scaling is in volts or amperes and the horizontal scaling is always *in units of time.*

Waveform: The path traced by a quantity, such as the voltage in Fig. 13.3, plotted as a function of some variable such as time (as above), position, degrees, radians, temperature, and so on.

Instantaneous value: The magnitude of a waveform at any instant of time; denoted by lowercase letters (e_1, e_2).

Peak amplitude: The maximum value of a waveform as measured from its *average,* or *mean,* value, denoted by uppercase letters (such as E_m for sources of voltage and V_m for the voltage drop across a load). For the waveform of Fig. 13.3 the average value is zero volts and E_m is as defined by the figure.

Peak value: The maximum instantaneous value of a function as measured from the zero-volt level. For the waveform of Fig. 13.3 the peak amplitude and peak value are the same, since the average value of the function is zero volts.

Peak-to-peak value: Denoted by $E_{p\text{-}p}$ or $V_{p\text{-}p}$, the full voltage between positive and negative peaks of the waveform, that is, the sum of the magnitude of the positive and negative peaks.

Periodic waveform: A waveform that continually repeats itself after the same time interval. The waveform of Fig. 13.3 is a periodic waveform.

Period (T): The time interval between successive repetitions of a periodic waveform (the period $T_1 = T_2 = T_3$ in Fig. 13.3), so long as successive *similar points* of the periodic waveform are used in determining T.

Cycle: The portion of a waveform contained in *one period* of time. The cycles within T_1, T_2, and T_3 of Fig. 13.3 may appear different in Fig. 13.4, but they are all bounded by one period of time and therefore satisfy the definition of a cycle.

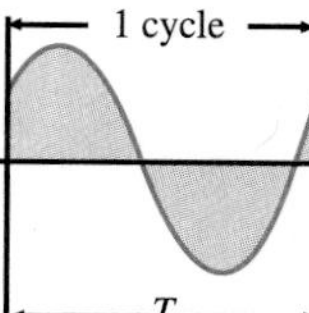

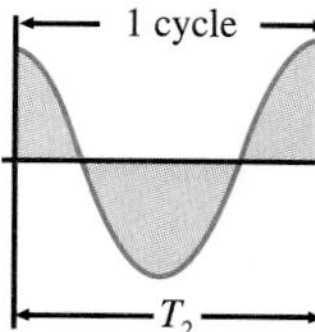

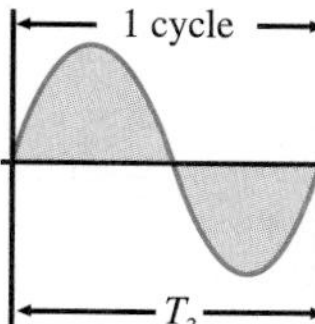

FIG. 13.4
Defining the cycle and period of a sinusoidal waveform.

Frequency (f): The number of cycles that occur in 1 s. The frequency of the waveform of Fig. 13.5(a) is 1 cycle per second, and for Fig. 13.5(b), 2½ cycles per second. If a waveform of similar shape had a period of 0.5 s [Fig. 13.5(c)], the frequency would be 2 cycles per second.

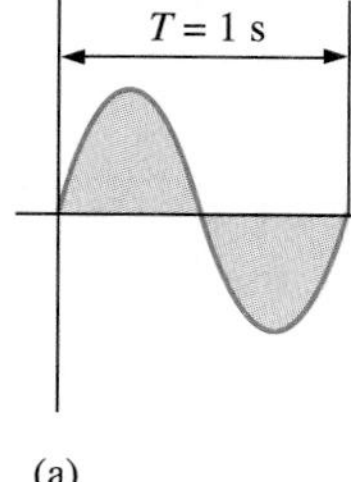

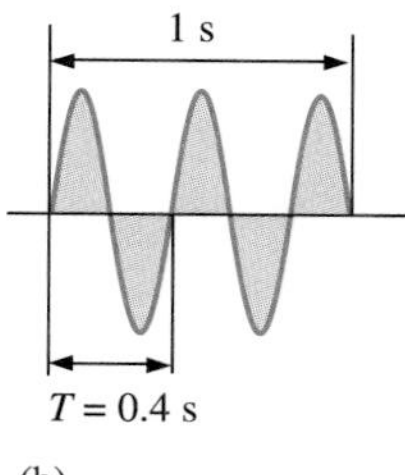

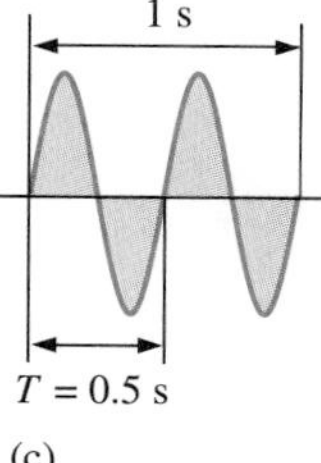

FIG. 13.5
Demonstrating the effect of a changing frequency on the period of a sinusoidal waveform.

The unit of measure for frequency is the **hertz** (Hz), where

$$1 \text{ hertz (Hz)} = 1 \text{ cycle per second (c/s)} \tag{13.1}$$

The unit *hertz* is derived from the surname of Heinrich Rudolph Hertz (Fig. 13.6), who did original research in the area of alternating currents and voltages and their effect on the basic R, L, and C elements.

In North America the frequency standard is 60 Hz, whereas for Europe it is mainly 50 Hz. Electric utilities monitor and very carefully regulate the frequency of their ac generating systems typically to within

LUMINARIES

German (Hamburg, Berlin, Karlsruhe) (1857–1894) Physicist Professor of Physics, Kalsruhe Polytechnic and University of Bonn

FIG. 13.6
Heinrich Rudolph Hertz

Encouraged by the earlier predictions of the English physicist James Clerk Maxwell, Heinrich Hertz produced *electromagnetic waves* in his laboratory at the Karlsruhe Polytechnic while in his early 30s. The basic *transmitter* and *receiver* were the first to broadcast and receive radio waves. He was able to measure the *wavelength* of the electromagnetic waves and confirmed that the *velocity of propagation* is in the same order of magnitude as light. In addition, he demonstrated that the *reflective* and *refractive* properties of electromagnetic waves are the same as those for heat and light waves. It was very unfortunate that this ingenious, industrious person passed away at the very early age of 37 due to a bone disease.

Courtesy of the Smithsonian Institution, Photo No. 666,606

a hundredth of a percent of the standard. This is because, as with all standards, any variation from the norm will cause difficulties. In 1993, Berlin, Germany, received all its power from Eastern European plants, whose output frequency was varying between 50.03 and 51 Hz. The result was that clocks were gaining as much as 4 minutes a day. Alarms went off too soon, VCRs clicked off before the end of the program, etc., requiring clocks to be continually reset. However, in 1994 when their power was linked with the rest of Europe, the precise standard of 50 Hz was reestablished and everyone was on time again.

Since the frequency is inversely related to the period—that is, as one increases, the other decreases by an equal amount—the two can be related by the following equation:

$$f = \frac{1}{T} \quad \text{(hertz, Hz)} \qquad (13.2)$$

or

$$T = \frac{1}{f} \quad \text{(seconds, s)} \qquad (13.3)$$

EXAMPLE 13.1

Find the period of a periodic waveform with a frequency of
a. 60 Hz
b. 1000 Hz

Solutions:

a. $T = \frac{1}{f} = \frac{1}{60\text{ Hz}} \cong 0.01667\text{ s, or } \mathbf{16.67\text{ ms}}$

(a recurring value since 60 Hz is so common)

b. $T = \frac{1}{f} = \frac{1}{1000\text{ Hz}} = 10^{-3}\text{ s} = \mathbf{1\text{ ms}}$

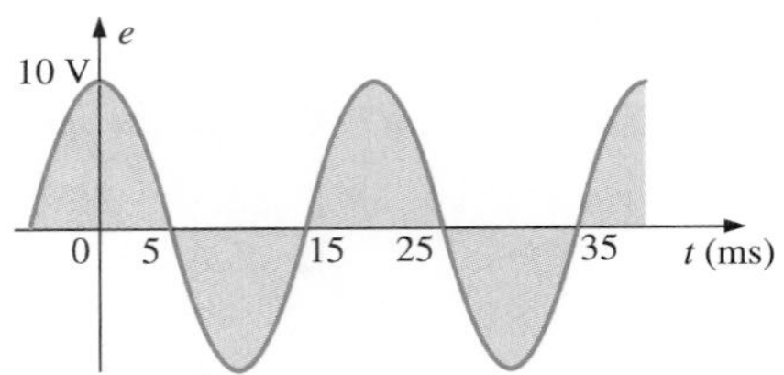

FIG. 13.7
Example 13.2.

EXAMPLE 13.2

Determine the frequency of the waveform of Fig. 13.7.

Solution: From the figure, $T = (25\text{ ms} - 5\text{ ms}) = 20\text{ ms}$ and

$$f = \frac{1}{T} = \frac{1}{20 \times 10^{-3}\text{ s}} = \mathbf{50\text{ Hz}}$$

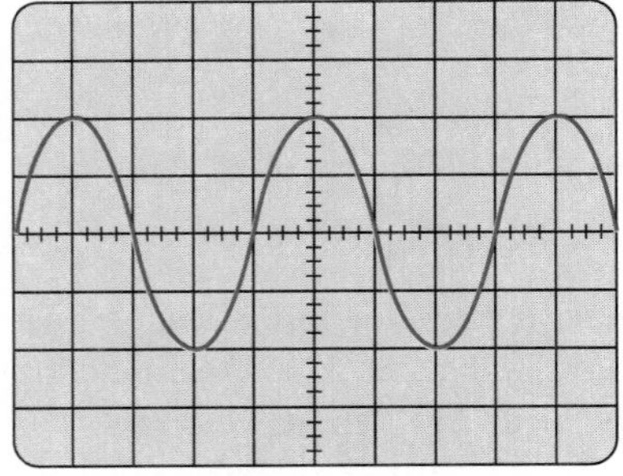

Vertical sensitivity = 0.1 V/div.
Horizontal sensitivity = 50 μs/div.

FIG. 13.8
Example 13.3.

EXAMPLE 13.3

The oscilloscope is an instrument that displays alternating waveforms such as those described above. A sinusoidal pattern appears on the oscilloscope of Fig. 13.8 with the indicated vertical and horizontal sensitivities. The vertical sensitivity defines the voltage associated with each vertical division of the display. Almost all oscilloscope screens are cut into a crosshatch pattern of lines separated by 1 cm in the vertical and horizontal directions. The horizontal sensitivity defines the time period associated with each horizontal division of the display.

For the pattern of Fig. 13.8 and the indicated sensitivities, determine the period, frequency, and peak value of the waveform.

continued

Solution: One cycle spans 4 divisions. The period is therefore

$$T = 4\ \text{div.}\left(\frac{50\ \mu\text{s}}{\text{div.}}\right) = \mathbf{200\ \mu s}$$

and the frequency is

$$f = \frac{1}{T} = \frac{1}{200 \times 10^{-6}\ \text{s}} = \mathbf{5\ kHz}$$

The vertical height above the horizontal axis includes 2 divisions. Therefore,

$$V_m = 2\ \text{div.}\left(\frac{0.1\ \text{V}}{\text{div.}}\right) = \mathbf{0.2\ V}$$

Defined Polarities and Direction

In the following analysis, we will need to establish a set of polarities for the sinusoidal ac voltage and a direction for the sinusoidal ac current. In each case, the polarity and current direction will be for an instant of time in the positive portion of the sinusoidal waveform. The polarities are indicated in Fig. 13.9 for a sinusoidal ac voltage source, and in Fig. 13.10 for a sinusoidal current source. The voltage and current are represented by lowercase letters to show that the quantities are time-dependent—their values change as a function of time. You will see the need for defining polarities and current direction when we consider multisource ac networks. Note that we didn't use the term *sinusoidal* before the phrase *ac networks.* This will continue to occur as we progress; *sinusoidal* is to be understood unless otherwise indicated.

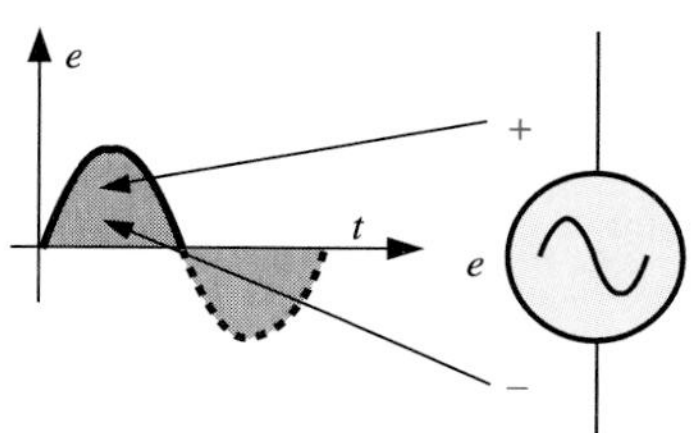

FIG. 13.9
Waveform polarities for sinusoidal ac voltage source.

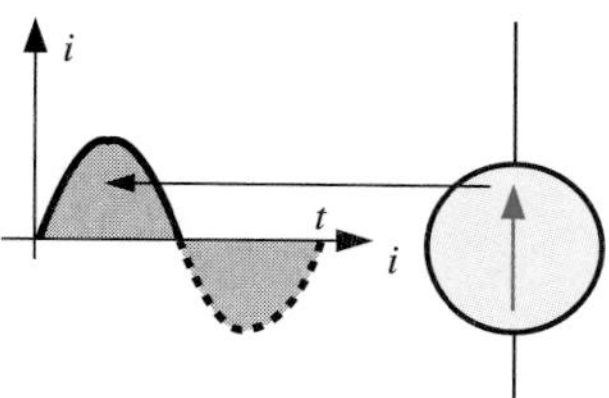

FIG. 13.10
Waveform polarities for sinusoidal ac current source.

13.3 THE SINE WAVE

The terms defined in the previous section can be applied to any type of periodic waveform, whether smooth or discontinuous. The sinusoidal waveform is of particular importance, however, since it lends itself readily to the mathematics and the physical phenomena associated with electric circuits. Consider the power of the following statement:

The sine wave is the only alternating waveform whose shape is unaffected by the response characteristics of R, L, and C elements.

In other words, if the voltage across (or current through) a resistor, coil, or capacitor is sinusoidal in nature, the resulting current (or voltage, respectively) for each will also have sinusoidal characteristics, as shown in Fig. 13.11. If a square wave or a triangular wave were applied, this would not be the case. The above statement also applies to the cosine

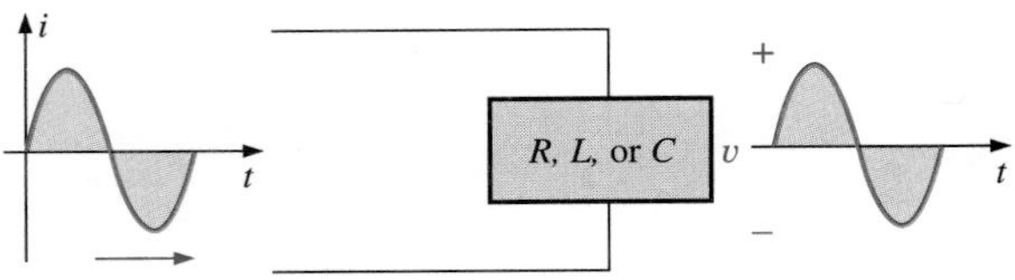

FIG. 13.11
The sine wave is the only alternating waveform whose shape is not altered by the response characteristics of a pure resistor, inductor, or capacitor.

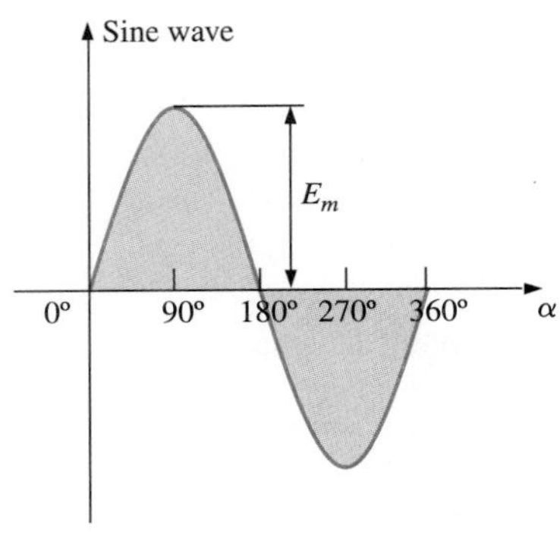

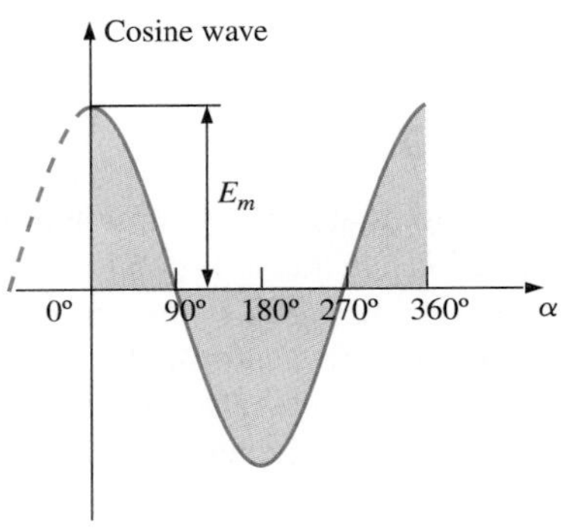

FIG. 13.12
Sine wave and cosine wave with the horizontal axis in degrees.

wave, since the waves differ only by a 90° shift on the horizontal axis, as shown in Fig. 13.12.

The unit of measurement for the horizontal axis of Fig. 13.12 is the *degree.* A second unit of measurement frequently used is the **radian** (rad). It is defined by a quadrant of a circle such as in Fig. 13.13, where the distance subtended on the circumference equals the radius of the circle.

If we define x as the number of intervals of r (the radius) around the circumference of the circle, then

$$C = 2\pi r = x \cdot r$$

and we find

$$x = 2\pi$$

Therefore, there are 2π rad around a 360° circle, as shown in Fig. 13.14, and

$$2\pi \text{ rad} = 360° \tag{13.4}$$

with

$$1 \text{ rad} = 57.296° \cong 57.3° \tag{13.5}$$

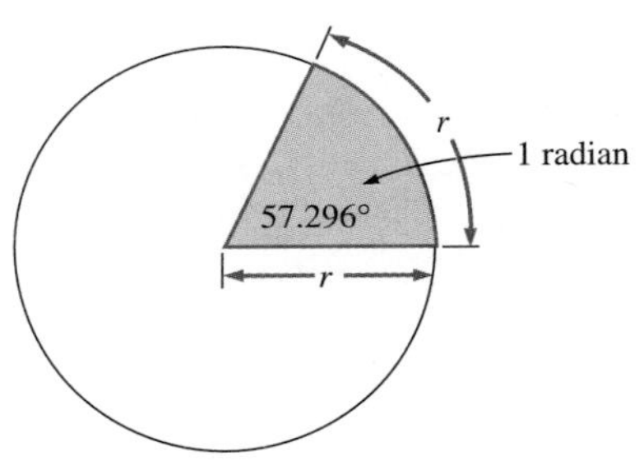

FIG. 13.13
Defining the radian.

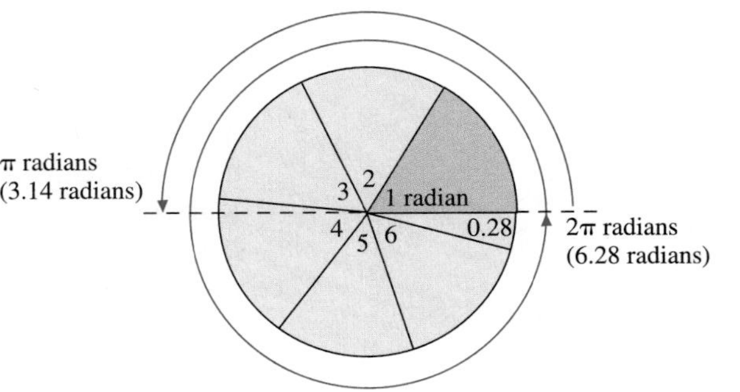

FIG. 13.14
There are 2π radians in one full circle of 360°.

A number of electrical formulas contain a multiplier of π. This is one reason it is sometimes better to measure angles in radians than in degrees.

The quantity π is the ratio of the circumference of a circle to its diameter.

The value of π built into scientific calculators is often correct to 10 digits or more:

$$\pi = 3.141\,592\,654 \cdots$$

Although the approximation $\pi \cong 3.14$ is often used, all the calculations in this text will use the π function as provided on all scientific calculators.

For 180° and 360°, the two units of measurement are related as shown in Fig. 13.14. The conversion equations between the two are the following:

$$\text{Radians} = \left(\frac{\pi}{180^\circ}\right) \times (\text{degrees}) \tag{13.6}$$

$$\text{Degrees} = \left(\frac{180^\circ}{\pi}\right) \times (\text{radians}) \tag{13.7}$$

Applying these equations, we find

$$\mathbf{90^\circ:}\ \text{Radians} = \frac{\pi}{180^\circ}(90^\circ) = \frac{\pi}{2}\ \textbf{rad}$$

$$\mathbf{30^\circ:}\ \text{Radians} = \frac{\pi}{180^\circ}(30^\circ) = \frac{\pi}{6}\ \textbf{rad}$$

$$\frac{\pi}{3}\ \textbf{rad:}\ \text{Degrees} = \frac{180^\circ}{\pi}\left(\frac{\pi}{3}\right) = \mathbf{60^\circ}$$

$$\frac{3\pi}{2}\ \textbf{rad:}\ \text{Degrees} = \frac{180^\circ}{\pi}\left(\frac{3\pi}{2}\right) = \mathbf{270^\circ}$$

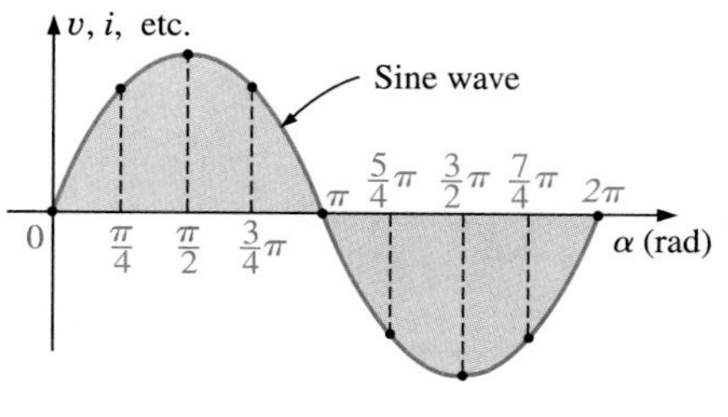

FIG. 13.15
Plotting a sine wave versus radians.

Using the radian as the unit of measurement for the horizontal axis, we still get a sine wave, as shown in Fig. 13.15.

It is of particular interest that the sinusoidal waveform can be derived from the length of the *vertical projection* of a radius vector rotating in a uniform circular motion about a fixed point. Starting as shown in Fig. 13.16(a) and plotting the amplitude (above and below zero) on the coordinates drawn to the right [Figs. 13.16(b) through (i)], we will trace a complete sinusoidal waveform after the radius vector has completed a 360° rotation about the centre.

The velocity with which the radius vector rotates about the centre, called the *angular velocity,* can be determined from the following equation:

$$\text{Angular velocity} = \frac{\text{distance (degrees or radians)}}{\text{time (seconds)}} \tag{13.8}$$

Substituting into Eq. (13.8) and assigning the Greek letter omega (ω) to the angular velocity, we have

$$\omega = \frac{\alpha}{t} \tag{13.9}$$

and

$$\alpha = \omega t \tag{13.10}$$

Since ω is usually provided in radians per second, the angle α obtained using Eq. (13.10) is usually in radians. If α is required in degrees, Eq. (13.7) must be applied. The importance of remembering the above will become obvious in the examples to follow.

Note: Physics and mathematics texts often use the Greek symbol θ to represent angular displacement (in radians), and α to represent angular

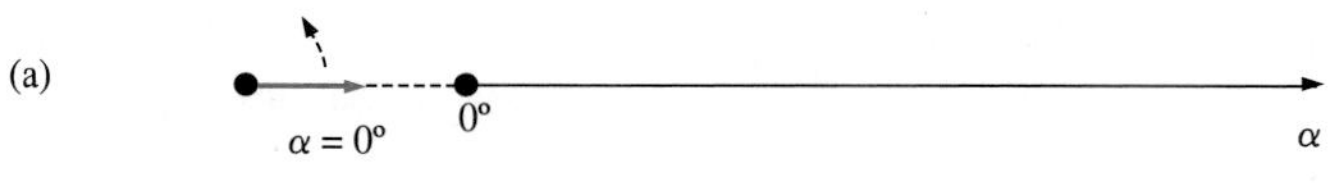

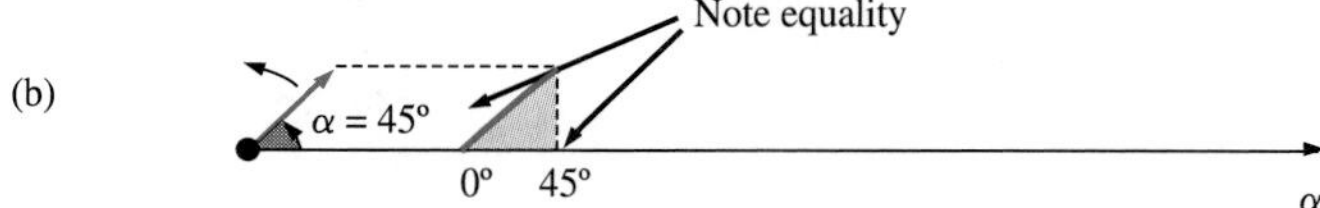

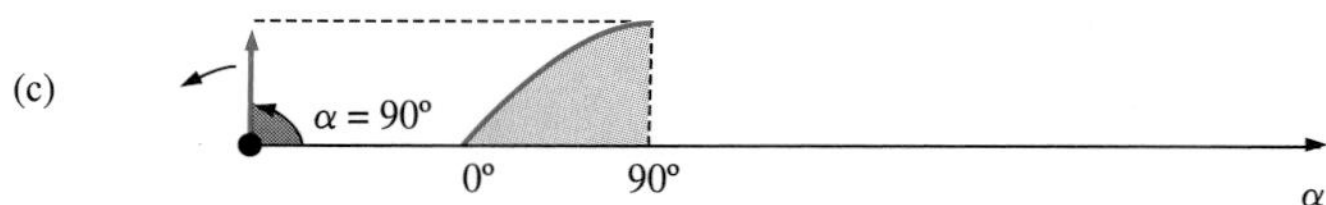

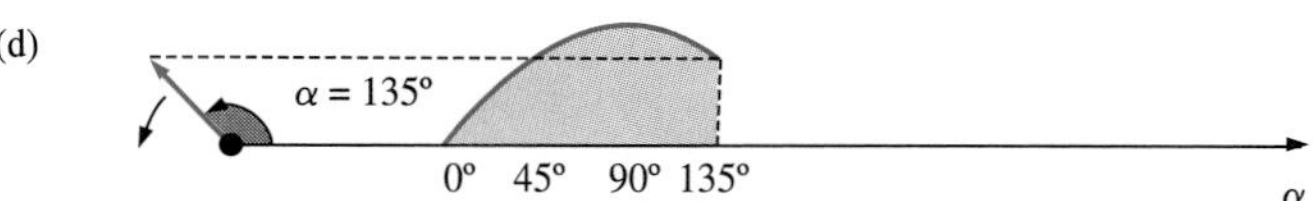

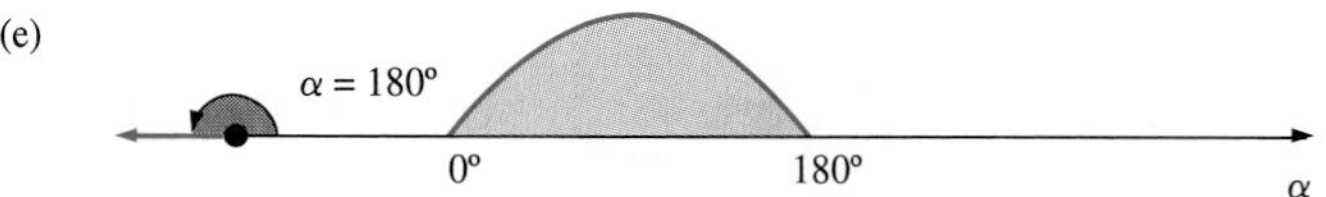

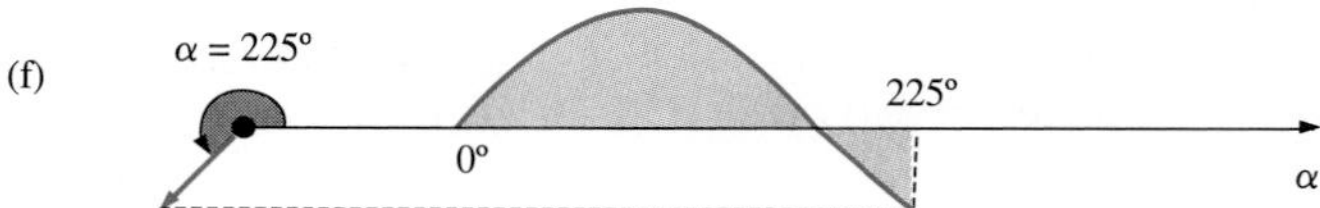

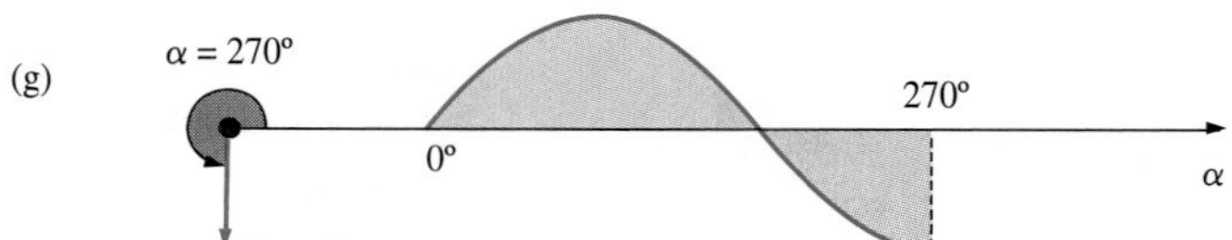

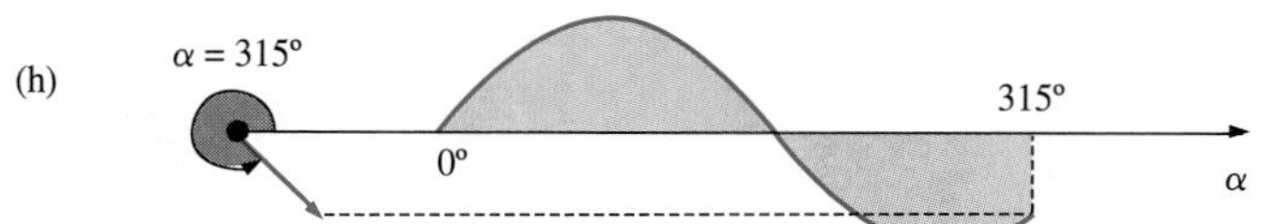

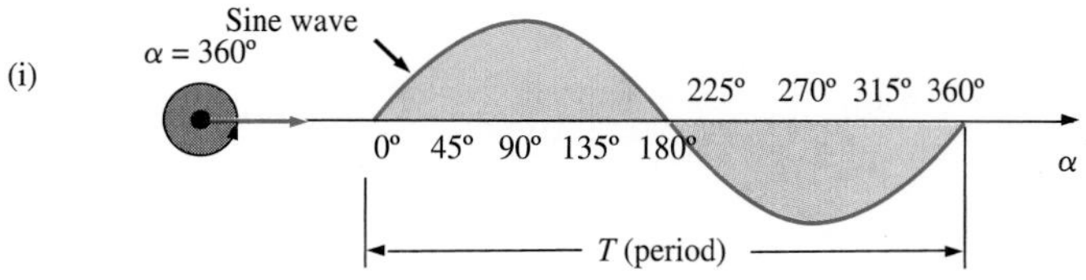

FIG. 13.16
Generating a sinusoidal waveform through the vertical projection of a rotating vector.

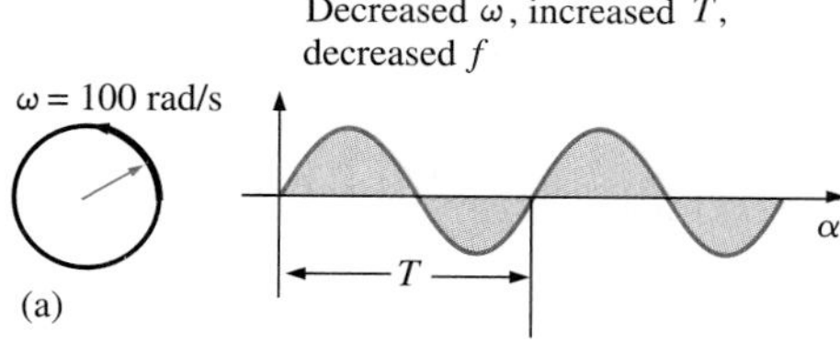

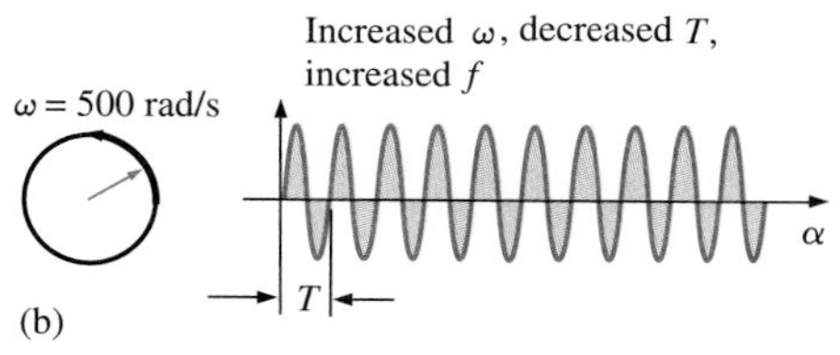

FIG. 13.17
Demonstrating the effect of ω on the frequency and period.

acceleration (rad/s^2). In this text, the α is used for angular displacement and θ (often in degrees) is used for phase shift.

In Fig. 13.16, the time required to complete one revolution is equal to the period (T) of the sinusoidal waveform of Fig. 13.16(i). The radians subtended in this time interval are 2π. Substituting, we have

$$\omega = \frac{2\pi}{T} \quad \text{(rad/s)} \tag{13.11}$$

In words, this equation states that the smaller the period of the sinusoidal waveform of Fig. 13.16(i), or the smaller the time interval before one complete cycle is generated, the greater must be the angular velocity of the rotating radius vector. We can now go one step further and apply the fact that the frequency of the generated waveform is inversely related to the period of the waveform; that is, $f = 1/T$. Thus,

$$\omega = 2\pi f \quad \text{(rad/s)} \tag{13.12}$$

This equation states that the higher the frequency of the generated sinusoidal waveform, the higher must be the angular velocity. Equations (13.11) and (13.12) are verified somewhat by Fig. 13.17, where for the same radius vector, $\omega = 100$ rad/s and 500 rad/s.

EXAMPLE 13.4

Determine the angular velocity of a sine wave having a frequency of 60 Hz.

Solution:

$$\omega = 2\pi f = (2\pi)(60 \text{ Hz}) \cong \mathbf{377\ rad/s}$$

This value of **377 rad/s** will appear frequently without further comment because it is the value for the angular velocity of the 60 Hz electrical systems used throughout North America.

EXAMPLE 13.5

Determine the frequency and period of the sine wave of Fig. 13.17(b).

Solution: Since $\omega = 2\pi/T$,

$$T = \frac{2\pi}{\omega} = \frac{2\pi \text{ rad}}{500 \text{ rad/s}} = \frac{2\pi \text{ rad}}{500 \text{ rad/s}} = \mathbf{12.57\ ms}$$

and

$$f = \frac{1}{T} = \frac{1}{12.57 \times 10^{-3} \text{ s}} = \mathbf{79.58\ Hz}$$

EXAMPLE 13.6

Given $\omega = 200$ rad/s, determine how long it will take the sinusoidal waveform to pass through an angle of 90°.

Solution: Eq. (13.10): $\alpha = \omega t$, and

$$t = \frac{\alpha}{\omega}$$

continued

However, α must be converted as $\pi/2$ (= 90°) since ω is in radians per second:

$$t = \frac{\alpha}{\omega} = \frac{\pi/2 \text{ rad}}{200 \text{ rad/s}} = \frac{\pi}{400 \text{ s}} = \mathbf{7.85 \text{ ms}}$$

EXAMPLE 13.7

Find the angle through which a sinusoidal waveform of 60 Hz will pass in a period of 5 ms.

Solution: Eq. (13.11): $\alpha = \omega t$, or

$$\alpha = 2\pi f t = (2\pi)(60 \text{ Hz})(5 \times 10^{-3} \text{s}) = \mathbf{1.885 \text{ rad}}$$

If you're not careful, you might be tempted to interpret the answer as 1.885°. However,

$$\alpha\ (°) = \frac{180°}{\pi \text{ rad}}(1.885 \text{ rad}) = \mathbf{108°}$$

13.4 GENERAL FORMAT FOR THE SINUSOIDAL VOLTAGE OR CURRENT

The basic mathematical format for the sinusoidal waveform is

$$A_m \sin \alpha \tag{13.13}$$

where A_m is the peak value of the waveform and α is the unit of measure for the horizontal axis, as shown in Fig. 13.18.

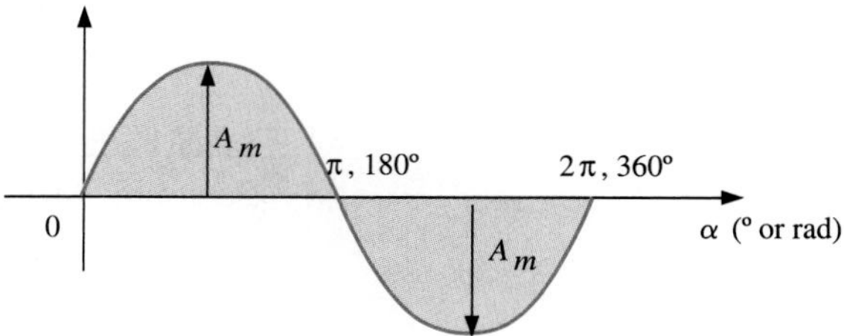

FIG. 13.18
Basic sinusoidal function.

The equation $\alpha = \omega t$ states that the angle α that the rotating vector of Fig. 13.16 passes through is determined by the angular velocity of the rotating vector and the length of time the vector rotates. For example, for a particular angular velocity (fixed ω), the longer the radius vector is permitted to rotate (that is, the greater the value of t), the greater the number of degrees or radians the vector will pass through. Relating this statement to the sinusoidal waveform, for a particular angular velocity, the longer the time, the greater the number of cycles shown. For a fixed time interval, the greater the angular velocity, the greater the number of cycles generated.

Due to Eq. (13.10), the general format of a sine wave can also be written

$$A_m \sin \omega t \tag{13.14}$$

with ωt as the horizontal unit of measure.

For electrical quantities such as current and voltage, the general format is

$$i = I_m \sin \omega t = I_m \sin \alpha$$
$$e = E_m \sin \omega t = E_m \sin \alpha$$

where the capital letters with the subscript m represent the amplitude and the lowercase letters i and e represent the instantaneous value of current or voltage, respectively, at any time t. This format is particularly important since it presents the sinusoidal voltage or current as a function of time, which is the horizontal scale for the oscilloscope. Recall that the horizontal sensitivity of a scope is in time per division and not degrees per centimetre.

EXAMPLE 13.8

Given $e = 5 \sin \alpha$, determine e at (a) $\alpha = 40°$ and (b) $\alpha = 0.8\pi$ rad.

Solutions:

a. For $\alpha = 40°$,

$$e = 5 \sin 40° = 5(0.6428) = \mathbf{3.214\ V}$$

b. For $\alpha = 0.8\pi$ rad,

$$e = 5 \sin (0.8\pi \text{ rad}) = 5(0.5878) = \mathbf{2.939\ V}$$

The angle at which a particular voltage level is attained can be determined by rearranging the equation

$$e = E_m \sin \alpha$$

in the following way:

$$\sin \alpha = \frac{e}{E_m}$$

which can be written

$$\alpha = \sin^{-1} \frac{e}{E_m} \tag{13.15}$$

Similarly, for a particular current level,

$$\alpha = \sin^{-1} \frac{i}{I_m} \tag{13.16}$$

The function $\sin^{-1}$ is available on all scientific calculators.

EXAMPLE 13.9

a. Determine the angle at which the magnitude of the sinusoidal function $v = 10 \sin 377t$ is 4 V. (Frequency is 60 Hz.)
b. Determine the time at which the magnitude is attained.

continued

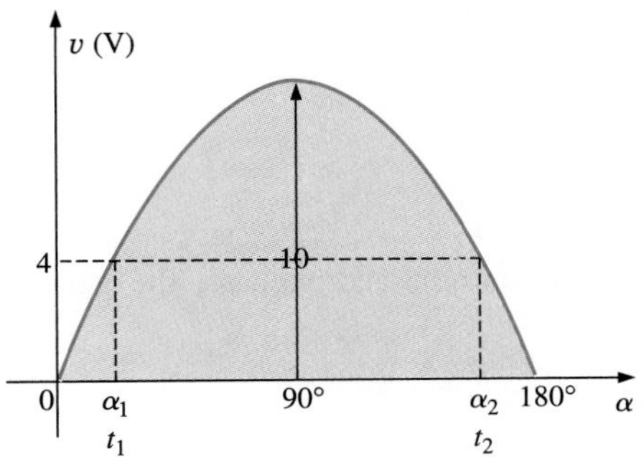

FIG. 13.19
Example 13.9.

Solutions:

a. Eq. (13.15):

$$\alpha_1 = \sin^{-1}\frac{v}{E_m} = \sin^{-1}\frac{4\text{ V}}{10\text{ V}} = \sin^{-1} 0.4 = \mathbf{23.578°}$$

Fig. 13.19 reveals that the magnitude of 4 V (positive) will be attained at two points between 0° and 180°. The second intersection is determined by

$$\alpha_2 = 180° - 23.578° = \mathbf{156.422°}$$

In general, therefore, keep in mind that Eqs. (13.15) and (13.16) will provide an angle with a magnitude between 0° and 90°.

b. Eq. (13.10): $\alpha = \omega t$, and so $t = \alpha/\omega$, with α in radians. Thus,

$$\alpha\text{ (rad)} = \frac{\pi}{180°}(23.578°) = 0.411\text{ rad}$$

and

$$t_1 = \frac{\alpha}{\omega} = \frac{0.411\text{ rad}}{377\text{ rad/s}} = \mathbf{1.09\text{ ms}}$$

For the second intersection,

$$\alpha\text{ (rad)} = \frac{\pi}{180°}(156.422°) = 2.73\text{ rad}$$

$$t_2 = \frac{\alpha}{\omega} = \frac{2.73\text{ rad}}{377\text{ rad/s}} = \mathbf{7.24\text{ ms}}$$

The sine wave can also be plotted against *time* on the horizontal axis. The time period for each interval can be determined from $t = \alpha/\omega$, but the most direct route is simply to find the period T from $T = 1/f$ and break it up into the required intervals. This technique will be demonstrated in Example 13.10.

Before reviewing the example, take special note of the relative simplicity of the mathematical equation that can represent a sinusoidal waveform. Any alternating waveform whose characteristics differ from those of the sine wave cannot be represented by a single term, but may require two, four, six, or perhaps an infinite number of terms to be represented accurately. Additional description of *nonsinusoidal waveforms* can be found in the CD-ROM supplement.

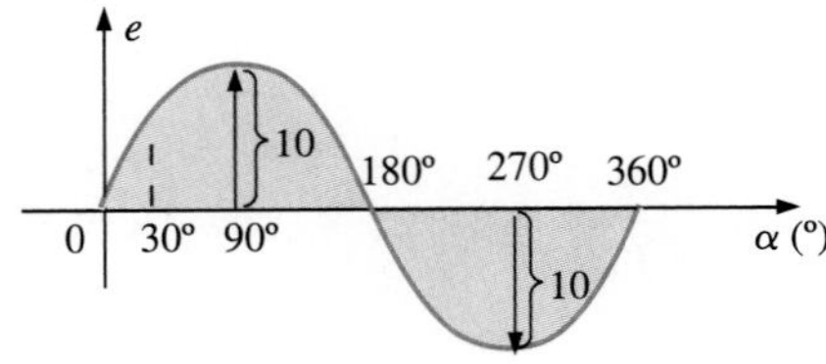

FIG. 13.20
Example 13.10, horizontal axis in degrees.

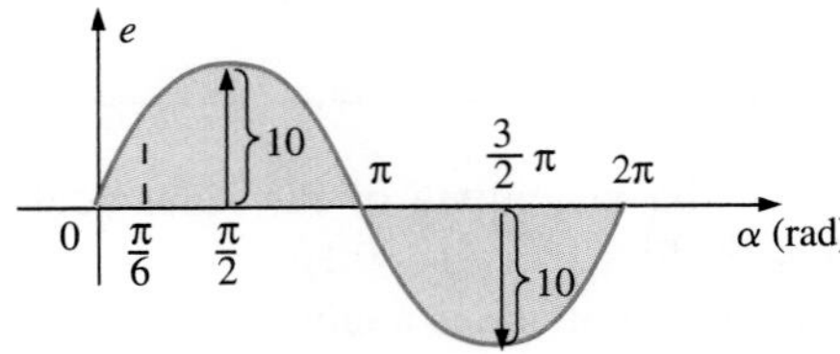

FIG. 13.21
Example 13.10, horizontal axis in radians.

EXAMPLE 13.10

Sketch $e = 10 \sin 314t$ with the horizontal axis in

a. angle (α) in degrees.
b. angle (α) in radians.
c. time (t) in seconds.

Solutions:

a. See Fig 13.20. (Note that no calculations are required.)
b. See Fig. 13.21. (Once the relationship between degrees and radians is understood, there is again no need for calculations.)
c. 360°: $T = \dfrac{2\pi}{\omega} = \dfrac{2\pi}{314} = 20\text{ ms}$

180°: $\dfrac{T}{2} = \dfrac{20\text{ ms}}{2} = 10\text{ ms}$

continued

$$90°: \frac{T}{4} = \frac{20 \text{ ms}}{4} = 5 \text{ ms}$$

$$30°: \frac{T}{12} = \frac{20 \text{ ms}}{12} = 1.67 \text{ ms}$$

See Fig. 13.22.

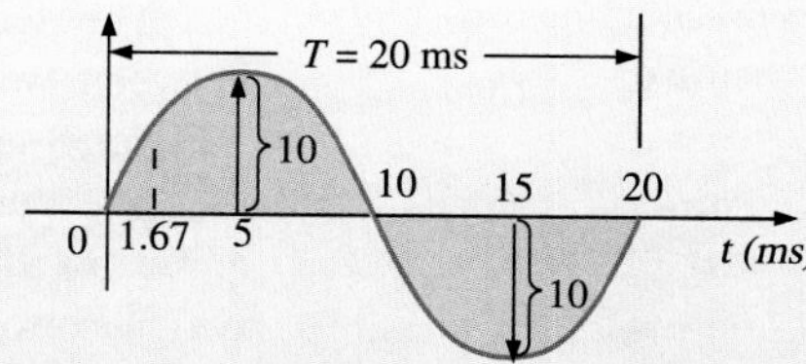

FIG. 13.22
Example 13.10, horizontal axis in milliseconds.

EXAMPLE 13.11

Given $i = 6 \times 10^{-3} \sin 1000t$, determine i at $t = 2$ ms.

Solution:

$$\alpha = \omega t = 1000t = (1000 \text{ rad/s})(2 \times 10^{-3} \text{ s}) = 2 \text{ rad}$$

$$\alpha\ (°) = \frac{180°}{\pi \text{ rad}}(2 \text{ rad}) = 114.59°$$

$$i = (6 \times 10^{-3})(\sin 114.59°) = (6 \text{ mA})(0.9093) = \mathbf{5.46 \text{ mA}}$$

13.5 PHASE RELATIONS

So far we have considered only sine waves that have maxima at $\pi/2$ and $3\pi/2$, with a zero value at 0, π, and 2π, as shown in Fig. 13.21. If the waveform is shifted to the right or left of 0°, the expression becomes

$$A_m \sin(\omega t \pm \theta) \qquad (13.17)$$

where θ is the angle in degrees or radians that the waveform has been shifted.

If the waveform passes through the horizontal axis with a *positive-going* (increasing with time) slope *before* 0°, as shown in Fig. 13.23, the expression is

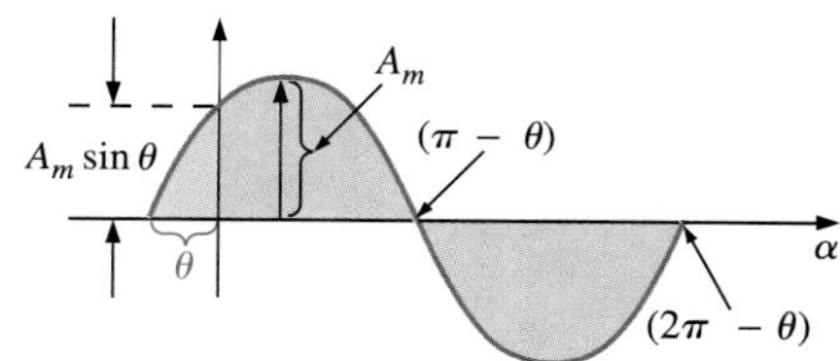

FIG. 13.23
Defining the phase shift for a sinusoidal function that crosses the horizontal axis with a positive slope before 0°.

$$A_m \sin(\omega t + \theta) \qquad (13.18)$$

At $\omega t = \alpha = 0°$, the magnitude is determined by $A_m \sin \theta$. If the waveform passes through the horizontal axis with a positive-going slope *after* 0°, as shown in Fig. 13.24, the expression is

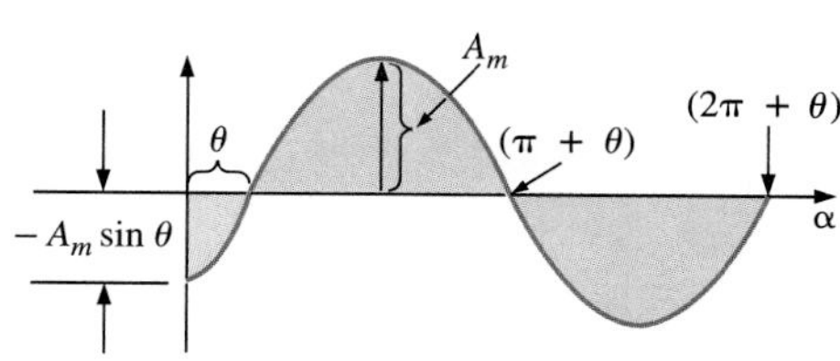

FIG. 13.24
Defining the phase shift for a sinusoidal function that crosses the horizontal axis with a positive slope after 0°.

$$A_m \sin(\omega t - \theta) \qquad (13.19)$$

And at $\omega t = \alpha = 0°$, the magnitude is $A_m \sin(-\theta)$, which, by a trigonometric identity, is $-A_m \sin \theta$.

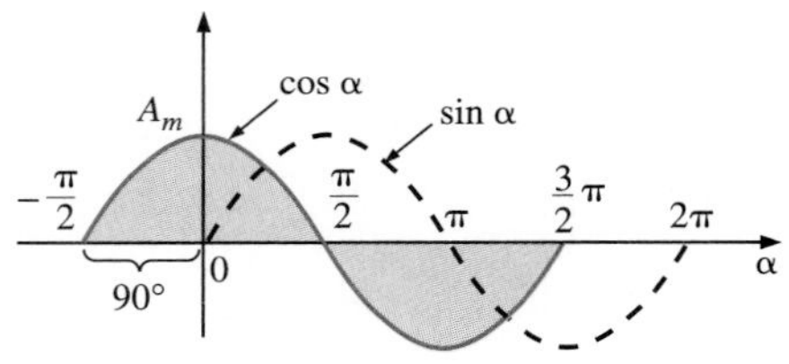

FIG. 13.25
Phase relationship between a sine wave and a cosine wave.

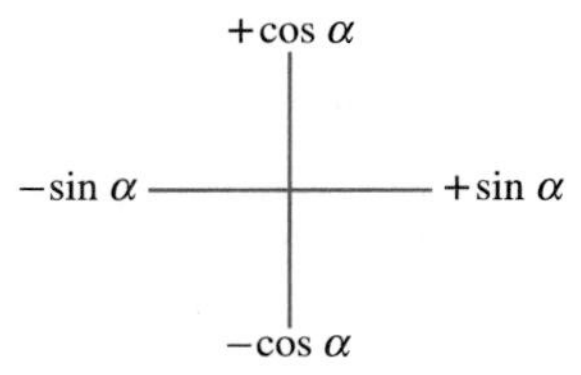

FIG. 13.26
Graphic tool for finding the relationship between specific sine and cosine functions.

If the waveform crosses the horizontal axis with a positive-going slope 90° ($\pi/2$) sooner, as shown in Fig. 13.25, it is called a *cosine wave;* that is,

$$\sin(\omega t + 90°) = \sin\left(\omega t + \frac{\pi}{2}\right) = \cos \omega t \qquad (13.20)$$

or

$$\sin \omega t = \cos(\omega t - 90°) = \cos\left(\omega t - \frac{\pi}{2}\right) \qquad (13.21)$$

The terms *lead* and *lag* are used to indicate the relationship between two sinusoidal waveforms of the *same frequency* plotted on the same set of axes. In Fig. 13.25, the cosine curve is said to *lead* the sine curve by 90°, and the sine curve is said to *lag* the cosine curve by 90°. The 90° is referred to as the phase angle between the two waveforms. In language commonly used, the waveforms are *out of phase* by 90°. Note that the phase angle between the two waveforms is measured between the two points on the horizontal axis through which each passes with the *same slope.* If both waveforms cross the axis at the same point with the same slope, they are *in phase.*

The geometric relationship between various forms of the sine and cosine functions can be derived from Fig. 13.26. For instance, starting at the $\sin \alpha$ position, we find that $\cos \alpha$ is an additional 90° in the counterclockwise direction. Therefore, $\cos \alpha = \sin(\alpha + 90°)$. For $-\sin \alpha$ we must travel 180° in the counterclockwise (or clockwise) direction so that $-\sin \alpha = \sin(\alpha \pm 180°)$, and so on, as listed below:

$$\begin{aligned} \cos \alpha &= \sin(\alpha + 90°) \\ \sin \alpha &= \cos(\alpha - 90°) \\ -\sin \alpha &= \sin(\alpha \pm 180°) \\ -\cos \alpha &= \sin(\alpha + 270°) = \sin(\alpha - 90°) \\ &\text{etc.} \end{aligned} \qquad (13.22)$$

In addition you should know that

$$\begin{aligned} \sin(-\alpha) &= -\sin \alpha \\ \cos(-\alpha) &= \cos \alpha \end{aligned} \qquad (13.23)$$

If a sinusoidal expression appears as

$$e = -E_m \sin \omega t$$

the negative sign is associated with the sine portion of the expression, not the peak value E_m. Except for convenience, the expression would be written

$$e = E_m(-\sin \omega t)$$

Since

$$-\sin \omega t = \sin(\omega t \pm 180°)$$

the expression can also be written

$$e = E_m \sin(\omega t \pm 180°)$$

revealing that a negative sign can be replaced by a 180° change in phase angle (+ or −); that is,

$$e = E_m \sin \omega t = E_m \sin(\omega t + 180°)$$
$$= E_m \sin(\omega t - 180°)$$

A plot of each will clearly show their equivalence. There are two correct mathematical representations for the functions.

The *phase relationship* between two waveforms indicates which one leads or lags, and by how many degrees or radians.

EXAMPLE 13.12

What is the phase relationship between the sinusoidal waveforms of each of the following sets?

a. $v = 10 \sin(\omega t + 30°)$
   $i = 5 \sin(\omega t + 70°)$

b. $i = 15 \sin(\omega t + 60°)$
   $v = 10 \sin(\omega t - 20°)$

c. $i = 2 \cos(\omega t + 10°)$
   $v = 3 \sin(\omega t - 10°)$

d. $i = -\sin(\omega t + 30°)$
   $v = 2 \sin(\omega t + 10°)$

e. $i = -2 \cos(\omega t - 60°)$
   $v = 3 \sin(\omega t - 150°)$

Solutions:

a. See Fig. 13.27.
   ***i* leads *v* by 40°, or *v* lags *i* by 40°.**

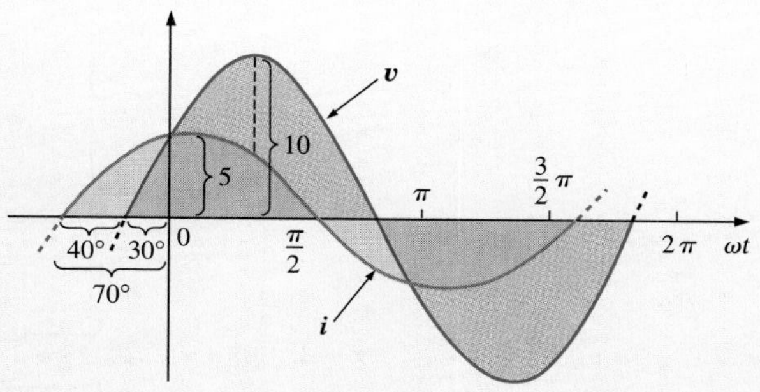

FIG. 13.27
Example 13.12, i leads v by 40°.

b. See Fig. 13.28.
   ***i* leads *v* by 80°, or *v* lags *i* by 80°.**

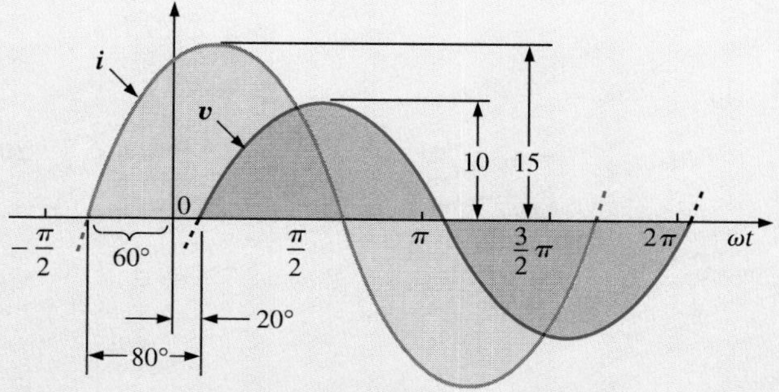

FIG. 13.28
Example 13.12, i leads v by 80°.

continued

c. See Fig. 13.29.

$$i = 2\cos(\omega t + 10^\circ) = 2\sin(\omega t + 10^\circ + 90^\circ)$$
$$= 2\sin(\omega t + 100^\circ)$$

***i* leads *v* by 110°, or *v* lags *i* by 110°.**

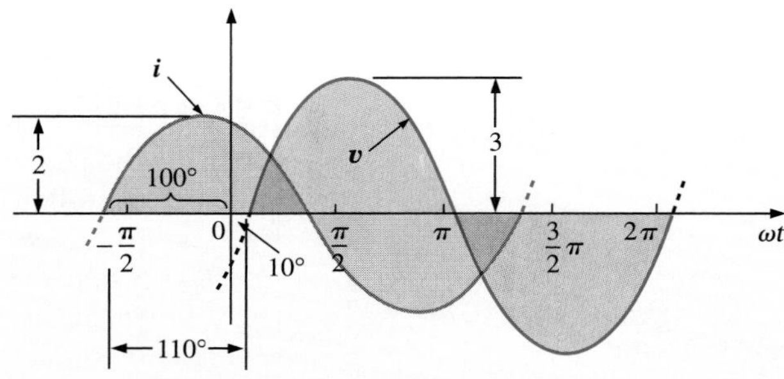

FIG. 13.29
Example 13.12, i leads v by 110°.

d. See Fig. 13.30.

$$-\sin(\omega t + 30^\circ) = \sin(\omega t + 30^\circ \overset{\text{Note}}{-} 180^\circ)$$
$$= \sin(\omega t - 150^\circ)$$

***v* leads *i* by 160°, or *i* lags *v* by 160°.**

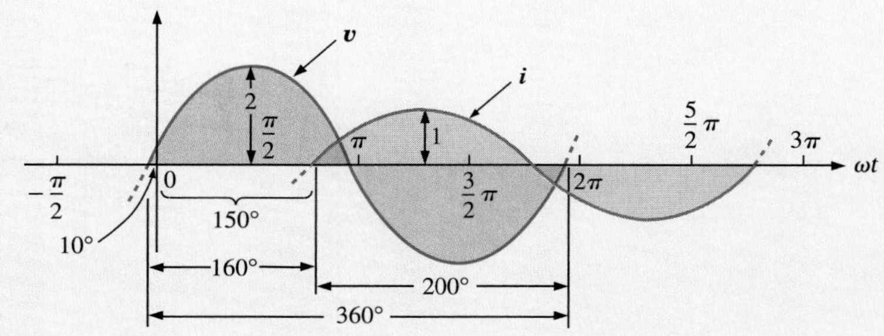

FIG. 13.30
Example 13.12, v leads i by 160°.

Or using

$$-\sin(\omega t + 30^\circ) = \sin(\omega t + 30^\circ \overset{\text{Note}}{+} 180^\circ)$$
$$= \sin(\omega t + 210^\circ)$$

***i* leads *v* by 200°, or *v* lags *i* by 200°.**

e. See Fig. 13.31.

$$i = -2\cos(\omega t - 60^\circ) = 2\cos(\omega t - 60^\circ \overset{\text{By choice}}{-} 180^\circ)$$
$$= 2\cos(\omega t - 240^\circ)$$

continued

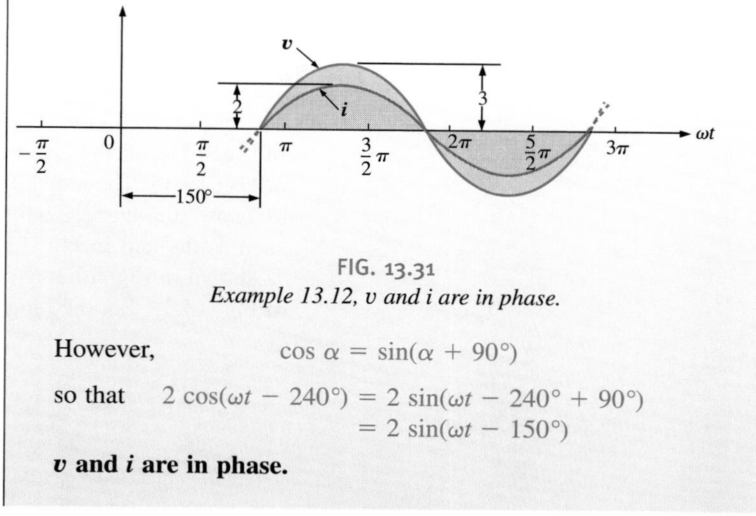

FIG. 13.31
Example 13.12, v and i are in phase.

However,
$$\cos \alpha = \sin(\alpha + 90°)$$

so that
$$2 \cos(\omega t - 240°) = 2 \sin(\omega t - 240° + 90°) = 2 \sin(\omega t - 150°)$$

***v* and *i* are in phase.**

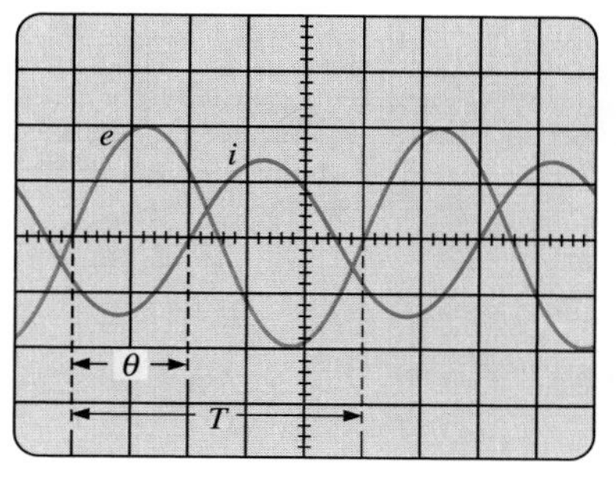

Vertical sensitivity = 2 V/div.
Horizontal sensitivity = 0.2 ms/div.

FIG. 13.32
Finding the phase angle between waveforms using a dual-trace oscilloscope.

Phase Measurements

The hookup procedure for using an oscilloscope to measure phase angles is covered in detail in Section 16.3. However, we can introduce the equation for determining the phase angle using Fig. 13.32. Note that each sinusoidal function has the same frequency, so either waveform can be used to determine the period. For the waveform chosen in Fig. 13.32, the period covers 5 divisions at 0.2 ms/div. The phase shift between the waveforms is 2 divisions. Since the full period represents a cycle of 360°, the following ratio [from which Eq. (13.24) can be derived] can be formed:

$$\frac{360°}{T\text{ (no. of div.)}} = \frac{\theta}{\text{phase shift (no. of div.)}}$$

and

$$\theta = \frac{\text{phase shift (no. of div.)}}{T\text{ (no. of div.)}} \times 360° \qquad (13.24)$$

Substituting into Eq. (13.24) will result in

$$\theta = \frac{(2\text{ div.})}{(5\text{ div.})} \times 360° = 144°$$

and e leads i by 144°.

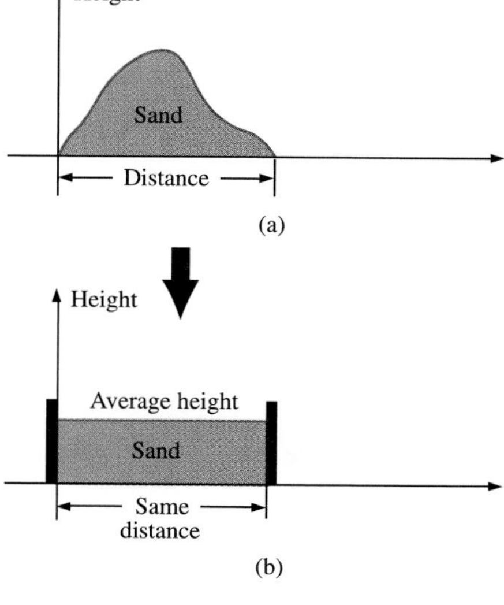

FIG. 13.33
Defining average value.

13.6 AVERAGE VALUE

Even though the concept of the *average value* is important in most technical fields, its true meaning is often misunderstood. In Fig. 13.33(a), for example, the average height of the sand may be required to determine the volume of sand available. The average height of the sand is the height obtained if the distance from one end to the other is maintained while the sand is levelled off, as shown in Fig. 13.33(b). The area under the mound of Fig. 13.33(a) will then equal the area under the rectangular shape of Fig. 13.33(b) as determined by $A = b \times h$. Of course, the depth (into the page) of the sand must be the same for Fig. 13.33(a) and 13.33(b) for the preceding conclusions to have any meaning.

In Fig. 13.33 the distance was measured from one end to the other. In Fig. 13.34(a) the distance extends beyond the end of the original pile of Fig. 13.33. The situation could be one where a landscaper would like to know the average height of the sand if spread out over a distance such as defined in Fig. 13.34(a). The result of an increased distance is as shown in Fig. 13.34(b). The average height has decreased compared to Fig. 13.33. So, the longer the distance, the lower the average height.

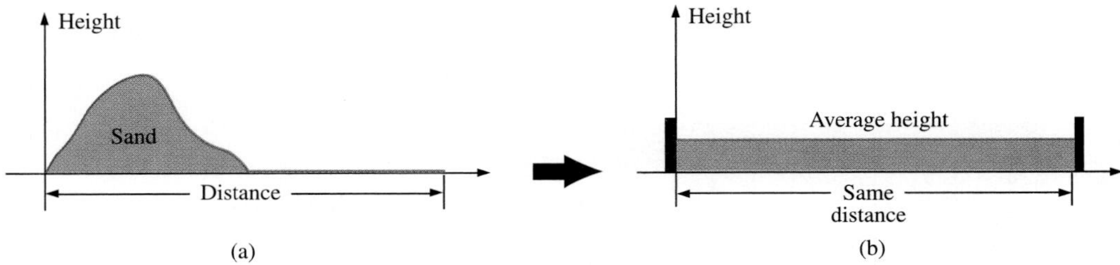

FIG. 13.34
Effect of distance (length) on average value.

If the distance parameter includes a depression, as shown in Fig. 13.35(a), some of the sand will be used to fill the depression, resulting in an even lower average height for the landscaper, as shown in Fig. 13.35(b). For a sinusoidal waveform, the depression would have the same shape as the mound of sand (over one full cycle), resulting in an average height of "ground level" (or zero volts for a sinusoidal voltage over one full period).

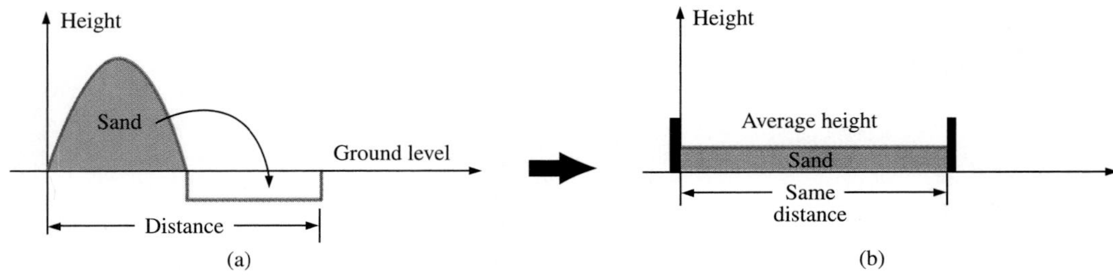

FIG. 13.35
Effect of depressions (negative excursions) on average value.

After travelling far by car, some drivers like to calculate their average speed for the entire trip. This is usually done by dividing the kilometres travelled by the hours required to drive that distance. For example, if a person travelled 360 km in 5 h, the average speed was 360 km/5 h, or 72 km/h. This same distance may have been travelled at various speeds for various intervals of time, as shown in Fig. 13.36.

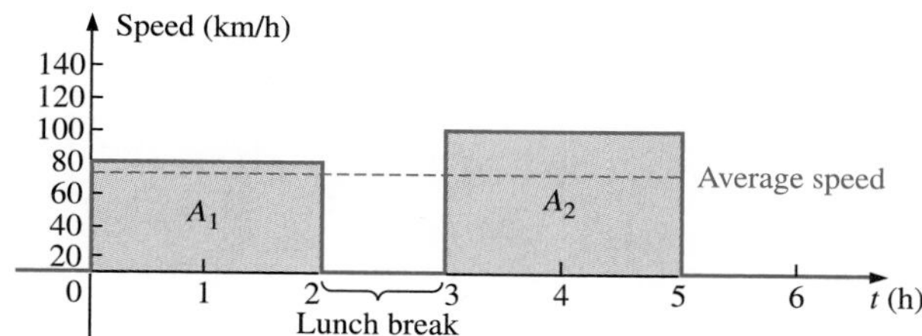

FIG. 13.36
Plotting speed versus time for a car trip.

By finding the total area under the curve for the 5 h and then dividing the area by 5 h (the total time for the trip), we obtain the same result of 72 km/h; that is,

$$\text{Average speed} = \frac{\text{area under curve}}{\text{length of curve}} \tag{13.25}$$

$$= \frac{A_1 + A_2}{5 \text{ h}}$$

$$= \frac{(80 \text{ km/h})(2 \text{ h}) + (100 \text{ km/h})(2 \text{ h})}{5 \text{ h}}$$

$$= \frac{360}{5} \text{ km/h}$$

$$= \mathbf{72 \text{ km/h}}$$

Equation (13.25) can be extended to include any variable quantity, such as current or voltage, if we let G represent the average value, as follows:

$$G \text{ (average value)} = \frac{\text{algebraic sum of areas}}{\text{length of curve}} \tag{13.26}$$

The *algebraic* sum of the areas must be determined, since some area contributions will be from below the horizontal axis. Areas above the axis will have a positive sign, and those below, a negative sign. A positive average value will be above the axis, and a negative value, below.

The average value of *any* current or voltage is the value indicated on a dc meter. In other words, over a complete cycle, the average value is the equivalent dc value. In the analysis of electronic circuits to be considered in a later course, both dc and ac sources of voltage will be applied to the same network. It will then be necessary to know or determine the dc (or average value) and ac components of the voltage or current in various parts of the system.

EXAMPLE 13.13

Determine the average value of the waveforms of Fig. 13.37.

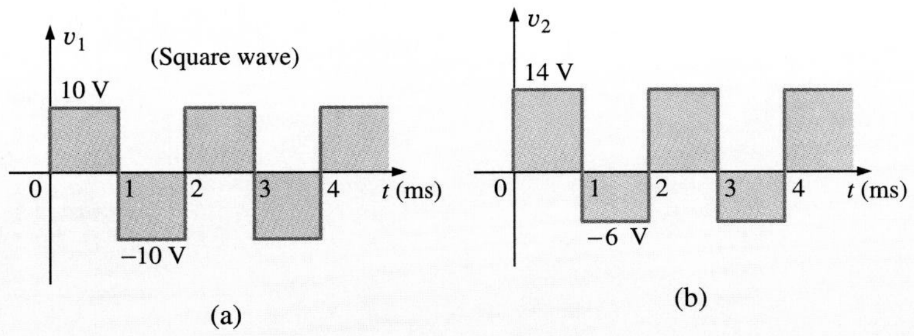

FIG. 13.37
Example 13.13.

Solutions:

a. You can see that the area above the axis equals the area below over one cycle, resulting in an average value of zero volts. Using Eq. (13.26):

$$G = \frac{(10\text{ V})(1\text{ ms}) - (10\text{ V})(1\text{ ms})}{2\text{ ms}}$$

$$= \frac{0}{2\text{ ms}} = \mathbf{0\ V}$$

b. Using Eq. (13.26):

$$G = \frac{(14\text{ V})(1\text{ ms}) - (6\text{ V})(1\text{ ms})}{2\text{ ms}}$$

$$= \frac{14\text{ V} - 6\text{ V}}{2} = \frac{8\text{ V}}{2} = \mathbf{4\ V}$$

as shown in Fig. 13.38.

The waveform of Fig. 13.37(b) is simply the square wave of Fig. 13.37(a) with a dc shift of 4 V; that is,

$$v_2 = v_1 + 4\text{ V}$$

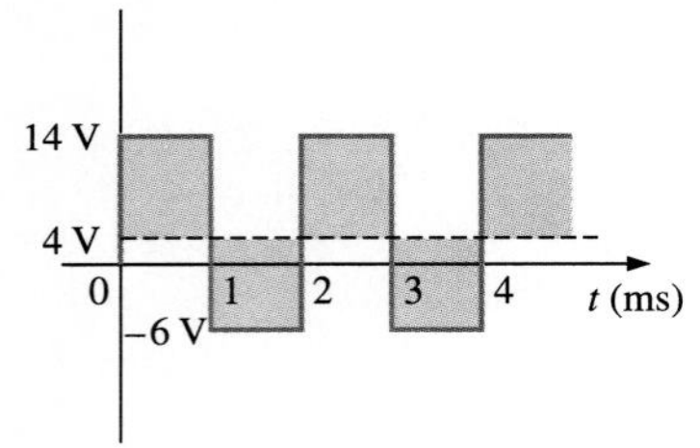

FIG. 13.38
Defining the average value for the waveform of Fig. 13.37(b).

EXAMPLE 13.14

Find the average values of the following waveforms over one full cycle:

a. Figure 13.39.

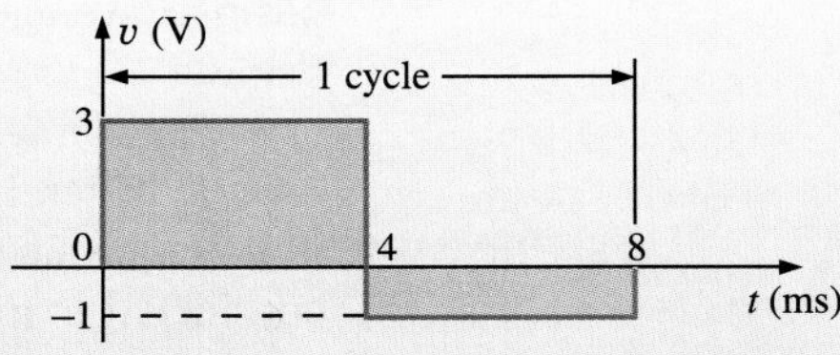

FIG. 13.39
Example 13.14, part (a).

continued

b. Figure 13.40.

i (A)
1 cycle
4
0
−2
−10
2
4
6
8
10
t (ms)

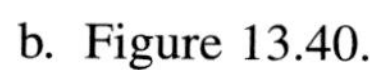

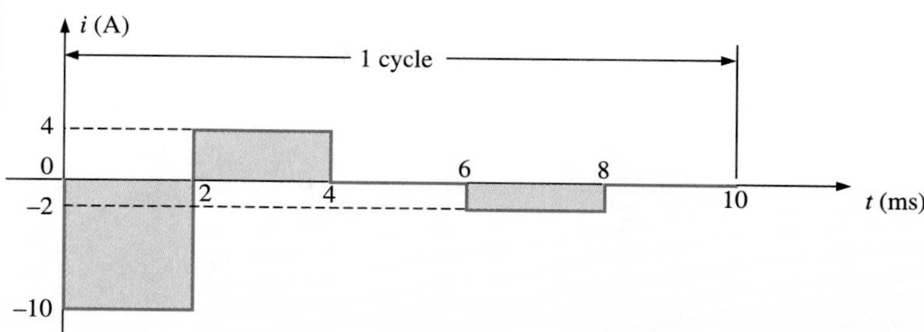

FIG. 13.40
Example 13.14, part (b).

Solutions:

a. $G = \dfrac{+(3\text{ V})(4\text{ ms}) - (1\text{ V})(4\text{ ms})}{8\text{ ms}} = \dfrac{12\text{ V} - 4\text{ V}}{8} = \mathbf{1\text{ V}}$

Note Fig. 13.41.

v_{av} (V)
1
8
t (ms)
0 1V
dc voltmeter (between 0 and 8 ms)

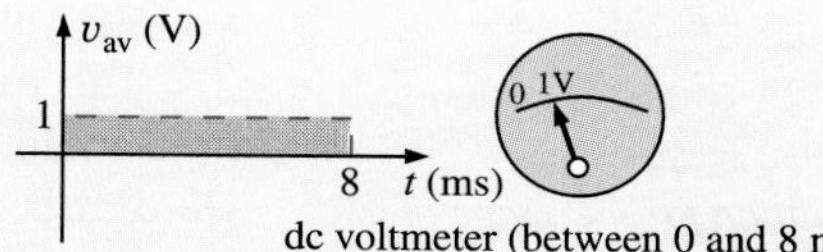

FIG. 13.41
The response of a dc meter to the waveform of Fig. 13.39.

b. $G = \dfrac{-(10\text{ A})(2\text{ ms}) + (4\text{ V})(2\text{ ms}) - (2\text{ V})(2\text{ ms})}{10\text{ ms}}$

$= \dfrac{-20\text{ V} + 8\text{ V} - 4\text{ V}}{10} = -\dfrac{16\text{ V}}{10} = \mathbf{-1.6\text{ V}}$

Note Fig. 13.42.

i_{av} (A)
0
−1.6
10
t (ms)
−1.6
dc ammeter (between 0 and 10 ms)

FIG. 13.42
The response of a dc meter to the waveform of Fig. 13.40.

We found the areas under the curves in this example by using a simple geometric formula. If we had a sine wave or any other unusual shape, however, we would have to find the area by some other means. We can get a good approximation of the area by trying to reproduce the original wave shape using a number of small rectangles or other familiar

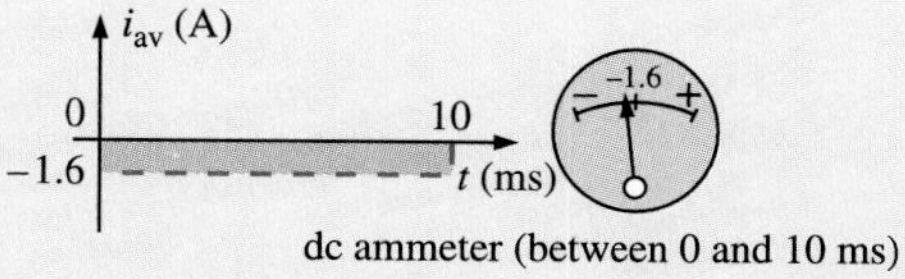

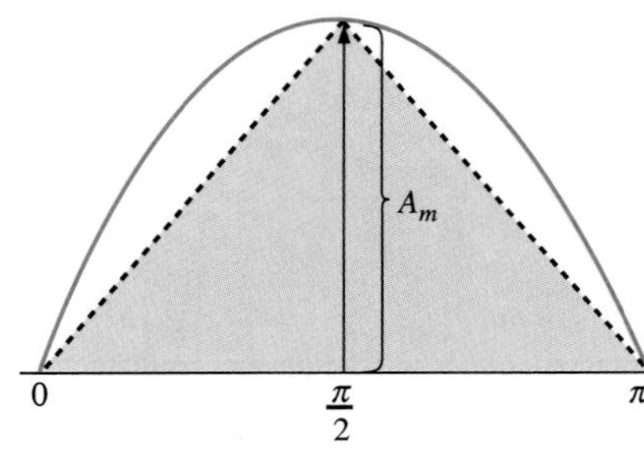

FIG. 13.43
Approximating the shape of the positive pulse of a sinusoidal waveform with two right triangles.

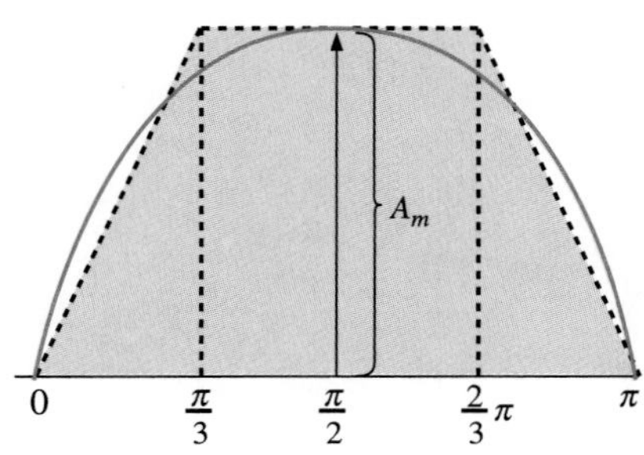

FIG. 13.44
A better approximation for the shape of the positive pulse of a sinusoidal waveform.

shapes, whose area we already know through simple geometric formulas. For example,

the area of the positive (or negative) pulse of a sine wave is $2A_m$.

Approximating this waveform by two triangles (Fig. 13.43), we obtain (using *area* = 1/2 *base* × *height* for the area of a triangle) a rough idea of the actual area:

$$\text{Area shaded} = 2\left(\frac{1}{2}bh\right) = 2\left[\left(\frac{1}{2}\right)\overset{b}{\left(\frac{\pi}{2}\right)}\overset{h}{(A_m)}\right] = \frac{\pi}{2}A_m$$
$$\cong 1.58A_m$$

A closer approximation might be a rectangle with two similar triangles (Fig. 13.44):

$$\text{Area} = A_m\frac{\pi}{3} + 2\left(\frac{1}{2}\,bh\right) = A_m\frac{\pi}{3} + \frac{\pi}{3}A_m = \frac{2}{3}\,\pi A_m$$
$$= 2.094A_m$$

which is certainly close to the actual area. If we used an infinite number of forms, an exact answer of $2A_m$ could be obtained. For irregular waveforms, this method can be especially useful if data such as the average value are desired.

The exact solution of $2A_m$ was derived by **integration**, a procedure included in calculus courses. Many other formulae in this text were also developed using the mathematical tools used in associated calculus courses. Integration is introduced in the following derivation only to illustrate its general usefulness in practical problem solving. It is not necessary for continuing through the rest of this text.

Finding the area under the positive pulse of a sine wave using integration, we have

$$\text{Area} = \int_0^{\pi} A_m \sin\alpha\, d\alpha$$

where $\int$ is the sign of integration, 0 and π are the limits of integration, $A_m \sin\alpha$ is the function to be integrated, and $d\alpha$ indicates that we are integrating with respect to α.

Integrating, we obtain

$$\begin{aligned}\text{Area} &= A_m[-\cos\alpha]_0^{\pi}\\ &= -A_m(\cos\pi - \cos 0°)\\ &= -A_m[-1 - (+1)] = -A_m(-2)\end{aligned}$$

$$\text{Area} = 2A_m \qquad (13.27)$$

Since we know the area under the positive (or negative) pulse, we can easily determine the average value of the positive (or negative) region of a sine wave pulse by applying Eq. (13.26):

$$G = \frac{2A_m}{\pi}$$

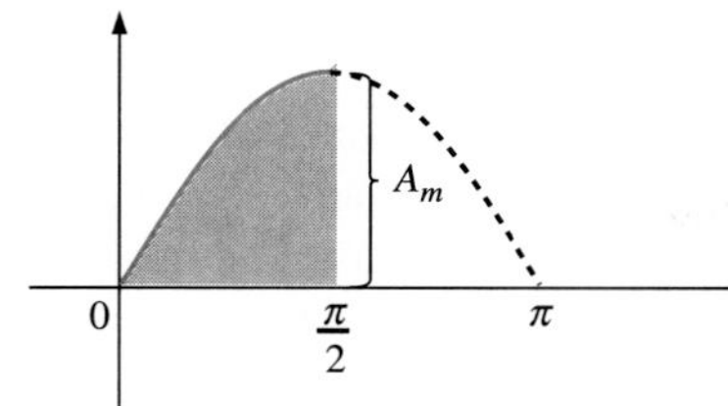

FIG. 13.45
Finding the average value of one-half the positive pulse of a sinusoidal waveform.

and

$$G = 0.637A_m \qquad (13.28)$$

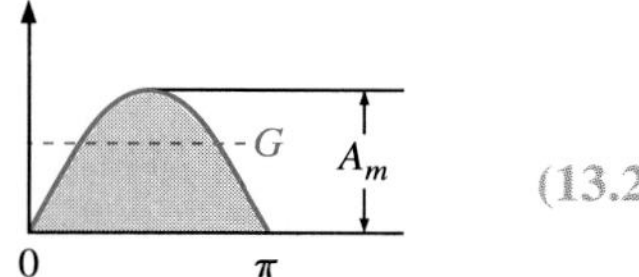

For the waveform of Fig. 13.45,

$$G = \frac{(2A_m/2)}{\pi/2} = \frac{2A_m}{\pi}$$ (average the same as for a full pulse)

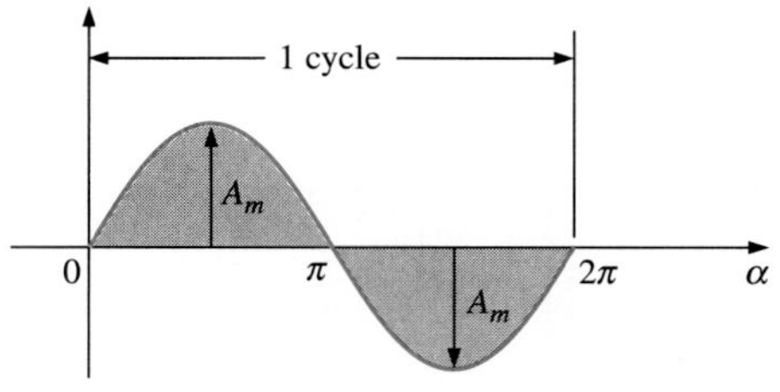

FIG. 13.46
Example 13.15.

EXAMPLE 13.15

Determine the average value of the sinusoidal waveform of Fig. 13.46.

Solution: By looking it is fairly obvious that

the average value of a pure sinusoidal waveform over one full cycle is zero.

Equation (13.26):

$$G = \frac{+2A_m - 2A_m}{2\pi} = \mathbf{0\ V}$$

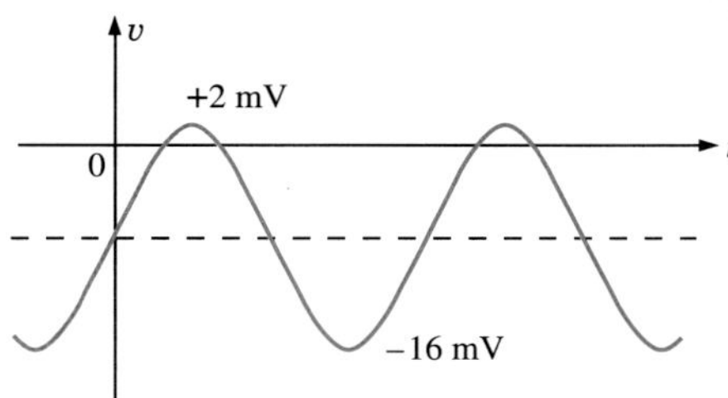

FIG. 13.47
Example 13.16.

EXAMPLE 13.16

Determine the average value of the waveform of Fig. 13.47.

Solution: The peak-to-peak value of the sinusoidal function is 16 mV + 2 mV = 18 mV. The peak amplitude of the sinusoidal waveform is, therefore, 18 mV/2 = 9 mV. Counting down 9 mV from 2 mV (or 9 mV up from −16 mV) results in an average or dc level of −7 mV, as noted by the dashed line of Fig. 13.47.

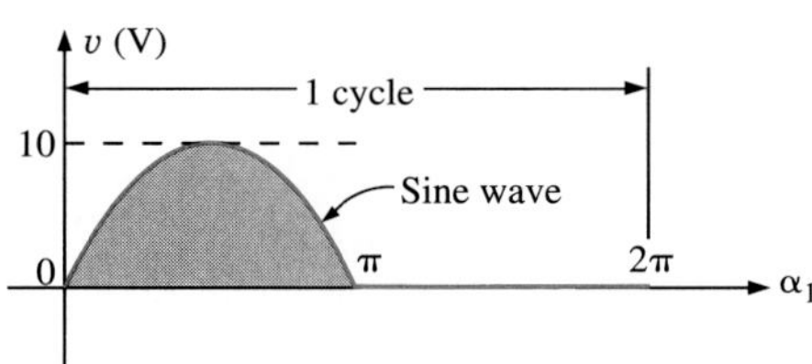

FIG. 13.48
Example 13.17.

EXAMPLE 13.17

Determine the average value of the waveform of Fig. 13.48.

Solution:

$$G = \frac{2A_m + 0}{2\pi} = \frac{2(10\ \text{V})}{2\pi} \cong \mathbf{3.18\ V}$$

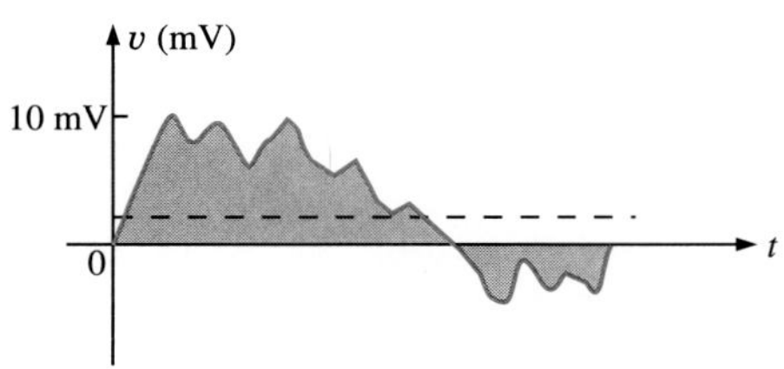

FIG. 13.49
Example 13.18.

EXAMPLE 13.18

For the waveform of Fig. 13.49 determine whether the average value is positive or negative and determine its approximate value.

Solution: From the appearance of the waveform, the average value is positive and around 2 mV. Occasionally, you will have to make judgments of this type.

Instrumentation

The dc level or average value of any waveform can be found using a digital multimeter (DMM) or an oscilloscope. For purely dc circuits, simply set the DMM on dc and read the voltage or current levels. Oscilloscopes are limited to voltage levels using the sequence of steps listed below:

1. First choose GND from the DC-GND-AC option list associated with each vertical channel. The GND option blocks any signal to which the oscilloscope probe may be connected from entering the oscilloscope and responds with just a horizontal line. Set the resulting line in the middle of the vertical axis on the horizontal axis, as shown in Fig. 13.50(a).

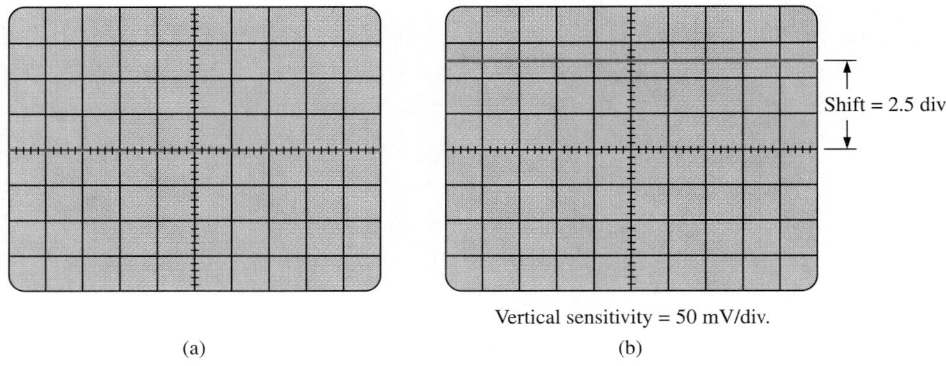

FIG. 13.50
Using the oscilloscope to measure dc voltages: (a) setting the GND condition; (b) the vertical shift resulting from a dc voltage when shifted to the DC option.

2. Apply the oscilloscope probe to the voltage to be measured (if not already connected) and switch to the DC option. If a dc voltage is present, the horizontal line will shift up or down, as demonstrated in Fig. 13.50(b). Multiplying the shift by the vertical sensitivity will result in the dc voltage. An upward shift is a positive voltage (higher potential at the red or positive lead of the oscilloscope), while a downward shift is a negative voltage (lower potential at the red or positive lead of the oscilloscope).

In general,

$$V_{dc} = (\text{vertical shift in div.}) \times (\text{vertical sensitivity in V/div.}) \quad (13.29)$$

For the waveform of Fig. 13.50(b),

$$V_{dc} = (2.5\text{ div.})(50\text{ mV/div.}) = \mathbf{125\text{ mV}}$$

The oscilloscope can also be used to measure the dc or average level of any waveform using the following sequence:

1. Using the GND option, reset the horizontal line to the middle of the screen.
2. Switch to AC (all dc components of the signal to which the probe is connected will be blocked from entering the oscilloscope—only the alternating, or changing, components will be displayed)

and note the location of some definitive point on the waveform, such as the bottom of the half-wave rectified waveform of Fig. 13.51(a); that is, note its position on the vertical scale. For the future, whenever you use the AC option, keep in mind that the computer will distribute the waveform above and below the horizontal axis such that the average value is zero; that is, the area above the axis will equal the area below.

3. Then switch to DC (to permit both the dc and ac components of the waveform to enter the oscilloscope) and note the shift in the chosen level of part 2, as shown in Fig. 13.51(b). Equation (13.29) can then be used to determine the dc or average value of the waveform. For the waveform of Fig. 13.51(b), the average value is about

$$V_{\text{average}} = V_{\text{dc}} = (0.9 \text{ div.})(5 \text{ V/div.}) = \mathbf{4.5\ V}$$

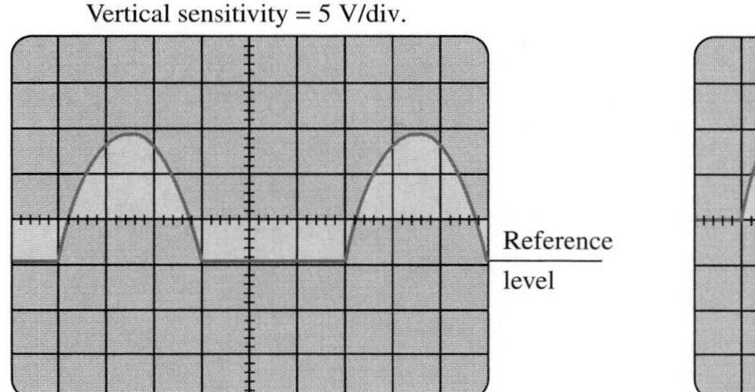

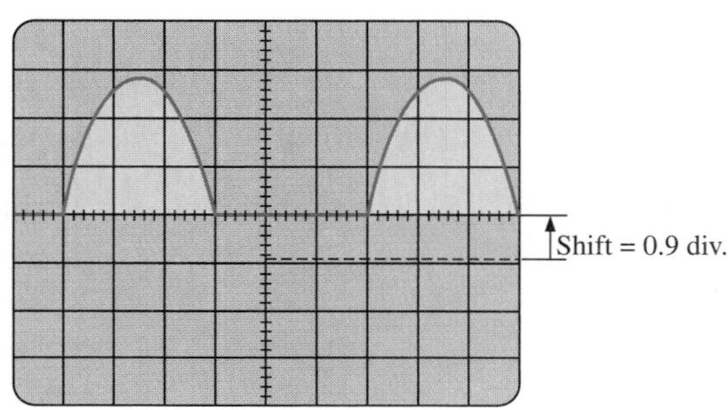

FIG. 13.51
Determining the average value of a nonsinusoidal waveform using the oscilloscope: (a) vertical channel on the ac mode; (b) vertical channel on the dc mode.

The procedure outlined above can be applied to any alternating waveform such as the one in Fig. 13.49. In some cases the average value may require moving the starting position of the waveform under the AC option to a different region of the screen or choosing a higher voltage scale. DMMs can read the average or dc level of any waveform by simply choosing the appropriate scale.

13.7 EFFECTIVE VALUES

Here we investigate the relation between ac and dc quantities involved in delivering power to a load. This will allow us to determine the amplitude of a sinusoidal ac current that will deliver the same power as a particular dc current.

While it appears that the net ac current in any one direction is zero (average value = 0), the resulting power delivered is not zero. Current of any magnitude through a resistor will deliver power to that resistor, *regardless* of direction. So, during both the positive and negative portions of a sinusoidal ac current, power is being delivered to the resistor at *each instant of time*. The power delivered at each instant will vary with the magnitude of the sinusoidal ac current, but there will be a net power flow over the full cycle. The net power flow will equal twice the flow delivered by either the positive or negative regions of the sinusoidal quantity.

A fixed relationship between ac and dc voltages and currents can be derived from the experimental setup shown in Fig. 13.52. A resistor in a water bath is connected by switches to a dc and an ac supply. If switch 1 is closed, a dc current *I*, determined by the resistance *R* and battery voltage *E*, will be established through the resistor *R*. The temperature reached by the water in a certain time is determined by the dc power dissipated in the form of heat by the resistor.

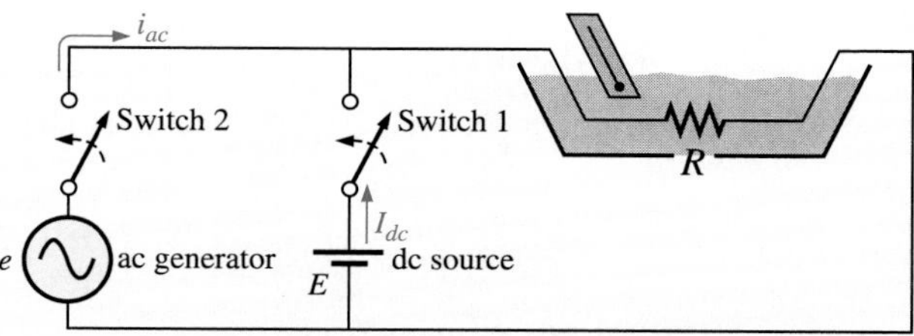

FIG. 13.52
An experimental setup to establish a relationship between dc and ac quantities.

If switch 2 is closed and switch 1 is opened, the ac current through the resistor will have a peak value of I_m. The temperature reached by the water is now determined by the ac power dissipated in the form of heat by the resistor. The ac input is varied until the same temperature is reached in the same time as with the dc input. When this is accomplished, the average electrical power delivered to the resistor *R* by the ac source is the same as that delivered by the dc source.

The power delivered by the ac supply at any instant of time is

$$P_{ac} = (i_{ac})^2R = (I_m \sin \omega t)^2R = (I_m^2 \sin^2\omega t)R$$

but

$$\sin^2\omega t = \frac{1}{2}(1 - \cos 2\omega t) \quad \text{(trigonometric identity)}$$

Therefore,

$$P_{ac} = I_m^2\left[\frac{1}{2}(1 - \cos 2\omega t)\right]R$$

and

$$P_{ac} = \frac{I_m^2R}{2} - \frac{I_m^2R}{2}\cos 2\omega t \tag{13.30}$$

The *average power* delivered by the ac source is just the first term, since the average value of a cosine wave is zero even though the wave may have twice the frequency of the original input current waveform. Equating the average power delivered by the ac generator to that delivered by the dc source,

$$P_{av(ac)} = P_{dc}$$

$$\frac{I_m^2R}{2} = I_{dc}^2R \text{ and } I_m = \sqrt{2}I_{dc}$$

or

$$I_{dc} = \frac{I_m}{\sqrt{2}} = 0.707I_m$$

continued

which, in words, states that

the equivalent dc value of a sinusoidal current or voltage is 1/$\sqrt{2}$, or 0.707 of its maximum value.

The equivalent dc value is called the effective value of the sinusoidal quantity.

In summary,

$$I_{\text{eq dc}} = I_{\text{eff}} = 0.707I_m \qquad (13.31)$$

or

$$I_m = \sqrt{2}I_{\text{eff}} = 1.414I_{\text{eff}} \qquad (13.32)$$

and

$$E_{\text{eff}} = 0.707E_m \qquad (13.33)$$

or

$$E_m = \sqrt{2}E_{\text{eff}} = 1.414E_{\text{eff}} \qquad (13.34)$$

As a simple numerical example, it would require an ac current with a peak value of $\sqrt{2}(10) = 14.14$ A to deliver the same power to the resistor in Fig. 13.52 as a dc current of 10 A. The effective value of any quantity plotted as a function of time can be found by using the following equation derived from the experiment just described:

$$I_{\text{eff}} = \sqrt{\frac{\int_0^T i^2(t)\,dt}{T}} \qquad (13.35)$$

or

$$I_{\text{eff}} = \sqrt{\frac{\text{area }(i^2(t))}{T}} \qquad (13.36)$$

which, in words, states that to find the effective value, the function $i(t)$ must first be squared. After $i(t)$ is squared, the area under the curve is found by integration. It is then divided by T, the length of the cycle or period of the waveform, to obtain the average or *mean* value of the squared waveform. The final step is to take the *square root* of the mean value. This procedure gives us another designation for the effective value, the *root-mean-square* (rms) value.

EXAMPLE 13.19

Find the effective values of the sinusoidal waveform in each part of Fig. 13.53.

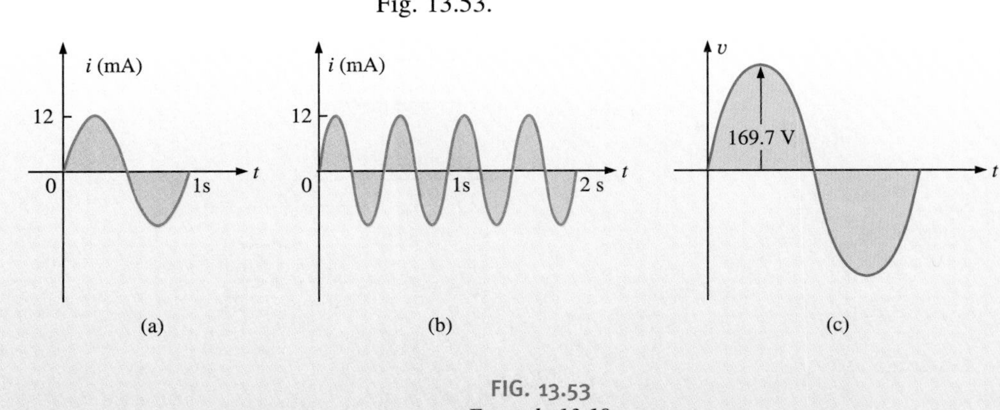

FIG. 13.53
Example 13.19.

continued

Solution: For part (a), $I_{\text{eff}} = 0.707(12 \times 10^{-3}\text{ A}) = \mathbf{8.484\text{ mA}}$. For part (b), again $I_{\text{eff}} = \mathbf{8.484\text{ mA}}$. Note that frequency did not change the effective value in (b) above as compared to (a). For part (c), $V_{\text{eff}} = 0.707(169.73\text{ V}) \cong \mathbf{120\text{ V}}$, the same as available from a home outlet.

EXAMPLE 13.20

The 120-V dc source of Fig. 13.54(a) delivers 3.6 W to the load. Determine the peak value of the applied voltage (E_m) and the current (I_m) if the ac source (Fig. 13.54(b)) is to deliver the same power to the load.

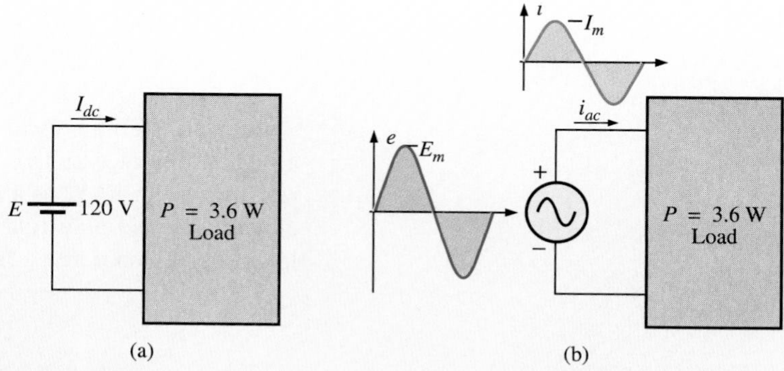

FIG. 13.54
Example 13.20.

Solution:

$$P_{\text{dc}} = V_{\text{dc}}I_{\text{dc}}$$

and

$$I_{\text{dc}} = \frac{P_{\text{dc}}}{V_{\text{dc}}} = \frac{3.6\text{ W}}{120\text{ V}} = 30\text{ mA}$$

$$I_m = \sqrt{2}I_{\text{dc}} = (1.414)(30\text{ mA}) = \mathbf{42.42\text{ mA}}$$

$$E_m = \sqrt{2}E_{\text{dc}} = (1.414)(120\text{ V}) = \mathbf{169.68\text{ V}}$$

EXAMPLE 13.21

Find the effective or rms value of the waveform of Fig. 13.55.

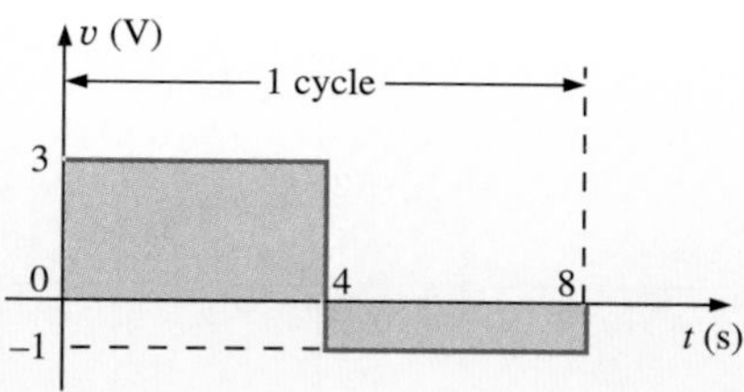

FIG. 13.55
Example 13.21.

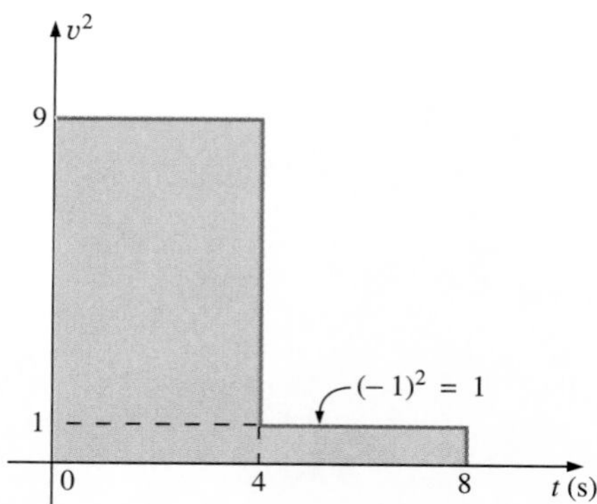

FIG. 13.56
The squared waveform of Fig. 13.55.

Solution:
v^2 (Fig. 13.56):

$$V_{\text{eff}} = \sqrt{\frac{(9)(4) + (1)(4)}{8}} = \sqrt{\frac{40}{8}} = \mathbf{2.236\text{ V}}$$

EXAMPLE 13.22

Calculate the effective value of the voltage of Fig. 13.57.

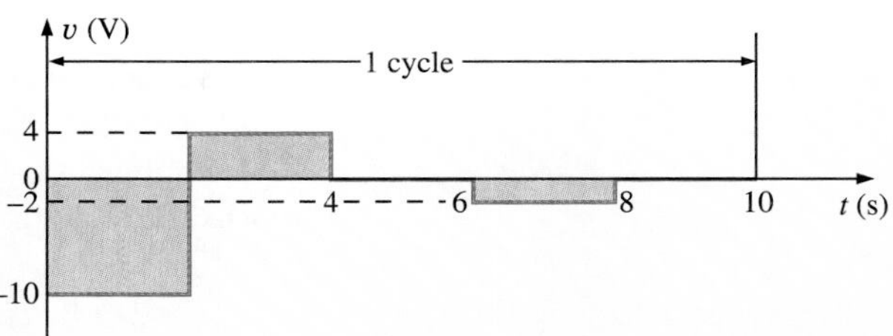

FIG. 13.57
Example 13.22.

Solution:

v^2 (Fig. 13.58):

$$V_{\text{eff}} = \sqrt{\frac{(100)(2) + (16)(2) + (4)(2)}{10}} = \sqrt{\frac{240}{10}}$$

$$= \mathbf{4.899\ V}$$

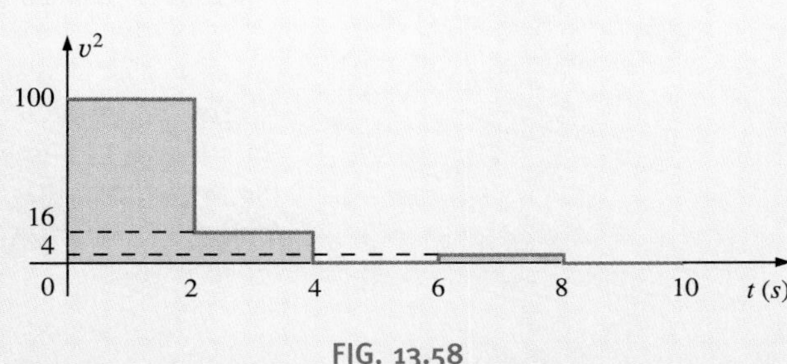

FIG. 13.58
The squared waveform of Fig. 13.57.

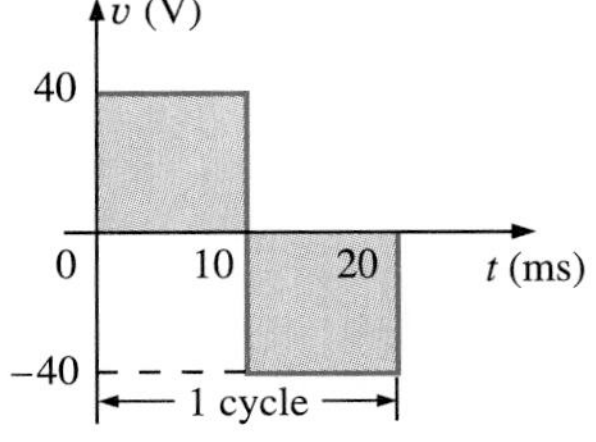

FIG. 13.59
Example 13.23.

EXAMPLE 13.23

Determine the average and effective values of the square wave of Fig. 13.59.

Solution: You can see that the average value is zero.

v^2 (Fig. 13.60):

$$V_{\text{eff}} = \sqrt{\frac{(1600)(10 \times 10^{-3}) + (1600)(10 \times 10^{-3})}{20 \times 10^{-3}}}$$

$$= \sqrt{\frac{32\,000 \times 10^{-3}}{20 \times 10^{-3}}} = \sqrt{1600}$$

$$V_{\text{eff}} = \mathbf{40\ V}$$

(the maximum value of the waveform of Fig. 13.60).

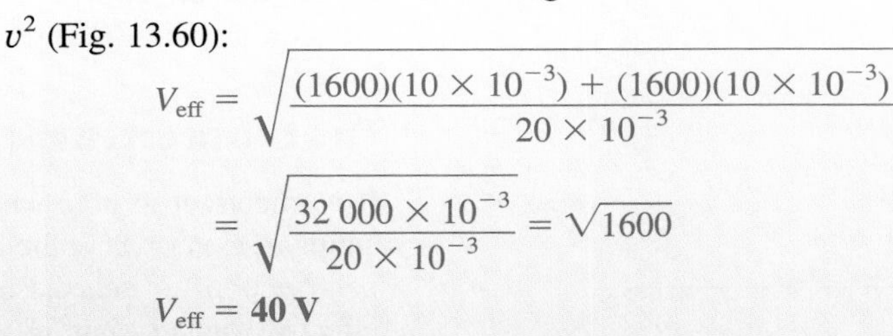

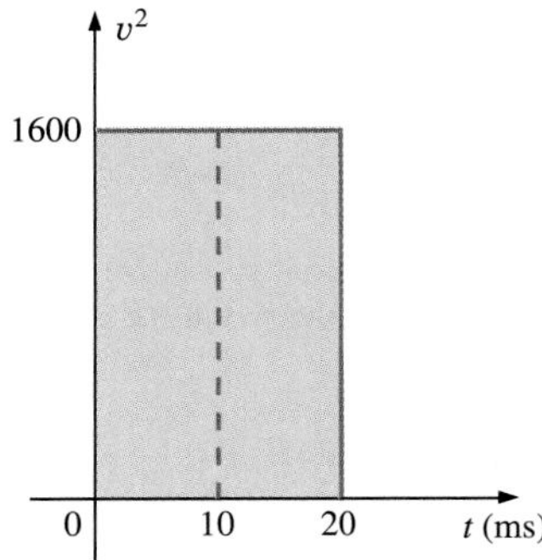

FIG. 13.60
The squared waveform of Fig. 13.59.

The waveforms in these examples are the same as those used in the examples on the average value. You might find it interesting to compare the effective and average values of these waveforms.

The effective values of sinusoidal quantities such as voltage or current will be represented by E and I. These symbols are the same as those used for dc voltages and currents. To avoid confusion, the peak value of a waveform will always have a subscript m associated with it: $I_m \sin \omega t$.

Caution: When finding the effective value of the positive pulse of a sine wave, note that the squared area is *not* simply $(2A_m)^2 = 4A^2_m$; it must be found by a completely new integration. This will always be the case for any waveform that is not rectangular.

A unique situation arises if a waveform has both a dc and an ac component that may be due to a source such as the one in Fig. 13.61. The combination appears frequently in the analysis of electronic networks where both dc and ac levels are present in the same system.

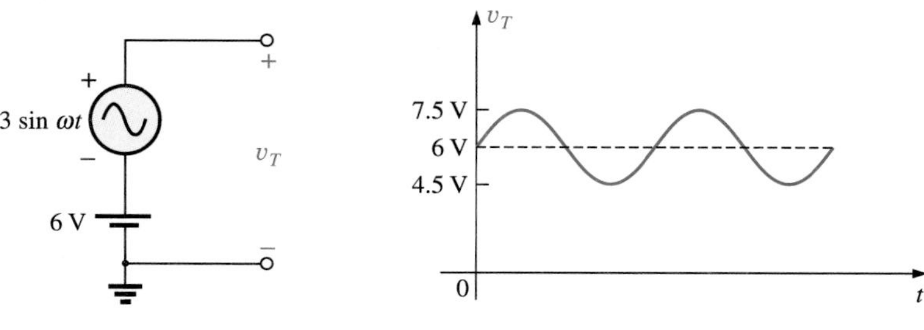

FIG. 13.61
Generation and display of a waveform having a dc and an ac component.

You might be tempted simply to assume that the effective value of the voltage v_T is the sum of the effective values of each component of the waveform; that is, $V_T(\text{eff}) = 0.7071\ (1.5\ \text{V}) + 6\ \text{V} = 1.06\ \text{V} + 6\ \text{V} = 7.06\ \text{V}$. However, the rms value is actually determined by

$$V_{\text{eff}} = \sqrt{V_{\text{dc}}^2 + V_{\text{ac rms}}^2} \qquad (13.37)$$

which for the above example is

$$V_{\text{eff}} = \sqrt{(6\ \text{V})^2 + (1.06\ \text{V})^2}$$
$$= \sqrt{37.124}\ \text{V}$$
$$\cong \mathbf{6.1\ V}$$

This result is noticeably less than that of the first assumption. Eq. (13.37) is derived in the supplementary section on non-sinusoidal circuits on the CD-ROM.

Instrumentation

It is important to note whether the DMM in use is a *true rms* meter or simply a meter where the average value is calibrated (as described in the next section) to indicate the rms level. A **true rms** meter will read the effective value of any waveform (such as Figs. 13.49 and 13.61) and is not limited to sinusoidal waveforms. Since the label *true rms* is normally not placed on the face of the meter, it is a good idea to check the manual if you are going to have waveforms that aren't purely sinusoidal.

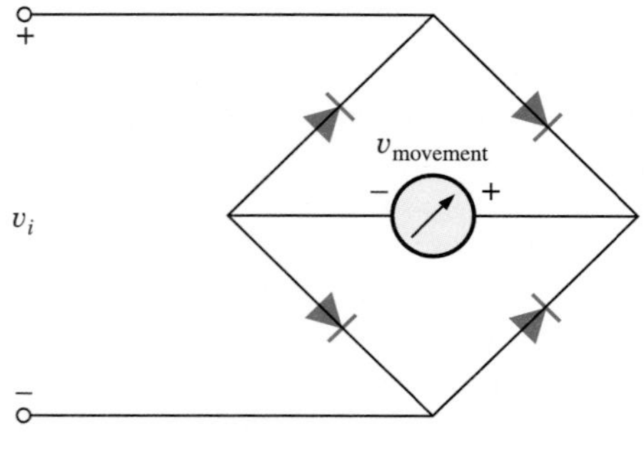

13.8 ac METERS AND INSTRUMENTS

The d'Arsonval movement (Chapter 7) used in dc meters can also be used to measure sinusoidal voltages and currents if the **bridge rectifier** of Fig. 13.62 is placed between the signal to be measured and the average reading movement.

FIG. 13.62
Full-wave bridge rectifier.

The bridge rectifier, made up of four **diodes** (electronic switches), will convert the input signal of zero average value to one with an average value sensitive to the peak value of the input signal. The conversion process is well described in most basic electronics texts. Basically, conduction is permitted through the diodes so as to convert the sinusoidal input of Fig. 13.63(a) to one that looks like Fig. 13.63(b). The negative portion of the input has been effectively "flipped over" by the bridge configuration. The resulting waveform of Fig. 13.63(b) is called a **full-wave rectified waveform**.

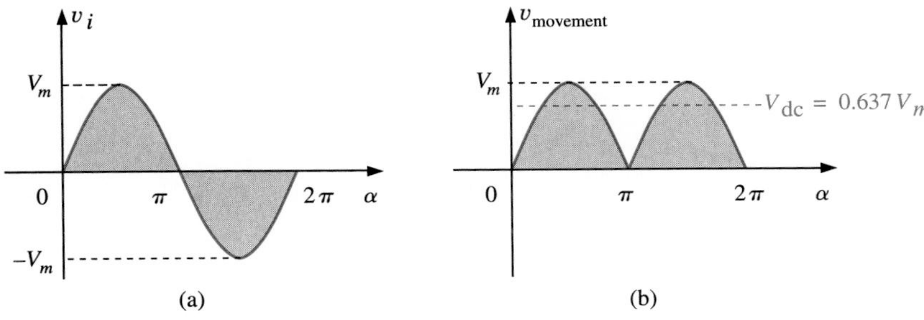

FIG. 13.63
(a) Sinusoidal input; (b) full-wave rectified signal.

The zero average value of Fig. 13.63(a) has been replaced by a pattern having an average value determined by

$$G = \frac{2V_m + 2V_m}{2\pi} = \frac{4V_m}{2\pi} = \frac{2V_m}{\pi} = 0.637V_m$$

The movement of the pointer will therefore be directly related to the peak value of the signal by the factor 0.637.

Forming the ratio between the rms and dc levels will result in

$$\frac{V_{\text{rms}}}{V_{\text{dc}}} = \frac{0.707V_m}{0.637V_m} \cong 1.11$$

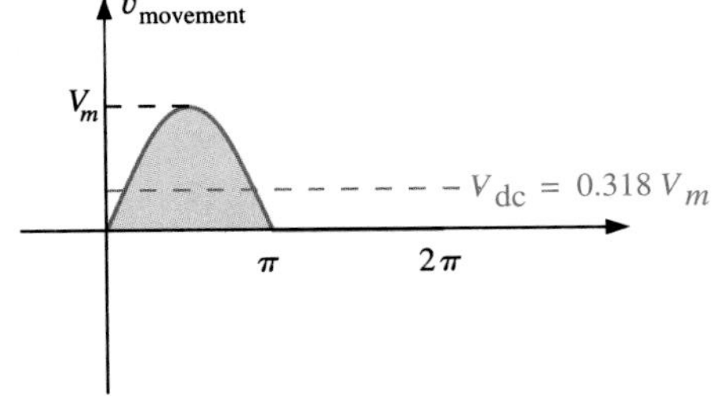

FIG. 13.64
Half-wave rectified signal.

revealing that the scale indication is 1.11 times the dc level measured by the movement; that is,

$$\text{Meter indication} = 1.11 \text{ (dc or average value)} \quad \textit{full-wave} \tag{13.38}$$

Some ac meters use a half-wave rectifier arrangement that results in the waveform of Fig. 13.64, which has half the average value of Fig. 13.63(b) over one full cycle. The result is

$$\text{Meter indication} = 2.22 \text{ (dc or average value)} \quad \textit{half-wave} \tag{13.39}$$

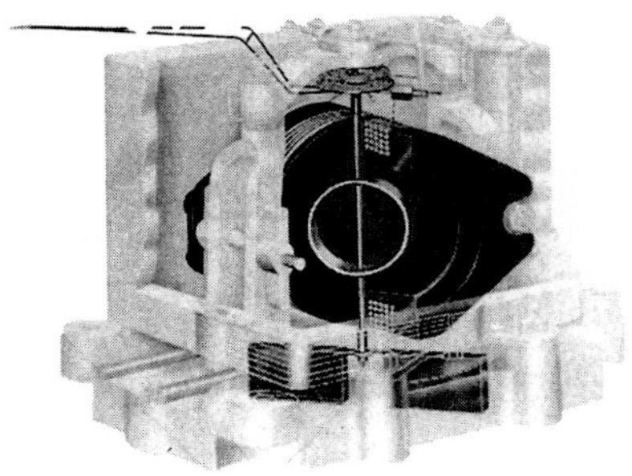

FIG. 13.65
Electrodynamometer movement. (Courtesy of Weston Instruments, Inc.)

A second movement, called the electrodynamometer movement (Fig. 13.65), can measure both ac and dc quantities without a change in internal circuitry. The movement can read the effective value of any ac waveform because a reversal in current direction reverses the fields of both the stationary and the movable coils—the net result is that the deflection of the pointer is always in the same direction.

The VOM, introduced in Chapter 2, can be used to measure both dc and ac voltages using a d'Arsonval movement and the proper switching networks. That is, when the meter is used for dc measurements, the dial setting will establish the proper series resistance for the chosen scale and permit the appropriate dc level to pass directly to the movement.

For ac measurements, the dial setting will introduce a network that uses a full- or half-wave rectifier to establish a dc level. As discussed above, each setting is properly calibrated to indicate the desired quantity on the face of the instrument.

EXAMPLE 13.24

Determine the reading of each meter for each situation of Fig. 13.66.

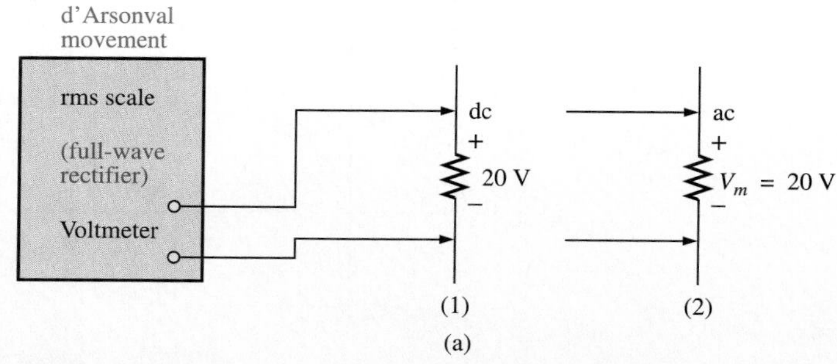

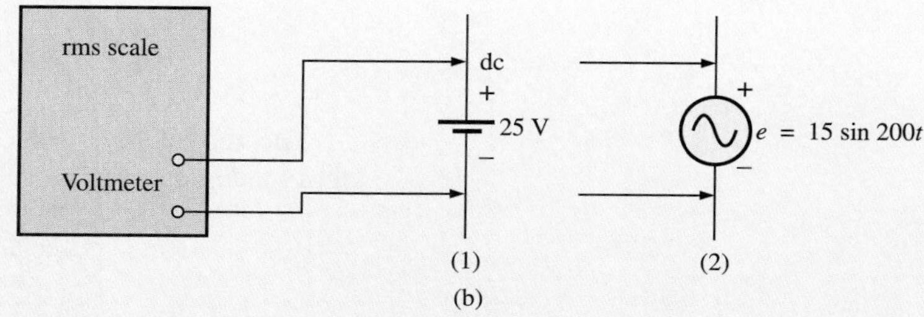

FIG. 13.66
Example 13.24.

Solution: For part (a), situation (1): By Eq. (13.38),

$$\text{Meter indication} = 1.11(20\text{ V}) = \mathbf{22.2\text{ V}}$$

For part (a), situation (2):

$$V_{\text{rms}} = 0.707V_m = 0.707(20\text{ V}) = \mathbf{14.14\text{ V}}$$

For part (b), situation (1):

$$V_{\text{rms}} = V_{\text{dc}} = \mathbf{25\text{ V}}$$

For part (b), situation (2):

$$V_{\text{rms}} = 0.707V_m = 0.707(15\text{ V}) \cong \mathbf{10.6\text{ V}}$$

Most DMMs use a full-wave rectification system to convert the input ac signal to one with an average value. In fact, for the DMM of Fig. 2.25, the same scale factor of Eq. (13.38) is used; that is, the aver-

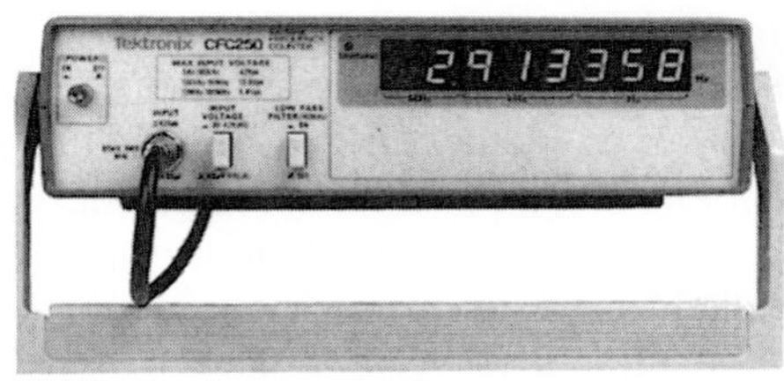

FIG. 13.67
Frequency counter. (Courtesy of Tektronix, Inc.)

age value is scaled up by a factor of 1.11 to obtain the rms value. In the digital meters, however, there are no moving parts such as in the d'Arsonval or electrodynamometer movements to display the signal level. Instead, the average value is sensed by a multiprocessor integrated circuit (IC), which in turn determines which digits should appear on the digital display.

Digital meters can also be used to measure nonsinusoidal signals, but the scale factor of each input waveform must first be known (normally provided by the manufacturer in the operator's manual). For instance, the scale factor for an average responding DMM on the ac rms scale will produce an indication for a square-wave input that is 1.11 times the peak value. For a triangular input, the response is 0.555 times the peak value. For a sine wave input, the response is 0.707 times the peak value.

For any instrument, it is always good practice to read (if only briefly) the operator's manual if you think you will use the instrument on a regular basis.

FIG. 13.68
Amp-Clamp®. (Courtesy of Simpson Instruments, Inc.)

For frequency measurements, the frequency counter of Fig. 13.67 provides a digital readout of sine, square, and triangular waves from 5 Hz to 100 MHz at input levels from 30 mV to 42 V. Note the relative simplicity of the panel and the high degree of accuracy available.

The Amp-Clamp® of Fig. 13.68 is an instrument that can measure alternating current in the ampere range without having to open the circuit. The loop is opened by squeezing the "trigger"; then it is placed around the current-carrying conductor. Through transformer action, the level of current in rms units will appear on the appropriate scale. The accuracy of this instrument is ±3% of full scale at 60 Hz, and its scales have maximum values ranging from 6 A to 300 A. The addition of two leads, as indicated in the figure, permits its use as both a voltmeter and an ohmmeter.

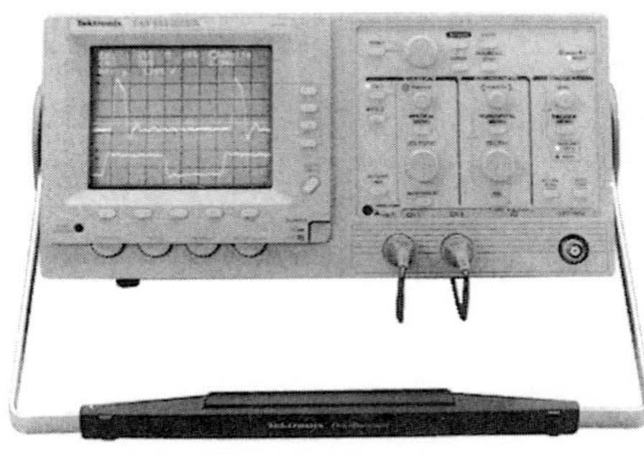

FIG. 13.69
Dual-channel oscilloscope. (Courtesy of Tektronix, Inc.)

One of the most versatile and important instruments in the electronics industry is the oscilloscope, which has already been introduced in this chapter. It provides a display of the waveform on a cathode-ray tube to allow you to detect irregularities and determine quantities such as magnitude, frequency, period, dc component, and so on. The analog oscilloscope of Fig. 13.69 can display two waveforms at the same time (dual-channel) using an innovative interface (front panel). It uses menu buttons to set the vertical and horizontal scales by choosing from selections appearing on the screen. Up to four measurement setups can be stored for future use.

If you are used to watching TV you might be confused when you first see an oscilloscope. You might, at least initially, assume that the oscilloscope is generating the waveform on the screen—much like a TV broadcast. However, it is important to clearly understand that

an oscilloscope only displays signals generated elsewhere and connected to the input terminals of the oscilloscope. The absence of an external signal will simply result in a horizontal line on the screen of the scope.

On most modern oscilloscopes, there is a switch or knob with the choice DC/GND/AC, as shown in Fig. 13.70(a), that is often ignored or treated too lightly in the early stages of scope use. The effect of each position is basically as shown in Fig. 13.70(b). In the DC mode the dc and ac components of the input signal can pass directly to the display. In the AC position the dc input is blocked by the capacitor, but the ac portion of the signal can pass through to the screen. In the GND position the input signal is prevented from reaching the scope display by a direct ground connection, which reduces the scope display to a single horizontal line.

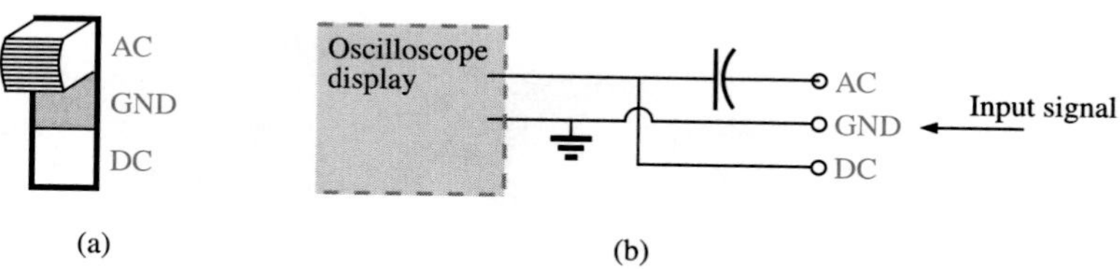

FIG. 13.70
AC-GND-DC switch for the vertical channel of an oscilloscope.

PROBLEMS

SECTION 13.2 Sinusoidal ac Voltage Characteristics and Definitions

1. For the periodic waveform of Fig. 13.71:
   a. Find the period T.
   b. How many cycles are shown?
   c. What is the frequency?
   *d. Determine the positive amplitude and peak-to-peak value (think!).

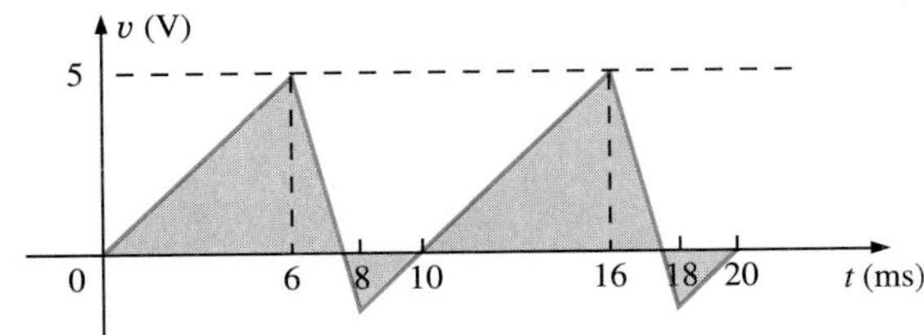

FIG. 13.71
Problem 1.

2. Repeat Problem 1 for the periodic waveform of Fig. 13.72.

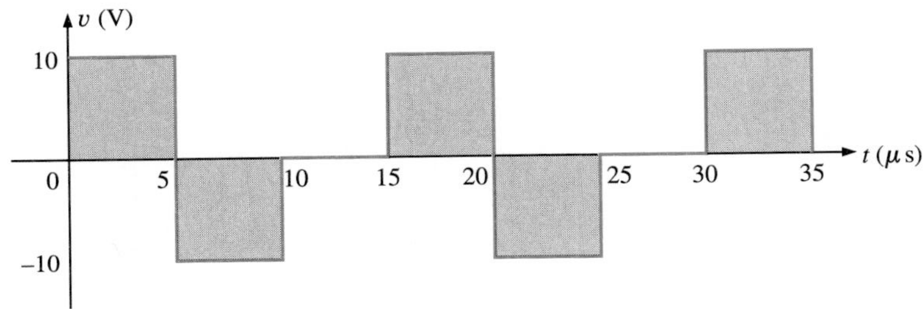

FIG. 13.72
Problems 2 and 47.

3. Determine the period and frequency of the sawtooth waveform of Fig. 13.73.

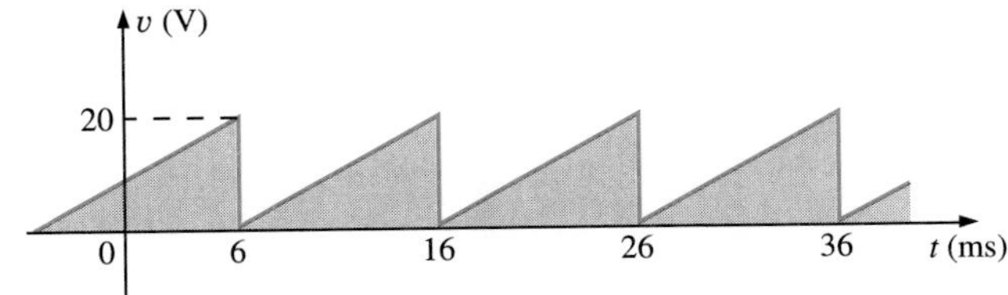

FIG. 13.73
Problems 3 and 48.

4. Find the period of a periodic waveform whose frequency is
   a. 400 Hz
   b. 36 kHz
   c. 27 MHz
   d. 2.45 GHz
5. Find the frequency of a repeating waveform whose period is
   a. 1/60 s
   b. 17.3 ms
   c. 37 μs
   d. 12.6 ns

6. Find the period of a sinusoidal waveform that completes
   a. 80 cycles in 24 ms
   b. 200 cycles in 17 μs
   c. 157 cycles in 3.2 s
   d. 6000 cycles in 25.7 ms
7. If a periodic waveform has a frequency of 50 Hz, how long (in seconds) will it take to complete
   a. 2 cycles?
   b. 14 cycles?
   c. 36.6 cycles?
   d. 1257 cycles?
8. What is the frequency of a periodic waveform that completes 42 cycles in
   a. 6 s?
   b. 15 s?
   c. 1.60 s?
   d. 360 ms?
9. Sketch a periodic square wave like the one in Fig. 13.72 with a frequency of 20 000 Hz and a peak value of 10 mV.
10. For the oscilloscope pattern of Fig. 13.74:
    a. Determine the peak amplitude.
    b. Find the period.
    c. Calculate the frequency.
    Redraw the oscilloscope pattern if a +25-mV dc level were added to the input waveform.

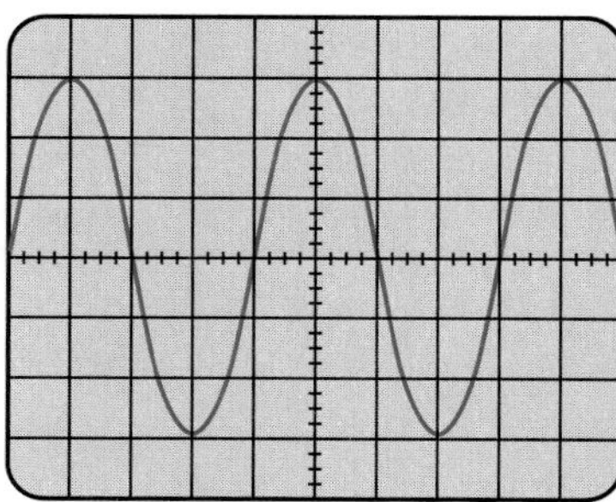

Vertical sensitivity = 50 mV/div.
Horizontal sensitivity = 10 μs/div.

FIG. 13.74
Problem 10.

SECTION 13.3 The Sine Wave

11. Convert the following degrees to radians:
    a. 45° b. 60°
    c. 120° d. 270°
    e. 178° f. 221°
12. Convert the following radians to degrees:
    a. $\pi/4$ b. $\pi/6$
    c. $\frac{1}{10}\pi$ d. $\frac{7}{6}\pi$
    e. 3π f. 0.55π
13. Find the angular velocity of a waveform with a period of
    a. 2 s b. 0.3 ms
    c. 4 μs d. $\frac{1}{25}$ s
14. Find the angular velocity of a waveform with a frequency of
    a. 50 Hz b. 600 Hz
    c. 2 kHz d. 0.004 MHz
15. Find the frequency and period of sine waves having an angular velocity of
    a. 754 rad/s b. 8.4 rad/s
    c. 6000 rad/s d. $\frac{1}{16}$ rad/s
16. Given $f = 60$ Hz, determine how long it will take the sinusoidal waveform to pass through an angle of
    a. 45°
    b. $\pi/3°$
    c. 180°
    d. $7\pi/2$
17. Determine the angular velocity of the waveform if it passes through an angle of 30° in
    a. 5 ms
    b. 12 μs
    c. 214 ns
    d. 0.7 s

SECTION 13.4 General Format for the Sinusoidal Voltage or Current

18. Find the amplitude and frequency of the following waves:
    a. $20 \sin 377t$ b. $5 \sin 754t$
    c. $10^6 \sin 10\,000t$ d. $0.001 \sin 942t$
    e. $-7.6 \sin 43.6t$ f. $(\frac{1}{42}) \sin 6.283t$

19. Sketch 5 sin $754t$ with the horizontal axis as
   a. angle in degrees.
   b. angle in radians.
   c. time in seconds.

20. Sketch 10^6 sin 10 000t with the horizontal axis as
   a. angle in degrees.
   b. angle in radians.
   c. time in seconds.

21. Sketch -7.6 sin 43.6t with the horizontal axis as
   a. angle in degrees.
   b. angle in radians.
   c. time in seconds.

22. If $e = 300 \sin 157t$, how long (in seconds) does it take this waveform to complete 1/2 cycle?

23. Given $i = 0.5 \sin \alpha$, determine i at
   a. $\alpha = 72°$
   b. $\alpha = \pi/6$ rad
   c. $\alpha = 193°$
   d. $\alpha = 3\pi/2$ rad

24. Given $v = 20 \sin \alpha$, determine v at
   a. $\alpha = 1.2\pi$
   b. $\alpha = -12.5°$
   c. $\alpha = 37.5°$
   d. $\alpha = 5\pi/4$

***25.** Given $v = 30 \times 10^{-3} \sin \alpha$, determine the angles at which v will be 6 mV.

***26.** If $v = 40$ V at $\alpha = 30°$ and $t = 1$ ms, determine the mathematical expression for the sinusoidal voltage.

SECTION 13.5 Phase Relations

27. Sketch $\sin(377t + 60°)$ with the horizontal axis as
   a. angle in degrees.
   b. angle in radians.
   c. time in seconds.

28. Sketch the following waveforms:
   a. $50 \sin(\omega t + 0°)$
   b. $-20 \sin(\omega t + 2°)$
   c. $5 \sin(\omega t + 60°)$
   d. $4 \cos \omega t$
   e. $2 \cos(\omega t + 10°)$
   f. $-5 \cos(\omega t + 20°)$

29. Find the phase relationship between the waveforms of each set:
   a. $v = 4 \sin(\omega t + 50°)$
   $i = 6 \sin(\omega t + 40°)$
   b. $v = 25 \sin(\omega t - 80°)$
   $i = 5 \times 10^{-3} \sin(\omega t - 10°)$
   c. $v = 0.2 \sin(\omega t - 60°)$
   $i = 0.1 \sin(\omega t + 20°)$
   d. $v = 200 \sin(\omega t - 210°)$
   $i = 25 \sin(\omega t - 60°)$

***30.** Repeat Problem 29 for the following sets:
   a. $v = 2 \cos(\omega t - 30°)$
   $i = 5 \sin(\omega t + 60°)$
   b. $v = -1 \sin(\omega t + 20°)$
   $i = 10 \sin(\omega t - 70°)$
   c. $v = -4 \cos(\omega t + 90°)$
   $i = -2 \sin(\omega t + 10°)$

31. Write the analytical expression for the waveforms of Fig. 13.75 with the phase angle in degrees.

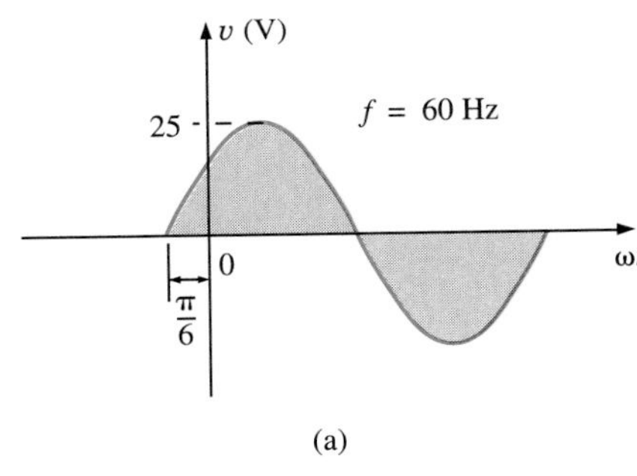

FIG. 13.75
Problem 31.

32. Repeat Problem 31 for the waveforms of Fig. 13.76.

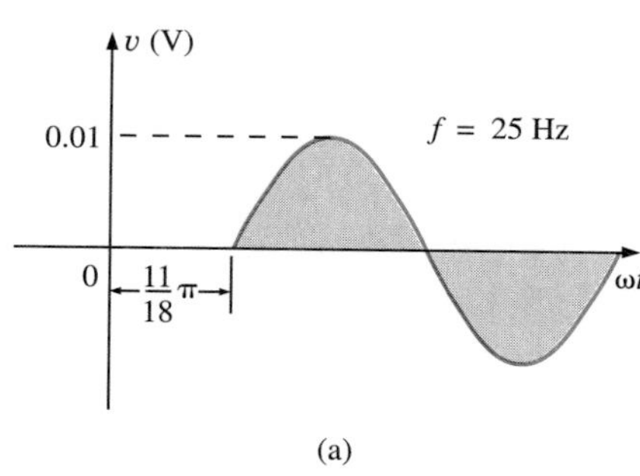

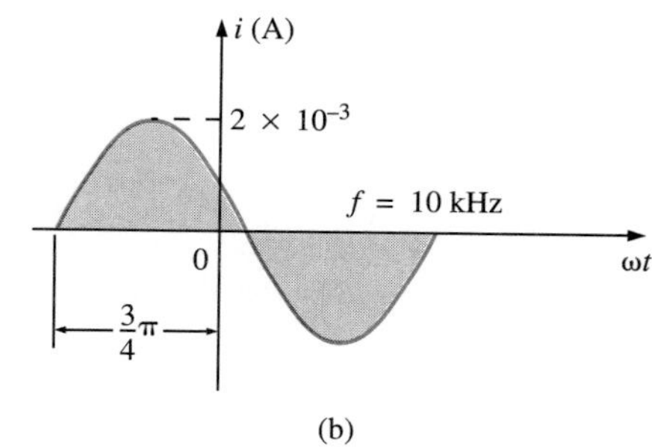

FIG. 13.76
Problem 32.

***33.** The sinusoidal voltage $v = 200 \sin(2\pi 1000t + 60°)$ is plotted in Fig. 13.77. Determine the time t_1.

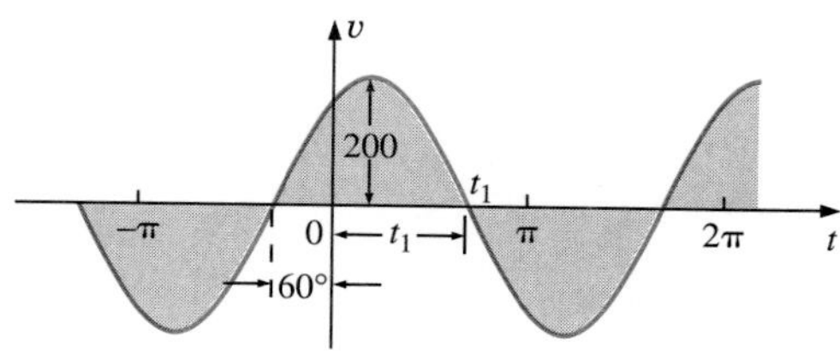

FIG. 13.77
Problem 33.

***34.** The sinusoidal current $i = 4 \sin(50\ 000t - 40°)$ is plotted in Fig. 13.78. Determine the time t_1.

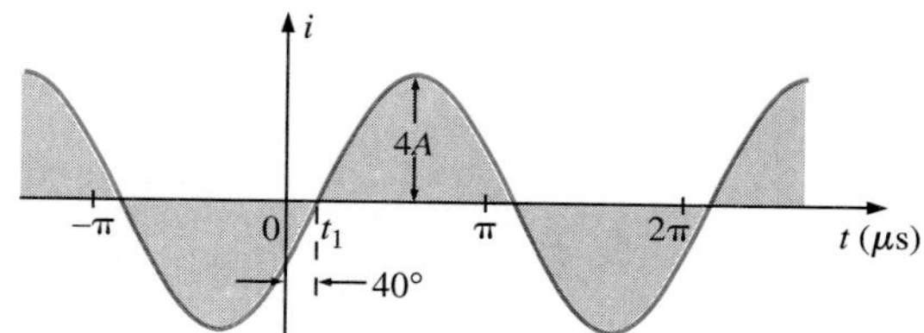

FIG. 13.78
Problem 34.

***35.** Determine the phase delay in milliseconds between the following two waveforms:

$$v = 60 \sin(1800t + 20°)$$
$$i = 1.2 \sin(1800t - 20°)$$

36. For the oscilloscope display of Fig. 13.79:

a. Determine the period of each waveform.
b. Determine the frequency of each waveform.
c. Find the rms value of each waveform.
d. Determine the phase shift between the two waveforms and which leads or lags.

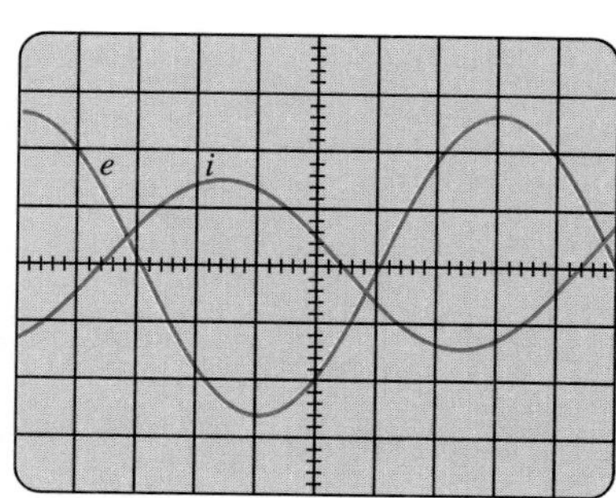

Vertical sensitivity = 0.5 V/div.
Horizontal sensitivity = 1 ms/div.

FIG. 13.79
Problem 36.

SECTION 13.6 Average Value

37. For the waveform of Fig. 13.80:

a. Determine the period.
b. Find the frequency.
c. Determine the average value.
d. Sketch the resulting oscilloscope display if the vertical channel is switched from DC to AC.

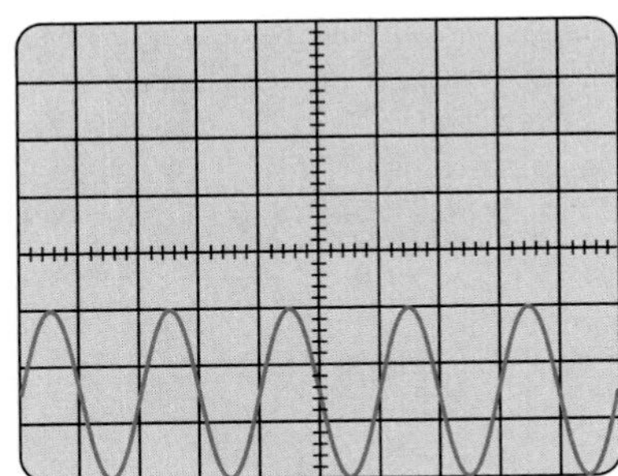

Vertical sensitivity = 10 mV/div.
Horizontal sensitivity = 0.2 ms/div.

FIG. 13.80
Problem 37.

38. Find the average value of the periodic waveforms of Fig. 13.81 over one full cycle.

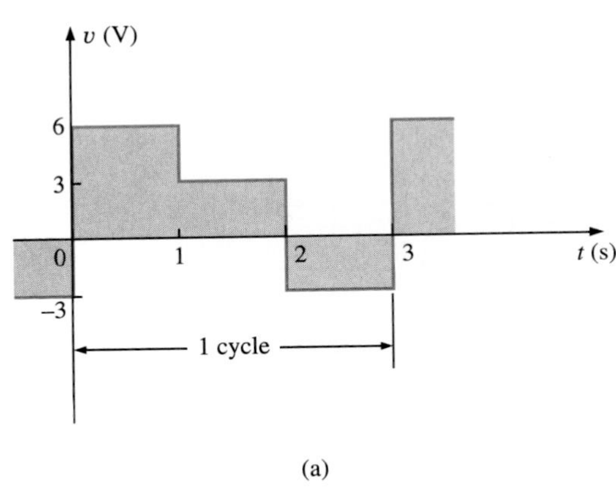

(a)

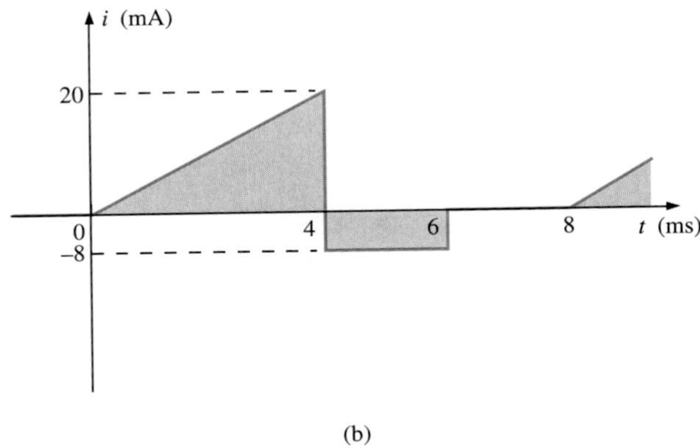

(b)

FIG. 13.81
Problem 38.

39. Find the average value of the periodic waveforms of Fig. 13.82 over one full cycle.

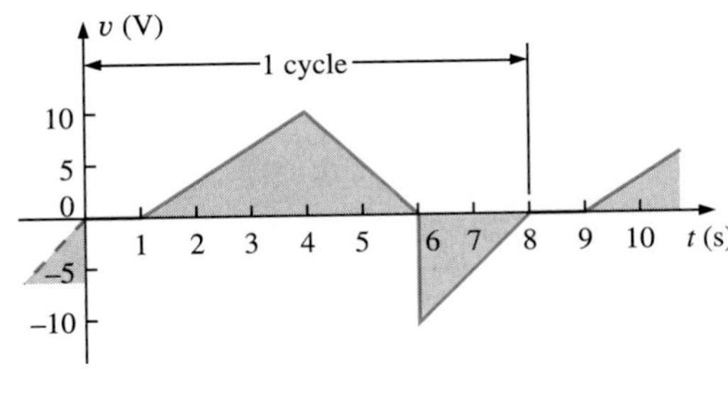

(a)

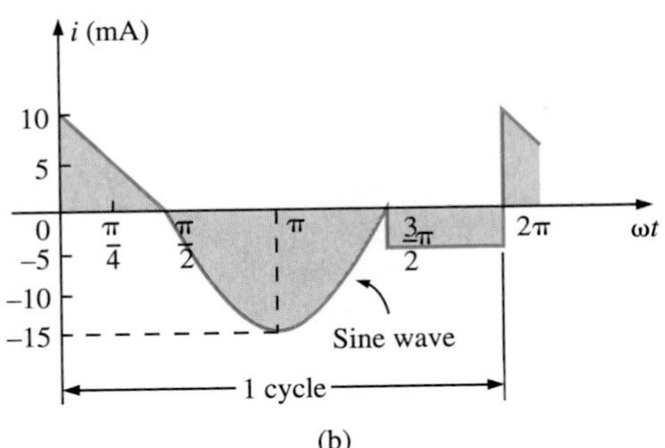

(b)

FIG. 13.82
Problem 39.

***40.** **a.** By the method of approximation, using familiar geometric shapes, find the area under the curve of Fig. 13.83 from zero to 10 s. Compare your solution with the actual area of 5 volt-seconds (V·s).

b. Find the average value of the waveform from zero to 10 s.

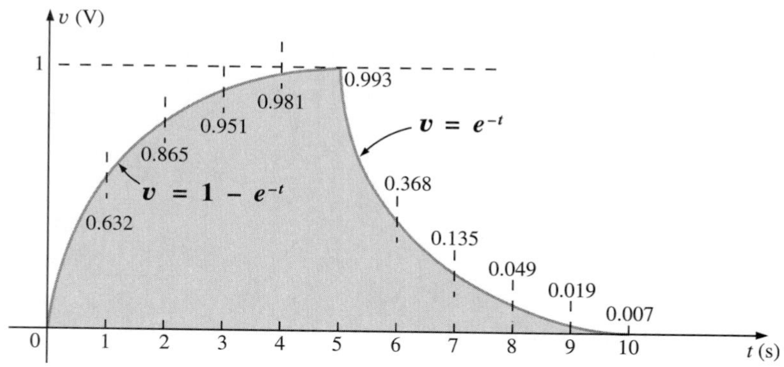

FIG. 13.83
Problem 40.

***41.** For the waveform of Fig. 13.84:

a. Determine the period.

b. Find the frequency.

c. Determine the average value.

d. Sketch the resulting oscilloscope display if the vertical channel is switched from DC to AC.

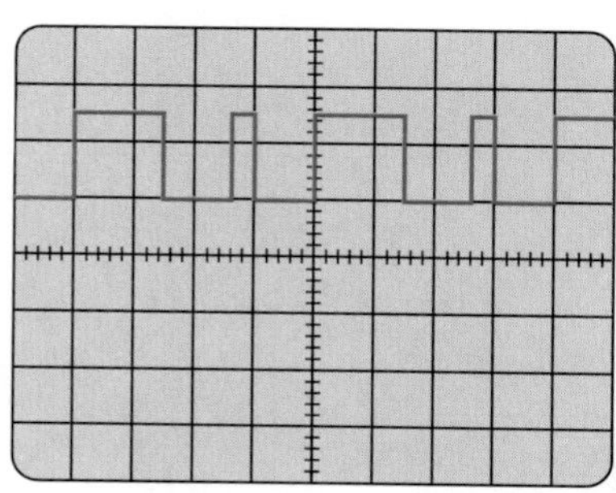

Vertical sensitivity = 10 mV/div.
Horizontal sensitivity = 10 μs/div.

FIG. 13.84
Problem 41.

SECTION 13.7 Effective Values

42. Find the effective values of the following sinusoidal waveforms:

a. $v - 26.3 \sin 754t$

b. $v = 0.06 \sin (70.7t - 24°)$

c. $i = 16.2 \text{ mA} \sin (600t + \pi/2)$

d. $i = 70.70 \sin (377t - 15°)$

43. Write the sinusoidal expressions for voltages and currents having the following effective values at a frequency of 60 Hz with zero phase shift:

a. 1.414 V
b. 70.7 V
c. 0.06 A
d. 24 μA

44. Find the effective value of the periodic waveform of Fig. 13.85 over one full cycle.

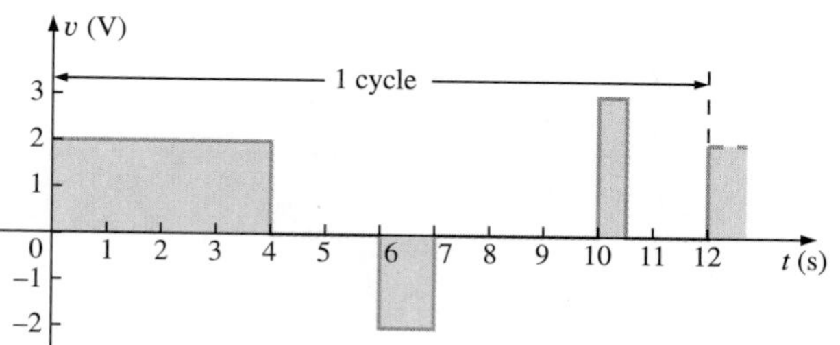

FIG. 13.85
Problem 44.

45. Find the effective value of the periodic waveform of Fig. 13.86 over one full cycle.

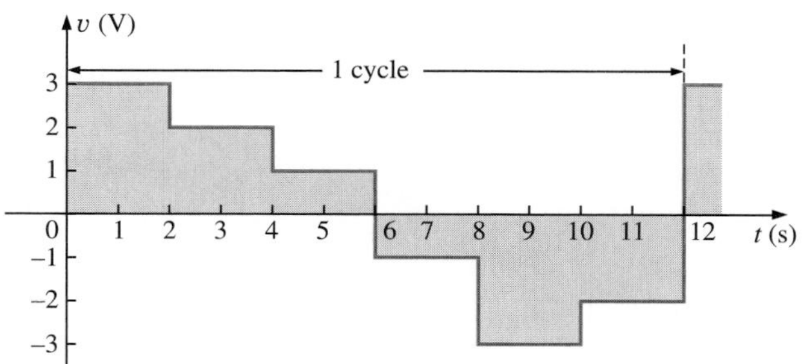

FIG. 13.86
Problem 45.

46. What are the average and effective values of the square wave of Fig. 13.87?

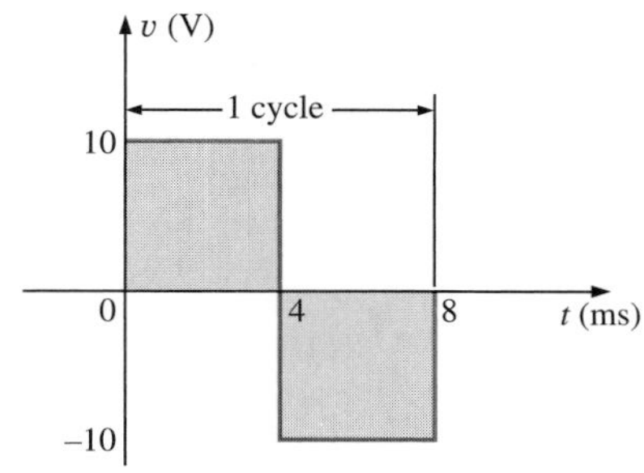

FIG. 13.87
Problem 46.

47. What are the average and effective values of the waveform of Fig. 13.72?

48. What is the average value of the waveform of Fig. 13.73?

49. For each waveform of Fig. 13.88 determine the period, frequency, average value, and effective value.

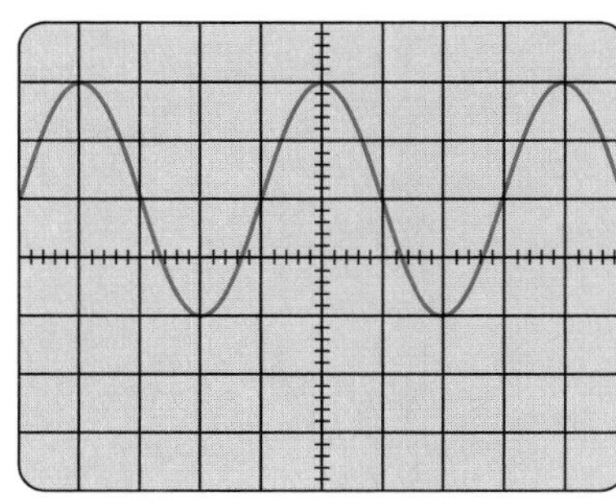

Vertical sensitivity = 20 mV/div.
Horizontal sensitivity = 10 μs/div.

(a)

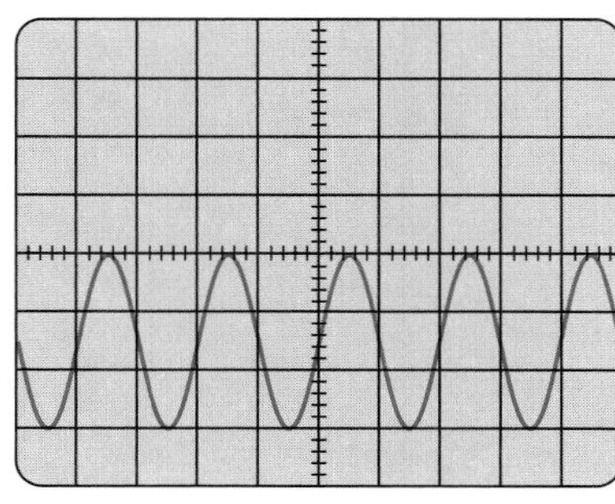

Vertical sensitivity = 0.2 V/div.
Horizontal sensitivity = 50 μs/div.

(b)

FIG. 13.88
Problem 49.

SECTION 13.8 ac Meters and Instruments

50. Determine the reading of the meter for each situation of Fig. 13.89.

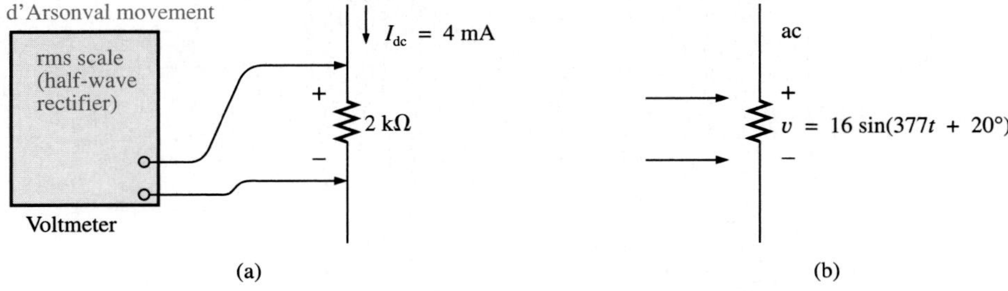

FIG. 13.89
Problem 50.

14

The Basic Elements and Phasors

Outline

Learning Outcomes

After completing this chapter you will be able to

- express ac sinusoidal voltages and currents in either rectangular form or polar form
- perform operations of addition, subtraction, multiplication, and division with complex numbers
- perform phasor calculations for ac sinusoidal voltages and currents at the same frequency
- determine the phase relationship between voltage and current for *R*, *L*, and *C* elements
- sketch a graph showing the response of the basic elements *R*, *L*, and *C* to an ac sinusoidal voltage or current
- calculate the instantaneous values of current and voltage given a simple *R-L* circuit or a simple *R-C* circuit
- calculate the equivalent resistance and reactance of an ac circuit given a voltage and a current
- calculate average power and power factor of an ac circuit given a voltage and a current

Electric utilities use complex numbers and phasor diagrams to monitor and measure power demand. This allows them to forecast demand and respond to sudden changes in the system.

14.1 INTRODUCTION

In this chapter, we will examine the response of the basic R, L, and C elements to a sinusoidal voltage and current, with special note of how frequency will affect the "opposing" characteristic of each element. We will then introduce phasor notation to establish a method of analysis that corresponds directly with of the methods, theorems, and concepts introduced in the dc chapters.

14.2 THE DERIVATIVE

The procedure of **differentiation**, which is a major component of any calculus course, is fundamental to the understanding of the response of the basic R, L, and C elements. Differentiation is introduced here only briefly to illustrate and help you understand the concepts of R, L, and C elements; it is not necessary for continuing through the rest of this text.

The derivative dx/dt is defined as the rate of change of a variable x with respect to time t. If there is no change in x for some time period, then $dx = 0$ (for that time period), and the derivative is zero ($dx/dt = 0$). Using more mathematical terms, the derivative dx/dt is actually the slope of the graph at any instant of time. Thus when the slope is zero, it means $dx/dt = 0$ and vice versa.

For the sinusoidal waveform shown in Fig. 14.1, x does not change at the instants of time when $\omega t = \pi/2$ and $3\pi/2$. This means that dx/dt is zero (that is, the slope is zero) only at the positive and negative peaks. Further examination of the sinusoidal waveform of Fig. 14.1 will also indicate that the greatest change in x occurs at the instants $\omega t = 0$, π, and 2π, and this is where the value of the derivative is at a maximum.

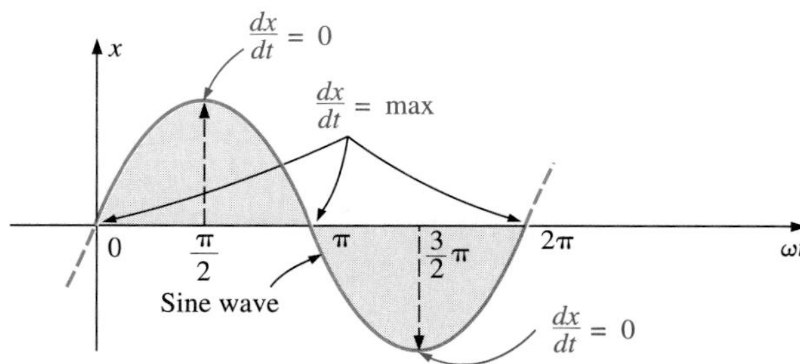

FIG. 14.1
Defining the points in a sinusoidal waveform that have maximum and minimum derivatives.

At $t = 0$, and 2π, x increases at its greatest rate, and dx/dt is given a positive sign since x increases with time. Similarly, at $t = \pi$, dx/dt decreases with time (at the same rate that it increases at $t = 0$ and 2π), so dx/dt is given a negative sign. The rate of change at 0, π, and 2π is the same, and so the magnitude of the derivative at these points is the same. For various values of ωt between these maximum and minimum points, the derivative will exist and have values that range from the minimum to the maximum. The first course in calculus will discuss the actual techniques of obtaining a derivative of a trigonometric function.

Fig. 14.2 is a plot of the derivative of the sine wave of Fig. 14.1 which shows that

the derivative of a sine wave is a cosine wave.

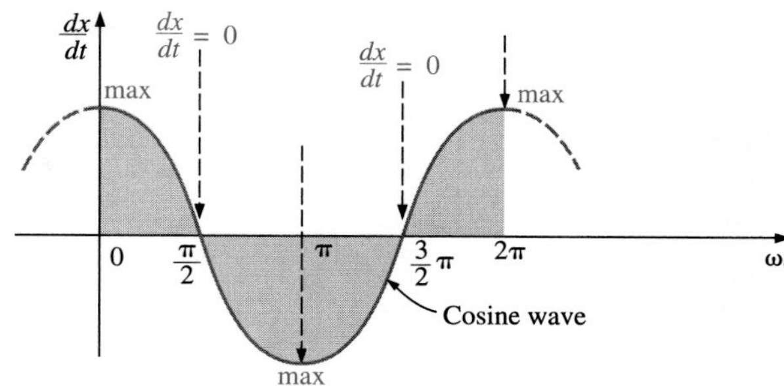

FIG. 14.2
Derivative of the sine wave of Fig. 14.1.

The peak value of the cosine wave is directly related to the frequency of the original waveform. The higher the frequency, the steeper the slope at the horizontal axis and the greater the value of dx/dt, as shown in Fig. 14.3 for two different frequencies.

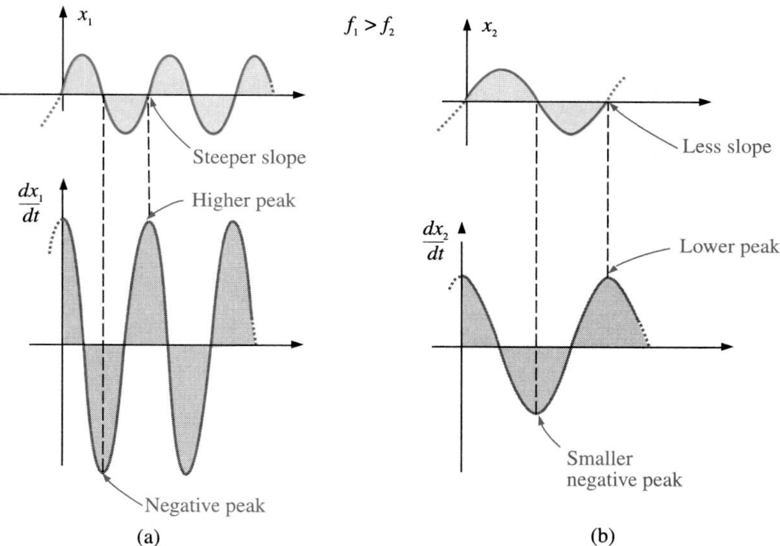

FIG. 14.3
Effect of frequency on the peak value of the derivative.

Note in Fig. 14.3 that even though both waveforms (x_1 and x_2) have the same peak value, the sinusoidal function with the higher frequency produces the larger peak value for the derivative. In addition, note that

the derivative of a sine wave has the same period and frequency as the original sinusoidal waveform.

For the sinusoidal voltage

$$e(t) = E_m \sin(\omega t \pm \theta)$$

the derivative can be found directly by differentiation (calculus) to produce the following:

$$\frac{d}{dt}e(t) = \omega E_m \cos(\omega t \pm \theta)$$
$$= 2\pi f E_m \cos(\omega t \pm \theta) \tag{14.1}$$

Note that the peak value of the derivative, $2\pi fE_m$, is a function of the frequency of $e(t)$ and the derivative of a sine wave is a cosine wave.

14.3 RESPONSE OF BASIC *R*, *L*, AND *C* ELEMENTS TO A SINUSOIDAL VOLTAGE OR CURRENT

Now that we are familiar with the characteristics of the derivative of a sinusoidal function, we can investigate the response of the basic elements *R*, *L*, and *C* to a sinusoidal voltage or current.

Resistor

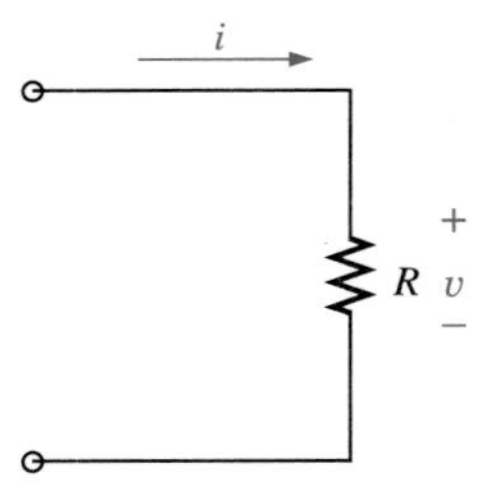

FIG. 14.4
Determining the sinusoidal response for a resistive element.

For power-line frequencies and frequencies up to a few hundred kilohertz, resistance is, for all practical purposes, unaffected by the frequency of the applied sinusoidal voltage or current. For this frequency region, the resistor *R* of Fig. 14.4 can be treated as a constant, and Ohm's law can be applied as follows. For $v = V_m \sin \omega t$,

$$i = \frac{v}{R} = \frac{V_m \sin \omega t}{R} = \frac{V_m}{R}\sin \omega t = I_m \sin \omega t$$

where
$$I_m = \frac{V_m}{R} \tag{14.2}$$

In addition, for a given *i*,

$$v = iR = (I_m \sin \omega t)R = I_m R \sin \omega t = V_m \sin \omega t$$

where
$$V_m = I_m R \tag{14.3}$$

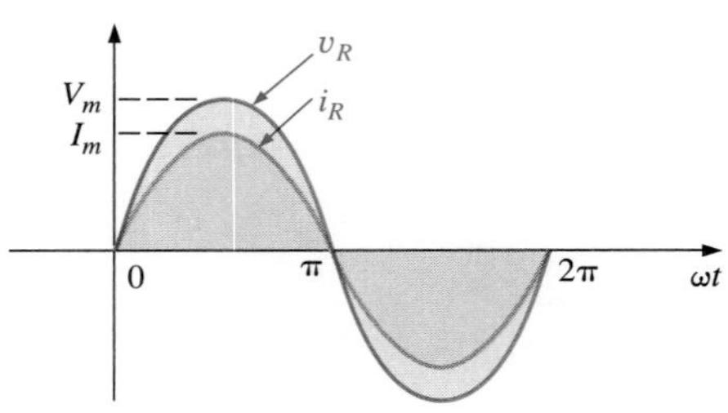

FIG. 14.5
The voltage and current of a resistive element are in phase.

The plot of *v* and *i* in Fig.14.5 reveals that for a purely resistive element, the voltage across and the current through the element are in phase with their peak values related by Ohm's law. This can be expressed simply as

for a resistor, v_R and i_R are in phase with each other.

Inductor

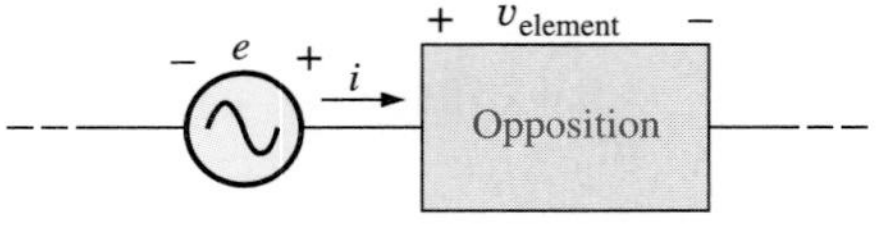

FIG. 14.6
Defining the opposition of an element to the flow of charge through the element.

For the series configuration of Fig. 14.6, the voltage $v_{element}$ of the boxed "opposition" element opposes the source *e* and thereby reduces the magnitude of the current *i*. The magnitude of the voltage across the element ($v_{element}$) is determined by the opposition of that element to the flow of charge, or current *i*.

For a resistor, we have found that the opposition is its resistance and that $v_{element}$ and *i* are determined by $v_{element} = iR$.

In Chapter 12, we found that the voltage across an inductor is directly related to the rate of change of current through the coil (the higher the frequency, the greater the magnitude of the voltage). We also found that the inductance of a coil determines the rate of change of the

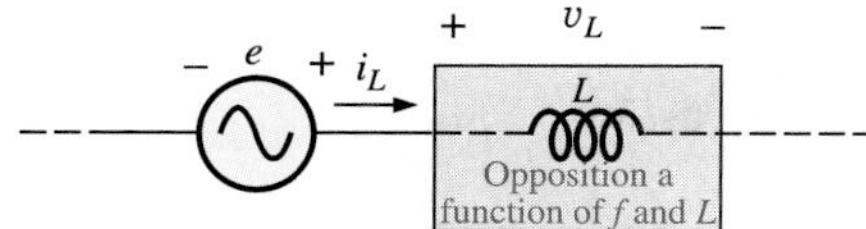

FIG. 14.7
Defining the parameters that determine the opposition of an inductive element to the flow of charge.

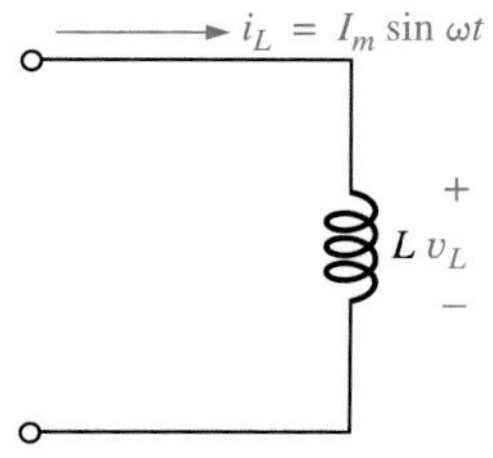

FIG. 14.8
Investigating the sinusoidal response of an inductive element.

flux linking a coil for a particular change in current through the coil (that is the higher the inductance, the greater the rate of change of the flux linkages, and the greater the resulting voltage across the coil). The inductive voltage, therefore, is directly related to the frequency and the inductance of the coil. For increasing values of f and L in Fig.14.7, the magnitude of v_L will increase, which suggests a directly related increase in levels of opposition in Fig.14.6.

We will now verify some of the preceding conclusions using a more mathematical approach and then define a few important quantities to be used in the sections and chapters to follow.

For the inductor of Fig. 14.8, we recall from Chapter 12 that

$$v_L = L\frac{di_L}{dt}$$

and, applying differentiation,

$$\frac{di_L}{dt} = \frac{d}{dt}(I_m \sin \omega t) = \omega I_m \cos \omega t$$

Therefore, $$v_L = L\frac{di_L}{dt} = L(\omega I_m \cos \omega t) = \omega L I_m \cos \omega t$$

or $$v_L = V_m \sin(\omega t + 90^\circ)$$

where $$V_m = \omega L I_m$$

Note that the peak value of v_L is directly related to ω $(= 2\pi f)$ and L as predicted in the discussion above.

A plot of v_L and i_L in Fig. 14.9 reveals that

for an inductor, v_L leads i_L by 90°, or i_L lags v_L by 90°.

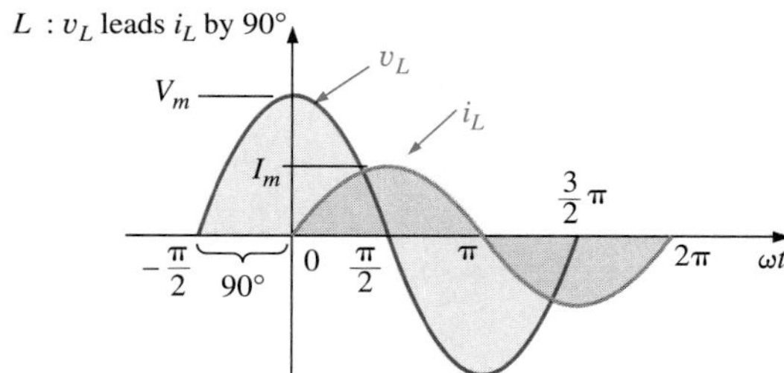

FIG. 14.9
For a pure inductor, the voltage across the coil leads the current through the coil by 90°.

If a phase angle is included in the sinusoidal expression for i_L, such as

$$i_L = I_m \sin(\omega t \pm \theta)$$

then $$v_L = \omega L I_m \sin(\omega t \pm \theta + 90^\circ)$$

The opposition established by an inductor in a sinusoidal ac network can now be found by applying Eq. (4.1):

$$\text{Effect} = \frac{\text{cause}}{\text{opposition}}$$

which, for our purposes, can be written

$$\text{Opposition} = \frac{\text{cause}}{\text{effect}}$$

Substituting values, we have

$$\text{Opposition} = \frac{V_m}{I_m} = \frac{\omega L I_m}{I_m} = \omega L$$

revealing that the opposition established by an inductor in an ac sinusoidal network is directly related to the product of the angular velocity ($\omega = 2\pi f$) and the inductance. This verifies our earlier conclusions.

The quantity ωL, called the *reactance* (from the word *reaction*) of an inductor, is symbolically represented by X_L and is measured in ohms; that is,

$$X_L = \omega L \qquad \text{(ohms, } \Omega) \qquad (14.4)$$

In an Ohm's law format, its magnitude can be determined from

$$X_L = \frac{V_m}{I_m} \qquad \text{(ohms, } \Omega) \qquad (14.5)$$

Inductive reactance is the opposition to the flow of current, which results in the continual interchange of energy between the source and the magnetic field of the inductor. In other words, inductive reactance, unlike resistance (which dissipates energy in the form of heat), does not dissipate electrical energy (ignoring the effects of the internal resistance of the inductor).

Capacitor

Let us now return to the series configuration of Fig. 14.6 and insert the capacitor as the element of interest. For the capacitor, however, we will determine i for a particular voltage across the element. When this approach reaches its conclusion, the relationship between the voltage and current will be known, and the opposing voltage (v_{element}) can be determined for any sinusoidal current i.

Our investigation of the inductor revealed that the inductive voltage across a coil opposes the instantaneous change in current through the coil. For capacitive networks, the voltage across the capacitor is limited by the rate at which charge can be deposited on, or released by, the plates of the capacitor during the charging and discharging phases, respectively. In other words, an instantaneous change in voltage across a capacitor is opposed by the fact that there is an element of time required to deposit charge on (or release charge from) the plates of a capacitor, and $V = Q/C$.

Since capacitance is a measure of the rate at which a capacitor will store charge on its plates,

for a particular change in voltage across the capacitor, the greater the value of capacitance, the greater the resulting capacitive current.

In addition, the fundamental equation relating the voltage across a capacitor to the current of a capacitor [$i = C(dv/dt)$] indicates that

for a particular capacitance the greater the rate of change of voltage across the capacitor, the greater the capacitive current.

An increase in frequency corresponds to an increase in the rate of change of voltage across the capacitor and to an increase in the current of the capacitor.

The current of a capacitor is therefore directly related to the frequency (or, more specifically, the angular velocity) and the capacitance of the capacitor. An increase in either quantity will result in an increase in the current of the capacitor. For the basic configuration of Fig. 14.10, however, we are interested in determining the opposition of the capacitor as compared to the resistance of a resistor or ωL for the inductor. Since an increase in current corresponds to a decrease in opposition, and i_C is proportional to ω and C, the opposition of a capacitor is inversely related to ω $(= 2\pi f)$ and C.

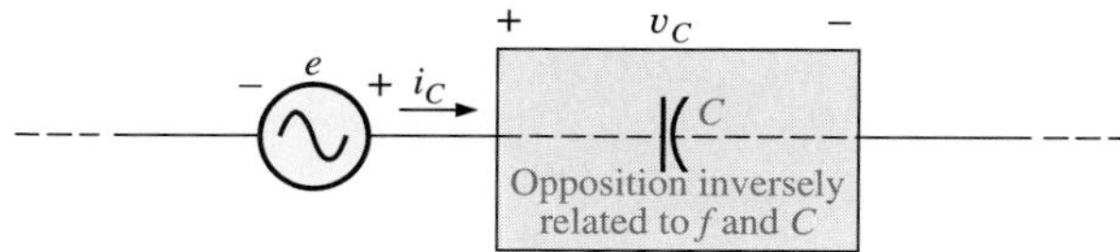

FIG. 14.10
Defining the parameters that determine the opposition of a capacitive element to the flow of the charge.

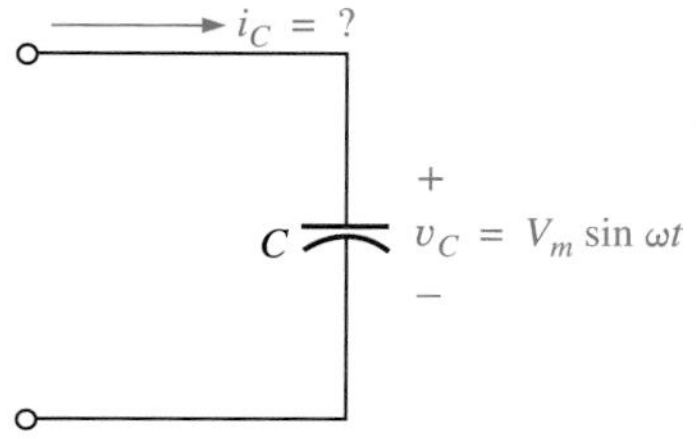

We will now verify, as we did for the inductor, some of the above conclusions using a more mathematical approach.

For the capacitor of Fig. 14.11, we recall from Chapter 10 that

$$i_C = C\frac{dv_C}{dt}$$

and, applying differentiation,

$$\frac{dv_C}{dt} = \frac{d}{dt}(V_m \sin \omega t) = \omega V_m \cos \omega t$$

FIG. 14.11
Investigating the sinusoidal response of a capacitive element.

Therefore,

$$i_C = C\frac{dv_C}{dt} = C(\omega V_m \cos \omega t) = \omega C V_m \cos \omega t$$

or $$i_C = I_m \sin(\omega t + 90°)$$

where $$I_m = \omega C V_m$$

Note that the peak value of i_C is directly related to ω $(= 2\pi f)$ and C, as predicted in the discussion above.

A plot of v_C and i_C in Fig. 14.12 reveals that

for a capacitor, i_C leads v_C by 90°, or v_C lags i_C by 90°.*

C: i_C leads v_C by 90°.

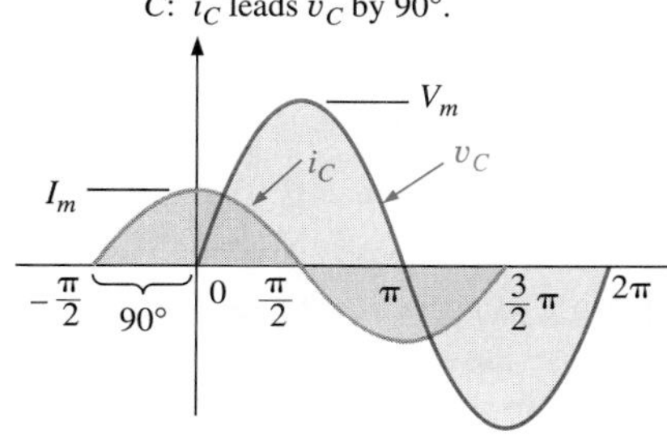

FIG. 14.12
The current of a purely capacitive element leads the voltage across the element by 90°.

If a phase angle is included in the sinusoidal expression for v_C, such as

$$v_C = V_m \sin(\omega t \pm \theta)$$

then $$i_C = \omega C V_m \sin(\omega t \pm \theta + 90°)$$

---

*A mnemonic phrase sometimes used to remember the phase relationship between the voltage and current of a coil and capacitor is "*ELI* the *ICE* man." Note that the *L* (inductor) has the *E* before the *I* (*e* leads *i* by 90°), and the *C* (capacitor) has the *I* before the *E* (*i* leads *e* by 90°).

Applying

$$\text{Opposition} = \frac{\text{cause}}{\text{effect}}$$

and substituting values, we obtain

$$\text{Opposition} = \frac{V_m}{I_m} = \frac{V_m}{\omega C V_m} = \frac{1}{\omega C}$$

which agrees with the results obtained above.

The quantity $1/\omega C$, called the *reactance* of a capacitor, is symbolically represented by X_C and is measured in ohms; that is,

$$X_C = \frac{1}{\omega C} \qquad \text{(ohms, } \Omega) \tag{14.6}$$

In an Ohm's law format, its magnitude can be determined from

$$X_C = \frac{V_m}{I_m} \qquad \text{(ohms, } \Omega) \tag{14.7}$$

Capacitive reactance is the opposition to the flow of charge, which results in the continual interchange of energy between the source and the electric field of the capacitor. Like the inductor, the capacitor does *not* dissipate energy in any form (ignoring the effects of the leakage resistance).

In the circuits just considered, the current was given in the inductive circuit, and the voltage in the capacitive circuit. This was done to avoid using integration in finding the unknown quantities. In the inductive circuit,

$$v_L = L\frac{di_L}{dt}$$

but

$$i_L = \frac{1}{L}\int v_L\, dt \tag{14.8}$$

In the capacitive circuit,

$$i_C = C\frac{dv_C}{dt}$$

but

$$v_C = \frac{1}{C}\int i_C\, dt \tag{14.9}$$

We will soon consider a method of analyzing ac circuits that permits us to solve for an unknown quantity with sinusoidal input without having to use direct integration or differentiation.

It is possible to determine whether a network with one or more elements is mainly capacitive or inductive by noting the phase relationship between the input voltage and current.

If the source current leads the applied voltage, the network is predominantly capacitive, and if the applied voltage leads the source current, it is predominantly inductive.

Since we now have an equation for the reactance of an inductor or capacitor, we do not need to use derivatives or integration in the examples to be considered. Simply applying Ohm's law, $I_m = E_m/X_L$ (or X_C), and keeping in mind the phase relationship between the voltage and current for each element, will be enough to complete the examples.

EXAMPLE 14.1

The voltage across a resistor is indicated. Find the sinusoidal expression for the current if the resistor is 10 Ω. Sketch the curves for v and i.

a. $v = 100 \sin 377t$
b. $v = 25 \sin(377t + 60°)$

Solutions:

a. Eq. (14.2): $I_m = \dfrac{V_m}{R} = \dfrac{100\text{ V}}{10\ \Omega} = 10\text{ A}$

(v and i are in phase), resulting in

$$i = \mathbf{10 \sin 377}t$$

The curves are sketched in Fig. 14.13.

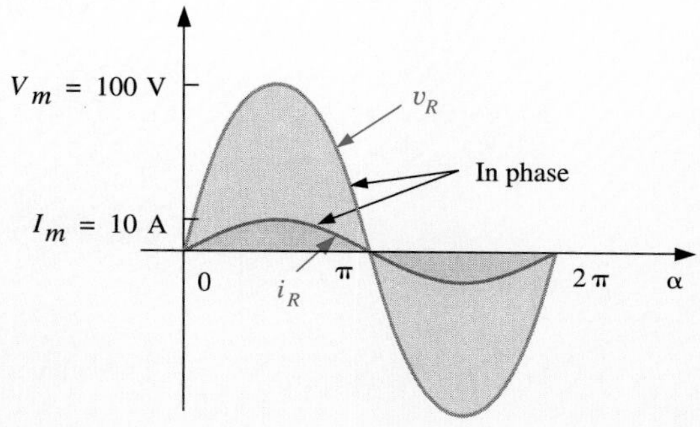

FIG. 14.13
Example 14.1(a).

b. Eq. (14.2): $I_m = \dfrac{V_m}{R} = \dfrac{25\text{ V}}{10\ \Omega} = 2.5\text{ A}$

(v and i are in phase), resulting in

$$i = \mathbf{2.5 \sin(377}t\mathbf{ + 60°)}$$

The curves are sketched in Fig. 14.14.

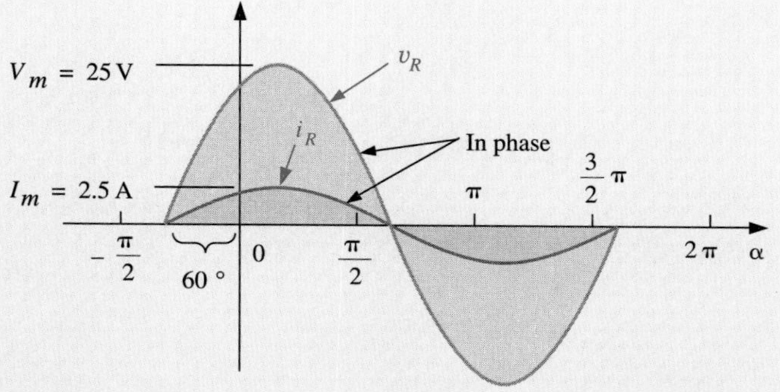

FIG. 14.14
Example 14.1(b).

EXAMPLE 14.2

The current through a 5-Ω resistor is given. Find the sinusoidal expression for the voltage across the resistor for $i = 40 \sin(377t + 30°)$.

Solution: Eq. (14.3): $V_m = I_m R = (40 \text{ A})(5 \ \Omega) = 200 \text{ V}$ (v and i are in phase), resulting in

$$v = \mathbf{200 \sin(377t + 30°)}$$

EXAMPLE 14.3

The current through a 0.1-H coil is provided. Find the sinusoidal expression for the voltage across the coil. Sketch the v and i curves.

a. $i = 10 \sin 377t$
b. $i = 7 \sin(377t - 70°)$

Solutions:

a. Eq. (14.4): $X_L = \omega L = (377 \text{ rad/s})(0.1 \text{ H}) = 37.7 \ \Omega$

Eq. (14.5): $V_m = I_m X_L = (10 \text{ A})(37.7 \ \Omega) = 377 \text{ V}$

and we know for a coil that v leads i by 90°. Therefore,

$$v = \mathbf{377 \sin(377t + 90°)}$$

The curves are sketched in Fig. 14.15.

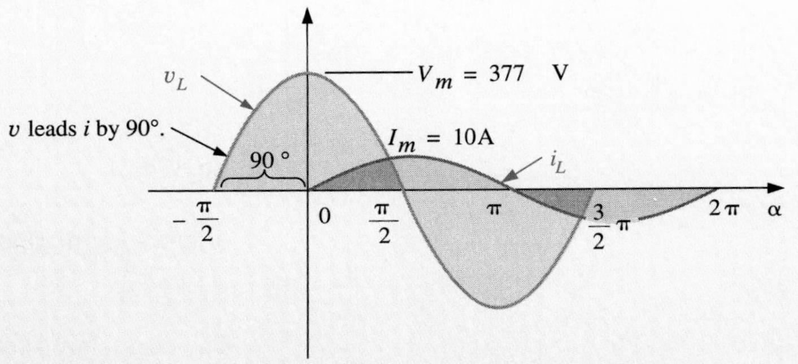

FIG. 14.15
Example 14.3(a).

b. X_L remains at 37.7 Ω.

$$V_m = I_m X_L = (7 \text{ A})(37.7 \ \Omega) = 263.9 \text{ V}$$

and we know for a coil that v leads i by 90°. Therefore,

$$v = 263.9 \sin(377t - 70° + 90°)$$

and

$$v = \mathbf{263.9 \sin(377t + 20°)}$$

continued

The curves are sketched in Fig. 14.16.

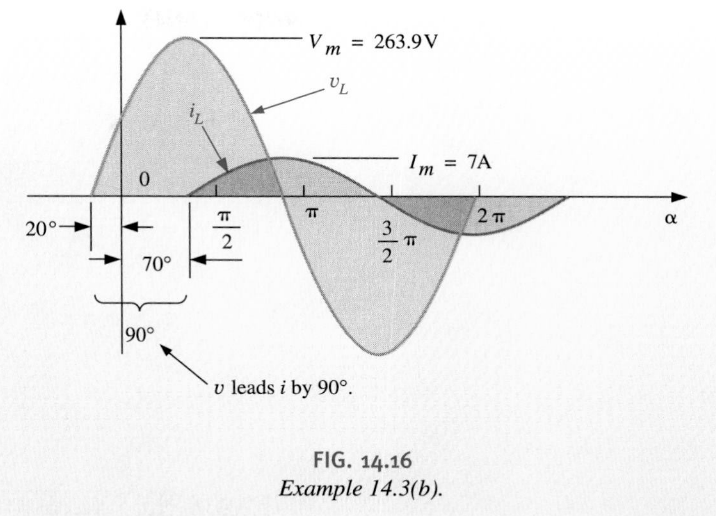

FIG. 14.16
Example 14.3(b).

EXAMPLE 14.4

The voltage across a 0.5-H coil is provided below. What is the sinusoidal expression for the current?

$$v = 100 \sin 20t$$

Solution:

$$X_L = \omega L = (20 \text{ rad/s})(0.5 \text{ H}) = 10 \ \Omega$$

$$I_m = \frac{V_m}{X_L} = \frac{100 \text{ V}}{10 \ \Omega} = 10 \text{ A}$$

and we know that i lags v by 90°. Therefore,

$$\mathbf{i = 10 \sin(20t - 90°)}$$

EXAMPLE 14.5

The voltage across a 1-μF capacitor is provided below. What is the sinusoidal expression for the current? Sketch the v and i curves.

$$v = 30 \sin 400t$$

Solution:

Eq. (14.6): $X_C = \dfrac{1}{\omega C} = \dfrac{1}{(400 \text{ rad/s})(1 \times 10^{-6} \text{ F})} = \dfrac{10^6 \ \Omega}{400} = 2500 \ \Omega$

Eq. (14.7): $I_m = \dfrac{V_m}{X_C} = \dfrac{30 \text{ V}}{2500 \ \Omega} = 0.0120 \text{ A} = 12 \text{ mA}$

and we know for a capacitor that i leads v by 90°. Therefore,

$$\mathbf{i = 12 \times 10^{-3} \sin(400t + 90°)}$$

continued

The curves are sketched in Fig. 14.17.

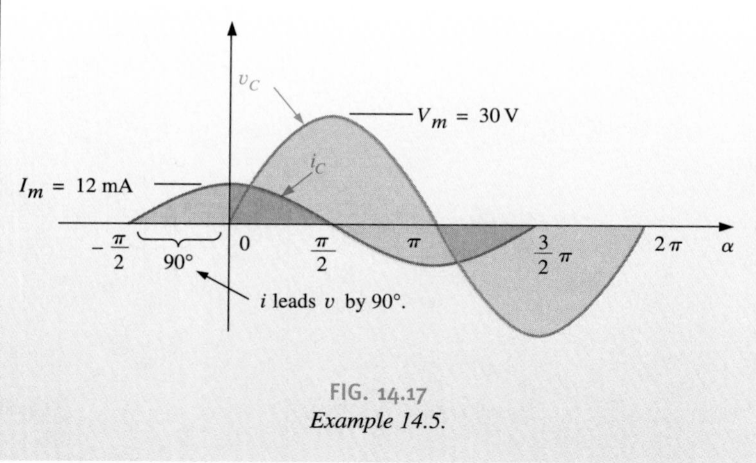

FIG. 14.17
Example 14.5.

EXAMPLE 14.6

The current through a 100-μF capacitor is given. Find the sinusoidal expression for the voltage across the capacitor.

$$i = 40 \sin(500t + 60°)$$

Solution:
$$X_C = \frac{1}{\omega C} = \frac{1}{(500 \text{ rad/s})(100 \times 10^{-6} \text{ F})} = \frac{10^6 \ \Omega}{5 \times 10^4} = \frac{10^2 \ \Omega}{5} = 20 \ \Omega$$

$$V_m = I_m X_C = (40 \text{ A})(20 \ \Omega) = 800 \text{ V}$$

and we know for a capacitor that v lags i by 90°. Therefore,

$$v = 800 \sin(500t + 60° - 90°)$$

and $$\mathbf{v = 800 \sin(500t - 30°)}$$

EXAMPLE 14.7

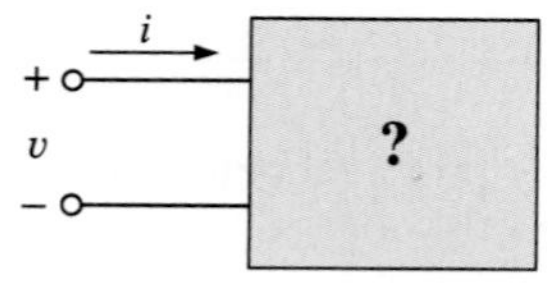

FIG. 14.18
Example 14.7.

For the following pairs of voltages and currents, determine whether the element involved is a capacitor, inductor, or resistor, and determine the value of C, L, or R, if sufficient data are provided (Fig. 14.18):

a. $v = 100 \sin(\omega t + 40°)$
   $i = 20 \sin(\omega t + 40°)$
b. $v = 1000 \sin(377t + 10°)$
   $i = 5 \sin(377t - 80°)$
c. $v = 500 \sin(157t + 30°)$
   $i = 1 \sin(157t + 120°)$
d. $v = 50 \cos(\omega t + 20°)$
   $i = 5 \sin(\omega t + 110°)$

Solutions:

a. Since v and i are *in phase,* the element is a *resistor,* and

$$R = \frac{V_m}{I_m} = \frac{100 \text{ V}}{20 \text{ A}} = \mathbf{5 \ \Omega}$$

continued

b. Since v *leads* i by 90°, the element is an *inductor,* and

$$X_L = \frac{V_m}{I_m} = \frac{1000\text{ V}}{5\text{ A}} = 200\ \Omega$$

so that $X_L = \omega L = 200\ \Omega$

$$\text{or } L = \frac{200\ \Omega}{\omega} = \frac{200\ \Omega}{377\text{ rad/s}} = \mathbf{0.531\ H}$$

c. Since i *leads* v by 90°, the element is a *capacitor,* and

$$X_C = \frac{V_m}{I_m} = \frac{500\text{ V}}{1\text{ A}} = 500\ \Omega$$

$$\text{so that } X_C = \frac{1}{\omega C} = 500\ \Omega$$

$$\text{or } C = \frac{1}{\omega 500\ \Omega} = \frac{1}{(157\text{ rad/s})(500\ \Omega)} = \mathbf{12.74\ \mu F}$$

d. $v = 50\cos(\omega t + 20°) = 50\sin(\omega t + 20° + 90°)$
$= 50\sin(\omega t + 110°)$

Since v and i are *in phase,* the element is a *resistor,* and

$$R = \frac{V_m}{I_m} = \frac{50\text{ V}}{5\text{ A}} = \mathbf{10\ \Omega}$$

dc, High-, and Low-Frequency Effects on *L* and *C*

For dc circuits, the frequency is zero, and the reactance of a coil is

$$X_L = 2\pi f L = 2\pi(0)L = 0\ \Omega$$

The use of the short-circuit equivalence for the inductor in dc circuits (Chapter 12) is now validated. At very high frequencies, $X_L\uparrow = 2\pi f\uparrow L$ is very large, and for some practical applications the inductor can be replaced by an open circuit. In equation form

$$X_L = 0\ \Omega \qquad (\text{dc}, f = 0\text{ Hz}) \tag{14.10}$$

and

$$X_L \cong \infty\ \Omega \qquad (f = \text{very high frequencies}) \tag{14.11}$$

The capacitor can be replaced by an open-circuit equivalence in dc circuits since $f = 0$, and

$$X_C = \frac{1}{2\pi f C} = \frac{1}{2\pi(0)C} = \infty\ \Omega$$

once again validating our previous action (Chapter 10). At very high frequencies, for finite capacitances,

$$X_C\downarrow = \frac{1}{2\pi f\uparrow C}$$

is very small, and for some practical applications the capacitor can be replaced by a short circuit. In equation form

$$X_C \cong \infty\ \Omega \qquad (\text{dc}, f = 0\text{ Hz}) \tag{14.12}$$

and $X_C \cong 0\ \Omega$ (f = very high frequencies) (14.13)

Table 14.1 reviews the preceding conclusions. The effects at very high frequencies are discussed further in Chapter 21.

TABLE 14.1
Effect of high and low frequencies on the circuit model of an inductor and capacitor.

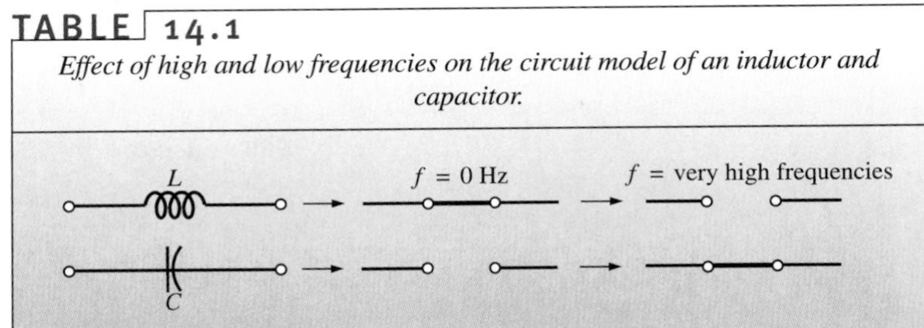

Phase Angle Measurements Between the Applied Voltage and Source Current

Now that you are familiar with phase relationships and understand how the elements affect the phase relationship between the applied voltage and resulting current, we can introduce the use of the oscilloscope to measure the phase angle. Recall from past discussions that the oscilloscope can be used only to display voltage levels versus time. However, now that we realize that the voltage across a resistor is in phase with the current through a resistor, we can consider the phase angle associated with the voltage across any resistor actually to be the phase angle of the current.

For example, suppose we want to find the phase angle introduced by the unknown system of Fig. 14.19(a). In Fig. 14.19(b) a resistor was added to the input leads, and the two channels of a dual trace (most modern oscilloscopes can display two signals at the same time) were connected as shown. One channel will display the input voltage v_i, and the other will display v_R, as shown in Fig. 14.19(c). However, as noted before, since v_R and i_R are in phase, the phase angle appearing in Fig. 14.19(c) is also the phase angle between v_i and i_i.

A *sensing* resistor can be added to determine the phase angle introduced by the system and can be used to determine the magnitude of the resulting current. (A **sensing** resistor is one with a magnitude that will not adversely affect the input characteristics of the system). We will leave the details of the connections that must be made and how the actual phase angle is determined for the laboratory experience.

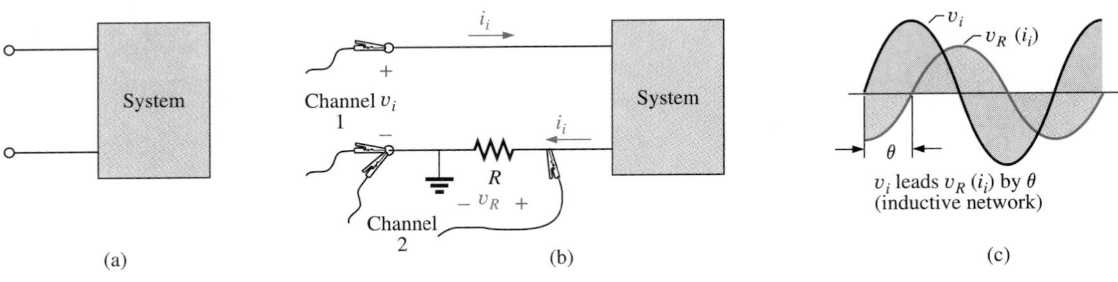

FIG. 14.19
Using an oscilloscope to determine the phase angle between the applied voltage and the source current.

14.4 FREQUENCY RESPONSE OF THE BASIC ELEMENTS

The analysis of Section 14.3 was limited to a particular applied frequency. What is the effect of varying the frequency on the level of opposition offered by a resistive, inductive, or capacitive element? We know from the last section that the inductive reactance increases with frequency while the capacitive reactance decreases. However, what is the pattern to this increase or decrease in opposition? Does it continue indefinitely on the same path? Since applied signals may have frequencies extending from a few hertz to megahertz, it is important to be aware of the effect of frequency on the opposition level.

R

So far we have assumed that the resistance of a resistor is independent of the applied frequency. However, in the real world each resistive element has stray capacitance and lead inductance that are sensitive to the applied frequency. However, the capacitive and inductive levels involved are usually so small that their real effect is not noticed until the megahertz range.

Although frequency does influence a resistor's resistance, we will ignore it for the frequency range we are considering. For these relatively low frequencies, we assume that resistance is independent of frequency, as shown in the graph of Fig. 14.20.

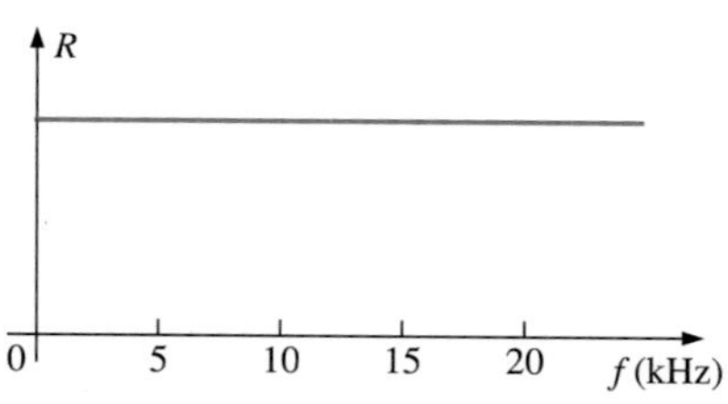

FIG. 14.20
R versus f for the range of interest.

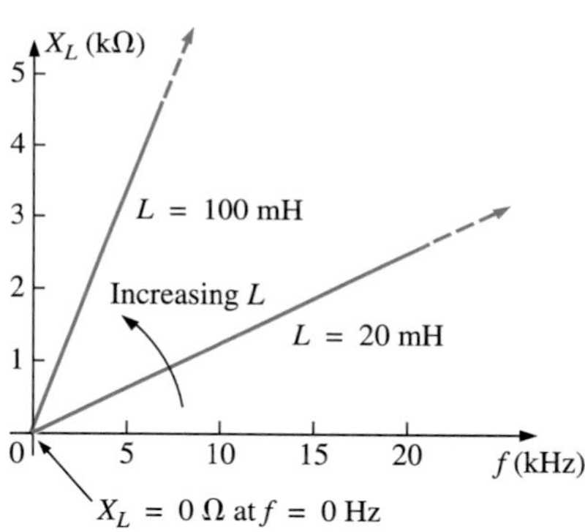

FIG. 14.21
X_L versus frequency.

L

For inductors, the equation

$$X_L = \omega L = 2\pi fL = 2\pi Lf$$

is directly related to the straight-line equation

$$y = mx + b = (2\pi L)f + 0$$

with a slope (m) of $2\pi L$ and a y-intercept (b) of zero. X_L is the y variable and f is the x variable, as shown in Fig. 14.21.

The larger the inductance, the greater the slope ($m = 2\pi L$) for the same frequency range, as shown in Fig. 14.21. Keep in mind, as Fig. 14.21 shows, that the opposition of an inductor at very low frequencies approaches that of a short circuit, while at high frequencies the reactance approaches that of an open circuit.

C

For the capacitor, the reactance equation

$$X_C = \frac{1}{2\pi fC}$$

can be written

$$X_C f = \frac{1}{2\pi C}$$

which matches the basic format of a hyperbola,

$$yx = k$$

with $y = X_C$, $x = f$, and the constant $k = 1/(2\pi C)$.

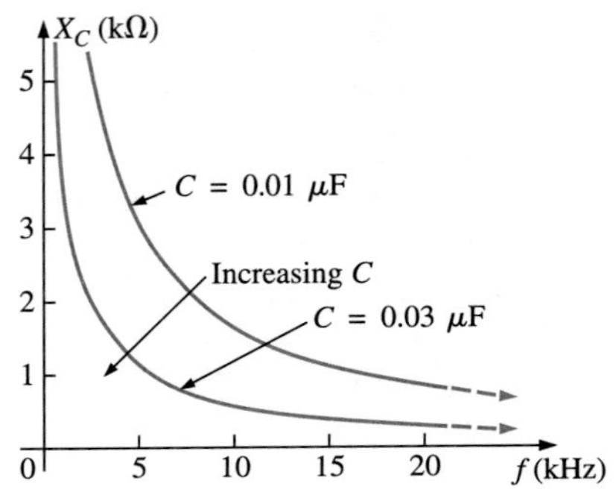

FIG. 14.22
X_C versus frequency.

At $f = 0$ Hz, the reactance of the capacitor is so large, as shown in Fig. 14.22, that it can be replaced by an open-circuit equivalent. As the frequency increases, the reactance decreases, until eventually a short-circuit equivalent would be appropriate. Note how an increase in capacitance causes the reactance to drop off more rapidly with frequency.

In summary, as the applied frequency increases,

a. the resistance of a resistor remains constant,
b. the reactance of an inductor increases linearly,
c. the reactance of a capacitor decreases nonlinearly.

EXAMPLE 14.8

At what frequency will the reactance of a 200-mH inductor match the resistance level of a 5-kΩ resistor?

Solution: The resistance remains constant at 5 kΩ for the frequency range of the inductor. Therefore,

$$R = 5000\ \Omega = X_L = 2\pi fL = 2\pi Lf$$
$$= 2\pi(200 \times 10^{-3}\ \text{H})f = 1.257f$$

and
$$f = \frac{5000\ \text{Hz}}{1.257} \cong \mathbf{3.98\ kHz}$$

EXAMPLE 14.9

At what frequency will an inductor of 5 mH have the same reactance as a capacitor of 0.1 μF?

Solution:

$$X_L = X_C$$
$$2\pi fL = \frac{1}{2\pi fC}$$
$$f^2 = \frac{1}{4\pi^2 LC}$$

and
$$f = \frac{1}{2\pi\sqrt{LC}} = \frac{1}{2\pi\sqrt{(5 \times 10^{-3}\ \text{H})(0.1 \times 10^{-6}\ \text{F})}}$$
$$= \frac{1}{2\pi\sqrt{5 \times 10^{-10}}} = \frac{1}{(2\pi)(2.236 \times 10^{-5})}$$
$$f = \frac{10^5\ \text{Hz}}{14.05} \cong \mathbf{7.12\ kHz}$$

You must also be aware that commercial inductors are not ideal elements. The terminal characteristics of an inductance will vary with several factors, such as frequency, temperature, and current.

Most inductors are built so that the larger the inductance, the lower the frequency at which parasitic factors become important. That is, for inductors in the milli-henry range (which is very typical), frequencies approaching 100 kHz can have an effect on the ideal characteristics of the element. For inductors in the micro-henry range, a frequency of 1 MHz may introduce negative effects. In general, the frequency of application for a coil becomes important at increasing frequencies. Inductors lose their ideal characteristics and in fact begin to act as capacitive elements with increasing losses at very high frequencies.

The capacitor, like the inductor, is not ideal at higher frequencies. In fact, there is a transition point where the characteristics of the capacitor will actually be inductive. In general, the frequency of application is important for capacitive elements because there comes a point with increasing frequency when the element will take on inductive characteristics. Also, the frequency of application defines the type of capacitor (or inductor) that should be applied: Electrolytics are limited to frequencies up to perhaps 10 kHz, while ceramic or mica can handle frequencies beyond 10 MHz.

The expected temperature range of operation can have an important impact on the type of capacitor chosen for a particular application. Electrolytics, tantalum, and some high-k ceramic capacitors are very sensitive to colder temperatures. With exposure and experience, you will learn the type of capacitor used for each application, and will be concerned only when you are dealing with very high frequencies, extreme temperatures, or very high currents or voltages.

14.5 AVERAGE POWER AND POWER FACTOR

For any load in a sinusoidal ac network, the voltage across the load and the current through the load will vary in a sinusoidal manner. The power to the load is determined by the product vi, but since both v and i vary with time, we need to determine what fixed value can be assigned to the power.

If we take the general case shown in Fig. 14.23 and use the following for v and i:

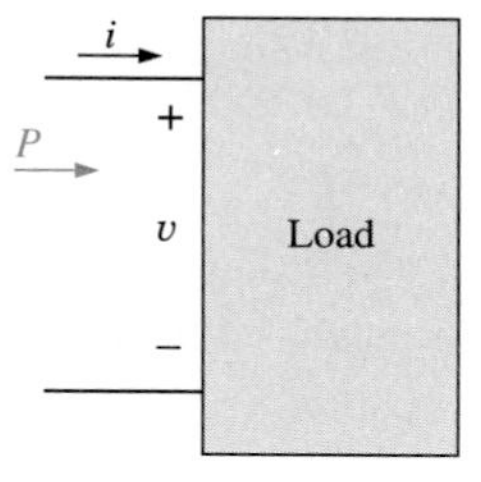

FIG. 14.23
Determining the power delivered in a sinusoidal ac network.

$$v = V_m \sin(\omega t + \theta_v)$$

$$i = I_m \sin(\omega t + \theta_i)$$

then the power is defined by

$$p = vi = V_m \sin(\omega t + \theta_v) I_m \sin(\omega t + \theta_i)$$

$$= V_m I_m \sin(\omega t + \theta_v) \sin(\omega t + \theta_i)$$

Using the trigonometric identity

$$\sin A \sin B = \frac{\cos(A - B) - \cos(A + B)}{2}$$

the function $\sin(\omega t + \theta_v) \sin(\omega t + \theta_i)$ becomes

$$\sin(\omega t + \theta_v) \sin(\omega t + \theta_i)$$

$$= \frac{\cos[(\omega t + \theta_v) - (\omega t + \theta_i)] - \cos[(\omega t + \theta_v) + (\omega t + \theta_i)]}{2}$$

$$= \frac{\cos(\theta_v - \theta_i) - \cos(2\omega t + \theta_v + \theta_i)}{2}$$

so that

$$p = \underbrace{\left[\frac{V_m I_m}{2}\cos(\theta_v - \theta_i)\right]}_{\text{Fixed value}} - \underbrace{\left[\frac{V_m I_m}{2}\cos(2\omega t + \theta_v + \theta_i)\right]}_{\text{Time-varying (function of } t)}$$

A plot of v, i, and p on the same set of axes is shown in Fig. 14.24.

Note that the second factor in the preceding equation is a cosine wave with an amplitude of $V_mI_m/2$, and a frequency twice that of the voltage or current. The average value of this term is zero over one cycle, producing no net transfer of energy in any one direction.

The first term in the preceding equation, however, has a constant magnitude (no time dependence) and therefore provides some net transfer of energy. This term is referred to as the *average power,* the reason for which is obvious from Fig. 14.24. The average power, or *real* power as it is sometimes called, is the power delivered to and dissipated by the load. It corresponds to the power calculations performed for dc networks. The angle $(\theta_v - \theta_i)$ is the phase angle between v and i. Since $\cos(-\alpha) = \cos \alpha$,

the magnitude of average power delivered is independent of whether v leads i or i leads v.

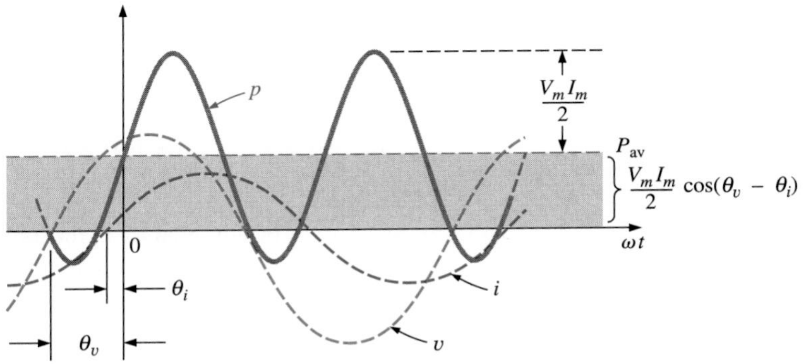

FIG. 14.24
Defining the average power for a sinusoidal ac network.

Defining θ as equal to $|\theta_v - \theta_i|$, where | | indicates that only the magnitude is important and the sign doesn't matter, we have

$$P = \frac{V_mI_m}{2}\cos\theta \qquad \text{(watts, W)} \qquad (14.14)$$

where P is the average power in watts. This equation can also be written

$$P = \left(\frac{V_m}{\sqrt{2}}\right)\left(\frac{I_m}{\sqrt{2}}\right)\cos\theta$$

or, since

$$V_{\text{eff}} = \frac{V_m}{\sqrt{2}} \text{ and } I_{\text{eff}} = \frac{I_m}{\sqrt{2}}$$

Eq. (14.14) becomes

$$P = V_{\text{eff}}I_{\text{eff}}\cos\theta \qquad (14.15)$$

Let us now apply Eqs. (14.14) and (14.15) to the basic R, L, and C elements.

Resistor

In a purely resistive circuit, since v and i are in phase, $|\theta_v - \theta_i| = \theta = 0°$, and $\cos\theta = \cos 0° = 1$, so that

$$P = \frac{V_m I_m}{2} = V_{\text{eff}} I_{\text{eff}} \qquad \text{(W)} \qquad (14.16)$$

Or, since $$I_{\text{eff}} = \frac{V_{\text{eff}}}{R}$$

then $$P = \frac{V_{\text{eff}}^2}{R} = I_{\text{eff}}^2 R \qquad \text{(W)} \qquad (14.17)$$

Inductor

In a purely inductive circuit, since v leads i by 90°, $|\theta_v - \theta_i| = \theta = |-90°\ | = 90°$. Therefore,

$$P = \frac{V_m I_m}{2} \cos 90° = \frac{V_m I_m}{2}(0) = \mathbf{0\ W}$$

The average power or power dissipated by the ideal inductor (no associated resistance) is zero watts.

Capacitor

In a purely capacitive circuit, since i leads v by 90°, $|\theta_v - \theta_i| = \theta = |-90°| = 90°$. Therefore,

$$P = \frac{V_m I_m}{2} \cos(90°) = \frac{V_m I_m}{2}(0) = \mathbf{0\ W}$$

The average power or power dissipated by the ideal capacitor (no associated resistance) is zero watts.

EXAMPLE 14.10

Find the average power dissipated in a network whose input current and voltage are the following:

$$i = 5 \sin(\omega t + 40°)$$
$$v = 10 \sin(\omega t + 40°)$$

Solution: Since v and i are in phase, the circuit appears to be purely resistive at the input terminals. Therefore,

$$P = \frac{V_m I_m}{2} = \frac{(10\text{ V})(5\text{ A})}{2} = \mathbf{25\ W}$$

or $$R = \frac{V_m}{I_m} = \frac{10\text{ V}}{5\text{ A}} = 2\ \Omega$$

and $$P = \frac{V_{\text{eff}}^2}{R} = \frac{[(0.707)(10\text{ V})]^2}{2} = \mathbf{25\ W}$$

or $$P = I_{\text{eff}}^2 R = [(0.707)(5\text{ A})]^2(2) = \mathbf{25\ W}$$

For the following example, the circuit consists of a combination of resistances and reactances producing phase angles between the input current and voltage different from 0° or 90°.

EXAMPLE 14.11

Determine the average power delivered to networks having the following input voltage and current:

a. $v = 100 \sin(\omega t + 40°)$
$i = 20 \sin(\omega t + 70°)$

b. $v = 150 \sin(\omega t - 70°)$
$i = 3 \sin(\omega t - 50°)$

Solutions:

a. $V_m = 100,\ \theta_v = 40°$
$I_m = 20,\ \theta_i = 70°$
$\theta = |\theta_v - \theta_i| = |40° - 70°| = |-30°| = 30°$

and

$$P = \frac{V_m I_m}{2} \cos\theta = \frac{(100\text{ V})(20\text{ A})}{2} \cos(30°) = (1000\text{ W})(0.866)$$
$$= \mathbf{866\ W}$$

b. $V_m = 150\text{ V},\ \theta_v = -70°$
$I_m = 3\text{ A},\ \theta_i = -50°$
$\theta = |\theta_v - \theta_i| = |-70° - (-50°)|$
$= |-70° + 50°| = |-20°| = 20°$

and

$$P = \frac{V_m I_m}{2} \cos\theta = \frac{(150\text{ V})(3\text{ A})}{2} \cos(20°) = (225\text{ W})(0.9397)$$
$$= \mathbf{211.43\ W}$$

Power Factor

In the equation $P = (V_m I_m/2)\cos\theta$ the factor that has significant control on the delivered power level is the $\cos\theta$. No matter how large the voltage or current, if $\cos\theta = 0$, the power is zero; if $\cos\theta = 1$, the power delivered is a maximum. Since it has such control, the expression is called **power factor** and is defined by

$$\text{Power factor} = F_p = \cos\theta \qquad (14.18)$$

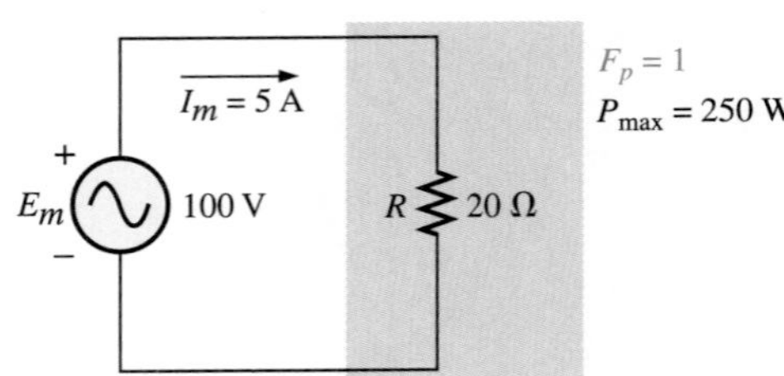

FIG. 14.25
Purely resistive load with $F_p = 1$.

For a purely resistive load such as the one shown in Fig. 14.25, the phase angle between v and i is 0° and $F_p = \cos\theta = \cos 0° = 1$. The power delivered is a maximum of $(V_m I_m/2)\cos\theta = ((100\text{ V})(5\text{ A})/2)\ (1) = 250$ W.

For a purely reactive load (inductive or capacitive) such as the one shown in Fig. 14.26, the phase angle between v and i is 90° and $F_p = \cos\theta = \cos 90° = 0$. The power delivered is then the minimum value of 0 watts, **even though the current has the same peak value** as the one in Fig. 14.25.

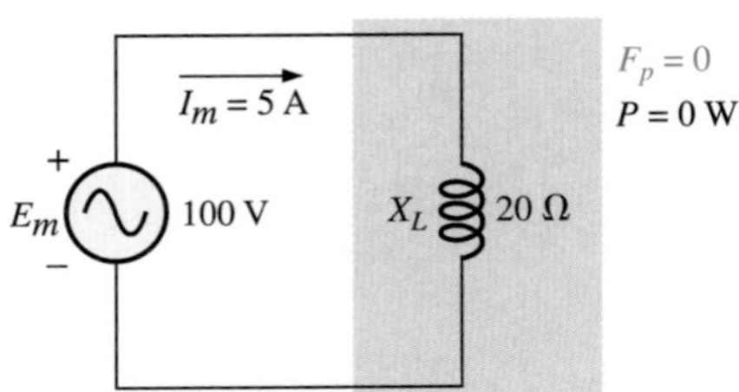

FIG. 14.26
Purely inductive load with $F_p = 0$.

For situations where the load is a combination of resistive and reactive elements, the power factor will vary between 0 and 1. The more resistive the total impedance, the closer the power factor is to 1; the more reactive the total impedance, the closer the power factor is to 0.

In terms of the average power and the terminal voltage and current,

$$F_p = \cos\theta = \frac{P}{V_{\text{eff}}I_{\text{eff}}} \tag{14.19}$$

The terms *leading* and *lagging* are often written together with the power factor. *They are defined by the current through the load.* If the current leads the voltage across a load, the load has a leading power factor. If the current lags the voltage across the load, the load has a lagging power factor. In other words,

capacitive networks have leading power factors, and inductive networks have lagging power factors.

The importance of the power factor on power distribution systems is examined in Chapter 19. In fact, one section deals only with power factor correction.

EXAMPLE 14.12

Determine the power factors of the following loads, and indicate whether they are leading or lagging:

a. Fig. 14.27(a)
b. Fig. 14.27(b)
c. Fig. 14.27(c)

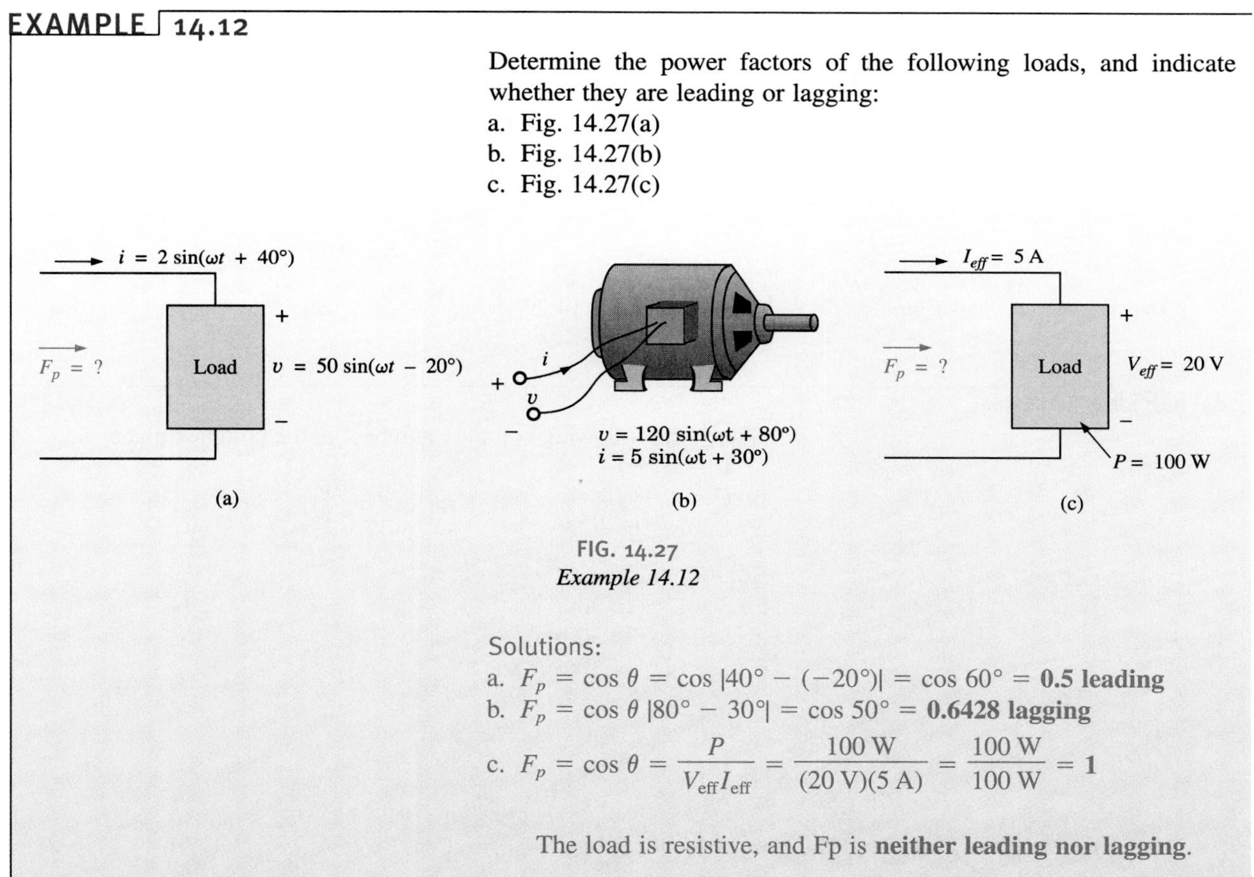

FIG. 14.27
Example 14.12

Solutions:

a. $F_p = \cos\theta = \cos|40° - (-20°)| = \cos 60° = \mathbf{0.5\ leading}$
b. $F_p = \cos\theta\ |80° - 30°| = \cos 50° = \mathbf{0.6428\ lagging}$
c. $F_p = \cos\theta = \dfrac{P}{V_{\text{eff}}I_{\text{eff}}} = \dfrac{100\text{ W}}{(20\text{ V})(5\text{ A})} = \dfrac{100\text{ W}}{100\text{ W}} = \mathbf{1}$

The load is resistive, and Fp is **neither leading nor lagging**.

14.6 COMPLEX NUMBERS

When we analyzed dc networks, we had to determine the algebraic sum of voltages and currents. We have to do the same when analyzing ac networks. However, the problem is how to determine the algebraic sum of two *sinusoidal* voltages or currents. One solution is to find the algebraic sum of the component waveforms on a point-to-point basis. This is a long and tedious process with relatively poor accuracy and a large opportunity for errors.

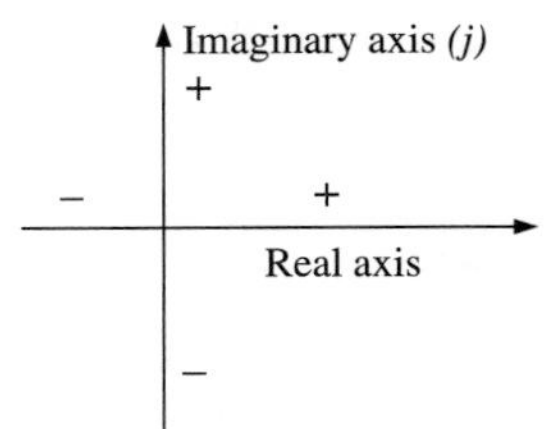

FIG. 14.28
Defining the real and imaginary axes of a complex plane.

This section introduces a system of *complex numbers* that provides a quick, direct, and accurate technique for finding the algebraic sum of sinusoidal waveforms. In the following chapters, we will extend the technique to allow you to analyze sinusoidal ac networks in a way that is similar to how we analyze dc networks. The methods and theorems described for dc networks can then be easily applied to sinusoidal ac networks.

A **complex number** represents a point in a two-dimensional plane located with reference to two distinct axes. The horizontal axis is called the **real** axis, while the vertical axis is called the **imaginary** axis (see Fig.14.28). For reasons that will be obvious later, the real axis is sometimes called the *resistance* axis and the imaginary axis, the *reactance* axis.

Complex numbers can be represented in two different forms: *rectangular* and *polar.* Each represents a point in the plane or a radius vector drawn from the origin to that point. In the **rectangular** form, a complex number has two parts: a *real component* and an *imaginary component.* The imaginary component is represented by the symbol j. In the **polar** form, a complex number also has two parts: a *magnitude* and an *associated angle*.

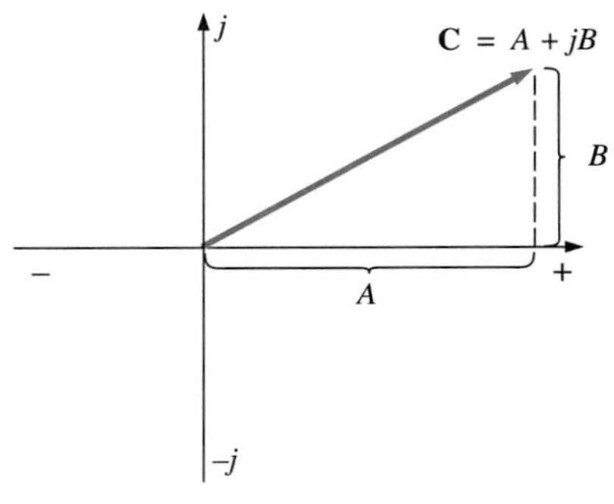

FIG. 14.29
Defining the rectangular form.

Rectangular Form

The format for the rectangular form is

$$\mathbf{C} = A + jB \tag{14.20}$$

as shown in Fig. 14.29.

EXAMPLE 14.13

Sketch the following complex numbers in the complex plane:

a. $\mathbf{C} = 3 + j4$
b. $\mathbf{C} = 0 - j6$
c. $\mathbf{C} = -10 - j20$

Solutions:

a. See Fig. 14.30(a).
b. See Fig. 14.30(b).
c. See Fig. 14.30(c).

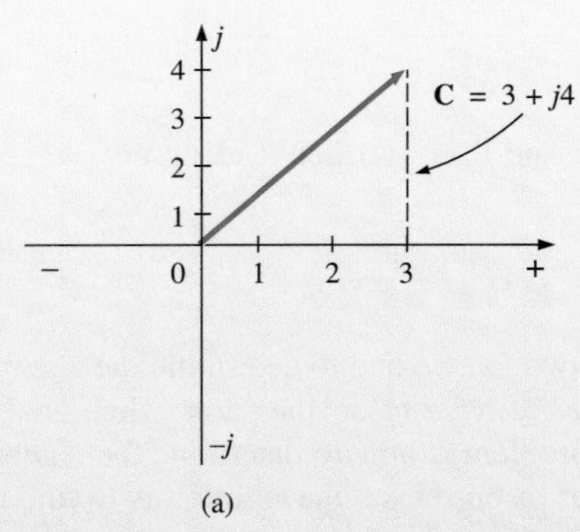

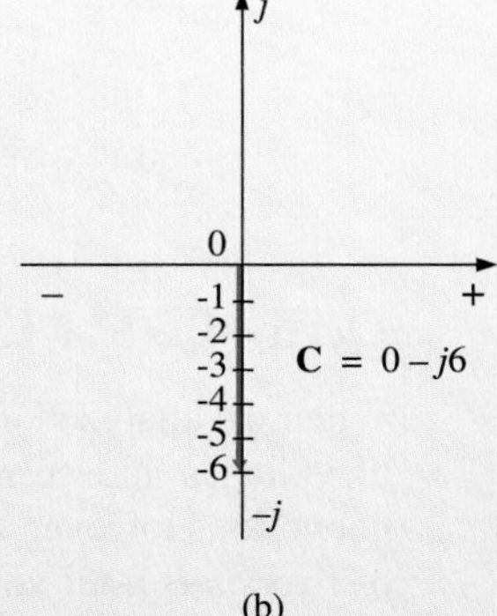

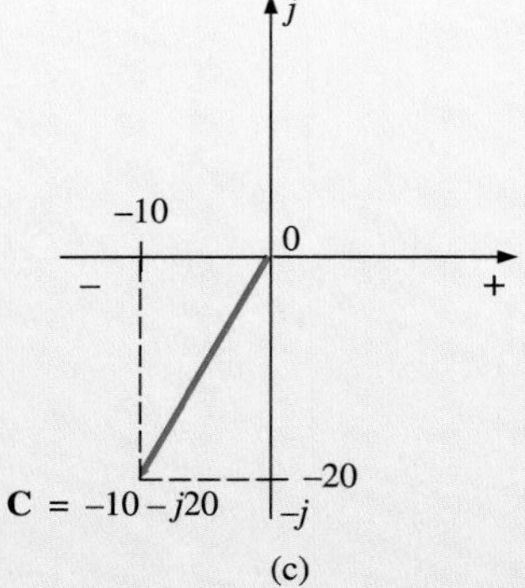

FIG. 14.30
Example 14.13

Polar Form

The format for the polar form is

$$\mathbf{C} = C \angle \theta \tag{14.21}$$

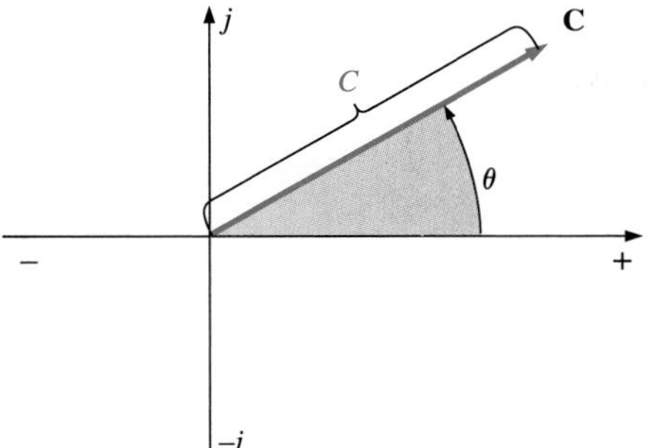

FIG. 14.31
Defining the polar form.

where C indicates magnitude only and θ is always measured counterclockwise from the *positive real axis,* as shown in Fig. 14.31. Angles measured in the clockwise direction from the positive real axis must have a negative sign associated with them.

In mathematics, the notation form of Eq. (14.21) is called *polar* form. When describing voltages and currents in electronics, this representation is called *phasor* form. Both terms mean the same thing. Phasors are discussed in more detail in Section 14.8.

A negative sign in front of the polar form has the effect shown in Fig. 14.32. Note that it results in a complex number directly opposite the complex number with a positive sign.

$$-\mathbf{C} = -C \angle \theta = C \angle \theta \pm \pi \tag{14.22}$$

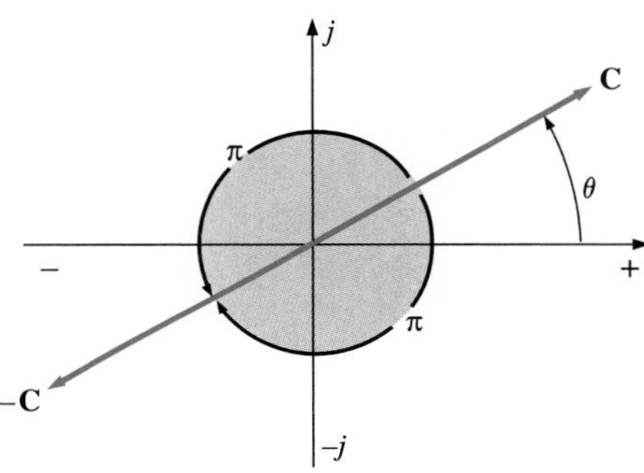

FIG. 14.32
Demonstrating the effect of a negative sign on the polar form.

EXAMPLE 14.14

Sketch the following complex numbers in the complex plane:

a. $\mathbf{C} = 5 \angle 30°$
b. $\mathbf{C} = 7 \angle -120°$
c. $\mathbf{C} = -4.2 \angle 60°$

Solutions:

a. See Fig. 14.33(a).
b. See Fig. 14.33(b).
c. See Fig. 14.33(c).

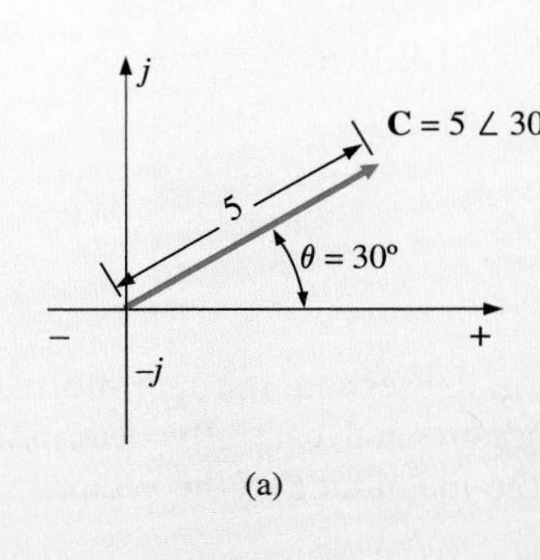

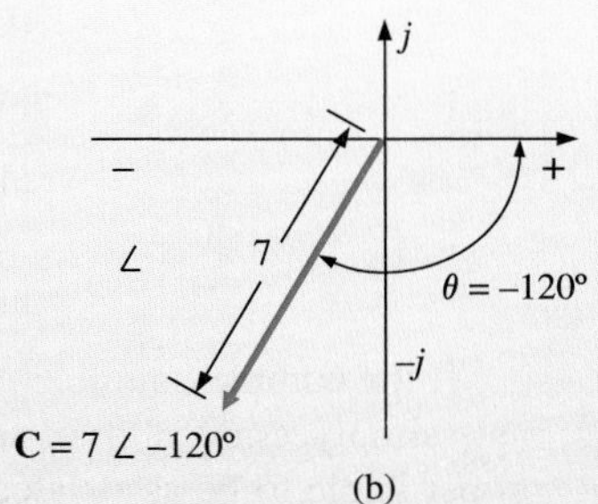

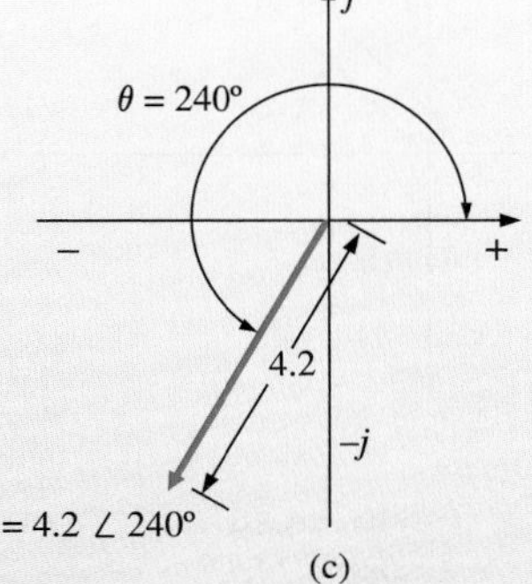

FIG. 14.33
Example 14.14.

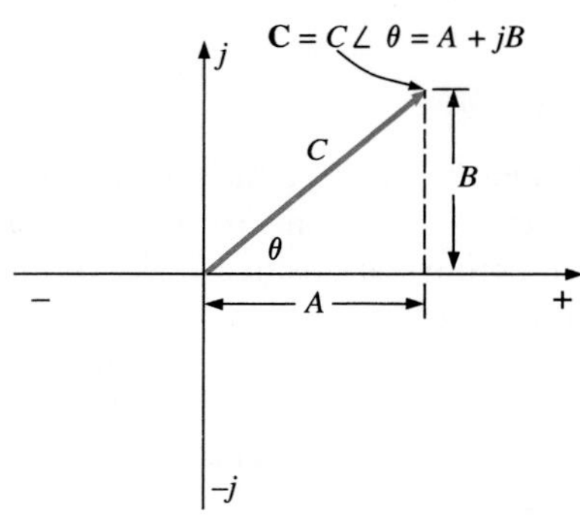

FIG. 14.34
Conversion between forms.

Conversion Between Forms

The rectangular and polar forms are related by the following four equations, using the notation of Fig. 14.34. To convert from rectangular to polar form, apply equations 14.23 and 14.24.

$$C = \sqrt{A^2 + B^2} \tag{14.23}$$

$$\theta = \tan^{-1}\frac{B}{A} \tag{14.24}$$

To convert from polar to rectangular form, apply equations 14.25 and 14.26.

$$A = C\cos\theta \tag{14.25}$$

$$B = C\sin\theta \tag{14.26}$$

Conversion between rectangular and polar forms can be done on most modern scientific calculators. Consult the owner's manual for your particular calculator to familiarize yourself with the steps and how to perform these conversions correctly.

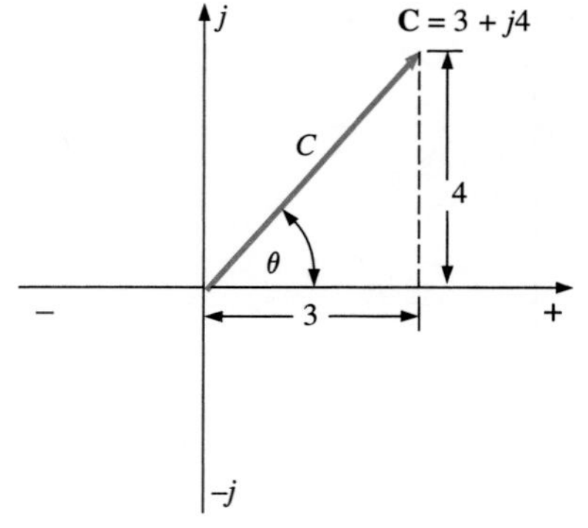

FIG. 14.35
Example 14.15.

EXAMPLE 14.15

Convert the following from rectangular to polar form:

$$\mathbf{C} = 3 + j\,4 \quad \text{(Fig. 14.35)}$$

Solution:

$$C = \sqrt{(3)^2 + (4)^2} = \sqrt{25} = 5$$

$$\theta = \tan^{-1}\left(\frac{4}{3}\right) = 53.13^\circ$$

and $\mathbf{C = 5\,\angle 53.13^\circ}$

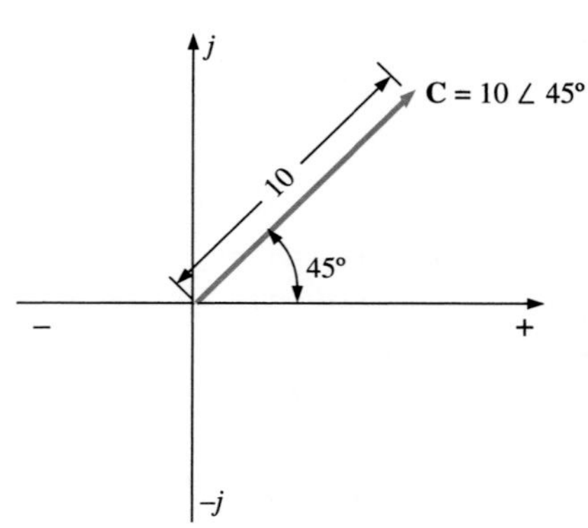

FIG. 14.36
Example 14.16.

EXAMPLE 14.16

Convert the following from polar to rectangular form:

$$\mathbf{C} = 10\angle 45^\circ \quad \text{(Fig. 14.36)}$$

Solution:

$$A = 10\cos 45^\circ = (10))(0.707) = 7.07$$

$$B = 10\sin 45^\circ = (10)(0.707) = 7.07$$

and $\mathbf{C = 7.07 + j\,7.07}$

If the complex number appears in the second, third, or fourth quadrant, simply convert it in that quadrant, and carefully determine the proper angle to be associated with the magnitude of the vector.

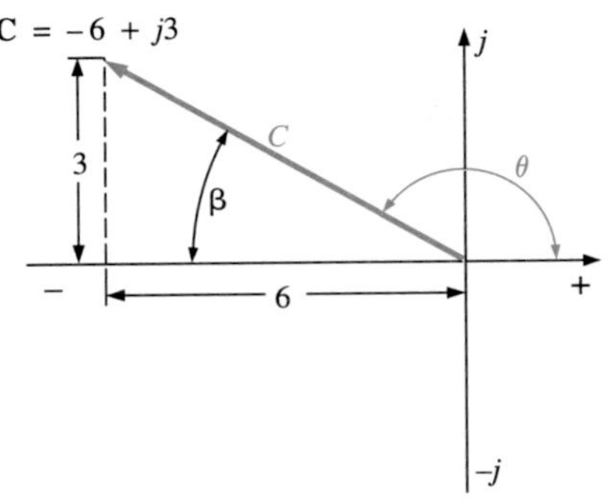

FIG. 14.37
Example 14.17.

EXAMPLE 14.17

Convert the following from rectangular to polar form:

$$\mathbf{C} = -6 + j\,3 \quad \text{(Fig. 14.37)}$$

Solution:

$$C = \sqrt{(6)^2 + (3)^2} = \sqrt{45} = 6.71$$

$$\beta = \tan^{-1}\left(\frac{3}{6}\right) = 26.57^\circ$$

$$\theta = 180 - 26.57^\circ = 153.43^\circ$$

and $$\mathbf{C} = \mathbf{6.71\,\angle 153.43^\circ}$$

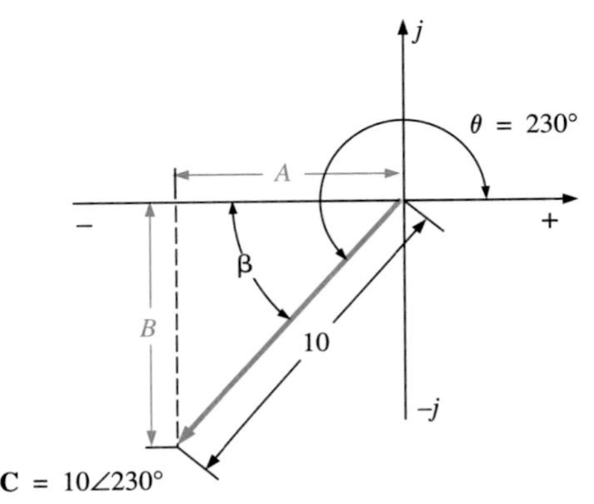

FIG. 14.38
Example 14.18.

EXAMPLE 14.18

Convert the following from polar to rectangular form:

$$\mathbf{C} = 10\,\angle 230^\circ \quad \text{(Fig. 14.38)}$$

Solution:

$$A = C\cos\beta = 10\cos(230^\circ - 180^\circ) = 10\cos 50^\circ$$
$$= (10)(0.6428) = 6.428$$
$$B = C\sin\beta = 10\sin 50^\circ = (10)(0.7660) = 7.660$$

and $$\mathbf{C} = \mathbf{-6.428 - j7.660}$$

14.7 MATHEMATICAL OPERATIONS WITH COMPLEX NUMBERS

The basic mathematical operations of addition, subtraction, multiplication, and division can be applied to complex numbers. A few basic rules and definitions must be understood before considering these operations.

Let us first examine the symbol j associated with imaginary numbers. By definition,

$$j = \sqrt{-1} \tag{14.27}$$

Therefore,

$$j^2 = -1 \tag{14.28}$$

and $$j^3 = j^2 j = -1j = -j$$

with $$j^4 = j^2 j^2 = (-1)(-1) = +1$$
$$j^5 = j$$

and so on. Also,

$$\frac{1}{j} = (1)\left(\frac{1}{j}\right) = \left(\frac{j}{j}\right)\left(\frac{1}{j}\right) = \frac{j}{j^2} = \frac{j}{-1}$$

and

$$\frac{1}{j} = -j \tag{14.29}$$

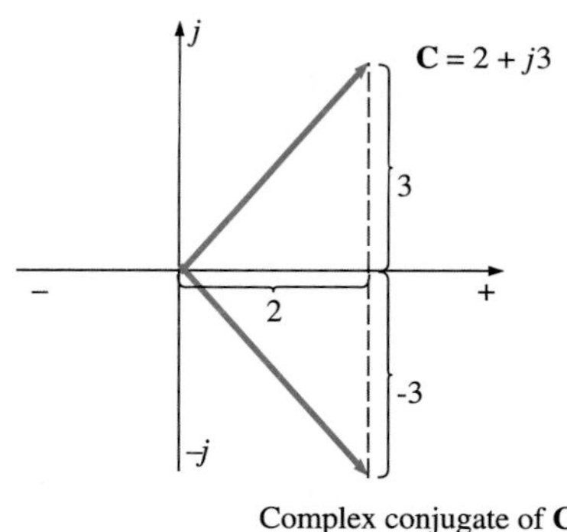

FIG. 14.39
Defining the complex conjugate of a complex number in rectangular form.

Complex Conjugate

The *conjugate* or *complex conjugate* of a complex number can be found by simply changing the sign of the imaginary part in the rectangular form or by using the negative of the angle of the polar form. For example, the conjugate of

$$\mathbf{C} = 2 + j3$$

is

$$2 - j3$$

as shown in Fig. 14.39. The conjugate of

$$\mathbf{C} = 2 \angle 30°$$

is

$$2 \angle -30°$$

as shown in Fig. 14.40.

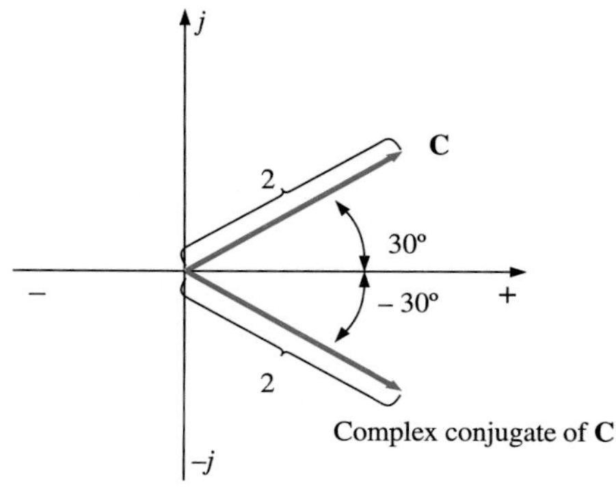

FIG. 14.40
Defining the complex conjugate of a complex number in polar form.

Reciprocal

The *reciprocal* of a complex number is 1 divided by the complex number. For example, the reciprocal of

$$\mathbf{C} = A + jB$$

is

$$\frac{1}{A + jB}$$

and of $C \angle \theta$,

$$\frac{1}{C \angle \theta}$$

We can now consider the four basic operations of *addition, subtraction, multiplication,* and *division* with complex numbers. These operations can be performed on many modern scientific calculators.

Addition

To add two or more complex numbers, simply add the real and imaginary parts separately. For example, if

$$\mathbf{C}_1 = \pm A_1 \pm jB_1 \text{ and } \mathbf{C}_2 = \pm A_2 \pm jB_2$$

then $$\mathbf{C}_1 + \mathbf{C}_2 = (\pm A_1 \pm A_2) + j(\pm B_1 \pm B_2) \quad (14.30)$$

You do not need to memorize the equation. Simply set one above the other and consider the real and imaginary parts separately, as shown in Example 14.19.

EXAMPLE 14.19

a. Add $\mathbf{C}_1 = 2 + j4$ and $\mathbf{C}_2 = 3 + j1$.
b. Add $\mathbf{C}_1 = 3 + j6$ and $\mathbf{C}_2 = -6 + j3$.

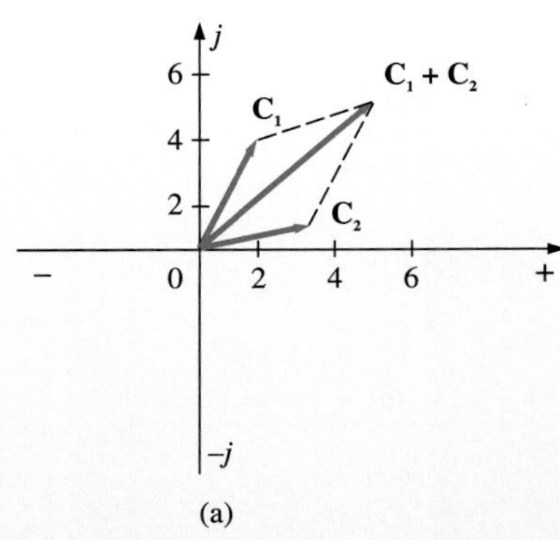

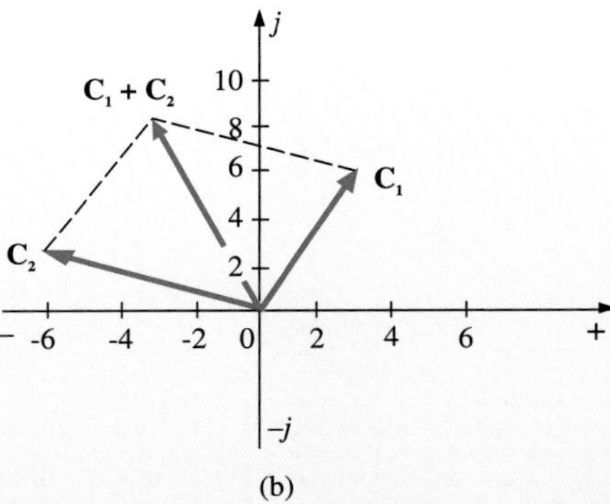

FIG. 14.41
Example 14.19.

Solutions:

a. By Eq. (14.30),

$$\mathbf{C}_1 + \mathbf{C}_2 = (2 + 3) + j(4 + 1) = \mathbf{5 + j5}$$

Note Fig. 14.41(a). An alternative method is

$$\begin{array}{c} 2 + j4 \\ \underline{3 + j1} \\ \downarrow \quad \downarrow \\ \mathbf{5 + j5} \end{array}$$

b. By Eq. (14.30),

$$\mathbf{C}_1 + \mathbf{C}_2 = (3 - 6) + j(6 + 3) = \mathbf{-3 + j9}$$

Note Fig. 14.41(b). An alternative method is

$$\begin{array}{c} 3 + j6 \\ \underline{-6 + j3} \\ \downarrow \quad \downarrow \\ \mathbf{-3 + j9} \end{array}$$

Subtraction

In subtraction, the real and imaginary parts are again considered separately. For example, if

$$\mathbf{C}_1 = \pm A_1 \pm jB_1 \text{ and } \mathbf{C}_2 = \pm A_2 \pm jB_2$$

then

$$\mathbf{C}_1 - \mathbf{C}_2 = [\pm A_2 - (\pm A_2)] + j[\pm B_1 - (\pm B_2)] \quad (14.31)$$

Again, you do not need to memorize the equation if you use the alternative method of Example 14.20.

EXAMPLE 14.20

a. Subtract $\mathbf{C}_2 = 1 + j4$ from $\mathbf{C}_1 = 4 + j6$.
b. Subtract $\mathbf{C}_2 = -2 + j5$ from $\mathbf{C}_1 = +3 + j3$.

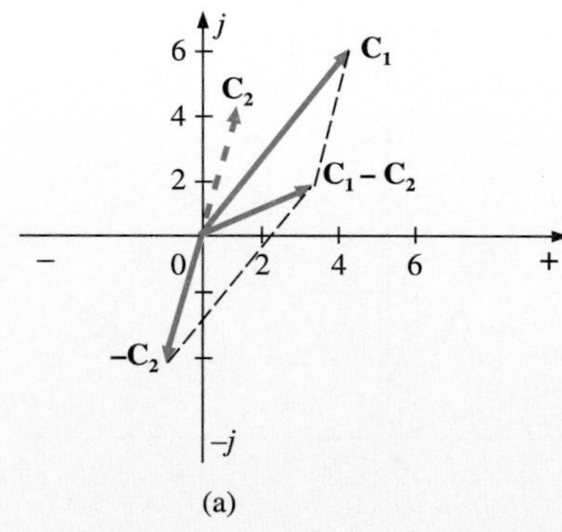

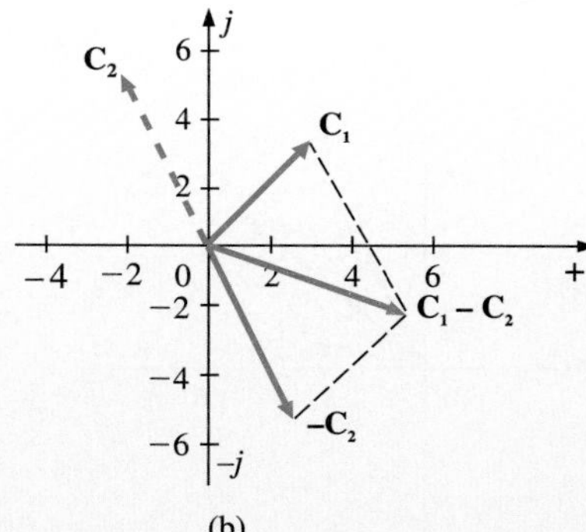

FIG. 14.42
Example 14.20.

Solutions:

a. By Eq. (14.31),

$$\mathbf{C}_1 - \mathbf{C}_2 = (4 - 1) + j(6 - 4) = \mathbf{3 + j2}$$

Note Fig. 14.42(a). An alternative method is

$$\begin{array}{r} 4 + j6 \\ \underline{-(1 + j4)} \\ \downarrow \quad \downarrow \\ \mathbf{3 + j2} \end{array}$$

b. By Eq. (14.31),

$$\mathbf{C}_1 - \mathbf{C}_2 = [3 - (-2)] + j(3 - 5) = \mathbf{5 - j2}$$

Note Fig. 14.42(b). An alternative method is

$$\begin{array}{r} 3 + j3 \\ \underline{-(-2 + j5)} \\ \downarrow \quad \downarrow \\ \mathbf{5 - j2} \end{array}$$

Note that $\mathbf{C}_1 - \mathbf{C}_2 = \mathbf{C}_1 + (-\mathbf{C}_2)$

Addition or subtraction cannot be performed in polar form unless the complex numbers have the same angle θ or differ only by multiples of 180°.

EXAMPLE 14.21

a. $2 \angle 45° + 3 \angle 45° = \mathbf{5 \angle 45°}$

Note Fig. 14.43(a). Or

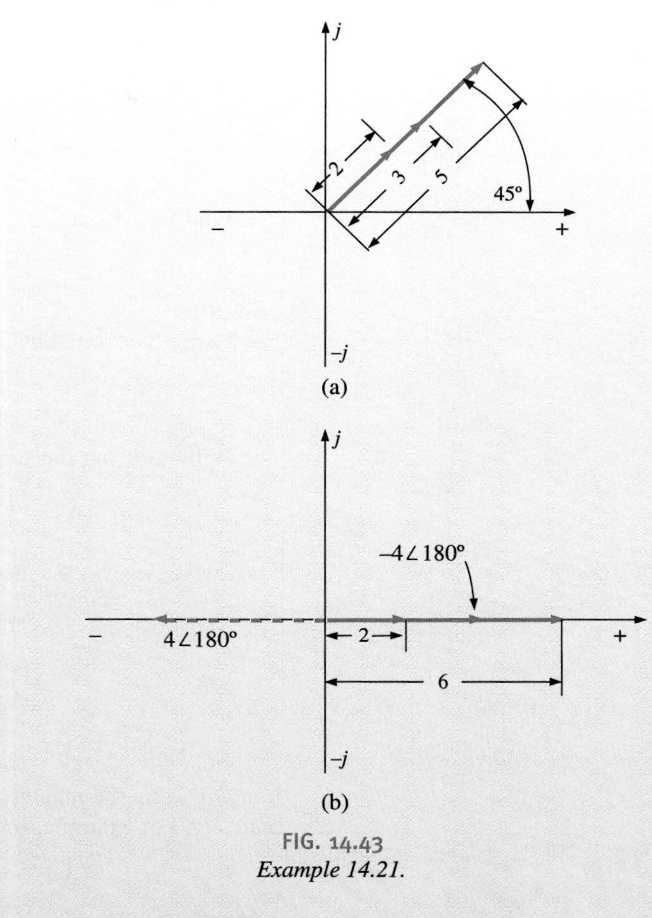

FIG. 14.43
Example 14.21.

b. $2 \angle 0° - 4 \angle 180° = \mathbf{6 \angle 0°}$

Note Fig. 14.43(b).

Multiplication

To multiply two complex numbers in *rectangular* form, multiply the real and imaginary parts of one in turn by the real and imaginary parts of the other. For example, if

$$\mathbf{C}_1 = A_1 + jB_1 \text{ and } \mathbf{C}_2 = A_2 + jB_2$$

then $\mathbf{C}_1 \cdot \mathbf{C}_2$:

$$\begin{array}{r} A_1 + jB_1 \\ A_2 + jB_2 \\ \hline A_1A_2 + jB_1A_2 \\ + jA_1B_2 + j^2B_1B_2 \\ \hline A_1A_2 + j(B_1A_2 + A_1B_2) + B_1B_2(-1) \end{array}$$

and
$$\mathbf{C}_1 \cdot \mathbf{C}_2 = (A_1A_2 - B_1B_2) + j\,(B_1A_2 + A_1B_2) \tag{14.32}$$

In Example 14.22(b), we obtain a solution without having to memorize Eq. (14.32). Simply carry along the *j* factor when multiplying each part of one vector with the real and imaginary parts of the other.

EXAMPLE 14.22

a. Find $\mathbf{C}_1 \cdot \mathbf{C}_2$ if

$$\mathbf{C}_1 = 2 + j3 \text{ and } \mathbf{C}_2 = 5 + j10$$

b. Find $\mathbf{C}_1 \cdot \mathbf{C}_2$ if

$$\mathbf{C}_1 = -2 - j3 \text{ and } \mathbf{C}_2 = +4 - j6$$

Solutions:

a. Using the format above, we have

$$\begin{aligned}\mathbf{C}_1 \cdot \mathbf{C}_2 &= [(2)(5) - (3)(10)] + j\,[(3)(5) + (2)(10)] \\ &= \mathbf{-20 + j35}\end{aligned}$$

b. Without using the format, we obtain

$$\begin{array}{l} -2 - j3 \\ +4 - j6 \\ \hline -8 - j12 \\ \quad + j12 + j^2 18 \\ \hline -8 + j(-12 + 12) - 18 \end{array}$$

and
$$\mathbf{C}_1 \cdot \mathbf{C}_2 = \mathbf{-26 = 26 \angle 180^\circ}$$

In *polar* form, the magnitudes are multiplied and the angles added algebraically. For example, for

$$\mathbf{C}_1 = C_1 \angle\theta_1 \text{ and } \mathbf{C}_2 = C_2 \angle\theta_2$$

we write

$$\mathbf{C}_1 \cdot \mathbf{C}_2 = C_1C_2 \angle\theta_1 + \theta_2 \tag{14.33}$$

EXAMPLE 14.23

a. Find $\mathbf{C}_1 \cdot \mathbf{C}_2$ if

$$\mathbf{C}_1 = 5 \angle 20^\circ \text{ and } \mathbf{C}_2 = 10 \angle 30^\circ$$

b. Find $\mathbf{C}_1 \cdot \mathbf{C}_2$ if

$$\mathbf{C}_1 = 2 \angle -40^\circ \text{ and } \mathbf{C}_2 = 7 \angle +120^\circ$$

Solutions:

a. $\mathbf{C}_1 \cdot \mathbf{C}_2 = (5 \angle 20^\circ)(10 \angle 30^\circ) = (5)(10) \angle 20^\circ + 30^\circ = \mathbf{50 \angle 50^\circ}$

b. $\mathbf{C}_1 \cdot \mathbf{C}_2 = (2 \angle -40^\circ)(7 \angle + 120^\circ) = (2)(7) \angle -40^\circ + 120^\circ$
$= \mathbf{14 \angle +80^\circ}$

To multiply a complex number in rectangular form by a real number, both the real part and the imaginary part must be multiplied by the real number. For example,

$$(10)(2 + j3) = 20 + j30$$

and $$50 \angle 0°(0 + j6) = j300 = 300 \angle 90°$$

Division

To divide two complex numbers in *rectangular* form, multiply the numerator and denominator by the conjugate of the denominator and collect the resulting real and imaginary parts. That is, if

$$\mathbf{C}_1 = A_1 + jB_1 \text{ and } \mathbf{C}_2 = A_2 + jB_2$$

then
$$\frac{\mathbf{C}_1}{\mathbf{C}_2} = \frac{(A_1 + jB_1)(A_2 - jB_2)}{(A_2 + jB_2)(A_2 - jB_2)}$$
$$= \frac{(A_1A_2 + B_1B_2) + j(A_2B_1 - A_1B_2)}{A_2^2 + B_2^2}$$

and
$$\frac{\mathbf{C}_1}{\mathbf{C}_2} = \frac{A_1A_2 + B_1B_2}{A_2^2 + B_2^2} + j\frac{A_2B_1 - A_1B_2}{A_2^2 + B_2^2} \qquad (14.34)$$

The equation does not have to be memorized if you use the steps above to obtain it. That is, first multiply the numerator by the complex conjugate of the denominator and separate the real and imaginary terms. Then divide each term by the sum of each term of the denominator squared.

EXAMPLE 14.24

a. Find $\mathbf{C}_1/\mathbf{C}_2$ if $\mathbf{C}_1 = 1 + j4$ and $\mathbf{C}_2 = 4 + j5$.
b. Find $\mathbf{C}_1/\mathbf{C}_2$ if $\mathbf{C}_1 = -4 - j8$ and $\mathbf{C}_2 = +6 - j1$.

Solutions:

a. By Eq. (14.34),

$$\frac{\mathbf{C}_1}{\mathbf{C}_2} = \frac{(1)(4) + (4)(5)}{4^2 + 5^2} + j\frac{(4)(4) - (1)(5)}{4^2 + 5^2}$$
$$= \frac{24}{41} + \frac{j11}{41} \cong \mathbf{0.585 + j\,0.268}$$

b. Using an alternative method, we obtain

$$\begin{array}{r} -4 - j8 \\ +6 + j1 \\ \hline -24 - j48 \\ -j4 - j^28 \\ \hline -24 - j52 + 8 = -16 - j52 \end{array}$$

$$\begin{array}{r} +6 - j1 \\ +6 + j1 \\ \hline 36 + j6 \\ -j6 - j^21 \\ \hline 36 + 0 + 1 = 37 \end{array}$$

and
$$\frac{\mathbf{C}_1}{\mathbf{C}_2} = \frac{-16}{37} - \frac{j52}{37} = \mathbf{-0.432 - j1.405}$$

To divide a complex number in rectangular form by a real number, both the real part and the imaginary part must be divided by the real number. For example,

$$\frac{8 + j10}{2} = 4 + j5$$

and

$$\frac{6.8 - j0}{2} = 3.4 - j0 = 3.4 \angle 0°$$

In *polar* form, division is accomplished by simply dividing the magnitude of the numerator by the magnitude of the denominator and subtracting the angle of the denominator from that of the numerator. That is, for

$$\mathbf{C}_1 = C_1 \angle \theta_1 \text{ and } \mathbf{C}_2 = C_2 \angle \theta_2$$

we write

$$\frac{\mathbf{C}_1}{\mathbf{C}_2} = \frac{C_1}{C_2} \angle \theta_1 - \theta_2 \qquad (14.35)$$

EXAMPLE 14.25

a. Find $\mathbf{C}_1/\mathbf{C}_2$ if $\mathbf{C}_1 = 15 \angle 10°$ and $\mathbf{C}_2 = 2 \angle 7°$.
b. Find $\mathbf{C}_1/\mathbf{C}_2$ if $\mathbf{C}_1 = 8 \angle 120°$ and $\mathbf{C}_2 = 16 \angle -50°$.

Solutions:

a. $\frac{\mathbf{C}_1}{\mathbf{C}_2} = \frac{15 \angle 10°}{2 \angle 7°} = \frac{15}{2} \angle 10° - 7° = \mathbf{7.5 \angle 3°}$

b. $\frac{\mathbf{C}_1}{\mathbf{C}_2} = \frac{8 \angle 120°}{16 \angle -50°} = \frac{8}{16} \angle 120° - (-50°) = \mathbf{0.5 \angle 170°}$

We obtain the *reciprocal* in the rectangular form by multiplying the numerator and denominator by the complex conjugate of the denominator:

$$\frac{1}{A + jB} = \left(\frac{1}{A + jB}\right)\left(\frac{A - jB}{A - jB}\right) = \frac{A - jB}{A^2 + B^2}$$

and

$$\frac{1}{A + jB} = \frac{A}{A^2 + B^2} - j\frac{B}{A^2 + B^2} \qquad (14.36)$$

In the polar form, the reciprocal is

$$\frac{1}{C \angle \theta} = \frac{1}{C} \angle -\theta \qquad (14.37)$$

A final example using the four basic operations follows.

EXAMPLE 14.26

Perform the following operations, leaving the answer in polar or rectangular form:

a. $\frac{(2 + j3) + (4 + j6)}{(7 + j7) - (3 - j3)} = \frac{(2 + 4) + j(3 + 6)}{(7 - 3) + j(7 + 3)}$

$$= \frac{(6 + j9)(4 - j10)}{(4 + j10)(4 - j10)}$$

continued

$$= \frac{[(6)(4) + (9)(10)] + j[(4)(9) - (6)(10)]}{4^2 + 10^2}$$

$$= \frac{114 - j24}{116} = \mathbf{0.983 - j0.207}$$

b. $$\frac{(50 \angle 30°)(5 + j5)}{10 \angle -20°} = \frac{(50 \angle 30°)(7.07 \angle 45°)}{10 \angle -20°} = \frac{353.5 \angle 75°}{10 \angle -20°}$$

$$= 35.35 \angle 75° - (-20°) = \mathbf{35.35 \angle 95°}$$

c. $$\frac{(2 \angle 20°)^2(3 + j4)}{8 - j6} = \frac{(2 \angle 20°)(2 \angle 20°)(5 \angle 53.13°)}{10 \angle -36.87°}$$

$$= \frac{(4 \angle 40°)(5 \angle 53.13°)}{10 \angle -36.87°} = \frac{20 \angle 93.13°}{10 \angle -36.87°}$$

$$= 2 \angle 93.13° - (-36.87°) = \mathbf{2.0 \angle 130°}$$

d. $$3 \angle 27° - 6 \angle -40° = (2.673 + j1.362) - (4.596 - j3.857)$$

$$= (2.673 - 4.596) + j(1.362 + 3.857)$$

$$= \mathbf{-1.923 + j5.219}$$

LUMINARIES

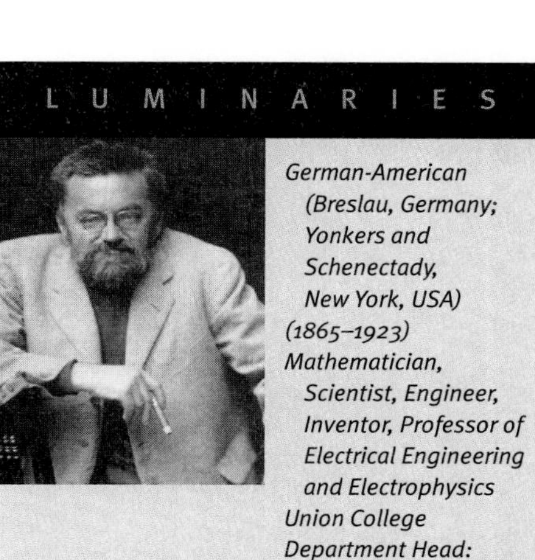

German-American (Breslau, Germany; Yonkers and Schenectady, New York, USA) (1865–1923) Mathematician, Scientist, Engineer, Inventor, Professor of Electrical Engineering and Electrophysics Union College Department Head: General Electric Co.

FIG. 14.44

Charles Proteus Steinmetz

Although he held 200 patents and was recognized worldwide for his contributions to the study of hysteresis losses and electrical transients, Charles Steinmetz is best known for his contribution to the study of ac networks. His "Symbolic Method of Alternating-current Calculations" provided an approach to the analysis of ac networks that removed much of the confusion and frustration experienced by engineers making the transition from dc to ac systems. His approach (which the phasor notation of this text is based on) permitted a direct analysis of ac systems using many of the theorems and methods of analysis developed for dc systems. In 1897 he wrote the epic work *Theory and Calculation of Alternating Current Phenomena*, which became the "bible" for practising engineers. He was fondly referred to as "The Doctor" at General Electric Company, where he worked for 30 years in a number of important roles. His recognition as a "multigifted genius" is supported by his friendships with Albert Einstein, Guglielmo Marconi, and Thomas A. Edison, among others. He was President of the American Institute of Electrical Engineers (AIEE) and the National Association of Corporation Schools and actively supported his local community (Schenectady) as president of the Board of Education and of the Commission on Parks and City Planning.

Courtesy of the Hall of History Foundation, Schenectady, New York

14.8 PHASORS

As noted earlier in this chapter, the addition of sinusoidal voltages and currents will frequently be required in the analysis of ac circuits (Fig. 14.44). One long but valid method of performing this operation is to place both sinusoidal waveforms on the same set of axes and add algebraically the magnitudes of each at every point along the horizontal axis, as shown for $c = a + b$ in Fig. 14.45. This, however, can be a long and slow process with limited accuracy. A shorter method uses the rotating radius vector first shown in Fig. 13.16. This *radius vector,* having a *constant magnitude* (length) with *one end fixed at the origin,* is called a *phasor* when applied to electric circuits. During its rotational development of the sine wave, the phasor will, at the instant $t = 0$, have the positions shown in Fig. 14.46(a) for each waveform in Fig. 14.46(b).

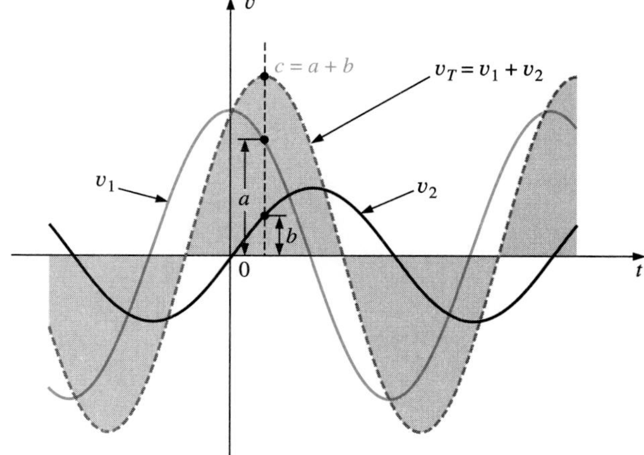

FIG. 14.45
Adding two sinusoidal waveforms on a point-by-point basis.

Note in Fig. 14.46(b) that v_2 passes through the horizontal axis at $t = 0$ s, requiring that the radius vector in Fig. 14.46(a) be on the horizontal axis to ensure a vertical projection of zero volts at $t = 0$ s. Its length in Fig. 14.46(a) is equal to the peak value of the sinusoid as required by the radius vector of Fig. 13.16. The other sinusoid has passed through 90° of its rotation by the time $t = 0$ s is reached and therefore has its maximum vertical projection as shown in Fig. 14.46(a). Since the vertical projection is a maximum, the peak value of the sinusoid that it will generate is also attained at $t = 0$ s, as shown in Fig. 14.46(b). Note also that $v_T = v_1$ at $t = 0$ s, since $v_2 = 0$ V at this instant.

It can be shown [see Fig. 14.46(a)] using the vector algebra described in Section 14.7 that

$$1 \text{ V} \angle 0° + 2 \text{ V} \angle 90° = 2.236 \text{ V} \angle 63.43°$$

In other words, if we convert v_1 and v_2 to the phasor form using

$$v = V_m \sin(\omega t \pm \theta) \Rightarrow V_m \angle \pm \theta$$

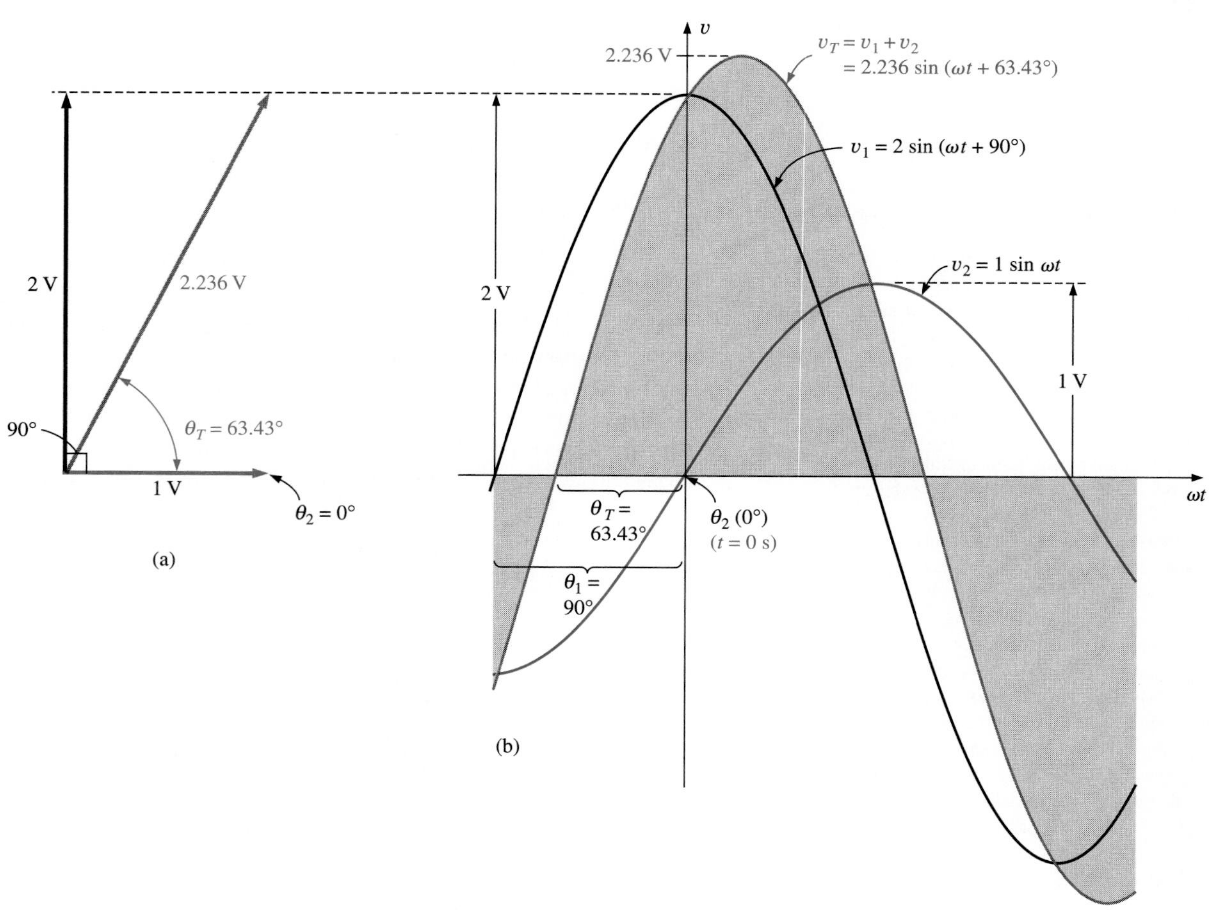

FIG. 14.46
(a) The phasor representation of the sinusoidal waveforms of Fig. 14.46(b); (b) finding the sum of two sinusoidal waveforms of v_1 and v_2.

and add them using complex number algebra, we can find the phasor form for v_T very easily. It can then be converted to the time domain and plotted on the same set of axes, as shown in Fig. 14.46(b). Figure 14.46(a), showing the magnitudes and relative positions of the various phasors, is called a *phasor diagram.* It is actually a "snapshot" of the rotating radius vectors at $t = 0$ s.

In the future, therefore, if you have to add two sinusoids, they should first be converted to the phasor domain and the sum found using complex algebra. The result can then be converted to the time domain.

The case of two sinusoidal functions having phase angles different from 0° and 90° appears in Fig. 14.47. Note again that the vertical height of the functions in Fig. 14.47(b) at $t = 0$ s is determined by the rotational positions of the radius vectors in Fig. 14.47(a).

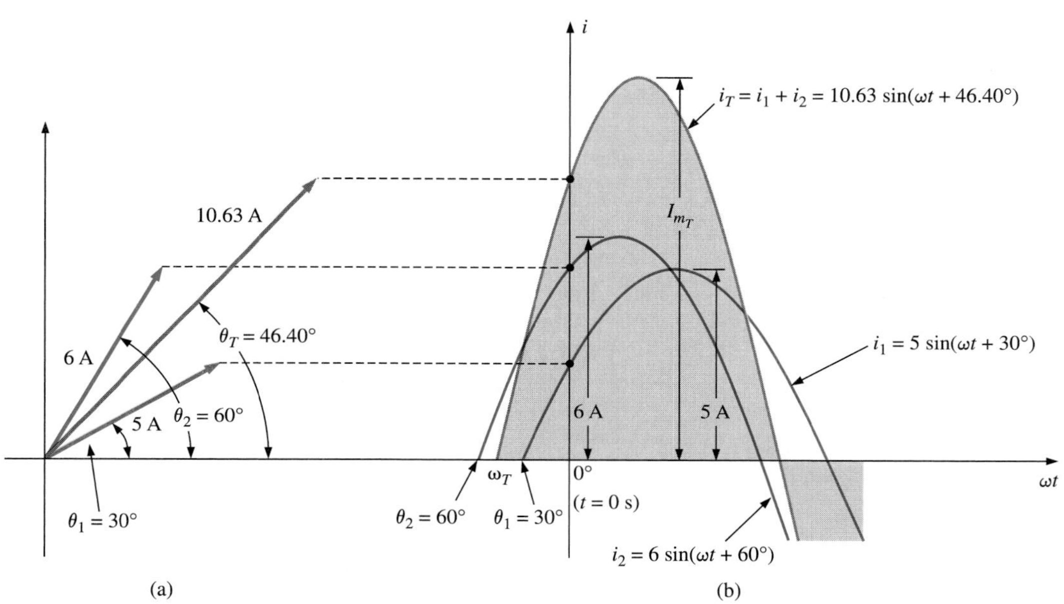

FIG. 14.47
Adding two sinusoidal currents with phase angles other than 90°.

Since the effective, rather than the peak, values are almost always used in the analysis of ac circuits, we will now redefine the phasor for the purposes of practicality and uniformity as having a magnitude equal to the *effective value* of the sine wave it represents. The angle associated with the phasor will remain as previously described—the phase angle.

In general, for all of the analyses to follow, the phasor form of a sinusoidal voltage or current will be

$$\mathbf{V} = V \angle \theta \text{ and } \mathbf{I} = I \angle \theta$$

where V and I are effective values and θ is the phase angle. In phasor notation, the sine wave is always the reference, and the frequency is not represented.

Phasor algebra for sinusoidal quantities is applicable only for waveforms having the same frequency.

EXAMPLE 14.27

Convert the following from the time to the phasor domain:

Time Domain	Phasor Domain
a. $\sqrt{2}(50)\sin\omega t$	$\mathbf{50 \angle 0^\circ}$
b. $69.6\sin(\omega t + 72^\circ)$	$(0.707)(69.6)\angle 72^\circ = \mathbf{49.21 \angle 72^\circ}$
c. $45\cos\omega t$	$(0.707)(45)\angle 90^\circ = \mathbf{31.82 \angle 90^\circ}$

EXAMPLE 14.28

Write the sinusoidal expression for the following phasors if the frequency is 60 Hz:

Phasor Domain	Time Domain
a. $\mathbf{I} = 10\angle 30^\circ$	$i = \sqrt{2}(10)\sin(2\pi 60t + 30^\circ)$ and $i = \mathbf{14.14\sin(377t + 30^\circ)}$
b. $\mathbf{V} = 115\angle -70^\circ$	$v = \sqrt{2}(115)\sin(377t - 70^\circ)$ and $v = \mathbf{162.6\sin(377t - 70^\circ)}$

EXAMPLE 14.29

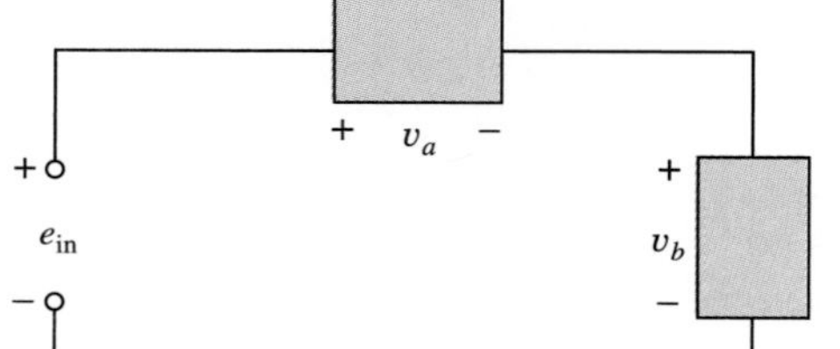

FIG. 14.48
Example 14.29.

Find the input voltage of the circuit of Fig. 14.48 if

$$\left.\begin{aligned} v_a &= 50\sin(377t + 30^\circ) \\ v_b &= 30\sin(377t + 60^\circ) \end{aligned}\right\} f = 60\text{ Hz}$$

Solution: Applying Kirchhoff's voltage law, we have

$$e_{in} = v_a + v_b$$

Converting from the time to the phasor domain gives

$$v_a = 50\sin(377t + 30^\circ) \Rightarrow \mathbf{V}_a = 35.35\text{ V}\angle 30^\circ$$
$$v_b = 30\sin(377t + 60^\circ) \Rightarrow \mathbf{V}_b = 21.21\text{ V}\angle 60^\circ$$

Converting from polar to rectangular form for addition gives

$$\mathbf{V}_a = 35.35\text{ V}\angle 30^\circ = 30.61\text{ V} + j17.68\text{ V}$$
$$\mathbf{V}_b = 21.21\text{ V}\angle 60^\circ = 10.61\text{ V} + j18.37\text{ V}$$

Then

$$\mathbf{E}_{in} = \mathbf{V}_a + \mathbf{V}_b = (30.61\text{ V} + j17.68\text{ V}) + (10.61\text{ V} + j18.37\text{ V}) = 41.22\text{ V} + j36.05\text{ V}$$

Converting from rectangular to polar form, we have

$$\mathbf{E}_{in} = 41.22\text{ V} + j36.05\text{ V} = 54.76\text{ V}\angle 41.17^\circ$$

Converting from the phasor to the time domain, we obtain

$$\mathbf{E}_{in} = 54.76\text{ V}\angle 41.17^\circ \Rightarrow e_{in} = \sqrt{2}(54.76)\sin(377t + 41.17^\circ)$$

and $$e_{in} = \mathbf{77.43\sin(377t + 41.17^\circ)}$$

continued

A plot of the three waveforms is shown in Fig. 14.49. Note that at each instant of time, the sum of the two waveforms does add up to e_{in}. At $t = 0$ ($\omega t = 0$), e_{in} is the sum of the two positive values, while at a value of ωt, almost midway between $\pi/2$ and π, the sum of the positive value of v_a and the negative value of v_b results in $e_{in} = 0$.

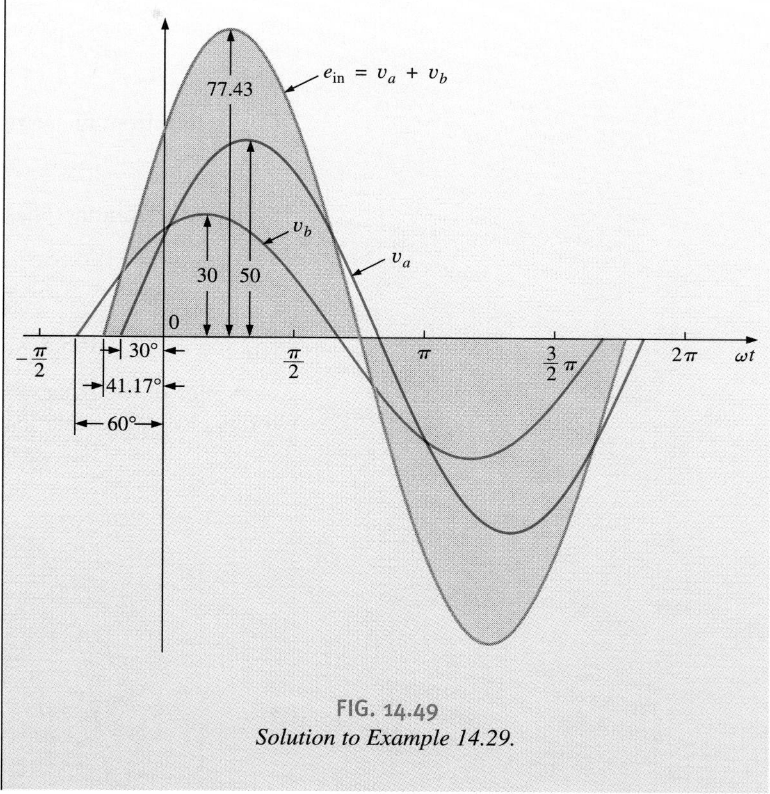

FIG. 14.49
Solution to Example 14.29.

EXAMPLE 14.30

Determine the current i_2 for the network of Fig. 14.50.

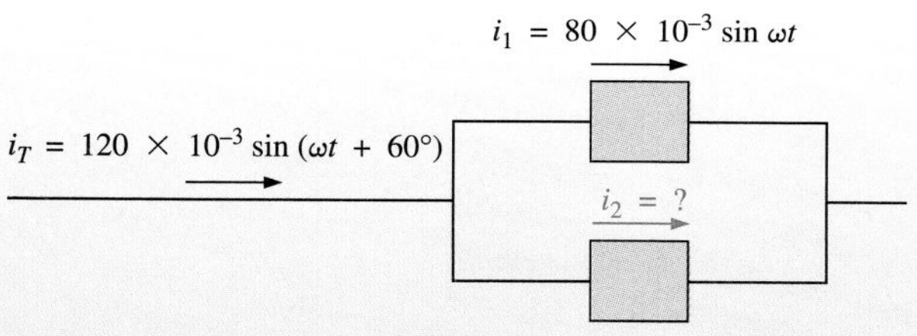

FIG. 14.50
Example 14.30.

Solution: Applying Kirchhoff's current law, we obtain

$$i_T = i_1 + i_2 \text{ or } i_2 = i_T - i_1$$

Converting from the time to the phasor domain gives

$$i_T = 120 \times 10^{-3} \sin(\omega t + 60°) \Rightarrow 84.84 \text{ mA } \angle 60°$$
$$i_1 = 80 \times 10^{-3} \sin \omega t \Rightarrow 56.56 \text{ mA } \angle 0°$$

continued

Converting from polar to rectangular form for subtraction gives

$$\mathbf{I}_T = 84.84 \text{ mA} \angle 60° = 42.42 \text{ mA} + j73.47 \text{ mA}$$
$$\mathbf{I}_1 = 56.56 \text{ mA} \angle 0° = 56.56 \text{ mA} + j0$$

Then

$$\mathbf{I}_2 = \mathbf{I}_T - \mathbf{I}_1$$
$$= (42.42 \text{ mA} + j73.47 \text{ mA}) - (56.56 \text{ mA} + j0)$$

and $\mathbf{I}_2 = -14.14 \text{ mA} + j73.47 \text{ mA}$

Converting from rectangular to polar form, we have

$$\mathbf{I}_2 = 74.82 \text{ mA} \angle 100.89°$$

Converting from the phasor to the time domain, we have

$$\mathbf{I}_2 = 74.82 \text{ mA} \angle 100.89° \Rightarrow$$
$$i_2 = \sqrt{2}(74.82 \times 10^{-3}) \sin(\omega t + 100.89°)$$

and $i_2 = \mathbf{105.8 \times 10^{-3} \sin(\omega t + 100.89°)}$

A plot of the three waveforms appears in Fig. 14.51. The waveforms clearly indicate that $i_T = i_1 + i_2$.

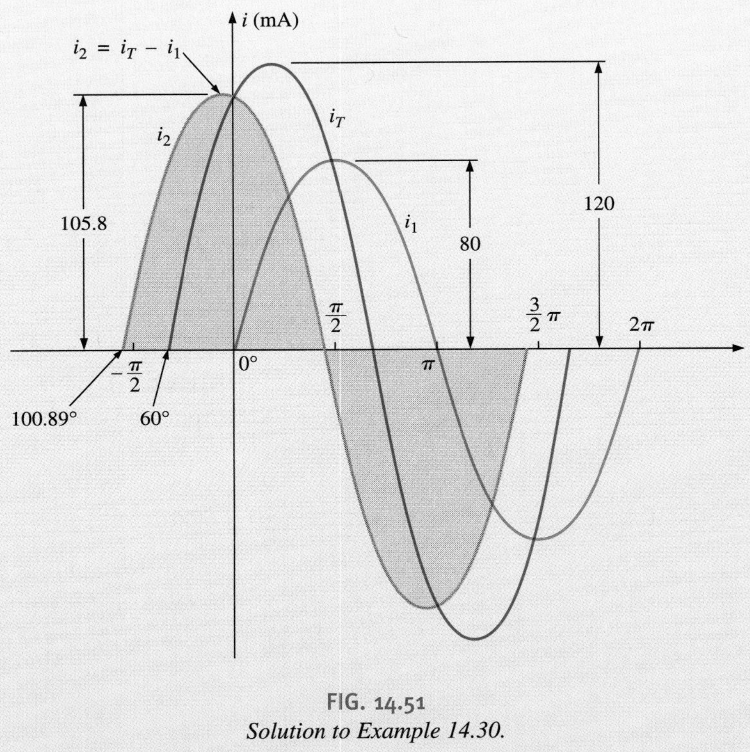

FIG. 14.51
Solution to Example 14.30.

PROBLEMS

SECTION 14.2 The Derivative

1. Plot the following waveform versus time showing one clear, complete cycle. Then determine the derivative of the waveform using Eq. (14.1), and sketch one complete cycle of the derivative directly under the original waveform. Compare the magnitude of the derivative at various points versus the slope of the original sinusoidal function.

$$v = 1 \sin 3.14t$$

2. Repeat Problem 1 for the following sinusoidal function and compare results. In particular, determine the frequency of the waveforms of Problems 1 and 2 and compare the magnitude of the derivative.

$$v = 1 \sin 15.71t$$

3. What is the derivative of each of the following sinusoidal expressions?
   a. $10 \sin 377t$ **b.** $0.6 \sin(754t + 20°)$
   c. $\sqrt{2}\, 20 \sin(157t - 20°)$ **d.** $-200 \sin(t + 180°)$

SECTION 14.3 Response of Basic *R, L,* and *C* Elements to a Sinusoidal Voltage or Current

4. The voltage across a 5-Ω resistor is as indicated. Find the sinusoidal expression for the current. In addition, sketch the v and i sinusoidal waveforms on the same axis.
   a. $150 \sin 377t$ **b.** $30 \sin(377t + 20°)$
   c. $40 \cos(\omega t + 10°)$ **d.** $-80 \sin(\omega t + 40°)$
5. The current through a 7-kΩ resistor is as indicated. Find the sinusoidal expression for the voltage. In addition, sketch the v and i sinusoidal waveforms on the same axis.
   a. $0.03 \sin 754t$
   b. $2 \times 10^{-3}\sin(400t - 120°)$
   c. $6 \times 10^{-6}\cos(\omega t - 2°)$
   d. $-0.004 \cos(\omega t - 90°)$
6. Determine the inductive reactance (in ohms) of a 2-H coil for
   a. dc
   and for the following frequencies:
   b. 25 Hz **c.** 60 Hz
   d. 2000 Hz **e.** 100 000 Hz
7. Determine the inductance of a coil that has a reactance of
   a. 20 Ω at f = 2 Hz.
   b. 1000 Ω at f = 60 Hz.
   c. 5280 Ω at f = 1000 Hz.
8. Determine the frequency at which a 10-H inductance has the following inductive reactances:
   a. 50 Ω **b.** 3770 Ω
   c. 15.7 kΩ **d.** 243 Ω
9. The current through a 20-Ω inductive reactance is given. What is the sinusoidal expression for the voltage? Sketch the v and i sinusoidal waveforms on the same axis.
   a. $i = 5 \sin \omega t$ **b.** $i = 0.4 \sin(\omega t + 60°)$
   c. $i = -6 \sin(\omega t - 30°)$ **d.** $i = 3 \cos(\omega t + 10°)$
10. The current through a 0.1-H coil is given. What is the sinusoidal expression for the voltage?
   a. $30 \sin 30t$
   b. $0.006 \sin 377t$
   c. $5 \times 10^{-6}\sin(400t + 20°)$
   d. $-4 \cos(20t - 70°)$
11. The voltage across a 50-Ω inductive reactance is given. What is the sinusoidal expression for the current? Sketch the v and i sinusoidal waveforms on the same axis.
   a. $50 \sin \omega t$ **b.** $30 \sin(\omega t + 20°)$
   c. $40 \cos(\omega t + 10°)$ **d.** $-80 \sin(377t + 40°)$
12. The voltage across a 0.2-H coil is given. What is the sinusoidal expression for the current?
   a. $1.5 \sin 60t$
   b. $0.016 \sin(t + 4°)$
   c. $-4.8 \sin(0.05t + 50°)$
   d. $9 \times 10^{-3} \cos(377t + 360°)$
13. Determine the capacitive reactance (in ohms) of a 5-μF capacitor for
   a. dc
   and for the following frequencies:
   b. 60 Hz **c.** 120 Hz
   d. 1800 Hz **e.** 24 000 Hz
14. Determine the capacitance in microfarads if a capacitor has a reactance of
   a. 250 Ω at f = 60 Hz.
   b. 55 Ω at f = 312 Hz.
   c. 10 Ω at f = 25 Hz.
15. Determine the frequency at which a 50-μF capacitor has the following capacitive reactances:
   a. 342 Ω **b.** 684 Ω
   c. 171 Ω **d.** 2000 Ω
16. The voltage across a 2.5-Ω capacitive reactance is given. What is the sinusoidal expression for the current? Sketch the v and i sinusoidal waveforms on the same axis.
   a. $100 \sin \omega t$ **b.** $0.4 \sin(\omega t + 20°)$
   c. $8 \cos(\omega t + 10°)$ **d.** $-70 \sin(\omega t + 40°)$
17. The voltage across a 1-μF capacitor is given. What is the sinusoidal expression for the current?
   a. $30 \sin 200t$ **b.** $90 \sin 377t$
   c. $-120 \sin(374t + 30°)$ **d.** $70 \cos(800t - 20°)$
18. The current through a 10-Ω capacitive reactance is given. Write the sinusoidal expression for the voltage. Sketch the v and i sinusoidal waveforms on the same axis.
   a. $i = 50 \sin \omega t$ **b.** $i = 40 \sin(\omega t + 60°)$
   c. $i = -6 \sin(\omega t - 30°)$ **d.** $i = 3 \cos(\omega t + 10°)$
19. The current through a 0.5-μF capacitor is given. What is the sinusoidal expression for the voltage?
   a. $0.20 \sin 300t$ **b.** $0.007 \sin 377t$
   c. $0.048 \cos 754t$ **d.** $0.08 \sin(1600t - 80°)$

*20. For the following pairs of voltages and currents, indicate whether the element involved is a capacitor, inductor, or resistor, and the value of C, L, or R if sufficient data are given:
   a. $v = 550 \sin(377t + 40°)$
   $i = 11 \sin(377t - 50°)$
   b. $v = 36 \sin(754t + 80°)$
   $i = 4 \sin(754t + 170°)$
   c. $v = 10.5 \sin(\omega t + 13°)$
   $i = 1.5 \sin(\omega t + 13°)$

*21. Repeat Problem 20 for the following pairs of voltages and currents:
   a. $v = 2000 \sin \omega t$
   $i = 5 \cos \omega t$
   b. $v = 80 \sin(157t + 150°)$
   $i = 2 \sin(157t + 60°)$
   c. $v = 35 \sin(\omega t - 20°)$
   $i = 7 \cos(\omega t - 110°)$

SECTION 14.4 Frequency Response of the Basic Elements

22. Plot X_L versus frequency for a 5-mH coil using a frequency range of zero to 100 kHz on a linear scale.

23. Plot X_C versus frequency for a 1-μF capacitor using a frequency range of zero to 10 kHz on a linear scale.

24. At what frequency will the reactance of a capacitor and resistance of a resistor be equal for a
   a. 1 μF capacitor and a 2 kΩ resistor?
   b. 3.3 μF capacitor and a 5.6 kΩ resistor?
   c. 0.15 μF capacitor and a 270 kΩ resistor?
   d. 0.68 μF capacitor and a 1.2 MΩ resistor?

25. Determine the required inductance of a coil that establishes an inductive reactance that matches a
   a. 10 kΩ resistor at a frequency of 5 kHz
   b. 270 kΩ resistor at a frequency of 400 kHz
   c. 1.8 kΩ resistor at a frequency of 3600 kHz
   d. 7.5 kΩ resistor at a frequency of 60 kHz

26. At what frequency will the reactance of a capacitor and an inductor be equal for a
   a. 1 μF capacitor and a 10 mH inductor?
   b. 3.3 μF capacitor and a 0.22 μH inductor?
   c. 0.15 μF capacitor and a 0.56 μH inductor?
   d. 0.68 μF capacitor and a 2.7 mH inductor?

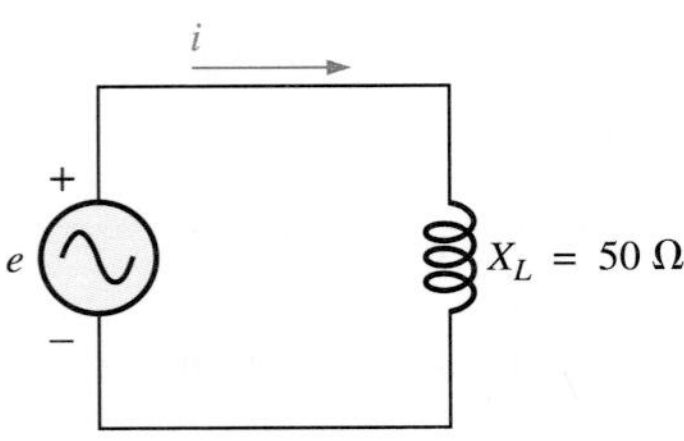

FIG. 14.53
Problem 35.

27. Determine the capacitance of a capacitor that establishes a capacitive reactance that matches a
   a. 2.0 mH inductor at a frequency of 50 kHz
   b. 0.82 μH inductor at a frequency of 600 kHz
   c. 0.27 mH inductor at a frequency of 1.6 kHz
   d. 1.5 mH inductor at a frequency of 7.2 kHz

SECTION 14.5 Average Power and Power Factor

28. Find the average power loss in watts for each set in Problem 20.

29. Find the average power loss in watts for each set in Problem 21.

*30. Find the average power loss and power factor for each of the circuits whose input current and voltage are as follows:
   a. $v = 60 \sin(\omega t + 30°)$
   $i = 15 \sin(\omega t + 60°)$
   b. $v = -50 \sin(\omega t - 20°)$
   $i = -2 \sin(\omega t + 40°)$
   c. $v = 50 \sin(\omega t + 80°)$
   $i = 3 \cos(\omega t + 20°)$
   d. $v = 75 \sin(\omega t - 5°)$
   $i = 0.08 \sin(\omega t - 35°)$

31. If the current through and voltage across an element are $i = 8 \sin(\omega t + 40°)$ and $v = 48 \sin(\omega t + 40°)$, respectively, compute the power by I^2R, $(V_m I_m/2) \cos \theta$, and $VI \cos \theta$, and compare answers.

32. A circuit dissipates 100 W (average power) at 150 V (effective input voltage) and 2 A (effective input current). What is the power factor? Repeat if the power is 0 W; 300 W.

*33. The power factor of a circuit is 0.5 lagging. The power delivered in watts is 500. If the input voltage is $50 \sin(\omega t + 10°)$, find the sinusoidal expression for the input current.

34. In Fig. 14.52, $e = 30 \sin(377t + 20°)$.
   a. What is the sinusoidal expression for the current?
   b. Find the power loss in the circuit.
   c. How long (in seconds) does it take the current to complete 6 cycles?

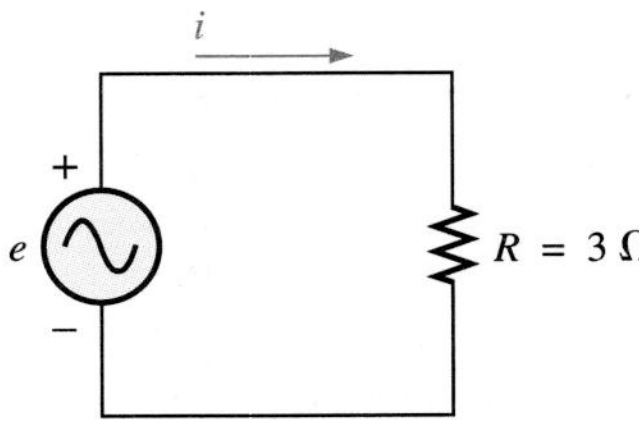

FIG. 14.52
Problem 34.

35. In Fig. 14.53, $e = 100 \sin(157t + 30°)$.
   a. Find the sinusoidal expression for i.
   b. Find the value of the inductance L.
   c. Find the average power loss by the inductor.

36. In Fig. 14.54, $i = 3 \sin(377t - 20°)$.
  a. Find the sinusoidal expression for e.
  b. Find the value of the capacitance C in microfarads.
  c. Find the average power loss in the capacitor.

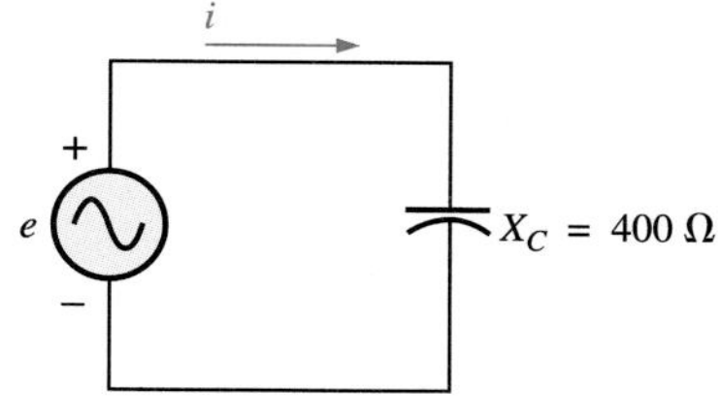

FIG. 14.54
Problem 36.

***37.** For the network of Fig. 14.55 and the applied signal:
  a. Determine i_1 and i_2.
  b. Find i_s.

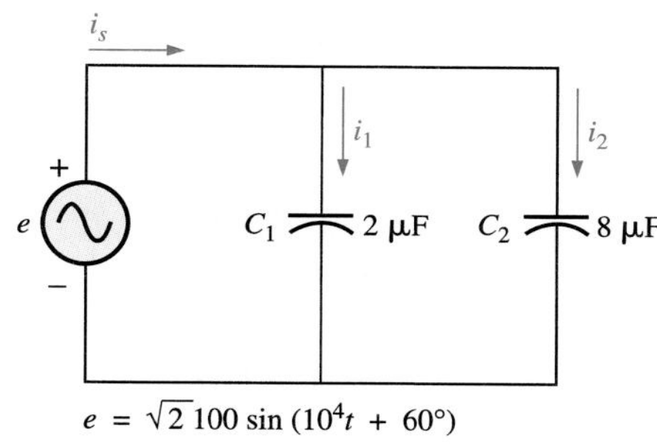

FIG. 14.55
Problem 37.

***38.** For the network of Fig. 14.56 and the applied source:
  a. Determine the source voltage v_s.
  b. Find the currents i_1 and i_2.

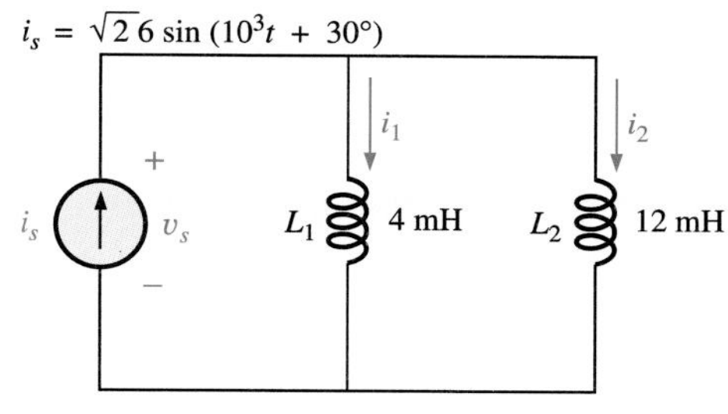

FIG. 14.56
Problem 38.

SECTION 14.6 Complex Numbers

39. Convert the following from rectangular to polar form:
  a. $4 + j3$ **b.** $2 + j2$
  c. $3.5 + j16$ **d.** $100 + j800$
  e. $1000 + j400$ **f.** $0.001 + j0.0065$
  g. $7.6 - j9$ **h.** $-8 + j4$
  i. $-15 - j60$ **j.** $+78 - j65$
  k. $-2400 + j3600$ **l.** $5 \times 10^{-3} - j25 \times 10^{-3}$

40. Convert the following from polar to rectangular form:
  a. $6 \angle 30°$ **b.** $40 \angle 80°$
  c. $7400 \angle 70°$ **d.** $4 \times 10^{-4} \angle 8°$
  e. $0.04 \angle 80°$ **f.** $0.0093 \angle 23°$
  g. $65 \angle 150°$ **h.** $1.2 \angle 135°$
  i. $500 \angle 200°$ **j.** $6320 \angle -35°$
  k. $7.52 \angle -125°$ **l.** $0.008 \angle 310°$

41. Convert the following from rectangular to polar form:

a. $1 + j15$ **b.** $60 + j5$
c. $0.01 + j0.3$ **d.** $100 - j2000$
e. $-5.6 + j86$ **f.** $-2.7 - j38.6$

42. Convert the following from polar to rectangular form:

a. $13 \angle 5°$ **b.** $160 \angle 87°$
c. $7 \times 10^{-6} \angle 2°$ **d.** $8.7 \angle 177°$
e. $76 \angle -4°$ **f.** $396 \angle +265°$

SECTION 14.7 Mathematical Operations with Complex Numbers

Perform the following operations.

43. Addition and subtraction (express your answers in rectangular form):

a. $(4.2 + j6.8) + (7.6 + j0.2)$
b. $(142 + j7) + (9.8 + j42) + (0.1 + j0.9)$
c. $(4 \times 10^{-6} + j76) + (7.2 \times 10^{-7} - j5)$
d. $(9.8 + j6.2) - (4.6 + j4.6)$
e. $(167 + j243) - (-42.3 - j68)$
f. $(-36.0 + j78) - (-4 - j6) + (10.8 - j72)$
g. $6 \angle 20° + 8 \angle 80°$
h. $42 \angle 45° + 62 \angle 60° - 70 \angle 120°$

44. Multiplication [express your answers in rectangular form for parts (a) through (d), and in polar form for parts (e) through (h)]:

a. $(2 + j3)(6 + j8)$
b. $(7.8 + j1)(4 + j2)(7 + j6)$
c. $(0.002 + j0.006)(-2 + j2)$
d. $(400 - j200)(-0.01 - j0.5)(-1 + j3)$
e. $(2 \angle 60°)(4 \angle 22°)$
f. $(6.9 \angle 8°)(7.2 \angle -72°)$
g. $0.002 \angle 120°)(0.5 \angle 200°)(40 \angle -60°)$
h. $(540 \angle -20°)(-5 \angle 180°)(6.2 \angle 0°)$

45. Division (express your answers in polar form):

a. $(42 \angle 10°)/(7 \angle 60°)$
b. $(0.006 \angle 120°)/(30 \angle -20°)$
c. $(4360 \angle -20°)/(40 \angle 210°)$
d. $(650 \angle -80°)/(8.5 \angle 360°)$
e. $(8 + j8)/(2 + j2)$
f. $(8 + j42)/(-6 + j60)$
g. $(0.05 + j0.25)/(8 - j60)$
h. $(-4.5 - j6)/(0.1 - j0.4)$

***46.** Perform the following operations (express your answers in rectangular form):

a. $\dfrac{(4 + j3) + (6 - j8)}{(3 + j3) - (2 + j3)}$

b. $\dfrac{8 \angle 60°}{(2 \angle 0°) + (100 + j100)}$

c. $\dfrac{(6 \angle 20°)(120 \angle -40°)(3 + j4)}{2 \angle -30°}$

d. $\dfrac{(0.4 \angle 60°)^2(300 \angle 40°)}{3 + j9}$

e. $\left(\dfrac{1}{(0.02 \angle 10°)^2}\right)\left(\dfrac{2}{j}\right)^3\left(\dfrac{1}{6^2 - j\sqrt{900}}\right)$

***47. a.** Determine a solution for x and y if

$$(x + j4) + (3x + jy) - j7 = 16 \angle 0°$$

b. Determine x if

$$(10 \angle 20°)(x \angle -60°) = 30.64 - j25.72$$

c. Determine a solution for x and y if

$$(5x + j10)(2 - jy) = 90 - j70$$

d. Determine θ if

$$\frac{80 \angle 0°}{20 \angle \theta} = 3.464 - j2$$

SECTION 14.8 Phasors

48. Express the following in phasor form:

a. $\sqrt{2}(100) \sin(\omega t + 30°)$
b. $\sqrt{2}(0.25) \sin(157t - 40°)$
c. $100 \sin(\omega t - 90°)$
d. $42 \sin(377t + 0°)$
e. $6 \times 10^{-6} \cos \omega t$
f. $3.6 \times 10^{-6} \cos(754t - 20°)$

49. Express the following phasor currents and voltages as sine waves if the frequency is 60 Hz:

a. $\mathbf{I} = 40 \text{ A} \angle 20°$ **b.** $\mathbf{V} = 120 \text{ V} \angle 0°$
c. $\mathbf{I} = 8 \times 10^{-3} \text{ A} \angle 120°$ **d.** $\mathbf{V} = 5 \text{ V} \angle 90°$
e. $\mathbf{I} = 1200 \text{ A} \angle -120°$ **f.** $\mathbf{V} = \dfrac{6000}{\sqrt{2}} \text{ V} \angle -180°$

50. For the system of Fig. 14.57, find the sinusoidal expression for the unknown voltage v_a if

$$e_{in} = 60 \sin(377t + 20°)$$
$$v_b = 20 \sin 377t$$

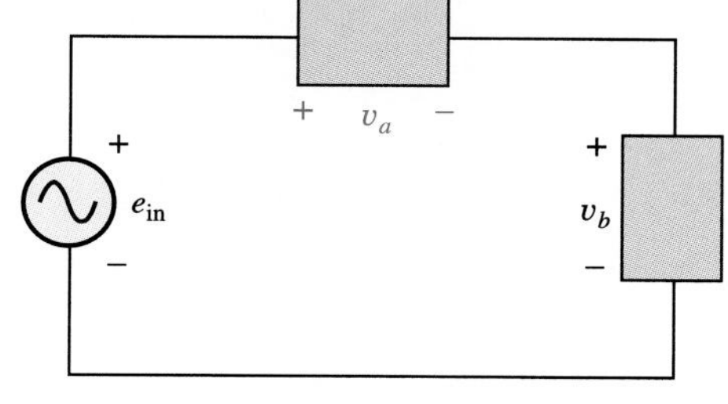

FIG. 14.57
Problem 50.

51. For the system of Fig. 14.58, find the sinusoidal expression for the unknown current i_1 if

$$i_s = 20 \times 10^{-6} \sin(\omega t + 90°)$$
$$i_2 = 6 \times 10^{-6} \sin(\omega t - 60°)$$

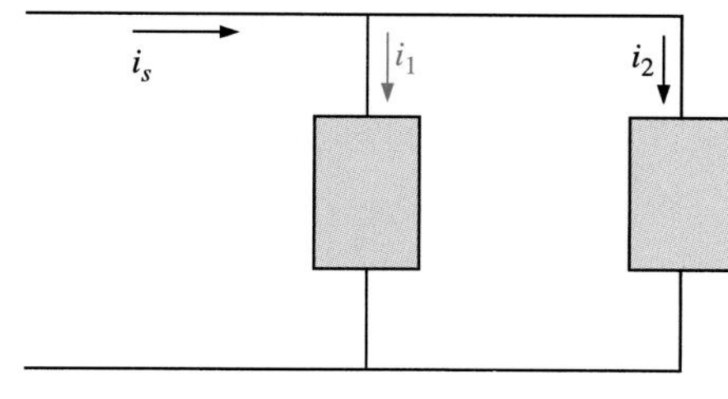

FIG. 14.58
Problem 51.

52. Find the sinusoidal expression for the applied voltage e for the system of Fig. 14.59 if

$$v_a = 60 \sin(\omega t + 30°)$$
$$v_b = 30 \sin(\omega t - 30°)$$
$$v_c = 40 \sin(\omega t + 120°)$$

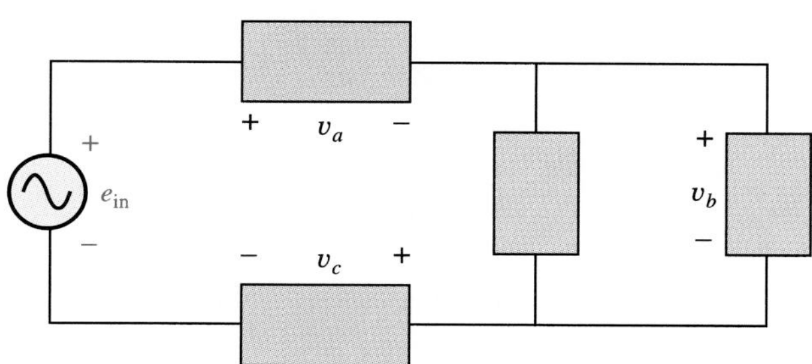

FIG. 14.59
Problem 52.

53. Find the sinusoidal expression for the current i_s for the system of Fig. 14.60 if

$$i_1 = 6 \times 10^{-3} \sin(377t + 180°)$$
$$i_2 = 8 \times 10^{-3} \sin 377t$$
$$i_3 = 2i_2$$

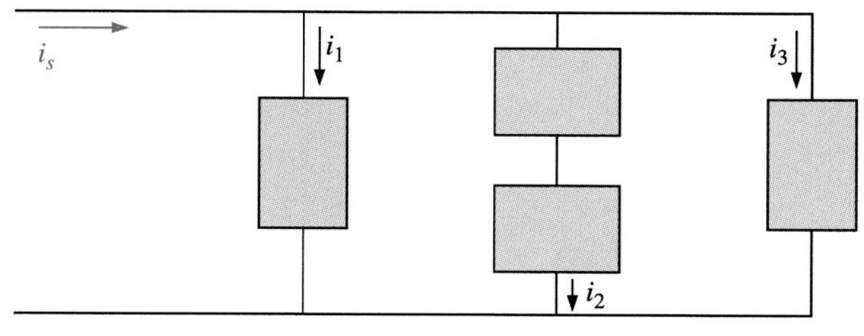

FIG. 14.60
Problem 53.

54. Repeat Problem 50 using the following values:

$$e_{in} = -15 \sin(377t + 30°)$$
$$v_b = 15 \sin 377t$$

55. Repeat Problem 50 using the following values:

$$e_{in} = 2.4 \sin(377t + 45°)$$
$$v_b = 3.6 \sin(377t - 12.5°)$$

56. Repeat Problem 51 using the following values:

$$i_s = 220 \sin(\omega t + 37°)$$
$$i_2 = 120 \sin(\omega t - 45°)$$

57. Repeat Problem 51 using the following values:

$$i_s = 7.4 \sin(\omega t + 14.3°)$$
$$i_2 = -10.4 \sin(\omega t - 17°)$$

15

Series and Parallel ac Circuits

Outline

Learning Outcomes

After completing this chapter you will be able to

- apply complex algebra to find the current and voltage through basic elements in a circuit
- calculate the equivalent resistance and reactance of a series ac circuit containing *R, L,* and *C* elements
- express the equivalent resistance and reactance in phasor notation for a series ac circuit
- sketch an impedance diagram for a series ac circuit
- apply Kirchhoff's voltage law to series ac circuits using phasor notation
- apply the voltage divider rule to ac circuits
- determine the frequency response of a series *R-C* circuit
- calculate the equivalent resistance and reactance of a parallel ac circuit containing *R, L,* and *C* elements
- express the equivalent resistance and reactance in phasor notation for a parallel ac network
- sketch an admittance diagram for a parallel ac network
- apply Kirchhoff's current law to parallel ac networks using phasor notation
- apply the current divider rule to ac networks
- determine the frequency response of a parallel *R-L* network

Series and parallel capactive, inductive, and resistive elements interact with each other making possible applications such as an electric machinery control panel.

15.1 INTRODUCTION

In this chapter, *phasor algebra* will be used to develop a quick, direct method for solving both the series and the parallel ac circuits. The close relationship between this method for solving for unknown quantities and the approach used for dc circuits will become obvious after a few simple examples are considered. Once this association is established, many of the rules (voltage divider rule, current divider rule, and so on) for dc circuits can be readily applied to ac circuits. Series ac circuits are analyzed in Sections 15.2 to 15.6; parallel ac circuits are analyzed in Sections 15.7 to 15.12.

15.2 IMPEDANCE AND THE PHASOR DIAGRAM

Resistive Elements

In Chapter 14, we found, for the purely resistive circuit of Fig. 15.1, that v and i were in phase, and the magnitude

$$I_m = \frac{V_m}{R} \quad \text{or} \quad V_m = I_m R$$

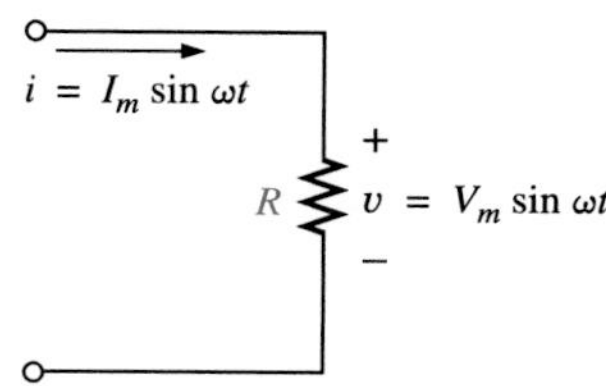

FIG. 15.1
Resistive ac circuit.

In phasor form,

$$v = V_m \sin \omega t \Rightarrow \mathbf{V} = V \angle 0^\circ$$

where $V = 0.707V_m$.

Applying Ohm's law and using phasor algebra, we have

$$\mathbf{I} = \frac{V \angle 0^\circ}{R \angle \theta_R} = \frac{V}{R} \angle 0^\circ - \theta_R$$

Since i and v are in phase, the angle associated with i also must be 0°. To satisfy this condition, θ_R must equal 0°. Substituting $\theta_R = 0°$, we find

$$\mathbf{I} = \frac{V \angle 0^\circ}{R \angle 0^\circ} = \frac{V}{R} \angle 0^\circ - 0^\circ = \frac{V}{R} \angle 0^\circ$$

so that in the time domain,

$$i = \sqrt{2}\left(\frac{V}{R}\right) \sin \omega t$$

The fact that $\theta_R = 0°$ will now be used in the following polar format to ensure the proper phase relationship between the voltage and current of a resistor.

$$\mathbf{Z}_R = R \angle 0^\circ \qquad (15.1)$$

The boldface roman quantity $\mathbf{Z}_R$, having both magnitude and an associated angle, is referred to as the **impedance** of a resistive element.

It is measured in ohms and is a measure of how much the element will "impede" the flow of charge through the network. The above format will prove to be a useful "tool" when the networks become more complex and phase relationships become less obvious. It is important to realize, however, that $\mathbf{Z}_R$ is *not a phasor,* even though the format $R \angle 0°$ is very similar to the phasor notation for sinusoidal currents and voltages. The term **phasor** is reserved for quantities that vary with time, and R and its associated angle of 0° are fixed, nonvarying quantities.

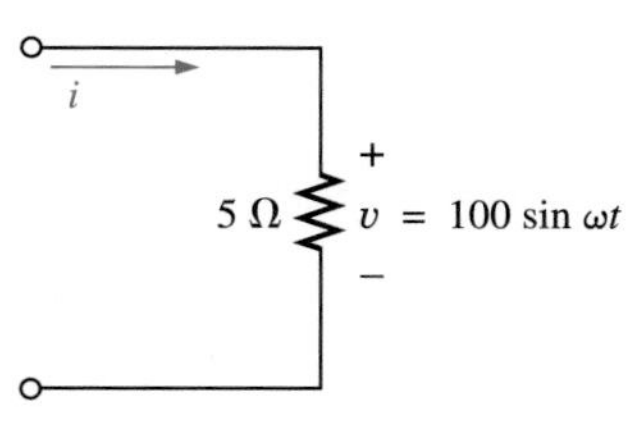

FIG. 15.2
Example 15.1.

EXAMPLE 15.1

Using complex algebra, find the current i for the circuit of Fig. 15.2. Sketch the waveforms of v and i.

Solution: Note Fig. 15.3:

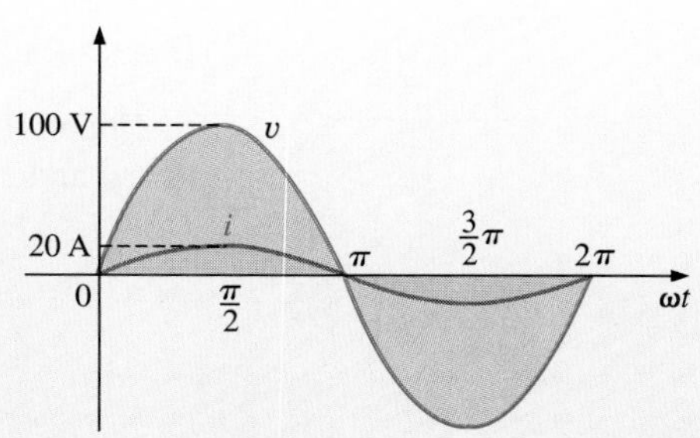

FIG. 15.3
Waveforms for Example 15.1.

$$v = 100 \sin \omega t \Rightarrow \text{phasor form } \mathbf{V} = 70.71 \text{ V} \angle 0°$$

$$\mathbf{I} = \frac{\mathbf{V}}{\mathbf{Z}_R} = \frac{V \angle \theta}{R \angle 0°} = \frac{70.71 \text{ V} \angle 0°}{5 \ \Omega \angle 0°} = 14.14 \text{ A} \angle 0°$$

and $$i = \sqrt{2}(14.14) \sin \omega t = \mathbf{20 \sin \omega t}$$

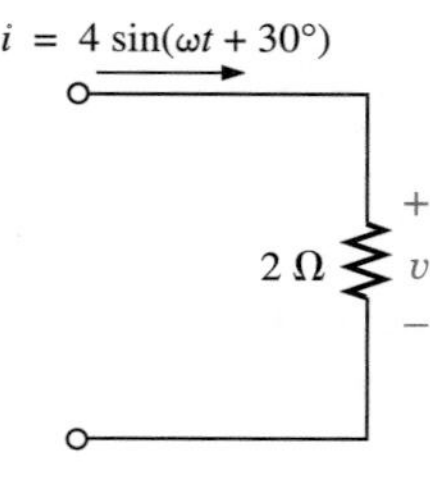

FIG. 15.4
Example 15.2.

EXAMPLE 15.2

Using complex algebra, find the voltage v for the circuit of Fig. 15.4. Sketch the waveforms of v and i.

Solution: Note Fig. 15.5:

$$i = 4 \sin(\omega t + 30°) \Rightarrow \text{phasor form } \mathbf{I} = 2.828 \text{ A} \angle 30°$$

$$\mathbf{V} = \mathbf{IZ}_R = (I \angle \theta)(R \angle 0°) = (2.828 \text{ A} \angle 30°)(2 \ \Omega \angle 0°)$$
$$= 5.656 \text{ V} \angle 30°$$

and $$v = \sqrt{2}(5.656) \sin(\omega t + 30°) = \mathbf{8.0 \sin(\omega t + 30°)}$$

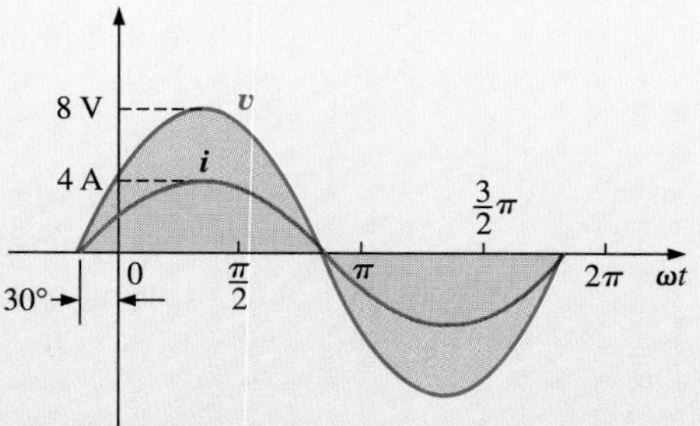

FIG. 15.5
Waveforms for Example 15.2.

It is often helpful when analyzing networks to have a **phasor diagram**, which shows at a glance the *magnitudes* and *phase relations* among the various quantities within the network. For example, the phasor diagrams of the circuits considered in the preceding examples would be as shown in Fig. 15.6. In both cases, it is immediately obvious that v and i are in phase since they both have the same phase angle.

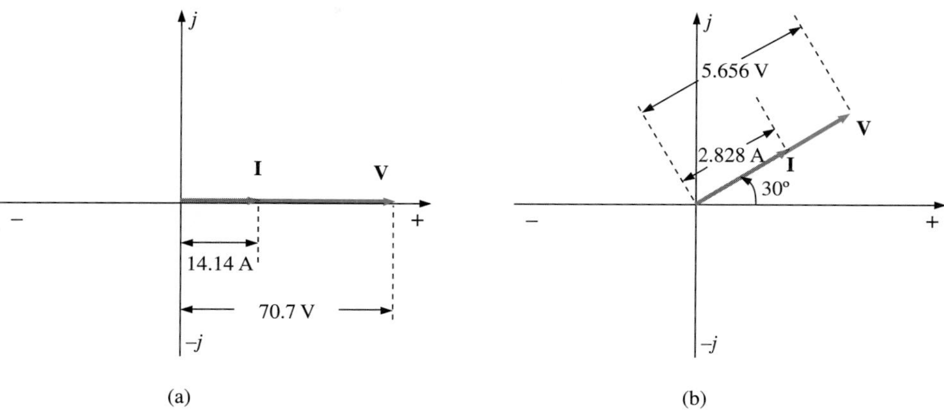

FIG. 15.6
Phasor diagrams for Examples 15.1 and 15.2.

Inductive Reactance

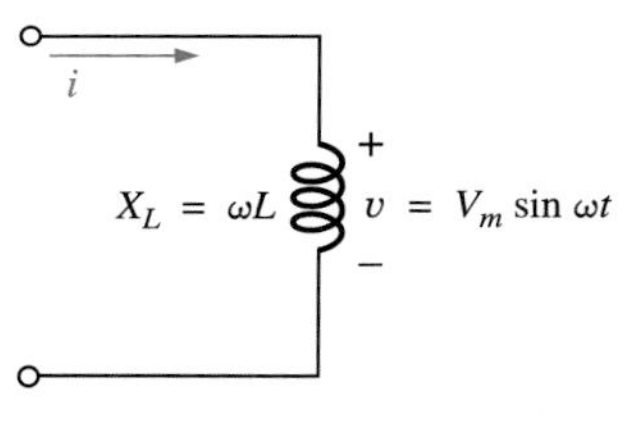

FIG. 15.7
Inductive ac circuit.

For the pure inductor of Fig. 15.7, you learned in Chapter 13 that the voltage leads the current by 90°, and that the reactance of the coil X_L is determined by ωL.

$$v = V_m \sin \omega t \Rightarrow \text{phasor form } \mathbf{V} = V \angle 0^\circ$$

By Ohm's law,

$$\mathbf{I} = \frac{V \angle 0^\circ}{X_L \angle \theta_L} = \frac{V}{X_L} \underline{/0^\circ - \theta_L}$$

Since v leads i by 90°, i must have an angle of $-90°$ associated with it. To satisfy this condition, θ_L must equal $+90°$. Substituting $\theta_L = 90°$, we obtain

$$\mathbf{I} = \frac{V \angle 0^\circ}{X_L \angle 90^\circ} = \frac{V}{X_L} \underline{/0^\circ - 90^\circ} = \frac{V}{X_L} \angle -90^\circ$$

so that in the time domain,

$$i = \sqrt{2}\left(\frac{V}{X_L}\right)\sin(\omega t - 90^\circ)$$

Note that $\theta_L = 90°$ will be used in the following polar format for inductive reactance to ensure the proper phase relationship between the voltage and current of an inductor.

$$\mathbf{Z}_L = X_L \angle 90^\circ \tag{15.2}$$

The boldface roman quantity $\mathbf{Z}_L$, having both magnitude and an associated angle, is referred to as the **impedance** of an inductive element. It is measured in ohms and is a measure of how much the inductive element will "control or impede" the level of current through the network (always keep in mind that inductive elements are storage devices and do not dissipate like resistors). The above format, like that defined for the resistive element, will prove to be a useful "tool" in the analysis of ac networks. Again, be aware that $\mathbf{Z}_L$ is not a phasor quantity, for the same reasons indicated for a resistive element.

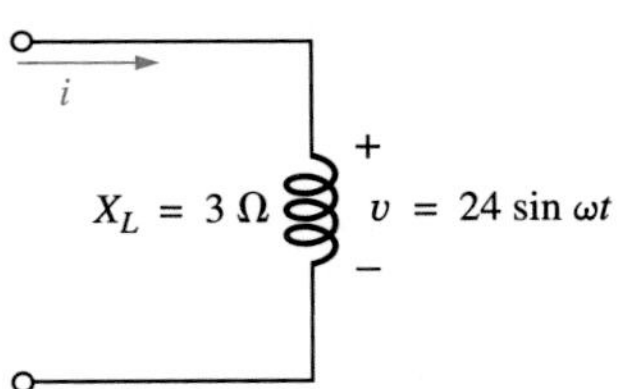

FIG. 15.8
Example 15.3.

EXAMPLE 15.3

Using complex algebra, find the current i for the circuit of Fig. 15.8. Sketch the v and i curves.

Solution: Note Fig. 15.9:

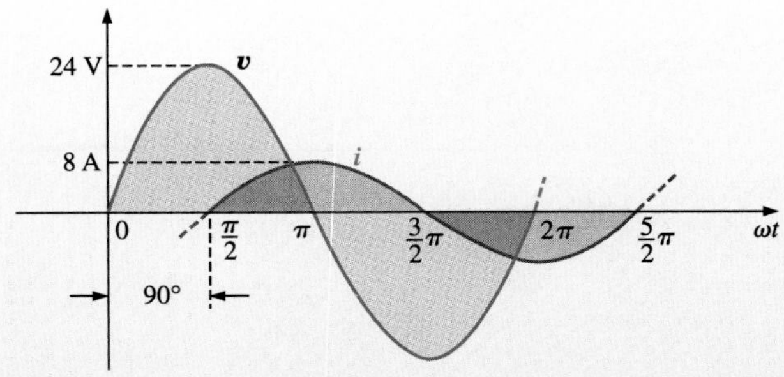

FIG. 15.9
Waveforms for Example 15.3.

$$v = 24 \sin \omega t \Rightarrow \text{phasor form } \mathbf{V} = 16.968\ \text{V} \angle 0^\circ$$

$$\mathbf{I} = \frac{\mathbf{V}}{\mathbf{Z}_L} = \frac{V \angle \theta}{X_L \angle 90^\circ} = \frac{16.968\ \text{V} \angle 0^\circ}{3\ \Omega \angle 90^\circ} = 5.656\ \text{A} \angle -90^\circ$$

and $\quad i = \sqrt{2}(5.656) \sin(\omega t - 90^\circ) = \mathbf{8.0 \sin(\omega t - 90^\circ)}$

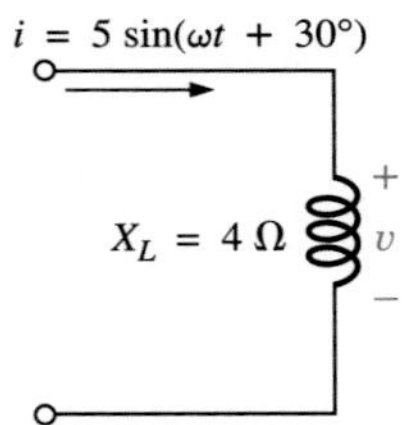

FIG. 15.10
Example 15.4.

EXAMPLE 15.4

Using complex algebra, find the voltage v for the circuit of Fig. 15.10. Sketch the v and i curves.

Solution: Note Fig. 15.11:

$$i = 5 \sin(\omega t + 30^\circ) \Rightarrow \text{phasor form } \mathbf{I} = 3.535\ \text{A} \angle 30^\circ$$

$$\mathbf{V} = \mathbf{I}\mathbf{Z}_L = (I \angle \theta)(X_L \angle 90^\circ) = (3.535\ \text{A} \angle 30^\circ)(4\ \Omega \angle +90^\circ)$$
$$= 14.140\ \text{V} \angle 120^\circ$$

and $\quad v = \sqrt{2}(14.140) \sin(\omega t + 120^\circ) = \mathbf{20 \sin(\omega t + 120^\circ)}$

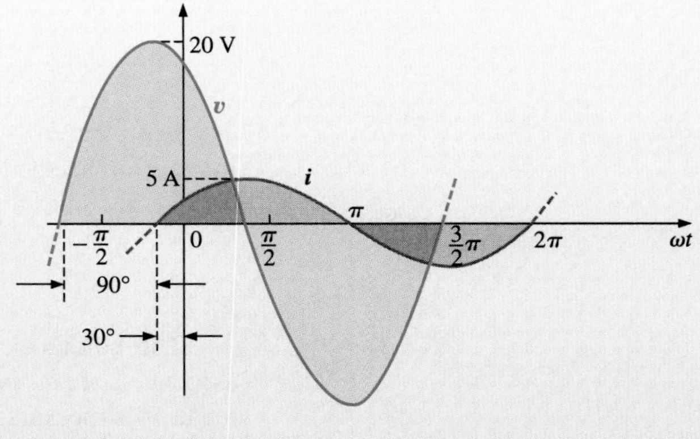

FIG. 15.11
Waveforms for Example 15.4.

The phasor diagrams for the two circuits of the preceding examples are shown in Fig. 15.12. Both indicate quite clearly that the voltage leads the current by 90°.

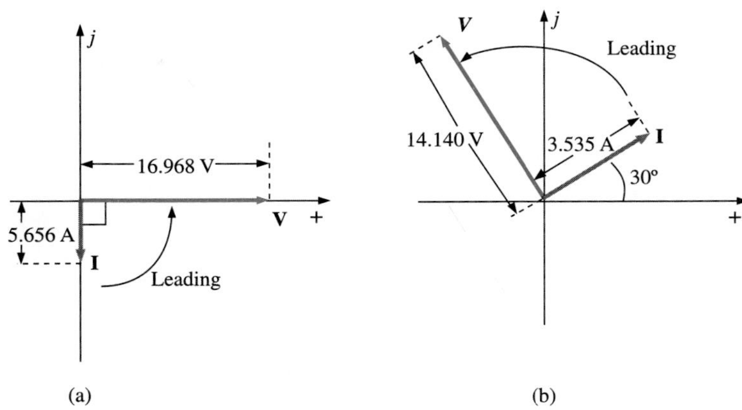

FIG. 15.12
Phasor diagrams for Examples 15.3 and 15.4.

Capacitive Reactance

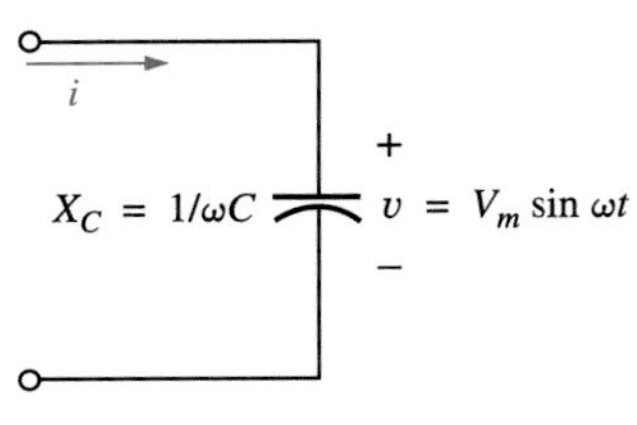

FIG. 15.13
Capacitive ac circuit.

For the pure capacitor of Fig. 15.13, you learned in Chapter 13 that the current leads the voltage by 90°, and that the reactance of the capacitor X_C is determined by $1/\omega C$.

$$v = V_m \sin \omega t \Rightarrow \text{phasor form } \mathbf{V} = V \angle 0^\circ$$

Applying Ohm's law and using phasor algebra, we find

$$\mathbf{I} = \frac{V \angle 0^\circ}{X_C \angle \theta_C} = \frac{V}{X_C} \underline{/0^\circ - \theta_C}$$

Since we know i leads v by 90°, i must have an angle of $+90°$ associated with it. To satisfy this condition, θ_C must equal $-90°$. Substituting $\theta_C = -90°$ gives us

$$\mathbf{I} = \frac{V \angle 0^\circ}{X_C \angle -90^\circ} = \frac{V}{X_C} \underline{/0^\circ - (-90^\circ)} = \frac{V}{X_C} \angle 90^\circ$$

so, in the time domain,

$$i = \sqrt{2}\left(\frac{V}{X_C}\right) \sin(\omega t + 90^\circ)$$

Note that $\theta_C = -90°$ will be used in the following polar format for capacitive reactance to ensure the proper phase relationship between the voltage and current of a capacitor.

$$\mathbf{Z}_C = X_C \angle -90^\circ \qquad (15.3)$$

The boldface roman quantity $\mathbf{Z}_C$, having both magnitude and an associated angle, is referred to as the **impedance** of a capacitive element. It is measured in ohms and is a measure of how much the capacitive element will "control or impede" the level of current through the network (always keep in mind that capacitive elements are storage devices and do not dissipate like resistors). The above format, like that defined for the resistive element, is a very useful "tool" in the analysis of ac networks. Again, be aware that $\mathbf{Z}_C$ is not a phasor quantity, for the same reasons indicated for a resistive element.

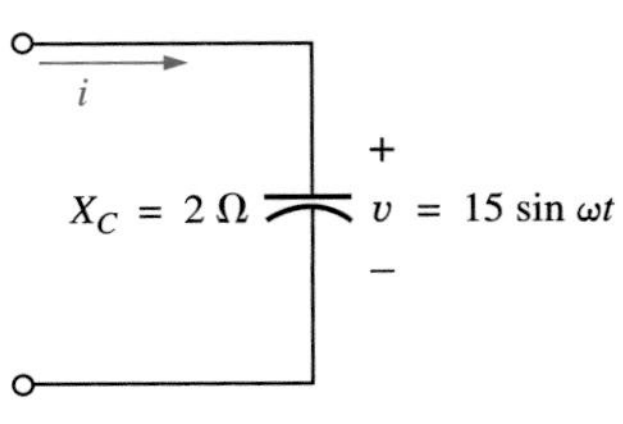

FIG. 15.14
Example 15.5.

EXAMPLE 15.5

Using complex algebra, find the current i for the circuit of Fig. 15.14. Sketch the v and i curves.

Solution: Note Fig. 15.15:

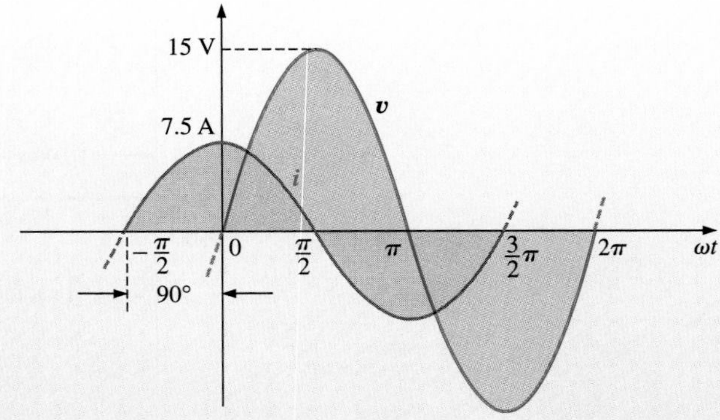

FIG. 15.15
Waveforms for Example 15.5.

$$v = 15 \sin \omega t \Rightarrow \text{phasor notation } \mathbf{V} = 10.605\text{ V} \angle 0^\circ$$

$$\mathbf{I} = \frac{\mathbf{V}}{\mathbf{Z}_C} = \frac{V \angle \theta}{X_C \angle -90^\circ} = \frac{10.605\text{ V} \angle 0^\circ}{2\ \Omega \angle -90^\circ} = 5.303\text{ A} \angle 90^\circ$$

and $\quad i = \sqrt{2}(5.303) \sin(\omega t + 90^\circ) = \mathbf{7.5 \sin(\omega t + 90^\circ)}$

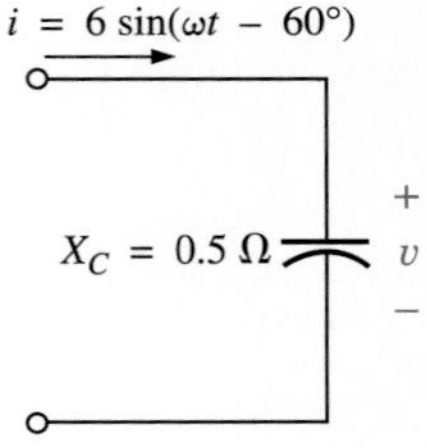

FIG. 15.16
Example 15.6.

EXAMPLE 15.6

Using complex algebra, find the voltage v for the circuit of Fig. 15.16. Sketch the v and i curves.

Solution: Note Fig. 15.17:

$$i = 6 \sin(\omega t - 60^\circ) \Rightarrow \text{phasor notation } \mathbf{I} = 4.242\text{ A} \angle -60^\circ$$

$$\mathbf{V} = \mathbf{I}\mathbf{Z}_C = (I \angle \theta)(X_C \angle -90^\circ) = (4.242\text{ A} \angle -60^\circ)(0.5\ \Omega \angle -90^\circ)$$
$$= 2.121\text{ V} \angle -150^\circ$$

and $\quad v = \sqrt{2}(2.121) \sin(\omega t - 150^\circ) = \mathbf{3.0 \sin(\omega t - 150^\circ)}$

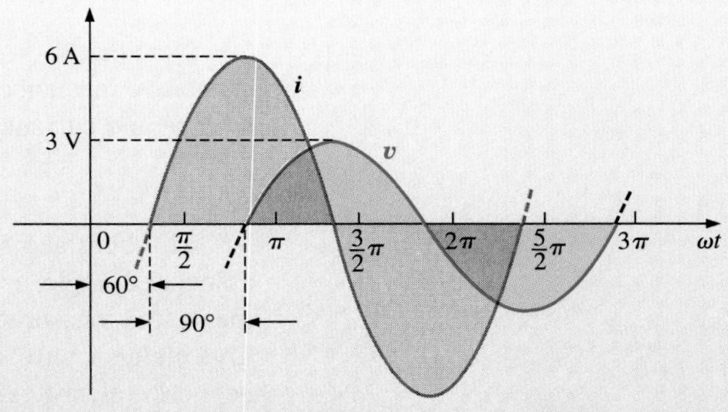

FIG. 15.17
Waveforms for Example 15.6.

The phasor diagrams for the two circuits of the preceding examples are shown in Fig. 15.18. Both indicate quite clearly that the current i leads the voltage v by 90°.

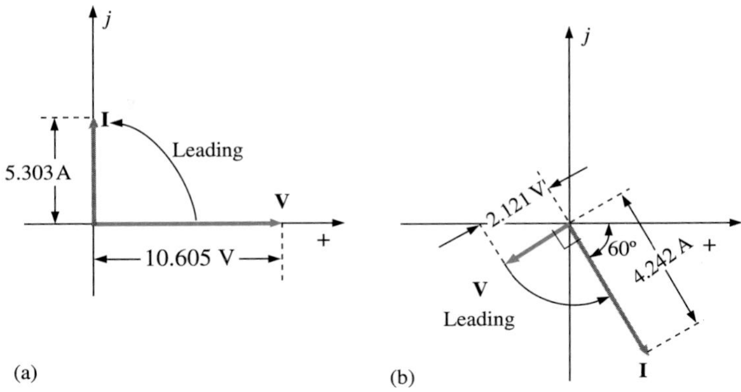

FIG. 15.18
Phasor diagrams for Examples 15.5 and 15.6.

Impedance Diagram

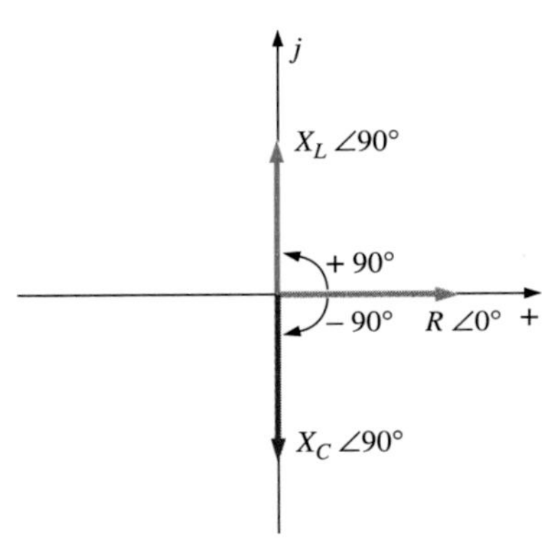

FIG. 15.19
Impedance diagram.

Now that an angle is associated with resistance, inductive reactance, and capacitive reactance, each can be placed on a complex plane diagram, as shown in Fig. 15.19. For any network, the resistance will *always* appear on the positive real axis, the inductive reactance on the positive imaginary axis, and the capacitive reactance on the negative imaginary axis. The result is an *impedance diagram* that can reflect the individual and total impedance levels of an ac network.

We will find in the sections and chapters to follow that networks combining different types of elements will have total impedances that extend from $-90°$ to $+90°$. If the total impedance has an angle of 0°, it is said to be resistive in nature. If it is closer to 90°, it is inductive in nature; and if it is closer to $-90°$, it is capacitive in nature.

Of course, for single-element networks the angle associated with the impedance will be the same as that of the resistive or reactive element, as revealed by Eqs. (15.1) to (15.3). It is important to remember that impedance, like resistance or reactance, is not a phasor quantity representing a time-varying function with a particular phase shift. It is simply an operating "tool" that is extremely useful in determining the magnitude and angle of quantities in a sinusoidal ac network.

Once the total impedance of a network is determined, its magnitude will define the resulting current level (through Ohm's law), and its angle will reveal whether the network is primarily inductive or capacitive or simply resistive.

For any configuration (series, parallel, series-parallel, etc.), the angle associated with the total impedance is the angle by which the applied voltage leads the source current. For inductive networks, θ_T will be positive, whereas for capacitive networks, θ_T will be negative.

15.3 SERIES CONFIGURATION

The overall properties of series ac circuits (Fig. 15.20) are the same as those for dc circuits. For instance, the total impedance of a system is the sum of the individual impedances:

$$\mathbf{Z}_T = \mathbf{Z}_1 + \mathbf{Z}_2 + \mathbf{Z}_3 + \cdots + \mathbf{Z}_N \tag{15.4}$$

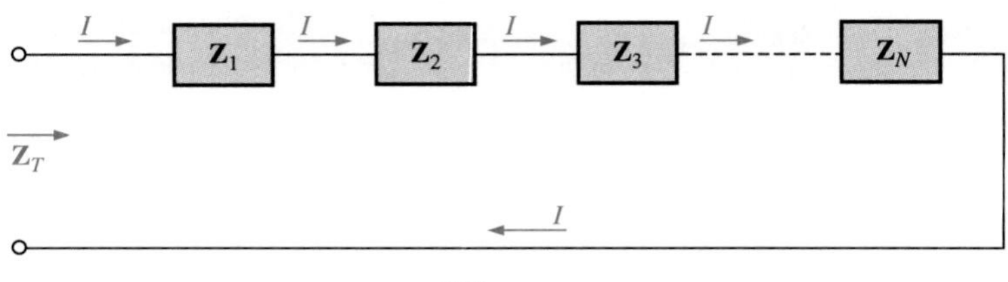

FIG. 15.20
Series impedances.

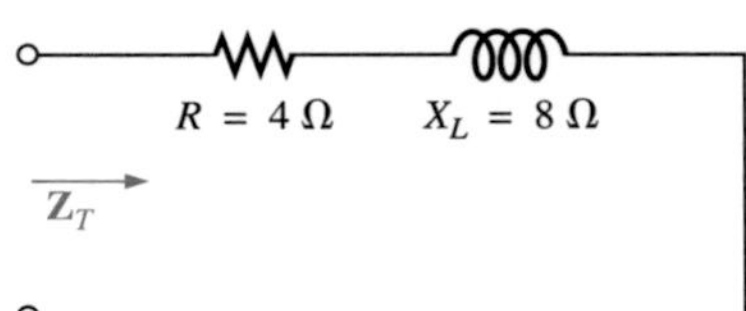

FIG. 15.21
Example 15.7.

EXAMPLE 15.7

Draw the impedance diagram for the circuit of Fig. 15.21 and find the total impedance.

Solution: As indicated by Fig. 15.22, the input impedance can be found graphically from the impedance diagram by properly scaling the real and imaginary axes and finding the length of the resultant vector Z_T and angle θ_T. Or, by using vector algebra, we obtain

$$\begin{aligned} \mathbf{Z}_T &= \mathbf{Z}_1 + \mathbf{Z}_2 \\ &= R \angle 0^\circ + X_L \angle 90^\circ \\ &= R + jX_L = 4\ \Omega + j8\ \Omega \\ \mathbf{Z}_T &= \mathbf{8.944\ \Omega \angle 63.43^\circ} \end{aligned}$$

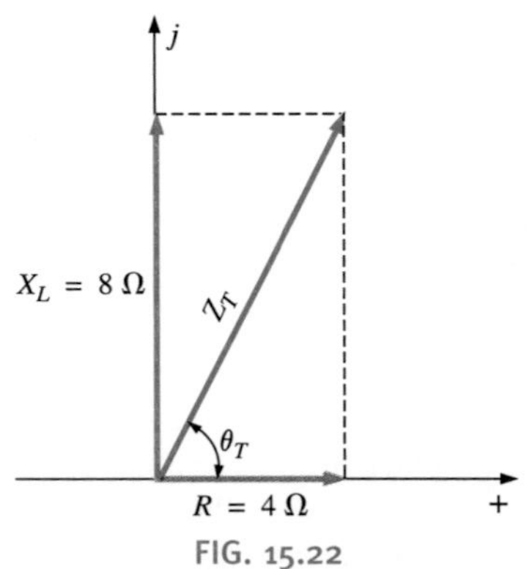

FIG. 15.22
Impedance diagram for Example 15.7.

EXAMPLE 15.8

Determine the input impedance to the series network of Fig. 15.23. Draw the impedance diagram.

Solution:

$$\begin{aligned} \mathbf{Z}_T &= \mathbf{Z}_1 + \mathbf{Z}_2 + \mathbf{Z}_3 \\ &= R \angle 0^\circ + X_L \angle 90^\circ + X_C \angle -90^\circ \\ &= R + jX_L - jX_C \\ &= R + j(X_L - X_C) = 6\ \Omega + j(10\ \Omega - 12\ \Omega) = 6\ \Omega - j2\ \Omega \\ \mathbf{Z}_T &= \mathbf{6.325\ \Omega \angle -18.43^\circ} \end{aligned}$$

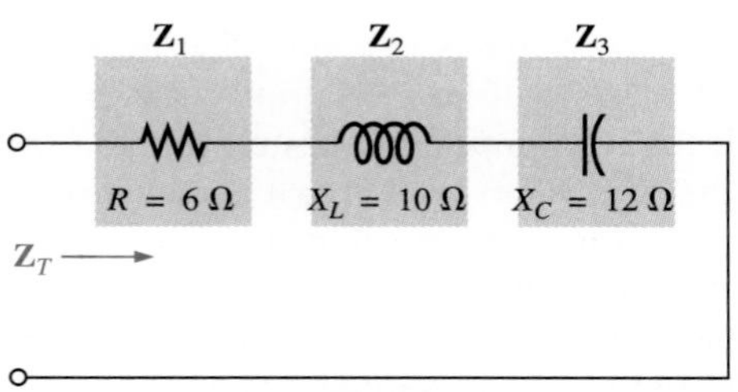

FIG. 15.23
Example 15.8

The impedance diagram appears in Fig. 15.24. Note that in this example, series inductive and capacitive reactances are in direct opposition. For the circuit of Fig. 15.23, if the inductive reactance were equal to the capacitive reactance, the input impedance would be purely resistive. We will have more to say about this particular condition in a later chapter.

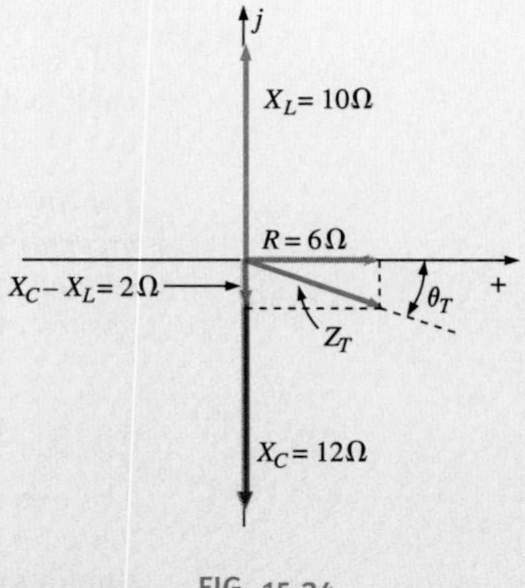

FIG. 15.24
Impedance diagram for Example 15.8.

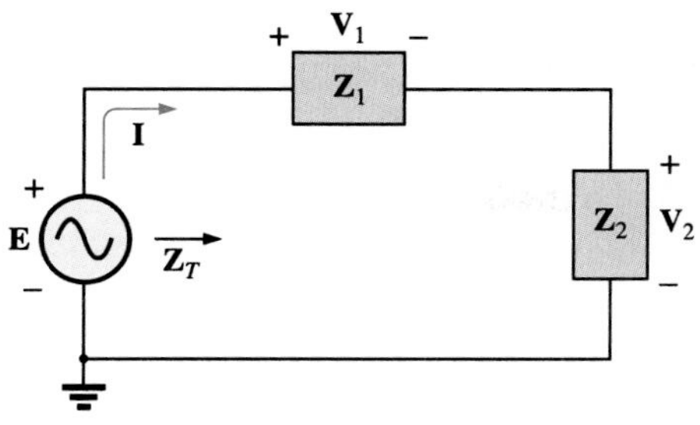

FIG. 15.25
Series ac circuit.

For the representative series ac network of Fig. 15.25 having two impedances, *the current is the same through each element* (as it was for the series dc circuits) and is determined by Ohm's law:

$$\mathbf{Z}_T = \mathbf{Z}_1 + \mathbf{Z}_2$$

and

$$\mathbf{I} = \frac{\mathbf{E}}{\mathbf{Z}_T} \tag{15.5}$$

The voltage across each element can then be found by another application of Ohm's law:

$$\mathbf{V}_1 = \mathbf{I}\mathbf{Z}_1 \tag{15.6a}$$

$$\mathbf{V}_2 = \mathbf{I}\mathbf{Z}_2 \tag{15.6b}$$

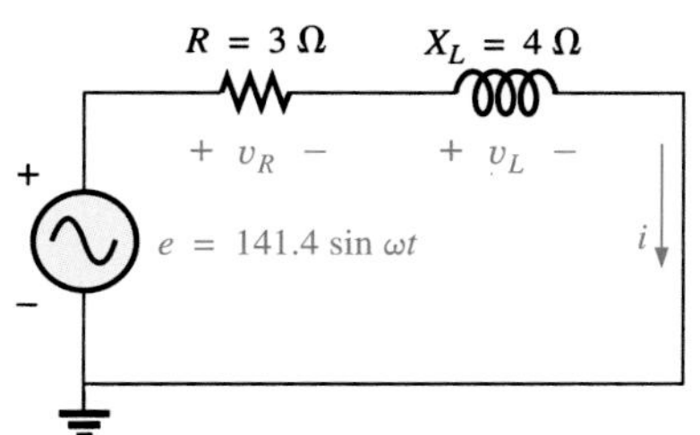

FIG. 15.26
Series R-L circuit.

Kirchhoff's voltage law can then be applied as it is for dc circuits (Section 5.4). However, keep in mind that we are now dealing with the algebraic manipulation of quantities that have both magnitude and direction.

$$\mathbf{E} - \mathbf{V}_1 - \mathbf{V}_2 = 0$$

or

$$\mathbf{E} = \mathbf{V}_1 + \mathbf{V}_2 \tag{15.7}$$

The power to the circuit can be determined by

$$P = EI \cos \theta_T \tag{15.8}$$

where θ_T is the phase angle between **E** and **I**.

Now that a general approach has been introduced, we will look in detail at the simplest of series configurations to further emphasize the similarities in the analysis of dc circuits. In many of the circuits to be considered, $3 + j4 = 5 \angle 53.13°$ and $4 + j3 = 5 \angle 36.87°$ will be used often to make the approach as clear as possible and to reduce mathematical complexity. Of course, the problems at the end of the chapter will provide plenty of experience with random values.

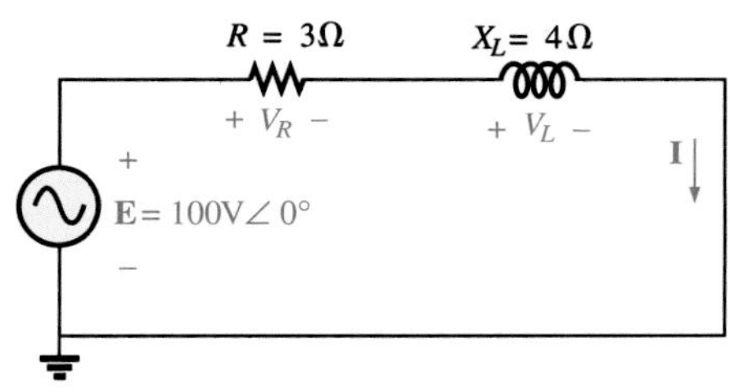

FIG. 15.27
Applying phasor notation to the network of Fig. 15.26.

R-L

Refer to Fig. 15.26.

Phasor notation:

$$e = 141.4 \sin \omega t \Rightarrow \mathbf{E} = 100\text{ V} \angle 0°$$

Note Fig. 15.27.

$\mathbf{Z}_T$:

$$\mathbf{Z}_T = \mathbf{Z}_1 + \mathbf{Z}_2 = 3\ \Omega \angle 0° + 4\ \Omega \angle 90° = 3\ \Omega + j4\ \Omega$$

and

$$\mathbf{Z}_T = \mathbf{5\ \Omega \angle 53.13°}$$

Impedance diagram: See Fig. 15.28.

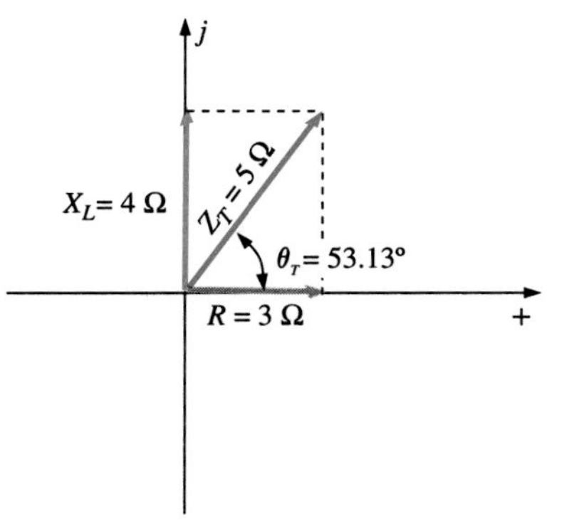

FIG. 15.28
Impedance diagram for the series R-L circuit of Fig. 15.26.

I:

$$\mathbf{I} = \frac{\mathbf{E}}{\mathbf{Z}_T} = \frac{100\text{ V} \angle 0°}{5\ \Omega \angle 53.13°} = \mathbf{20\text{ A} \angle -53.13°}$$

$\mathbf{V}_R$, $\mathbf{V}_L$:

Ohm's law:

$$\mathbf{V}_R = \mathbf{I}\mathbf{Z}_R = (20\text{ A}\angle -53.13°)(3\ \Omega\angle 0°) = \mathbf{60\ V\angle -53.13°}$$

$$\mathbf{V}_L = \mathbf{I}\mathbf{Z}_L = (20\text{ A}\angle -53.13°)(4\ \Omega\angle 90°) = \mathbf{80\ V\angle 36.87°}$$

Kirchhoff's voltage law:

$$\Sigma_{\circlearrowleft} \mathbf{V} = \mathbf{E} - \mathbf{V}_R - \mathbf{V}_L = 0$$

or

$$\mathbf{E} = \mathbf{V}_R + \mathbf{V}_L$$

In rectangular form,

$$\mathbf{V}_R = 60\text{ V}\angle -53.13° = 36\text{ V} - j48\text{ V}$$

$$\mathbf{V}_L = 80\text{ V}\angle +36.87° = 64\text{ V} + j48\text{ V}$$

and

$$\mathbf{E} = \mathbf{V}_R + \mathbf{V}_L = (36\text{ V} - j48\text{ V}) + (64\text{ V} + j48\text{ V}) = 100\text{ V} + j0 = 100\text{ V}\angle 0°$$

as applied.

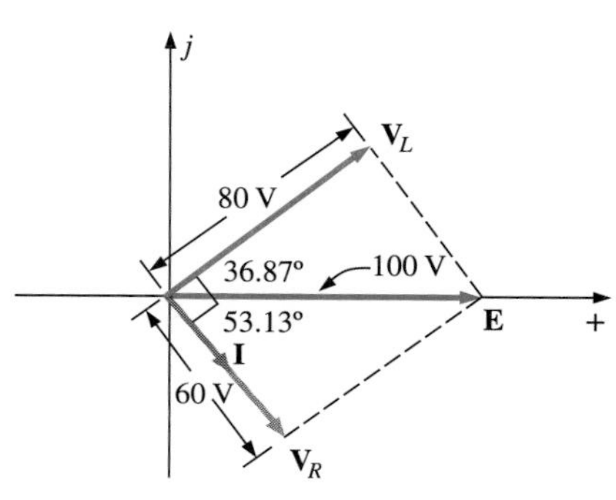

FIG. 15.29
Phasor diagram for the series R-L circuit of Fig. 15.26.

Phasor diagram: Note that for the phasor diagram of Fig. 15.29, **I** is in phase with the voltage across the resistor and lags the voltage across the inductor by 90°.

Power: The total power in watts delivered to the circuit is

$$P_T = EI\cos\theta_T = (100\text{ V})(20\text{ A})\cos 53.13° = (2000\text{ W})(0.6) = \mathbf{1200\ W}$$

where E and I are effective values and θ_T is the phase angle between E and I, or

$$P_T = I^2R = (20\text{ A})^2(3\ \Omega) = (400)(3) = \mathbf{1200\ W}$$

where I is the effective value; or, finally,

$$P_T = P_R + P_L = V_RI\cos\theta_R + V_LI\cos\theta_L = (60\text{ V})(20\text{ A})\cos 0° + (80\text{ V})(20\text{ A})\cos 90° = 1200\text{ W} + 0 = \mathbf{1200\ W}$$

where θ_R is the phase angle between $\mathbf{V}_R$ and **I**, and θ_L is the phase angle between $\mathbf{V}_L$ and **I**.

Power factor: The power factor F_p of the circuit is cos 53.13° = **0.6 lagging,** where 53.13° is the phase angle between **E** and **I.**

If we write the basic power equation $P = EI\cos\theta$ as follows:

$$\cos\theta = \frac{P}{EI}$$

where E and I are the input quantities and P is the power delivered to the network, and then perform the following substitutions from the basic series ac circuit:

$$\cos\theta = \frac{P}{EI} = \frac{I^2R}{EI} = \frac{IR}{E} = \frac{R}{E/I} = \frac{R}{Z_T}$$

we find

$$F_p = \cos\theta_T = \frac{R}{Z_T} \tag{15.9}$$

Fig. 15.28 also indicates that θ is the impedance angle θ_T as written in Eq. (15.9), further supporting the fact that the impedance angle θ_T is also the phase angle between the input voltage and current for a series ac circuit. To determine the power factor, you only have to form the ratio of the total resistance to the magnitude of the input impedance. For this case,

$$F_p = \cos\theta = \frac{R}{Z_T} = \frac{3\ \Omega}{5\ \Omega} = \mathbf{0.6\ lagging}$$

as found above.

R-C

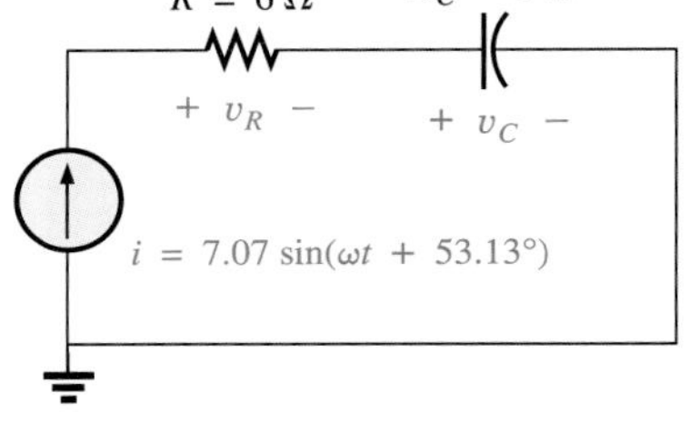

FIG. 15.30
Series R-C ac circuit.

Refer to Fig. 15.30.

Phasor notation:

$$i = 7.07 \sin(\omega t + 53.13^\circ) \Rightarrow \mathbf{I} = 5\ \text{A} \angle 53.13^\circ$$

Note Fig. 15.31.

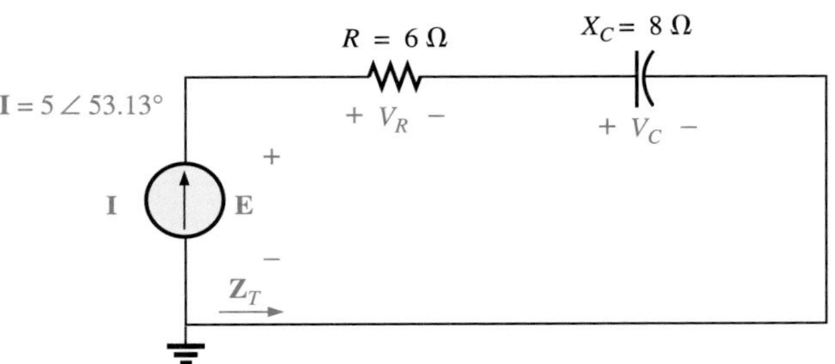

FIG. 15.31
Applying phasor notation to the circuit of Fig. 15.30.

$\mathbf{Z}_T$:

$$\mathbf{Z}_T = \mathbf{Z}_1 + \mathbf{Z}_2 = 6\ \Omega \angle 0^\circ + 8\ \Omega \angle -90^\circ = 6\ \Omega - j8\ \Omega$$

and

$$\mathbf{Z}_T = \mathbf{10\ \Omega \angle -53.13^\circ}$$

Impedance diagram: As shown in Fig. 15.32.

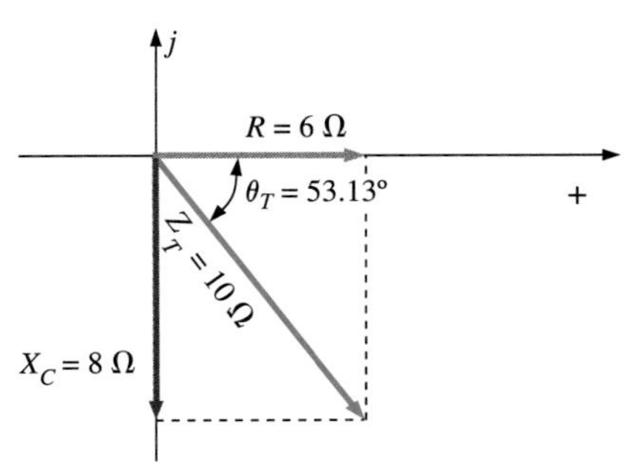

FIG. 15.32
Impedance diagram for the series R-C circuit of Fig. 15.30.

E:

$$\mathbf{E} = \mathbf{I}\mathbf{Z}_T = (5\ \text{A} \angle 53.13^\circ)(10\ \Omega \angle -53.13^\circ) = \mathbf{50\ V \angle 0^\circ}$$

$\mathbf{V}_R$, $\mathbf{V}_C$:

$$\begin{aligned}\mathbf{V}_R &= \mathbf{I}\mathbf{Z}_R = (I \angle \theta)(R \angle 0^\circ) = (5\ \text{A} \angle 53.13^\circ)(6\ \Omega \angle 0^\circ) \\ &= \mathbf{30\ V \angle 53.13^\circ}\end{aligned}$$

$$\begin{aligned}\mathbf{V}_C &= \mathbf{I}\mathbf{Z}_C = (I \angle \theta)(X_C \angle -90^\circ) = (5\ \text{A} \angle 53.13^\circ)(8\ \Omega \angle -90^\circ) \\ &= \mathbf{40\ V \angle -36.87^\circ}\end{aligned}$$

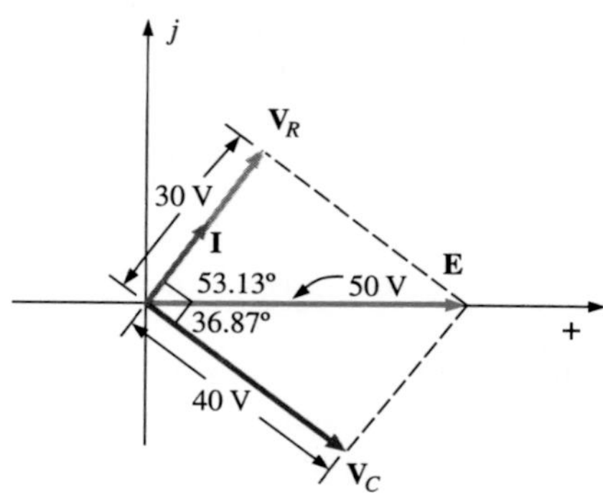

FIG. 15.33
Phasor diagram for the series R-C circuit of Fig. 15.30.

Kirchhoff's voltage law:

$$\Sigma_{\circlearrowright} \mathbf{V} = \mathbf{E} - \mathbf{V}_R - \mathbf{V}_C = 0$$

or

$$\mathbf{E} = \mathbf{V}_R + \mathbf{V}_C$$

which can be verified by vector algebra as demonstrated for the *R-L* circuit.

Phasor diagram: Note on the phasor diagram of Fig. 15.33 that the current **I** is in phase with the voltage across the resistor and leads the voltage across the capacitor by 90°.

Time domain: In the time domain,

$$e = \sqrt{2}(50) \sin \omega t = \mathbf{70.70 \sin \omega t}$$
$$v_R = \sqrt{2}(30) \sin(\omega t + 53.13°) = \mathbf{42.42 \sin(\omega t + 53.13°)}$$
$$v_C = \sqrt{2}(40) \sin(\omega t - 36.87°) = \mathbf{56.56 \sin(\omega t - 36.87°)}$$

A plot of all of the voltages and the current of the circuit appears in Fig. 15.34. Note again that i and v_R are in phase and that v_C lags i by 90°.

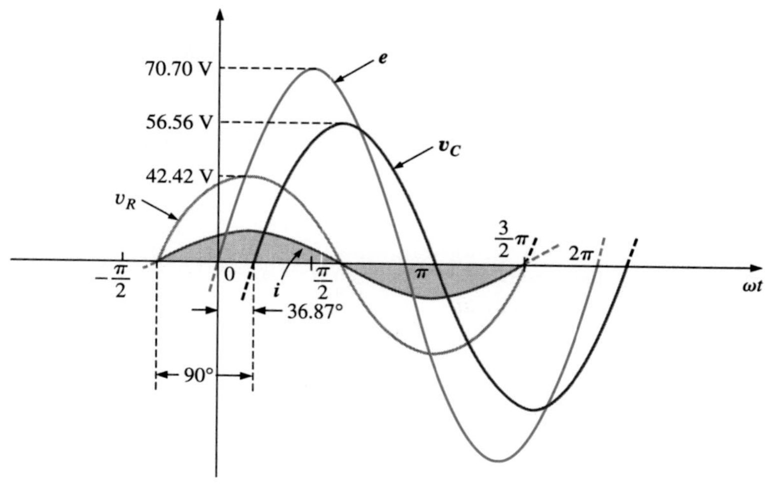

FIG. 15.34
Waveforms for the series R-C circuit of Fig. 15.30.

Power: The total power in watts delivered to the circuit is

$$P_T = EI \cos \theta_T = (50 \text{ V})(5 \text{ A}) \cos 53.13° = (250)(0.6) = \mathbf{150 \text{ W}}$$

or

$$P_T = I^2R = (5 \text{ A})^2(6 \ \Omega) = (25)(6) = \mathbf{150 \text{ W}}$$

or, finally,

$$\begin{aligned} P_T = P_R + P_C &= V_RI \cos \theta_R + V_CI \cos \theta_C \\ &= (30 \text{ V})(5 \text{ A}) \cos 0° + (40 \text{ V})(5 \text{ A}) \cos 90° \\ &= 150 \text{ W} + 0 \\ &= \mathbf{150 \text{ W}} \end{aligned}$$

Power factor: The power factor of the circuit is

$$F_p = \cos \theta = \cos 53.13° = \mathbf{0.6 \text{ leading}}$$

Using Eq. (15.9), we obtain

$$F_p = \cos\theta = \frac{R}{Z_T} = \frac{6\ \Omega}{10\ \Omega}$$
$$= \mathbf{0.6\ leading}$$

as determined above.

R-L-C

Refer to Fig. 15.35.

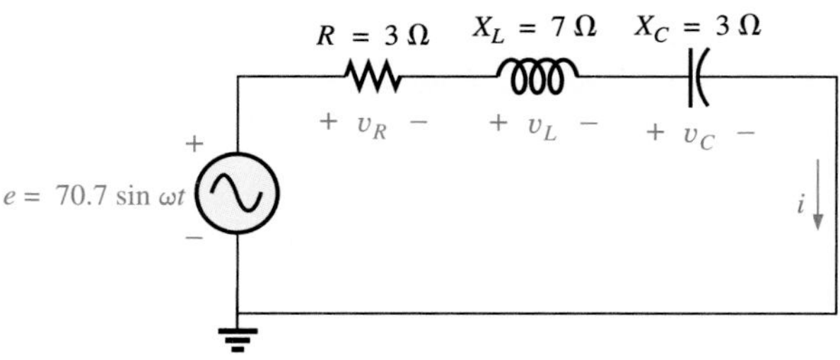

FIG. 15.35
Series R-L-C ac circuit.

Phasor notation: As shown in Fig. 15.36.

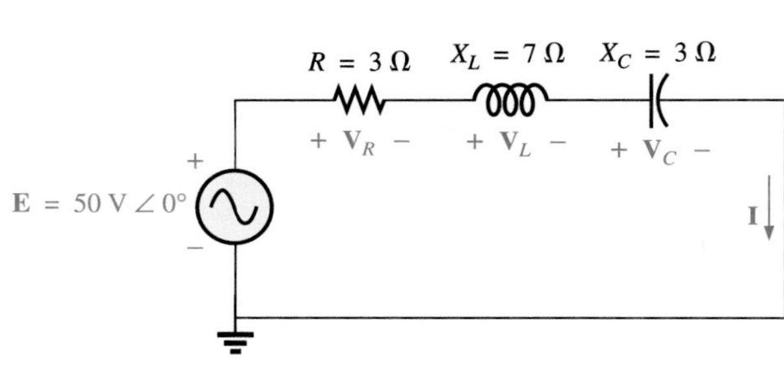

FIG. 15.36
Applying phasor notation to the circuit of Fig. 15.35.

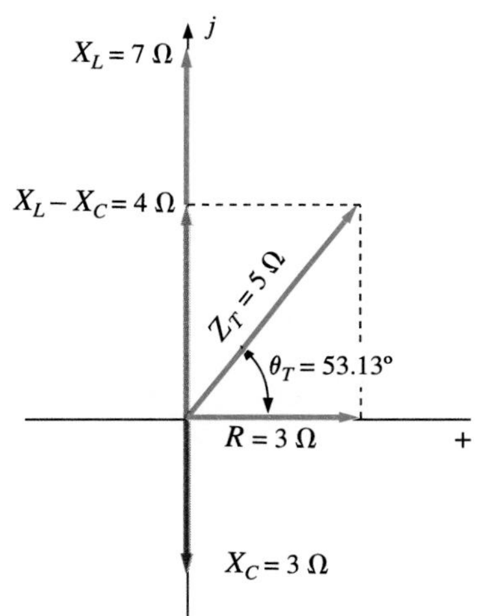

FIG. 15.37
Impedance diagram for the series R-L-C circuit of Fig. 15.35.

$\mathbf{Z}_T$:

$$\mathbf{Z}_T = \mathbf{Z}_1 + \mathbf{Z}_2 + \mathbf{Z}_3 = R\ \angle 0^\circ + X_L\ \angle 90^\circ + X_C\ \angle -90^\circ$$
$$= 3\ \Omega + j\,7\ \Omega - j\,3\ \Omega = 3\ \Omega + j\,4\ \Omega$$

and
$$\mathbf{Z}_T = \mathbf{5\ \Omega\ \angle 53.13^\circ}$$

Impedance diagram: As shown in Fig. 15.37.

I:

$$\mathbf{I} = \frac{\mathbf{E}}{\mathbf{Z}_T} = \frac{50\ \text{V}\ \angle 0^\circ}{5\ \Omega\ \angle 53.13^\circ} = \mathbf{10\ A\ \angle -53.13^\circ}$$

$\mathbf{V}_R$, $\mathbf{V}_L$, $\mathbf{V}_C$:

$$\mathbf{V}_R = \mathbf{I}\mathbf{Z}_R = (I\ \angle\theta)(R\ \angle 0^\circ) = (10\ \text{A}\ \angle -53.13^\circ)(3\ \Omega\ \angle 0^\circ)$$
$$= \mathbf{30\ V\ \angle -53.13^\circ}$$

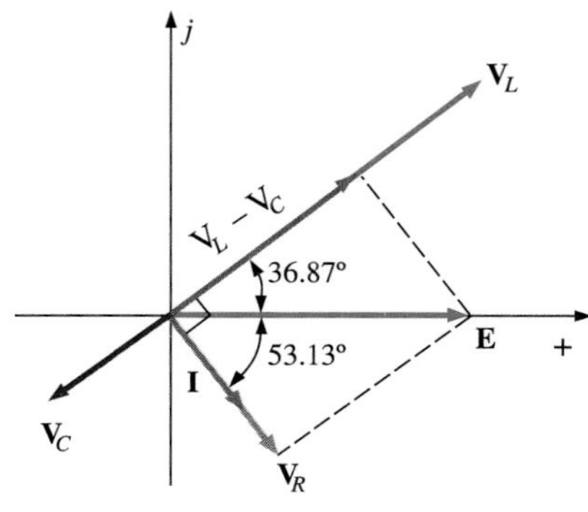

FIG. 15.38
Phasor diagram for the series R-L-C circuit of Fig. 15.35.

$$\mathbf{V}_L = \mathbf{I}\mathbf{Z}_L = (I \angle\theta)(X_L \angle 90^\circ) = (10 \text{ A} \angle -53.13^\circ)(7\ \Omega \angle 90^\circ)$$
$$= \mathbf{70\ V \angle 36.87^\circ}$$
$$\mathbf{V}_C = \mathbf{I}\mathbf{Z}_C = (I \angle\theta)(X_C \angle -90^\circ) = (10 \text{ A} \angle -53.13^\circ)(3\ \Omega \angle -90^\circ)$$
$$= \mathbf{30\ V \angle -143.13^\circ}$$

Kirchhoff's voltage law:

$$\Sigma_{\circlearrowleft} \mathbf{V} = \mathbf{E} - \mathbf{V}_R - \mathbf{V}_L - \mathbf{V}_C = 0$$

or

$$\mathbf{E} = \mathbf{V}_R + \mathbf{V}_L + \mathbf{V}_C$$

which can also be verified through vector algebra.

Phasor diagram: The phasor diagram of Fig. 15.38 indicates that the current **I** is in phase with the voltage across the resistor, lags the voltage across the inductor by 90°, and leads the voltage across the capacitor by 90°.

Time domain:

$$i = \sqrt{2}(10) \sin(\omega t - 53.13^\circ) = \mathbf{14.14 \sin(\omega t - 53.13^\circ)}$$
$$v_R = \sqrt{2}(30) \sin(\omega t - 53.13^\circ) = \mathbf{42.42 \sin(\omega t - 53.13^\circ)}$$
$$v_L = \sqrt{2}(70) \sin(\omega t + 36.87^\circ) = \mathbf{98.98 \sin(\omega t + 36.87^\circ)}$$
$$v_C = \sqrt{2}(30) \sin(\omega t - 143.13^\circ) = \mathbf{42.42 \sin(\omega t - 143.13^\circ)}$$

A plot of all the voltages and the current of the circuit appears in Fig. 15.39.

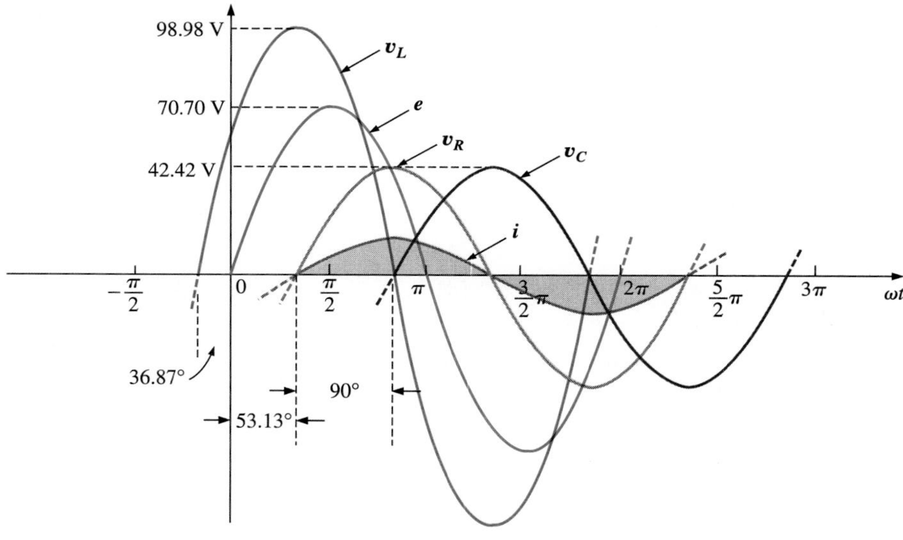

FIG. 15.39
Waveforms for the series R-L circuit of Fig. 15.35.

Power: The total power in watts delivered to the circuit is

$$P_T = EI \cos\theta_T = (50 \text{ V})(10 \text{ A}) \cos 53.13^\circ = (500)(0.6) = \mathbf{300\ W}$$

or

$$P_T = I^2R = (10 \text{ A})^2(3\ \Omega) = (100)(3) = \mathbf{300\ W}$$

or

$$P_T = P_R + P_L + P_C$$
$$= V_RI \cos\theta_R + V_LI \cos\theta_L + V_CI \cos\theta_C$$
$$= (30 \text{ V})(10 \text{ A}) \cos 0^\circ + (70 \text{ V})(10 \text{ A}) \cos 90^\circ + (30 \text{ V})(10 \text{ A}) \cos 90^\circ$$
$$= (30 \text{ V})(10 \text{ A}) + 0 + 0 = \mathbf{300\ W}$$

Power factor: The power factor of the circuit is

$$F_p = \cos \theta_T = \cos 53.13° = \mathbf{0.6 \text{ lagging}}$$

Using Eq. (15.9), we obtain

$$F_p = \cos \theta = \frac{R}{Z_T} = \frac{3\ \Omega}{5\ \Omega} = \mathbf{0.6 \text{ lagging}}$$

15.4 VOLTAGE DIVIDER RULE

The basic format for the *voltage divider rule* in ac circuits is exactly the same as that for dc circuits:

$$\mathbf{V}_x = \frac{\mathbf{Z}_x\mathbf{E}}{\mathbf{Z}_T} \qquad (15.10)$$

where $\mathbf{V}_x$ is the voltage across one or more elements in series that have total impedance $\mathbf{Z}_x$, $\mathbf{E}$ is the total voltage appearing across the series circuit, and $\mathbf{Z}_T$ is the total impedance of the series circuit.

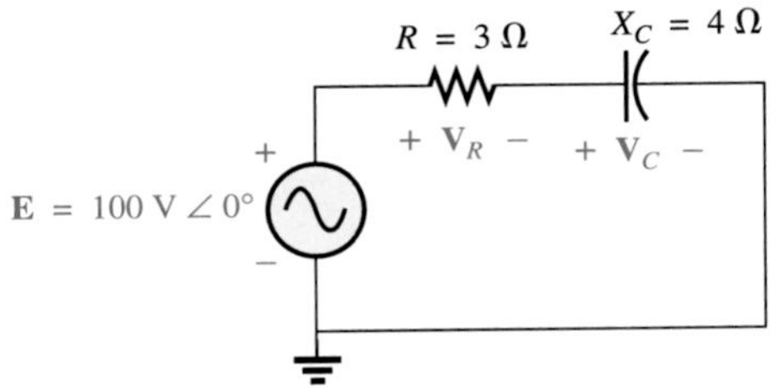

FIG. 15.40
Example 15.9.

EXAMPLE 15.9

Using the voltage divider rule, find the voltage across each element of the circuit of Fig. 15.40.

Solution:

$$\mathbf{V}_C = \frac{\mathbf{Z}_C\mathbf{E}}{\mathbf{Z}_C + \mathbf{Z}_R} = \frac{(4\ \Omega \angle -90°)(100\ \text{V} \angle 0°)}{4\ \Omega \angle -90° + 3\ \Omega \angle 0°} = \frac{400 \angle -90°}{3 - j4}$$

$$= \frac{400 \angle -90°}{5 \angle -53.13°} = \mathbf{80\ V \angle -36.87°}$$

$$\mathbf{V}_R = \frac{\mathbf{Z}_R\mathbf{E}}{\mathbf{Z}_C + \mathbf{Z}_R} = \frac{(3\ \Omega \angle 0°)(100\ \text{V} \angle 0°)}{5\ \Omega \angle -53.13°} = \frac{300 \angle 0°}{5 \angle -53.13°}$$

$$= \mathbf{60\ V \angle +53.13°}$$

EXAMPLE 15.10

Using the voltage divider rule, find the unknown voltages $\mathbf{V}_R$, $\mathbf{V}_L$, $\mathbf{V}_C$, and $\mathbf{V}_1$ for the circuit of Fig. 15.41.

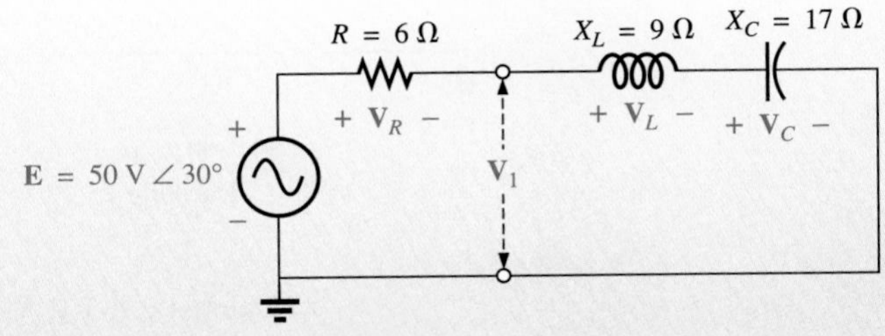

FIG. 15.41
Example 15.10.

continued

Solution:

$$\mathbf{V}_R = \frac{\mathbf{Z}_R\mathbf{E}}{\mathbf{Z}_R + \mathbf{Z}_L + \mathbf{Z}_C} = \frac{(6\ \Omega\ \angle 0^\circ)(50\ \text{V}\ \angle 30^\circ)}{6\ \Omega\ \angle 0^\circ + 9\ \Omega\ \angle 90^\circ + 17\ \Omega\ \angle -90^\circ}$$

$$= \frac{300\ \angle 30^\circ}{6 + j\,9 - j\,17} = \frac{300\ \angle 30^\circ}{6 - j\,8}$$

$$= \frac{300\ \angle 30^\circ}{10\ \angle -53.13^\circ} = \mathbf{30\ V\ \angle 83.13^\circ}$$

$$\mathbf{V}_L = \frac{\mathbf{Z}_L\mathbf{E}}{\mathbf{Z}_T} = \frac{(9\ \Omega\ \angle 90^\circ)(50\ \text{V}\ \angle 30^\circ)}{10\ \Omega\ \angle -53.13^\circ} = \frac{450\ \text{V}\angle 120^\circ}{10\ \angle -53.13^\circ}$$

$$= \mathbf{45\ V\ \angle 173.13^\circ}$$

$$\mathbf{V}_C = \frac{\mathbf{Z}_C\mathbf{E}}{\mathbf{Z}_T} = \frac{(17\ \Omega\ \angle -90^\circ)(50\ \text{V}\ \angle 30^\circ)}{10\ \Omega\ \angle -53.13^\circ} = \frac{850\ \text{V}\angle -60^\circ}{10\ \angle -53^\circ}$$

$$= \mathbf{85\ V\ \angle -6.87^\circ}$$

$$\mathbf{V}_1 = \frac{(\mathbf{Z}_L + \mathbf{Z}_C)\mathbf{E}}{\mathbf{Z}_T} = \frac{(9\ \Omega\ \angle 90^\circ + 17\ \Omega\ \angle -90^\circ)(50\ \text{V}\ \angle 30^\circ)}{10\ \Omega\ \angle -53.13^\circ}$$

$$= \frac{(8\ \angle -90^\circ)(50\ \angle 30^\circ)}{10\ \angle -53.13^\circ}$$

$$= \frac{400\ \angle -60^\circ}{10\ \angle -53.13^\circ} = \mathbf{40\ V\ \angle -6.87^\circ}$$

EXAMPLE 15.11

For the circuit of Fig. 15.42:

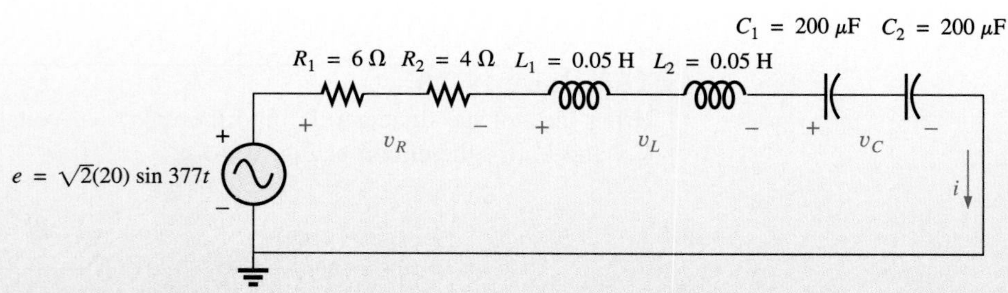

FIG. 15.42
Example 15.11.

a. Calculate $\mathbf{I}$, $\mathbf{V}_R$, $\mathbf{V}_L$, and $\mathbf{V}_C$ in phasor form.
b. Calculate the total power factor.
c. Calculate the average power delivered to the circuit.
d. Draw the phasor diagram.

continued

e. Obtain the phasor sum of $\mathbf{V}_R$, $\mathbf{V}_L$, and $\mathbf{V}_C$, and show that it equals the input voltage **E**.
f. Find $\mathbf{V}_R$ and $\mathbf{V}_C$ using the voltage divider rule.

Solutions:

a. Combining common elements and finding the reactance of the inductor and capacitor, we obtain

$$R_T = 6\ \Omega + 4\ \Omega = 10\ \Omega$$
$$L_T = 0.05\ \text{H} + 0.05\ \text{H} = 0.1\ \text{H}$$
$$C_T = \frac{200\ \mu\text{F}}{2} = 100\ \mu\text{F}$$

$$X_L = \omega L = (377\ \text{rad/s})(0.1\ \text{H}) = 37.70\ \Omega$$

$$X_C = \frac{1}{\omega C} = \frac{1}{(377\ \text{rad/s})(100 \times 10^{-6}\ \text{F})} = \frac{10^6\ \Omega}{37\ 700} = 26.53\ \Omega$$

Redrawing the circuit using phasor notation gives us Fig. 15.43.

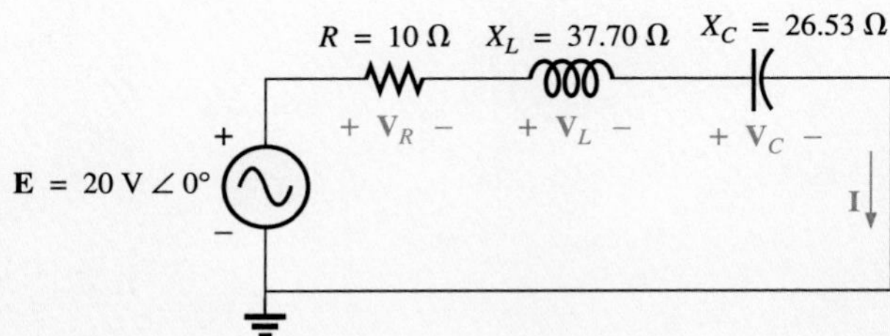

FIG. 15.43
Applying phasor notation to the circuit of Fig. 15.42.

For the circuit of Fig. 15.43,

$$\begin{aligned}\mathbf{Z}_T &= R\angle 0^\circ + X_L\angle 90^\circ + X_C\angle -90^\circ\\ &= 10\ \Omega + j\,37.70\ \Omega - j\,26.53\ \Omega\\ &= 10\ \Omega + j\,11.17\ \Omega = \mathbf{15\ \Omega\ \angle 48.16^\circ}\end{aligned}$$

The current **I** is

$$\mathbf{I} = \frac{\mathbf{E}}{\mathbf{Z}_T} = \frac{20\ \text{V}\ \angle 0^\circ}{15\ \Omega\ \angle 48.16^\circ} = \mathbf{1.33\ A\ \angle -48.16^\circ}$$

The voltage across the resistor, inductor, and capacitor can be found using Ohm's law:

$$\begin{aligned}\mathbf{V}_R = \mathbf{IZ}_R &= (I\angle\theta)(R\angle 0^\circ) = (1.33\ \text{A}\ \angle -48.16^\circ)(10\ \Omega\ \angle 0^\circ)\\ &= \mathbf{13.30\ V\ \angle -48.16^\circ}\end{aligned}$$

$$\begin{aligned}\mathbf{V}_L = \mathbf{IZ}_L &= (I\angle\theta)(X_L\angle 90^\circ) = (1.33\ \text{A}\ \angle -48.16^\circ)(37.70\ \Omega\ \angle 90^\circ)\\ &= \mathbf{50.14\ V\ \angle 41.84^\circ}\end{aligned}$$

$$\begin{aligned}\mathbf{V}_C = \mathbf{IZ}_C &= (I\angle\theta)(X_C\angle -90^\circ) = (1.33\ \text{A}\ \angle -48.16^\circ)(26.53\ \Omega\angle -90^\circ)\\ &= \mathbf{35.28\ V\ \angle -138.16^\circ}\end{aligned}$$

b. The total power factor, determined by the angle between the applied voltage **E** and the resulting current **I**, is 48.16°:

$$F_p = \cos\theta = \cos 48.16^\circ = \mathbf{0.667\ lagging}$$

continued

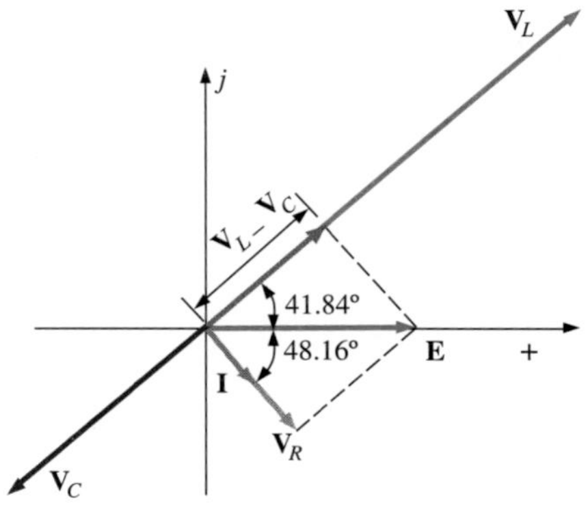

FIG. 15.44
Phasor diagram for the circuit of Fig. 15.42.

or $$F_p = \cos\theta = \frac{R}{Z_T} = \frac{10\ \Omega}{15\ \Omega} = \mathbf{0.667\ lagging}$$

c. The total power in watts delivered to the circuit is

$$P_T = EI\cos\theta = (20\text{ V})(1.33\text{ A})(0.667) = \mathbf{17.74\ W}$$

d. The phasor diagram appears in Fig. 15.44.
e. The phasor sum of $\mathbf{V}_R$, $\mathbf{V}_L$, and $\mathbf{V}_C$ is

$$\begin{aligned}\mathbf{E} &= \mathbf{V}_R + \mathbf{V}_L + \mathbf{V}_C\\ &= 13.30\text{ V}\angle -48.16° + 50.14\text{ V}\angle 41.84° + 35.28\text{ V}\\ &\quad \angle -138.16°\\ \mathbf{E} &= 13.30\text{ V}\angle -48.16° + 14.86\text{ V}\angle 41.84°\end{aligned}$$

Therefore,

$$E = \sqrt{(13.30\text{ V})^2 + (14.86\text{ V})^2} = \mathbf{20\ V}$$

and $$\theta_E = \mathbf{0°} \quad \text{(from phasor diagram)}$$

and $$\mathbf{E} = 20\angle 0°$$

f. $$\mathbf{V}_R = \frac{\mathbf{Z}_R\mathbf{E}}{\mathbf{Z}_T} = \frac{(10\ \Omega\angle 0°)(20\text{ V}\angle 0°)}{15\ \Omega\angle 48.16°} = \frac{200\text{ V}\angle 0°}{15\angle 48.16°}$$

$$= \mathbf{13.3\ V\angle -48.16°}$$

$$\mathbf{V}_C = \frac{\mathbf{Z}_C\mathbf{E}}{\mathbf{Z}_T} = \frac{(26.5\ \Omega\angle -90°)(20\text{ V}\angle 0°)}{15\ \Omega\angle 48.16°} = \frac{530.6\text{ V}\angle -90°}{15\angle 48.16°}$$

$$= \mathbf{35.37\ V\angle -138.16°}$$

15.5 FREQUENCY RESPONSE OF THE *R-C* CIRCUIT

We have limited our analysis of series circuits so far to a particular frequency. We will now examine the effect of frequency on the response of an *R-C* series configuration such as the one in Fig. 15.45. The magnitude of the source is fixed at 10 V, but the frequency range of analysis will extend from zero to 20 kHz.

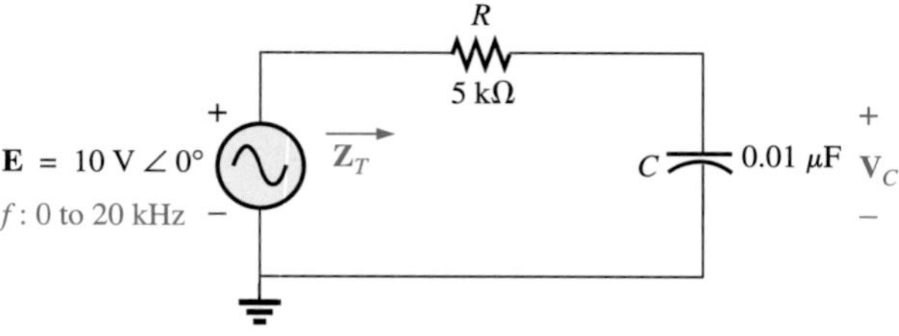

FIG. 15.45
Determining the frequency response of a series R-C circuit.

$\mathbf{Z}_T$:

Let us first determine how the impedance of the circuit $\mathbf{Z}_T$ will vary with frequency for the specified frequency range of interest. Before getting into specifics, however, let us first develop a sense for what we should expect by noting the impedance-versus-frequency curve of each element, as drawn in Fig. 15.46.

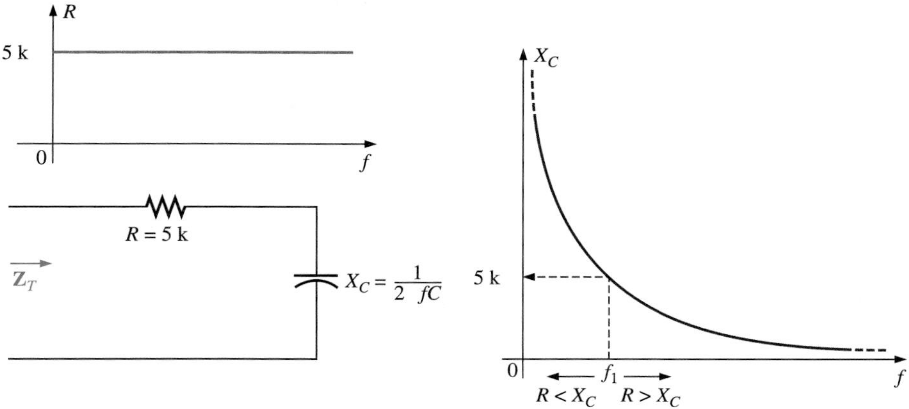

FIG. 15.46
The frequency response of the individual elements of a series R-C circuit.

At low frequencies the reactance of the capacitor will be quite high and considerably more than the level of the resistance R, suggesting that the total impedance will be primarily capacitive in nature. At high frequencies the reactance X_C will drop below the $R = 5\text{-k}\Omega$ level and the network will start to shift toward one of a purely resistive nature (at 5 kΩ). The frequency at which $X_C = R$ can be determined in the following manner:

$$X_C = \frac{1}{2\pi f_1 C} = R$$

and
$$f_1 = \frac{1}{2\pi RC} \qquad (X_C = R) \tag{15.11}$$

which for the network of interest is

$$f_1 = \frac{1}{2\pi(5\text{ k}\Omega)(0.01\ \mu\text{F})} \cong \mathbf{3183.1\ Hz}$$

For frequencies less than f_1, $X_C > R$, and for frequencies greater than f_1, $R > X_C$, as shown in Fig. 15.46.

Now for the details. The total impedance is determined by the following equation:

$$Z_T = R - jX_C$$

and
$$\mathbf{Z}_T = Z_T \angle\theta_T = \sqrt{R^2 + X_C^2} \angle -\tan^{-1}\frac{X_C}{R} \tag{15.12}$$

The magnitude and angle of the total impedance can now be found at any frequency of interest by simply substituting into Eq. (15.12). The presence of the capacitor suggests that we start from a low frequency (100 Hz) and then choose a few evenly spaced values of f up to 20 kHz.

f = 100 Hz:

$$X_C = \frac{1}{2\pi fC} = \frac{1}{2\pi(100\text{ Hz})(0.01\ \mu\text{F})} = 159.16\text{ k}\Omega$$

and $Z_T = \sqrt{R^2 + X_C^2} = \sqrt{(5\text{ k}\Omega)^2 + (159.16\text{ k}\Omega)^2} = 159.24\text{ k}\Omega$

with $$\theta_T = -\tan^{-1}\frac{X_C}{R} = -\tan^{-1}\frac{159.16\text{ k}\Omega}{5\text{ k}\Omega} = -\tan^{-1}31.83$$
$$= -88.2°$$

and $$\mathbf{Z}_T = \mathbf{159.24\ k\Omega\ \angle{-88.2°}}$$

which is very close to $\mathbf{Z}_C$ = 159.16 kΩ ∠−90° if the circuit were purely capacitive (R = 0 Ω). Our assumption that the circuit is primarily capacitive at low frequencies is therefore confirmed.

f = 1 kHz:

$$X_C = \frac{1}{2\pi fC} = \frac{1}{2\pi(1\text{ kHz})(0.01\ \mu\text{F})} = 15.92\text{ k}\Omega$$

and $Z_T = \sqrt{R^2 + X_C^2} = \sqrt{(5\text{ k}\Omega)^2 + (15.92\text{ k}\Omega)^2} = 16.69\text{ k}\Omega$

with $\theta_T = -\tan^{-1}\dfrac{X_C}{R} = -\tan^{-1}\dfrac{15.92\text{ k}\Omega}{5\text{ k}\Omega} = -\tan^{-1}3.18 = -72.54°$

and $\mathbf{Z}_T = \mathbf{16.69\ k\Omega\ \angle{-72.54°}}$

A noticeable drop in the magnitude has occurred and the impedance angle has dropped almost 17° from the purely capacitive level.

Continuing:

f = 5 kHz: $\mathbf{Z}_T = \mathbf{5.93\ k\Omega\ \angle{-32.48°}}$

f = 10 kHz: $\mathbf{Z}_T = \mathbf{5.25\ k\Omega\ \angle{-17.66°}}$

f = 15 kHz: $\mathbf{Z}_T = \mathbf{5.11\ k\Omega\ \angle{-11.98°}}$

f = 20 kHz: $\mathbf{Z}_T = \mathbf{5.06\ k\Omega\ \angle{-9.04°}}$

Note how close the magnitude of Z_T at f = 20 kHz is to the resistance level of 5 kΩ. In addition, note how the phase angle is approaching that associated with a pure resistive network (0°).

A plot of Z_T versus frequency in Fig. 15.47 completely supports our assumption based on the curves of Fig. 15.46. The plot of θ_T versus frequency in Fig. 15.48 further suggests that the total impedance changed from one of a capacitive nature (θ_T = −90°) to one with resistive characteristics (θ_T = 0°).

V_C:

Applying the voltage divider rule to determine the voltage across the capacitor in phasor form gives us

$$\mathbf{V}_C = \frac{\mathbf{Z}_C\mathbf{E}}{\mathbf{Z}_R + \mathbf{Z}_C}$$
$$= \frac{(X_C\angle{-90°})(E\angle 0°)}{R - jX_C} = \frac{X_C E\angle{-90°}}{R - jX_C}$$
$$= \frac{X_C E\angle{-90°}}{\sqrt{R^2 + X_C^2}\ \angle{-\tan^{-1}X_C/R}}$$

or $$\mathbf{V}_C = V_C\angle\theta_C = \frac{X_C E}{\sqrt{R^2 + X_C^2}}\angle{-90° + \tan^{-1}(X_C/R)}$$

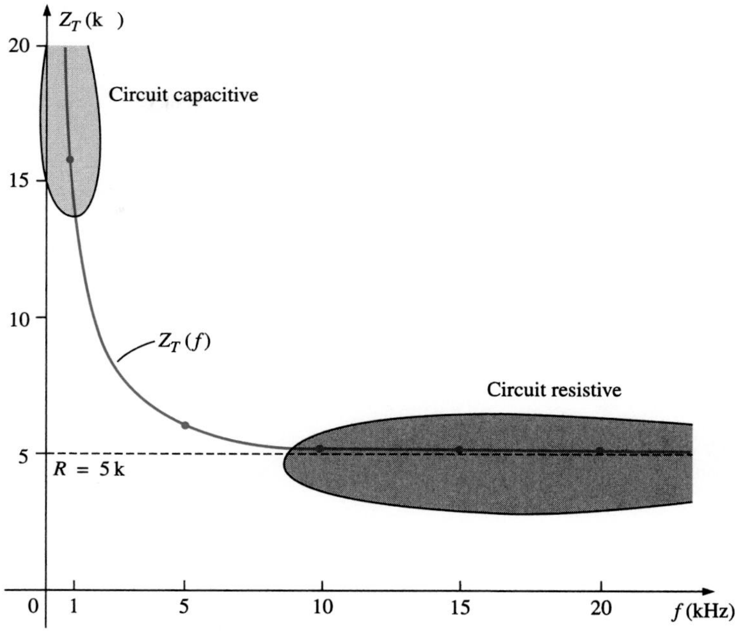

FIG. 15.47
The magnitude of the input impedance versus frequency for the circuit of Fig. 15.45.

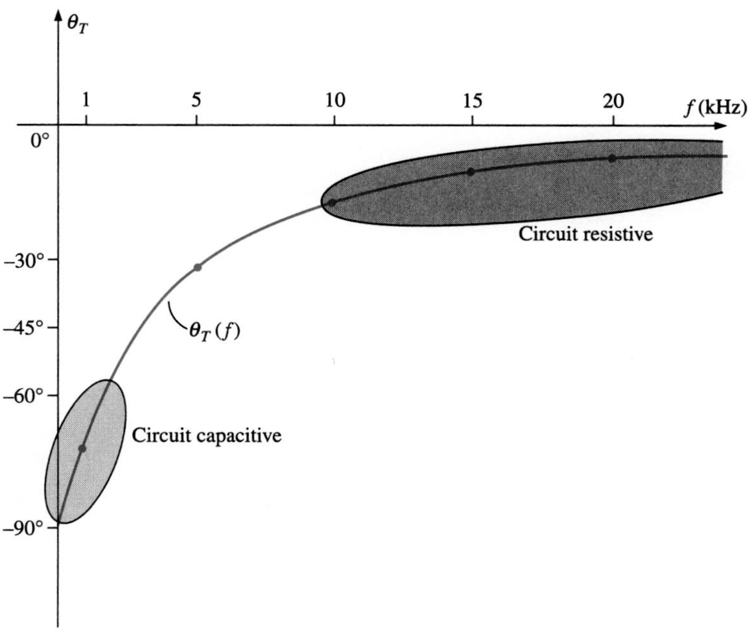

FIG. 15.48
The phase angle of the input impedance versus frequency for the circuit of Fig. 15.45.

The magnitude of $\mathbf{V}_C$ is therefore determined by

$$V_C = \frac{X_C E}{\sqrt{R^2 + X_C^2}} \tag{15.13}$$

and the phase angle θ_C by which $\mathbf{V}_C$ leads $\mathbf{E}$ is given by

$$\theta_C = -90^\circ + \tan^{-1}\frac{X_C}{R} = -\tan^{-1}\frac{R}{X_C} \tag{15.14}$$

To determine the frequency response, X_C must be calculated for each frequency of interest and inserted into Eqs. (15.13) and (15.14).

The first step in any analysis of an ac circuit is to consider the dc conditions, that is, the case when $f = 0$ Hz.

f = 0 Hz:

$$X_C = \frac{1}{2\pi(0)C} = \frac{1}{0} \Rightarrow \text{very large value}$$

Applying the open-circuit equivalent for the capacitor based on the above calculation will result in the following:

$$\mathbf{V}_C = \mathbf{E} = 10 \text{ V} \angle 0^\circ$$

If we apply Eq. (15.13), we find

$$X_C^2 >> R^2$$

and $$\sqrt{R^2 + X_C^2} \cong \sqrt{X_C^2} = X_C$$

and $$V_C = \frac{X_C E}{\sqrt{R^2 + X_C^2}} = \frac{X_C E}{X_C} = E$$

with $$\theta_C = -\tan^{-1}\frac{R}{X_C} = -\tan^{-1} 0 = 0^\circ$$

verifying the above conclusions.

f = 1 kHz:

Applying Eq. (15.13):

$$X_C = \frac{1}{2\pi fC} = \frac{1}{(2\pi)(1 \times 10^3 \text{ Hz})(0.01 \times 10^{-6} \text{ F})} \cong \mathbf{15.92\ k\Omega}$$

$$\sqrt{R^2 + X_C^2} = \sqrt{(5 \text{ k}\Omega)^2 + (15.92 \text{ k}\Omega)^2} \cong 16.69 \text{ k}\Omega$$

and $$V_C = \frac{X_C E}{\sqrt{R^2 + X_C^2}} = \frac{(15.92 \text{ k}\Omega)(10)}{16.69 \text{ k}\Omega} = \mathbf{9.54\ V}$$

Applying Eq. (15.14):

$$\theta_C = -\tan^{-1}\frac{R}{X_C} = -\tan^{-1}\frac{5 \text{ k}\Omega}{15.9 \text{ k}\Omega}$$

$$= -\tan^{-1} 0.314 = \mathbf{-17.46^\circ}$$

and $$\mathbf{V}_C = \mathbf{9.53\ V \angle -17.46^\circ}$$

As expected, the high reactance of the capacitor at low frequencies has resulted in the major part of the applied voltage appearing across the capacitor.

If we plot the phasor diagrams for $f = 0$ Hz and $f = 1$ kHz, as shown in Fig. 15.49, we find that $\mathbf{V}_C$ is beginning a clockwise rotation with an increase in frequency that will increase the angle θ_C and decrease the phase angle between **I** and **E.** Recall that for a purely

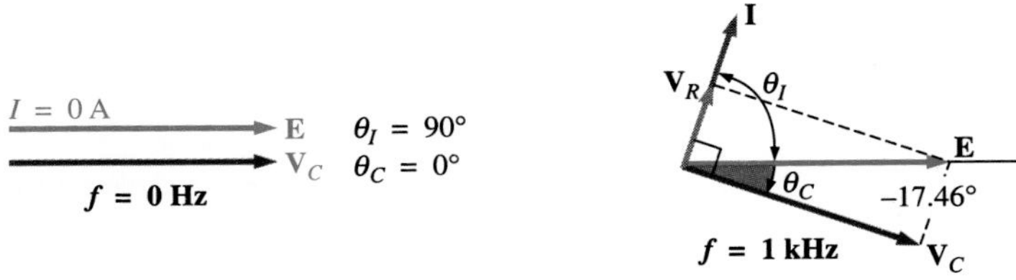

FIG. 15.49
The phasor diagram for the circuit of Fig. 15.45 for $f = 0$ Hz and 1 kHz.

capacitive network, **I** leads **E** by 90°. As the frequency increases, therefore, the capacitive reactance is decreasing, and eventually $R >> X_C$ with $\theta_C = -90°$, and the angle between **I** and **E** will approach 0°. Keep in mind as we proceed through the other frequencies that θ_C is the phase angle between $\mathbf{V}_C$ and **E** and that the magnitude of the angle by which **I** leads **E** is determined by

$$|\theta_I| = 90° - |\theta_C| \qquad (15.15)$$

$f = 5$ kHz:

Applying Eq. (15.13):

$$X_C = \frac{1}{2\pi fC} = \frac{1}{(2\pi)(5 \times 10^3 \text{ Hz})(0.01 \times 10^{-6} \text{ F})} \cong \mathbf{3.18\ k\Omega}$$

Note the dramatic drop in X_C from 1 kHz to 5 kHz. In fact, X_C is now less than the resistance R of the network, and the phase angle determined by $\tan^{-1}(X_C/R)$ must be less than 45°. Here,

$$V_C = \frac{X_C E}{\sqrt{R^2 + X_C^2}} = \frac{(3.18\text{ k}\Omega)(10\text{ V})}{\sqrt{(5\text{ k}\Omega)^2 + (3.18\text{ k}\Omega)^2}} = \mathbf{5.37\ V}$$

with $\theta_C = -\tan^{-1}\frac{R}{X_C} = -\tan^{-1}\frac{5\text{ k}\Omega}{3.2\text{ k}\Omega}$

$$= -\tan^{-1} 1.56 = \mathbf{-57.38°}$$

$f = 10$ kHz:

$$X_C \cong \mathbf{1.59\ k\Omega},\ V_C = \mathbf{3.03\ V},\ \theta_C = \mathbf{-72.34°}$$

$f = 15$ kHz:

$$X_C \cong \mathbf{1.06\ k\Omega},\ V_C = \mathbf{2.07\ V},\ \theta_C = \mathbf{-78.02°}$$

$f = 20$ kHz:

$$X_C \cong \mathbf{795.78\ \Omega},\ V_C = \mathbf{1.57\ V},\ \theta_C = \mathbf{-80.96°}$$

The phasor diagrams for $f = 5$ kHz and $f = 20$ kHz appear in Fig. 15.50 to show the continuing rotation of the $\mathbf{V}_C$ vector.

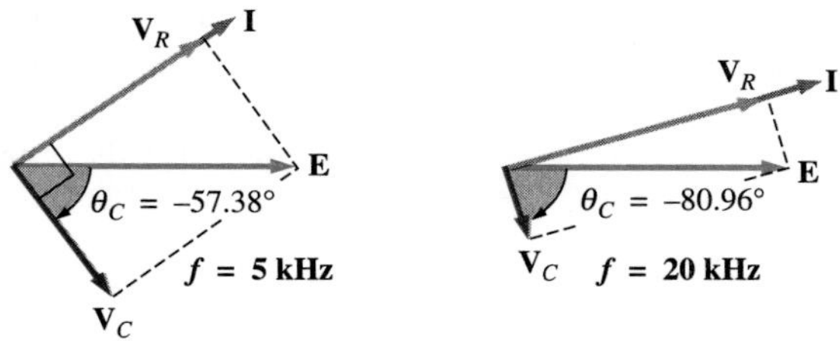

FIG. 15.50
The phasor diagram for the circuit of Fig. 15.45 for $f = 5$ kHz and 20 kHz.

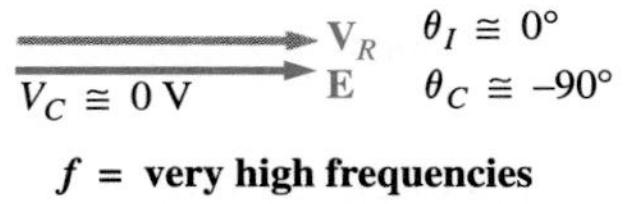

$\mathbf{V}_R$ $\theta_I \cong 0°$
$V_C \cong 0$ V E $\theta_C \cong -90°$
f = **very high frequencies**

FIG. 15.51
The phasor diagram for the circuit of Fig. 15.45 at very high frequencies.

Note also from Figs. 15.49 and 15.50 that the vector $\mathbf{V}_R$ and the current **I** have grown in magnitude with the reduction in the capacitive reactance. Eventually, at very high frequencies X_C will approach zero ohms and the short-circuit equivalent can be applied, resulting in $V_C \cong 0$ V and $\theta_C \cong -90°$, and producing the phasor diagram of Fig. 15.51. The network is then resistive and the phase angle between **I** and **E** is essentially zero degrees, and V_R and I are their maximum values.

A plot of V_C versus frequency appears in Fig. 15.52. At low frequencies $X_C >> R$ and V_C is very close to E in magnitude. As the applied frequency increases, X_C decreases in magnitude along with V_C as V_R captures more of the applied voltage. A plot of θ_C versus frequency is provided in Fig. 15.53. At low frequencies the phase angle between $\mathbf{V}_C$ and **E** is very small since $\mathbf{V}_C \cong \mathbf{E}$. Recall that if two phasors are equal, they must have the same angle. As the applied frequency increases, the network becomes more resistive and the phase angle between $\mathbf{V}_C$ and **E** approaches 90°.

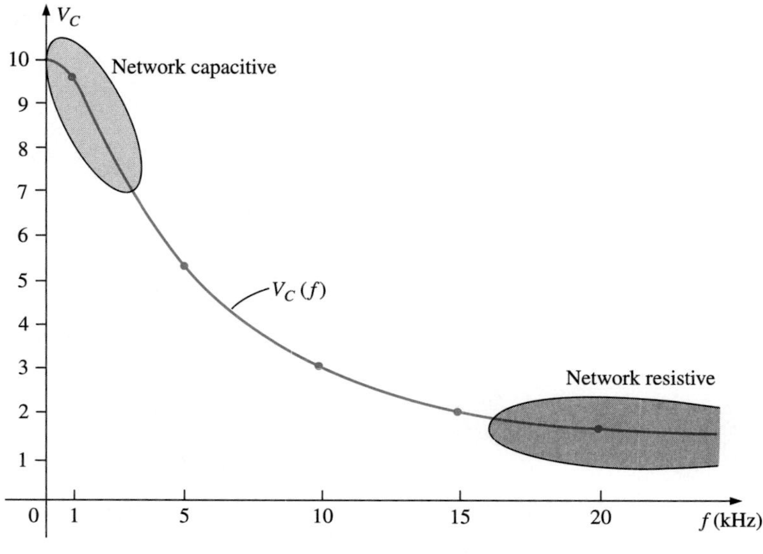

FIG. 15.52
The magnitude of the voltage V_C versus frequency for the circuit of Fig. 15.45.

A plot of V_R versus frequency would approach E volts from zero volts with an increase in frequency, but remember $V_R \neq E - V_C$ due to the vector relationship. The phase angle between **I** and **E** could be plotted directly from Fig. 15.53 using Eq. (15.15).

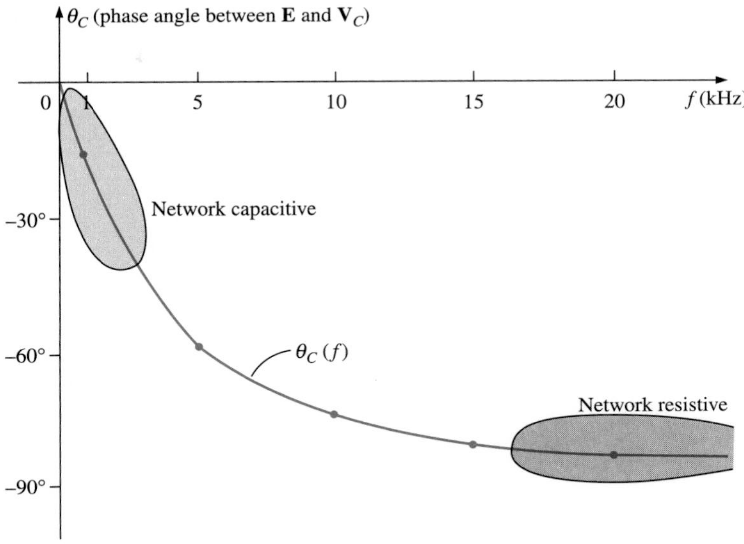

FIG. 15.53
The phase angle between **E** *and* $\mathbf{V}_C$ *versus frequency for the circuit of Fig. 15.45.*

From this analysis, you should note that any network connected across the capacitor will receive the greatest potential level at low frequencies and be effectively *shorted out* at very high frequencies. Similarly, the analysis of a series *R-L* circuit would proceed in much the same way, except that X_L and V_L will increase with frequency and the angle between **I** and **E** will approach 90° (voltage leading the current) rather than 0°. If V_L were plotted versus frequency, $\mathbf{V}_L$ would approach **E**, and X_L would eventually attain a level at which the *open-circuit* equivalent would be appropriate.

The analysis of this section will be extended in Chapter 21 to a much wider frequency range using a logarithmic axis for frequency. We will show there that an *R-C* circuit like the one in Fig.15.45 can be used as a filter to determine which frequencies will have the greatest impact on the stage to follow.

15.6 SUMMARY—SERIES ac CIRCUITS

The following is a review of important conclusions that can be made from the discussion and examples of the previous sections. The list is not all-inclusive, but it does emphasize some of the conclusions that should be carried forward in the future analysis of ac systems.

For series ac circuits with reactive elements:

1. ***The total impedance will be frequency dependent.***
2. ***The impedance of any one element can be greater than the total impedance of the network.***

3. ***The inductive and capacitive reactances are always in direct opposition on an impedance diagram.***
4. ***Depending on the frequency applied, the same circuit can be either predominantly inductive or capacitive.***
5. ***At lower frequencies the capacitive elements will usually have the most impact on the total impedance, while at high frequencies the inductive elements will usually have the most impact.***
6. ***The magnitude of the voltage across any one element can be greater than the applied voltage.***
7. ***The magnitude of the voltage across an element as compared to the other elements of the circuit is directly related to the magnitude of its impedance; that is, the larger the impedance of an element, the larger the magnitude of the voltage across the element.***
8. ***The voltages across an inductor or capacitor are always in direct opposition on a phasor diagram.***
9. ***The current is always in phase with the voltage across the resistive elements, lags the voltage across all the inductive elements by 90°, and leads the voltage across all the capacitive elements by 90°.***
10. ***The larger the resistive element of a circuit compared to the net reactive impedance, the closer the power factor is to unity.***

15.7 ADMITTANCE AND SUSCEPTANCE

The discussion for *parallel ac circuits* will be very similar to the one for dc circuits. In dc circuits, **conductance** (G) was defined as being equal to $1/R$. The total conductance of a parallel circuit was then found by adding the conductance of each branch. The total resistance R_T is simply $1/G_T$.

In ac circuits, we define **admittance** ($\mathbf{Y}$) as being equal to $1/\mathbf{Z}$. The unit of measure for admittance as defined by the SI system is **siemens**, which has the symbol S. Admittance is a measure of how well an ac circuit will *admit*, or allow, current to flow in the circuit. The larger its value, therefore, the heavier the current flow for the same applied potential. The total admittance of a circuit can also be found by finding the sum of the parallel admittances. The total impedance $\mathbf{Z}_T$ of the circuit is then $1/\mathbf{Y}_T$; that is, for the network of Fig. 15.54:

$$\mathbf{Y}_T = \mathbf{Y}_1 + \mathbf{Y}_2 + \mathbf{Y}_3 + \cdots + \mathbf{Y}_N \tag{15.16}$$

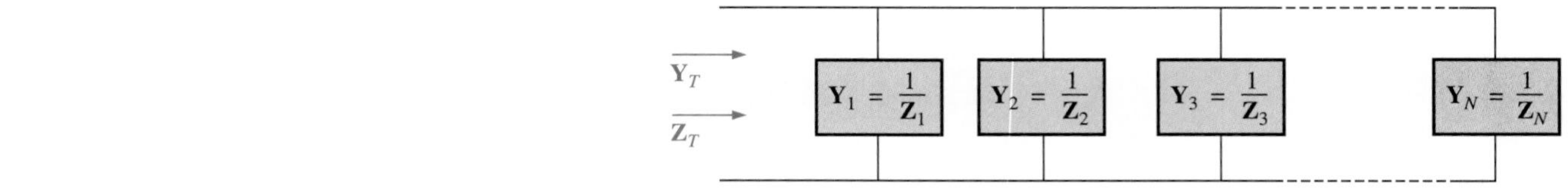

FIG. 15.54
Parallel ac network.

or, since $\mathbf{Z} = 1/\mathbf{Y}$,

$$\frac{1}{\mathbf{Z}_T} = \frac{1}{\mathbf{Z}_1} + \frac{1}{\mathbf{Z}_2} + \frac{1}{\mathbf{Z}_3} + \cdots + \frac{1}{\mathbf{Z}_N} \tag{15.17}$$

For two impedances in parallel,

$$\frac{1}{\mathbf{Z}_T} = \frac{1}{\mathbf{Z}_1} + \frac{1}{\mathbf{Z}_2}$$

If the manipulations used in Chapter 6 to find the total resistance of two parallel resistors are now applied, the following similar equation will result:

$$\mathbf{Z}_T = \frac{\mathbf{Z}_1\mathbf{Z}_2}{\mathbf{Z}_1 + \mathbf{Z}_2} \qquad (15.18)$$

For three parallel impedances,

$$\mathbf{Z}_T = \frac{\mathbf{Z}_1\mathbf{Z}_2\mathbf{Z}_3}{\mathbf{Z}_1\mathbf{Z}_2 + \mathbf{Z}_2\mathbf{Z}_3 + \mathbf{Z}_1\mathbf{Z}_3} \qquad (15.19)$$

As pointed out in the introduction to this section, conductance is the reciprocal of resistance, and

$$\mathbf{Y}_R = \frac{1}{\mathbf{Z}_R} = \frac{1}{R \angle 0°} = G \angle 0° \qquad (15.20)$$

The reciprocal of reactance ($1/X$) is called **susceptance** and is a measure of how *susceptible* an element is to the passage of current through it. Susceptance is also measured in *siemens* and is represented by the capital letter B.

For the inductor,

$$\mathbf{Y}_L = \frac{1}{\mathbf{Z}_L} = \frac{1}{X_L \angle 90°} = \frac{1}{X_L} \angle -90° \qquad (15.21)$$

Defining $$B_L = \frac{1}{X_L} \quad \text{(siemens, S)} \qquad (15.22)$$

we have $$\mathbf{Y}_L = B_L \angle -90° \qquad (15.23)$$

Note that for inductance, an increase in frequency or inductance will result in a decrease in susceptance or, correspondingly, in admittance.

For the capacitor,

$$\mathbf{Y}_C = \frac{1}{\mathbf{Z}_C} = \frac{1}{X_C \angle -90°} = \frac{1}{X_C} \angle 90° \qquad (15.24)$$

Defining $$B_C = \frac{1}{X_C} \quad \text{(siemens, S)} \qquad (15.25)$$

we have $$\mathbf{Y}_C = B_C \angle 90° \qquad (15.26)$$

For the capacitor, therefore, an increase in frequency or capacitance will result in an increase in its susceptibility.

For parallel ac circuits, the *admittance diagram* is used with the three admittances, represented as shown in Fig. 15.55.

Note in Fig. 15.55 that the conductance (like resistance) is on the positive real axis, whereas inductive and capacitive susceptances are in direct opposition on the imaginary axis.

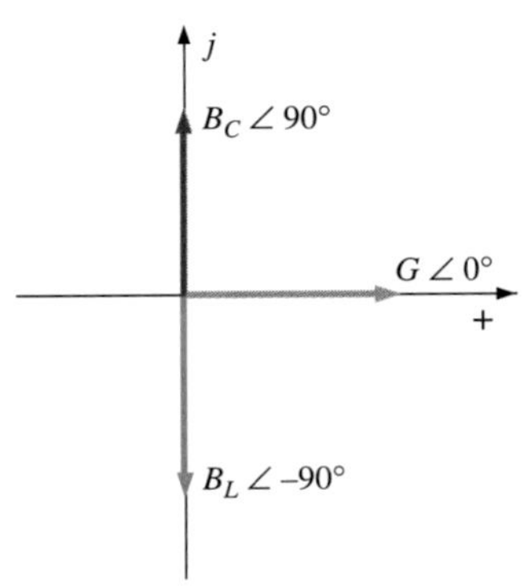

FIG. 15.55
Admittance diagram.

For any configuration (series, parallel, series-parallel, etc.), the angle associated with the total admittance is the angle by which the source current leads the applied voltage. For inductive networks, θ_T is negative, whereas for capacitive networks, θ_T is positive.

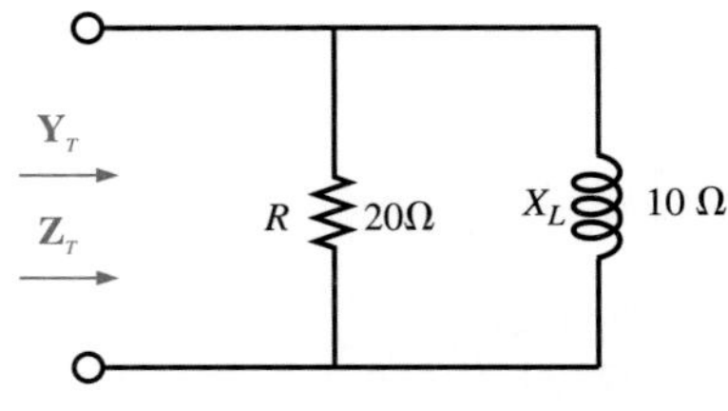

FIG. 15.56
Example 15.12.

EXAMPLE 15.12

For the network of Fig. 15.56:

a. Find the admittance of each parallel branch.
b. Determine the input admittance.
c. Calculate the input impedance.
d. Draw the admittance diagram.

Solutions:

a. $\mathbf{Y}_R = G\angle 0° = \dfrac{1}{R}\angle 0° = \dfrac{1}{20\ \Omega}\angle 0°$

$= \mathbf{0.05\ S\angle 0° = 0.05\ S} + \boldsymbol{j}\ \mathbf{0}$

$\mathbf{Y}_L = B_L\angle -90° = \dfrac{1}{X_L}\angle -90° = \dfrac{1}{10\ \Omega}\angle -90°$

$= \mathbf{0.1\ S\angle -90° = 0} - \boldsymbol{j}\ \mathbf{0.1\ S}$

b. $\mathbf{Y}_T = \mathbf{Y}_R + \mathbf{Y}_L = (0.05\ \text{S} + j\ 0) + (0 - j\ 0.1\ \text{S})$

$= \mathbf{0.05\ S} - \boldsymbol{j}\ \mathbf{0.1\ S} = \mathbf{0.112S\ \angle\ -63.43°}$

c. $\mathbf{Z}_T = \dfrac{1}{\mathbf{Y}_T} = \dfrac{1}{0.112\ \text{S}\angle -63.43°}$

$= \mathbf{8.93\ \Omega\ \angle 63.43°}$

or Eq. (15.17):

$$\mathbf{Z}_T = \frac{\mathbf{Z}_R\mathbf{Z}_L}{\mathbf{Z}_R + \mathbf{Z}_L} = \frac{(20\ \Omega\angle 0°)(10\ \Omega\angle 90°)}{20\ \Omega + j\ 10\ \Omega}$$

$$= \frac{200\ \Omega\angle 90°}{22.36\angle 26.57°} = \mathbf{8.93\ \Omega\ \angle 63.43°}$$

d. The admittance diagram appears in Fig. 15.57.

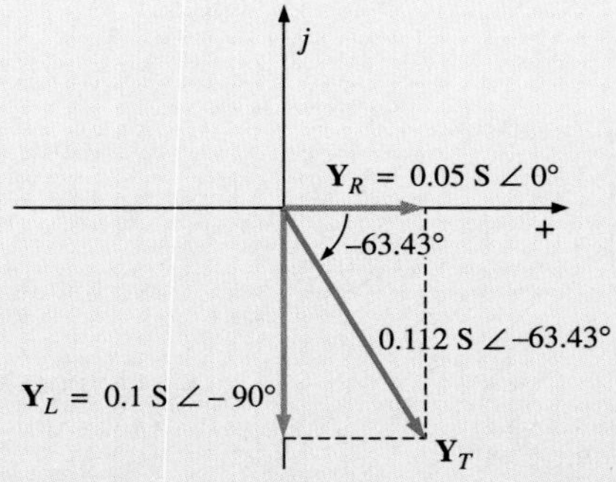

FIG. 15.57
Admittance diagram for the network of Fig. 15.56.

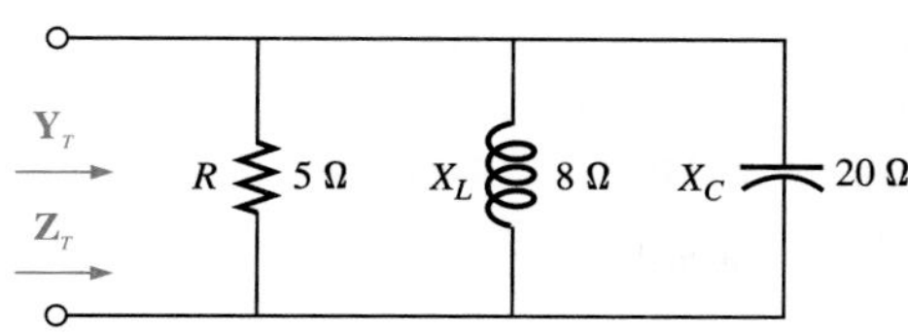

FIG. 15.58
Example 15.13.

EXAMPLE 15.13

Repeat Example 15.12 for the parallel network of Fig. 15.58.

Solutions:

a. $$\mathbf{Y}_R = G \angle 0° = \frac{1}{R} \angle 0° = \frac{1}{5\ \Omega} \angle 0°$$
$$= \mathbf{0.2\ S \angle 0° = 0.2\ S + j\ 0}$$
$$\mathbf{Y}_L = B_L \angle -90° = \frac{1}{X_L} \angle -90° = \frac{1}{8\ \Omega} \angle -90°$$
$$= \mathbf{0.125\ S \angle -90° = 0 - j\ 0.125\ S}$$
$$\mathbf{Y}_C = B_C \angle 90° = \frac{1}{X_C} \angle 90° = \frac{1}{20\ \Omega} \angle 90°$$
$$= \mathbf{0.050\ S \angle +90° = 0 + j\ 0.050\ S}$$

b. $$\mathbf{Y}_T = \mathbf{Y}_R + \mathbf{Y}_L + \mathbf{Y}_C$$
$$= (0.2\ S + j\ 0) + (0 - j\ 0.125\ S) + (0 + j\ 0.050\ S)$$
$$= 0.2\ S - j\ 0.075\ S = \mathbf{0.2136\ S \angle -20.56°}$$

c. $$\mathbf{Z}_T = \frac{1}{0.2136\ S \angle -20.56°} = \mathbf{4.68\ \Omega \angle 20.56°}$$

or

$$\mathbf{Z}_T = \frac{\mathbf{Z}_R\mathbf{Z}_L\mathbf{Z}_C}{\mathbf{Z}_R\mathbf{Z}_L + \mathbf{Z}_L\mathbf{Z}_C + \mathbf{Z}_R\mathbf{Z}_C}$$
$$= \frac{(5\ \Omega \angle 0°)(8\ \Omega \angle 90°)(20\ \Omega \angle -90°)}{(5\ \Omega \angle 0°)(8\ \Omega \angle 90°) + (8\ \Omega \angle 90°)(20\ \Omega \angle -90°) + (5\ \Omega \angle 0°)(20\ \Omega \angle -90°)}$$
$$= \frac{800\ \Omega \angle 0°}{40 \angle 90° + 160 \angle 0° + 100 \angle -90°}$$
$$= \frac{800\ \Omega}{160 + j\ 40 - j\ 100} = \frac{800\ \Omega}{160 - j\ 60}$$
$$= \frac{800\ \Omega}{170.88 \angle -20.56°}$$
$$= \mathbf{4.68\ \Omega \angle 20.56°}$$

d. The admittance diagram appears in Fig. 15.59.

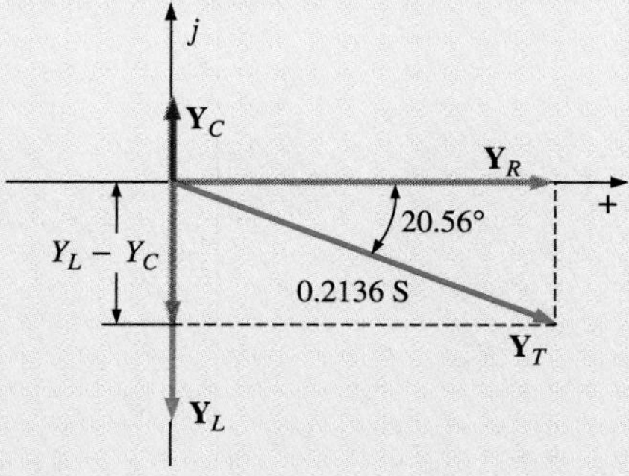

FIG. 15.59
Admittance diagram for the network of Fig. 15.58.

Often, the inverse relationship $\mathbf{Y}_T = 1/\mathbf{Z}_T$ or $\mathbf{Z}_T = 1/\mathbf{Y}_T$ will require that we divide the number 1 by a complex number having a real and an imaginary part. This division, if not performed in the polar form, requires that we multiply the numerator and denominator by the conjugate of the denominator, as follows:

$$\mathbf{Y}_T = \frac{1}{\mathbf{Z}_T} = \frac{1}{4\,\Omega + j\,6\,\Omega} = \left(\frac{1}{4\,\Omega + j\,6\,\Omega}\right)\left(\frac{(4\,\Omega - j\,6\,\Omega)}{(4\,\Omega - j\,6\,\Omega)}\right) = \frac{4 - j\,6}{4^2 + 6^2}$$

and

$$\mathbf{Y}_T = \frac{4}{52}\,\text{S} - j\,\frac{6}{52}\,\text{S}$$

To avoid this long and slow task each time we want to find the reciprocal of a complex number in rectangular form, a format can be developed using the following complex number, which is symbolic of any impedance or admittance in the first or fourth quadrant:

$$\frac{1}{a_1 \pm j\,b_1} = \left(\frac{1}{a_1 \pm j\,b_1}\right)\left(\frac{a_1 \mp j\,b_1}{a_1 \mp j\,b_1}\right) = \frac{a_1 \mp j\,b_1}{a_1^2 + b_1^2}$$

or

$$\frac{1}{a_1 \pm j\,b_1} = \frac{a_1}{a_1^2 + b_1^2} \mp j\,\frac{b_1}{a_1^2 + b_1^2} \tag{15.27}$$

Note that the denominator is simply the sum of the squares of each term. The sign is inverted between the real and imaginary parts. A few examples will develop some familiarity with the use of this equation.

EXAMPLE 15.14

Find the admittance of each set of series elements in Fig. 15.60.

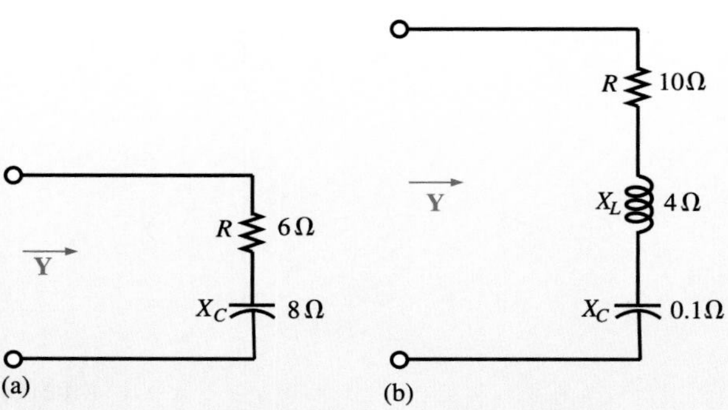

FIG. 15.60
Example 15.14.

Solutions:

a. $\mathbf{Z} = R - j\,X_C = 6\,\Omega - j\,8\,\Omega$

Eq. (15.27): $$\mathbf{Y} = \frac{1}{6\,\Omega - j\,8\,\Omega} = \frac{6}{(6)^2 + (8)^2} + j\,\frac{8}{(6)^2 + (8)^2}$$

$$= \frac{\mathbf{6}}{\mathbf{100}}\,\mathbf{S} + j\,\frac{\mathbf{8}}{\mathbf{100}}\,\mathbf{S}$$

b. $\mathbf{Z} = 10\,\Omega + j\,4\,\Omega + (-j\,0.1\,\Omega) = 10\,\Omega + j\,3.9\,\Omega$

Eq. (15.27):

$$\mathbf{Y} = \frac{1}{\mathbf{Z}} = \frac{1}{10\,\Omega + j\,3.9\,\Omega} = \frac{10}{(10)^2 + (3.9)^2} - j\,\frac{3.9}{(10)^2 + (3.9)^2}$$

$$= \frac{10}{115.21} - j\,\frac{3.9}{115.21} = \mathbf{0.087\,S - j\,0.034\,S}$$

15.8 PARALLEL ac NETWORKS

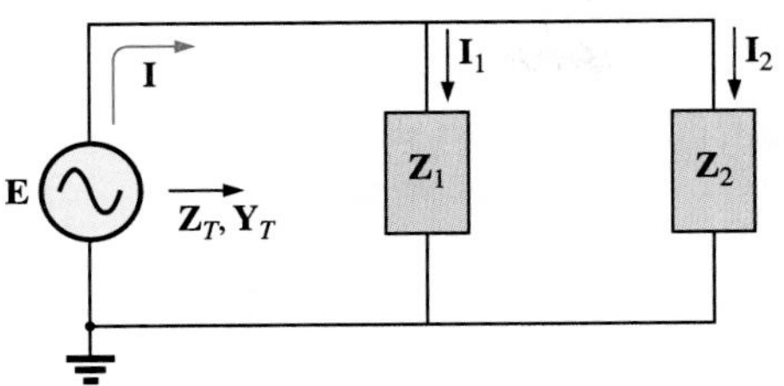

FIG. 15.61
Parallel ac network.

For the representative parallel ac network of Fig. 15.61, the total impedance or admittance is determined as described in the previous section, and the source current is determined by Ohm's law as follows:

$$\mathbf{I} = \frac{\mathbf{E}}{\mathbf{Z}_T} = \mathbf{E}\mathbf{Y}_T \tag{15.28}$$

Since the voltage is the same across parallel elements, the current through each branch can then be found through another application of Ohm's law:

$$\mathbf{I}_1 = \frac{\mathbf{E}}{\mathbf{Z}_1} = \mathbf{E}\mathbf{Y}_1 \tag{15.29a}$$

$$\mathbf{I}_2 = \frac{\mathbf{E}}{\mathbf{Z}_2} = \mathbf{E}\mathbf{Y}_2 \tag{15.29b}$$

Kirchhoff's current law can then be applied as it was for dc networks (Section 6.5). However, keep in mind that we are now dealing with the algebraic manipulation of quantities that have both magnitude and direction.

$$\mathbf{I} - \mathbf{I}_1 - \mathbf{I}_2 = 0$$

or

$$\mathbf{I} = \mathbf{I}_1 + \mathbf{I}_2 \tag{15.30}$$

The power to the network can be determined by

$$P = EI\cos\theta_T \tag{15.31}$$

where θ_T is the phase angle between **E** and **I**.

Let us now look at a few examples carried out in great detail for the first one.

R-L

Refer to Fig. 15.62.

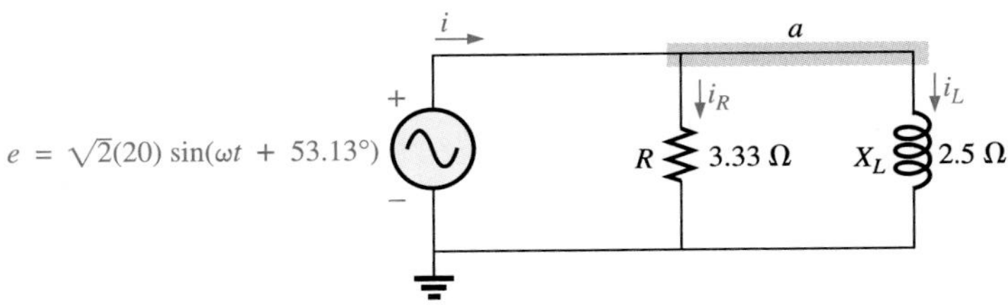

FIG. 15.62
Parallel R-L network.

Phasor notation: As shown in Fig. 15.63.

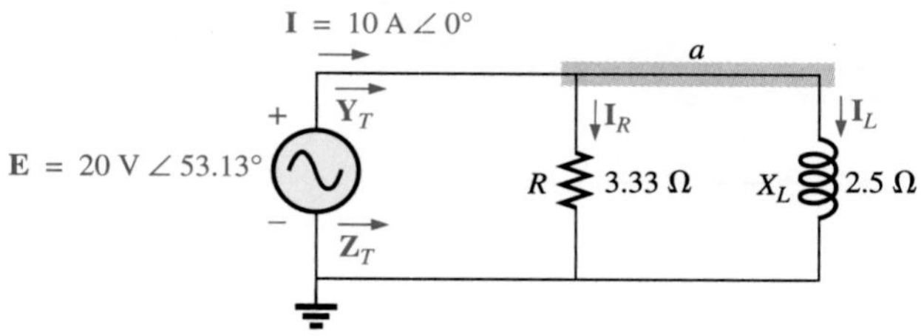

FIG. 15.63
Applying phasor notation to the network of Fig. 15.62.

$\mathbf{Y}_T$, $\mathbf{Z}_T$:

$$\mathbf{Y}_T = \mathbf{Y}_R + \mathbf{Y}_L$$
$$= G \angle 0^\circ + B_L \angle -90^\circ = \frac{1}{3.33\ \Omega} \angle 0^\circ + \frac{1}{2.5\ \Omega} \angle -90^\circ$$
$$= 0.3\text{ S} \angle 0^\circ + 0.4\text{ S} \angle -90^\circ = 0.3\text{ S} - j\,0.4\text{ S}$$
$$= \mathbf{0.5\ S} \angle \mathbf{-53.13^\circ}$$
$$\mathbf{Z}_T = \frac{1}{\mathbf{Y}_T} = \frac{1}{0.5\text{ S} \angle -53.13^\circ} = \mathbf{2\ \Omega} \angle \mathbf{53.13^\circ}$$

Admittance diagram: As shown in Fig. 15.64.

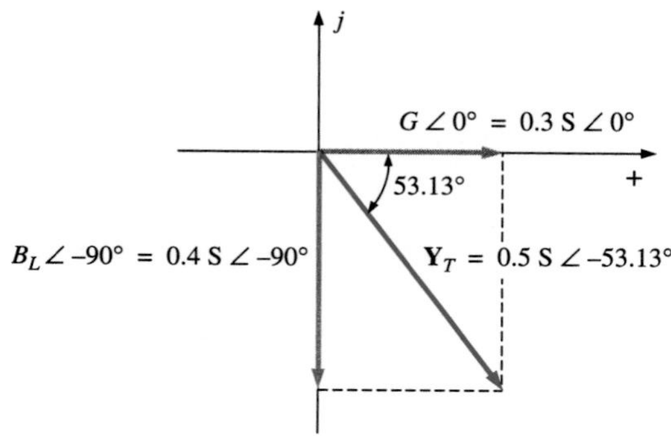

FIG. 15.64
Admittance diagram for the parallel R-L network of Fig. 15.62.

I:

$$\mathbf{I} = \frac{\mathbf{E}}{\mathbf{Z}_T} = \mathbf{E}\mathbf{Y}_T = (20\text{ V} \angle 53.13^\circ)(0.5\text{ S} \angle -53.13^\circ) = \mathbf{10\ A} \angle \mathbf{0^\circ}$$

$\mathbf{I}_R$, $\mathbf{I}_L$:

$$\mathbf{I}_R = \frac{E \angle \theta}{R \angle 0^\circ} = (E \angle \theta)(G \angle 0^\circ)$$
$$= (20\text{ V} \angle 53.13^\circ)(0.3\text{ S} \angle 0^\circ) = \mathbf{6\ A} \angle \mathbf{53.13^\circ}$$

$$\mathbf{I}_L = \frac{E \angle \theta}{X_L \angle 90^\circ} = (E \angle \theta)(B_L \angle -90^\circ)$$
$$= (20 \text{ V} \angle 53.13^\circ)(0.4 \text{ S} \angle -90^\circ)$$
$$= \mathbf{8 \text{ A} \angle -36.87^\circ}$$

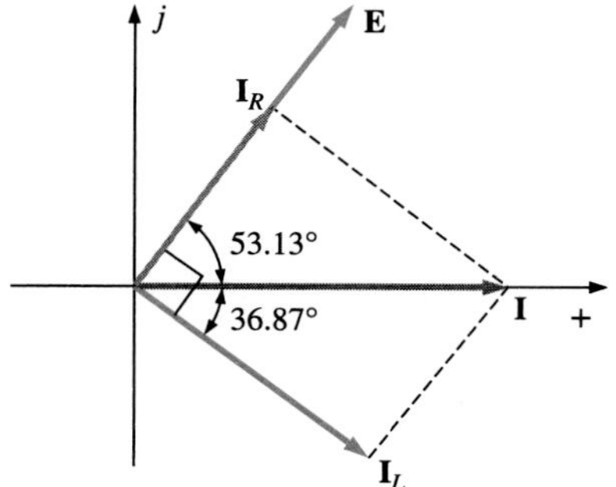

FIG. 15.65
Phasor diagram for the parallel R-L network of Fig. 15.62.

Kirchhoff's current law: At node *a*,

$$\mathbf{I} - \mathbf{I}_R - \mathbf{I}_L = 0$$

or

$$\mathbf{I} = \mathbf{I}_R + \mathbf{I}_L$$
$$10 \text{ A} \angle 0^\circ = 6 \text{ A} \angle 53.13^\circ + 8 \text{ A} \angle -36.87^\circ$$
$$10 \text{ A} \angle 0^\circ = (3.60 \text{ A} + j\,4.80 \text{ A}) + (6.40 \text{ A} - j\,4.80 \text{ A}) = 10 \text{ A} + j\,0$$

and $\mathbf{10 \text{ A} \angle 0^\circ = 10 \text{ A} \angle 0^\circ}$ (checks)

Phasor diagram: The phasor diagram of Fig. 15.65 indicates that the applied voltage **E** is in phase with the current $\mathbf{I}_R$ and leads the current $\mathbf{I}_L$ by 90°.

Power: The total power in watts delivered to the circuit is

$$P_T = EI \cos \theta_T$$
$$= (20 \text{ V})(10 \text{ A}) \cos 53.13^\circ = (200 \text{ W})(0.6)$$
$$= \mathbf{120 \text{ W}}$$

or $$P_T = I^2 R = \frac{V_R^2}{R} = V_R^2 G = (20 \text{ V})^2(0.3 \text{ S}) = \mathbf{120 \text{ W}}$$

or, finally,

$$P_T = P_R + P_L = EI_R \cos \theta_R + EI_L \cos \theta_L$$
$$= (20 \text{ V})(6 \text{ A}) \cos 0^\circ + (20 \text{ V})(8 \text{ A}) \cos 90^\circ = 120 \text{ W} + 0$$
$$= \mathbf{120 \text{ W}}$$

Power factor: The power factor of the circuit is

$$F_p = \cos \theta_T = \cos 53.13^\circ = \mathbf{0.6 \text{ lagging}}$$

or, through an analysis similar to the one done for a series ac circuit,

$$\cos \theta_T = \frac{P}{EI} = \frac{E^2/R}{EI} = \frac{EG}{I} = \frac{G}{I/V} = \frac{G}{Y_T}$$

and $$\boldsymbol{F_p = \cos \theta_T = \frac{G}{Y_T}} \qquad (15.32)$$

where G and Y_T are the magnitudes of the total conductance and admittance of the parallel network. For this case,

$$F_p = \cos \theta_T = \frac{0.3 \text{ S}}{0.5 \text{ S}} = \mathbf{0.6 \text{ lagging}}$$

Impedance approach: The current **I** can also be found by first finding the total impedance of the network:

$$\mathbf{Z}_T = \frac{\mathbf{Z}_R \mathbf{Z}_L}{\mathbf{Z}_R + \mathbf{Z}_L} = \frac{(3.33\ \Omega \angle 0^\circ)(2.5\ \Omega \angle 90^\circ)}{3.33\ \Omega \angle 0^\circ + 2.5\ \Omega \angle 90^\circ}$$

$$= \frac{8.325 \angle 90^\circ}{4.164 \angle 36.87^\circ} = \mathbf{2\ \Omega \angle 53.13^\circ}$$

And then, using Ohm's law, we obtain

$$\mathbf{I} = \frac{\mathbf{E}}{\mathbf{Z}_T} = \frac{20\text{ V}\angle 53.13^\circ}{2\ \Omega\angle 53.13^\circ} = \mathbf{10\text{ A}\angle 0^\circ}$$

R-C

Refer to Fig. 15.66.

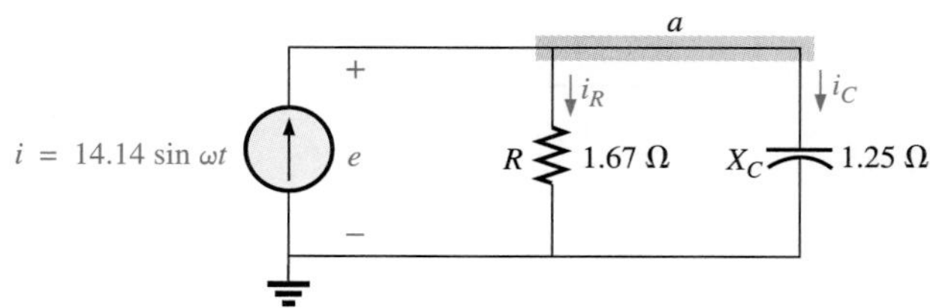

FIG. 15.66
Parallel R-C network.

Phasor notation: As shown in Fig. 15.67.

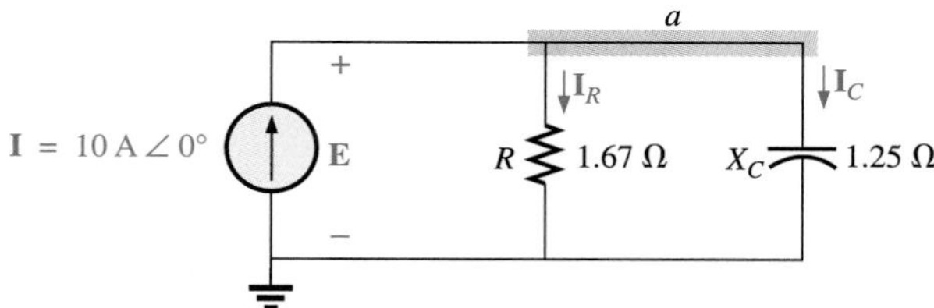

FIG. 15.67
Applying phasor notation to the network of Fig. 15.66.

$\mathbf{Y}_T$, $\mathbf{Z}_T$:

$$\mathbf{Y}_T = \mathbf{Y}_R + \mathbf{Y}_C = G\angle 0^\circ + B_C\angle 90^\circ = \frac{1}{1.67\ \Omega}\angle 0^\circ + \frac{1}{1.25\ \Omega}\angle 90^\circ$$

$$= 0.6\text{ S}\angle 0^\circ + 0.8\text{ S}\angle 90^\circ = 0.6\text{ S} + j\,0.8\text{ S} = \mathbf{1.0\text{ S}\angle 53.13^\circ}$$

$$\mathbf{Z}_T = \frac{1}{\mathbf{Y}_T} = \frac{1}{1.0\text{ S}\angle 53.13^\circ} = \mathbf{1\ \Omega\angle -53.13^\circ}$$

Admittance diagram: As shown in Fig. 15.68.

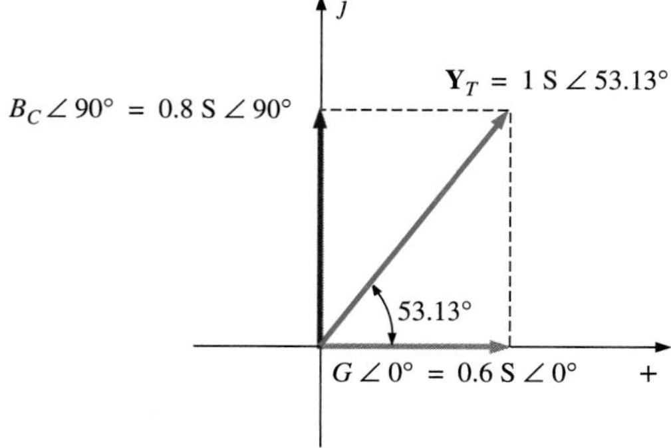

FIG. 15.68
Admittance diagram for the parallel R-C network of Fig. 15.66.

E:

$$\mathbf{E} = \mathbf{I}\mathbf{Z}_T = \frac{\mathbf{I}}{\mathbf{Y}_T} = \frac{10\text{ A}\angle 0^\circ}{1\text{ S}\angle 53.13^\circ} = \mathbf{10\text{ V}\angle -53.13^\circ}$$

$\mathbf{I}_R$, $\mathbf{I}_C$:

$$\begin{aligned}\mathbf{I}_R &= (E\angle\theta)(G\angle 0^\circ)\\ &= (10\text{ V}\angle -53.13^\circ)(0.6\text{ S}\angle 0^\circ) = \mathbf{6\text{ A}\angle -53.13^\circ}\\ \mathbf{I}_C &= (E\angle\theta)(B_C\angle 90^\circ)\\ &= (10\text{ V}\angle -53.13^\circ)(0.8\text{ S}\angle 90^\circ) = \mathbf{8\text{ A}\angle 36.87^\circ}\end{aligned}$$

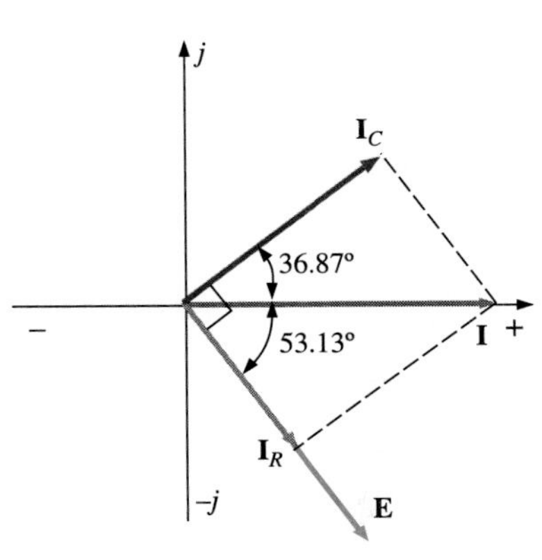

FIG. 15.69
Phasor diagram for the parallel R-C network of Fig. 15.66.

Kirchhoff's current law: At node *a*,

$$\mathbf{I} - \mathbf{I}_R - \mathbf{I}_C = 0$$

or

$$\mathbf{I} = \mathbf{I}_R + \mathbf{I}_C$$

which can also be verified (as for the *R-L* network) through vector algebra.

Phasor diagram: The phasor diagram of Fig. 15.69 indicates that **E** is in phase with the current through the resistor $\mathbf{I}_R$ and lags the capacitive current $\mathbf{I}_C$ by 90°.

Time domain:

$$\begin{aligned}e &= \sqrt{2}(10)\sin(\omega t - 53.13^\circ) = \mathbf{14.14\sin(\omega t - 53.13^\circ)}\\ i_R &= \sqrt{2}(6)\sin(\omega t - 53.13^\circ) = \mathbf{8.48\sin(\omega t - 53.13^\circ)}\\ i_C &= \sqrt{2}(8)\sin(\omega t + 36.87^\circ) = \mathbf{11.31\sin(\omega t + 36.87^\circ)}\end{aligned}$$

A plot of all of the currents and the voltage appears in Fig. 15.70. Note that e and i_R are in phase and e lags i_C by 90°.

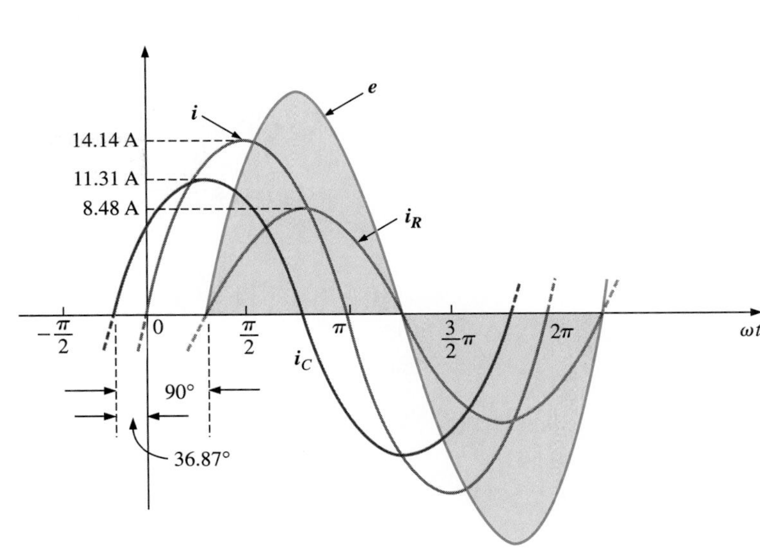

FIG. 15.70
Waveforms for the parallel R-C network of Fig. 15.66.

Power:

$$\begin{aligned}P_T &= EI\cos\theta = (10\text{ V})(10\text{ A})\cos 53.13^\circ = (10)^2(0.6)\\ &= \mathbf{60\text{ W}}\end{aligned}$$

or $$P_T = E^2G = (10\text{ V})^2(0.6\text{ S}) = \mathbf{60\text{ W}}$$

or, finally,

$$\begin{aligned}P_T = P_R + P_C &= EI_R \cos\theta_R + EI_C \cos\theta_C\\ &= (10\text{ V})(6\text{ A})\cos 0° + (10\text{ V})(8\text{ A})\cos 90°\\ &= \mathbf{60\text{ W}}\end{aligned}$$

Power factor: The power factor of the circuit is

$$F_p = \cos 53.13° = \mathbf{0.6\text{ leading}}$$

Using Eq. (15.32), we have

$$F_p = \cos\theta_T = \frac{G}{Y_T} = \frac{0.6\text{ S}}{1.0\text{ S}} = \mathbf{0.6\text{ leading}}$$

Impedance approach: The voltage **E** can also be found by first finding the total impedance of the circuit:

$$\begin{aligned}\mathbf{Z}_T = \frac{\mathbf{Z}_R\mathbf{Z}_C}{\mathbf{Z}_R + \mathbf{Z}_C} &= \frac{(1.67\ \Omega\angle 0°)(1.25\ \Omega\angle -90°)}{1.67\ \Omega\angle 0° + 1.25\ \Omega\angle -90°}\\ &= \frac{2.09\angle -90°}{2.09\angle -36.81°} = \mathbf{1\ \Omega\angle -53.19°}\end{aligned}$$

and then, using Ohm's law, we find

$$\mathbf{E} = \mathbf{IZ}_T = (10\text{ A}\angle 0°)(1\ \Omega\angle -53.19°) = \mathbf{10\text{ V}\angle -53.19°}$$

R-L-C

Refer to Fig. 15.71.

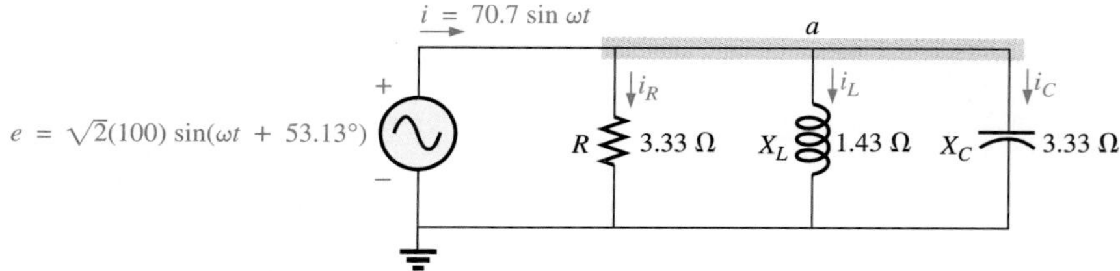

FIG. 15.71
Parallel R-L-C ac network.

Phasor notation: As shown in Fig. 15.72.

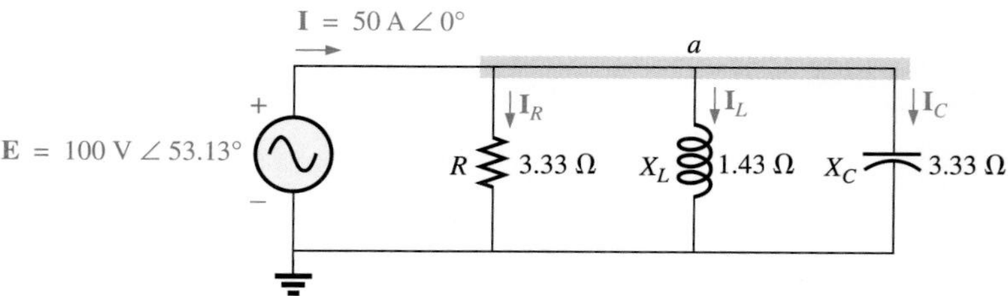

FIG. 15.72
Applying phasor notation to the network of Fig. 15.71.

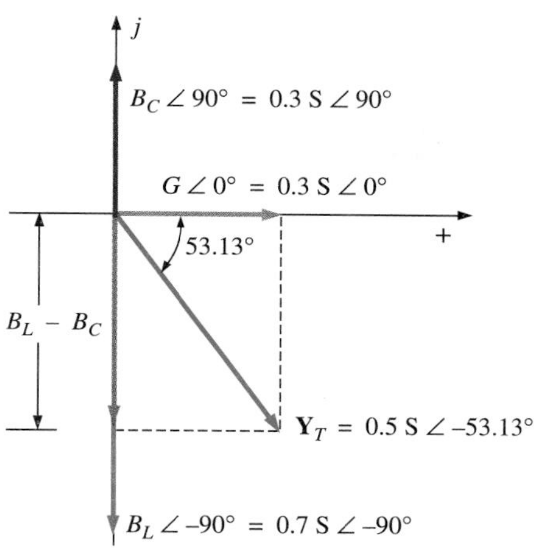

FIG. 15.73
Admittance diagram for the parallel R-L-C network of Fig. 15.71.

$\mathbf{Y}_T$, $\mathbf{Z}_T$:

$$\begin{aligned}\mathbf{Y}_T &= \mathbf{Y}_R + \mathbf{Y}_L + \mathbf{Y}_C = G \angle 0° + B_L \angle -90° + B_C \angle 90° \\ &= \frac{1}{3.33\ \Omega} \angle 0° + \frac{1}{1.43\ \Omega} \angle -90° + \frac{1}{3.33\ \Omega} \angle 90° \\ &= 0.3\text{ S} \angle 0° + 0.7\text{ S} \angle -90° + 0.3\text{ S} \angle 90° \\ &= 0.3\text{ S} - j\,0.7\text{ S} + j\,0.3\text{ S} \\ &= 0.3\text{ S} - j\,0.4\text{ S} = \mathbf{0.5\ S \angle -53.13°}\end{aligned}$$

$$\mathbf{Z}_T = \frac{1}{\mathbf{Y}_T} = \frac{1}{0.5\text{ S} \angle -53.13°} = \mathbf{2\ \Omega \angle 53.13°}$$

Admittance diagram: As shown in Fig. 15.73.

$\mathbf{I}$:

$$\mathbf{I} = \frac{\mathbf{E}}{\mathbf{Z}_T} = \mathbf{E}\mathbf{Y}_T = (100\text{ V} \angle 53.13°)(0.5\text{ S} \angle -53.13°) = \mathbf{50\ A \angle 0°}$$

$\mathbf{I}_R$, $\mathbf{I}_L$, $\mathbf{I}_C$:

$$\begin{aligned}\mathbf{I}_R &= (E \angle \theta)(G \angle 0°) \\ &= (100\text{ V} \angle 53.13°)(0.3\text{ S} \angle 0°) = \mathbf{30\ A \angle 53.13°} \\ \mathbf{I}_L &= (E \angle \theta)(B_L \angle -90°) \\ &= (100\text{ V} \angle 53.13°)(0.7\text{ S} \angle -90°) = \mathbf{70\ A \angle -36.87°} \\ \mathbf{I}_C &= (E \angle \theta)(B_C \angle 90°) \\ &= (100\text{ V} \angle 53.13°)(0.3\text{ S} \angle +90°) = \mathbf{30\ A \angle 143.13°}\end{aligned}$$

Kirchhoff's current law: At node *a*,

$$\mathbf{I} - \mathbf{I}_R - \mathbf{I}_L - \mathbf{I}_C = 0$$

or

$$\mathbf{I} = \mathbf{I}_R + \mathbf{I}_L + \mathbf{I}_C$$

Phasor diagram: The phasor diagram of Fig. 15.74 indicates that the impressed voltage **E** is in phase with the current $\mathbf{I}_R$ through the resistor, leads the current $\mathbf{I}_L$ through the inductor by 90°, and lags the current $\mathbf{I}_C$ of the capacitor by 90°.

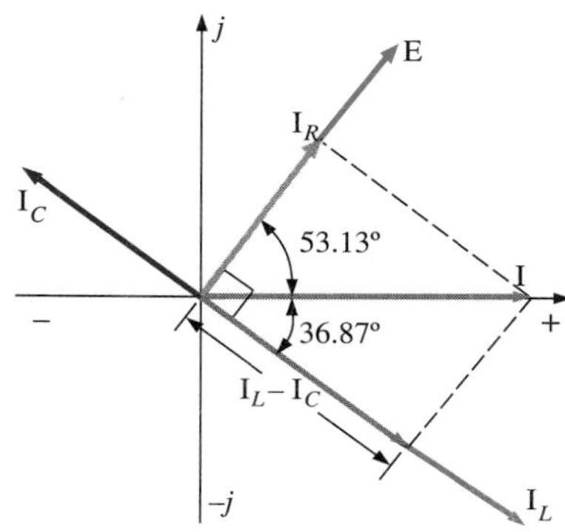

FIG. 15.74
Phasor diagram for the parallel R-L-C network of Fig. 15.71.

Time domain:

$$i = \sqrt{2}(50) \sin \omega t = \mathbf{70.70 \sin \omega t}$$
$$i_R = \sqrt{2}(30) \sin(\omega t + 53.13°) = \mathbf{42.42 \sin(\omega t + 53.13°)}$$
$$i_L = \sqrt{2}(70) \sin(\omega t - 36.87°) = \mathbf{98.98 \sin(\omega t - 36.87°)}$$
$$i_C = \sqrt{2}(30) \sin(\omega t + 143.13°) = \mathbf{42.42 \sin(\omega t + 143.13°)}$$

A plot of all of the currents and the impressed voltage appears in Fig. 15.75.

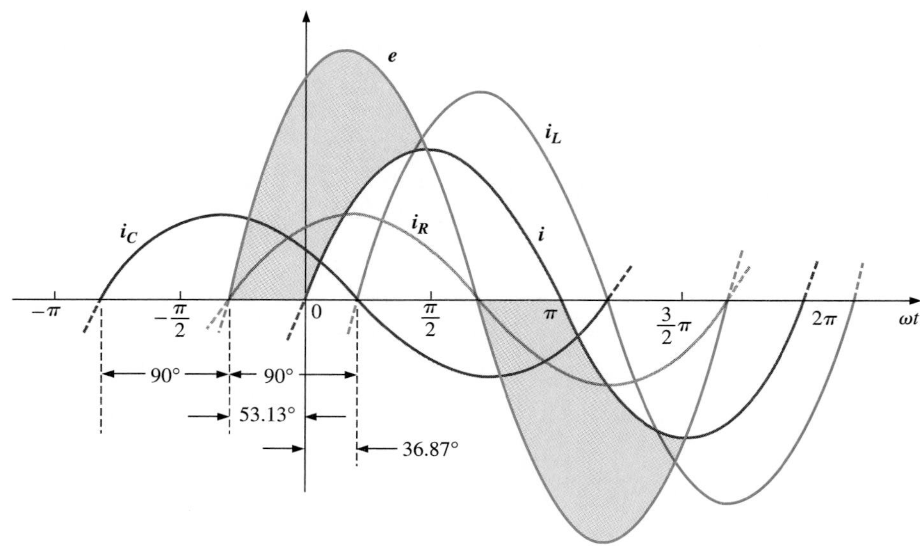

FIG. 15.75
Waveforms for the parallel R-L-C network of Fig. 15.71.

Power: The total power in watts delivered to the circuit is

$$P_T = EI \cos \theta = (100 \text{ V})(50 \text{ A}) \cos 53.13° = (5000)(0.6)$$
$$= \mathbf{3000 \text{ W}}$$

or
$$P_T = E^2G = (100 \text{ V})^2(0.3 \text{ S}) = \mathbf{3000 \text{ W}}$$

or, finally,

$$P_T = P_R + P_L + P_C$$
$$= EI_R \cos \theta_R + EI_L \cos \theta_L + EI_C \cos \theta_C$$
$$= (100 \text{ V})(30 \text{ A}) \cos 0° + (100 \text{ V})(70 \text{ A}) \cos 90° + (100 \text{ V})(30 \text{ A}) \cos 90°$$
$$= 3000 \text{ W} + 0 + 0$$
$$= \mathbf{3000 \text{ W}}$$

Power factor: The power factor of the circuit is

$$F_p = \cos \theta_T = \cos 53.13° = \mathbf{0.6 \text{ lagging}}$$

Using Eq. (15.32), we obtain

$$F_p = \cos \theta_T = \frac{G}{Y_T} = \frac{0.3 \text{ S}}{0.5 \text{ S}} = \mathbf{0.6 \text{ lagging}}$$

Impedance approach: The input current **I** can also be determined by first finding the total impedance as follows:

$$\mathbf{Z}_T = \frac{\mathbf{Z}_R\mathbf{Z}_L\mathbf{Z}_C}{\mathbf{Z}_R\mathbf{Z}_L + \mathbf{Z}_L\mathbf{Z}_C + \mathbf{Z}_R\mathbf{Z}_C} = \mathbf{2\ \Omega\ \angle 53.13^\circ}$$

and, applying Ohm's law, we obtain

$$\mathbf{I} = \frac{\mathbf{E}}{\mathbf{Z}_T} = \frac{100\text{ V}\ \angle 53.13^\circ}{2\ \Omega\ \angle 53.13^\circ} = \mathbf{50\ A\ \angle 0^\circ}$$

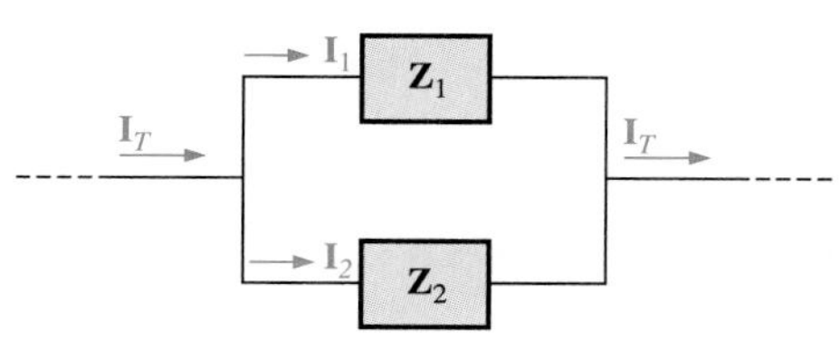

FIG. 15.76
Applying the current divider rule.

15.9 CURRENT DIVIDER RULE

The basic format for the *current divider rule* in ac circuits is exactly the same as that for dc circuits; that is, for two parallel branches with impedances $\mathbf{Z}_1$ and $\mathbf{Z}_2$ as shown in Fig. 15.76,

$$\mathbf{I}_1 = \frac{\mathbf{Z}_2\mathbf{I}_T}{\mathbf{Z}_1 + \mathbf{Z}_2} \quad \text{or} \quad \mathbf{I}_2 = \frac{\mathbf{Z}_1\mathbf{I}_T}{\mathbf{Z}_1 + \mathbf{Z}_2} \tag{15.33}$$

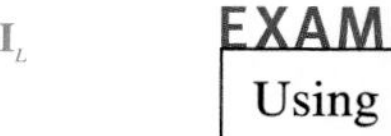
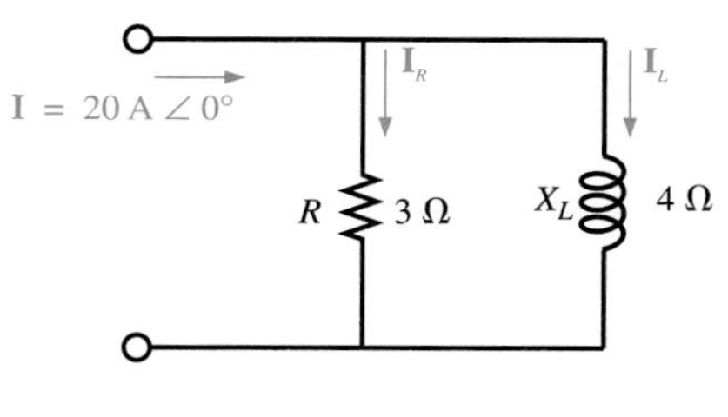

FIG. 15.77
Example 15.15.

EXAMPLE 15.15

Using the current divider rule, find the current through each impedance of Fig. 15.77.

Solution:

$$\mathbf{I}_R = \frac{\mathbf{Z}_L\mathbf{I}_T}{\mathbf{Z}_R + \mathbf{Z}_L} = \frac{(4\ \Omega\ \angle 90^\circ)(20\text{ A}\ \angle 0^\circ)}{3\ \Omega\ \angle 0^\circ + 4\ \Omega\ \angle 90^\circ} = \frac{80\text{ A}\ \angle 90^\circ}{5\ \angle 53.13^\circ}$$

$$= \mathbf{16\ A\ \angle 36.87^\circ}$$

$$\mathbf{I}_L = \frac{\mathbf{Z}_R\mathbf{I}_T}{\mathbf{Z}_R + \mathbf{Z}_L} = \frac{(3\ \Omega\ \angle 0^\circ)(20\text{ A}\ \angle 0^\circ)}{5\ \Omega\ \angle 53.13^\circ} = \frac{60\text{ A}\ \angle 0^\circ}{5\ \angle 53.13^\circ}$$

$$= \mathbf{12\ A\ \angle -53.13^\circ}$$

EXAMPLE 15.16

Using the current divider rule, find the current through each parallel branch of Fig. 15.78.

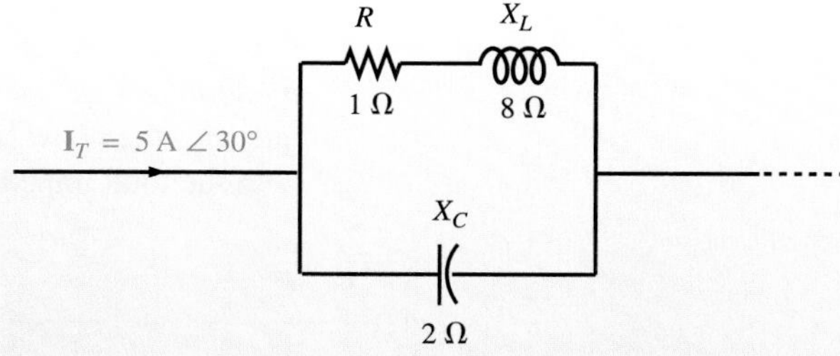

FIG. 15.78
Example 15.16.

Solution:

$$\mathbf{I}_{R\text{-}L} = \frac{\mathbf{Z}_C\mathbf{I}_T}{\mathbf{Z}_C + \mathbf{Z}_{R\text{-}L}} = \frac{(2\ \Omega\ \angle -90^\circ)(5\text{ A}\ \angle 30^\circ)}{-j\,2\ \Omega + 1\ \Omega + j\,8\ \Omega} = \frac{10\text{ A}\angle -60^\circ}{1 + j\,6}$$

$$= \frac{10\text{ A}\angle -60^\circ}{6.083\ \angle 80.54^\circ} \cong \mathbf{1.644\ A\ \angle -140.54^\circ}$$

$$\mathbf{I}_C = \frac{\mathbf{Z}_{R\text{-}L}\mathbf{I}_T}{\mathbf{Z}_{R\text{-}L} + \mathbf{Z}_C} = \frac{(1\ \Omega + j\,8\ \Omega)(5\text{ A}\ \angle 30^\circ)}{6.08\ \Omega\ \angle 80.54^\circ}$$

continued

$$= \frac{(8.06 \angle 82.87^\circ)(5 \text{ A} \angle 30^\circ)}{6.08 \angle 80.54^\circ} = \frac{40.30 \text{ A} \angle 112.87^\circ}{6.083 \angle 80.54^\circ}$$

$$= \mathbf{6.625 \text{ A} \angle 32.33^\circ}$$

15.10 FREQUENCY RESPONSE OF THE PARALLEL *R-L* NETWORK

In Section 15.5 the frequency response of a series *R-C* circuit was analyzed. Let us now note the impact of frequency on the total impedance and inductive current for the parallel *R-L* network of Fig. 15.79 for a frequency range of 0 through 40 kHz.

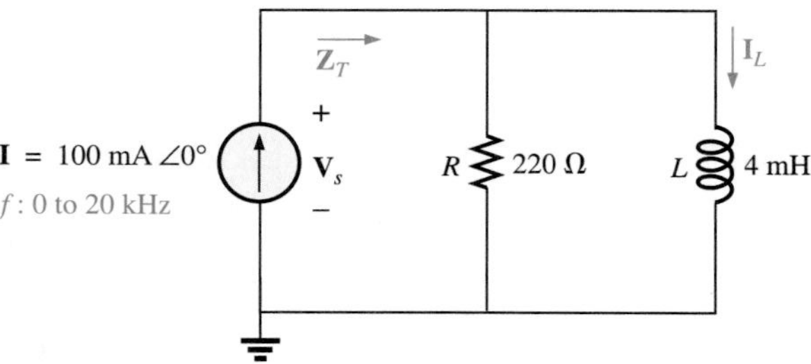

FIG. 15.79
Determining the frequency response of a parallel R-L network.

$\mathbf{Z}_T$:

Before getting into specifics, let us first develop a "sense" for the impact of frequency on the network of Fig. 15.79 by noting the impedance-versus-frequency curves of the individual elements, as shown in Fig. 15.80. The fact that the elements are now in parallel means that we should consider their characteristics in a different way than we did for the series *R-C* circuit of Section 15.5. Recall that for parallel elements, the element with the smallest impedance will have the greatest impact on the total impedance at that frequency. In Fig. 15.80, for example, X_L is very small at low frequencies compared to R, establishing X_L as the predominant factor in this frequency range. In other words, at low frequencies the network will be primarily inductive and the angle associated with the total impedance will be close to 90°, as with a pure inductor. As

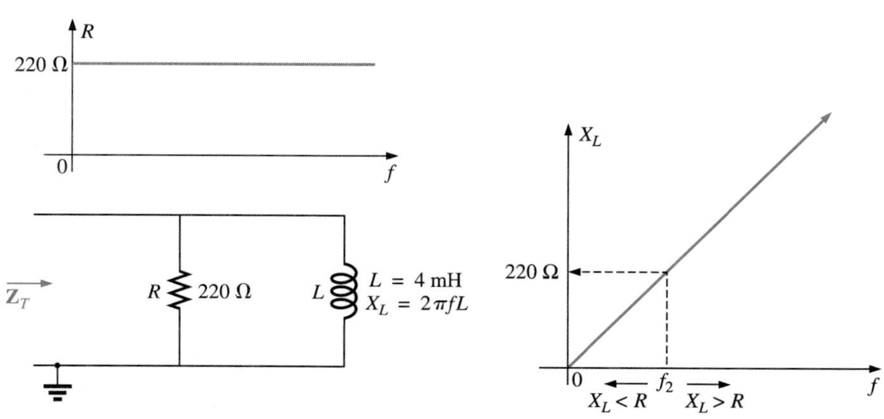

FIG. 15.80
The frequency response of the individual elements of a parallel R-L network.

the frequency increases, X_L will increase until it equals the impedance of the resistor (220 Ω). We can find the frequency at which this situation occurs in the following way:

$$X_L = 2\pi f_2 L = R$$

and

$$f_2 = \frac{R}{2\pi L} \tag{15.34}$$

which for the network of Fig. 15.79 is

$$f_2 = \frac{R}{2\pi L} = \frac{220\ \Omega}{2\pi(4 \times 10^{-3}\ \text{H})}$$

$$\cong \mathbf{8.75\ kHz}$$

which falls within the frequency range of interest.

For frequencies less than f_2, $X_L < R$, and for frequencies greater than f_2, $X_L > R$, as shown in Fig. 15.80. A general equation for the total impedance in vector form can be developed as follows:

$$\mathbf{Z}_T = \frac{\mathbf{Z}_R\mathbf{Z}_L}{\mathbf{Z}_R + \mathbf{Z}_L}$$

$$= \frac{(R \angle 0°)(X_L \angle 90°)}{R + jX_L} = \frac{RX_L \angle 90°}{\sqrt{R^2 + X_L^2} \angle \tan^{-1} X_L/R}$$

and

$$\mathbf{Z}_T = \frac{RX_L}{\sqrt{R^2 + X_L^2}} \angle 90° - \tan^{-1} X_L/R$$

so that

$$Z_T = \frac{RX_L}{\sqrt{R^2 + X_L^2}} \tag{15.35}$$

and

$$\theta_T = 90° - \tan^{-1}\frac{X_L}{R} = \tan^{-1}\frac{R}{X_L} \tag{15.36}$$

The magnitude and angle of the total impedance can now be found at any frequency of interest simply by substituting Eqs. (15.35) and (15.36).

f = 1 kHz:

$$X_L = 2\pi fL = 2\pi(1\ \text{kHz})(4 \times 10^{-3}\ \text{H}) = 25.12\ \Omega$$

and

$$Z_T = \frac{RX_L}{\sqrt{R^2 + X_L^2}} = \frac{(220\ \Omega)(25.12\ \Omega)}{\sqrt{(220\ \Omega)^2 + (25.12\Omega)^2}} = \mathbf{24.96\ \Omega}$$

with

$$\theta_T = \tan^{-1}\frac{R}{X_L} = \tan^{-1}\frac{220\ \Omega}{25.12\ \Omega}$$

$$= \tan^{-1} 8.76 = 83.49°$$

and

$$\mathbf{Z}_T = \mathbf{24.96\ \Omega \angle 83.49°}$$

This value is very close to $X_L = 25.12\ \Omega \angle 90°$, which it would be if the network were purely inductive ($R = \infty\ \Omega$). Our assumption that the network is primarily inductive at low frequencies is therefore confirmed.

Continuing:

$f = 5 \text{ kHz}$: $\mathbf{Z}_T = \mathbf{109.1\ \Omega \angle 60.23^\circ}$
$f = 10 \text{ kHz}$: $\mathbf{Z}_T = \mathbf{165.5\ \Omega \angle 41.21^\circ}$
$f = 15 \text{ kHz}$: $\mathbf{Z}_T = \mathbf{189.99\ \Omega \angle 30.28^\circ}$
$f = 20 \text{ kHz}$: $\mathbf{Z}_T = \mathbf{201.53\ \Omega \angle 23.65^\circ}$
$f = 30 \text{ kHz}$: $\mathbf{Z}_T = \mathbf{211.19\ \Omega \angle 16.27^\circ}$
$f = 40 \text{ kHz}$: $\mathbf{Z}_T = \mathbf{214.91\ \Omega \angle 12.35^\circ}$

At $f = 40$ kHz, note how closely the magnitude of Z_T has approached the resistance level of 220 Ω and how the associated angle with the total impedance is approaching zero degrees. The result is a network with terminal characteristics that are becoming more and more resistive as the frequency increases, which further confirms the earlier conclusions developed by the curves of Fig. 15.80.

Plots of Z_T versus frequency in Fig. 15.81 and θ_T in Fig. 15.82 clearly reveal the transition from an inductive network to one that has

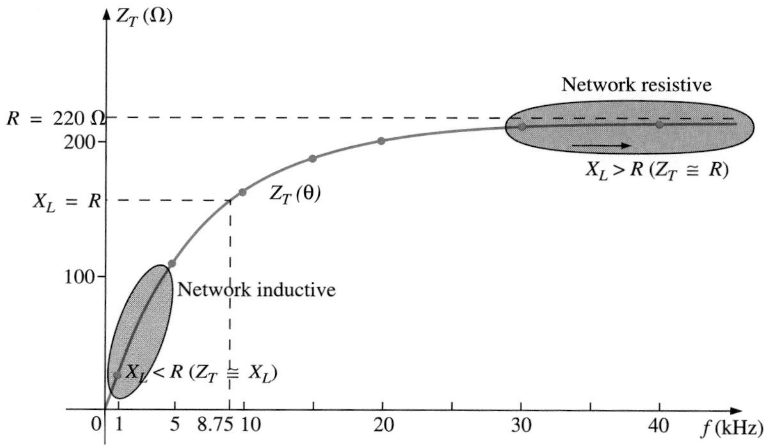

FIG. 15.81
The magnitude of the input impedance versus frequency for the network of Fig. 15.79.

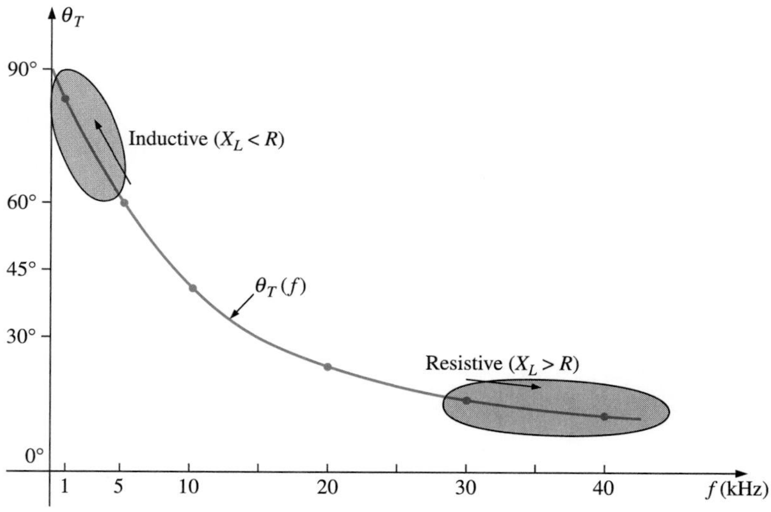

FIG. 15.82
The phase angle of the input impedance versus frequency for the network of Fig. 15.79.

resistive characteristics. Note that the transition frequency of 8.75 kHz occurs right in the middle of the knee of the curves for both Z_T and θ_T.

A review of Figs. 15.47 and 15.81 will reveal that a series *R-C* and a parallel *R-L* network will have an impedance level that approaches the resistance of the network at high frequencies. The capacitive circuit approaches the level from above, whereas the inductive network does the same from below. For the series *R-L* circuit and the parallel *R-C* network, the total impedance will begin at the resistance level and then display the characteristics of the reactive elements at high frequencies.

$\mathbf{I}_L$:

Applying the current divider rule to the network of Fig. 15.79 will result in the following:

$$\mathbf{I}_L = \frac{\mathbf{Z}_R\mathbf{I}}{\mathbf{Z}_R + \mathbf{Z}_L}$$

$$= \frac{(R \angle 0°)(I \angle 0°)}{R + j X_L} = \frac{RI \angle 0°}{\sqrt{R^2 + X_L^2} \angle \tan^{-1} X_L/R}$$

and $$\mathbf{I}_L = I_L \angle \theta_L = \frac{RI}{\sqrt{R^2 + X_L^2}} \angle -\tan^{-1} X_L/R$$

The magnitude of I_L is therefore determined by

$$I_L = \frac{RI}{\sqrt{R^2 + X_L^2}} \tag{15.37}$$

and the phase angle θ_L, by which $\mathbf{I}_L$ leads $\mathbf{I}$, is given by

$$\theta_L = -\tan^{-1} \frac{X_L}{R} \tag{15.38}$$

Because θ_L is always negative, the magnitude of θ_L is, in reality, the angle by which $\mathbf{I}_L$ lags $\mathbf{I}$.

To begin our analysis, let us first consider the case of $f = 0$ Hz (dc conditions).

$f = 0$ Hz:

$$X_L = 2\pi f L = 2\pi(0 \text{ Hz})L = 0\ \Omega$$

Applying the short-circuit equivalent for the inductor in Fig. 15.79 would result in

$$\mathbf{I}_L = \mathbf{I} = 100 \text{ mA} \angle 0°$$

as shown in Figs. 15.83 and 15.84.

f = 1 kHz:

Applying Eq. (15.37):

$$X_L = 2\pi fL = 2\pi(1\text{ kHz})(4\text{ mH}) = 25.12\ \Omega$$

and
$$\sqrt{R^2 + X_L^2} = \sqrt{(220\ \Omega)^2 + (25.12\ \Omega)^2} = 221.43\ \Omega$$

and
$$I_L = \frac{RI}{\sqrt{R^2 + X_L^2}} = \frac{(220\ \Omega)(100\text{ mA})}{221.43\ \Omega} = \mathbf{99.35\text{ mA}}$$

with

$$\theta_L = \tan^{-1}\frac{X_L}{R} = -\tan^{-1}\frac{25.12\ \Omega}{220\ \Omega} = -\tan^{-1}0.114 = \mathbf{-6.51°}$$

and
$$\mathbf{I}_L = \mathbf{99.35\text{ mA}\ \angle -6.51°}$$

The result is a current $\mathbf{I}_L$ that is still very close to the source current **I** in both magnitude and phase.

Continuing:

$$f = 5\text{ kHz: } \mathbf{I}_L = \mathbf{86.84\text{ mA}\ \angle -29.72°}$$
$$f = 10\text{ kHz: } \mathbf{I}_L = \mathbf{65.88\text{ mA}\ \angle -48.79°}$$
$$f = 15\text{ kHz: } \mathbf{I}_L = \mathbf{50.43\text{ mA}\ \angle -59.72°}$$
$$f = 20\text{ kHz: } \mathbf{I}_L = \mathbf{40.11\text{ mA}\ \angle -66.35°}$$
$$f = 30\text{ kHz: } \mathbf{I}_L = \mathbf{28.02\text{ mA}\ \angle -73.73°}$$
$$f = 40\text{ kHz: } \mathbf{I}_L = \mathbf{21.38\text{ mA}\ \angle -77.65°}$$

The plot of the magnitude of I_L versus frequency shown in Fig. 15.83 reveals that the current through the coil dropped from its maximum of 100 mA to almost 20 mA at 40 kHz. As the reactance of the coil increased with frequency, more of the source current chose the lower-resistance path of the resistor. The magnitude of the phase angle between $\mathbf{I}_L$ and **I** is approaching 90° with an increase in frequency, as shown in Fig. 15.84, leaving its initial value of zero degrees at f = 0 Hz far behind.

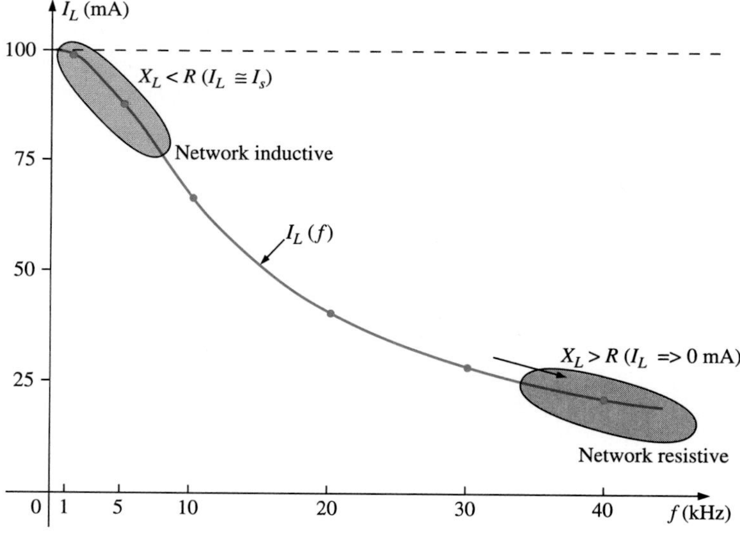

FIG. 15.83
The magnitude of the current $\mathbf{I}_L$ versus frequency for the parallel R-L network of Fig. 15.79.

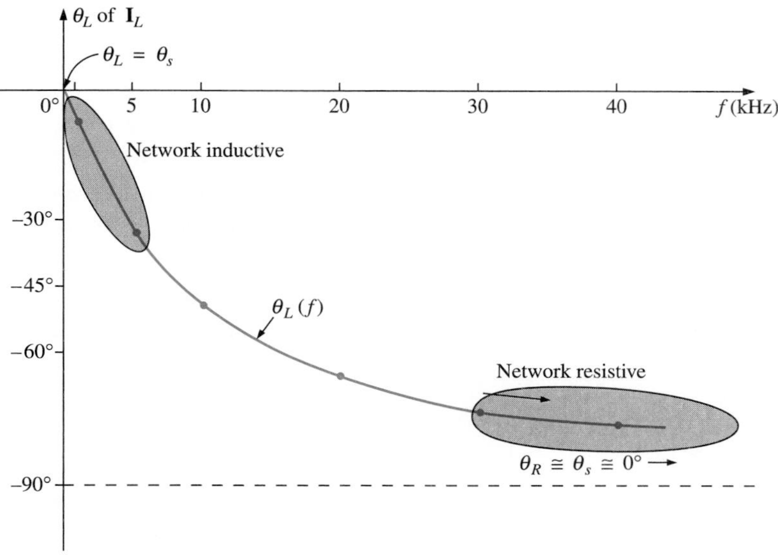

FIG. 15.84
The phase angle of the current $\mathbf{I}_L$ versus frequency for the parallel R-L network of Fig. 15.79.

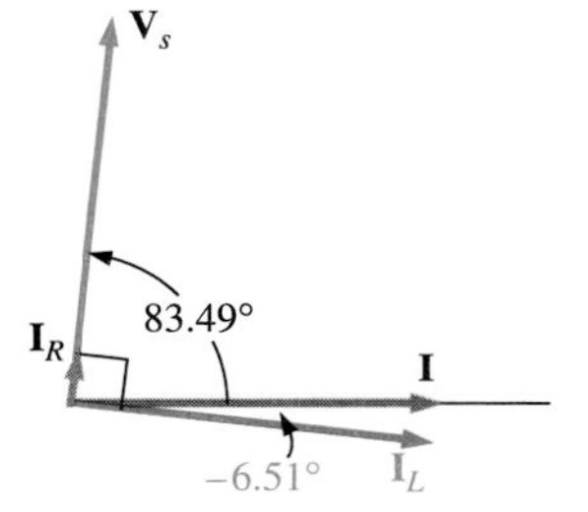

FIG. 15.85
The phasor diagram for the parallel R-L network of Fig. 15.79 at f = 1 kHz.

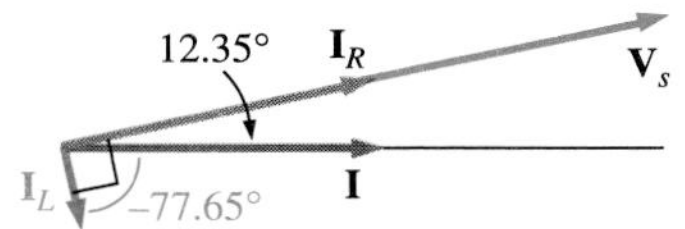

FIG. 15.86
The phasor diagram for the parallel R-L network of Fig. 15.79 at f = 40 kHz.

At $f = 1$ kHz the phasor diagram of the network appears as shown in Fig. 15.85. First note that the magnitude and phase angle of $\mathbf{I}_L$ are very close to those of $\mathbf{I}$. Since the voltage across a coil must lead the current through a coil by 90°, the voltage $\mathbf{V}_s$ appears as shown. The voltage across a resistor is in phase with the current through the resistor, resulting in the direction of $\mathbf{I}_R$ shown in Fig. 15.85. Of course, at this frequency $R > X_L$, and the current I_R is relatively small in magnitude.

At $f = 40$ kHz the phasor diagram changes to that appearing in Fig. 15.86. Note that now $\mathbf{I}_R$ and $\mathbf{I}$ are close in magnitude and phase because $X_L > R$. The magnitude of $\mathbf{I}_L$ has dropped to very low levels and the phase angle associated with $\mathbf{I}_L$ is approaching −90°. The network is now more "resistive" as compared to its "inductive" characteristics at low frequencies.

The analysis of a parallel *R-C* or *R-L-C* network would proceed in much the same way, with the inductive impedance predominating at low frequencies and the capacitive reactance predominating at high frequencies.

15.11 SUMMARY—PARALLEL ac NETWORKS

The following is a review of important conclusions that can be made from the discussion and examples of the previous sections. The list is not all inclusive, but it does emphasize some of the conclusions that should be carried forward in the future analysis of ac systems.

For parallel ac networks with reactive elements:

1. ***The total admittance (impedance) will be frequency dependent.***
2. ***The impedance of any one element can be less than the total impedance (for dc circuits recall that the total resistance must always be less than the smallest parallel resistor).***
3. ***The inductive and capacitive susceptances are in direct opposition on an admittance diagram.***
4. ***Depending on the frequency applied, the same network can be either predominantly inductive or capacitive.***
5. ***At lower frequencies the inductive elements will usually have the most impact on the total impedance, while at high frequencies the capacitive elements will usually have the most impact.***
6. ***The magnitude of the current through any one branch can be greater than the source current.***
7. ***The magnitude of the current through an element, as compared to the other elements of the network, is directly related to the magnitude of its impedance; that is, the smaller the impedance of an element, the larger the magnitude of the current through the element.***
8. ***The current through an inductor is always in direct opposition to the current through a capacitor on a phasor diagram.***
9. ***The applied voltage is always in phase with the current through the resistive elements, leads the voltage across all the inductive elements by 90°, and lags the current through all capacitive elements by 90°.***
10. ***The smaller the resistive element of a network compared to the net reactive susceptance, the closer the power factor is to unity.***

PROBLEMS

SECTION 15.2 Impedance and the Phasor Diagram

1. Express the impedances of Fig. 15.87 in both polar and rectangular form.

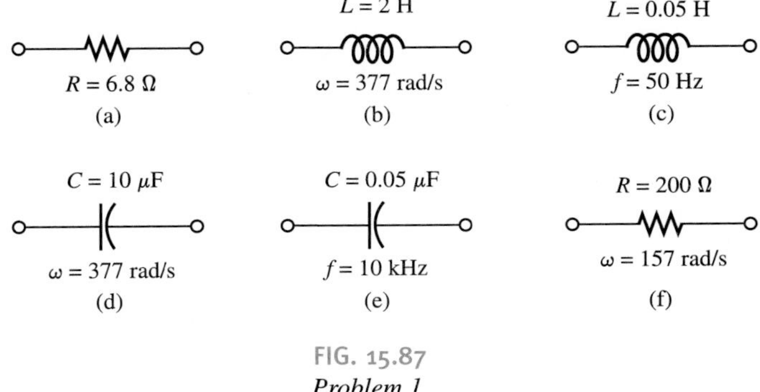

FIG. 15.87
Problem 1.

2. Find the current i for the elements of Fig. 15.88 using complex algebra.

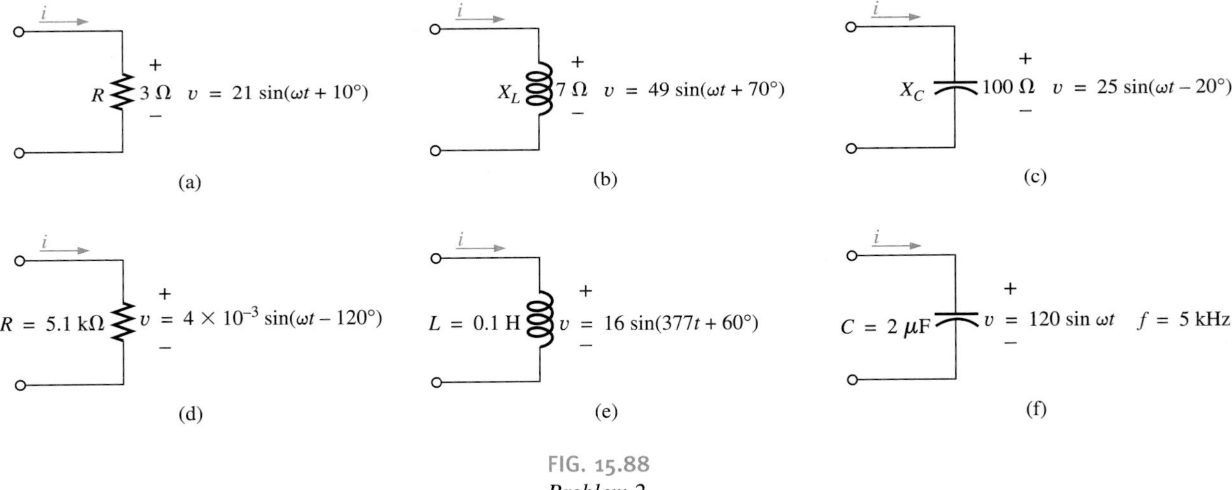

FIG. 15.88
Problem 2.

3. Find the voltage v for the elements of Fig. 15.89 using complex algebra.

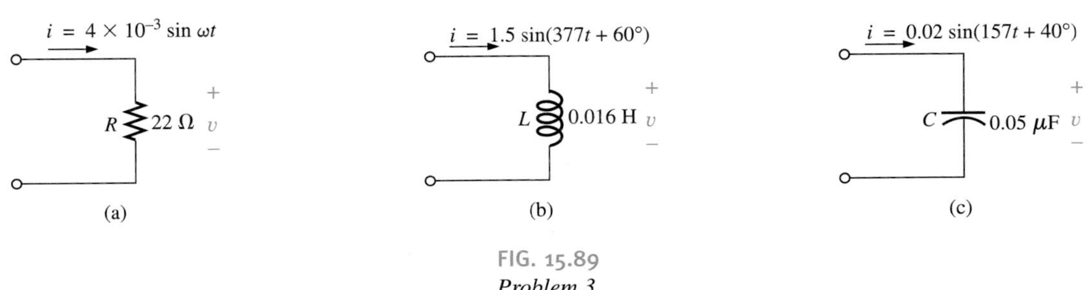

FIG. 15.89
Problem 3.

SECTION 15.3 Series Configuration

4. Calculate the total impedance of the circuits of Fig. 15.90. Express your answer in rectangular and polar form, and draw the impedance diagram.

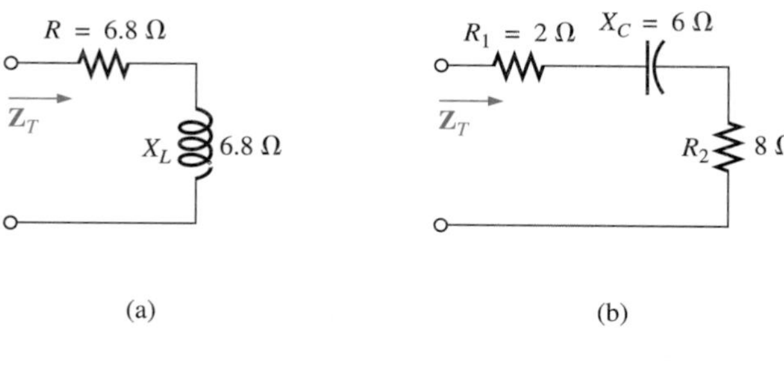

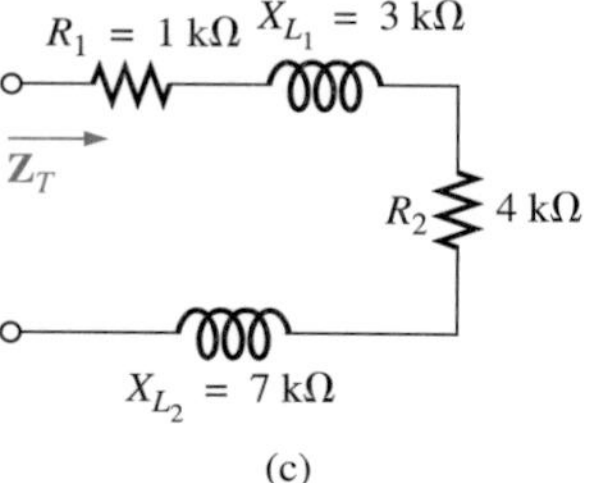

FIG. 15.90
Problem 4.

5. Calculate the total impedance of the circuits of Fig. 15.91. Express your answer in rectangular and polar form, and draw the impedance diagram.

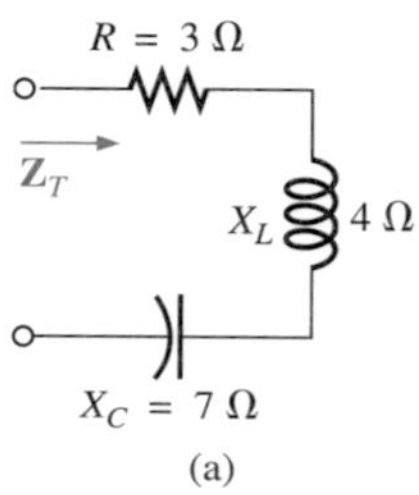

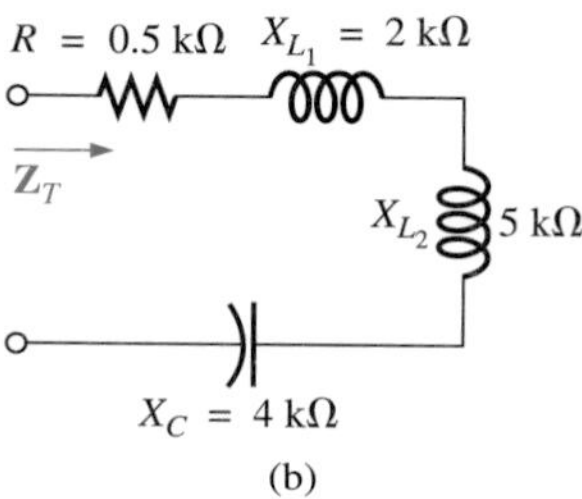

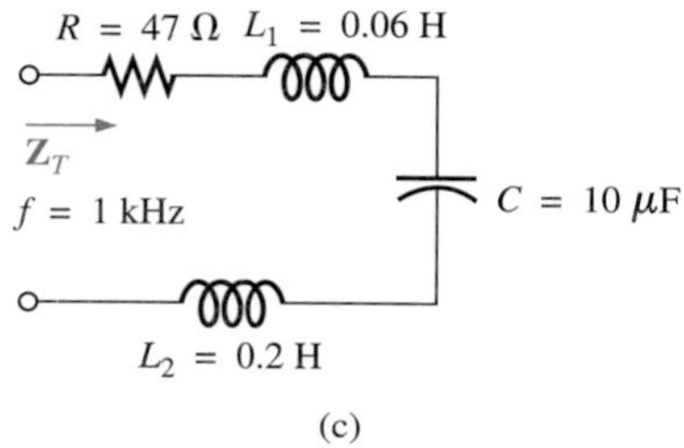

FIG. 15.91
Problem 5.

6. Find the type and impedance in ohms of the series circuit elements that must be in the closed container of Fig. 15.92 for the indicated voltages and currents to exist at the input terminals. (Find the simplest series circuit that will satisfy the indicated conditions.)

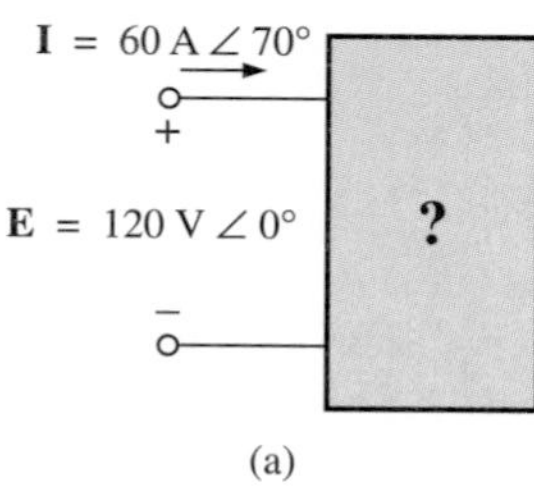

I = 20 mA ∠ 40°
+
E = 80 V ∠ 320°
?
–
(b)

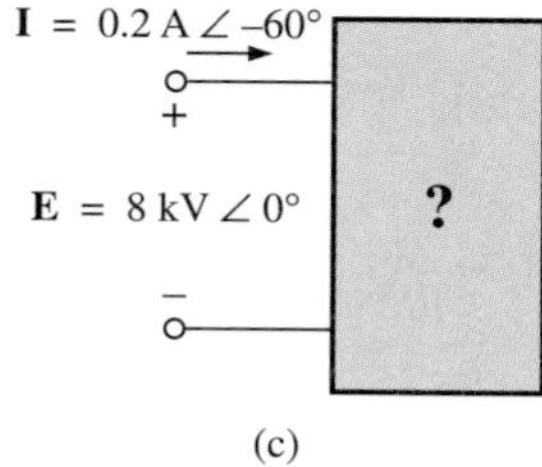

FIG. 15.92
Problems 6 and 28.

7. For the circuit of Fig. 15.93:

a. Find the total impedance $\mathbf{Z}_T$ in polar form.
b. Draw the impedance diagram.
c. Find the current $\mathbf{I}$ and the voltages $\mathbf{V}_R$ and $\mathbf{V}_L$ in phasor form.
d. Draw the phasor diagram of the voltages $\mathbf{E}$, $\mathbf{V}_R$, and $\mathbf{V}_L$, and the current $\mathbf{I}$.
e. Verify Kirchhoff's voltage law around the closed loop.
f. Find the average power delivered to the circuit.
g. Find the power factor of the circuit and indicate whether it is leading or lagging.
h. Find the sinusoidal expressions for the voltages and current if the frequency is 60 Hz.

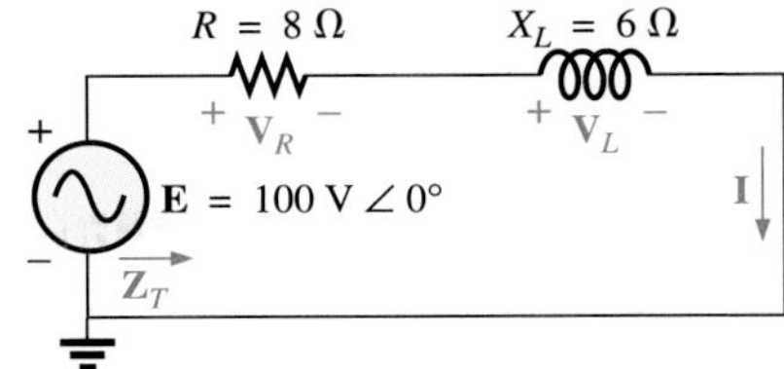

FIG. 15.93
Problem 7.

8. Repeat Problem 7 for the circuit of Fig. 15.94, replacing $\mathbf{V}_L$ with $\mathbf{V}_C$ in parts (c) and (d).

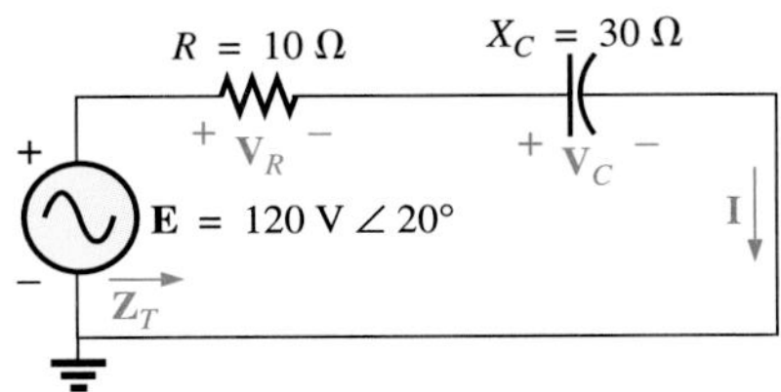

FIG. 15.94
Problem 8.

9. Given the network of Fig. 15.95:

a. Determine $\mathbf{Z}_T$.
b. Find $\mathbf{I}$.
c. Calculate $\mathbf{V}_R$ and $\mathbf{V}_C$.
d. Find P and F_p.

10. Repeat Problem 9 using a frequency of $f = 2.7$ Hz.

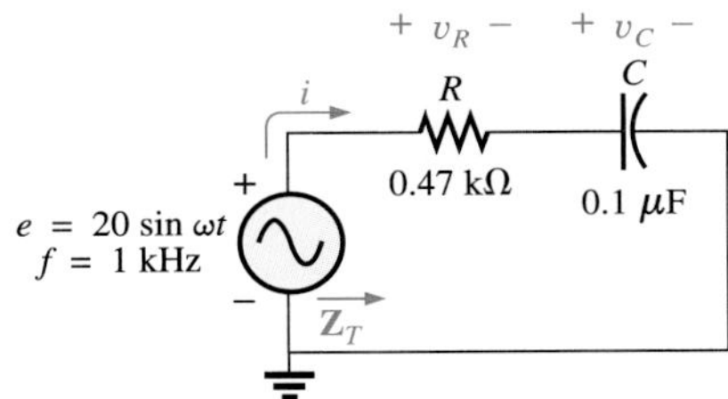

FIG. 15.95
Problem 9 and 10.

11. For the circuit of Fig. 15.96:

a. Find the total impedance $\mathbf{Z}_T$ in polar form.
b. Draw the impedance diagram.
c. Find the value of C in microfarads and L in henries.
d. Find the current $\mathbf{I}$ and the voltages $\mathbf{V}_R$, $\mathbf{V}_L$, and $\mathbf{V}_C$ in phasor form.
e. Draw the phasor diagram of the voltages $\mathbf{E}$, $\mathbf{V}_R$, $\mathbf{V}_L$, and $\mathbf{V}_C$, and the current $\mathbf{I}$.
f. Verify Kirchhoff's voltage law around the closed loop.
g. Find the average power delivered to the circuit.
h. Find the power factor of the circuit and indicate whether it is leading or lagging.
i. Find the sinusoidal expressions for the voltages and current.
j. Plot the waveforms for the voltages and current on the same set of axes.

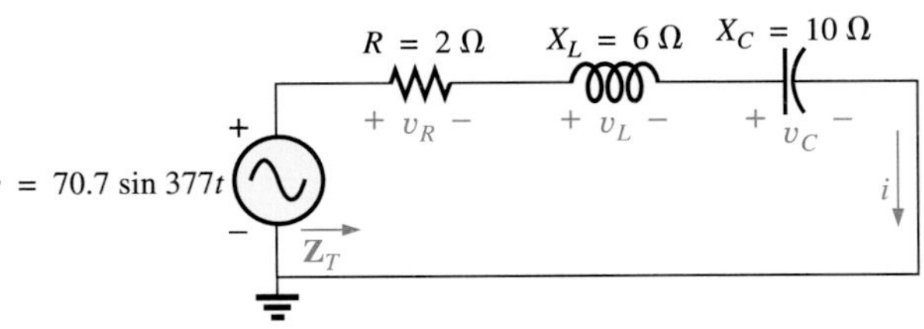

FIG. 15.96
Problem 11.

12. Repeat Problem 11 for the circuit of Fig. 15.97.

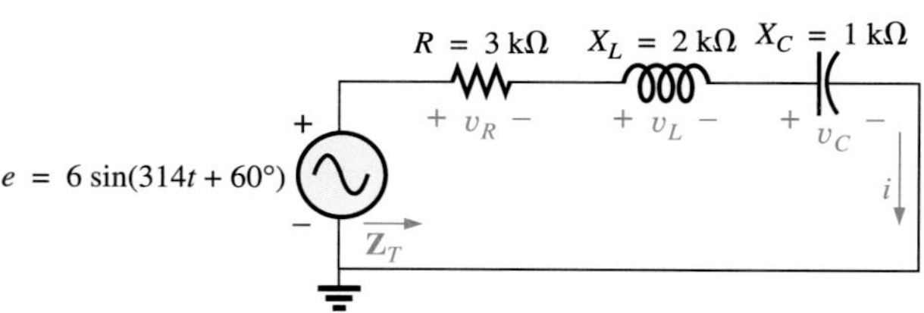

FIG. 15.97
Problem 12.

13. Using the oscilloscope reading of Fig. 15.98, determine the resistance R.

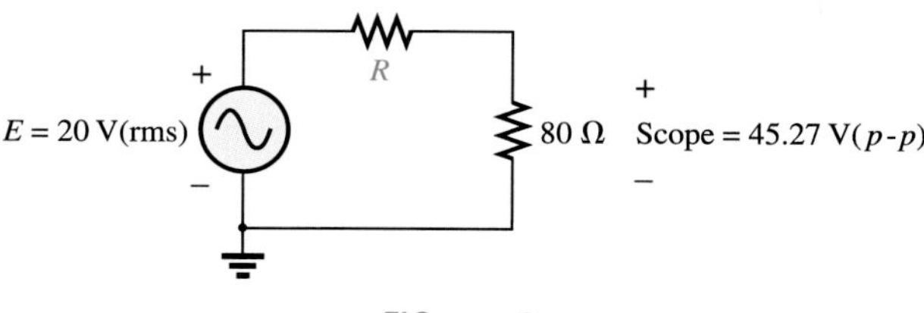

FIG. 15.98
Problem 13.

14. Using the DMM current reading and the oscilloscope measurement of Fig. 15.99:

a. Determine the inductance L.

b. Find the resistance R.

c. How do these results change if the system is operated at a frequency of = 12.5 kHz?

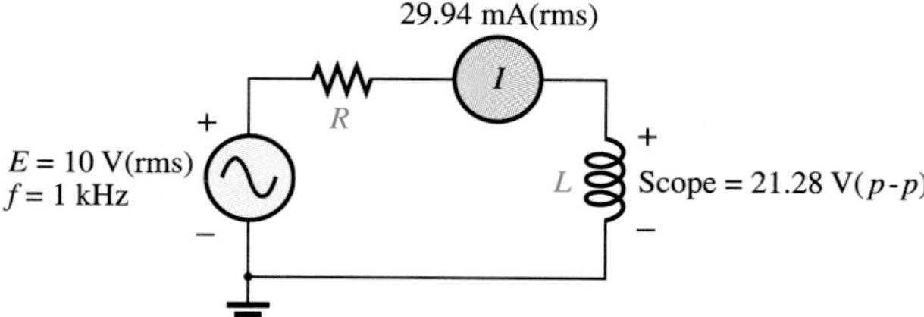

FIG. 15.99
Problem 14.

15. Using the oscilloscope reading of Fig. 15.100, determine the capacitance C.

16. Repeat Problem 15 using a frequency of $f = 2.7$ Hz and a resistance of 6.4 kΩ.

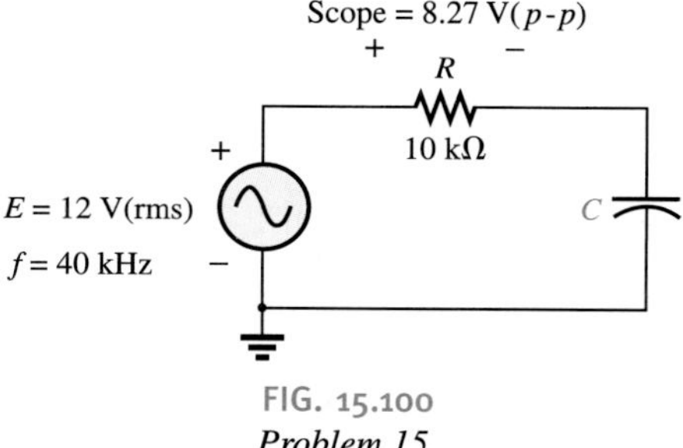

FIG. 15.100
Problem 15.

SECTION 15.4 Voltage Divider Rule

17. Calculate the voltages $\mathbf{V}_1$ and $\mathbf{V}_2$ for the circuit of Fig. 15.101 in phasor form using the voltage divider rule.

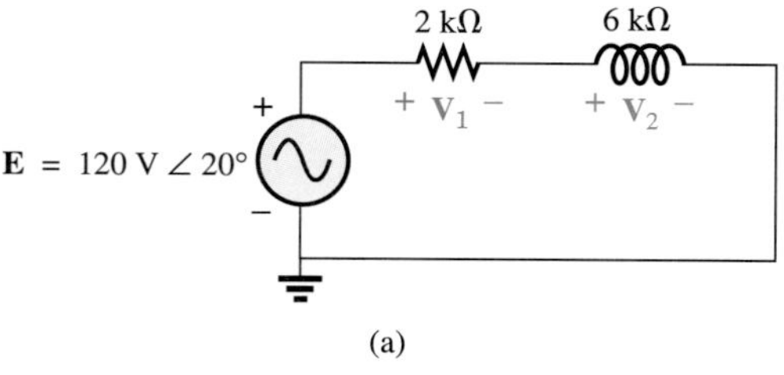

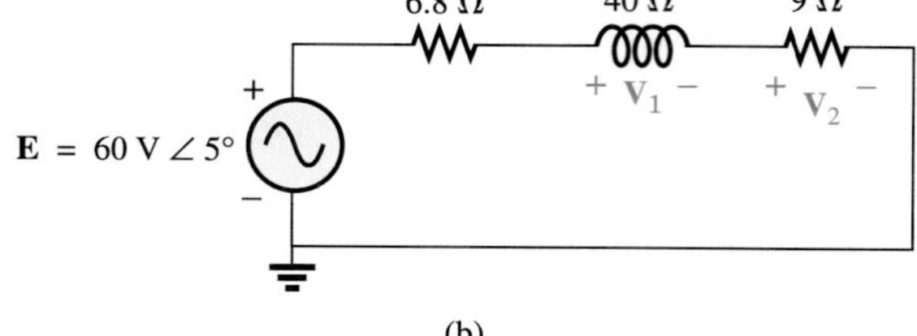

FIG. 15.101
Problem 17.

18. Calculate the voltages $\mathbf{V}_1$ and $\mathbf{V}_2$ for the circuit of Fig. 15.102 in phasor form using the voltage divider rule.

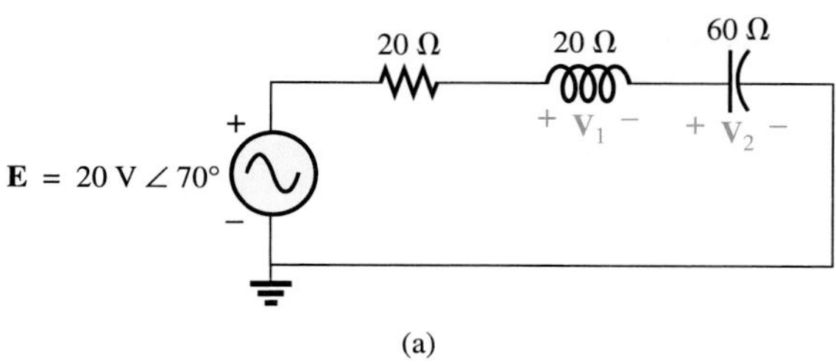

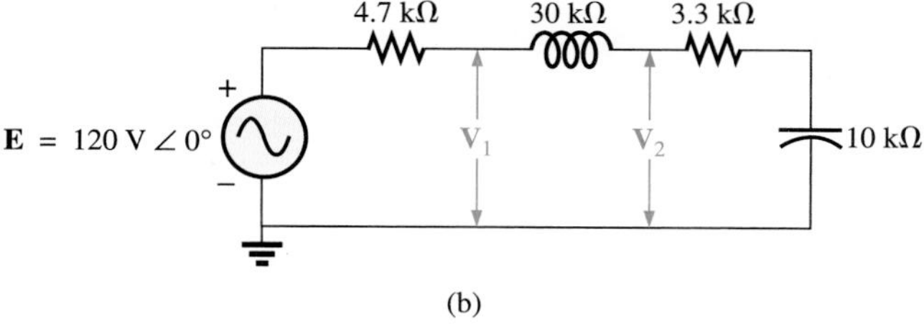

FIG. 15.102
Problem 18.

19. For the circuit of Fig. 15.103:

a. Determine $\mathbf{I}$, $\mathbf{V}_R$, and $\mathbf{V}_C$ in phasor form.

b. Calculate the total power factor and indicate whether it is leading or lagging.

c. Calculate the average power delivered to the circuit.

d. Draw the impedance diagram.

e. Draw the phasor diagram of the voltages $\mathbf{E}$, $\mathbf{V}_R$, and $\mathbf{V}_C$, and the current $\mathbf{I}$.

f. Find the voltages $\mathbf{V}_R$ and $\mathbf{V}_C$ using the voltage divider rule, and compare them with the results of part (a) above.

g. Draw the equivalent series circuit of the above as far as the total impedance and the current i are concerned.

20. Repeat Problem 19 if the capacitance is changed to 1000 μF.

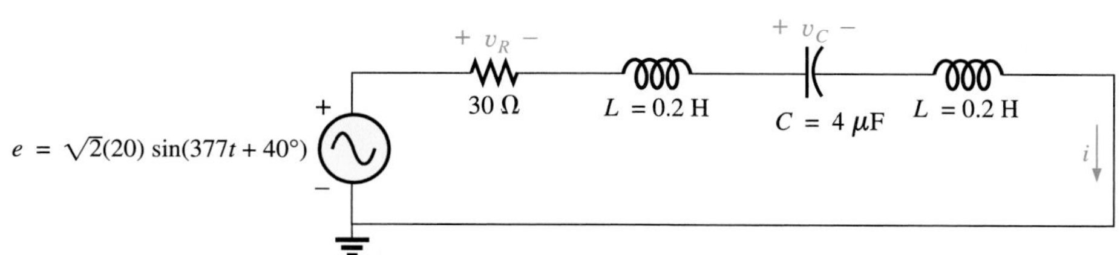

FIG. 15.103
Problems 19 and 20.

21. An electrical load has a power factor of 0.8 lagging. It dissipates 8 kW at a voltage of 200 V. Calculate the impedance of this load in rectangular coordinates.

22. Find the series element or elements that must be in the enclosed container of Fig. 15.104 to satisfy the following conditions:

a. Average power to circuit = 300 W.

b. Circuit has a lagging power factor.

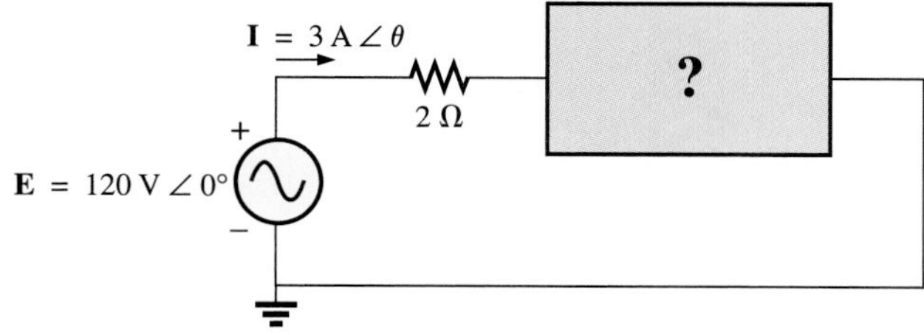

FIG. 15.104
Problem 22.

SECTION 15.5 Frequency Response of the *R-C* Circuit

23. For the circuit of Fig. 15.105:

a. Plot Z_T and θ_T versus frequency for a frequency range of zero to 20 kHz.

b. Plot V_L versus frequency for the frequency range of part (a).

c. Plot θ_L versus frequency for the frequency range of part (a).

d. Plot V_R versus frequency for the frequency range of part (a).

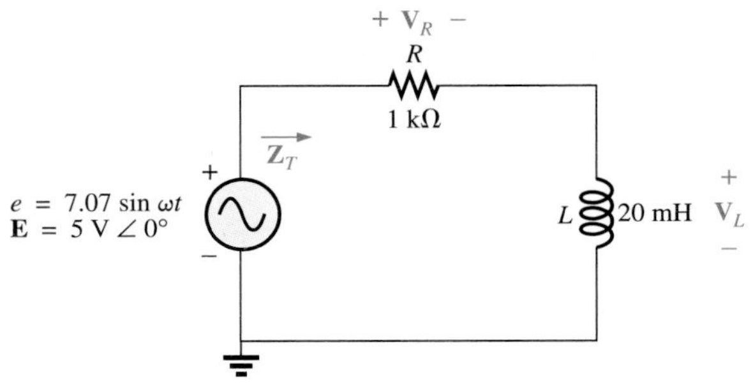

FIG. 15.105
Problem 23.

24. For the circuit of Fig. 15.106:

a. Plot Z_T and θ_T versus frequency for a frequency range of zero to 10 kHz.

b. Plot V_C versus frequency for the frequency range of part (a).

c. Plot θ_C versus frequency for the frequency range of part (a).

d. Plot V_R versus frequency for the frequency range of part (a).

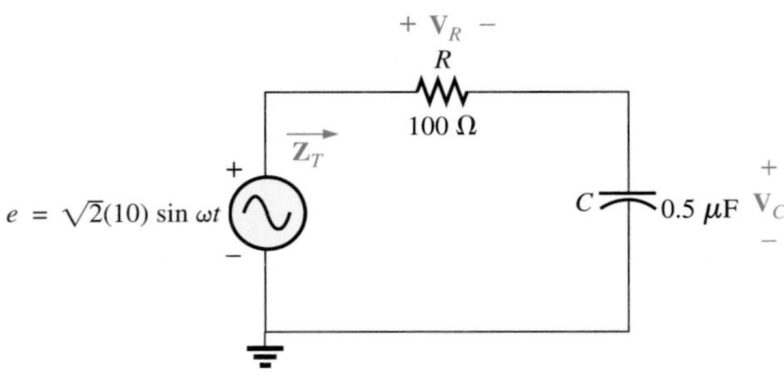

FIG. 15.106
Problem 24.

25. For the series *R-L-C* circuit of Fig. 15.107:

a. Plot Z_T and θ_T versus frequency for a frequency range of zero to 20 kHz in increments of 1 kHz.

b. Plot V_C (magnitude only) versus frequency for the same frequency range of part (a).

c. Plot I (magnitude only) versus frequency for the same frequency range of part (a).

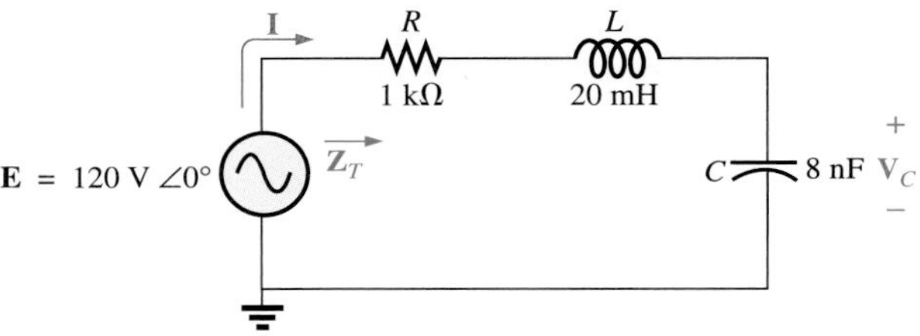

FIG. 15.107
Problem 25.

SECTION 15.7 Admittance and Susceptance

26. Find the total admittance and impedance of the circuits of Fig. 15.108. Identify the values of conductance and susceptance, and draw the admittance diagram.

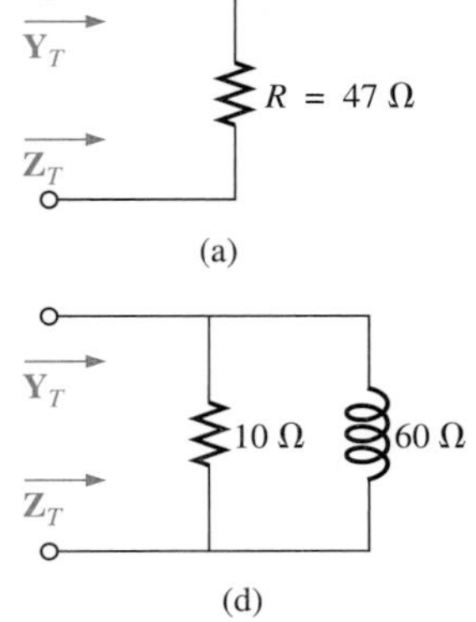

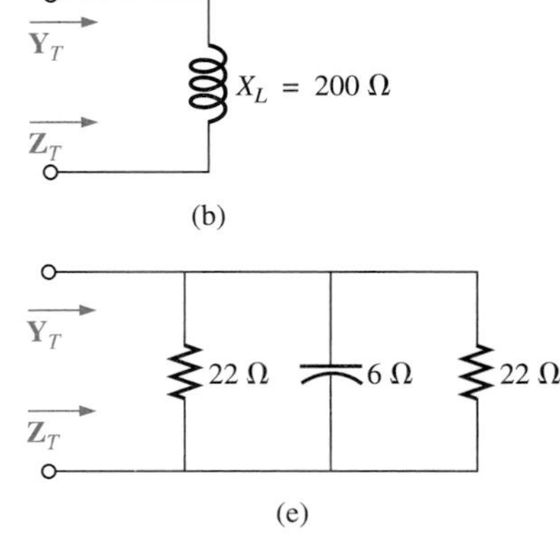

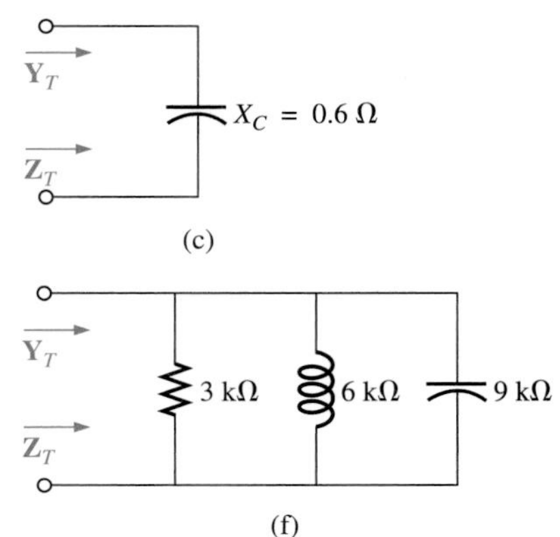

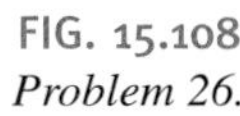

FIG. 15.108
Problem 26.

27. Find the total admittance and impedance of the circuits of Fig. 15.109. Identify the values of conductance and susceptance, and draw the admittance diagram.

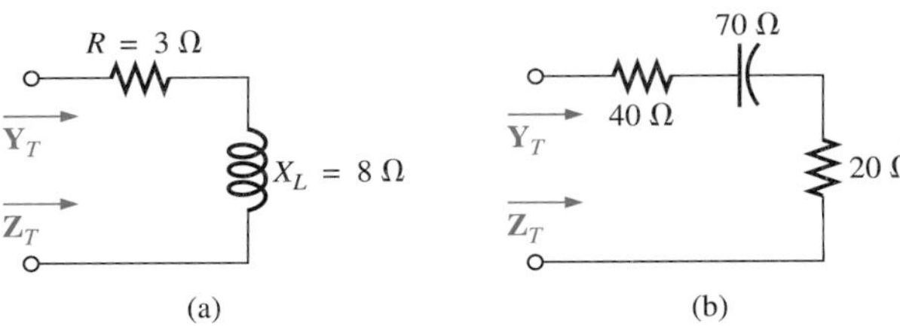

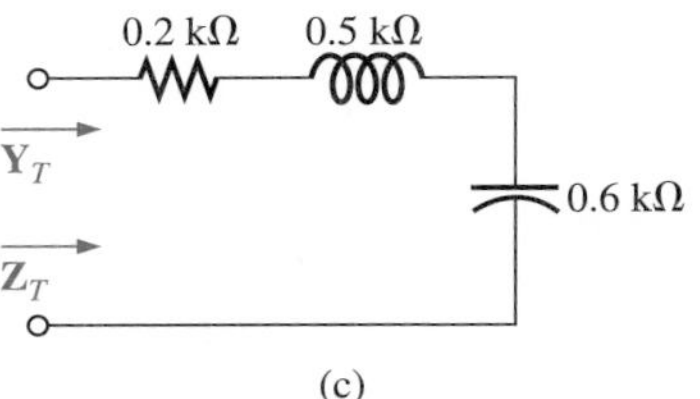

FIG. 15.109
Problem 27.

SECTION 15.8 Parallel ac Networks

28. Repeat Problem 6 for the **parallel** circuit elements that must be in the closed container for the same voltage and current to exist at the input terminals. (Find the simplest parallel circuit that will satisfy the conditions indicated.)

29. For the circuit of Fig. 15.110:

a. Find the total admittance $\mathbf{Y}_T$ in polar form.
b. Draw the admittance diagram.
c. Find the voltage **E** and the currents $\mathbf{I}_R$ and $\mathbf{I}_L$ in phasor form.
d. Draw the phasor diagram of the currents $\mathbf{I}_s$, $\mathbf{I}_R$, and $\mathbf{I}_L$ and the voltage **E**.
e. Verify Kirchhoff's current law at one node.
f. Find the average power delivered to the circuit.
g. Find the power factor of the circuit and indicate whether it is leading or lagging.
h. Find the sinusoidal expressions for the currents and voltage if the frequency is 60 Hz.

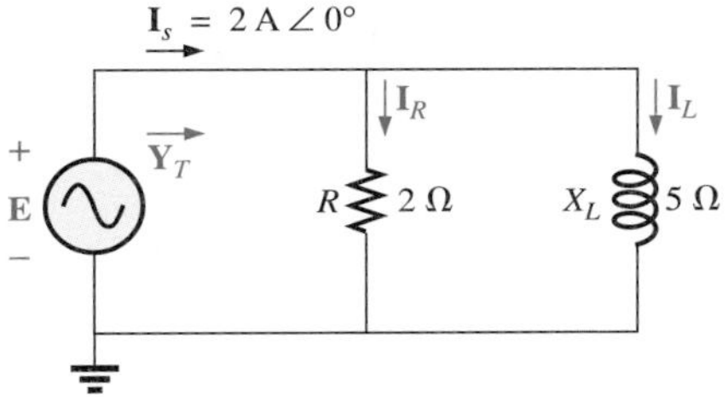

FIG. 15.110
Problem 29.

30. Repeat Problem 29 for the circuit of Fig. 15.111, replacing $\mathbf{I}_L$ with $\mathbf{I}_C$ in parts (c) and (d).

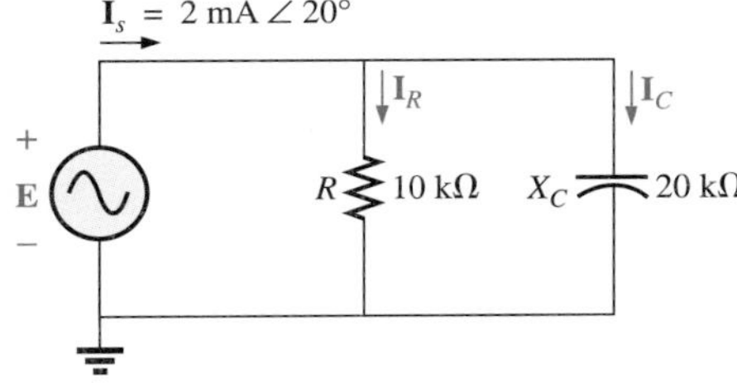

FIG. 15.111
Problem 30.

31. Repeat Problem 29 for the circuit of Fig. 15.112, replacing **E** with $\mathbf{I}_s$ in part (c).

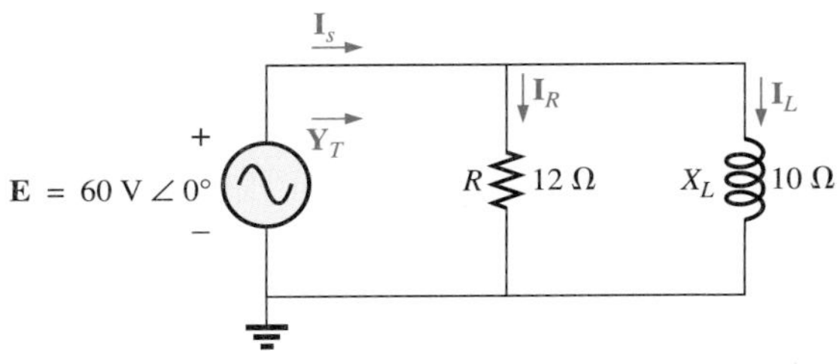

FIG. 15.112
Problem 31.

32. For the circuit of Fig. 15.113:

a. Find the total admittance $\mathbf{Y}_T$ in polar form.
b. Draw the admittance diagram.
c. Find the value of C in microfarads and L in henries.
d. Find the voltage **E** and currents $\mathbf{I}_R$, $\mathbf{I}_L$, and $\mathbf{I}_C$ in phasor form.
e. Draw the phasor diagram of the currents $\mathbf{I}_s$, $\mathbf{I}_R$, $\mathbf{I}_L$, and $\mathbf{I}_C$, and the voltage **E.**
f. Verify Kirchhoff's current law at one node.
g. Find the average power delivered to the circuit.
h. Find the power factor of the circuit and indicate whether it is leading or lagging.
i. Find the sinusoidal expressions for the currents and voltage.

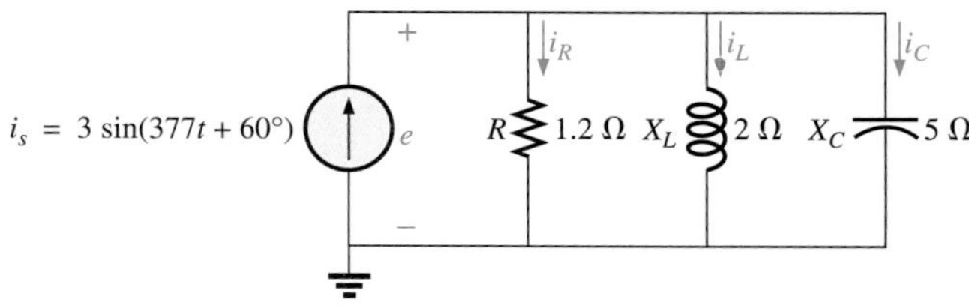

FIG. 15.113
Problem 32.

33. Repeat Problem 32 for the circuit of Fig. 15.114.

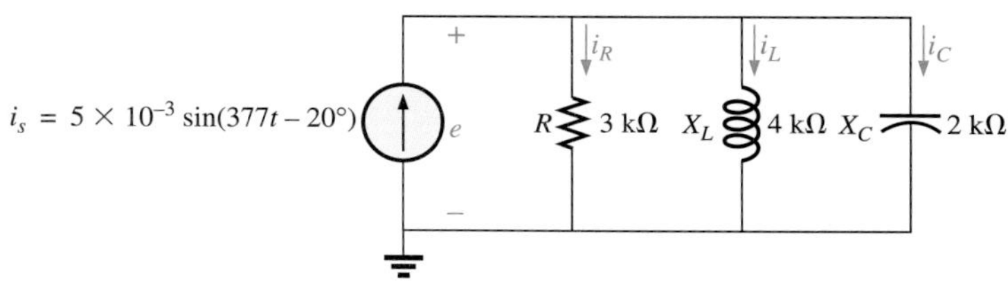

FIG. 15.114
Problem 33.

34. Repeat Problem 32 for the circuit of Fig. 15.115, replacing e with i_s in part (d).

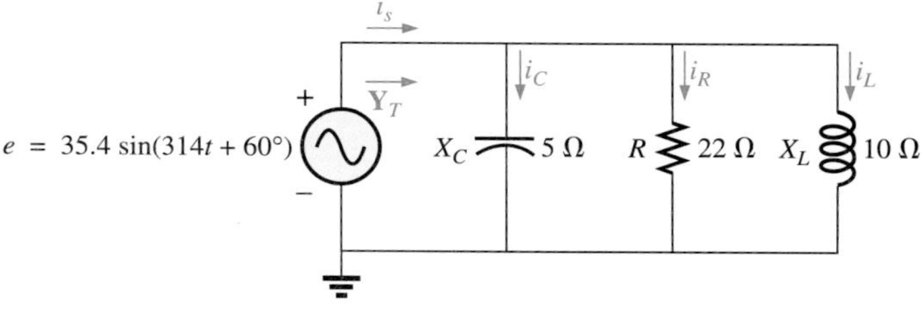

FIG. 15.115
Problem 34.

SECTION 15.9 Current Divider Rule

35. Calculate the currents $\mathbf{I}_1$ and $\mathbf{I}_2$ of Fig. 15.116 in phasor form using the current divider rule.

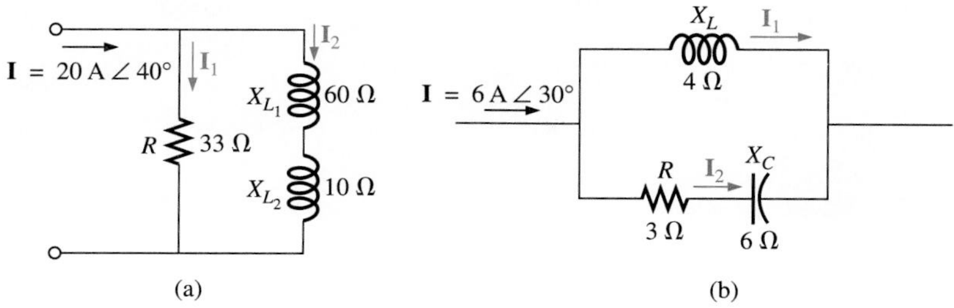

FIG. 15.116
Problem 35.

SECTION 15.10 Frequency Response of the Parallel *R-L* Network

36. For the parallel *R-C* network of Fig. 15.117:

a. Plot Z_T and θ_T versus frequency for a frequency range of zero to 20 kHz.

b. Plot V_C versus frequency for the frequency range of part (a).

c. Plot I_R versus frequency for the frequency range of part (a).

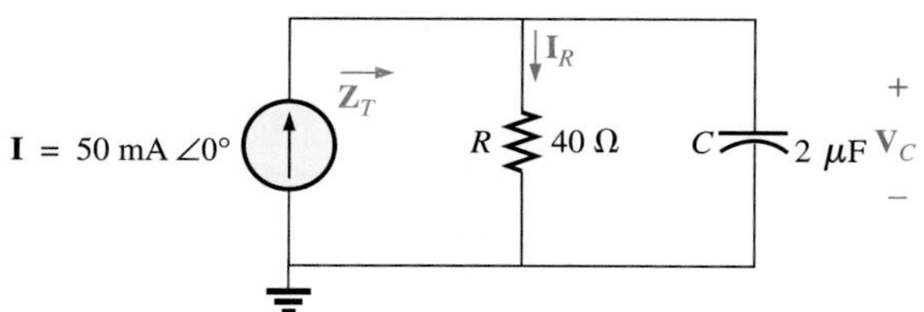

FIG. 15.117
Problems 36 and 37.

37. Plot Y_T and θ_T (of $\mathbf{Y}_T = Y_T \angle\theta_T$) for a frequency range of zero to 20 kHz for the network of Fig. 15.117.

38. For the parallel *R-L* network of Fig. 15.118:

a. Plot Z_T and θ_T versus frequency for a frequency range of zero to 10 kHz.

b. Plot I_L versus frequency for the frequency range of part (a).

c. Plot I_R versus frequency for the frequency range of part (a).

39. Plot Y_T and θ_T (of $\mathbf{Y}_T = Y_T \angle\theta_T$) for a frequency range of zero to 10 kHz for the network of Fig. 15.118.

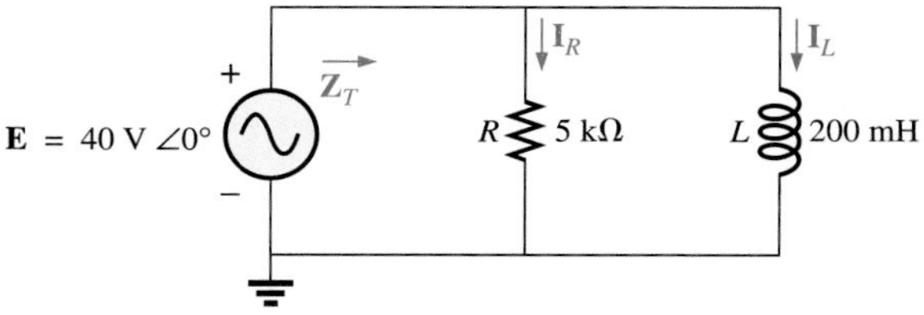

FIG. 15.118
Problems 38 and 39.

40. For the parallel *R-L-C* network of Fig. 15.119:

a. Plot Y_T and θ_T (of $\mathbf{Y}_T = Y_T \angle\theta_T$) for a frequency range of zero to 20 kHz.

b. Repeat part (a) for Z_T and θ_T (of $\mathbf{Z}_T = Z_T \angle\theta_T$).

c. Plot V_C versus frequency for the frequency range of part (a).

d. Plot I_L versus frequency for the frequency range of part (a).

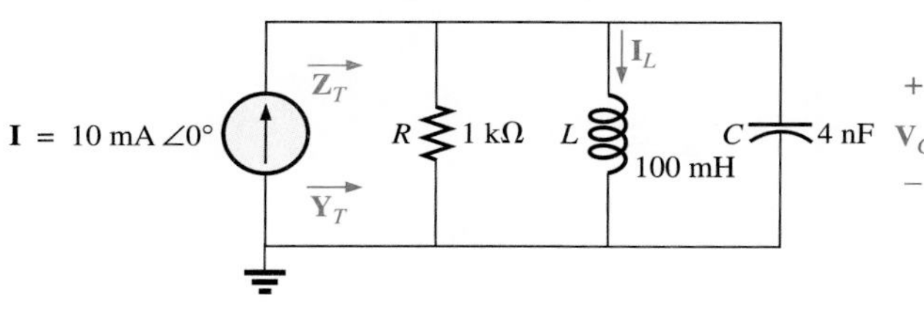

FIG. 15.119
Problem 40.

16 Series-Parallel ac Networks

Outline

Learning Outcomes

After completing this chapter, you will be able to

- redraw series-parallel elements of a single energy source ac network using equivalent block impedances
- reduce the equivalent block impedance network to an equivalent fundamental structure of a single energy source plus *R-C* or *R-L* elements
- determine the source current requirements for a series-parallel network with a single voltage source
- determine the total impedance of a series-parallel network
- verify that the results are reasonable by considering the energy source and elements in the system

The switching yard of an electrical distribution system contains a network of conductors with resistance, transformers with inductive properties, and power factor correction elements with capacitive properties.

16.1 INTRODUCTION

In the last chapter we examined ac series and parallel networks containing *R, L,* and *C* elements. The concept of combining these into an equivalent total impedance is presented in this chapter. It will prove to be very useful for simplifying and analyzing more complicated networks. In addition, there is a practical section discussing how to obtain values from an oscilloscope.

The chapter continues by using the basic concepts of the previous chapter to develop a technique for solving series-parallel ac networks. The measurements of voltage, current, and phase angle permit you to find key values needed for calculating the equivalent total impedance of a system.

The circuits discussed in this chapter have only one source of energy, either potential or current. A brief review of the methods of dc network analysis in Chapter 7 may be helpful since the approach to analyzing ac networks here will be quite similar. Ac networks with two or more sources will be considered in Chapters 17 and 18, also using methods previously described for dc circuits.

Steps in Analyzing Series-Parallel ac Networks

In general, when working with series-parallel ac networks, consider the following approach:

Step 1. Redraw the network, using block impedances to combine obvious sets of series or parallel elements.

This will reduce the network to one that clearly shows the basic structure of the system.

Step 2. Study the problem and make a brief mental sketch of the overall approach you plan to use.

Doing this may result in shortcuts that will save both time and energy. In some cases a lengthy, drawn out analysis may not be necessary. A single application of a fundamental law of circuit analysis may result in the desired solution.

Step 3. Consider each branch involved in your method independently before tying them together in series-parallel combinations.

After the overall approach has been determined, it is usually best to work back from the obvious series and parallel combinations to the source to determine the total impedance of the network. The source current can then be determined, and the path back to specific unknowns can be defined. As you progress back to the source, continually define those unknowns that have not been lost in the reduction process. It will save time when you have to work back through the network to find specific quantities.

Step 4. When you have arrived at a solution, check to see that it is reasonable.

Consider the magnitudes of the energy source and of the elements in the circuit. If they do not seem reasonable, either solve the network using another approach, or check over your work very carefully. At this point a computer solution can be very helpful for checking.

16.2 EQUIVALENT CIRCUITS

In a series ac circuit, the total impedance of two or more elements in series is often equivalent to an impedance that can be achieved with fewer elements of different values, the elements and their values being determined by the frequency applied. This is also true for parallel circuits. For the circuit of Fig. 16.1,

$$\mathbf{Z}_T = \frac{\mathbf{Z}_C\mathbf{Z}_L}{\mathbf{Z}_C + \mathbf{Z}_L} = \frac{(5\ \Omega \angle -90^\circ)(10\ \Omega \angle 90^\circ)}{5\ \Omega \angle -90^\circ + 10\ \Omega \angle 90^\circ} = \frac{50 \angle 0^\circ}{5 \angle 90^\circ}$$
$$= 10\ \Omega \angle -90^\circ$$

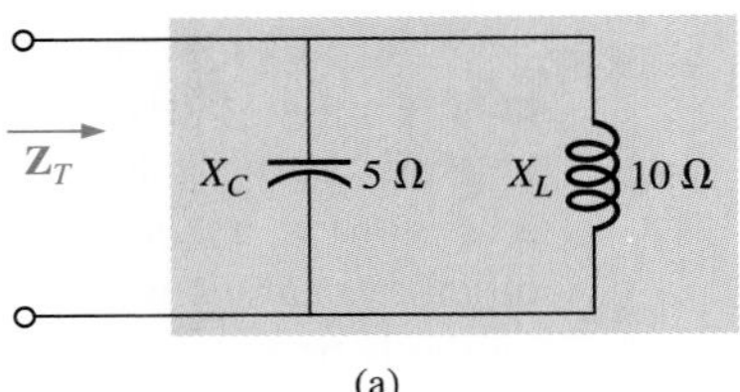

(a)

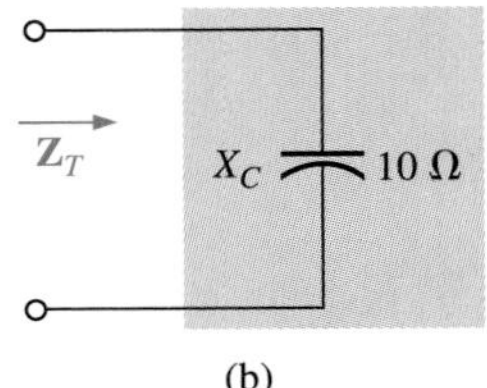

(b)

FIG. 16.1
Defining the equivalence between two networks at a specific frequency.

The total impedance at the frequency applied is equivalent to a capacitor with a reactance of 10 Ω, as shown in Fig. 16.1(b). Always keep in mind that this equivalence is true only at the applied frequency. If the frequency changes, the reactance of each element changes, and the equivalent circuit will change—perhaps from capacitive to inductive in the above example.

Another interesting development appears if the impedance of a parallel circuit, such as the one of Fig. 16.2(a), is found in rectangular form. In this case,

$$\mathbf{Z}_T = \frac{\mathbf{Z}_L\mathbf{Z}_R}{\mathbf{Z}_L + \mathbf{Z}_R} = \frac{(4\ \Omega \angle 90^\circ)(3\ \Omega \angle 0^\circ)}{4\ \Omega \angle 90^\circ + 3\ \Omega \angle 0^\circ}$$
$$= \frac{12 \angle 90^\circ}{5 \angle 53.13^\circ} = 2.40\ \Omega \angle 36.87^\circ$$
$$= 1.920\ \Omega + j\,1.440\ \Omega$$

which is the impedance of a series circuit with a resistor of 1.92 Ω and an inductive reactance of 1.44 Ω, as shown in Fig. 16.2(b).

The current **I** will be the same in each circuit of Fig. 16.1 or Fig. 16.2 if the same input voltage **E** is applied. For a parallel circuit of one resistive element and one reactive element, the series circuit with the same input impedance will always be composed of one resistive and one reactive element. The impedance of each element of the series circuit will be different from that of the parallel circuit, but the reactive elements will always be of the same type; that is, an *R-L* circuit and an *R-C* parallel circuit will have an equivalent *R-L* and *R-C* series circuit, respectively. The same is true when converting from a series to a parallel circuit. In the discussion to follow, keep in mind that

the term equivalent refers only to the fact that for the same applied potential, the same impedance and input current will result.

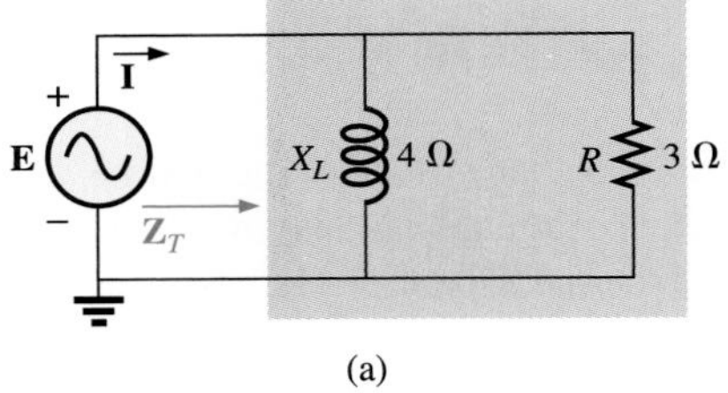

(a)

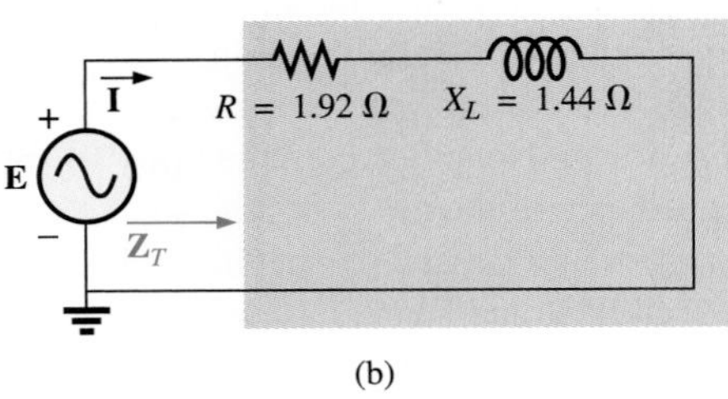

(b)

FIG. 16.2
Finding the series equivalent circuit for a parallel R-L network.

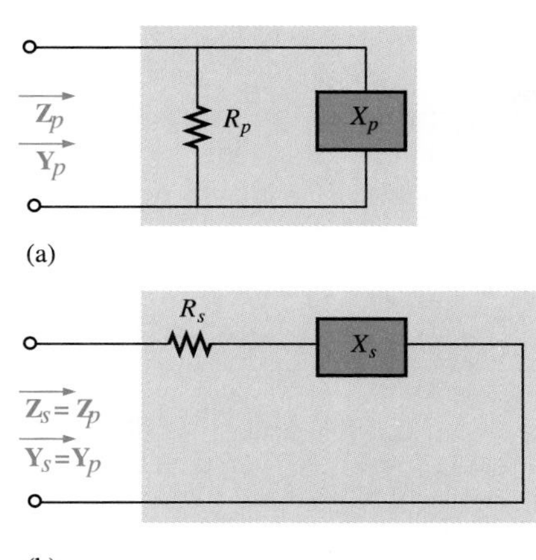

FIG. 16.3
Defining the parameters of equivalent series and parallel networks.

To formulate the equivalence between the series and parallel circuits, the equivalent series circuit for a resistor and reactance in parallel can be found by determining the total impedance of the circuit in rectangular form; that is, for the circuit of Fig. 16.3(a),

$$\mathbf{Y}_p = \frac{1}{R_p} + \frac{1}{\pm j X_p} = \frac{1}{R_P} \mp j\frac{1}{X_p}$$

and

$$\mathbf{Z}_p = \frac{1}{\mathbf{Y}_p} = \frac{1}{(1/R_p) \pm j\,(1/X_p)}$$

$$= \frac{1/R_p}{(1/R_p)^2 + (1/X_p)^2} \pm j\frac{1/X_p}{(1/R_p)^2 + (1/X_p)^2}$$

Multiplying the numerator and denominator of each term by $R_p^2X_p^2$ results in

$$\mathbf{Z}_p = \frac{R_pX_p^2}{X_p^2 + R_p^2} \pm j\frac{R_p^2X_p}{X_p^2 + R_p^2}$$

$$= R_s \pm j\,X_s \quad \text{[Fig. 16.3(b)]}$$

and

$$R_s = \frac{R_pX_p^2}{X_p^2 + R_p^2} \tag{16.1}$$

with

$$X_s = \frac{R_p^2X_p}{X_p^2 + R_p^2} \tag{16.2}$$

For the network of Fig. 16.2,

$$R_s = \frac{R_pX_p^2}{X_p^2 + R_p^2} = \frac{(3\ \Omega)(4\ \Omega)^2}{(4\ \Omega)^2 + (3\ \Omega)^2} = \frac{48\ \Omega}{25} = \mathbf{1.920\ \Omega}$$

and

$$X_s = \frac{R_p^2\,X_p}{X_p^2 + R_p^2} = \frac{(3\ \Omega)^2(4\ \Omega)}{(4\ \Omega)^2 + (3\ \Omega)^2} = \frac{36\ \Omega}{25} = \mathbf{1.440\ \Omega}$$

which agrees with the previous result.

The equivalent parallel circuit for a circuit with a resistor and reactance in series can be found by simply finding the total admittance of the system in rectangular form; that is, for the circuit of Fig. 16.3(b),

$$\mathbf{Z}_s = R_s \pm j\,X_s$$

$$\mathbf{Y}_s = \frac{1}{\mathbf{Z}_s} = \frac{1}{R_s \pm j\,X_s} = \frac{R_s}{R_s^2 + X_s^2} \mp j\frac{X_s}{R_s^2 + X_s^2}$$

$$= G_p \mp j\,B_p = \frac{1}{R_p} \mp j\frac{1}{X_p} \quad \text{[Fig. 15.89(a)]}$$

or

$$R_p = \frac{R_s^2 + X_s^2}{R_s} \tag{16.3}$$

with

$$X_p = \frac{R_s^2 + X_s^2}{X_s} \tag{16.4}$$

For the above example,

$$R_p = \frac{R_s^2 + X_s^2}{R_s} = \frac{(1.92\ \Omega)^2 + (1.44\ \Omega)^2}{1.92\ \Omega} = \frac{5.76\ \Omega}{1.92} = \mathbf{3.0\ \Omega}$$

and

$$X_p = \frac{R_s^2 + X_s^2}{X_s} = \frac{5.76\ \Omega}{1.44} = \mathbf{4.0\ \Omega}$$

as shown in Fig. 16.2(a).

EXAMPLE 16.1

Determine the series equivalent circuit for the network of Fig. 16.4.

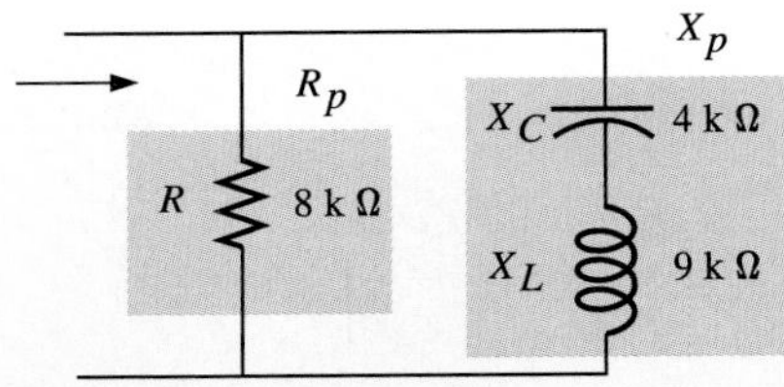

FIG. 16.4
Example 16.1

Solution:

$$R_p = 8\ \text{k}\Omega$$

$$X_p\ (\text{resultant}) = |X_L - X_C| = |9\ \text{k}\Omega - 4\ \text{k}\Omega| = 5\ \text{k}\Omega$$

and

$$R_s = \frac{R_p X_p^2}{X_p^2 + R_p^2} = \frac{(8\ \text{k}\Omega)(5\ \text{k}\Omega)^2}{(5\ \text{k}\Omega)^2 + (8\ \text{k}\Omega)^2} = \frac{200\ \text{k}\Omega}{89} = \mathbf{2.247\ k\Omega}$$

with

$$X_s = \frac{R_p^2 X_p}{X_p^2 + R_p^2} = \frac{(8\ \text{k}\Omega)^2(5\ \text{k}\Omega)}{(5\ \text{k}\Omega)^2 + (8\ \text{k}\Omega)^2} = \frac{320\ \text{k}\Omega}{89} = \mathbf{3.596\ k\Omega}\ \ \textbf{(inductive)}$$

The equivalent series circuit appears in Fig. 16.5.

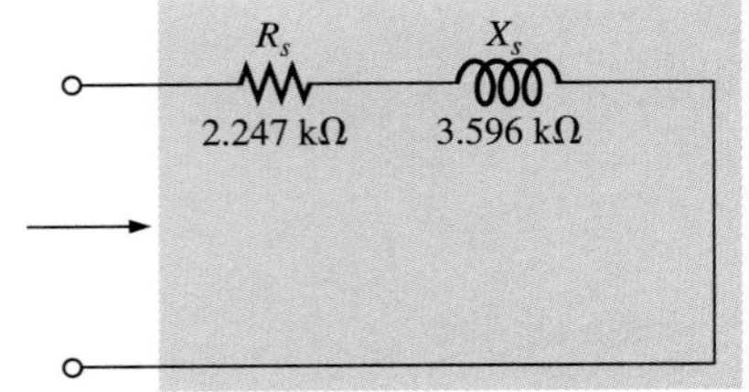

FIG. 16.5
The equivalent series circuit for the parallel network of Fig. 16.4.

EXAMPLE 16.2

For the network of Fig. 16.6:

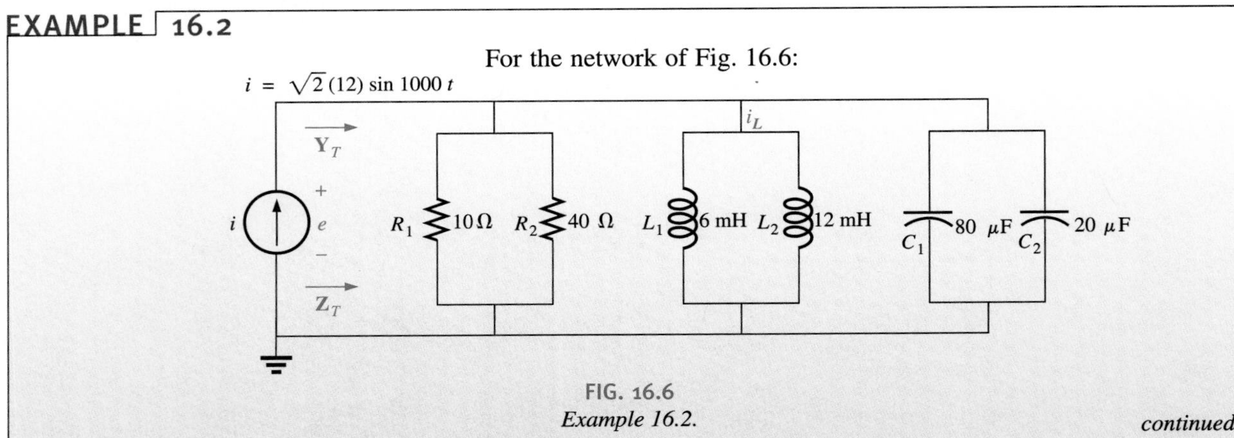

FIG. 16.6
Example 16.2.

continued

a. Determine $\mathbf{Y}_T$.
b. Sketch the admittance diagram.
c. Find **E** and $\mathbf{I}_L$.
d. Compute the power factor of the network and the power delivered to the network.
e. Determine the equivalent series circuit as far as the terminal characteristics of the network are concerned.
f. Using the equivalent circuit developed in part (e), calculate **E** and compare it with the result of part (c).
g. Determine the power delivered to the network and compare it with the solution of part (d).
h. Determine the equivalent parallel network from the equivalent series circuit and calculate the total admittance $\mathbf{Y}_T$. Compare the result with the solution of part (a).

Solutions:

a. Combining common elements and finding the reactance of the inductor and capacitor, we obtain

$$R_T = 10\ \Omega \parallel 40\ \Omega = 8\ \Omega$$
$$L_T = 6\text{ mH} \parallel 12\text{ mH} = 4\text{ mH}$$
$$C_T = 80\ \mu\text{F} + 20\ \mu\text{F} = 100\ \mu\text{F}$$
$$X_L = \omega L = (1000\text{ rad/s})(4\text{ mH}) = 4\ \Omega$$
$$X_C = \frac{1}{\omega C} = \frac{1}{(1000\text{ rad/s})(100\ \mu\text{F})} = 10\ \Omega$$

The network is redrawn in Fig. 16.7 with phasor notation. The total admittance is

$$\begin{aligned}\mathbf{Y}_T &= \mathbf{Y}_R + \mathbf{Y}_L + \mathbf{Y}_C \\ &= G\angle 0^\circ + B_L\angle -90^\circ + B_C\angle +90^\circ \\ &= \frac{1}{8\ \Omega}\angle 0^\circ + \frac{1}{4\ \Omega}\angle -90^\circ + \frac{1}{10\ \Omega}\angle +90^\circ \\ &= 0.125\text{ S}\angle 0^\circ + 0.25\text{ S}\angle -90^\circ + 0.1\text{ S}\angle +90^\circ \\ &= 0.125\text{ S} - j\,0.25\text{ S} + j\,0.1\text{ S} \\ &= 0.125\text{ S} - j\,0.15\text{ S} = \mathbf{0.195\ S\angle -50.194^\circ}\end{aligned}$$

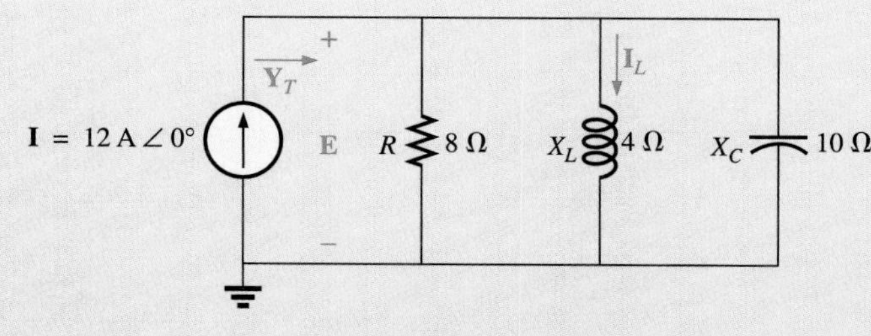

FIG. 16.7
Applying phasor notation to the network of Fig. 16.6.

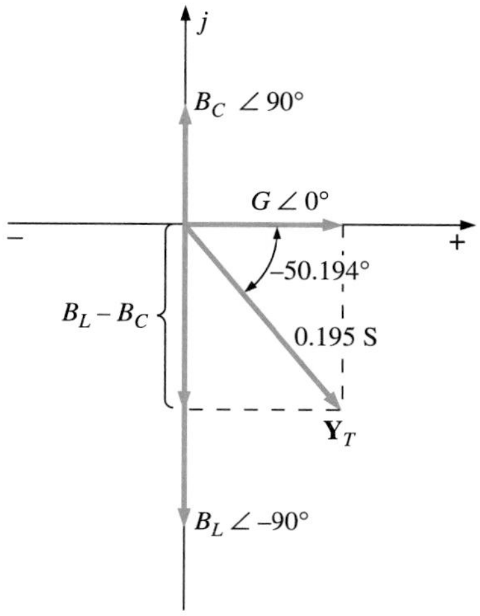

b. See Fig. 16.8.

FIG. 16.8
Admittance diagram for the parallel R-L-C network of Fig. 16.6.

continued

c. $\mathbf{E} = \mathbf{I}\mathbf{Z}_T = \dfrac{\mathbf{I}}{\mathbf{Y}_T} = \dfrac{12\text{ A}\angle 0^\circ}{0.195\text{ S}\angle -50.194^\circ} = \mathbf{61.538\text{ V}\angle 50.194^\circ}$

$$\mathbf{I}_L = \frac{\mathbf{V}_L}{\mathbf{Z}_L} = \frac{\mathbf{E}}{\mathbf{Z}_L} = \frac{61.538\text{ V}\angle 50.194^\circ}{4\,\Omega\angle 90^\circ} = \mathbf{15.385\text{ A}\angle -39.81^\circ}$$

d. $F_p = \cos\theta = \dfrac{G}{Y_T} = \dfrac{0.125\text{ S}}{0.195\text{ S}} = \mathbf{0.641}$ lagging (**E** leads **I**)

$$P = EI\cos\theta = (61.538\text{ V})(12\text{ A})\cos 50.194^\circ$$
$$= \mathbf{472.75\text{ W}}$$

e. $\mathbf{Z}_T = \dfrac{1}{\mathbf{Y}_T} = \dfrac{1}{0.195\text{ S}\angle -50.194^\circ} = 5.128\,\Omega\angle +50.194^\circ$

$$= 3.283\,\Omega + j3.939\,\Omega$$
$$= R + jX_L$$

$$X_L = 3.939\,\Omega = \omega L$$

and $L = \dfrac{3.939\,\Omega}{\omega} = \dfrac{3.939\,\Omega}{1000\text{ rad/s}} = \mathbf{3.939\text{ mH}}$

The series equivalent circuit appears in Fig. 16.9.

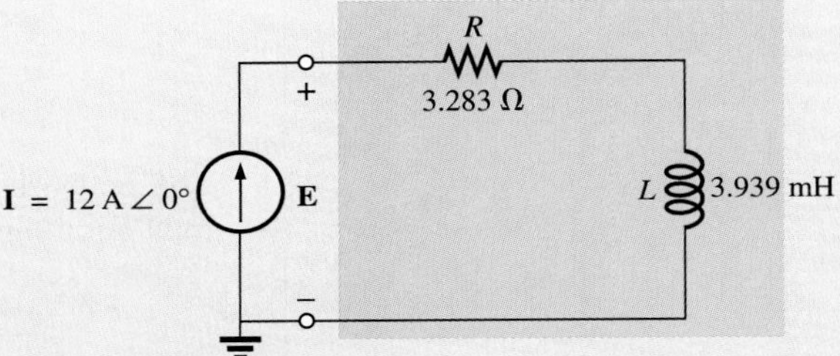

FIG. 16.9
Series equivalent circuit for the parallel R-L-C network of Fig. 16.6 with $\omega = 1000$ rad/s.

f. $\mathbf{E} = \mathbf{I}\mathbf{Z}_T = (12\text{ A}\angle 0^\circ)(5.128\,\Omega\angle 50.194^\circ)$
$= \mathbf{61.536\text{ V}\angle 50.194^\circ}$, as above

g. $P = I^2R = (12\text{ A})^2(3.283\,\Omega) = \mathbf{472.75\text{ W}}$ (as above)

h. $R_p = \dfrac{R_s^2 + X_s^2}{R_s} = \dfrac{(3.283\,\Omega)^2 + (3.939\,\Omega)^2}{3.283\,\Omega} = \mathbf{8\,\Omega}$

$$X_p = \frac{R_s^2 + X_s^2}{X_s} = \frac{(3.283\,\Omega)^2 + (3.939\,\Omega)^2}{3.939\,\Omega} = \mathbf{6.675\,\Omega}$$

continued

The parallel equivalent circuit appears in Fig. 16.10.

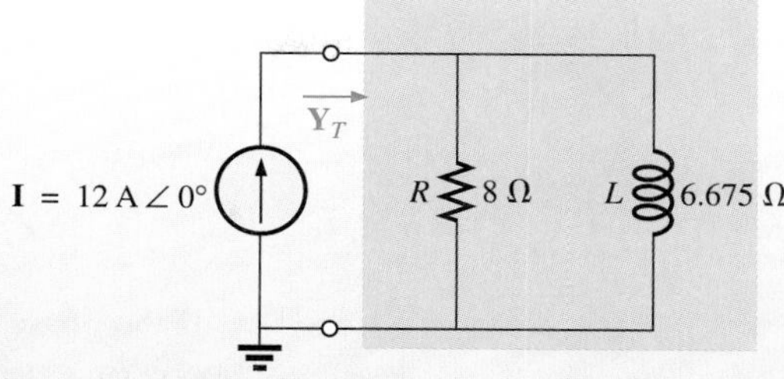

FIG. 16.10
Parallel equivalent of the circuit of Fig. 16.9.

$$\mathbf{Y}_T = G\angle 0^\circ + B_L\angle -90^\circ = \frac{1}{8\,\Omega}\angle 0^\circ + \frac{1}{6.675\,\Omega}\angle -90^\circ$$
$$= 0.125\text{ S}\angle 0^\circ + 0.15\text{ S}\angle -90^\circ$$
$$= 0.125\text{ S} - j\,0.15\text{ S} = \mathbf{0.195\text{ S}\angle -50.194^\circ} \quad \text{(as above)}$$

16.3 PHASE MEASUREMENTS: DUAL-TRACE OSCILLOSCOPE

The phase shift between the voltages of a network or between the voltages and currents of a network can be found using a **dual-trace** (two signals displayed at the same time) **oscilloscope**. Phase-shift measurements can also be performed using a single-trace oscilloscope by properly interpreting the resulting Lissajous patterns obtained on the screen. This second approach, however, will be left for the laboratory experience.

In Fig. 16.11 channel 1 of the dual-trace oscilloscope is hooked up to display the applied voltage e. Channel 2 is connected to display the voltage across the inductor v_L. The important fact is that the ground of the

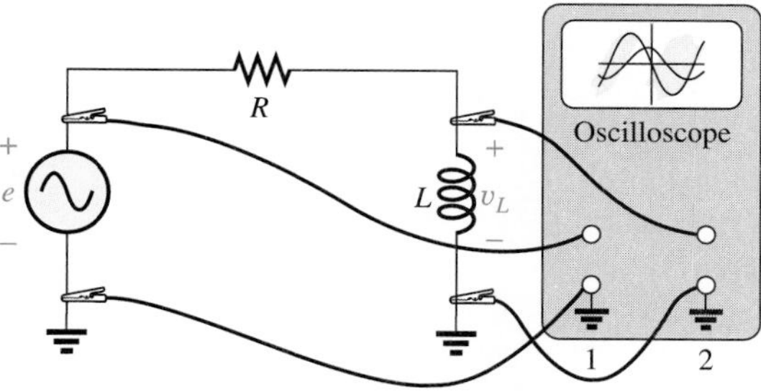

FIG. 16.11
Determining the phase relationship between e and v_L.

circuit is connected to the ground of the oscilloscope for both channels. In other words, there is only one common ground for the circuit and oscilloscope. The resulting waveforms may appear as shown in Fig. 16.12.

For the chosen horizontal sensitivity, each waveform of Fig. 16.12 has a period T defined by eight horizontal divisions, and the phase angle between the two waveforms is defined by $1\frac{1}{2}$ divisions. Using the fact that each period of a sinusoidal waveform covers 360°, the following ratios can be set up to determine the phase angle θ.

$$\frac{8 \text{ divisions}}{360°} = \frac{1.5 \text{ divisions}}{\theta}$$

and
$$\theta = \left(\frac{1.5}{8}\right)360° = \mathbf{67.5°}$$

In general,

$$\theta = \frac{(\text{Div. for } \theta)}{(\text{Div. for } T)} \times 360° \qquad (16.5)$$

If the phase relationship between e and v_R is required, the oscilloscope *must not* be hooked up as shown in Fig. 16.13. Points a and b have a common ground that will establish a zero-volt drop between the two points; this drop will have the same effect as a short-circuit connection between a and b. The resulting short circuit will "short out" the inductive element, and the current will increase due to the drop in impedance for the circuit. A dangerous situation can arise if the inductive element has a high impedance and the resistor has a relatively low impedance. The current, controlled solely by the resistance R, could jump to dangerous levels and damage the equipment.

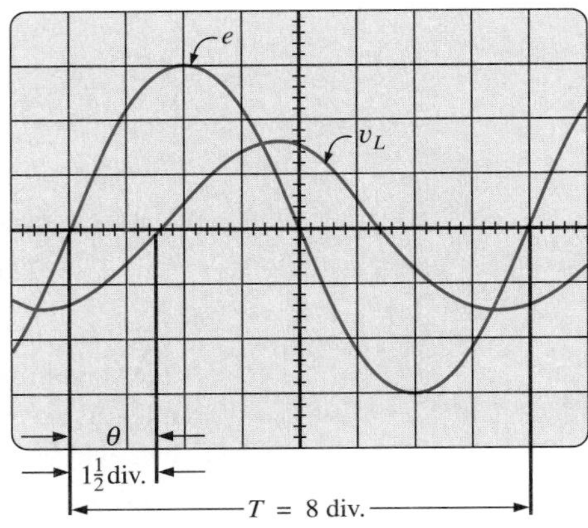

FIG. 16.12
Determining the phase angle between e and v_L.

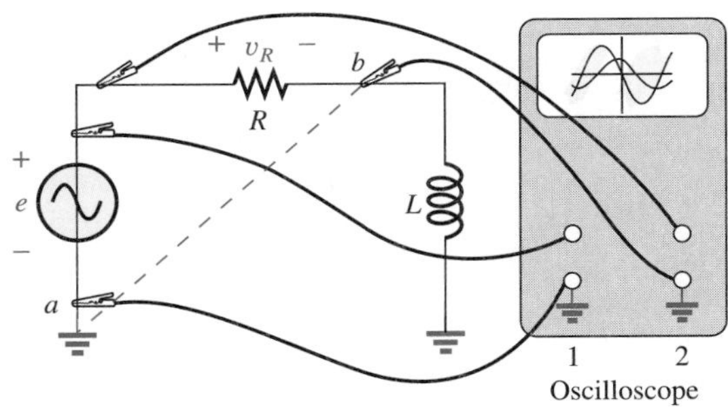

FIG. 16.13
An improper phase-measurement connection.

The phase relationship between e and v_R can be determined by simply interchanging the positions of the coil and resistor or by introducing a sensing resistor, as shown in Fig. 16.14. A sensing resistor is exactly that, introduced to "sense" a quantity without degrading the behaviour of the network. In other words, the sensing resistor must be small

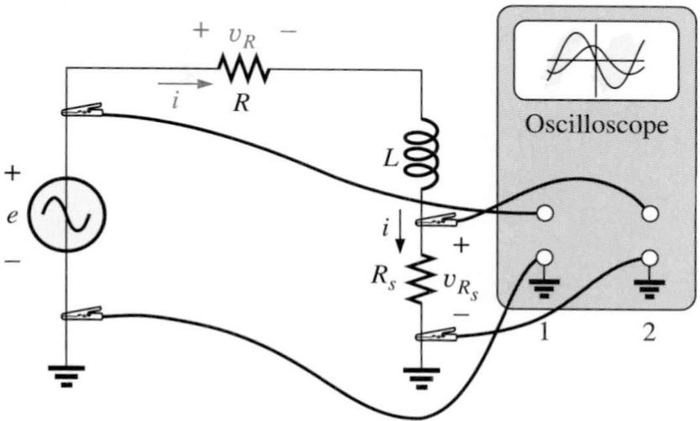

FIG. 16.14
Determining the phase relationship between e and v_R or e and i using a sensing resistor.

enough compared to the other impedances of the network not to cause a significant change in the voltage and current levels or phase relationships. Note that the sensing resistor is introduced in a way that will result in one end being connected to the common ground of the network. In Fig. 16.14, channel 2 will display the voltage v_{R_s}, which is in phase with the current i. However, the current i is also in phase with the voltage v_R across the resistor R. The net result is that the voltages v_{R_s} and v_R are in phase and the phase relationship between e and v_R can be determined from the waveforms e and v_{R_s}. Since v_{R_s} and i are in phase, the above procedure will also determine the phase angle between the applied voltage e and the source current i. If the magnitude of R_s is small enough compared to R or X_L, the phase measurements of Fig. 16.11 can be performed with R_s in place. That is, channel 2 can be connected to the top of the inductor and to ground, and the effect of R_s can be ignored. In the above application, the sensing resistor will not reveal the magnitude of the voltage v_R but simply the phase relationship between e and v_R.

For the parallel network of Fig. 16.15, the phase relationship between two of the branch currents, i_R and i_L, can be determined using a sensing resistor, as shown in the figure. Channel 1 will display the voltage v_R and channel 2 will display the voltage v_{R_s}. Since v_R is in

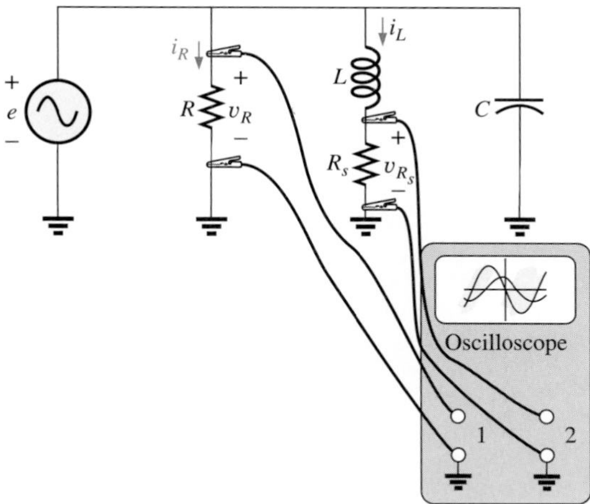

FIG. 16.15
Determining the phase relationship between i_R and i_L.

phase with i_R and v_{R_s} is in phase with the current i_L, the phase relationship between v_R and v_{R_s} will be the same as that between i_R and i_L. In this case, the magnitudes of the current levels can be determined using Ohm's law and the resistance levels R and R_s, respectively.

If the phase relationship between e and i_s of Fig. 16.15 is required, a sensing resistor can be used, as shown in Fig. 16.16.

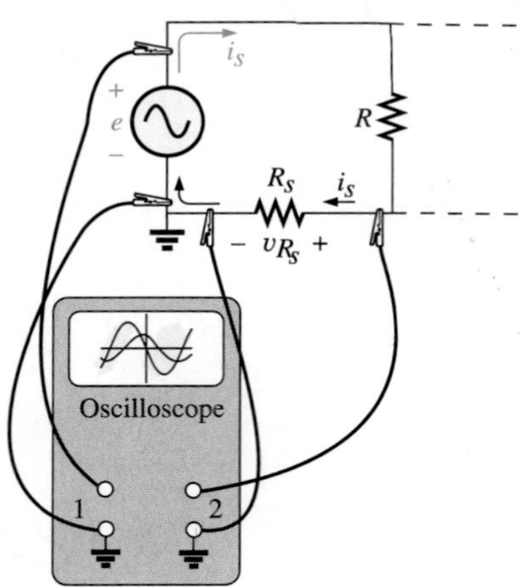

FIG. 16.16
Determining the phase relationship between e and i_s.

In general, therefore, for dual-trace measurements of phase relationships, be particularly careful of the grounding arrangement and make full use of the in-phase relationship between the voltage and current of a resistor.

16.4 ILLUSTRATIVE EXAMPLES

EXAMPLE 16.3

For the network of Fig. 16.17:

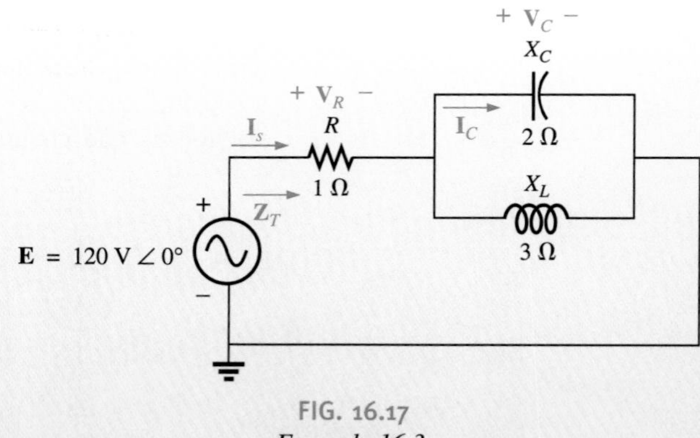

FIG. 16.17
Example 16.3.

a. Calculate $\mathbf{Z}_T$.
b. Determine $\mathbf{I}_s$.
c. Calculate $\mathbf{V}_R$ and $\mathbf{V}_C$.
d. Find $\mathbf{I}_C$.
e. Compute the power delivered.
f. Find F_p of the network.

continued

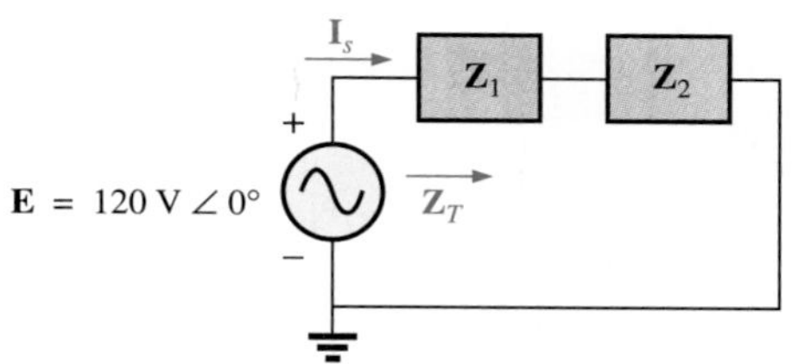

FIG. 16.18
Network of Fig. 16.17 after assigning the block impedances.

Solutions:

a. As suggested in the introduction, the network has been redrawn with block impedances, as shown in Fig. 16.18. The impedance $\mathbf{Z}_1$ is simply the resistor R of 1 Ω, and $\mathbf{Z}_2$ is the parallel combination of X_C and X_L. The network now clearly reveals that it is basically a series circuit, suggesting a direct path toward the total impedance and the source current.

As noted in the introduction, for many such problems you must first work back to the source to find the total impedance and then the source current. When the unknown quantities are found in terms of these subscripted impedances, the numerical values can then be substituted to find the magnitude and phase angle of the unknown. In other words, try to find the desired solution solely in terms of the subscripted impedances before substituting numbers. This approach will usually make the chosen path toward a solution clearer while saving time and preventing careless calculation errors. Note also in Fig. 16.18 that all the unknown quantities except $\mathbf{I}_C$ have been preserved, meaning that Fig. 16.18 can be used to determine these quantities rather than having to return to the more complex network of Fig. 16.17.

The total impedance is defined by

$$\mathbf{Z}_T = \mathbf{Z}_1 + \mathbf{Z}_2$$

with

$$\mathbf{Z}_1 = R\angle 0° = 1\,\Omega\angle 0°$$

$$\mathbf{Z}_2 = \mathbf{Z}_C \parallel \mathbf{Z}_L = \frac{(X_C\angle -90°)(X_L\angle 90°)}{-j X_C + j X_L} = \frac{(2\,\Omega\angle -90°)(3\,\Omega\angle 90°)}{-j\,2\,\Omega + j\,3\,\Omega}$$

$$= \frac{6\,\Omega\angle 0°}{j\,1} = \frac{6\,\Omega\angle 0°}{1\angle 90°} = 6\,\Omega\angle -90°$$

and

$$\mathbf{Z}_T = \mathbf{Z}_1 + \mathbf{Z}_2 = 1\,\Omega - j\,6\,\Omega = \mathbf{6.08\,\Omega\angle -80.54°}$$

b. $$\mathbf{I}_s = \frac{\mathbf{E}}{\mathbf{Z}_T} = \frac{120\text{ V}\angle 0°}{6.08\,\Omega\angle -80.5°} = \mathbf{19.74\text{ A}\angle 80.54°}$$

c. Referring to Fig. 16.18, we find that $\mathbf{V}_R$ and $\mathbf{V}_C$ can be found by directly applying Ohm's law:

$$\mathbf{V}_R = \mathbf{I}_s\mathbf{Z}_1 = (19.74\text{ A}\angle 80.54°)(1\,\Omega\angle 0°) = \mathbf{19.74\text{ V}\angle 80.54°}$$

$$\mathbf{V}_C = \mathbf{I}_s\mathbf{Z}_2 = (19.74\text{ A}\angle 80.54°)(6\,\Omega\angle -90°)$$

$$= \mathbf{118.44\text{ V}\angle -9.46°}$$

d. Now that $\mathbf{V}_C$ is known, the current $\mathbf{I}_C$ can also be found using Ohm's law.

$$\mathbf{I}_C = \frac{\mathbf{V}_C}{\mathbf{Z}_C} = \frac{118.44\text{ V}\angle -9.46°}{2\,\Omega\angle -90°} = \mathbf{59.22\text{ A}\angle 80.54°}$$

e. $P_{del} = I_s^2 R = (19.74\text{ A})^2(1\,\Omega) = \mathbf{389.67\text{ W}}$

f. $F_p = \cos\theta = \cos 80.54° = \mathbf{0.164\ leading}$

continued

The fact that the total impedance has a negative phase angle (revealing that $\mathbf{I}_s$ leads $\mathbf{E}$) clearly shows that the network is capacitive in nature and therefore has a leading power factor. The fact that the network is capacitive can be determined from the original network by first realizing that, for the parallel L-C elements, the smaller impedance predominates and results in an R-C network.

EXAMPLE 16.4

For the network of Fig. 16.19(a):

a. If $\mathbf{I}$ is 50 A $\angle 30°$, calculate $\mathbf{I}_1$ using the current divider rule.
b. Repeat part (a) for $\mathbf{I}_2$.
c. Verify Kirchhoff's current law at one node.

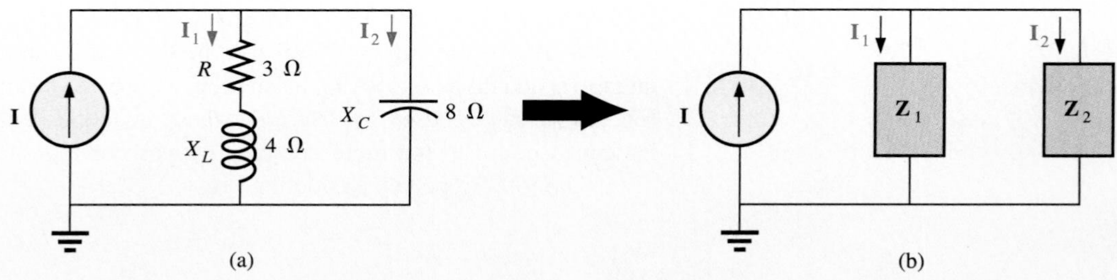

FIG. 16.19
Example 16.4.

Solutions:

a. Redrawing the circuit as in Fig. 16.19(b), we have

$$\mathbf{Z}_1 = R + jX_L = 3\ \Omega + j\,4\ \Omega = 5\ \Omega\ \angle 53.13°$$
$$\mathbf{Z}_2 = -jX_C = -j\,8\ \Omega = 8\ \Omega\ \angle -90°$$

Using the current divider rule gives us

$$\mathbf{I}_1 = \frac{\mathbf{Z}_2\mathbf{I}}{\mathbf{Z}_2 + \mathbf{Z}_1} = \frac{(8\ \Omega\ \angle -90°)(50\text{ A}\ \angle 30°)}{(-j\,8\ \Omega) + (3\ \Omega + j\,4\ \Omega)} = \frac{400\ \angle -60°}{3 - j\,4}$$
$$= \frac{400\ \angle -60°}{5\ \angle -53.13°} = \mathbf{80\ A\ \angle -6.87°}$$

b. $$\mathbf{I}_2 = \frac{\mathbf{Z}_1\mathbf{I}}{\mathbf{Z}_2 + \mathbf{Z}_1} = \frac{(5\ \Omega\ \angle 53.13°)(50\text{ A}\ \angle 30°)}{5\ \Omega\ \angle -53.13°} = \frac{250\ \angle 83.13°}{5\ \angle -53.13°}$$
$$= \mathbf{50\ A\ \angle 136.26°}$$

c.
$$\mathbf{I} = \mathbf{I}_1 + \mathbf{I}_2$$
$$50\text{ A}\ \angle 30° = 80\text{ A}\ \angle -6.87° + 50\text{ A}\ \angle 136.26°$$
$$= (79.43 - j\,9.57) + (-36.12 + j\,34.57)$$
$$= 43.31 + j\,25.0$$
$$50\text{ A}\ \angle 30° = 50\text{ A}\ \angle 30° \quad \text{(checks)}$$

EXAMPLE 16.5

For the network of Fig. 16.20(a):

a. Calculate the voltage $\mathbf{V}_C$ using the voltage divider rule.
b. Calculate the current $\mathbf{I}_s$.

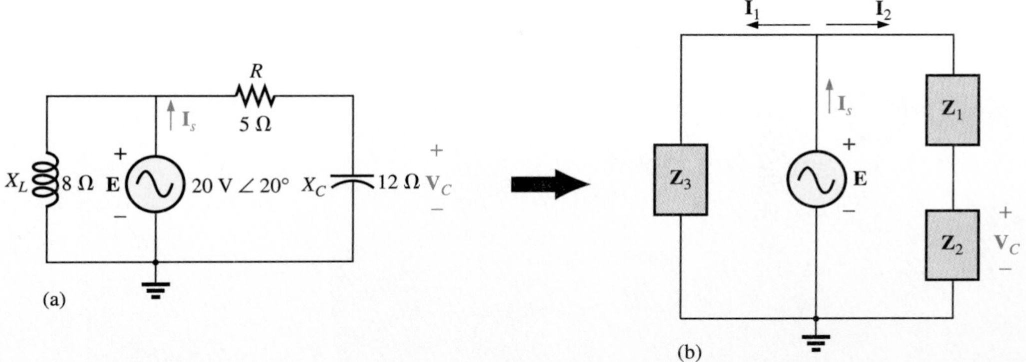

FIG. 16.20
Example 16.5.

Solutions:

a. The network is redrawn as shown in Fig. 16.20(b), with

$$\mathbf{Z}_1 = 5\ \Omega = 5\ \Omega \angle 0^\circ$$
$$\mathbf{Z}_2 = -j\,12\ \Omega = 12\ \Omega \angle -90^\circ$$
$$\mathbf{Z}_3 = +j\,8\ \Omega = 8\ \Omega \angle 90^\circ$$

Since $\mathbf{V}_C$ is desired, we will not combine R and X_C into a single block impedance. Note also how Fig. 16.20(b) clearly reveals that **E** is the total voltage across the series combination of $\mathbf{Z}_1$ and $\mathbf{Z}_2$, permitting the use of the voltage divider rule to calculate $\mathbf{V}_C$. In addition, note that all the currents necessary to determine $\mathbf{I}_s$ have been preserved in Fig. 16.20(b), so there is no need ever to return to the network of Fig. 16.20(a)—everything is defined by Fig. 16.20(b).

$$\mathbf{V}_C = \frac{\mathbf{Z}_2\mathbf{E}}{\mathbf{Z}_1 + \mathbf{Z}_2} = \frac{(12\ \Omega \angle -90^\circ)(20\ \text{V} \angle 20^\circ)}{5\ \Omega - j\,12\ \Omega} = \frac{240\ \text{V} \angle -70^\circ}{13 \angle -67.38^\circ}$$
$$= \mathbf{18.46\ V \angle -2.62^\circ}$$

b. $$\mathbf{I}_1 = \frac{\mathbf{E}}{\mathbf{Z}_3} = \frac{20\ \text{V} \angle 20^\circ}{8\ \Omega \angle 90^\circ} = 2.5\ \text{A} \angle -70^\circ$$

$$\mathbf{I}_2 = \frac{\mathbf{E}}{\mathbf{Z}_1 + \mathbf{Z}_2} = \frac{20\ \text{V} \angle 20^\circ}{13\ \Omega \angle -67.38^\circ} = 1.54\ \text{A} \angle 87.38^\circ$$

and

$$\mathbf{I}_s = \mathbf{I}_1 + \mathbf{I}_2$$
$$= 2.5\ \text{A} \angle -70^\circ + 1.54\ \text{A} \angle 87.38^\circ$$
$$= (0.86 - j\,2.35) + (0.07 + j\,1.54)$$
$$\mathbf{I}_s = 0.93 - j\,0.81 = \mathbf{1.23\ A \angle -41.05^\circ}$$

EXAMPLE 16.6

For Fig. 16.21(a):

a. Calculate the current $\mathbf{I}_s$.
b. Find the voltage $\mathbf{V}_{ab}$.

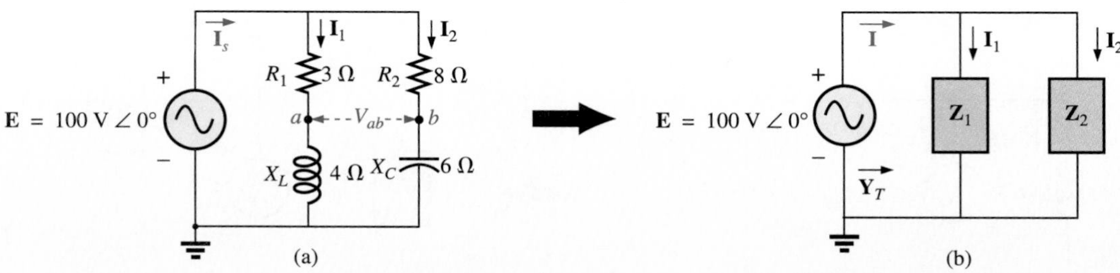

FIG. 16.21
Example 16.6.

Solutions:

a. Redrawing the circuit as in Fig. 16.21(b), we obtain

$$\mathbf{Z}_1 = R_1 + j X_L = 3\ \Omega + j\,4\ \Omega = 5\ \Omega\ \angle 53.13^\circ$$
$$\mathbf{Z}_2 = R_2 - j X_C = 8\ \Omega - j\,6\ \Omega = 10\ \Omega\ \angle -36.87^\circ$$

Here the voltage $\mathbf{V}_{ab}$ is lost in the redrawn network, but the currents $\mathbf{I}_1$ and $\mathbf{I}_2$ remain defined for future calculations necessary to determine $\mathbf{V}_{ab}$. Figure 16.21(b) shows that the total impedance can be found using the equation for two parallel impedances:

$$\mathbf{Z}_T = \frac{\mathbf{Z}_1\mathbf{Z}_2}{\mathbf{Z}_1 + \mathbf{Z}_2} = \frac{(5\ \Omega\ \angle 53.13^\circ)(10\ \Omega\ \angle -36.87^\circ)}{(3\ \Omega + j\,4\ \Omega) + (8\ \Omega - j\,6\ \Omega)}$$
$$= \frac{50\ \Omega\ \angle 16.26^\circ}{11 - j\,2} = \frac{50\ \Omega\ \angle 16.26^\circ}{11.18\ \angle -10.30^\circ}$$
$$= \mathbf{4.472\ \Omega\ \angle 26.56^\circ}$$

and $$\mathbf{I}_s = \frac{\mathbf{E}}{\mathbf{Z}_T} = \frac{100\ \text{V}\ \angle 0^\circ}{4.472\ \Omega\ \angle 26.56^\circ} = \mathbf{22.36\ A\ \angle -26.56^\circ}$$

b. By Ohm's law,

$$\mathbf{I}_1 = \frac{\mathbf{E}}{\mathbf{Z}_1} = \frac{100\ \text{V}\ \angle 0^\circ}{5\ \Omega\ \angle 53.13^\circ} = \mathbf{20\ A\ \angle -53.13^\circ}$$
$$\mathbf{I}_2 = \frac{\mathbf{E}}{\mathbf{Z}_2} = \frac{100\ \text{V}\ \angle 0^\circ}{10\ \Omega\ \angle -36.87^\circ} = \mathbf{10\ A\ \angle 36.87^\circ}$$

Returning to Fig. 16.21(a), we have

$$\mathbf{V}_{R_1} = \mathbf{I}_1\mathbf{Z}_{R_1} = (20\ \text{A}\ \angle -53.13^\circ)(3\ \Omega\ \angle 0^\circ) = \mathbf{60\ V\ \angle -53.13^\circ}$$
$$\mathbf{V}_{R_2} = \mathbf{I}_1\mathbf{Z}_{R_2} = (10\ \text{A}\ \angle +36.87^\circ)(8\ \Omega\ \angle 0^\circ) = \mathbf{80\ V\ \angle +36.87^\circ}$$

Instead of using the two steps just shown, $\mathbf{V}_{R_1}$ or $\mathbf{V}_{R_2}$ could have been determined in one step using the voltage divider rule:

$$\mathbf{V}_{R_1} = \frac{(3\ \Omega\ \angle 0^\circ)(100\ \text{V}\ \angle 0^\circ)}{3\ \Omega\ \angle 0^\circ + 4\ \Omega\ \angle 90^\circ} = \frac{300\ \text{V}\ \angle 0^\circ}{5\ \angle 53.13^\circ} = \mathbf{60\ V\ \angle -53.13^\circ}$$

continued

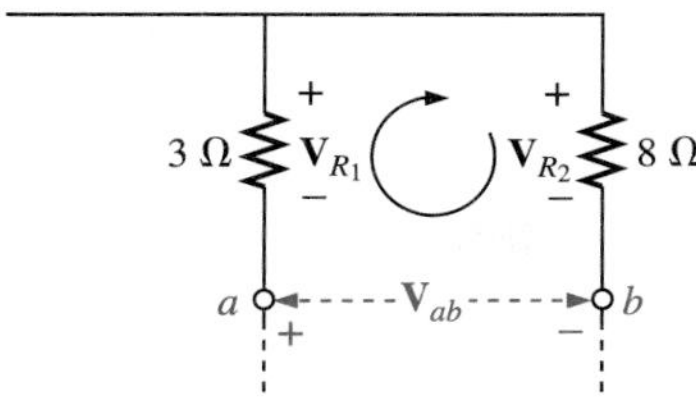

FIG. 16.22
Determining the voltage V_{ab} for the network of Fig. 16.21(a).

To find $\mathbf{V}_{ab}$, Kirchhoff's voltage law must be applied around the loop (Fig.16.22) consisting of the 3-Ω and 8-Ω resistors. By Kirchhoff's voltage law,

$$\mathbf{V}_{ab} + \mathbf{V}_{R_1} - \mathbf{V}_{R_2} = 0$$

or

$$\begin{aligned}\mathbf{V}_{ab} &= \mathbf{V}_{R_2} - \mathbf{V}_{R_1}\\ &= 80\text{ V}\angle 36.87^\circ - 60\text{ V}\angle -53.13^\circ\\ &= (64 + j\,48) - (36 - j\,48)\\ &= 28 + j\,96\\ \mathbf{V}_{ab} &= \mathbf{100\text{ V}\angle 73.74^\circ}\end{aligned}$$

EXAMPLE 16.7

The network of Fig. 16.23(a) is common in *transistor* networks. The transistor equivalent circuit includes a current source **I** and an output impedance R_o. The resistor R_C is a biasing resistor to establish specific dc conditions, and the resistor R_i represents the loading of the next stage. The coupling capacitor is designed to be an open circuit for dc and as low an impedance as possible for the frequencies of interest to ensure that $\mathbf{V}_L$ is a maximum value. The frequency range includes the entire audio (hearing) spectrum from 100 Hz to 20 kHz. This example shows that for the full audio range, the effect of the capacitor can be ignored. It acts as a dc blocking agent but permits the ac to pass through with little disturbance.

a. Find $\mathbf{V}_L$ for the network of Fig. 16.23(a) at a frequency of 100 Hz.
b. Repeat part (a) at a frequency of 20 kHz.
c. Compare the results of parts (a) and (b).

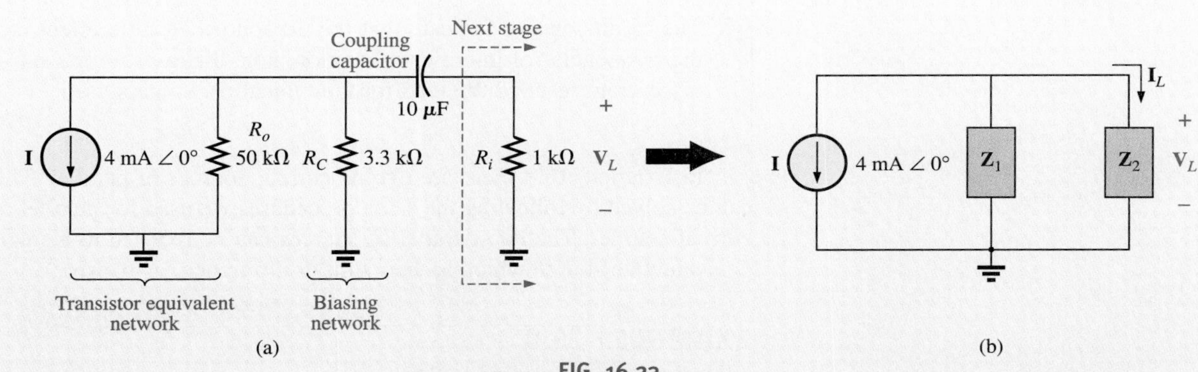

FIG. 16.23
Basic transistor amplifier.

Solutions:

a. The network is redrawn with subscripted impedances in Fig. 16.23(b).

$$\mathbf{Z}_1 = 50\text{ k}\Omega\angle 0^\circ \parallel 3.3\text{ k}\Omega\angle 0^\circ = 3.096\text{ k}\Omega\angle 0^\circ$$
$$\mathbf{Z}_2 = R_i - j\,X_C$$

At $f = 100$ Hz: $X_C = \dfrac{1}{2\pi fC} = \dfrac{1}{2\pi(100\text{ Hz})(10\ \mu\text{F})} = 159.16\ \Omega$

and $\mathbf{Z}_2 = 1\text{ k}\Omega - j\,159.16\ \Omega$

continued

Current divider rule:

$$\mathbf{I}_L = \frac{\mathbf{Z}_1\mathbf{I}}{\mathbf{Z}_1 + \mathbf{Z}_2} = \frac{(3.096\text{ k}\Omega \angle 0°)(4\text{ mA} \angle 0°)}{3.096\text{ k}\Omega + 1\text{ k}\Omega - j\,159.16\ \Omega}$$

$$= \frac{12.384\text{ A} \angle 0°}{4096 - j\,159.16} = \frac{12.384\text{ A} \angle 0°}{4099 \angle -2.225°}$$

$$= 3.021\text{ mA} \angle 2.225°$$

and

$$\mathbf{V}_L = \mathbf{I}_L\mathbf{Z}_R$$
$$= (3.021\text{ mA} \angle 2.225°)(1\text{ k}\Omega \angle 0°)$$
$$= \mathbf{3.021\text{ V} \angle 2.225°}$$

b. At $f = 20$ kHz: $X_C = \dfrac{1}{2\pi fC} = \dfrac{1}{2\pi(20\text{ kHz})(10\ \mu\text{F})} = 0.796\ \Omega$

Note the dramatic change in X_C with frequency. The higher the frequency, the more closely X_C approximates a short circuit.

$$\mathbf{Z}_2 = 1\text{ k}\Omega - j\,0.796\ \Omega$$

Current divider rule:

$$\mathbf{I}_L = \frac{\mathbf{Z}_1\mathbf{I}}{\mathbf{Z}_1 + \mathbf{Z}_2} = \frac{(3.096\text{ k}\Omega \angle 0°)(4\text{ mA} \angle 0°)}{3.096\text{ k}\Omega + 1\text{ k}\Omega - j\,0.796\ \Omega}$$

$$= \frac{12.384\text{ A} \angle 0°}{4096 - j\,0.796\ \Omega} = \frac{12.384\text{ A} \angle 0°}{4096 \angle -0.011°}$$

$$= 3.023\text{ mA} \angle 0.011°$$

and

$$\mathbf{V}_L = \mathbf{I}_L\mathbf{Z}_R$$
$$= (3.023\text{ mA} \angle 0.011°)(1\text{ k}\Omega \angle 0°)$$
$$= \mathbf{3.023\text{ V} \angle 0.011°}$$

c. The results clearly indicate that the capacitor had little effect on the frequencies of interest. In addition, note that most of the supply current reached the load for the typical parameters used.

In Example 16.8 there are two ac current sources in parallel. This can be solved by following the same procedures we used for parallel dc current sources. The two parallel ac sources can be reduced to a single equivalent ac source which is their sum or difference (as phasors).

EXAMPLE 16.8

For the network of Fig. 16.24:

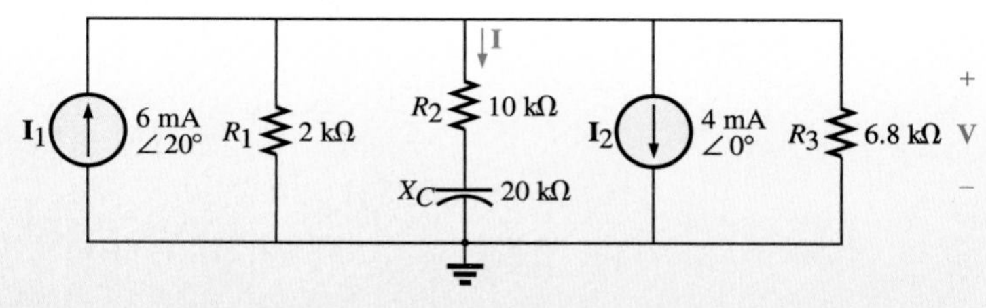

FIG. 16.24
Example 16.8.

a. Determine the current **I**.

b. Find the voltage **V**.

continued

Solutions:

a. Reducing the parallel ac current sources to an eqivalent current source by using their phasor sum (or difference), we get:

$$\begin{aligned}\mathbf{I}_T &= 6\text{ mA }\angle 20^\circ - 4\text{ mA }\angle 0^\circ \\ &= 5.638\text{ mA} + j\,2.052\text{ mA} - 4\text{ mA} \\ &= 1.638\text{ mA} + j\,2.052\text{ mA} \\ &= 2.626\text{ mA }\angle 51.402^\circ\end{aligned}$$

Redrawing the network using block impedances will result in the network of Fig. 16.25 where

$$\mathbf{Z}_1 = 2\text{ k}\Omega\angle 0^\circ \parallel 6.8\text{ k}\Omega\angle 0^\circ = 1.545\text{ k}\Omega\angle 0^\circ$$

and $\mathbf{Z}_2 = 10\text{ k}\Omega - j\,20\text{ k}\Omega = 22.361\text{ k}\Omega\angle -63.435^\circ$

Note that **I** and **V** are still defined in Fig. 16.25.

Current divider rule:

$$\mathbf{I} = \frac{\mathbf{Z}_1\mathbf{I}_T}{\mathbf{Z}_1+\mathbf{Z}_2} = \frac{(1.545\text{ k}\Omega\angle 0^\circ)(2.626\text{ mA}\angle 51.402^\circ)}{1.545\text{ k}\Omega + 10\text{ k}\Omega - j\,20\text{ k}\Omega}$$

$$= \frac{4.057\text{ A}\angle 51.402^\circ}{11.545\times 10^3 - j\,20\times 10^3} = \frac{4.057\text{ A}\angle 51.402^\circ}{23.093\times 10^3\angle -60.004^\circ}$$

$$= \mathbf{0.176\text{ mA}\angle 111.406^\circ}$$

b. $$\begin{aligned}\mathbf{V} &= \mathbf{I}\mathbf{Z}_2 \\ &= (0.176\text{ mA}\angle 111.406^\circ)(22.36\text{ k}\Omega\angle -63.435^\circ) \\ &= \mathbf{3.936\text{ V}\angle 47.971^\circ}\end{aligned}$$

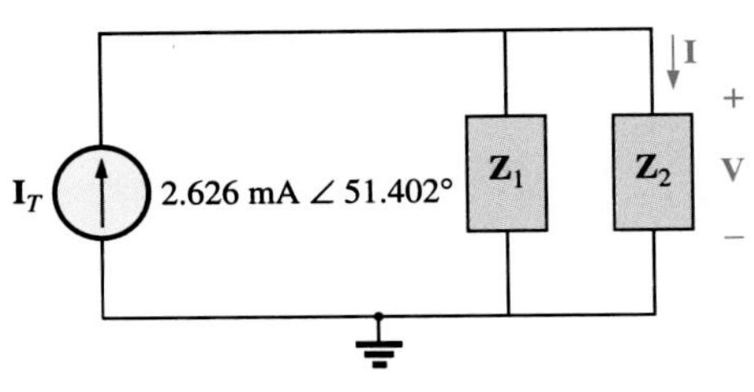

FIG. 16.25
Network of Fig. 16.24 after the subscripted impedances are assigned.

EXAMPLE 16.9

For the network of Fig. 16.26:

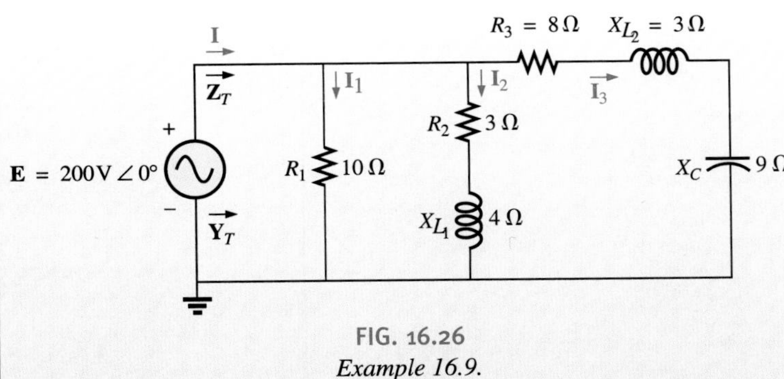

FIG. 16.26
Example 16.9.

a. Compute **I**.
b. Find $\mathbf{I}_1$, $\mathbf{I}_2$, and $\mathbf{I}_3$.
c. Verify Kirchhoff's current law by showing that

$$\mathbf{I} = \mathbf{I}_1 + \mathbf{I}_2 + \mathbf{I}_3$$

d. Find the total impedance of the circuit.

continued

Solutions:

a. Redrawing the circuit as in Fig. 16.27 reveals a strictly parallel network where

$$\mathbf{Z}_1 = R_1 = 10\ \Omega \angle 0°$$
$$\mathbf{Z}_2 = R_2 + j X_{L_1} = 3\ \Omega + j\,4\ \Omega$$
$$\mathbf{Z}_3 = R_3 + j X_{L_2} - j X_C = 8\ \Omega + j\,3\ \Omega - j\,9\ \Omega = 8\ \Omega - j\,6\ \Omega$$

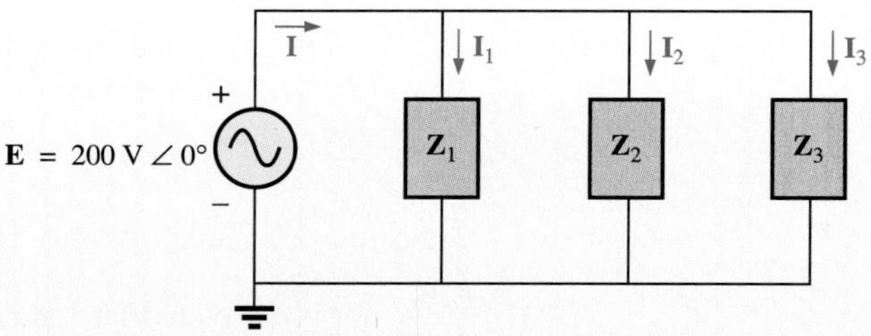

FIG. 16.27
Network of Fig. 16.26 after the subscripted impedances are assigned.

The total admittance is

$$\mathbf{Y}_T = \mathbf{Y}_1 + \mathbf{Y}_2 + \mathbf{Y}_3$$
$$= \frac{1}{\mathbf{Z}_1} + \frac{1}{\mathbf{Z}_2} + \frac{1}{\mathbf{Z}_3} = \frac{1}{10\ \Omega} + \frac{1}{3\ \Omega + j\,4\ \Omega} + \frac{1}{8\ \Omega - j\,6\ \Omega}$$
$$= 0.1\ \text{S} + \frac{1}{5\ \Omega \angle 53.13°} + \frac{1}{10\ \Omega \angle -36.87°}$$
$$= 0.1\ \text{S} + 0.2\ \text{S} \angle -53.13° + 0.1\ \text{S} \angle 36.87°$$
$$= 0.1\ \text{S} + 0.12\ \text{S} - j\,0.16\ \text{S} + 0.08\ \text{S} + j\,0.06\ \text{S}$$
$$= 0.3\ \text{S} - j\,0.1\ \text{S} = 0.316\ \text{S} \angle -18.435°$$

The current **I**:

$$\mathbf{I} = \mathbf{E}\mathbf{Y}_T = (200\ \text{V} \angle 0°)(0.316\ \text{S} \angle -18.435°)$$
$$= \mathbf{63.2\ A \angle -18.435°}$$

b. Since the voltage is the same across parallel branches,

$$\mathbf{I}_1 = \frac{\mathbf{E}}{\mathbf{Z}_1} = \frac{200\ \text{V} \angle 0°}{10\ \Omega \angle 0°} = \mathbf{20\ A \angle 0°}$$
$$\mathbf{I}_2 = \frac{\mathbf{E}}{\mathbf{Z}_2} = \frac{200\ \text{V} \angle 0°}{5\ \Omega \angle 53.13°} = \mathbf{40\ A \angle -53.13°}$$
$$\mathbf{I}_3 = \frac{\mathbf{E}}{\mathbf{Z}_3} = \frac{200\ \text{V} \angle 0°}{10\ \Omega \angle -36.87°} = \mathbf{20\ A \angle +36.87°}$$

c.
$$\mathbf{I} = \mathbf{I}_1 + \mathbf{I}_2 + \mathbf{I}_3$$
$$60 - j\,20 = 20 \angle 0° + 40 \angle -53.13° + 20 \angle +36.87°$$
$$= (20 + j\,0) + (24 - j\,32) + (16 + j\,12)$$
$$60 - j\,20 = 60 - j\,20 \quad \text{(checks)}$$

d.
$$\mathbf{Z}_T = \frac{1}{\mathbf{Y}_T} = \frac{1}{0.316\ \text{S} \angle -18.435°}$$
$$= \mathbf{3.165\ \Omega \angle 18.435°}$$

EXAMPLE 16.10

For the network of Fig. 16.28(a):

a. Calculate the total impedance $\mathbf{Z}_T$.
b. Compute **I**.
c. Find the total power factor.
d. Calculate $\mathbf{I}_1$ and $\mathbf{I}_2$.
e. Find the average power delivered to the circuit.

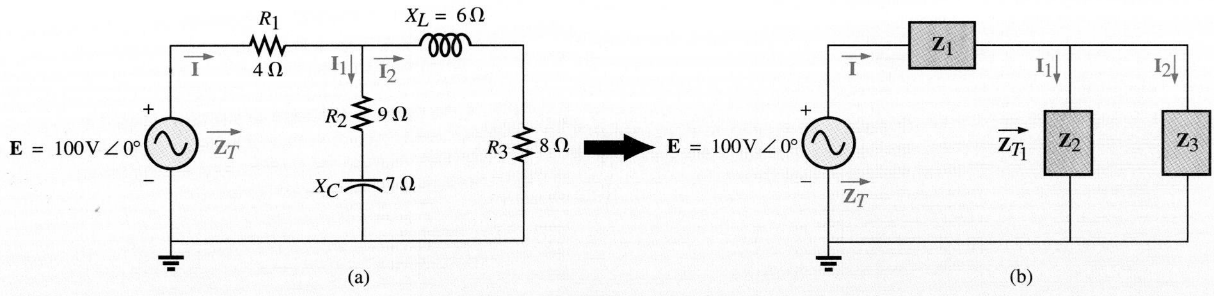

FIG. 16.28
Example 16.10.

Solutions:

a. Redrawing the circuit as in Fig. 16.28(b), we have

$$\mathbf{Z}_1 = R_1 = 4\ \Omega \angle 0^\circ$$
$$\mathbf{Z}_2 = R_2 - j X_C = 9\ \Omega - j\,7\ \Omega = 11.40\ \Omega \angle -37.87^\circ$$
$$\mathbf{Z}_3 = R_3 + j X_L = 8\ \Omega + j\,6\ \Omega = 10\ \Omega \angle +36.87^\circ$$

Notice that all the desired quantities were conserved in the redrawn network. The total impedance:

$$\mathbf{Z}_T = \mathbf{Z}_1 + \mathbf{Z}_{T_1}$$
$$= \mathbf{Z}_1 + \frac{\mathbf{Z}_2\mathbf{Z}_3}{\mathbf{Z}_2 + \mathbf{Z}_3}$$
$$= 4\ \Omega + \frac{(11.4\ \Omega \angle -37.87^\circ)(10\ \Omega \angle 36.87^\circ)}{(9\ \Omega - j\,7\ \Omega) + (8\ \Omega + j\,6\ \Omega)}$$
$$= 4\ \Omega + \frac{114\ \Omega \angle -1.00^\circ}{17.03 \angle -3.37^\circ} = 4\ \Omega + 6.69\ \Omega \angle 2.37^\circ$$
$$= 4\ \Omega + 6.68\ \Omega + j\,0.28\ \Omega = 10.68\ \Omega + j\,0.28\ \Omega$$
$$\mathbf{Z}_T = \mathbf{10.684\ \Omega \angle 1.5^\circ}$$

Calculator Cautions 1. Here is an opportunity to demonstrate the versatility of the calculator. For the above operation, you must be aware of the priority of mathematical operations. Some calculators require the operations to be performed in the same order as they would be performed longhand, while calculator models require data to be entered in the RPN order.

2. In addition, rounding off intermediate results, as done in Example 16.10, may give a slightly different answer than carrying all digits throughout. Carrying all the digits gives an answer of 10.692 Ω ∠ 1.481° instead of the 10.684 Ω ∠ 1.5° shown in the example above. The difference between the two is only about 0.1%.

continued

3. In all cases, consult your physics or calculus text, and your calculator owner's manual for a more complete discussion of:
a. how rounding off and the number of significant figures in a calculation can affect your results.
b. the correct order for entering data and operations.

c. $$\mathbf{I} = \frac{\mathbf{E}}{\mathbf{Z}_T} = \frac{100 \text{ V} \angle 0^\circ}{10.684\ \Omega \angle 1.5^\circ} = \mathbf{9.36\ A \angle -1.5^\circ}$$

d. $$F_p = \cos\theta_T = \frac{R}{Z_T} = \frac{10.68\ \Omega}{10.684\ \Omega} \cong \mathbf{1}$$

(essentially resistive, which is interesting, considering the complexity of the network).

e. Current divider rule:

$$\mathbf{I}_2 = \frac{\mathbf{Z}_1\mathbf{I}}{\mathbf{Z}_2 + \mathbf{Z}_3} = \frac{(11.40\ \Omega \angle -37.87^\circ)(9.36\text{ A} \angle -1.5^\circ)}{(9\ \Omega - j\,7\ \Omega) + (8\ \Omega + j\,6\ \Omega)}$$

$$= \frac{106.7\text{ A} \angle -39.37^\circ}{17 - j\,1} = \frac{106.7\text{ A} \angle -39.37^\circ}{17.03 \angle -3.37^\circ}$$

$$\mathbf{I}_2 = \mathbf{6.27\ A \angle -36^\circ}$$

Applying Kirchhoff's current law (rather than the current divider rule again) gives us

$$\mathbf{I}_1 = \mathbf{I} - \mathbf{I}_2$$

or
$$\mathbf{I} = \mathbf{I}_1 - \mathbf{I}_2$$
$$= (9.36\text{ A} \angle -1.5^\circ) - (6.27\text{ A} \angle -36^\circ)$$
$$= (9.36\text{ A} - j\,0.25\text{ A}) - (5.07\text{ A} - j\,3.69\text{ A})$$
$$\mathbf{I}_1 = 4.29\text{ A} + j\,3.44\text{ A} = \mathbf{5.5\ A \angle 38.72^\circ}$$

f. $$P_T = EI\cos\theta_T$$
$$= (100\text{ V})(9.36\text{ A})\cos 1.5^\circ$$
$$= (936)(0.99966)$$
$$P_T = \mathbf{935.68\ W}$$

LUMINARIES

Hugh Le Caine (1914–1977) (Member of the Canadian Science and Engineering Hall of Fame) Physicist, inventor, composer and pioneer in the design of electronic instruments.

FIG. 16.29
Hugh Le Caine

From a very early age, Hugh Le Caine dreamed of inventing new musical instruments. He pursued his interests in both music and electronics throughout his life. From an early age he studied music (particularly piano) and in his spare time experimented with electronics. He studied engineering at Queen's University in the late 1940s before accepting a job at the National Research Council in Ottawa in 1954. For over 20 years at the NRC, Hugh Le Caine continued to develop and refine new and innovative electronic musical instruments and equipment. Many of his inventions are preserved in the National Science and Technology Museum in Ottawa. One of them that he completed as early as 1948 was called the "Sackbut." With its touch sensitivity and electronic sound production it is considered be the first synthesizer.
His visionary work represents a true marrying of art and science: the aesthetics of musical composition with the practical demonstration of electronic principles in the instruments he designed and built.

16.5 LADDER NETWORKS

Ladder networks were discussed in some detail in Chapter 7. This section will simply apply the first method described in Section 7.3 to the general sinusoidal ac ladder network of Fig. 16.30. The current $\mathbf{I}_6$ is desired.

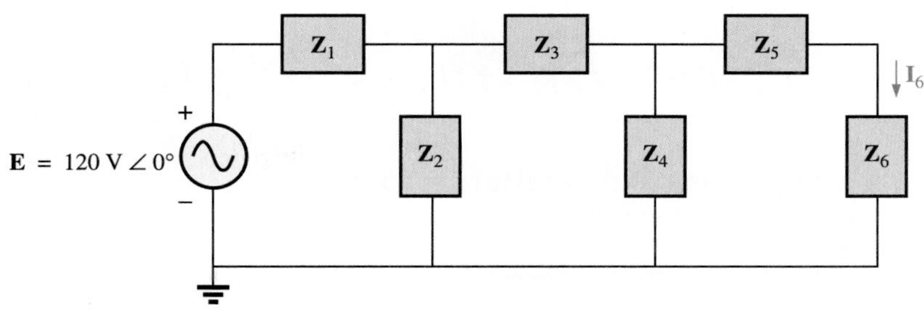

FIG. 16.30
Ladder network.

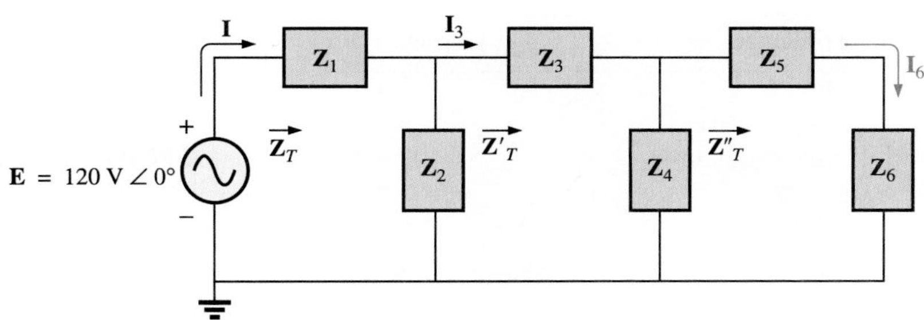

FIG. 16.31
Defining an approach to the analysis of ladder networks.

Impedances $\mathbf{Z}_T$, $\mathbf{Z}'_T$, and $\mathbf{Z}''_T$ and currents $\mathbf{I}_1$ and $\mathbf{I}_3$ are defined in Fig. 16.31:

$$\mathbf{Z}''_T = \mathbf{Z}_5 + \mathbf{Z}_6$$

and

$$\mathbf{Z}'_T = \mathbf{Z}_3 + \mathbf{Z}_4 \parallel \mathbf{Z}''_T$$

with

$$\mathbf{Z}_T = \mathbf{Z}_1 + \mathbf{Z}_2 \parallel \mathbf{Z}'_T$$

Then

$$\mathbf{I} = \frac{\mathbf{E}}{\mathbf{Z}_T}$$

and

$$\mathbf{I}_3 = \frac{\mathbf{Z}_2\mathbf{I}}{\mathbf{Z}_2 + \mathbf{Z}'_T}$$

with

$$\mathbf{I}_6 = \frac{\mathbf{Z}_4\mathbf{I}_3}{\mathbf{Z}_4 + \mathbf{Z}''_T}$$

Practical Business Applications

Seriously series and parallel parables

The next time you find yourself sitting in your kitchen enjoying a great cappuccino or mocha java, take a look around at the various appliances there. If your home is typical, you'll likely find a fridge, stove, coffee maker, microwave oven, toaster, etc.

Have you ever wondered how much electricity all these appliances use? To find out, just take a look at the electrical ratings sticker on the back or bottom of each item. The current ratings might look something like the following:

Appliance	Current Rating
Fridge	6 A
Coffee maker	4 A
Microwave oven	10 A
Stove	24 A
Toaster	8 A
Total	52 A

Wow! It doesn't take long to realize that kitchens use up a lot of current. So how are the various appliances hooked up? Are they in series, parallel, or both?

The answer is: it's a bit of each. Residential wiring in North America is standardized around a 240/120 V ac system called the Edison three-wire distribution system. That is, the local utility provides what is called a 2-phase circuit. This essentially consists of two 120 V ac circuits (each circuit is 180 degrees out of phase with respect to the other) as shown in the diagram.

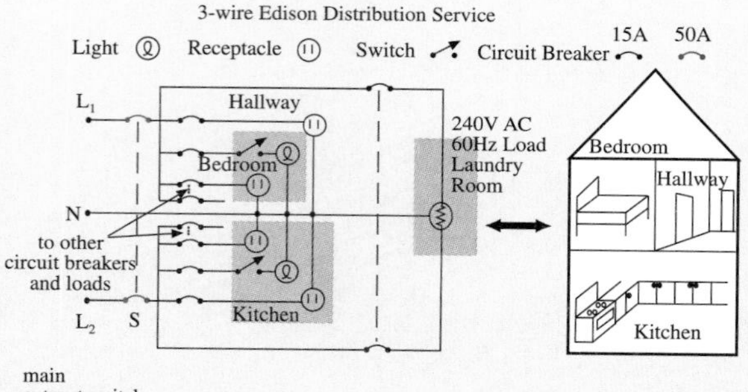

Now, take a look at your electrical panels, usually located in the basement. The fuses or circuit breakers are located in a distribution panel that divides each of these two phases into multiple parallel branch circuits. Each fuse or breaker is in series with one branch circuit.

You'll also notice that most of the fuses or circuit breakers are rated for a maximum of 15 A. That's because the standard wiring used throughout your home is American Wire Gauge (AWG) #14 copper, which is designed to safely handle this amount. Thus, one branch circuit is not enough to supply the whole kitchen. The only way to be able to use several appliances at once is to install more than one circuit.

To meet the needs of a typical kitchen, often four (or more) circuits are used. The stove needs two dedicated circuits by itself because it takes up to 30 A. Even if it's not in your kitchen, your dryer uses another two circuits because it also takes up to 30 A. Each kitchen wall outlet is usually on a separate circuit. In other rooms, several outlets will be connected in parallel to a single branch circuit.

Spreading the loads over more than one room is, in fact, an electrical building code requirement. How it's done depends on the electrician who wired up your house. It's generally safe to assume that we don't use all the rooms at once. Thus, it's very unlikely that we will ever overload any one circuit. If we do overload a circuit, you just simply go downstairs and replace the fuse or reset the circuit breaker.

Oh yes, don't forget to bring a flashlight down with you. It's usually dark down there and, as you know, we never play with electricity in the dark, right? So educate yourself about electricity and you'll be enlightened.

GFI (Ground fault interupter)

GFI outlets work just like regular electrical outlets except that they provide an increased measure of safety. Electrical codes require that GFI outlets be used for new outlets installed anywhere water or dampness could result in serious electrical shock injury, such as bathrooms, pools, marinas, etc. The outlet looks like any other except that it has a reset button and a test button in the center of the unit. A GFI outlet will shut off the power much more quickly than the main panel breaker. You may still feel a shock with a GFI outlet, but it will shut off the power so quickly (in a few milliseconds) that a person in normal health should not receive any serious electrical injury. (For more information see the CD-ROM.)

Coax cable

Coax cable can be found almost everywhere. It is used to connect a wide variety of equipment, from audio and video, medical instrumentation, to computers and more. So why is it so popular? The primary purpose of coax is to provide a noise-free communication channel between two connection points—a direct link in its purest form. Normal communication signals over standard two-wire connections can be affected by all types of radio frequency interference (RFI) from, for example, cellular phones, pagers, two-way radios (police, fire, ambulance, taxi, etc), radio and TV stations, etc. Compared to standard wire conductors, coax displays a lower loss of signal in transmission and has much improved high-frequency transmission characteristics.

The solid inner conductor of the cable is surrounded by a polyethylene dielectric (insulator). This insulator is covered by a braided copper or aluminum shield and finally covered with a (typically black or white coloured) waterproof outer plastic jacket. The braided outer shield reduces the effects of the fields established by any currents that may pass through it. Instead of a single continuous coating, the braiding breaks up the induced magnetic fields into smaller opposing components that effectively cancel each other out, providing the desired noise suppression.

Coax cable is often called RF (radio frequency) cable, referring to its most common use for carrying RF signals such as those provided by your local cable company. However, coax cable is suitable for a very wide range of frequencies, from dc (0 Hz) through the GHz range. In general, coax cable is used whenever there is a need to ensure that the transmitted data is undisturbed by any surrounding noise. Coax cable protects the signal inside the cable, and will not "leak" or transmit the internal signal to the surroundings. (For more information, see the CD-ROM.)

PROBLEMS

SECTION 16.2 Equivalent Circuits

1. For the series circuits of Fig. 16.32, find a parallel circuit that will have the same total impedance ($\mathbf{Z}_T$).

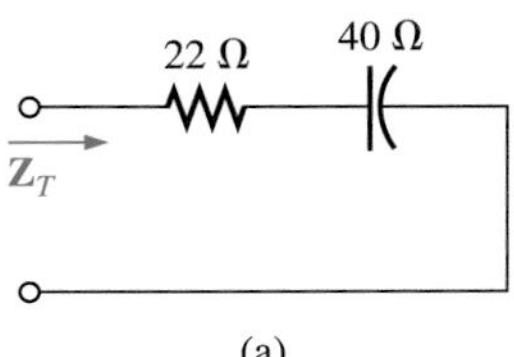

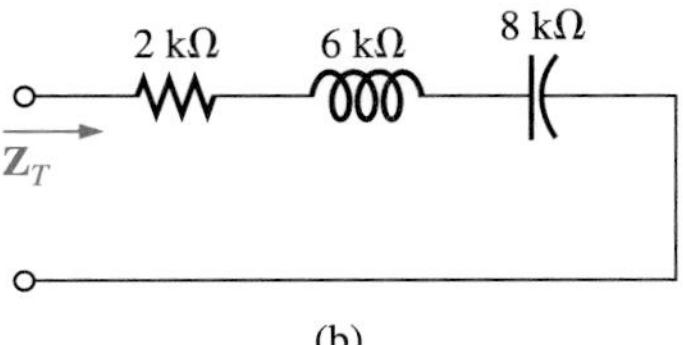

FIG. 16.32
Problem 1.

2. For the parallel circuits of Fig. 16.33, find a series circuit that will have the same total impedance.

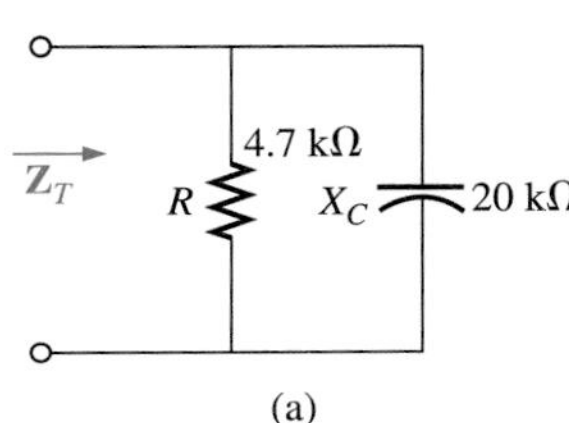

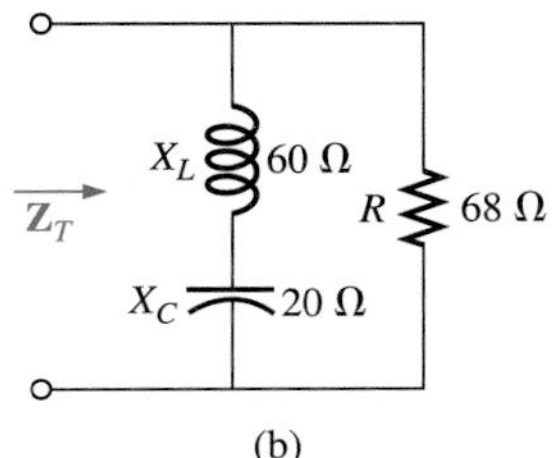

FIG. 16.33
Problem 2.

3. For the network of Fig. 16.34:
a. Calculate **E**, $\mathbf{I}_R$, and $\mathbf{I}_L$ in phasor form.
b. Calculate the total power factor and indicate whether it is leading or lagging.
c. Calculate the average power delivered to the circuit.
d. Draw the admittance diagram.
e. Draw the phasor diagram of the currents $\mathbf{I}_s$, $\mathbf{I}_R$, and $\mathbf{I}_L$, and the voltage **E**.
f. Find the current $\mathbf{I}_C$ for each capacitor using only Kirchhoff's current law.
g. Find the series circuit of one resistive and reactive element that will have the same impedance as the original circuit.

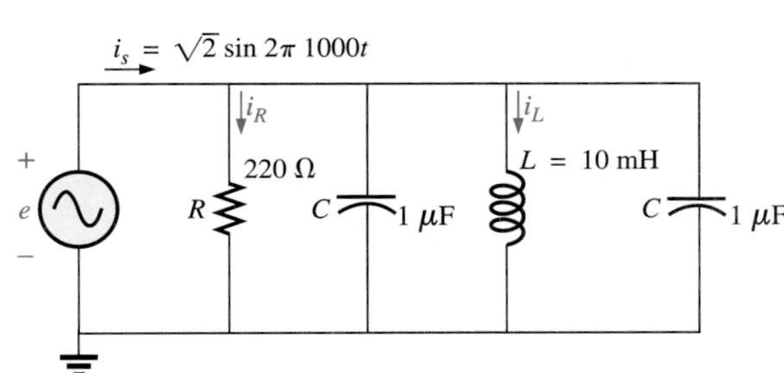

FIG. 16.34
Problems 3 and 4.

***4.** Repeat Problem 3 if the inductance is changed to 1 H.

5. Find the element or elements that must be in the closed container of Fig. 16.35 to satisfy the following conditions. (Find the simplest parallel circuit that will satisfy the indicated conditions.)
a. Average power to the circuit = 3000 W.
b. Circuit has a lagging power factor.

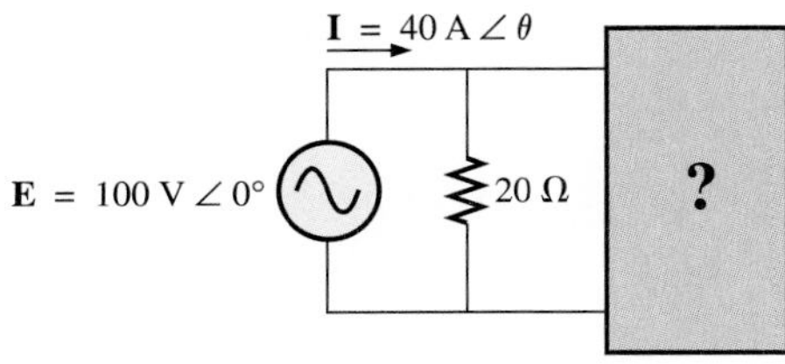

FIG. 16.35
Problem 5.

SECTION 16.3 Phase Measurements: Dual-Trace Oscilloscope

6. For the circuit of Fig. 16.36, determine the phase relationship between the following using a dual-trace oscilloscope. The circuit can be reconstructed differently for each part but do not use sensing resistors. Show all connections on a redrawn diagram.

a. e and v_C
b. e and i_s
c. e and v_L

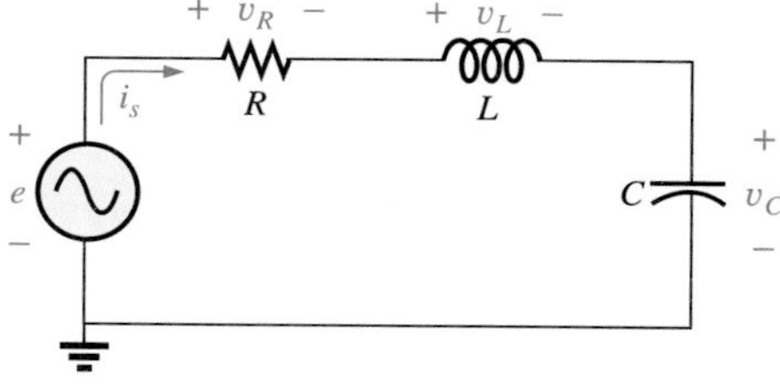

FIG. 16.36
Problem 6.

7. For the network of Fig. 16.37, determine the phase relationship between the following using a dual-trace oscilloscope. The network must remain as constructed in Fig. 16.36 but sensing resistors can be introduced. Show all connections on a redrawn diagram.

a. e and v_{R_2}
b. e and i_s
c. i_L and i_C

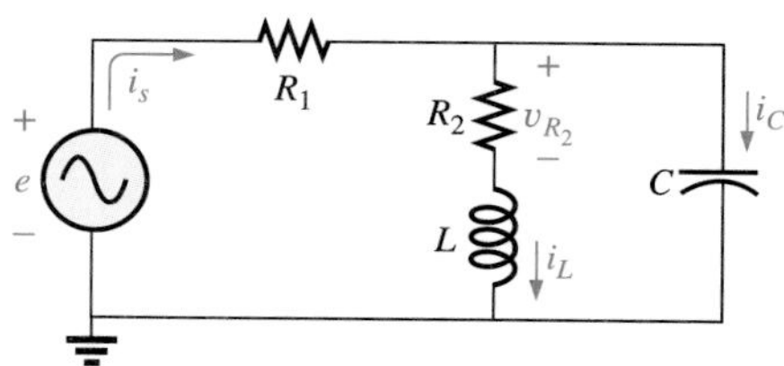

FIG. 16.37
Problem 7.

8. For the oscilloscope traces of Fig. 16.38:

a. Determine the phase relationship between the waveforms and indicate which one leads or lags.
b. Determine the peak-to-peak and rms values of each waveform.
c. Find the frequency of each waveform.

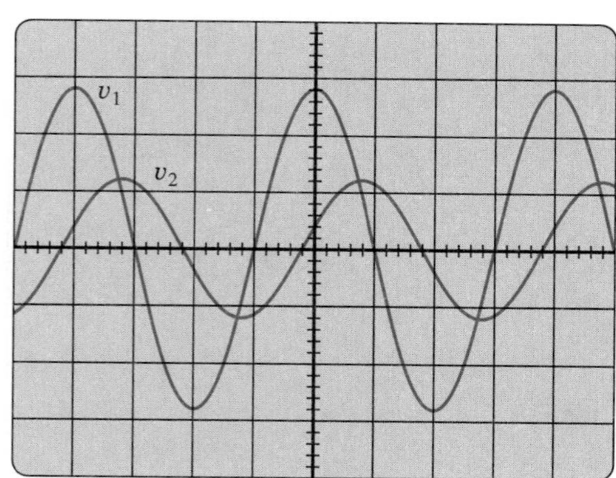

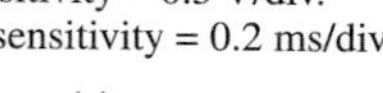

Vertical sensitivity = 0.5 V/div.
Horizontal sensitivity = 0.2 ms/div.

(a)

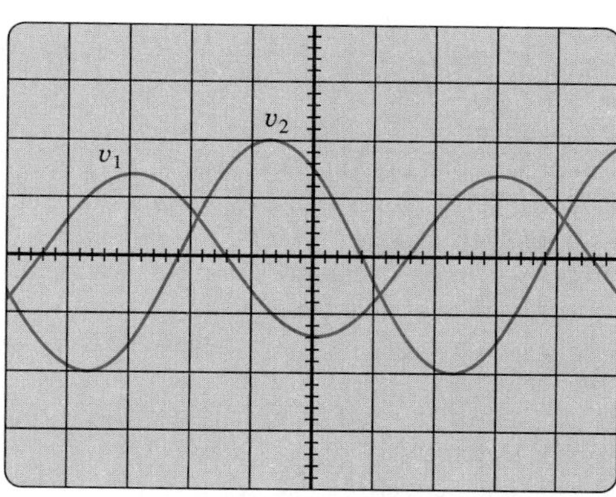

Vertical sensitivity = 2 V/div.
Horizontal sensitivity = 10 μs/div.

(b)

FIG. 16.38
Problem 8.

SECTION 16.4 Illustrative Examples

9. For the series-parallel network of Fig. 16.39:
   a. Calculate $\mathbf{Z}_T$.
   b. Determine $\mathbf{I}$.
   c. Determine $\mathbf{I}_1$.
   d. Find $\mathbf{I}_2$ and $\mathbf{I}_3$.
   e. Find $\mathbf{V}_L$.

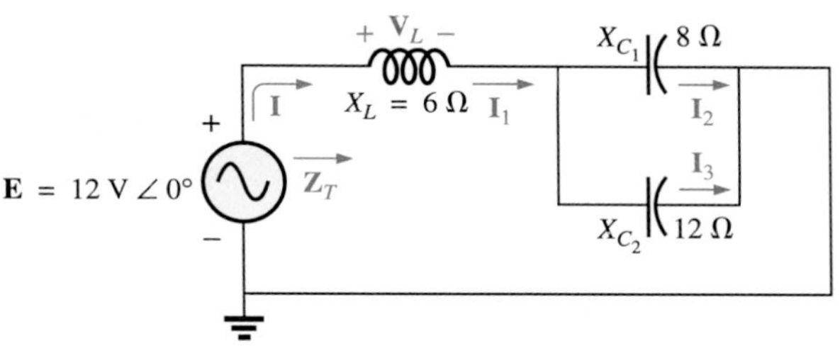

FIG. 16.39
Problem 9.

10. For the network of Fig. 16.40:
    a. Find the total impedance $\mathbf{Z}_T$.
    b. Determine the current $\mathbf{I}_s$.
    c. Calculate $\mathbf{I}_C$ using the current divider rule.
    d. Calculate $\mathbf{V}_L$ using the voltage divider rule.

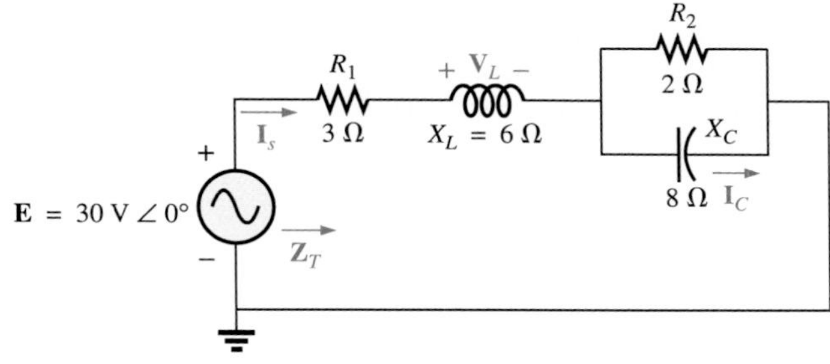

FIG. 16.40
Problem 10.

11. For the network of Fig. 16.41:
    a. Find the total impedance $\mathbf{Z}_T$ and the total admittance $\mathbf{Y}_T$.
    b. Find the current $\mathbf{I}_s$.
    c. Calculate $\mathbf{I}_2$ using the current divider rule.
    d. Calculate $\mathbf{V}_C$.
    e. Calculate the average power delivered to the network.

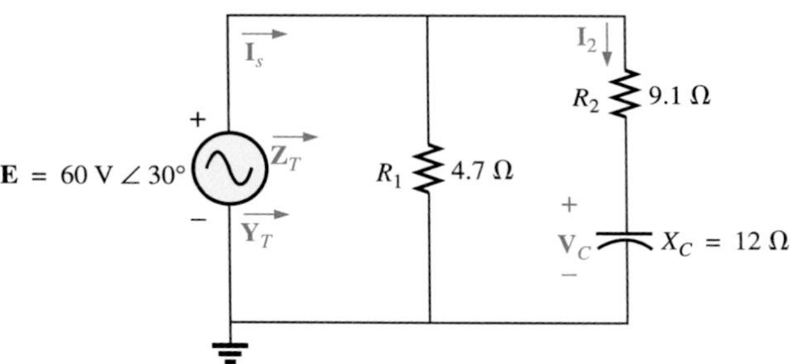

FIG. 16.41
Problem 11.

12. For the network of Fig. 16.42:
    a. Find the total impedance $\mathbf{Z}_T$.
    b. Calculate the voltage $\mathbf{V}_2$ and the current $\mathbf{I}_L$.
    c. Find the power factor of the network.

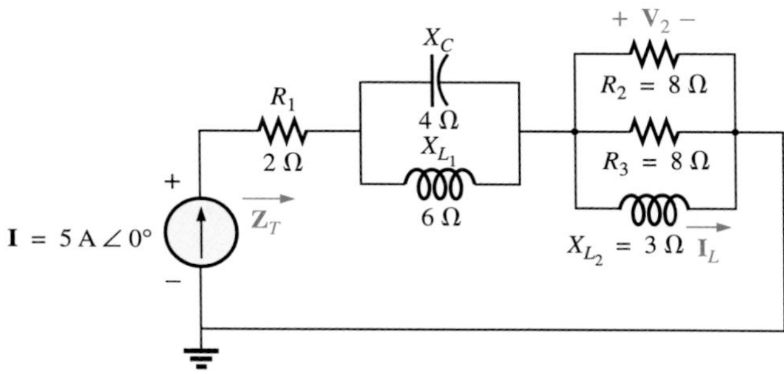

FIG. 16.42
Problem 12.

13. For the network of Fig. 16.43:

a. Find the current **I**.

b. Find the voltage $\mathbf{V}_C$.

c. Find the average power delivered to the network.

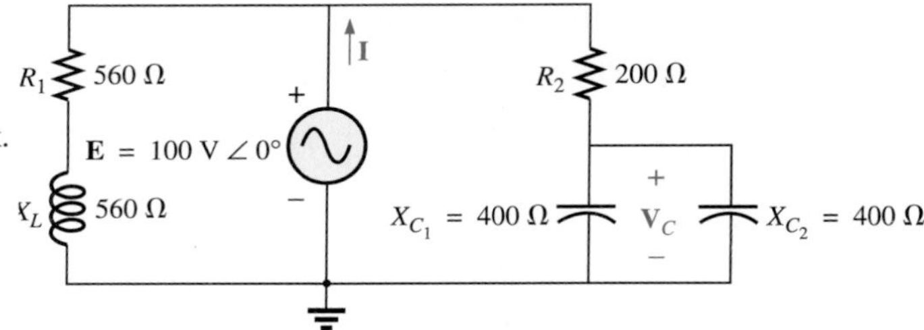

FIG. 16.43
Problem 13.

***14.** For the network of Fig. 16.44:

a. Find the current $\mathbf{I}_1$.

b. Calculate the voltage $\mathbf{V}_C$ using the voltage divider rule.

c. Find the voltage $\mathbf{V}_{ab}$.

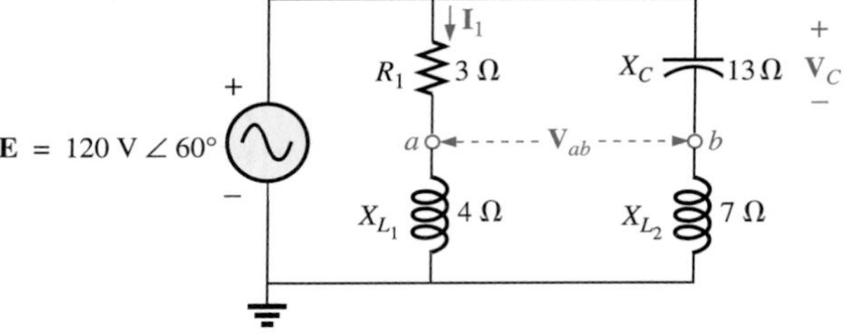

FIG. 16.44
Problem 14.

***15.** For the network of Fig. 16.45:

a. Find the current $\mathbf{I}_1$.

b. Find the voltage $\mathbf{V}_1$.

c. Calculate the average power delivered to the network.

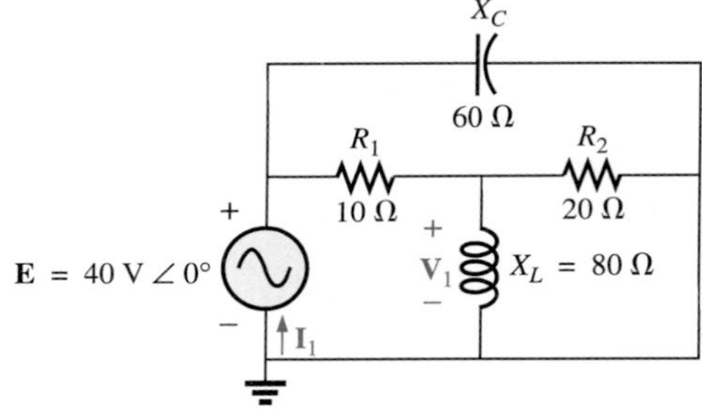

FIG. 16.45
Problem 15.

16. For the network of Fig. 16.46:

a. Find the total impedance $\mathbf{Z}_T$ and the admittance $\mathbf{Y}_T$.

b. Find the currents $\mathbf{I}_1$, $\mathbf{I}_2$, and $\mathbf{I}_3$.

c. Verify Kirchhoff's current law by showing that $\mathbf{I}_s = \mathbf{I}_1 + \mathbf{I}_2 + \mathbf{I}_3$.

d. Find the power factor of the network and indicate whether it is leading or lagging.

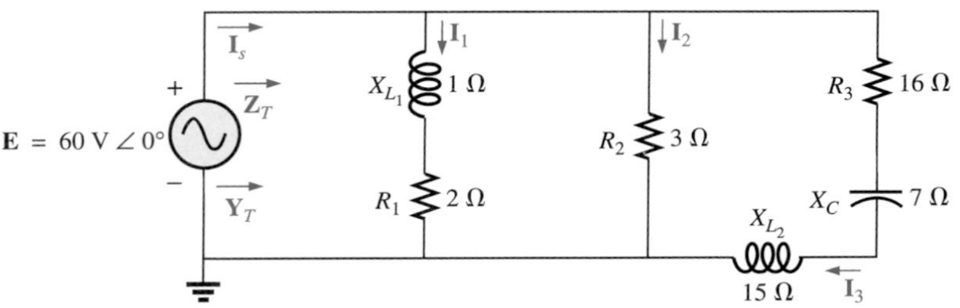

FIG. 16.46
Problem 16.

*17. For the network of Fig. 16.47:
   a. Find the total admittance $\mathbf{Y}_T$.
   b. Find the voltages $\mathbf{V}_1$ and $\mathbf{V}_2$.
   c. Find the current $\mathbf{I}_3$.

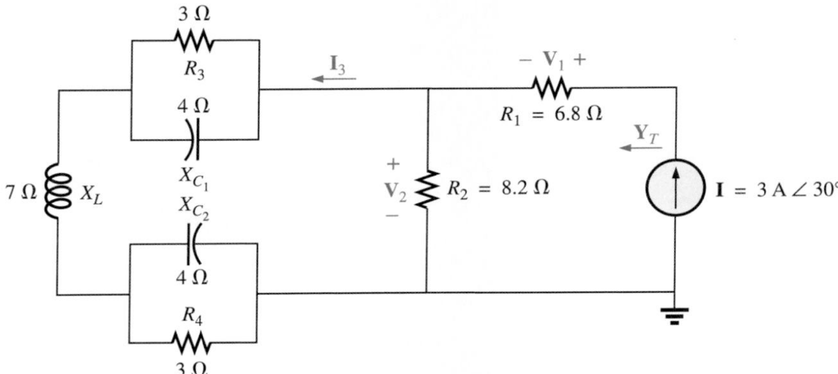

FIG. 16.47
Problem 17.

*18. For the network of Fig. 16.48:
   a. Find the total impedance $\mathbf{Z}_T$ and the admittance $\mathbf{Y}_T$.
   b. Find the source current $\mathbf{I}_s$ in phasor form.
   c. Find the currents $\mathbf{I}_1$ and $\mathbf{I}_2$ in phasor form.
   d. Find the voltages $\mathbf{V}_1$ and $\mathbf{V}_{ab}$ in phasor form.
   e. Find the average power delivered to the network.
   f. Find the power factor of the network and indicate whether it is leading or lagging.

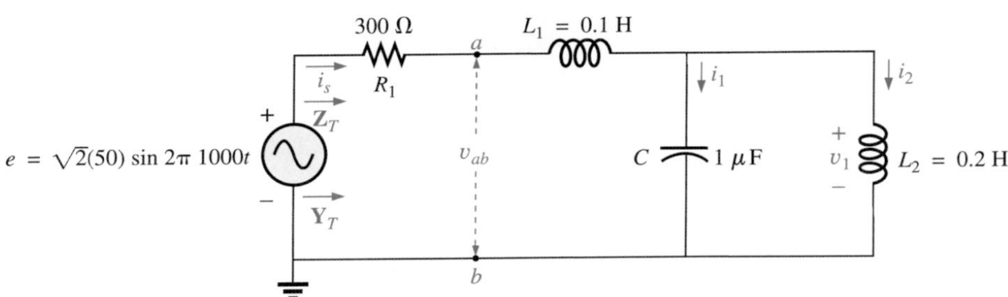

FIG. 16.48
Problem 18.

*19. Find the current $\mathbf{I}$ for the network of Fig. 16.49.

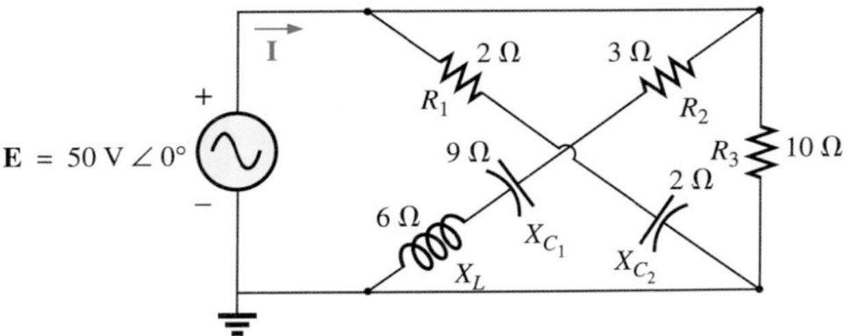

FIG. 16.49
Problem 19.

SECTION 16.5 Ladder Networks

20. Find the current $\mathbf{I}_5$ for the network of Fig. 16.50. Note the effect of one reactive element on the resulting calculations.

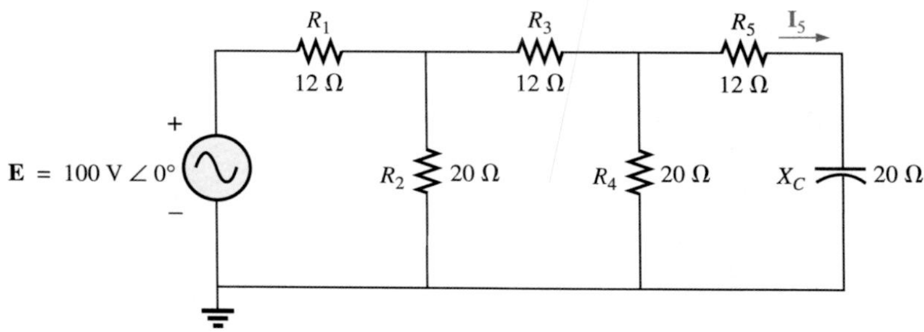

FIG. 16.50
Problem 20.

21. Find the average power delivered to R_4 in Fig. 16.51.

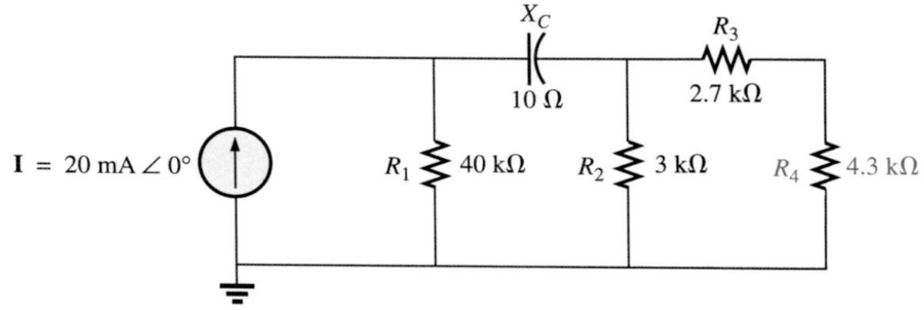

FIG. 16.51
Problem 21.

22. Find the current $\mathbf{I}_1$ for the network of Fig. 16.52.

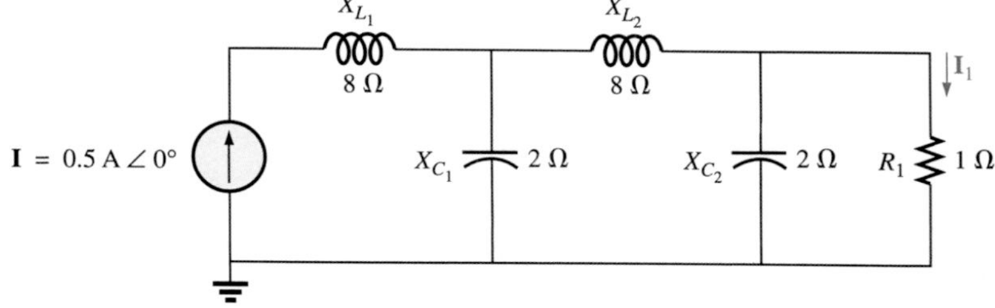

FIG. 16.52
Problem 22.

17

Methods of Analysis and Selected Topics (ac)

Outline

Learning Outcomes

After completing this chapter, you will be able to

- distinguish between independent and dependent sources in a network with two or more energy sources
- connect an independent voltage source to an independent current source and vice versa
- connect a dependent voltage source to a dependent current source and vice versa
- apply Kirchhoff's voltage law to the loops of an ac network using impedance and phasors (instead of resistance and real numbers)
- apply Kirchhoff's current law at a junction of an ac network using impedance and phasors (instead of resistance and real numbers)
- perform mesh analysis on ac networks with two or more sources that are not in series or parallel
- perform nodal analysis on ac networks with two or more sources that are not in series or parallel
- use mesh and nodal analysis techniques to analyze ac bridge networks
- perform Δ–Y and Y–Δ conversions on ac circuits

The central control room of an electric utility is staffed 24 hours a day, every day of the year, to monitor and analyze the ever-changing combination of power sources and power consumers.

17.1 INTRODUCTION

For networks with two or more sources that are not in series or parallel, the methods described in the last two chapters cannot be applied. Instead, methods such as mesh analysis or nodal analysis must be used. Since these methods were discussed in detail for dc circuits in Chapter 8, this chapter will consider the variations required to apply these methods to ac circuits. Dependent sources will also be introduced for both mesh and modal analysis.

The branch-current method will not be discussed again because it falls within the framework of mesh analysis. In addition to the methods mentioned above, the bridge network and Δ-Y, Y-Δ conversions will also be discussed for ac circuits.

Before we examine these topics, however, we must consider the subject of independent and controlled sources.

17.2 INDEPENDENT VERSUS DEPENDENT (CONTROLLED) SOURCES

In the previous chapters, each source appearing in the analysis of dc or ac networks was an **independent source**, such as E and I (or **E** and **I**) in Fig. 17.1.

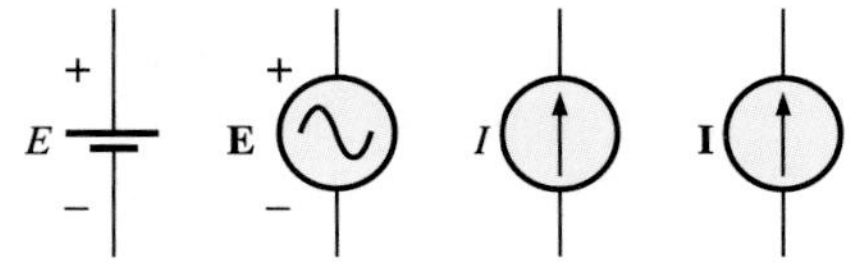

FIG. 17.1
Independent sources.

The term independent specifies that the magnitude of the source is independent of the network to which it is applied and that it displays its terminal characteristics even if completely isolated.

A dependent or controlled source is one whose magnitude is determined (or controlled) by a current or voltage of the system in which it appears.

There are currently two symbols used for controlled sources. One simply uses the independent symbol with an indication of the controlling element, as shown in Fig. 17.2. In Fig. 17.2(a), the magnitude and phase of the voltage are controlled by a voltage **V** elsewhere in the system, with the magnitude further controlled by the constant k_1. In Fig. 17.2(b), the magnitude and phase of the current source are controlled by

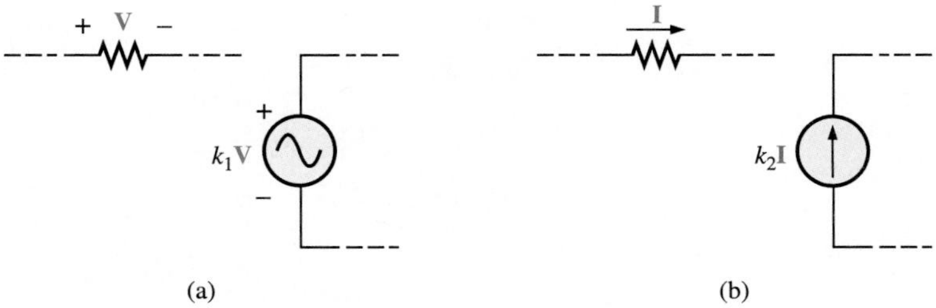

(a)

(b)

FIG. 17.2
Controlled or dependent sources.

a current **I** elsewhere in the system, with the magnitude further controlled by the constant k_2. Recently, some electronic engineers have chosen to distinguish dependent sources from independent ones by replacing the circle symbol with a diamond, as in Fig. 17.3. Others continue to use the symbols of Fig. 17.2, especially for modelling such devices as the transistor and FET. While you should be aware of both, this text will use the symbols of Fig. 17.3.

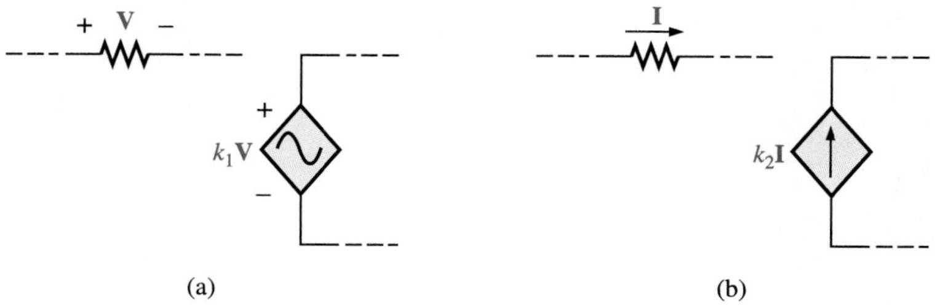

FIG. 17.3
Special notation for controlled or dependent sources.

Possible combinations for controlled sources are indicated in Fig. 17.4. Note that the magnitude of current sources or voltage sources can be controlled by a voltage and a current, respectively. Unlike with the independent source, isolation such that **V** or **I** = 0 in Fig. 17.4(a) will result in the short-circuit or open-circuit equivalent as indicated in Fig. 17.4(b). Note that the type of representation under these conditions is controlled by whether it is a current source or a voltage source, not by the controlling agent (**V** or **I**).

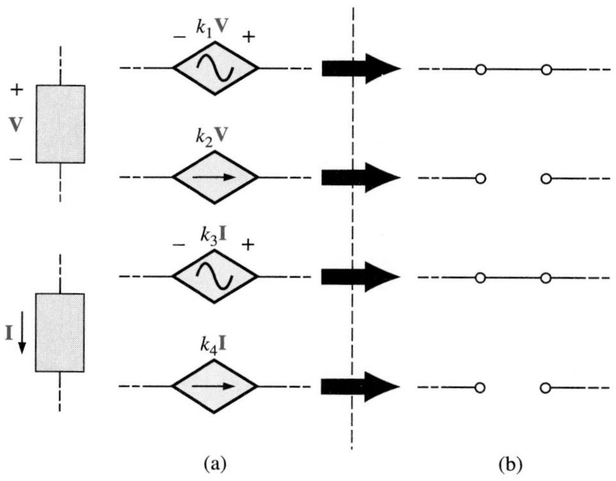

FIG. 17.4
Conditions of $V = 0$ V and $I = 0$ A for a dependent source.

17.3 SOURCE CONVERSIONS

When applying the methods to be discussed, it may be necessary to convert a current source to a voltage source, or a voltage source to a current source. This can be done in much the same way as for dc circuits, except now we will be dealing with phasors and impedances instead of just real numbers and resistors.

Independent Sources

In general, the format for converting one type of independent source to another is as shown in Fig. 17.5.

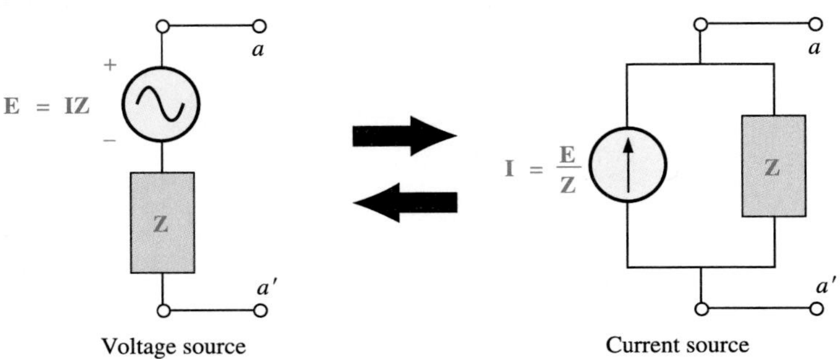

FIG. 17.5
Source conversion.

EXAMPLE 17.1

Convert the voltage source of Fig. 17.6(a) to a current source.

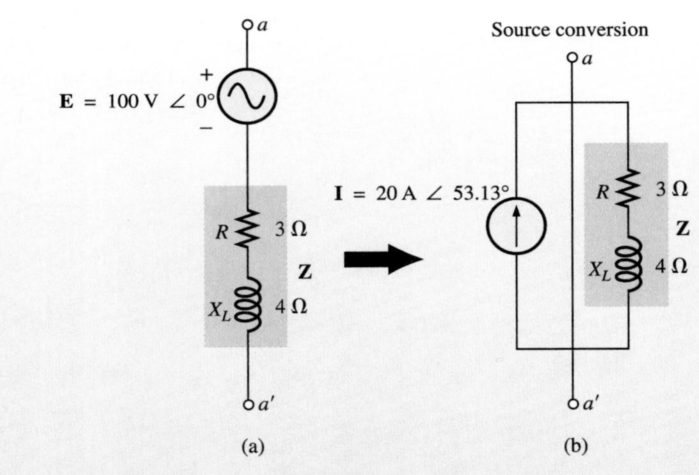

FIG. 17.6
Example 17.1.

Solution:

$$\mathbf{I} = \frac{\mathbf{E}}{\mathbf{Z}} = \frac{100\text{ V}\angle 0^\circ}{5\ \Omega\angle 53.13^\circ}$$
$$= \mathbf{20\ A\angle -53.13^\circ} \quad \text{[Fig. 17.6(b)]}$$

EXAMPLE 17.2

Convert the current source of Fig. 17.7(a) to a voltage source.

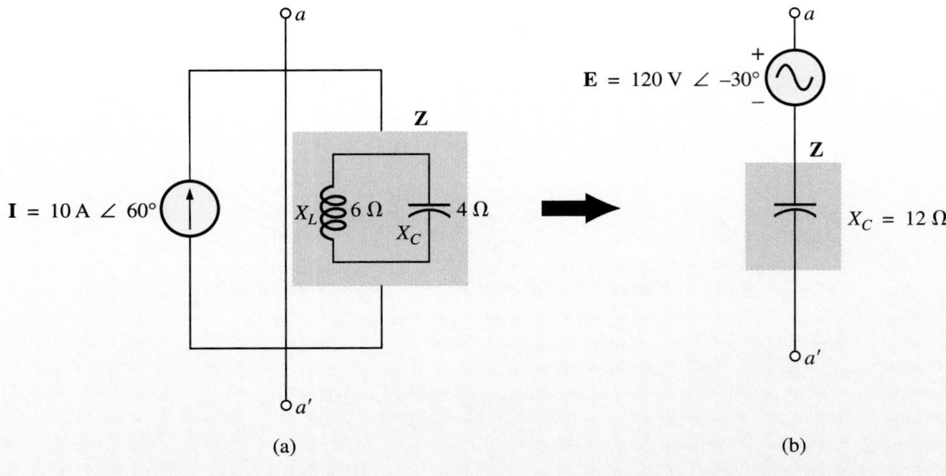

FIG. 17.7
Example 17.2.

Solution:

$$\mathbf{Z} = \frac{\mathbf{Z}_C\mathbf{Z}_L}{\mathbf{Z}_C + \mathbf{Z}_L} = \frac{(X_C \angle -90^\circ)(X_L \angle 90^\circ)}{-j X_C + j X_L}$$

$$= \frac{(4\ \Omega \angle -90^\circ)(6\ \Omega \angle 90^\circ)}{-j\,4\ \Omega + j\,6\ \Omega} = \frac{24\ \Omega \angle 0^\circ}{2 \angle 90^\circ}$$

$$= \mathbf{12\ \Omega \angle -90^\circ} \quad \text{[Fig. 17.7(b)]}$$

$$\mathbf{E} = \mathbf{IZ} = (10\ \text{A} \angle 60^\circ)(12\ \Omega \angle -90^\circ)$$

$$= \mathbf{120\ V \angle -30^\circ} \quad \text{[Fig. 17.7(b)]}$$

Dependent Sources

For dependent sources, the direct conversion of Fig. 17.5 can be applied if the controlling variable (**V** or **I** in Fig. 17.4) is not determined by a portion of the network to which the conversion is to be applied. For example, in Figs. 17.8 and 17.9, **V** and **I**, respectively, are controlled by an external portion of the network. Conversions of the other kind, where **V** and **I** are controlled by a portion of the network to be converted, will be considered in Sections 18.3 and 18.4.

EXAMPLE 17.3

Convert the voltage source of Fig. 17.8(a) to a current source.

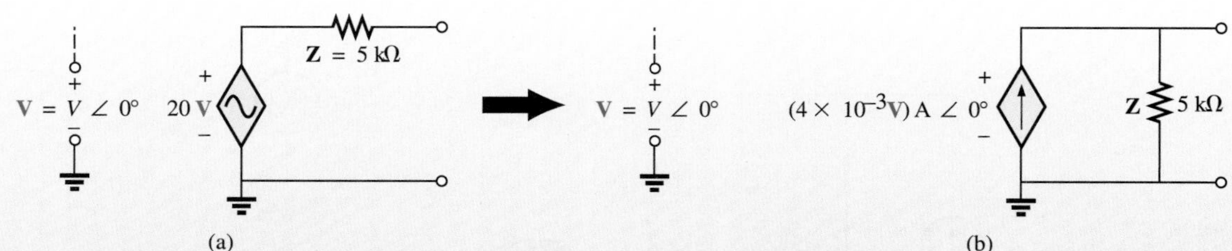

FIG. 17.8
Source conversion with a voltage-controlled voltage source.

Solution:

$$\mathbf{I} = \frac{\mathbf{E}}{\mathbf{Z}} = \frac{(20\mathbf{V})\ \mathrm{V}\ \angle 0^\circ}{5\ \mathrm{k}\Omega\ \angle 0^\circ}$$
$$= \mathbf{(4 \times 10^{-3}\ V)\ A\ \angle 0^\circ} \quad \text{[Fig. 17.8(b)]}$$

EXAMPLE 17.4

Convert the current source of Fig. 17.9(a) to a voltage source.

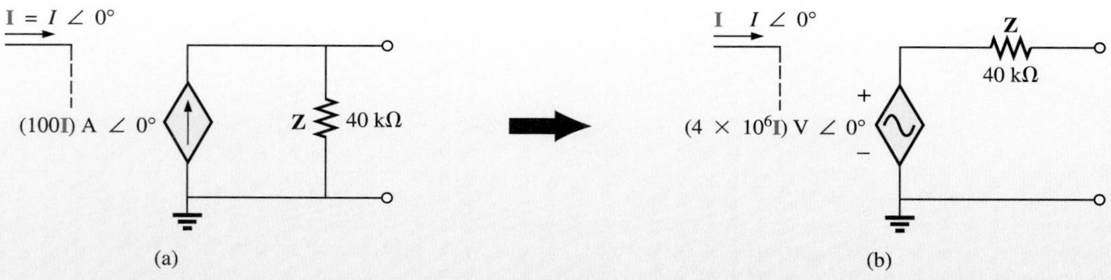

FIG. 17.9
Source conversion with a current-controlled current source.

Solution:

$$\mathbf{E} = \mathbf{IZ} = [(100\mathbf{I})\ \mathrm{A}\ \angle 0^\circ][40\ \mathrm{k}\Omega\ \angle 0^\circ]$$
$$= \mathbf{(4 \times 10^{6}I)\ V\ \angle 0^\circ} \quad \text{[Fig. 17.9(b)]}$$

17.4 MESH ANALYSIS

General Approach

Independent Voltage Sources Before examining how to apply this method to ac networks, you should first review the appropriate sections on mesh analysis in Chapter 8, since this section will only refer to the general conclusions of Chapter 8.

The general approach to mesh analysis for independent sources includes the same sequence of steps found in Chapter 8. In fact, throughout this section the only change from the dc coverage will be to substitute impedance for resistance and admittance for conductance in the general procedure.

1. Assign a distinct current in the clockwise direction to each independent closed loop of the network.

   (It is not absolutely necessary to choose the clockwise direction for each loop current. However, it eliminates the need to have to choose a direction for each application. Any direction can be chosen for each loop current with no loss in accuracy as long as the remaining steps are followed properly.)
2. Indicate the polarities within each loop for each impedance as determined by the assumed direction of loop current for that loop.
3. Apply Kirchhoff's voltage law around each closed loop in the clockwise direction. Again, the clockwise direction was chosen to establish uniformity and prepare us for the format approach to follow.
   a. If an impedance has two or more assumed currents through it, the total current will be composed of parts. The total current through the impedance is the assumed current of the loop in which Kirchhoff's voltage law is being applied, ***plus*** the assumed currents of the other loops passing through in the same direction, ***minus*** the assumed currents passing through in the opposite direction.
   b. The polarity of a voltage source is unaffected by the direction of the assigned loop currents.
4. Solve the resulting simultaneous linear equations for the assumed loop currents.

This technique is applied as above for all networks with *independent sources*, or for networks with **dependent sources** *where the controlling variable is not a part of the network under investigation.* If the controlling variable is part of the network being examined, a method we will describe shortly must be applied.

EXAMPLE 17.5

Using the general approach to mesh analysis, find the current $\mathbf{I}_1$ in Fig. 17.10.

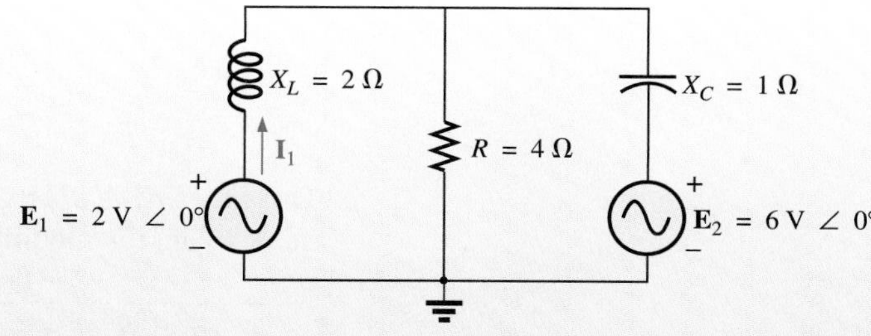

FIG. 17.10
Example 17.5.

Solution: When applying these methods to ac circuits, it is good practice to represent the resistors and reactances (or combinations of them) by subscripted impedances. When the total solution is found in terms of these subscripted impedances, the numerical values can be substituted to find the unknown quantities.

continued

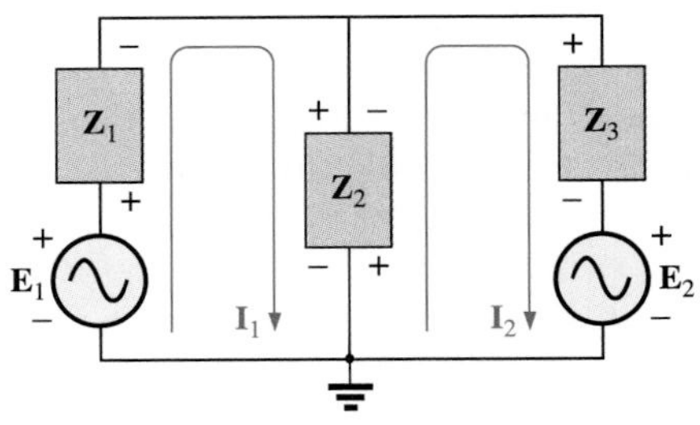

FIG. 17.11
Assigning the mesh currents and subscripted impedances for the network of Fig. 17.10.

The network is redrawn in Fig. 17.11 with subscripted impedances:

$$\mathbf{Z}_1 = +j\,X_L = +j\,2\ \Omega \qquad \mathbf{E}_1 = 2\text{ V}\ \angle 0^\circ$$
$$\mathbf{Z}_2 = R = 4\ \Omega \qquad \mathbf{E}_2 = 6\text{ V}\ \angle 0^\circ$$
$$\mathbf{Z}_3 = -j\,X_C = -j\,1\ \Omega$$

Steps 1 and 2 are as indicated in Fig. 17.11.

Step 3:

$$+\mathbf{E}_1 - \mathbf{I}_1\mathbf{Z}_1 - \mathbf{Z}_2(\mathbf{I}_1 - \mathbf{I}_2) = 0$$
$$-\mathbf{Z}_2(\mathbf{I}_2 - \mathbf{I}_1) - \mathbf{I}_2\mathbf{Z}_3 - \mathbf{E}_2 = 0$$

or

$$\mathbf{E}_1 - \mathbf{I}_1\mathbf{Z}_1 - \mathbf{I}_1\mathbf{Z}_2 + \mathbf{I}_2\mathbf{Z}_2 = 0$$
$$-\mathbf{I}_2\mathbf{Z}_2 + \mathbf{I}_1\mathbf{Z}_2 - \mathbf{I}_2\mathbf{Z}_3 - \mathbf{E}_2 = 0$$

so that

$$\mathbf{I}_1(\mathbf{Z}_1 + \mathbf{Z}_2) - \mathbf{I}_2\mathbf{Z}_2 = \mathbf{E}_1$$
$$\mathbf{I}_2(\mathbf{Z}_2 + \mathbf{Z}_3) - \mathbf{I}_1\mathbf{Z}_2 = -\mathbf{E}_2$$

which are rewritten as

$$\mathbf{I}_1(\mathbf{Z}_1 + \mathbf{Z}_2) - \mathbf{I}_2\mathbf{Z}_2 \qquad = \mathbf{E}_1$$
$$-\mathbf{I}_1\mathbf{Z}_2 \qquad + \mathbf{I}_2(\mathbf{Z}_2 + \mathbf{Z}_3) = -\mathbf{E}_2$$

Step 4: Using determinants, we obtain

$$\mathbf{I}_1 = \frac{\begin{vmatrix} \mathbf{E}_1 & -\mathbf{Z}_2 \\ -\mathbf{E}_2 & \mathbf{Z}_2 + \mathbf{Z}_3 \end{vmatrix}}{\begin{vmatrix} \mathbf{Z}_1 + \mathbf{Z}_2 & -\mathbf{Z}_2 \\ -\mathbf{Z}_2 & \mathbf{Z}_2 + \mathbf{Z}_3 \end{vmatrix}}$$

$$= \frac{\mathbf{E}_1(\mathbf{Z}_2 + \mathbf{Z}_3) - \mathbf{E}_2(\mathbf{Z}_2)}{(\mathbf{Z}_1 + \mathbf{Z}_2)(\mathbf{Z}_2 + \mathbf{Z}_3) - (\mathbf{Z}_2)^2}$$

$$= \frac{(\mathbf{E}_1 - \mathbf{E}_2)\mathbf{Z}_2 + \mathbf{E}_1\mathbf{Z}_3}{\mathbf{Z}_1\mathbf{Z}_2 + \mathbf{Z}_1\mathbf{Z}_3 + \mathbf{Z}_2\mathbf{Z}_3}$$

Substituting numerical values gives us

$$\mathbf{I}_1 = \frac{(2\text{ V} - 6\text{ V})(4\ \Omega) + (2\text{ V})(-j\,1\ \Omega)}{(+j\,2\ \Omega)(4\ \Omega) + (+j\,2\ \Omega)(-j\,2\ \Omega) + (4\ \Omega)(-j\,2\ \Omega)}$$

$$= \frac{-16 - j\,2}{j\,8 - j^2 2 - j\,4} = \frac{-16 - j\,2}{2 + j\,4} = \frac{16.12\text{ A}\ \angle -172.87^\circ}{4.47\ \angle 63.43^\circ}$$

$$= \mathbf{3.61\ A\ \angle -236.30^\circ} \text{ or } \mathbf{3.61\ A\ \angle 123.70^\circ}$$

Dependent Voltage Sources For dependent voltage sources, the procedure is modified as follows:

1. **Steps 1 and 2** are the same as those applied for independent voltage sources.
2. **Step 3** is modified: Treat each dependent source like an independent source when Kirchhoff's voltage law is applied to each independent loop. However, once the equation is written, substitute the equation for the controlling quantity to ensure that the unknowns are limited solely to the chosen mesh currents.
3. **Step 4** is as before.

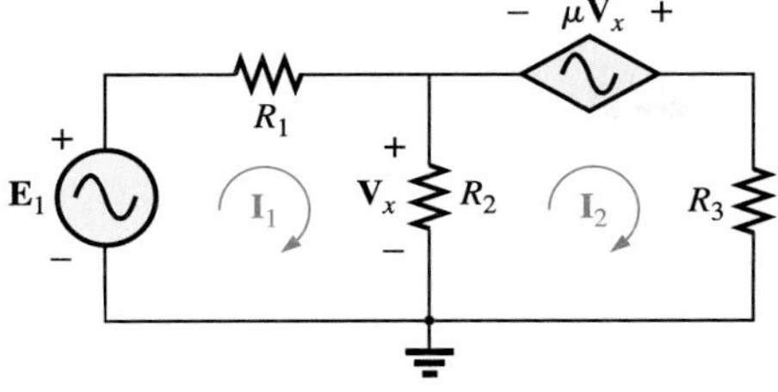

FIG. 17.12
Applying mesh analysis to a network with a voltage-controlled voltage source.

EXAMPLE 17.6

Write the mesh currents for the network of Fig. 17.12 having a dependent voltage source.

Solution:

Steps 1 and 2 are defined on Fig. 17.12.

Step 3: $E_1 - I_1R_1 - R_2(I_1 - I_2) = 0$

$$R_2(I_2 - I_1) + \mu V_x - I_2R_3 = 0$$

with $V_x = (I_1 - I_2)R_2$

The result is two equations and two unknowns.

$$E_1 - I_1R_1 - R_2(I - I_2) = 0$$
$$R_2(I_2 - I_1) + \mu R_2(I_1 - I_2) - I_2R_3 = 0$$

Independent Current Sources For independent current sources, the procedure is modified as follows:

1. **Steps 1 and 2** are the same as those applied for independent sources.
2. **Step 3** is modified: Treat each current source as an open circuit (recall the *supermesh* from Chapter 8), and write the mesh equations for each remaining independent path. Then relate the chosen mesh currents to the dependent sources to ensure that the unknowns of the final equations are limited simply to the mesh currents.
3. **Step 4** is as before.

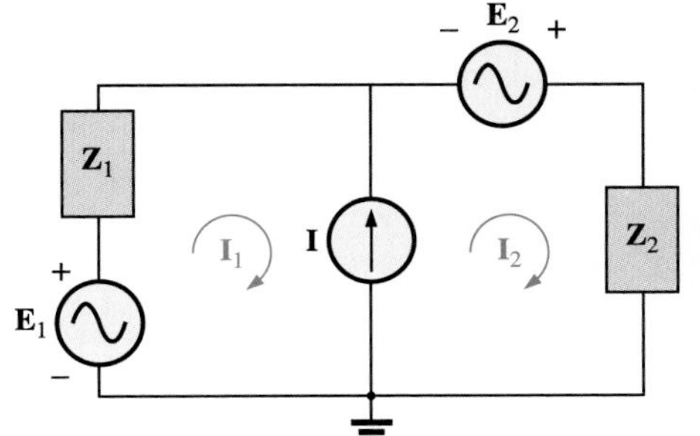

FIG. 17.13
Applying mesh analysis to a network with an independent current source.

EXAMPLE 17.7

Write the mesh currents for the network of Fig. 17.13 having an independent current source.

Solution:

Steps 1 and 2 are defined on Fig. 17.13.

Step 3: $E_1 - I_1Z_1 + E_2 - I_2Z_2 = 0$ (only remaining independent path)

with $I_1 + I = I_2$

The result is two equations and two unknowns.

Dependent Current Sources For dependent current sources, the procedure is modified as follows:

1. **Steps 1 and 2** are the same as those applied for independent sources.
2. **Step 3** is modified: The procedure is essentially the same as that applied for independent current sources, except now the dependent sources have to be defined in terms of the chosen mesh currents to ensure that the final equations have only mesh currents as the unknown quantities.
3. **Step 4** is as before.

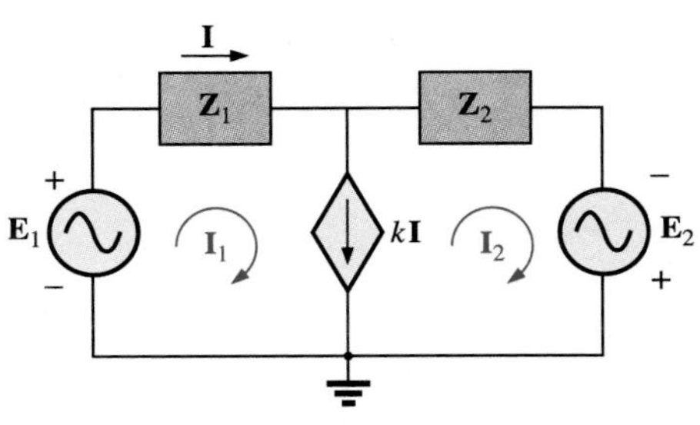

FIG. 17.14
Applying mesh analysis to a network with a current-controlled current source.

EXAMPLE 17.8

Write the mesh currents for the network of Fig. 17.14 having a dependent current source.

Solution:

Steps 1 and 2 are defined on Fig. 17.14.

Step 3: $\mathbf{E}_1 - \mathbf{I}_1\mathbf{Z}_1 - \mathbf{I}_2\mathbf{Z}_2 + \mathbf{E}_2 = 0$

with $\quad k\mathbf{I} = \mathbf{I}_1 - \mathbf{I}_2 \quad$ and $\quad \mathbf{I} = \mathbf{I}_1$

so that $k\mathbf{I}_1 = \mathbf{I}_1 - \mathbf{I}_2 \quad$ or $\quad \mathbf{I}_2 = \mathbf{I}_1(1 - k)$

The result is two equations and two unknowns.

Format Approach

The format approach was introduced in Section 8.8. The steps for applying this method are repeated here with changes for using it in ac circuits:

1. Assign a loop current to each independent closed loop (as in the previous section) in a clockwise direction.
2. The number of required equations is equal to the number of chosen independent closed loops. Column 1 of each equation is formed by simply summing the impedance values of those impedances through which the loop current of interest passes and multiplying the result by that loop current.
3. We must now consider the mutual terms that are always subtracted from the terms in the first column. It is possible to have more than one mutual term if the loop current of interest has an element in common with more than one other loop current. Each mutual term is the product of the mutual impedance and the other loop current passing through the same element.
4. The column to the right of the equals sign is the algebraic sum of the voltage sources through which the loop current of interest passes. Positive signs are assigned to sources of voltage with a polarity such that the loop current passes from the negative to the positive terminal. A negative sign is assigned to those potentials for which the reverse is true.
5. Solve resulting simultaneous equations for the desired loop currents.

The technique is applied as above for all networks with independent sources or networks with dependent sources where the controlling variable is not a part of the network under investigation. If the controlling variable is part of the network being examined, you must be extra careful when applying the above steps.

EXAMPLE 17.9

Using the format approach to mesh analysis, find the current $\mathbf{I}_2$ in Fig. 17.15(a).

continued

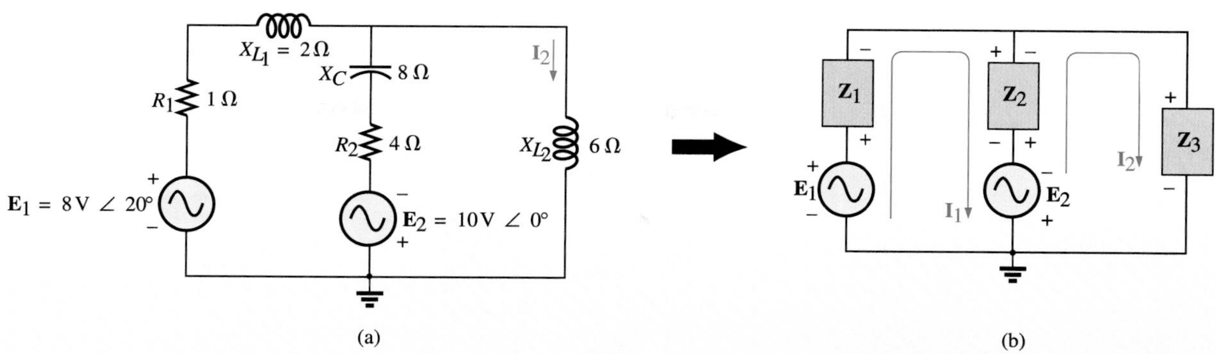

FIG. 17.15
For Example 17.9, assign the mesh currents and subscripted impedances.

Solution: The network is redrawn in Fig. 17.15(b):

$$\mathbf{Z}_1 = R_1 + jX_{L_1} = 1\,\Omega + j\,2\,\Omega \qquad \mathbf{E}_1 = 8\text{ V}\angle 20°$$
$$\mathbf{Z}_2 = R_2 - jX_C = 4\,\Omega - j\,8\,\Omega \qquad \mathbf{E}_2 = 10\text{ V}\angle 0°$$
$$\mathbf{Z}_3 = +jX_{L_2} = +j\,6\,\Omega$$

Note how substitution of the subscripted impedances reduces the complexity of the problem.

Step 1 is as indicated in Fig. 17.15(b).

Steps 2 to 4:

$$\mathbf{I}_1(\mathbf{Z}_1 + \mathbf{Z}_2) - \mathbf{I}_2\mathbf{Z}_2 = \mathbf{E}_1 + \mathbf{E}_2$$
$$\mathbf{I}_2(\mathbf{Z}_2 + \mathbf{Z}_3) - \mathbf{I}_1\mathbf{Z}_2 = -\mathbf{E}_2$$

which are rewritten as

$$\begin{aligned} \mathbf{I}_1(\mathbf{Z}_1 + \mathbf{Z}_2) - \mathbf{I}_2\mathbf{Z}_2 \qquad\qquad &= \mathbf{E}_1 + \mathbf{E}_2 \\ -\mathbf{I}_1\mathbf{Z}_2 \qquad + \mathbf{I}_2(\mathbf{Z}_2 + \mathbf{Z}_3) &= -\mathbf{E}_2 \end{aligned}$$

Step 5: Using determinants, we have

$$\mathbf{I}_2 = \frac{\begin{vmatrix} \mathbf{Z}_1 + \mathbf{Z}_2 & \mathbf{E}_1 + \mathbf{E}_2 \\ -\mathbf{Z}_2 & -\mathbf{E}_2 \end{vmatrix}}{\begin{vmatrix} \mathbf{Z}_1 + \mathbf{Z}_2 & -\mathbf{Z}_2 \\ -\mathbf{Z}_2 & \mathbf{Z}_2 + \mathbf{Z}_3 \end{vmatrix}}$$
$$= \frac{-(\mathbf{Z}_1 + \mathbf{Z}_2)\mathbf{E}_2 + \mathbf{Z}_2(\mathbf{E}_1 + \mathbf{E}_2)}{(\mathbf{Z}_1 + \mathbf{Z}_2)(\mathbf{Z}_2 + \mathbf{Z}_3) - \mathbf{Z}_2^2}$$
$$= \frac{\mathbf{Z}_2\mathbf{E}_1 - \mathbf{Z}_1\mathbf{E}_2}{\mathbf{Z}_1\mathbf{Z}_2 + \mathbf{Z}_1\mathbf{Z}_3 + \mathbf{Z}_2\mathbf{Z}_3}$$

Substituting numerical values gives us

$$\mathbf{I}_2 = \frac{(4\,\Omega - j\,8\,\Omega)(8\text{ V}\angle 20°) - (1\,\Omega + j\,2\,\Omega)(10\text{ V}\angle 0°)}{(1\,\Omega + j\,2\,\Omega)(4\,\Omega - j\,8\,\Omega) + (1\,\Omega + j\,2\,\Omega)(+j\,6\,\Omega) + (4\,\Omega - j\,8\,\Omega)(+j\,6\,\Omega)}$$
$$= \frac{(4 - j\,8)(7.52 + j\,2.74) - (10 + j\,20)}{20 + (j\,6 - 12) + (j\,24 + 48)}$$
$$= \frac{(52.0 - j\,49.20) - (10 + j\,20)}{56 + j\,30} = \frac{42.0 - j\,69.20}{56 + j\,30} = \frac{80.95\text{ A}\angle -58.74°}{63.53\angle 28.18°}$$
$$= \mathbf{1.27\text{ A}\angle -86.92°}$$

EXAMPLE 17.10

Write the mesh equations for the network shown in Fig. 17.16. Do not solve.

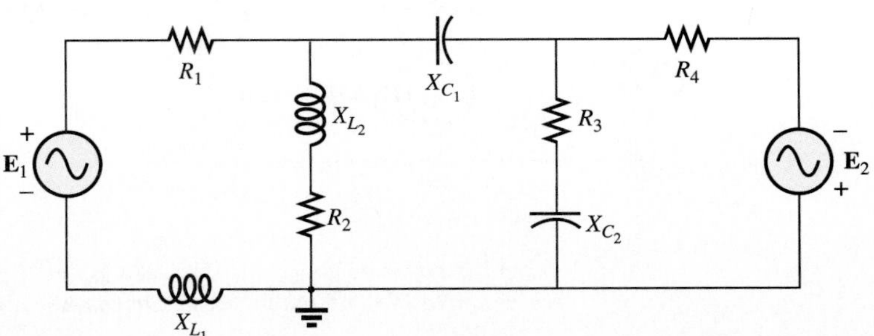

FIG. 17.16
Example 17.10.

Solution: The network is redrawn in Fig. 17.17. Again note how using subscripted impedances reduces the complexity and makes the problem clearer:

$$\mathbf{Z}_1 = R_1 + j X_{L_1} \quad \mathbf{Z}_4 = R_3 - j X_{C_2}$$
$$\mathbf{Z}_2 = R_2 + j X_{L_2} \quad \mathbf{Z}_5 = R_4$$
$$\mathbf{Z}_3 = j X_{C_1}$$

and

$$\mathbf{I}_1(\mathbf{Z}_1 + \mathbf{Z}_2) - \mathbf{I}_2\mathbf{Z}_2 = \mathbf{E}_1$$
$$\mathbf{I}_2(\mathbf{Z}_2 + \mathbf{Z}_3 + \mathbf{Z}_4) - \mathbf{I}_1\mathbf{Z}_2 - \mathbf{I}_3\mathbf{Z}_4 = 0$$
$$\mathbf{I}_3(\mathbf{Z}_4 + \mathbf{Z}_5) - \mathbf{I}_2\mathbf{Z}_4 = \mathbf{E}_2$$

or

$$\begin{aligned}
&\mathbf{I}_1(\mathbf{Z}_1 + \mathbf{Z}_2) - \mathbf{I}_2(\mathbf{Z}_2) &&+ 0 &&= \mathbf{E}_1\\
&\mathbf{I}_1\mathbf{Z}_2 - \mathbf{I}_2(\mathbf{Z}_2 + \mathbf{Z}_3 + \mathbf{Z}_4) &&+ \mathbf{I}_3(\mathbf{Z}_4) &&= 0\\
&0 - \mathbf{I}_2(\mathbf{Z}_4) &&+ \mathbf{I}_3(\mathbf{Z}_4 + \mathbf{Z}_5) &&= \mathbf{E}_2
\end{aligned}$$

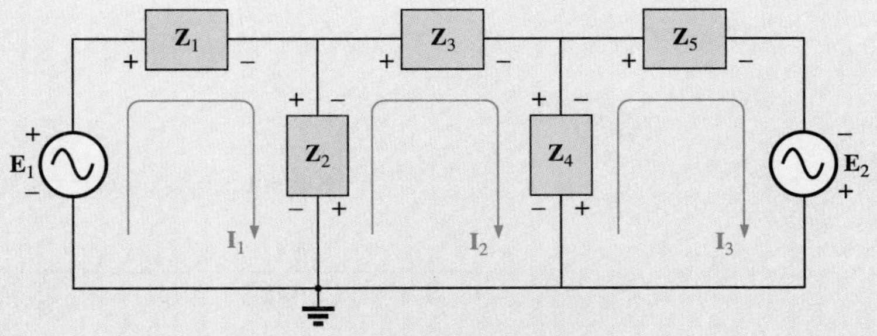

FIG. 17.17
Assigning the mesh currents and subscripted impedances for the network of Fig. 17.16.

EXAMPLE 17.11

Using the format approach, write the mesh equations for the network of Fig. 17.18(a).

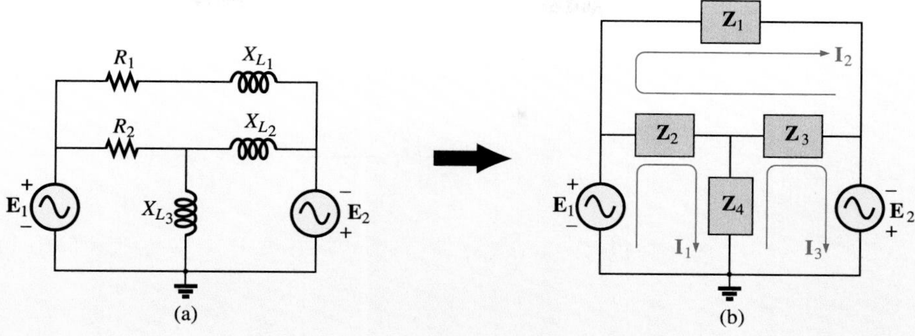

FIG. 17.18
Example 17.11

Solution: The network is redrawn as shown in Fig. 17.18(b), where

$$\mathbf{Z}_1 = R_1 + j X_{L_1} \quad \mathbf{Z}_3 = j X_{L_2}$$
$$\mathbf{Z}_2 = R_2 \quad \mathbf{Z}_4 = j X_{L_3}$$

and

$$\mathbf{I}_1(\mathbf{Z}_2 + \mathbf{Z}_4) - \mathbf{I}_2\mathbf{Z}_2 - \mathbf{I}_3\mathbf{Z}_4 = \mathbf{E}_1$$
$$\mathbf{I}_2(\mathbf{Z}_1 + \mathbf{Z}_2 + \mathbf{Z}_3) - \mathbf{I}_1\mathbf{Z}_2 - \mathbf{I}_3\mathbf{Z}_3 = 0$$
$$\mathbf{I}_3(\mathbf{Z}_3 + \mathbf{Z}_4) - \mathbf{I}_2\mathbf{Z}_3 - \mathbf{I}_1\mathbf{Z}_4 = \mathbf{E}_2$$

or

$$\begin{array}{llll} \mathbf{I}_1(\mathbf{Z}_2 + \mathbf{Z}_4) & - \mathbf{I}_2\mathbf{Z}_2 & - \mathbf{I}_3\mathbf{Z}_4 & = \mathbf{E}_1 \\ -\mathbf{I}_1\mathbf{Z}_2 & + \mathbf{I}_2(\mathbf{Z}_1 + \mathbf{Z}_2 + \mathbf{Z}_3) & - \mathbf{I}_3\mathbf{Z}_3 & = 0 \\ -\mathbf{I}_1\mathbf{Z}_4 & - \mathbf{I}_2\mathbf{Z}_3 & + \mathbf{I}_3(\mathbf{Z}_3 + \mathbf{Z}_4) & = \mathbf{E}_2 \end{array}$$

Note the symmetry *about* the diagonal axis; that is, note the location of $-\mathbf{Z}_2$, $-\mathbf{Z}_4$, and $-\mathbf{Z}_3$ off the diagonal.

17.5 NODAL ANALYSIS

General Approach

Independent Sources Before examining how to apply the method to ac networks, you should review the appropriate sections on nodal analysis in Chapter 8, since this section will cover only the general conclusions of Chapter 8.

The basic steps are the following:

1. Determine the number of nodes within the network.
2. Pick a reference node and label each remaining node with a subscripted value of voltage: $\mathbf{V}_1$, $\mathbf{V}_2$, and so on.
3. Apply Kirchhoff's current law at each node except the reference. Assume that all unknown currents leave the node for each application of Kirchhoff's current law.
4. Solve the resulting equations for the nodal voltages.

A few examples will refresh your memory about the content of Chapter 8 and the general approach to a nodal analysis solution.

EXAMPLE 17.12

Determine the voltage across the inductor for the network of Fig. 17.19.

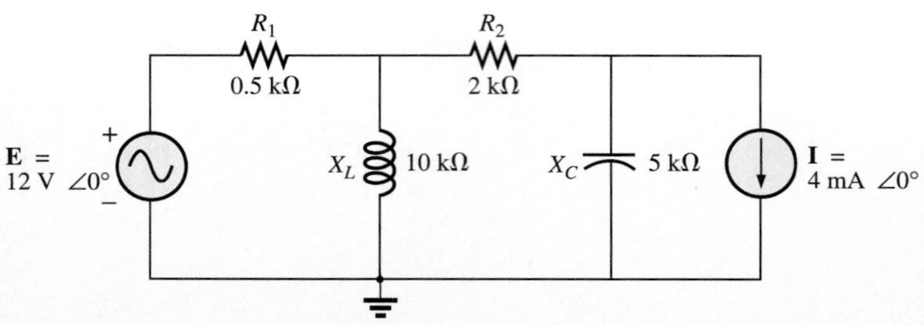

FIG. 17.19
Example 17.12.

Solution:

Steps 1 and 2 are as indicated in Fig. 17.20.

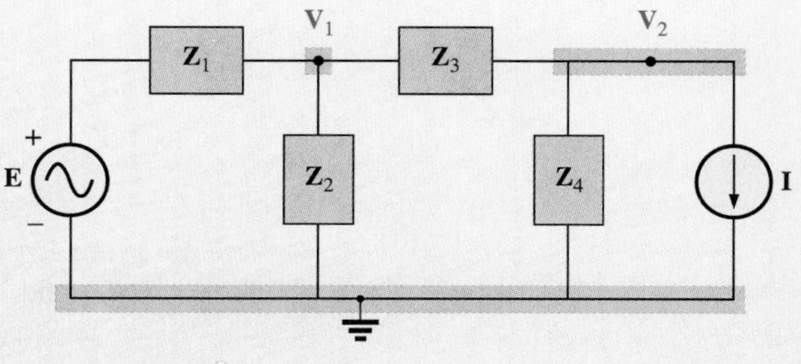

FIG. 17.20
Assigning the nodal voltages and subscripted impedances to the network of Fig. 17.19.

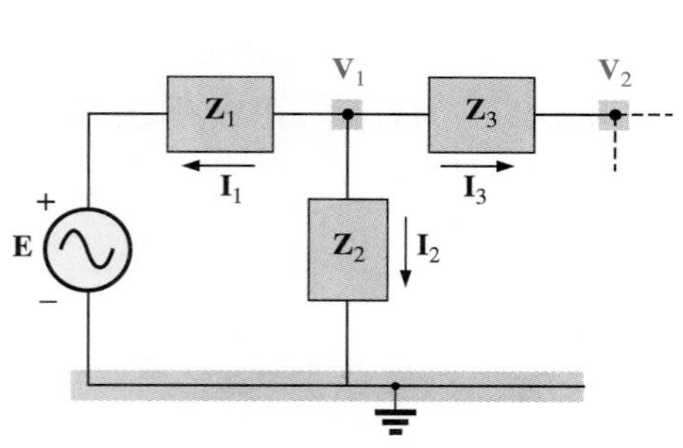

FIG. 17.21
Applying Kirchhoff's current law to the node $\mathbf{V}_1$ of Fig. 17.20.

Step 3: Note Fig. 17.21 for the application of Kirchhoff's current law to node $\mathbf{V}_1$.

$$\Sigma \mathbf{I}_i = \Sigma \mathbf{I}_o$$

$$0 = \mathbf{I}_1 + \mathbf{I}_2 + \mathbf{I}_3$$

$$\frac{\mathbf{V}_1 - \mathbf{E}}{\mathbf{Z}_1} + \frac{\mathbf{V}_1}{\mathbf{Z}_2} + \frac{\mathbf{V}_1 - \mathbf{V}_2}{\mathbf{Z}_3} = 0$$

Rearranging terms:

$$\mathbf{V}_1\left[\frac{1}{\mathbf{Z}_1} + \frac{1}{\mathbf{Z}_2} + \frac{1}{\mathbf{Z}_3}\right] - \mathbf{V}_2\left[\frac{1}{\mathbf{Z}_3}\right] = \frac{\mathbf{E}_1}{\mathbf{Z}_1} \qquad (17.1)$$

Note Fig. 17.22 for the application of Kirchhoff's current law to node $\mathbf{V}_2$.

$$0 = \mathbf{I}_3 + \mathbf{I}_4 + \mathbf{I}$$

$$\frac{\mathbf{V}_2 - \mathbf{V}_1}{\mathbf{Z}_3} + \frac{\mathbf{V}_2}{\mathbf{Z}_4} + \mathbf{I} = 0$$

continued

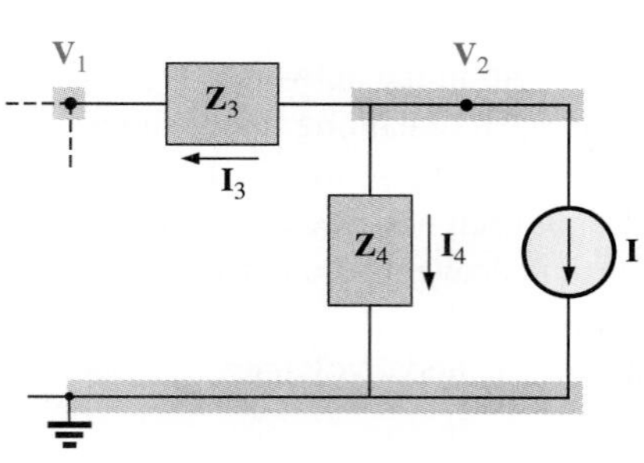

FIG. 17.22
Applying Kirchhoff's current law to the node $\mathbf{V}_2$ of Fig. 17.20.

Rearranging terms:

$$\mathbf{V}_2\left[\frac{1}{\mathbf{Z}_3}+\frac{1}{\mathbf{Z}_4}\right]-\mathbf{V}_1\left[\frac{1}{\mathbf{Z}_3}\right]=-\mathbf{I} \qquad (17.2)$$

Grouping equations:

$$\mathbf{V}_1\left[\frac{1}{\mathbf{Z}_1}+\frac{1}{\mathbf{Z}_2}+\frac{1}{\mathbf{Z}_3}\right]-\mathbf{V}_2\left[\frac{1}{\mathbf{Z}_3}\right]=\frac{\mathbf{E}}{\mathbf{Z}_1}$$

$$\mathbf{V}_1\left[\frac{1}{\mathbf{Z}_3}\right]-\mathbf{V}_2\left[\frac{1}{\mathbf{Z}_3}+\frac{1}{\mathbf{Z}_4}\right]=\mathbf{I}$$

$$\frac{1}{\mathbf{Z}_1}+\frac{1}{\mathbf{Z}_2}+\frac{1}{\mathbf{Z}_3}=\frac{1}{0.5\text{ k}\Omega}+\frac{1}{j\,10\text{ k}\Omega}+\frac{1}{2\text{ k}\Omega}=2.5\text{ mS}\angle-2.29^\circ$$

$$\frac{1}{\mathbf{Z}_3}+\frac{1}{\mathbf{Z}_4}=\frac{1}{2\text{ k}\Omega}+\frac{1}{-j\,5\text{ k}\Omega}=0.539\text{ mS}\angle 21.80^\circ$$

and

$$\mathbf{V}_1[2.5\text{ mS}\angle-2.29^\circ]-\mathbf{V}_2[0.5\text{ mS}\angle 0^\circ]=24\text{ mA}\angle 0^\circ$$

$$\mathbf{V}_1[0.5\text{ mS}\angle 0^\circ]-\mathbf{V}_2[0.539\text{ mS}\angle 21.80^\circ]=4\text{ mA}\angle 0^\circ$$

with

$$\mathbf{V}_1=\frac{\begin{vmatrix}24\text{ mA}\angle 0^\circ & -0.5\text{ mS}\angle 0^\circ\\ 4\text{ mA}\angle 0^\circ & -0.539\text{ mS}\angle 21.80^\circ\end{vmatrix}}{\begin{vmatrix}2.5\text{ mS}\angle-2.29^\circ & -0.5\text{ mS}\angle 0^\circ\\ 0.5\text{ mS}\angle 0^\circ & -0.539\text{ mS}\angle 21.80^\circ\end{vmatrix}}$$

$$=\frac{(24\text{ mA}\angle 0^\circ)(-0.539\text{ mS}\angle 21.80^\circ)+(0.5\text{ mS}\angle 0^\circ)(4\text{ mA}\angle 0^\circ)}{(2.5\text{ mS}\angle-2.29^\circ)(-0.539\text{ mS}\angle 21.80^\circ)+(0.5\text{ mS}\angle 0^\circ)(0.5\text{ mS}\angle 0^\circ)}$$

$$=\frac{-12.94\times 10^{-6}\text{ V}\angle 21.80^\circ+2\times 10^{-6}\text{ V}\angle 0^\circ}{-1.348\times 10^{-6}\angle 19.51^\circ+0.25\times 10^{-6}\angle 0^\circ}$$

$$=\frac{-(12.01+j\,4.81)\times 10^{-6}\text{ V}+2\times 10^{-6}\text{ V}}{-(1.271+j\,0.45)\times 10^{-6}+0.25\times 10^{-6}}$$

$$=\frac{-10.01\text{ V}-j\,4.81\text{ V}}{-1.021-j\,0.45}=\frac{11.106\text{ V}\angle-154.33^\circ}{1.116\angle-156.21^\circ}$$

$$\mathbf{V}_1=\mathbf{9.95\text{ V}\angle 1.88^\circ}$$

Dependent Current Sources For dependent current sources, the procedure is modified as follows:

1. **Steps 1 and 2** are the same as those applied for independent sources.
2. **Step 3** is modified: Treat each dependent current source like an independent source when Kirchhoff's current law is applied to each defined node. However, once the equations are established, substitute the equation for the controlling quantity to ensure that the unknowns are limited solely to the chosen nodal voltages.
3. **Step 4** is as before.

EXAMPLE 17.13

Write the nodal equations for the network of Fig. 17.23 having a dependent current source.

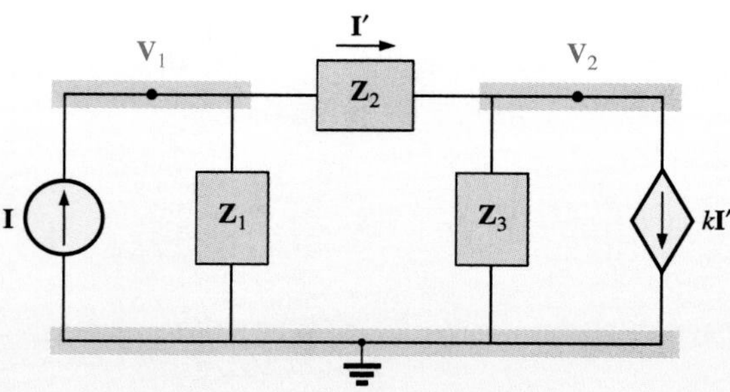

FIG. 17.23
Applying nodal analysis to a network with a current-controlled current source.

Solution:

Steps 1 and 2 are as defined in Fig. 17.23.

Step 3: At node $\mathbf{V}_1$,

$$\mathbf{I} = \mathbf{I}_1 + \mathbf{I}_2$$

$$\frac{\mathbf{V}_1}{\mathbf{Z}_1} + \frac{\mathbf{V}_1 - \mathbf{V}_2}{\mathbf{Z}_2} - \mathbf{I} = 0$$

and

$$\mathbf{V}_1\left[\frac{1}{\mathbf{Z}_1} + \frac{1}{\mathbf{Z}_2}\right] - \mathbf{V}_2\left[\frac{1}{\mathbf{Z}_2}\right] = \mathbf{I}$$

At node $\mathbf{V}_2$,

$$\mathbf{I}_2 + \mathbf{I}_3 + k\mathbf{I}' = 0$$

$$\frac{\mathbf{V}_2 - \mathbf{V}_1}{\mathbf{Z}_2} + \frac{\mathbf{V}_2}{\mathbf{Z}_3} + k\left[\frac{\mathbf{V}_1 - \mathbf{V}_2}{\mathbf{Z}_2}\right] = 0$$

and

$$\mathbf{V}_1\left[\frac{1-k}{\mathbf{Z}_2}\right] - \mathbf{V}_2\left[\frac{1-k}{\mathbf{Z}_2} + \frac{1}{\mathbf{Z}_3}\right] = 0$$

resulting in two equations and two unknowns.

Independent Voltage Sources Between Assigned Nodes

For independent voltage sources between assigned nodes, the procedure is modified as follows:

1. **Steps 1 and 2** are the same as those applied for independent sources.
2. **Step 3** is modified: Treat each source between defined nodes as a short circuit (recall the *supernode* classification of Chapter 8), and write the nodal equations for each remaining independent node. Then relate the chosen nodal voltages to the independent voltage source to ensure that the unknowns of the final equations are limited solely to the nodal voltages.
3. **Step 4** is as before.

EXAMPLE 17.14

Write the nodal equations for the network of Fig. 17.24 having an independent source between two assigned nodes.

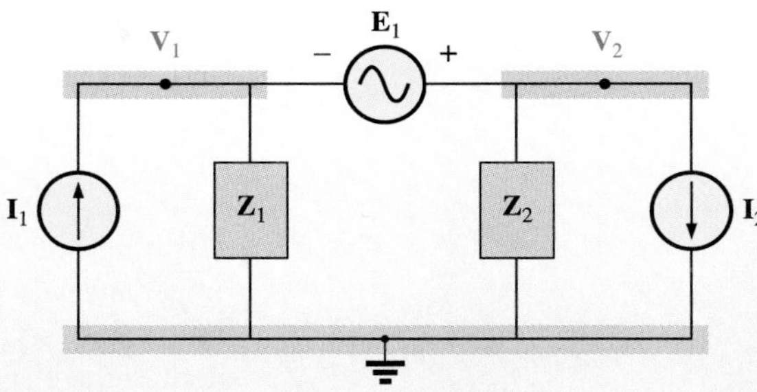

FIG. 17.24
Applying nodal analysis to a network with an independent voltage source between defined nodes.

Solution:

Steps 1 and 2 are defined in Fig. 17.24.

Step 3: Replacing the independent source **E** with a short-circuit equivalent results in a supernode that will generate the following equation when Kirchhoff's current law is applied to node $\mathbf{V}_1$:

$$\mathbf{I}_1 = \frac{\mathbf{V}_1}{\mathbf{Z}_1} + \frac{\mathbf{V}_2}{\mathbf{Z}_2} + \mathbf{I}_2$$

with
$$\mathbf{V}_2 - \mathbf{V}_1 = \mathbf{E}$$

and we have two equations and two unknowns.

Dependent Voltage Sources Between Defined Nodes

For dependent voltage sources between defined nodes, the procedure is modified as follows:

1. **Steps 1 and 2** are the same as those applied for independent voltage sources.
2. **Step 3** is modified: The procedure is essentially the same as that applied for independent voltage sources, except now the dependent sources have to be defined in terms of the chosen nodal voltages to ensure that the final equations have only nodal voltages as their unknown quantities.
3. **Step 4** is as before.

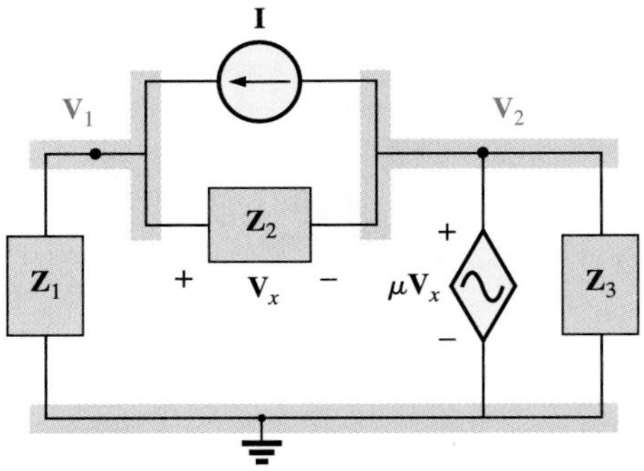

FIG. 17.25
Applying nodal analysis to a network with a voltage-controlled voltage source.

EXAMPLE 17.15

Write the nodal equations for the network of Fig. 17.25 having a dependent voltage source between two defined nodes.

Solution:

Steps 1 and 2 are defined in Fig. 17.25.

Step 3: Replacing the dependent source $\mu\mathbf{V}_x$ with a short-circuit equivalent will result in the following equation when Kirchhoff's

continued

current law is applied at node $\mathbf{V}_1$:

$$\mathbf{I} = \mathbf{I}_1 + \mathbf{I}_2$$

$$\frac{\mathbf{V}_1}{\mathbf{Z}_1} + \frac{(\mathbf{V}_1 - \mathbf{V}_2)}{\mathbf{Z}_2} - \mathbf{I} = 0$$

and

$$\mathbf{V}_2 = \mu\mathbf{V}_x = \mu[\mathbf{V}_1 - \mathbf{V}_2]$$

or

$$\mathbf{V}_2 = \frac{\mu}{1 + \mu}\mathbf{V}_1$$

resulting in two equations and two unknowns. Note that because the impedance $\mathbf{Z}_3$ is in parallel with a voltage source, it does not appear in the analysis. It will, however, affect the current through the dependent voltage source.

Format Approach

If you examine closely Eqs. (17.1) and (17.2) in Example 17.12 you will see that they are the same equations that would have been obtained using the format approach introduced in Chapter 8. Recall that the approach required that the voltage source first be converted to a current source, but the writing of the equations was quite direct and minimized any chances of an error due to a lost sign or missing term.

The sequence of steps required to apply the **format approach** is the following:

1. Choose a reference node and assign a subscripted voltage label to the $(N - 1)$ remaining independent nodes of the network.
2. The number of equations required for a complete solution is equal to the number of subscripted voltages $(N - 1)$. Column 1 of each equation is formed by summing the admittances tied to the node of interest and multiplying the result by that subscripted nodal voltage.
3. The mutual terms are always subtracted from the terms of the first column. It is possible to have more than one mutual term if the nodal voltage of interest has an element in common with more than one other nodal voltage. Each mutual term is the product of the mutual admittance and the other nodal voltage tied to that admittance.
4. The column to the right of the equals sign is the algebraic sum of the current sources tied to the node of interest. A current source is assigned a positive sign if it supplies current to a node, and a negative sign if it draws current from the node.
5. Solve resulting simultaneous equations for the desired nodal voltages. The comments made for mesh analysis regarding independent and dependent sources apply here also.

EXAMPLE 17.16

Using the format approach to nodal analysis, find the voltage across the 4-Ω resistor in Fig. 17.26.

continued

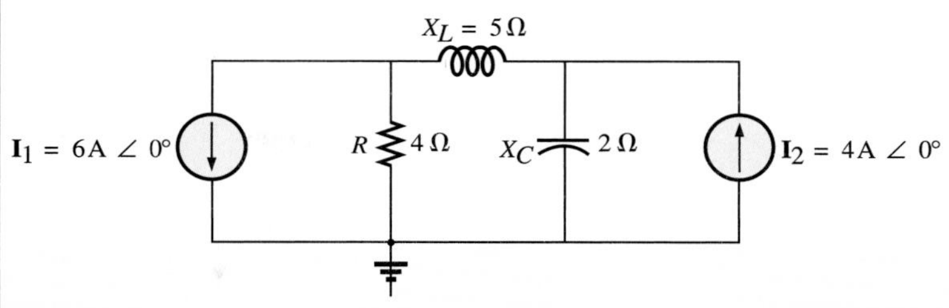

FIG. 17.26
Example 17.16.

Solution:

Step 1: Choosing nodes (Fig. 17.27) and writing the nodal equations, we have

$$\mathbf{Z}_1 = R = 4\ \Omega \qquad \mathbf{Z}_2 = j X_L = j\,5\ \Omega \qquad \mathbf{Z}_3 = -j X_C = -j\,2\ \Omega$$

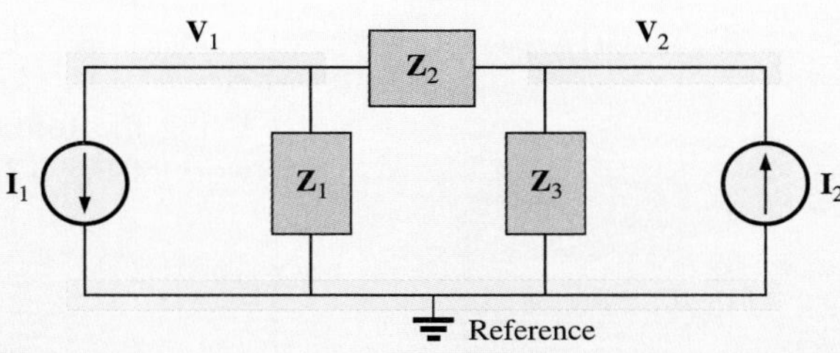

FIG. 17.27
Assigning the nodal voltages and subscripted impedances for the network of Fig. 17.26.

Steps 2, 3, and 4:

$$\begin{aligned}\mathbf{V}_1(\mathbf{Y}_1 + \mathbf{Y}_2) - \mathbf{V}_2(\mathbf{Y}_2) &= -\mathbf{I}_1\\ \mathbf{V}_2(\mathbf{Y}_3 + \mathbf{Y}_2) - \mathbf{V}_1(\mathbf{Y}_2) &= +\mathbf{I}_2\end{aligned}$$

or

$$\begin{aligned}\mathbf{V}_1(\mathbf{Y}_1 + \mathbf{Y}_2) - \mathbf{V}_2(\mathbf{Y}_2) \qquad\qquad &= -\mathbf{I}_1\\ -\mathbf{V}_1(\mathbf{Y}_2) \qquad + \mathbf{V}_2(\mathbf{Y}_3 + \mathbf{Y}_2) &= +\mathbf{I}_2\end{aligned}$$

$$\mathbf{Y}_1 = \frac{1}{\mathbf{Z}_1} \qquad \mathbf{Y}_2 = \frac{1}{\mathbf{Z}_2} \qquad \mathbf{Y}_3 = \frac{1}{\mathbf{Z}_3}$$

Step 5: Using determinants gives us

$$\mathbf{V}_1 = \frac{\begin{vmatrix} -\mathbf{I}_1 & -\mathbf{Y}_2 \\ +\mathbf{I}_2 & \mathbf{Y}_3 + \mathbf{Y}_2 \end{vmatrix}}{\begin{vmatrix} \mathbf{Y}_1 + \mathbf{Y}_2 & -\mathbf{Y}_2 \\ -\mathbf{Y}_2 & \mathbf{Y}_3 + \mathbf{Y}_2 \end{vmatrix}}$$

$$= \frac{-(\mathbf{Y}_3 + \mathbf{Y}_2)\mathbf{I}_1 + \mathbf{I}_2\mathbf{Y}_2}{(\mathbf{Y}_1 + \mathbf{Y}_2)(\mathbf{Y}_3 + \mathbf{Y}_2) - \mathbf{Y}_2^2}$$

$$= \frac{-(\mathbf{Y}_3 + \mathbf{Y}_2)\mathbf{I}_1 + \mathbf{I}_2\mathbf{Y}_2}{\mathbf{Y}_1\mathbf{Y}_3 + \mathbf{Y}_2\mathbf{Y}_3 + \mathbf{Y}_1\mathbf{Y}_2}$$

continued

Substituting numerical values, we have

$$\mathbf{V}_1 = \frac{-[(1/-j\,2\,\Omega) + (1/j\,5\,\Omega)]6\text{ A}\angle 0^\circ + 4\text{ A}\angle 0^\circ(1/j\,5\,\Omega)}{(1/4\,\Omega)(1/-j\,2\,\Omega) + (1/j\,5\,\Omega)(1/-j\,2\,\Omega) + (1/4\,\Omega)(1/j\,5\,\Omega)}$$

$$= \frac{-(+j\,0.5 - j\,0.2)6\angle 0^\circ + 4\angle 0^\circ(-j\,0.2)}{(1/-j\,8) + (1/10) + (1/j\,20)}$$

$$= \frac{(-0.3\angle 90^\circ)(6\angle 0^\circ) + (4\angle 0^\circ)(0.2\angle -90^\circ)}{j\,0.125 + 0.1 - j\,0.05}$$

$$= \frac{-1.8\angle 90^\circ + 0.8\angle -90^\circ}{0.1 + j\,0.075}$$

$$= \frac{2.6\text{ V}\angle -90^\circ}{0.125\angle 36.87^\circ}$$

$$= \mathbf{20.80\text{ V}\angle -126.87^\circ}$$

EXAMPLE 17.17

Using the format approach, write the nodal equations for the network of Fig. 17.28.

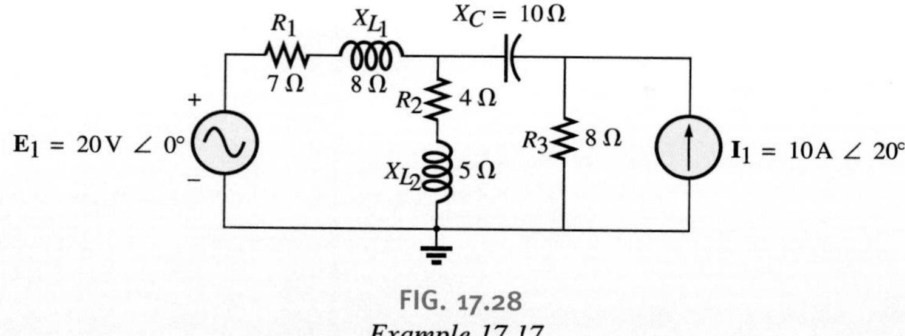

FIG. 17.28
Example 17.17.

Solution:

Step 1: The circuit is redrawn in Fig. 17.29, where

$$\mathbf{Z}_1 = R_1 + j\,X_{L_1} = 7\,\Omega + j\,8\,\Omega \qquad \mathbf{E}_1 = 20\text{ V}\angle 0^\circ$$
$$\mathbf{Z}_2 = R_2 + j\,X_{L_2} = 4\,\Omega + j\,5\,\Omega \qquad \mathbf{I}_1 = 10\text{ A}\angle 20^\circ$$
$$\mathbf{Z}_3 = -j\,X_C = -j\,10\,\Omega$$
$$\mathbf{Z}_4 = R_3 = 8\,\Omega$$

Converting the voltage source to a current source and choosing nodes, we obtain Fig. 17.30. Note the neat appearance of the network using the subscripted impedances. Working directly with Fig. 17.28 would be more difficult and could produce errors.

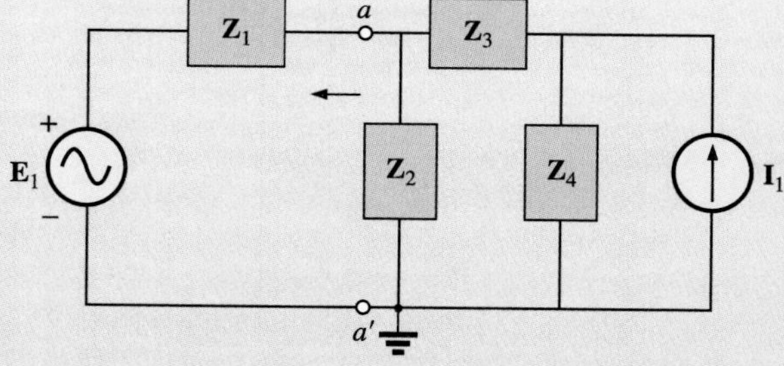

FIG. 17.29
Assigning the subscripted impedances for the network of Fig. 17.28.

continued

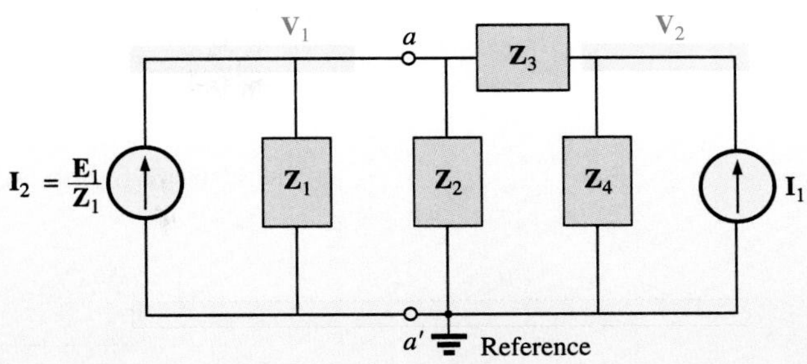

FIG. 17.30
Converting the voltage source of Fig. 17.29 to a current source and defining the nodal voltages.

Steps 2, 3, and 4: Write the nodal equations:

$$\mathbf{V}_1(\mathbf{Y}_1 + \mathbf{Y}_2 + \mathbf{Y}_3) - \mathbf{V}_2(\mathbf{Y}_3) = +\mathbf{I}_2$$
$$\mathbf{V}_2(\mathbf{Y}_3 + \mathbf{Y}_4) - \mathbf{V}_1(\mathbf{Y}_3) = +\mathbf{I}_1$$

$$\mathbf{Y}_1 = \frac{1}{\mathbf{Z}_1} \quad \mathbf{Y}_2 = \frac{1}{\mathbf{Z}_2} \quad \mathbf{Y}_3 = \frac{1}{\mathbf{Z}_3} \quad \mathbf{Y}_4 = \frac{1}{\mathbf{Z}_4}$$

which are rewritten as

$$\mathbf{V}_1(\mathbf{Y}_1 + \mathbf{Y}_2 + \mathbf{Y}_3) - \mathbf{V}_2(\mathbf{Y}_3) = +\mathbf{I}_2$$
$$-\mathbf{V}_1(\mathbf{Y}_3) \quad + \mathbf{V}_2(\mathbf{Y}_3 + \mathbf{Y}_4) = +\mathbf{I}_1$$

EXAMPLE 17.18

Write the nodal equations for the network of Fig. 17.31(a). Do not solve.

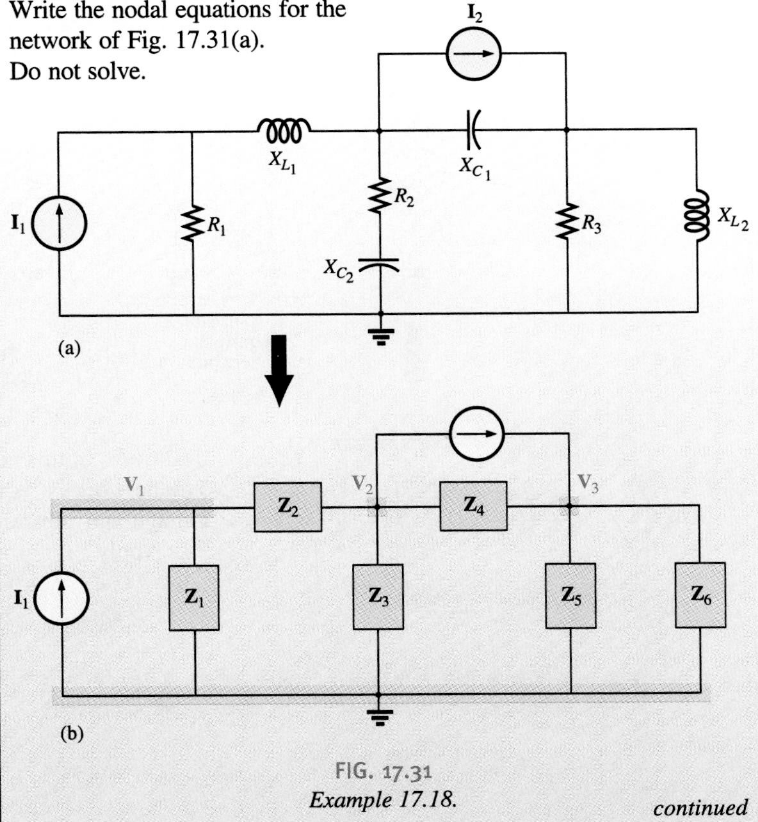

FIG. 17.31
Example 17.18.

continued

Solution: Choose nodes [Fig. 17.31(b)]:

$$\mathbf{Z}_1 = R_1 \qquad \mathbf{Z}_2 = jX_{L_1} \qquad \mathbf{Z}_3 = R_2 - jX_{C_2}$$
$$\mathbf{Z}_4 = -jX_{C_1} \qquad \mathbf{Z}_5 = R_3 \qquad \mathbf{Z}_6 = jX_{L_2}$$

and write the nodal equations:

$$\mathbf{V}_1(\mathbf{Y}_1 + \mathbf{Y}_2) - \mathbf{V}_2(\mathbf{Y}_2) = +\mathbf{I}_1$$
$$\mathbf{V}_2(\mathbf{Y}_2 + \mathbf{Y}_3 + \mathbf{Y}_4) - \mathbf{V}_1(\mathbf{Y}_2) - \mathbf{V}_3(\mathbf{Y}_4) = -\mathbf{I}_2$$
$$\mathbf{V}_3(\mathbf{Y}_4 + \mathbf{Y}_5 + \mathbf{Y}_6) - \mathbf{V}_2(\mathbf{Y}_4) = +\mathbf{I}_2$$

which are rewritten as

$$\begin{aligned}
\mathbf{V}_1(\mathbf{Y}_1 + \mathbf{Y}_2) &- \mathbf{V}_2(\mathbf{Y}_2) &+ 0 &= +\mathbf{I}_1\\
-\mathbf{V}_1(\mathbf{Y}_2) &+ \mathbf{V}_2(\mathbf{Y}_2 + \mathbf{Y}_3 + \mathbf{Y}_4) &- \mathbf{V}_3(\mathbf{Y}_4) &= -\mathbf{I}_2\\
0 &- \mathbf{V}_2(\mathbf{Y}_4) &+ \mathbf{V}_3(\mathbf{Y}_4 + \mathbf{Y}_5 + \mathbf{Y}_6) &= +\mathbf{I}_2
\end{aligned}$$

$$\mathbf{Y}_1 = \frac{1}{R_1} \qquad \mathbf{Y}_2 = \frac{1}{jX_{L_1}} \qquad \mathbf{Y}_3 = \frac{1}{R_2 - jX_{C_2}}$$
$$\mathbf{Y}_4 = \frac{1}{-jX_{C_1}} \qquad \mathbf{Y}_5 = \frac{1}{R_3} \qquad \mathbf{Y}_6 = \frac{1}{jX_{L_2}}$$

Note the symmetry about the diagonal for this example and the ones preceding it in this section.

EXAMPLE 17.19

Apply nodal analysis to the network of Fig. 17.32(a). Determine the voltage $\mathbf{V}_L$.

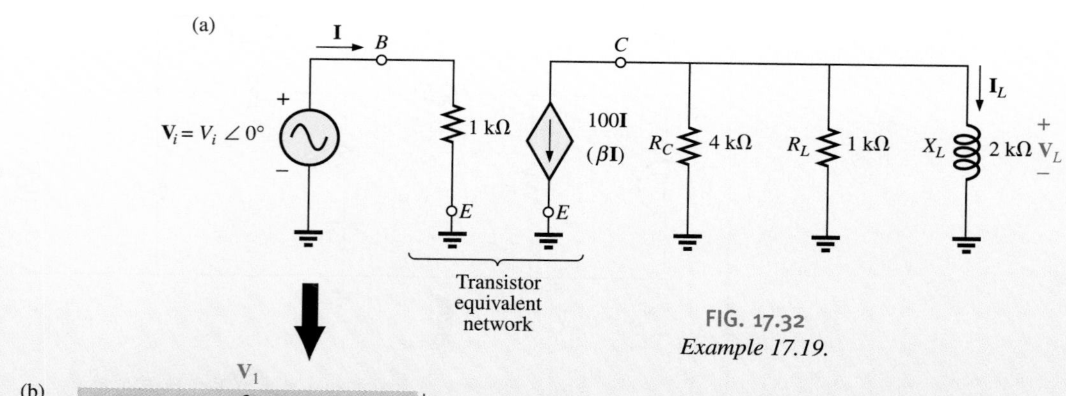

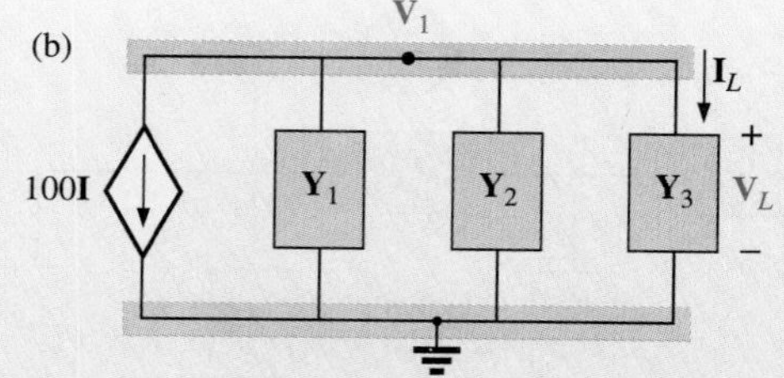

FIG. 17.32
Example 17.19.

Solution: In this case there is no need for a source conversion. The network is redrawn in Fig. 17.32(b) with the chosen node voltage and subscripted impedances.

Apply the format approach:

$$\mathbf{Y}_1 = \frac{1}{\mathbf{Z}_1} = \frac{1}{4\text{ k}\Omega} = 0.25\text{ mS}\angle 0^\circ = G_1\angle 0^\circ$$

$$\mathbf{Y}_2 = \frac{1}{\mathbf{Z}_2} = \frac{1}{1\text{ k}\Omega} = 1\text{ mS}\angle 0^\circ = G_2\angle 0^\circ$$

continued

$$\mathbf{Y}_3 = \frac{1}{\mathbf{Z}_3} = \frac{1}{2\text{ k}\Omega \angle 90^\circ} = 0.5\text{ mS} \angle -90^\circ$$

$$= -j\,0.5\text{ mS} = -j\,B_L$$

$$\mathbf{V}_1\text{: } (\mathbf{Y}_1 + \mathbf{Y}_2 + \mathbf{Y}_3)\mathbf{V}_1 = -100\mathbf{I}$$

and
$$\mathbf{V}_i = \frac{-100\mathbf{I}}{\mathbf{Y}_1 + \mathbf{Y}_2 + \mathbf{Y}_3}$$

$$= \frac{-100\mathbf{I}}{0.25\text{ mS} + 1\text{ mS} - j\,0.5\text{ mS}}$$

$$= \frac{-100 \times 10^3\mathbf{I}}{1.25 - j\,0.5} = \frac{-100 \times 10^3\mathbf{I}}{1.3463 \angle -21.80^\circ}$$

$$= -74.28 \times 10^3\mathbf{I} \angle 21.80^\circ$$

$$= -74.28 \times 10^3\left(\frac{\mathbf{V}_i}{1\text{ k}\Omega}\right) \angle 21.80^\circ$$

$$\mathbf{V}_1 = \mathbf{V}_L = \mathbf{-(74.28V}_i\mathbf{)\text{ V}\angle 21.80^\circ}$$

17.6 BRIDGE NETWORKS (ac)

The basic bridge configuration was discussed in some detail in Section 8.11 for dc networks. We now continue to examine bridge networks by considering those that have reactive components and a sinusoidal ac voltage or current applied.

We will first analyze various familiar forms of the bridge network using *mesh analysis* and *nodal analysis* (the *format approach*). The balance conditions will be investigated throughout the section.

A. Mesh Analysis Apply mesh analysis to the network of Fig. 17.33(a).

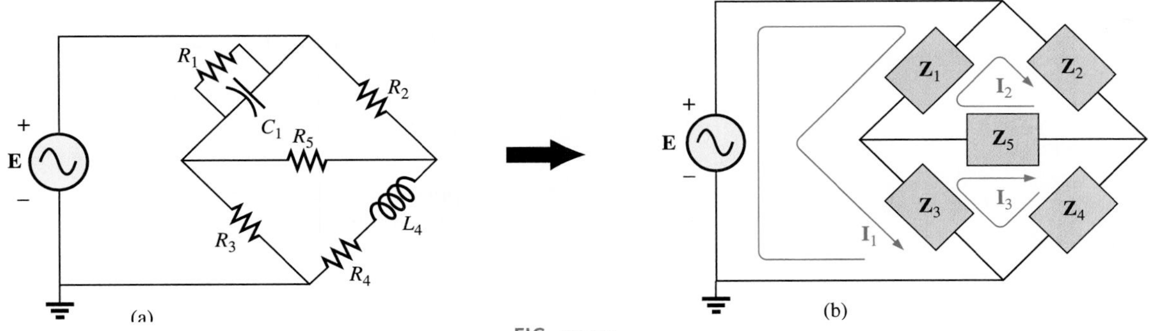

FIG. 17.33
Maxwell bridge.

The network is redrawn in Fig. 17.33(b), where

$$\mathbf{Z}_1 = \frac{1}{\mathbf{Y}_1} = \frac{1}{G_1 + j\,B_C} = \frac{G_1}{G_1^2 + B_C^2} - j\,\frac{B_C}{G_1^2 + B_C^2}$$

$$\mathbf{Z}_2 = R_2 \quad \mathbf{Z}_3 = R_3 \quad \mathbf{Z}_4 = R_4 + j\,X_L \quad \mathbf{Z}_5 = R_5$$

Applying the format approach:

$$(\mathbf{Z}_1 + \mathbf{Z}_3)\mathbf{I}_1 - (\mathbf{Z}_1)\mathbf{I}_2 - (\mathbf{Z}_3)\mathbf{I}_3 = \mathbf{E}$$
$$(\mathbf{Z}_1 + \mathbf{Z}_2 + \mathbf{Z}_5)\mathbf{I}_2 - (\mathbf{Z}_1)\mathbf{I}_1 - (\mathbf{Z}_5)\mathbf{I}_3 = 0$$
$$(\mathbf{Z}_3 + \mathbf{Z}_4 + \mathbf{Z}_5)\mathbf{I}_3 - (\mathbf{Z}_3)\mathbf{I}_1 - (\mathbf{Z}_5)\mathbf{I}_2 = 0$$

which are rewritten as

$$\begin{array}{llll} \mathbf{I}_1(\mathbf{Z}_1 + \mathbf{Z}_3) & - \mathbf{I}_2\mathbf{Z}_1 & - \mathbf{I}_3\mathbf{Z}_3 & = \mathbf{E} \\ -\mathbf{I}_1\mathbf{Z}_1 & + \mathbf{I}_2(\mathbf{Z}_1 + \mathbf{Z}_2 + \mathbf{Z}_5) & - \mathbf{I}_3\mathbf{Z}_5 & = 0 \\ -\mathbf{I}_1\mathbf{Z}_3 & - \mathbf{I}_2\mathbf{Z}_5 & + \mathbf{I}_3(\mathbf{Z}_3 + \mathbf{Z}_4 + \mathbf{Z}_5) & = 0 \end{array}$$

Note the symmetry about the diagonal of the above equations. For a balanced condition, such as discussed in part C, *balanced Hay bridge*, the value of $\mathbf{I}_{\mathbf{Z}_5}$ must be set to zero ($\mathbf{I}_{\mathbf{Z}_5}$ = 0 A), and

$$\mathbf{I}_{\mathbf{Z}_5} = \mathbf{I}_2 - \mathbf{I}_3 = 0$$

From the above equations,

$$\mathbf{I}_2 = \frac{\begin{vmatrix} \mathbf{Z}_1 + \mathbf{Z}_3 & \mathbf{E} & -\mathbf{Z}_3 \\ -\mathbf{Z}_1 & 0 & -\mathbf{Z}_5 \\ -\mathbf{Z}_3 & 0 & (\mathbf{Z}_3 + \mathbf{Z}_4 + \mathbf{Z}_5) \end{vmatrix}}{\begin{vmatrix} \mathbf{Z}_1 + \mathbf{Z}_3 & -\mathbf{Z}_1 & -\mathbf{Z}_3 \\ -\mathbf{Z}_1 & (\mathbf{Z}_1 + \mathbf{Z}_2 + \mathbf{Z}_5) & -\mathbf{Z}_5 \\ -\mathbf{Z}_3 & -\mathbf{Z}_5 & (\mathbf{Z}_3 + \mathbf{Z}_4 + \mathbf{Z}_5) \end{vmatrix}}$$

$$= \frac{\mathbf{E}(\mathbf{Z}_1\mathbf{Z}_3 + \mathbf{Z}_1\mathbf{Z}_4 + \mathbf{Z}_1\mathbf{Z}_5 + \mathbf{Z}_3\mathbf{Z}_5)}{\Delta}$$

where Δ signifies the determinant of the denominator (or coefficients). Similarly,

$$\mathbf{I}_3 = \frac{\mathbf{E}(\mathbf{Z}_1\mathbf{Z}_3 + \mathbf{Z}_3\mathbf{Z}_2 + \mathbf{Z}_1\mathbf{Z}_5 + \mathbf{Z}_3\mathbf{Z}_5)}{\Delta}$$

and

$$\mathbf{I}_{\mathbf{Z}_5} = \mathbf{I}_2 - \mathbf{I}_3 = \frac{\mathbf{E}(\mathbf{Z}_1\mathbf{Z}_4 - \mathbf{Z}_3\mathbf{Z}_2)}{\Delta}$$

For $\mathbf{I}_{\mathbf{Z}_5}$ = 0, the following must be satisfied (for a finite Δ not equal to zero):

$$\mathbf{Z}_1\mathbf{Z}_4 = \mathbf{Z}_3\mathbf{Z}_2 \qquad \mathbf{I}_{\mathbf{Z}_5} = 0 \tag{17.3}$$

This condition will be analyzed in greater depth in part B, *nodal analysis*, and part C, *balanced Hay bridge*.

B. Nodal Analysis Applying nodal analysis to the network of Fig. 17.34(a) will result in the configuration of Fig. 17.34(b), where

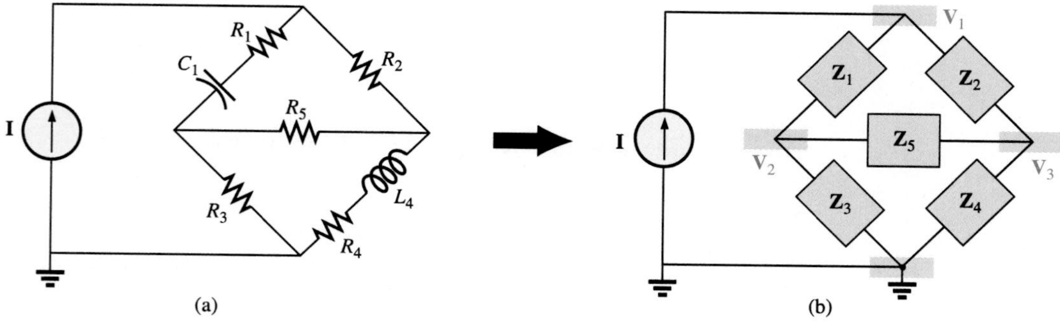

FIG. 17.34
Hay bridge.

$$\mathbf{Y}_1 = \frac{1}{\mathbf{Z}_1} = \frac{1}{R_1 - jX_C} \qquad \mathbf{Y}_2 = \frac{1}{\mathbf{Z}_2} = \frac{1}{R_2}$$

$$\mathbf{Y}_3 = \frac{1}{\mathbf{Z}_3} = \frac{1}{R_3} \qquad \mathbf{Y}_4 = \frac{1}{\mathbf{Z}_4} = \frac{1}{R_4 + jX_L} \qquad \mathbf{Y}_5 = \frac{1}{R_5}$$

and

$$\begin{aligned}(\mathbf{Y}_1 + \mathbf{Y}_2)\mathbf{V}_1 - (\mathbf{Y}_1)\mathbf{V}_2 - (\mathbf{Y}_2)\mathbf{V}_3 &= \mathbf{I}\\ (\mathbf{Y}_1 + \mathbf{Y}_3 + \mathbf{Y}_5)\mathbf{V}_2 - (\mathbf{Y}_1)\mathbf{V}_1 - (\mathbf{Y}_5)\mathbf{V}_3 &= 0\\ (\mathbf{Y}_2 + \mathbf{Y}_4 + \mathbf{Y}_5)\mathbf{V}_3 - (\mathbf{Y}_2)\mathbf{V}_1 - (\mathbf{Y}_5)\mathbf{V}_2 &= 0\end{aligned}$$

which are rewritten as

$$\begin{array}{llll}\mathbf{V}_1(\mathbf{Y}_1 + \mathbf{Y}_2) & - \mathbf{V}_2\mathbf{Y}_1 & - \mathbf{V}_3\mathbf{Y}_2 & = \mathbf{I}\\ -\mathbf{V}_1\mathbf{Y}_1 & + \mathbf{V}_2(\mathbf{Y}_1 + \mathbf{Y}_3 + \mathbf{Y}_5) & - \mathbf{V}_3\mathbf{Y}_5 & = 0\\ -\mathbf{V}_1\mathbf{Y}_2 & - \mathbf{V}_2\mathbf{Y}_5 & + \mathbf{V}_3(\mathbf{Y}_2 + \mathbf{Y}_4 + \mathbf{Y}_5) & = 0\end{array}$$

Again, note the symmetry about the diagonal axis. For balance, $\mathbf{V}_{\mathbf{Z}_5} = 0$ V, and

$$\mathbf{V}_{\mathbf{Z}_5} = \mathbf{V}_2 - \mathbf{V}_3 = 0$$

From the above equations,

$$\mathbf{V}_2 = \frac{\begin{vmatrix}\mathbf{Y}_1 + \mathbf{Y}_2 & \mathbf{I} & -\mathbf{Y}_2\\ -\mathbf{Y}_1 & 0 & -\mathbf{Y}_5\\ -\mathbf{Y}_2 & 0 & (\mathbf{Y}_2 + \mathbf{Y}_4 + \mathbf{Y}_5)\end{vmatrix}}{\begin{vmatrix}\mathbf{Y}_1 + \mathbf{Y}_2 & -\mathbf{Y}_1 & -\mathbf{Y}_2\\ -\mathbf{Y}_1 & (\mathbf{Y}_1 + \mathbf{Y}_3 + \mathbf{Y}_5) & -\mathbf{Y}_5\\ -\mathbf{Y}_2 & -\mathbf{Y}_5 & (\mathbf{Y}_2 + \mathbf{Y}_4 + \mathbf{Y}_5)\end{vmatrix}}$$

$$= \frac{\mathbf{I}(\mathbf{Y}_1\mathbf{Y}_3 + \mathbf{Y}_1\mathbf{Y}_4 + \mathbf{Y}_1\mathbf{Y}_5 + \mathbf{Y}_3\mathbf{Y}_5)}{\Delta}$$

Similarly,

$$\mathbf{V}_3 = \frac{\mathbf{I}(\mathbf{Y}_1\mathbf{Y}_3 + \mathbf{Y}_3\mathbf{Y}_2 + \mathbf{Y}_1\mathbf{Y}_5 + \mathbf{Y}_3\mathbf{Y}_5)}{\Delta}$$

Note the similarities between the above equations and those obtained for mesh analysis. Then

$$\mathbf{V}_{\mathbf{Z}_5} = \mathbf{V}_2 - \mathbf{V}_3 = \frac{\mathbf{I}(\mathbf{Y}_1\mathbf{Y}_4 - \mathbf{Y}_3\mathbf{Y}_2)}{\Delta}$$

For $\mathbf{V}_{\mathbf{Z}_5} = 0$, the following must be satisfied for a finite Δ not equal to zero:

$$\mathbf{Y}_1\mathbf{Y}_4 = \mathbf{Y}_3\mathbf{Y}_2 \qquad \mathbf{V}_{\mathbf{Z}_5} = 0 \tag{17.4}$$

However, substituting $\mathbf{Y}_1 = 1/\mathbf{Z}_1$, $\mathbf{Y}_2 = 1/\mathbf{Z}_2$, $\mathbf{Y}_3 = 1/\mathbf{Z}_3$, and $\mathbf{Y}_4 = 1/\mathbf{Z}_4$, we have

$$\frac{1}{\mathbf{Z}_1\mathbf{Z}_4} = \frac{1}{\mathbf{Z}_3\mathbf{Z}_2}$$

or

$$\mathbf{Z}_1\mathbf{Z}_4 = \mathbf{Z}_3\mathbf{Z}_2 \qquad \mathbf{V}_{\mathbf{Z}_5} = 0$$

corresponding with Eq. (17.3) obtained earlier.

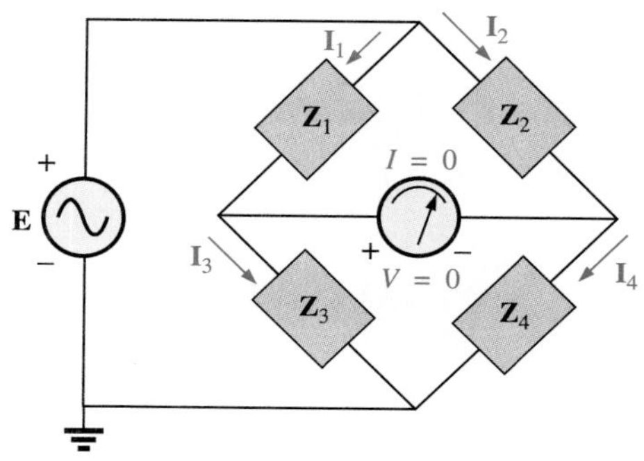

FIG. 17.35
Investigating the balance criteria for an ac bridge configuration.

C. Balanced Hay Bridge Let us now investigate the balance criteria in more detail by considering the network of Fig. 17.35, where it is specified that **I**, **V** = 0.

Since **I** = 0,

$$\mathbf{I}_1 = \mathbf{I}_3 \tag{17.5a}$$

and

$$\mathbf{I}_2 = \mathbf{I}_4 \tag{17.5b}$$

In addition, for **V** = 0,

$$\mathbf{I}_1\mathbf{Z}_1 = \mathbf{I}_2\mathbf{Z}_2 \tag{17.5c}$$

and

$$\mathbf{I}_3\mathbf{Z}_3 = \mathbf{I}_4\mathbf{Z}_4 \tag{17.5d}$$

Substituting the preceding current relations into Eq. (17.5d), we have

$$\mathbf{I}_1\mathbf{Z}_3 = \mathbf{I}_2\mathbf{Z}_4$$

and

$$\mathbf{I}_2 = \frac{\mathbf{Z}_3}{\mathbf{Z}_4}\mathbf{I}_1$$

Substituting this relationship for $\mathbf{I}_2$ into Eq. (17.5c) results in

$$\mathbf{I}_1\mathbf{Z}_1 = \left(\frac{\mathbf{Z}_3}{\mathbf{Z}_4}\mathbf{I}_1\right)\mathbf{Z}_2$$

and

$$\mathbf{Z}_1\mathbf{Z}_4 = \mathbf{Z}_2\mathbf{Z}_3$$

as obtained earlier.

Rearranging, we have

$$\frac{\mathbf{Z}_1}{\mathbf{Z}_3} = \frac{\mathbf{Z}_2}{\mathbf{Z}_4} \tag{17.6}$$

corresponding with Eq. (8.4) for dc resistive networks.

For the network of Fig. 17.34(a), which is referred to as a **Hay bridge** when $\mathbf{Z}_5$ is replaced by a sensitive galvanometer,

$$\mathbf{Z}_1 = R_1 - jX_C$$
$$\mathbf{Z}_2 = R_2$$
$$\mathbf{Z}_3 = R_3$$
$$\mathbf{Z}_4 = R_4 + jX_L$$

This particular network is used for measuring the resistance and inductance of coils in which the resistance is a small fraction of the reactance X_L.

Substitute into Eq. (17.6) in the following form:

$$\mathbf{Z}_2\mathbf{Z}_3 = \mathbf{Z}_4\mathbf{Z}_1$$
$$R_2R_3 = (R_4 + jX_L)(R_1 - jX_C)$$

or

$$R_2R_3 = R_1R_4 + j(R_1X_L - R_4X_C) + X_CX_L$$

so that

$$R_2R_3 + j0 = (R_1R_4 + X_CX_L) + j(R_1X_L - R_4X_C)$$

For the equations to be equal, *the real and imaginary parts must be equal.* Therefore, for a balanced Hay bridge,

$$R_2R_3 = R_1R_4 + X_CX_L \tag{17.7a}$$

and $$0 = R_1X_L - R_4X_C \tag{17.7b}$$

or substituting $$X_L = \omega L \quad \text{and} \quad X_C = \frac{1}{\omega C}$$

we have $$X_CX_L = \left(\frac{1}{\omega C}\right)(\omega L) = \frac{L}{C}$$

and $$R_2R_3 = R_1R_4 + \frac{L}{C}$$

with $$R_1\omega L = \frac{R_4}{\omega C}$$

Solving for R_4 in the last equation gives us

$$R_4 = \omega^2 LCR_1$$

and substituting into the previous equation, we have

$$R_2R_3 = R_1(\omega^2 LCR_1) + \frac{L}{C}$$

Multiply through by C and factor:

$$CR_2R_3 = L(\omega^2C^2R_1^2 + 1)$$

and $$L = \frac{CR_2R_3}{1 + \omega^2C^2R_1^2} \tag{17.8a}$$

With additional algebra this results in:

$$R_4 = \frac{\omega^2C^2R_1R_2R_3}{1 + \omega^2C^2R^2{}_1} \tag{17.8b}$$

Equations (17.7) and (17.8) are the balance conditions for the Hay bridge. Note that each is frequency dependent. For different frequencies, the resistive and capacitive elements must vary for a particular coil to achieve balance. For a coil placed in the Hay bridge as shown in Fig. 17.34, the resistance and inductance of the coil can be determined by Eqs. (17.8a) and (17.8b) when balance is achieved.

The bridge of Fig. 17.33 is referred to as a **Maxwell bridge** when $\mathbf{Z}_5$ is replaced by a sensitive galvanometer. This setup is used for inductance measurements when the resistance of the coil is large enough not to require a Hay bridge.

Application of Eq. (17.6) in the form:

$$\mathbf{Z}_2\mathbf{Z}_3 = \mathbf{Z}_4\mathbf{Z}_1$$

and substituting

$$\mathbf{Z}_1 = R_1 \angle 0° \parallel X_{C_1} \angle -90° = \frac{(R_1 \angle 0°)(X_{C_1} \angle -90°)}{R_1 - jX_{C_1}}$$

$$= \frac{R_1X_{C_1} \angle -90°}{R_1 - jX_{C_1}} = \frac{-jR_1X_{C_1}}{R_1 - jX_{C_1}}$$

$$\mathbf{Z}_2 = R_2$$

$$\mathbf{Z}_3 = R_3$$

and $$\mathbf{Z}_4 = R_4 + jX_{L_4}$$

we have $$(R_2)(R_3) = (R_4 + j\,X_{L_4})\left(\frac{-j\,R_1X_{C_1}}{R_1 - j\,X_{C_1}}\right)$$

$$R_2R_3 = \frac{-j\,R_1R_4X_{C_1} + R_1X_{C_1}X_{L_4}}{R_1 - j\,X_{C_1}}$$

or $$(R_2R_3)(R_1 - j\,X_{C_1}) = R_1X_{C_1}X_{L_4} - j\,R_1R_4X_{C_1}$$

and $$R_1R_2R_3 - j\,R_2R_3X_{C_1} = R_1X_{C_1}X_{L_4} - j\,R_1R_4X_{C_1}$$

so that for balance

$$\cancel{R_1}R_2R_3 = \cancel{R_1}X_{C_1}X_{L_4}$$

$$R_2R_3 = \left(\frac{1}{\cancel{2\pi} f C_1}\right)(\cancel{2\pi} f L_4)$$

and $$L_4 = C_1R_2R_3 \qquad (17.9)$$

and $$R_2R_3\cancel{X_{C_1}} = R_1R_4\cancel{X_{C_1}}$$

so that $$R_4 = \frac{R_2R_3}{R_1} \qquad (17.10)$$

Note the absence of frequency in Eqs. (17.9) and (17.10).

One remaining popular bridge is the **capacitance comparison bridge** of Fig. 17.36. An unknown capacitance and its associated resistance can be determined using this bridge. Application of Eq. (17.6) will give the following results:

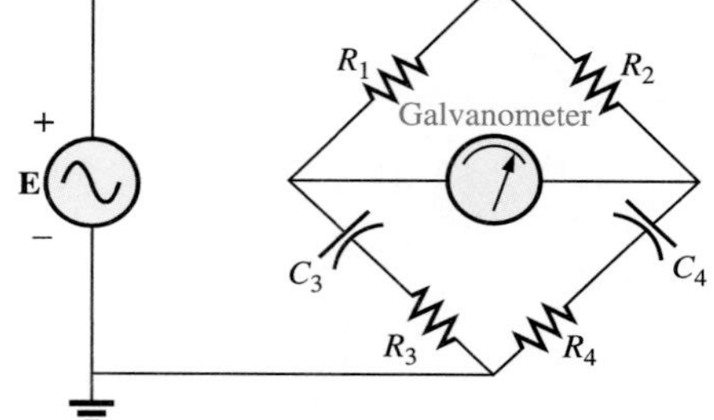

FIG. 17.36
Capacitance comparison bridge.

$$C_4 = C_3\frac{R_1}{R_2} \qquad (17.11)$$

$$R_4 = \frac{R_2R_3}{R_1} \qquad (17.12)$$

The derivation of these equations will appear as a problem at the end of the chapter.

17.7 Δ-Y, Y-Δ CONVERSIONS

Refer back to Section 8.12 to recall the circuit configurations, Y and Δ. Also recall the equivalence of Y with T, and Δ with π. The relations established there for resistances in a dc circuit are exactly the same for *impedances in ac circuits*. The general diagram for the **Δ-Y and Y-Δ conversions** is shown in Fig. 17.37. The equations for expressing impedances in the Y in terms of those for the Δ are:

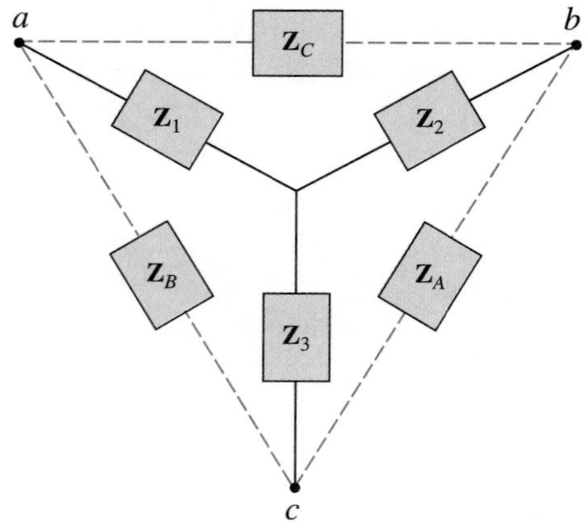

FIG. 17.37
Δ-Y configuration.

$$\mathbf{Z}_1 = \frac{\mathbf{Z}_B\mathbf{Z}_C}{\mathbf{Z}_A + \mathbf{Z}_B + \mathbf{Z}_C} \qquad (17.13)$$

$$\mathbf{Z}_2 = \frac{\mathbf{Z}_A\mathbf{Z}_C}{\mathbf{Z}_A + \mathbf{Z}_B + \mathbf{Z}_C} \qquad (17.14)$$

$$\mathbf{Z}_3 = \frac{\mathbf{Z}_A\mathbf{Z}_B}{\mathbf{Z}_A + \mathbf{Z}_B + \mathbf{Z}_C} \qquad (17.15)$$

For the impedances of the Δ in terms of those for the Y, the equations are

$$\mathbf{Z}_B = \frac{\mathbf{Z}_1\mathbf{Z}_2 + \mathbf{Z}_1\mathbf{Z}_3 + \mathbf{Z}_2\mathbf{Z}_3}{\mathbf{Z}_2} \tag{17.16}$$

$$\mathbf{Z}_A = \frac{\mathbf{Z}_1\mathbf{Z}_2 + \mathbf{Z}_1\mathbf{Z}_3 + \mathbf{Z}_2\mathbf{Z}_3}{\mathbf{Z}_1} \tag{17.17}$$

$$\mathbf{Z}_C = \frac{\mathbf{Z}_1\mathbf{Z}_2 + \mathbf{Z}_1\mathbf{Z}_3 + \mathbf{Z}_2\mathbf{Z}_3}{\mathbf{Z}_3} \tag{17.18}$$

Note that each impedance of the Y is equal to the product of the impedances in the two closest branches of the Δ, divided by the sum of the impedances in the Δ.

Further, the value of each impedance of the Δ is equal to the sum of the possible product combinations of the impedances of the Y, divided by the impedances of the Y farthest from the impedance to be determined.

Equations 17.13 to 17.18 are used to make the Δ-Y and Y-Δ conversions shown in Fig. 17.38.

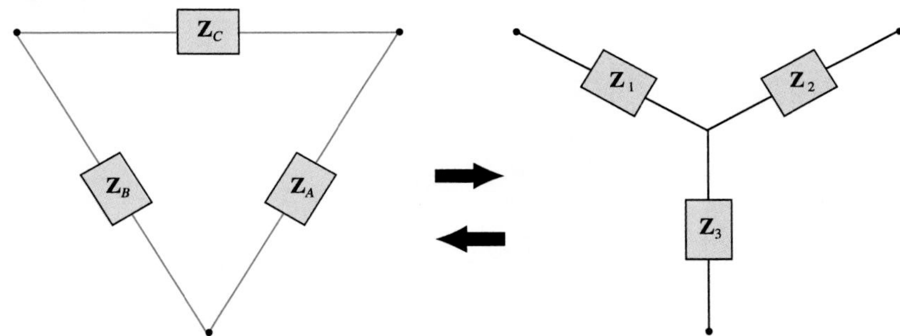

FIG. 17.38
Δ-Y and Y-Δ conversions.

Helpful Hint: If you do not immediately see the type of connectivity in a network, and you do not know how to analyze it, it is sometimes helpful to redraw it. Note that the Y-Δ network of Fig. 17.38 can be redrawn as a T-π network shown in Fig. 17.39 and vice versa. The rest of the discussion in the chapter will assume that this point is clear.

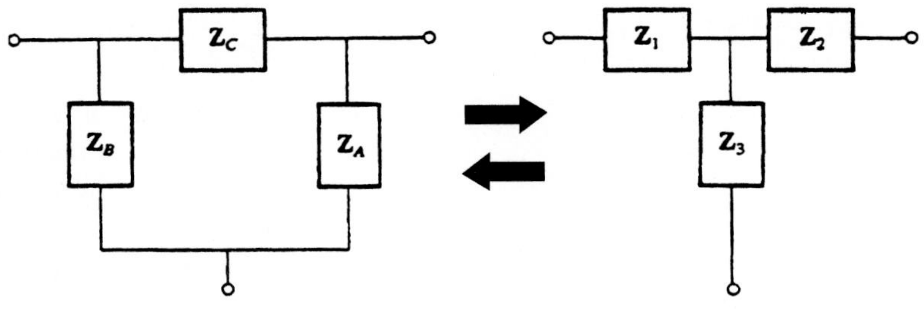

FIG. 17.39
The T and π configurations

In the study of dc networks, we found that if all of the resistors of the Δ or Y were the same, we could convert from one to the other using the equation

$$R_\Delta = 3R_Y \text{ or } R_Y = \frac{R_\Delta}{3}$$

For ac networks,

$$\mathbf{Z}_\Delta = 3\mathbf{Z}_Y \text{ or } \mathbf{Z}_Y = \frac{\mathbf{Z}_\Delta}{3} \tag{17.19}$$

Be careful when using this simplified form. It is not sufficient for all the impedances of the Δ or Y to be of the same magnitude: *The angle associated with each must also be the same.*

EXAMPLE 17.20

Find the total impedance Z_T of the network of Fig. 17.40.

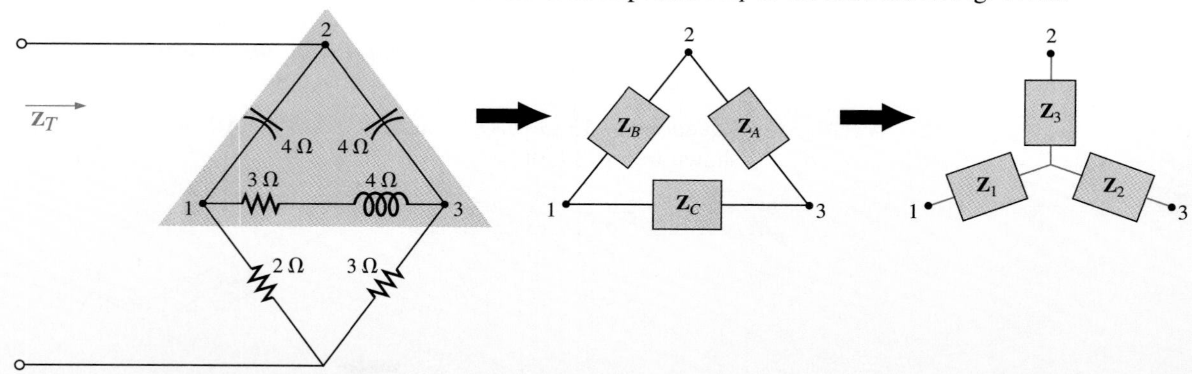

FIG. 17.40
Converting the upper Δ of a bridge configuration to a Y.

Solution:

$$\mathbf{Z}_B = -j\,4 \quad \mathbf{Z}_A = -j\,4 \quad \mathbf{Z}_C = 3 + j\,4$$

$$\mathbf{Z}_1 = \frac{\mathbf{Z}_B\mathbf{Z}_C}{\mathbf{Z}_A + \mathbf{Z}_B + \mathbf{Z}_C} = \frac{(-j\,4\,\Omega)(3\,\Omega + j\,4\,\Omega)}{(-j\,4\,\Omega) + (-j\,4\,\Omega) + (3\,\Omega + j\,4\,\Omega)}$$

$$= \frac{(4\,\angle -90^\circ)(5\,\angle 53.13^\circ)}{3 - j\,4} = \frac{20\,\angle -36.87^\circ}{5\,\angle -53.13^\circ}$$

$$= 4\,\Omega\,\angle 16.13^\circ = 3.84\,\Omega + j\,1.11\,\Omega$$

$$\mathbf{Z}_2 = \frac{\mathbf{Z}_A\mathbf{Z}_C}{\mathbf{Z}_A + \mathbf{Z}_B + \mathbf{Z}_C} = \frac{(-j\,4\,\Omega)(3\,\Omega + j\,4\,\Omega)}{5\,\Omega\,\angle -53.13^\circ}$$

$$= 4\,\Omega\,\angle 16.13^\circ = 3.84\,\Omega + j\,1.11\,\Omega$$

Recall from the study of dc circuits that if two branches of the Y or Δ were the same, the corresponding Δ or Y, respectively, would also have two similar branches. In this example, $\mathbf{Z}_A = \mathbf{Z}_B$. Therefore, $\mathbf{Z}_1 = \mathbf{Z}_2$, and

$$\mathbf{Z}_3 = \frac{\mathbf{Z}_A\mathbf{Z}_B}{\mathbf{Z}_A + \mathbf{Z}_B + \mathbf{Z}_C} = \frac{(-j\,4\,\Omega)(-j\,4\,\Omega)}{5\,\Omega\,\angle -53.13^\circ}$$

$$= \frac{16\,\Omega\,\angle -180^\circ}{5\,\angle -53.13^\circ} = 3.2\,\Omega\,\angle -126.87^\circ = -1.92\,\Omega - j\,2.56\,\Omega$$

continued

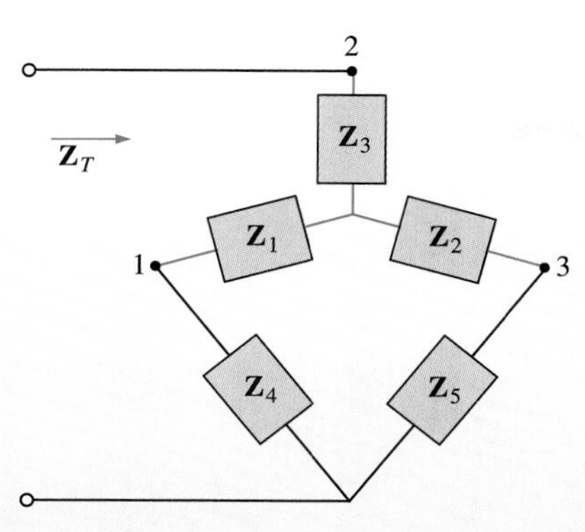

FIG. 17.41
The network of Fig. 17.38 following the substitution of the Y configuration.

Replace the Δ by the Y (Fig. 17.41):

$$\mathbf{Z}_1 = 3.84\ \Omega + j\,1.11\ \Omega \qquad \mathbf{Z}_2 = 3.84\ \Omega + j\,1.11\ \Omega$$
$$\mathbf{Z}_3 = -1.92\ \Omega - j\,2.56\ \Omega \quad \mathbf{Z}_4 = 2\ \Omega \quad \mathbf{Z}_5 = 3\ \Omega$$

Impedances $\mathbf{Z}_1$ and $\mathbf{Z}_4$ are in series:

$$\mathbf{Z}_{T_1} = \mathbf{Z}_1 + \mathbf{Z}_4 = 3.84\ \Omega + j\,1.11\ \Omega + 2\ \Omega = 5.84\ \Omega + j\,1.11\ \Omega$$
$$= 5.94\ \Omega \angle 10.76^\circ$$

Impedances $\mathbf{Z}_2$ and $\mathbf{Z}_5$ are in series:

$$\mathbf{Z}_{T_2} = \mathbf{Z}_2 + \mathbf{Z}_5 = 3.84\ \Omega + j\,1.11\ \Omega + 3\ \Omega = 6.84\ \Omega + j\,1.11\ \Omega$$
$$= 6.93\ \Omega \angle 9.22^\circ$$

Impedances $\mathbf{Z}_{T_1}$ and $\mathbf{Z}_{T_2}$ are in parallel:

$$\mathbf{Z}_{T_3} = \frac{\mathbf{Z}_{T_1}\mathbf{Z}_{T_2}}{\mathbf{Z}_{T_1} + \mathbf{Z}_{T_2}} = \frac{(5.94\ \Omega \angle 10.76^\circ)(6.93\ \Omega \angle 9.22^\circ)}{5.84\ \Omega + j\,1.11\ \Omega + 6.84\ \Omega + j\,1.11\ \Omega}$$

$$= \frac{41.16\ \Omega \angle 19.98^\circ}{12.68 + j\,2.22} = \frac{41.16\ \Omega \angle 19.98^\circ}{12.87 \angle 9.93^\circ} = 3.198\ \Omega \angle 10.05^\circ$$

$$= 3.15\ \Omega + j\,0.56\ \Omega$$

Impedances $\mathbf{Z}_3$ and $\mathbf{Z}_{T_3}$ are in series. Therefore,

$$\mathbf{Z}_T = \mathbf{Z}_3 + \mathbf{Z}_{T_3} = -1.92\ \Omega - j\,2.56\ \Omega + 3.15\ \Omega + j\,0.56\ \Omega$$
$$= 1.23\ \Omega - j\,2.0\ \Omega = \mathbf{2.35\ \Omega \angle -58.41^\circ}$$

EXAMPLE 17.21

Using both the Δ-Y and Y-Δ transformations, find the total impedance $\mathbf{Z}_T$ for the network of Fig. 17.42.

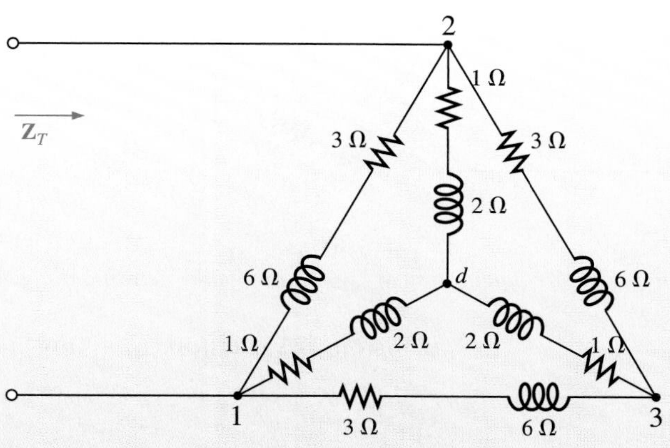

FIG. 17.42
Example 17.21.

Solution: *Using the Δ-Y transformation,* we obtain Fig. 17.43. In this case, since both systems are balanced (same impedance in each

continued

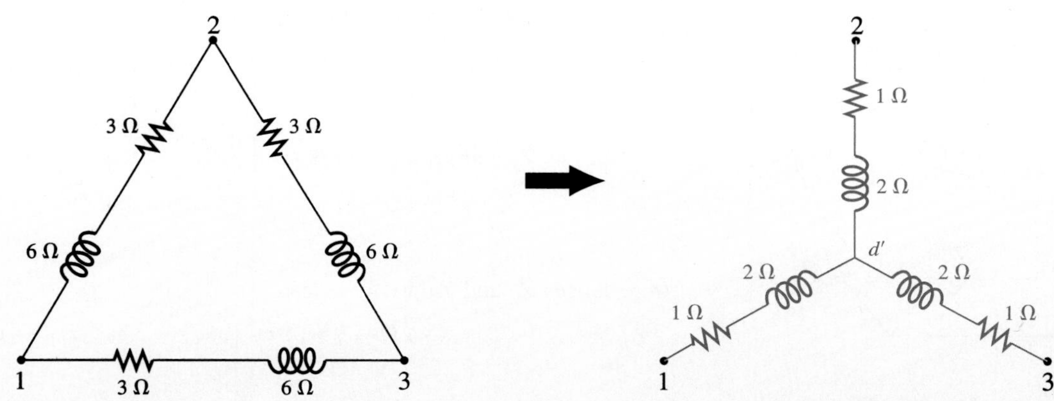

FIG. 17.43
Converting a Δ configuration to a Y configuration.

branch), the centre point d' of the transformed Δ will be the same as point d of the original Y:

$$\mathbf{Z}_{Y} = \frac{\mathbf{Z}_{\Delta}}{3} = \frac{3\ \Omega + j\,6\ \Omega}{3} = 1\ \Omega + j\,2\ \Omega$$

and (Fig. 17.44)

$$\mathbf{Z}_{T} = 2\left(\frac{1\ \Omega + j\,2\ \Omega}{2}\right) = \mathbf{1\ \Omega + j\,2\ \Omega}$$

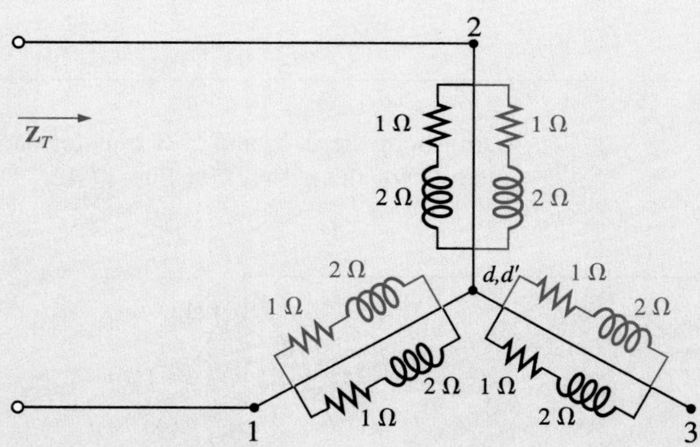

FIG. 17.44
Substituting the Y configuration of Fig. 17.43 into the network of Fig. 17.42.

Using the Y-Δ transformation (Fig. 17.45), we obtain

$$\mathbf{Z}_{\Delta} = 3\mathbf{Z}_{Y} = 3(1\ \Omega + j\,2\ \Omega) = 3\ \Omega + j\,6\ \Omega$$

continued

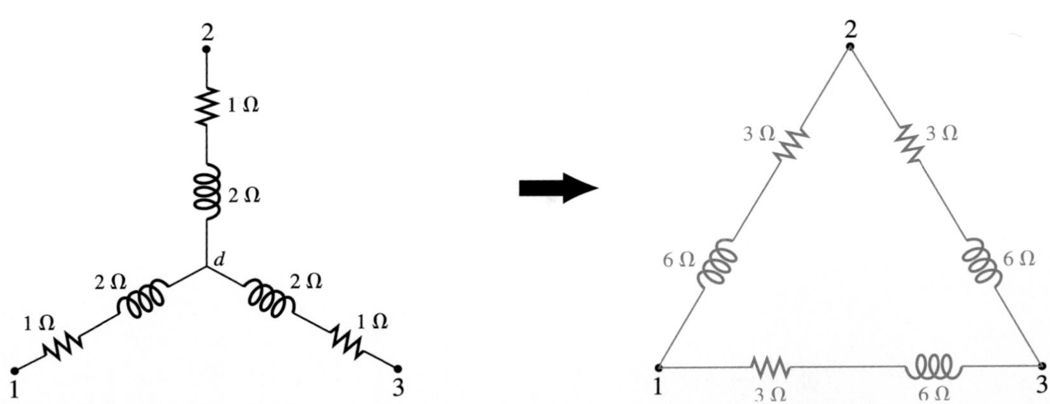

FIG. 17.45
Converting the Y configuration of Fig. 17.42 to a Δ.

Each resulting parallel combination in Fig. 17.46 will have the following impedance:

$$\mathbf{Z}' = \frac{3\ \Omega + j\,6\ \Omega}{2} = 1.5\ \Omega + j\,3\ \Omega$$

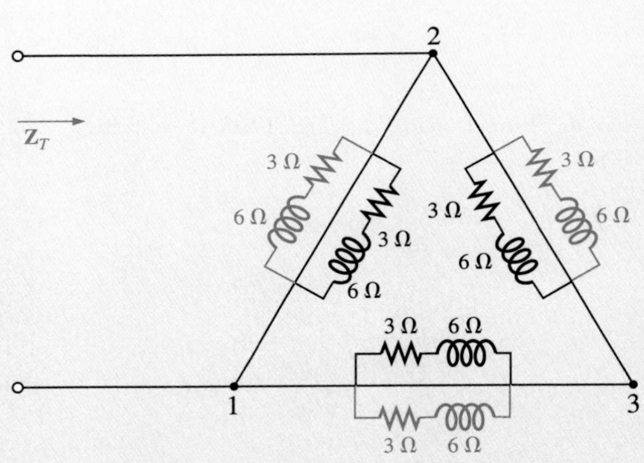

FIG. 17.46
Substituting the Δ configuration of Fig. 17.45 into the network of Fig. 17.42.

and
$$\mathbf{Z}_T = \frac{\mathbf{Z}'(2\mathbf{Z}')}{\mathbf{Z}' + 2\mathbf{Z}'} = \frac{2(\mathbf{Z}')^2}{3\mathbf{Z}'} = \frac{2\mathbf{Z}'}{3}$$
$$= \frac{2(1.5\ \Omega + j\,3\ \Omega)}{3} = \mathbf{1\ \Omega + j\,2\ \Omega}$$

which agrees with the previous result.

PROBLEMS

SECTION 17.2 Independent Versus Dependent (Controlled) Sources

1. Discuss, in your own words, the difference between a controlled and an independent source.

SECTION 17.3 Source Conversions

2. Convert the voltage sources of Fig. 17.47 to current sources.

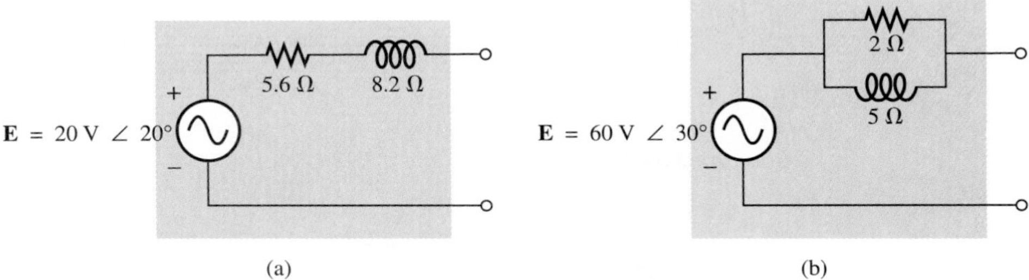

FIG. 17.47
Problem 2.

3. Convert the current sources of Fig. 17.48 to voltage sources.

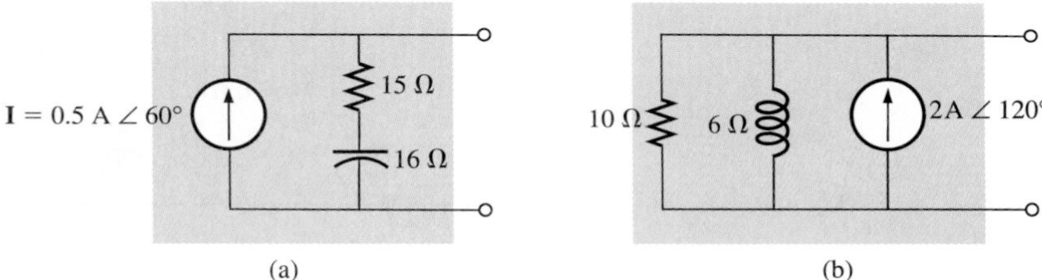

FIG. 17.48
Problem 3.

4. Convert the voltage source of Fig. 17.49(a) to a current source and the current source of Fig. 17.49(b) to a voltage source.

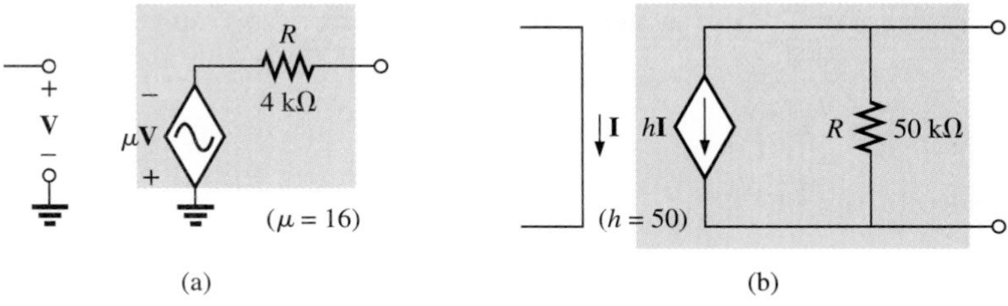

FIG. 17.49
Problem 4.

SECTION 17.4 Mesh Analysis

5. Write the mesh equations for the networks of Fig. 17.50. Determine the current through the resistor R.

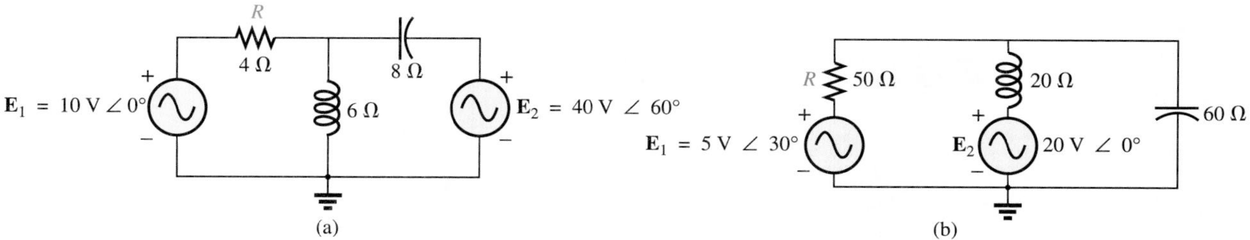

FIG. 17.50
Problem 5.

6. Write the mesh equations for the networks of Fig. 17.51. Determine the current through the resistor R_1.

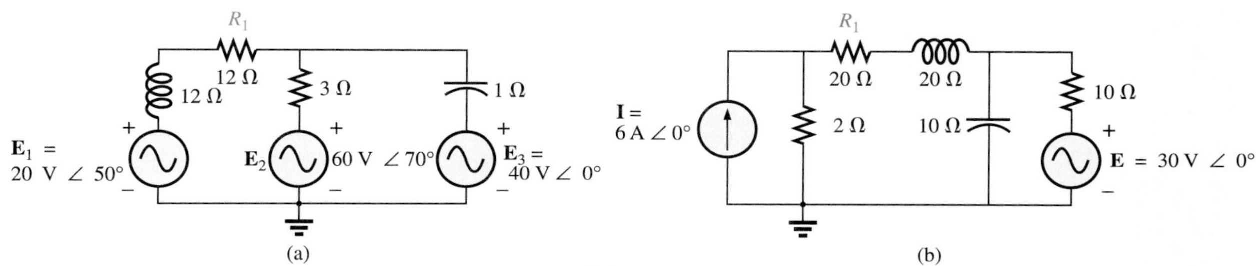

FIG. 17.51
Problems 6 and 16.

*7. Write the mesh equations for the networks of Fig. 17.52. Determine the current through the resistor R_1.

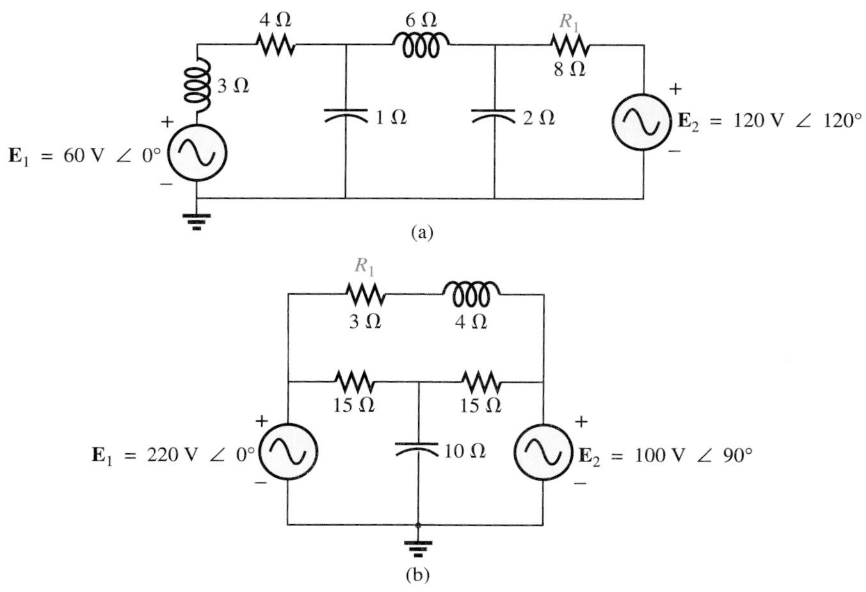

FIG. 17.52
Problems 7 and 17.

*8. Write the mesh equations for the networks of Fig. 17.53. Determine the current through the resistor R_1.

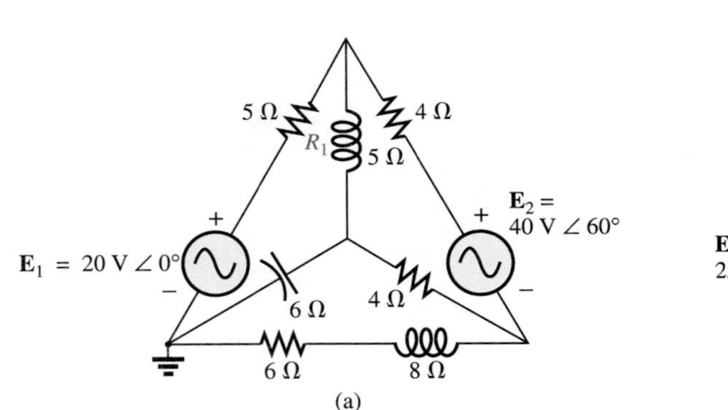

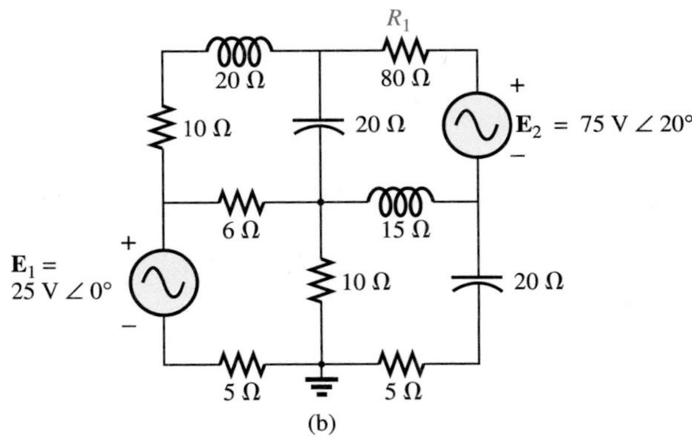

FIG. 17.53
Problems 8, 18, and 19.

9. Using mesh analysis, determine the current $\mathbf{I}_L$ (in terms of **V**) for the network of Fig. 17.54.

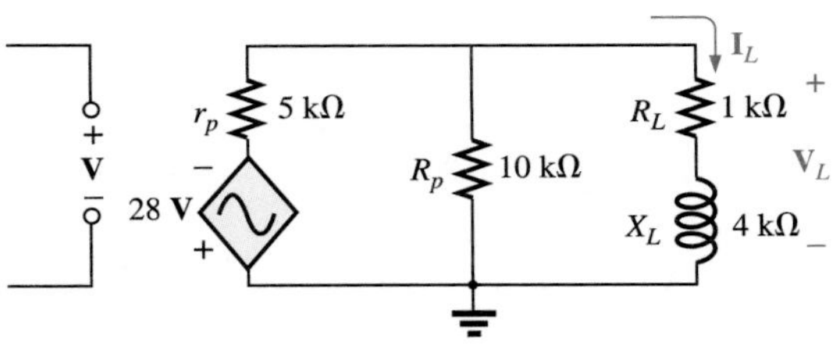

FIG. 17.54
Problem 9.

*10. Using mesh analysis, determine the current $\mathbf{I}_L$ (in terms of **I**) for the network of Fig. 17.55.

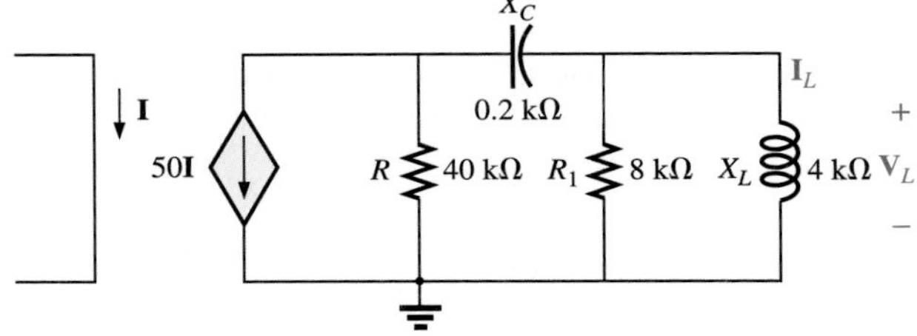

FIG. 17.55
Problem 10.

*11. Write the mesh equations for the network of Fig. 17.56 and determine the current through the 1-kΩ and 2-kΩ resistors.

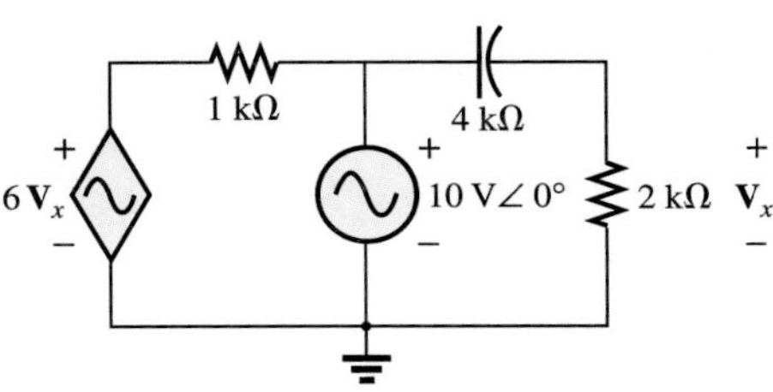

FIG. 17.56
Problem 11.

*12. Write the mesh equations for the network of Fig. 17.57 and determine the current through the 10-kΩ resistor.

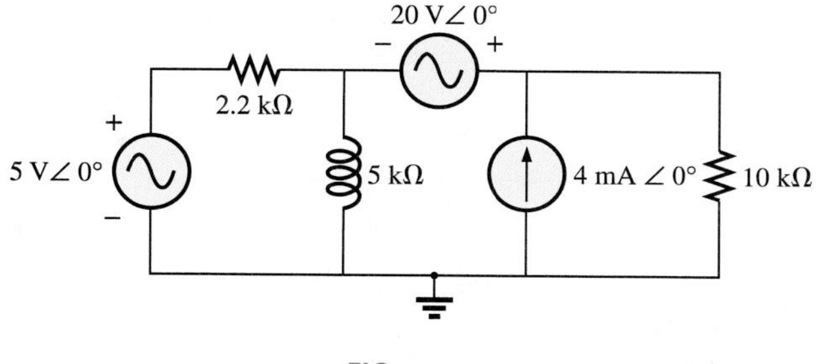

FIG. 17.57
Problem 12.

*13. Write the mesh equations for the network of Fig. 17.58 and determine the current through the inductive element.

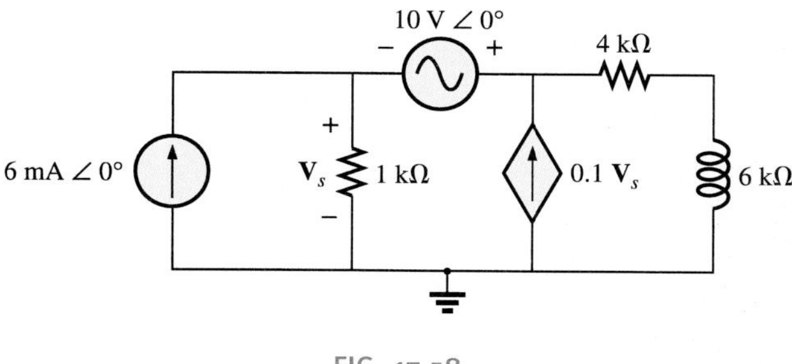

FIG. 17.58
Problem 13.

SECTION 17.5 Nodal Analysis

14. Determine the nodal voltages for the networks of Fig. 17.59.

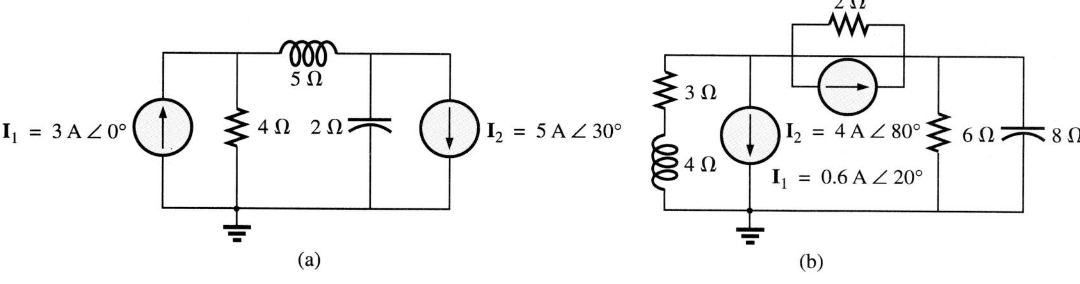

FIG. 17.59
Problem 14.

15. Determine the nodal voltages for the networks of Fig. 17.60.

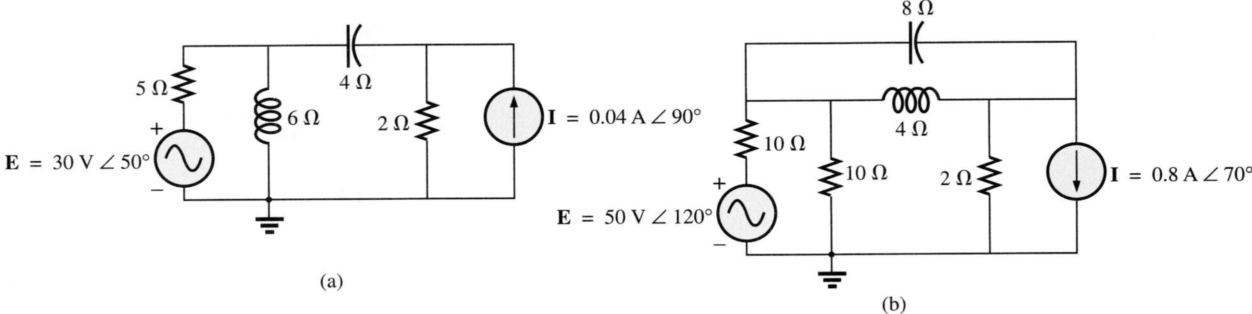

FIG. 17.60
Problem 15.

16. Determine the nodal voltages for the network of Fig. 17.51(b).

17. Determine the nodal voltages for the network of Fig. 17.52(b).

***18.** Determine the nodal voltages for the network of Fig. 17.53(a).

***19.** Determine the nodal voltages for the network of Fig. 17.53(b).

***20.** Determine the nodal voltages for the networks of Fig. 17.61.

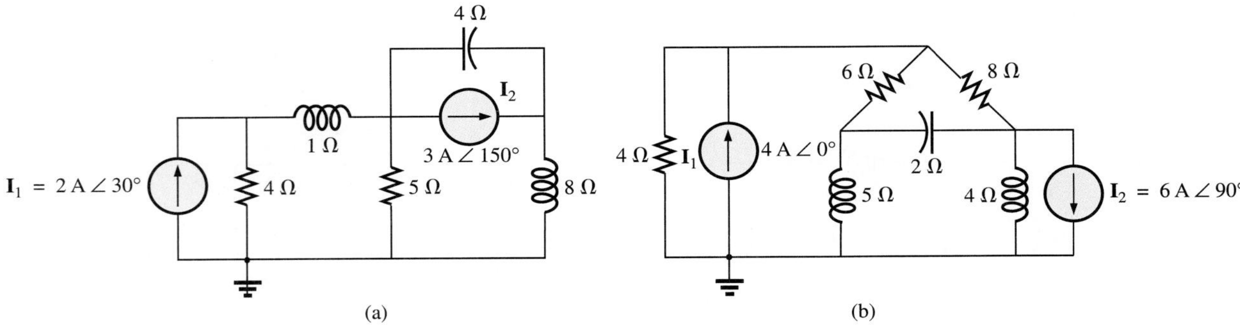

FIG. 17.61
Problem 20.

***21.** Write the nodal equations for the network of Fig. 17.62 and find the voltage across the 1-kΩ resistor.

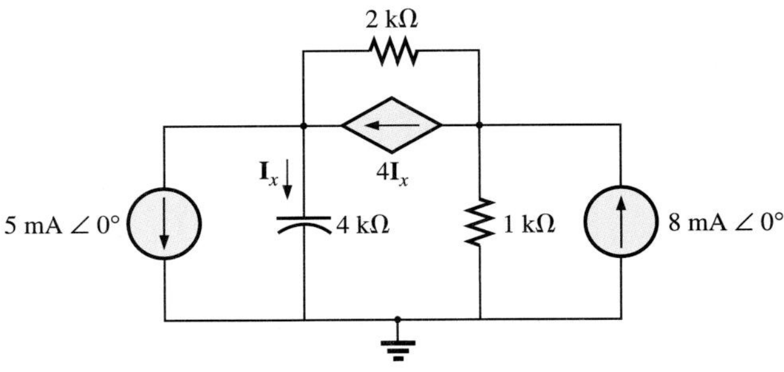

FIG. 17.62
Problem 21.

*22. Write the nodal equations for the network of Fig. 17.63 and find the voltage across the capacitive element.

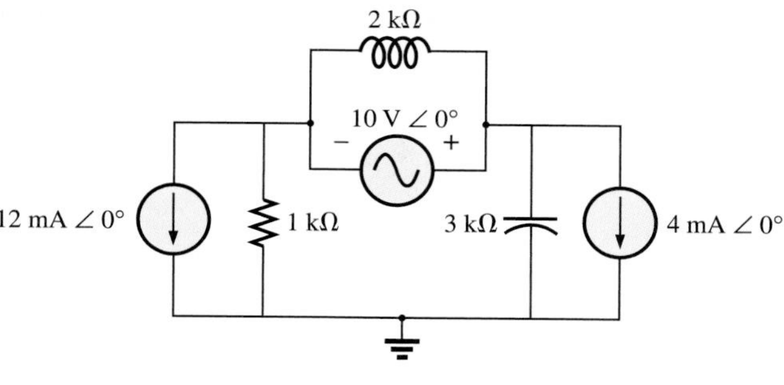

FIG. 17.63
Problem 22.

*23. Write the nodal equations for the network of Fig. 17.64 and find the voltage across the 2-kΩ resistor.

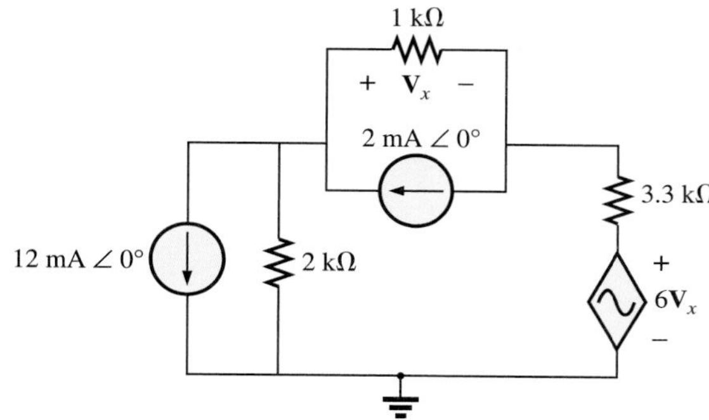

FIG. 17.64
Problem 23.

*24. Write the nodal equations for the network of Fig. 17.65 and find the voltage across the 2-kΩ resistor.

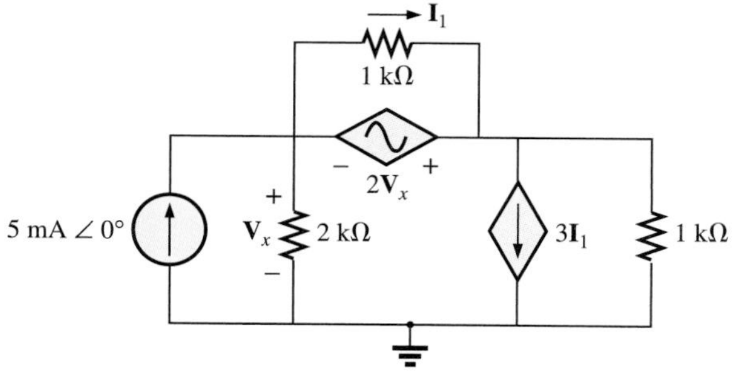

FIG. 17.65
Problem 24.

*25. For the network of Fig. 17.66, determine the voltage $\mathbf{V}_L$ in terms of the voltage $\mathbf{E}_i$.

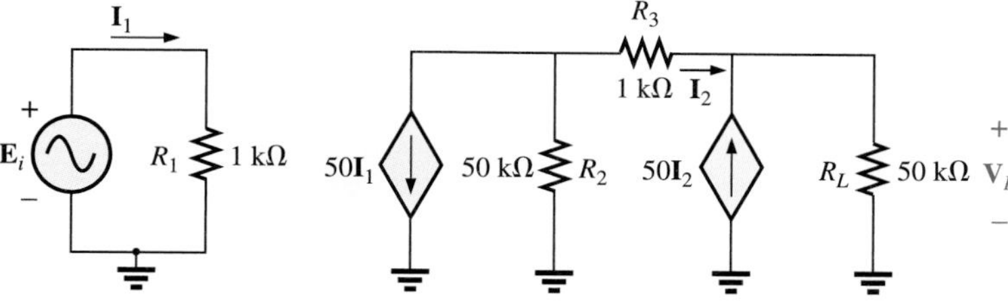

FIG. 17.66
Problem 25.

SECTION 17.6 Bridge Networks (ac)

26. For the bridge network of Fig. 17.67:

a. Is the bridge balanced?

b. Using mesh analysis, determine the current through the capacitive reactance.

c. Using nodal analysis, determine the voltage across the capacitive reactance.

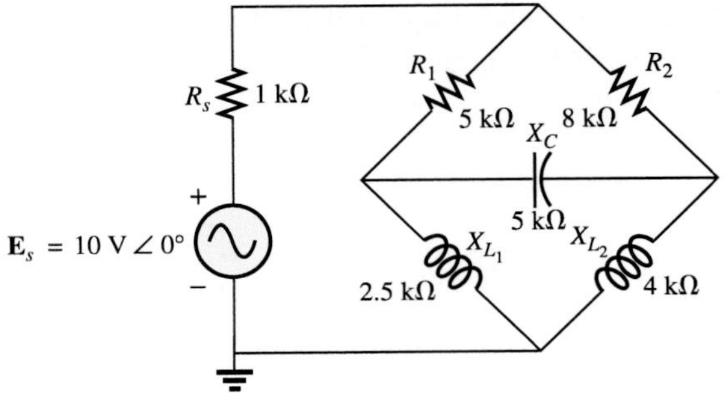

FIG. 17.67
Problem 26.

27. For the bridge network of Fig. 17.68:

a. Is the bridge balanced?

b. Using mesh analysis, determine the current through the capacitive reactance.

c. Using nodal analysis, determine the voltage across the capacitive reactance.

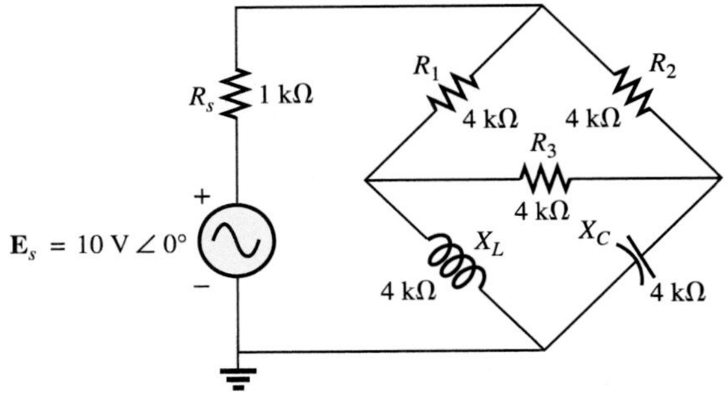

FIG. 17.68
Problem 27.

28. The Hay bridge of Fig. 17.69 is balanced. Using Eq. (17.3), determine the unknown inductance L_x and resistance R_x.

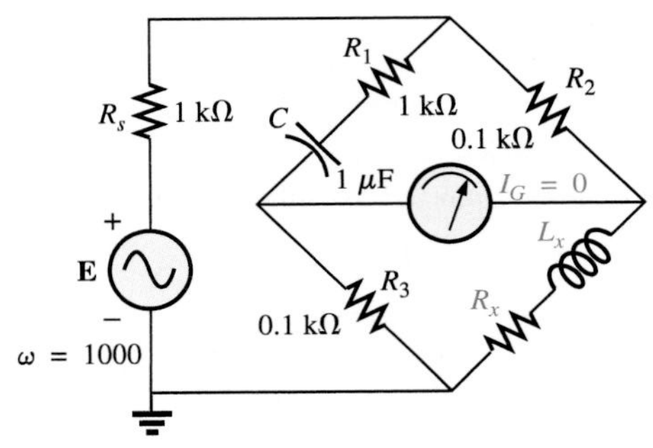

FIG. 17.69
Problem 28.

29. Determine whether the Maxwell bridge of Fig. 17.70 is balanced (ω = 1000 rad/s).

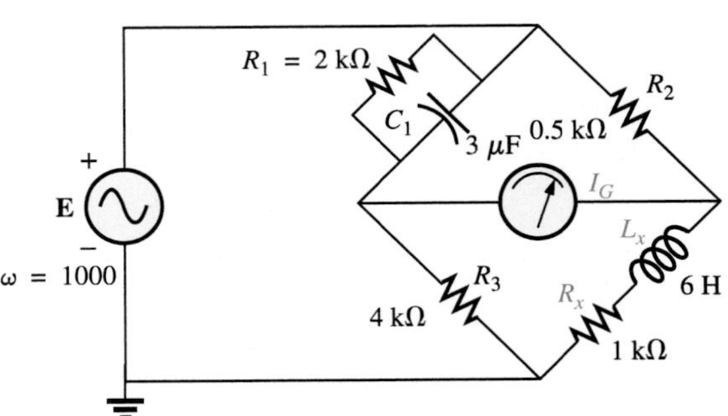

FIG. 17.70
Problem 29.

30. Derive the balance equations (17.11) and (17.12) for the capacitance comparison bridge.

31. Determine the balance equations for the inductance bridge of Fig. 17.71.

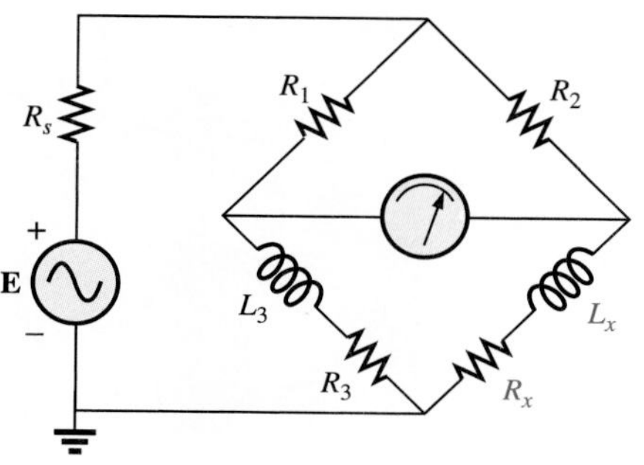

FIG. 17.71
Problem 31.

SECTION 17.7 Δ-Y, Y-Δ Conversions

32. Using the Δ-Y or Y-Δ conversion, determine the current **I** for the networks of Fig. 17.72.

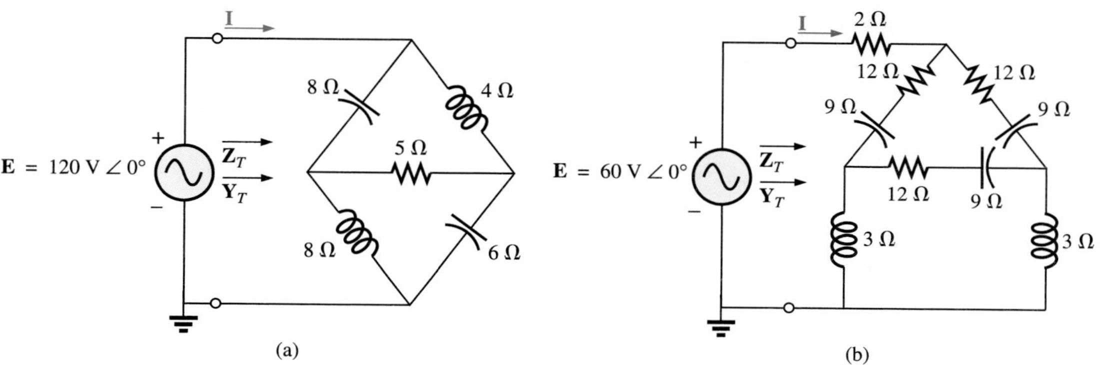

FIG. 17.72
Problem 32.

33. Using the Δ-Y or Y-Δ conversion, determine the current **I** for the networks of Fig. 17.73. (**E** = 100 V ∠0° in each case.)

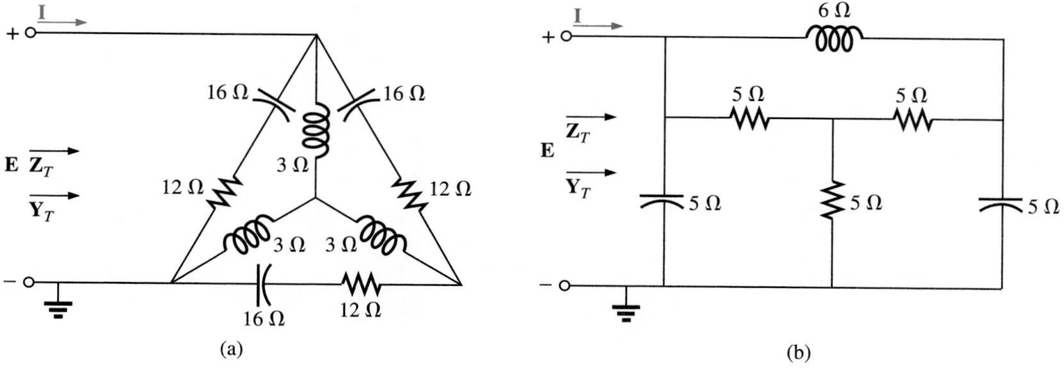

FIG. 17.73
Problem 33.

18

Network Theorems (ac)

Outline

Learning Outcomes

After completing this chapter you will be able to

- analyze ac circuits and networks with two or more sources at the same frequency
- simplify the analysis of ac circuits and networks using the theorem of superposition
- simplify the analysis of ac circuits and networks using Thévenin's theorem
- simplify the analysis of ac circuits and networks using Norton's theorem
- determine the output power obtained from ac circuits or networks using the maximum power transfer theorem

After the Great Ice Storm natural disaster in Eastern Canada power networks had to be analyzed.

18.1 INTRODUCTION

This chapter deals with **ac networks**. Its sections are given in the same order as in Chapter 9, which deals with dc networks. Reviewing each theorem in Chapter 9 before beginning this chapter will be very helpful because many of the comments there will not be repeated here.

Some sections have been divided into two parts: (a) independent sources and (b) dependent sources to help you to develop confidence in distinguishing between different types of sources, as well as in using the various theorems with ac networks. Theorems to be considered in detail include:

the superposition theorem
Thévenin's theorem
Norton's theorem
the maximum power theorem

The substitution and reciprocity theorems and Millman's theorem are not discussed in detail here because reviewing Chapter 9 will enable you to apply them to sinusoidal ac networks easily. Some problems have been included at the end of the chapter for your self study.

18.2 SUPERPOSITION THEOREM (ac)

You will recall from Chapter 9 that the superposition theorem made it unnecessary to solve simultaneous linear equations by considering the effects of each source independently. To consider the effects of each source, we had to remove the remaining sources. This was done by setting voltage sources to zero (short-circuit representation) and current sources to zero (open-circuit representation). The current through, or voltage across, a portion of the network produced by each source was then added algebraically to find the total solution for the current or voltage.

The only variation in applying this method to ac networks with independent sources is that we will now be working with impedances and phasors instead of just resistors and real numbers.

Note: The superposition theorem is not applicable to power effects in ac networks since we are still dealing with a nonlinear relationship. It can be applied to networks with sources of different frequencies only if the total response for *each* frequency is found independently and the results are expanded in a nonsinusoidal expression.

The superposition theorem is often applied to electronic systems where the dc and ac analyses are treated separately; the total solution is then the sum of the two. This is an important application of the superposition theorem because the impact of the reactive elements changes dramatically in response to the two types of independent sources. In addition, the dc analysis of an electronic system often defines important parameters for the ac analysis.

Independent Sources (superposition theorem)

Examples 18.1 through 18.3 examine networks using only independent voltage or current sources. They follow the same order as the dc analysis in Chapter 9. Example 18.4 demonstrates the impact of the applied source on the general configuration of the network.

EXAMPLE 18.1

Using the superposition theorem, find the current **I** through the 4-Ω reactance (X_{L_2}) of Fig. 18.1(a).

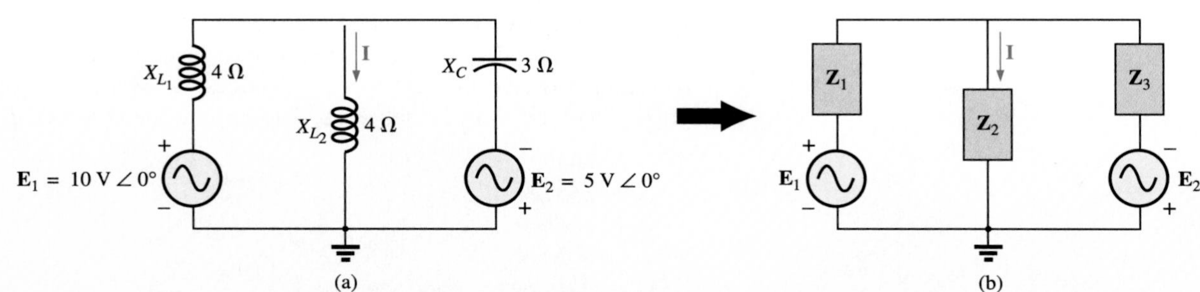

FIG. 18.1
Example 18.1.

Solution: For the redrawn circuit (Fig. 18.1(b)),

$$\mathbf{Z}_1 = +j\,X_{L_1} = j\,4\ \Omega$$
$$\mathbf{Z}_2 = +j\,X_{L_2} = j\,4\ \Omega$$
$$\mathbf{Z}_3 = -j\,X_C = -j\,3\ \Omega$$

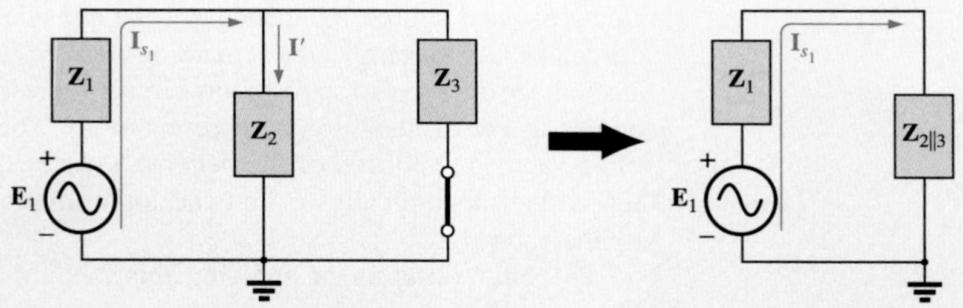

FIG. 18.2
Determining the effect of the voltage source $\mathbf{E}_1$ on the current **I** *of the network of Fig. 18.1.*

Considering the effects of the voltage source $\mathbf{E}_1$ (Fig. 18.2), we have

$$\mathbf{Z}_{2\|3} = \frac{\mathbf{Z}_2\mathbf{Z}_3}{\mathbf{Z}_2 + \mathbf{Z}_3} = \frac{(j\,4\ \Omega)(-j\,3\ \Omega)}{j\,4\ \Omega - j\,3\ \Omega} = \frac{12\ \Omega}{j} = -j\,12\ \Omega$$
$$= 12\ \Omega\ \angle -90^\circ$$

$$I_{s_1} = \frac{\mathbf{E}_1}{\mathbf{Z}_{2\|3} + \mathbf{Z}_1} = \frac{10\ \text{V}\ \angle 0^\circ}{-j\,12\ \Omega + j\,4\ \Omega} = \frac{10\ \text{V}\ \angle 0^\circ}{8\ \Omega\ \angle -90^\circ}$$
$$= 1.25\ \text{A}\ \angle 90^\circ$$

and

$$\mathbf{I}' = \frac{\mathbf{Z}_3\mathbf{I}_{s_1}}{\mathbf{Z}_2 + \mathbf{Z}_3} \quad \text{(current divider rule)}$$

$$= \frac{(-j\,3\ \Omega\,)(\,j\,1.25\ \text{A})}{j\,4\ \Omega - j\,3\ \Omega} = \frac{3.75\ \text{A}}{j\,1} = 3.75\ \text{A}\ \angle -90^\circ$$

continued

Considering the effects of the voltage source $\mathbf{E}_2$ (Fig. 18.3), we have

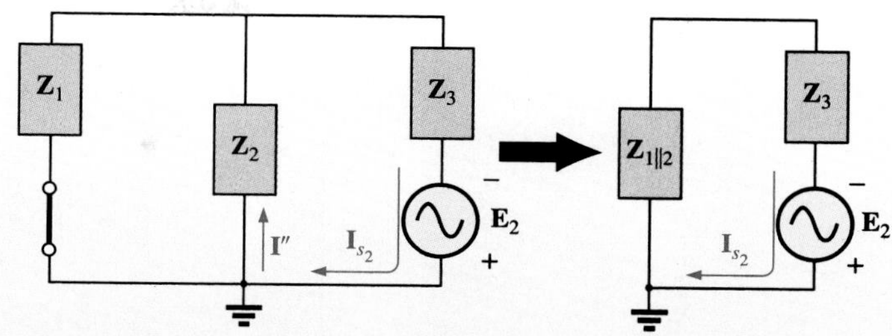

FIG. 18.3
Determining the effect of the voltage source $\mathbf{E}_2$ on the current $\mathbf{I}$ of the network of Fig. 18.1.

$$\mathbf{Z}_{1\|2} = \frac{\mathbf{Z}_1}{N} = \frac{j\,4\,\Omega}{2} = j\,2\,\Omega$$

$$\mathbf{I}_{s_2} = \frac{\mathbf{E}_2}{\mathbf{Z}_{1\|2} + \mathbf{Z}_3} = \frac{5\text{ V}\angle 0^\circ}{j\,2\,\Omega - j\,3\,\Omega} = \frac{5\text{ V}\angle 0^\circ}{1\,\Omega\angle -90^\circ} = 5\text{ A}\angle 90^\circ$$

and
$$\mathbf{I}'' = \frac{\mathbf{I}_{s_2}}{2} = 2.5\text{ A}\angle 90^\circ$$

The resulting current through the 4-Ω reactance X_{L_2} (Fig. 18.4) is

$$\begin{aligned}\mathbf{I} &= \mathbf{I}' - \mathbf{I}''\\ &= 3.75\text{ A}\angle -90^\circ - 2.50\text{ A}\angle 90^\circ = -j\,3.75\text{ A} - j\,2.50\text{ A}\\ &= -j\,6.25\text{ A}\\ \mathbf{I} &= \mathbf{6.25\text{ A}\angle -90^\circ}\end{aligned}$$

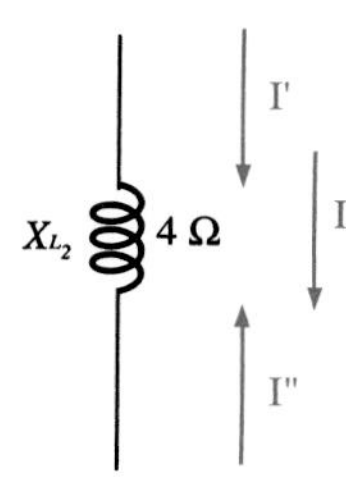

FIG. 18.4
Determining the resulting current for the network of Fig. 18.1.

EXAMPLE 18.2

Using superposition, find the current $\mathbf{I}$ through the 6-Ω resistor of Fig. 18.5(a).

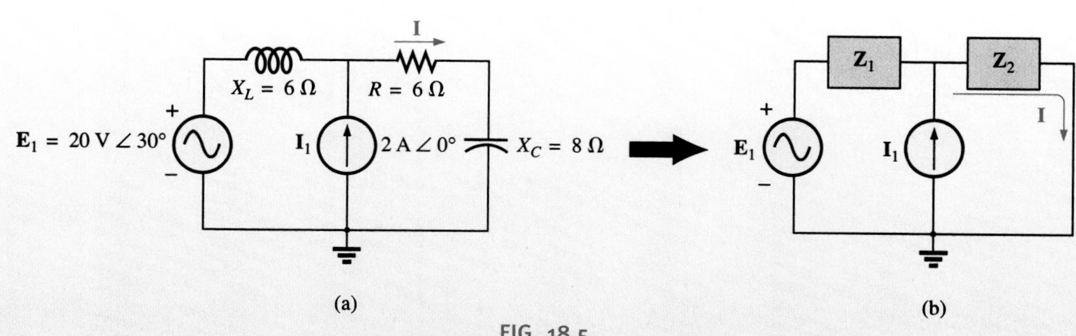

FIG. 18.5
Examples 18.2 and 18.3.

Solution: For the redrawn circuit (Fig. 18.5(b)),

$$\mathbf{Z}_1 = j\,6\,\Omega \quad \mathbf{Z}_2 = 6 - j\,8\,\Omega$$

continued

Consider the effects of the current source [Fig. 18.6(a)]. Applying the current divider rule, we have

$$\mathbf{I}' = \frac{\mathbf{Z}_1\mathbf{I}_1}{\mathbf{Z}_1 + \mathbf{Z}_2} = \frac{(j\,6\,\Omega)(2\,\text{A})}{j\,6\,\Omega + 6\,\Omega - j\,8\,\Omega} = \frac{j\,12\,\text{A}}{6 - j\,2}$$

$$= \frac{12\,\text{A}\,\angle 90^\circ}{6.32\,\angle -18.43^\circ}$$

$$\mathbf{I}' = 1.9\,\text{A}\,\angle 108.43^\circ$$

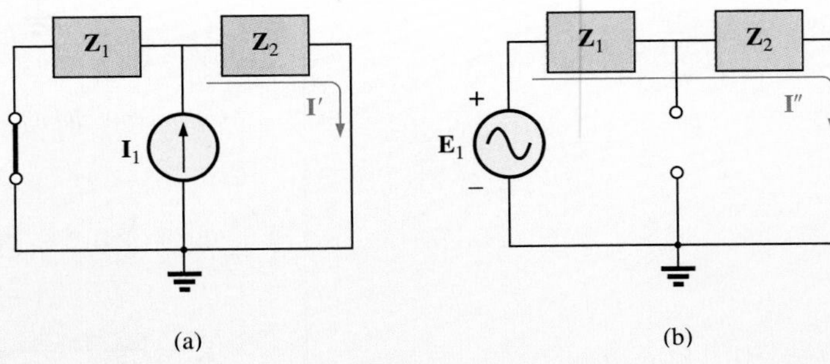

FIG. 18.6
(a) Determining the effect of the current source $\mathbf{I}_1$ *on the current* **I** *of the network of Fig. 18.5.*
(b) Determining the effect of the voltage source $\mathbf{E}_1$ *on the current* **I** *of the network of Fig. 18.5.*

Consider the effects of the voltage source [Fig. 18.6(b)]. Applying Ohm's law gives us

$$\mathbf{I}'' = \frac{\mathbf{E}_1}{\mathbf{Z}_T} = \frac{\mathbf{E}_1}{\mathbf{Z}_1 + \mathbf{Z}_2} = \frac{20\,\text{V}\,\angle 30^\circ}{6.32\,\Omega\,\angle -18.43^\circ}$$

$$= 3.16\,\text{A}\,\angle 48.43^\circ$$

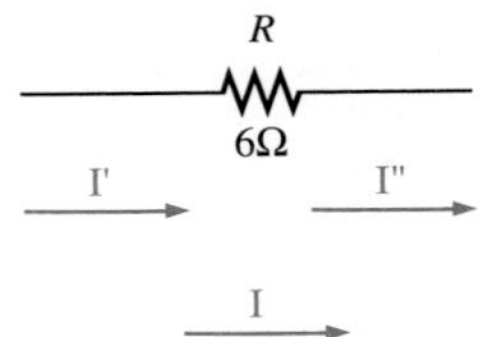

FIG. 18.7
Determining the resulting current **I** *for the network of Fig. 18.5.*

The total current through the 6-Ω resistor (Fig. 18.7) is

$$\mathbf{I} = \mathbf{I}' + \mathbf{I}''$$

$$= 1.9\,\text{A}\,\angle 108.43^\circ + 3.16\,\text{A}\,\angle 48.43^\circ$$

$$= (-0.60\,\text{A} + j\,1.80\,\text{A}) + (2.10\,\text{A} + j\,2.36\,\text{A})$$

$$= 1.50\,\text{A} + j\,4.16\,\text{A}$$

$$\mathbf{I} = \mathbf{4.42\,A\,\angle 70.2^\circ}$$

EXAMPLE 18.3

Using superposition, find the voltage across the 6-Ω resistor in Fig. 18.5. Check the results against $\mathbf{V}_{6\Omega} = \mathbf{I}(6\,\Omega)$, where **I** is the current found through the 6-Ω resistor in the previous example.

continued

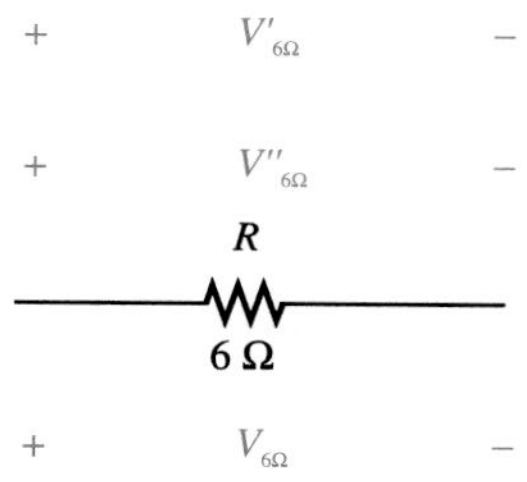

Solution: For the current source,

$$\mathbf{V}'_{6\Omega} = \mathbf{I}'(6\ \Omega) = (1.9\ \text{A}\ \angle 108.43°)(6\ \Omega) = 11.4\ \text{V}\ \angle 108.43°$$

For the voltage source,

$$\mathbf{V}''_{6\Omega} = \mathbf{I}''(6) = (3.16\ \text{A}\ \angle 48.43°)(6\ \Omega) = 18.96\ \text{V}\ \angle 48.43°$$

The total voltage across the 6-Ω resistor (Fig. 18.8) is

$$\begin{aligned}\mathbf{V}_{6\Omega} &= \mathbf{V}'_{6\Omega} + \mathbf{V}''_{6\Omega}\\ &= 11.4\ \text{V}\ \angle 108.43° + 18.96\ \text{V}\ \angle 48.43°\\ &= (-3.60\ \text{V} + j\,10.82\ \text{V}) + (12.58\ \text{V} + j\,14.18\ \text{V})\\ &= 8.98\ \text{V} + j\,25.0\ \text{V}\\ \mathbf{V}_{6\Omega} &= \mathbf{26.5\ V}\ \angle\mathbf{70.2°}\end{aligned}$$

Checking the result, we have

$$\begin{aligned}\mathbf{V}_{6\Omega} &= \mathbf{I}(6\ \Omega) = (4.42\ \text{A}\ \angle 70.2°)(6\ \Omega)\\ &= \mathbf{26.5\ V}\ \angle\mathbf{70.2°} \quad \text{(checks)}\end{aligned}$$

FIG. 18.8
Determining the resulting voltage $\mathbf{V}_{6\Omega}$ for the network of Fig. 18.5.

EXAMPLE 18.4

For the network of Fig. 18.9, determine the sinusoidal expression for the voltage v_3 using superposition.

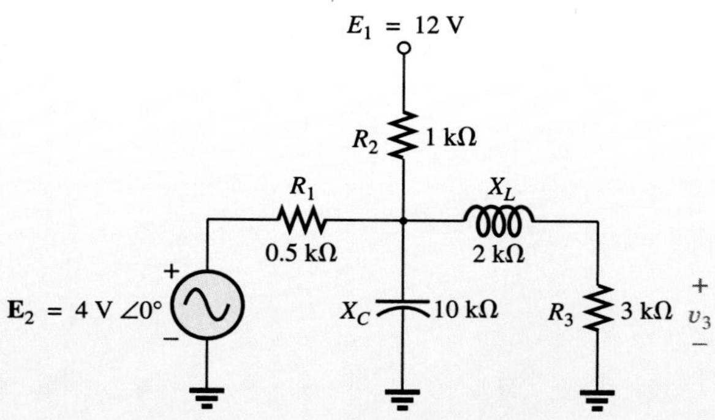

FIG. 18.9
Example 18.4.

Solution:

dc analysis: For the dc source, recall that for dc analysis, in the steady state the capacitor can be replaced by an open-circuit equivalent and the inductor by a short-circuit equivalent. The result is the network of Fig. 18.10.

The resistors R_1 and R_3 are then in parallel and the voltage V_3 can be determined using the voltage divider rule:

$$R' = R_1 \parallel R_3 = 0.5\ \text{k}\Omega \parallel 3\ \text{k}\Omega = 0.429\ \text{k}\Omega$$

and
$$\begin{aligned}V_3 &= \frac{R'E_1}{R' + R_2}\\ &= \frac{(0.429\ \text{k}\Omega)(12\ \text{V})}{0.429\ \text{k}\Omega + 1\ \text{k}\Omega} = \frac{5.148\ \text{V}}{1.429}\\ V_3 &\cong \mathbf{3.6\ V}\end{aligned}$$

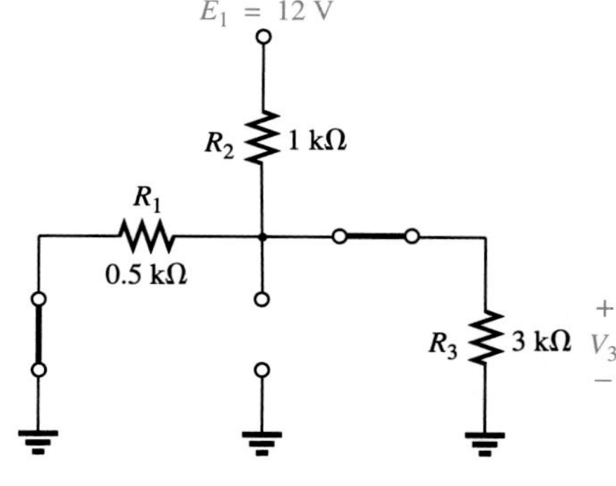

FIG. 18.10
Determining the effect of the dc voltage source E_1 on the voltage v_3 of the network of Fig. 18.9.

continued

ac analysis: The dc source is set to zero and the network is redrawn, as shown in Fig. 18.11(a).

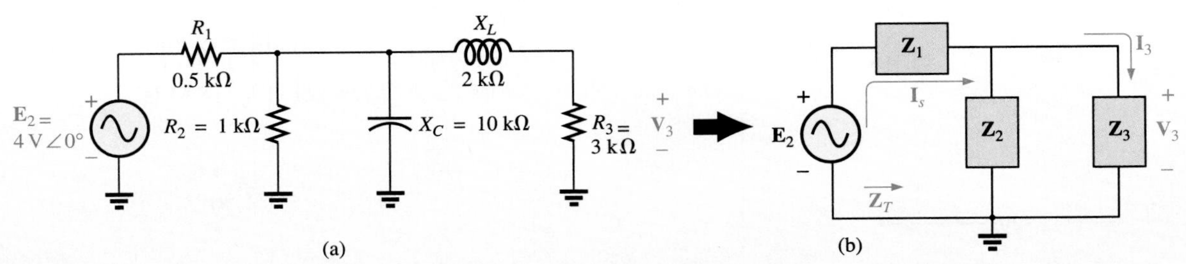

FIG. 18.11
(a) Redrawing the network of Fig. 18.9 to determine the effect of the ac voltage source $\mathbf{E}_2$*.*
(b) Assigning the subscripted impedances to the network.

The block impedances are then defined as in Fig. 18.11(b) and series-parallel techniques are applied as follows:

$$\mathbf{Z}_1 = 0.5\ \text{k}\Omega\ \angle 0^\circ$$

$$\mathbf{Z}_2 = (R_2\ \angle 0^\circ \parallel (X_C\ \angle -90^\circ)$$

$$= \frac{(1\ \text{k}\Omega\ \angle 0^\circ)(10\ \text{k}\Omega\ \angle -90^\circ)}{1\ \text{k}\Omega - j\,10\ \text{k}\Omega} = \frac{10\ \text{k}\Omega\ \angle -90^\circ}{10.05\ \angle -84.29^\circ}$$

$$= 0.995\ \text{k}\Omega\ \angle -5.71^\circ$$

$$\mathbf{Z}_3 = R_3 + j\,X_L = 3\ \text{k}\Omega + j\,2\ \text{k}\Omega = 3.61\ \text{k}\Omega\ \angle 33.69^\circ$$

and

$$\mathbf{Z}_T = \mathbf{Z}_1 + \mathbf{Z}_2 \parallel \mathbf{Z}_3$$

$$= 0.5\ \text{k}\Omega + (0.995\ \text{k}\Omega\ \angle -5.71^\circ) \parallel (3.61\ \text{k}\Omega\ \angle 33.69^\circ)$$

$$= 1.312\ \text{k}\Omega\ \angle 1.57^\circ$$

$$\mathbf{I}_s = \frac{\mathbf{E}_2}{\mathbf{Z}_T} = \frac{4\ \text{V}\ \angle 0^\circ}{1.312\ \text{k}\Omega\ \angle 1.57^\circ} = 3.05\ \text{mA}\ \angle -1.57^\circ$$

Current divider rule:

$$\mathbf{I}_3 = \frac{\mathbf{Z}_2\mathbf{I}_s}{\mathbf{Z}_2 + \mathbf{Z}_3} = \frac{(0.995\ \text{k}\Omega\ \angle -5.71^\circ)(3.05\ \text{mA}\ \angle -1.57^\circ)}{0.995\ \text{k}\Omega\ \angle -5.71^\circ + 3.61\ \text{k}\Omega\ \angle 33.69^\circ}$$

$$= 0.686\ \text{mA}\ \angle -32.74^\circ$$

with

$$\mathbf{V}_3 = (I_3\ \angle \theta)(R_3\ \angle 0^\circ)$$

$$= (0.686\ \text{mA}\ \angle -32.74^\circ)(3\ \text{k}\Omega\ \angle 0^\circ)$$

$$= \mathbf{2.06\ V\ \angle -32.74^\circ}$$

The total solution:

$$v_3 = v_3\,(\text{dc}) + v_3\,(\text{ac})$$

$$= 3.6\ \text{V} + 2.06\ \text{V}\ \angle -32.74^\circ$$

$$v_3 = \mathbf{3.6 + 2.91\,\sin(\omega t - 32.74^\circ)}$$

The result is a sinusoidal voltage having a peak value of 2.91 V riding on an average value of 3.6 V, as shown in Fig. 18.12.

continued

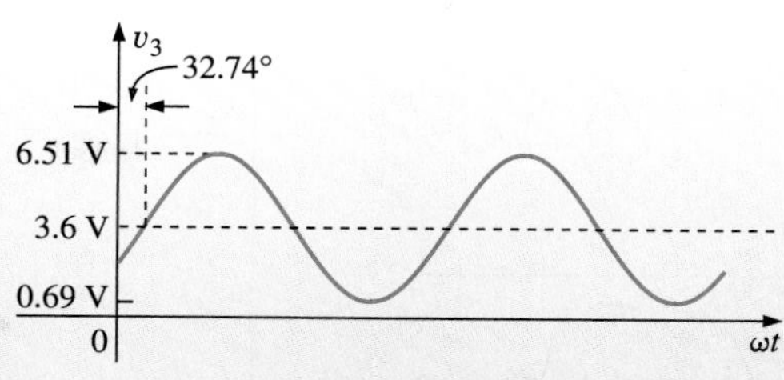

FIG. 18.12
The resulting voltage v_3 for the network of Fig. 18.9.

Dependent Sources (superposition theorem)

There are two cases to consider for **dependent sources**:

Case 1: Where *the controlling variable is* **not** *determined by the network to which the superposition theorem is to be applied.* For Case 1, the theorem is applied in basically the same way as for independent sources. Example 18.5 illustrates that the solution obtained will simply be in terms of the controlling variables.

EXAMPLE 18.5

Using the superposition theorem, determine the current $\mathbf{I}_2$ for the network of Fig. 18.13(a). The quantities μ and h are constants.

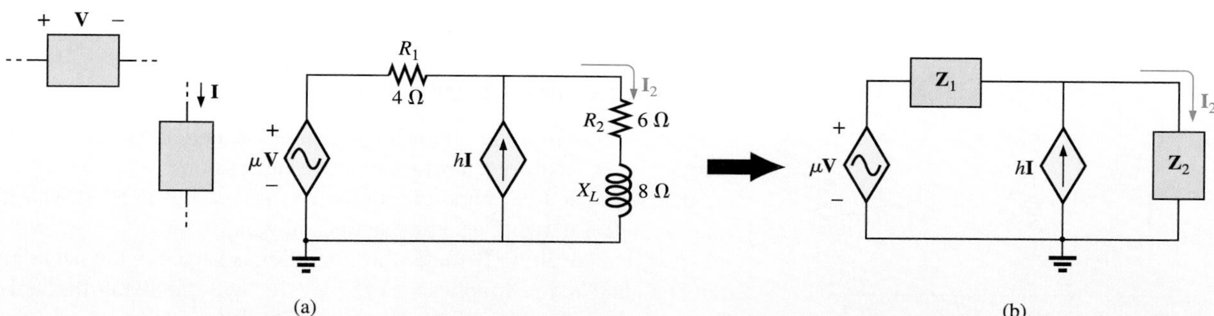

FIG. 18.13
Example 18.5.

Solution: With a portion of the system redrawn [Fig. 18.13(b)],

$$\mathbf{Z}_1 = R_1 = 4\ \Omega \quad \mathbf{Z}_2 = R_2 + jX_L = 6 + j\,8\ \Omega$$

For the voltage source (Fig. 18.14),

$$\mathbf{I}' = \frac{\mu\mathbf{V}}{\mathbf{Z}_1 + \mathbf{Z}_2} = \frac{\mu\mathbf{V}}{4\ \Omega + 6\ \Omega + j\,8\ \Omega} = \frac{\mu\mathbf{V}}{10\ \Omega + j\,8\ \Omega}$$

$$= \frac{\mu\mathbf{V}}{12.8\ \Omega\ \angle 38.66^\circ} = 0.078\ \mu\mathbf{V}/\Omega\ \angle -38.66^\circ$$

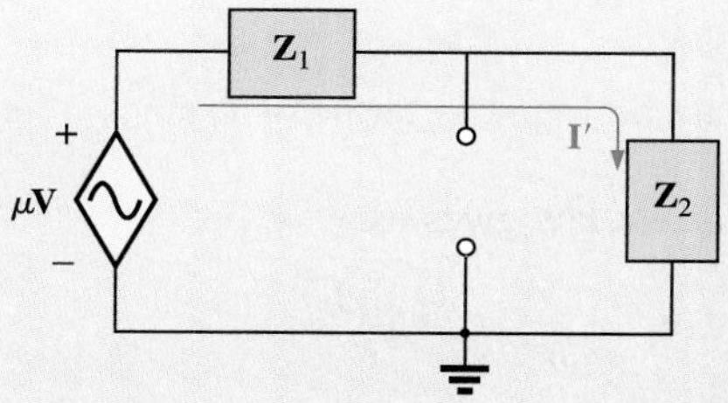

FIG. 18.14
Determining the effect of the voltage-controlled voltage source on the current $\mathbf{I}_2$ for the network of Fig. 18.13.

For the current source (Fig. 18.15),

$$\mathbf{I}'' = \frac{\mathbf{Z}_1(h\mathbf{I})}{\mathbf{Z}_1 + \mathbf{Z}_2} = \frac{(4\ \Omega)(h\mathbf{I})}{12.8\ \Omega\ \angle 38.66^\circ} = 4(0.078)h\mathbf{I}\ \angle -38.66^\circ$$

$$= 0.312h\mathbf{I}\ \angle -38.66^\circ$$

continued

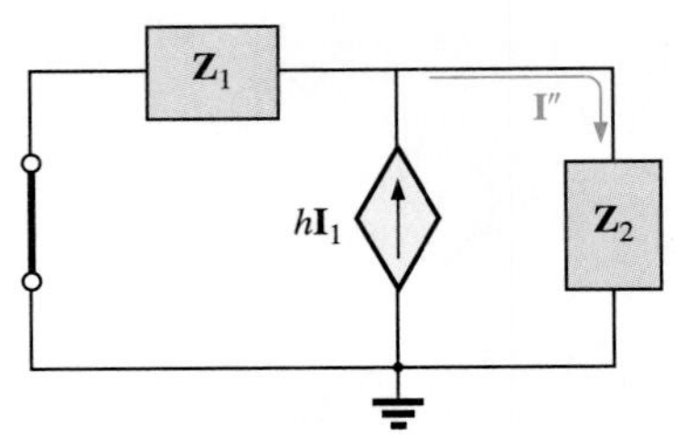

FIG. 18.15
Determining the effect of the current-controlled current source on the current $\mathbf{I}_2$ for the network of Fig. 18.13.

The current $\mathbf{I}_2$ is

$$\mathbf{I}_2 = \mathbf{I}' + \mathbf{I}'' = 0.078\ \mu\mathbf{V}/\Omega\ \angle -38.66^\circ + 0.312h\mathbf{I}\ \angle \mathbf{-38.66^\circ}$$

For $\mathbf{V} = 10\text{ V}\ \angle 0^\circ$, $\mathbf{I} = 20\text{ mA}\ \angle 0^\circ$, $\mu = 20$, $h = 100$,

$$\mathbf{I}_2 = 0.078(20)(10\text{ V}\ \angle 0^\circ)/\Omega\ \angle -38.66^\circ + 0.312(100)(20\text{ mA}\ \angle 0^\circ)\ \angle -38.66^\circ$$
$$= 15.60\text{ A}\ \angle -38.66^\circ + 0.62\text{ A}\ \angle -38.66^\circ$$
$$\mathbf{I}_2 = \mathbf{16.22\ A\ \angle -38.66^\circ}$$

Case 2: Where *the controlling variable* **is** *determined by the network to which the superposition theorem is to be applied.* In this case, the dependent source cannot be set to zero unless the controlling variable is also zero. Example 18.6 illustrates this case where the superposition theorem is applied for each independent source and each dependent source not having a controlling variable in the portions of the network under investigation.

Recall from the discussion in Chapter 17 that:

dependent sources are not actually sources of energy in the sense that if all independent sources are removed from a system, then all currents and voltages must be zero.

EXAMPLE 18.6

Determine the current $\mathbf{I}_L$ through the resistor R_L of Fig. 18.16.

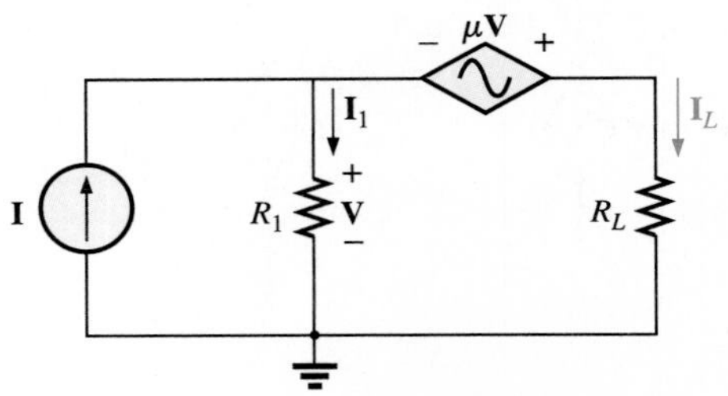

FIG. 18.16
Example 18.6.

Solution: Note that the controlling variable **V** is determined by the network to be analyzed. From the above discussions, you know that the dependent source cannot be set to zero unless **V** is zero. If we set **I** to zero, the network lacks a source of voltage, and $\mathbf{V} = 0$ with $\mu\mathbf{V} = 0$. The resulting $\mathbf{I}_L$ under this condition is zero. So, the network must be analyzed as it appears in Fig. 18.16, with the result that neither source can be eliminated, as is normally done using the superposition theorem.

Applying Kirchhoff's voltage law, we have

$$\mathbf{V}_L = \mathbf{V} + \mu\mathbf{V} = (1 + \mu)\mathbf{V}$$

and
$$\mathbf{I}_L = \frac{\mathbf{V}_L}{R_L} = \frac{(1 + \mu)\mathbf{V}}{R_L}$$

The result, however, must be found in terms of **I**, since **V** and $\mu\mathbf{V}$ are only dependent variables.

Applying Kirchhoff's current law gives us

$$\mathbf{I} = \mathbf{I}_1 + \mathbf{I}_L = \frac{\mathbf{V}}{R_1} + \frac{(1 + \mu)\mathbf{V}}{R_L}$$

and
$$\mathbf{I} = \mathbf{V}\left(\frac{1}{R_1} + \frac{1 + \mu}{R_L}\right)$$

or
$$\mathbf{V} = \frac{\mathbf{I}}{(1/R_1) + [(1 + \mu)/R_L]}$$

continued

Substituting into the above results in

$$\mathbf{I}_L = \frac{(1+\mu)\mathbf{V}}{R_L} = \frac{(1+\mu)}{R_L}\left(\frac{\mathbf{I}}{(1/R_1) + [(1+\mu)/R_L]}\right)$$

Therefore,
$$\mathbf{I}_L = \frac{(1+\mu)R_1\mathbf{I}}{R_L + (1+\mu)R_1}$$

18.3 THÉVENIN'S THEOREM (ac)

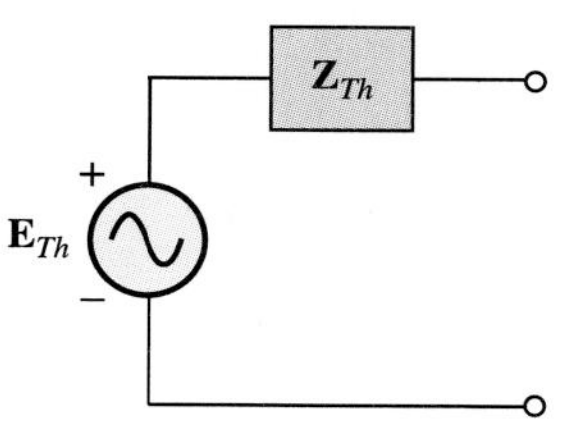

FIG. 18.17
Thévenin equivalent circuit for ac networks.

Thévenin's theorem, as stated for sinusoidal ac circuits, is changed only to include the term *impedance* instead of *resistance*; that is,

any two-terminal linear ac network can be replaced with an equivalent circuit consisting of a voltage source and an impedance in series, as shown in Fig. 18.17.

Caution: Since the reactances of a circuit are often dependent, the Thévenin circuit found for a particular network is applicable only at *one* frequency.

As in Section 18.2, independent and dependent sources will be treated separately.

Independent Sources (Thévenin's theorem)

The following steps are repeated here for convenience. They are essentially same as in Section 9.3. The only change is to replace the term *resistance* with *impedance.* Example 18.7 applies these steps to a Thévenin equivalent circuit for a network with an external *resistor.* Example 18.8 applies the steps to a network with an external *power supply.*

1. Remove the portion of the network across which the Thévenin equivalent circuit is to be found.
2. Mark (○, •, and so on) the terminals of the remaining two-terminal network.
3. Calculate Z_{Th} by first setting all voltage and current sources to zero (short circuit and open circuit, respectively) and then finding the resulting impedance between the two marked terminals.
4. Calculate E_{Th} by first replacing the voltage and current sources and then finding the open-circuit voltage between the marked terminals.
5. Draw the Thévenin equivalent circuit with the portion of the circuit previously removed replaced between the terminals of the Thévenin equivalent circuit.

EXAMPLE 18.7

Find the Thévenin equivalent circuit for the network external to resistor R in Fig. 18.18.

continued

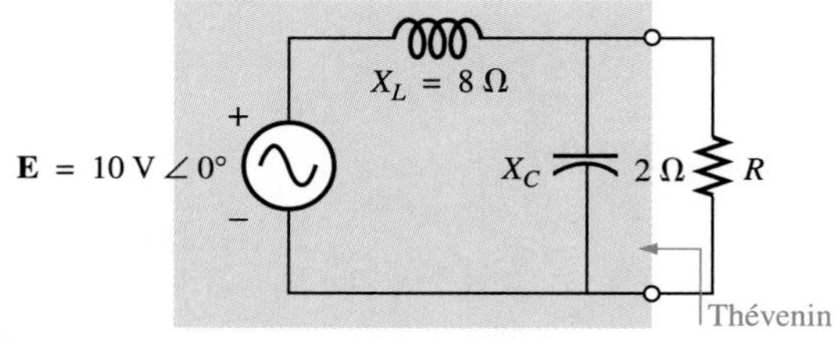

FIG. 18.18
Example 18.7.

Solution:

Steps 1 and 2 (Fig. 18.19):

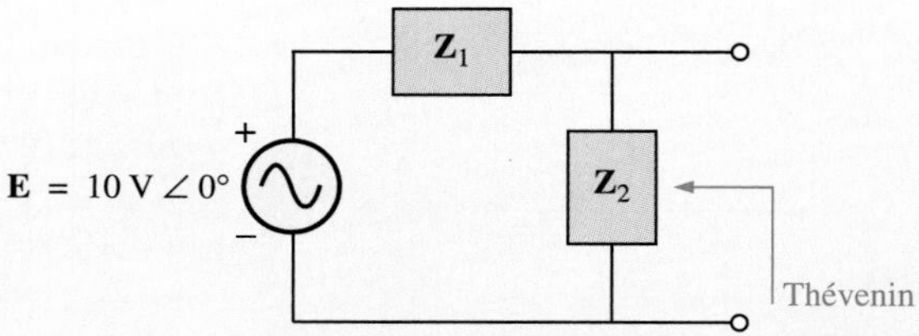

FIG. 18.19
Assigning the subscripted impedances to the network of Fig. 18.18.

$$\mathbf{Z}_1 = jX_L = j\,8\ \Omega \quad \mathbf{Z}_2 = -jX_C = -j\,2\ \Omega$$

Step 3 (Fig. 18.20):

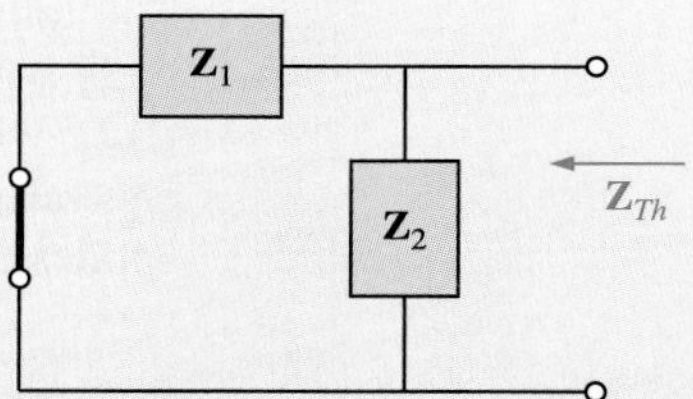

FIG. 18.20
Determining the Thévenin impedance for the network of Fig. 18.18.

$$\mathbf{Z}_{Th} = \frac{\mathbf{Z}_1\mathbf{Z}_2}{\mathbf{Z}_1 + \mathbf{Z}_2} = \frac{(j\,8\ \Omega)(-j\,2\ \Omega)}{j\,8\ \Omega - j\,2\ \Omega} = \frac{-j^2 16\ \Omega}{j\,6} = \frac{16\ \Omega}{6\ \angle 90^\circ}$$
$$= \mathbf{2.67\ \Omega\ \angle -90^\circ}$$

Step 4 (Fig. 18.21):

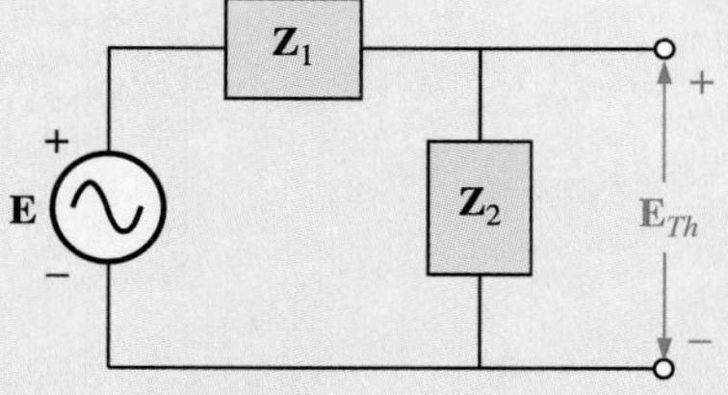

FIG. 18.21
Determining the open-circuit Thévenin voltage for the network of Fig. 18.18.

continued

$$\mathbf{E}_{Th} = \frac{\mathbf{Z}_2\mathbf{E}}{\mathbf{Z}_1 + \mathbf{Z}_2} \quad \text{(voltage divider rule)}$$

$$= \frac{(-j\,2\,\Omega)(10\text{ V})}{j\,8\,\Omega - j\,2\,\Omega} = \frac{-j\,20\text{ V}}{j\,6} = \mathbf{3.33\text{ V} \angle -180^\circ}$$

Step 5: The Thévenin equivalent circuit is shown in Fig. 18.22.

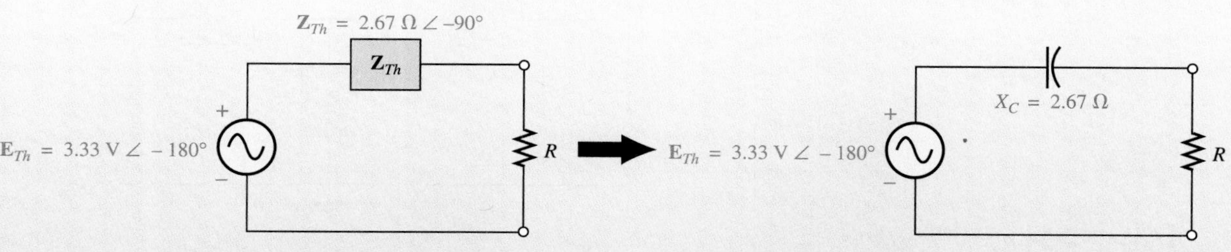

FIG. 18.22
The Thévenin equivalent circuit for the network of Fig. 18.18.

EXAMPLE 18.8

Find the Thévenin equivalent circuit for the network external to branch *a-a′* in Fig. 18.23.

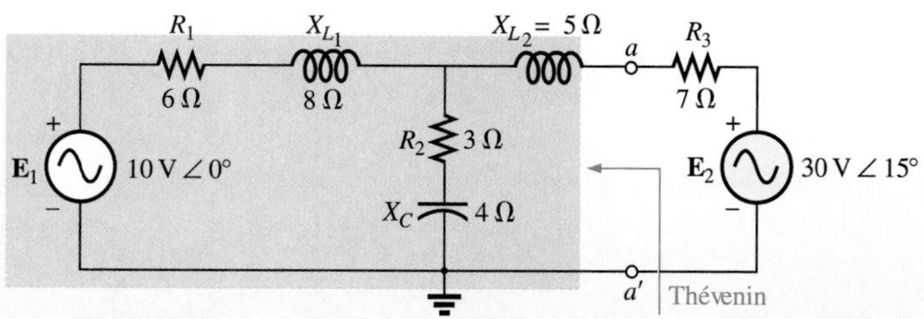

FIG. 18.23
Example 18.8.

Solution:

Steps 1 and 2 (Fig. 18.24): Note the reduced complexity with subscripted impedances:

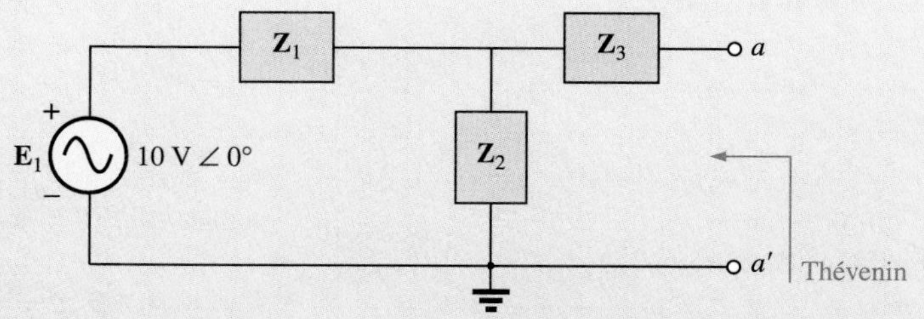

FIG. 18.24
Assigning the subscripted impedances to the network of Fig. 18.23.

continued

$$\mathbf{Z}_1 = R_1 + jX_{L_1} = 6\ \Omega + j\,8\ \Omega$$
$$\mathbf{Z}_2 = R_2 - jX_C = 3\ \Omega - j\,4\ \Omega$$
$$\mathbf{Z}_3 = +jX_{L_2} = j\,5\ \Omega$$

Step 3 (Fig. 18.25):

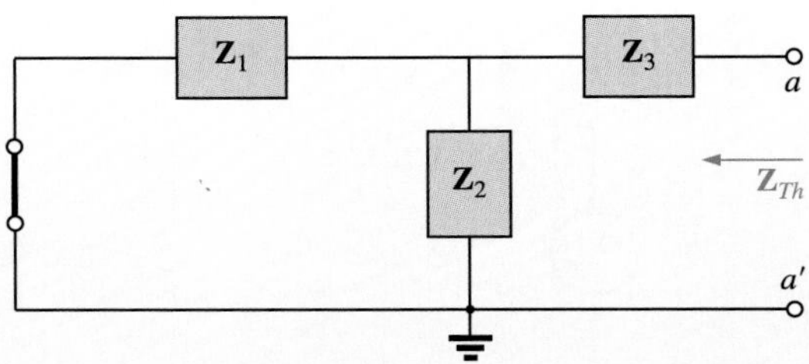

FIG. 18.25
Determining the Thévenin impedance for the network of Fig. 18.23.

$$\mathbf{Z}_{Th} = \mathbf{Z}_3 + \frac{\mathbf{Z}_1\mathbf{Z}_2}{\mathbf{Z}_1 + \mathbf{Z}_2} = j\,5\ \Omega + \frac{(10\ \Omega \angle 53.13^\circ)(5\ \Omega \angle -53.13^\circ)}{(6\ \Omega + j\,8\ \Omega) + (3\ \Omega - j\,4\ \Omega)}$$
$$= j\,5 + \frac{50 \angle 0^\circ}{9 + j\,4} = j\,5 + \frac{50 \angle 0^\circ}{9.85 \angle 23.96^\circ}$$
$$= j\,5 + 5.08 \angle -23.96^\circ = j\,5 + 4.64 - j\,2.06$$
$$\mathbf{Z}_{Th} = \mathbf{4.64\ \Omega + j\,2.94\ \Omega = 5.49\ \Omega \angle 32.36^\circ}$$

Step 4 (Fig. 18.26): Since *a-a′* is an open circuit, $\mathbf{I}_{\mathbf{Z}_3} = 0$. Then $\mathbf{E}_{Th}$ is the voltage drop across $\mathbf{Z}_2$:

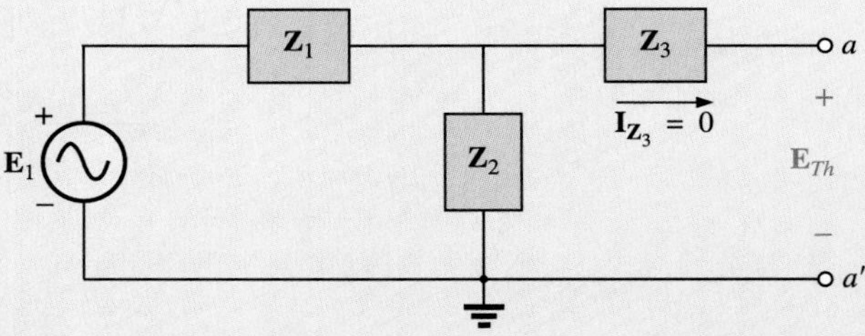

FIG. 18.26
Determining the open-circuit Thévenin voltage for the network of Fig. 18.23.

$$\mathbf{E}_{Th} = \frac{\mathbf{Z}_2\mathbf{E}}{\mathbf{Z}_2 + \mathbf{Z}_1} \quad \text{(voltage divider rule)}$$
$$= \frac{(5\ \Omega \angle -53.13^\circ)(10\ \text{V} \angle 0^\circ)}{9.85\ \Omega \angle 23.96^\circ}$$
$$\mathbf{E}_{Th} = \frac{50\ \text{V} \angle -53.13^\circ}{9.85 \angle 23.96^\circ} = \mathbf{5.08\ V \angle -77.09^\circ}$$

continued

Step 5: The Thévenin equivalent circuit is shown in Fig. 18.27.

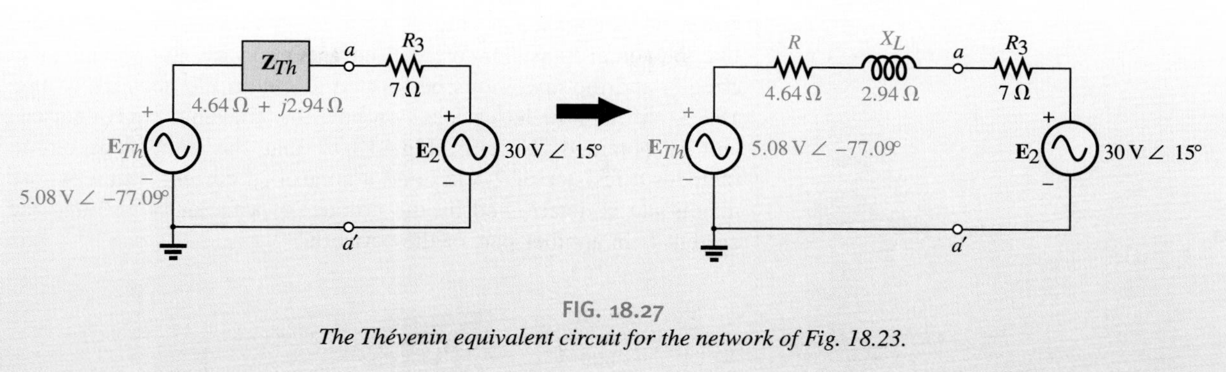

FIG. 18.27
The Thévenin equivalent circuit for the network of Fig. 18.23.

Example 18.9 demonstrates the use of superposition to separate the dc and ac analyses. The fact that the controlling variable in this analysis is not in the part of the network connected directly to the terminals of interest allows us to analyze the network in the same way as in the previous two examples for independent sources.

EXAMPLE 18.9

Determine the Thévenin equivalent circuit for the transistor network external to the resistor R_L in the following network (Fig. 18.28) and then determine $\mathbf{V}_L$.

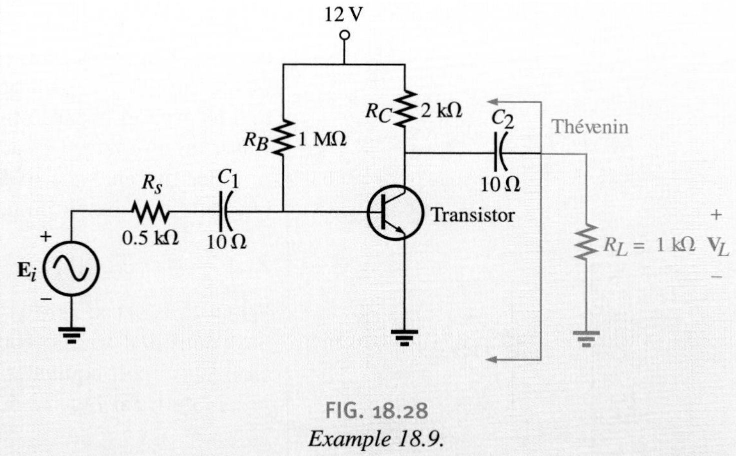

FIG. 18.28
Example 18.9.

Solution: Applying superposition.

dc conditions:

Substituting the open-circuit equivalent for the coupling capacitor C_2 will isolate the dc source and the resulting currents from the load resistor. The result is for dc conditions that $V_L = 0$ V. Although the output dc voltage is zero, the application of the dc voltage is important to the basic operation of the transistor in a number of important ways, one of which is to determine the parameters of the "equivalent circuit" to appear in the ac analysis to follow.

continued

ac conditions:
For the ac analysis, an equivalent circuit is substituted for the transistor, as established by the dc conditions above, that will behave like the actual transistor. We will have more to say about equivalent circuits and the operations performed to obtain the network of Fig. 18.29, but for now let us focus on how the Thévenin equivalent circuit is obtained. Note in Fig. 18.29 that the equivalent circuit includes a resistor of 2.3 kΩ and a controlled current source whose magnitude is determined by the product of a factor of 100 and the current I_1 in another part of the network.

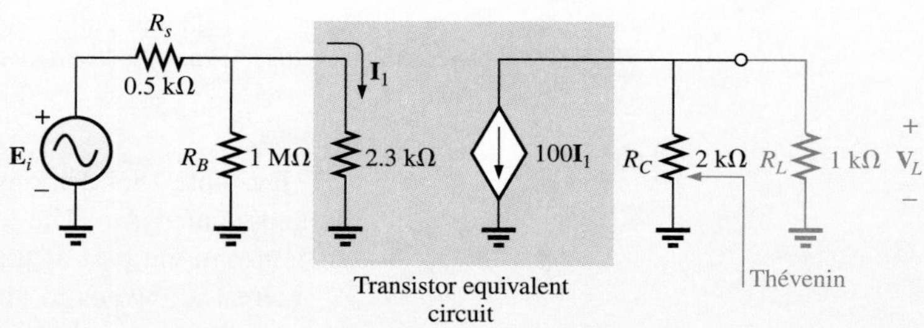

FIG. 18.29
The ac equivalent network for the transistor amplifier of Fig. 18.28.

Note in Fig. 18.29 the absence of the coupling capacitors for the ac analysis. In general, coupling capacitors are designed to be open circuits for dc and short circuits for ac analysis. The short-circuit equivalent is valid because the other impedances in series with the coupling capacitors are so much larger in magnitude that the effect of the coupling capacitors can be ignored. Both R_B and R_C are now tied to ground because the dc source was set to zero volts (superposition) and replaced by a short-circuit equivalent to ground.

For the analysis to follow, the effect of the resistor R_B will be ignored since it is so much larger than the parallel 2.3-kΩ resistor.

$\mathbf{Z}_{Th}$:

When $\mathbf{E}_i$ is set to zero volts, the current $\mathbf{I}_1$ will be zero amperes, and the controlled source $100\mathbf{I}_1$ will be zero amperes also. The result is an open-circuit equivalent for the source, as appears in Fig. 18.30.

Note from Fig. 18.30 that

$$\mathbf{Z}_{Th} = 2\ \mathrm{k\Omega}$$

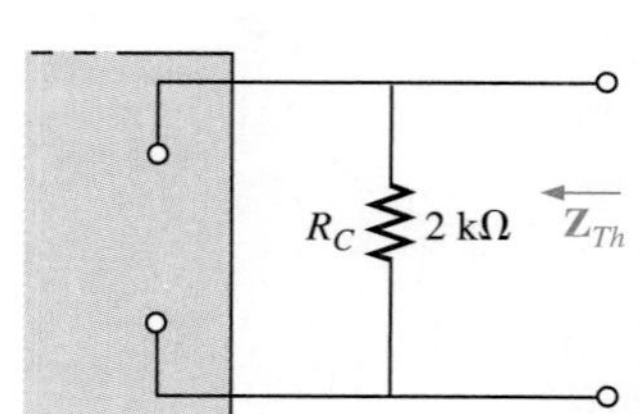

FIG. 18.30
Determining the Thévenin impedance for the network of Fig. 18.29.

$\mathbf{E}_{Th}$:

For $\mathbf{E}_{Th}$ the current $\mathbf{I}_1$ of Fig. 18.29 will be

$$\mathbf{I}_1 = \frac{\mathbf{E}_i}{R_s + 2.3\ \mathrm{k\Omega}} = \frac{\mathbf{E}_i}{0.5\ \mathrm{k\Omega} + 2.3\ \mathrm{k\Omega}} = \frac{\mathbf{E}_i}{2.8\ \mathrm{k\Omega}}$$

and
$$100\mathbf{I}_1 = (100)\left(\frac{\mathbf{E}_i}{2.8\ \mathrm{k\Omega}}\right) = 35.71 \times 10^{-3}/\Omega\ \mathbf{E}_i$$

Referring to Fig. 18.31, we find that

$$\begin{aligned}\mathbf{E}_{Th} &= -(100\mathbf{I}_1)R_C\\ &= -(35.71 \times 10^{-3}/\Omega\ \mathbf{E}_i)(2 \times 10^3\ \Omega)\\ \mathbf{E}_{Th} &= -71.42\mathbf{E}_i\end{aligned}$$

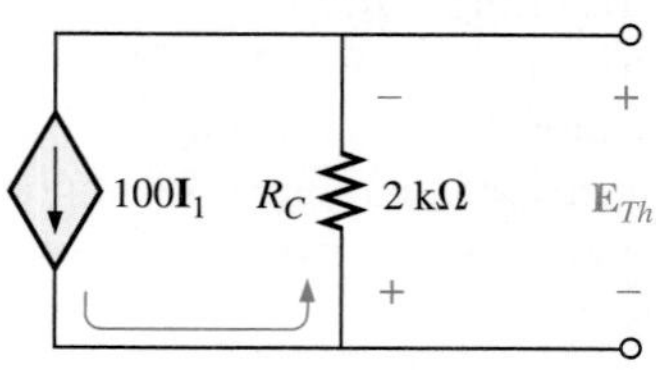

FIG. 18.31
Determining the Thévenin voltage for the network of Fig. 18.29.

continued

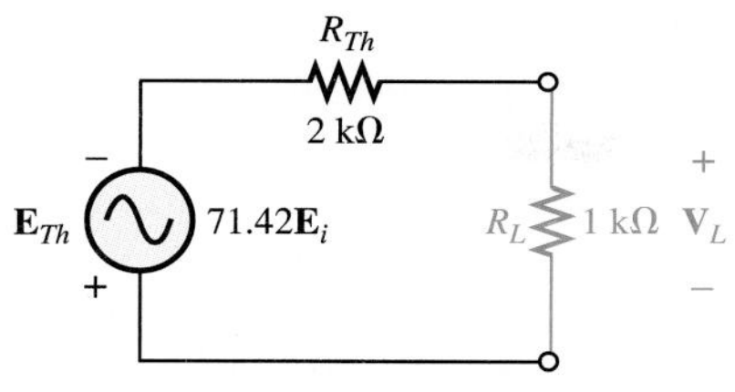

FIG. 18.32
The Thévenin equivalent circuit for the network of Fig. 18.29.

The Thévenin equivalent circuit appears in Fig. 18.32 with the original load R_L.

$\mathbf{V}_L$:

The output voltage $\mathbf{V}_L$:

$$\mathbf{V}_L = \frac{-R_L\mathbf{E}_{Th}}{R_L + \mathbf{Z}_{Th}} = \frac{-(1\text{ k}\Omega)(71.42\mathbf{E}_i)}{1\text{ k}\Omega + 2\text{ k}\Omega}$$

and
$$\mathbf{V}_L = \mathbf{-23.81E}_i$$

revealing that the output voltage is 23.81 times the applied voltage with a phase shift of 180° due to the minus sign.

Dependent Sources (Thévenin's theorem)

We will present three methods for applying Thévenin's theorem to solve dependent sources.

Case 1: For dependent sources with a *controlling variable* **not** *in the network under investigation*, the same procedure can be applied that was used for independent sources earlier. (This is like Case 1 in section 18.2, which also deals with dependent sources).

Case 2: For dependent sources (as with Case 2 in Section 18.2) where the *controlling variable* **is** *part of the network to which the theorem is to be applied*, another approach must be used. This method can also be applied to any dc or sinusoidal ac network. Consider the Thévenin equivalent circuit of Fig. 18.33(a).

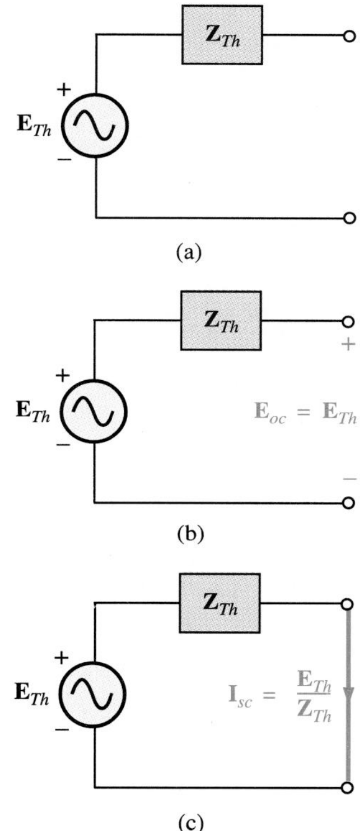

FIG. 18.33
Defining an alternative approach for determining the Thévenin impedance.

As shown in Fig. 18.33(b), the open-circuit terminal voltage ($\mathbf{E}_{oc}$) of the Thévenin equivalent circuit is the Thévenin equivalent voltage; that is,

$$\mathbf{E}_{oc} = \mathbf{E}_{Th} \tag{18.1}$$

If the terminals are short circuited as in Fig. 18.33(c), the resulting short-circuit current is determined by

$$\mathbf{I}_{sc} = \frac{\mathbf{E}_{Th}}{\mathbf{Z}_{Th}} \tag{18.2}$$

or rearranged, we get the Thévenin equivalent impedance,

$$\mathbf{Z}_{Th} = \frac{\mathbf{E}_{Th}}{\mathbf{I}_{sc}}$$

$$\mathbf{Z}_{Th} = \frac{\mathbf{E}_{oc}}{\mathbf{I}_{sc}} \tag{18.3}$$

Case 3: There is a third approach to the Thévenin equivalent circuit. The Thévenin voltage is found as in the previous two methods. However, the Thévenin *impedance* is obtained by applying a source of voltage to the terminals of interest and determining the source current as indicated in Fig. 18.34. For this method, the source voltage of the original network is set to zero. The Thévenin impedance is then determined by the following equation:

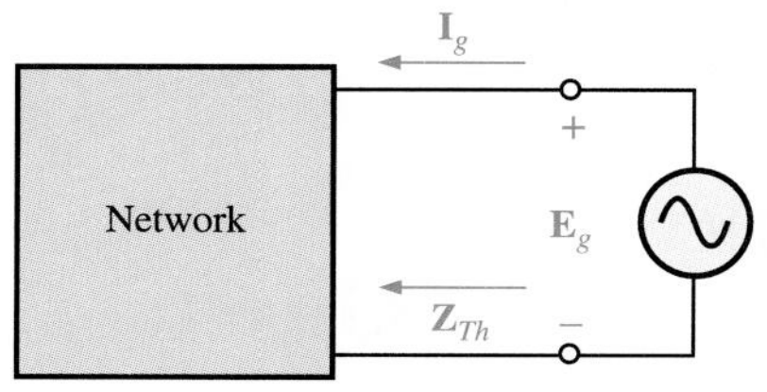

FIG. 18.34
Determining $\mathbf{Z}_{Th}$ using the approach $\mathbf{Z}_{Th} = \mathbf{E}_g/\mathbf{I}_g$.

$$\mathbf{Z}_{Th} = \frac{\mathbf{E}_g}{\mathbf{I}_g} \tag{18.4}$$

Note that for all three methods, $\mathbf{E}_{Th} = \mathbf{E}_{oc}$, but the Thévenin impedance is found in different ways. The following examples will illustrate each of these methods.

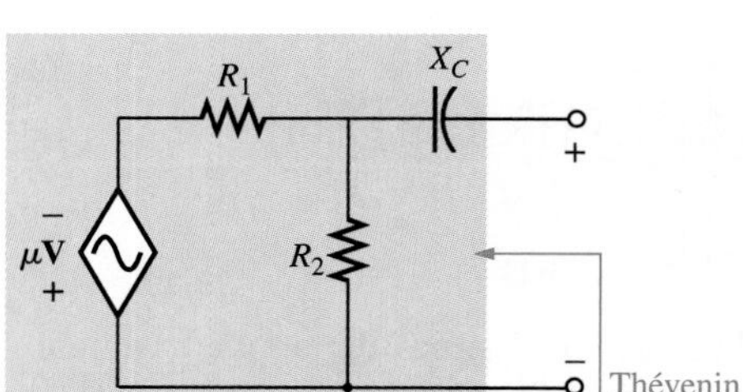

FIG. 18.35
Example 18.10.

EXAMPLE 18.10

Using each of the three techniques described in this section, determine the Thévenin equivalent circuit for the network of Fig. 18.35.

Solution: Since for each approach the Thévenin voltage is found in exactly the same way, it will be determined first. From Fig. 18.40, where $\mathbf{I}_{X_C} = 0$,

Due to the polarity for **V** and defined terminal polarities

$$\mathbf{V}_{R_1} = \mathbf{E}_{Th} = \mathbf{E}_{oc} = -\frac{R_2(\mu\mathbf{V})}{R_1 + R_2} = -\frac{\mu R_2 \mathbf{V}}{R_1 + R_2}$$

The following three methods for determining the Thévenin impedance appear in the order in which they were introduced in this section.

Method 1 (Fig. 18.36):

$$\mathbf{Z}_{Th} = R_1 \parallel R_2 - j\,X_C$$

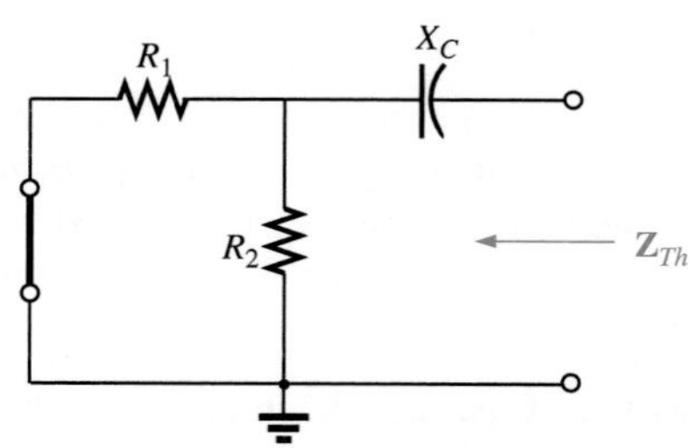

FIG. 18.36
Determining the Thévenin impedance for the network of Fig. 18.35.

Method 2 (Fig. 18.37): Converting the voltage source to a current source (Fig. 18.38), we have (current divider rule)

$$\mathbf{I}_{sc} = \frac{-(R_1 \parallel R_2)\dfrac{\mu\mathbf{V}}{R_1}}{(R_1 \parallel R_2) - j\,X_C} = \frac{-\dfrac{R_1R_2}{R_1 + R_2}\left(\dfrac{\mu\mathbf{V}}{R_1}\right)}{(R_1 \parallel R_2) - j\,X_C}$$

$$= \frac{\dfrac{-\mu R_2\mathbf{V}}{R_1 + R_2}}{(R_1 \parallel R_2) - j\,X_C}$$

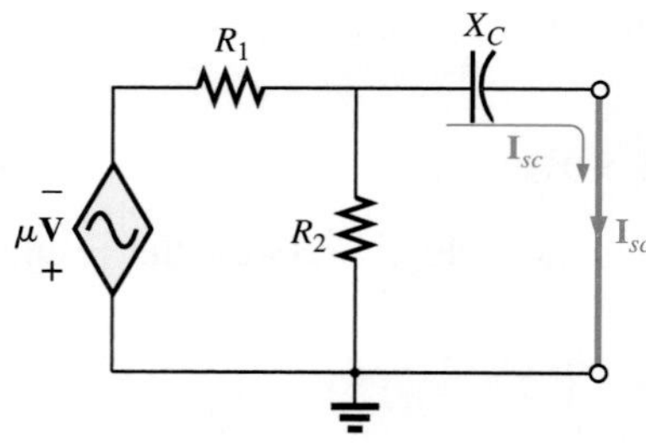

FIG. 18.37
Determining the short-circuit current for the network of Fig. 18.35.

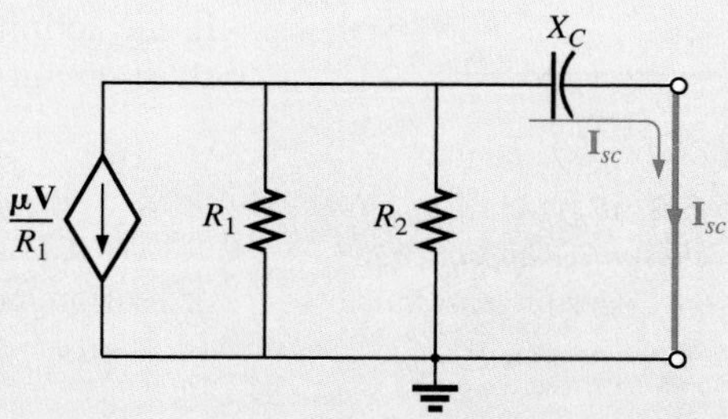

FIG. 18.38
Converting the voltage source of Fig. 18.37 to a current source.

and

$$\mathbf{Z}_{Th} = \frac{\mathbf{E}_{oc}}{\mathbf{I}_{sc}} = \frac{\dfrac{-\mu R_2\mathbf{V}}{R_1 + R_2}}{\dfrac{\dfrac{-\mu R_2\mathbf{V}}{R_1 + R_2}}{(R_1 \parallel R_2) - j\,X_C}} = \frac{1}{\dfrac{1}{(R_1 \parallel R_2) - j\,X_C}}$$

$$= R_1 \parallel R_2 - j\,X_C$$

continued

Method 3 (Fig. 18.39):

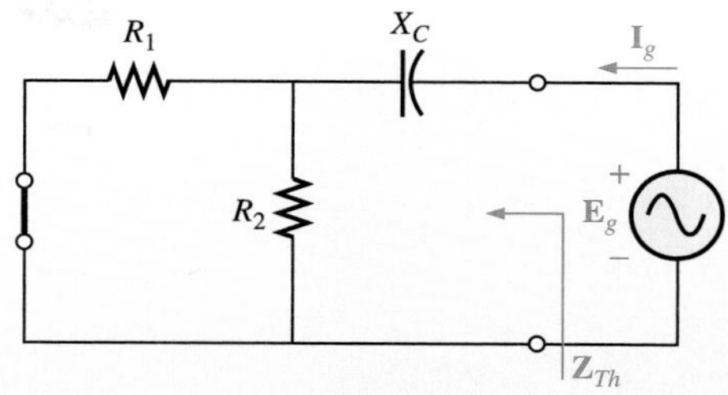

FIG. 18.39
Determining the Thévenin impedance for the network of Fig. 18.35 using the approach $\mathbf{Z}_{Th} = \mathbf{E}_g/\mathbf{I}_g$.

$$\mathbf{I}_g = \frac{\mathbf{E}_g}{(R_1 \parallel R_2) - j X_C}$$

and
$$\mathbf{Z}_{Th} = \frac{\mathbf{E}_g}{\mathbf{I}_g} = R_1 \parallel R_2 - j X_C$$

In each case, the Thévenin impedance is the same. The resulting Thévenin equivalent circuit is shown in Fig. 18.40.

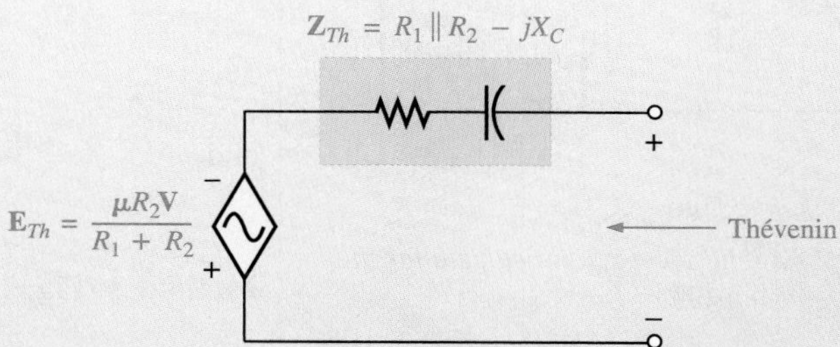

FIG. 18.40
The Thévenin equivalent circuit for the network of Fig. 18.35.

EXAMPLE 18.11

Repeat Example 18.10 for the network of Fig. 18.41.

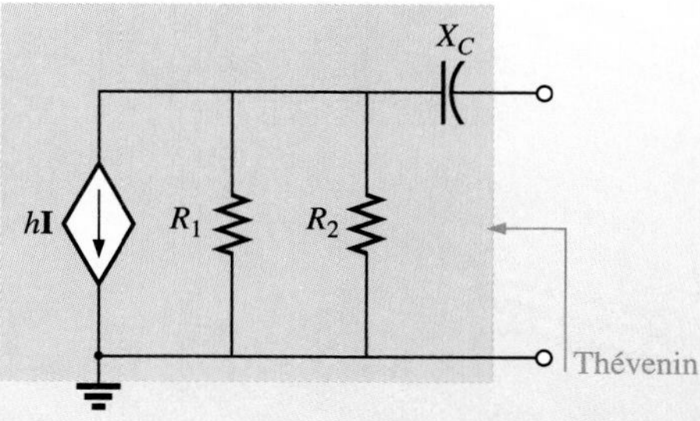

FIG. 18.41
Example 18.11.

continued

Solution: From Fig. 18.41, $\mathbf{E}_{Th}$ is

$$\mathbf{E}_{Th} = \mathbf{E}_{oc} = -h\mathbf{I}(R_1 \parallel R_2) = -\frac{hR_1R_2\mathbf{I}}{R_1 + R_2}$$

Method 1 (Fig. 18.42):

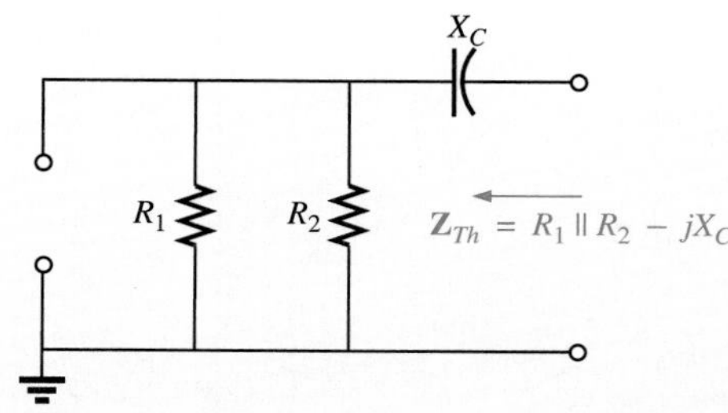

FIG. 18.42
Determining the Thévenin impedance for the network of Fig. 18.41.

$$\mathbf{Z}_{Th} = \boldsymbol{R_1 \parallel R_2 - j\,X_C}$$

Note the similarity between this solution and the one in the previous example.

Method 2 (Fig. 18.43):

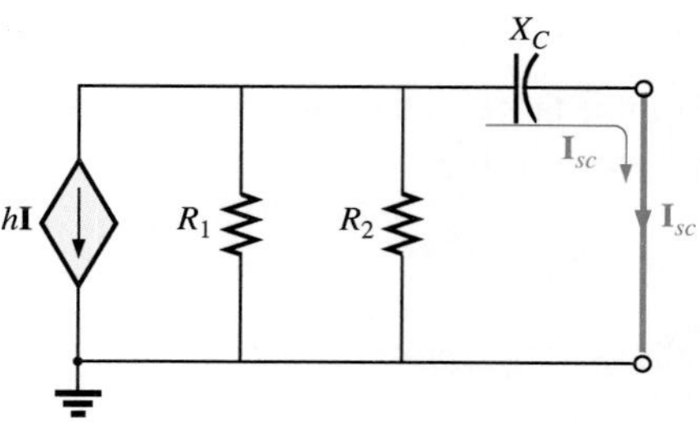

FIG. 18.43
Determining the short-circuit current for the network of Fig. 18.41.

$$\mathbf{I}_{sc} = \frac{-(R_1 \parallel R_2)h\mathbf{I}}{(R_1 \parallel R_2) - j\,X_C}$$

and
$$\mathbf{Z}_{Th} = \frac{\mathbf{E}_{oc}}{\mathbf{I}_{sc}} = \frac{-h\mathbf{I}(R_1 \parallel R_2)}{\dfrac{-(R_1 \parallel R_2)h\mathbf{I}}{(R_1 \parallel R_2) - j\,X_C}} = \boldsymbol{R_1 \parallel R_2 - j\,X_C}$$

Method 3 (Fig. 18.44):

$$\mathbf{I}_g = \frac{\mathbf{E}_g}{(R_1 \parallel R_2) - j\,X_C}$$

and
$$\mathbf{Z}_{Th} = \frac{\mathbf{E}_g}{\mathbf{I}_g} = \boldsymbol{R_1 \parallel R_2 - j\,X_C}$$

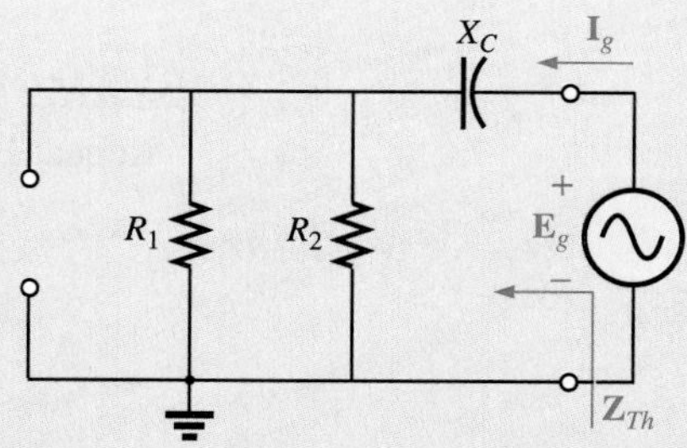

FIG. 18.44
Determining the Thévenin impedance using the approach $\mathbf{Z}_{Th} = \mathbf{E}_g/\mathbf{I}_g$.

The next example (18.12) has a dependent source that will not allow you to use Method 1 as described above. However, we will apply all three methods, so that the incorrect assumption can be exposed and the remaining two correct methods compared.

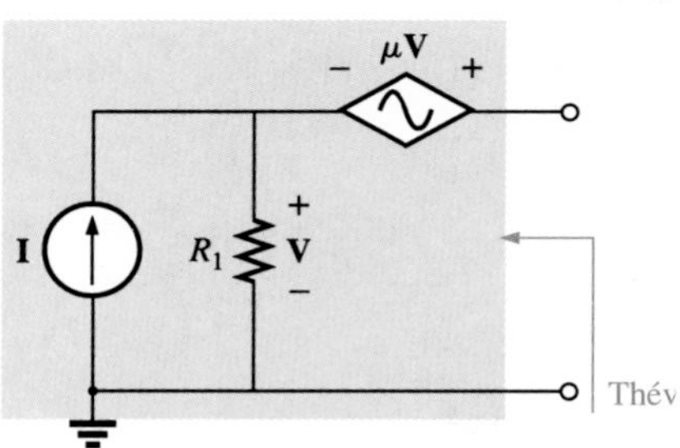

FIG. 18.45
Example 18.12.

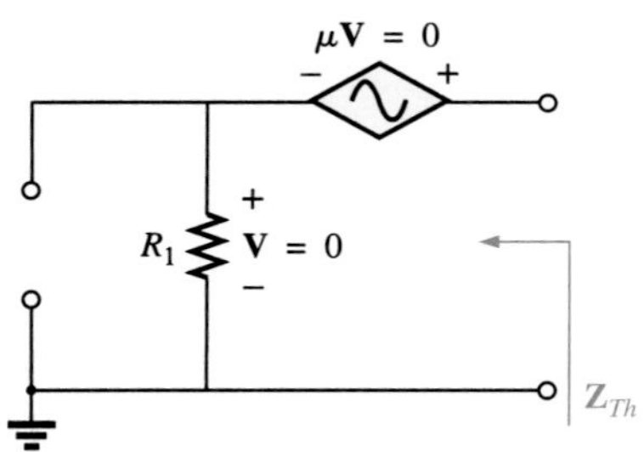

FIG. 18.46
Determining $\mathbf{Z}_{Th}$ incorrectly.

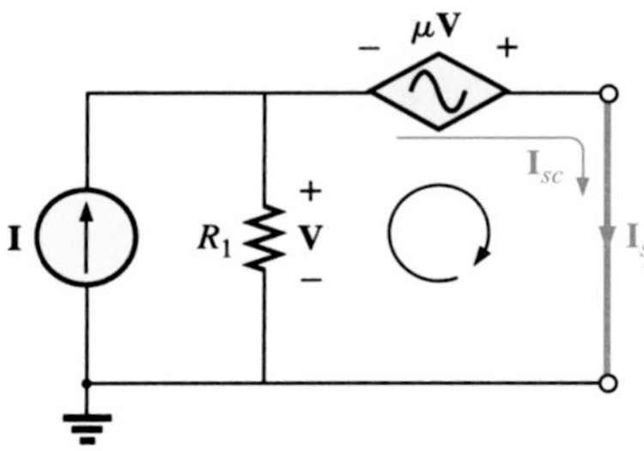

FIG. 18.47
Determining $\mathbf{I}_{sc}$ for the network of Fig. 18.45.

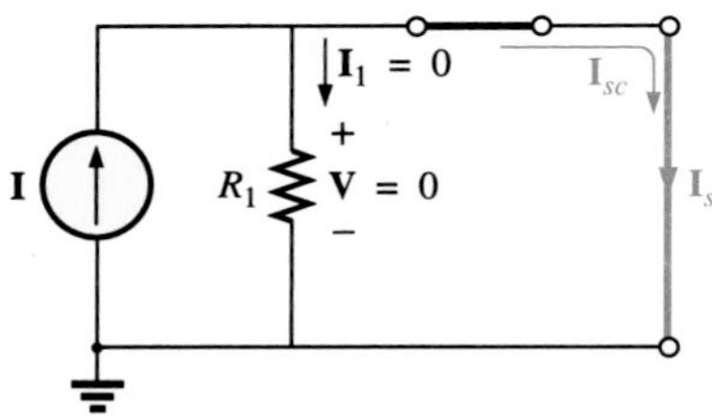

FIG. 18.48
Substituting $\mathbf{V} = 0$ into the network of Fig. 18.47.

EXAMPLE 18.12

For the network of Fig. 18.45 (introduced in Example 18.6), determine the Thévenin equivalent circuit between the indicated terminals using each method described in this section. Compare your results.

Solution: First, using Kirchhoff's voltage law, $\mathbf{E}_{Th}$ (which is the same for each method) is written

$$\mathbf{E}_{Th} = \mathbf{V} + \mu\mathbf{V} = (1 + \mu)\mathbf{V}$$

However,

$$\mathbf{V} = \mathbf{I}R_1$$

so

$$\mathbf{E}_{Th} = (1 + \mu)\mathbf{I}R_1$$

$\mathbf{Z}_{Th}$:

Method 1 (Fig. 18.46): Since $\mathbf{I} = 0$, $\mathbf{V}$ and $\mu\mathbf{V} = 0$, and

$$\cancel{\mathbf{Z}_{Th} = R_1} \quad \textit{(incorrect)}$$

Method 2 (Fig. 18.47): Kirchhoff's voltage law around the indicated loop gives us

$$\mathbf{V} + \mu\mathbf{V} = 0$$

and

$$\mathbf{V}(1 + \mu) = 0$$

Since μ is a positive constant, the above equation can be satisfied only when $\mathbf{V} = 0$. Substituting this result into Fig. 18.47 gives us the configuration of Fig. 18.48, and

$$\mathbf{I}_{sc} = \mathbf{I}$$

with

$$\mathbf{Z}_{Th} = \frac{\mathbf{E}_{oc}}{\mathbf{I}_{sc}} = \frac{(1 + \mu)\mathbf{I}R_1}{\mathbf{I}} = (1 + \mu)R_1 \quad \textit{(correct)}$$

Method 3 (Fig. 18.49):

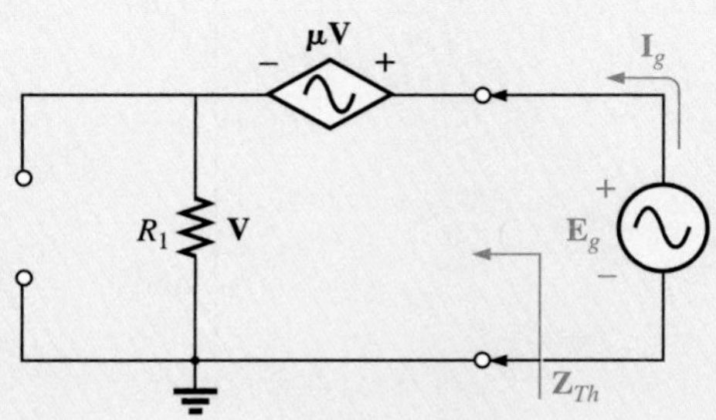

FIG. 18.49
Determining $\mathbf{Z}_{Th}$ using the approach $\mathbf{Z}_{Th} = \mathbf{E}_g/\mathbf{I}_g$.

continued

$$\mathbf{E}_g = \mathbf{V} + \mu\mathbf{V} = (1 + \mu)\mathbf{V}$$

or
$$\mathbf{V} = \frac{\mathbf{E}_g}{1 + \mu}$$

and
$$\mathbf{I}_g = \frac{\mathbf{V}}{R_1} = \frac{\mathbf{E}_g}{(1 + \mu)R_1}$$

and
$$\mathbf{Z}_{Th} = \frac{\mathbf{E}_g}{\mathbf{I}_g} = (1 + \mu)R_1 \quad \textit{(correct)}$$

The Thévenin equivalent circuit appears in Fig. 18.50, and

$$\mathbf{I}_L = \frac{(1 + \mu)R_1\mathbf{I}}{R_L + (1 + \mu)R_1}$$

which compares with the result of Example 18.6.

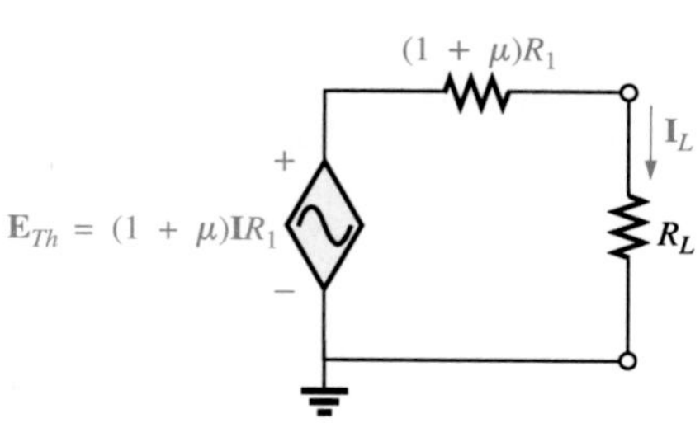

FIG. 18.50
The Thévenin equivalent circuit for the network of Fig. 18.45.

The network of Fig. 18.51 is the basic configuration of the transistor equivalent circuit applied most frequently today (although some other texts in electronics will use the circle rather than the diamond outline for the source). Needless to say, it is necessary to know its characteristics and to be good at using it. Note that there is a controlled voltage and current source, each controlled by variables in the configuration.

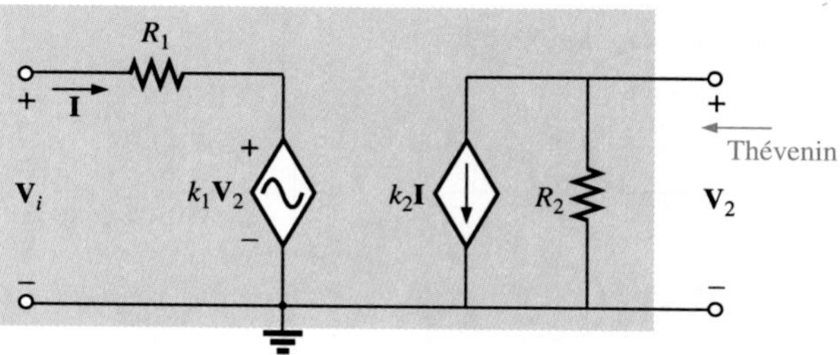

FIG. 18.51
Example 18.13: Transistor equivalent network.

EXAMPLE 18.13

Determine the Thévenin equivalent circuit for the indicated terminals of the network of Fig. 18.51.

Solution: Apply the second method introduced in this section.

$\mathbf{E}_{Th}$:

$$\mathbf{E}_{oc} = \mathbf{V}_2$$

$$\mathbf{I} = \frac{\mathbf{V}_i - k_1\mathbf{V}_2}{R_1} = \frac{\mathbf{V}_i - k_1\mathbf{E}_{oc}}{R_1}$$

and
$$\mathbf{E}_{oc} = -k_2\mathbf{I}R_2 = -k_2R_2\left(\frac{\mathbf{V}_i - k_1\mathbf{E}_{oc}}{R_1}\right)$$

$$= \frac{-k_2R_2\mathbf{V}_i}{R_1} + \frac{k_1k_2R_2\mathbf{E}_{oc}}{R_1}$$

continued

or
$$\mathbf{E}_{oc}\left(1 - \frac{k_1k_2R_2}{R_1}\right) = \frac{-k_2R_2\mathbf{V}_i}{R_1}$$

and
$$\mathbf{E}_{oc}\left(\frac{R_1 - k_1k_2R_2}{R_1}\right) = \frac{-k_2R_2\mathbf{V}_i}{R_1}$$

so
$$\mathbf{E}_{oc} = \frac{-k_2R_2\mathbf{V}_i}{R_1 - k_1k_2R_2} = \mathbf{E}_{Th} \qquad (18.5)$$

$\mathbf{I}_{sc}$:

For the network of Fig. 18.52, where
$$\mathbf{V}_2 = 0 \quad k_1\mathbf{V}_2 = 0 \quad \mathbf{I} = \frac{\mathbf{V}_i}{R_1}$$

and
$$\mathbf{I}_{sc} = -k_2\mathbf{I} = \frac{-k_2\mathbf{V}_i}{R_1}$$

so
$$\mathbf{Z}_{Th} = \frac{\mathbf{E}_{oc}}{\mathbf{I}_{sc}} = \frac{\dfrac{-k_2R_2\mathbf{V}_i}{R_1-k_1k_2R_2}}{\dfrac{-k_2\mathbf{V}_i}{R_1}} = \frac{R_1R_2}{R_1-k_1k_2R_2}$$

and
$$\mathbf{Z}_{Th} = \frac{R_2}{1 - \dfrac{k_1k_2R_2}{R_1}} \qquad (18.6)$$

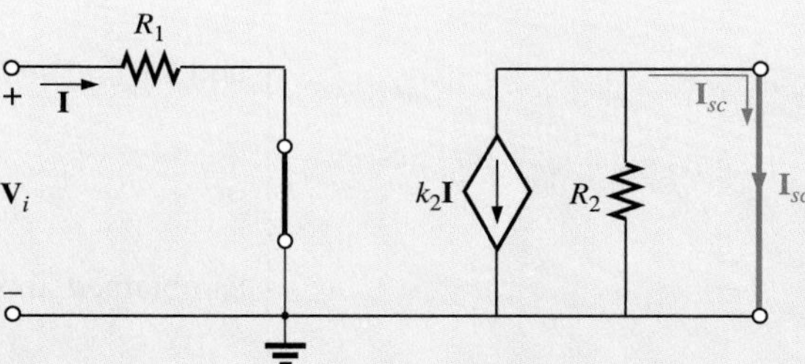

FIG. 18.52
Determining $\mathbf{I}_{sc}$ for the network of Fig. 18.51.

The approximation $k_1 \cong 0$ is often applied. Then the Thévenin voltage and impedance are

$$\mathbf{E}_{Th} = \frac{-k_2R_2\mathbf{V}_i}{R_1} \qquad k_1 = 0 \qquad (18.7)$$

$$\mathbf{Z}_{Th} = R_2 \qquad k_1 = 0 \qquad (18.8)$$

continued

Apply $\mathbf{Z}_{Th} = \mathbf{E}_g/\mathbf{I}_g$ to the network of Fig. 18.53, where

$$\mathbf{I} = \frac{-k_1\mathbf{V}_2}{R_1}$$

But

$$\mathbf{V}_2 = \mathbf{E}_g$$

so

$$\mathbf{I} = \frac{-k_1\mathbf{E}_g}{R_1}$$

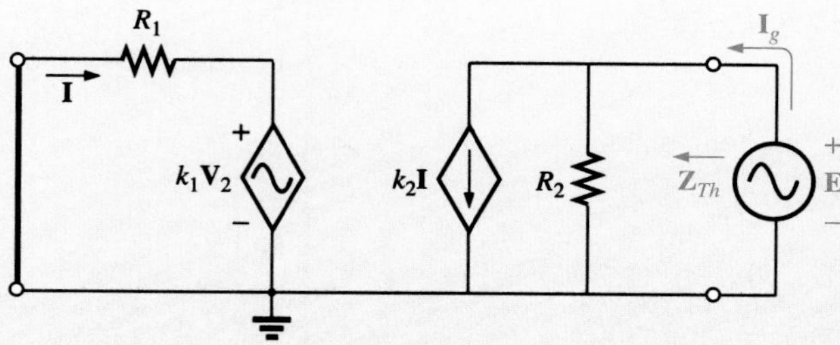

FIG. 18.53
Determining $\mathbf{Z}_{Th}$ using the procedure $\mathbf{Z}_{Th} = \mathbf{E}_g/\mathbf{I}_g$.

Applying Kirchhoff's current law, we have

$$\mathbf{I}_g = k_2\mathbf{I} + \frac{\mathbf{E}_g}{R_2} = k_2\left(-\frac{k_1\mathbf{E}_g}{R_1}\right) + \frac{\mathbf{E}_g}{R_2}$$

$$= \mathbf{E}_g\left(\frac{1}{R_2} - \frac{k_1k_2}{R_1}\right)$$

and

$$\frac{\mathbf{I}_g}{\mathbf{E}_g} = \frac{R_1 - k_1k_2R_2}{R_1R_2}$$

or

$$\mathbf{Z}_{Th} = \frac{\mathbf{E}_g}{\mathbf{I}_g} = \frac{R_1R_2}{R_1 - k_1k_2R_2}$$

as obtained above.

The last two methods presented in this section were applied only to networks in which the magnitudes of the controlled sources were dependent on a variable within the network for which the Thévenin equivalent circuit was to be obtained. Understand that both of these methods can also be applied to any dc or sinusoidal ac network containing only independent sources or dependent sources of the other kind.

18.4 NORTON'S THEOREM (ac)

The three methods described for Thévenin's theorem will each be altered so that they can be used with Norton's theorem. Since the Thévenin and Norton impedances are the same for a particular network, certain parts of the discussion will be quite similar to those in the

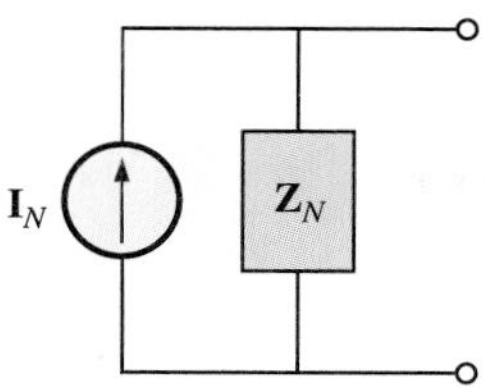

FIG. 18.54
The Norton equivalent circuit for ac networks.

previous section. We will first consider independent sources and the approach developed in Chapter 9, followed by dependent sources and the new techniques developed for Thévenin's theorem.

You will recall from Chapter 9 that Norton's theorem allows us to replace any two-terminal linear bilateral ac network with an equivalent circuit consisting of a current source and impedance, as in Fig. 18.54.

Caution: Since the reactances of a circuit are frequency dependent, the Norton equivalent circuit found for a particular network applies only at *one* frequency. This is the same condition as for the Thévenin equivalent circuit.

Independent Sources (Norton's theorem)

The procedure outlined below to find the Norton equivalent of a sinusoidal ac network is changed (from that in Chapter 9) in only one way: the term *resistance* has been replaced with the term *impedance.*

1. Remove the portion of the network across which the Norton equivalent circuit is to be found.
2. Mark (○, •, and so on) the terminals of the remaining two-terminal network.
3. Calculate Z_N by first setting all voltage and current sources to zero (short circuit and open circuit, respectively) and then finding the resulting impedance between the two marked terminals.
4. Calculate I_N by first replacing the voltage and current sources and then finding the short-circuit current between the marked terminals.
5. Draw the Norton equivalent circuit with the portion of the circuit previously removed replaced between the terminals of the Norton equivalent circuit.

The Norton and Thévenin equivalent circuits can be found from each other by using the source transformation shown in Fig. 18.55. The source transformation is applicable for any Thévenin or Norton equivalent circuit determined from a network with any combination of independent or dependent sources.

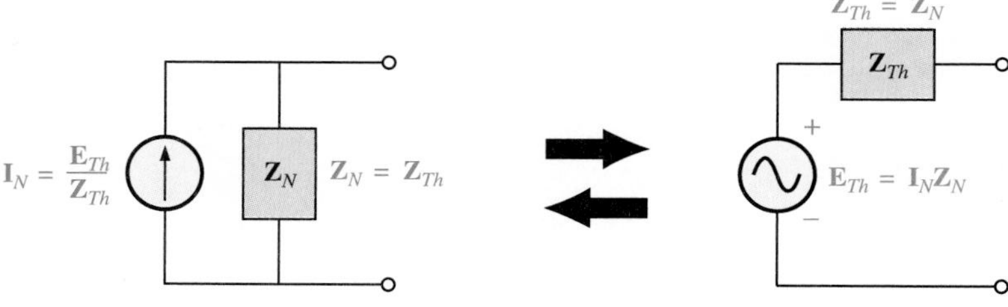

FIG. 18.55
Conversion between the Thévenin and Norton equivalent circuits.

EXAMPLE 18.14

Determine the Norton equivalent circuit for the network external to the 6-Ω resistor of Fig. 18.56.

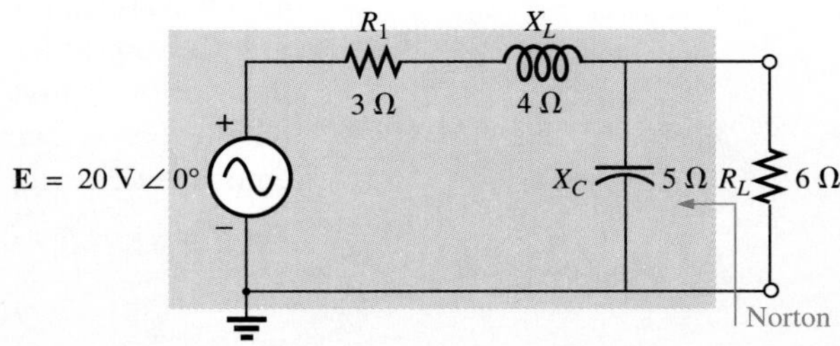

FIG. 18.56
Example 18.14.

Solution:

Steps 1 and 2 [Fig. 18.57(a)]:

$$\mathbf{Z}_1 = R_1 + j\,X_L = 3\,\Omega + j\,4\,\Omega = 5\,\Omega\,\angle 53.13^\circ$$

$$\mathbf{Z}_2 = -j\,X_C = -j\,5\,\Omega$$

Step 3 [Fig. 18.57(b)]:

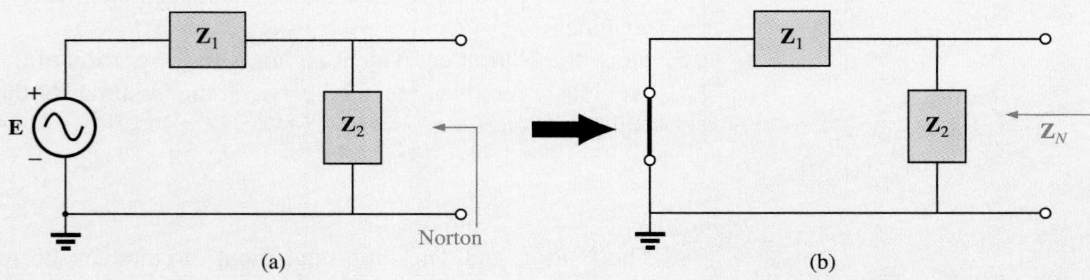

FIG. 18.57
(a) Assigning the subscripted impedances to the network of Fig. 18.56.
(b) Determining the Norton impedance for the network of Fig. 18.56.

$$\mathbf{Z}_N = \frac{\mathbf{Z}_1\mathbf{Z}_2}{\mathbf{Z}_1 + \mathbf{Z}_2} = \frac{(5\,\Omega\,\angle 53.13^\circ)(5\,\Omega\,\angle -90^\circ)}{3\,\Omega + j\,4\,\Omega - j\,5\,\Omega} = \frac{25\,\Omega\,\angle -36.87^\circ}{3 - j\,1}$$

$$= \frac{25\,\Omega\,\angle -36.87^\circ}{3.16\,\angle -18.43^\circ} = 7.91\,\Omega\,\angle -18.44^\circ = \mathbf{7.50\,\Omega - j\,2.50\,\Omega}$$

continued

Step 4 (Fig. 18.58):

$$\mathbf{I}_N = \mathbf{I}_1 = \frac{\mathbf{E}}{\mathbf{Z}_1} = \frac{20 \text{ V} \angle 0^\circ}{5 \,\Omega \angle 53.13^\circ} = \mathbf{4 \text{ A} \angle -53.13^\circ}$$

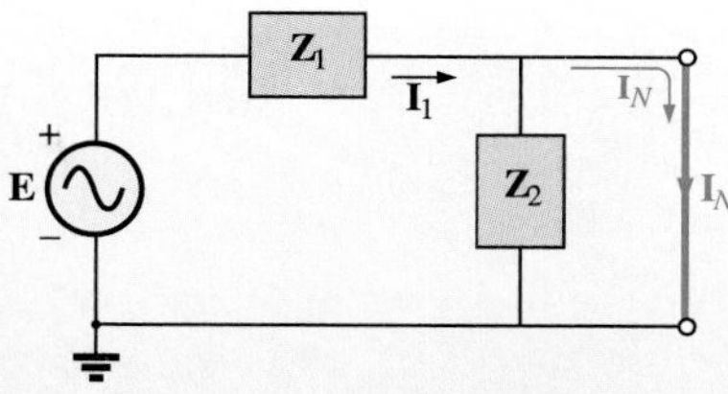

FIG. 18.58
Determining $\mathbf{I}_N$ for the network of Fig. 18.56.

Step 5: The Norton equivalent circuit is shown in Fig. 18.59.

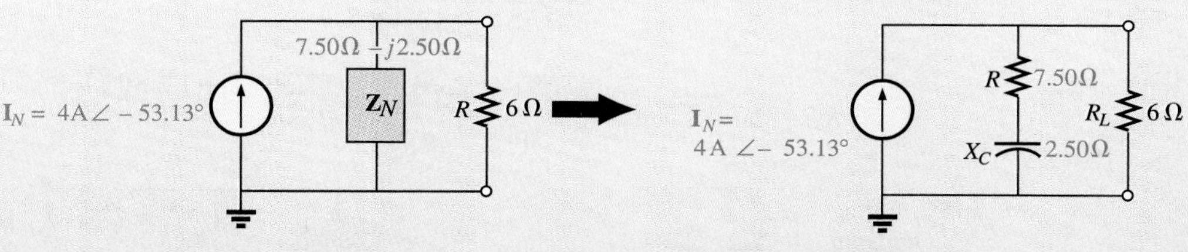

FIG. 18.59
The Norton equivalent circuit for the network of Fig. 18.56.

EXAMPLE 18.15

Find the Norton equivalent circuit for the network external to the 7-Ω capacitive reactance in Fig. 18.60.

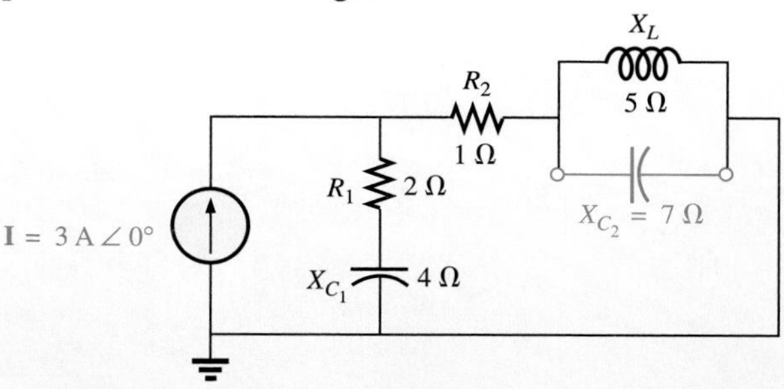

FIG. 18.60
Examples 18.15 and 18.16.

Solution: **Steps 1 and 2** (Fig. 18.61):

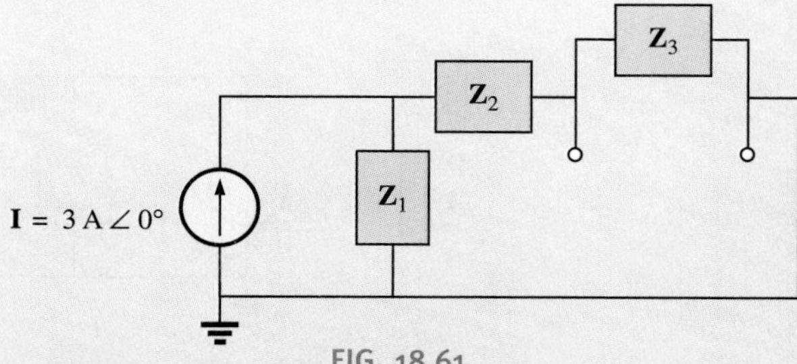

FIG. 18.61
Assigning the subscripted impedances to the network of Fig. 18.60.

continued

$$\mathbf{Z}_1 = R_1 - j X_{C_1} = 2\ \Omega - j\,4\ \Omega$$
$$\mathbf{Z}_2 = R_2 = 1\ \Omega$$
$$\mathbf{Z}_3 = +j X_L = j\,5\ \Omega$$

Step 3 (Fig. 18.62):

$$\mathbf{Z}_N = \frac{\mathbf{Z}_3(\mathbf{Z}_1 + \mathbf{Z}_2)}{\mathbf{Z}_3 + (\mathbf{Z}_1 + \mathbf{Z}_2)}$$

$$\mathbf{Z}_1 + \mathbf{Z}_2 = 2\ \Omega - j\,4\ \Omega + 1\ \Omega = 3\ \Omega - j\,4\ \Omega = 5\ \Omega \angle -53.13^\circ$$

$$\mathbf{Z}_N = \frac{(5\ \Omega \angle 90^\circ)(5\ \Omega \angle -53.13^\circ)}{j\,5\ \Omega + 3\ \Omega - j\,4\ \Omega} = \frac{25\ \Omega \angle 36.87^\circ}{3 + j\,1}$$

$$= \frac{25\ \Omega \angle 36.87^\circ}{3.16 \angle +18.43^\circ}$$

$$\mathbf{Z}_N = 7.91\ \Omega \angle 18.44^\circ = \mathbf{7.50\ \Omega + j\,2.50\ \Omega}$$

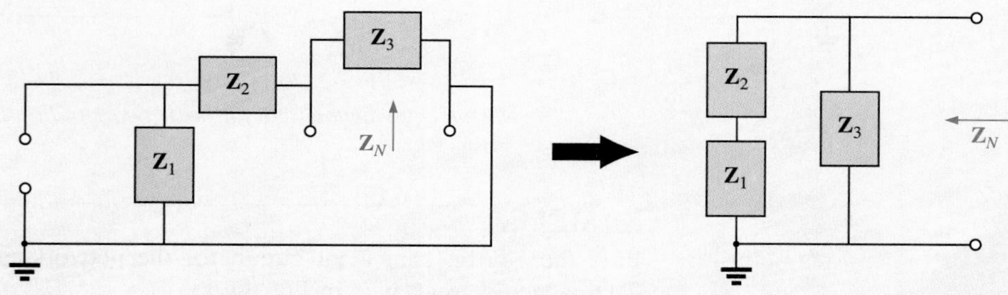

FIG. 18.62
Finding the Norton impedance for the network of Fig. 18.60.

Step 4 (Fig. 18.63):

$$\mathbf{I}_N = \mathbf{I}_1 = \frac{\mathbf{Z}_1 \mathbf{I}}{\mathbf{Z}_1 + \mathbf{Z}_2} \quad \text{(current divider rule)}$$

$$= \frac{(2\ \Omega - j\,4\ \Omega)(3\ \text{A})}{3\ \Omega - j\,4\ \Omega} = \frac{6\ \text{A} - j\,12\ \text{A}}{5 \angle -53.13^\circ} = \frac{13.4\ \text{A} \angle -63.43^\circ}{5 \angle -53.13^\circ}$$

$$\mathbf{I}_N = \mathbf{2.68\ A \angle -10.3^\circ}$$

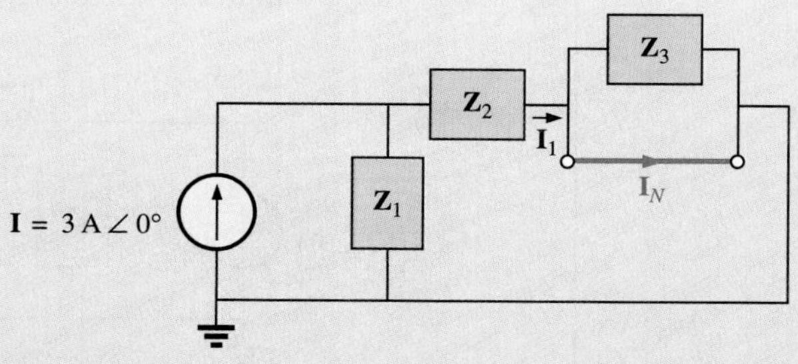

FIG. 18.63
Determining $\mathbf{I}_N$ for the network of Fig. 18.60.

continued

Step 5: The Norton equivalent circuit is shown in Fig. 18.64.

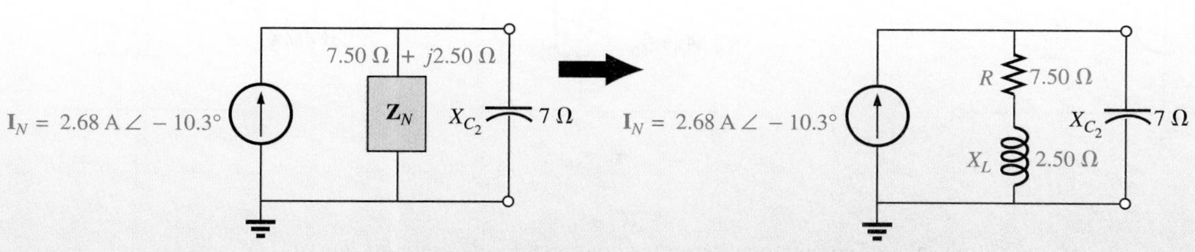

FIG. 18.64
The Norton equivalent circuit for the network of Fig. 18.60.

EXAMPLE 18.16

Find the Thévenin equivalent circuit for the network external to the 7-Ω capacitive reactance in Fig. 18.60.

Solution: Using the conversion between sources [Fig. 18.65(a)], we obtain

$$\mathbf{Z}_{Th} = \mathbf{Z}_N = \mathbf{7.50\ \Omega + j\ 2.50\ \Omega}$$

$$\mathbf{E}_{Th} = \mathbf{I}_N\mathbf{Z}_N = (2.68\text{ A}\angle -10.3^\circ)(7.91\ \Omega\angle 18.44^\circ)$$

$$= \mathbf{21.2\ V\angle 8.14^\circ}$$

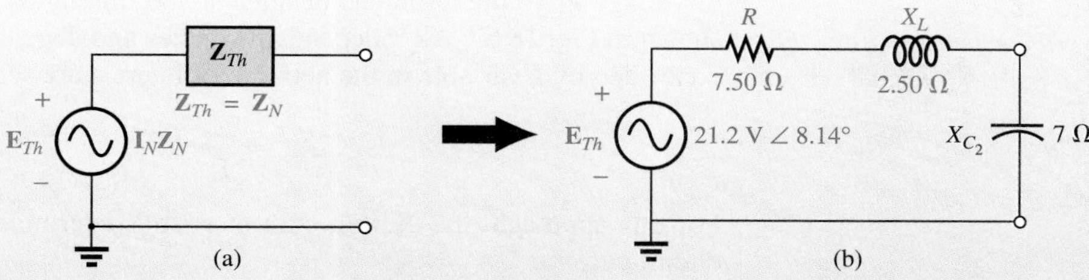

FIG. 18.65
(a) Determining the Thévenin equivalent circuit for the Norton equivalent of Fig. 18.64.
(b) The Thévenin equivalent circuit for the network of Fig. 18.60.

The Thévenin equivalent circuit is shown in Fig. 18.65(b).

Dependent Sources (Norton's theorem)

As stated for Thévenin's theorem, *dependent sources in which the controlling variable is not determined by the network* for which the Norton equivalent circuit is to be found do not alter the procedure outlined above.

For dependent sources of the other kind, one of the two following procedures must be applied. Both of these procedures can also be applied to networks with any combination of independent sources and dependent sources not controlled by the network under investigation.

The Norton equivalent circuit appears in Fig. 18.66(a). In Fig. 18.66(b), we find that

$$\mathbf{I}_{sc} = \mathbf{I}_N \qquad (18.9)$$

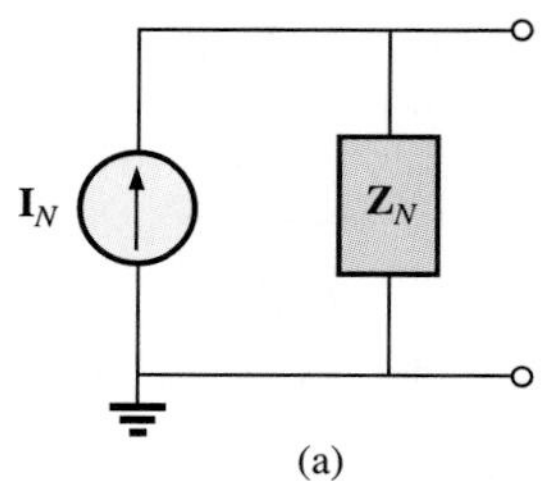

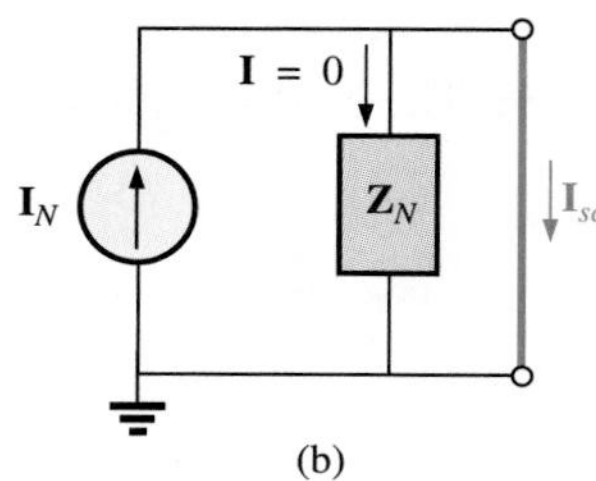

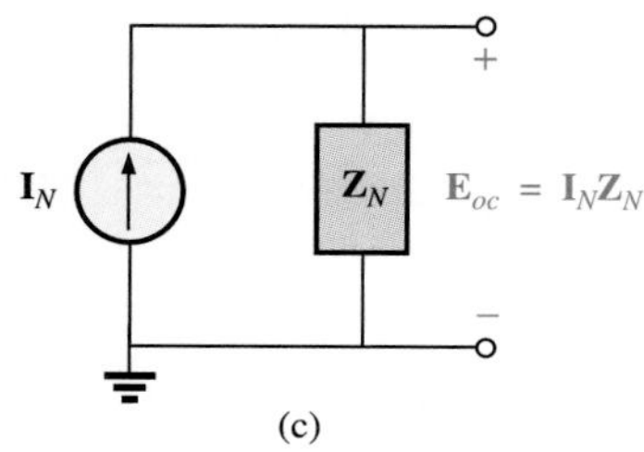

FIG. 18.66
Defining an alternative approach for determining $\mathbf{Z}_N$.

and in Fig. 18.66(c) that

$$\mathbf{E}_{oc} = \mathbf{I}_N \mathbf{Z}_N$$

Or, rearranging, we have

$$\mathbf{Z}_N = \frac{\mathbf{E}_{oc}}{\mathbf{I}_N}$$

and

$$\mathbf{Z}_N = \frac{\mathbf{E}_{oc}}{\mathbf{I}_{sc}} \tag{18.10}$$

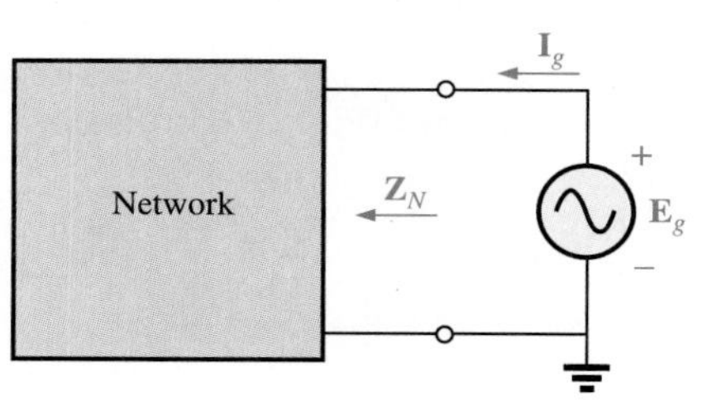

FIG. 18.67
Determining the Norton impedance using the approach $\mathbf{Z}_N = \mathbf{E}_g/\mathbf{I}_g$.

The Norton impedance can also be determined by applying a source of voltage $\mathbf{E}_g$ to the terminals of interest and finding the resulting $\mathbf{I}_g$, as shown in Fig. 18.67. All independent sources and dependent sources not controlled by a variable in the network of interest are set to zero, and

$$\mathbf{Z}_N = \frac{\mathbf{E}_g}{\mathbf{I}_g} \tag{18.11}$$

For this approach, the Norton current is still determined by the short-circuit current.

EXAMPLE 18.17

Using each method described for dependent sources, find the Norton equivalent circuit for the network of Fig. 18.68(a).

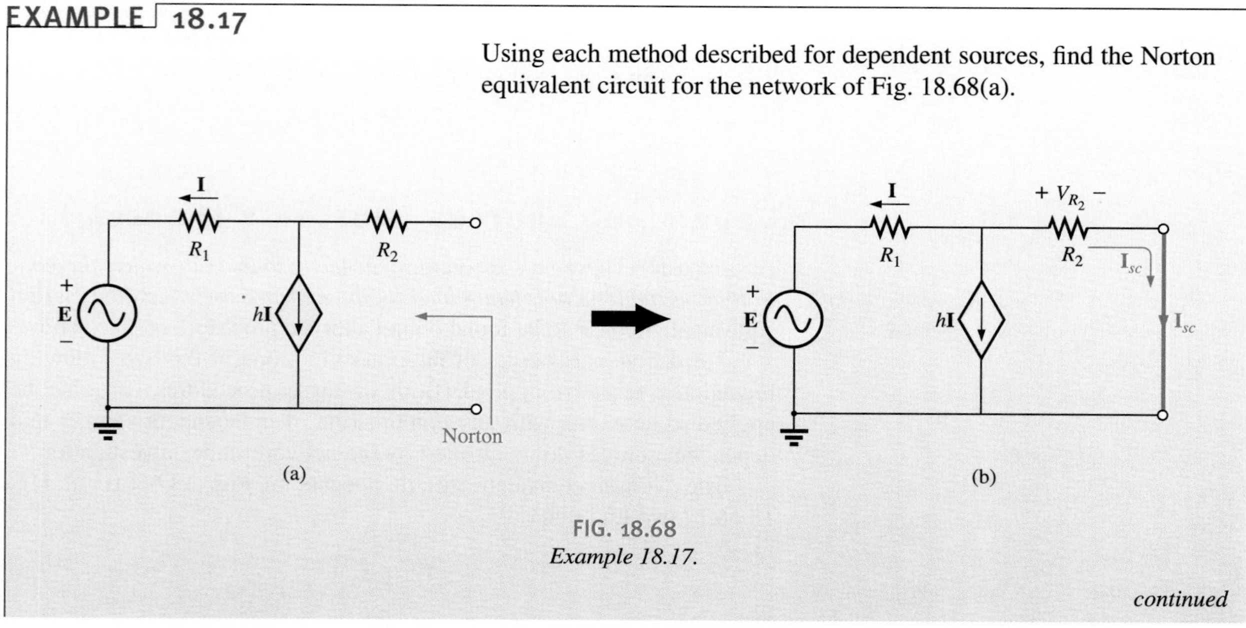

FIG. 18.68
Example 18.17.

continued

Solution:

$\mathbf{I}_N$:

For each method, $\mathbf{I}_N$ is determined in the same way. From Fig. 18.68(b), using Kirchhoff's current law, we have

$$0 = \mathbf{I} + h\mathbf{I} + \mathbf{I}_{sc}$$

or

$$\mathbf{I}_{sc} = -(1 + h)\mathbf{I}$$

Applying Kirchhoff's voltage law gives us

$$\mathbf{E} + \mathbf{I}R_1 - \mathbf{I}_{sc}R_2 = 0$$

and

$$\mathbf{I}R_1 = \mathbf{I}_{sc}R_2 - \mathbf{E}$$

or

$$\mathbf{I} = \frac{\mathbf{I}_{sc}R_2 - \mathbf{E}}{R_1}$$

so

$$\mathbf{I}_{sc} = -(1 + h)\mathbf{I} = -(1 + h)\left(\frac{\mathbf{I}_{sc}R_2 - \mathbf{E}}{R_1}\right)$$

or

$$R_1\mathbf{I}_{sc} = -(1 + h)\mathbf{I}_{sc}R_2 + (1 + h)\mathbf{E}$$

$$\mathbf{I}_{sc}[R_1 + (1 + h)R_2] = (1 + h)\mathbf{E}$$

$$\mathbf{I}_{sc} = \frac{(1 + h)\mathbf{E}}{R_1 + (1 + h)R_2} = \mathbf{I}_N$$

$\mathbf{Z}_N$:

Method 1: $\mathbf{E}_{oc}$ is determined from the network of Fig. 18.69. By Kirchhoff's current law,

$$0 = \mathbf{I} + h\mathbf{I} \text{ or } \mathbf{I}(h + 1) = 0$$

For h a positive constant, **I** must equal zero to satisfy the above. Therefore,

$$\mathbf{I} = 0 \text{ and } h\mathbf{I} = 0$$

and

$$\mathbf{E}_{oc} = \mathbf{E}$$

with

$$\mathbf{Z}_N = \frac{\mathbf{E}_{oc}}{\mathbf{I}_{sc}} = \frac{\mathbf{E}}{\dfrac{(1 + h)\mathbf{E}}{R_1 + (1 + h)R_2}} = \frac{R_1 + (1 + h)R_2}{(1 + h)}$$

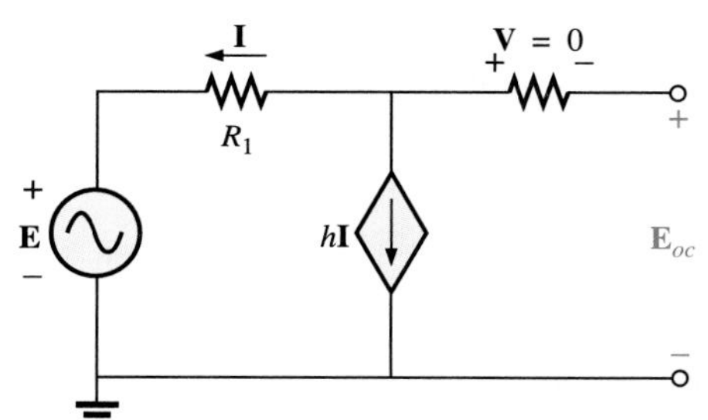

FIG. 18.69
Determining $\mathbf{E}_{oc}$ for the network of Fig. 18.68.

Method 2: Note Fig. 18.70. By Kirchhoff's current law,

$$\mathbf{I}_g = \mathbf{I} + h\mathbf{I} = (1 + h)\mathbf{I}$$

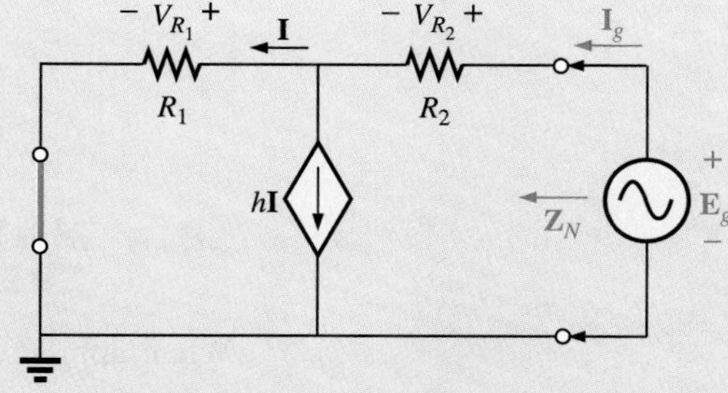

FIG. 18.70
Determining the Norton impedance using the approach $\mathbf{Z}_N = \mathbf{E}_g/\mathbf{E}_g$.

continued

By Kirchhoff's voltage law,

$$\mathbf{E}_g - \mathbf{I}_g R_2 - \mathbf{I} R_1 = 0$$

or

$$\mathbf{I} = \frac{\mathbf{E}_g - \mathbf{I}_g R_2}{R_1}$$

Substituting, we have

$$\mathbf{I}_g = (1 + h)\mathbf{I} = (1 + h)\left(\frac{\mathbf{E}_g - \mathbf{I}_g R_2}{R_1}\right)$$

and

$$\mathbf{I}_g R_1 = (1 + h)\mathbf{E}_g - (1 + h)\mathbf{I}_g R_2$$

so

$$\mathbf{E}_g(1 + h) = \mathbf{I}_g[R_1 + (1 + h)R_2]$$

or

$$\mathbf{Z}_N = \frac{\mathbf{E}_g}{\mathbf{I}_g} = \frac{R_1 + (1 + h)R_2}{1 + h}$$

which agrees with the above.

EXAMPLE 18.18

Find the Norton equivalent circuit for the network configuration of Fig. 18.51.

Solution: By source conversion,

$$\mathbf{I}_N = \frac{\mathbf{E}_{Th}}{\mathbf{Z}_{Th}} = \frac{\dfrac{-k_2 R_2 \mathbf{V}_i}{R_1 - k_1 k_2 R_2}}{\dfrac{R_1 R_2}{R_1 - k_1 k_2 R_2}}$$

and

$$\mathbf{I}_N = \frac{-k_2 \mathbf{V}_i}{R_1} \tag{18.12}$$

which is $\mathbf{I}_{sc}$ as determined in Example 18.13, and

$$\mathbf{Z}_N = \mathbf{Z}_{Th} = \frac{R_2}{1 - \dfrac{k_1 k_2 R_2}{R_1}} \tag{18.13}$$

For $k_1 \cong 0$, we have

$$\mathbf{I}_N = \frac{-k_2 \mathbf{V}_i}{R_1} \qquad k_1 = 0 \tag{18.14}$$

$$\mathbf{Z}_N = R_2 \qquad k_1 = 0 \tag{18.15}$$

18.5 MAXIMUM POWER TRANSFER THEOREM (ac)

When applied to ac circuits, the **maximum power transfer theorem** states that

maximum power will be delivered to a load when the load impedance is the conjugate of the Thévenin impedance across its terminals.

That is, for Fig. 18.71, for maximum power transfer to the load,

$$Z_L = Z_{Th} \text{ and } \theta_L = -\theta_{Th_Z} \tag{18.16}$$

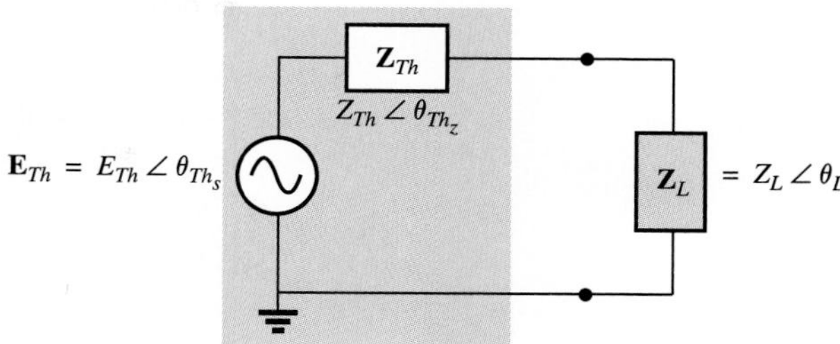

FIG. 18.71
Defining the conditions for maximum power transfer to a load.

or, in rectangular form,

$$R_L = R_{Th} \text{ and } \pm j\,X_{\text{load}} = \mp j\,X_{Th} \tag{18.17}$$

The conditions just mentioned will make the total impedance of the circuit appear purely resistive, as indicated in Fig. 18.72:

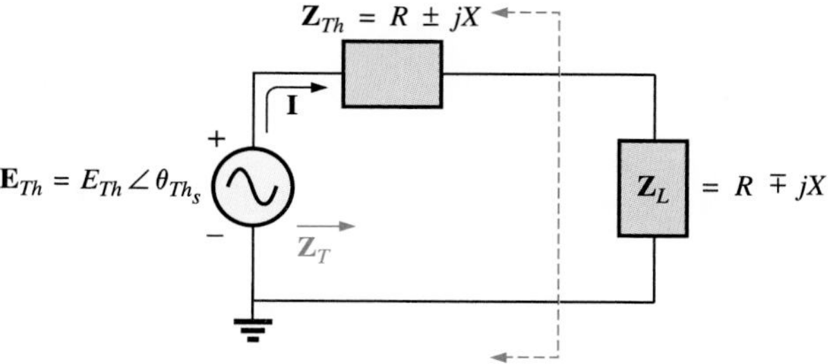

FIG. 18.72
Conditions for maximum power transfer to $\mathbf{Z}_L$.

$$\mathbf{Z}_T = (R \pm j\,X) + (R \mp j\,X)$$

and

$$\mathbf{Z}_T = 2R \tag{18.18}$$

Since the circuit is purely resistive, the power factor (recall Section 14.5) of the circuit under maximum power conditions is 1; that is,

$$F_p = 1 \qquad \text{(maximum power transfer)} \tag{18.19}$$

Power factor is discussed further in Section 19.3. The magnitude of the current **I** of Fig. 18.72 is

$$I = \frac{E_{Th}}{Z_T} = \frac{E_{Th}}{2R}$$

The maximum power to the load is

$$P_{\text{max}} = I^2 R = \left(\frac{E_{Th}}{2R}\right)^2 R$$

and

$$P_{\text{max}} = \frac{E_{Th}^2}{4R} \tag{18.20}$$

EXAMPLE 18.19

Find (a) the load impedance ($\mathbf{Z}_L$) in Fig. 18.73 for maximum power to the load, and (b) the maximum power (P_{max}).

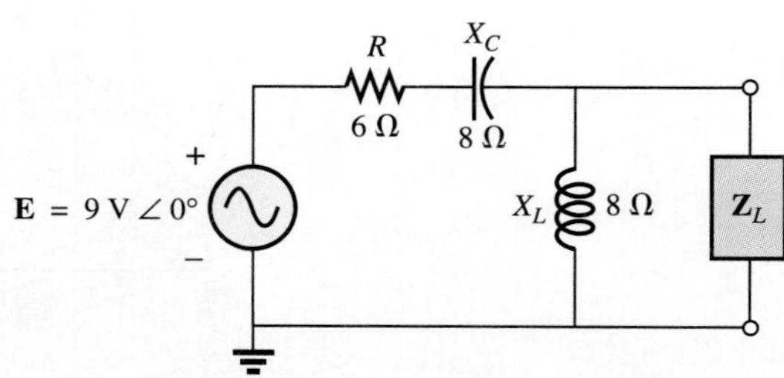

FIG. 18.73
Example 18.19.

Solutions: (a) Determine $\mathbf{Z}_{Th}$ [Fig. 18.74(a)]:

$$\mathbf{Z}_1 = R - jX_C = 6\ \Omega - j\,8\ \Omega = 10\ \Omega\ \angle -53.13^\circ$$

$$\mathbf{Z}_2 = +jX_L = j\,8\ \Omega$$

$$\mathbf{Z}_{Th} = \frac{\mathbf{Z}_1\mathbf{Z}_2}{\mathbf{Z}_1 + \mathbf{Z}_2} = \frac{(10\ \Omega\ \angle -53.13^\circ)(8\ \Omega\ \angle 90^\circ)}{6\ \Omega - j\,8\ \Omega + j\,8\ \Omega} = \frac{80\ \Omega\ \angle 36.87^\circ}{6\ \angle 0^\circ}$$

$$= 13.33\ \Omega\ \angle 36.87^\circ = 10.66\ \Omega + j\,8\ \Omega$$

and $\qquad \mathbf{Z}_L = 13.3\ \Omega\ \angle -36.87^\circ = \mathbf{10.66\ \Omega - j\,8\ \Omega}$

(b) To find the maximum power, we must first find $\mathbf{E}_{Th}$ [Fig. 18.74(b)], as follows:

$$\mathbf{E}_{Th} = \frac{\mathbf{Z}_2\mathbf{E}}{\mathbf{Z}_2 + \mathbf{Z}_1} \quad \text{(voltage divider rule)}$$

$$= \frac{(8\ \Omega\ \angle 90^\circ)(9\ \text{V}\ \angle 0^\circ)}{j\,8\ \Omega + 6\ \Omega - j\,8\ \Omega} = \frac{72\ \text{V}\ \angle 90^\circ}{6\ \angle 0^\circ} = 12\ \text{V}\ \angle 90^\circ$$

Then $\qquad P_{max} = \dfrac{E_{Th}^2}{4R} = \dfrac{(12\ \text{V})^2}{4(10.66\ \Omega)} = \dfrac{144}{42.64} = \mathbf{3.38\ W}$

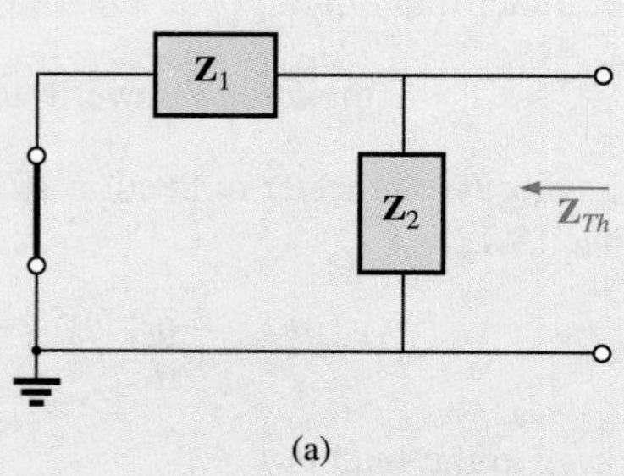

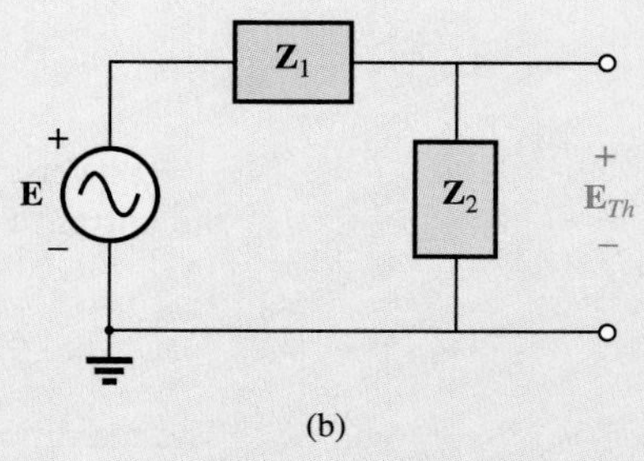

FIG. 18.74
Determining (a) $\mathbf{Z}_{Th}$ and (b) $\mathbf{E}_{Th}$ for the network external to the load in Fig. 18.73.

EXAMPLE 18.20

Find (a) the load impedance ($\mathbf{Z}_L$) in Fig. 18.75 for maximum power to the load, and (b) the maximum power (P_{max}).

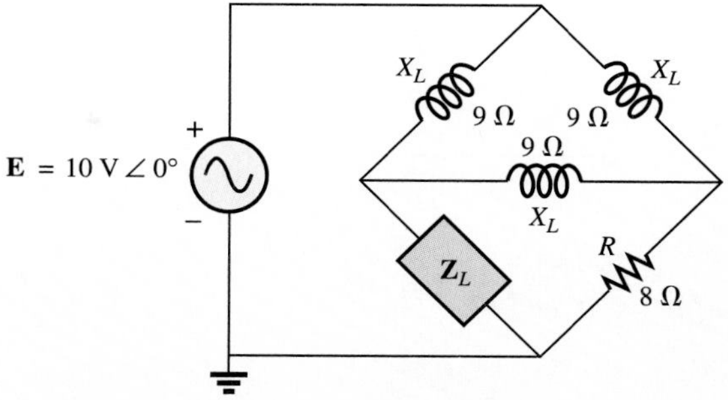

FIG. 18.75
Example 18.20.

Solutions: (a) First we must find $\mathbf{Z}_{Th}$ (Fig. 18.76).

$$\mathbf{Z}_1 = +j X_L = j\,9\,\Omega \quad \mathbf{Z}_2 = R = 8\,\Omega$$

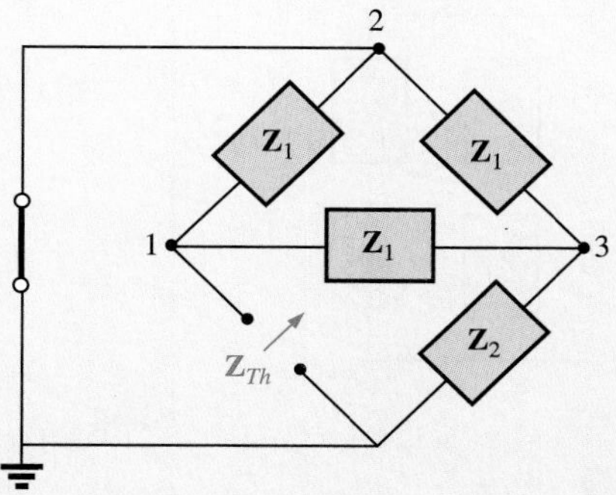

FIG. 18.76
Defining the subscripted impedances for the network of Fig. 18.75.

Converting from a Δ to a Y (Fig. 18.77), we have

$$\mathbf{Z}'_1 = \frac{\mathbf{Z}_1}{3} = j\,3\,\Omega \quad \mathbf{Z}_2 = 8\,\Omega$$

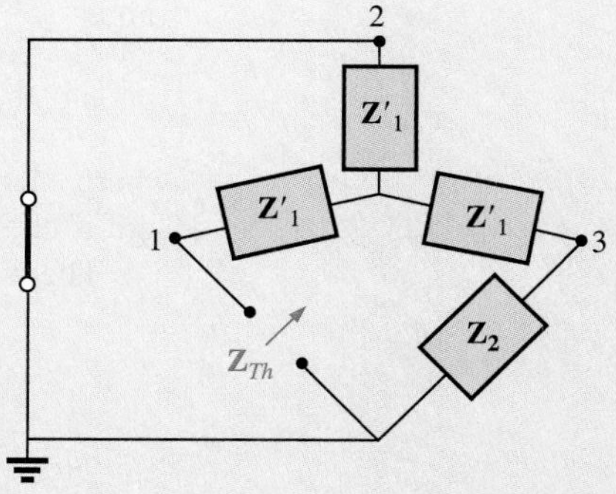

FIG. 18.77
Substituting the Y equivalent for the upper Δ configuration of Fig. 18.76.

continued

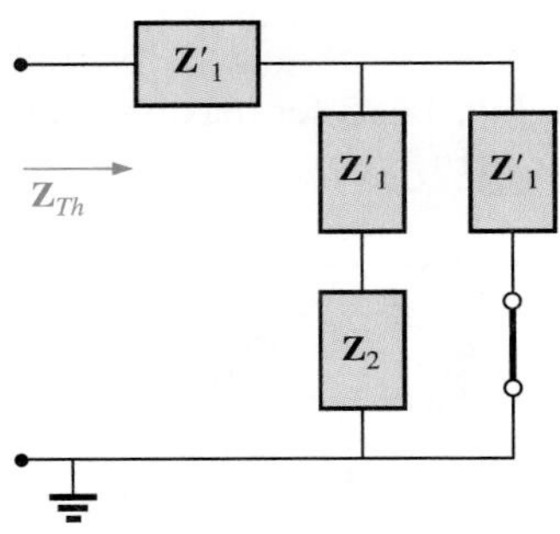

FIG. 18.78
Determining $\mathbf{Z}_{Th}$ for the network of Fig. 18.75.

The redrawn circuit (Fig. 18.78) shows

$$\mathbf{Z}_{Th} = \mathbf{Z}'_1 + \frac{\mathbf{Z}'_1(\mathbf{Z}'_1 + \mathbf{Z}_2)}{\mathbf{Z}'_1 + (\mathbf{Z}'_1 + \mathbf{Z}_2)}$$

$$= j\,3\;\Omega + \frac{3\;\Omega\;\angle 90^\circ(j\,3\;\Omega + 8\;\Omega)}{j\,6\;\Omega + 8\;\Omega}$$

$$= j\,3 + \frac{(3\;\angle 90^\circ)(8.54\;\angle 20.56^\circ)}{10\;\angle 36.87^\circ}$$

$$= j\,3 + \frac{25.62\;\angle 110.56^\circ}{10\;\angle 36.87^\circ} = j\,3 + 2.56\;\angle 73.69^\circ$$

$$= j\,3 + 0.72 + j\,2.46$$

$$\mathbf{Z}_{Th} = 0.72\;\Omega + j\,5.46\;\Omega$$

and $$\mathbf{Z}_L = \mathbf{0.72\;\Omega - j\,5.46\;\Omega}$$

(b) For $\mathbf{E}_{Th}$, use the modified circuit of Fig. 18.79 with the voltage source replaced in its original position. Since $I_1 = 0$, $\mathbf{E}_{Th}$ is the voltage across the series impedance of $\mathbf{Z}'_1$ and $\mathbf{Z}_2$. Using the voltage divider rule gives us

$$\mathbf{E}_{Th} = \frac{(\mathbf{Z}'_1 + \mathbf{Z}_2)\mathbf{E}}{\mathbf{Z}'_1 + \mathbf{Z}_2 + \mathbf{Z}'_1} = \frac{(j\,3\;\Omega + 8\;\Omega)(10\;\text{V}\;\angle 0^\circ)}{8\;\Omega + j\,6\;\Omega}$$

$$= \frac{(8.54\;\angle 20.56^\circ)(10\;\text{V}\;\angle 0^\circ)}{10\;\angle 36.87^\circ}$$

$$\mathbf{E}_{Th} = 8.54\;\text{V}\;\angle -16.31^\circ$$

and $$P_{\max} = \frac{E_{Th}^2}{4R} = \frac{(8.54\;\text{V})^2}{4(0.72\;\Omega)} = \frac{72.93}{2.88}\;\text{W}$$
$$= \mathbf{25.32\;W}$$

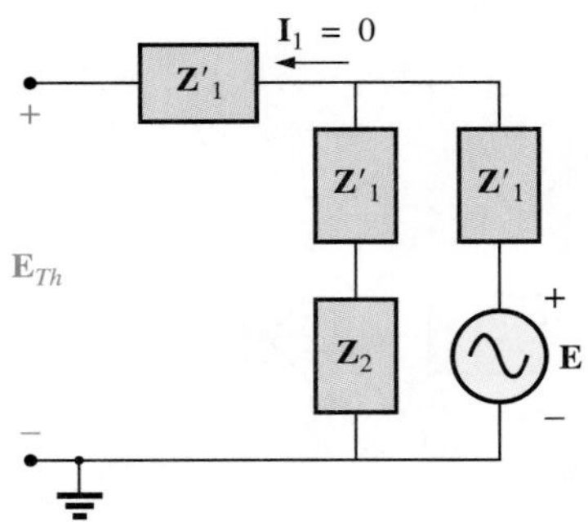

FIG. 18.79
Finding the Thévenin voltage for the network of Fig. 18.75.

If the load resistance is adjustable but the magnitude of the load reactance cannot be set equal to the magnitude of the Thévenin reactance, then the maximum power *that can be delivered* to the load will occur when the load reactance is made as close to the Thévenin reactance as possible and the load resistance is set to the following value:

$$R_L = \sqrt{R_{Th}^2 + (X_{Th} + X_{\text{load}})^2} \qquad (18.21)$$

where each reactance carries a positive sign if inductive and a negative sign if capacitive.

The power delivered will be determined by

$$P = E_{Th}^2/4R_{\text{av}} \qquad (18.22)$$

where $$R_{\text{av}} = \frac{R_{Th} + R_L}{2} \qquad (18.23)$$

The derivation of the above equations is given in Appendix E of the text. The following example demonstrates their use.

EXAMPLE 18.21

For the network of Fig. 18.80:

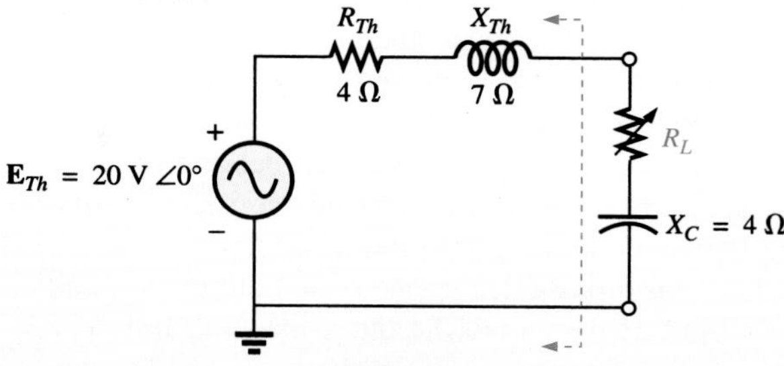

FIG. 18.80
Example 18.21.

a. Determine the value of R_L for maximum power to the load if the load reactance is fixed at 4 Ω.
b. Find the power delivered to the load under the conditions of part (a).
c. Find the maximum power to the load if the load reactance is made adjustable to any value, and compare the result to part (b) above.

Solutions:

a. Eq. (18.21): $R_L = \sqrt{R_{Th}^2 + (X_{Th} + X_{load})^2}$

$$= \sqrt{(4\ \Omega)^2 + (7\ \Omega - 4\ \Omega)^2}$$

$$= \sqrt{16 + 9} = \sqrt{25}$$

$$R_L = \mathbf{5\ \Omega}$$

b. Eq. (18.22): $P = \dfrac{E_{Th}^2}{4R_{av}}$

$$= \frac{(20\ \text{V})^2}{4(4.5\ \Omega)} = \frac{400}{18}\ \text{W}$$

$$\cong \mathbf{22.22\ W}$$

Eq. (18.23): $R_{av} = \dfrac{R_{Th} + R_L}{2} = \dfrac{4\ \Omega + 5\ \Omega}{2}$

$$= \mathbf{4.5\ \Omega}$$

c. For $\mathbf{Z}_L = 4\ \Omega - j\,7\ \Omega$,

$$P_{max} = \frac{E_{Th}^2}{4R_{Th}} = \frac{(20\ \text{V})^2}{4(4\ \Omega)}$$

$$= \mathbf{25\ W}$$

exceeding the result of part (b) by 2.78 W.

Practical Business Application

No fried chips for me, please!

Years ago, a keen electrical student set out to make his small home-stereo into a powerful rock concert hall. The idea that seemed to make sense at the time was to wire up lots of extra speakers in parallel and to place these speakers uniformly all around the room. Something like an early surround-sound concept.

After installing three speakers on both the left and right output of the stereo, he did a listening test. The student sat in the very middle of the room and listened to music first with only one speaker, then with the three speakers connected to each stereo output. Can you guess what the result was?

Initially, when the additional speakers where hooked up, the student thought his stereo had suddenly blown up. Or maybe his ears where fooling him. After trying every possible arrangement he could think of, he gave up in frustration. He found that every time he used more than one speaker, the music volume in the room was actually less than with only one speaker. "But how could this be?" he wondered. "Doesn't having more speakers in parallel mean you'll get more volume?" The answer is: Nope.

This student hadn't yet learned about the maximum power transfer theorem. Here's a simple circuit diagram of an ideal voltage source in series with the internal impedance of a real source, which is in series with a load, the speaker:

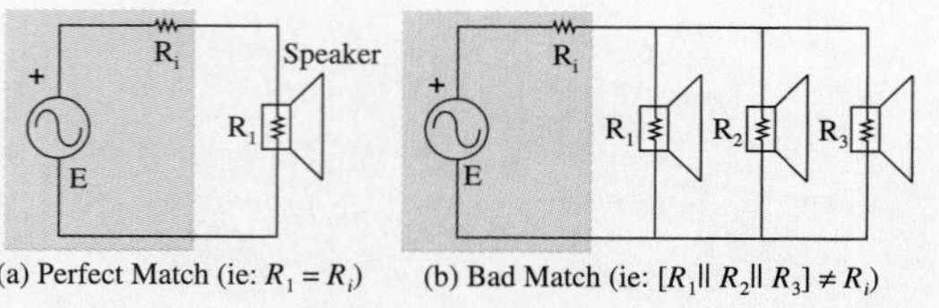

(a) Perfect Match (ie: $R_1 = R_i$) (b) Bad Match (ie: $[R_1 \| R_2 \| R_3] \neq R_i$)

A simplified circuit diagram can be used to illustrate the problem. Let's start by assuming a constant ideal voltage source of E volts. For a typical stereo system, the speakers each have resistance $R = 8\ \Omega$, and the internal impedance of the stereo is designed to match this ($R_i = 8\ \Omega$). Thus, the net voltage delivered to this speaker, E_1, is:

$$E_1 = E R_1/(R_1 + R_i) = E/2$$

And the power, P_1, is:

$$P_1 = (E_1)^2/R_1 = E^2/(4R_1) = E^2/32$$

By putting three speakers in parallel instead, the total load impedance, R_2, drops to:

$$R_2 = R_1/3$$

The net voltage, E_2, and power, P_2, delivered to all the speakers are:

$$E_2 = E R_2/(R_2 + \mathrm{Ri}) = \mathrm{E}/4$$

$$\mathrm{P}_2 = (\mathrm{E}_2)^2/\mathrm{R}_2 = 3\mathrm{E}^2/(16\mathrm{R}_1)$$

Got it? The power has dropped by one sixteenth. That's not exactly the three times improvement our young student was expecting. By the way, did you know that the student could very well have blown up his stereo's output section?

That's right! Because if you put three speakers in parallel, you'll be attaching one third of the resistance to the output section of your amplifier inside your stereo system. According to Ohm's law, with constant voltage and reduced resistance, what must the current become?

Violà! Up it goes. And so, by playing loud rock music while the integrated circuit chips inside your stereo can't handle the higher current, you'll be roasting your stereo—and frying your chips.

Rock On Dude! And try to think of a way to connect four speakers to avoid this problem.

PROBLEMS

SECTION 18.2 Superposition Theorem

1. Using superposition, determine the current through the inductance X_L for each network of Fig. 18.82.

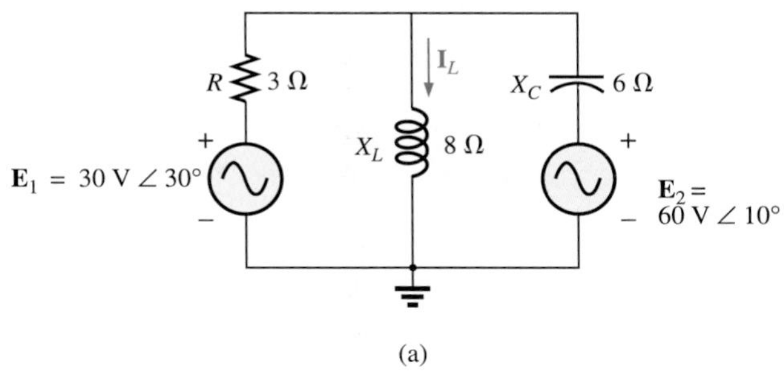

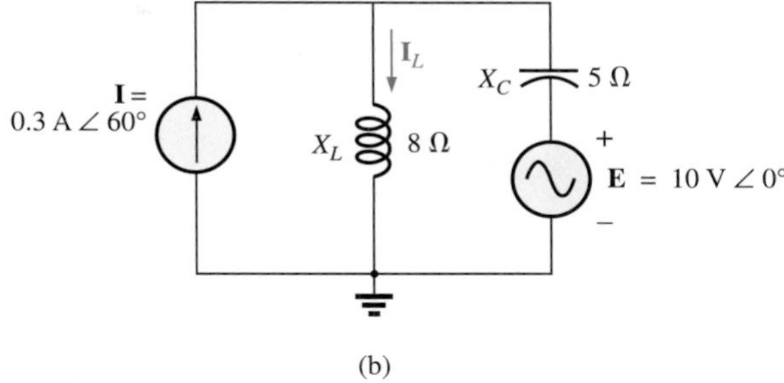

FIG. 18.82
Problem 1.

*2. Using superposition, determine the current $\mathbf{I}_L$ for each network of Fig. 18.83.

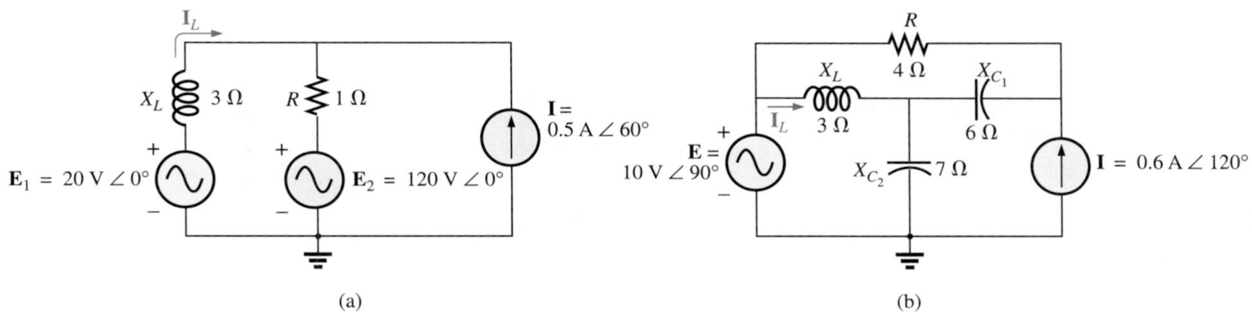

FIG. 18.83
Problem 2.

*3. Using superposition, find the sinusoidal expression for the current i for the network of Fig. 18.84.

4. Using superposition, find the sinusoidal expression for the voltage v_C for the network of Fig. 18.85.

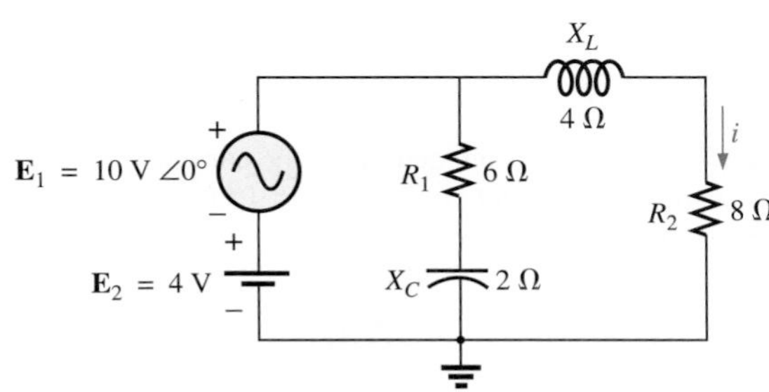

FIG. 18.84
Problems 3, 15, 30, and 42.

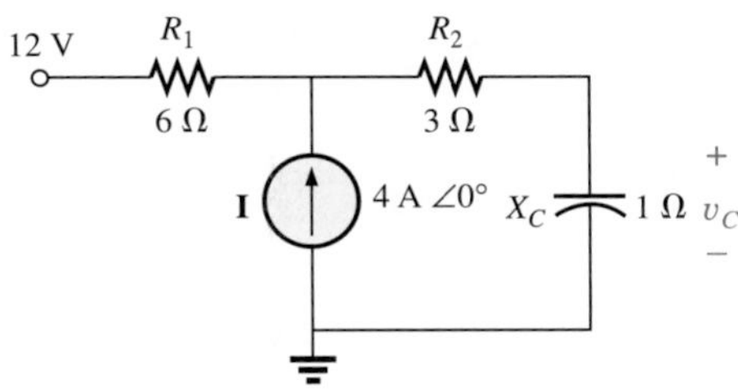

FIG. 18.85
Problems 4, 16, 31, and 43.

***5.** Using superposition, find the current **I** for the network of Fig. 18.86.

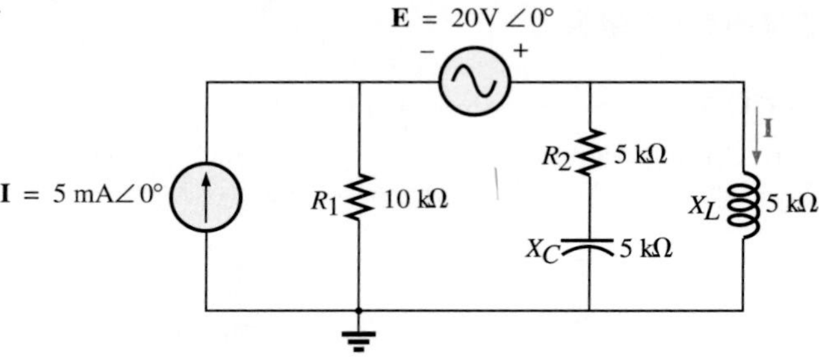

FIG. 18.86
Problems 5, 17, 32, and 44.

6. Using superposition, determine the current $\mathbf{I}_L$ ($h = 100$) for the network of Fig. 18.87.

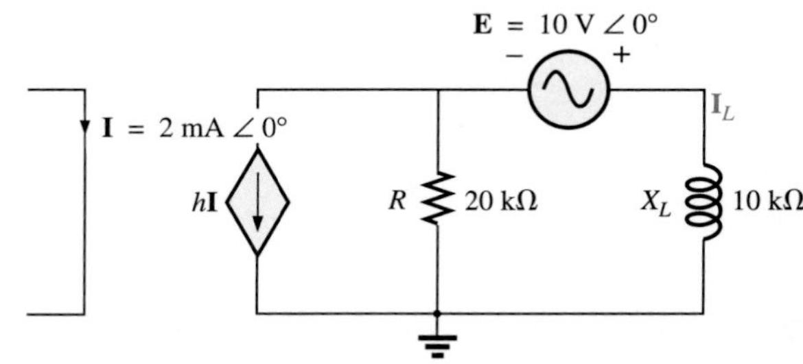

FIG. 18.87
Problems 6 and 20.

7. Using superposition, for the network of Fig. 18.88, determine the voltage $\mathbf{V}_L$ ($\mu = 20$).

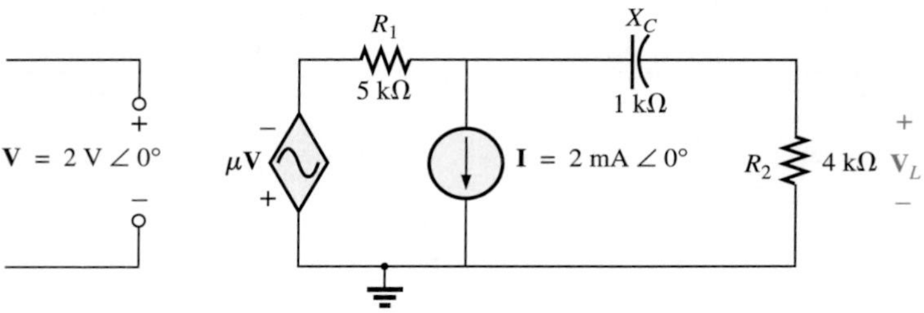

FIG. 18.88
Problems 7, 21, and 35.

***8.** Using superposition, determine the current $\mathbf{I}_L$ for the network of Fig. 18.89 ($\mu = 20$; $h = 100$).

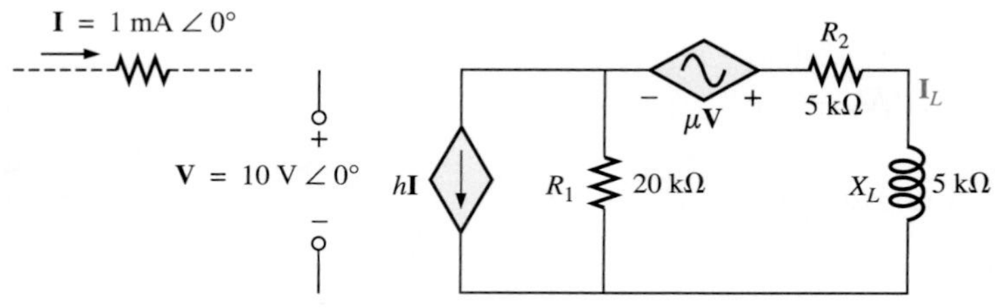

FIG. 18.89
Problems 8, 22, and 36.

***9.** Determine $\mathbf{V}_L$ for the network of Fig. 18.90 ($h = 50$).

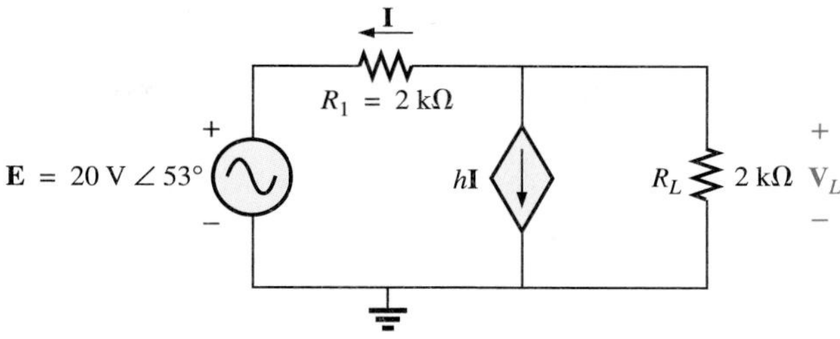

FIG. 18.90
Problems 9 and 23.

***10.** Calculate the current **I** for the network of Fig. 18.91.

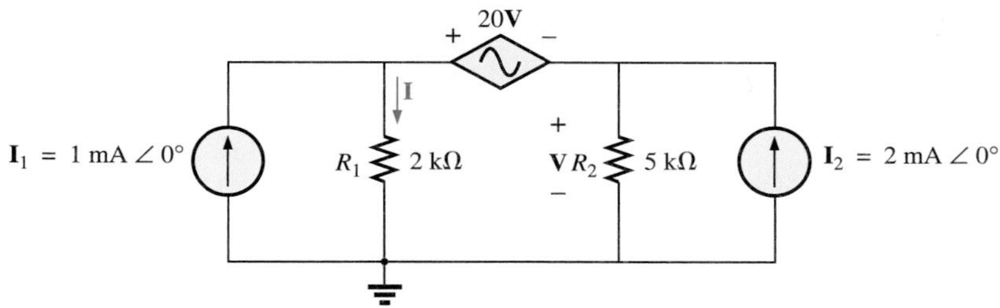

FIG. 18.91
Problems 10, 24, and 38.

11. Find the voltage $\mathbf{V}_s$ for the network of Fig. 18.92.

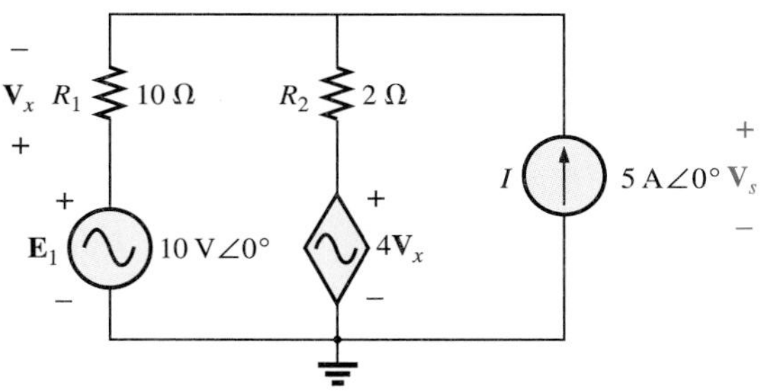

FIG. 18.92
Problem 11.

SECTION 18.3 Thévenin's Theorem

12. Find the Thévenin equivalent circuit for the portions of the networks of Fig. 18.93 external to the elements between points *a* and *b*.

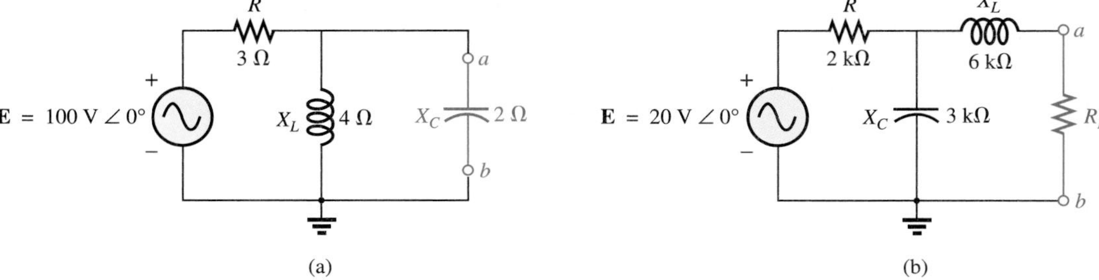

FIG. 18.93
Problems 12 and 26.

*13. Find the Thévenin equivalent circuit for the portions of the networks of Fig. 18.94 external to the elements between points a and b.

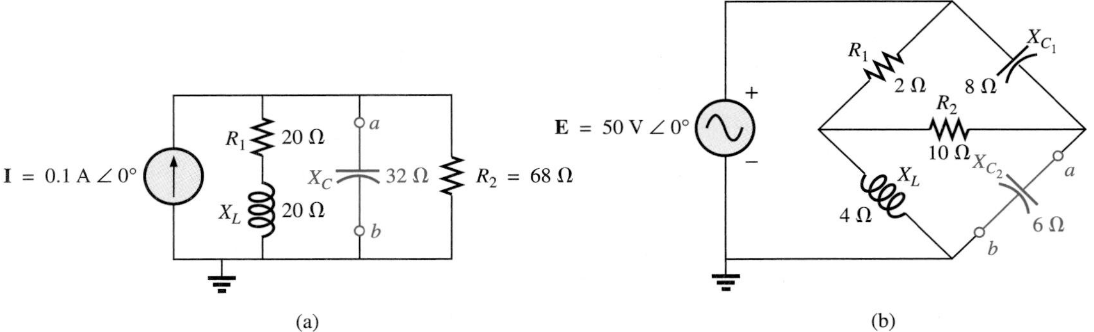

FIG. 18.94
Problems 13 and 27.

*14. Find the Thévenin equivalent circuit for the portions of the networks of Fig. 18.95 external to the elements between points a and b.

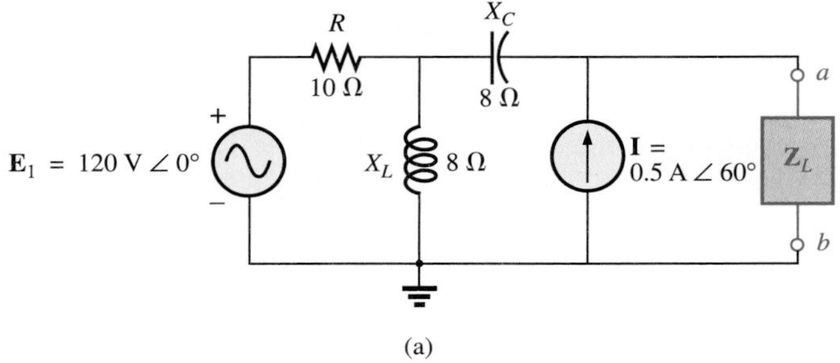

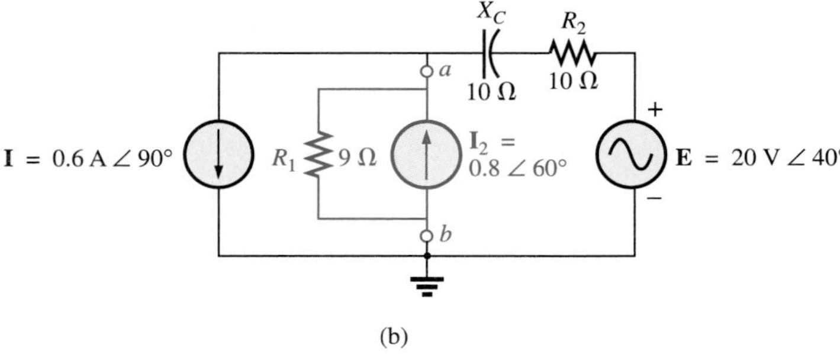

FIG. 18.95
Problems 14 and 28.

*15. **a.** Find the Thévenin equivalent circuit for the network external to the resistor R_2 in Fig. 18.84.
b. Using the results of part (a), determine the current i of the same figure.

16. a. Find the Thévenin equivalent circuit for the network external to the capacitor of Fig. 18.85.
b. Using the results of part (a), determine the voltage $\mathbf{V}_C$ for the same figure.

*17. **a.** Find the Thévenin equivalent circuit for the network external to the inductor of Fig. 18.86.
b. Using the results of part (a), determine the current **I** of the same figure.

18. Determine the Thévenin equivalent circuit for the network external to the 5-kΩ inductive reactance of Fig. 18.96 (in terms of **V**).

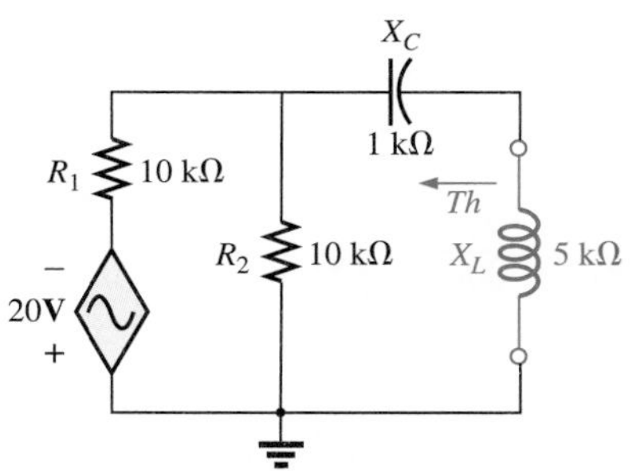

FIG. 18.96
Problems 18 and 33.

19. Determine the Thévenin equivalent circuit for the network external to the 4-kΩ inductive reactance of Fig. 18.97 (in terms of **I**).

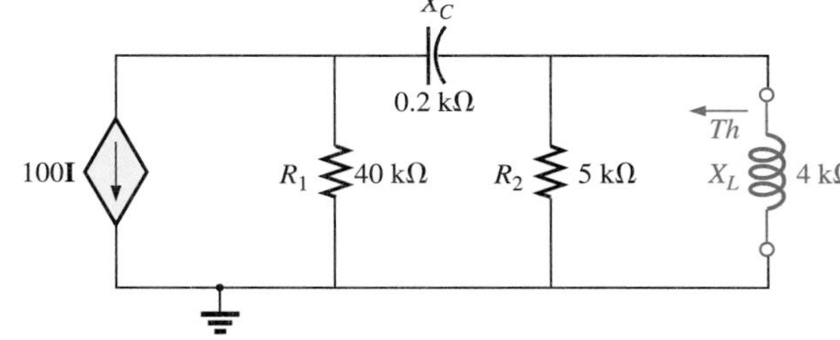

FIG. 18.97
Problems 19 and 34.

20. Find the Thévenin equivalent circuit for the network external to the 10-kΩ inductive reactance of Fig. 18.87.

21. Determine the Thévenin equivalent circuit for the network external to the 4-kΩ resistor of Fig. 18.88.

***22.** Find the Thévenin equivalent circuit for the network external to the 5-kΩ inductive reactance of Fig. 18.89.

***23.** Determine the Thévenin equivalent circuit for the network external to the 2-kΩ resistor of Fig. 18.90.

***24.** Find the Thévenin equivalent circuit for the network external to the resistor R_1 of Fig. 18.91.

***25.** Find the Thévenin equivalent circuit for the network to the left of terminals a-a' of Fig. 18.98.

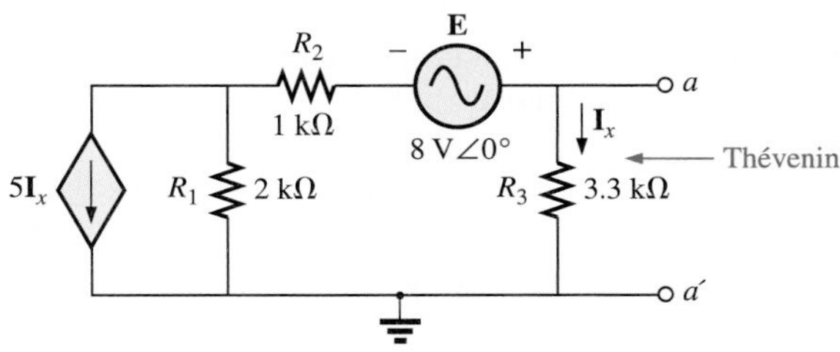

FIG. 18.98
Problem 25.

SECTION 18.4 Norton's Theorem

26. Find the Norton equivalent circuit for the network external to the elements between a and b for the networks of Fig. 18.93.

27. Find the Norton equivalent circuit for the network external to the elements between a and b for the networks of Fig. 18.94.

28. Find the Norton equivalent circuit for the network external to the elements between a and b for the networks of Fig. 18.95.

***29.** Find the Norton equivalent circuit for the portions of the networks of Fig. 18.99 external to the elements between points *a* and *b*.

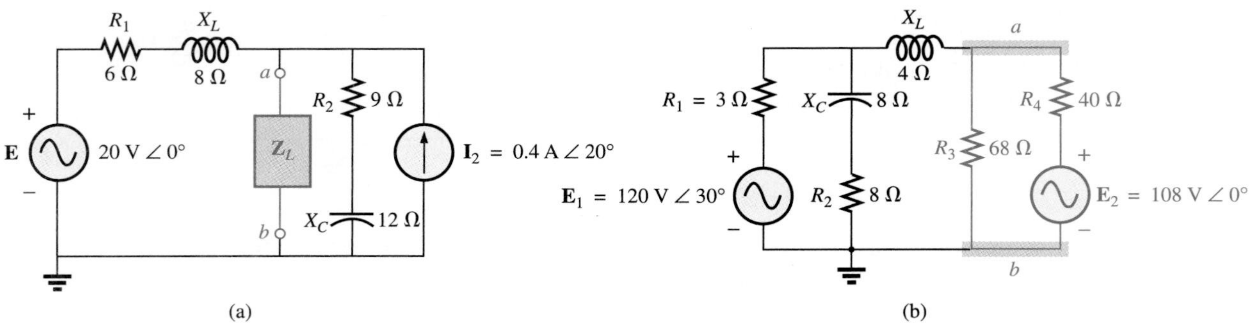

FIG. 18.99
Problem 29.

***30. a.** Find the Norton equivalent circuit for the network external to the resistor R_2 in Fig. 18.84.
b. Using the results of part (a), determine the current **I** of the same figure.

***31. a.** Find the Norton equivalent circuit for the network external to the capacitor of Fig. 18.85.
b. Using the results of part (a), determine the voltage $\mathbf{V}_C$ for the same figure.

***32. a.** Find the Norton equivalent circuit for the network external to the inductor of Fig. 18.86.
b. Using the results of part (a), determine the current **I** of the same figure.

33. Determine the Norton equivalent circuit for the network external to the 5-kΩ inductive reactance of Fig. 18.96.

34. Determine the Norton equivalent circuit for the network external to the 4-kΩ inductive reactance of Fig. 18.97.

35. Find the Norton equivalent circuit for the network external to the 4-kΩ resistor of Fig. 18.88.

***36.** Find the Norton equivalent circuit for the network external to the 5-kΩ inductive reactance of Fig. 18.89.

***37.** For the network of Fig. 18.100, find the Norton equivalent circuit for the network external to the 2-kΩ resistor.

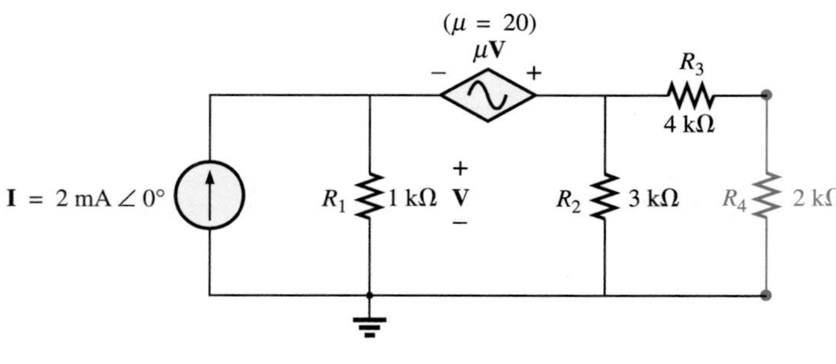

FIG. 18.100
Problem 37.

*38. Find the Norton equivalent circuit for the network external to the $\mathbf{I}_1$ current source of Fig. 18.91.

SECTION 18.5 Maximum Power Transfer Theorem

39. Find the load impedance $\mathbf{Z}_L$ for the networks of Fig. 18.101 for maximum power to the load, and find the maximum power to the load.

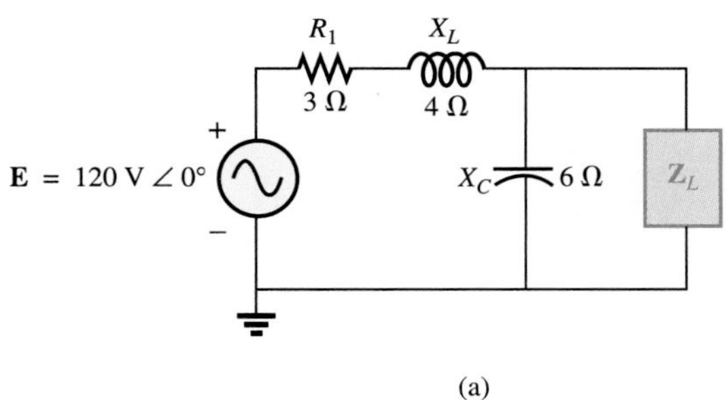

(a)

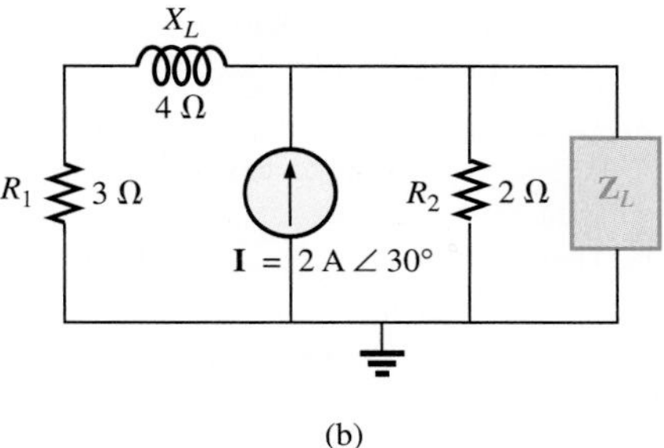

(b)

FIG. 18.101
Problem 39.

*40. Find the load impedance $\mathbf{Z}_L$ for the networks of Fig. 18.102 for maximum power to the load, and find the maximum power to the load.

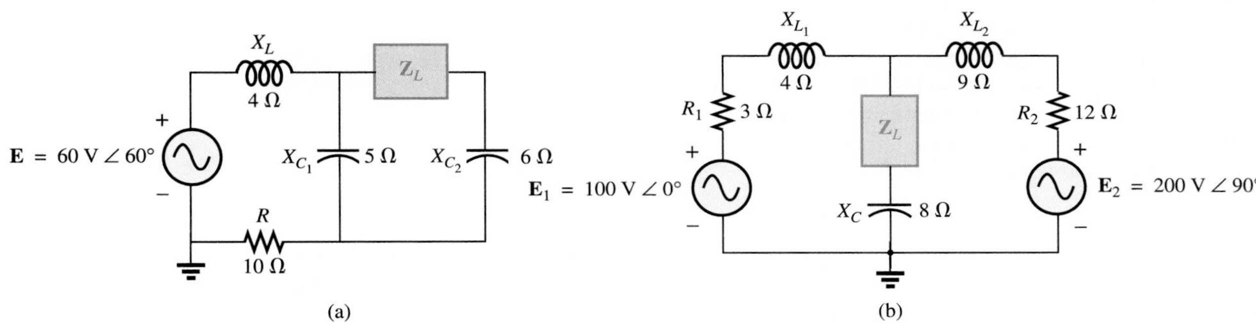

FIG. 18.102
Problem 40.

41. Find the load impedance R_L for the network of Fig. 18.103 for maximum power to the load, and find the maximum power to the load.

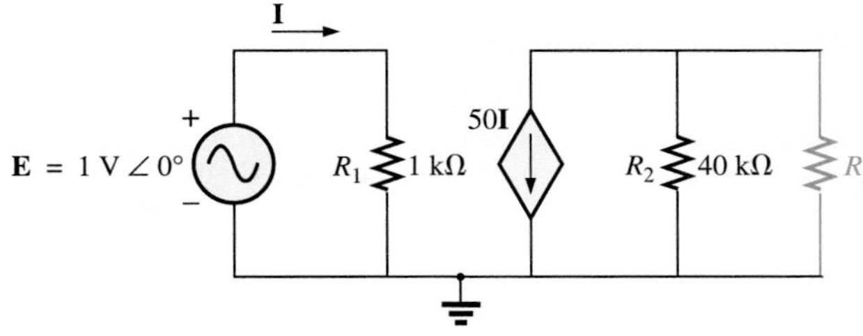

FIG. 18.103
Problem 41.

*42. **a.** Determine the load impedance to replace the resistor R_2 of Fig. 18.84 to ensure maximum power to the load.

b. Using the results of part (a), determine the maximum power to the load.

*43. **a.** Determine the load impedance to replace the capacitor X_C of Fig. 18.85 to ensure maximum power to the load.

b. Using the results of part (a), determine the maximum power to the load.

*44. **a.** Determine the load impedance to replace the inductor X_L of Fig. 18.86 to ensure maximum power to the load.

b. Using the results of part (a), determine the maximum power to the load.

45. **a.** For the network of Fig. 18.104, determine the value of R_L that will result in maximum power to the load.

b. Using the results of part (a), determine the maximum power delivered.

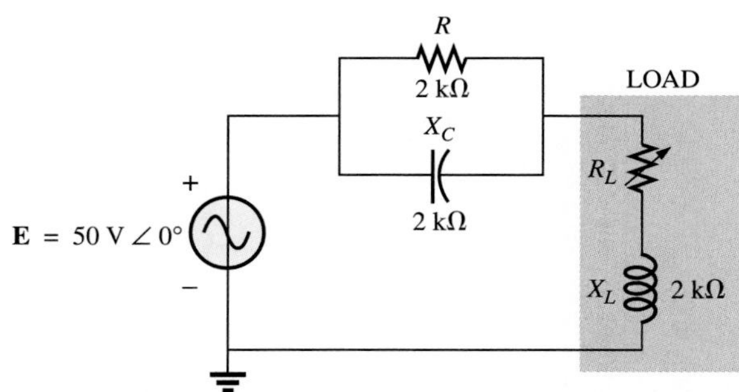

FIG. 18.104
Problem 45.

*46. **a.** For the network of Fig. 18.105, determine the level of capacitance that will ensure maximum power to the load if the range of capacitance is limited to 1 to 5 nF.

b. Using the results of part (a), determine the value of R_L that will ensure maximum power to the load.

c. Using the results of parts (a) and (b), determine the maximum power to the load.

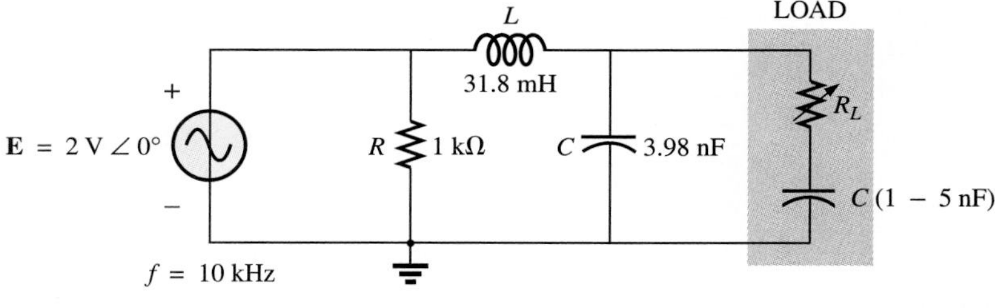

FIG. 18.105
Problem 46.

Substitution, Reciprocity, and Millman's Theorems

47. For the network of Fig. 18.106, determine two equivalent branches through the substitution theorem for the branch *a-b*.

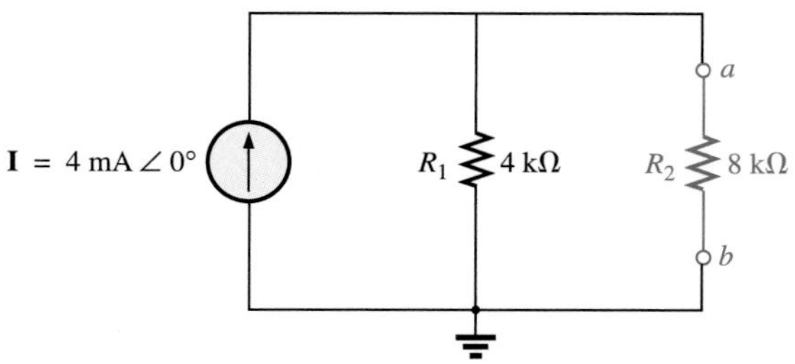

FIG. 18.106
Problem 47.

48. **a.** For the network of Fig. 18.107(a), find the current **I**.
b. Repeat part (a) for the network of Fig. 18.107(b).
c. How do the results of parts (a) and (b) compare?

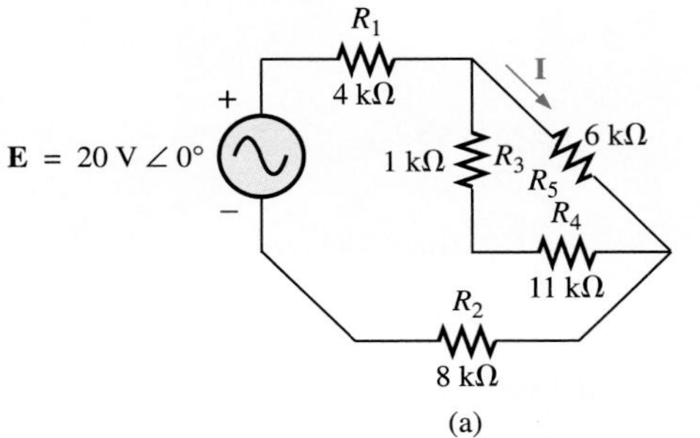

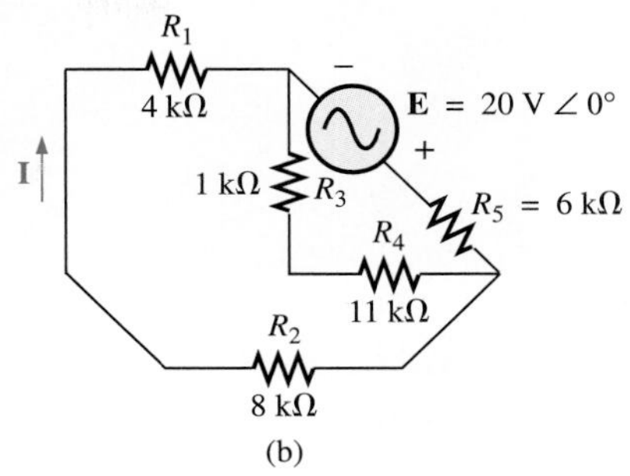

FIG. 18.107
Problem 48.

49. Using Millman's theorem, determine the current through the 4-kΩ capacitive reactance of Fig. 18.108.

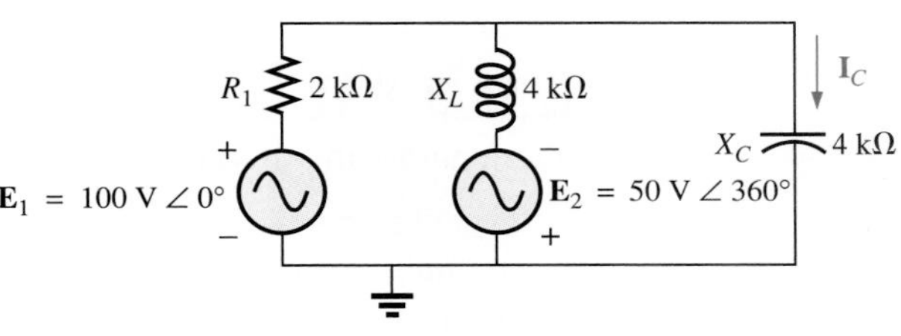

FIG. 18.108
Problem 49.

19 Power (ac)

Outline

Learning Outcomes

After completing this chapter, you will be able to

- describe and predict the flow of energy in resistors, inductors, and capacitors
- define and describe the differences between average, reactive, and apparent power
- determine all values in a load using its rated values
- calculate the average power, reactive power, apparent power, and power factor for any circuit or load
- draw the power triangle for any circuit or load
- explain the effect of a low power factor in an industrial load
- analyze and correct an industrial load with a poor power factor
- determine the effective resistance by measurement

Reactance is a property that causes excessive current and voltage fluctuation. This chapter shows how to minimize current and optimize circuit performance.

19.1 INTRODUCTION

The discussion of power in Chapter 14 included only the average power delivered to an ac network. We will now examine the total power equation in a slightly different form and introduce two additional types of power: **apparent** and **reactive**.

For any system such as Fig. 19.1, the power delivered to a load at any instant is defined by the product of the applied voltage and the resulting current; that is,

$$p = vi$$

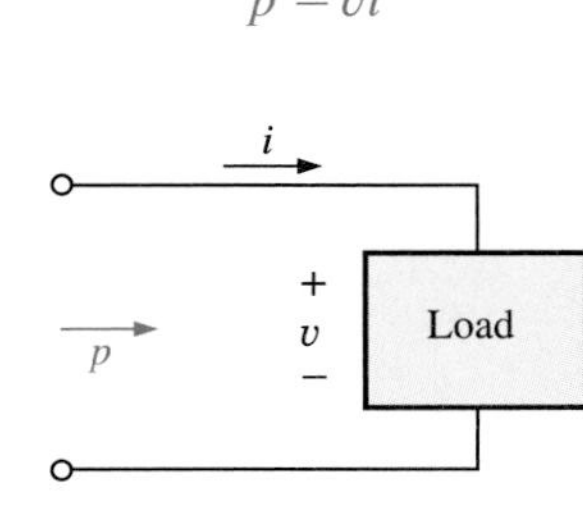

FIG. 19.1
Defining the power delivered to a load.

In this case, since v and i are sinusoidal quantities, let us establish a general case where

$$v = V_m \sin(\omega t + \theta)$$

and
$$i = I_m \sin \omega t$$

The chosen v and i include all possibilities because, if the load is purely resistive, $\theta = 0°$. If the load is purely inductive or capacitive, $\theta = 90°$ or $\theta = -90°$, respectively. For a network that is primarily inductive, θ is positive (v leads i), and for a network that is primarily capacitive θ is negative (i leads v).

Substituting the above equations for v and i into the power equation will result in

$$p = V_m I_m \sin \omega t \sin(\omega t + \theta)$$

The graph of this equation, made by multiplying the instantaneous values of v and i, is shown in Fig. 19.2.

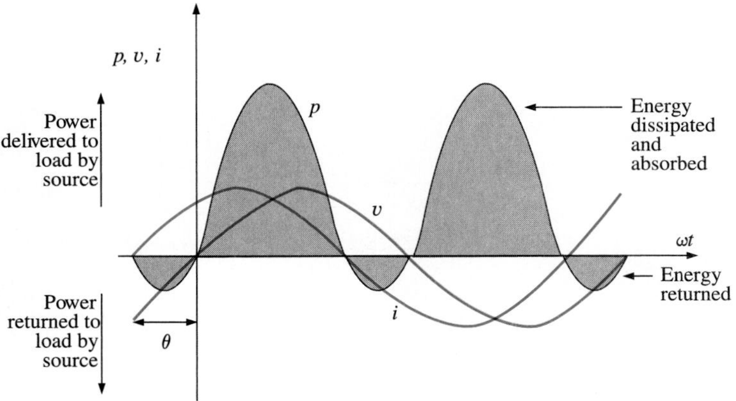

FIG. 19.2
Power vs. time for a general load.

If we now apply a number of trigonometric identities, the following form for the power equation will result:

$$p = VI\cos\theta(1 - \cos 2\omega t) + VI\sin\theta(\sin 2\omega t) \qquad (19.1)$$

where V and I are now effective values. The conversion from peak values V_m and I_m to effective values resulted from the operations performed using the trigonometric identities.

It would seem at first that nothing has been gained by putting the equation in this form. However, we will show how useful the form of Eq. (19.1) is in the following sections.

If Eq. (19.1) is expanded to the form

$$p = \underbrace{VI\cos\theta}_{\text{Average}} - \underbrace{VI\cos\theta}_{\text{Peak}}\underbrace{\cos 2\omega t}_{2x} + \underbrace{VI\sin\theta}_{\text{Peak}}\underbrace{\sin 2\omega t}_{2x}$$

we can make several points. First, the average power still appears as an isolated term that is time independent. Second, both terms that follow vary at a frequency twice that of the applied voltage or current with peak values having a very similar format. Third, the $VI\cos\theta$ term represents the components of V and I that are in phase, indicating power to a resistance. Fourth, the $VI\sin\theta$ term represents the components of V and I that are 90° out of phase, indicating power to an inductance or capacitance.

To ensure completeness and order in presentation, we will treat each basic element (R, L, and C) separately.

19.2 RESISTIVE CIRCUIT

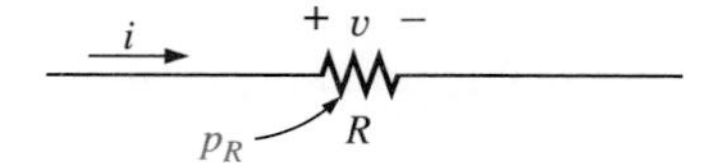

FIG. 19.3
Determining the power delivered to a purely resistive load.

For a purely resistive circuit (such as the one in Fig. 19.3), v and i are in phase, and $\theta = 0°$, as shown in Fig. 19.4. Substituting $\theta = 0°$ into Eq. (19.1), we obtain

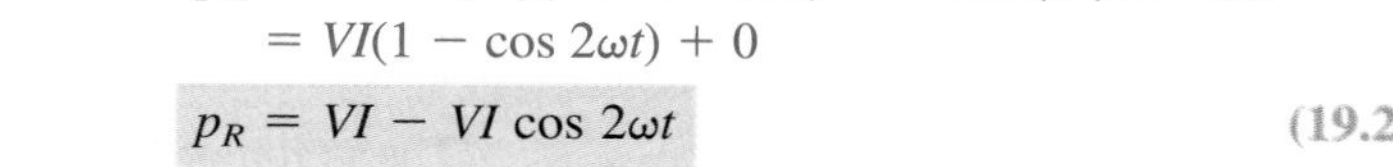

$$p_R = VI\cos(0°)(1 - \cos 2\omega t) + VI\sin(0°)\sin 2\omega t$$
$$= VI(1 - \cos 2\omega t) + 0$$

or

$$p_R = VI - VI\cos 2\omega t \qquad (19.2)$$

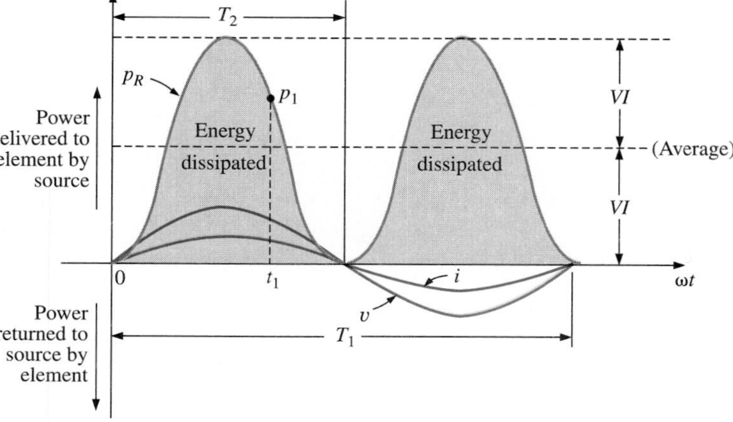

FIG. 19.4
Power versus time for a purely resistive load.

where VI is the average or dc term and $-VI \cos 2\omega t$ is a negative cosine wave with twice the frequency of either input quantity (v or i) and a peak value of VI.

Plotting the waveform for p_R (Fig. 19.4), we see that

$$T_1 = \text{period of input quantities, } v \text{ and } i$$

$$T_2 = \text{period of power curve } p_R$$

Note that in Fig. 19.4 the power curve passes through two cycles about its average value of VI for each cycle of either v or i ($T_1 = 2T_2$ or $f_2 = 2f_1$). Also, since the peak and average values of the power curve are the same, the curve is always above the horizontal axis. This means that

the total power delivered to a resistor will be dissipated in the form of heat.

The power returned to the source is represented by the portion of the curve below the axis, which is zero in this case. The power dissipated at any instant of time t_1 by the resistor can be found by simply substituting the time t_1 into Eq. (19.2) to find p_1, as indicated in Fig. 19.4. The average power from Eq. (19.2), or Fig. 19.4, is VI; or, as a summary,

$$P = VI = \frac{V_m I_m}{2} = I^2R = \frac{V^2}{R} \qquad \text{(watts, W)} \qquad (19.3)$$

as derived in Chapter 14. Other names for power to a resistance are *active* and *real* power.

The energy dissipated by the resistor (W_R) over one full cycle of the applied voltage (Fig. 19.4) can be found using the following equation:

$$W = Pt$$

where P is the average value and t is the period of the applied voltage; that is,

$$W_R = VIT_1 \qquad \text{(joules, J)} \qquad (19.4)$$

or, since $T_1 = 1/f_1$,

$$W_R = \frac{VI}{f_1} \qquad \text{(joules, J)} \qquad (19.5)$$

19.3 APPARENT POWER

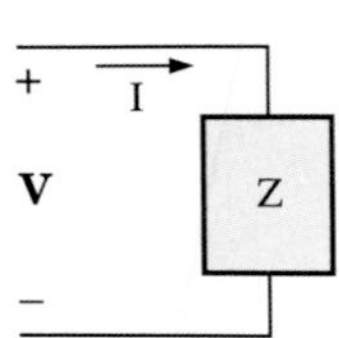

FIG. 19.5
Defining the apparent power to a load.

From our analysis of dc networks (and resistive elements above), it would seem *apparent* that the power delivered to the load of Fig. 19.5 is simply determined by the product of the applied voltage and current, with no concern for the components of the load; that is, $P = VI$. However, we found in Chapter 14 that the power factor ($\cos \theta$) of the load will have a significant effect on the number of watts dissipated. Although the product of the voltage and current is not always the number of watts delivered, it is a very useful power rating for describing and analyzing sinusoidal ac networks and for the maximum rating of many electrical components and systems. It is called the **apparent power** and its symbol is S. Since it is simply the product of voltage and current, its units are *volt-amperes*, abbreviated VA. Its magnitude is determined by

$$S = VI \qquad \text{(volt-amperes, VA)} \qquad (19.6)$$

or, since

$$V = IZ \text{ and } I = \frac{V}{Z}$$

then $$S = I^2Z \quad \text{(volt-amperes, VA)} \tag{19.7}$$

and $$S = \frac{V^2}{Z} \quad \text{(volt-amperes, VA)} \tag{19.8}$$

The average power to the load of Fig. 19.5 is

$$P = VI \cos \theta$$

However, $$S = VI$$

Therefore, $$P = S \cos \theta \quad \text{(watts, W)} \tag{19.9}$$

and the power factor of a system F_p is

$$F_p = \cos \theta = \frac{P}{S} \quad \text{(unitless)} \tag{19.10}$$

The power factor of a circuit, therefore, is the ratio of the average power to the apparent power. For a purely resistive circuit, we have

$$P = VI = S$$

and $$F_p = \cos \theta = \frac{P}{S} = 1$$

In general, power equipment is rated in volt-amperes (VA) or in kilovolt-amperes (kVA) and not in watts. By knowing the volt-ampere rating and the rated voltage of a device, we can easily determine the *maximum* current rating. For example, a device rated at 10 kVA at 200 V has a maximum current rating of $I =$ 10 000 VA/200 V $=$ 50 A when operated under rated conditions. The volt-ampere rating of a piece of equipment is equal to the wattage rating only when the F_p is 1. It is therefore a maximum power dissipation rating. This condition exists only when the total impedance of a system $Z \angle\theta$ is such that $\theta = 0°$.

The exact current demand of a device, when used under normal operating conditions, could also be determined if the wattage rating and power factor were given instead of the volt-ampere rating. However, the power factor is sometimes not available, or it may vary with the load.

The reason for rating some electrical equipment in volt-amperes rather than in watts can be described using the configuration of Fig. 19.6. The load has an apparent power rating of 750 VA at 150 V, and a

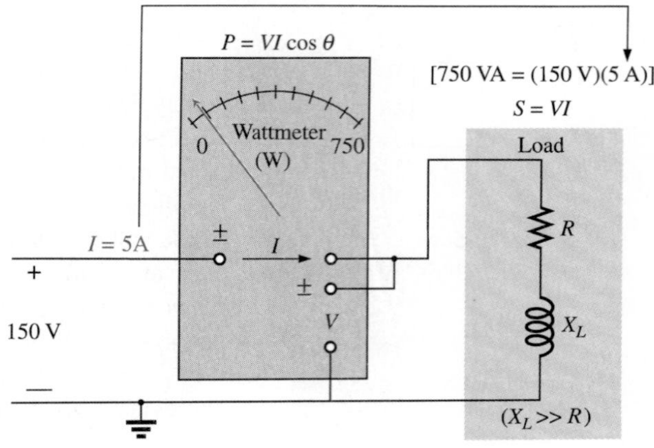

FIG. 19.6
Demonstrating the reason for rating a load in VA rather than W.

power factor of $F_P = 0.1$. Therefore the power is $P = (750\text{ V})(0.1) = 75\text{ W}$. The current is determined by the VA rating, not the wattage rating. The current is $(750\text{ VA})/(150\text{ V}) = 5\text{ A}$. Therefore we use the wattmeter's 0–150 V and 0–5 A inputs. If we had used the load's wattage rating to determine the wattmeter's current input, we could have selected a current input coil that was too small, causing damage to the coil. In other words, the wattmeter reading is an indication of the watts dissipated and may not reflect the magnitude of the current drawn. In fact, if the load were purely reactive, the wattmeter reading would be zero even if the elements were being damaged by a high current level.

19.4 INDUCTIVE CIRCUIT AND REACTIVE POWER

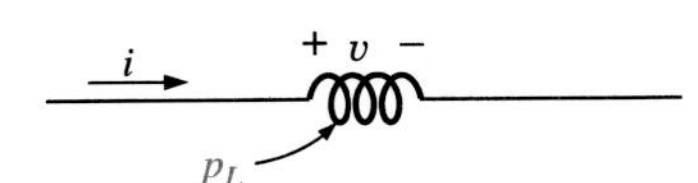

FIG. 19.7
Defining the power level for a purely inductive load.

For a purely inductive circuit (such as that in Fig. 19.7), v leads i by 90°, as shown in Fig. 19.8. Therefore, in Eq. (19.1), $\theta = 90°$. Substituting $\theta = 90°$ into Eq. (19.1) gives us

$$p_L = VI\cos(90°)(1 - \cos 2\omega t) + VI\sin(90°)(\sin 2\omega t)$$

$$= 0 + VI\sin 2\omega t$$

or

$$p_L = VI\sin 2\omega t \qquad (19.11)$$

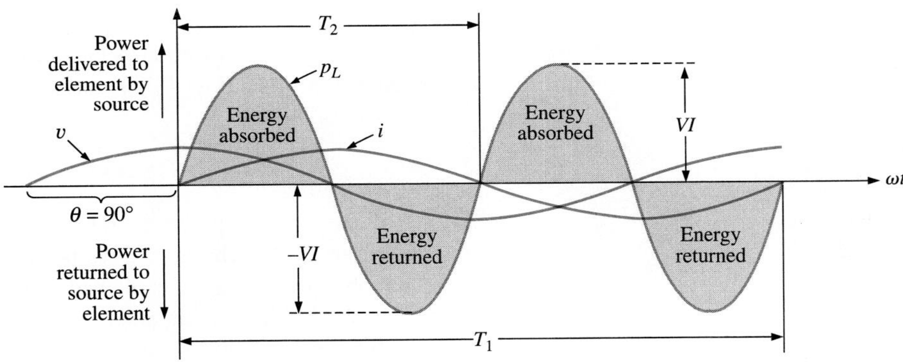

FIG. 19.8
The power curve for a purely inductive load.

where $VI\sin 2\omega t$ is a sine wave with twice the frequency of either input quantity (v or i) and a peak value of VI. Note the absence of an average or constant term in the equation.

Plotting the waveform for p_L (Fig. 19.8), we obtain

$$T_1 = \text{period of either input quantity}$$

$$T_2 = \text{period of } p_L \text{ curve}$$

Note that over one full cycle of p_L (T_2), the area above the horizontal axis in Fig. 19.8 is exactly equal to the area below the axis. This indicates that over a full cycle of p_L, the power delivered by the source to the inductor is exactly equal to the power returned to the source by the inductor.

The net flow of power to a pure (ideal) inductor over a full cycle (that is, the average power) is zero. No energy is lost in the transaction.

The power absorbed or returned by the inductor at any instant of time t_1 can be found simply by substituting t_1 into Eq. (19.11). The peak value of the curve VI is defined as the *reactive power* associated with a pure inductor.

In general, the reactive power associated with any circuit is defined as $VI \sin \theta$, a factor appearing in the second term of Eq. (19.1). Note that it is the peak value of that term of the total power equation that produces no net transfer of energy. The symbol for reactive power is Q, and its unit of measure is the *volt-ampere reactive* (VAR). The Q is derived from the quadrature (90°) relationship between the various powers, to be discussed in detail in a later section. Therefore,

$$Q = VI \sin \theta \qquad \text{(volt-ampere-reactive, VAR)} \qquad (19.12)$$

where θ is the phase angle between V and I.

For the inductor,

$$Q_L = VI \quad \text{(VAR)} \qquad (19.13)$$

or, since $V = IX_L$ or $I = V/X_L$,

$$Q_L = I^2 X_L \quad \text{(VAR)} \qquad (19.14)$$

or

$$Q_L = \frac{V^2}{X_L} \quad \text{(VAR)} \qquad (19.15)$$

The apparent power associated with an inductor is $S = VI$, and the average power is $P = 0$, as noted in Fig. 19.8. The power factor is therefore

$$F_p = \cos \theta = \frac{P}{S} = \frac{0}{VI} = 0$$

If the average power is zero, and the energy supplied is returned within one cycle, why is reactive power of any significance? The reason is not obvious but can be explained using the curve of Fig. 19.8. At every instant of time along the power curve that the curve is above the axis (positive), energy must be supplied to the inductor, even though it will be returned during the negative portion of the cycle. This power requirement during the positive portion of the cycle means that the generating plant must provide this energy during that interval. Therefore, the effect of reactive elements such as the inductor can be to raise the power requirement of the generating plant, even though the reactive power is not dissipated but simply "borrowed." The increased power demand during these intervals is a cost factor that must be passed on to the industrial consumer. In fact, most larger users of electrical energy pay for the apparent power demand rather than the watts dissipated since the volt-amperes used are sensitive to the reactive power requirement (see Section 19.6). In other words, the closer the power factor of an industrial outfit is to 1, the more efficient is the plant's operation, since it is limiting its use of "borrowed" power.

The energy stored by the inductor during the positive portion of the cycle (Fig. 19.8) is equal to the energy returned during the negative portion and can be determined using the equation

$$W = Pt$$

where P is the average value for the interval and t is the associated interval of time.

Recall from Section 13.6 that the average value of the positive portion of a sinusoid equals 2(peak value/π) and $t = T_2/2$. Therefore,

$$W_L = \left(\frac{2VI}{\pi}\right) \times \left(\frac{T_2}{2}\right)$$

and
$$W_L = \frac{VIT_2}{\pi} \qquad \text{(joules, J)} \qquad (19.16)$$

or, since $T_2 = 1/f_2$, where f_2 is the frequency of the p_L curve, we have

$$W_L = \frac{VI}{\pi f_2} \qquad \text{(joules, J)} \qquad (19.17)$$

Since the frequency f_2 of the power curve is twice that of the input quantity, if we substitute the frequency f_1 of the input voltage or current, Eq. (19.17) becomes

$$W_L = \frac{VI}{\pi(2f_1)} = \frac{VI}{\omega_1}$$

However,
$$V = IX_L = I\omega_1 L$$

so that
$$W_L = \frac{(I\omega_1 L)I}{\omega_1}$$

and
$$W_L = LI^2 \qquad \text{(joules, J)} \qquad (19.18)$$

providing an equation for the energy stored or released by the inductor in one half-cycle of the applied voltage in terms of the inductance and effective value of the current squared.

19.5 CAPACITIVE CIRCUIT AND REACTIVE POWER

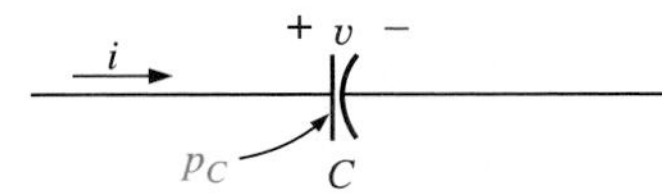

FIG. 19.9
Defining the power level for a purely capacitive load.

For a purely capacitive circuit (such as the one in Fig. 19.9), i leads v by 90°, as shown in Fig. 19.10. Therefore, in Eq. (19.1), $\theta = -90°$. Substituting $\theta = -90°$ into Eq. (19.1), we obtain

$$\begin{aligned} p_C &= VI\cos(-90°)(1 - \cos 2\omega t) + VI\sin(-90°)(\sin 2\omega t) \\ &= 0 - VI\sin 2\omega t \end{aligned}$$

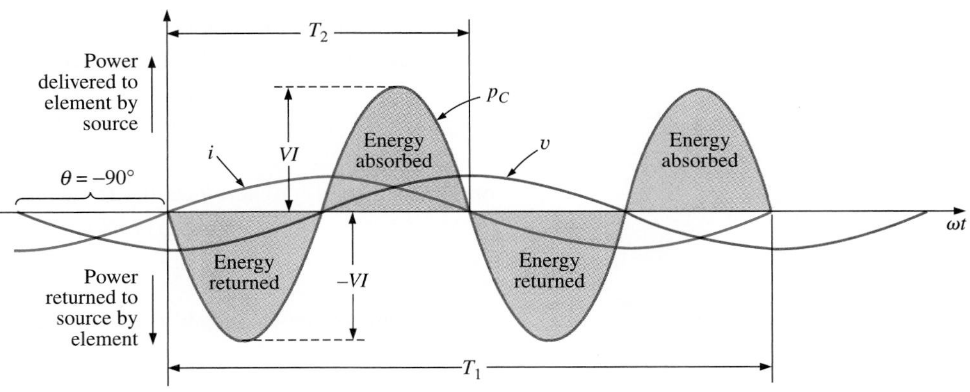

FIG. 19.10
The power curve for a purely capacitive load.

or

$$p_C = -VI \sin 2\omega t \qquad (19.19)$$

where $-VI \sin 2\omega t$ is a negative sine wave with twice the frequency of either input (v or i) and a peak value of VI. Again, note the absence of an average or constant term.

Plotting the waveform for p_C (Fig. 19.10) gives us

$$T_1 = \text{period of either input quantity}$$
$$T_2 = \text{period of } p_C \text{ curve}$$

Note that the same situation exists here for the p_C curve as existed for the p_L curve. The power delivered by the source to the capacitor is exactly equal to that returned to the source by the capacitor over one full cycle.

The net flow of power to the pure (ideal) capacitor is zero over a full cycle,

and no energy is lost in the transaction. The power absorbed or returned by the capacitor at any instant of time t_1 can be found by substituting t_1 into Eq. (19.19).

The reactive power associated with the capacitor is equal to the peak value of the p_C curve, as follows:

$$Q_C = VI \quad \text{(VAR)} \qquad (19.20)$$

but, since $V = IX_C$ and $I = V/X_C$, the reactive power to the capacitor can also be written

$$Q_C = I^2X_C \quad \text{(VAR)} \qquad (19.21)$$

and

$$Q_C = \frac{V^2}{X_C} \quad \text{(VAR)} \qquad (19.22)$$

The apparent power associated with the capacitor is

$$S = VI \quad \text{(volt-amperes, VA)} \qquad (19.23)$$

and the average power is $P = 0$, as noted from Eq. (19.19) or Fig. 19.10. The power factor is, therefore,

$$F_p = \cos\theta = \frac{P}{S} = \frac{0}{VI} = 0$$

The energy stored by the capacitor during the positive portion of the cycle (Fig. 19.10) is equal to the energy returned during the negative portion and can be determined using the equation $W = Pt$.

Proceeding as we did for the inductor, we can show that

$$W_C = \frac{VIT_2}{\pi} \quad \text{(joules, J)} \qquad (19.24)$$

or, since $T_2 = 1/f_2$, where f_2 is the frequency of the p_C curve,

$$W_C = \frac{VI}{\pi f_2} \quad \text{(joules, J)} \qquad (19.25)$$

In terms of the frequency f_1 of the input quantities v and i,

$$W_C = \frac{VI}{\pi(2f_1)} = \frac{VI}{\omega_1} = \frac{V(V\omega_1 C)}{\omega_1}$$

and

$$W_C = CV^2 \quad \text{(joules, J)} \qquad (19.26)$$

providing an equation for the energy stored or released by the capacitor in one half-cycle of the applied voltage in terms of the capacitance and effective value of the voltage squared.

19.6 THE POWER TRIANGLE

The three quantities *average power, apparent power,* and *reactive power* can be related in the vector domain by

$$\mathbf{S} = \mathbf{P} + \mathbf{Q} \tag{19.27}$$

with

$$\mathbf{P} = P \angle 0^\circ \quad \mathbf{Q}_L = Q_L \angle 90^\circ \quad \mathbf{Q}_C = Q_C \angle -90^\circ$$

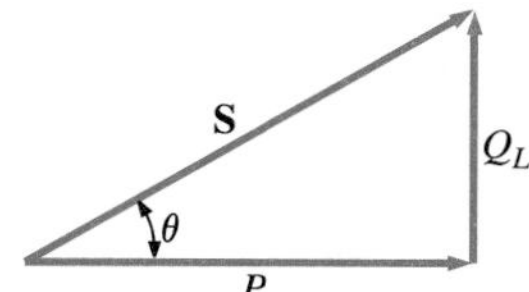

FIG. 19.11
Power diagram for inductive loads.

For an inductive load, the **phasor power S,** as it is often called, is defined by

$$\mathbf{S} = P + j\,Q_L$$

as shown in Fig. 19.11.

The 90° shift in Q_L from P is the source of another term for reactive power: *quadrature power.*

For a capacitive load, the phasor power **S** is defined by

$$\mathbf{S} = P - j\,Q_C$$

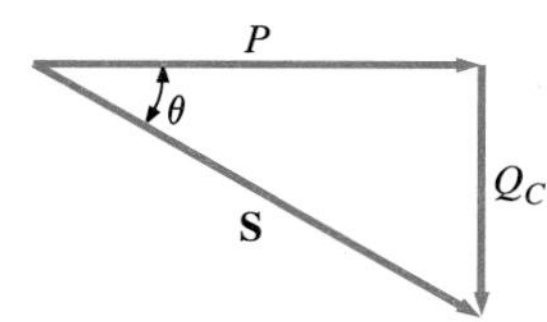

FIG. 19.12
Power diagram for capacitive loads.

as shown in Fig. 19.12.

If a network has both capacitive and inductive elements, the reactive component of the power triangle will be determined by the *difference* between the reactive power delivered to each. If $Q_L > Q_C$, the resulting power triangle will be similar to Fig. 19.11. If $Q_C > Q_L$, the resulting power triangle will be similar to Fig. 19.12.

We can show that the total reactive power is the difference between the reactive powers of the inductive and capacitive elements by considering Eqs. (19.11) and (19.19). Using these equations, the reactive power delivered to each reactive element has been plotted for a series L-C circuit on the same set of axes in Fig. 19.13. The reactive elements were chosen such that $X_L > X_C$. Note that the power curve for each is exactly 180° out of phase. The curve for the resulting reactive power is therefore determined by the algebraic resultant of the two at each

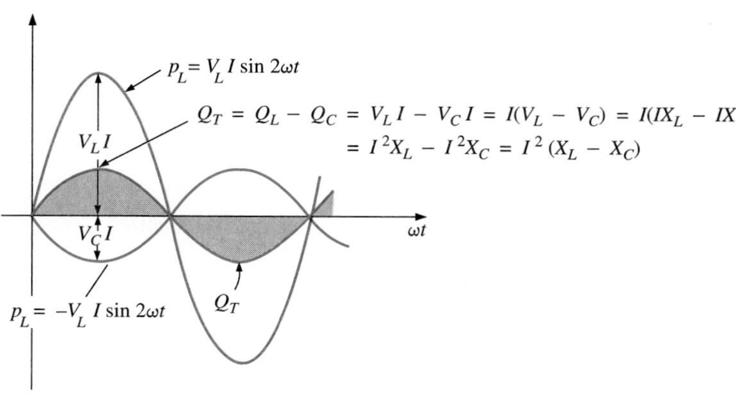

FIG. 19.13
Demonstrating why the net reactive power is the difference between that delivered to inductive and capacitive elements.

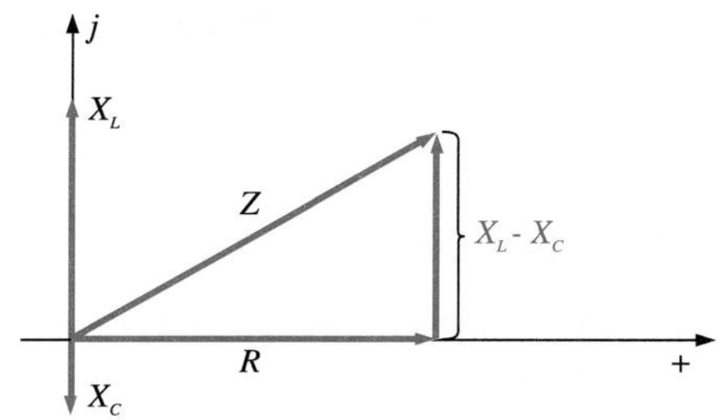

FIG. 19.14
Impedance diagram for a series R-L-C circuit.

instant of time. Since the reactive power is defined as the peak value, the reactive component of the power triangle is as indicated in Fig. 19.13: $I^2(X_L - X_C)$.

Additional verification comes from first considering the impedance diagram of a series *R-L-C* circuit (Fig. 19.14). If we multiply each radius vector by the current squared (I^2), we obtain the results shown in Fig. 19.15, which is the power triangle for a predominantly inductive circuit.

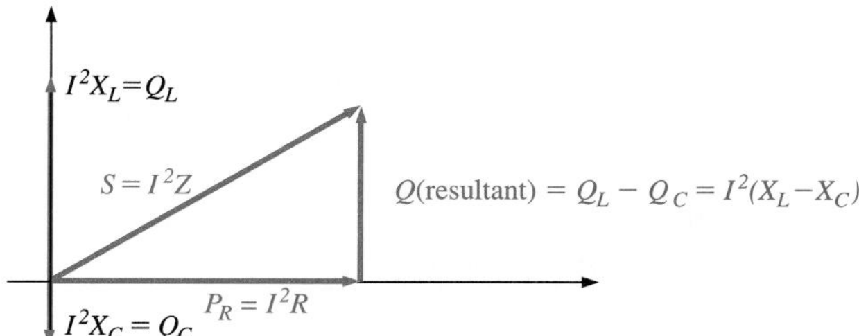

FIG. 19.15
The result of multiplying each vector of Fig. 19.14 by I^2 for a series R-L-C circuit.

Since the reactive power and average power are always angled 90° to each other, the three powers are related by the Pythagorean theorem; that is,

$$S^2 = P^2 + Q^2 \tag{19.28}$$

Therefore, the third power can always be found if the other two are known.

It is particularly interesting that the equation

$$\mathbf{S} = \mathbf{VI}^* \tag{19.29}$$

will provide the vector form of the apparent power of a system. Here, **V** is the voltage across the system and $\mathbf{I}^*$ is the complex conjugate of the current.

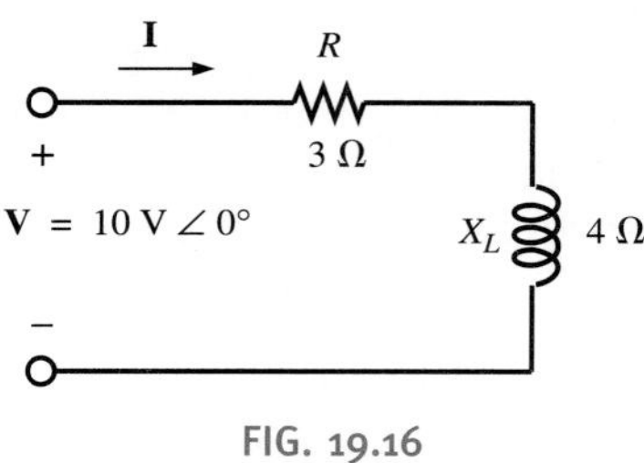

FIG. 19.16
Demonstrating Eq. (19.29).

Consider, for example, the simple *R-L* circuit of Fig. 19.16, where

$$\mathbf{I} = \frac{\mathbf{V}}{\mathbf{Z}_T} = \frac{10\text{ V}\angle 0°}{3\,\Omega + j\,4\,\Omega} = \frac{10\text{ V}\angle 0°}{5\,\Omega\angle 53.13°} = 2\text{ A}\angle -53.13°$$

The real power (the term *real* coming from the positive real axis of the complex plane) is

$$P = I^2R = (2\text{ A})^2(3\,\Omega) = 12\text{ W}$$

and the reactive power is

$$Q_L = I^2X_L = (2\text{ A})^2(4\,\Omega) = 16\text{ VAR}$$

with $\quad \mathbf{S} = P + j\,Q_L = 12\text{ W} + j\,16\text{ VAR} = 20\text{ VA}\angle 53.13°$

as shown in Fig. 19.17. Applying Eq. (19.29) gives us

$$\mathbf{S} = \mathbf{VI}^* = (10\text{ V}\angle 0°)(2\text{ A}\angle +53.13°) = 20\text{ VA}\angle 53.13°$$

as obtained above.

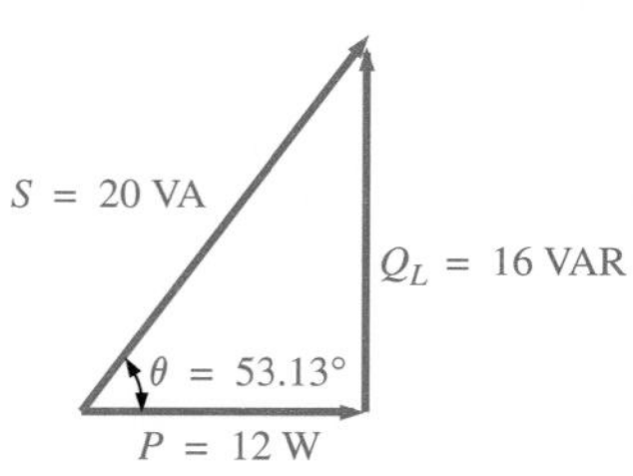

FIG. 19.17
The power triangle for the circuit of Fig. 19.16.

The angle θ associated with **S** in Figs. 19.11, 19.12, and 19.17 is the power-factor angle of the network. Since

$$P = VI \cos \theta$$

or

$$P = S \cos \theta$$

then

$$F_p = \cos \theta = \frac{P}{S} \qquad (19.30)$$

19.7 THE TOTAL P, Q, AND S

The total number of watts, volt-amperes reactive, and volt-amperes, and the power factor of any system can be found using the following procedure:

1. Find the real power and reactive power for each branch of the circuit.
2. The total real power of the system (P_T) is then the sum of the average power delivered to each branch.
3. The total reactive power (Q_T) is the difference between the reactive power of the inductive loads and that of the capacitive loads.
4. The total apparent power is $S_T = \sqrt{P_T^2 + Q_T^2}$
5. The total power factor is P_T/S_T.

There are two important points in the above procedure. First, the total apparent power must be determined from the total average and reactive powers and *cannot* be determined from the apparent powers of each branch. Second, and more important, it is *not necessary* to consider the series-parallel arrangement of branches. In other words, the total real, reactive, or apparent power doesn't depend on whether the loads are in series, parallel, or series-parallel. The following examples will demonstrate how easily all the quantities of interest can be found.

EXAMPLE 19.1

Find the total number of watts, volt-amperes reactive, and volt-amperes, and the power factor F_p of the network in Fig. 19.18. Draw the power triangle and find the current in phasor form.

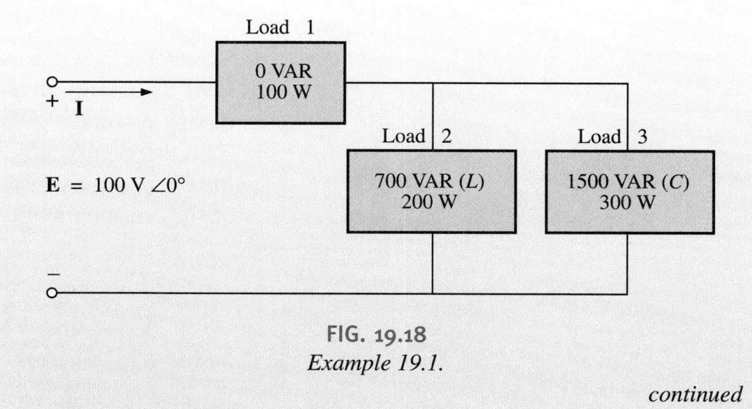

FIG. 19.18
Example 19.1.

continued

Solution: Use a table:

Load	W	VAR	VA
1	100	0	100
2	200	700 (L)	$\sqrt{(200)^2 + (700)^2} = 728.0$
3	300	1500 (C)	$\sqrt{(300)^2 + (1500)^2} = 1529.71$
	$P_T =$ **600**	$Q_T =$ **800 (C)**	$S_T = \sqrt{(600)^2 + (800)^2} =$ **1000**
	Total power dissipated	Resultant reactive power of network	(Note that $S_T \neq$ sum of S for each branch: $1000 \neq 100 + 728 + 1529.71$)

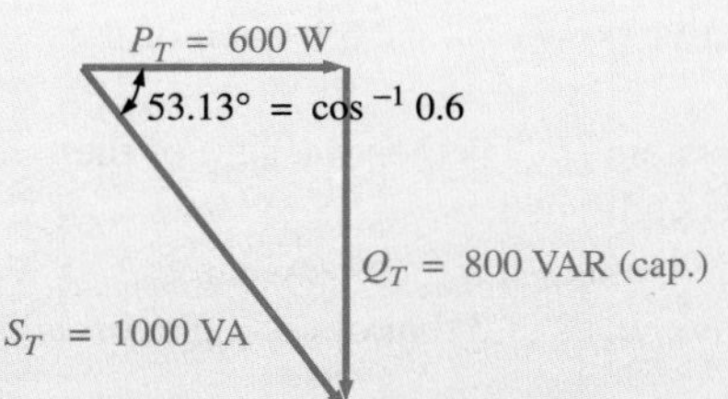

FIG. 19.19
Power triangle for Example 19.1.

$$F_p = \frac{P_T}{S_T} = \frac{600\text{ W}}{1000\text{ VA}} = \mathbf{0.6\ leading\ (C)}$$

The power triangle is shown in Fig. 19.19.

Since $S_T = VI = 1000$ VA, $I = 1000$ VA/100 V $= 10$ A; and since θ of $\cos\theta = F_p$ is the angle between the input voltage and current,

$$\mathbf{I = 10\ A\ \angle{+53.13°}}$$

The plus sign is associated with the phase angle since the circuit is predominantly capacitive.

EXAMPLE 19.2

a. Find the total number of watts, volt-amperes reactive, and volt-amperes, and the power factor F_p for the network of Fig. 19.20.

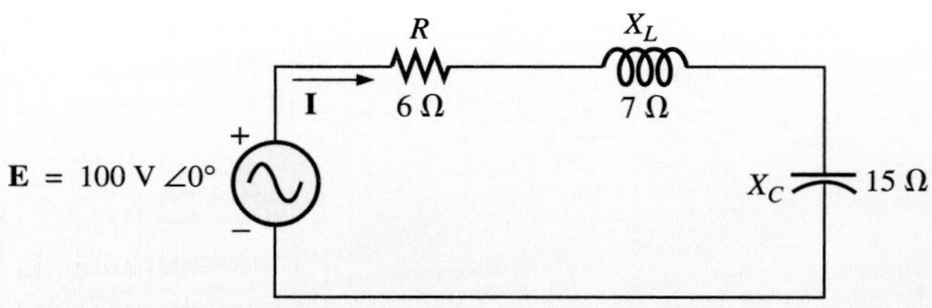

FIG. 19.20
Example 19.2.

b. Sketch the power triangle.
c. Find the energy dissipated by the resistor over one full cycle of the input voltage if the frequency of the input quantities is 60 Hz.
d. Find the energy stored in, or returned by, the capacitor or inductor over one half-cycle of the power curve for each if the frequency of the input quantities is 60 Hz.

Solutions:

a. $$\mathbf{I} = \frac{\mathbf{E}}{\mathbf{Z}_T} = \frac{100\text{ V}\angle 0°}{6\ \Omega + j\,7\ \Omega - j\,15\ \Omega} = \frac{100\text{ V}\angle 0°}{10\ \Omega\angle -53.13°}$$
$$= 10\text{ A}\angle 53.13°$$
$$\mathbf{V}_R = (10\text{ A}\angle 53.13°)(6\ \Omega\angle 0°) = 60\text{ V}\angle 53.13°$$

continued

$$\mathbf{V}_L = (10\text{ A}\angle 53.13°)(7\ \Omega\angle 90°) = 70\text{ V}\angle 143.13°$$
$$\mathbf{V}_C = (10\text{ A}\angle 53.13°)(15\ \Omega\angle -90°) = 150\text{ V}\angle -36.87°$$

$$P_T = EI\cos\theta = (100\text{ V})(10\text{ A})\cos 53.13° = \mathbf{600\ W}$$
$$= I^2R = (10\text{ A})^2(6\ \Omega) = \mathbf{600\ W}$$
$$= \frac{V_R^2}{R} = \frac{(60\text{ V})^2}{6} = \mathbf{600\ W}$$

$$S_T = EI = (100\text{ V})(10\text{ A}) = \mathbf{1000\ VA}$$
$$= I^2Z_T = (10\text{ A})^2(10\ \Omega) = \mathbf{1000\ VA}$$
$$= \frac{E^2}{Z_T} = \frac{(100\text{ V})^2}{10\ \Omega} = \mathbf{1000\ VA}$$

$$Q_T = EI\sin\theta = (100\text{ V})(10\text{ A})\sin 53.13° = \mathbf{800\ VAR}$$
$$= Q_C - Q_L$$
$$= I^2(X_C - X_L) = (10\text{ A})^2(15\ \Omega - 7\ \Omega) = \mathbf{800\ VAR}$$
$$Q_T = \frac{V_C^2}{X_C} - \frac{V_L^2}{X_L} = \frac{(150\text{ V})^2}{15\ \Omega} - \frac{(70\text{ V})^2}{7\ \Omega}$$
$$= 1500\text{ VAR} - 700\text{ VAR} = \mathbf{800\ VAR}$$

$$F_p = \frac{P_T}{S_T} = \frac{600\text{ W}}{1000\text{ VA}} = \mathbf{0.6\ leading\ (cap.)}$$

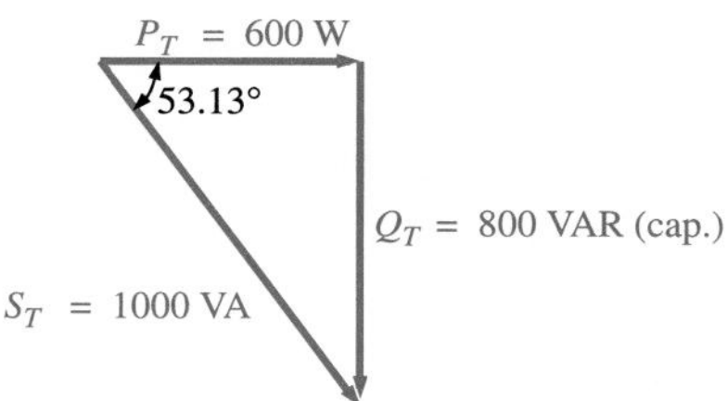

FIG. 19.21
Power triangle for Example 19.2.

b. The power triangle is as shown in Fig. 19.21.

c. $W_R = \dfrac{V_RI}{f_1} = \dfrac{(60\text{ V})(10\text{ A})}{60\text{ Hz}} = \mathbf{10\ J}$

d. $W_L = \dfrac{V_LI}{\omega_1} = \dfrac{(70\text{ V})(10\text{ A})}{(2\pi)(60\text{ Hz})} = \dfrac{700\text{ J}}{377} = \mathbf{1.86\ J}$

$W_C = \dfrac{V_CI}{\omega_1} = \dfrac{(150\text{ V})(10\text{ A})}{377\text{ rad/s}} = \dfrac{1500\text{ J}}{377} = \mathbf{3.98\ J}$

EXAMPLE 19.3

For the system of Fig. 19.22,

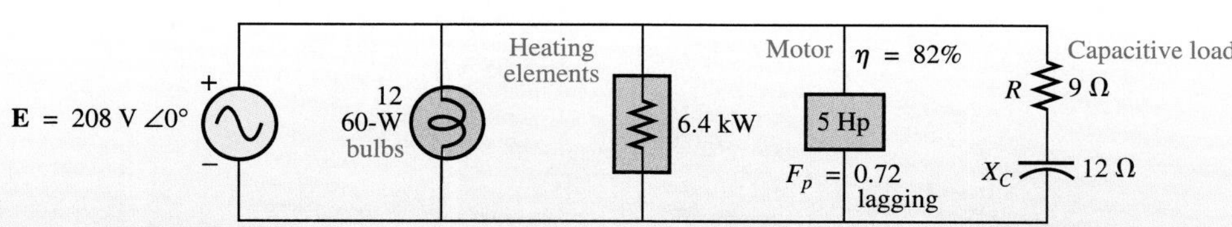

FIG. 19.22
Example 19.3.

a. Find the average power, apparent power, reactive power, and F_p for each branch.
b. Find the total number of watts, volt-amperes reactive, and volt-amperes, and the power factor of the system. Sketch the power triangle.
c. Find the source current *I*.

continued

Solutions:

a. **Bulbs:**

Total dissipation of applied power

$$P_1 = 12(60\text{ W}) = \mathbf{720\text{ W}}$$
$$Q_1 = \mathbf{0\text{ VAR}}$$
$$S_1 = P_1 = \mathbf{720\text{ VA}}$$
$$F_{p_1} = \mathbf{1}$$

Heating elements:

Total dissipation of applied power

$$P_2 = \mathbf{6.4\text{ kW}}$$
$$Q_2 = \mathbf{0\text{ VAR}}$$
$$S_2 = P_2 = \mathbf{6.4\text{ kVA}}$$
$$F_{p_2} = \mathbf{1}$$

Motor:

$$\eta = \frac{P_o}{P_i} \rightarrow P_i = \frac{P_o}{\eta} = \frac{5(746\text{ W})}{0.82} = \mathbf{4548.78\text{ W}} = P_3$$

$F_p = \mathbf{0.72\text{ lagging}}$

$$P_3 = S_3 \cos\theta \rightarrow S_3 = \frac{P_3}{\cos\theta} = \frac{4548.78\text{ W}}{0.72} = \mathbf{6317.75\text{ VA}}$$

Also, $\theta = \cos^{-1} 0.72 = 43.95°$, so that

$$Q_3 = S_3 \sin\theta = (6317.75\text{ VA})(\sin 43.95°)$$
$$= (6317.75\text{ VA})(0.694) = \mathbf{4384.71\text{ VAR}}$$

Capacitive load:

$$\mathbf{I} = \frac{\mathbf{E}}{\mathbf{Z}} = \frac{208\text{ V}\angle 0°}{9\ \Omega - j\,12\ \Omega} = \frac{208\text{ V}\angle 0°}{15\ \Omega\angle -53.13°} = 13.87\text{ A}\angle 53.13°$$

$$P_4 = I^2R = (13.87\text{ A})^2 \cdot 9\ \Omega = \mathbf{1731.39\text{ W}}$$

$$Q_4 = I^2X_C = (13.87\text{ A})^2 \cdot 12\ \Omega = \mathbf{2308.52\text{ VAR}}$$

$$S_4 = \sqrt{P_4^2 + Q_4^2} = \sqrt{(1731.39\text{ W})^2 + (2308.52\text{ VAR})^2}$$
$$= \mathbf{2885.65\text{ VA}}$$

$$F_p = \frac{P_4}{S_4} = \frac{1731.39\text{ W}}{2885.65\text{ VA}} = \mathbf{0.6\text{ leading}}$$

b. $P_T = P_1 + P_2 + P_3 + P_4$

$$= 720\text{ W} + 6400\text{ W} + 4548.78\text{ W} + 1731.39\text{ W}$$
$$= \mathbf{13\,400.17\text{ W}}$$

$$Q_T = \pm Q_1 \pm Q_2 \pm Q_3 \pm Q_4$$
$$= 0 + 0 + 4384.71\text{ VAR} - 2308.52\text{ VAR}$$
$$= \mathbf{2076.19\text{ VAR (ind.)}}$$

$$S_T = \sqrt{P_T^2 + Q_T^2} = \sqrt{(13\,400.17\text{ W})^2 + (2076.19\text{ VAR})^2}$$
$$= \mathbf{13\,560.06\text{ VA}}$$

continued

$$F_p = \frac{P_T}{S_T} = \frac{13.4 \text{ kW}}{13\,560.06 \text{ VA}} = \mathbf{0.988 \text{ lagging}}$$

$$\theta = \cos^{-1} 0.988 = 8.89°$$

Note Fig. 19.23.

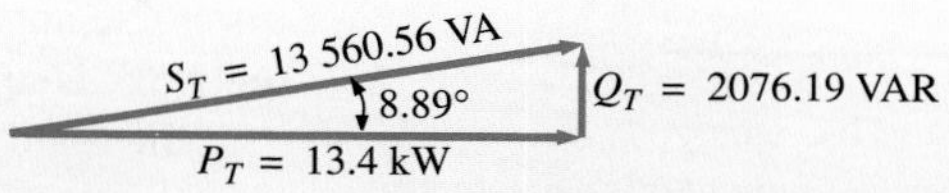

S_T = 13 560.56 VA, 8.89°, Q_T = 2076.19 VAR, P_T = 13.4 kW

FIG. 19.23
Power triangle for Example 19.3.

c. $S_T = EI \rightarrow I = \dfrac{S_T}{E} = \dfrac{13\,559.89 \text{ VA}}{208 \text{ V}} = 65.19 \text{ A}$

Lagging power factor: **E** leads **I** by 8.89° and

$$\mathbf{I = 65.19 \text{ A} \angle -8.89°}$$

EXAMPLE 19.4

An electrical device is rated 5 kVA, 100 V at a 0.6 power-factor lag. What is the impedance of the device in rectangular coordinates?

Solution:

$$S = EI = 5000 \text{ VA}$$

Therefore,
$$I = \frac{5000 \text{ VA}}{100 \text{ V}} = 50 \text{ A}$$

For $F_p = 0.6$, we have

$$\theta = \cos^{-1} 0.6 = 53.13°$$

Since the power factor is lagging, the circuit is predominantly inductive, and **I** lags **E.** Or, for $\mathbf{E} = 100 \text{ V} \angle 0°$,

$$\mathbf{I} = 50 \text{ A} \angle -53.13°$$

However,

$$\mathbf{Z}_T = \frac{\mathbf{E}}{\mathbf{I}} = \frac{100 \text{ V} \angle 0°}{50 \text{ A} \angle -53.13°} = 2\ \Omega \angle 53.13° = \mathbf{1.2\ \Omega\ + j\,1.6\ \Omega}$$

which is the impedance of the circuit of Fig. 19.24.

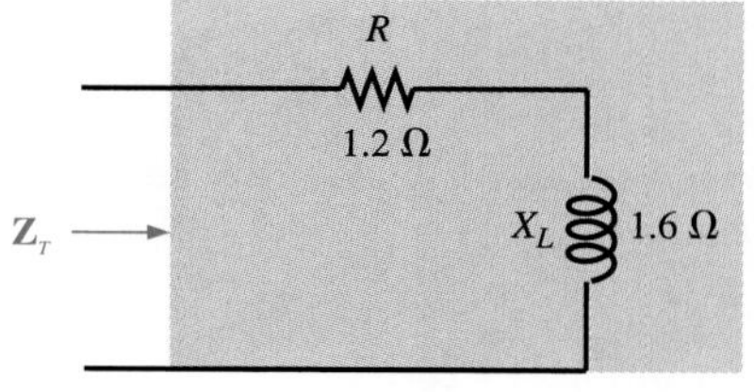

FIG. 19.24
Example 19.4.

19.8 POWER-FACTOR CORRECTION

The design of any power transmission system is very sensitive to the magnitude of the current in the lines as determined by the applied loads. Increased currents result in increased power losses (by a squared factor since $P = I^2R$) in the transmission lines due to the resistance of the lines. Heavier currents also require larger conductors, increasing the amount of copper needed for the system, and, quite obviously, they require increased generating capacities by the utility company.

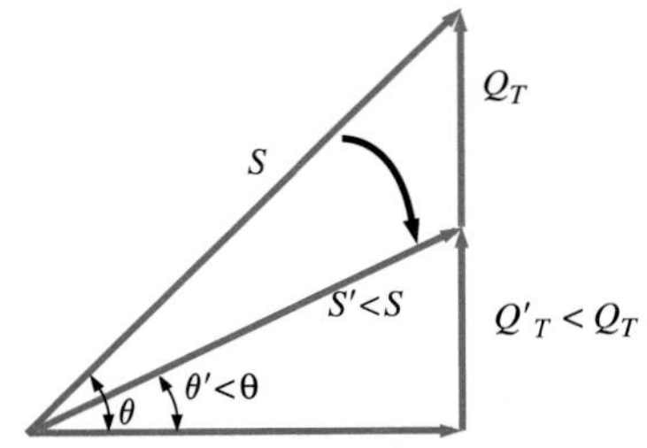

FIG. 19.25
Demonstrating the impact of power-factor correction on the power triangle of a network.

Every effort must therefore be made to keep current levels at a minimum. Since the line voltage of a transmission system is fixed, the apparent power is directly related to the current level. In turn, the smaller the net apparent power, the smaller the current drawn from the supply. Minimum current is therefore drawn from a supply when $S = P$ and $Q_T = 0$. Note the effect of decreasing levels of Q_T on the length (and magnitude) of S in Fig. 19.25 for the same real power. Note also that the power-factor angle approaches zero degrees and F_p approaches 1, revealing that the network is appearing more and more resistive at the input terminals.

In reality, it is difficult to keep the power factor at $F_P = 1.0$. So, electrical utility companies have a rate structure that accepts a power factor as low as 90% ($F_P \geq 0.9$). However, for $F_P < 0.9$, they introduce penalties that can be very expensive.

The process of introducing reactive elements to bring the power factor closer to unity is called **power-factor correction**. Since most loads are inductive, the process normally involves introducing elements with capacitive terminal characteristics having the sole purpose of improving the power factor.

In Fig. 19.26(a), for instance, an inductive load is drawing a current I_L that has a real and an imaginary component. In Fig. 19.26(b) a capacitive load was added in parallel with the original load to raise the power factor of the total system to the unity power-factor level. Note that by placing all the elements in parallel, the load still receives the same terminal voltage and draws the same current I_L. In other words, the load is unaware of and unconcerned about whether it is hooked up as shown in Fig. 19.26(a) or Fig. 19.26(b).

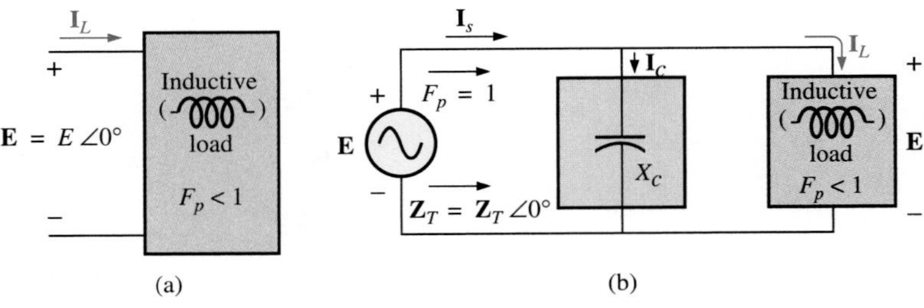

FIG. 19.26
Demonstrating the impact of a capacitive element on the power factor of a network.

Solving for the source current in Fig. 19.26(b):

$$\begin{aligned}\mathbf{I}_s &= \mathbf{I}_C + \mathbf{I}_L \\ &= -j\,I_C + (I_L + j\,I'_L) \\ &= I_L + j\,(I'_L - I_C)\end{aligned}$$

If X_C is chosen such that $I_C = I'_L$,

$$\mathbf{I}_s = I_L + j\,(0) = I_L \angle 0°$$

The result is a source current whose magnitude is simply equal to the real part of the load current, which can be considerably less than the magnitude of the load current of Fig. 19.26(a). In addition, since the phase angle associated with both the applied voltage and the source current is the same, the system appears "resistive" at the input terminals and all the power supplied is absorbed, creating maximum efficiency for a generating utility.

EXAMPLE 19.5

A 5-hp motor with a 0.6 lagging power factor and an efficiency of 92% is connected to a 208-V, 60-Hz supply.

a. Establish the power triangle for the load.
b. Determine the power factor capacitor that must be placed in parallel with the load to raise the power factor to unity.
c. Determine the change in supply current from the uncompensated to the compensated system.
d. Find the network equivalent of the above and verify the conclusions.

Solutions:

a. Since 1 hp = 746 W,

$$P_o = 5 \text{ hp} = 5(746 \text{ W}) = 3730 \text{ W}$$

and P_i (drawn from the line) $= \dfrac{P_o}{\eta} = \dfrac{3730 \text{ W}}{0.92} = 4054.35 \text{ W}$

Also, $F_P = \cos\theta = 0.6$

and $\theta = \cos^{-1} 0.6 = 53.13°$

Applying $\tan\theta = \dfrac{Q_L}{P_i}$

we obtain $Q_L = P_i \tan\theta = (4054.35 \text{ W}) \tan 53.13°$
$= 5405.8 \text{ VAR}$

and

$$S = \sqrt{P_i^2 + Q_L^2} = \sqrt{(4054.35 \text{ W})^2 + (5405.8 \text{ VAR})^2}$$
$$= 6757.25 \text{ VA}$$

The power triangle appears in Fig. 19.27.

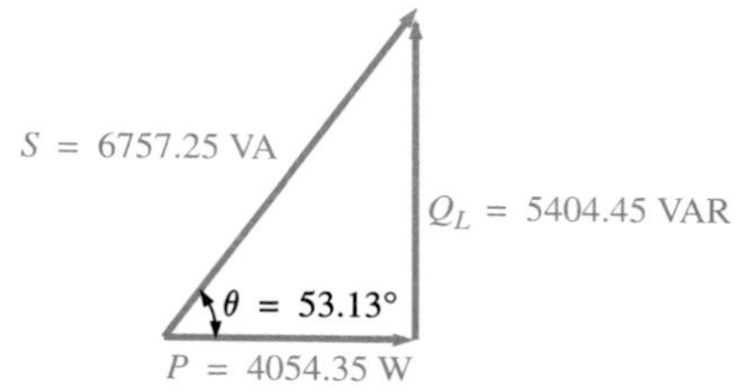

S = 6757.25 VA
Q_L = 5404.45 VAR
θ = 53.13°
P = 4054.35 W

FIG. 19.27
Initial power triangle for the load of Example 19.5.

b. A net unity power-factor level is established by introducing a capacitive reactive power level of 5405.8 VAR to balance Q_L. Since the capacitor is connected in parallel, use

$$Q_C = \frac{V^2}{X_C}$$

then $X_C = \dfrac{V^2}{Q_C} = \dfrac{(208 \text{ V})^2}{5405.8 \text{ VAR}} = 8\ \Omega$

and $C = \dfrac{1}{2\pi f X_C} = \dfrac{1}{(2\pi)(60 \text{ Hz})(8\ \Omega)} = \mathbf{331.6\ \mu F}$

c. **At 0.6F_p,**

$$S = VI = 6757.25 \text{ VA}$$

and $I = \dfrac{S}{V} = \dfrac{6757.25 \text{ VA}}{208 \text{ V}} = \mathbf{32.49\ A}$

At unity F_p,

$$S = VI = 4054.35 \text{ VA}$$

and $I = \dfrac{S}{V} = \dfrac{4054.35 \text{ VA}}{208 \text{ V}} = \mathbf{19.49\ A}$

producing a 40% reduction in supply current.

continued

d. For the motor, the angle by which the applied voltage leads the current is

$$\theta = \cos^{-1} 0.6 = 53.13°$$

and $P = EI_m \cos \theta = 4054.35$ W, from above, so that

$$I_m = \frac{P}{E \cos \theta} = \frac{4054.35 \text{ W}}{(208 \text{ V})(0.6)} = \mathbf{32.49\ A} \quad \text{(as above)}$$

resulting in

$$\mathbf{I}_m = 32.49 \text{ A} \angle -53.13°$$

Therefore,

$$\mathbf{Z}_m = \frac{\mathbf{E}}{\mathbf{I}_m} = \frac{208 \text{ V} \angle 0°}{32.49 \text{ A} \angle -53.13°} = 6.4\ \Omega \angle 53.13°$$
$$= 3.84\ \Omega + j\,5.12\ \Omega$$

as shown in Fig. 19.28(a).

The equivalent parallel load is determined from

$$\mathbf{Y} = \frac{1}{\mathbf{Z}} = \frac{1}{6.4\ \Omega \angle 53.13°}$$
$$= 0.156 \text{ S} \angle -53.13° = 0.094 \text{ S} - j\,0.125 \text{ S}$$
$$= \frac{1}{10.64\ \Omega} + \frac{1}{j\,8\ \Omega}$$

as shown in Fig. 19.28(b).

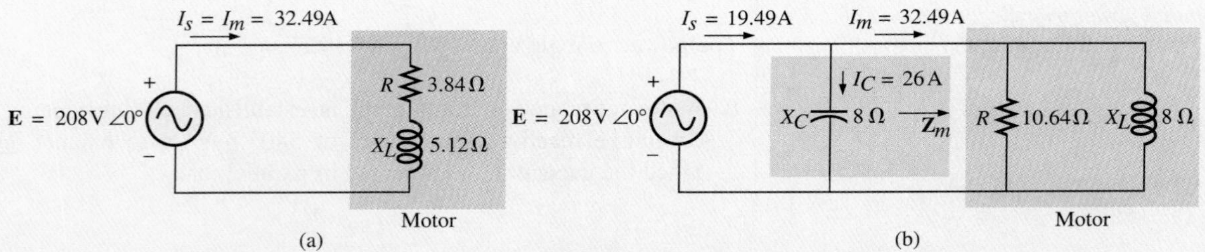

FIG. 19.28
Demonstrating the impact of power-factor corrections on the source current.

It is now clear that the effect of the 8-Ω inductive reactance can be compensated for by a parallel capacitive reactance of 8 Ω using a power factor correction capacitor of 332 μF.

Since

$$\mathbf{Y}_T = \frac{1}{-j\,X_C} + \frac{1}{R} + \frac{1}{+j\,X_L} = \frac{1}{R}$$

$$I_s = EY_T = E\left(\frac{1}{R}\right) = (208 \text{ V})\left(\frac{1}{10.64\ \Omega}\right) = \mathbf{19.54\ A} \quad \text{as above}$$

In addition, the magnitude of the capacitive current can be determined as follows:

$$I_C = \frac{E}{X_C} = \frac{208 \text{ V}}{8\ \Omega} = \mathbf{26\ A}$$

EXAMPLE 19.6

A small industrial plant has a 10-kW heating load and a 20-kVA inductive load due to a bank of induction motors. The heating elements are considered purely resistive ($F_p = 1$) and the induction motors have a lagging power factor of 0.7. If the supply is 600 V at 60 Hz, determine the capacitive element required to raise the power factor to 0.95.

Solution: For the induction motors,

$$S = VI = 20 \text{ kVA}$$
$$P = VI\cos\theta = (20 \times 10^3 \text{ VA})(0.7) = 14 \times 10^3 \text{ W}$$
$$\theta = \cos^{-1} 0.7 \cong 45.6°$$

and

$$Q_L = VI\sin\theta = (20 \times 10^3 \text{ VA})(0.714) = 14.28 \times 10^3 \text{ VAR}$$

The power triangle for the total system appears in Fig. 19.29.

Note the addition of real powers and the resulting S_T:

$$S_T = \sqrt{(24 \text{ kW})^2 + (14.28 \text{ kVAR})^2} = 27.93 \text{ kVA}$$

with
$$I = \frac{S_T}{V} = \frac{27.93 \text{ kVA}}{600 \text{ V}} = 46.55 \text{ A}$$

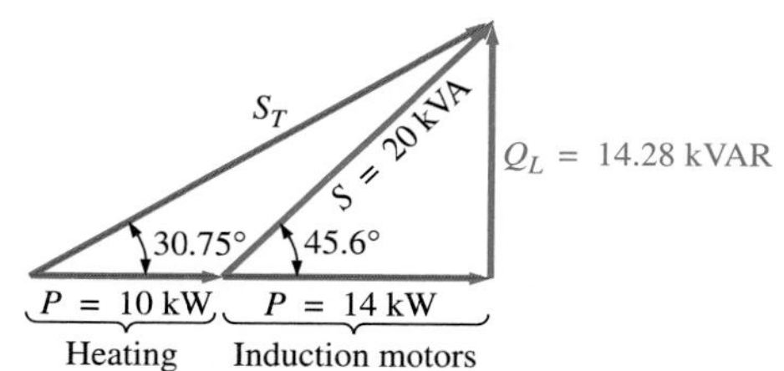

FIG. 19.29
Initial power triangle for the load of Example 19.6.

The desired power factor of 0.95 results in an angle between S and P of

$$\theta = \cos^{-1} 0.95 = 18.19°$$

changing the power triangle to that of Fig. 19.30:

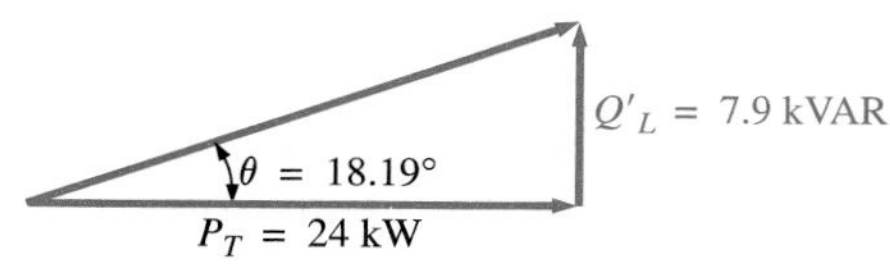

FIG. 19.30
Power triangle for the load of Example 19.6 after raising the power factor to 0.95.

$$\text{with } \tan\theta = \frac{Q'_L}{P_T} \rightarrow Q'_L = P_T\tan\theta = (24 \times 10^3 \text{ W})(\tan 18.19°)$$
$$= (24 \times 10^3 \text{ W})(0.329) = 7.9 \text{ kVAR}$$

The inductive reactive power must therefore be reduced by

$$Q_L - Q'_L = 14.28 \text{ kVAR} - 7.9 \text{ kVAR} = 6.38 \text{ kVAR}$$

Therefore, $Q_C = 6.38$ kVAR, and using

$$Q_C = \frac{V^2}{X_C}$$

we obtain

$$X_C = \frac{V^2}{Q_C} = \frac{(600 \text{ V})^2}{6.38 \times 10^3 \text{ VAR}} = 56.43\ \Omega$$

and
$$C = \frac{1}{2\pi f X_C} = \frac{1}{(2\pi)(60 \text{ Hz})(47\ \Omega)} = \mathbf{56.43\ \mu F}$$

19.9 WATTMETERS AND POWER-FACTOR METERS

The wattmeter was introduced in Section 4.4, along with its movement and terminal connections. The same meter can be used to measure the power in a dc or ac network using the same connection strat-

FIG. 19.31
Digital wattmeter. (Courtesy of Yokogawa Corporation of America.)

egy; in fact, it can be used to measure the wattage of any network with a periodic or nonperiodic input.

The digital display wattmeter of Fig. 19.31 employs a sophisticated electronic package to sense the voltage and current levels and, through the use of an analog-to-digital conversion unit, display the proper digits. It is capable of providing a digital readout for distorted nonsinusoidal waveforms and can provide the phase power, total power, apparent power, reactive power, and power factor.

When using a wattmeter, the operator must take care not to exceed its current, voltage, or wattage rating. The product of the voltage and current ratings may or may not equal the wattage rating. In the high-power-factor wattmeter, the product of the voltage and current ratings is usually equal to the wattage rating, or at least 80% of it. For a low-power-factor wattmeter, the product of the current and voltage ratings is much greater than the wattage rating. For obvious reasons, the low-power-factor meter is used only in circuits with low power factors (total impedance highly reactive). Typical ratings for high-power-factor (HPF) and low-power-factor (LPF) meters are shown in Table 19.1. Meters of both high and low power factors have an accuracy of 0.5% to 1% of full scale.

TABLE 19.1

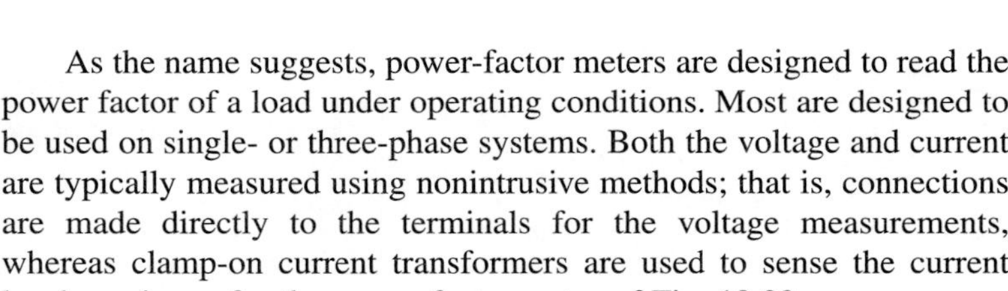

Meter	Current Ratings	Voltage Ratings	Wattage Ratings
HPF	2.5 A	150 V	1500/750/375
	5.0 A	300 V	
LPF	2.5 A	150 V	300/150/75
	5.0 A	300 V	

FIG. 19.32
Clamp-on power-factor meter. (Courtesy of the AEMC Corporation.)

As the name suggests, power-factor meters are designed to read the power factor of a load under operating conditions. Most are designed to be used on single- or three-phase systems. Both the voltage and current are typically measured using nonintrusive methods; that is, connections are made directly to the terminals for the voltage measurements, whereas clamp-on current transformers are used to sense the current level, as shown for the power-factor meter of Fig. 19.32.

Once the power factor is known, most power-factor meters come with a set of tables that will help define the power-factor capacitor that should be used to improve the power factor. Power-factor capacitors are typically rated in kVARs, with typical ratings extending from 1 to 25 kVARs at 240 V and 1 to 50 kVARs at 480 V or 600 V.

19.10 EFFECTIVE RESISTANCE

The resistance of a conductor as determined by the equation $R = \rho(l/A)$ is often called the *dc*, *ohmic*, or *geometric* resistance. It is a constant quantity determined only by the material used and its physical dimensions. In ac circuits, the actual resistance of a conductor (called the *effective* resistance) differs from the dc resistance because of the varying currents and voltages that introduce effects not present in dc circuits.

These effects include radiation losses, skin effect, eddy currents, and hysteresis losses. The first two effects apply to any network, while the second two are concerned with the additional losses introduced by the presence of ferromagnetic materials in a changing magnetic field.

Experimental Procedure

The effective resistance of an ac circuit cannot be measured by the ratio V/I since this ratio is now the impedance of a circuit that may have both resistance and reactance. The effective resistance can be found, however, by using the power equation $P = I^2R$, where

$$R_{\text{eff}} = \frac{P}{I^2} \tag{19.31}$$

A wattmeter and an ammeter are therefore necessary for measuring the effective resistance of an ac circuit.

There is another method, using an oscilloscope to measure the angle between the voltage and the current. The ratio V/I gives $Z \angle \theta$, which can then be resolved into R_{eff} and X.

Radiation Losses

We will now examine the various losses in greater detail. The radiation loss is the loss of energy in the form of electromagnetic waves during the transfer of energy from one element to another. This loss in energy requires that the input power be larger to establish the same current I, causing R to increase as determined by Eq. (19.31). At a frequency of 60 Hz, the effects of radiation losses can be completely ignored. However, at radio frequencies, this is an important effect and may in fact become the main effect in an electromagnetic device such as an antenna.

Skin Effect

The explanation of skin effect requires using some basic concepts previously described. Remember from Chapter 11 that a magnetic field exists around every current-carrying conductor (Fig. 19.33). Since the amount of charge flowing in ac circuits changes with time, the magnetic field surrounding the moving charge (current) also changes. Recall also that a wire placed in a changing magnetic field will have an induced voltage across its terminals as determined by Faraday's law $e = N \times (d\phi/dt)$. The higher the frequency of the changing flux as determined by an alternating current, the greater the induced voltage will be.

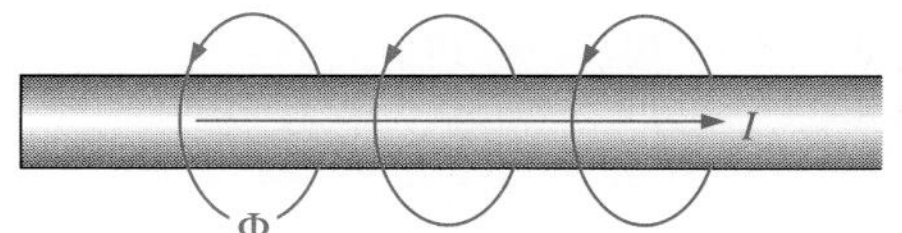

FIG. 19.33
Demonstrating the skin effect on the effective resistance of a conductor.

For a conductor carrying alternating current, the changing magnetic field surrounding the wire links the wire itself, developing within the wire an induced voltage that opposes the original flow of charge or current. These effects are stronger at the centre of the conductor than at the surface because the centre is linked by the changing flux inside the wire as well as that outside the wire. As the frequency of the applied signal increases, the flux linking the wire will change at a greater rate. An increase in frequency will therefore increase the counter-induced voltage at the centre of the wire to the point where the current will, for all practical purposes, flow on the surface of the conductor. At 60 Hz, the skin effect is almost unnoticeable. However, at radio frequencies the skin effect is so strong that conductors are frequently made hollow because the centre part is relatively ineffective. The skin effect, therefore, reduces the effective area through which the current can flow, and causes the resistance of the conductor, given by the equation $R\uparrow = \rho(l/A\downarrow)$, to increase.

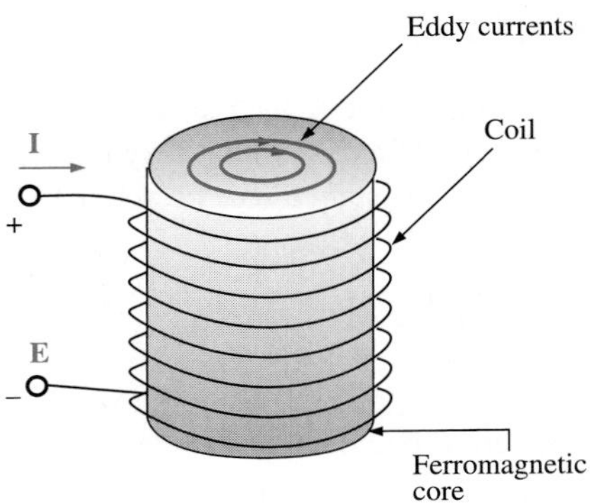

FIG. 19.34
Defining the eddy current losses of a ferromagnetic core.

Hysteresis and Eddy Current Losses

As mentioned earlier, hysteresis and eddy current losses will appear when a ferromagnetic material is placed in the region of a changing magnetic field. To describe eddy current losses in greater detail, we will consider the effects of an alternating current passing through a coil wrapped around a ferromagnetic core. As the alternating current passes through the coil, it will develop a changing magnetic flux Φ linking both the coil and the core that will develop an induced voltage within the core as determined by Faraday's law. This induced voltage and the geometric resistance of the core $R_C = \rho(l/A)$ cause currents to be developed within the core, $i_{\text{core}} = (e_{\text{ind}}/R_C)$, called *eddy currents*. The currents flow in circular paths, as shown in Fig. 19.34, changing direction with the applied ac potential.

The eddy current losses are determined by

$$P_{\text{eddy}} = i^2_{\text{eddy}} R_{\text{core}}$$

The magnitude of these losses is determined primarily by the type of core used. If the core is nonferromagnetic—and has a high resistivity like wood or air—the eddy current losses can be ignored. In terms of the frequency of the applied signal and the magnetic field strength produced, the eddy current loss is proportional to the square of the frequency times the square of the magnetic field strength:

$$P_{\text{eddy}} \propto f^2 B^2$$

Eddy current losses can be reduced if the core is constructed of thin, laminated sheets of ferromagnetic material insulated from one another and aligned parallel to the magnetic flux. Such construction reduces the magnitude of the eddy currents by placing more resistance in their path.

Hysteresis losses can be effectively reduced by injecting small amounts of silicon into the magnetic core, constituting some 2% or 3% of the total composition of the core. This must be done carefully, however, because too much silicon makes the core brittle and difficult to machine into the shape desired.

EXAMPLE 19.7

a. An air-core coil is connected to a 120-V, 60-Hz source as shown in Fig. 19.35. The current is found to be 5 A, and a wattmeter reading of 75 W is observed. Find the effective resistance and the inductance of the coil.

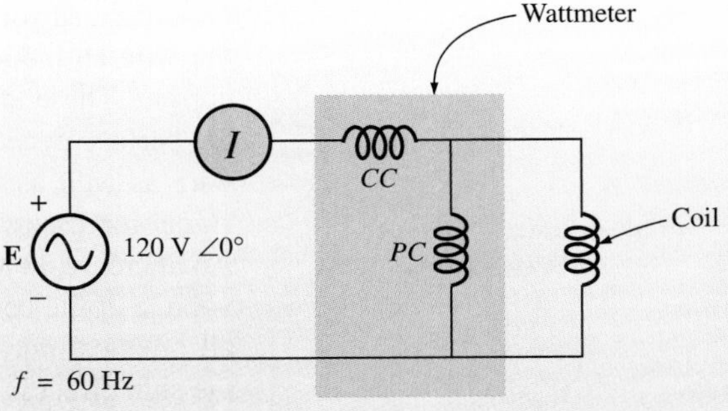

FIG. 19.35
The basic components required to determine the effective resistance and inductance of the coil.

continued

b. A brass core is then inserted in the coil; the ammeter reads 4 A, and the wattmeter reads 80 W. Calculate the effective resistance of the core. To what do you attribute the increase in value over that of part (a)?

c. If a solid iron core is inserted in the coil, the current is found to be 2 A, and the wattmeter reads 52 W. Calculate the resistance and the inductance of the coil. Compare these values to those of part (a), and account for the changes.

Solutions:

a.
$$R = \frac{P}{I^2} = \frac{75\text{ W}}{(5\text{ A})^2} = \mathbf{3\ \Omega}$$

$$Z_T = \frac{E}{I} = \frac{120\text{ V}}{5\text{ A}} = 24\ \Omega$$

$$X_L = \sqrt{Z_T^2 - R^2} = \sqrt{(24\ \Omega)^2 - (3\ \Omega)^2} = 23.81\ \Omega$$

and
$$X_L = 2\pi f L$$

or
$$L = \frac{X_L}{2\pi f} = \frac{23.81\ \Omega}{377\text{ rad/s}} = \mathbf{63.16\ mH}$$

b.
$$R = \frac{P}{I^2} = \frac{80\text{ W}}{(4\text{ A})^2} = \frac{80\ \Omega}{16} = \mathbf{5\ \Omega}$$

The brass core has less reluctance than the air core. Therefore, a greater magnetic flux density B will be created in it. Since $P_{\text{eddy}} \propto f^2B^2$, and $P_{\text{hys}} \propto f^1B^n$, as the flux density increases, the core losses and the effective resistance increase.

c.
$$R = \frac{P}{I^2} = \frac{52\text{ W}}{(2\text{ A})^2} = \frac{52\ \Omega}{4} = \mathbf{13\ \Omega}$$

$$Z_T = \frac{E}{I} = \frac{120\text{ V}}{2\text{ A}} = 60\ \Omega$$

$$X_L = \sqrt{Z_T^2 - R^2} = \sqrt{(60\ \Omega)^2 - (13\ \Omega)^2} = 58.57\ \Omega$$

$$L = \frac{X_L}{2\pi f} = \frac{58.57\ \Omega}{377\text{ rad/s}} = \mathbf{155.36\ mH}$$

The iron core has less reluctance than the air or brass cores. Therefore, a greater magnetic flux density B will be developed in the core. Again, since $P_{\text{eddy}} \propto f^2B^2$, and $P_{\text{hys}} \propto f^1B^n$, the increased flux density will cause the core losses and the effective resistance to increase.

Since the inductance L is related to the change in flux by the equation $L = N(d\phi/di)$, the inductance will be greater for the iron core because the changing flux linking the core will increase.

Practical Business Applications

Show me the money

Now for a short lesson in economics. Why, you may ask? Because it's necessary. You see, college graduates in the new millennium will not only have to understand technology but also demonstrate (at least at arm's length) an ability for doing business.

This doesn't mean that everyone's got to be a budding entrepreneur, just that they should appreciate making, spending, and saving money. You can take short courses almost anywhere on making money and, since nobody needs to learn how to spend money, we'll focus on saving money.

If you're studying electrical engineering technology, you'll likely get a job in some company that either manufactures electrical products or offers commercial electrical services. That company may be a heavy or light power consumer, depending on what it does.

Let's suppose that a large industrial company hires you to check its electrical utility bill every month. The following meter readings are taken at the end of your first month:

kWh consumption	146.2 MWh
peak kW demand	241 kW
kW demand	233 kW
kVA demand	250 kVA

Also, suppose the local power authority that supplies your company has the following rate schedule:

Energy	1st 450 kWh @ 22.3¢/kWh Additional kWh @ 8.9¢/kWh	Next 12 MWh @ 17.1¢/kWh
Power	1st 240 kW @ free Additional kW @ $12.05/kW	

The final bill has two parts to it—the energy consumption cost and the power demand cost. The first thing you should do is calculate the month's energy consumption cost. This is the easy part and should be done as follows:

$$\begin{aligned}\text{Cost} &= (450\text{ kWh})(\$0.223/\text{kWh}) + (12\text{ MWh})(\$0.171/\text{kWh})\\ &\quad + [146.2\text{ MWh} - (12\text{ MWh} + 450\text{ KWh})](\$0.089/\text{kWh})\\ &= \$100.35 + \$2052.00 + \$11\,903.75\\ &= \$14\,056.10\end{aligned}$$

The next thing to do is determine the month's power demand cost. This gets trickier because now your understanding of power factor will determine how well you do in your new job.

You see, the kW demand portion of the bill requires that you compare the kW demand that was read from a kW demand meter with a calculation involving the power factor. Recall that the real power, P, in an electric circuit or system is the power factor ($\cos\theta$) times the apparent power, S.

Most power authorities use 0.9 as the magic number for the power factor value (PF) in the calculation. That's because the power authority figures if you keep the power factor between 0.9 leading and 0.9 lagging, that's close enough to unity that no penalty is charged. The authority will charge the higher of either the metered kW demand or the real power demand calculation. So, in our example we get the following:

$$\text{kW demand} = 233\text{ kW}$$

$$\begin{aligned}\text{Real power demand} &= \text{PF} \times S\\ &= 0.9 \times \text{kVA demand}\\ &= 0.9 \times 250\text{ kVA}\\ &= 225\text{ kW}\end{aligned}$$

But, from the power authority's rate schedule, the first 240 kW are free, so there is no charge for the power demand this month. However, what if the kW demand value that was metered was 267 kW, which is greater than the maximum allowable value of 240 kW? Then you would be charged the difference between 267 kW and 240 kW times the kW demand rate of $12.05/kW, or $325.35.

Also, the calculation of real power demand could turn out to be higher than the kW demand value metered and the maximum allowable rate. Then, in this case, you would be charged the difference between the calculated real power demand and the maximum allowable rate times the kW demand rate.

Finally, we can now compute the company's total electric utility bill:

$$\begin{aligned}\text{Total cost} &= \text{Energy cost} + \text{power cost}\\ &= \$14\,056.10 + \text{free}\\ &= \$14\,056.10\end{aligned}$$

So, you see, it's important to know how power factor and power factor correction works because some day you may need to know this in your job. You can probably see that if you don't have a firm grasp of how real power is calculated from power factor values and metered values, you could make a serious error that could cost your employer lots of money. So when she says to you, "Show me the money!", you'd better be sure you can.

PROBLEMS

SECTIONS 19.1 through 19.7

1. For the battery of bulbs (purely resistive) appearing in Fig. 19.36:
   a. Determine the total power dissipation.
   b. Calculate the total reactive and apparent power.
   c. Find the source current I_s.
   d. Calculate the resistance of each bulb for the specified operating conditions.
   e. Determine the currents I_1 and I_2.

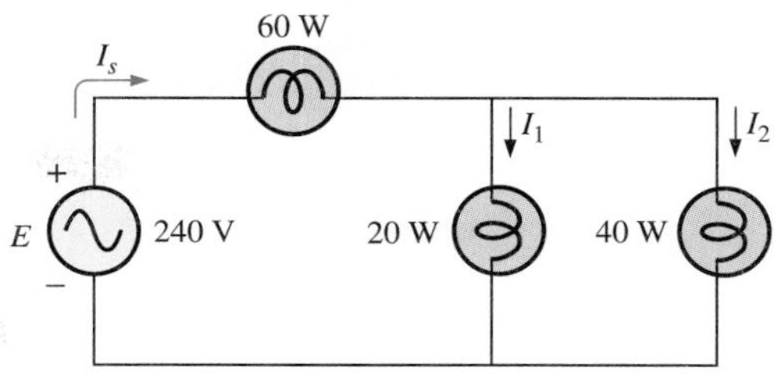

FIG. 19.36
Problem 1.

2. For the network of Fig. 19.37:
   a. Find the average power delivered to each element.
   b. Find the reactive power for each element.
   c. Find the apparent power for each element.
   d. Find the total number of watts, volt-amperes reactive, and volt-amperes, and the power factor F_p of the circuit.
   e. Sketch the power triangle.
   f. Find the energy dissipated by the resistor over one full cycle of the input voltage.
   g. Find the energy stored or returned by the capacitor and the inductor over one half-cycle of the power curve for each.

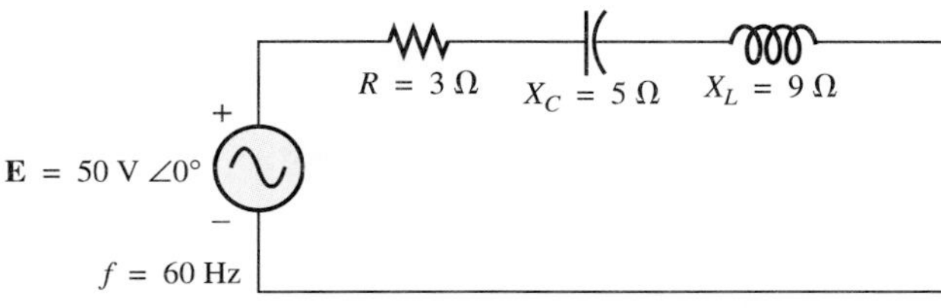

FIG. 19.37
Problem 2.

3. For the system of Fig. 19.38:
   a. Find the total number of watts, volt-amperes reactive, and volt-amperes, and the power factor F_p.
   b. Draw the power triangle.
   c. Find the current $\mathbf{I}_s$.

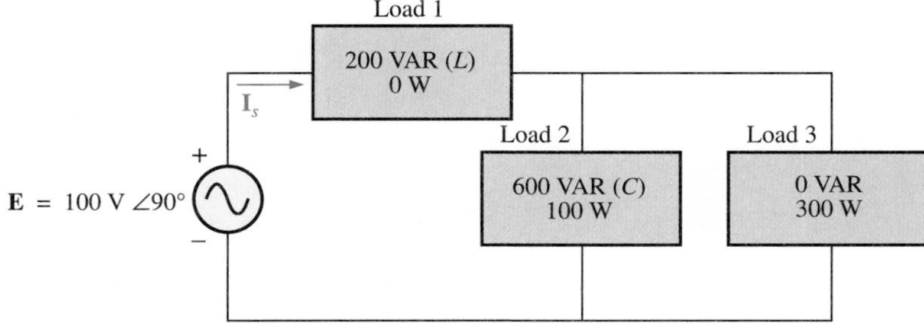

FIG. 19.38
Problem 3.

4. For the system of Fig. 19.39:
 a. Find the P, Q, S, and F_p for load 3.
 c. Find the magnitude and angle of the current in load 3.
 d. Find the combined P, Q, S, F_p, and impedance angle for loads 1 and 2.
 e. Find the magnitude and angle of the current into loads 1 and 2.
 f. Using the results of (c) and (e), find current I_s.

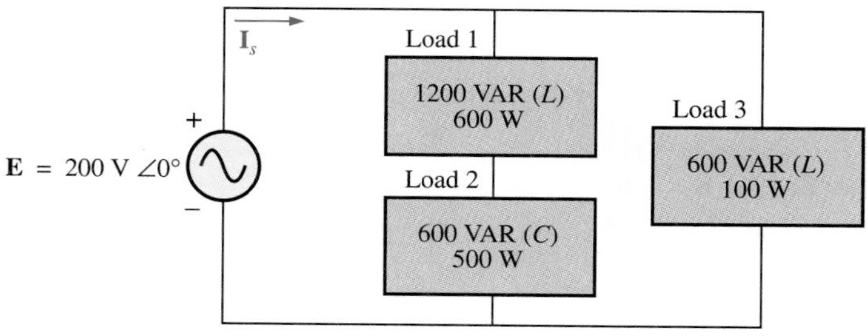

FIG. 19.39
Problem 4.

5. For the system of Fig. 19.40:
 a. Find P_T, Q_T, and S_T.
 b. Find the power factor F_p.
 c. Draw the power triangle.
 d. Find $\mathbf{I}_s$.

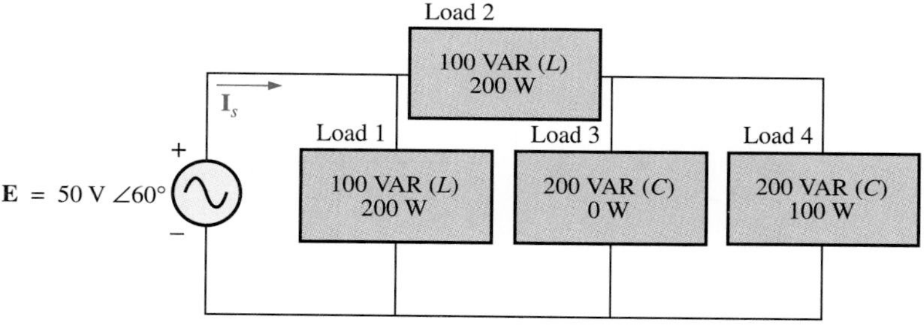

FIG. 19.40
Problem 5.

6. For the circuit of Fig. 19.41:
 a. Find the average, reactive, and apparent power for the 20-Ω resistor.
 b. Repeat part (a) for the 10-Ω inductive reactance.
 c. Find the total number of watts, volt-amperes reactive, and volt-amperes, and the power factor F_p.
 d. Find the current $\mathbf{I}_s$.

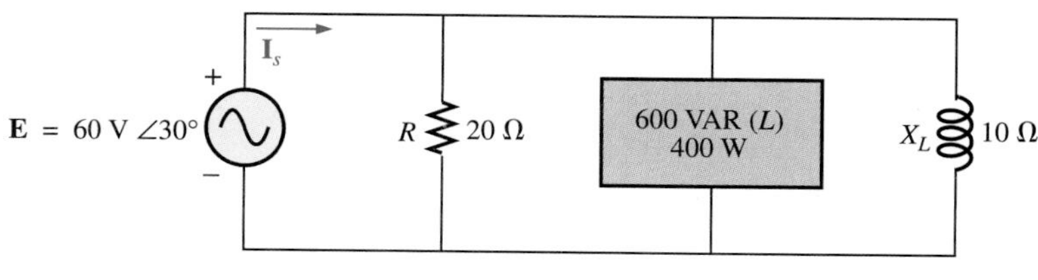

FIG. 19.41
Problem 6.

7. For the network of Fig. 19.42:

a. Find the average power delivered to each element.
b. Find the reactive power for each element.
c. Find the apparent power for each element.
d. Find P_T, Q_T, S_T, and F_p for the system.
e. Sketch the power triangle.
f. Find $\mathbf{I}_s$.

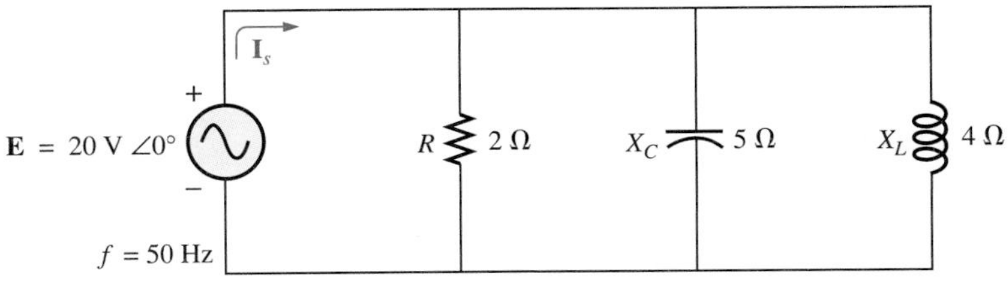

FIG. 19.42
Problem 7.

8. For the circuit of Fig. 19.43, the total power factor is known to be 0.9 leading (capacitive).

a. Sketch the power diagram and identify each part.
b. Find the average power P and reactive power Q_L for R and L.
c. Find the total apparent power S of the circuit.
d. Find the reactive power of the capacitor.
e. Find the ohms of the capacitor.

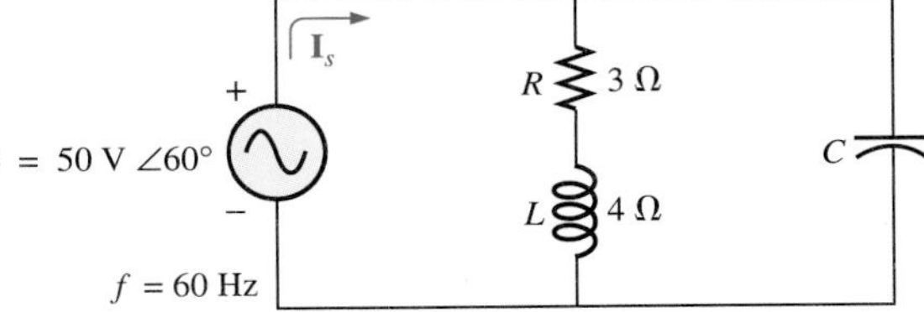

FIG. 19.43
Problem 8.

***9.** For the network of Fig. 19.44:

a. Find the average power delivered to each element.
b. Find the reactive power for each element.
c. Find the apparent power for each element.
d. Find the total number of watts, volt-amperes reactive, and volt-amperes, and the power factor F_p of the circuit.
e. Sketch the power triangle.
f. Find the energy dissipated by the resistor over one full cycle of the input voltage.
g. Find the energy stored or returned by the capacitor and the inductor over one half-cycle of the power curve for each.

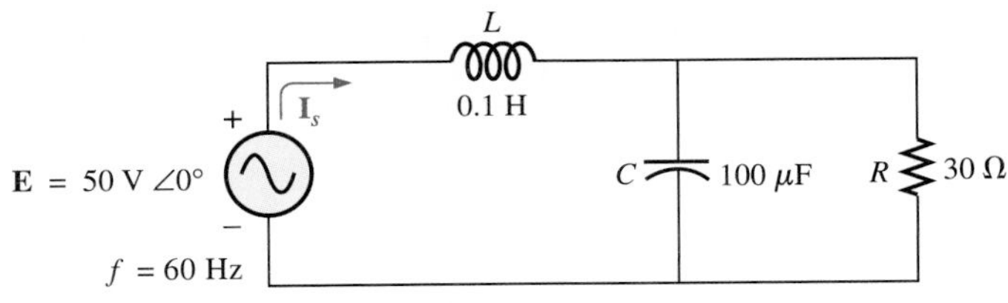

FIG. 19.44
Problem 9.

10. An electrical motor is rated 10 hp, 200 V at a 0.7 lagging power factor, and has an efficiency of 80%.

a. Find the average power into the motor in watts.

b. Find the reactive and apparent power to the motor.

c. Find the impedance of the motor in rectangular coordinates.

11. An electrical system is rated 5 kVA, 120 V, at a 0.8 lagging power factor.

a. Determine the impedance of the system in rectangular coordinates.

b. Find the average power delivered to the system.

***12.** For the system of Fig. 19.45:

a. Find the total number of watts, volt-amperes reactive, and volt-amperes, and F_p.

b. Find the current $\mathbf{I}_s$.

c. Draw the power triangle.

d. Find the type elements and their impedance in ohms within each electrical box. (Assume that all elements of a load are in series.)

e. Verify that the result of part (b) is correct by finding the current $\mathbf{I}_s$ using only the input voltage **E** and the results of part (d). Compare the value of $\mathbf{I}_s$ with that obtained for part (b).

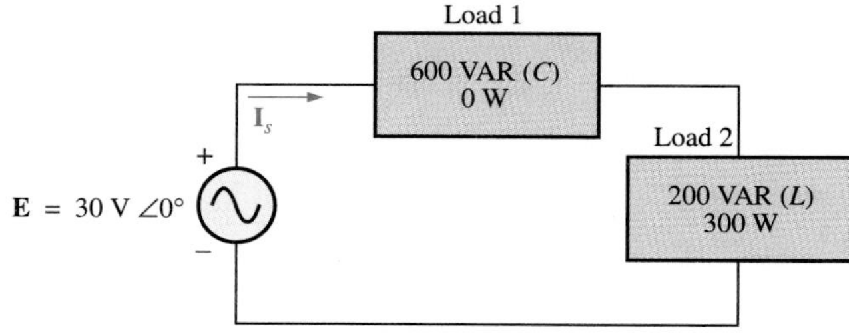

FIG. 19.45
Problem 12.

***13.** Repeat Problem 12 for the system of Fig. 19.46:

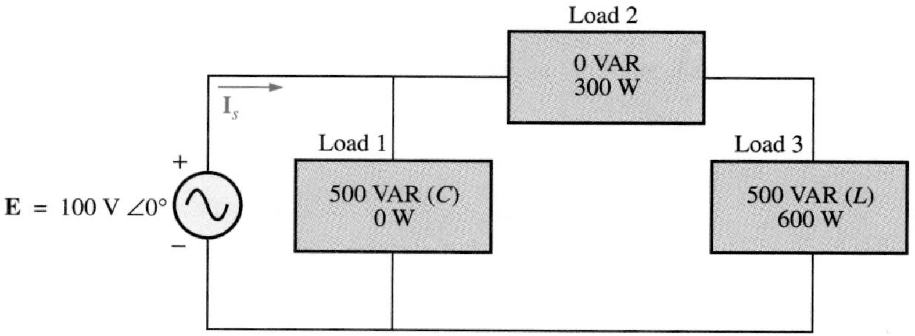

FIG. 19.46
Problem 13.

***14.** For the circuit of Fig. 19.47:

a. Find the total number of watts, volt-amperes reactive, and volt-amperes, and F_p.

b. Find the current $\mathbf{I}_s$.

c. Find the type elements and their impedance in each box. (Assume that the elements within each box are in series.)

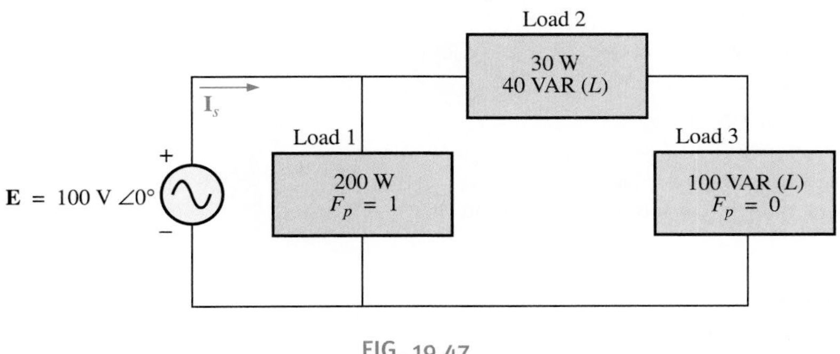

FIG. 19.47
Problem 14.

15. For the circuit of Fig. 19.48:

a. Find the total number of watts, volt-amperes reactive, and volt-amperes, and F_p.

b. Find the voltage **E**.

c. Find the type elements and their impedance in each box. (Assume that the elements within each box are in series.)

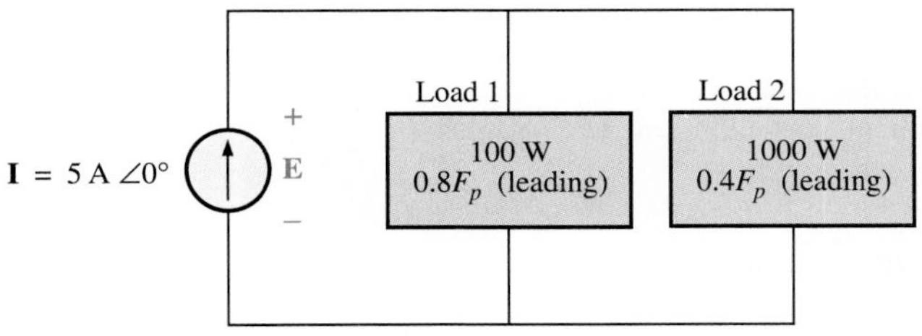

FIG. 19.48
Problem 15.

SECTION 19.8 Power-Factor Correction

***16.** The lighting and motor loads of a small factory establish a 10-kVA power demand at a 0.7 lagging power factor on a 208-V, 60-Hz supply.

a. Establish the power triangle for the load.

b. Determine the power-factor capacitor that must be placed in parallel with the load to raise the power factor to unity.

c. Determine the change in supply current from the uncompensated to compensated system.

d. Repeat parts (b) and (c) if the power factor is increased to 0.9.

17. The load on a 120-V, 60-Hz supply is 5 kW (resistive), 8 kVAR (inductive), and 2 kVAR (capacitive).
   a. Find the total kilovolt-amperes.
   b. Determine the F_p of the combined loads.
   c. Find the current drawn from the supply.
   d. Calculate the capacitance necessary to establish a unity power factor.
   e. Find the current drawn from the supply at unity power factor and compare it to the uncompensated level.

18. The loading of a factory on a 600-V, 60-Hz system includes:

   20-kW heating (unity power factor)
   10-kW (P_i) induction motors (0.7 lagging power factor)
   5-kW lighting (0.85 lagging power factor)

   a. Draw the power triangle for the total loading on the supply.
   b. Determine the size of capacitor (Q_c) to correct the power factor to unity.
   c. Determine the resulting line current after correction to unity and state the % change.
   d. Determine the size of capacitor (Q_c) to correct the power factor to 0.9 lagging.
   e. Determine the resulting line current after correction to 0.9 lagging and state the % change.
   f. Compare the improvement in current in going from 0.9 lagging to unity power factor to the required increase in the size of the capacitor. Comment.

SECTION 19.9 Wattmeters and Power-Factor Meters

19. **a.** A wattmeter is connected with its current coil as shown in Fig. 19.49 and the potential coil across points *f*-*g*. What does the wattmeter read?
   b. Repeat part (a) with the potential coil (*PC*) across *a*-*b*, *b*-*c*, *a*-*c*, *a*-*d*, *c*-*d*, *d*-*e*, and *f*-*e*.

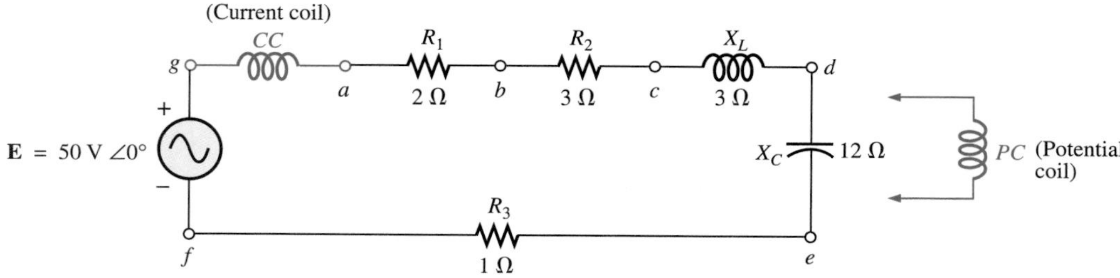

FIG. 19.49
Problem 19.

20. It is desired to determine the internal characteristics of the load in Fig. 19.50 using only a voltmeter, ammeter, and a wattmeter. The readings are:

$$V = 120\text{ V} \qquad I = 5.0\text{ A} \qquad P = 400\text{W}$$

a. Calculate the apparent power S and power factor F_p.
b. Calculate the series resistance.
c. Calculate the series reactance. (Based on the measurements alone, are you able to determine if the load is inductive or capacitive?)

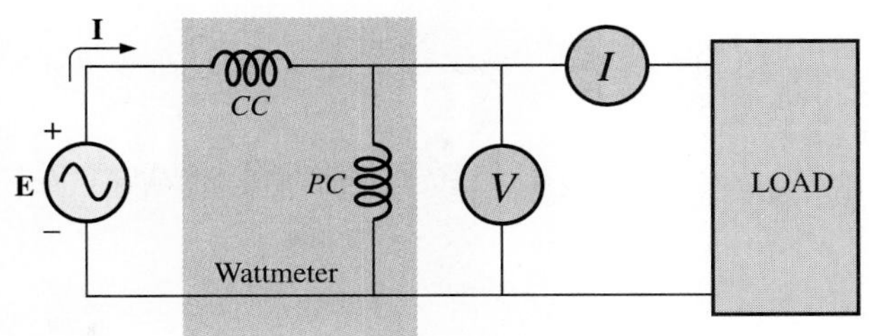

FIG. 19.50
Problem 20.

SECTION 19.10 Effective Resistance

21. **a.** An air-core coil is connected to a 200-V, 60-Hz source. The current is found to be 4 A, and a wattmeter reading of 80 W is observed. Find the effective resistance and the inductance of the coil.
b. A brass core is inserted in the coil. The ammeter reads 3 A, and the wattmeter reads 90 W. Calculate the effective resistance of the core. Explain the increase over the value of part (a).
c. If a solid iron core is inserted in the coil, the current is found to be 2 A, and the wattmeter reads 60 W. Calculate the resistance and inductance of the coil. Compare these values to the values of part (a), and account for the changes.

22. **a.** The inductance of an air-core coil is 0.08 H, and the effective resistance is 4 Ω when a 60-V, 50-Hz source is connected across the coil. Find the current passing through the coil and the reading of a wattmeter across the coil.
b. If a brass core is inserted in the coil, the effective resistance increases to 7 Ω, and the wattmeter reads 30 W. Find the current passing through the coil and the inductance of the coil.
c. If a solid iron core is inserted in the coil, the effective resistance of the coil increases to 10 Ω, and the current decreases to 1.7 A. Find the wattmeter reading and the inductance of the coil.
d. For parts (a), (b), and (c), find the angle between the voltage and current.
e. Sketch the circuit and show how you could use an oscilloscope to measure the angle between the voltage and current.

20 Resonance

Outline

Learning Outcomes

After completing this chapter, you will be able to

- explain the meanings of resonance, bandwidth, quality factor (Q), and selectivity
- describe the relationships among resonance, bandwidth, quality factor (Q), and selectivity
- calculate the resonant and bandwidth frequencies for a given series or parallel *R-L-C* circuit
- calculate voltage, current, and power in resonant circuits at and near resonance
- draw the response graphs of resonant circuits at and near resonance
- determine what happens to resonance, bandwidth, and impedance when using practical coils and lossy circuit elements
- design a series or parallel resonant circuit to meet given specifications

This chapter shows how inductance and capacitance are used to tune desired radio signals.

20.1 INTRODUCTION

This chapter will introduce the very important *resonant* (or *tuned*) circuit, which is fundamental to the operation of a wide variety of electrical and electronic systems in use today. The resonant circuit is a combination of R, L, and C elements having a frequency response characteristic similar to the one appearing in Fig. 20.1. Note in the figure that the response is a maximum for the frequency f_r, decreasing to the right and left of this frequency. In other words, for a particular range of frequencies the response will be near or equal to the maximum. The frequencies to the far left or right have very low voltage or current levels and, for all practical purposes, have little effect on the system's response.

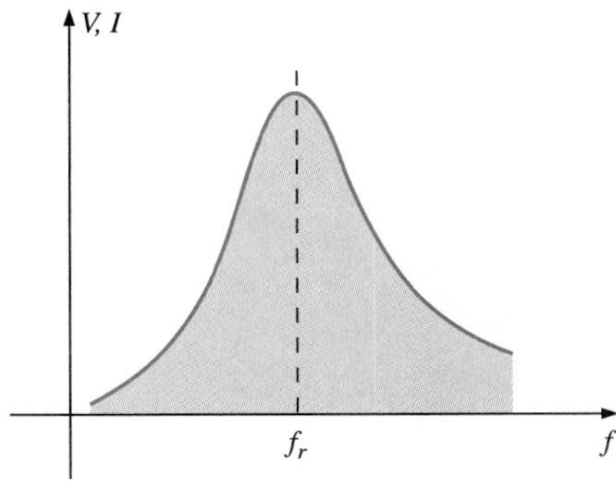

FIG. 20.1
Resonance curve.

Radio and television receivers have a response curve for each broadcast station of the type indicated in Fig. 20.1. When a receiver is set (or tuned) to a particular station, it is set on or near the frequency f_r of Fig. 20.1. Stations transmitting at frequencies to the far right or left of this resonant frequency are not carried through with significant power to affect the program of interest. The tuning process (setting the dial to f_r) as described above is the reason for the terminology **tuned circuit**. When the response is at or near the maximum, the circuit is said to be in a state of **resonance**.

The concept of resonance is not limited to electrical or electronic systems. If mechanical impulses are applied to a mechanical system at the proper frequency, the system will enter a state of resonance in which sustained vibrations of very large amplitude will develop. The frequency at which this occurs is called the **natural frequency** of the system. The classic example of this effect was the Tacoma Narrows Bridge built in 1940 over Puget Sound in Washington State. Four months after the bridge, with its suspended span of 850 m, was completed, a 68-km/h pulsating gale set the bridge into oscillations at its natural frequency. The amplitude of the oscillations increased to the point where the main span broke up and fell into the water below. It has since been replaced by the new Tacoma Narrows Bridge, completed in 1950.

The resonant electrical circuit *must* have both inductance and capacitance. In addition, resistance will always be present due either to the lack of ideal elements or to the control offered on the shape of the resonance curve. When resonance occurs due to the application of the proper frequency (f_r), the energy absorbed by one reactive element is the same as that released by another reactive element within the system. In other words, energy pulsates from one reactive element to the other.

Therefore, once an ideal (pure C, L) system has reached a state of resonance, it requires no further reactive power, since it is self-sustaining. In a practical circuit, there is some resistance associated with the reactive elements that will result in the eventual "damping" of the oscillations between reactive elements.

There are two types of resonant circuits: **series** and **parallel**. Sections 20.2 to 20.7 cover series resonant circuits, and Sections 20.8 to 20.12 cover parallel resonant circuits.

20.2 SERIES RESONANT CIRCUIT

A resonant circuit (series or parallel) must have both inductive and capacitive elements. A resistive element will always be present due to the internal resistance of the source (R_s), the internal resistance of the inductor (R_l), and any added resistance to control the shape of the response curve (R_{design}). The basic configuration for the **series resonant circuit** appears in Fig. 20.2(a) with the resistive elements listed above. The cleaner appearance of Fig. 20.2(b) is a result of combining the series resistive elements into one total value. That is,

$$R = R_s + R_l + R_d \qquad (20.1)$$

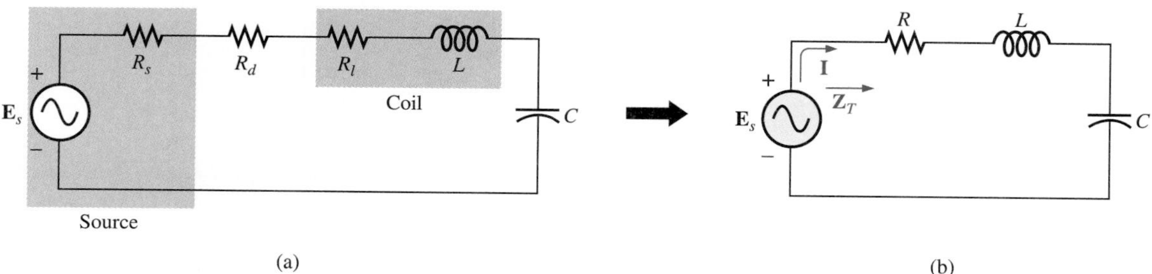

FIG. 20.2
Series resonant circuit.

The total impedance of this network at any frequency is determined by

$$\mathbf{Z}_T = R + jX_L - jX_C = R + j(X_L - X_C)$$

The resonant conditions described in the introduction will occur when

$$X_L = X_C \qquad (20.2)$$

removing the reactive component from the total impedance equation. The total impedance at resonance is then simply

$$\mathbf{Z}_{T_s} = R \qquad (20.3)$$

representing the minimum value of $\mathbf{Z}_T$ at any frequency. The subscript s will be used to indicate series resonant conditions.

The resonant frequency can be determined in terms of the inductance (in henries) and capacitance (in farads) by examining the defining equation for resonance [Eq. (20.2)]:

$$X_L = X_C$$

Substituting results in

$$\omega L = \frac{1}{\omega C} \quad \text{and} \quad \omega^2 = \frac{1}{LC}$$

and $$\omega_s = \frac{1}{\sqrt{LC}} \tag{20.4}$$

or $$f_s = \frac{1}{2\pi\sqrt{LC}} \quad \text{(hertz, Hz)} \tag{20.5}$$

Take note that the series resonant frequency is independent of the resistance in the circuit.

The current through the circuit at resonance is

$$\mathbf{I} = \frac{E \angle 0°}{R \angle 0°} = \frac{E}{R} \angle 0°$$

which you will note is the maximum current for the circuit of Fig. 20.2 for an applied voltage **E** since $\mathbf{Z}_T$ is a minimum value. Consider also that *the input voltage and current are in phase at resonance.*

Since the current is the same through the capacitor and inductor, the voltage across each is equal in magnitude but 180° out of phase at resonance:

$$\left.\begin{aligned} \mathbf{V}_L &= (I \angle 0°)(X_L \angle 90°) = IX_L \angle 90° \\ \mathbf{V}_C &= (I \angle 0°)(X_C \angle -90°) = IX_C \angle -90° \end{aligned}\right\} \begin{matrix}180° \\ \text{out of} \\ \text{phase}\end{matrix}$$

and, since $X_L = X_C$, the magnitude of V_L equals V_C at resonance; that is,

$$V_{L_s} = V_{C_s} \tag{20.6}$$

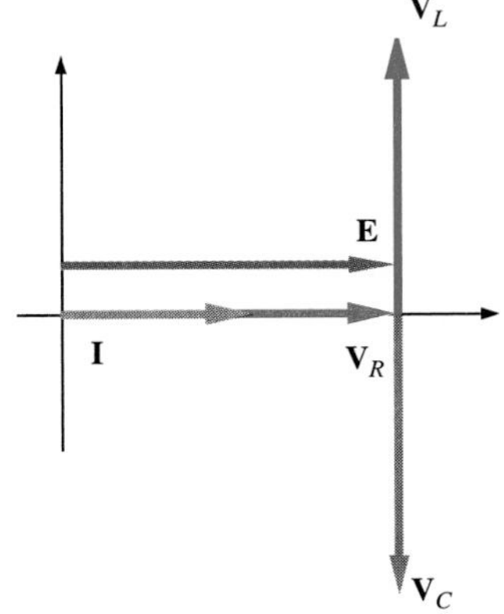

FIG. 20.3
Phasor diagram for the series resonant circuit at resonance.

Figure 20.3, a phasor diagram of the voltages and current, clearly indicates that the voltage across the resistor at resonance is the input voltage and **E,** and **I,** and $\mathbf{V}_R$ are in phase at resonance.

The average power to the resistor at resonance is equal to I^2R, and the reactive power to the capacitor and inductor are I^2X_C and I^2X_L, respectively.

The power triangle at resonance (Fig. 20.4) shows that the total apparent power is equal to the average power dissipated by the resistor since $Q_L = Q_C$. The power factor of the circuit at resonance is

$$F_p = \cos\theta = \frac{P}{S}$$

and $$F_{p_s} = 1 \tag{20.7}$$

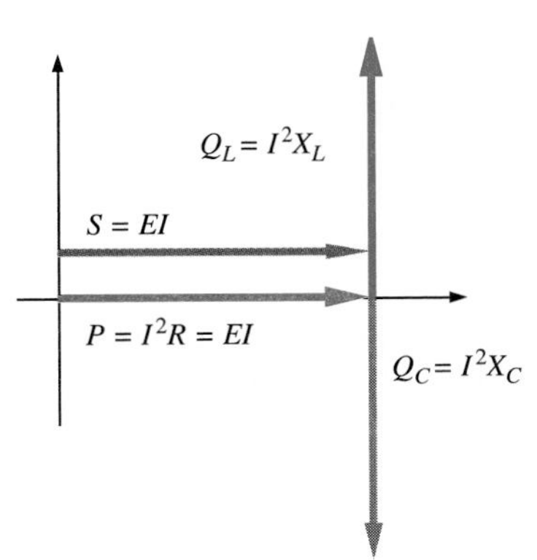

FIG. 20.4
Power triangle for the series resonant circuit at resonance.

Plotting the power curves of each element on the same set of axes (Fig. 20.5), we note that, even though the total reactive power at any instant is equal to zero (note $t = t'$), energy is still being absorbed and released by the inductor and capacitor at resonance.

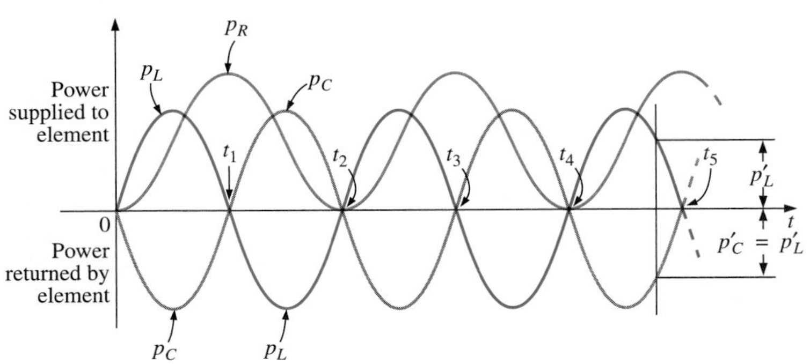

FIG. 20.5
Power curves at resonance for the series resonant circuit.

A closer examination reveals that the energy absorbed by the inductor from time 0 to t_1 is the same as the energy released by the capacitor from 0 to t_1. The reverse occurs from t_1 to t_2, and so on. Therefore, the total apparent power continues to be equal to the average power, even though the inductor and capacitor are absorbing and releasing energy. This condition occurs only at resonance. The slightest change in frequency introduces a reactive component into the power triangle, which will increase the apparent power of the system above the average power dissipation, and resonance will no longer exist.

20.3 THE QUALITY FACTOR (Q)

The **quality factor** Q of a series resonant circuit is defined as the ratio of the reactive power of either the inductor or the capacitor to the average power of the resistor at resonance; that is,

$$Q_s = \frac{\text{reactive power}}{\text{average power}} \tag{20.8}$$

The quality factor is also an indication of how much energy is placed in storage (continual transfer from one reactive element to the other) as compared to that dissipated. The lower the level of dissipation for the same reactive power, the larger the Q_s factor and the more concentrated and intense the region of resonance.

Substituting for an inductive reactance in Eq. (20.8) at resonance gives us

$$Q_s = \frac{I^2X_L}{I^2R}$$

and

$$Q_s = \frac{X_L}{R} = \frac{\omega_s L}{R} \tag{20.9}$$

If the resistance R is just the resistance of the coil (R_l), we can speak of the Q of the coil, where

$$Q_{\text{coil}} = Q_l = \frac{X_L}{R_l} \qquad R = R_l \tag{20.10}$$

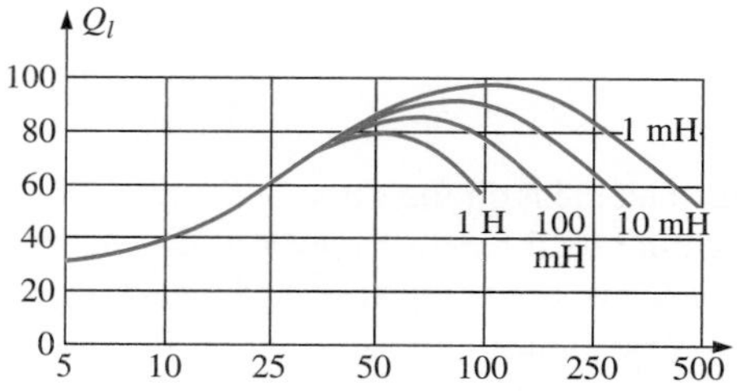

FIG. 20.6
Q_l versus frequency for a series of inductors of similar construction.

Since the quality factor of a coil is typically the information provided by manufacturers of inductors, it is often given the symbol Q without an associated subscript. It would appear from Eq. (20.10) that Q_l will increase linearly with frequency since $X_L = 2\pi fL$. That is, if the frequency doubles, then Q_l will also increase by a factor of 2. This is approximately true for the low range to the midrange of frequencies such as shown for the coils of Fig. 20.6. Unfortunately, however, as the frequency increases, the effective resistance of the coil will also increase, due primarily to skin effect phenomena, and the resulting Q_l will decrease. In addition, the capacitive effects between the windings will increase, further reducing the Q_l of the coil. For this reason, Q_l must be specified for a particular frequency or frequency range. For wide frequency applications, a plot of Q_l versus frequency is often provided. The maximum Q_l for most commercially available coils is less than 200, with most having a maximum near 100. Note in Fig. 20.6 that for coils of the same type, Q_l drops off more quickly for higher levels of inductance.

If we substitute

$$\omega_s = 2\pi f_s$$

and then

$$f_s = \frac{1}{2\pi\sqrt{LC}}$$

into Eq. (20.9), we have

$$Q_s = \frac{\omega_s L}{R} = \frac{2\pi f_s L}{R} = \frac{2\pi}{R}\left(\frac{1}{2\pi\sqrt{LC}}\right)L$$

$$= \frac{L}{R}\left(\frac{1}{\sqrt{LC}}\right) = \left(\frac{\sqrt{L}}{\sqrt{L}}\right)\frac{L}{R\sqrt{LC}}$$

and

$$Q_s = \frac{1}{R}\sqrt{\frac{L}{C}} \tag{20.11}$$

providing Q_s in terms of the circuit parameters.

For series resonant circuits used in communication systems, Q_s is usually greater than 1. By applying the voltage divider rule to the circuit of Fig. 20.2, we obtain

$$V_L = \frac{X_L E}{Z_T} = \frac{X_L E}{R} \quad \text{(at resonance)}$$

and

$$V_{L_s} = Q_s E \tag{20.12}$$

or

$$V_C = \frac{X_C E}{Z_T} = \frac{X_C E}{R}$$

and

$$V_{C_s} = Q_s E \tag{20.13}$$

Since Q_s is usually greater than 1, the voltage across the capacitor or inductor of a series resonant circuit can be significantly greater than the input voltage. In fact, in many cases the Q_s is so high that careful design and handling (including adequate insulation) are necessary for the voltage across the capacitor and inductor.

In the circuit of Fig. 20.7, for example, which is in the state of resonance,

$$Q_s = \frac{X_L}{R} = \frac{480\ \Omega}{6\ \Omega} = 80$$

and

$$V_L = V_C = Q_s E = (80)(10\text{ V}) = \mathbf{800\ V}$$

which is certainly a potential of significant magnitude.

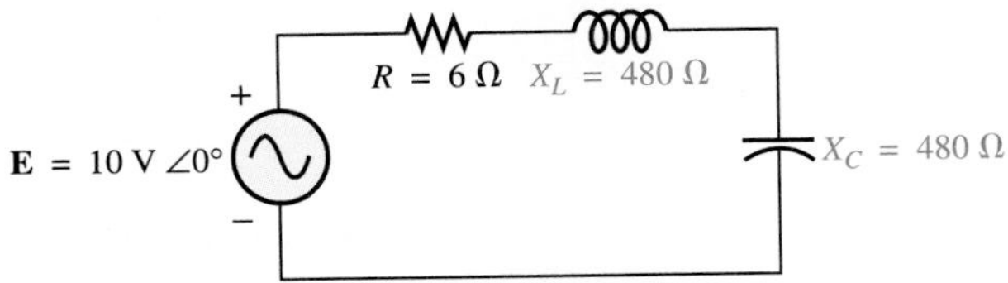

FIG. 20.7
High-Q series resonant circuit.

20.4 Z_T VERSUS FREQUENCY

The **total impedance** of the series *R-L-C* circuit of Fig. 20.2 at any frequency is determined by

$$\mathbf{Z}_T = R + jX_L - jX_C \text{ or } \mathbf{Z}_T = R + j(X_L - X_C)$$

The magnitude of the impedance $\mathbf{Z}_T$ versus frequency is determined by

$$Z_T = \sqrt{R^2 + (X_L - X_C)^2}$$

The total-impedance-versus-frequency curve for the series resonant circuit of Fig. 20.2 can be found by applying the impedance-versus-frequency curve for each element of the equation just derived, written in the following form:

$$Z_T(f) = \sqrt{[R(f)]^2 + [X_L(f) - X_C(f)]^2} \tag{20.14}$$

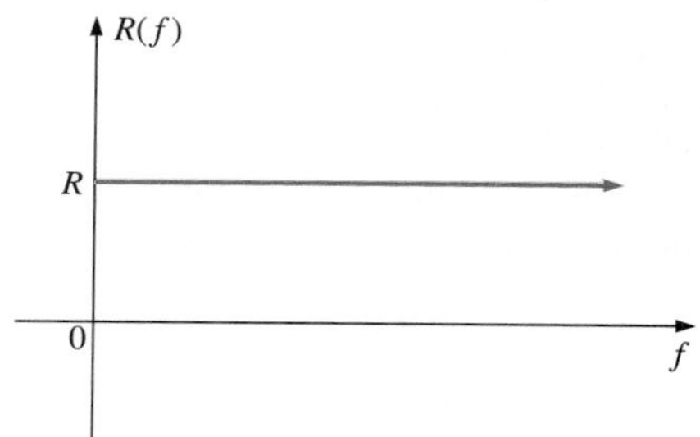

FIG. 20.8
Resistance versus frequency.

where $Z_T(f)$ means the total impedance as a *function* of frequency. For the frequency range of interest, we will assume that the resistance R does not change with frequency, resulting in the plot of Fig. 20.8. The curve for the inductance, as determined by the reactance equation, is a straight line intersecting the origin with a slope equal to the inductance of the coil. The mathematical expression for any straight line in a two-dimensional plane is given by

$$y = mx + b$$

Thus, for the coil,

$$\begin{array}{cccccc} X_L = 2\pi fL + 0 = & (2\pi L) & (f) & + 0 \\ \downarrow & \downarrow & \downarrow & \downarrow \\ y = & m & \cdot\ x & +\ b \end{array}$$

(where $2\pi L$ is the slope), producing the results shown in Fig. 20.9.

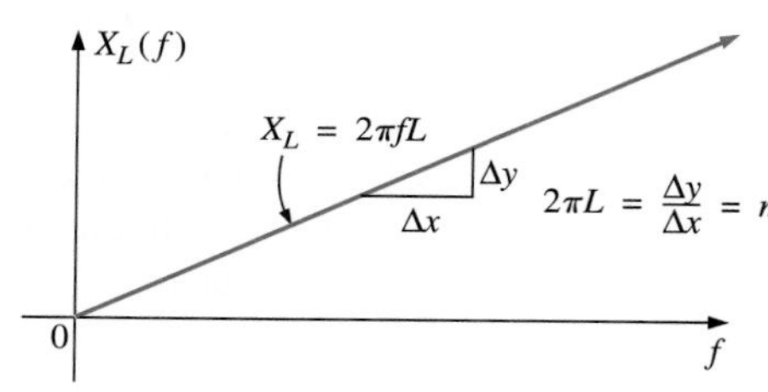

FIG. 20.9
Inductive reactance versus frequency.

For the capacitor,

$$X_C = \frac{1}{2\pi fC} \quad \text{or} \quad X_C f = \frac{1}{2\pi C}$$

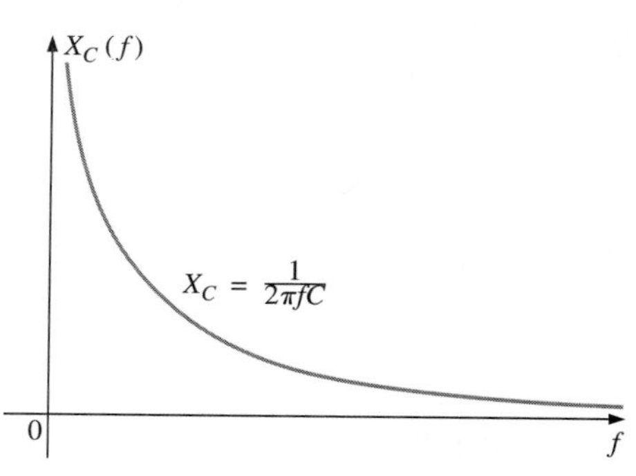

FIG. 20.10
Capacitive reactance versus frequency.

which becomes $yx = k$, the equation for a hyperbola, where

$$y \text{ (variable)} = X_C$$
$$x \text{ (variable)} = f$$
$$k \text{ (constant)} = \frac{1}{2\pi C}$$

The hyperbolic curve for $X_C(f)$ is plotted in Fig. 20.10. In particular, note its very large magnitude at low frequencies and its rapid drop-off as the frequency increases.

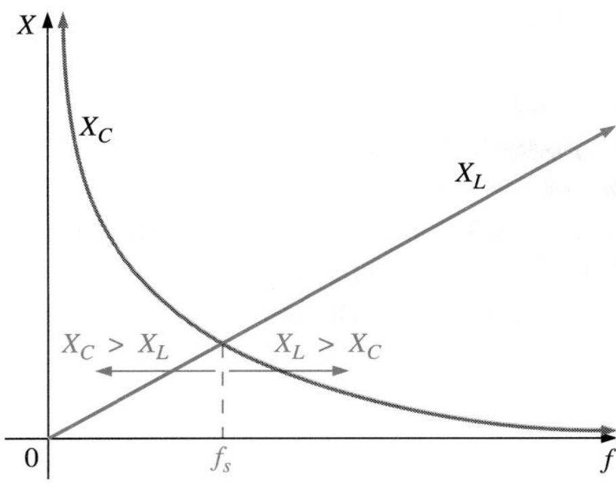

FIG. 20.11
Placing the frequency response of the inductive and capacitive reactance of a series R-L-C circuit on the same axis.

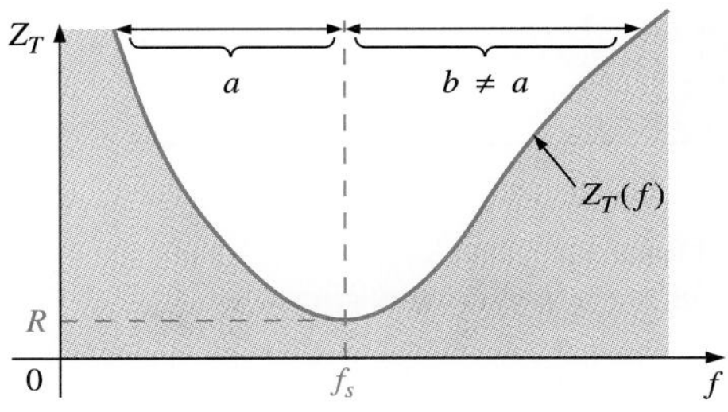

FIG. 20.12
Z_T versus frequency for the series resonant circuit.

If we place Figs. 20.9 and 20.10 on the same axis, we obtain the curves of Fig. 20.11. The condition of resonance is now clearly defined by the point of intersection, where $X_L = X_C$. For frequencies less than f_s, it is also quite clear that the network is primarily capacitive ($X_C > X_L$). For frequencies above the resonant condition, $X_L > X_C$ and the network is inductive.

Applying

$$Z_T(f) = \sqrt{[R(f)]^2 + [X_L(f) - X_C(f)]^2}$$

$$= \sqrt{[R(f)]^2 + [X(f)]^2}$$

to the curves of Fig. 20.11, where $X(f) = X_L(f) - X_C(f)$, we obtain the curve for $Z_T(f)$ as shown in Fig. 20.12. The minimum impedance occurs at the resonant frequency and is equal to the resistance R. Note that the curve is not symmetrical about the resonant frequency (especially at higher values of Z_T).

The phase angle associated with the total impedance is

$$\theta = \tan^{-1}\frac{(X_L - X_C)}{R} \tag{20.15}$$

For the $\tan^{-1} x$ function (resulting when $X_L > X_C$), the larger x is, the larger the angle θ (closer to 90°). However, for regions where $X_C > X_L$ you must also be aware that

$$\tan^{-1}(-x) = -\tan^{-1} x \tag{20.16}$$

At low frequencies, $X_C > X_L$ and θ will approach −90° (capacitive), as shown in Fig. 20.13, whereas at high frequencies, $X_L > X_C$ and θ will approach 90°. In general, therefore, for a series resonant circuit:

$f < f_s$: network capacitive, **I** leads **E**
$f > f_s$: network inductive, **E** leads **I**
$f = f_s$: network resistive, **E** and **I** are in phase

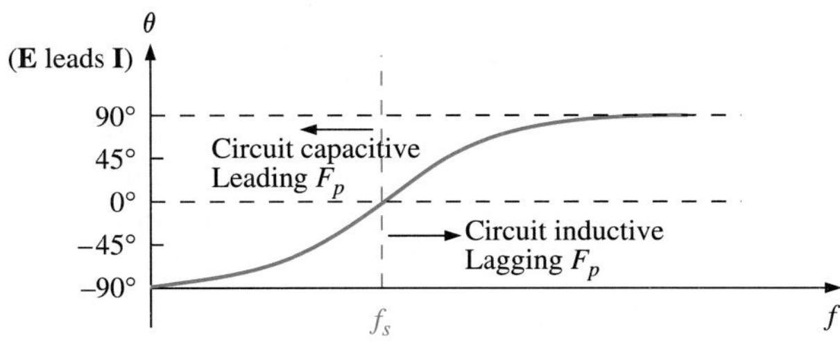

FIG. 20.13
Phase plot for the series resonant circuit.

20.5 SELECTIVITY

If we now plot the magnitude of the current $I = E/Z_T$ versus frequency for a *fixed* applied voltage E, we obtain the curve shown in Fig. 20.14, which rises from zero to a maximum value of E/R (where Z_T is a minimum) and then drops toward zero (as Z_T increases) at a slower rate than it rose to its peak value. The curve is actually the inverse of the impedance-versus-frequency curve. Since the Z_T curve is not absolutely symmetrical about the resonant frequency, the curve of the current versus frequency has the same property.

There is a definite range of frequencies at which the current is near its maximum value and the impedance is at a minimum. Those frequencies corresponding to 0.707 of the maximum current are called the **band frequencies**, **cutoff frequencies**, or **half-power frequencies**. They are indicated by f_1 and f_2 in Fig. 20.14. The range of frequencies between the two is referred to as the **bandwidth** (abbreviated *BW*) of the resonant circuit.

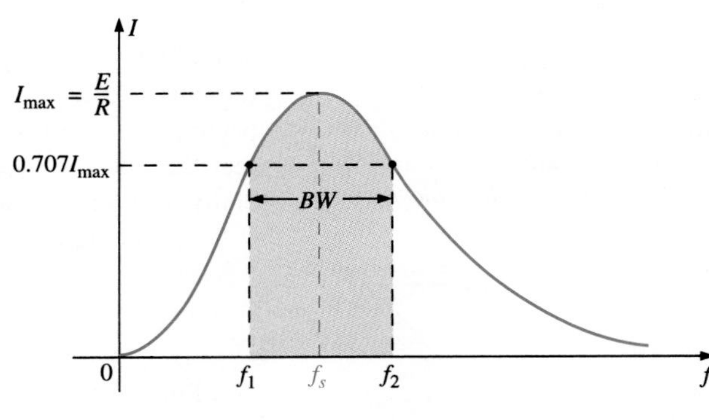

FIG. 20.14
I versus frequency for the series resonant circuit.

Half-power frequencies are those frequencies at which the power delivered is one-half the power delivered at the resonant frequency; that is,

$$P_{HPF} = \frac{1}{2}P_{max} \tag{20.17}$$

The above condition is derived using the fact that

$$P_{max} = I^2_{max}R$$

and $$P_{HPF} = I^2R = (0.707I_{max})^2R = (0.5)(I^2_{max}R) = \frac{1}{2}P_{max}$$

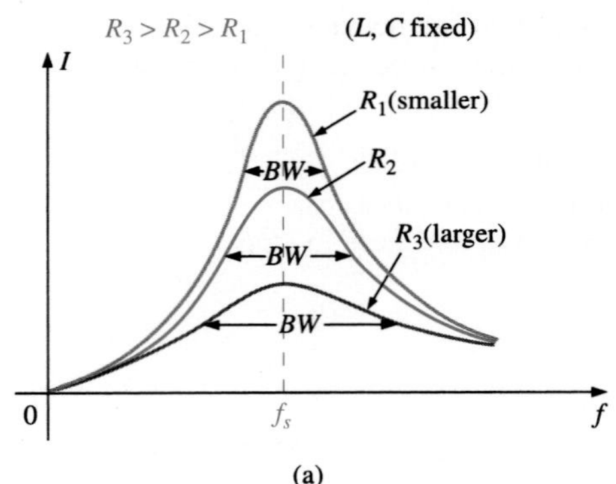

(a)

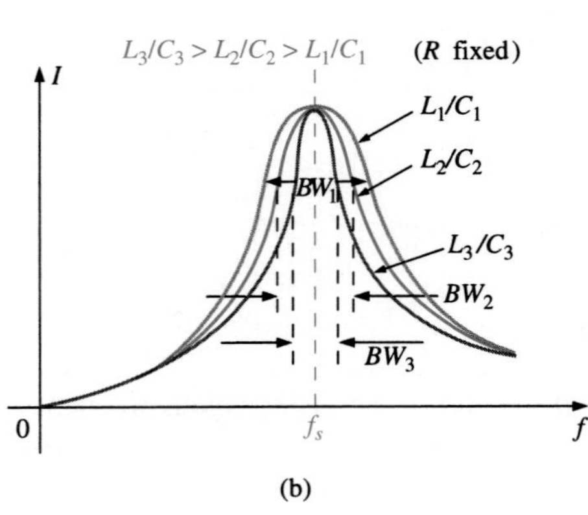

(b)

FIG. 20.15
Effect of R, L, and C on the selectivity curve for the series resonant circuit.

Since the resonant circuit is adjusted to select a band of frequencies, the curve of Fig. 20.14 is called the **selectivity curve**. The term comes from the fact that one must be *selective* in choosing the frequency to ensure that it is in the bandwidth. The smaller the bandwidth, the higher the selectivity. The shape of the curve, as shown in Fig. 20.15, depends on each element of the series *R-L-C* circuit. If the resistance is made smaller with a fixed inductance and capacitance, the bandwidth decreases and the selectivity increases. Similarly, if the ratio *L/C* increases with fixed resistance, the bandwidth again decreases with an increase in selectivity.

In terms of Q_s, if *R* is larger for the same X_L, then Q_s is less, as determined by the equation $Q_s = \omega_s L/R$.

A small Q_s, therefore, is associated with a resonant curve having a large bandwidth and a small selectivity, while a large Q_s indicates the opposite.

For circuits where $Q_s \geq 10$, a widely accepted approximation is that the resonant frequency bisects the bandwidth and that the resonant curve is symmetrical about the resonant frequency.

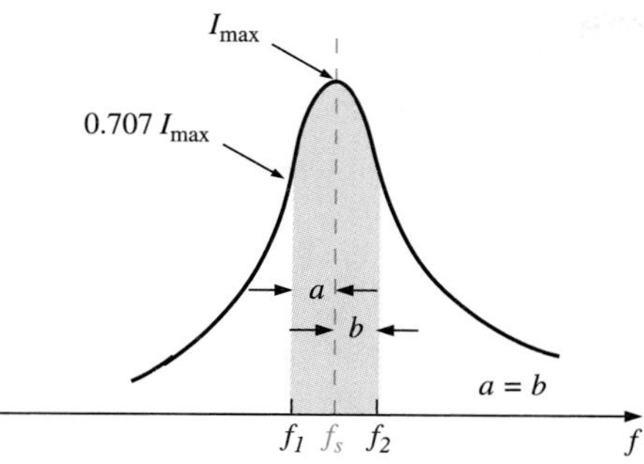

FIG. 20.16
Approximate series resonance curve for $Q_s \geq 10$.

These conditions are shown in Fig. 20.16, indicating that the cutoff frequencies are then equidistant from the resonant frequency.

For any Q_s, the preceding case is not true. The cutoff frequencies f_1 and f_2 can be found for the general case (any Q_s) by first using the fact that a drop in current to 0.707 of its resonant value corresponds to an increase in impedance equal to $1/0.707 = \sqrt{2}$ times the resonant value, which is R.

Substituting $\sqrt{2}R$ into the equation for the magnitude of Z_T, we find that

$$Z_T = \sqrt{R^2 + (X_L - X_C)^2}$$

becomes

$$\sqrt{2}R = \sqrt{R^2 + (X_L - X_C)^2}$$

or, squaring both sides, that

$$2R^2 = R^2 + (X_L - X_C)^2$$

and

$$R^2 = (X_L - X_C)^2$$

Taking the square root of both sides gives us

$$R = X_L - X_C \text{ or } R - X_L + X_C = 0$$

Let us first consider the case where $X_L > X_C$, which relates to f_2 or ω_2. Substituting $\omega_2 L$ for X_L and $1/\omega_2 C$ for X_C and bringing both quantities to the left of the equals sign, we have

$$R - \omega_2 L + \frac{1}{\omega_2 C} = 0 \text{ or } R\omega_2 - \omega_2^2 L + \frac{1}{C} = 0$$

which can be written

$$\omega_2^2 - \frac{R}{L}\omega_2 - \frac{1}{LC} = 0$$

Solving the quadratic, we have

$$\omega_2 = \frac{-(-R/L) \pm \sqrt{[-(R/L)]^2 - [-(4/LC)]}}{2}$$

and

$$\omega_2 = +\frac{R}{2L} \pm \frac{1}{2}\sqrt{\frac{R^2}{L^2} + \frac{4}{LC}}$$

with

$$f_2 = \frac{1}{2\pi}\left[\frac{R}{2L} + \frac{1}{2}\sqrt{\left(\frac{R}{L}\right)^2 + \frac{4}{LC}}\right] \quad \text{(Hz)} \qquad (20.18)$$

The negative sign in front of the second factor was dropped because $(1/2)\sqrt{(R/L)^2 + 4/LC}$ is always greater than $R/(2L)$. If we didn't drop it, there would be a negative solution for the radian frequency ω.

If we repeat the same procedure for $X_C > X_L$, which relates to ω_1 or f_1 such that $Z_T = \sqrt{R^2 + (X_C - X_L)^2}$, the solution f_1 becomes

$$f_1 = \frac{1}{2\pi}\left[-\frac{R}{2L} + \frac{1}{2}\sqrt{\left(\frac{R}{L}\right)^2 + \frac{4}{LC}}\right] \quad \text{(Hz)} \qquad (20.19)$$

The bandwidth (BW) is

$$BW = f_2 - f_1 = \text{Eq. (20.18)} - \text{Eq. (20.19)}$$

continued

and
$$BW = f_2 - f_1 = \frac{R}{2\pi L} \qquad (20.20)$$

Substituting $R/L = \omega_s/Q_s$ from $Q_s = \omega_s L/R$ and $1/2\pi = f_s/\omega_s$ from $\omega_s = 2\pi f_s$ gives us

$$BW = \frac{R}{2\pi L} = \left(\frac{1}{2\pi}\right)\left(\frac{R}{L}\right) = \left(\frac{f_s}{\omega_s}\right)\left(\frac{\omega_s}{Q_s}\right)$$

or
$$BW = \frac{f_s}{Q_s} \qquad (20.21)$$

which is a very convenient form, since it relates the bandwidth to the Q_s of the circuit. As mentioned earlier, Eq. (20.21) verifies that the larger the Q_s, the smaller the bandwidth, and vice versa.

Written in a slightly different form, Eq. (20.21) becomes

$$\frac{f_2 - f_1}{f_s} = \frac{1}{Q_s} \qquad (20.22)$$

The ratio $(f_2 - f_1)/f_s$ is sometimes called the **fractional bandwidth**, providing an indication of the width of the bandwidth as compared to the resonant frequency.

It can also be shown by mathematically manipulating the relevant equations that the resonant frequency is related to the geometric mean of the band frequencies; that is,

$$f_s = \sqrt{f_1 f_2} \qquad (20.23)$$

20.6 V_R, V_L, AND V_C

Plotting the *magnitude* (effective value) of the voltages $\mathbf{V}_R$, $\mathbf{V}_L$, and $\mathbf{V}_C$ and the current **I** versus *frequency* for the series resonant circuit on the same set of axes, we obtain the curves shown in Fig. 20.17. Note that the V_R curve has the same shape as the I curve and a peak value equal to the magnitude of the input voltage E. The V_C curve builds up slowly at first from a value equal to the input voltage since the reactance of the

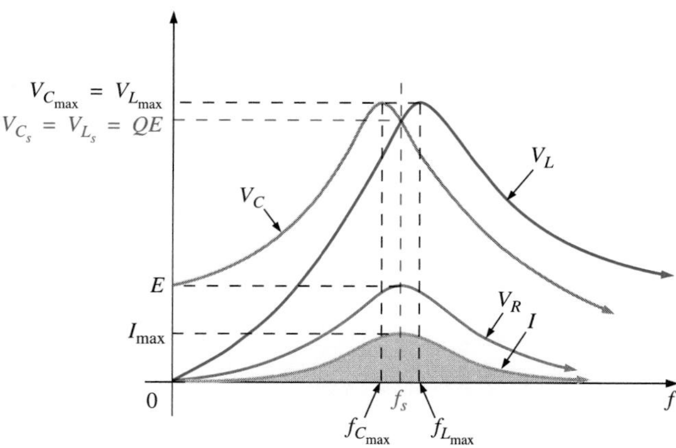

FIG. 20.17
V_R, V_L, V_C, and I versus frequency for a series resonant circuit.

capacitor is infinite (open circuit) at zero frequency and the reactance of the inductor is zero (short circuit) at this frequency. As the frequency increases, $1/\omega C$ of the equation

$$V_C = IX_C = (I)\left(\frac{1}{\omega C}\right)$$

becomes smaller, but I increases at a rate faster than that at which $1/\omega C$ drops. Therefore, V_C rises and will continue to rise due to the quickly rising current, until the frequency nears resonance. As it approaches the resonant condition, the rate of change of I decreases. When this occurs, the factor $1/\omega C$, which decreased as the frequency rose, will overcome the rate of change of I, and V_C will start to drop. The peak value will occur at a frequency just before resonance. After resonance, both V_C and I drop in magnitude, and V_C approaches zero.

The higher the Q_s of the circuit, the closer $f_{C_{max}}$ will be to f_s, and the closer $V_{C_{max}}$ will be to Q_sE. For circuits with $Q_s \geq 10$, $f_{C_{max}} \cong f_s$, and $V_{C_{max}} \cong Q_s E$.

The curve for V_L increases steadily from zero to the resonant frequency since both quantities ωL and I of the equation $V_L = IX_L = (I)(\omega L)$ increase over this frequency range. At resonance, I has reached its maximum value, but ωL is still rising. Therefore, V_L will reach its maximum value after resonance. After reaching its peak value, the voltage V_L will drop toward E since the drop in I will overcome the rise in ωL. It approaches E because X_L will eventually be infinite, and X_C will be zero.

As Q_s of the circuit increases, the frequency $f_{L_{max}}$ drops toward f_s, and $V_{L_{max}}$ approaches $Q_s E$. For circuits with $Q_s \geq 10$, $f_{L_{max}} \cong f_s$, and $V_{L_{max}} \cong Q_s E$.

The V_L curve has a greater magnitude than the V_C curve for any frequency above resonance, and the V_C curve has a greater magnitude than the V_L curve for any frequency below resonance. This again verifies that the series *R-L-C* circuit is predominantly capacitive from zero to the resonant frequency and predominantly inductive for any frequency above resonance.

For the condition $Q_s \geq 10$, the curves of Fig. 20.17 will appear as shown in Fig. 20.18. Note that they each peak (on an approximate basis) at the resonant frequency and have a similar shape.

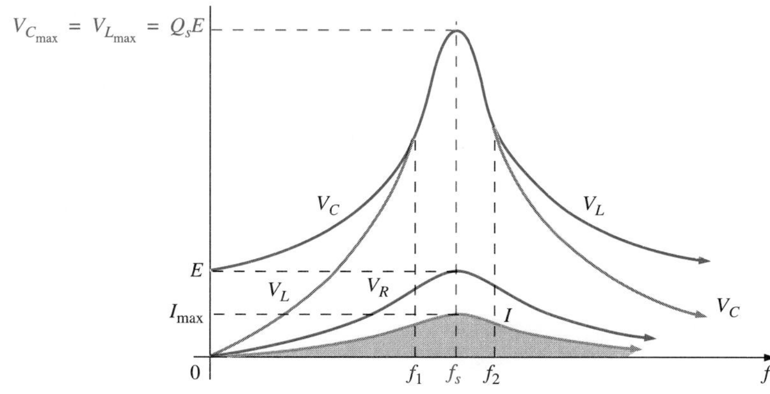

FIG. 20.18
V_R, V_L, V_C, and I for a series resonant circuit where $Q_s \geq 10$.

In review,

1. V_C and V_L are at their maximum values at or near resonance (depending on Q_s).
2. At very low frequencies, V_C is very close to the source voltage and V_L is very close to zero volts, whereas at very high frequencies, V_L approaches the source voltage and V_C approaches zero volts.
3. Both V_R and I peak at the resonant frequency and have the same shape.

20.7 EXAMPLES (SERIES RESONANCE)

EXAMPLE 20.1

a. For the series resonant circuit of Fig. 20.19, find **I**, $\mathbf{V}_R$, $\mathbf{V}_L$, and $\mathbf{V}_C$ at resonance.

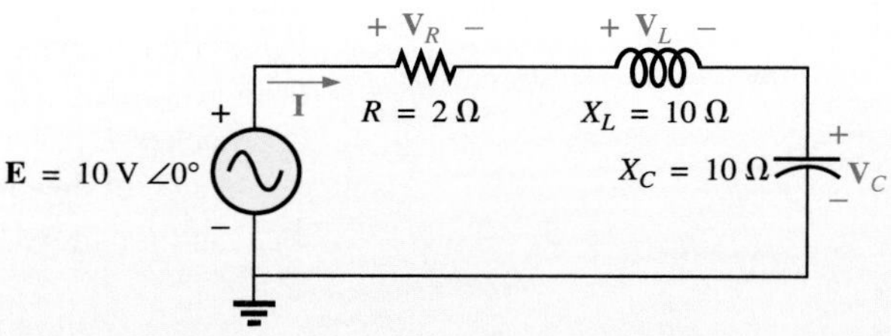

FIG. 20.19
Example 20.1.

b. What is the Q_s of the circuit?
c. If the resonant frequency is 5000 Hz, find the bandwidth.
d. What is the power dissipated in the circuit at the half-power frequencies?

Solutions:

a. $\mathbf{Z}_{T_s} = R = 2\ \Omega$

$$\mathbf{I} = \frac{\mathbf{E}}{\mathbf{Z}_{T_s}} = \frac{10\ \text{V}\ \angle 0°}{2\ \Omega\ \angle 0°} = \mathbf{5\ A\ \angle 0°}$$

$$\mathbf{V}_R = \mathbf{E} = 10\ \text{V}\ \angle 0°$$

$$\mathbf{V}_L = (I\ \angle 0°)(X_L\ \angle 90°) = (5\ \text{A}\ \angle 0°)(10\ \Omega\ \angle 90°) = \mathbf{50\ V\ \angle 90°}$$

$$\mathbf{V}_C = (I\ \angle 0°)(X_C\ \angle -90°) = (5\ \text{A}\ \angle 0°)(10\ \Omega\ \angle -90°) = \mathbf{50\ V\ \angle -90°}$$

b. $$Q_s = \frac{X_L}{R} = \frac{10\ \Omega}{2\ \Omega} = \mathbf{5}$$

c. $$BW = f_2 - f_1 = \frac{f_s}{Q_s} = \frac{5000\ \text{Hz}}{5} = \mathbf{1000\ Hz}$$

d. $$P_{\text{HPF}} = \frac{1}{2}P_{\text{max}} = \frac{1}{2}I_{\text{max}}^2 R = \left(\frac{1}{2}\right)(5\ \text{A})^2(2\ \Omega) = \mathbf{25\ W}$$

EXAMPLE 20.2

The bandwidth of a series resonant circuit is 400 Hz.

a. If the resonant frequency is 4000 Hz, what is the value of Q_s?
b. If $R = 10\ \Omega$, what is the value of X_L at resonance?
c. Find the inductance L and capacitance C of the circuit.

Solutions:

a. $BW = \dfrac{f_s}{Q_s}$ or $Q_s = \dfrac{f_s}{BW} = \dfrac{4000\text{ Hz}}{400\text{ Hz}} = \mathbf{10}$

b. $Q_s = \dfrac{X_L}{R}$ or $X_L = Q_sR = (10)(10\ \Omega) = \mathbf{100\ \Omega}$

c. $X_L = 2\pi f_sL$ or $L = \dfrac{X_L}{2\pi f_s} = \dfrac{100\ \Omega}{2\pi(4000\text{ Hz})} = \mathbf{3.98\text{ mH}}$

$$X_C = \frac{1}{2\pi f_sC} \text{ or } C = \frac{1}{2\pi f_sX_C} = \frac{1}{2\pi(4000\text{ Hz})(100\ \Omega)}$$
$$= \mathbf{0.398\ \mu F}$$

EXAMPLE 20.3

A series *R-L-C* circuit has a series resonant frequency of 12 000 Hz.

a. If $R = 5\ \Omega$ and X_L at resonance is 300 Ω, find the bandwidth.
b. Find the cutoff frequencies.

Solutions:

a. $Q_s = \dfrac{X_L}{R} = \dfrac{300\ \Omega}{5\ \Omega} = 60$

$$BW = \frac{f_s}{Q_s} = \frac{12\,000\text{ Hz}}{60} = \mathbf{200\text{ Hz}}$$

b. Since $Q_s \geq 10$, the bandwidth is bisected by f_s. Therefore,

$$f_2 = f_s + \frac{BW}{2} = 12\,000\text{ Hz} + 100\text{ Hz} = \mathbf{12\,100\text{ Hz}}$$

and $f_1 = 12\,000\text{ Hz} - 100\text{ Hz} = \mathbf{11\,900\text{ Hz}}$

EXAMPLE 20.4

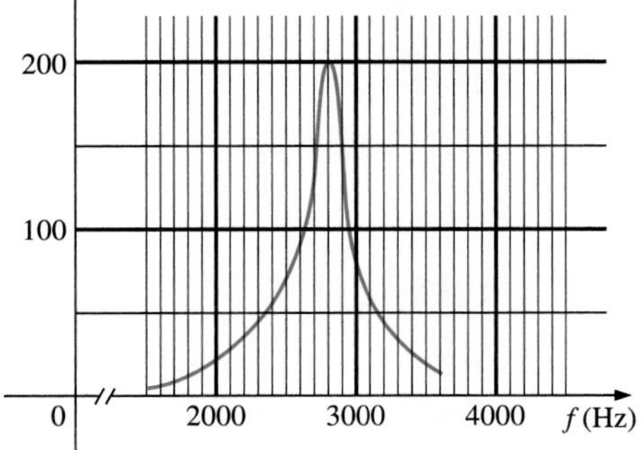

FIG. 20.20
Example 20.4.

a. Determine the Q_s and bandwidth for the response curve of Fig. 20.20.
b. For $C = 101.5$ nF, determine L and R for the series resonant circuit.
c. Determine the applied voltage.

Solutions:

a. The resonant frequency is 2800 Hz. At 0.707 times the peak value,

$$BW = \mathbf{200\text{ Hz}}$$

and $$Q_s = \frac{f_s}{BW} = \frac{2800\text{ Hz}}{200\text{ Hz}} = \mathbf{14}$$

b. $f_s = \dfrac{1}{2\pi\sqrt{LC}}$ or $L = \dfrac{1}{4\pi^2f_s^2C}$

continued

$$= \frac{1}{4\pi^2(2.8 \times 10^3 \text{ Hz})^2(101.5 \times 10^{-9} \text{ F})}$$

$$= \mathbf{31.832 \text{ mH}}$$

$$Q_s = \frac{X_L}{R} \text{ or } R = \frac{X_L}{Q_s} = \frac{2\pi(2800 \text{ Hz})(31.832 \times 10^{-3} \text{ H})}{14}$$

$$= \mathbf{40 \ \Omega}$$

c. $I_{max} = \frac{E}{R}$ or $E = I_{max}R$

and $E = (200 \text{ mA})(40 \ \Omega) = \mathbf{8 \text{ V}}$

EXAMPLE 20.5

A series *R-L-C* circuit is designed to resonate at $\omega_s = 10^5$ rad/s, have a bandwidth of $0.15\omega_s$, and draw 16 W from a 120-V source at resonance.

a. Determine the value of *R*.
b. Find the bandwidth in hertz.
c. Find the nameplate values of *L* and *C*.
d. What is the Q_s of the circuit?
e. Determine the fractional bandwidth.

Solutions:

a. $P = \frac{E^2}{R}$ and $R = \frac{E^2}{P} = \frac{(120 \text{ V})^2}{16 \text{ W}} = \mathbf{900 \ \Omega}$

b. $f_s = \frac{\omega_s}{2\pi} = \frac{10^5 \text{ rad/s}}{2\pi} = 15{,}915.49 \text{ Hz}$

$$BW = 0.15f_s = 0.15(15{,}915.49 \text{ Hz}) = \mathbf{2387.32 \text{ Hz}}$$

c. Eq. (20.20):

$$BW = \frac{R}{2\pi L} \text{ and } L = \frac{R}{2\pi BW} = \frac{900 \ \Omega}{2\pi(2387.32 \text{ Hz})} = \mathbf{60 \text{ mH}}$$

$$f_s = \frac{1}{2\pi\sqrt{LC}} \text{ and } C = \frac{1}{4\pi^2 f_s^2 L} = \frac{1}{4\pi^2(15{,}915.49 \text{ Hz})^2(60 \times 10^{-3} \text{ H})}$$

$$= \mathbf{1.67 \text{ nF}}$$

d. $Q_s = \frac{X_L}{R} = \frac{2\pi f_s L}{R} = \frac{2\pi(15{,}915.49 \text{ Hz})(60 \text{ mH})}{900 \ \Omega} = \mathbf{6.67}$

e. $\frac{f_2 - f_1}{f_s} = \frac{BW}{f_s} = \frac{1}{Q_s} = \frac{1}{6.67} = \mathbf{0.15}$

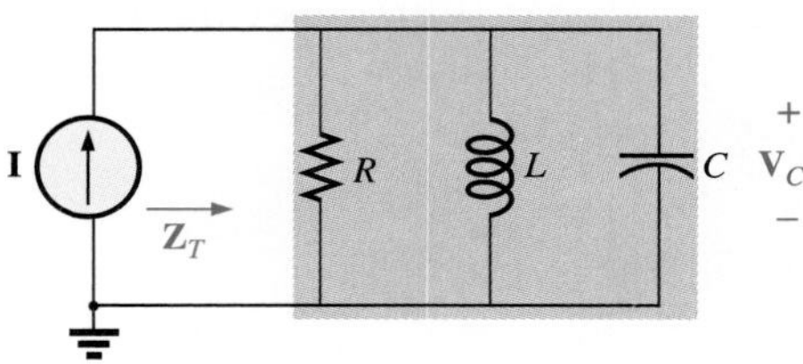

FIG. 20.21
Ideal parallel resonant network.

20.8 PARALLEL RESONANT CIRCUIT

The basic format of the series resonant circuit is a series *R-L-C* combination in series with an applied voltage source. The **parallel resonant circuit** has the basic configuration of Fig. 20.21, a parallel *R-L-C* combination in parallel with an applied current source.

For the series circuit, the impedance was a minimum at resonance, producing a significant current that resulted in a high output voltage for $\mathbf{V}_C$ and $\mathbf{V}_L$. For the parallel resonant circuit, the impedance is relatively high at resonance, producing a significant voltage for $\mathbf{V}_C$ and $\mathbf{V}_L$ through the Ohm's law relationship ($\mathbf{V}_C = \mathbf{IZ}_T$). For the network of Fig. 20.21, resonance will occur when $X_L = X_C$, and the resonant frequency will have the same format obtained for series resonance.

If the practical equivalent of Fig. 20.21 had the format of Fig. 20.21, the analysis would be as direct and clear as the one we did for series resonance. However, in the practical world, the internal resistance of the coil must be placed in series with the inductor, as shown in Fig. 20.22. The resistance R_l can no longer be included in a simple series or parallel combination with the source resistance and any other resistance added for design purposes. Even though R_l is usually relatively small in magnitude compared with other resistance and reactance levels of the network, it does have an important impact on the parallel resonant condition, as we will demonstrate in the sections to follow. In other words, the network of Fig. 20.21 is an ideal situation that can be assumed only for specific network conditions.

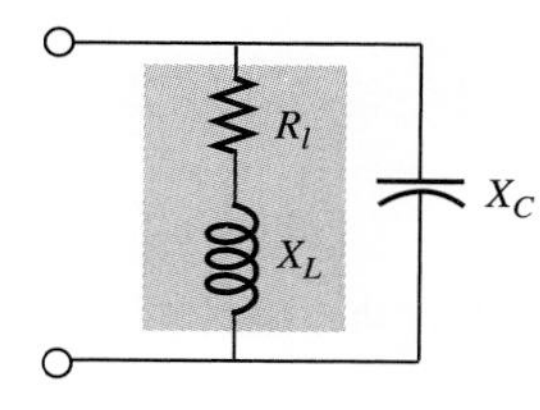

FIG. 20.22
Practical parallel L-C network.

First we have to find a parallel network equivalent (at the terminals) for the series *R-L* branch of Fig. 20.22 using the technique introduced in Section 15.7. That is,

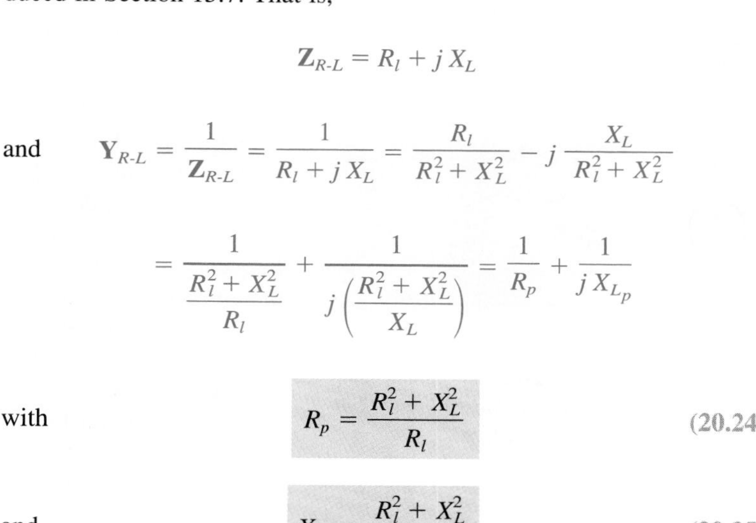

$$\mathbf{Z}_{R\text{-}L} = R_l + j\,X_L$$

and
$$\mathbf{Y}_{R\text{-}L} = \frac{1}{\mathbf{Z}_{R\text{-}L}} = \frac{1}{R_l + j\,X_L} = \frac{R_l}{R_l^2 + X_L^2} - j\,\frac{X_L}{R_l^2 + X_L^2}$$

$$= \frac{1}{\dfrac{R_l^2 + X_L^2}{R_l}} + \frac{1}{j\left(\dfrac{R_l^2 + X_L^2}{X_L}\right)} = \frac{1}{R_p} + \frac{1}{j\,X_{L_p}}$$

with
$$R_p = \frac{R_l^2 + X_L^2}{R_l} \qquad (20.24)$$

and
$$X_{L_p} = \frac{R_l^2 + X_L^2}{X_L} \qquad (20.25)$$

as shown in Fig. 20.23.

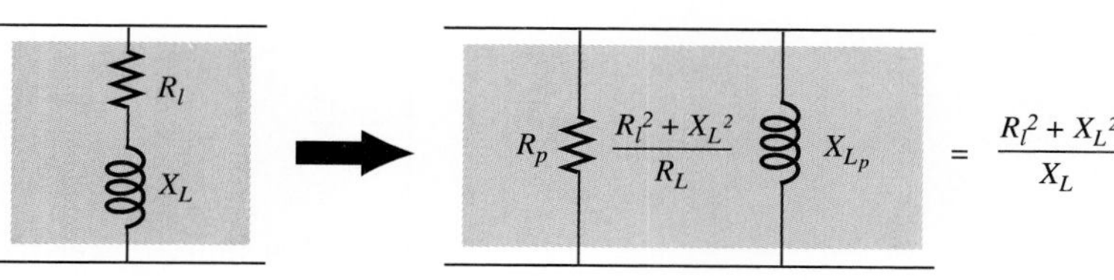

FIG. 20.23
Equivalent parallel network for a series R-L combination.

Redrawing the network of Fig. 20.22 with the equivalent of Fig. 20.23 and a practical current source having an internal resistance R_s will result in the network of Fig. 20.24.

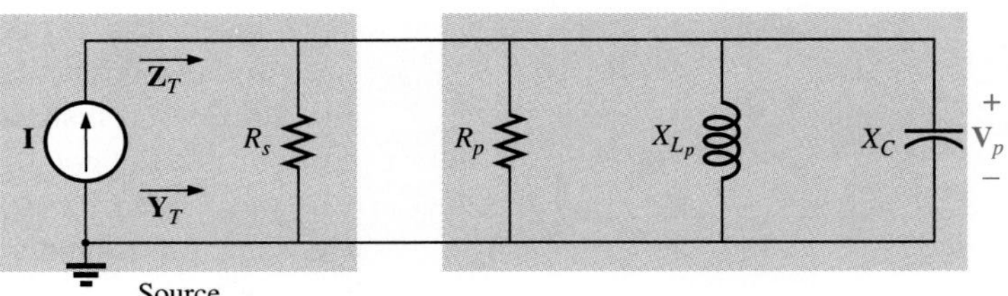

FIG. 20.24
Substituting the equivalent parallel network for the series R-L combination of Fig. 20.22.

If we define the parallel combination of R_s and R_p by the notation

$$R = R_s \parallel R_p \qquad (20.26)$$

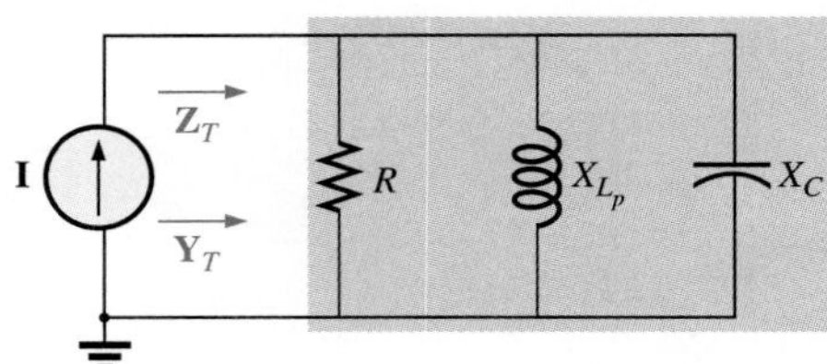

FIG. 20.25
Substituting $R = R_s \parallel R_p$ for the network of Fig. 20.24.

the network of Fig. 20.25 will result. It has the same format as the ideal configuration of Fig. 20.21.

We can now define the resonance conditions for the practical parallel resonant configuration. Recall for series resonance that the resonant frequency was the frequency at which the impedance was a minimum, the current a maximum, and the input impedance purely resistive, and the network had a unity power factor. For parallel networks, since the resistance R_p in our equivalent model is frequency dependent, the frequency at which maximum V_C is obtained is not the same as required for the unity power factor characteristic. Since both conditions are often used to define the resonant state, the frequency at which each occurs will be designated by different subscripts.

Unity Power Factor, f_p

For the network of Fig. 20.25,

$$\mathbf{Y}_T = \frac{1}{\mathbf{Z}_1} + \frac{1}{\mathbf{Z}_2} + \frac{1}{\mathbf{Z}_3} = \frac{1}{R} + \frac{1}{j\,X_{L_p}} + \frac{1}{-j\,X_C}$$

$$= \frac{1}{R} - j\left(\frac{1}{X_{L_p}}\right) + j\left(\frac{1}{X_C}\right)$$

and

$$\mathbf{Y}_T = \frac{1}{R} + j\left(\frac{1}{X_C} - \frac{1}{X_{L_p}}\right) \qquad (20.27)$$

For unity power factor, the reactive component must be zero as defined by

continued

$$\frac{1}{X_C} - \frac{1}{X_{L_p}} = 0$$

Therefore,

$$\frac{1}{X_C} = \frac{1}{X_{L_p}}$$

and

$$X_{L_p} = X_C \tag{20.28}$$

Substituting for X_{L_p} gives us

$$\frac{R_l^2 + X_L^2}{X_L} = X_C \tag{20.29}$$

The resonant frequency, f_p, can now be determined from Eq. (20.29) as follows:

$$R_l^2 + X_L^2 = X_C X_L = \left(\frac{1}{\omega C}\right)\omega L = \frac{L}{C}$$

or

$$X_L^2 = \frac{L}{C} - R_l^2$$

with

$$2\pi f_p L = \sqrt{\frac{L}{C} - R_l^2}$$

and

$$f_p = \frac{1}{2\pi L}\sqrt{\frac{L}{C} - R_l^2}$$

Multiplying the top and bottom of the factor within the square-root sign by C/L produces

$$f_p = \frac{1}{2\pi L}\sqrt{\frac{1 - R_l^2(C/L)}{C/L}} = \frac{1}{2\pi L\sqrt{C/L}}\sqrt{1 - \frac{R_l^2 C}{L}}$$

and

$$f_p = \frac{1}{2\pi\sqrt{LC}}\sqrt{1 - \frac{R_l^2 C}{L}} \tag{20.30}$$

or

$$f_p = f_s\sqrt{1 - \frac{R_l^2 C}{L}} \tag{20.31}$$

where f_p is the resonant frequency of a parallel resonant circuit (for $F_p = 1$) and f_s is the resonant frequency as determined by $X_L = X_C$ for series resonance. Note that unlike a series resonant circuit, the resonant frequency f_p is a function of resistance (in this case R_l). Note also, however, the absence of the source resistance R_s in Eqs. (20.30) and (20.31). Since the factor $\sqrt{1 - (R_l^2 C/L)}$ is less than one, f_p is less than f_s. Recognize also that as the magnitude of R_l approaches zero, f_p rapidly approaches f_s.

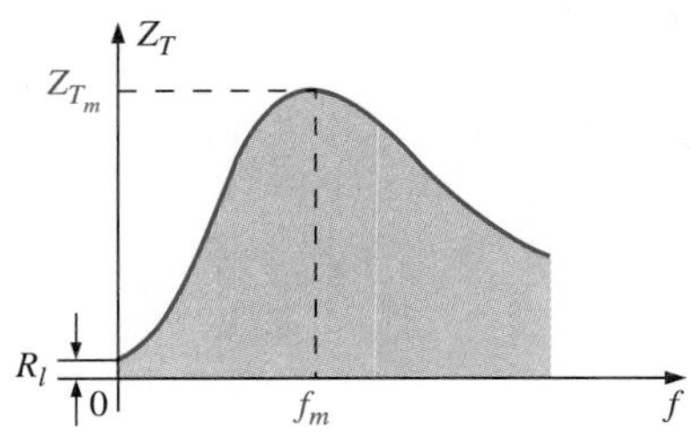

FIG. 20.26
Z_T versus frequency for the parallel resonant circuit.

Maximum Impedance, f_m

At $f = f_p$ the input impedance of a parallel resonant circuit will be near its maximum value but not quite its maximum value due to the frequency dependence of R_p. The frequency at which maximum impedance will occur is defined by f_m and is slightly more than f_p, as demonstrated in Fig. 20.26. The frequency f_m is determined by differentiating (calculus) the general equation for Z_T with respect to frequency and then determining the frequency at which the resulting equation is equal to zero. The algebra is quite extensive and complicated and will not be included here. The resulting equation, however, is the following:

$$f_m = f_s \sqrt{1 - \frac{1}{4}\left(\frac{R_l^2 C}{L}\right)} \quad (20.32)$$

Note the similarities with Eq. (20.31). Since the square root factor of Eq. (20.32) is always more than the similar factor of Eq. (20.31), f_m is always closer to f_s and more than f_p. In general,

$$f_s > f_m > f_p \quad (20.33)$$

Once f_m is determined, the network of Fig. 20.25 can be used to determine the magnitude and phase angle of the total impedance at the resonance condition simply by substituting $f = f_m$ and performing the required calculations. That is,

$$Z_{T_m} = R \parallel X_{L_p} \parallel X_C \Big|_{(f = f_m)} \quad (20.34)$$

20.9 SELECTIVITY CURVE FOR PARALLEL RESONANT CIRCUITS

The Z_T-versus-frequency curve of Fig. 20.26 clearly reveals that a parallel resonant circuit exhibits maximum impedance at resonance (f_m), unlike the series resonant circuit, which experiences minimum resistance levels at resonance. Note also that Z_T is approximately R_l at $f =$ 0 Hz since $Z_T = R_s \parallel R_l \cong R_l$.

Since the current I of the current source is constant for any value of Z_T or frequency, the voltage across the parallel circuit will have the same shape as the total impedance Z_T, as shown in Fig. 20.27.

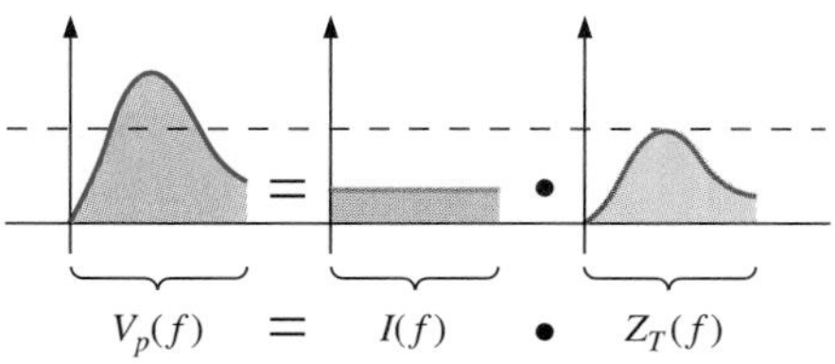

FIG. 20.27
Defining the shape of the $V_p(f)$ curve.

For the parallel circuit, the resonance curve of interest is that of the voltage V_C across the capacitor. The reason for this interest in V_C comes from electronic considerations that often place the capacitor at the input to another stage of a network.

Since the voltage across parallel elements is the same,

$$V_C = V_p = IZ_T \tag{20.35}$$

The resonant value of V_C is therefore determined by the value of Z_{T_m} and the magnitude of the current source I.

The quality factor of the parallel resonant circuit continues to be determined by the ratio of the reactive power to the real power. That is,

$$Q_p = \frac{V_p^2/X_{L_p}}{V_p^2/R}$$

where $R = R_s \| R_p$, and V_p is the voltage across the parallel branches. The result is

$$Q_p = \frac{R}{X_{L_p}} = \frac{R_s \| R_p}{X_{L_p}} \tag{20.36a}$$

or since $X_{L_p} = X_C$ at resonance,

$$Q_p = \frac{R_s \| R_p}{X_C} \tag{20.36b}$$

For the ideal current source ($R_s = \infty\ \Omega$) or when R_s is sufficiently large compared to R_p, we can make the following approximation:

$$R = R_s \| R_p \cong R_p$$

and

$$Q_p = \frac{R_s \| R_p}{X_{L_p}} = \frac{R_p}{X_{L_p}} = \frac{(R_l^2 + X_L^2)/R_l}{(R_l^2 + X_L^2)/X_L}$$

so that

$$Q_p = \frac{X_L}{R_l} = Q_l \qquad (R_s \gg R_p) \tag{20.37}$$

which is simply the quality factor Q_l of the coil.

In general, the bandwidth is still related to the resonant frequency and the quality factor by

$$BW = f_2 - f_1 = \frac{f_r}{Q_p} \tag{20.38}$$

The cutoff frequencies f_1 and f_2 can be determined using the equivalent network of Fig. 20.25 and the unity power condition for resonance. The half-power frequencies are defined by the condition that the output voltage is 0.707 times the maximum value. However, for parallel resonance with a current source driving the network, the frequency response for the driving point impedance is the same as that for the output voltage. This similarity allows us to define each cutoff frequency as the frequency at which the input impedance is 0.707 times its maximum value. Since the maximum value is the equivalent resistance R of Fig. 20.25, the cutoff frequencies will be associated with an impedance equal to $0.707R$ or $(1/\sqrt{2})R$.

Setting the input impedance for the network of Fig. 20.25 equal to this value will result in the following relationship:

$$\mathbf{Z} = \frac{1}{\dfrac{1}{R} + j\left(\omega C - \dfrac{1}{\omega L}\right)} = 0.707R$$

which can be written as

$$\mathbf{Z} = \frac{1}{\dfrac{1}{R}\left[1 + jR\left(\omega C - \dfrac{1}{\omega L}\right)\right]} = \frac{R}{\sqrt{2}}$$

or

$$\frac{R}{1 + jR\left(\omega C - \dfrac{1}{\omega L}\right)} = \frac{R}{\sqrt{2}}$$

and finally

$$\frac{1}{1 + jR\left(\omega C - \dfrac{1}{\omega L}\right)} = \frac{1}{\sqrt{2}}$$

The only way the equality can be satisfied is if the magnitude of the imaginary term on the bottom left is equal to 1 because the magnitude of $1 + j\,1$ must be equal to $\sqrt{2}$.

The following relationship, therefore, defines the cutoff frequencies for the system:

$$R\left(\omega C - \frac{1}{\omega L}\right) = 1$$

Substituting $\omega = 2\pi f$ and rearranging will result in the following quadratic equation:

$$f^2 - \frac{f}{2\pi RC} - \frac{1}{4\pi^2 LC} = 0$$

having the form

$$af^2 + bf + c = 0$$

with

$$a = 1,\ b = -\frac{1}{2\pi RC},\ \text{and } c = -\frac{1}{4\pi^2 LC}$$

Substituting into the equation:

$$f = \frac{-b \pm \sqrt{b^2 - 4ac}}{2a}$$

results in the following after careful mathematical manipulations:

$$f_1 = \frac{1}{4\pi C}\left[\frac{1}{R} - \sqrt{\frac{1}{R^2} + \frac{4C}{L}}\right] \tag{20.39a}$$

$$f_2 = \frac{1}{4\pi C}\left[\frac{1}{R} + \sqrt{\frac{1}{R^2} + \frac{4C}{L}}\right] \tag{20.39b}$$

Since the term in the brackets of Eq. (20.39a) will always be negative, simply associate f_1 with the magnitude of the result.

The effect of R_l, L, and C on the shape of the parallel resonance curve, as shown in Fig. 20.28 for the input impedance, is quite similar to their effect on the series resonance curve in Fig. 20.15. Whether or not R_l is zero, the parallel resonant circuit will frequently appear in a network schematic as shown in Fig. 20.22.

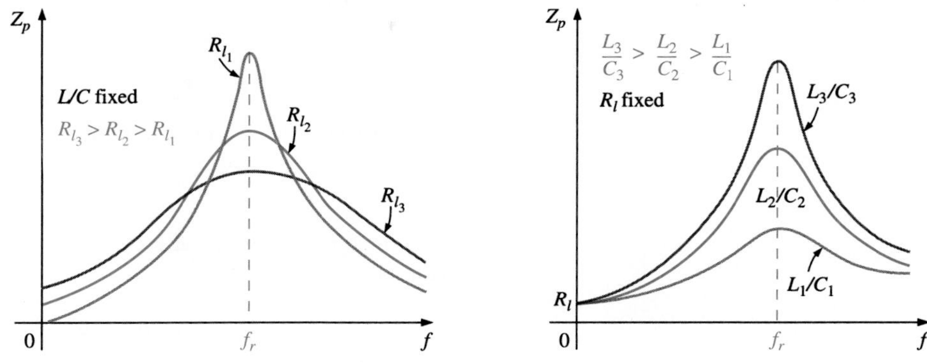

FIG. 20.28
Effect of R_l, L, and C on the parallel resonance curve.

At resonance, an increase in R_l or a decrease in the ratio L/C will result in a decrease in the resonant impedance, with a corresponding increase in the current. The bandwidth of the resonance curves is given by Eq. (20.38). For increasing R_l or decreasing L (or L/C for constant C), the bandwidth will increase as shown in Fig. 20.28.

At low frequencies, the capacitive reactance is quite high, and the inductive reactance is low. Since the elements are in parallel, the total impedance at low frequencies will therefore be inductive. At high frequencies, the reverse is true, and the network is capacitive. At resonance (f_p), the network appears resistive. These facts lead to the phase plot of Fig. 20.29. Note that it is the inverse of the one for the series resonant circuit, because at low frequencies the series resonant circuit was capacitive and at high frequencies it was inductive.

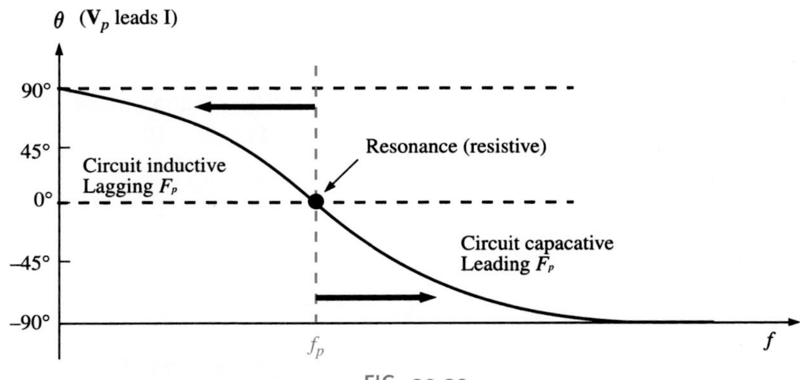

FIG. 20.29
Phase plot for the parallel resonant circuit.

20.10 EFFECT OF $Q_L \geq 10$

The previous section may suggest that analyzing parallel resonant circuits is much more complex than analyzing series resonant circuits.

Fortunately, however, this is not the case since, for most parallel resonant circuits, the quality factor of the coil Q_l is large enough to permit a number of approximations that simplify the required analysis.

Inductive Reactance, X_{L_p}

If we expand X_{L_p} as

$$X_{L_p} = \frac{R_l^2 + X_L^2}{X_L} = \frac{R_l^2(X_L)}{X_L(X_L)} + X_L = \frac{X_L}{Q_l^2} + X_L$$

then, for $Q_l \geq 10$, $X_L/Q_l^2 \cong 0$ compared to X_L and

$$X_{L_p} \cong X_L \quad (Q_l \geq 10) \tag{20.40}$$

and since resonance is defined by $X_{L_p} = X_C$, the resulting condition for resonance is reduced to:

$$X_L \cong X_C \quad (Q_l \geq 10) \tag{20.41}$$

Resonant Frequency, f_p (Unity Power Factor)

We can rewrite the factor R_l^2C/L of Eq. (20.31) as

$$\frac{R_l^2C}{L} = \frac{1}{\dfrac{L}{R_l^2C}} = \frac{1}{\dfrac{(\omega)}{(\omega)}\dfrac{L}{R_l^2C}} = \frac{1}{\dfrac{\omega L}{R_l^2\omega C}} = \frac{1}{\dfrac{X_LX_C}{R_l^2}}$$

and substitute Eq. (20.40) ($X_L = X_C$):

$$\frac{1}{\dfrac{X_LX_C}{R_l^2}} = \frac{1}{\dfrac{X_L^2}{R_l^2}} = \frac{1}{Q_l^2}$$

Equation (20.31) then becomes

$$f_p = f_s\sqrt{1 - \frac{1}{Q_l^2}} \quad (Q_l \geq 10) \tag{20.42}$$

clearly revealing that as Q_l increases, f_p becomes closer and closer to f_s.

For $Q_l \geq 10$,

$$1 - \frac{1}{Q_l^2} \cong 1$$

and

$$f_p \cong f_s = \frac{1}{2\pi\sqrt{LC}} \quad (Q_l \geq 10) \tag{20.43}$$

Resonant Frequency, f_m (Maximum V_C)

Using the equivalency $R_l^2C/L = 1/Q_l^2$ derived for Eq. (20.42), Eq. (20.32) will take on the following form:

$$f_m \cong f_s\sqrt{1 - \frac{1}{4}\left(\frac{1}{Q_l^2}\right)} \quad (Q_l \geq 10) \tag{20.44}$$

The fact that the negative term under the square root will always be less than that appearing in the equation for f_p reveals that f_m will always be closer to f_s than f_p.

For $Q_l \geq 10$ the negative term becomes very small and can be ignored, leaving:

$$f_m \cong f_s = \frac{1}{2\pi\sqrt{LC}} \quad (Q_l \geq 10) \tag{20.45}$$

In total, therefore, for $Q_l \geq 10$,

$$f_p \cong f_m \cong f_s \quad (Q_l \geq 10) \tag{20.46}$$

R_p

$$R_p = \frac{R_l^2 + X_L^2}{R_l} = R_l + \frac{X_L^2}{R_l}\left(\frac{R_l}{R_l}\right) = R_l + \frac{X_L^2}{R_l^2}R_l$$

$$= R_l + Q_l^2 R_l = (1 + Q_l^2)R_l$$

For $Q_l \geq 10$, $1 + Q_l^2 \cong Q_l^2$ and

$$R_p \cong Q_l^2 R_l \quad (Q_l \geq 10) \tag{20.47}$$

Applying these approximations to the network of Fig. 20.24 will result in the approximate equivalent network for $Q_l \geq 10$ of Fig. 20.30, which certainly looks a lot cleaner.

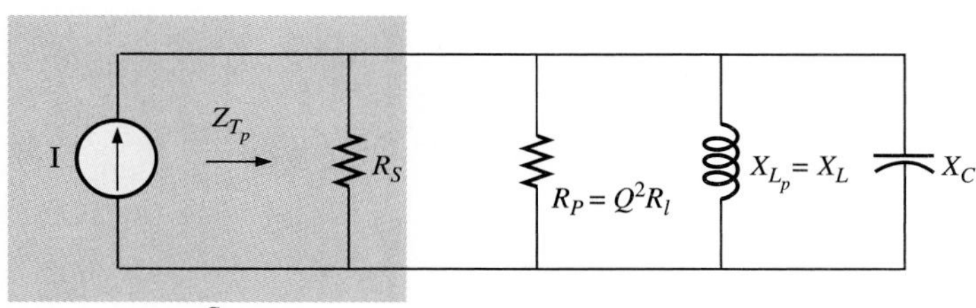

FIG. 20.30
Approximate equivalent circuit for $Q_l \geq 10$.

Substituting $Q_l = \dfrac{X_L}{R_l}$ into Eq. (20.47)

$$R_p \cong Q_l^2 R_l = \frac{X_L^2}{R_l^2}R_l = \frac{X_L^2}{R_l} = \frac{X_L X_C}{R_l} = \frac{2\pi f L}{R_l(2\pi f C)}$$

and

$$R_p \cong \frac{L}{R_l C} \quad (Q_l \geq 10) \tag{20.48}$$

Z_{T_p}

The total impedance at resonance is now defined by:

$$Z_{T_p} \cong R_s \parallel R_p = R_s \parallel Q_l^2 R_l \quad (Q_l \geq 10) \tag{20.49}$$

For an ideal current source ($R_s = \infty\ \Omega$) or if $R_s \gg R_p$, the equation reduces to

$$Z_{T_p} \cong Q_l^2 R_l \quad (Q_l \geq 10,\ R_s \gg R_p) \tag{20.50}$$

Q_p

The quality factor is now defined by

$$Q_p = \frac{R}{X_{L_p}} \cong \frac{R_s \parallel Q_l^2 R_l}{X_L} \tag{20.51}$$

Therefore, R_s does have an impact on the quality factor of the network and the shape of the resonant curves.

If an ideal current source ($R_s = \infty\ \Omega$) is used or if $R_s \gg R_p$,

$$Q_p \cong \frac{R_s \parallel Q_l^2 R_l}{X_L} = \frac{Q_l^2 R_l}{X_L} = \frac{Q_l^2}{X_L/R_l} = \frac{Q_l^2}{Q_l}$$

and

$$Q_p \cong Q_l \quad (Q_l \geq 10,\ R_s \gg R_p) \tag{20.52}$$

BW

The bandwidth defined by f_p is

$$BW = f_2 - f_1 = \frac{f_p}{Q_p} \tag{20.53}$$

By substituting Q_p from above and performing a few algebraic manipulations, it can be shown that

$$BW = f_2 - f_1 \cong \frac{1}{2\pi}\left[\frac{R_l}{L} + \frac{1}{R_s C}\right] \tag{20.54}$$

revealing the impact of R_s on the resulting bandwidth. Of course, if $R_s = \infty\ \Omega$ (ideal current source):

$$BW = f_2 - f_1 \cong \frac{R_l}{2\pi L} \quad (R_s = \infty\ \Omega) \tag{20.55}$$

I_L and I_C

A portion of Fig. 20.30 is reproduced in Fig. 20.31, with I_T defined as shown.

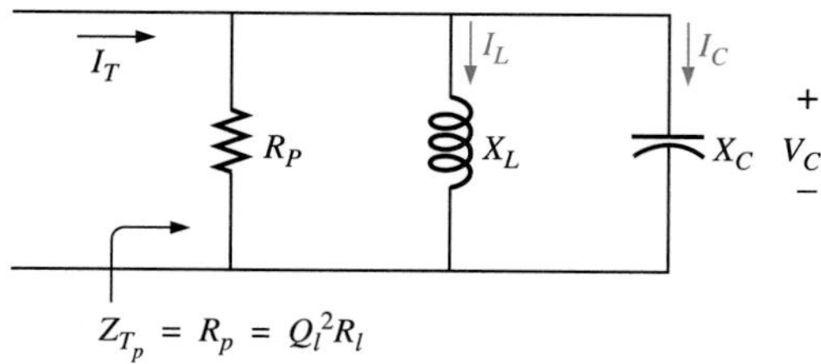

FIG. 20.31
Establishing the relationship between I_C and I_L and the current I_T.

As indicated, Z_{T_p} at resonance is $Q_l^2R_l$. The voltage across the parallel network is, therefore,

$$V_C = V_L = V_R = I_T Z_{T_p} = I_T Q_l^2 R_l$$

The magnitude of the current I_C can then be determined using Ohm's law, as follows:

$$I_C = \frac{V_C}{X_C} = \frac{I_T Q_l^2 R_l}{X_C}$$

Substituting $X_C = X_L$ when $Q_l \geq 10$,

$$I_C = \frac{I_T Q_l^2 R_l}{X_L} = I_T \frac{Q_l^2}{\dfrac{X_L}{R_l}} = I_T \frac{Q_l^2}{Q_l}$$

and

$$I_C \cong Q_l I_T \quad (Q_l \geq 10) \tag{20.56}$$

revealing that the capacitive current is Q_l times the magnitude of the current entering the parallel resonant circuit. For large Q_l, the current I_C can be significant.

A similar derivation results in

$$I_L \cong Q_l I_T \quad (Q_l \geq 10) \tag{20.57}$$

Conclusions

The equations resulting from applying the condition $Q_l \geq 10$ are much easier to apply than those obtained earlier. It is, therefore, a condition that should be checked early in an analysis to determine which approach to apply. Although the condition $Q_l \geq 10$ was applied throughout, many of the equations are still good approximations for $Q_l < 10$. For instance, if $Q_l = 5$, $X_{L_p} = (X_L/Q_l^2) + X_l = (X_L/25) + X_L = 1.04X_L$, which is very close to X_L. In fact, for $Q_l = 2$, $X_{L_p} = (X_L/4) + X_L = 1.25X_L$, which is not exactly X_L, but it is only 25% off. In general, be aware that the approximate equations can be applied with good accuracy with $Q_l < 10$. The smaller the level of Q_l, however, the less valid the approximation. The approximate equations are certainly valid for a range of values of $Q_l < 10$ if you need a rough approximation to the actual response rather than one that is accurate to the hundredths place.

20.11 SUMMARY TABLE

To limit any confusion resulting from the introduction of f_p and f_m and an approximate approach dependent on Q_l, we have developed Summary Table 20.1. You can always use the equations for any Q_l, but knowing how to apply the approximate equations defined by Q_l will help you greatly in the long run.

TABLE 20.1

Parallel resonant circuit ($f_s = 1/(2\pi\sqrt{LC})$).

	Any Q_l	$Q_l \geq 10$	$Q_l \geq 10$, $R_s \gg Q_l^2 R_l$
f_p	$f_s\sqrt{1 - \frac{R_l^2 C}{L}}$	f_s	f_s
f_m	$f_s\sqrt{1 - \frac{1}{4}\left[\frac{R_l^2 C}{L}\right]}$	f_s	f_s
Z_{T_p}	$R_s \parallel R_p = R_s \parallel \left(\frac{R_l^2 + X_L^2}{R_l}\right)$	$R_s \parallel Q_l^2 R_l$	$Q_l^2 R_l$
Z_{T_m}	$R_s \parallel \mathbf{Z}_{R-L} \parallel \mathbf{Z}_C$	$R_s \parallel Q_l^2 R_l$	$Q_l^2 R_l$
Q_p	$\frac{Z_{T_p}}{X_{L_p}} = \frac{Z_{T_p}}{X_C}$	$\frac{Z_{T_p}}{X_L} = \frac{Z_{T_p}}{X_C}$	Q_l
BW	$\frac{f_p}{Q_p}$ or $\frac{f_m}{Q_p}$	$\frac{f_p}{Q_p} = \frac{f_s}{Q_p}$	$\frac{f_p}{Q_l} = \frac{f_s}{Q_l}$
I_L, I_C	Network analysis	$I_L = I_C = Q_l I_T$	$I_L = I_C = Q_l I_T$

For the future, the analysis of a parallel resonant network might proceed as follows:

1. Determine f_s to obtain some idea of the resonant frequency. Recall that for most situations, f_s, f_m, and f_p will be relatively close to each other.
2. Calculate an approximate Q_l using f_s from above and compare to the condition $Q_l \geq 10$. If satisfied, use the approximate approach unless a high degree of accuracy is required.
3. If Q_l is less than 10, the approximate approach can be applied, but you must understand that the smaller the level of Q_l, the less accurate the solution. However, considering the typical variations from nameplate values for many of our components and that a resonant frequency to the tenths place is rarely required, the use of the approximate approach for many practical situations is usually quite valid.

20.12 EXAMPLES (PARALLEL RESONANCE)

EXAMPLE 20.6

Given the parallel network of Fig. 20.32 composed of "ideal" elements:

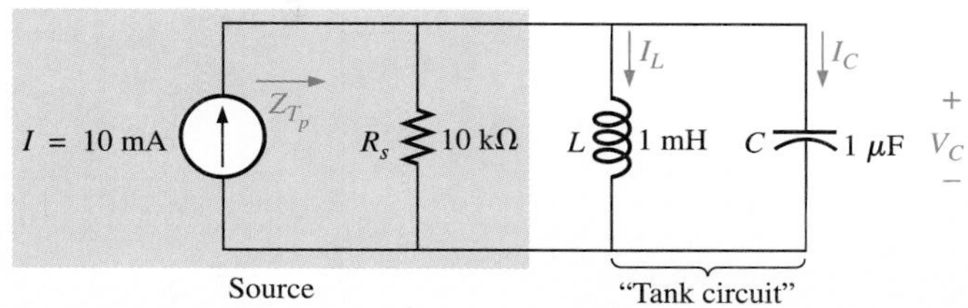

FIG. 20.32
Example 20.6.

a. Determine the resonant frequency f_p.
b. Find the total impedance at resonance.
c. Calculate the quality factor, bandwidth, and cutoff frequencies f_1 and f_2 of the system.
d. Find the voltage V_C at resonance.
e. Determine the currents I_L and I_C at resonance.

Solutions:

a. The fact that R_l is zero ohms results in a very high Q_l $(= X_L/R_l)$, allowing you to use the following equation for f_p.

$$f_p = f_s = \frac{1}{2\pi\sqrt{LC}} = \frac{1}{2\pi\sqrt{(1\text{ mH})(1\ \mu\text{F})}}$$
$$= \mathbf{5.03\ kHz}$$

b. For the parallel reactive elements:

$$\mathbf{Z}_L \parallel \mathbf{Z}_C = \frac{(X_L \angle 90°)(X_C \angle -90°)}{+j(X_L - X_C)}$$

but $X_L = X_C$ at resonance, resulting in a zero in the denominator of the equation and a very high impedance that can be approximated by an open circuit. Therefore,

$$Z_{T_p} = R_s \parallel \mathbf{Z}_L \parallel \mathbf{Z}_C = R_s = \mathbf{10\ k\Omega}$$

c. $$Q_p = \frac{R_s}{X_{L_p}} = \frac{R_s}{2\pi f_p L} = \frac{10\text{ k}\Omega}{2\pi(5.03\text{ kHz})(1\text{ mH})} = \mathbf{316.41}$$

$$BW = \frac{f_p}{Q_p} = \frac{5.03\text{ kHz}}{316.41} = \mathbf{15.90\ Hz}$$

Eq. (20.39a):

$$f_1 = \frac{1}{4\pi C}\left[\frac{1}{R} - \sqrt{\frac{1}{R^2} + \frac{4C}{L}}\right]$$

continued

$$= \frac{1}{4\pi(1\ \mu F)}\left[\frac{1}{10\ k\Omega} - \sqrt{\frac{1}{(10\ k\Omega)^2} + \frac{4(1\ \mu F)}{1\ mH}}\right]$$

$= \mathbf{5.025\ kHz}$ (ignoring the negative sign)

Eq. (20.39b):

$$f_2 = \frac{1}{4\pi C}\left[\frac{1}{R} + \sqrt{\frac{1}{R^2} + \frac{4C}{L}}\right]$$

$$= \mathbf{5.041\ kHz}$$

d. $V_C = IZ_{T_p} = (10\ mA)(10\ k\Omega) = \mathbf{100\ V}$

e. $I_L = \dfrac{V_L}{X_L} = \dfrac{V_C}{2\pi f_p L} = \dfrac{100\ V}{2\pi(5.03\ kHz)(1\ mH)} = \dfrac{100\ V}{31.6\ \Omega} = \mathbf{3.16\ A}$

$I_C = \dfrac{V_C}{X_C} = \dfrac{100\ V}{31.6\ \Omega} = \mathbf{3.16\ A}\ (= Q_p I)$

This example demonstrates the impact of R_s on the calculations associated with parallel resonance. The source impedance is the only factor to limit the input impedance and the level of V_C.

EXAMPLE 20.7

For the parallel resonant circuit of Fig. 20.33 with $R_s = \infty\ \Omega$:

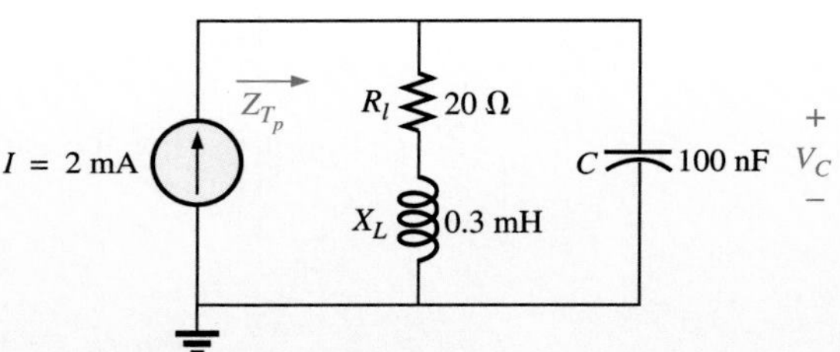

FIG. 20.33
Example 20.7.

a. Determine f_s, f_m, and f_p and compare their levels.
b. Calculate the maximum impedance and the magnitude of the voltage V_C at f_m.
c. Determine the quality factor Q_p.
d. Calculate the bandwidth.
e. Compare the above results with those obtained using the equations associated with $Q_l \geq 10$.

Solutions:

a. $f_s = \dfrac{1}{2\pi\sqrt{LC}} = \dfrac{1}{2\pi\sqrt{(0.3\ mH)(100\ nF)}} = \mathbf{29\ 057.58\ Hz}$

$$f_m = f_s\sqrt{1 - \frac{1}{4}\left[\frac{R_l^2 C}{L}\right]}$$

continued

$$= (29\,057.58 \text{ Hz})\sqrt{1 - \frac{1}{4}\left[\frac{(20\ \Omega)^2(100 \text{ nF})}{0.3 \text{ mH}}\right]}$$

$$= \mathbf{28\,569.19\ Hz}$$

$$f_p = f_s\sqrt{1 - \frac{R_l^2 C}{L}} = (29\,057.58 \text{ Hz})\sqrt{1 - \left[\frac{(20\ \Omega)^2(100 \text{ nF})}{0.3 \text{ mH}}\right]}$$

$$= \mathbf{27\,051.14\ Hz}$$

Both f_m and f_p are less than f_s, as predicted. In addition, f_m is closer to f_s than f_p, as expected. f_m is about 0.5 kHz less than f_s, and f_p is about 2 kHz less. The differences among f_s, f_m, and f_p suggest a low-Q network.

b. $\mathbf{Z}_{T_m} = (R_l + j\,X_L) \parallel -j\,X_C$ at $f = f_m$

$$X_L = 2\pi f_m L = 2\pi(28\,569.19 \text{ Hz})(0.3 \text{ mH}) = 53\,852\ \Omega$$

$$X_C = \frac{1}{2\pi f_m C} = \frac{1}{2\pi(28\,569.19 \text{ Hz})(100 \text{ nF})} = 55\,709\ \Omega$$

$$R_l + j\,X_L = 20\ \Omega + j\,53.852\ \Omega = 57.446\ \Omega\ \angle 69.626^\circ$$

$$\mathbf{Z}_{T_m} = \frac{(57.446\ \Omega\ \angle 69.626^\circ)(55.709\ \Omega\ \angle -90^\circ)}{20\ \Omega + j\,53.852\ \Omega - j\,55.709\ \Omega}$$

$$= \mathbf{159.34\ \Omega\ \angle -15.069^\circ}$$

$$V_{C_{max}} = IZ_{T_m} = (2 \text{ mA})(159.34\ \Omega) = \mathbf{318.68\ mV}$$

c. $R_s = \infty\ \Omega$; therefore,

$$Q_p = \frac{R_s \parallel R_p}{X_{L_p}} = \frac{R_p}{X_{L_p}} = Q_l = \frac{X_L}{R_l}$$

$$= \frac{2\pi(27\,051.14 \text{ Hz})(0.3 \text{ mH})}{20\ \Omega} = \frac{50.990\ \Omega}{20\ \Omega} = \mathbf{2.55}$$

The low Q confirms our conclusion of part (a). The differences among f_s, f_m, and f_p will be significantly less for higher-Q networks.

d. $$BW = \frac{f_p}{Q_p} = \frac{27\,051.14 \text{ Hz}}{2.55} = \mathbf{10\,608.29\ Hz}$$

e. For $Q_l \geq 10$, $f_m = f_p = f_s = \mathbf{29\,057.28\ Hz}$

$$Q_p = Q_l = \frac{2\pi f_s L}{R_l} = \frac{2\pi(29\,057.58 \text{ Hz})(0.3 \text{ mH})}{20\ \Omega} = \mathbf{2.739}$$

(versus 2.55 above)

$$Z_{T_p} = Q_l^2 R_l = (2.739)^2 \cdot 20\ \Omega = \mathbf{150.04\ \Omega\ \angle 0^\circ}$$

(versus 159.34 Ω ∠−15.069° above)

$$V_{C_{max}} = IZ_{T_p} = (2 \text{ mA})(150.04\ \Omega) = \mathbf{300.08\ mV}$$

(versus 318.68 mV above)

$$BW = \frac{f_p}{Q_p} = \frac{29\,057.58 \text{ Hz}}{2.739} = \mathbf{10\,608.83\ Hz}$$

(versus 10 608.29 Hz above)

continued

The results reveal that, even for a relatively low Q system, the approximate solutions are still in the ball park compared to those obtained using the full equations. The primary difference is between f_s and f_p (about 7%), with the difference between f_s and f_m at less than 2%. For the future, using f_s to determine Q_l will certainly provide a measure of Q_l that can be used to determine whether the approximate approach is appropriate.

EXAMPLE 20.8

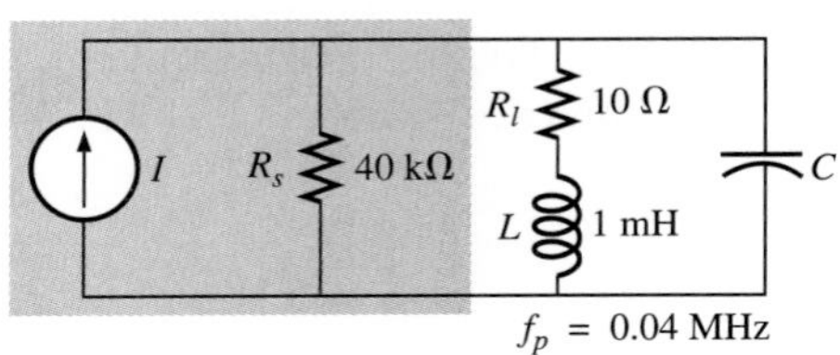

FIG. 20.34
Example 20.8.

For the network of Fig. 20.34 with f_p provided:

a. Determine Q_l.
b. Determine R_p.
c. Calculate Z_{T_p}.
d. Find C at resonance.
e. Find Q_p.
f. Calculate the BW and cutoff frequencies.

Solutions:

a. $$Q_l = \frac{X_L}{R_l} = \frac{2\pi f_p L}{R_l} = \frac{2\pi(0.04 \text{ MHz})(1 \text{ mH})}{10\ \Omega} = \mathbf{25.12}$$

b. $Q_l \geq 10$. Therefore,

$$R_p \cong Q_l^2 R_l = (25.12)^2(10\ \Omega) = \mathbf{6.31\ k\Omega}$$

c. $Z_{T_p} = R_s \parallel R_p = 40\text{ k}\Omega \parallel 6.31\text{ k}\Omega = \mathbf{5.45\ k\Omega}$

d. $Q_l \geq 10$. Therefore,

$$f_p \cong \frac{1}{2\pi\sqrt{LC}}$$

and $$C = \frac{1}{4\pi^2 f^2 L} = \frac{1}{4\pi^2(0.04 \text{ MHz})^2(1 \text{ mH})} = \mathbf{15.83\ nF}$$

e. $Q_l \geq 10$. Therefore,

$$Q_p = \frac{Z_{T_p}}{X_L} = \frac{R_s \parallel Q_l^2 R_l}{2\pi f_p L} = \frac{5.45\text{ k}\Omega}{2\pi(0.04 \text{ MHz})(1 \text{ mH})} = \mathbf{21.68}$$

f. $$BW = \frac{f_p}{Q_p} = \frac{0.04 \text{ MHz}}{21.68} = \mathbf{1.85\ kHz}$$

$$f_1 = \frac{1}{4\pi C}\left[\frac{1}{R} - \sqrt{\frac{1}{R^2} + \frac{4C}{L}}\right]$$

$$= \frac{1}{4\pi(15.9 \text{ mF})}\left[\frac{1}{5.45\text{ k}\Omega} - \sqrt{\frac{1}{(5.45\text{ k}\Omega)^2} + \frac{4(15.9 \text{ mF})}{1 \text{ mH}}}\right]$$

$$= 5.005 \times 10^6[183.486 \times 10^{-6} - 7.977 \times 10^{-3}]$$

$$= 5.005 \times 10^6[-7.794 \times 10^{-3}]$$

$$= \mathbf{39.009\ kHz} \quad \text{(ignoring the negative sign)}$$

$$f_2 = \frac{1}{4\pi C}\left[\frac{1}{R} + \sqrt{\frac{1}{R^2} + \frac{4C}{L}}\right]$$

continued

$$= 5.005 \times 10^6[183.486 \times 10^{-6} + 7.977 \times 10^{-3}]$$
$$= 5.005 \times 10^6[8.160 \times 10^{-3}]$$
$$= \mathbf{40.843\ kHz}$$

Note that $f_2 - f_1 = 40.843\ \text{kHz} - 39.009\ \text{kHz} = 1.834\ \text{kHz}$, confirming our solution for the bandwidth above. Note also that the bandwidth is not symmetrical about the resonant frequency, with 991 Hz below and 843 Hz above.

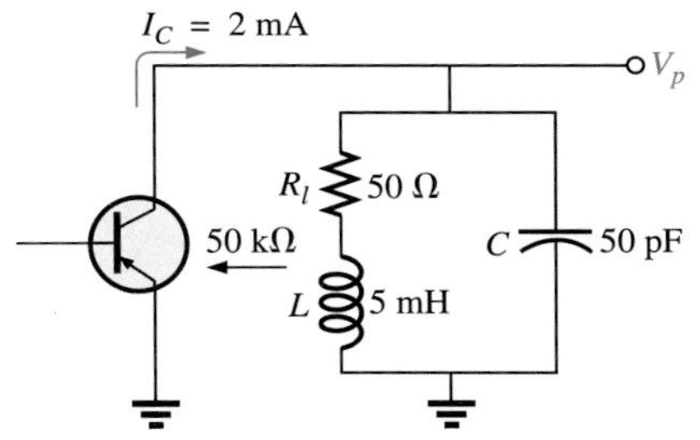

FIG. 20.35
Example 20.9.

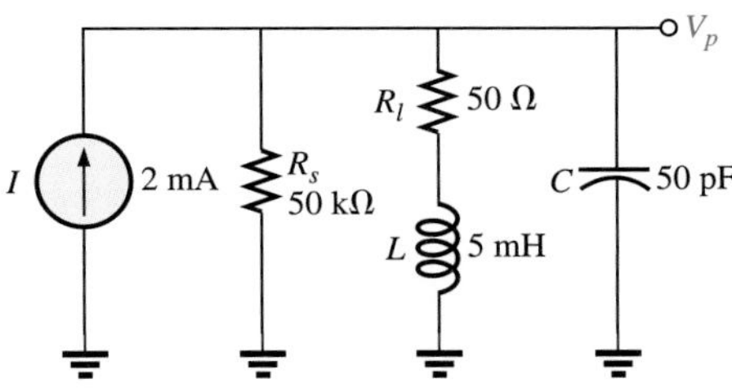

FIG. 20.36
Equivalent network for the transistor configuration of Fig. 20.35.

EXAMPLE 20.9

The equivalent network for the transistor configuration of Fig. 20.35 is provided in Fig. 20.36.

a. Find f_p.
b. Determine Q_p.
c. Calculate the BW.
d. Determine V_p at resonance.
e. Sketch the curve of V_C versus frequency.

Solutions:

a. $$f_s = \frac{1}{2\pi\sqrt{LC}} = \frac{1}{2\pi\sqrt{(5\ \text{mH})(50\ \text{pF})}} = 318.31\ \text{kHz}$$

$$X_L = 2\pi f_s L = 2\pi(318.31\ \text{kHz})(5\ \text{mH}) = 10\ \text{k}\Omega$$

$$Q_l = \frac{X_L}{R_l} = \frac{10\ \text{k}\Omega}{50\ \Omega} = 200 > 10$$

Therefore, $f_p = f_s =$ **318.31 kHz.** Using Eq. (20.31) would result in $\cong$ 318.5 kHz.

b. $$Q_p = \frac{R_s \| R_p}{X_L}$$

$$R_p = Q_l^2 R_l = (200)^2 50\ \Omega = 2\ \text{M}\Omega$$

$$Q_p = \frac{50\ \text{k}\Omega \| 2\ \text{M}\Omega}{10\ \text{k}\Omega} = \frac{48.78\ \text{k}\Omega}{10\ \text{k}\Omega} = \mathbf{4.88}$$

Note the drop in Q from $Q_l = 200$ to $Q_p = 4.88$ due to R_s.

c. $$BW = \frac{f_p}{Q_p} = \frac{318.31\ \text{kHz}}{4.88} = \mathbf{65.23\ kHz}$$

On the other hand,

$$BW = \frac{1}{2\pi}\left(\frac{R_l}{L} + \frac{1}{R_s C}\right) = \frac{1}{2\pi}\left[\frac{50\ \Omega}{5\ \text{mH}} + \frac{1}{(50\ \text{k}\Omega)(50\ \text{pF})}\right]$$
$$= \mathbf{65.25\ kHz}$$

compares very well with the above solution.

d. $V_p = IZ_{T_p} = (2\ \text{mA})(R_s \| R_p) = (2\ \text{mA})(48.78\ \text{k}\Omega) =$ **97.56 V**

continued

e. See Fig. 20.37.

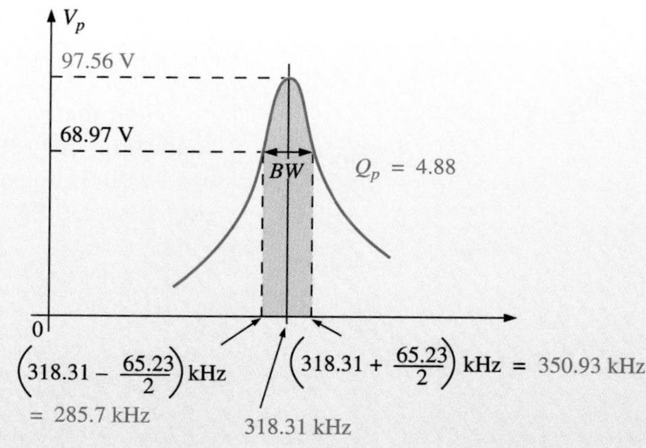

FIG. 20.37
Resonance curve for the network of Fig. 20.36.

EXAMPLE 20.10

Repeat Example 20.9, but ignore the effects of R_s and compare results.

Solutions:

a. f_p is the same, **318.31 kHz.**

b. For $R_s = \infty\ \Omega$

$$Q_p = Q_l = \mathbf{200} \quad \text{(versus 4.88)}$$

c. $BW = \dfrac{f_p}{Q_p} = \dfrac{318.31 \text{ kHz}}{200} = \mathbf{1.592\ kHz} \quad \text{(versus 65.23 kHz)}$

d. $Z_{T_p} = R_p = 2\ \text{M}\Omega \quad \text{(versus 48.78 k}\Omega)$
$V_p = IZ_{T_p} = (2\ \text{mA})(2\ \text{M}\Omega) = \mathbf{4000\ V} \quad \text{(versus 97.56 V)}$

These results clearly reveal that the source resistance can have a significant impact on the response characteristics of a parallel resonant circuit.

EXAMPLE 20.11

Design a parallel resonant circuit to have the response curve of Fig. 20.38 using a 1-mH, 10-Ω inductor, and a current source with an internal resistance of 40 kΩ.

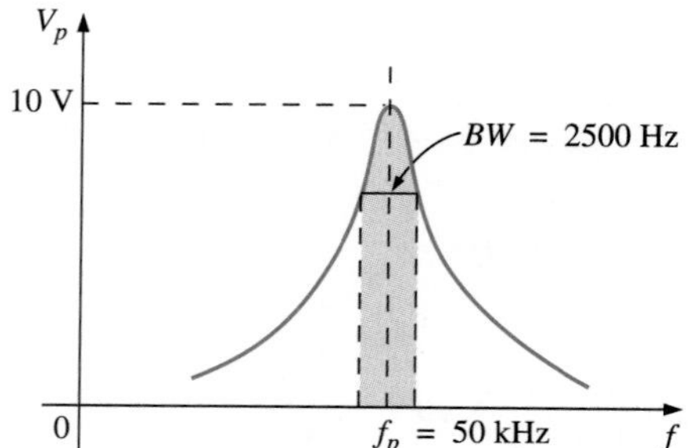

FIG. 20.38
Example 20.11.

Solution:

$$BW = \frac{f_p}{Q_p}$$

Therefore,

$$Q_p = \frac{f_p}{BW} = \frac{50\ 000 \text{ Hz}}{2500 \text{ Hz}} = \mathbf{20}$$

continued

$$X_L = 2\pi f_p L = 2\pi(50 \text{ kHz})(1 \text{ mH}) = 314\ \Omega$$

and
$$Q_l = \frac{X_L}{R_l} = \frac{314\ \Omega}{10\ \Omega} = \mathbf{31.4}$$

$$R_p = Q_l^2 R = (31.4)^2(10\ \Omega) = \mathbf{9859.6\ \Omega}$$

$$Q_p = \frac{R}{X_L} = \frac{R_s \parallel 9859.6\ \Omega}{314\ \Omega} = 20 \quad \text{(from above)}$$

so that
$$\frac{(R_s)(9859.6)}{R_s + 9859.6} = 6280$$

resulting in
$$R_s = 17.298\text{ k}\Omega$$

However, the source resistance was given as 40 kΩ. We must therefore add a parallel resistor (R') that will reduce the 40 kΩ to approximately 17.298 kΩ; that is,

$$\frac{(40\text{ k}\Omega)(R')}{40\text{ k}\Omega + R'} = 17.298\text{ k}\Omega$$

Solving for R':

$$R' = \mathbf{30.481\text{ k}\Omega}$$

The closest commercial value is **30 kΩ**. At resonance, $X_L = X_C$, and

$$X_C = \frac{1}{2\pi f_p C}$$

$$C = \frac{1}{2\pi f_p X_C} = \frac{1}{2\pi(50\text{ kHz})(314\ \Omega)}$$

and
$$C \cong \mathbf{0.01\ \mu F} \quad \text{(commercially available)}$$

$$\begin{aligned} Z_{T_p} &= R_s \parallel Q_l^2 R_l \\ &= 17.298\text{ k}\Omega \parallel 9859.6\ \Omega \\ &= 6.28\text{ k}\Omega \end{aligned}$$

with
$$V_p = IZ_{T_p}$$

and
$$I = \frac{V_p}{Z_{T_p}} = \frac{10\text{ V}}{6.28\text{ k}\Omega} \cong \mathbf{1.6\ mA}$$

The network appears in Fig. 20.39.

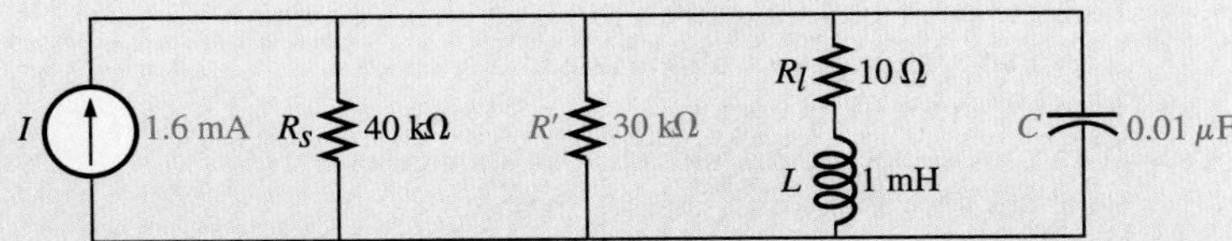

FIG. 20.39
Network designed to meet the criteria of Fig. 20.38.

PROBLEMS

SECTIONS 20.2 THROUGH 20.7 Series Resonance

1. Find the resonant ω_s and f_s for the series circuit with the following parameters:
   a. $R = 10\ \Omega$, $L = 1$ H, $C = 16\ \mu$F
   b. $R = 300\ \Omega$, $L = 0.5$ H, $C = 0.16\ \mu$F
   c. $R = 20\ \Omega$, $L = 0.28$ mH, $C = 7.46\ \mu$F
2. For the series circuit of Fig. 20.40:
   a. Find the value of X_C for resonance.
   b. Determine the total impedance of the circuit at resonance.
   c. Find the magnitude of the current I.
   d. Calculate the voltages V_R, V_L, and V_C at resonance. How are V_L and V_C related? How does V_R compare to the applied voltage E?
   e. What is the quality factor of the circuit? Is it a high- or low-Q circuit?
   f. What is the power dissipated by the circuit at resonance?

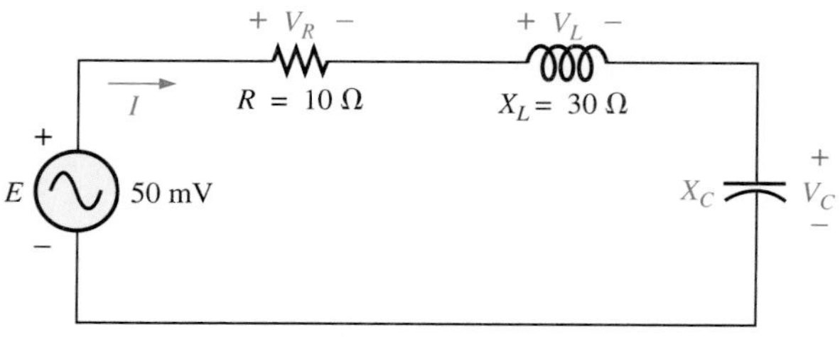

FIG. 20.40
Problem 2.

3. For the series circuit of Fig. 20.41:
   a. Find the value of X_L for resonance.
   b. Determine the magnitude of the current I at resonance.
   c. Find the voltages V_R, V_L, and V_C at resonance and compare their magnitudes.
   d. Determine the quality factor of the circuit. Is it a high- or low-Q circuit?
   e. If the resonant frequency is 5 kHz, determine the value of L and C.
   f. Find the bandwidth of the response if the resonant frequency is 5 kHz.
   g. What are the low and high cutoff frequencies?

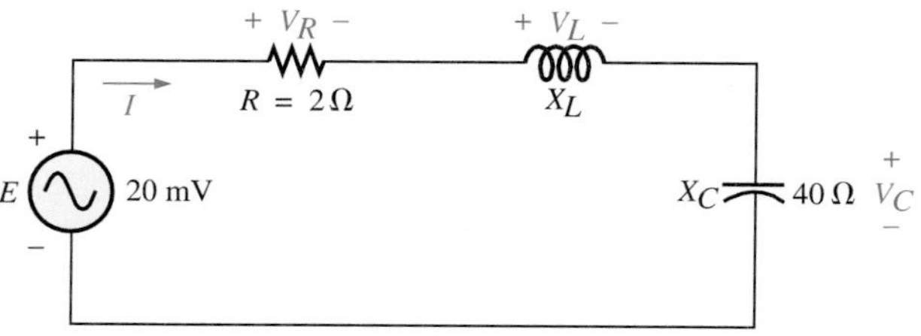

FIG. 20.41
Problem 3.

4. For the circuit of Fig. 20.42:
   a. Find the value of L in millihenries if the resonant frequency is 1800 Hz.
   b. Calculate X_L and X_C. How do they compare?
   c. Find the magnitude of the current I_{rms} at resonance.
   d. Find the power dissipated by the circuit at resonance.
   e. What is the apparent power delivered to the system at resonance?
   f. What is the power factor of the circuit at resonance?
   g. Calculate the Q of the circuit and the resulting bandwidth.
   h. Find the cutoff frequencies and calculate the power dissipated by the circuit at these frequencies.

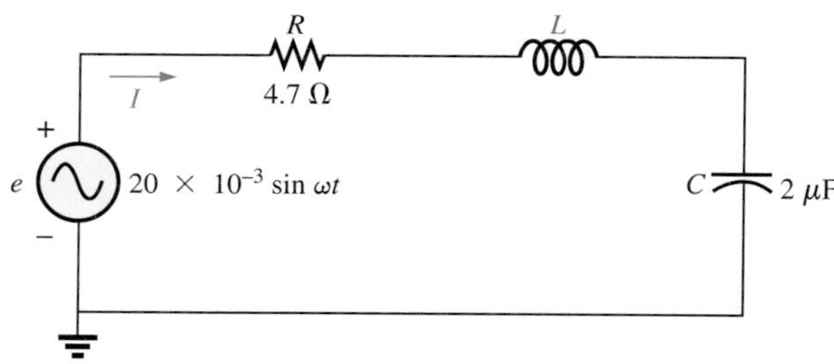

FIG. 20.42
Problem 4.

5. a. Find the bandwidth of a series resonant circuit having a resonant frequency of 6000 Hz and a Q_s of 15.
   b. Find the cutoff frequencies.
   c. If the resistance of the circuit at resonance is 3 Ω, what are the values of X_L and X_C in ohms?
   d. What is the power dissipated at the half-power frequencies if the maximum current flowing through the circuit is 0.5 A?

6. A series circuit has a resonant frequency of 10 kHz. The resistance of the circuit is 5 Ω, and X_C at resonance is 200 Ω.

a. Find the bandwidth.
b. Find the cutoff frequencies.
c. Find Q_s.
d. If the input voltage is 30 V ∠0°, find the voltage across the coil and capacitor in phasor form.
e. Find the power dissipated at resonance.

7. a. The bandwidth of a series resonant circuit is 200 Hz. If the resonant frequency is 2000 Hz, what is the value of Q_s for the circuit?
b. If $R = 2\ \Omega$, what is the value of X_L at resonance?
c. Find the value of L and C at resonance.
d. Find the cutoff frequencies.

8. The cutoff frequencies of a series resonant circuit are 5600 Hz and 6000 Hz. The current at 5600 Hz was measured to be 7.07 mA with 20 mV applied to the circuit.

a. Assuming that Q_s is >10, find the resonant frequency f_s.
b. Find the bandwidth of the circuit.
c. Calculate Q_s. Was your assumption in (a) a good one?
d. If the applied voltage is also 20mV at resonance, does the current increase or decrease, and to what value?
e. Find the resistance of the circuit at resonance.
f. Find the values of L and C.

***9.** Design a series resonant circuit with an input voltage of 5 V ∠0° to have the following specifications:

a. A peak current of 500 mA at resonance
b. A bandwidth of 120 Hz
c. A resonant frequency of 8400 Hz

Find the value of L and C and the cutoff frequencies.

***10.** A series resonant circuit is made of a 0.16 H inductor in series with a 0.1 μF capacitor. The inductor has resistance and the capacitor is ideal. While at resonance, the voltage across the capacitor is measured as 15 volts using an ideal voltmeter with 1 volt applied to the circuit.

a. Find the resonant frequency.
b. Find Q_s of the coil.
c. Find the resistance of the coil.
d. Find the current at resonance and at the cutoff frequencies.
e. If the voltmeter across the capacitor has a resistance of 400 kohms, determine what effect this has on the results of (a), (b), (c), and (d) above.

***11.** Design a series resonant circuit to have a bandwidth of 500 Hz using an 800 μH coil with a Q_s of 25 at the resonant frequency. Determine the values of the coil resistance, the capacitance, the resonant frequency and the half-power frequencies.

***12.** A series resonant circuit will resonate at a frequency of 1 MHz with a fractional bandwidth of 0.2. If the quality factor of the coil at resonance is 12.5 and its inductance is 100 μH, determine

a. The resistance of the coil.
b. The additional resistance required to establish the indicated fractional bandwidth.
c. The required value of capacitance.

SECTIONS 20.8 through 20.12 Parallel Resonance

13. For the "ideal" parallel resonant circuit of Fig. 20.43:

a. Determine the resonant frequency (f_p).
b. Find the voltage V_C at resonance.
c. Determine the currents I_L and I_C at resonance.
d. Find Q_p.

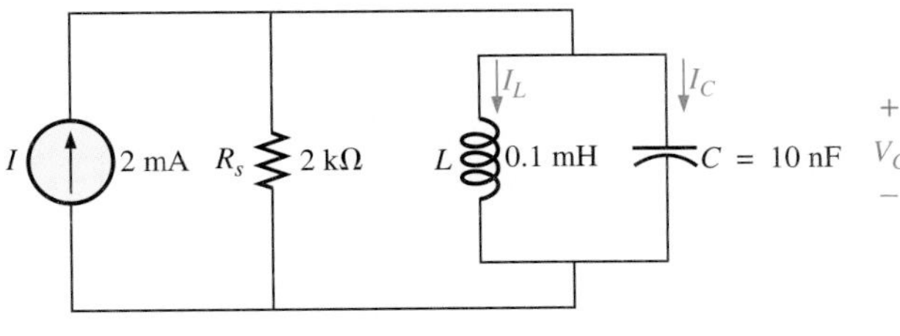

FIG. 20.43
Problem 13.

14. For the parallel resonant network of Fig. 20.44:
 a. Calculate f_s.
 b. Determine Q_l using $f = f_s$. Can the approximate approach be applied?
 c. Determine f_p and f_m.
 d. Calculate X_L and X_C using f_p. How do they compare?
 e. Find the total impedance at resonance (f_p).
 f. Calculate V_C at resonance (f_p).
 g. Determine Q_p and the BW using f_p.
 h. Calculate I_L and I_C at f_p.

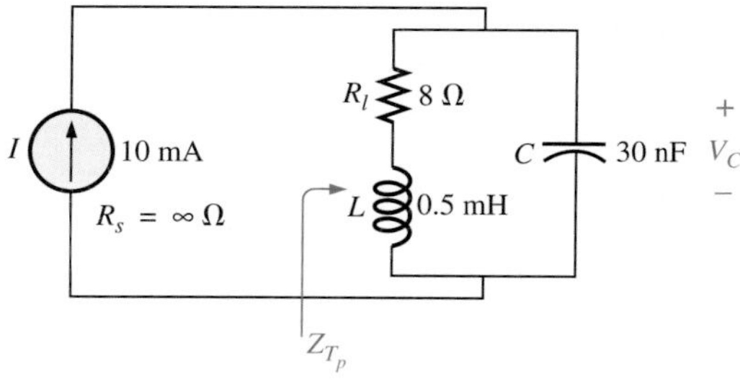

FIG. 20.44
Problem 14.

15. Repeat Problem 14 for the network of Fig. 20.45.

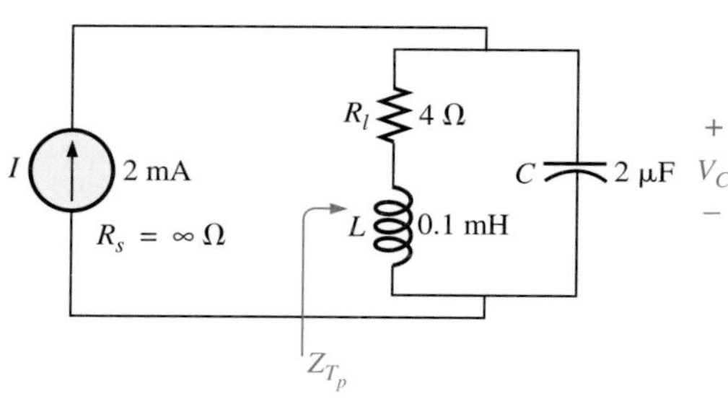

FIG. 20.45
Problem 15.

16. For the network of Fig. 20.46:
 a. Find the value of X_C at resonance (f_p).
 b. Find the total impedance Z_{T_p} at resonance (f_p).
 c. Find the currents I_L and I_C at resonance (f_p).
 d. If the resonant frequency is 20 000 Hz, find the value of L and C at resonance.
 e. Find Q_p and BW.

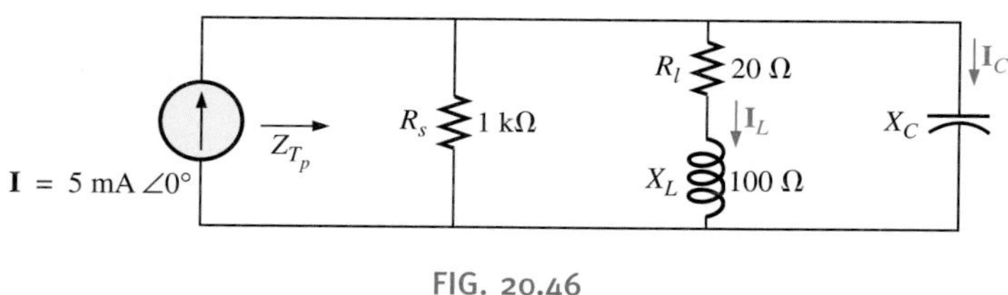

FIG. 20.46
Problem 16.

17. Repeat Problem 16 for the network of Fig. 20.47.

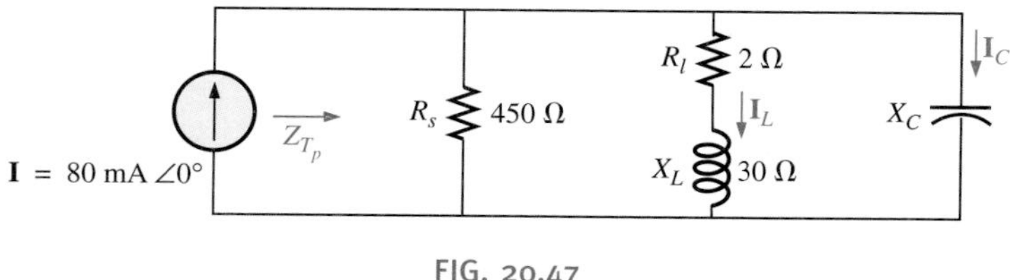

FIG. 20.47
Problem 17.

18. For the network of Fig. 20.48:

a. Find the resonant frequencies f_s, f_p, and f_m. What do the results suggest about the Q_p of the network?

b. Find the value of X_L and X_C at resonance (f_p). How do they compare?

c. Find the impedance Z_{T_p} at resonance (f_p).

d. Calculate Q_p and the *BW*.

e. Find the magnitude of currents I_L and I_C at resonance (f_p).

f. Calculate the voltage V_C at resonance (f_p).

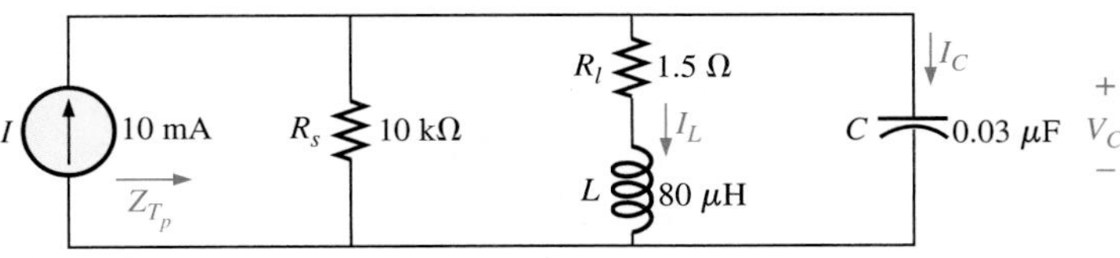

FIG. 20.48
Problem 18.

***19.** Repeat Problem 18 for the network of Fig. 20.49.

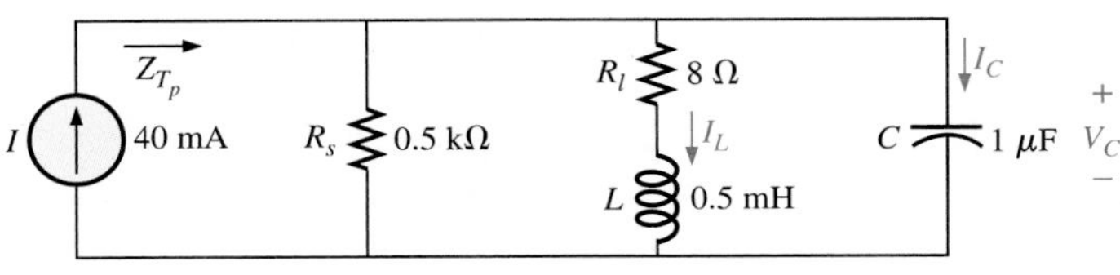

FIG. 20.49
Problem 19.

20. It is desired that the impedance Z_T of the high-Q circuit of Fig. 20.50 to 50 kΩ∠0° at resonance (f_p).

a. Find the value of X_L.

b. Find the value of X_C.

c. If the voltage across the capacitor, V_C, is 50 V, calculate the current from the supply.

d. If the 50 ohm resistor is increased to 100 ohms, predict if the current from the supply will increase or decrease while V_C remains 50 V.

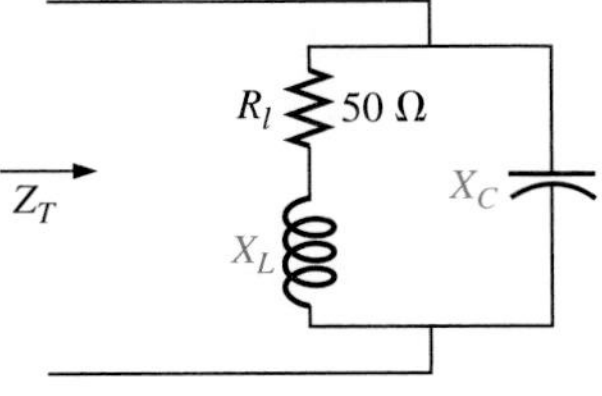

FIG. 20.50
Problem 20.

21. For the network of Fig. 20.51:

a. Find f_p.

b. Calculate the magnitude of V_C at resonance (f_p).

c. Determine the power absorbed at resonance.

d. Find the *BW*.

e. Find the current into the coil and capacitor combination.

f. Find the impedance seen by the 40 kohm resistor.

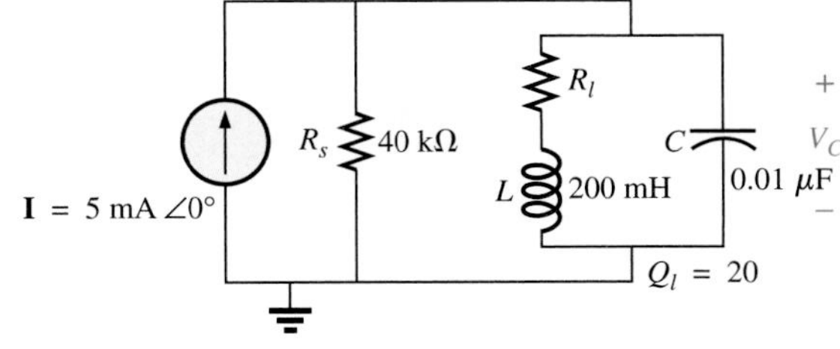

FIG. 20.51
Problem 21.

***22.** For the network of Fig. 20.52:

a. Find the value of X_L for resonance.

b. Find Q_l.

c. Find the resonant frequency (f_p) if the bandwidth is 1 kHz.

d. Find the maximum value of the voltage V_C.

e. Sketch the curve of V_C versus frequency. Indicate its peak value, resonant frequency, and band frequencies.

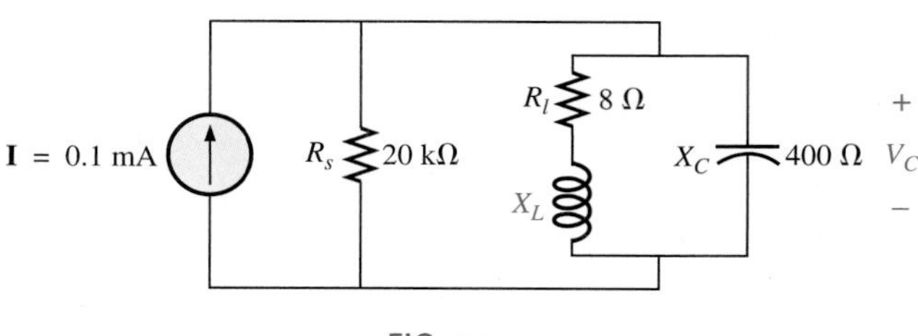

FIG. 20.52
Problem 22.

***23.** Repeat Problem 22 for the network of Fig. 20.53.

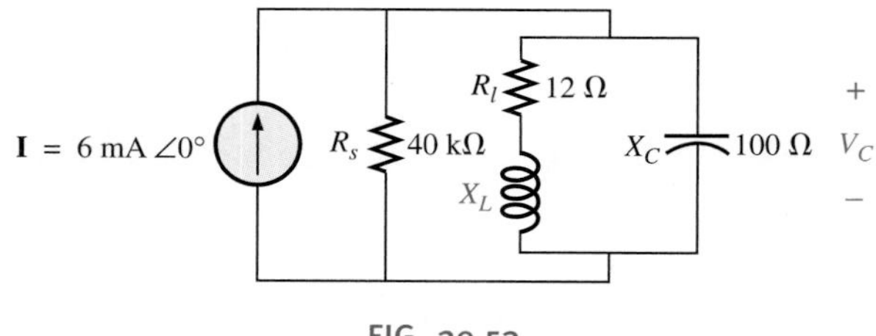

FIG. 20.53
Problem 23.

***24.** For the network of Fig. 20.54:

a. Find f_s, f_p, and f_m.

b. Determine Q_l and Q_p at f_p after a source conversion is performed.

c. Find the input impedance Z_{T_p}.

d. Find the magnitude of the voltage V_C.

e. Calculate the bandwidth using f_p.

f. Determine the magnitude of the currents I_C and I_L.

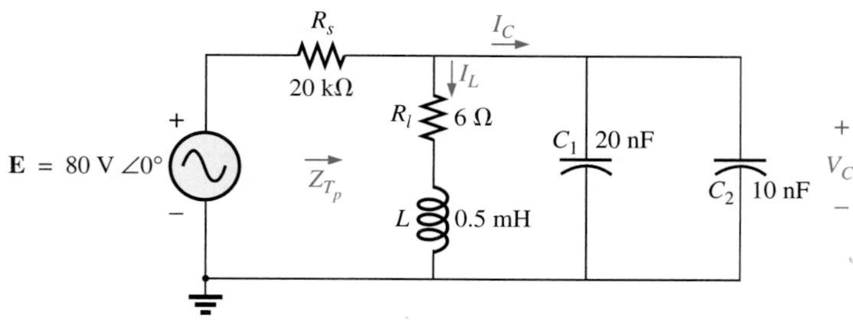

FIG. 20.54
Problem 24.

*25. For the network of Fig. 20.55, the following are specified:

$$f_p = 20 \text{ kHz}$$
$$BW = 1.8 \text{ kHz}$$
$$L = 2 \text{ mH}$$
$$Q_l = 80$$

Find R_s and C.

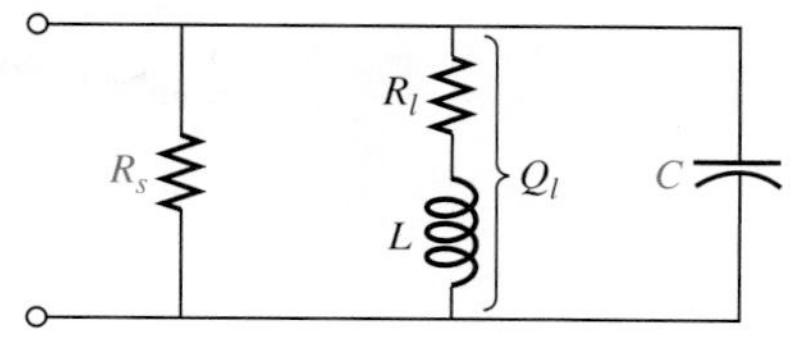

FIG. 20.55
Problem 25.

*26. Design the network of Fig. 20.56 to have the following characteristics:

a. a bandwidth of 500 Hz
b. $Q_p = 30$
c. $V_{C_{max}} = 1.8$ V

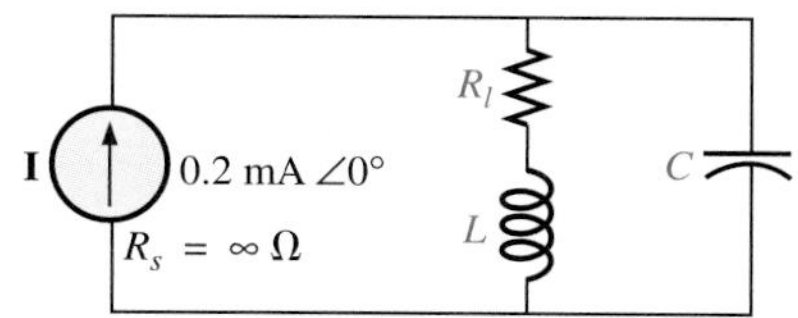

FIG. 20.56
Problem 26.

*27. For the parallel resonant circuit of Fig. 20.57:

a. Determine the resonant frequency.
b. Find the total impedance at resonance.
c. Find Q_p.
d. Calculate the bandwidth.
e. Repeat parts (a) through (d) for $L = 20\ \mu$H and $C = 20$ nF.
f. Repeat parts (a) through (d) for $L = 0.4$ mH and $C = 1$ nF.
g. For the network of Fig. 20.57 and the parameters of parts (e) and (f), determine the ratio L/C.
h. Do your results confirm the conclusions of Fig. 20.28 for changes in the L/C ratio?

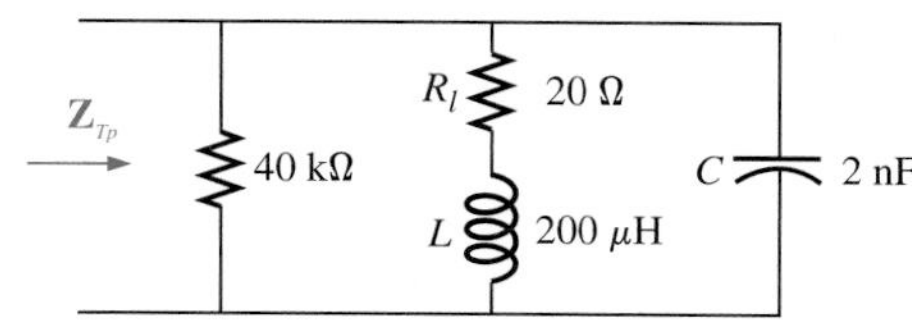

FIG. 20.57
Problem 27.

21 Decibels, Filters, and Bode Plots

Outline

Learning Outcomes

After completing this chapter you will be able to

- calculate and specify power and voltage gain as a ratio and in decibels
- explain why devices respond to increased levels of power in a logarithmic fashion
- be able to design and analyze *R-C* and *R-L* low and high pass band filter for a specified cutoff frequency
- design and analyze simple band pass and band stop filters
- sketch the gain and phase shift versus frequency graphs for simple low, high, pass, and stop band filters
- describe and design simple double tuned filters
- draw the Bode response of gain and phase shift versus frequency graphs for simple low, high, pass, and stop band filters

This chapter introduces circuits that pass desired signals while filtering out interfering signals and noise.

21.1 INTRODUCTION

Up to this point we have looked at methods used to analyze circuits regardless of their configuration, frequency, or application. Our objective has been to determine any electrical quantity (current, voltage, or power) at any instant, in any element, in many different circuits.

Now we will consider the circuit as a **system** made up of an input and an output as shown in Fig. 21.1. We will compare the output to the input in terms of the *voltage gain* and *power gain* (or **attenuation**) *A*, and the phase shift. In particular, we will consider the category of circuits known as **filters**, and see how to represent them both mathematically and graphically.

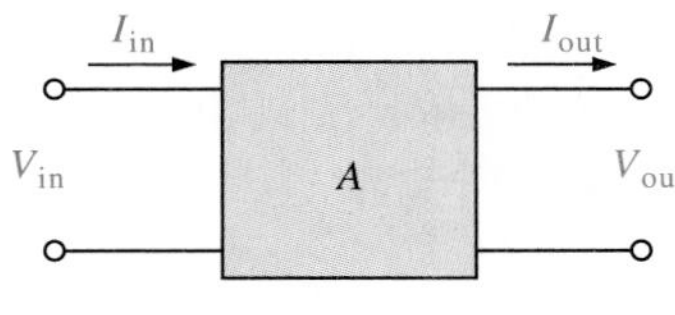

FIG. 21.1
General circuit with gain A.

21.2 DECIBELS

Power Gain

Two levels of power can be compared using a unit of measure called the **bel**, which is defined by the following equation:

$$\text{B} = \log_{10} \frac{P_2}{P_1} \quad \text{(bels)} \tag{21.1}$$

However, to provide a unit of measure of *less* magnitude, a **decibel** is defined, where

$$1 \text{ bel} = 10 \text{ decibels (dB)} \tag{21.2}$$

The result is the following important equation, which compares power levels P_2 and P_1 in decibels.

$$\text{dB} = 10 \log_{10} \frac{P_2}{P_1} \quad \text{(decibels, dB)} \tag{21.3}$$

If the power levels are equal ($P_2 = P_1$), there is no change in power level, and dB $= 0$. If there is an increase in power level ($P_2 > P_1$), the resulting decibel level is positive. If there is a decrease in power level ($P_2 < P_1$), the resulting decibel level will be negative.

For the special case of $P_2 = 2P_1$, the gain in decibels is

$$\text{dB} = 10 \log_{10} \frac{P_2}{P_1} = 10 \log_{10} 2 = \mathbf{3\ dB}$$

Therefore, for a speaker system, a 3-dB increase in output would require that the power level be doubled. In the audio industry, it is a generally accepted rule that an increase in sound level is accomplished with 3-dB increments in the output level. In other words, a 1-dB increase is barely detectable and a 2-dB increase, just discernible. A 3-dB increase normally results in a readily detectable increase in sound level. An additional increase in the sound level is normally accomplished by simply increasing the output level another 3 dB. If an 8-W system were in use, a 3-dB increase would require a 16-W output, whereas an additional increase of 3 dB (a total of 6 dB) would require a 32-W system, as demonstrated by the calculations below:

$$\text{dB} = 10 \log_{10} \frac{P_2}{P_1} = 10 \log_{10} \frac{16}{8} = 10 \log_{10} 2 = \mathbf{3\ dB}$$

$$\text{dB} = 10 \log_{10} \frac{P_2}{P_1} = 10 \log_{10} \frac{32}{8} = 10 \log_{10} 4 = \mathbf{6\ dB}$$

For $P_2 = 10P_1$,

$$\text{dB} = 10 \log_{10} \frac{P_2}{P_1} = 10 \log_{10} 10 = 10(1) = \mathbf{10\ dB}$$

resulting in the unique situation where the power gain has the same magnitude as the decibel level.

For some applications, a reference level is established to allow decibel levels to be compared from one situation to another. For communication systems, a commonly applied reference level is

$$P_{\text{ref}} = 1\ \text{mW} \quad (\text{across a } 600\text{-}\Omega \text{ load})$$

Equation (21.3) is then typically written as

$$\text{dB}_m = 10 \log_{10} \frac{P}{1\ \text{mW}}\bigg|_{600\ \Omega} \tag{21.4}$$

Note the subscript m to indicate that the decibel level is determined with a reference level of 1 mW.

In particular, for $P = 40$ mW,

$$\text{dB}_m = 10 \log_{10} \frac{40\ \text{mW}}{1\ \text{mW}} = 10 \log_{10} 40 = 10(1.6) = \mathbf{16\ dB_m}$$

whereas for $P = 4$ W,

$$\text{dB}_m = 10 \log_{10} \frac{4000\ \text{mW}}{1\ \text{mW}} = 10 \log_{10} 4000 = 10(3.6) = \mathbf{36\ dB_m}$$

Even though the power level has increased by a factor of 4000 mW/40 mW $= 100$, the dB_m increase is limited to 20 dB_m. In time, the significance of dB_m levels of 16 dB_m and 36 dB_m will generate an immediate appreciation regarding the power levels involved. An increase of 20 dB_m will also be associated with a significant gain in power levels.

Voltage Gain

Decibels are also used to provide a comparison between voltage levels. Substituting the basic power equations $P_2 = V_2^2/R_2$ and $P_1 = V_1^2/R_1$ into Eq. (21.3) will result in

$$\text{dB} = 10 \log_{10} \frac{P_2}{P_1} = 10 \log_{10} \frac{V_2^2/R_2}{V_1^2/R_1}$$

$$= 10 \log_{10} \frac{V_2^2/V_1^2}{R_2/R_1} = 10 \log_{10}\left(\frac{V_2}{V_1}\right)^2 - 10 \log_{10}\left(\frac{R_2}{R_1}\right)$$

and
$$\text{dB} = 20 \log_{10} \frac{V_2}{V_1} - 10 \log_{10} \frac{R_2}{R_1}$$

For the situation where $R_2 = R_1$, a condition normally assumed when comparing voltage levels on a decibel basis, the second term of the preceding equation will drop out ($\log_{10} 1 = 0$) and

$$\text{dB}_v = 20 \log_{10} \frac{V_2}{V_1} \quad (\text{dB}) \tag{21.5}$$

Note the subscript v to define the decibel level obtained.

EXAMPLE 21.1

Find the voltage gain in dB of a system where the applied signal is 2 mV and the output voltage is 1.2 V.

Solution:

$$\text{dB}_v = 20 \log_{10} \frac{V_o}{V_i} = 20 \log_{10} \frac{1.2\text{ V}}{2\text{ mV}} = 20 \log_{10} 600 = \mathbf{55.56\ dB}$$

for a voltage gain $A_v = V_o/V_i$ of 600.

EXAMPLE 21.2

If a system has a voltage gain of 36 dB, find the applied voltage if the output voltage is 6.8 V.

Solution:

$$\text{dB}_v = 10 \log_{10} \frac{V_o}{V_i}$$

$$36 = 20 \log_{10} \frac{V_o}{V_i}$$

$$1.8 = \log_{10} \frac{V_o}{V_i}$$

From the antilogarithm: $$\frac{V_o}{V_i} = 63.096$$

and $$V_i = \frac{V_o}{63.096} = \frac{6.8\text{ V}}{63.096} = \mathbf{107.77\ mV}$$

TABLE 21.1

V_o/V_i	$\text{dB}_v = 20 \log_{10}(V_o/V_i)$
1	0 dB
2	6 dB
10	20 dB
20	26 dB
100	40 dB
1 000	60 dB
100 000	100 dB

Table 21.1 compares the magnitude of specific gains to the resulting decibel level. In particular, note that when voltage levels are compared, doubling the level results in a change of 6 dB rather than 3 dB as obtained for power levels.

In addition, note that an increase in gain from 1 to 100 000 results in a change in decibels that can easily be plotted on a single graph. Also note that doubling the gain (from 1 to 2 and 10 to 20) results in a 6-dB increase in the decibel level, while a change of 10 to 1 (from 1 to 10, 10 to 100, and so on) always results in a 20-dB decrease in the decibel level.

The Human Auditory Response

One of the most common uses of the decibel scale is in the communication and entertainment industries. The human ear does not respond in a linear way to changes in source power level; that is, a doubling of the audio power level from 1/2 W to 1 W does not result in a doubling of the loudness level for the human ear. In addition, a change from 5 W to 10 W will be received by the ear as the same change in sound intensity as experienced from 1/2 W to 1 W. In other words, the ratio between levels is the same in each case (1 W/0.5 W = 10 W/5 W = 2). The ear, therefore, responds in a logarithmic fashion to changes in audio power levels.

To establish a basis for comparison between audio levels, a reference level of 0.0002 microbar (μbar) was chosen, where 1 μbar is equal to the sound pressure of 0.1 Pa (one-tenth of a pascal), or about 1 millionth of the normal atmospheric pressure at sea level. The 0.0002-μbar level is the threshold level of hearing. Using this reference level, the sound pressure level in decibels is defined by the following equation:

$$\text{dB}_s = 20 \log_{10} \frac{P}{0.0002\ \mu\text{bar}} \tag{21.6}$$

where P is the sound pressure in microbars.

The decibel levels of Fig. 21.2 are defined by Eq. (21.6). Meters designed to measure audio levels are calibrated to the levels defined by Eq. (21.6) and shown in Fig. 21.2.

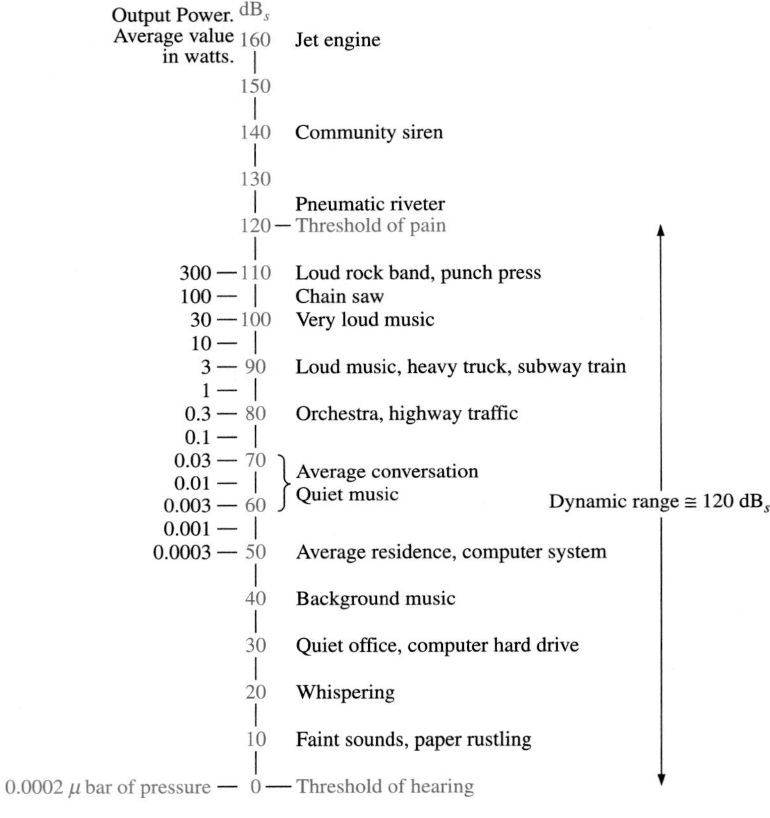

FIG. 21.2
Typical sound levels and their decibel levels.

A common question about audio levels is how much the power level of an acoustical source must be increased to double the sound level received by the human ear. The question is not as simple as it first seems due to considerations such as the frequency content of the sound, the acoustical conditions of the surrounding area, the physical characteristics of the surrounding medium, and—of course—the unique characteristics of the human ear. However, a general conclusion can be formulated that has practical value if we note the power levels of an acoustical source appearing to the left of Fig. 21.2. Each power level is associated with a particular decibel level, and a change of 10 dB in the scale corresponds with an increase or decrease in power by a factor of 10. For instance, a change from 90 dB to 100 dB is associated with a change in wattage from 3 W to 30 W. Through experimentation it has been found on an average basis that the loudness level will double for every 10-dB change in audio level—a conclusion somewhat verified by the examples to the right of Fig. 21.2. Using the fact that a 10-dB change corresponds with a tenfold increase in power level supports the following conclusion (on an approximate basis):

To double the sound level received by the human ear, the power rating of the acoustical source (in watts) must be increased by a factor of 10.

In other words, doubling the sound level available from a 1-W acoustical source would require moving up to a 10-W source.

Instrumentation

A number of modern VOMs and DMMs have a dB scale designed to provide an indication of power ratios referenced to a standard level of 1 mW at 600 Ω. Since the reading is accurate only if the load has a characteristic impedance of 600 Ω, the 1-mW, 600 Ω reference level is normally printed somewhere on the face of the meter, as shown in Fig. 21.3. The dB scale is usually calibrated to the lowest ac scale of the meter. In other

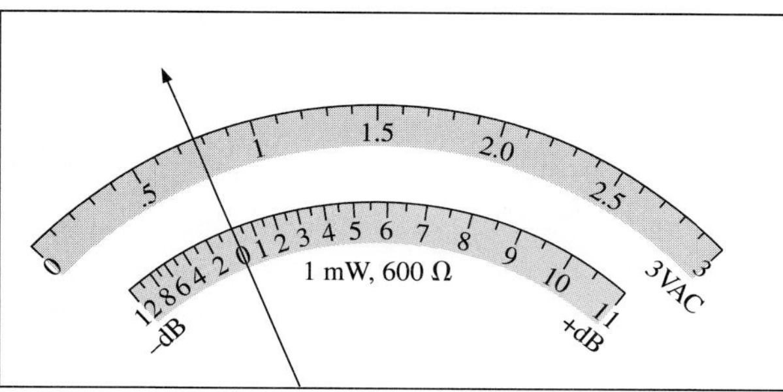

FIG. 21.3
Defining the relationship between a dB scale referenced to 1 mW, 600 Ω and a 3 V rms voltage scale.

words, when making the dB measurement, choose the lowest ac voltage scale, but read the dB scale. If a higher voltage scale is chosen, a correction factor must be used that is sometimes printed on the face of the meter but always available in the meter manual. If the impedance is other than 600 Ω or not purely resistive, other correction factors must be used that are normally included in the meter manual.

Using the basic power equation $P = V^2/R$ will reveal that 1 mW across a 600-Ω load is the same as applying 0.775 V rms across a 600-Ω load: that is, $V = \sqrt{PR} = \sqrt{(1\text{ mW})(600\ \Omega)} = 0.775$ V. The result is that an analog display will have 0 dB [defining the reference point of 1 mW, $\text{dB} = 10 \log_{10} P_2/P_1 = 10 \log_{10}$ (1 mW/1 mW(ref) = 0 dB] and 0.775 V rms on the same pointer projection, as shown in Fig. 21.3. A voltage of 2.5 V across a 600-Ω load would result in a dB level of $\text{dB} = 20 \log_{10} V_2/V_1 = 20 \log_{10} 2.5\text{ V}/0.775 = 10.17$ dB, resulting in 2.5 V and 10.17 dB appearing along the same pointer projection. A voltage of less than 0.775 V, such as 0.5 V, will result in a dB level of $\text{dB} = 20 \log_{10} V_2/V_1 = 20 \log_{10} 0.5\text{ V}/0.775\text{ V} = -3.8$ dB, as is also shown on the scale of Fig. 21.3. Although a reading of 10 dB will reveal that the power level is 10 times the reference, don't assume that a reading of 5 dB means the output level is 5 mW. The 10:1 ratio is a special one in logarithmic circles. For the 5-dB level, the power level must be found using the antilogarithm (3.126), which reveals that the power level associated with 5 dB is about 3.1 times the reference, or 3.1 mW. A conversion table is usually provided in the manual for such conversions.

21.3 FILTERS

Any combination of passive (*R, L,* and *C*) and/or active (transistors or operational amplifiers) elements designed to select or reject a band of frequencies is called a *filter.* In communication systems, filters are used to pass frequencies containing the desired information and reject the remaining frequencies. In stereo systems, filters can be used to isolate particular bands of frequencies for increased or decreased emphasis by the output acoustical system (amplifier, speaker, etc.). Filters are used to filter out any unwanted frequencies, commonly called *noise,* due to the nonlinear characteristics of some electronic devices or signals picked up from the surrounding medium. In general, there are two classifications of filters:

1. *Passive filters* are filters made of series or parallel combinations of *R, L,* and *C* elements.
2. *Active filters* are filters that use active devices such as transistors and operational amplifiers in combination with *R, L,* and *C* elements.

Since this text deals only with passive devices, we will only analyze passive filters in this chapter. In addition, we will examine only the most fundamental forms in the next few sections. The subject of filters is a very broad one that continues to receive extensive research support from industry and the government as new communication systems are developed to meet the demands of increased volume and speed. There are courses and texts that deal only with analyzing and designing filter systems that can become quite complex and sophisticated. In general, however, all filters belong to the four broad categories of *low-pass, high-pass, pass-band,* and *stop-band,* as shown on the next page in Fig. 21.4. For each form there are critical frequencies that define the regions of pass- and stop-bands (often called *reject* bands). Any frequency in the pass-band will pass through to the next stage with at least 70.7% of the maximum output voltage. Recall the use of the 0.707 level to define the bandwidth of a series or parallel resonant circuit (both with the general shape of the pass-band filter).

For some stop-band filters, the stop-band is defined by conditions other than the 0.707 level. In fact, for many stop-band filters, the condition that $V_o = 1/1000V_{\max}$ (corresponding with -60 dB in the discussion to follow) is used to define the stop-band region, with the pass-band continuing to be defined by the 0.707-V level. The resulting frequencies between the two regions are then called the *transition frequencies* and establish the *transition region.*

At least one example of each filter of Fig. 21.4 will be discussed in some detail in the sections to follow. Take particular note of how simple some of the designs are.

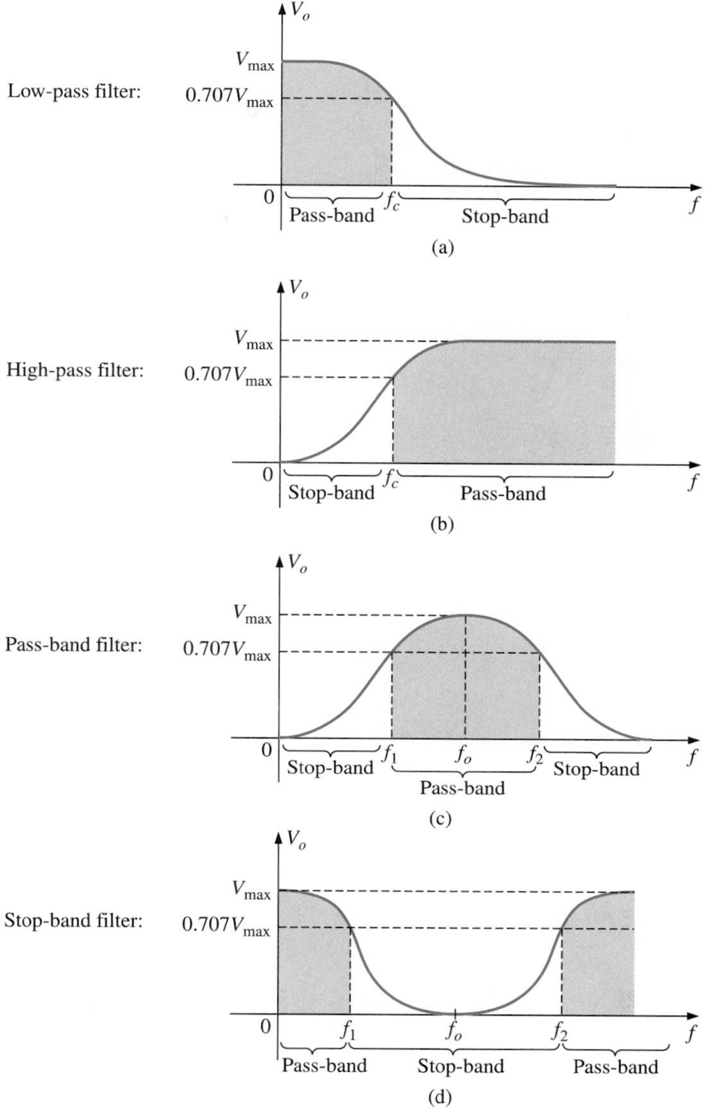

FIG. 21.4
Defining the four broad categories of filters.

21.4 *R-C* LOW-PASS FILTER

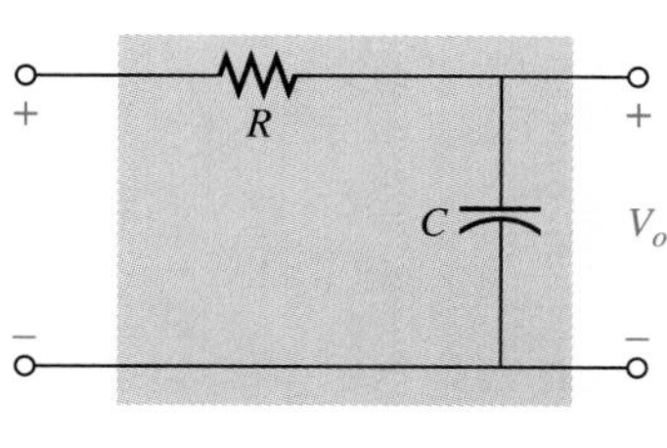

FIG. 21.5
Low-pass filter.

The *R-C* filter, incredibly simple in design, can be used as a low-pass or high-pass filter. If the output is taken off the capacitor, as shown in Fig. 21.5, it will respond as a low-pass filter. If the positions of the resistor and capacitor are interchanged and the output is taken off the resistor, it will respond as a high-pass filter.

A glance at Fig. 21.4(a) reveals that the circuit should behave in a way that will result in a high-level output for low frequencies and a declining level for frequencies above the critical value. Let us first examine the network at the frequency extremes of $f = 0$ Hz and very high frequencies to test the response of the circuit.

At $f = 0$ Hz,

$$X_C = \frac{1}{2\pi fC} = \infty\ \Omega$$

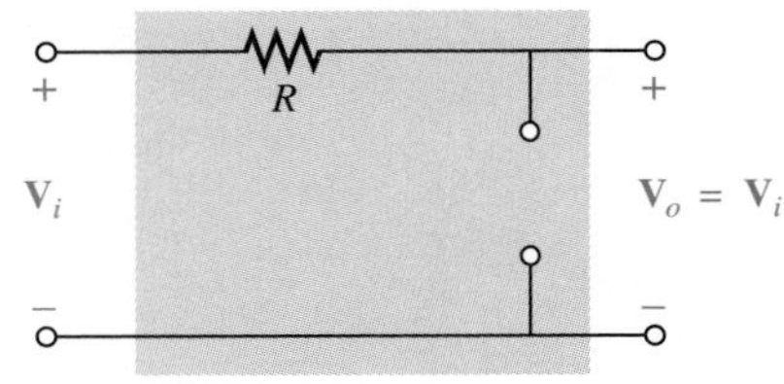

FIG. 21.6
R-C low-pass filter at low frequencies.

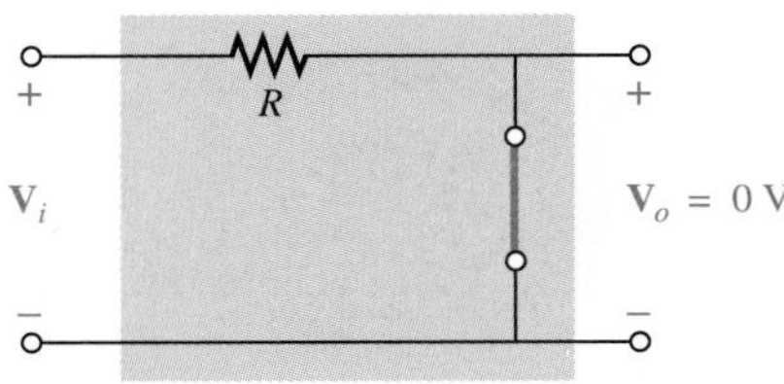

FIG. 21.7
R-C low-pass filter at high frequencies.

and the open-circuit equivalent can be substituted for the capacitor, as shown in Fig. 21.6, resulting in $\mathbf{V}_o = \mathbf{V}_i$.

At very high frequencies, the reactance is

$$X_C = \frac{1}{2\pi fC} \cong 0\ \Omega$$

and the short-circuit equivalent can be substituted for the capacitor, as shown in Fig. 21.7, resulting in $\mathbf{V}_o = 0$ V.

A plot of the magnitude of V_o versus frequency will result in the curve of Fig. 21.8. Our next goal is now clear: Find the frequency at which the transition takes place from a pass-band to a stop-band.

For filters, a *normalized plot* is used more often than the plot of V_o versus frequency of Fig. 21.8.

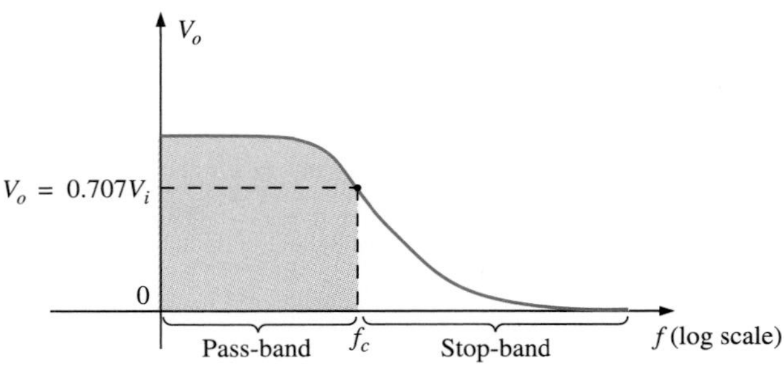

FIG. 21.8
V_o versus frequency for a low-pass R-C filter.

Normalization is a process whereby a quantity such as voltage, current, or impedance is divided by a quantity of the same unit of measure to establish a dimensionless level of a specific value or range.

A **normalized plot** in the filter domain can be obtained by dividing the plotted quantity such as V_o of Fig. 21.8 by the applied voltage V_i for the frequency range of interest. This is the gain A indicated in Fig. 21.1. When referring to voltage gain it is usually expressed as A_v. Since the maximum value of V_o for the low-pass filter of Fig. 21.8 is V_i, each level of V_o in Fig. 21.8 is divided by the level of V_i. The result is the plot of $A_v = V_o/V_i$ of Fig. 21.9. Note that the maximum value is 1 and the cutoff frequency is defined at the 0.707 level.

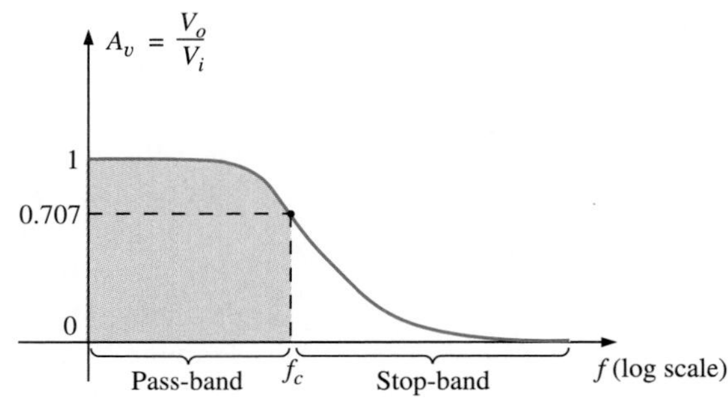

FIG. 21.9
Normalized plot of Fig. 21.8.

At any intermediate frequency, the output voltage $\mathbf{V}_o$ of Fig. 21.5 can be determined using the voltage divider rule:

$$\mathbf{V}_o = \frac{X_C \angle -90° \mathbf{V}_i}{R - jX_C}$$

or

$$\mathbf{A}_v = \frac{\mathbf{V}_o}{\mathbf{V}_i} = \frac{X_C \angle -90°}{R - jX_C} = \frac{X_C \angle -90°}{\sqrt{R^2 + X_C^2} \angle -\tan^{-1}(X_C/R)}$$

and

$$\mathbf{A}_v = \frac{\mathbf{V}_o}{\mathbf{V}_i} = \frac{X_C}{\sqrt{R^2 + X_C^2}} \angle -90° + \tan^{-1}\left(\frac{X_C}{R}\right)$$

The magnitude of the ratio V_o/V_i is therefore determined by

$$A_v = \frac{V_o}{V_i} = \frac{X_C}{\sqrt{R^2 + X_C^2}} = \frac{1}{\sqrt{\left(\frac{R}{X_C}\right)^2 + 1}} \qquad (21.7)$$

and the phase angle is determined by

$$\theta = -90° + \tan^{-1}\frac{X_C}{R} = -\tan^{-1}\frac{R}{X_C} \qquad (21.8)$$

For the special frequency at which $X_C = R$, the magnitude becomes

$$A_v = \frac{V_o}{V_i} = \frac{1}{\sqrt{\left(\frac{R}{X_C}\right)^2 + 1}} = \frac{1}{\sqrt{1+1}} = \frac{1}{\sqrt{2}} = 0.707$$

which defines the critical or cutoff frequency of Fig. 21.9.

The frequency at which $X_C = R$ is determined by

$$\frac{1}{2\pi f_c C} = R$$

and

$$f_c = \frac{1}{2\pi RC} \qquad (21.9)$$

The impact of Eq. (21.9) is greater than its relative simplicity. For any low-pass filter, the application of any frequency less than f_c will result in an output voltage V_o that is at least 70.7% of the maximum. For any frequency above f_c, the output is less than 70.7% of the applied signal.

Solving for $\mathbf{V}_o$ and substituting $\mathbf{V}_i = V_i \angle 0°$ gives

$$\mathbf{V}_o = \left[\frac{X_C}{\sqrt{R^2 + X_C^2}} \angle\theta\right] \mathbf{V}_i = \left[\frac{X_C}{\sqrt{R^2 + X_C^2}} \angle\theta\right] V_i \angle 0°$$

and

$$\mathbf{V}_o = \frac{X_C V_i}{\sqrt{R^2 + X_C^2}} \angle\theta$$

The angle θ is, therefore, the angle by which $\mathbf{V}_o$ leads $\mathbf{V}_i$. Since $\theta = -\tan^{-1} R/X_C$ is always negative (except at $f = 0$ Hz), it is clear that $\mathbf{V}_o$ will always lag $\mathbf{V}_i$, which explains why the network of Fig. 21.5 is called a *lagging network.*

At high frequencies, X_C is very small and R/X_C is quite large, resulting in $\theta = -\tan^{-1} R/X_C$ approaching $-90°$.

At low frequencies, X_C is quite large and R/X_C is very small, resulting in θ approaching $0°$.

At $X_C = R$, or $f = f_c$, $-\tan^{-1} R/X_C = -\tan^{-1} 1 = -45°$.

A plot of θ versus frequency results in the phase plot of Fig. 21.10.

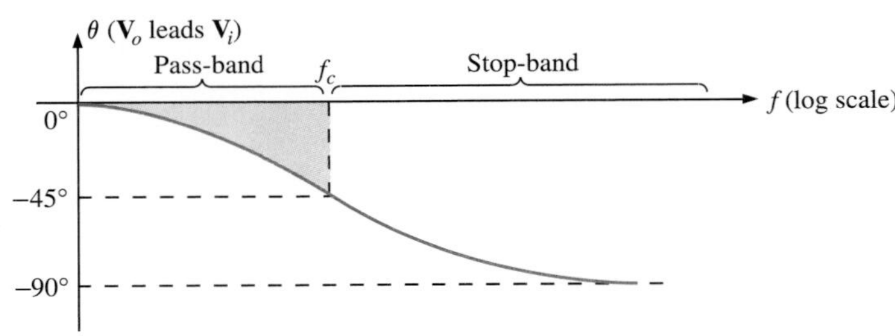

FIG. 21.10
Angle by which $\mathbf{V}_o$ *leads* $\mathbf{V}_i$.

The plot is of $\mathbf{V}_o$ leading $\mathbf{V}_i$, but since the phase angle is always negative, the phase plot of Fig. 21.11 ($\mathbf{V}_o$ lagging $\mathbf{V}_i$) is more appropriate. Note that a change in sign requires that the vertical axis be changed to the angle by which $\mathbf{V}_o$ lags $\mathbf{V}_i$. In particular, note that the phase angle between $\mathbf{V}_o$ and $\mathbf{V}_i$ is less than 45° in the pass-band and approaches 0° at lower frequencies. In summary, for the low-pass *R-C* filter of Fig. 21.8:

$$f_c = \frac{1}{2\pi RC}$$

For $\quad f < f_c, \quad V_o > 0.707V_i$

whereas for $\quad f > f_c, \quad V_o < 0.707V_i$

At f_c, $\quad \mathbf{V}_o$ lags $\mathbf{V}_i$ by 45°.

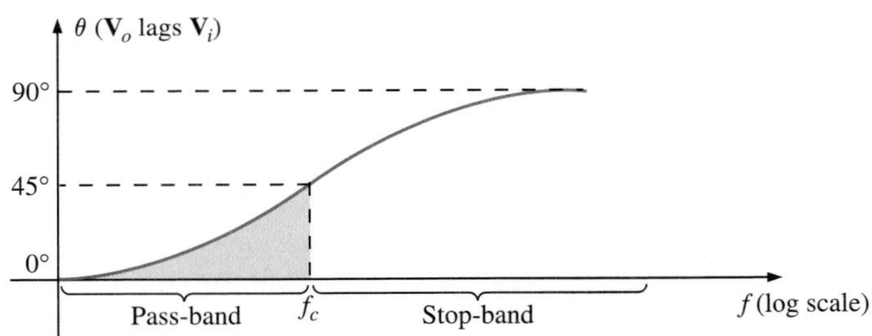

FIG. 21.11
Angle by which $\mathbf{V}_o$ *lags* $\mathbf{V}_i$.

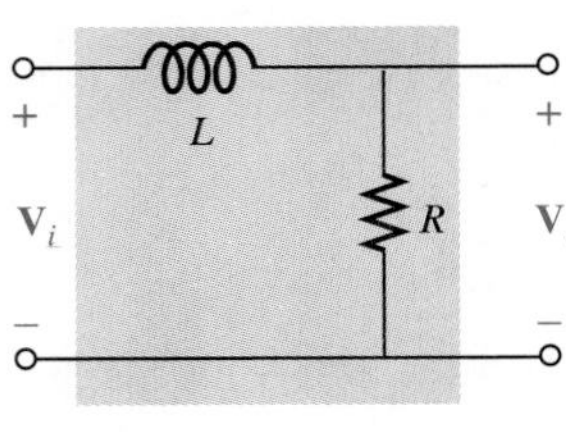

FIG. 21.12
Low-pass R-L filter.

The low-pass filter response of Fig. 21.4(a) can also be obtained using the *R-L* combination of Fig. 21.12 with

$$f_c = \frac{R}{2\pi L} \qquad (21.10)$$

In general, however, the *R-C* combination is more popular due to the smaller size of capacitive elements and the nonlinearities associated with inductive elements. Analyzing the low-pass *R-L* will be left as an exercise for you.

EXAMPLE 21.3

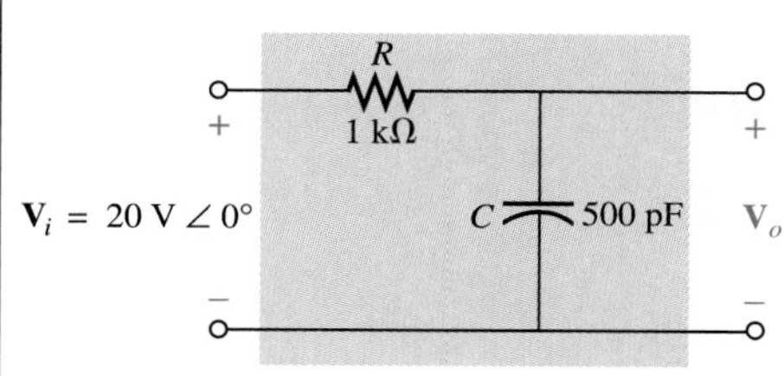

FIG. 21.13
Example 21.3.

a. Sketch the output voltage V_o versus frequency for the low-pass *R-C* filter of Fig. 21.13.
b. Determine the voltage V_o at $f = 100$ kHz and 1 MHz and compare the results to the results obtained from the curve of part (a).
c. Sketch the normalized gain $A_v = V_o/V_i$.

Solutions:

a. Equation (21.9):

$$f_c = \frac{1}{2\pi RC} = \frac{1}{2\pi(1\text{ k}\Omega)(500\text{ pF})} = \mathbf{318.31\text{ kHz}}$$

At f_c, $V_o = 0.707(20\text{ V}) = 14.14$ V. See Fig. 21.14.

b. Eq. (21.7):

$$V_o = \frac{V_i}{\sqrt{\left(\frac{R}{X_C}\right)^2 + 1}}$$

At $f = 100$ kHz:

$$X_C = \frac{1}{2\pi fC} = \frac{1}{2\pi(100\text{ kHz})(500\text{ pF})} = 3.18\text{ k}\Omega$$

and

$$V_o = \frac{20\text{ V}}{\sqrt{\left(\frac{1\text{ k}\Omega}{3.18\text{ k}\Omega}\right)^2 + 1}} = \mathbf{19.08\text{ V}}$$

At $f = 1$ MHz:

$$X_C = \frac{1}{2\pi fC} = \frac{1}{2\pi(1\text{ MHz})(500\text{ pF})} = 0.32\text{ k}\Omega$$

and

$$V_o = \frac{20\text{ V}}{\sqrt{\left(\frac{1\text{ k}\Omega}{0.32\text{ k}\Omega}\right)^2 + 1}} = \mathbf{6.1\text{ V}}$$

Both levels are verified by Fig. 21.14.

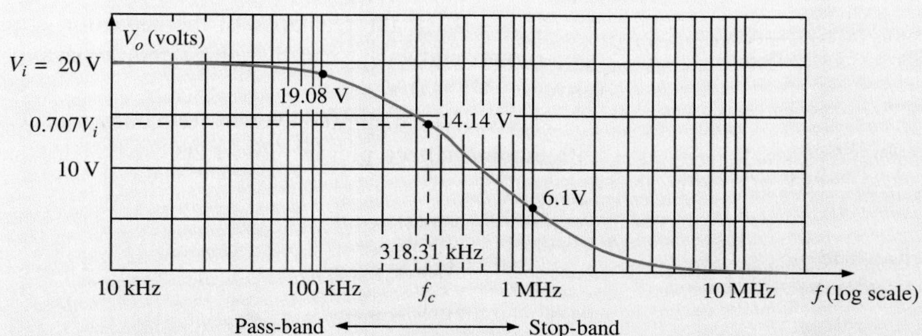

FIG. 21.14
Frequency response for the low-pass R-C network of Fig. 21.13.

continued

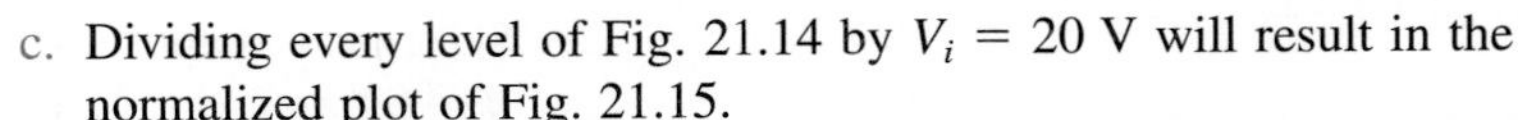

c. Dividing every level of Fig. 21.14 by $V_i = 20$ V will result in the normalized plot of Fig. 21.15.

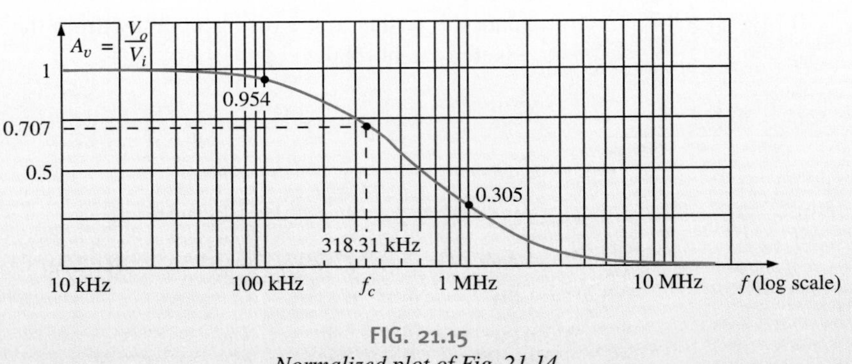

FIG. 21.15
Normalized plot of Fig. 21.14.

21.5 *R-C* HIGH-PASS FILTER

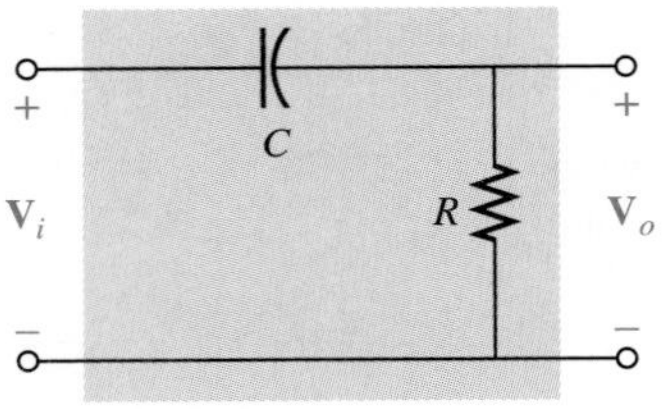

FIG. 21.16
High-pass filter.

As noted early in Section 21.4, a high-pass *R-C* filter can be constructed by simply reversing the positions of the capacitor and resistor, as shown in Fig. 21.16.

At very high frequencies the reactance of the capacitor is very small and the short-circuit equivalent can be substituted, as shown in Fig. 21.17. The result is $\mathbf{V}_o = \mathbf{V}_i$.

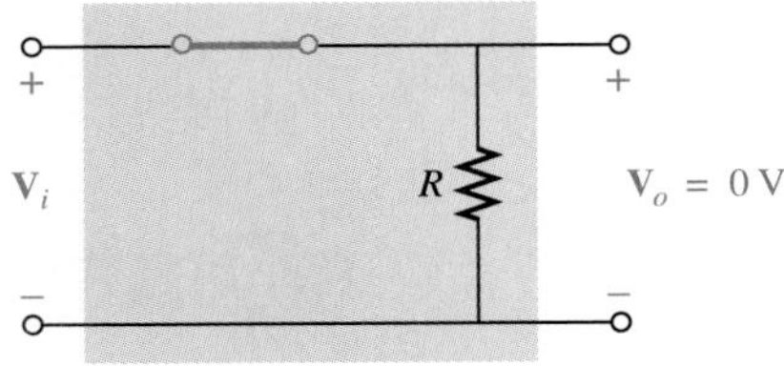

FIG. 21.17
R-C high-pass filter at very high frequencies.

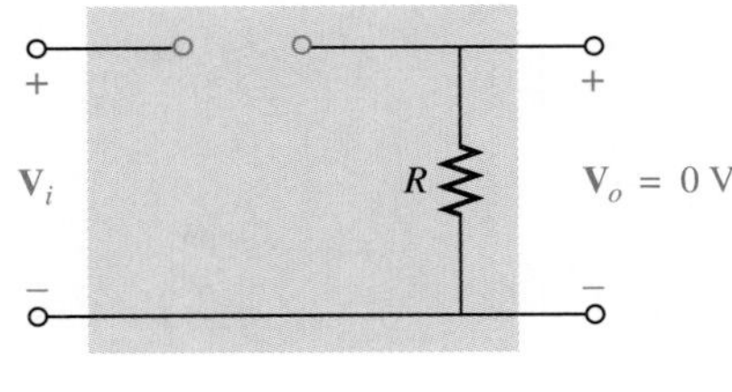

FIG. 21.18
R-C high-pass filter at $f = 0$ Hz.

At $f = 0$ Hz, the reactance of the capacitor is quite high, and the open-circuit equivalent can be substituted, as shown in Fig. 21.18. In this case, $\mathbf{V}_o = 0$ V.

A plot of the magnitude versus frequency is provided in Fig. 21.19, with the normalized plot in Fig. 21.20.

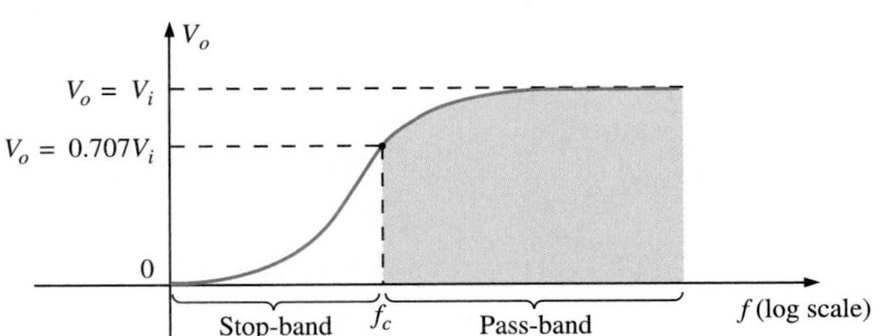

FIG. 21.19
V_o versus frequency for a high-pass R-C filter.

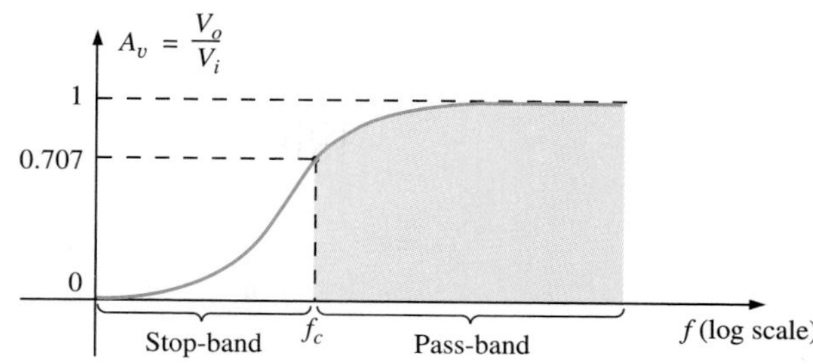

FIG. 21.20
Normalized plot of Fig. 21.19.

At any intermediate frequency, the output voltage can be determined using the voltage divider rule:

$$\mathbf{V}_o = \frac{R \angle 0° \, \mathbf{V}_i}{R - j X_C}$$

or

$$\frac{\mathbf{V}_o}{\mathbf{V}_i} = \frac{R \angle 0°}{R - j X_C} = \frac{R \angle 0°}{\sqrt{R^2 + X_C^2} \angle -\tan^{-1}(X_C/R)}$$

and

$$\frac{\mathbf{V}_o}{\mathbf{V}_i} = \frac{R}{\sqrt{R^2 + X_C^2}} \angle \tan^{-1}(X_C/R)$$

The magnitude of the ratio $\mathbf{V}_o/\mathbf{V}_i$ is therefore determined by

$$A_v = \frac{V_o}{V_i} = \frac{R}{\sqrt{R^2 + X_C^2}} = \frac{1}{\sqrt{1 + \left(\frac{X_C}{R}\right)^2}} \tag{21.11}$$

and the phase angle θ, by

$$\theta = \tan^{-1} \frac{X_C}{R} \tag{21.12}$$

For the frequency at which $X_C = R$, the magnitude becomes

$$\frac{V_o}{V_i} = \frac{1}{\sqrt{1 + \left(\frac{X_C}{R}\right)^2}} = \frac{1}{\sqrt{1+1}} = \frac{1}{\sqrt{2}} = 0.707$$

as shown in Fig. 21.20.

The frequency at which $X_C = R$ is determined by

$$X_C = \frac{1}{2\pi f_c C} = R$$

and

$$f_c = \frac{1}{2\pi RC} \tag{21.13}$$

For the high-pass *R-C* filter, applying any frequency greater than f_c will result in an output voltage V_o that is at least 70.7% of the magnitude of the input signal. For any frequency below f_c, the output is less than 70.7% of the applied signal.

For the phase angle, high frequencies result in small values of X_C, and the ratio X_C/R will approach zero with $\tan^{-1}(X_C/R)$ approaching 0°, as shown in Fig. 21.21. At low frequencies, the ratio X_C/R becomes quite large and $\tan^{-1}(X_C/R)$ approaches 90°. For the case $X_C = R$, $\tan^{-1}(X_C/R) = \tan^{-1} 1 = 45°$. Assigning a phase angle of 0° to $\mathbf{V}_i$ such that $\mathbf{V}_i = V_i \angle 0°$, the phase angle associated with $\mathbf{V}_o$ is θ, resulting in $\mathbf{V}_o = V_o \angle \theta$ and revealing that θ is the angle by which $\mathbf{V}_o$ leads $\mathbf{V}_i$. Since the angle θ is the angle by which $\mathbf{V}_o$ leads $\mathbf{V}_i$ throughout the frequency range of Fig. 21.21, the high-pass R-C filter is referred to as a *leading network*.

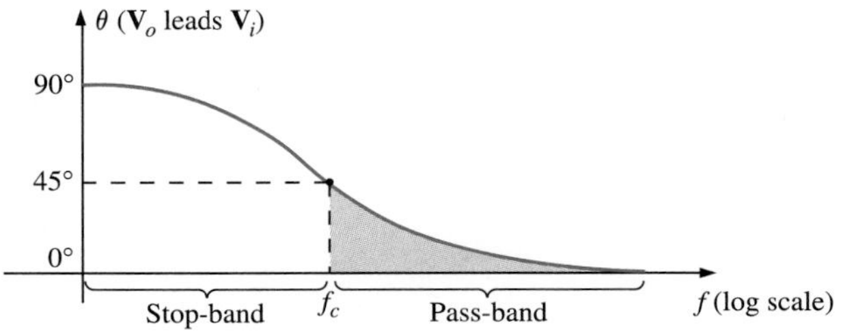

FIG. 21.21
Phase-angle response for the high-pass R-C filter.

In summary, for the high-pass R-C filter:

$$f_c = \frac{1}{2\pi RC}$$

For $\quad f < f_c, \quad V_o < 0.707V_i$

whereas for $\quad f > f_c, \quad V_o > 0.707V_i$

At f_c, $\quad \mathbf{V}_o$ leads $\mathbf{V}_i$ by 45°.

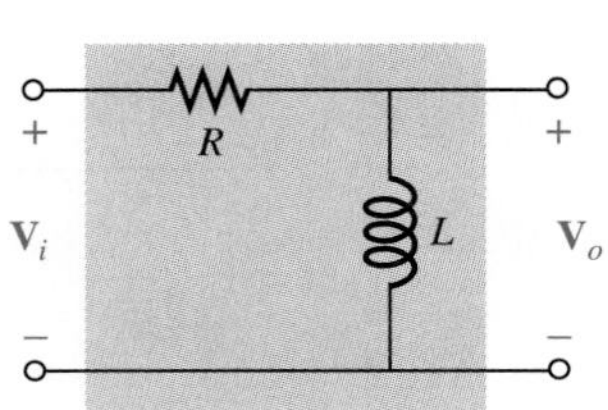

FIG. 21.22
High-pass R-L filter.

The high-pass filter response of Fig. 21.20 can also be obtained using the same elements of Fig. 21.12 but interchanging their positions, as shown in Fig. 21.22. Once again, the analysis is left as an exercise for you.

EXAMPLE 21.4

Given $R = 20\ \text{k}\Omega$ and $C = 1200$ pF:

a. Sketch the normalized plot if the filter is used as both a high- and a low-pass filter.
b. Sketch the phase plot for both filters of part (a).
c. Determine the magnitude and phase of $\mathbf{A}_v = \mathbf{V}_o/\mathbf{V}_i$ at $f = \frac{1}{2}f_c$ for the high-pass filter.

Solutions:

a. $f_c = \dfrac{1}{2\pi RC} = \dfrac{1}{(2\pi)(20\ \text{k}\Omega)(1200\ \text{pF})}$

$= \mathbf{6631.46\ Hz}$

The normalized plots appear in Fig. 21.23.

b. The phase plots appear in Fig. 21.24.

c. $f = \dfrac{1}{2}f_c = \dfrac{1}{2}(6631.46\ \text{Hz}) = 3315.73\ \text{Hz}$

$X_C = \dfrac{1}{2\pi fC} = \dfrac{1}{(2\pi)(3315.73\ \text{Hz})(1200\ \text{pF})}$

$\cong 40\ \text{k}\Omega$

continued

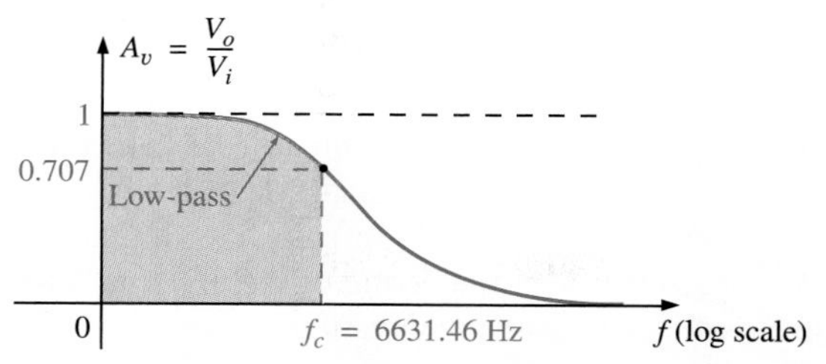

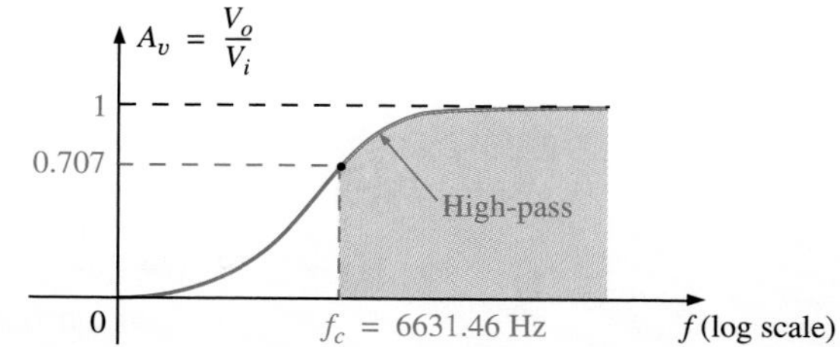

FIG. 21.23
Normalized plots for a low-pass and high-pass filter using the same elements.

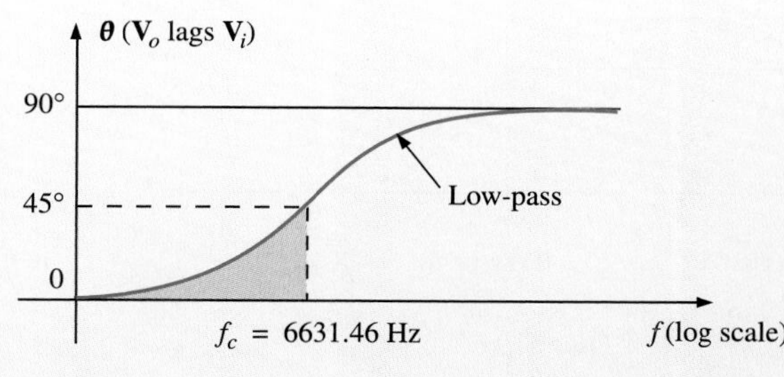

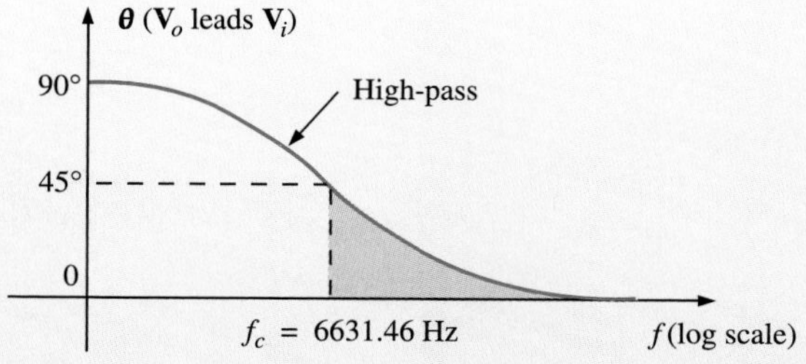

FIG. 21.24
Phase plots for a low-pass and high-pass filter using the same elements.

$$A_v = \frac{V_o}{V_i} = \frac{1}{\sqrt{1 + \left(\frac{X_C}{R}\right)^2}} = \frac{1}{\sqrt{1 + \left(\frac{40\ \text{k}\Omega}{20\ \text{k}\Omega}\right)^2}} = \frac{1}{\sqrt{1 + (2)^2}}$$

$$= \frac{1}{\sqrt{5}} = 0.4472$$

$$\theta = \tan^{-1}\frac{X_C}{R} = \tan^{-1}\frac{40\ \text{k}\Omega}{20\ \text{k}\Omega} = \tan^{-1} 2 = 63.43°$$

and
$$\mathbf{A}_v = \frac{\mathbf{V}_o}{\mathbf{V}_i} = \mathbf{0.4472\ \angle 63.43°}$$

21.6 PASS-BAND FILTERS

There are a number of methods to establish the pass-band characteristic of Fig. 21.4(c). One method uses both a low-pass and high-pass filter in cascade, as shown in Fig. 21.25.

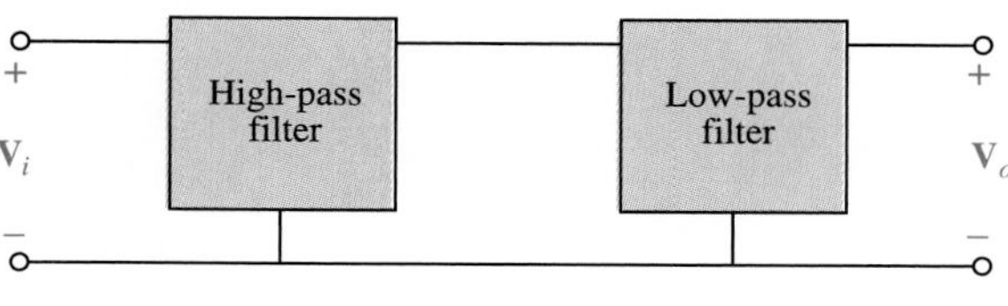

FIG. 21.25
Pass-band filter.

The components are chosen to establish a cutoff frequency for the high-pass filter that is lower than the critical frequency of the low-pass filter, as shown in Fig. 21.26. A frequency f_1 may pass through the low-pass filter but have little effect on V_o due to the reject characteristics of the high-pass filter. A frequency f_2 may pass through the high-pass filter but be prohibited from reaching the high-pass filter by the low-pass characteristics. A frequency f_o near the centre of the pass-band will pass through both filters with very little degeneration.

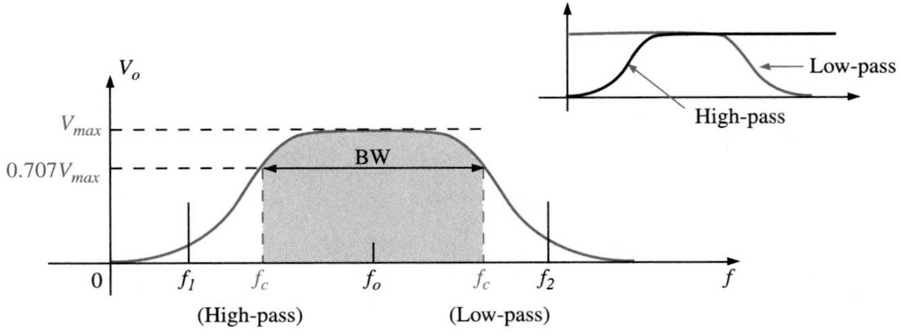

FIG. 21.26
Pass-band characteristics.

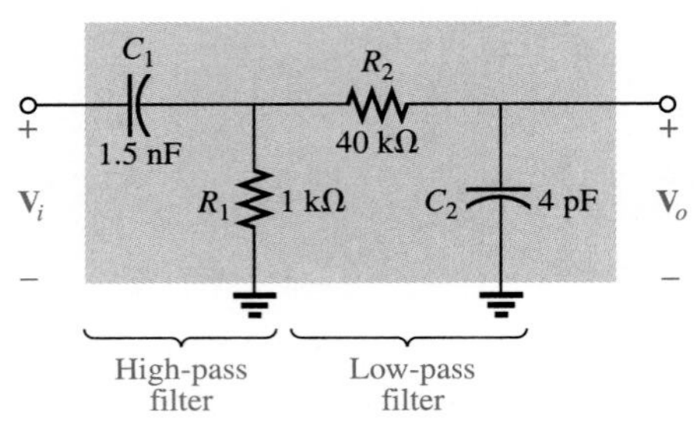

FIG. 21.27
Pass-band filter.

The network of Fig. 21.27 will generate the characteristics of Fig. 21.26. When used separately, each filter has its own critical or cutoff frequency where the output voltage is $V_o = 0.707\ V_i$. Each has its own pass-band where the output voltage $V_o = V_i$ is its highest possible value, V_{max}. However, when the two filters are cascaded, the output of the high pass filter is loaded down and V_o will drop lower than value V_i in the pass-band. Also, each circuit will cause a shift in the other's critical frequency. Through proper design, the level of V_o can be made to be very near V_i in the pass band, and the effect on the cutoff frequencies kept small. In addition, as the critical frequencies of each filter are brought closer to one another to increase the quality factor of the response curve, the maximum values within the pass band will continue to drop. For this reason, the bandwidth is defined where the output is 0.707 of the actual maximum output voltage V_o.

EXAMPLE 21.5

For the pass-band filter of Fig. 21.27:

a. Determine the critical frequencies for the low- and high-pass filters.
b. Find the maximum output V_o in volts and in dB relative to V_i. Then, using the critical frequencies found in (a), sketch the response characteristics.
c. Determine the actual value of V_o at the high-pass critical frequency calculated in part (a). Then compare it to the level resulting by multiplying the actual highest value of V_o in the pass band by 0.707.
d. If R_2 had been 1 kΩ, calculate the value of C_2 that would have the same low-pass critical frequency.
e. Using the values of R_2 and C_2 in part (d), calculate the value of V_o at the band-pass frequency used in part (c).

Solutions:

a. High-pass filter:

$$f_c = \frac{1}{2\pi R_1 C_1} = \frac{1}{2\pi(1\ \text{k}\Omega)(1.5\ \text{nF})} = \mathbf{106.1\ kHz}$$

continued

Low-pass filter:

$$f_c = \frac{1}{2\pi R_2 C_2} = \frac{1}{2\pi(40\ \text{k}\Omega)(4\ \text{pF})} = \mathbf{994.72\ kHz}$$

b. To find the maximum output voltage in the mid-region of the pass-band requires the exact frequency where it occurs. Using f_o = 500 kHz, near the centre of the mid-region, an analysis of the network reveals that $\mathbf{V}_o = 0.8720\ V_i\angle -14.69°$.

If we use the geometric mean frequency (Eq. 20.23), then

$$f_o = \sqrt{(106.1\ \text{kHz})(994.72\ \text{kHz})} = 324.9\ \text{kHz}$$

Analysis of the network at 324.9 kHz reveals that $\mathbf{V}_o = 0.9014\ V_i \angle 0.00°$, which is higher than at 500 kHz. The fact that the output is in phase with the input is a good indication that the effect of the capacitors is minimal, and that we are near the frequency where the output is maximum. The bandwidth is therefore defined when V_o is at a level of $0.707(0.9014\ V_i) = 0.6373\ V_i$. Fig. 21.28 indicates these values.

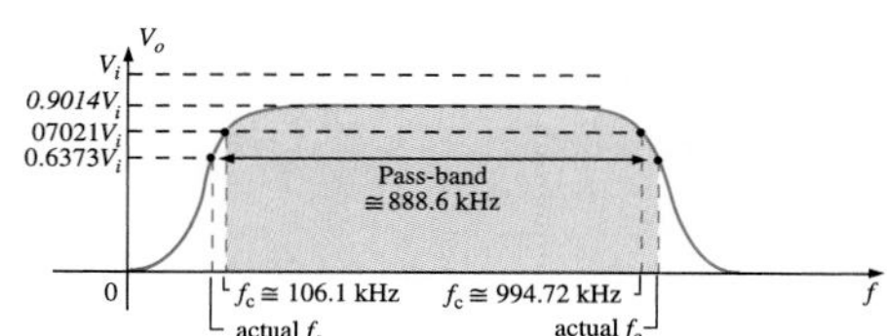

FIG. 21.28
Pass-band characteristics for the filter of Fig. 21.27.

Also at f_o = 324.9 kHz,

$$A_{v_{dB}} = 20 \log_{10} \frac{V_o}{V_i} = 20 \log_{10} 0.9014 \frac{V_o}{V_i} = -0.9\ \text{dB}$$

This means the output in the mid-band is down 0.9 dB from V_i due to the loading effect.

At the cutoff frequency f_c,

$$A_{v_{dB}} = 20 \log_{10} \frac{V_o}{V_i} = 20 \log_{10} 0.6373 \frac{V_o}{V_i} = -3.9\ \text{dB}$$

or down 3.9 dB from V_i. Note that the difference between the levels at f_c and f_o is -3.0 dB.

c. At f = 994.72 kHz,

$$X_{C_1} = \frac{1}{2\pi f C_1} \cong 107\ \Omega$$

and

$$X_{C_2} = \frac{1}{2\pi f C_2} = R_2 = 40\ \text{k}\Omega$$

resulting in the network of Fig. 21.29.

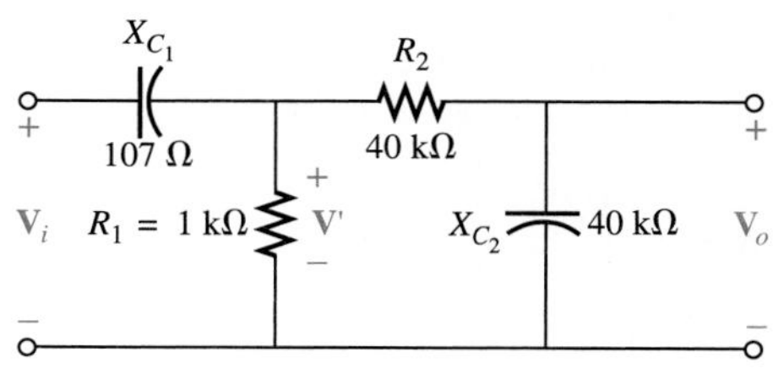

FIG. 21.29
Network of Fig. 21.27 at f = 994.72 kHz.

The parallel combination of $R_1 \parallel (R_2 - jX_{C_2})$ is $0.9876\ \Omega\angle -0.71^0$, which is close to the value of R_1. This is an indication the loading, while present, is not too severe.

Then,

$$V' = (0.9876\ \Omega\angle -0.71^0)\left(\frac{V_i}{0.9876\ \text{k}\Omega\angle -0.71° - j(0.107\ \text{k}\Omega)}\right) =$$

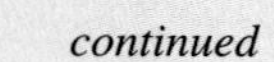

$0.9929\ \text{k}\Omega\angle 6.18°\mathbf{V}_i$

continued

With

$$\mathbf{V}_o = (40\ \text{k}\Omega\angle - 90^0)\left(\frac{0.9929V_i\angle 6.18^0}{40\ \text{k}\Omega - \text{j}(40\ \text{k}\Omega)}\right)$$

$$\mathbf{V}_o = 0.7021\ V_i \text{ at } f = 994.72\ \text{kHz}$$

But since the bandwidth is defined at 0.6373 V_i, the upper cutoff frequency will be higher than 994.72 kHz, as shown in Fig. 21.28. The lower cutoff frequency is affected also. These shifts were caused by the loading effect.

d. Low-pass filter

$$C_2 = \frac{1}{(2\pi f_c R_2)} = \frac{1}{(2\pi(994.72\ \text{kHz})(1.0\ \text{k}\Omega))} = 160\ \text{pF}$$

While $R_2 = 1.0\ \text{k}\Omega$ and $C_2 = 160$ pF has the same characteristic as the 40 kΩ and 4.0 pF, they will cause severe loading when cascaded as seen in part (e).

e. At a frequency of 324.9 kHz, the parallel combination of $R_1 \parallel (R_2 - jX_{c_2})$ is $0.8808\ \text{k}\Omega\angle -15.05°$. Compared to part (c), the loading is quite severe.

Then,

$$\mathbf{V}' = (0.8808\ \text{k}\Omega\angle - 15.05^0)\left(\frac{\mathbf{V}_i}{(0.8506\ \text{k}\Omega - j0.2287\ \text{k}\Omega) - j0.3266\ \text{k}\Omega}\right)$$

$$\mathbf{V}' = \frac{0.8808\ \text{k}\Omega\angle - 15.05^0}{1.0158\ \text{k}\Omega\angle - 33.14^0}\mathbf{V}_i = 0.8671\ \text{k}\Omega\angle 18.09^0\ \mathbf{V}_i$$

With

$$\mathbf{V}_o = \frac{3.062\ \text{k}\Omega\angle - 90^0}{1.0\ \text{k}\Omega - j(3.062\ \text{k}\Omega)}\mathbf{V}_i$$

Giving

$V_o = 0.8243\ \text{k}\Omega\angle 0.01°\mathbf{V}_i$ at $f = 324.9$ kHz (a drop of 1.7 dB due to loading).

The critical frequencies would be defined where $V_o = 0.707(0.8243\ V_i) = 0.5828\ V_i$ or 4.7 dB down.

The pass-band response can also be obtained using the series and parallel resonant circuits discussed in Chapter 20. In each case, however, V_o will not be equal to V_i in the pass-band, but a frequency range in which V_o will be equal to or greater than $0.707V_{\max}$ can be defined.

For the series resonant circuit of Fig. 21.30, $X_L = X_C$ at resonance and

$$V_{o_{\max}} = \frac{R}{R + R_l}V_i \Bigg|_{f = f_s} \tag{21.14}$$

and

$$f_s = \frac{1}{2\pi\sqrt{LC}} \tag{21.15}$$

with

$$Q_s = \frac{X_L}{R + R_l} \tag{21.16}$$

and

$$BW = \frac{f_s}{Q_s} \tag{21.17}$$

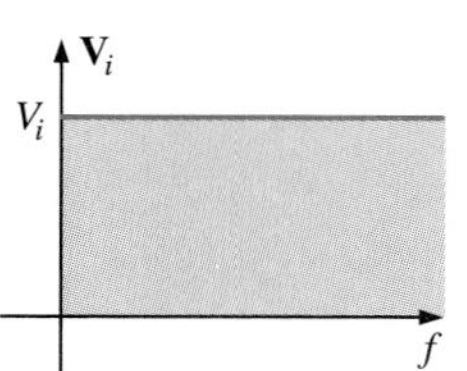

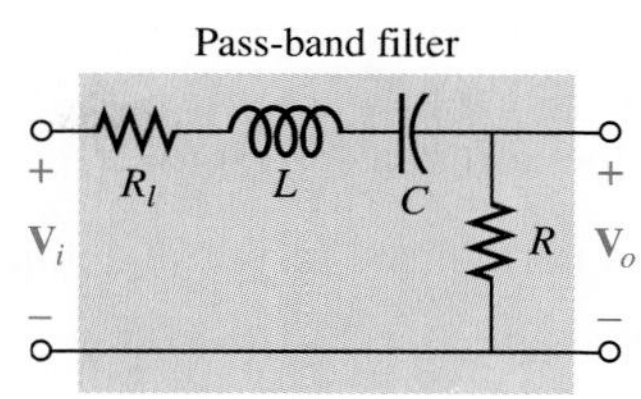

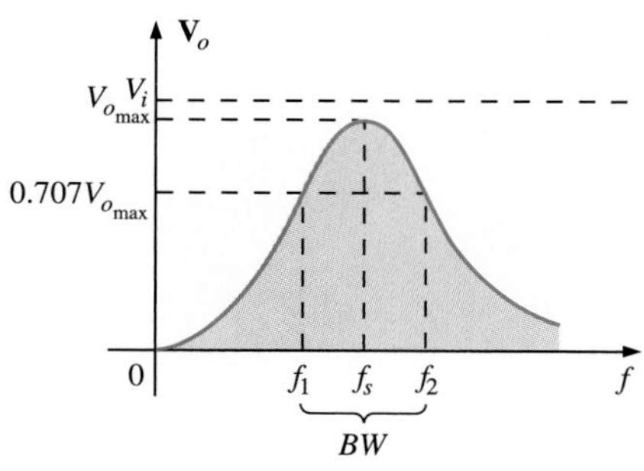

FIG. 21.30
Series resonant pass-band filter.

For the parallel resonant circuit of Fig. 21.31, Z_{T_p} is a maximum value at resonance and

$$V_{o_{max}} = \frac{Z_{T_p}V_i}{Z_{T_p} + R} \quad f = f_p \tag{21.18}$$

with

$$Z_{T_p} = Q_l^2 R_l \quad Q_l \geq 10 \tag{21.19}$$

and

$$f_p = \frac{1}{2\pi\sqrt{LC}} \quad Q_l \geq 10 \tag{21.20}$$

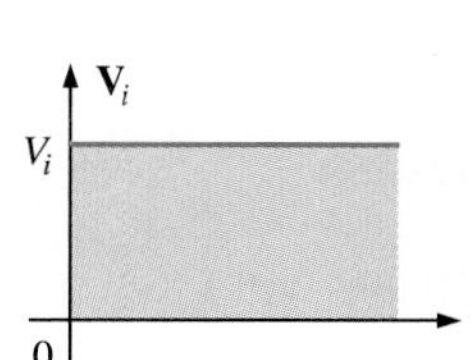

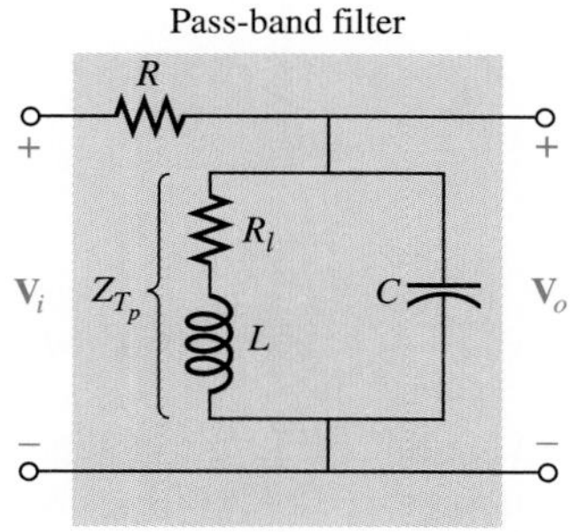

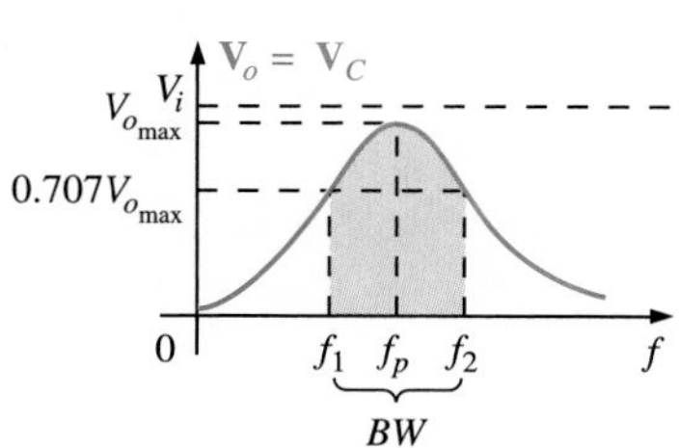

FIG. 21.31
Parallel resonant pass-band filter.

For the parallel resonant circuit

$$Q_p = \frac{X_L}{R_l} \tag{21.21}$$

and

$$BW = \frac{f_p}{Q_p} \tag{21.22}$$

As a first approximation that is acceptable for most practical applications, you can assume that the resonant frequency bisects the bandwidth.

EXAMPLE 21.6

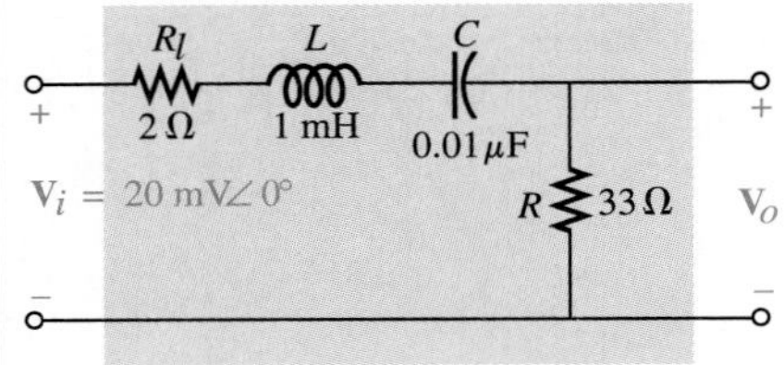

FIG. 21.32
Series resonant pass-band filter for Example 21.6.

a. Determine the frequency response for the voltage V_o for the series circuit of Fig. 21.32.
b. Plot the normalized response $A_v = V_o/V_i$.
c. Plot a normalized response defined by $A'_v = \dfrac{A_v}{A_{v_{\max}}}$.

Solutions:

a.
$$f_s = \frac{1}{2\pi\sqrt{LC}} = \frac{1}{2\pi\sqrt{(1\ \text{mH})(0.01\ \mu\text{F})}} = \mathbf{50\ 329.21\ Hz}$$

$$Q_s = \frac{X_L}{R + R_l} = \frac{2\pi(50\ 329.21\ \text{Hz})(1\ \text{mH})}{33\ \Omega + 2\ \Omega} = \mathbf{9.04}$$

$$BW = \frac{f_s}{Q_s} = \frac{50\ 329.21\ \text{Hz}}{9.04} = \mathbf{5.57\ kHz}$$

At resonance:

$$V_{o_{\max}} = \frac{RV_i}{R + R_l} = \frac{33\ \Omega(V_i)}{33\ \Omega + 2\ \Omega} = 0.943V_i = 0.943(20\ \text{mV})$$
$$= \mathbf{18.86\ mV}$$

At the cutoff frequencies:

$$V_o = (0.707)(0.943V_i) = 0.667V_i = 0.667(20\ \text{mV})$$
$$= \mathbf{13.34\ mV}$$

Note Fig. 21.33.

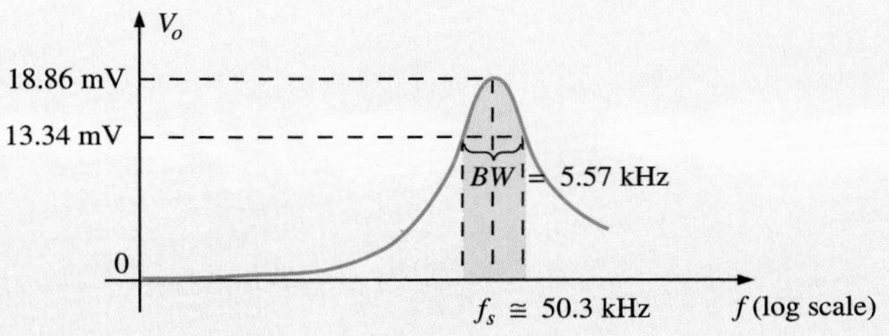

FIG. 21.33
Pass-band response for the network.

b. Dividing all levels of Fig. 21.33 by V_i = 20 mV will result in the normalized plot of Fig. 21.34(a).

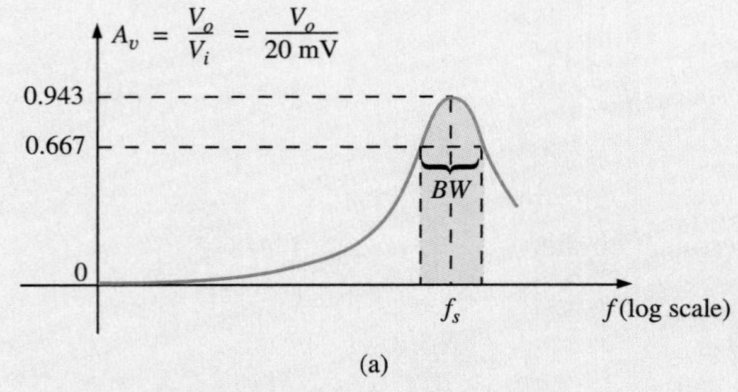

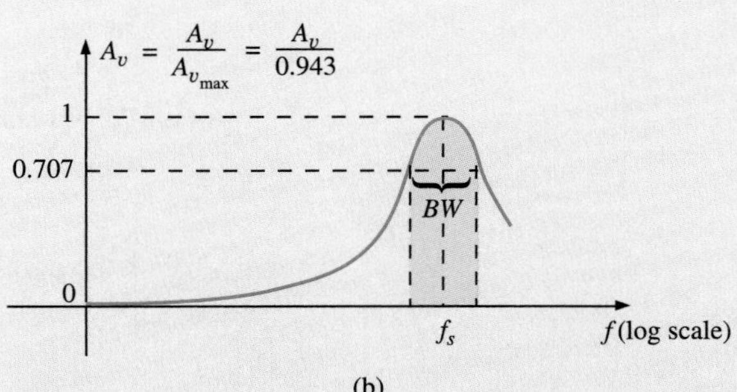

FIG. 21.34
Normalized plots for the pass-band filter of Fig. 21.32.

c. Dividing all levels of Fig. 21.34(a) by $A_{v_{\max}}$ = 0.943 will result in the normalized plot of Fig. 21.34(b).

21.7 STOP-BAND FILTERS

Stop-band filters can also be constructed using a low-pass and a high-pass filter. However, rather than the cascaded configuration used for the pass-band filter, a parallel arrangement is required, as shown in Fig. 21.35. A low frequency f_1 can pass through the low-pass filter and a higher frequency f_2 can use the parallel path, as shown in Figs. 21.35 and 21.36. However, a frequency such as f_o in the reject-band is higher than the low-pass critical frequency and lower than the high-pass critical frequency, and is therefore prevented from contributing to the levels of V_o above $0.707V_{max}$.

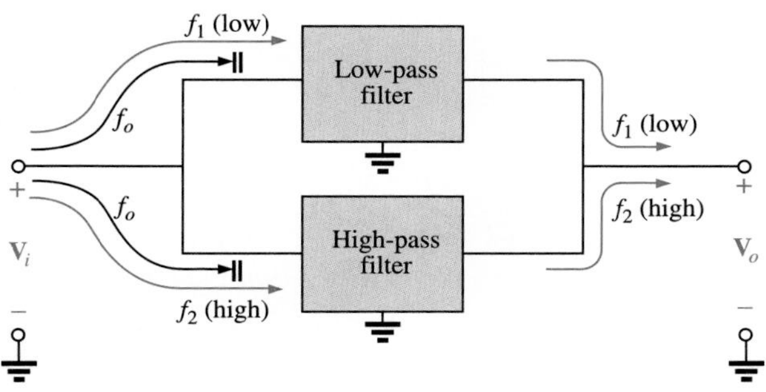

FIG. 21.35
Stop-band filter.

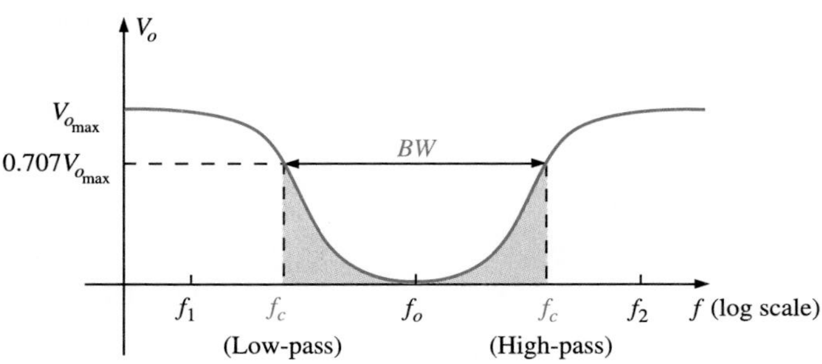

FIG. 21.36
Stop-band characteristics.

Since the characteristics of a stop-band filter are the inverse of the pattern of the pass-band filters, we can use the fact that at any frequency the sum of the magnitudes of the two waveforms to the right of the equals sign in Fig. 21.37 will equal the applied voltage V_i.

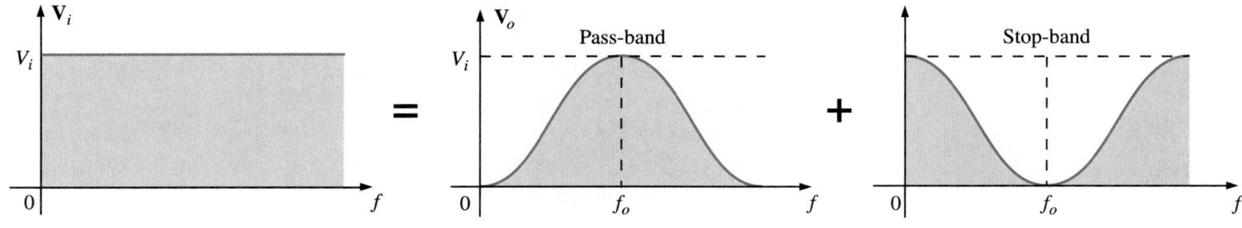

FIG. 21.37
Demonstrating how an applied signal of fixed magnitude can be broken down into a pass-band and stop-band response curve.

For the pass-band filters of Figs. 21.30 and 21.31, therefore, if we take the output off the other series elements as shown in Figs. 21.38 and 21.39, a stop-band characteristic will be obtained, as required by Kirchhoff's voltage law.

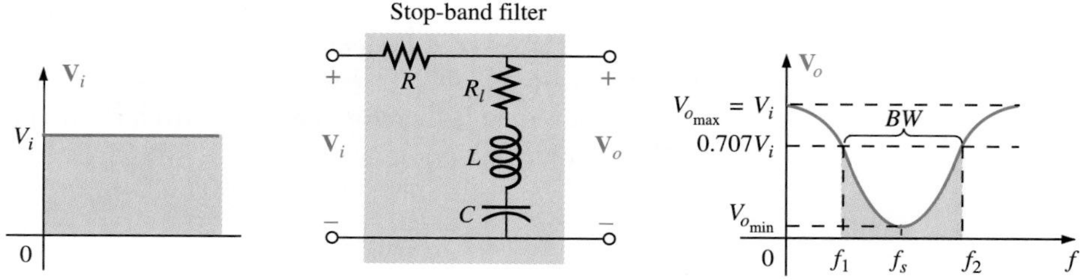

FIG. 21.38
Stop-band filter using a series resonant circuit.

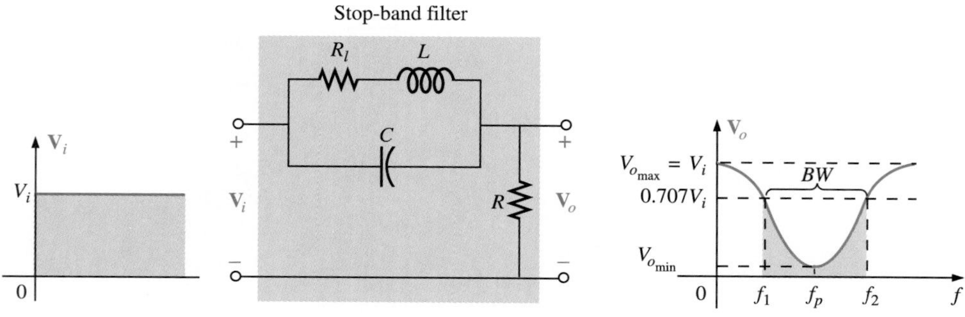

FIG. 21.39
Stop-band filter using a parallel resonant network.

For the series resonant circuit of Fig. 21.38, Eqs. (21.15) to (21.17) still apply, but now at resonance,

$$V_{o_{min}} = \frac{R_l V_i}{R_l + R} \tag{21.23}$$

For the parallel resonant circuit of Fig. 21.39, Eqs. (21.19) to (21.22) still apply, but now at resonance,

$$V_{o_{min}} = \frac{RV_i}{R + Z_{T_p}} \tag{21.24}$$

The maximum value of V_o for the series resonant circuit is V_i at the low end due to the open-circuit equivalent for the capacitor and V_i at the high end due to the high impedance of the inductive element.

For the parallel resonant circuit, at $f = 0$ Hz, the coil can be replaced by a short-circuit equivalent and the capacitor can be replaced by its open circuit and $V_o = RV_i/(R + R_l)$. At the high-frequency end, the capacitor approaches a short-circuit equivalent, and V_o increases toward V_i.

21.8 DOUBLE-TUNED FILTERS

Some network configurations display both a pass-band and a stop-band characteristic, as shown in Fig. 21.40. These circuits are called double tuned circuits. For the network of Fig. 21.40(a), the parallel resonant circuit will establish a stop-band for the range of frequencies not permitted to establish a significant V_L. The greater part of the applied voltage will appear across the parallel resonant circuit for this frequency range due to its very high impedance compared with R_L. For the pass-band, the parallel resonant circuit is designed to be capacitive (inductive if L_s is replaced by C_s). The inductance L_s is chosen to cancel the effects of the resulting net capacitive reactance at the resonant pass-band frequency of the tank circuit, thereby acting as a series resonant circuit. The applied voltage will then appear across R_L at this frequency.

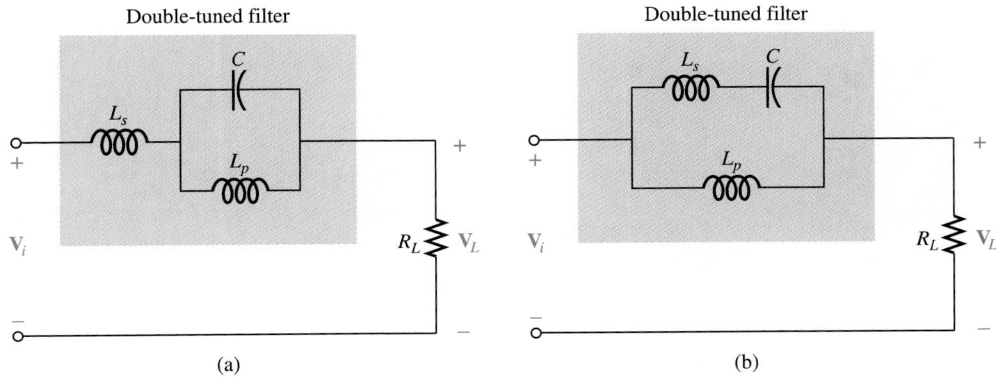

FIG. 21.40
Double-tuned networks.

For the network of Fig. 21.40(b), the series resonant circuit will still determine the pass-band, acting as a very low impedance across the parallel inductor at resonance. At the desired stop-band resonant frequency, the series resonant circuit is capacitive. The inductance L_p is chosen to establish parallel resonance at the resonant stop-band frequency. The high impedance of the parallel resonant circuit will result in a very low load voltage V_L.

For rejected frequencies below the pass-band, the networks should appear as shown in Fig. 21.40. For the reverse situation, L_s in Fig. 21.40(a) and L_p in Fig. 21.40(b) are replaced by capacitors.

EXAMPLE 21.7

For the network of Fig. 21.40(b), determine L_s and L_p for a capacitance C of 500 pF if a frequency of 200 kHz is to be rejected and a frequency of 600 kHz accepted.

Solution: For series resonance, we have

$$f_s = \frac{1}{2\pi\sqrt{LC}}$$

continued

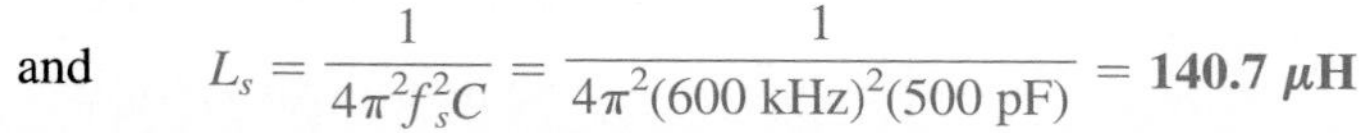

and $$L_s = \frac{1}{4\pi^2 f_s^2 C} = \frac{1}{4\pi^2(600\text{ kHz})^2(500\text{ pF})} = \mathbf{140.7\ \mu H}$$

At 200 kHz,

$$X_{L_s} = \omega L = 2\pi f_s L_s = (2\pi)(200\text{ kHz})(140.7\ \mu\text{H}) = 176.8\ \Omega$$

and $$X_C = \frac{1}{\omega C} = \frac{1}{(2\pi)(200\text{ kHz})(500\text{ pF})} = 1591.5\ \Omega$$

For the series elements,

$$j\,(X_{L_s} - X_C) = j\,(176.8\ \Omega - 1591.5\ \Omega) = -j\,1414.7\ \Omega = -j\,X'_C$$

At parallel resonance ($Q_l \geq 10$ assumed),

$$X_{L_p} = X'_C$$

and $$L_p = \frac{X_{L_p}}{\omega} = \frac{1414.7\ \Omega}{(2\pi)(200\text{ kHz})} = \mathbf{1.13\ mH}$$

LUMINARIES

American (Madison, Wis.; Summit, N.J.; Cambridge, Mass.) (1905–1981)
V.P. at Bell Laboratories
Professor of Systems Engineering, Harvard University

FIG. 21.41
Hendrik Wade Bode

In his early years at Bell Laboratories, Hendrik Bode was involved with *electric filter* and *equalizer* design. He then transferred to the Mathematics Research Group, where he specialized in research on electrical networks theory and its application to long distance communication facilities. In 1946 he was awarded the Presidential Certificate of Merit for his work in electronic fire control devices. In addition to publishing of the book *Network Analysis and Feedback Amplifier Design* in 1945, which is considered a classic in its field, he has received 25 patents in electrical engineering and systems design. When he retired, Bode was elected Gordon McKay Professor of Systems Engineering at Harvard University. He was a fellow of the IEEE and American Academy of Arts and Sciences.

Courtesy of AT&T Archives

21.9 BODE PLOTS

There is a technique for sketching the frequency response of such factors as filters, amplifiers, and systems on a decibel scale that can save a great deal of time and effort and can provide an excellent way to compare decibel levels at different frequencies.

The curves obtained for the magnitude and/or phase angle versus frequency are called Bode plots (Fig. 21.41). Through the use of straight-line segments called idealized Bode plots, the frequency response of a system can be found efficiently and accurately.

To ensure that you clearly understand the derivation of the method, we will examine the first network to be analyzed in some detail. We will treat the second network in a shorthand way, and finally, we'll introduce a method for quickly determining the response.

High-Pass *R-C* Filter

Let us start by reexamining the high-pass filter of Fig. 21.42. We chose the high-pass filter as our starting point since the frequencies of primary interest are at the low end of the frequency spectrum.

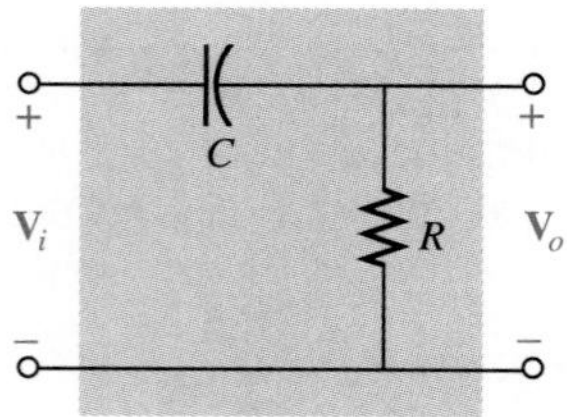

FIG. 21.42
High-pass filter.

The voltage gain of the system is given by:

$$\mathbf{A}_v = \frac{\mathbf{V}_o}{\mathbf{V}_i} = \frac{R}{R - jX_C} = \frac{1}{1 - j\dfrac{X_C}{R}} = \frac{1}{1 - j\dfrac{1}{2\pi fCR}}$$

$$= \frac{1}{1 - j\left(\dfrac{1}{2\pi RC}\right)\dfrac{1}{f}}$$

If we substitute $$f_c = \frac{1}{2\pi RC} \qquad (21.25)$$

continued

which we recognize as the cutoff frequency of earlier sections, we obtain

$$\mathbf{A}_v = \frac{1}{1 - j\,(f_c/f)} \tag{21.26}$$

The analysis to follow will show that the ability to reformat the gain to one having the general characteristics of Eq. (21.26) is critical to the application of the Bode technique. Different configurations will result in variations of the format of Eq. (21.26), but the desired similarities will become obvious as we progress through the material.

In magnitude and phase form:

$$\mathbf{A}_v = \frac{\mathbf{V}_o}{\mathbf{V}_i} = A_v \angle\theta = \frac{1}{\sqrt{1 + (f_c/f)^2}} \angle\tan^{-1}(f_c/f) \tag{21.27}$$

providing an equation for the magnitude and phase of the high-pass filter in terms of the frequency levels.

Using Eq. (21.5),

$$A_{v_{\text{dB}}} = 20 \log_{10} A_v$$

and, substituting the magnitude component of Eq. (21.27),

$$A_{v_{\text{dB}}} = 20 \log_{10} \frac{1}{\sqrt{1 + (f_c/f)^2}} = \underbrace{20 \log_{10} 1}_{0} - 20 \log_{10} \sqrt{1 + (f_c/f)^2}$$

and
$$\mathbf{A}_{v\text{dB}} = -20 \log_{10} \sqrt{1 + \left(\frac{f_c}{f}\right)^2}$$

Recognizing that $\log_{10}\sqrt{x} = \log_{10} x^{1/2} = \frac{1}{2}\log_{10} x$,

we have
$$A_{v_{\text{dB}}} = -\frac{1}{2}(20) \log_{10}\left[1 + \left(\frac{f_c}{f}\right)^2\right]$$

$$= -10 \log_{10}\left[1 + \left(\frac{f_c}{f}\right)^2\right]$$

For frequencies where $f \ll f_c$ or $(f_c/f)^2 \gg 1$,

$$1 + \left(\frac{f_c}{f}\right)^2 \cong \left(\frac{f_c}{f}\right)^2$$

and
$$A_{v_{\text{dB}}} = -10 \log_{10}\left(\frac{f_c}{f}\right)^2$$

but
$$\log_{10} x^2 = 2 \log_{10} x$$

resulting in
$$A_{v_{\text{dB}}} = -20 \log_{10} \frac{f_c}{f}$$

However, logarithms are such that

$$-\log_{10} b = +\log_{10} \frac{1}{b}$$

and substituting $b = f_c/f$ we have

$$A_{v_{\text{dB}}} = +20 \log_{10} \frac{f}{f_c} \qquad (f \ll f_c) \tag{21.28}$$

First note the similarities between Eq. (21.28) and the basic equation for gain in decibels: $A_{dB} = 20 \log_{10} V_o/V_i$. The comments about changes in decibel levels due to changes in V_o/V_i can therefore be applied here also, except now a change in frequency by a 2:1 ratio will result in a 6-dB change in gain. A change in frequency by a 10:1 ratio will result in a 20-dB change in gain.

Two frequencies separated by a 2:1 ratio are said to be an octave apart.

For Bode plots, a change in frequency by one octave will result in a 6-dB change in gain.

Two frequencies separated by a 10:1 ratio are said to be a decade apart.

For Bode plots, a change in frequency by one decade will result in a 20-dB change in gain.

You may wonder about all the mathematical development to obtain an equation that initially appears confusing and of limited value. As specified, Eq. (21.28) is accurate only for frequency levels much less than f_c.

First, realize that you will not have to repeat the mathematical development of Eq. (21.28) for each configuration you work with. Second, the equation itself is rarely applied but simply used, as we will show, to define a straight line on a log plot that permits sketching the frequency response of a system with minimum effort and high accuracy.

To plot Eq. (21.28), consider the following levels of increasing frequency:

$$\text{For } f = f_c/10,\ f/f_c = 0.1 \text{ and } +20 \log_{10} 0.1 = -20 \text{ dB}$$
$$\text{For } f = f_c/4,\ f/f_c = 0.25 \text{ and } +20 \log_{10} 0.25 = -12 \text{ dB}$$
$$\text{For } f = f_c/2,\ f/f_c = 0.5 \text{ and } +20 \log_{10} 0.5 = -6 \text{ dB}$$
$$\text{For } f = f_c,\ f/f_c = 1 \text{ and } +20 \log_{10} 1 = 0 \text{ dB}$$

Note from these equations that as the frequency of interest approaches f_c, the dB gain becomes less negative and approaches the final normalized value of 0 dB. The positive sign in front of Eq. (21.28) can therefore be interpreted to mean that the dB gain will have a positive slope with an increase in frequency. A plot of these points on a log scale will result in the straight-line segment of Fig. 21.43 to the left of f_c.

For the future, note that the resulting plot is a straight line intersecting the 0-dB line at f_c. It increases to the right at a rate of +6 dB per octave or +20 dB per decade. In other words, once f_c is determined,

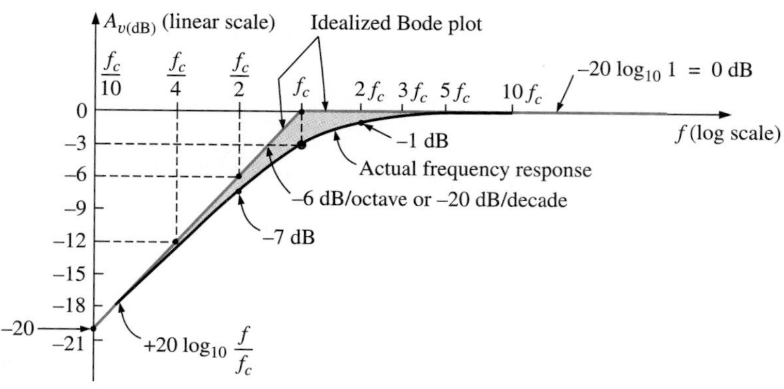

FIG. 21.43
Idealized Bode plot for the low-frequency region.

find $f_c/2$ and a plot point exists at -6 dB (or find $f_c/10$ and a plot point exists at -20 dB).

Bode plots are straight-line segments because the dB change per decade or octave is constant.

The actual response will approach an asymptote (straight-line segment) defined by $A_{v_{dB}} = 0$ dB since at high frequencies

$$f \gg f_c \text{ and } f_c/f \cong 0$$

with
$$A_{v_{dB}} = 20\log_{10}\frac{1}{\sqrt{1+(f_c/f)^2}} = 20\log_{10}\frac{1}{\sqrt{1+0}}$$
$$= 20\log_{10}1 = 0\text{ dB}$$

The two asymptotes defined above will intersect at f_c, as shown in Fig. 21.43, forming an envelope for the actual frequency response.

At $f = f_c$, the cutoff frequency,

$$A_{v_{dB}} = 20\log_{10}\frac{1}{\sqrt{1+(f_c/f)^2}} = 20\log_{10}\frac{1}{\sqrt{1+1}} = 20\log_{10}\frac{1}{\sqrt{2}}$$
$$= \mathbf{-3\ dB}$$

At $f = 2f_c$,

$$A_{v_{dB}} = -20\log_{10}\sqrt{1+\left(\frac{f_c}{2f_c}\right)^2} = -20\log_{10}\sqrt{1+\left(\frac{1}{2}\right)^2}$$
$$= -20\log_{10}\sqrt{1.25} = \mathbf{-1\ dB}$$

as shown in Fig. 21.43.

At $f = f_c/2$,

$$A_{v_{dB}} = -20\log_{10}\sqrt{1+\left(\frac{f_c}{f_c/2}\right)^2} = -20\log_{10}\sqrt{1+(2)^2}$$
$$= -20\log_{10}\sqrt{5}$$
$$= \mathbf{-7\ dB}$$

separating the idealized Bode plot from the actual response by 7 dB $-$ 6 dB = 1 dB, as shown in Fig. 21.43.

Reviewing the above,

at $f = f_c$ the actual response curve is 3 dB down from the idealized Bode plot, whereas at $f = 2f_c$ and $f_c/2$, the actual response curve is 1 dB down from the asymptotic response.

The phase response can also be sketched using straight-line asymptotes by considering a few critical points in the frequency spectrum.

Equation (21.27) specifies the phase response (the angle by which $\mathbf{V}_o$ leads $\mathbf{V}_i$) by

$$\theta = \tan^{-1}\frac{f_c}{f} \tag{21.29}$$

For frequencies well below f_c ($f \ll f_c$), $\theta = \tan^{-1}(f_c/f)$ approaches 90° and for frequencies well above f_c ($f \gg f_c$), $\theta = \tan^{-1}(f_c/f)$ will approach 0°, as discovered in earlier sections of the chapter. At $f = f_c$, $\theta = \tan^{-1}(f_c/f) = \tan^{-1} 1 = 45°$.

Defining $f \ll f_c$ for $f = f_c/10$ (and less) and $f \gg f_c$ for $f = 10f_c$ (and more), we can define an asymptote at $\theta = 90°$ for $f \ll f_c/10$, an asymptote at $\theta = 0°$ for $f \gg 10f_c$, and an asymptote from $f_c/10$ to $10f_c$ that passes through $\theta = 45°$ at $f = f_c$.

The asymptotes defined above all appear in Fig. 21.44. Again, the Bode plot for Eq.(21.29) is a straight line because the change in phase angle will be 45° for every tenfold change in frequency.

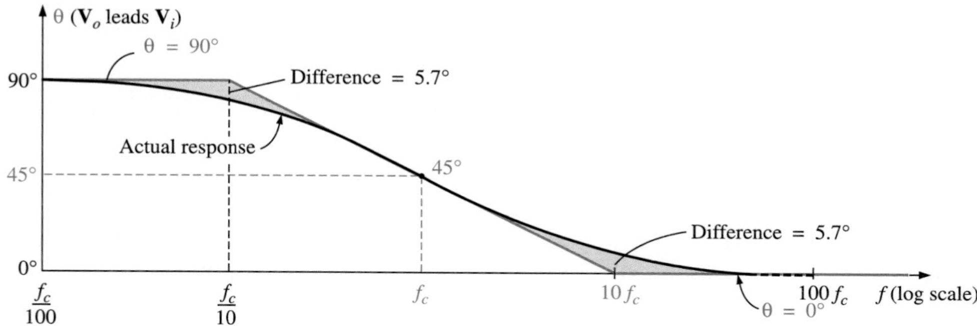

FIG. 21.44
Phase response for a high-pass R-C filter.

Substituting $f = f_c/10$ into Eq. (21.29),

$$\theta = \tan^{-1}\left(\frac{f_c}{f_c/10}\right) = \tan^{-1} 10 = \mathbf{84.29°}$$

for a difference of $90° - 84.29° \cong 5.7°$ from the idealized response.

Substituting $f = 10f_c$

$$\theta = \tan^{-1}\left(\frac{f_c}{10f_c}\right) = \tan^{-1}\frac{1}{10} \cong \mathbf{5.7°}$$

In summary, therefore,

at $f = f_c$, $\theta = 45°$, whereas at $f = f_c/10$ and $10f_c$, the difference between the actual phase response and the asymptotic plot is 5.7°.

EXAMPLE 21.8

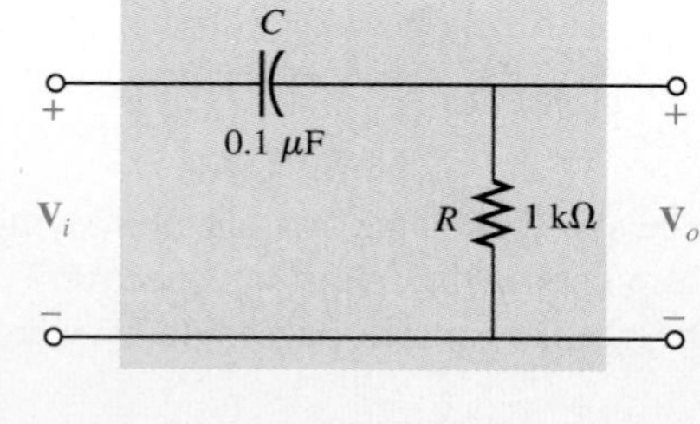

FIG. 21.45
Example 21.8.

a. Sketch $A_{v_{dB}}$ versus frequency for the high-pass *R-C* filter of Fig. 21.45.
b. Determine the decibel level at $f = 1$ kHz.
c. Sketch the phase response versus frequency on a log scale.

Solutions:

a. $$f_c = \frac{1}{2\pi RC} = \frac{1}{(2\pi)(1\text{ k}\Omega)(0.1\ \mu\text{F})} = 1591.55\text{ Hz}$$

The frequency f_c is identified on the log scale as shown in Fig. 21.46. A straight line is then drawn from f_c with a slope that will intersect -20 dB at $f_c/10 = 159.15$ Hz or -6 dB at $f_c/2 = 795.77$ Hz. A second asymptote is drawn from f_c to higher frequencies at 0 dB. The actual response curve can then be drawn through the -3-dB level at f_c approaching the two asymptotes of Fig. 21.46. Note the 1-dB difference between the actual response and the idealized Bode plot at $f = 2f_c$ and $0.5f_c$.

continued

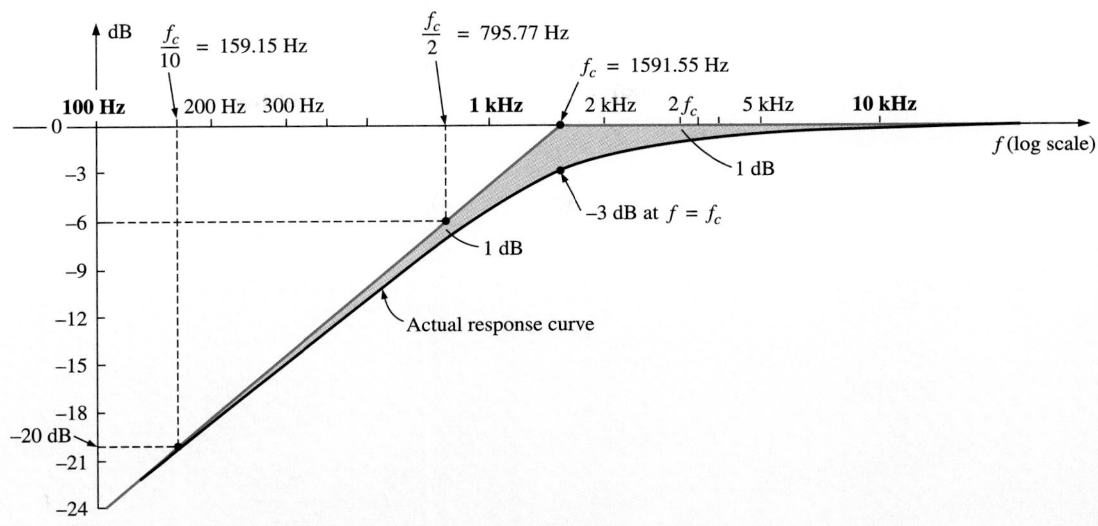

FIG. 21.46
Frequency response for the high-pass filter of Fig. 21.45.

Note in the preceding solution that there was no need to use Eq. (21.28) or to perform any extensive mathematical manipulations.

b. Eq. (21.26):

$$|A_{v\text{dB}}| = 20 \log_{10} \frac{1}{\sqrt{1 + \left(\frac{f_c}{f}\right)^2}} = 20 \log_{10} \frac{1}{\sqrt{1 + \left(\frac{1591.55 \text{ Hz}}{1000}\right)^2}}$$

$$= 20 \log_{10} \frac{1}{\sqrt{1 + (1.592)^2}} = 20 \log_{10} 0.5318 = \mathbf{-5.49\ dB}$$

as verified by Fig. 21.46.

c. See Fig. 21.47. Note that $\theta = 45°$ at $f = f_c = 1591.55$ Hz, and the difference between the straight-line segment and the actual response is 5.7° at $f = f_c/10 = 159.2$ Hz and $f = 10f_c = 15\,923.6$ Hz.

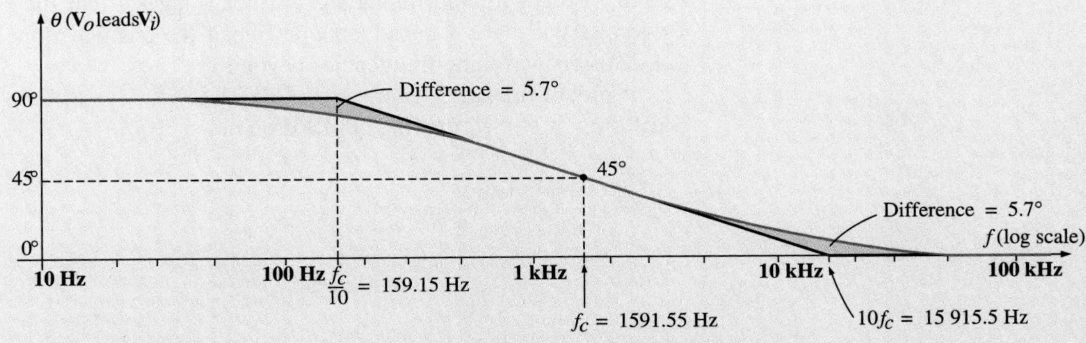

FIG. 21.47
Phase plot for the high-pass R-C filter.

Low-Pass *R-C* Filter

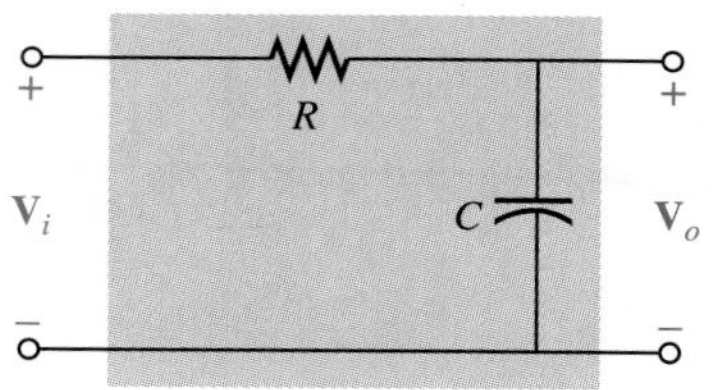

FIG. 21.48
Low-pass filter.

For the low-pass filter of Fig. 21.48,

$$\mathbf{A}_v = \frac{\mathbf{V}_o}{\mathbf{V}_i} = \frac{-jX_C}{R - jX_C} = \frac{1}{\dfrac{R}{-jX_C} + 1}$$

$$= \frac{1}{1 + j\dfrac{R}{X_C}} = \frac{1}{1 + j\dfrac{R}{\dfrac{1}{2\pi fC}}} = \frac{1}{1 + j\dfrac{f}{\dfrac{1}{2\pi RC}}}$$

and

$$\mathbf{A}_v = \frac{1}{1 + j\,(f/f_c)} \tag{21.30}$$

with

$$f_c = \frac{1}{2\pi RC} \tag{21.31}$$

as defined earlier.

Note that now the sign of the imaginary component in the denominator is positive and f_c appears in the denominator of the frequency ratio rather than in the numerator, as in the case of f_c for the high-pass filter.

In terms of magnitude and phase,

$$\mathbf{A}_v = \frac{\mathbf{V}_o}{\mathbf{V}_i} = A_v \angle\theta = \frac{1}{\sqrt{1 + (f/f_c)^2}} \angle -\tan^{-1}(f/f_c) \tag{21.32}$$

An analysis similar to that performed for the high-pass filter will result in

$$A_{v_{\text{dB}}} = -20\log_{10}\frac{f}{f_c} \qquad (f \gg f_c) \tag{21.33}$$

Note in particular that the equation is exact only for frequencies much greater than f_c, but a plot of Eq. (21.33) does provide an asymptote that performs the same function as the asymptote derived for the high-pass filter. In addition, note that it is exactly the same as Eq. (21.28), except for the minus sign, which suggests that the resulting Bode plot will have a negative slope [recall the positive slope for Eq. (21.28)] for increasing frequencies beyond f_c.

A plot of Eq. (21.33) appears in Fig. 21.49 for $f_c = 1$ kHz. Note the 6-dB drop at $f = 2f_c$ and the 20-dB drop at $f = 10f_c$.

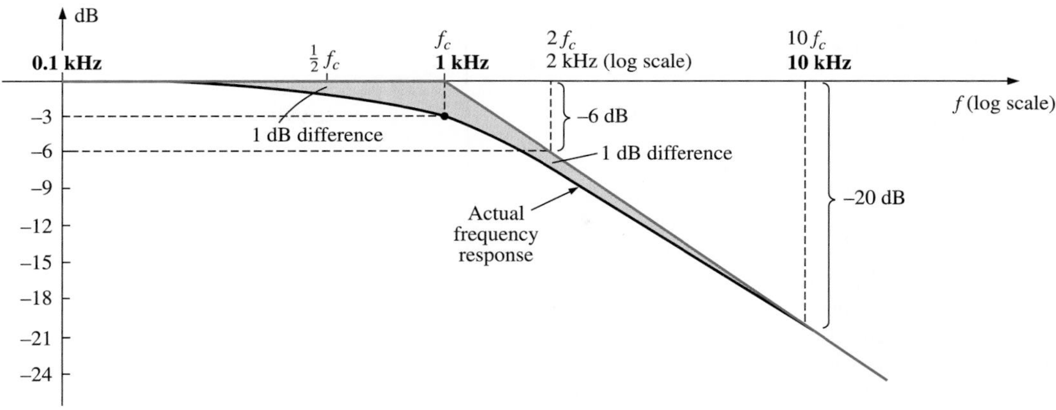

FIG. 21.49
Bode plot for the high-frequency region of a low-pass R-C filter.

At $f \gg f_c$, the phase angle $\theta = -\tan^{-1}(f/f_c)$ approaches $-90°$, whereas for frequencies $f \ll f_c$, $\theta = -\tan^{-1}(f/f_c)$ approaches 0°. At $f = f_c$, $\theta = -\tan^{-1} 1 = -45°$, establishing the plot of Fig. 21.50. Note again the 45° change in phase angle for each tenfold increase in frequency.

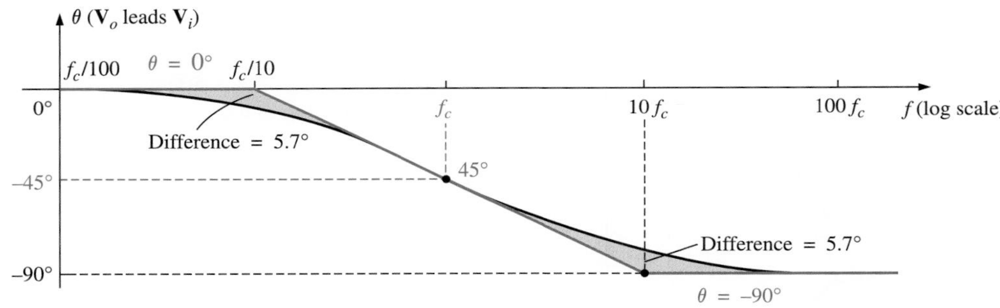

FIG. 21.50
Phase plot for a low-pass R-C filter.

Even though the preceding analysis relates only to the *R-C* combination, the results obtained will apply to networks that are much more complicated. One good example is the high- and low-frequency response of a standard transistor configuration. There are some capacitive elements in a practical transistor network that will affect the low-frequency response and others that will affect the high-frequency response. In the absence of the capacitive elements, the frequency response of a transistor would ideally stay level at the midband value. However, the coupling capacitors at low frequencies and the bypass and parasitic capacitors at high frequencies will define a bandwidth for numerous transistor configurations. In the low-frequency region, specific capacitors and resistors will form an *R-C* combination that will define a low cutoff frequency. There are then other elements and capacitors forming a second *R-C* combination that will define a high cutoff frequency. Once the cutoff frequencies are known, the -3-dB points are set and the bandwidth of the system can be determined.

Practical Business Applications

Rock on, dude!

Ever been to a rock concert to catch a great band only to find the sound system was lousy? Well, it's not necessarily that the sound system is bad—the sound technician may not have done the job well. You might become one of these sound technicians once you graduate from college. You've got to make sure you use the knowledge you've gained so that others don't have to put up with a lousy rock concert too.

So why was the sound so bad? First, what that sound technician should have done is characterize the concert hall, that is, do a thorough sound check. Technically, this is a sound absorption test to see what kind of frequency response the concert hall has.

There are a number of different ways to generate the frequency response of a rock hall. If you're giving a concert in a renowned place, the rock-hall management will probably already have a recently updated one. The bare minimum is one that was done when the concert hall was first constructed. The rock-hall frequency response might look something like the blue curve in the graph shown.

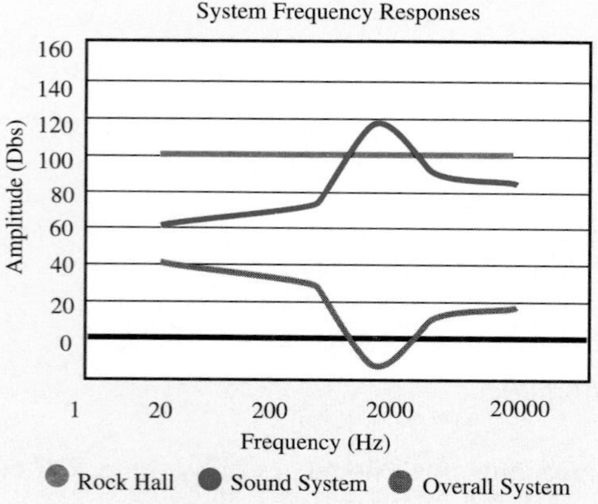

Your job is to match the rock-hall frequency response curve with a suitable sound-system frequency response curve like the one shown in green on the graph. The result will be the overall frequency response curve shown in red on the graph. You can probably see that this is just the superposition of the rock-hall and the sound-system frequency responses. So, your goal is to adjust the sound system somehow to produce this frequency response characteristic. But how do you do that?

The answer is to use an adjustable multiple bandpass/bandstop filter or, in other words, an audio graphic equalizer. You can buy these with a wide variety of features and controls but usually you'll get at least four centre frequencies with a range of probably 40 dB amplitude adjustment. Suppose this is the type of graphic equalizer that we have.

So, to produce the sound-system frequency response shown in green, we pass the rock band's audio output through the graphic equalizer. The various different frequency bands amplify (add amplitude to) or attenuate (remove amplitude from) the rock band's audio output to produce the flat red overall system frequency response. The following table shows this:

		Frequency Response (dB)		
Band (Hz)	Frequency	*Blue Curve* Rock Hall	*Red Curve* Sound System	*Green Curve* Overall System
1	20	60	40	100
2	200	70	30	100
3	2 000	110	–10	100
4	20 000	80	20	100

Therefore, to get the sound system frequency response shown in green, we need a graphic equalizer characteristic that looks something like this:

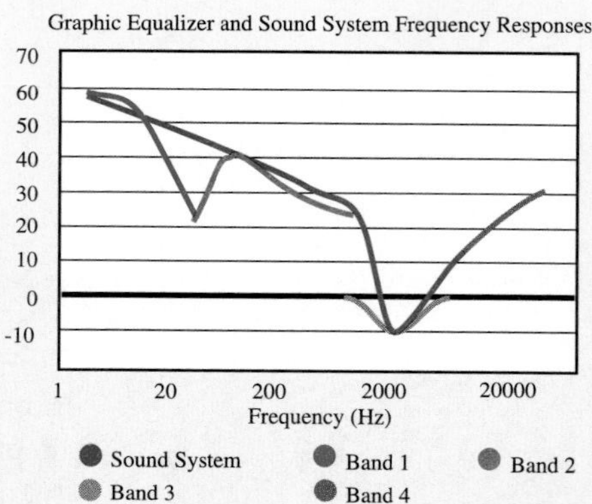

Looking at this graph, we see that each band contributes a small range of frequencies to the overall sound system frequency response. Band 1 is just a simple lowpass filter with a cutoff frequency around 10 Hz. Band 2 is a bandpass filter providing about 30 dB amplification centred around 200 Hz. Band 3 is a bandstop filter attenuating 10 dB at around 2000 Hz. Finally, band 4 is a highpass filter with cutoff frequency around 20 000 Hz.

Together, they flatten the rock-hall frequency response so that when you listen to the rock concert, it's music to your ears. Literally!

PROBLEMS

SECTION 21.2 Decibels

1. **a.** Determine the number of bels that relate power levels of $P_2 = 280$ mW and $P_1 = 4$ mW.
   b. Determine the number of decibels for the power levels of part (a) and compare results.
2. A power level of 100 W is 6 dB above what power level?
3. If a 2-W speaker is replaced by one with a 40-W output, what is the increase in decibel level?
4. Determine the dB_m level for an output power of 120 mW.
5. Find the dB_v gain of an amplifier that raises the voltage level from 0.1 mV to 8.4 V.
6. Find the output voltage of an amplifier if the applied voltage is 20 mV and a dB_v gain of 22 dB is attained.
7. If the sound pressure level is increased from 0.001 μbar to 0.016 μbar, what is the increase in dB_s level?
8. What is the required increase in acoustical power to raise a sound level from that of quiet music to very loud music? Use Fig. 21.2.
9. **a.** Using semilog paper, plot X_L versus frequency for a 10-mH coil and a frequency range of 100 Hz to 1 MHz. Choose the best vertical scaling for the range of X_L.
   b. Repeat part (a) using log-log graph paper. Compare to the results of part (a). Which plot is more informative?
   c. Using semilog paper, plot X_C versus frequency for a 1-μF capacitor and a frequency range of 10 Hz to 100 kHz. Again choose the best vertical scaling for the range of X_C.
   d. Repeat part (a) using log-log graph paper. Compare to the results of part (c). Which plot is more informative?
10. **a.** For the meter of Fig. 21.3, find the power delivered to a load for an 8-dB reading.
    b. Repeat part (a) for a −5-dB reading.

SECTION 21.4 *R-C* Low-Pass Filter

11. For the *R-C* low-pass filter of Fig. 21.51:
    a. Sketch $A_v = V_o/V_i$ versus frequency using a log scale for the frequency axis. Determine $A_v = V_o/V_i$ at $0.1f_c$, $0.5f_c$, f_c, $2f_c$, and $10f_c$.
    b. Sketch the phase plot of θ versus frequency, where θ is the angle by which $\mathbf{V}_o$ leads $\mathbf{V}_i$. Determine θ at $f = 0.1f_c$, $0.5f_c$, f_c, $2f_c$, and $10f_c$.

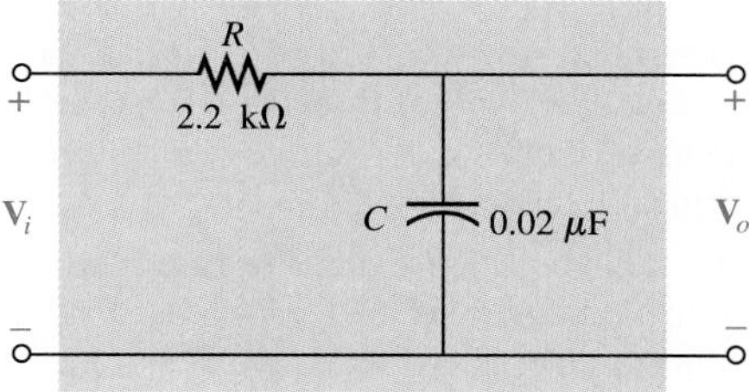

FIG. 21.51
Problem 11.

*12. For the network of Fig. 21.52:
  a. Determine V_o at a frequency one octave above the critical frequency.
  b. Determine V_o at a frequency one decade below the critical frequency.
  c. Do the levels of parts (a) and (b) verify the expected frequency plot of V_o versus frequency for the filter?

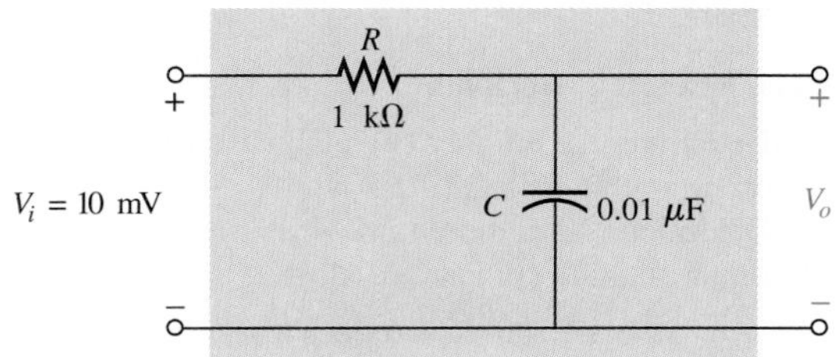

FIG. 21.52
Problem 12.

13. Design an *R-C* low-pass filter to have a cutoff frequency of 500 Hz using a resistor of 1.2 kΩ. Then sketch the resulting magnitude and phase plot for a frequency range of $0.1f_c$ to $10f_c$.

14. For the low-pass filter of Fig. 21.53:
  a. Determine f_c.
  b. Find $A_v = V_o/V_i$ at $f = 0.1f_c$ and compare to the maximum value of 1 for the low-frequency range.
  c. Find $A_v = V_o/V_i$ at $f = 10f_c$ and compare to the minimum value of 0 for the high-frequency range.
  d. Determine the frequency at which $A_v = 0.01$ or $V_o = \frac{1}{100} V_i$.

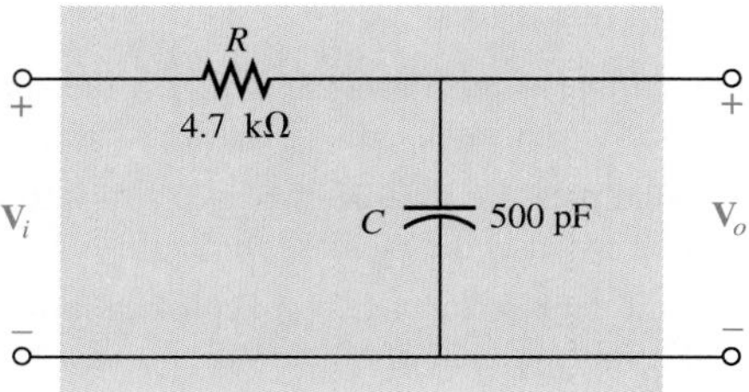

FIG. 21.53
Problem 14.

SECTION 21.5 *R-C* High-Pass Filter

15. For the *R-C* high-pass filter of Fig. 21.54:
  a. Sketch $A_v = V_o/V_i$ versus frequency using a log scale for the frequency axis. Determine $A_v = V_o/V_i$ at f_c, one octave above and below f_c, and one decade above and below f_c.
  b. Sketch the phase plot of θ versus frequency, where θ is the angle by which $\mathbf{V}_o$ leads $\mathbf{V}_i$. Determine θ at the same frequencies noted in part (a).

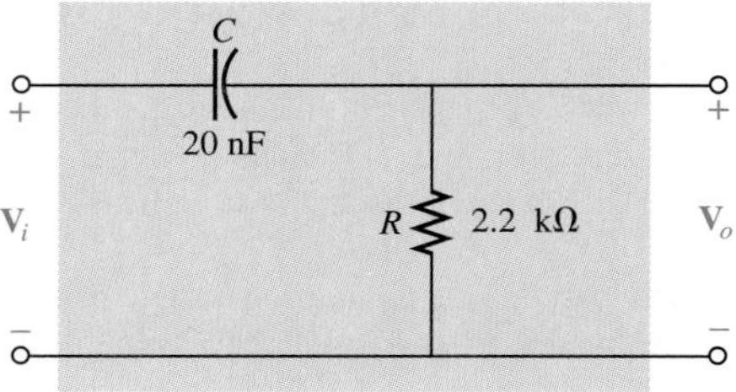

FIG. 21.54
Problem 15.

16. For the network of Fig. 21.55:
  a. Determine $A_v = V_o/V_i$ at $f = f_c$ for the high-pass filter.
  b. Determine $A_v = V_o/V_i$ at two octaves above f_c. Is the rise in V_o significant from the $f = f_c$ level?
  c. Determine $A_v = V_o/V_i$ at two decades above f_c. Is the rise in V_o significant from the $f = f_c$ level?
  d. If $V_i = 10$ mV, what is the power delivered to R at the critical frequency?

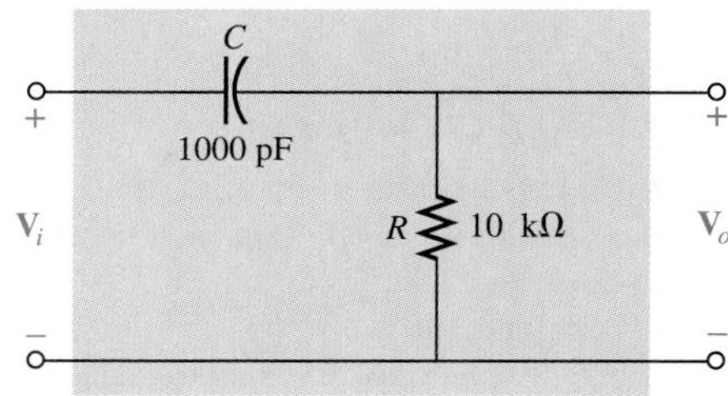

FIG. 21.55
Problem 16.

17. Design a high-pass *R-C* filter to have a cutoff or corner frequency of 2 kHz, given a capacitor of 0.1 μF. Choose the closest commercial value for R and then recalculate the resulting corner frequency. Sketch the normalized gain $A_v = V_o/V_i$ for a frequency range of $0.1f_c$ to $10f_c$.

18. For the high-pass filter of Fig. 21.56:

a. Determine f_c.

b. Find $A_v = V_o/V_i$ at $f = 0.01f_c$ and compare to the minimum level of 0 for the low-frequency region.

c. Find $A_v = V_o/V_i$ at $f = 100f_c$ and compare to the maximum level of 1 for the high-frequency region.

d. Determine the frequency at which $V_o = \frac{1}{2}V_i$.

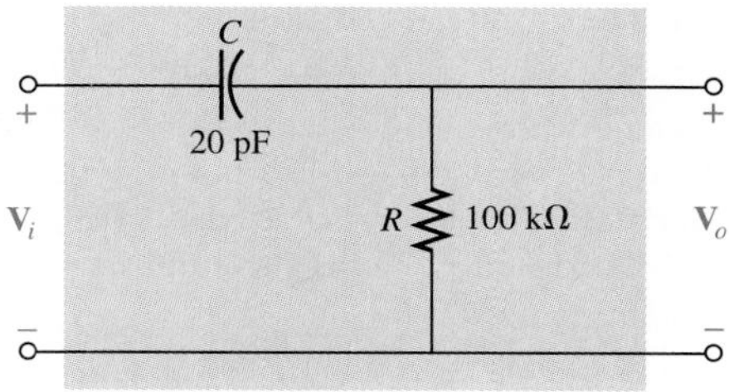

FIG. 21.56
Problem 18.

SECTION 21.6 Pass-Band Filters

19. For the pass-band filter of Fig. 21.57:

a. Sketch the frequency response of $A_v = V_o/V_i$ against a log scale extending from 10 Hz to 10 kHz.

b. What are the bandwidth and the centre frequency?

***20.** Design a pass-band filter such as the one in Fig. 21.57 to have a low cutoff frequency of 4 kHz and a high cutoff frequency of 80 kHz.

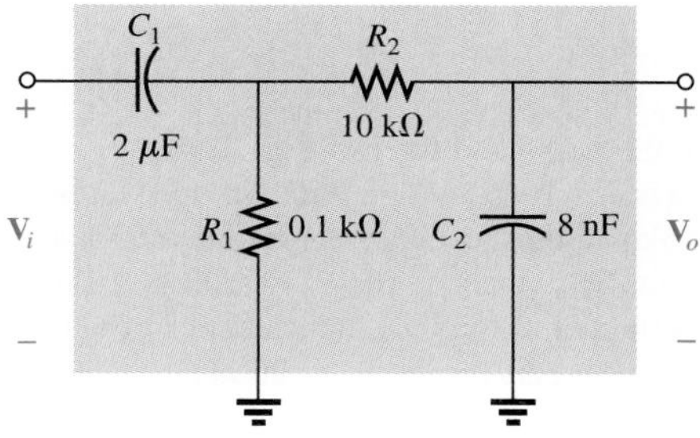

FIG. 21.57
Problem 19 and 20.

21. For the pass-band filter of Fig. 21.58:

a. Determine f_s.

b. Calculate Q_s and the bandwidth for $\mathbf{V}_o$.

c. Sketch $A_v = V_o/V_i$ for a frequency range of 1 kHz to 1 MHz.

d. Find the magnitude of V_o at $f = f_s$ and the cutoff frequencies.

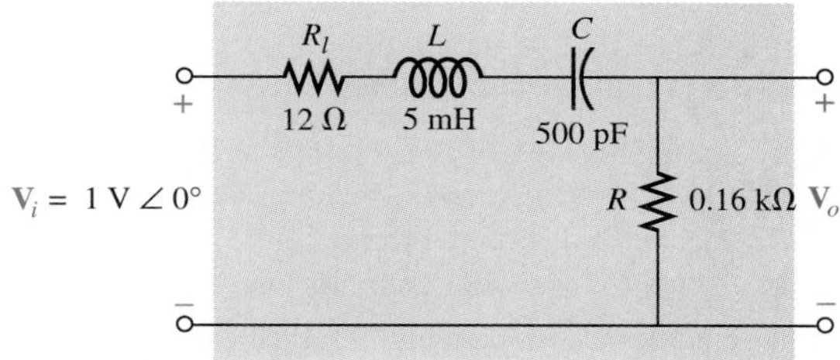

FIG. 21.58
Problem 21.

22. For the pass-band filter of Fig. 21.59:

a. Determine the frequency response of $A_v = V_o/V_i$ for a frequency range of 100 Hz to 1 MHz.

b. Find the quality factor Q_p and the bandwidth of the response.

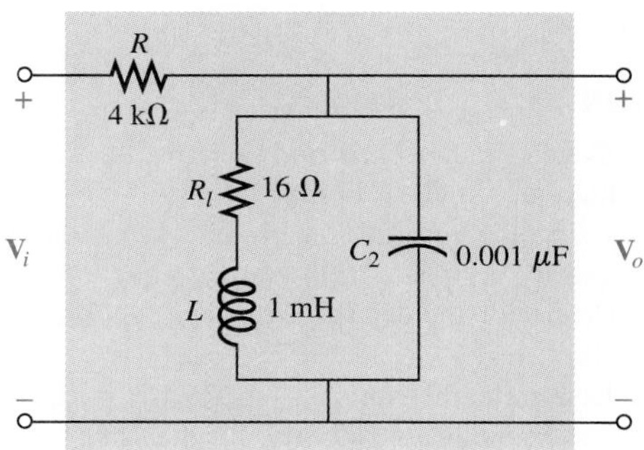

FIG. 21.59
Problem 22.

SECTION 21.7 Stop-Band Filters

*23. For the stop-band filter of Fig. 21.60:
  a. Determine Q_s.
  b. Find the bandwidth and half-power frequencies.
  c. Sketch the frequency characteristics of $A_v = V_o/V_i$.
  d. What is the effect on the curve of part (c) if a load of 2 kΩ is applied?

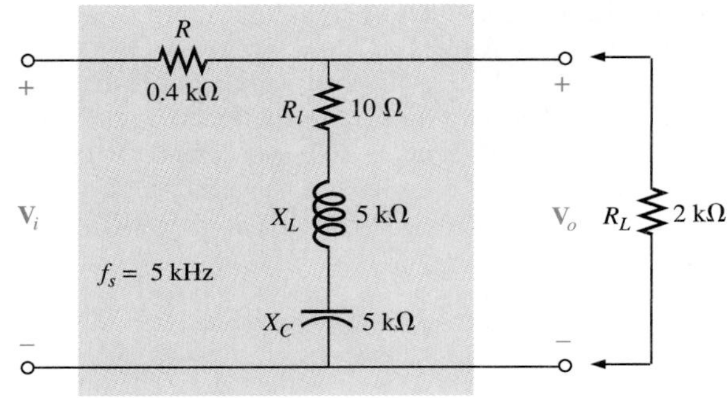

FIG. 21.60
Problem 23.

*24. For the pass-band filter of Fig. 21.61:
  a. Determine Q_p ($R_L = \infty$ Ω, an open circuit).
  b. Sketch the frequency characteristics of $A_v = V_o/V_i$.
  c. Find Q_p (loaded) for R_L = 100 kΩ, and indicate the effect of R_L on the characteristics of part (b).
  d. Repeat part (c) for R_L = 20 kΩ.

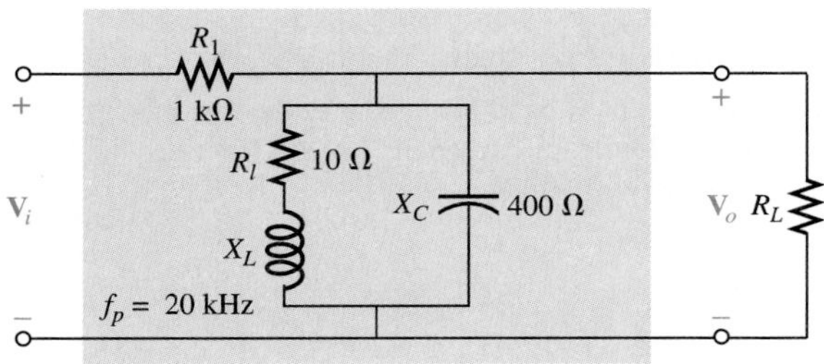

FIG. 21.61
Problem 24.

SECTION 21.8 Double-Tuned Filter

25. a. For the network of Fig. 21.40(a), if L_p = 400 μH ($Q >$ 10), L_s = 60 μH, and C = 120 pF, determine the rejected and accepted frequencies.
  b. Sketch the response curve for part (a).

26. a. For the network of Fig. 21.40(b), if the rejected frequency is 30 kHz and the accepted is 100 kHz, determine the values of L_s and L_p ($Q >$ 10) for a capacitance of 200 pF.
  b. Sketch the response curve for part (a).

SECTION 21.9 Bode Plots

27. a. Sketch the idealized Bode plot for $A_v = V_o/V_i$ for the high-pass filter of Fig. 21.62.
  b. Using the results of part (a), sketch the actual frequency response for the same frequency range.
  c. Determine the decibel level at f_c, $\frac{1}{2}f_c$, $2f_c$, $\frac{1}{10}f_c$, and $10f_c$.
  d. Determine the gain $A_v = V_o/V_i$ as $f = f_c$, $\frac{1}{2}f_c$, and $2f_c$.
  e. Sketch the phase response for the same frequency range.

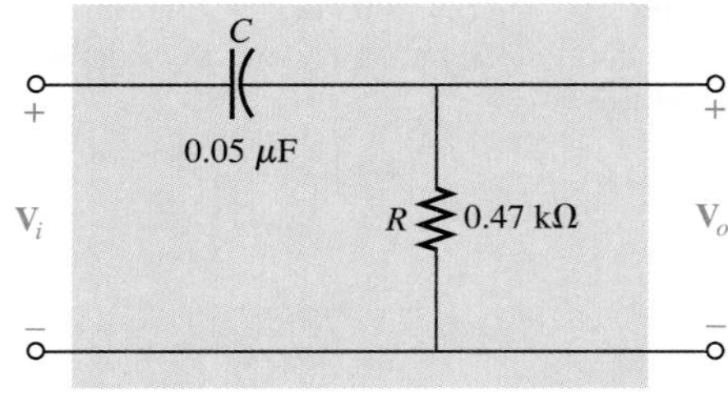

FIG. 21.62
Problem 27.

*28. **a.** Sketch the response of the magnitude of V_o (in terms of V_i) versus frequency for the high-pass filter of Fig. 21.63.

b. Using the results of part (a), sketch the response $A_v = V_o/V_i$ for the same frequency range.

c. Sketch the idealized Bode plot.

d. Sketch the actual response, indicating the dB difference between the idealized and actual response at $f = f_c$, $0.5f_c$, and $2f_c$.

e. Determine $A_{v_{dB}}$ at $f = 1.5f_c$ from the plot of part (d) and then determine the corresponding magnitude of $A_v = V_o/V_i$.

f. Sketch the phase response for the same frequency range (the angle by which $\mathbf{V}_o$ leads $\mathbf{V}_i$).

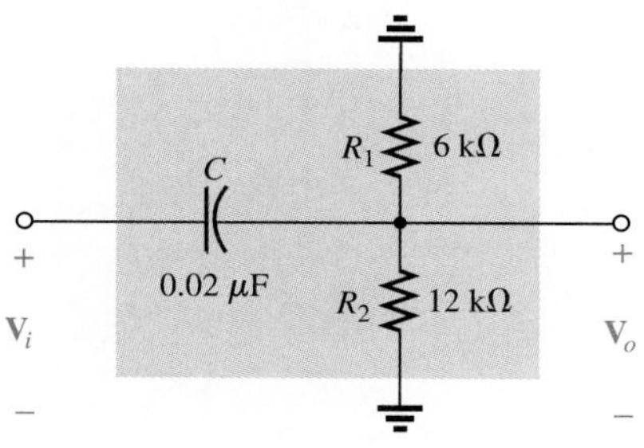

FIG. 21.63
Problem 28.

29. **a.** Sketch the idealized Bode plot for $A_v = V_o/V_i$ for the low-pass filter of Fig. 21.64.

b. Using the results of part (a), sketch the actual frequency response for the same frequency range.

c. Determine the decibel level at f_c, $\frac{1}{2}f_c$, $2f_c$, $\frac{1}{10}f_c$, and $10f_c$.

d. Determine the gain $A_v = V_o/V_i$ at $f = f_c$, $\frac{1}{2}f_c$, and $2f_c$.

e. Sketch the phase response for the same frequency range.

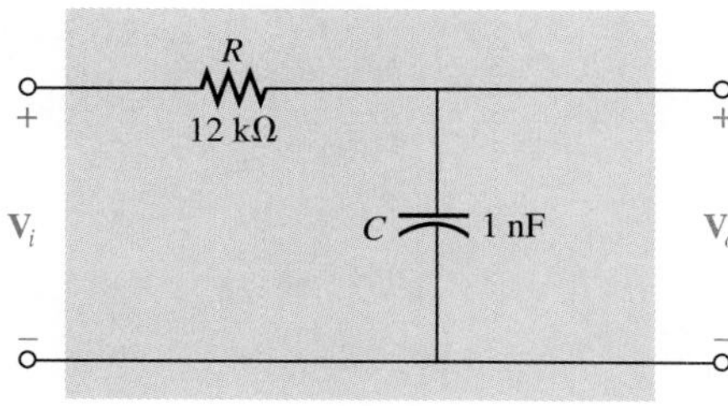

FIG. 21.64
Problem 29.

*30. **a.** Sketch the response of the magnitude of V_o (in terms of V_i) versus frequency for the low-pass filter of Fig. 21.65.

b. Using the results of part (a), sketch the response $A_v = V_o/V_i$ for the same frequency range.

c. Sketch the idealized Bode plot.

d. Sketch the actual response indicating the dB difference between the idealized and actual response at $f = f_c$, $0.5f_c$, and $2f_c$.

e. Determine $A_{v_{dB}}$ at $f = 0.25f_c$ from the plot of part (d) and then determine the corresponding magnitude of $A_v = V_o/V_i$.

f. Sketch the phase response for the same frequency range (the angle by which $\mathbf{V}_o$ leads $\mathbf{V}_i$).

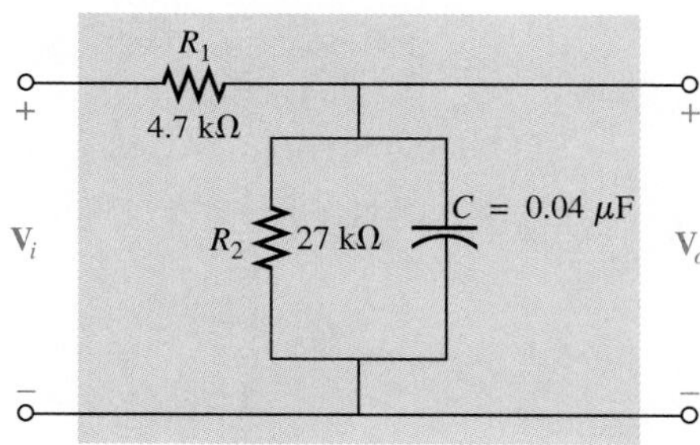

FIG. 21.65
Problem 30.

22

Transformers

Outline

Learning Outcomes

After completing this chapter you will be able to

- describe mutual inductance and calculate the effect of coupled flux on total inductance and induced voltage.
- determine the polarity of induced voltage using the transformer dot convention.
- calculate the primary and secondary voltages using the transformer turns ratio.
- calculate the impedance transferred from one side of a transformer to the other and use it to match impedances.
- model the equivalent circuit of a practical iron core transformer and calculate the effect of non-ideal characteristics on voltage transfer.
- describe the different types of transformers and their applications.

This chapter shows how transformers are able to change voltage levels, isolate circuits from one another, and change the value of impedances.

22.1 INTRODUCTION

Chapter 11 discussed the factors that affect flux developed in a magnetic circuit by a coil, and Chapter 12 discussed the self inductance of a coil. We shall now examine the mutual flux and mutual inductance that exists between coils of the same or different dimensions. Mutual inductance is a phenomenon basic to the operation of the *transformer,* an electrical device used today in almost every field of electrical engineering. This device plays an integral part in power distribution systems and can be found in many electronic circuits and measuring instruments. In this chapter, we will discuss three of the basic applications of a transformer: to build up or step down the voltage or current, to act as an impedance matching device, and to isolate (no physical connection) one portion of a circuit from another. In addition, we will introduce the **dot convention** and will consider the transformer equivalent circuit. The chapter will conclude with a word about writing mesh equations for a network with mutual inductance.

22.2 MUTUAL INDUCTANCE

A transformer is constructed of two coils placed so that the changing flux developed by one will link the other, as shown in Fig. 22.1. This

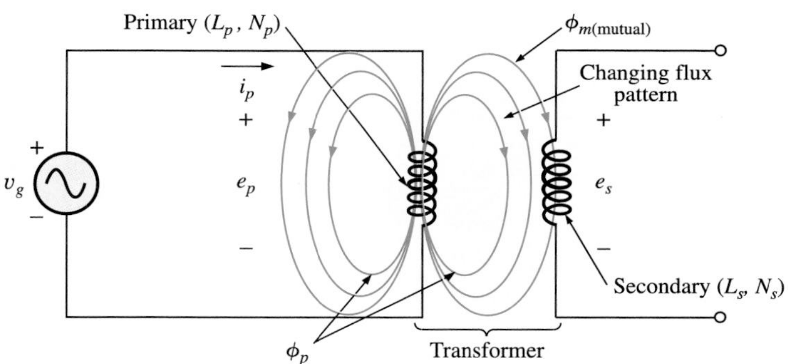

FIG. 22.1
Defining the components of a transformer.

will result in an induced voltage across each coil. To distinguish between the coils, we will apply the transformer convention that

***the coil to which the source is applied is called the* primary*, and the coil to which the load is applied is called the* secondary.**

For the primary of the transformer of Fig. 22.1, an application of Faraday's law [Eq. (12.1)] will result in

$$e_p = N_p \frac{d\phi_p}{dt} \quad \text{(volts, V)} \tag{22.1}$$

revealing that the voltage induced across the primary is directly related to the number of turns in the primary and the rate of change of magnetic flux linking the primary coil. Or, from Eq. (12.5),

$$e_p = L_p \frac{di_p}{dt} \quad \text{(volts, V)} \tag{22.2}$$

revealing that the induced voltage across the primary is also directly related to the self-inductance of the primary and the rate of change of current through the primary winding.

The magnitude of e_s, the voltage induced across the secondary, is determined by

$$e_s = N_s \frac{d\phi_m}{dt} \quad \text{(volts, V)} \tag{22.3}$$

where N_s is the number of turns in the secondary winding and ϕ_m is the portion of the primary flux ϕ_p that links the secondary winding.

If all of the flux linking the primary links the secondary, then

$$\phi_m = \phi_p$$

and

$$e_s = N_s \frac{d\phi_p}{dt} \quad \text{(volts, V)} \tag{22.4}$$

The **coefficient of coupling** (k) between two coils is determined by

$$k \text{ (coefficient of coupling)} = \frac{\phi_m}{\phi_p} \tag{22.5}$$

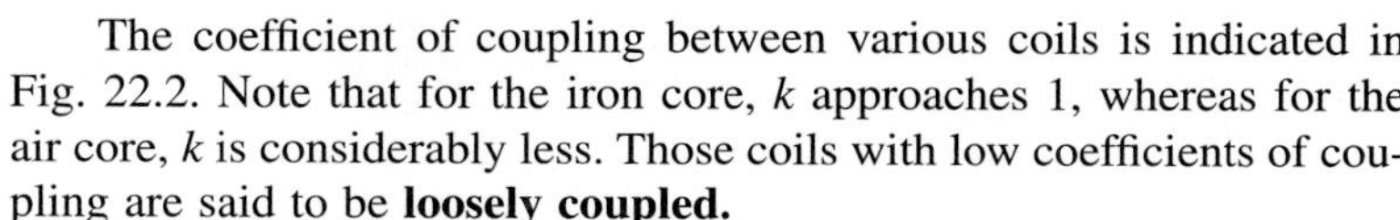

Since the maximum level of ϕ_m is ϕ_p, the coefficient of coupling between two coils can never be greater than 1.

The coefficient of coupling between various coils is indicated in Fig. 22.2. Note that for the iron core, k approaches 1, whereas for the air core, k is considerably less. Those coils with low coefficients of coupling are said to be **loosely coupled.**

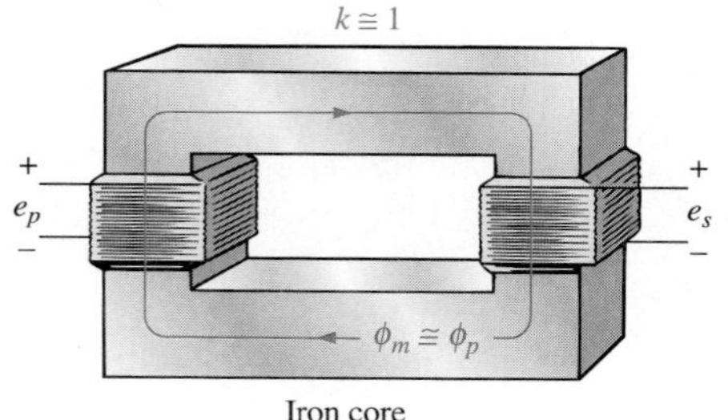

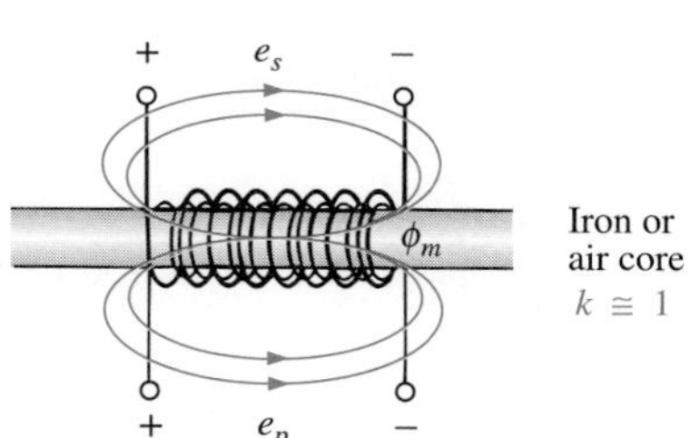

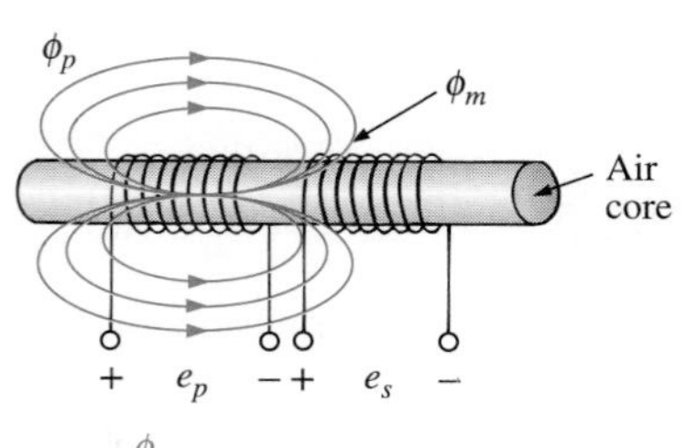

FIG. 22.2
Windings having different coefficients of coupling.

For the secondary, we have

$$e_s = N_s \frac{d\phi_m}{dt} = N_s \frac{dk\phi_p}{dt}$$

and

$$e_s = kN_s \frac{d\phi_p}{dt} \quad \text{(volts, V)} \tag{22.6}$$

The mutual inductance between the two coils of Fig. 22.1 is determined by

$$M = N_s \frac{d\phi_m}{di_p} \quad \text{(henries, H)} \tag{22.7}$$

or

$$M = N_p \frac{d\phi_p}{di_s} \quad \text{(henries, H)} \tag{22.8}$$

Note in the above equations that the symbol for mutual inductance is the capital letter M and that its unit of measurement, like that of self-inductance, is the *henry.* In words, Equations (22.7) and (22.8) state that the

mutual inductance ***between two coils is proportional to the instantaneous change in flux linking one coil due to an instantaneous change in current through the other coil.***

In terms of the inductance of each coil and the coefficient of coupling, the mutual inductance is determined by

$$M = k\sqrt{L_p L_s} \quad \text{(henries, H)} \tag{22.9}$$

The greater the coefficient of coupling (greater flux linkages), or the greater the inductance of either coil, the higher the mutual inductance between the coils. Relate this fact to the configurations of Fig. 22.2.

The secondary voltage e_s can also be found in terms of the mutual inductance if we rewrite Eq. (22.3) as

$$e_s = N_s\left(\frac{d\phi_m}{di_p}\right)\left(\frac{di_p}{dt}\right)$$

and, since $M = N_s(d\phi_m/di_p)$, it can also be written

$$e_s = M\frac{di_p}{dt} \quad \text{(volts, V)} \tag{22.10}$$

Similarly,

$$e_p = M\frac{di_s}{dt} \quad \text{(volts, V)} \tag{22.11}$$

EXAMPLE 22.1

For the transformer in Fig. 22.3:

a. Find the mutual inductance M.
b. Find the induced voltage e_p if the flux ϕ_p changes at the rate of 450 mWb/s.
c. Find the induced voltage e_s for the same rate of change indicated in part (b).
d. Find the induced voltages e_p and e_s if the current i_p changes at the rate of 0.2 A/ms.

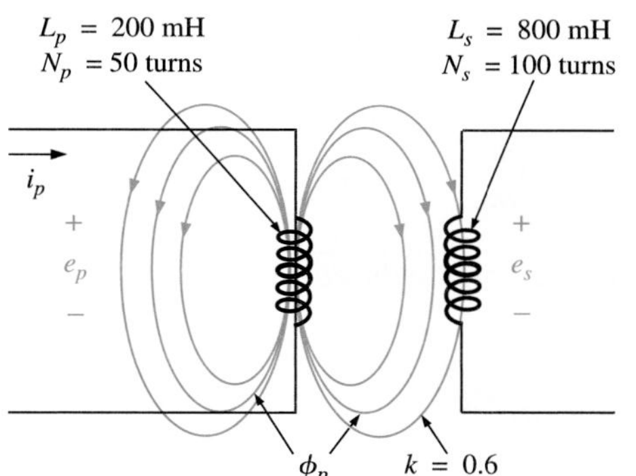

FIG. 22.3
Example 22.1.

Solution:

a. $M = k\sqrt{L_p L_s} = 0.6\sqrt{(200 \text{ mH})(800 \text{ mH})}$
$= 0.6\sqrt{16 \times 10^{-2}} = (0.6)(400 \times 10^{-3}) = \mathbf{240\ mH}$

b. $e_p = N_p\frac{d\phi_p}{dt} = (50)(450 \text{ mWb/s}) = \mathbf{22.5\ V}$

c. $e_s = kN_s\frac{d\phi_p}{dt} = (0.6)(100)(450 \text{ mWb/s}) = \mathbf{27\ V}$

d. $e_p = L_p\frac{di_p}{dt} = (200 \text{ mH})(0.2 \text{ A/ms}) = (200 \text{ mH})(200 \text{ A/s}) = \mathbf{40\ V}$

$e_s = M\frac{di_p}{dt} = (240 \text{ mH})(200 \text{ A/s}) = \mathbf{48\ V}$

22.3 SERIES CONNECTION OF MUTUALLY COUPLED COILS

In Chapter 12, we found that the total inductance of series isolated coils was determined simply by the sum of the inductances. For two coils

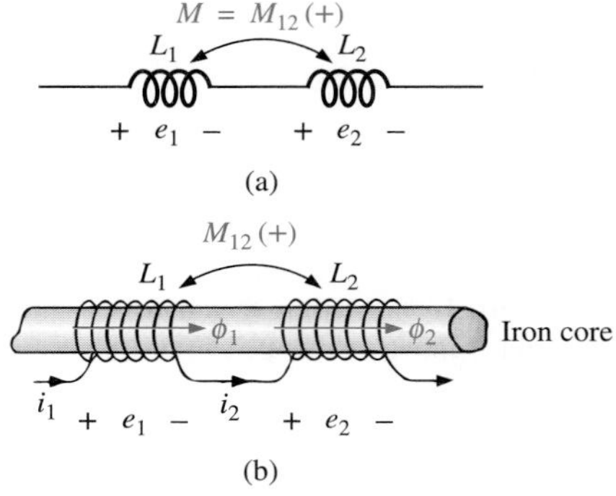

FIG. 22.4
Mutually coupled coils connected in series.

that are connected in series but also share the same flux linkages, such as those in Fig. 22.4(a), a mutual term is introduced that will alter the total inductance of the series combination. The physical picture of how the coils are connected is indicated in Fig. 22.4(b). An iron core is included, although the equations to be developed are for any two mutually coupled coils with any value of coefficient of coupling k. When referring to the voltage induced across the inductance L_1 (or L_2) due to the change in flux linkages of the inductance L_2 (or L_1, respectively), the mutual inductance is represented by M_{12}. This type of subscript notation is particularly important when there are two or more mutual terms.

Due to the presence of the mutual term, the induced voltage e_1 is composed of that due to the self-inductance L_1 and that due to the mutual inductance M_{12}. That is,

$$e_1 = L_1 \frac{di_1}{dt} + M_{12} \frac{di_2}{dt}$$

However, since $i_1 = i_2 = i$,

$$e_1 = L_1 \frac{di}{dt} + M_{12} \frac{di}{dt}$$

or

$$e_1 = (L_1 + M_{12})\frac{di}{dt} \quad \text{(volts, V)} \tag{22.12}$$

and, similarly,

$$e_2 = (L_2 + M_{12})\frac{di}{dt} \quad \text{(volts, V)} \tag{22.13}$$

For the series connection, the total induced voltage across the series coils, represented by e_T, is

$$e_T = e_1 + e_2 = (L_1 + M_{12})\frac{di}{dt} + (L_2 + M_{12})\frac{di}{dt}$$

or

$$e_T = (L_1 + L_2 + M_{12} + M_{12})\frac{di}{dt}$$

and the total effective inductance is

$$L_{T(+)} = L_1 + L_2 + 2M_{12} \quad \text{(henries, H)} \tag{22.14}$$

The subscript (+) was included to indicate that the mutual terms have a positive sign and are added to the self-inductance values to determine the total inductance. If the coils were wound such as shown in Fig. 22.5, where ϕ_1 and ϕ_2 are in opposition, the induced voltages due to the mutual terms would oppose that due to the self-inductance, and the total inductance would be determined by

$$L_{T(-)} = L_1 + L_2 - 2M_{12} \quad \text{(henries, H)} \tag{22.15}$$

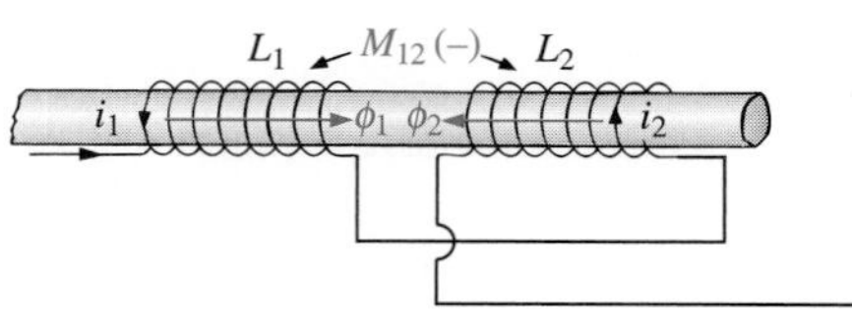

FIG. 22.5
Mutually coupled coils connected in series with negative mutual inductance.

Through Eqs. (22.14) and (22.15), the mutual inductance can be determined by

$$M_{12} = \frac{1}{4}(L_{T(+)} - L_{T(-)}) \qquad (22.16)$$

Equation (22.16) is very effective in determining the mutual inductance between two coils. It states that the mutual inductance is equal to one-quarter the difference between the total inductance with a positive and negative mutual effect.

From the preceding, it should be clear that the mutual inductance will directly affect the magnitude of the voltage induced across a coil since it will determine the net inductance of the coil. Additional examination reveals that the sign of the mutual term for each coil of a coupled pair is the same. For $L_{T(+)}$ they were both positive, and for $L_{T(-)}$ they were both negative. On a network schematic where it is inconvenient to indicate the windings and the flux path, a system of dots is employed that will determine whether the mutual terms are to be positive or negative. The dot convention is shown in Fig. 22.6 for the series coils of Figs. 22.4 and 22.5.

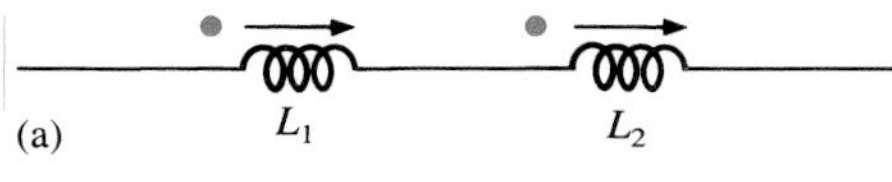

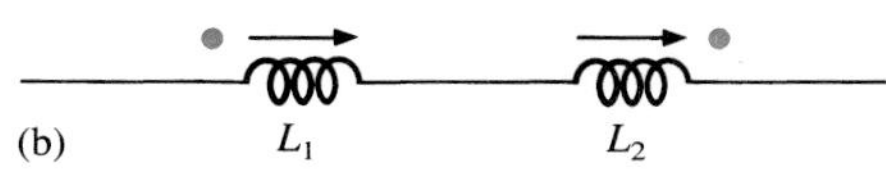

FIG. 22.6
Dot convention for the series coils of (a) Fig. 22.4 and (b) Fig. 22.5.

If the current through *each* of the mutually coupled coils is going away from (or toward) the dot as it *passes through the coil,* the mutual term will be positive, as shown for the case in Fig. 22.6(a). If the arrow indicating current direction through the coil is leaving the dot for one coil and entering the dot for the other, the mutual term is negative.

A few possibilities for mutually coupled transformer coils are indicated in Fig. 22.7. The sign of M is indicated for each. When determining the sign, be sure to examine the current direction within the coil itself. In Fig. 22.7(e), one direction was indicated outside for one coil and through for the other. It initially might appear that the sign should be positive since both currents enter the dot, but the current *through* coil 1 is leaving the dot; hence a negative sign is in order.

The dot convention also reveals the polarity of the *induced* voltage across the mutually coupled coil. If the reference direction for the current *in* a coil leaves the dot, the polarity at the dot for the induced voltage of the mutually coupled coil is positive. In Fig. 22.7(a) and (b), the polarity at the dots of the induced voltages is positive. Fig. 22.7(c), the polarity at the dot of the right-hand coil is negative, while the polarity at the dot of the left-hand coil is positive, since the current enters the dot (within the coil) of the right-hand coil. The comments for Fig. 22.7(c) can also be applied to Fig. 22.7(d).

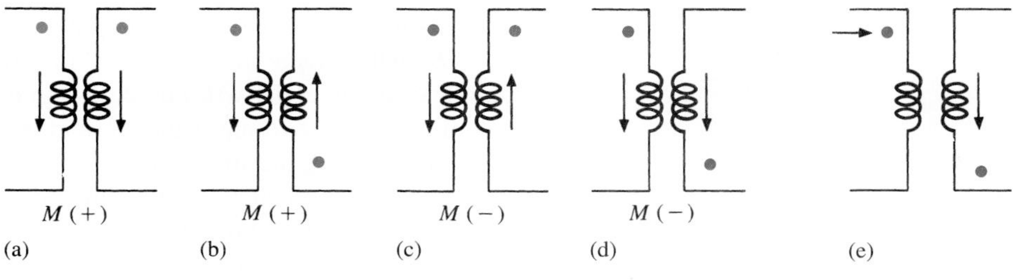

FIG. 22.7
Defining the sign of M for mutually coupled transformer coils.

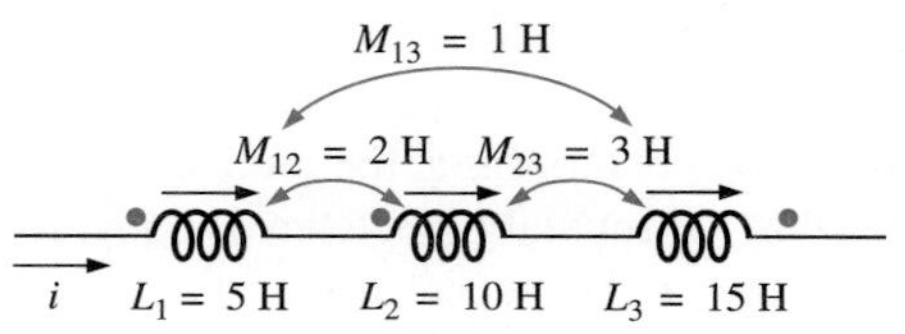

FIG. 22.8
Example 22.2.

EXAMPLE 22.2

Find the total inductance of the series coils of Fig. 22.8.

Solution:

Current vectors leave dot.

Coil 1: $L_1 + M_{12} - M_{13}$

One current vector enters dot, while one leaves.

Coil 2: $L_2 + M_{12} - M_{23}$

Coil 3: $L_3 - M_{23} - M_{13}$

and

$$L_T = (L_1 + M_{12} - M_{13}) + (L_2 + M_{12} - M_{23}) + (L_3 - M_{23} - M_{13})$$
$$= L_1 + L_2 + L_3 + 2M_{12} - 2M_{23} - 2M_{13}$$

Substituting values, we find

$$L_T = 5\text{ H} + 10\text{ H} + 15\text{ H} + 2(2\text{ H}) - 2(3\text{ H}) - 2(1\text{ H})$$
$$= 34\text{ H} - 8\text{ H} = \mathbf{26\text{ H}}$$

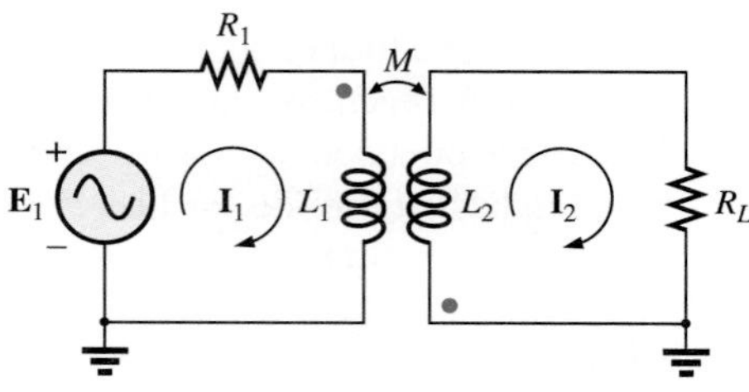

FIG. 22.9
Example 22.3.

EXAMPLE 22.3

Write the mesh equations for the transformer network in Fig. 22.9.

Solution:

For each coil, the mutual term is positive, and the sign of M in $\mathbf{X}_m = \omega M \angle 90°$ is positive, as determined by the direction of $\mathbf{I}_1$ and $\mathbf{I}_2$. Thus,

$$\mathbf{E}_1 - \mathbf{I}_1 R_1 - \mathbf{I}_1 X_{L_1} \angle 90° - \mathbf{I}_2 X_m \angle 90° = 0$$

or

$$\mathbf{E}_1 - \mathbf{I}_1(R_1 + j X_{L_1}) - \mathbf{I}_2 X_m \angle 90° = 0$$

For the other loop,

$$-\mathbf{I}_2 X_{L_2} \angle 90° - \mathbf{I}_1 X_m \angle 90° - \mathbf{I}_2 R_L = 0$$

or

$$\mathbf{I}_2(R_L + j X_{L_2}) + \mathbf{I}_1 X_m \angle 90° = 0$$

22.4 THE IRON-CORE TRANSFORMER

An iron-core transformer under loaded conditions is shown in Fig. 22.10. The iron core will serve to increase the coefficient of coupling between the coils by increasing the mutual flux ϕ_m. Recall from Chapter 11 that magnetic flux lines will always take the path of least reluctance, which in this case is the iron core.

We will assume in the analyses to follow in this chapter that all of the flux linking coil 1 will link coil 2. In other words, the coefficient of coupling is its maximum value, 1, and $\phi_m = \phi_p = \phi_s$. In addition, we will first analyze the transformer from an ideal viewpoint; that is, we will neglect losses such as the geometric or dc resistance of the coils, the leakage reactance due to the flux linking either coil that forms no part of ϕ_m, and the hysteresis and eddy current losses. This is not to convey the impression, however, that we will be far from the actual operation of a transformer. Most transformers manufactured today can

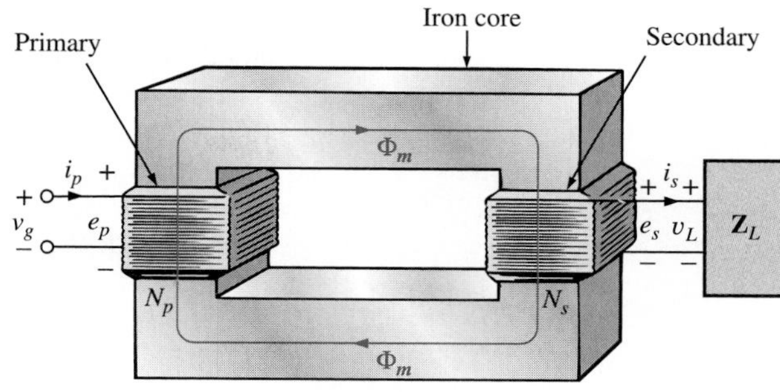

FIG. 22.10
Iron-core transformer.

be considered almost ideal. The equations we will develop under ideal conditions will be, in general, a first approximation to the actual response, which will never be off by more than a few percentage points. The losses will be considered in greater detail in Section 22.6.

When the current i_p through the primary circuit of the iron-core transformer is a maximum, the flux ϕ_m linking both coils is also a maximum. In fact, the magnitude of the flux is directly proportional to the current through the primary windings. Therefore, the two are in phase, and for sinusoidal inputs, the magnitude of the flux will vary as a sinusoid also. That is, if

$$i_p = \sqrt{2}I_p \sin \omega t$$

then

$$\phi_m = \Phi_m \sin \omega t$$

The induced voltage across the primary due to a sinusoidal input can be determined by Faraday's law:

$$e_p = N_p \frac{d\phi_p}{dt} = N_p \frac{d\phi_m}{dt}$$

Substituting for ϕ_m gives us

$$e_p = N_p \frac{d}{dt}(\Phi_m \sin \omega t)$$

and differentiating, we obtain

$$e_p = \omega N_p \Phi_m \cos \omega t$$

or

$$e_p = \omega N_p \Phi_m \sin(\omega t + 90°)$$

indicating that the induced voltage e_p leads the current through the primary coil by 90°.

The effective value of e_p is

$$E_p = \frac{\omega N_p \Phi_m}{\sqrt{2}} = \frac{2\pi f N_p \Phi_m}{\sqrt{2}}$$

and

$$E_p = 4.44 f N_p \Phi_m \qquad (22.17)$$

which is an equation for the effective value of the voltage across the primary coil in terms of the frequency of the input current or voltage, the number of turns of the primary, and the maximum value of the magnetic flux linking the primary.

For the case under discussion, where the flux linking the secondary equals that of the primary, if we repeat the procedure just described for the induced voltage across the secondary, we get

$$E_s = 4.44fN_s\Phi_m \tag{22.18}$$

Dividing Eq. (22.17) by Eq. (22.18), as follows:

$$\frac{E_p}{E_s} = \frac{4.44fN_p\Phi_m}{4.44fN_s\Phi_m}$$

we obtain

$$\frac{E_p}{E_s} = \frac{N_p}{N_s} \tag{22.19}$$

revealing an important relationship for transformers:

The ratio of the magnitudes of the induced voltages is the same as the ratio of the corresponding turns.

If we consider that

$$e_p = N_p\frac{d\phi_m}{dt} \quad \text{and} \quad e_s = N_s\frac{d\phi_m}{dt}$$

and divide one by the other, that is,

$$\frac{e_p}{e_s} = \frac{N_p(d\phi_m/dt)}{N_s(d\phi_m/dt)}$$

then

$$\frac{e_p}{e_s} = \frac{N_p}{N_s}$$

The *instantaneous* values of e_1 and e_2 are therefore related by a constant determined by the turns ratio. Since their instantaneous magnitudes are related by a constant, the induced voltages are in phase, and Equation (22.19) can be changed to include phasor notation; that is,

$$\frac{\mathbf{E}_p}{\mathbf{E}_s} = \frac{N_p}{N_s} \tag{22.20}$$

or, since $\mathbf{V}_g = \mathbf{E}_1$ and $\mathbf{V}_L = \mathbf{E}_2$ for the ideal situation,

$$\frac{\mathbf{V}_g}{\mathbf{V}_L} = \frac{N_p}{N_s} \tag{22.21}$$

The ratio N_p/N_s, usually represented by the lowercase letter a, is referred to as the **transformation ratio**:

$$a = \frac{N_p}{N_s} \tag{22.22}$$

If $a < 1$, the transformer is called a **step-up transformer** since the voltage $E_s > E_p$; that is,

$$\frac{E_p}{E_s} = \frac{N_p}{N_s} = a \quad \text{or} \quad E_s = \frac{E_p}{a}$$

and, if $a < 1$,

$$E_s > E_p$$

If $a > 1$, the transformer is called a **step-down transformer** since $E_s < E_p$; that is,

$$E_p = aE_s$$

and, if $a > 1$, then $$E_p > E_s$$

EXAMPLE 22.4

For the iron-core transformer of Fig. 22.11:

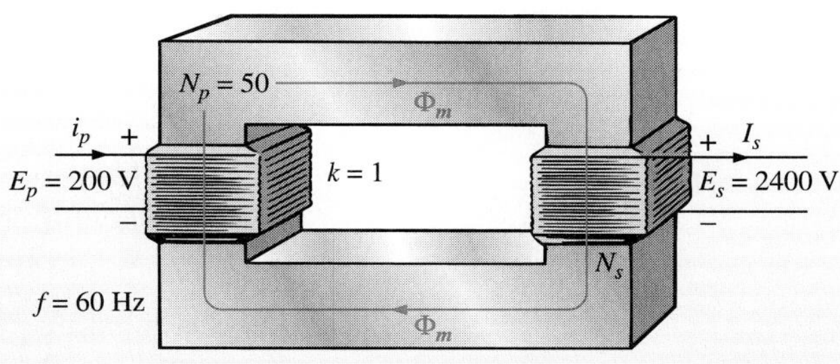

FIG. 22.11
Example 22.4.

a. Find the maximum flux Φ_m.
b. Find the secondary turns N_s.

Solution:

a. $E_p = 4.44N_pf\Phi_m$

Therefore, $$\Phi_m = \frac{E_p}{4.44N_pf} = \frac{200 \text{ V}}{(4.44)(50 \text{ t})(60 \text{ Hz})}$$

and $$\Phi_m = \mathbf{15.02 \text{ mWb}}$$

b. $\dfrac{E_p}{E_s} = \dfrac{N_p}{N_s}$

Therefore, $$N_s = \frac{N_pE_s}{E_p} = \frac{(50 \text{ t})(2400 \text{ V})}{200 \text{ V}}$$
$$= \mathbf{600 \text{ turns}}$$

The induced voltage across the secondary of the transformer of Fig. 22.10 will establish a current i_s through the load Z_L and the secondary windings. This current and the turns N_s will develop an mmf N_si_s that would not be present under no-load conditions since $i_s = 0$ and $N_si_s = 0$. Under loaded or unloaded conditions, however, the net ampere-turns on the core produced by both the primary and the secondary must remain unchanged for the same flux ϕ_m to be established in the core. The flux ϕ_m must remain the same to have the same induced voltage across the primary and to balance the voltage impressed across the primary. In order to counteract the mmf of the secondary, which is tending to change ϕ_m, an additional current must flow in the primary. This current is called the *load component of the primary current* and is represented by the notation i'_p.

For the balanced or equilibrium condition,

$$N_pi'_p = N_si_s$$

The total current in the primary under loaded conditions is

$$i_p = i'_p + i_{\phi_m}$$

where i_{ϕ_m} is the current in the primary necessary to establish the flux ϕ_m. For most practical applications, $i'_p > i_{\phi_m}$. For our analysis, we will assume $i_p \cong i'_p$, so

$$N_p i_p = N_s i_s$$

Since the instantaneous values of i_p and i_s are related by the turns ratio, the phasor quantities $\mathbf{I}_p$ and $\mathbf{I}_s$ are also related by the same ratio:

$$N_p \mathbf{I}_p = N_s \mathbf{I}_s$$

or

$$\frac{\mathbf{I}_p}{\mathbf{I}_s} = \frac{N_s}{N_p} \qquad (22.23)$$

The primary and secondary currents of a transformer are therefore related by the inverse ratios of the turns.

Keep in mind that Equation (22.23) holds true only if we neglect the effects of i_{ϕ_m}. Otherwise, the magnitudes of $\mathbf{I}_p$ and $\mathbf{I}_s$ are not related by the turns ratio, and $\mathbf{I}_p$ and $\mathbf{I}_s$ are not in phase.

For the step-up transformer, $a < 1$, and the current in the secondary, $I_s = aI_p$, is less in magnitude than that in the primary. For a step-down transformer, the reverse is true.

22.5 REFLECTED IMPEDANCE AND POWER

In the previous section we found that

$$\frac{\mathbf{V}_g}{\mathbf{V}_L} = \frac{N_p}{N_s} = a \quad \text{and} \quad \frac{\mathbf{I}_p}{\mathbf{I}_s} = \frac{N_s}{N_p} = \frac{1}{a}$$

Dividing the first by the second, we have

$$\frac{\mathbf{V}_g/\mathbf{V}_L}{\mathbf{I}_p/\mathbf{I}_s} = \frac{a}{1/a}$$

or

$$\frac{\mathbf{V}_g/\mathbf{I}_p}{\mathbf{V}_L/\mathbf{I}_s} = a^2 \quad \text{and} \quad \frac{\mathbf{V}_g}{\mathbf{I}_p} = a^2\frac{\mathbf{V}_L}{\mathbf{I}_s}$$

However, since

$$\mathbf{Z}_p = \frac{\mathbf{V}_g}{\mathbf{I}_p} \quad \text{and} \quad \mathbf{Z}_L = \frac{\mathbf{V}_L}{\mathbf{I}_s}$$

then

$$\mathbf{Z}_p = a^2\mathbf{Z}_L \qquad (22.24)$$

which in words states that the impedance of the primary circuit of an ideal transformer is the transformation ratio squared times the impedance of the load. If a transformer is used, therefore, an impedance can be made to appear larger or smaller at the primary by placing it in the secondary of a step-down ($a > 1$) or step-up ($a < 1$) transformer, respectively. Note that if the load is capacitive or inductive, the **reflected impedance** will also be capacitive or inductive.

For the ideal iron-core transformer,

$$\frac{E_p}{E_s} = a = \frac{I_s}{I_p}$$

or $$E_pI_p = E_sI_s \tag{22.25}$$

and $$P_{in} = P_{out} \quad \text{(ideal conditions)} \tag{22.26}$$

EXAMPLE 22.5

For the iron-core transformer of Fig. 22.12:

a. Find the magnitude of the current in the primary and the impressed voltage across the primary.
b. Find the input resistance of the transformer.

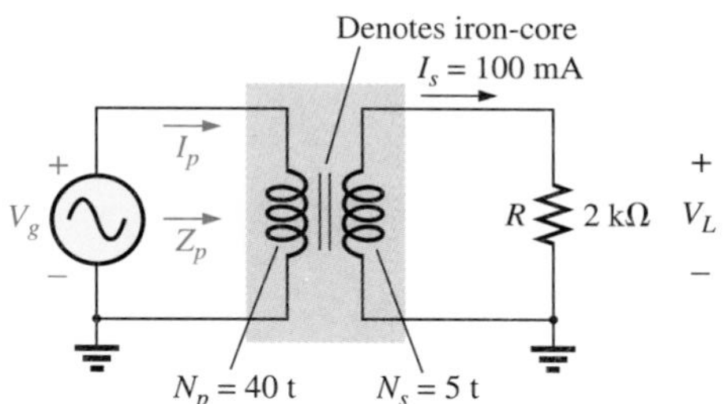

FIG. 22.12
Example 22.5.

Solution:

a. $\dfrac{I_p}{I_s} = \dfrac{N_s}{N_p}$

$$I_p = \frac{N_s}{N_p}I_s = \left(\frac{5\text{ t}}{40\text{ t}}\right)(0.1\text{ A}) = \mathbf{12.5\text{ mA}}$$

$$V_L = I_sZ_L = (0.1\text{ A})(2\text{ k}\Omega) = 200\text{ V}$$

Also, $$\frac{V_g}{V_L} = \frac{N_p}{N_s}$$

$$V_g = \frac{N_p}{N_s}V_L = \left(\frac{40\text{ t}}{5\text{ t}}\right)(200\text{ V}) = \mathbf{1600\text{ V}}$$

b. $Z_p = a^2Z_L$

$$a = \frac{N_p}{N_s} = 8$$

$$Z_p = (8)^2(2\text{ k}\Omega) = R_p = \mathbf{128\text{ k}\Omega}$$

EXAMPLE 22.6

For the residential supply appearing in Fig. 22.13, determine (assuming a totally resistive load) the following:

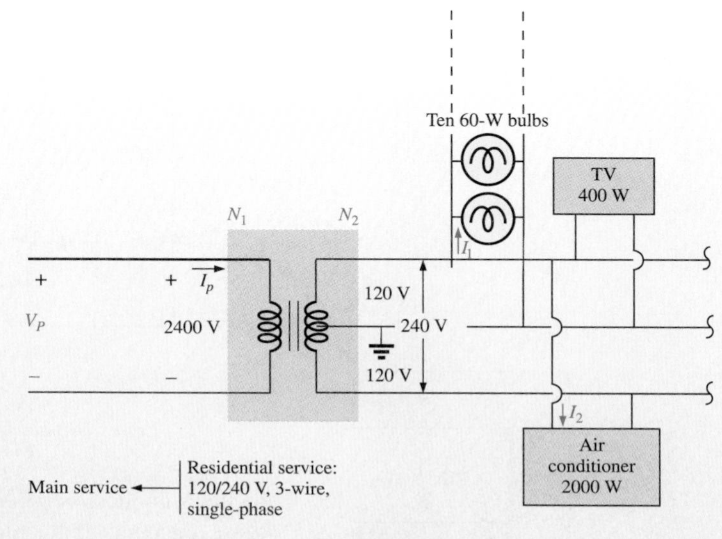

FIG. 22.13
Single-phase residential supply.

continued

a. the magnitude of I_1 and I_2.
b. the total secondary power P_T.
c. the turns ratio a.
d. the total resistance on the secondary.
e. the power into the primary.
f. the primary current I_P.
g. the primary resistance.

Solution:

a. $P_1 = 600 \text{ W} = VI_1 = (120\text{v})I_1$

and $I_1 = 5\text{A}$

$P_2 = 2000 \text{ W} = VI_2 = (240\text{v})I_2$

and $I_2 = 8.33\text{A}$

b. P_T on secondary $= (10)(60 \text{ W}) + 400 \text{ W} + 2000 \text{ W}$

$= 600 \text{ W} + 400 \text{ W} + 2000 \text{ W} = 3000 \text{ W}$

c. $a = N_p/N_s = V_p/V_s = 2400 \text{ V}/240 \text{ V} = 10$

d. $R_s = V_s^2/P_T = 240^2/3000 = 19.2 \text{ } \Omega$

e. $P_{in} = P_{out} = 3000 \text{ W}$

f. $I_p = P_p/V_p = P_{in}/V_p = 3000/2400 = 1.25\text{A}$

g. $R_p = V_p/I_p = 2400 \text{ V}/1.25 \text{ A} = 1920 \text{ } \Omega$

also $Z_p = V_p^2/P_T = 2400^2/3000 = 1920 \text{ } \Omega$

also $Z_p = a^2 Z_s = (10^2)(19.2) = 1920 \text{ } \Omega$

22.6 EQUIVALENT CIRCUIT (IRON-CORE TRANSFORMER)

For the nonideal or practical iron-core transformer, the equivalent circuit appears as in Fig. 22.14. As indicated, part of this equivalent circuit

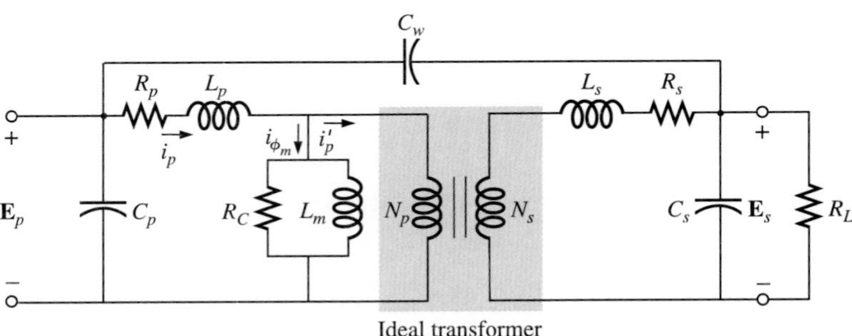

FIG. 22.14
Equivalent circuit for the practical iron-core transformer.

includes an ideal transformer. The remaining elements of Fig. 22.14 are those elements that contribute to the nonideal characteristics of the device. The resistances R_p and R_s are simply the dc or geometric resistance of the primary and secondary windings, respectively. For the primary and secondary coils of a transformer, there is a small amount of flux that links each coil but does not pass through the core, as shown in Fig. 22.15 for the primary winding. This **leakage flux**, representing a

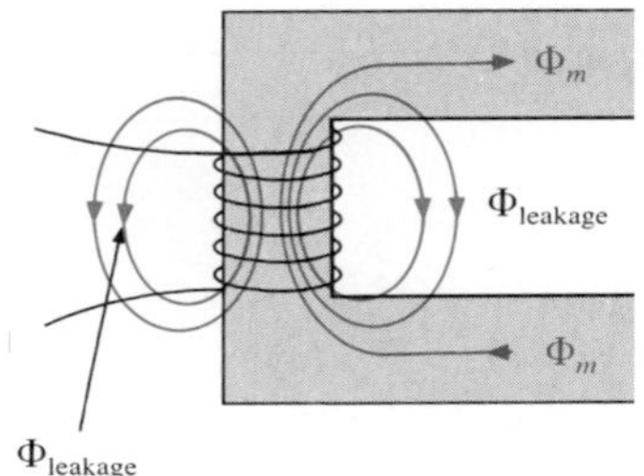

FIG. 22.15
Identifying the leakage flux of the primary.

definite loss in the system, is represented by an inductance L_p in the primary circuit and an inductance L_s in the secondary.

The resistance R_c represents the hysteresis and eddy current losses (core losses) within the core due to an ac flux through the core. The inductance L_m (magnetizing inductance) is the inductance associated with the magnetization of the core, that is, the establishing of the flux Φ_m in the core. The capacitances C_p and C_s are the lumped capacitances of the primary and secondary circuits, respectively, and C_w represents the equivalent lumped capacitances between the windings of the transformer.

Since i'_p is normally considerably larger than i_{ϕ_m} (the magnetizing current), we will ignore i_{ϕ_m} for the moment (set it equal to zero), resulting in the absence of R_c and L_m in the reduced equivalent circuit of Fig. 22.16. The capacitances C_p, C_w, and C_s do not appear in the equivalent circuit of Fig. 22.16 since their reactance at typical operating frequencies will not appreciably affect the transfer characteristics of the transformer.

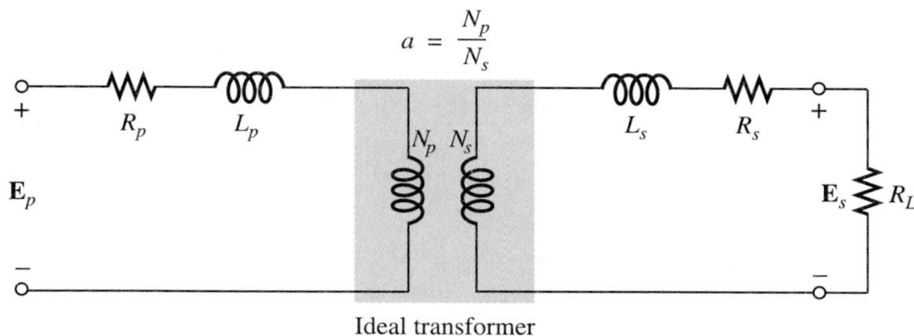

FIG. 22.16
Reduced equivalent circuit for the nonideal iron-core transformer.

If we now reflect the secondary circuit through the ideal transformer using Eq. (22.24), as shown in Fig. 22.17(a), we will have the load and generator voltage in the same continuous circuit. The total resistance and inductive reactance of the primary circuit are determined by

$$R_{\text{equivalent}} = R_e = R_p + a^2R_s \tag{22.27}$$

and

$$X_{\text{equivalent}} = X_e = X_p + a^2X_s \tag{22.28}$$

which result in the useful equivalent circuit of Fig. 22.17(b). The load voltage can be obtained directly from the circuit of Fig. 22.17(b) through the voltage divider rule:

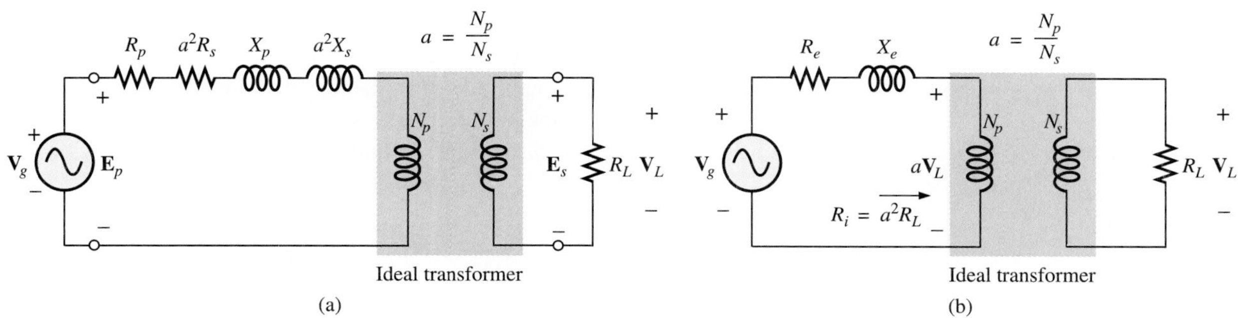

FIG. 22.17
Reflecting the secondary circuit into the primary side of the iron-core transformer.

$$a\mathbf{V}_L = \frac{(R_i)\mathbf{V}_g}{(R_e + R_i) + j X_e}$$

and

$$\mathbf{V}_L = \frac{a^2 R_L \mathbf{V}_g}{(R_e + a^2 R_L) + j X_e} \tag{22.29}$$

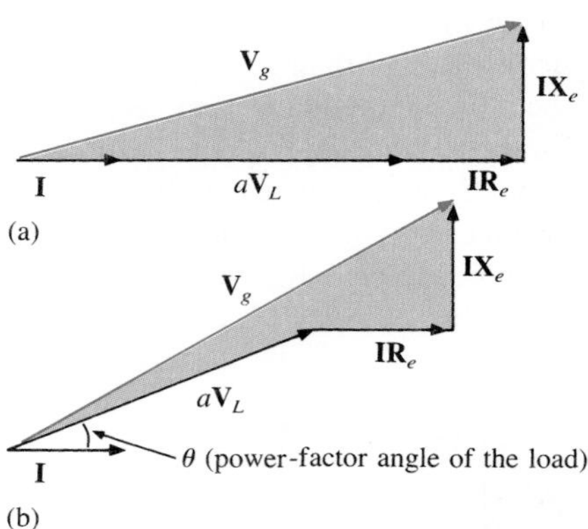

FIG. 22.18
Phasor diagram for the iron-core transformer with (a) unity power-factor load (resistive) and (b) lagging power-factor load (inductive).

The network of Fig. 22.17(b) will also allow us to calculate the generator voltage necessary to establish a particular load voltage. The voltages across the elements of Fig. 22.17(b) have the phasor relationship indicated in Fig. 22.18(a). Note that the current is the reference phasor for drawing the phasor diagram. That is, the voltages across the resistive elements are *in phase* with the current phasor, while the voltage across the equivalent inductance leads the current by 90°. The primary voltage, by Kirchhoff's voltage law, is then the phasor sum of these voltages, as indicated in Fig. 22.18(a). For an inductive load, the phasor diagram appears in Fig. 22.18(b). Note that $a\mathbf{V}_L$ leads $\mathbf{I}$ by the power-factor angle of the load. The remainder of the diagram is then similar to that for a resistive load. (The phasor diagram for a capacitive load will be left to the reader as an exercise.)

The effect of R_e and X_e on the magnitude of $\mathbf{V}_g$ for a particular $\mathbf{V}_L$ is obvious from Eq. (22.29) or Fig. 22.18. For increased values of R_e or X_e, an increase in $\mathbf{V}_g$ is required for the same load voltage. For R_e and $X_e = 0$, $\mathbf{V}_L$ and $\mathbf{V}_g$ are simply related by the turns ratio.

EXAMPLE 22.7

For a transformer having the equivalent circuit of Fig. 22.19:

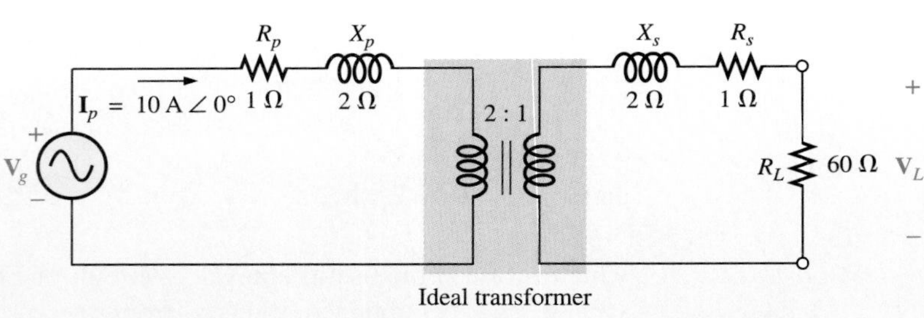

FIG. 22.19
Example 22.7.

a. Determine R_e and X_e.
b. Determine the magnitude of the voltages V_L and V_g.
c. Determine the magnitude of the voltage V_g to establish the same load voltage in part (b) if R_e and $X_e = 0\ \Omega$. Compare with the result of part (b).

Solution:

a. $R_e = R_p + a^2 R_s = 1\ \Omega + (2)^2(1\ \Omega) = \mathbf{5\ \Omega}$

$X_e = X_p + a^2 X_s = 2\ \Omega + (2)^2(2\ \Omega) = \mathbf{10\ \Omega}$

b. The transformed equivalent circuit appears in Fig. 22.20.

$$aV_L = (I_p)(a^2 R_L) = 2400\text{ V}$$

continued

Thus,

$$V_L = \frac{2400\ \text{V}}{a} = \frac{2400\ \text{V}}{2} = \mathbf{1200\ V}$$

and

$$\begin{aligned}\mathbf{V}_g &= \mathbf{I}_p(R_e + a^2R_L + j\,X_e)\\ &= 10\ \text{A}(5\ \Omega + 240\ \Omega + j\,10\ \Omega) = 10\ \text{A}(245\ \Omega + j\,10\ \Omega)\\ \mathbf{V}_g &= 2450\ \text{V} + j\,100\ \text{V} = \mathbf{2452.04\ V\ \angle 2.34^\circ}\end{aligned}$$

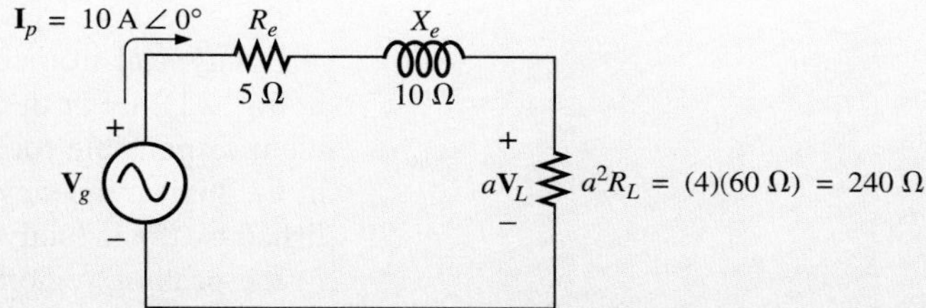

FIG. 22.20
Transformed equivalent circuit of Fig. 22.19.

c. For R_e and $X_e = 0$, $V_g = aV_L = (2)(1200\ \text{V}) = 2400\ \text{V}$.
Therefore, it is necessary to increase the generator voltage by 52.04 V (due to R_e and X_e) to obtain the same load voltage.

22.7 FREQUENCY CONSIDERATIONS

For certain frequency ranges, the effect of some parameters in the equivalent circuit of the iron-core transformer of Fig. 22.14 should not be ignored. Since it is convenient to consider a low-, mid-, and high-frequency region, the equivalent circuits for each will now be introduced and briefly examined.

For the low-frequency region, the series reactance ($2\pi fL$) of the primary and secondary leakage reactances can be ignored since they are small in magnitude. The magnetizing inductance must be included, however, since it appears in parallel with the secondary reflected circuit, and small impedances in a parallel network can have a dramatic impact on the terminal characteristics. The resulting equivalent network for the low-frequency region is provided in Fig. 22.21(a). As the frequency decreases, the reactance of the magnetizing inductance will reduce in magnitude, causing a reduction in the voltage across the secondary circuit. For $f = 0$ Hz, L_m is ideally a short circuit, and $V_L = 0$. As the frequency increases, the reactance of L_m will eventually be sufficiently large compared with the reflected secondary impedance to be neglected. The mid-frequency reflected equivalent circuit will then appear as shown in Fig. 22.21(b). Note the absence of reactive elements, resulting in an *in-phase* relationship between load and generator voltages.

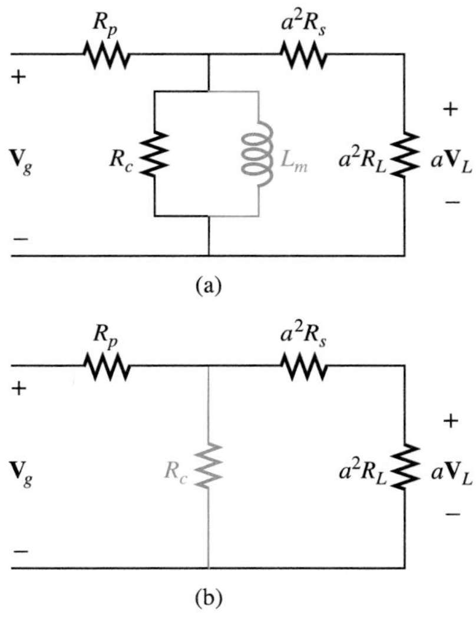

FIG. 22.21
(a) Low-frequency reflected equivalent circuit; (b) mid-frequency reflected circuit.

For higher frequencies, the capacitive elements and primary and secondary leakage reactances must be considered, as shown in Fig. 22.22. For discussion purposes, the effects of C_w and C_s appear as a lumped capacitor C in the reflected network of Fig. 22.22; C_p does not appear since the effect of C will predominate. As the frequency of interest increases, the capacitive reactance ($X_C = 1/2\pi fC$) will decrease to the point that it will have a shorting effect across the secondary circuit of the transformer, causing V_L to decrease in magnitude.

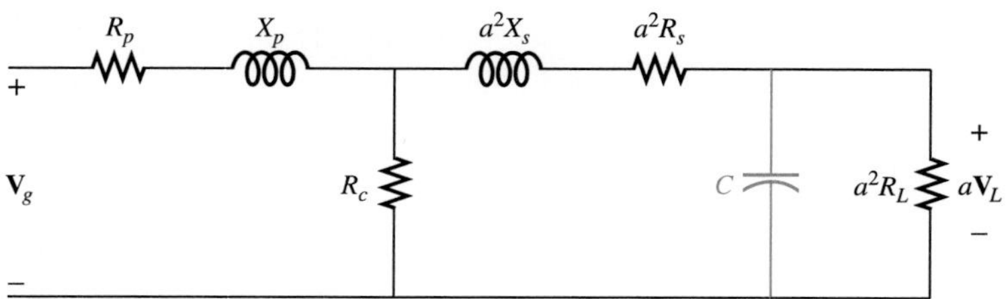

FIG. 22.22
High-frequency reflected equivalent circuit.

A typical iron-core transformer-frequency response curve appears in Fig. 22.23. For the low- and high-frequency regions, the primary element responsible for the drop-off is indicated. The peaking that occurs in the high-frequency region is due to the series resonant circuit established by the inductive and capacitive elements of the equivalent circuit. In the peaking region, the series resonant circuit is in, or near, its resonant or tuned state.

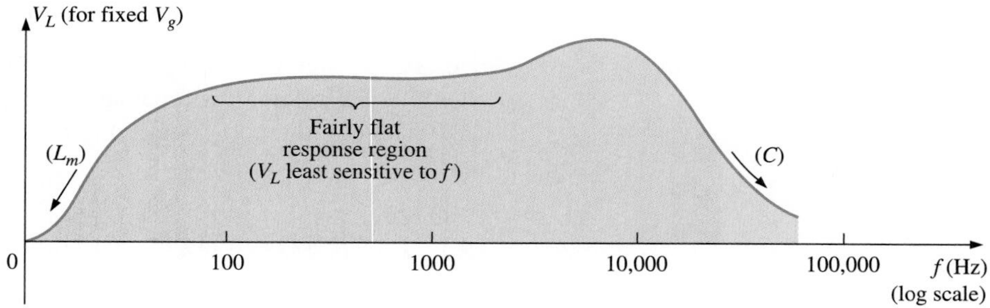

FIG. 22.23
Transformer-frequency response curve.

22.8 AIR-CORE TRANSFORMER

As the name implies, the air-core transformer does not have a ferromagnetic core to link the primary and secondary coils. Rather, the coils are placed sufficiently close to have a mutual inductance that will establish the desired transformer action. In Fig. 22.24, current direction and polarities have been defined for the air-core transformer. Note the presence of a mutual inductance term M, which will be positive in this case, as determined by the dot convention.

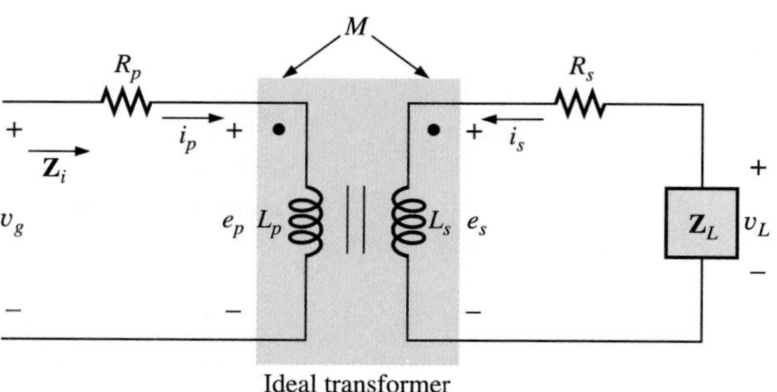

FIG. 22.24
Air-core transformer equivalent circuit.

From past analysis in this chapter, we now know that

$$e_p = L_p \frac{di_p}{dt} + M \frac{di_s}{dt} \tag{22.30}$$

for the primary circuit.

We found in Chapter 12 that for the pure inductor, with no mutual inductance present, the mathematical relationship

$$v_1 = L \frac{di_1}{dt}$$

resulted in the following useful form of the voltage across an inductor:

$$\mathbf{V}_1 = \mathbf{I}_1 X_L \angle 90^\circ \quad \text{where } X_L = \omega L$$

Similarly, it can be shown, for a mutual inductance, that

$$v_1 = M \frac{di_2}{dt}$$

will result in

$$\mathbf{V}_1 = \mathbf{I}_2 X_m \angle 90^\circ \quad \text{where } X_m = \omega M \tag{22.31}$$

Equation (22.30) can then be written (using phasor notation) as

$$\mathbf{E}_p = \mathbf{I}_p X_{L_p} \angle 90^\circ + \mathbf{I}_s X_m \angle 90^\circ \tag{22.32}$$

and
$$\mathbf{V}_g = \mathbf{I}_p R_p \angle 0^\circ + \mathbf{I}_p X_{L_p} \angle 90^\circ + \mathbf{I}_s X_m \angle 90^\circ$$

or
$$\mathbf{V}_g = \mathbf{I}_p (R_p + j\,X_{L_p}) + \mathbf{I}_s X_m \angle 90^\circ \tag{22.33}$$

For the secondary circuit,

$$\mathbf{E}_s = \mathbf{I}_s X_{L_s} \angle 90^\circ + \mathbf{I}_p X_m \angle 90^\circ \tag{22.34}$$

and
$$\mathbf{V}_L = \mathbf{I}_s R_s \angle 0^\circ + \mathbf{I}_s X_{L_s} \angle 90^\circ + \mathbf{I}_p X_m \angle 90^\circ$$

or
$$\mathbf{V}_L = \mathbf{I}(R_s + j\,X_{L_s}) + \mathbf{I}_p X_m \angle 90^\circ \tag{22.35}$$

Substituting
$$\mathbf{V}_L = -\mathbf{I}_s \mathbf{Z}_L$$

into Eq. (22.35) results in

$$0 = \mathbf{I}_s (R_s + j\,X_{L_s} + \mathbf{Z}_L) + \mathbf{I}_p X_m \angle 90^\circ$$

Solving for $\mathbf{I}_s$, we have

$$\mathbf{I}_s = \frac{-\mathbf{I}_p X_m \angle 90^\circ}{R_s + j\,X_{L_s} + \mathbf{Z}_L}$$

and, substituting into Eq. (22.33), we obtain

$$\mathbf{V}_g = \mathbf{I}_p (R_p + j\,X_{L_p}) + \left(\frac{-\mathbf{I}_p X_m \angle 90^\circ}{R_s + j\,X_{L_s} + \mathbf{Z}_L}\right) X_m \angle 90^\circ$$

Thus, the input impedance is

$$\mathbf{Z}_i = \frac{\mathbf{V}_g}{\mathbf{I}_p} = R_p + j\,X_{L_p} - \frac{(X_m \angle 90^\circ)^2}{R_s + j\,X_{L_s} + \mathbf{Z}_L}$$

or, defining

$$\mathbf{Z}_p = R_p + j\,X_{L_p} \quad \mathbf{Z}_s = R_s + j\,X_{L_s} \text{ and } X_m \angle 90^\circ = +j\,\omega M$$

we have

$$\mathbf{Z}_i = \mathbf{Z}_p - \frac{(+j\,\omega M)^2}{\mathbf{Z}_s + \mathbf{Z}_L}$$

and

$$\mathbf{Z}_i = \mathbf{Z}_p - \frac{(\omega M)^2}{\mathbf{Z}_s + \mathbf{Z}_L} \tag{22.36}$$

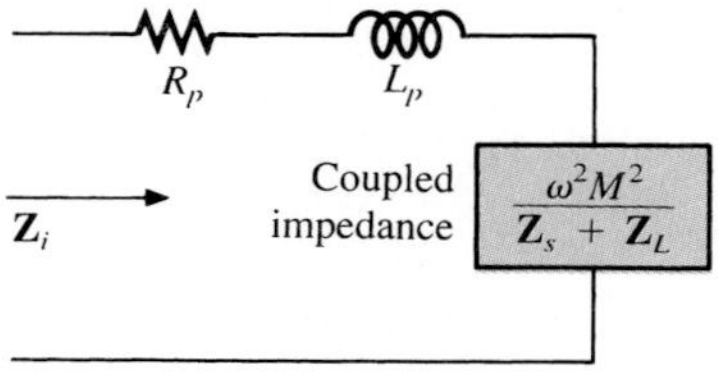

FIG. 22.25
Input characteristics for the air-core transformer.

The term $(\omega M)^2/(\mathbf{Z}_s + \mathbf{Z}_L)$ is called the *coupled impedance,* and it is independent of the sign of M since it is squared in the equation. Consider also that since $(\omega M)^2$ is a constant with 0° phase angle, if the load $\mathbf{Z}_L$ is resistive, the resulting coupled impedance term will appear capacitive due to division of $(\mathbf{Z}_s + R_L)$ into $(\omega M)^2$. This resulting capacitive reactance will oppose the series primary inductance L_p, causing a reduction in $\mathbf{Z}_i$. Including the effect of the mutual term, the input impedance to the network will appear as shown in Fig. 22.25.

EXAMPLE 22.8

Determine the input impedance to the air-core transformer in Fig. 22.26.

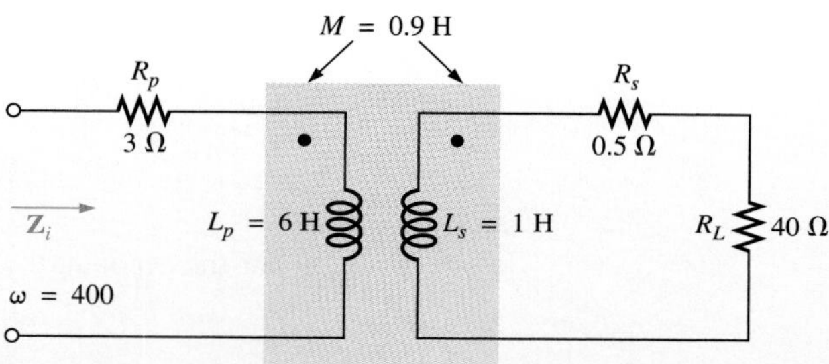

FIG. 22.26
Example 22.8.

Solution:

$$\mathbf{Z}_i = \mathbf{Z}_p + \frac{(\omega M)^2}{\mathbf{Z}_s + \mathbf{Z}_L}$$

$$= R_p + j\,X_{L_p} + \frac{(\omega M)^2}{R_s + j\,X_{L_s} + R_L}$$

$$= 3\ \Omega + j\,2.4\ \text{k}\Omega + \frac{[(400\ \text{rad/s})(0.9\ \text{H})]^2}{0.5\ \Omega + j\,400\ \Omega + 40\ \Omega}$$

$$\cong j\,2.4\ \text{k}\Omega + \frac{129.6 \times 10^3\ \Omega}{40.5 + j\,400}$$

$$= j\,2.4\ \text{k}\Omega + \underbrace{322.4\ \Omega \angle -84.22^\circ}_{\text{capacitive}}$$

$$= j\,2.4\ \text{k}\Omega + (0.0325\ \text{k}\Omega - j\,0.3208\ \text{k}\Omega)$$

$$= 0.0325\ \text{k}\Omega + j\,(2.40 - 0.3208)\ \text{k}\Omega$$

and $\mathbf{Z}_i = R_i + j\,X_{L_i} = \mathbf{32.5\ \Omega + j\,2079\ \Omega = 2079.25\ \Omega \angle 89.10^\circ}$

22.9 IMPEDANCE MATCHING, ISOLATION, AND DISPLACEMENT

Transformers can be particularly useful when you are trying to ensure that a load receives maximum power from a source. Recall that maximum power is transferred to a load when its impedance is a match with the internal resistance of the supply. Even if a perfect match is unattainable, the closer the load matches the internal resistance, the greater the power to the load and the more efficient the system. Unfortunately, unless it is planned as part of the design, most loads are not a close match with the internal impedance of the supply. However, transformers have a unique relationship between their primary and secondary impedances that can be put to good use in the impedance matching process. Example 22.9 will demonstrate the significant difference in the power delivered to the load with and without an impedance matching transformer.

EXAMPLE 22.9

a. The source impedance for the supply of Fig. 22.27(a) is 512 Ω, which is a poor match with the 8-Ω input impedance of the speaker. One can expect only that the power delivered to the speaker will be significantly less than the maximum possible level. Determine the power to the speaker under the conditions of Fig. 22.27(a).

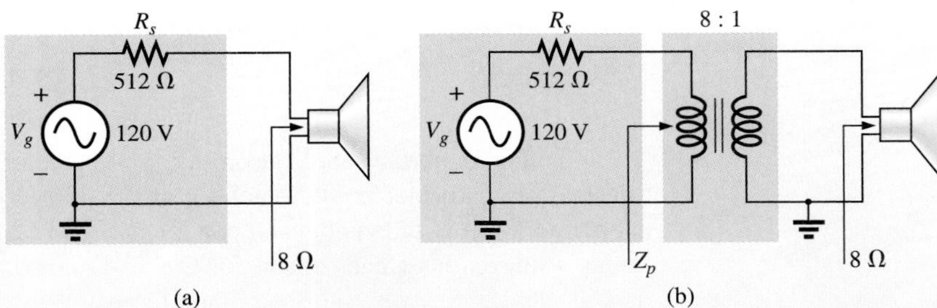

FIG. 22.27
Example 22.9.

b. In Fig. 22.27(b), an audio impedance matching transformer was introduced between the speaker and the source, and it was designed to ensure maximum power to the 8-Ω speaker. Determine the input impedance of the transformer and the power delivered to the speaker.

c. Compare the power delivered to the speaker under the conditions of parts (a) and (b).

Solution:

a. The source current:

$$I_s = \frac{E}{R_T} = \frac{120\text{ V}}{512\ \Omega + 8\ \Omega} = \frac{120\text{ V}}{520\ \Omega} = 230.8\text{ mA}$$

The power to the speaker:

$$P = I^2R = (230.8\text{ mA})^2 \cdot 8\ \Omega = \mathbf{426.15\ mW} \cong \mathbf{0.43\ W}$$

continued

or less than half a watt.

b. $Z_p = a^2 Z_L$

$$a = \frac{N_p}{N_s} = \frac{8}{1} = 8$$

and $Z_p = (8)^2 8\ \Omega = \mathbf{512\ \Omega}$

which matches that of the source. Maximum power transfer conditions have been established, and the source current is now determined by:

$$I_s = \frac{E}{R_T} = \frac{120\text{ V}}{512\ \Omega + 512\ \Omega} = \frac{120\text{ V}}{1024\ \Omega} = 117.19\text{ mA}$$

The power to the primary (which equals that to the secondary for the ideal transformer) is:

$$P = I^2 R = (117.19\text{ mA})^2\, 512\ \Omega = \mathbf{7.032\ W}$$

The result is not in milliwatts, as obtained above, and exceeds 7 W, which is a significant improvement.

c. Comparing levels, 7.032 W/426.15 mW = 16.5, or more than 16 times the power delivered to the speaker using the impedance matching transformer.

Another important application of the impedance matching capabilities of a transformer is the matching of the 300-Ω twin line transmission line from a television antenna to the 75-Ω input impedance of today's televisions (ready-made for the 75-Ω coaxial cable), as shown in Fig. 22.28. A match must be made to ensure the strongest signal to the television receiver.

Using the equation $Z_p = a^2 Z_L$ we find:

$$300\ \Omega = a^2 75\ \Omega$$

and
$$a = \sqrt{\frac{300\ \Omega}{75\ \Omega}} = \sqrt{4} = \mathbf{2}$$

with $N_p : N_s = 2 : 1$ (a step-down transformer)

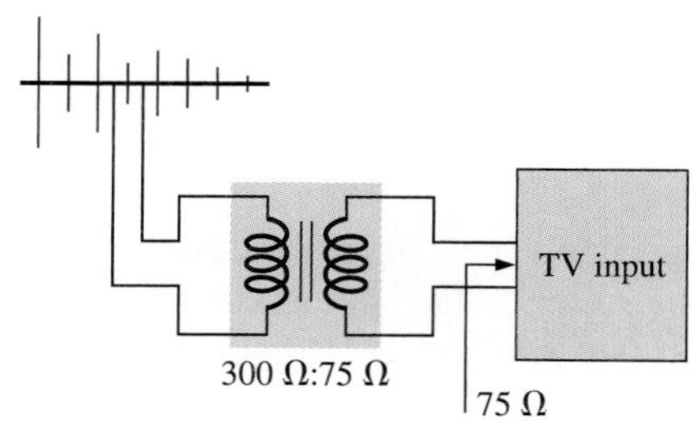

FIG. 22.28
Television impedance matching transformer.

EXAMPLE 22.10

Impedance matching transformers are also quite evident in public address systems, such as the one appearing in the 70.7-V system of Fig. 22.29. Although the system has only one set of output terminals, up to four speakers can be connected to this system (the number is a function of the chosen system). Each 8-Ω speaker is connected to the 70.7-V line through a 10-W audio-matching transformer (defining the frequency range of linear operation).

continued

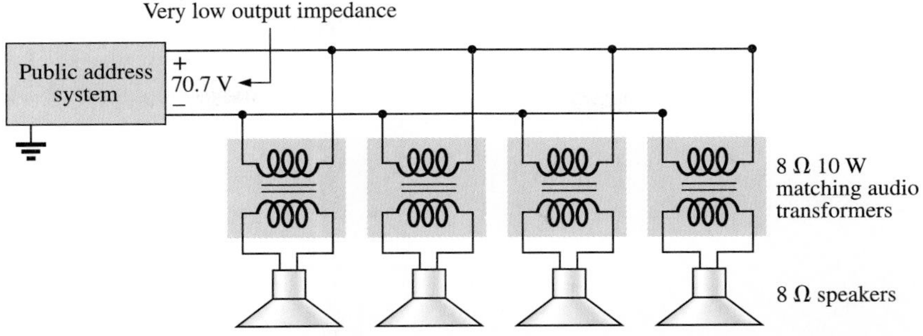

FIG. 22.29
Public address system.

a. If each speaker of Fig. 22.29 can receive 10 W of power, what is the maximum power drain on the source?
b. For each speaker, determine the impedance seen at the input side of the transformer if each is operating under its full 10 W of power.
c. Determine the turns ratio of the transformers.
d. At 10 W, what are the speaker voltage and current?
e. What is the load seen by the source with one, two, three, or four speakers connected?

Solution:

a. Ideally, the primary power equals the power delivered to the load, resulting in a maximum of **40 W** from the supply.

b. The power at the primary:

$$P_p = V_p I_p = (70.7\text{ V})\, I_p = 10\text{ W}$$

and
$$I_p = \frac{10\text{ W}}{70.7\text{ V}} = 141.4\text{ mA}$$

so that
$$Z_p = \frac{V_p}{I_p} = \frac{70.7\text{ V}}{141.4\text{ mA}} = \mathbf{500\ \Omega}$$

c. $Z_p = a^2 Z_L \Rightarrow a = \sqrt{\dfrac{Z_p}{Z_L}} = \sqrt{\dfrac{500\ \Omega}{8\ \Omega}} = \sqrt{62.5} = \mathbf{7.91 \cong 8:1}$

d. $V_s = V_L = \dfrac{V_p}{a} = \dfrac{70.7\text{ V}}{7.91} = \mathbf{8.94\ V \cong 9\ V}$

e. All the speakers are in parallel. Therefore,

One speaker: $R_T = \mathbf{500\ \Omega}$

Two speakers: $R_T = \dfrac{500\ \Omega}{2} = \mathbf{250\ \Omega}$

Three speakers: $R_T = \dfrac{500\ \Omega}{3} = \mathbf{167\ \Omega}$

Four speakers: $R_T = \dfrac{500\ \Omega}{4} = \mathbf{125\ \Omega}$

Even though the load seen by the source will vary with the number of speakers connected, the source impedance is so low (compared to the lowest load of 125 Ω) that the terminal voltage of 70.7 V is essentially constant. This is not the case where the desired result is to match the load to the input impedance; rather, it was to ensure 70.7 V at each primary, no matter how many speakers were connected, and to limit the current drawn from the supply.

The transformer is frequently used to isolate one portion of an electrical system from another. *Isolation* implies the absence of any direct physical connection. As a first example of its use as an isolation device, consider the measurement of line voltages on the order of 40 000 V (Fig. 22.30).

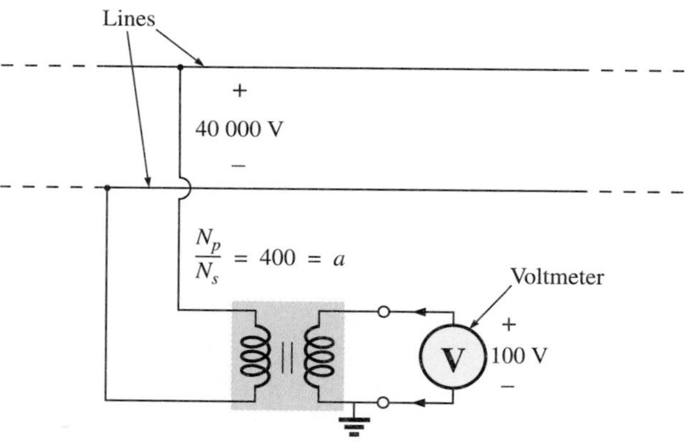

FIG. 22.30
Isolating a high-voltage line from the point of measurement.

To apply a voltmeter across 40 000 V would obviously be a dangerous task due to the possibility of physical contact with the lines when making the necessary connections. By including a transformer in the transmission system as original equipment, one can bring the potential down to a safe level for measurement purposes and can determine the line voltage using the turns ratio. Therefore, the transformer will serve both to isolate and to step down the voltage.

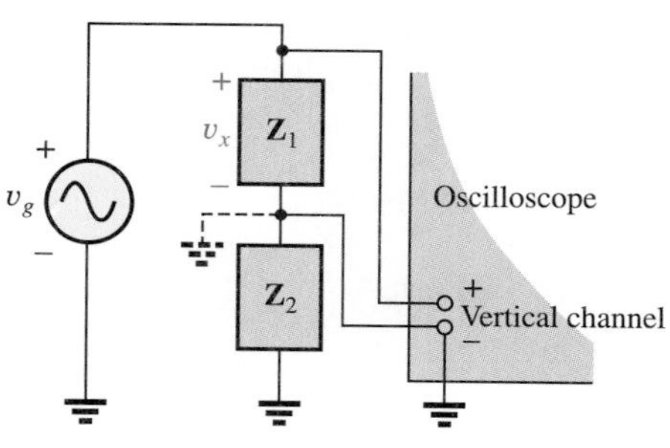

FIG. 22.31
Demonstrating the shorting effect introduced by the grounded side of the vertical channel of an oscilloscope.

As a second example, consider the application of the voltage v_x to the vertical input of the oscilloscope (a measuring instrument) in Fig. 22.31. If the connections are made as shown, and if the generator and oscilloscope have a common ground, the impedance $\mathbf{Z}_2$ has been effectively shorted out of the circuit by the ground connection of the oscilloscope. The input voltage to the oscilloscope will therefore be meaningless as far as the voltage v_x is concerned. In addition, if $\mathbf{Z}_2$ is the current-limiting impedance in the circuit, the current in the circuit may rise to a level that will cause severe damage to the circuit. If a transformer is used as shown in Fig. 22.32, this problem will be eliminated, and the input voltage to the oscilloscope will be v_x.

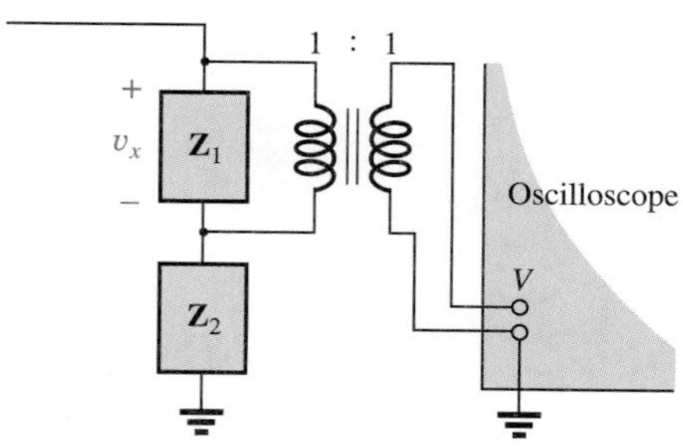

FIG. 22.32
Correcting the situation of Fig. 22.31 using an isolation transformer.

The linear variable differential transformer (LVDT) is a sensor that can reveal displacement using transformer effects. In its simplest form, the LVDT has a central winding and two secondary windings, as shown in Fig. 22.33(a). A ferromagnetic core inside the windings is free to move as dictated by some external force. A constant, low-level ac voltage is applied to the primary, and the output voltage is the difference between the voltages induced in the secondaries. If the core is in the position shown in Fig. 22.33(b), a relatively large voltage will be induced across the secondary winding labeled coil 1, and a relatively small voltage will be induced across the secondary winding labeled coil 2 (essentially an air-core transformer for this position). The result is a relatively large secondary output voltage. If the core is in the position shown in Fig. 22.33(c), the flux linking each coil is the

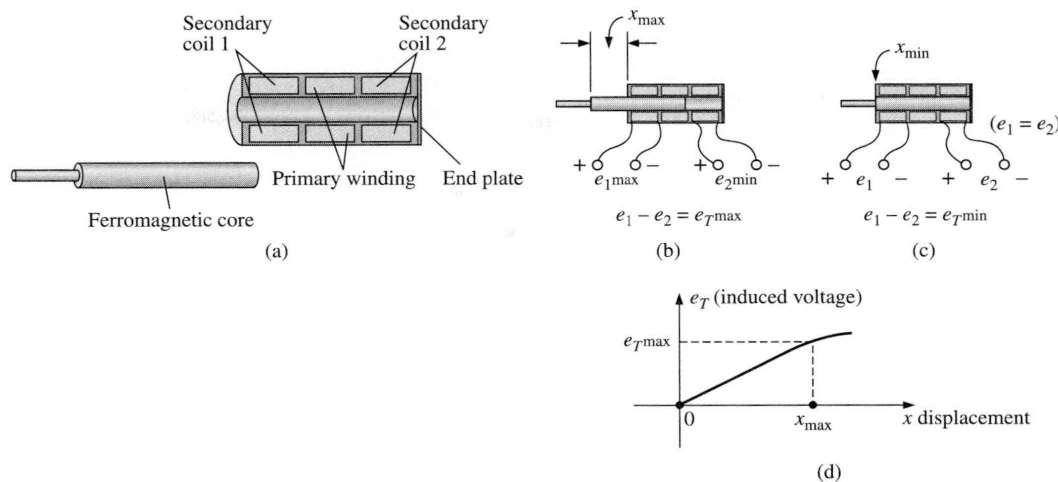

FIG. 22.33
LVDT transformer: (a) construction; (b) maximum displacement; (c) minimum displacement; (d) graph of induced voltage versus displacement.

same, and the output voltage (being the difference) will be quite small. In total, therefore, the position of the core can be related to the secondary voltage, and a position-versus-voltage graph can be developed as shown in Fig. 22.33(d). Due to the nonlinearity of the *B-H* curve, the curve becomes somewhat nonlinear if the core is moved too far out of the unit.

22.10 NAMEPLATE DATA

A typical iron-core power transformer rating, included in the **nameplate data** for the transformer, might be the following:

5 kVA 2000/100 V 60 Hz

The 2000 V or the 100 V can be either the primary or the secondary voltage; that is, if 2000 V is the primary voltage, then 100 V is the secondary voltage, and vice versa. The 5 kVA is the apparent power ($S = VI$) rating of the transformer. If the secondary voltage is 100 V, then the maximum load current is

$$I_L = \frac{S}{V_L} = \frac{5000\text{ VA}}{100\text{ V}} = 50\text{ A}$$

and if the secondary voltage is 2000 V, then the maximum load current is

$$I_L = \frac{S}{V_L} = \frac{5000\text{ VA}}{2000\text{ V}} = 2.5\text{ A}$$

The transformer is rated in terms of the apparent power rather than the average, or real, power for the reason demonstrated by the circuit of Fig. 22.34. Since the current through the load is greater than that determined by the apparent power rating, the transformer may be permanently damaged. Note, however, that since the load is purely capacitive, the average power to the load is zero. The wattage rating would therefore be meaningless regarding the ability of this load to damage the transformer.

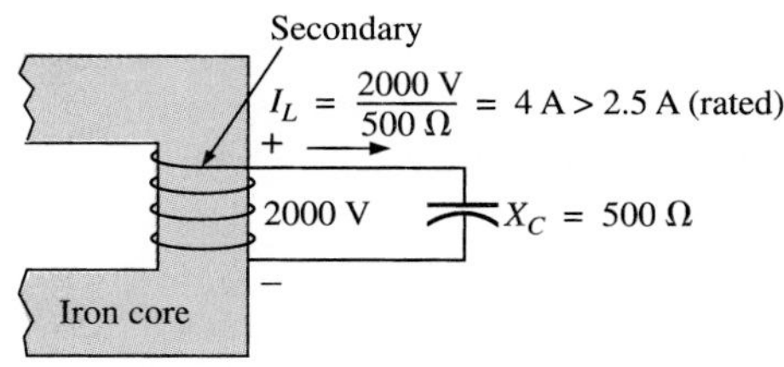

FIG. 22.34
Demonstrating why transformers are rated in kVA rather than kW.

The transformation ratio of the transformer under discussion can be either of two values. If the secondary voltage is 2000 V, the transformation ratio is $a = N_p/N_s = V_g/V_L = 100\text{ V}/2000\text{ V} = 1/20$, and the transformer is a step-up transformer. If the secondary voltage is 100 V, the transformation ratio is $a = N_p/N_s = V_g/V_L = 2000\text{ V}/100\text{ V} = 20$, and the transformer is a step-down transformer.

The rated primary current can be determined simply by applying Eq. (22.23):

$$I_p = \frac{I_s}{a}$$

which is equal to $[2.5\text{ A}/(1/20)] = 50\text{ A}$ if the secondary voltage is 2000 V, and $(50\text{ A}/20) = 2.5\text{ A}$ if the secondary voltage is 100 V.

To explain the necessity for including the frequency in the nameplate data, consider Eq. (22.7):

$$E_p = 4.44 f_p N_p \Phi_m$$

and the magnetization curve for the iron core of the transformer (Fig. 22.35).

The point of operation on the magnetization curve for most transformers is at the knee of the curve. If the frequency of the applied signal should drop, and N_p and E_p remain the same, then Φ_m must increase in magnitude, as determined by Eq. (22.7):

$$\Phi_m\uparrow = \frac{E_p}{4.44 f_p \downarrow N_p}$$

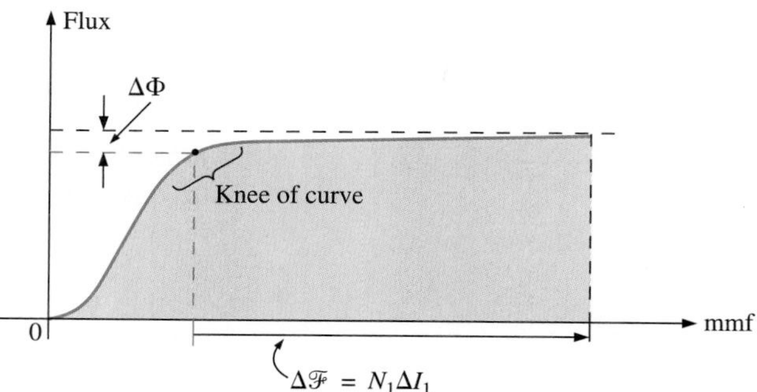

FIG. 22.35
Demonstrating why the frequency of application is important for transformers.

Due to the nonlinearity of the curve, as the flux increases, as shown in Fig. 22.35, the current increases also, but at a greater rate. The resulting ΔI could cause a very high current in the primary, resulting in possible damage to the transformer.

Magnetization curves are often plotted as the flux density B vs. the magnetic intensity H instead of flux vs. mmf as shown here. Refer to the expanded material in Chapter 11 of the CD ROM for further discussion of magnetic intensity H and the B-H curve.

22.11 TYPES OF TRANSFORMERS

Transformers are available in many different shapes and sizes. Some of the more common types include the power transformer, audio transformer, IF (intermediate-frequency) transformer, and RF (radio-frequency) transformer. Each is designed to fulfill a particular requirement in a specific area of application. The symbols for some of the basic types of transformers are shown in Fig. 22.36.

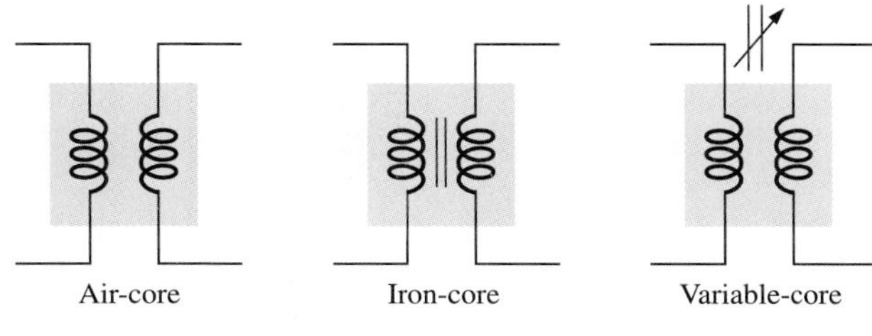

FIG. 22.36
Transformer symbols.

The method of construction varies from one transformer to another. Two of the many different ways in which the primary and secondary coils can be wound around an iron core are shown in Fig. 22.37. In either case, the core is made of laminated sheets of ferromagnetic material separated by an insulator to reduce the eddy current losses. The sheets themselves will also contain a small percentage of silicon to increase the electrical resistivity of the material and further reduce the eddy current losses.

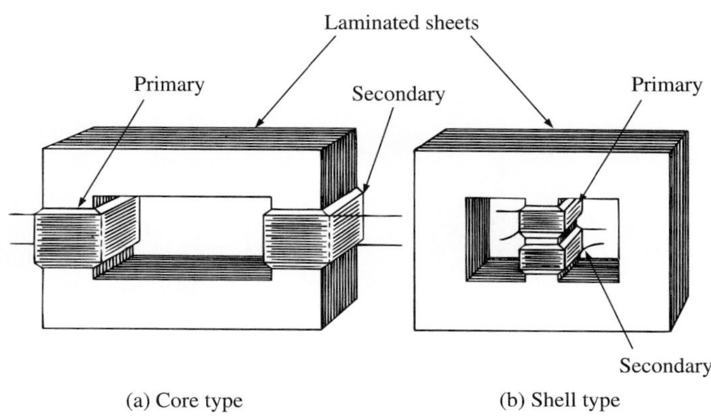

FIG. 22.37
Types of ferromagnetic core construction.

FIG. 22.38
Split bobbin, low-profile power transformer. (Courtesy of Microtran Company, Inc.)

A variation of the core-type transformer appears in Fig. 22.38. This transformer is designed for low-profile (the 2.5-VA size has a maximum height of only 0.65 in.) applications in power, control, and instrumentation applications. There are actually two transformers on the same core, with the primary and secondary of each wound side by side. The schematic representation appears in the same figure. Each set of terminals on the left can accept 115 V at 50 or 60 Hz, whereas each side of the output will provide 230 V at the same frequency. Note the dot convention, as described earlier in the chapter.

The **autotransformer** [Fig. 22.39(b)] is a type of power transformer that, instead of employing the two-circuit principle (complete isolation between coils), has one winding common to both the input and the output circuits. The induced voltages are related to the turns ratio in the same manner as that described for the two-circuit transformer. If the proper connection is used, a two-circuit power transformer can be employed as an autotransformer. The advantage of using it as an autotransformer is that a larger apparent power can be transformed. This can be demonstrated by the two-circuit transformer of Fig. 22.39(a), shown in Fig. 22.39(b) as an autotransformer.

For the two-circuit transformer, note that $S = (\frac{1}{20}\text{ A})(120\text{ V}) = 6$

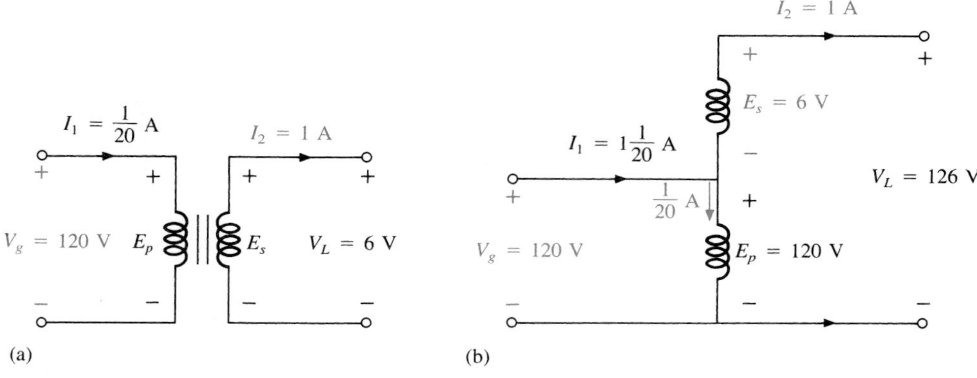

FIG. 22.39
(a) Two-circuit transformer; (b) autotransformer.

VA, whereas for the autotransformer, $S = (1\frac{1}{20}\text{ A})(120\text{ V}) = 126$ VA, which is many times that of the two-circuit transformer. Note also that the current and voltage of each coil are the same as those for the two-circuit configuration. The disadvantage of the autotransformer is obvious: loss of the isolation between the primary and secondary circuits.

A pulse transformer designed for printed-circuit applications where high-amplitude, long-duration pulses must be transferred without saturation appears in Fig. 22.40. Turns ratios are available from 1 : 1 to 5 : 1 at maximum line voltages of 240 V rms at 60 Hz. The upper unit is for printed-circuit applications with isolated dual primaries, whereas the lower unit is the bobbin variety with a single primary winding.

Two miniature ($\frac{1}{4}$ in. by $\frac{1}{4}$ in.) transformers with plug-in or insulated leads appear in Fig. 22.41, along with their schematic representations. Power ratings of 100 mW or 125 mW are available with a variety of turns ratios, such as 1 : 1, 5 : 1, 9.68 : 1, and 25 : 1.

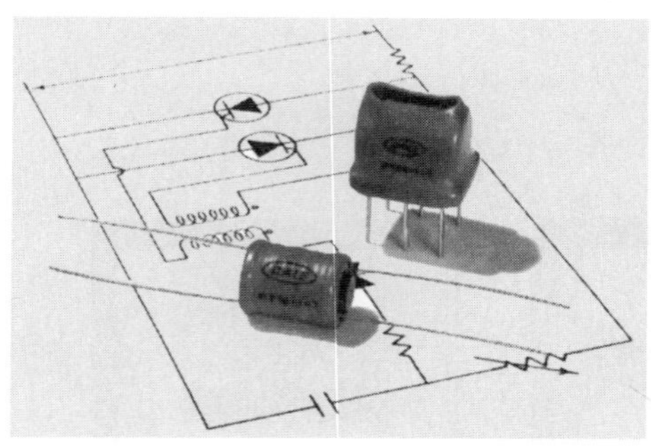

FIG. 22.40
Pulse transformers. (Courtesy of DALE Electronics, Inc.)

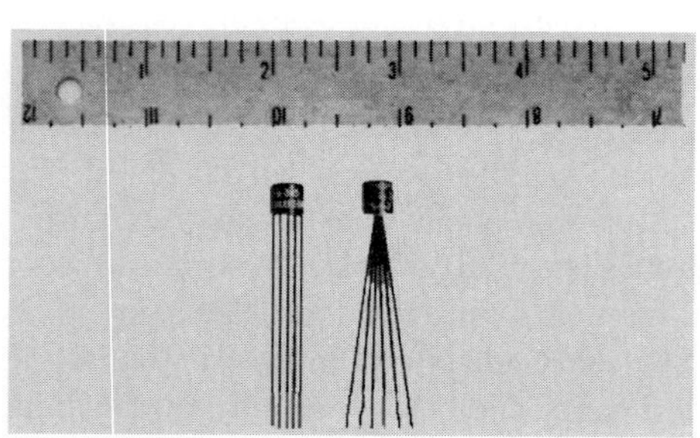

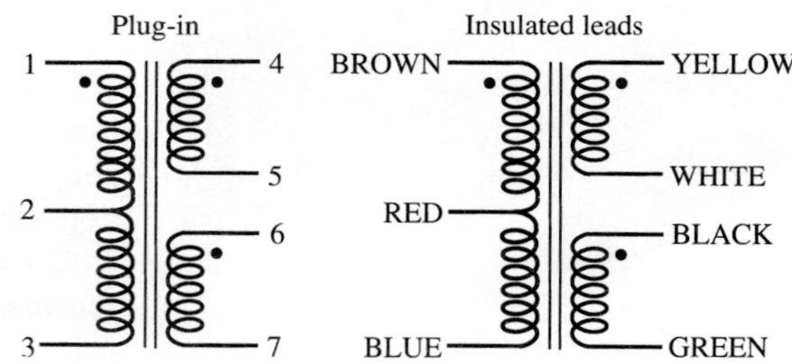

FIG. 22.41
Miniature transformers. (Courtesy of PICO Electronics, Inc.)

22.12 TAPPED AND MULTIPLE-LOAD TRANSFORMERS

For the **center-tapped** (primary) **transformer** of Fig. 22.42, where the voltage from the center tap to either outside lead is defined as $E_p/2$, the relationship between E_p and E_s is

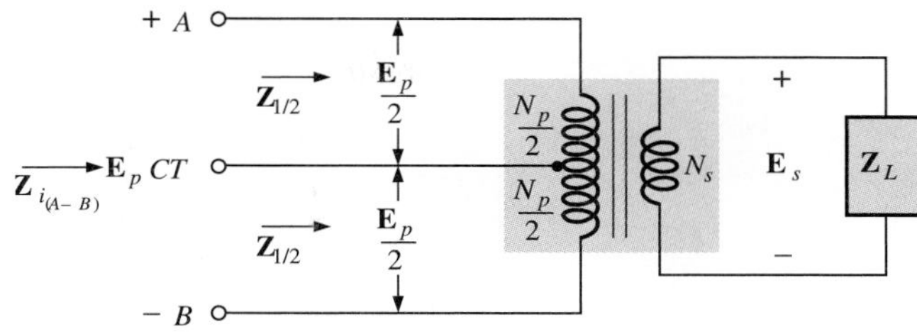

FIG. 22.42
Ideal transformer with a center-tapped primary.

$$\frac{\mathbf{E}_p}{\mathbf{E}_s} = \frac{N_p}{N_s} \qquad (22.37)$$

For each half-section of the primary,

$$\mathbf{Z}_{1/2} = \left(\frac{N_p/2}{N_s}\right)^2 \mathbf{Z}_L = \frac{1}{4}\left(\frac{N_p}{N_s}\right)^2 \mathbf{Z}_L$$

with
$$\mathbf{Z}_{i_{(A-B)}} = \left(\frac{N_p}{N_s}\right)^2 \mathbf{Z}_L$$

Therefore,
$$\mathbf{Z}_{1/2} = \frac{1}{4}\mathbf{Z}_i \qquad (22.38)$$

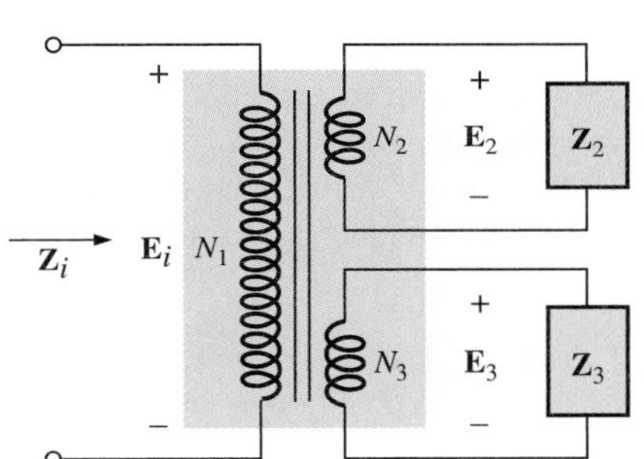

FIG. 22.43
Ideal transformer with multiple loads.

For the **multiple-load transformer** of Fig. 22.43, the following equations apply:

$$\frac{\mathbf{E}_i}{\mathbf{E}_2} = \frac{N_1}{N_2} \quad \frac{\mathbf{E}_1}{\mathbf{E}_3} = \frac{N_1}{N_3} \quad \frac{\mathbf{E}_2}{\mathbf{E}_3} = \frac{N_2}{N_3} \qquad (22.39)$$

The total input impedance can be determined by first noting that, for the ideal transformer, the power delivered to the primary is equal to the power dissipated by the load; that is,

$$P_1 = P_{L_2} + P_{L_3}$$

and, for resistive loads ($\mathbf{Z}_i = R_i$, $\mathbf{Z}_2 = R_2$, and $\mathbf{Z}_3 = R_3$),

$$\frac{E_i^2}{R_i} = \frac{E_2^2}{R_2} + \frac{E_3^2}{R_3}$$

or, since
$$E_2 = \frac{N_2}{N_1}E_i \text{ and } E_3 = \frac{N_3}{N_1}E_1$$

then
$$\frac{E_i^2}{R_i} = \frac{[(N_2/N_1)E_i]^2}{R_2} + \frac{[(N_3/N_1)E_i]^2}{R_3}$$

and
$$\frac{E_i^2}{R_i} = \frac{E_i^2}{(N_1/N_2)^2 R_2} + \frac{E_i^2}{(N_1/N_3)^2 R_3}$$

Thus,
$$\frac{1}{R_i} = \frac{1}{(N_1/N_2)^2 R_2} + \frac{1}{(N_1/N_3)^2 R_3} \qquad (22.40)$$

indicating that the load resistances are reflected in parallel.

For the configuration of Fig. 22.44, with E_2 and E_3 defined as shown, Equations (22.39) and (22.40) are applicable.

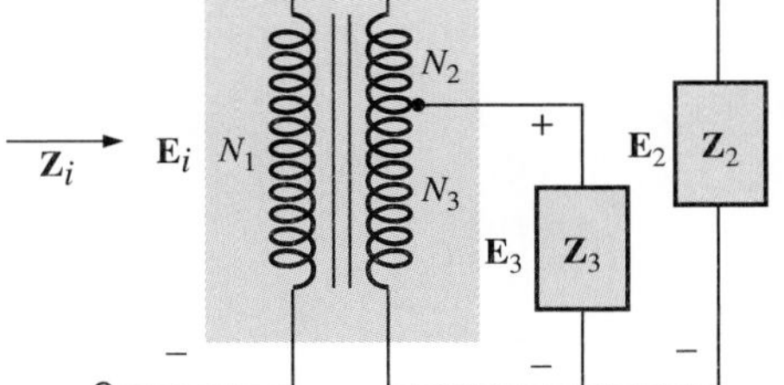

FIG. 22.44
Ideal transformer with a tapped secondary and multiple loads.

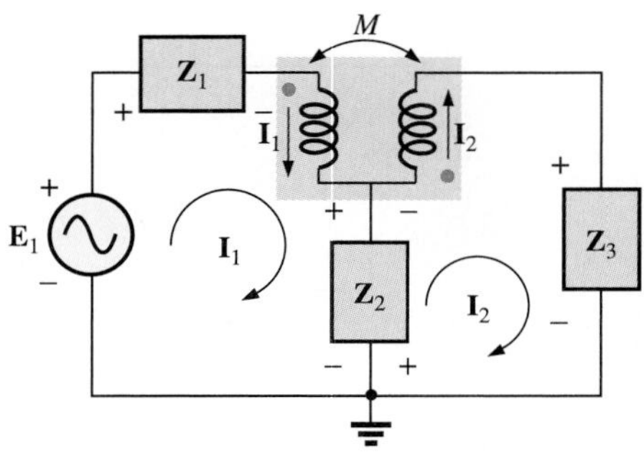

FIG. 22.45
Applying mesh analysis to magnetically coupled coils.

22.13 NETWORKS WITH MAGNETICALLY COUPLED COILS

For multiloop networks with magnetically coupled coils, the mesh-analysis approach is most frequently applied. A firm understanding of the dot convention discussed earlier should make the writing of the equations quite direct and free of errors. Before writing the equations for any particular loop, first determine whether the mutual term is positive or negative, keeping in mind that it will have the same sign as that for the other magnetically coupled coil. For the two-loop network of Fig. 22.45, for example, the mutual term has a positive sign since the current through each coil leaves the dot. For the primary loop,

$$\mathbf{E}_1 - \mathbf{I}_1\mathbf{Z}_1 - \mathbf{I}_1\mathbf{Z}_{L_1} - \mathbf{I}_2\mathbf{Z}_m - \mathbf{Z}_2(\mathbf{I}_1 - \mathbf{I}_2) = 0$$

where M of $\mathbf{Z}_m = \omega M \angle 90°$ is positive, and

$$\mathbf{I}_1(\mathbf{Z}_1 + \mathbf{Z}_{L_1} + \mathbf{Z}_2) - \mathbf{I}_2(\mathbf{Z}_2 - \mathbf{Z}_m) = \mathbf{E}_1$$

Note in the above that the mutual impedance was treated as if it were an additional inductance in series with the inductance L_1 having a sign determined by the dot convention and the voltage across which is determined by the current in the magnetically coupled loop.

For the secondary loop,

$$-\mathbf{Z}_2(\mathbf{I}_2 - \mathbf{I}_1) - \mathbf{I}_2\mathbf{Z}_{L_2} - \mathbf{I}_1\mathbf{Z}_m - \mathbf{I}_2\mathbf{Z}_3 = 0$$

or

$$\mathbf{I}_2(\mathbf{Z}_2 + \mathbf{Z}_{L_2} + \mathbf{Z}_3) - \mathbf{I}_1(\mathbf{Z}_2 - \mathbf{Z}_m) = 0$$

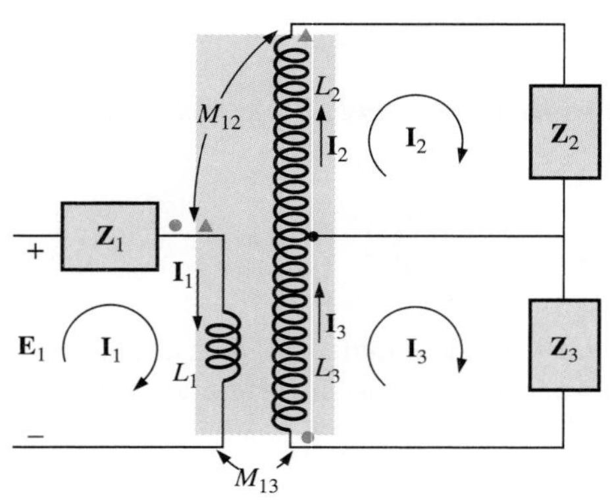

FIG. 22.46
Applying mesh analysis to a network with two magnetically coupled coils.

For the network of Fig. 22.46, we find a mutual term between L_1 and L_2 and L_1 and L_3, labeled M_{12} and M_{13}, respectively.

For the coils with the dots (L_1 and L_3), since each current through the coils leaves the dot, M_{13} is positive for the chosen direction of I_1 and I_3. However, since the current I_1 leaves the dot through L_1, and I_2 enters the dot through coil L_2, M_{12} is negative. Consequently, for the input circuit,

$$\mathbf{E}_1 - \mathbf{I}_1\mathbf{Z}_1 - \mathbf{I}_1\mathbf{Z}_{L_1} - \mathbf{I}_2(-\mathbf{Z}_{m_{12}}) - \mathbf{I}_3\mathbf{Z}_{m_{13}} = 0$$

or

$$\mathbf{E}_1 - \mathbf{I}_1(\mathbf{Z}_1 + \mathbf{Z}_{L_1}) + \mathbf{I}_2\mathbf{Z}_{m_{12}} - \mathbf{I}_3\mathbf{Z}_{m_{13}} = 0$$

For loop 2,

$$-\mathbf{I}_2\mathbf{Z}_2 - \mathbf{I}_2\mathbf{Z}_{L_2} - \mathbf{I}_1(-\mathbf{Z}_{m_{12}}) = 0$$
$$-\mathbf{I}_1\mathbf{Z}_{m_{12}} + \mathbf{I}_2(\mathbf{Z}_2 + \mathbf{Z}_{L_2}) = 0$$

and for loop 3,

$$-\mathbf{I}_3\mathbf{Z}_3 - \mathbf{I}_3\mathbf{Z}_{L_3} - \mathbf{I}_1\mathbf{Z}_{m_{13}} = 0$$

or

$$\mathbf{I}_1\mathbf{Z}_{m_{13}} + \mathbf{I}_3(\mathbf{Z}_3 + \mathbf{Z}_{L_3}) = 0$$

In determinant form,

$$\begin{aligned}
\mathbf{I}_1(\mathbf{Z}_1 + \mathbf{Z}_{L_1}) &- \mathbf{I}_2\mathbf{Z}_{m_{12}} &+ \mathbf{I}_3\mathbf{Z}_{m_{13}} &= \mathbf{E}_1\\
-\mathbf{I}_1\mathbf{Z}_{m_{12}} &+ \mathbf{I}_2(\mathbf{Z}_2 + \mathbf{Z}_{L_{12}}) &+ 0 &= 0\\
\mathbf{I}_1\mathbf{Z}_{m_{13}} &+ 0 &+ \mathbf{I}_3(\mathbf{Z}_3 + \mathbf{Z}_{13}) &= 0
\end{aligned}$$

PROBLEMS

SECTION 22.2 Mutual Inductance

1. For the air-core transformer of Fig. 22.47:
   a. Find the value of L_s if the mutual inductance M is equal to 80 mH.
   b. Find the induced voltages e_p and e_s if the flux linking the primary coil changes at the rate of 0.08 Wb/s.
   c. Find the induced voltages e_p and e_s if the current i_p changes at the rate of 0.3 A/ms.
2. a. Repeat Problem 1 if k is changed to 1.
   b. Repeat Problem 1 if k is changed to 0.2.
   c. Compare the results of parts (a) and (b).
3. Repeat Problem 1 for $k = 0.9$, $N_p = 300$ turns, and $N_s = 25$ turns.

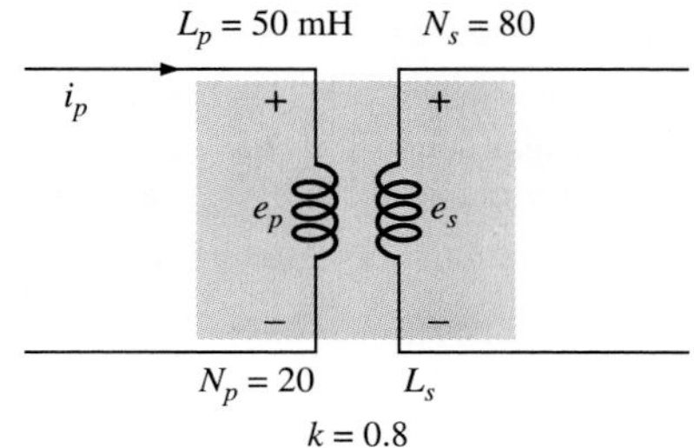

FIG. 22.47
Problems 1, 2, and 3.

SECTION 22.3 Series Connection of Mutually Coupled Coils

4. Determine the total inductance of the series coils of Fig. 22.48.
5. Determine the total inductance of the series coils of Fig. 22.49.

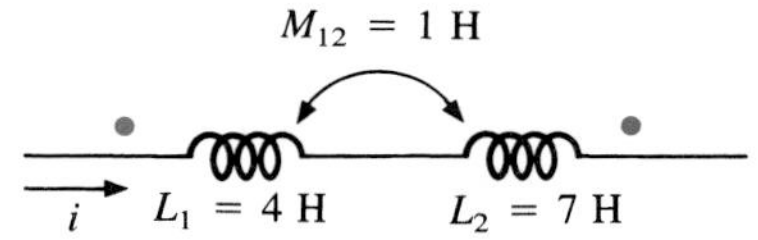

FIG. 22.48
Problem 4.

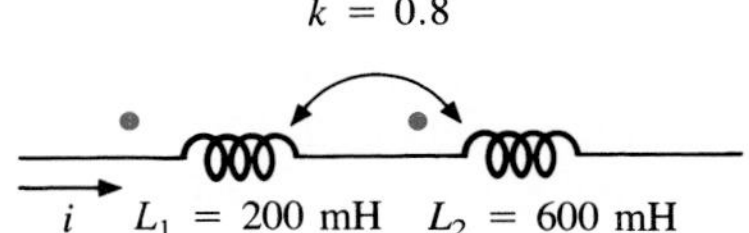

FIG. 22.49
Problem 5.

6. Determine the total inductance of the series coils of Fig. 22.50.
7. Write the mesh equations for the network of Fig. 22.51.

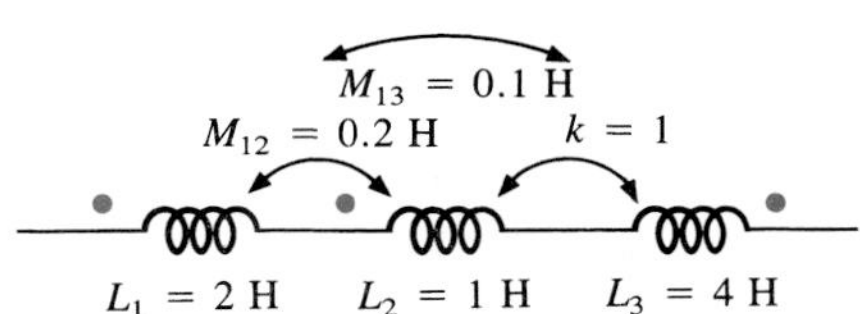

FIG. 22.50
Problem 6.

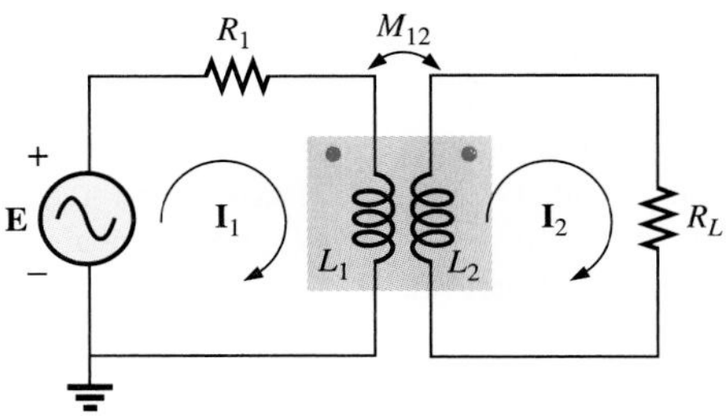

FIG. 22.51
Problem 7.

SECTION 22.4 The Iron-Core Transformer

8. For the iron-core transformer ($k = 1$) of Fig. 22.52:
  a. Find the magnitude of the induced voltage E_s.
  b. Find the maximum flux Φ_m.

9. Repeat Problem 8 for $N_p = 240$ and $N_s = 30$.

10. Find the applied voltage of an iron-core transformer if the secondary voltage is 240 V, and $N_p = 60$ with $N_s = 720$.

11. If the maximum flux passing through the core of Problem 8 is 12.5 mWb, find the frequency of the input voltage.

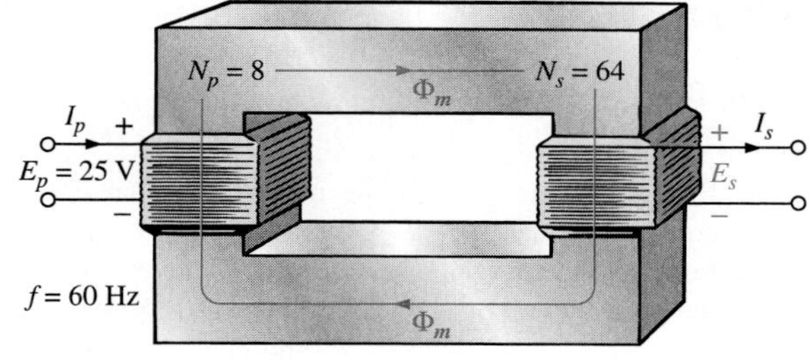

FIG. 22.52
Problems 8, 9, and 11.

SECTION 22.5 Reflected Impedance and Power

12. For the iron-core transformer of Fig. 22.53:
  a. Find the magnitude of the current I_L and the voltage V_L if $a = 1/5$, $I_p = 2$ A, and $Z_L = 2\text{-}\Omega$ resistor.
  b. Find the input resistance for the data specified in part (a).

13. Find the input impedance for the iron-core transformer of Fig. 22.53 if $a = 2$, $I_p = 4$ A, and $V_g = 1600$ V.

14. Find the voltage V_g and the current I_p if the input impedance of the iron-core transformer of Fig. 22.53 is 4 Ω, and $V_L = 1200$ V and $a = 1/4$.

15. If $V_L = 240$ V, $Z_L = 20\text{-}\Omega$ resistor, $I_p = 0.05$ A, and $N_s = 50$, find the number of turns in the primary circuit of the iron-core transformer of Fig. 22.53.

16. **a.** If $N_p = 400$, $N_s = 1200$, and $V_g = 100$ V, find the magnitude of I_p for the iron-core transformer of Fig. 22.53 if $Z_L = 9\ \Omega + j\ 12\ \Omega$.
  b. Find the magnitude of the voltage V_L and the current I_L for the conditions of part (a).

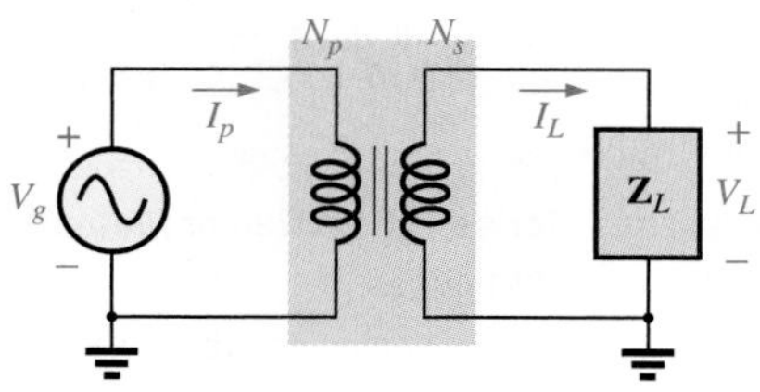

FIG. 22.53
Problems 12 through 16.

SECTION 22.6 Equivalent Circuit (Iron-Core Transformer)

17. For the transformer of Fig. 22.54, determine
  a. the equivalent resistance R_e.
  b. the equivalent reactance X_e.
  c. the equivalent circuit reflected to the primary.
  d. the primary current for $\mathbf{V}_g = 50\ \text{V}\ \angle 0°$.
  e. the load voltage V_L.
  f. the phasor diagram of the reflected primary circuit.
  g. the new load voltage if we assume the transformer to be ideal with a 4 : 1 turns ratio. Compare the result with that of part (e).

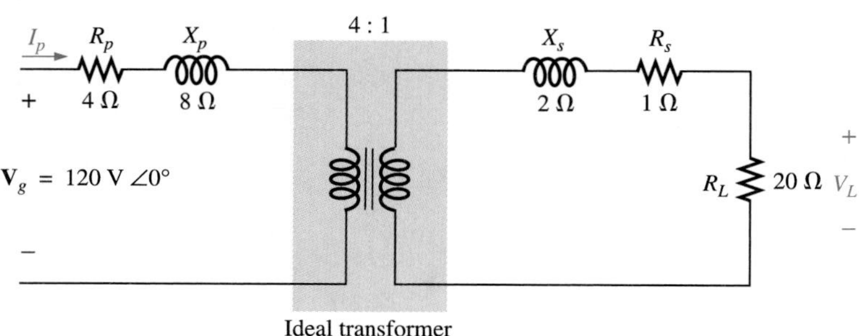

FIG. 22.54
Problems 17, 30, and 31.

18. For the transformer of Fig. 22.55, if the resistive load is replaced by an inductive reactance of 20 Ω:
   a. Determine the total reflected primary impedance.
   b. Calculate the primary current.
   c. Determine the voltage across R_e and X_e, and find the reflected load.
   d. Draw the phasor diagram.

19. Repeat Problem 18 for a capacitive load having a reactance of 20 Ω.

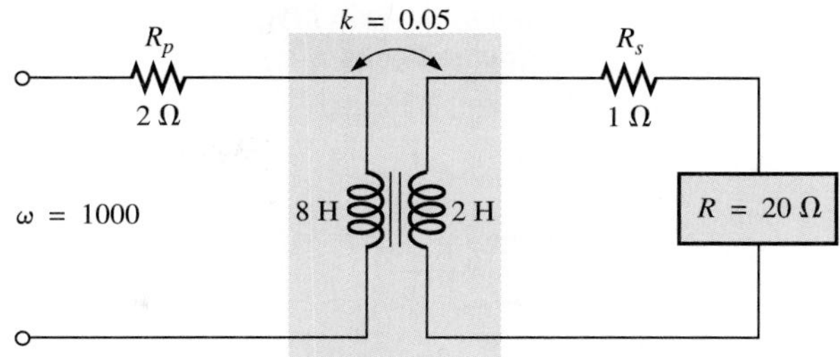

FIG. 22.55
Problems 18, 19, and 32.

SECTION 22.7 Frequency Considerations

20. Discuss in your own words the frequency characteristics of the transformer. Employ the applicable equivalent circuit and frequency characteristics appearing in this chapter.

SECTION 22.8 Air-Core Transformer

21. Determine the input impedance to the air-core transformer of Fig. 22.56. Sketch the reflected primary network.

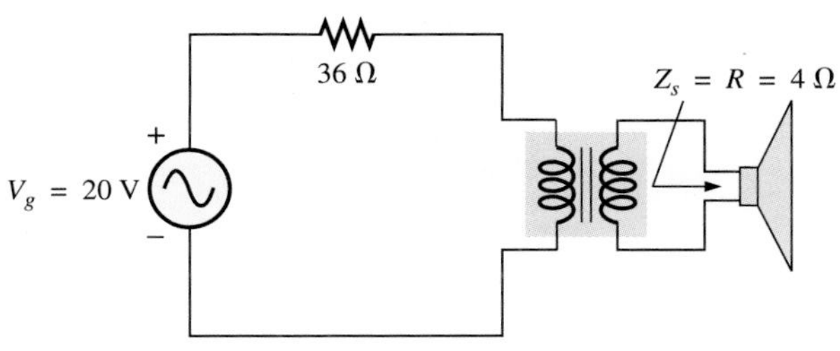

FIG. 22.56
Problems 21 and 22.

SECTION 22.9 Impedance Matching, Isolation, and Displacement

22. a. For the circuit of Fig. 22.56, find the transformation ratio required to deliver maximum power to the speaker.
   b. Find the maximum power delivered to the speaker.

SECTION 22.10 Nameplate Data

23. An ideal transformer is rated 10 kVA, 2400/120 V, 60 Hz.
   a. Find the transformation ratio if the 120 V is the secondary voltage.
   b. Find the current rating of the secondary if the 120 V is the secondary voltage.
   c. Find the current rating of the primary if the 120 V is the secondary voltage.
   d. Repeat parts (a) through (c) if the 2400 V is the secondary voltage.

SECTION 22.11 Types of Transformers

24. Determine the primary and secondary voltages and currents for the autotransformer of Fig. 22.57.

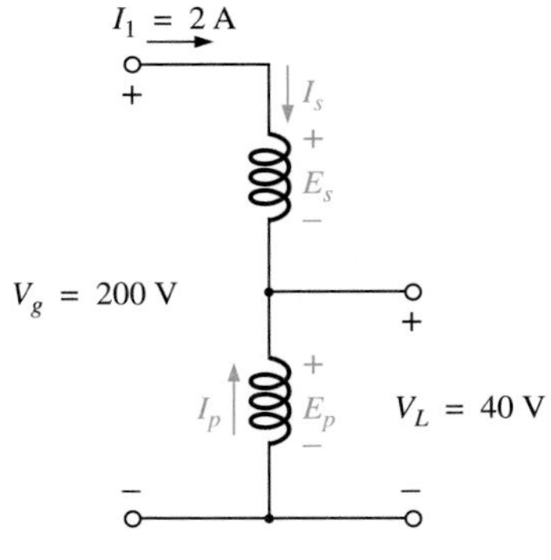

FIG. 22.57
Problem 24.

SECTION 22.12 Tapped and Multiple-Load Transformers

25. For the center-tapped transformer of Fig. 22.42 where $N_p = 100$, $N_s = 25$, $Z_L = R\ \angle 0° = 5\ \Omega\ \angle 0°$, and $\mathbf{E}_p = 100\ \text{V}\ \angle 0°$:
   a. Determine the load voltage and current.
   b. Find the impedance $\mathbf{Z}_i$.
   c. Calculate the impedance $\mathbf{Z}_{1/2}$.

26. For the multiple-load transformer of Fig. 22.43 where $N_1 = 90$, $N_2 = 15$, $N_3 = 45$, $\mathbf{Z}_2 = R_2\ \angle 0° = 8\ \Omega\ \angle 0°$, $\mathbf{Z}_3 = R_L\ \angle 0° = 5\ \Omega\ \angle 0°$, and $\mathbf{E}_i = 60\ \text{V}\ \angle 0°$:
   a. Determine the load voltages and currents.
   b. Calculate $\mathbf{Z}_1$.

27. For the multiple-load transformer of Fig. 22.44 where $N_1 = 120$, $N_2 = 40$, $N_3 = 30$, $\mathbf{Z}_2 = R_2\ \angle 0° = 12\ \Omega\ \angle 0°$, $\mathbf{Z}_3 = R_3\ \angle 0° = 10\ \Omega\ \angle 0°$, and $\mathbf{E}_1 = 120\ \text{V}\ \angle 60°$:
   a. Determine the load voltages and currents.
   b. Calculate $\mathbf{Z}_1$.

SECTION 22.13 Networks with Magnetically Coupled Coils

28. Write the mesh equations for the network of Fig. 22.58.

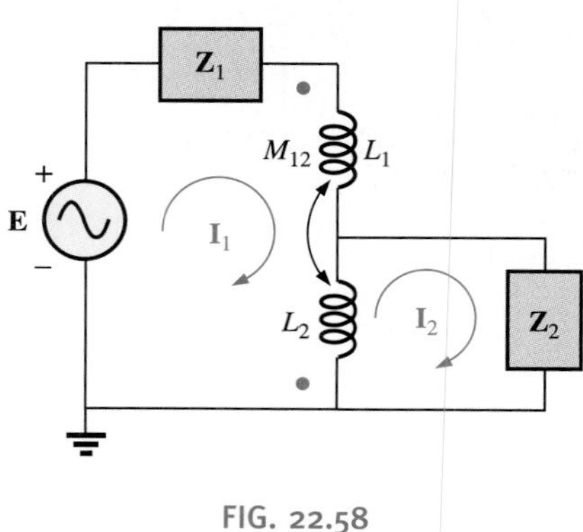

FIG. 22.58
Problem 28.

29. Write the mesh equations for the network of Fig. 22.59.

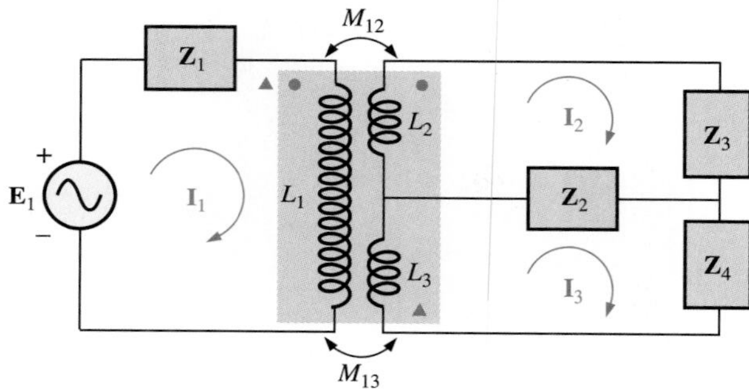

FIG. 22.59
Problem 29.

Appendices A–F

Appendix A
Determinants

Appendix B
Logarithms

Appendix C
The Greek Alphabet

Appendix D
Magnetic Parameter Conversions

Appendix E
Maximum Power Transfer Conditions

Appendix F
Answers to Selected Problems

Appendix A

Determinants

Determinants are used to solve systems of simultaneous equations. This procedure is particularly helpful when you have more than two simultaneous equations (in more than two unknowns). Determinants are recommended for solving complicated networks.

Consider the following equations, where x and y are the unknown variables and a_1, a_2, b_1, b_2, c_1, and c_2 are constants:

Col. 1	Col. 2	Col. 3
$a_1x +$	$b_1y =$	c_1
$a_2x +$	$b_2y =$	c_2

(A.1a)

(A.1b)

It is certainly possible to solve for one variable in Eq. (A.1a) and substitute into Eq. (A.1b). That is, solving for x in Eq. (A.1a),

$$x = \frac{c_1 - b_1y}{a_1}$$

and substituting the result in Eq. (A.1b),

$$a_2\left(\frac{c_1 - b_1y}{a_1}\right) + b_2y = c_2$$

It is now possible to solve for y, since it is the only variable remaining, and then substitute into either equation for x. This is acceptable for two equations, but it becomes a very slow and lengthy process for three or more simultaneous equations.

Using **determinants** to solve for x and y requires establishing the following formats for each variable:

Col. 1, Col. 2

$$x = \frac{\begin{vmatrix} c_1 & b_1 \\ c_2 & b_2 \end{vmatrix}}{\begin{vmatrix} a_1 & b_1 \\ a_2 & b_2 \end{vmatrix}} \qquad y = \frac{\begin{vmatrix} a_1 & c_1 \\ a_2 & c_2 \end{vmatrix}}{\begin{vmatrix} a_1 & b_1 \\ a_2 & b_2 \end{vmatrix}} \tag{A.2}$$

First note that only constants appear within the vertical brackets and that the denominator of each is the same. In fact, the denominator is simply the coefficients of x and y in the same arrangement as in Eqs. (A.1a) and (A.1b). When solving for x, the coefficients of x in the numerator are replaced by the constants to the right of the equals sign in Eqs. (A.1a) and (A.1b), whereas the coefficients of the y variable are simply repeated. When solving for y, the y coefficients in the numerator are replaced by the constants to the right of the equal sign and the coefficients of x are repeated.

Each configuration in the numerator and denominator of Eqs. (A.2) is referred to as a *determinant* (D), which can be evaluated numerically in the following way:

$$\text{Determinant} = D = \begin{array}{c} \text{Col. Col.} \\ 1 \quad 2 \\ \hline \begin{vmatrix} a_1 & b_1 \\ a_2 & b_2 \end{vmatrix} \end{array} = a_1b_2 - a_2b_1 \tag{A.3}$$

The expanded value is obtained by first multiplying the top left element by the bottom right and then subtracting the product of the lower left and upper right elements. This particular determinant is referred to as a *second-order* determinant, since it contains two rows and two columns.

It is important to remember when using determinants to place columns of the equations, as indicated in Eqs. (A.1a) and (A.1b), in the same order within the determinant configuration. That is, since a_1 and a_2 are in column 1 of Eqs. (A.1a) and (A.1b), they must be in column 1 of the determinant. (The same is true for b_1 and b_2.)

Expanding the entire expression for x and y, we have the following:

$$x = \frac{\begin{vmatrix} c_1 & b_1 \\ c_2 & b_2 \end{vmatrix}}{\begin{vmatrix} a_1 & b_1 \\ a_2 & b_2 \end{vmatrix}} = \frac{c_1b_2 - c_2b_1}{a_1b_2 - a_2b_1} \tag{A.4a}$$

$$y = \frac{\begin{vmatrix} a_1 & c_1 \\ a_2 & c_2 \end{vmatrix}}{\begin{vmatrix} a_1 & b_1 \\ a_2 & b_2 \end{vmatrix}} = \frac{a_1c_2 - a_2c_1}{a_1b_2 - a_2b_1} \tag{A.4b}$$

EXAMPLE A.1

Evaluate the following determinants:

a. $\begin{vmatrix} 2 & 2 \\ 3 & 4 \end{vmatrix} = (2)(4) - (3)(2) = 8 - 6 = \mathbf{2}$

b. $\begin{vmatrix} 4 & -1 \\ 6 & 2 \end{vmatrix} = (4)(2) - (6)(-1) = 8 + 6 = \mathbf{14}$

c. $\begin{vmatrix} 0 & -2 \\ -2 & 4 \end{vmatrix} = (0)(4) - (-2)(-2) = 0 - 4 = \mathbf{-4}$

d. $\begin{vmatrix} 0 & 0 \\ 3 & 10 \end{vmatrix} = (0)(10) - (3)(0) = \mathbf{0}$

EXAMPLE A.2

Solve for x and y:

$$\begin{aligned} 2x + y &= 3 \\ 3x + 4y &= 2 \end{aligned}$$

continued

Solution:

$$x = \frac{\begin{vmatrix} 3 & 1 \\ 2 & 4 \end{vmatrix}}{\begin{vmatrix} 2 & 1 \\ 3 & 4 \end{vmatrix}} = \frac{(3)(4) - (2)(1)}{(2)(4) - (3)(1)} = \frac{12 - 2}{8 - 3} = \frac{10}{5} = \mathbf{2}$$

$$y = \frac{\begin{vmatrix} 2 & 3 \\ 3 & 2 \end{vmatrix}}{5} = \frac{(2)(2) - (3)(3)}{5} = \frac{4 - 9}{5} = \frac{-5}{5} = \mathbf{-1}$$

Check:

$$\begin{aligned} 2x + y &= (2)(2) + (-1) \\ &= 4 - 1 = 3 \quad \text{(checks)} \\ 3x + 4y &= (3)(2) + (4)(-1) \\ &= 6 - 4 = 2 \quad \text{(checks)} \end{aligned}$$

EXAMPLE A.3

Solve for x and y:

$$\begin{aligned} -x + 2y &= 3 \\ 3x - 2y &= -2 \end{aligned}$$

Solution: In this example, note the effect of the minus sign and the use of parentheses to ensure the proper sign is obtained for each product:

$$x = \frac{\begin{vmatrix} 3 & 2 \\ -2 & -2 \end{vmatrix}}{\begin{vmatrix} -1 & 2 \\ 3 & -2 \end{vmatrix}} = \frac{(3)(-2) - (-2)(2)}{(-1)(-2) - (3)(2)}$$

$$= \frac{-6 + 4}{2 - 6} = \frac{-2}{-4} = \mathbf{\frac{1}{2}}$$

$$y = \frac{\begin{vmatrix} -1 & 3 \\ 3 & -2 \end{vmatrix}}{-4} = \frac{(-1)(-2) - (3)(3)}{-4}$$

$$= \frac{2 - 9}{-4} = \frac{-7}{-4} = \mathbf{\frac{7}{4}}$$

EXAMPLE A.4

Solve for x and y:

$$\begin{aligned} x &= 3 - 4y \\ 20y &= -1 + 3x \end{aligned}$$

Solution: In this case, the equations must first be placed in the format of Eqs. (A.1a) and (A.1b):

$$\begin{aligned} x + 4y &= 3 \\ -3x + 20y &= -1 \end{aligned}$$

continued

$$x = \frac{\begin{vmatrix} 3 & 4 \\ -1 & 20 \end{vmatrix}}{\begin{vmatrix} 1 & 4 \\ -3 & 20 \end{vmatrix}} = \frac{(3)(20) - (-1)(4)}{(1)(20) - (-3)(4)}$$

$$= \frac{60 + 4}{20 + 12} = \frac{64}{32} = \mathbf{2}$$

$$y = \frac{\begin{vmatrix} 1 & 3 \\ -3 & -1 \end{vmatrix}}{32} = \frac{(1)(-1) - (-3)(3)}{32}$$

$$= \frac{-1 + 9}{32} = \frac{8}{32} = \frac{\mathbf{1}}{\mathbf{4}}$$

Determinants are not used only for solving two simultaneous equations; determinants can be applied to any number of simultaneous linear equations. First we will examine a shorthand method that applies to third-order determinants only, since most of the problems in the text are limited to this level of difficulty. We will then investigate the general procedure for solving any number of simultaneous equations.

Consider the three following simultaneous equations:

Col. 1		Col. 2		Col. 3		Col. 4
a_1x	$+$	b_1y	$+$	c_1z	$=$	d_1
a_2x	$+$	b_2y	$+$	c_2z	$=$	d_2
a_3x	$+$	b_3y	$+$	c_3z	$=$	d_3

in which x, y, and z are the variables, and $a_{1,2,3}$, $b_{1,2,3}$, $c_{1,2,3}$ and $d_{1,2,3}$ are constants.

The determinant configuration for x, y, and z can be found in much the same way as we did for two simultaneous equations. That is, to solve for x, find the determinant in the numerator by replacing column 1 with the elements to the right of the equals sign. The denominator is the determinant of the coefficients of the variables (the same applies to y and z). Again, the denominator is the same for each variable.

$$x = \frac{\begin{vmatrix} d_1 & b_1 & c_1 \\ d_2 & b_2 & c_2 \\ d_3 & b_3 & c_3 \end{vmatrix}}{D}, \qquad y = \frac{\begin{vmatrix} a_1 & d_1 & c_1 \\ a_2 & d_2 & c_2 \\ a_3 & d_3 & c_3 \end{vmatrix}}{D}, \qquad z = \frac{\begin{vmatrix} a_1 & b_1 & d_1 \\ a_2 & b_2 & d_2 \\ a_3 & b_3 & d_3 \end{vmatrix}}{D}$$

where $$D = \begin{vmatrix} a_1 & b_1 & c_1 \\ a_2 & b_2 & c_2 \\ a_3 & b_3 & c_3 \end{vmatrix}$$

A shorthand method for evaluating the third-order determinant consists simply of repeating the first two columns of the determinant to the right of the determinant and then summing the products along specific diagonals as shown on the next page:

$$D = \begin{vmatrix} a_1 & b_1 & c_1 \\ a_2 & b_2 & c_2 \\ a_3 & b_3 & c_3 \end{vmatrix} \begin{matrix} a_1 & b_1 \\ a_2 & b_2 \\ a_3 & b_3 \end{matrix}$$

4(−) 5(−) 6(−)

1(+) 2(+) 3(+)

The products of the diagonals 1, 2, and 3 are positive and have the following magnitudes:

$$+a_1b_2c_3 + b_1c_2a_3 + c_1a_2b_3$$

The products of the diagonals 4, 5, and 6 are negative and have the following magnitudes:

$$-a_3b_2c_1 - b_3c_2a_1 - c_3a_2b_1$$

The total solution is the sum of the diagonals 1, 2, and 3 minus the sum of the diagonals 4, 5, and 6:

$$+(a_1b_2c_3 + b_1c_2a_3 + c_1a_2b_3) - (a_3b_2c_1 + b_3c_2a_1 + c_3a_2b_1) \quad \text{(A.5)}$$

Warning: **This method of expansion is good only for third-order determinants!** It cannot be applied to fourth-order and higher systems.

EXAMPLE A.5

Evaluate the following determinant:

$$\begin{vmatrix} 1 & 2 & 3 \\ -2 & 1 & 0 \\ 0 & 4 & 2 \end{vmatrix} \longrightarrow \begin{vmatrix} 1 & 2 & 3 \\ -2 & 1 & 0 \\ 0 & 4 & 2 \end{vmatrix} \begin{matrix} 1 & 2 \\ -2 & 1 \\ 0 & 4 \end{matrix}$$

(−) (−) (−)

(+) (+) (+)

Solution:

$$[(1)(1)(2) + (2)(0)(0) + (3)(-2)(4)]$$
$$-[(0)(1)(3) + (4)(0)(1) + (2)(-2)(2)]$$
$$= (2 + 0 - 24) - (0 + 0 - 8) = (-22) - (-8)$$
$$= -22 + 8 = \mathbf{-14}$$

EXAMPLE A.6

Solve for x, y, and z:

$$1x + 0y - 2z = -1$$
$$0x + 3y + 1z = +2$$
$$1x + 2y + 3z = 0$$

Solution:

$$x = \frac{\begin{vmatrix} -1 & 0 & -2 \\ 2 & 3 & 1 \\ 0 & 2 & 3 \end{vmatrix} \begin{matrix} -1 & 0 \\ 2 & 3 \\ 0 & 2 \end{matrix}}{\begin{vmatrix} 1 & 0 & -2 \\ 0 & 3 & 1 \\ 1 & 2 & 3 \end{vmatrix} \begin{matrix} 1 & 0 \\ 0 & 3 \\ 1 & 2 \end{matrix}}$$

continued

$$= \frac{[(-1)(3)(3) + (0)(1)(0) + (-2)(2)(2)] - [(0)(3)(-2) + (2)(1)(-1) + (3)(2)(0)]}{[(1)(3)(3) + (0)(1)(1) + (-2)(0)(2)] - [(1)(3)(-2) + (2)(1)(1) + (3)(0)(0)]}$$

$$= \frac{(-9 + 0 - 8) - (0 - 2 + 0)}{(9 + 0 + 0) - (-6 + 2 + 0)}$$

$$= \frac{-17 + 2}{9 + 4} = -\mathbf{\frac{15}{13}}$$

$$y = \frac{\begin{vmatrix} 1 & -1 & -2 \\ 0 & 2 & 1 \\ 1 & 0 & 3 \end{vmatrix} \begin{matrix} 1 & -1 \\ 0 & 2 \\ 1 & 0 \end{matrix}}{13}$$

$$= \frac{[(1)(2)(3) + (-1)(1)(1) + (-2)(0)(0)] - [(1)(2)(-2) + (0)(1)(1) + (3)(0)(-1)]}{13}$$

$$= \frac{(6 - 1 + 0) - (-4 + 0 + 0)}{13}$$

$$= \frac{5 + 4}{13} = \mathbf{\frac{9}{13}}$$

$$z = \frac{\begin{vmatrix} 1 & 0 & -1 \\ 0 & 3 & 2 \\ 1 & 2 & 0 \end{vmatrix} \begin{matrix} 1 & 0 \\ 0 & 3 \\ 1 & 2 \end{matrix}}{13}$$

$$= \frac{[(1)(3)(0) + (0)(2)(1) + (-1)(0)(2)] - [(1)(3)(-1) + (2)(2)(1) + (0)(0)(0)]}{13}$$

$$= \frac{(0 + 0 + 0) - (-3 + 4 + 0)}{13}$$

$$= \frac{0 - 1}{13} = -\mathbf{\frac{1}{13}}$$

or from $0x + 3y + 1z = +2$,

$$z = 2 - 3y = 2 - 3\left(\frac{9}{13}\right) = \frac{26}{13} - \frac{27}{13} = -\mathbf{\frac{1}{13}}$$

Check:

$$\left.\begin{aligned} 1x + 0y - 2z &= -1 \\ 0x + 3y + 1z &= +2 \\ 1x + 2y + 3z &= 0 \end{aligned}\right\} \quad \left.\begin{aligned} -\frac{15}{13} + 0 + \frac{2}{13} &= -1 \\ 0 + \frac{27}{13} + \frac{-1}{13} &= +2 \\ -\frac{15}{13} + \frac{18}{13} + \frac{-3}{13} &= 0 \end{aligned}\right\} \quad \begin{aligned} -\frac{13}{13} &= -1 \ \checkmark \\ \frac{26}{13} &= +2 \ \checkmark \\ -\frac{18}{13} + \frac{18}{13} &= 0 \ \checkmark \end{aligned}$$

The general approach to third-order or higher determinants requires that the determinant be expanded in the following form. There is more than one expansion that will generate the correct result, but this form is typically used when the material is first introduced.

$$D = \begin{vmatrix} a_1 & b_1 & c_1 \\ a_2 & b_2 & c_2 \\ a_3 & b_3 & c_3 \end{vmatrix} = \underset{\substack{\uparrow \\ \text{Multiplying factor}}}{a_1} \underbrace{\left(+ \underbrace{\begin{vmatrix} b_2 & c_2 \\ b_3 & c_3 \end{vmatrix}}_{\text{Minor}} \right)}_{\text{Cofactor}} + \underset{\substack{\uparrow \\ \text{Multiplying factor}}}{b_1} \underbrace{\left(- \underbrace{\begin{vmatrix} a_2 & c_2 \\ a_3 & c_3 \end{vmatrix}}_{\text{Minor}} \right)}_{\text{Cofactor}} + \underset{\substack{\uparrow \\ \text{Multiplying factor}}}{c_1} \underbrace{\left(+ \underbrace{\begin{vmatrix} a_2 & b_2 \\ a_3 & b_3 \end{vmatrix}}_{\text{Minor}} \right)}_{\text{Cofactor}}$$

This expansion was obtained by multiplying the elements of the first row of D by their corresponding cofactors. The first row doesn't have to be used as the multiplying factors. In fact, any *row* or *column* (not diagonals) may be used to expand a third-order determinant.

The sign of each cofactor is determined by the position of the multiplying factors (a_1, b_1, and c_1 in this case) as in the following standard format:

$$\begin{vmatrix} + \rightarrow & - & + \\ \downarrow & & \\ - & + & - \\ + & - & + \end{vmatrix}$$

Note that the proper sign for each element can be obtained by simply assigning the upper left element a positive sign and then changing sign as you move horizontally or vertically to the next position.

For the determinant D, the elements would have the following signs:

$$\begin{vmatrix} a_1^{(+)} & b_1^{(-)} & c_1^{(+)} \\ a_2^{(-)} & b_2^{(+)} & c_2^{(-)} \\ a_3^{(+)} & b_3^{(-)} & c_3^{(+)} \end{vmatrix}$$

The minors associated with each multiplying factor are obtained by covering up the row and column in which the multiplying factor is located and writing a second-order determinant to include the remaining elements in the same relative positions that they have in the third-order determinant.

Consider the cofactors associated with a_1 and b_1 in the expansion of D. The sign is positive for a_1 and negative for b_1 as determined by the standard format. Following the procedure outlined above, we can find the minors of a_1 and b_1 as follows:

$$a_{1(\text{minor})} = \begin{vmatrix} \not{a}_1 & \not{b}_1 & \not{c}_1 \\ \not{a}_2 & b_2 & c_2 \\ \not{a}_3 & b_3 & c_3 \end{vmatrix} = \begin{vmatrix} b_2 & c_2 \\ b_3 & c_3 \end{vmatrix}$$

$$b_{1(\text{minor})} = \begin{vmatrix} \not{a}_1 & \not{b}_1 & \not{c}_1 \\ a_2 & \not{b}_2 & c_2 \\ a_3 & \not{b}_3 & c_3 \end{vmatrix} = \begin{vmatrix} a_2 & c_2 \\ a_3 & c_3 \end{vmatrix}$$

We pointed out that any row or column may be used to expand the third-order determinant, and the same result will still be obtained. Using the first column of D, we obtain the expansion

$$D = \begin{vmatrix} a_1 & b_1 & c_1 \\ a_2 & b_2 & c_2 \\ a_3 & b_3 & c_3 \end{vmatrix} = a_1\left(+\begin{vmatrix} b_2 & c_2 \\ b_3 & c_3 \end{vmatrix}\right) + a_2\left(-\begin{vmatrix} b_1 & c_1 \\ b_3 & c_3 \end{vmatrix}\right) + a_3\left(+\begin{vmatrix} b_1 & c_1 \\ b_2 & c_2 \end{vmatrix}\right)$$

The proper choice of row or column can often reduce the amount of work required to expand the third-order determinant. For example, in the following determinants, the first column and third row, respectively, would reduce the number of cofactors in the expansion:

$$D = \begin{vmatrix} 2 & 3 & -2 \\ 0 & 4 & 5 \\ 0 & 6 & 7 \end{vmatrix} = 2\left(+\begin{vmatrix} 4 & 5 \\ 6 & 7 \end{vmatrix}\right) + 0 + 0 = 2(28 - 30)$$

$$= \mathbf{-4}$$

$$D = \begin{vmatrix} 1 & 4 & 7 \\ 2 & 6 & 8 \\ 2 & 0 & 3 \end{vmatrix} = 2\left(+\begin{vmatrix} 4 & 7 \\ 6 & 8 \end{vmatrix}\right) + 0 + 3\left(+\begin{vmatrix} 1 & 4 \\ 2 & 6 \end{vmatrix}\right)$$

$$= 2(32 - 42) + 3(6 - 8) = 2(-10) + 3(-2)$$
$$= \mathbf{-26}$$

EXAMPLE A.7

Expand the following third-order determinants:

a. $$D = \begin{vmatrix} 1 & 2 & 3 \\ 3 & 2 & 1 \\ 2 & 1 & 3 \end{vmatrix} = 1\left(+\begin{vmatrix} 2 & 1 \\ 1 & 3 \end{vmatrix}\right) + 3\left(-\begin{vmatrix} 2 & 3 \\ 1 & 3 \end{vmatrix}\right) + 2\left(+\begin{vmatrix} 2 & 3 \\ 2 & 1 \end{vmatrix}\right)$$

$$= 1[6 - 1] + 3[-(6 - 3)] + 2[2 - 6]$$
$$= 5 + 3(-3) + 2(-4)$$
$$= 5 - 9 - 8$$
$$= \mathbf{-12}$$

b. $$D = \begin{vmatrix} 0 & 4 & 6 \\ 2 & 0 & 5 \\ 8 & 4 & 0 \end{vmatrix} = 0 + 2\left(-\begin{vmatrix} 4 & 6 \\ 4 & 0 \end{vmatrix}\right) + 8\left(+\begin{vmatrix} 4 & 6 \\ 0 & 5 \end{vmatrix}\right)$$

$$= 0 + 2[-(0 - 24)] + 8[(20 - 0)]$$
$$= 0 + 2(24) + 8(20)$$
$$= 48 + 160$$
$$= \mathbf{208}$$

Appendix B
Logarithms

The use of **logarithms** in industry is so extensive that it is absolutely necessary that you understand their purpose and use. At first, logarithms often appear vague and mysterious due to the mathematical operations required to find the *logarithm* and *antilogarithm* using the longhand table approach that is usually taught in mathematics courses. However, almost all of today's scientific calculators have the common and natural log functions, eliminating the complexity of applying logarithms and allowing us to concentrate on the basic characteristics of the function.

Basic Relationships

Let us first examine the relationship between the variables of the logarithmic function. The mathematical expression

$$N = (b)^x$$

states that the number N is equal to the base b taken to the power x. A few examples:

$$100 = (10)^2$$
$$27 = (3)^3$$
$$54.6 = (e)^4 \quad \text{where } e = 2.7183$$

If you were asked to find the power x to satisfy the equation

$$1200 = (10)^x$$

the value of x could be determined using logarithms as follows:

$$x = \log_{10} 1200 = \mathbf{3.079}$$

revealing that

$$10^{3.079} = 1200$$

Note that the logarithm was taken to the base 10—the number to be taken to the power of x. There is no limitation to the numerical value of the base except that tables and calculators are designed to handle either a base of 10 (common logarithm, [log]) or base $e = 2.7183$ (natural logarithm, [ln]). In review, therefore,

$$\text{If } N = (b)^x, \text{ then } x = \log_b N. \tag{B.1}$$

The base to be used depends on the area of application. If a conversion from one base to the other is required, the following equation can be applied:

$$\log_e x = 2.3 \log_{10} x \tag{B.2}$$

In this section we will concentrate solely on the common logarithm. However, a number of the conclusions also apply to natural logarithms.

Some Areas of Application

The following is a short list of the most common applications of the logarithmic function.

1. The use of logarithms permits plotting the response of a system for a range of values that may otherwise be impossible or too complex with a linear scale (see Chapter 21).
2. Levels of power, voltage, and the like can be compared without dealing with very large or small numbers that often hide the true impact of the difference in magnitudes.
3. There are a number of systems that respond to outside stimuli in a nonlinear logarithmic manner. The result is a mathematical model that permits a direct calculation of the response of the system to a particular input signal.
4. The response of a cascaded or compound system can be rapidly determined using logarithms if the gain of each stage is known on a logarithmic basis. This characteristic will be demonstrated in an example to follow.

Graphs

Graph paper is available in the *semilog* and *log-log* varieties. Semilog paper has only one log scale, with the other a linear scale. Both scales of log-log paper are log scales. A section of semilog paper appears in Fig. B.1. Note the linear (even-spaced-interval) vertical scaling and the repeating intervals of the log scale at multiples of 10.

The spacing of the log scale is determined by taking the common log (base 10) of the number. The scaling starts with 1, since $\log_{10} 1 = 0$. The distance between 1 and 2 is determined by $\log_{10} 2 = 0.3010$, or approximately 30% of the full distance of a log interval, as shown on the graph. The distance between 1 and 3 is determined by $\log_{10} 3 = 0.4771$, or about 48% of the full width. For future reference, keep in mind that almost 50% of the width of one log interval is represented by a 3 rather than by the 5 of a linear scale. In addition, note that the number 5 is about 70% of the full width, and 8 is about 90%. Remembering the percentage of full width of the lines 2, 3, 5, and 8 will be particularly useful when the various lines of a log plot are left unnumbered.

Since

$$\begin{aligned} \log_{10} 1 &= 0 \\ \log_{10} 10 &= 1 \\ \log_{10} 100 &= 2 \\ \log_{10} 1000 &= 3 \quad \text{(etc.)} \end{aligned}$$

the spacing between 1 and 10, 10 and 100, 100 and 1000, and so on, will be the same as shown in Figs. B.1 and B.2.

Note in Figs. B.1 and B.2 how the log scale becomes compressed at the high end of each interval. With increasing frequency levels assigned to each interval, a single graph can provide a frequency plot extending from 1 Hz to 1 MHz, as shown in Fig. B.2, with particular reference to the 30%, 50%, 70%, and 90% levels of each interval.

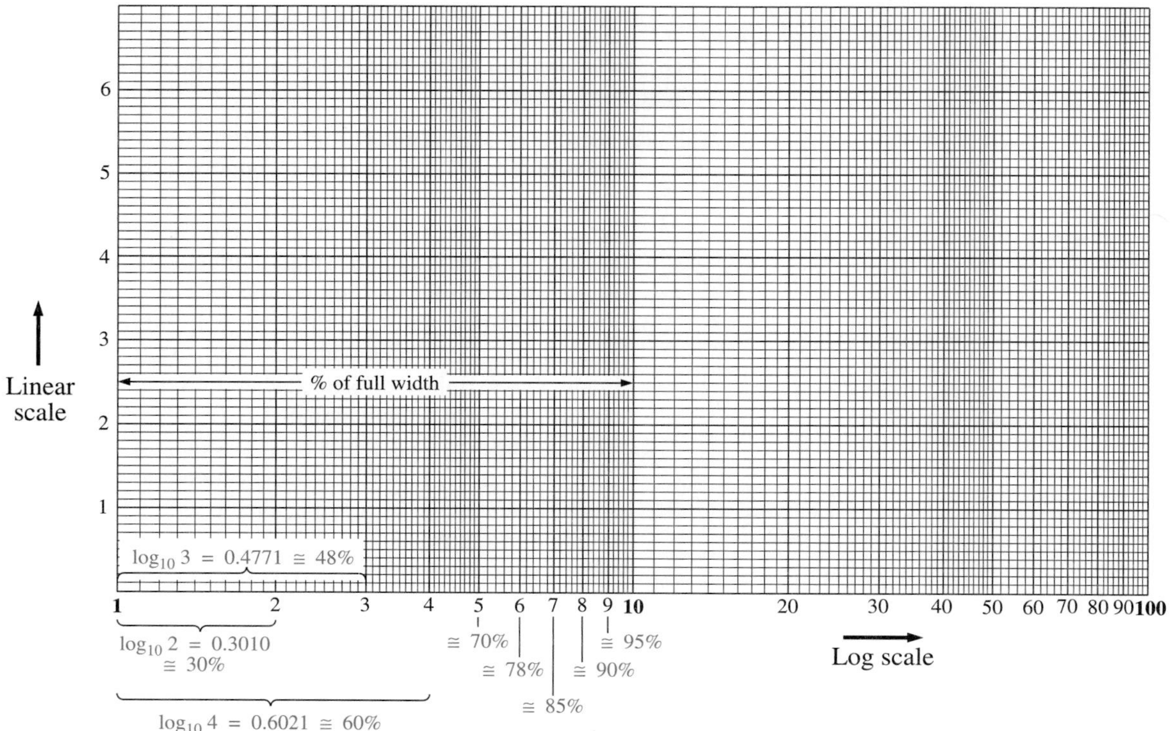

FIG. B.1
Semilog graph paper.

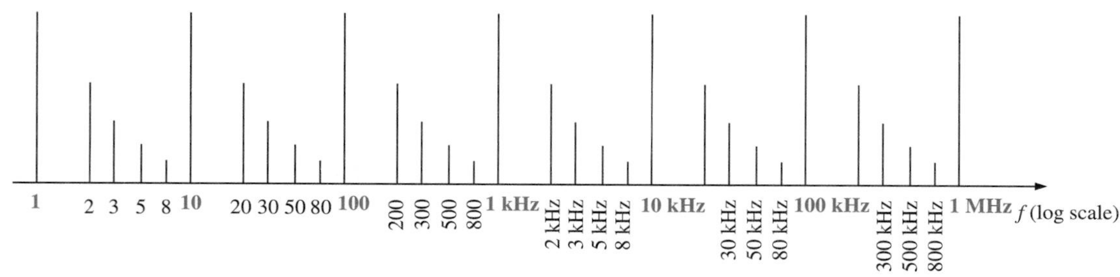

FIG. B.2
Frequency log scale.

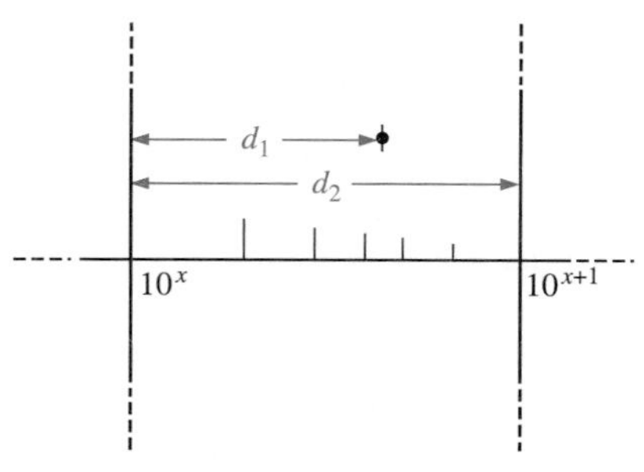

FIG. B.3
Finding a value on a log plot.

On many log plots the tick marks for most of the intermediate levels are left off because of space constraints. The following equation can be used to determine the logarithmic level at a particular point between known levels using a ruler or simply estimating the distances. The parameters are defined by Fig. B.3.

$$\text{Value} = 10^x \times 10^{d_1/d_2} \qquad \text{(B.3)}$$

The derivation of Eq. (B.3) is simply an extension of the details regarding distance appearing on Fig. B.1.

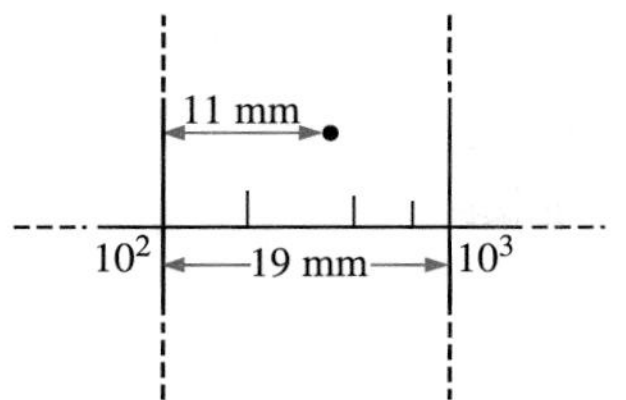

FIG. B.4
Example B.1.

EXAMPLE B.1

Determine the value of the point appearing on the logarithmic plot of Fig. B.4 using the measurements made by a ruler (linear).

Solution:

$$\frac{d_1}{d_2} = \frac{11\text{ mm}}{19\text{ mm}} = 0.579$$

Using a calculator:

$$10^{d_1/d_2} = 10^{0.579} = 3.793$$

Applying Eq. (B.3):

$$\begin{aligned}\text{Value} &= 10^x \times 10^{d_1/d_2} = 10^2 \times 3.793 \\ &= \mathbf{379.3}\end{aligned}$$

Properties of Logarithms

There are a few characteristics of logarithms that should be emphasized.

1. ***The common or natural logarithm of the number 1 is 0.***

$$\log_{10} 1 = 0 \tag{B.4}$$

just as $10^x = 1$ requires that $x = 0$.

2. ***The log of any number less than 1 is a negative number.***

$$\log_{10} \tfrac{1}{2} = \log_{10} 0.5 = -0.3$$

$$\log_{10} \tfrac{1}{10} = \log_{10} 0.1 = -1$$

3. ***The log of the product of two numbers is the sum of the logs of the numbers.***

$$\log_{10} ab = \log_{10} a + \log_{10} b \tag{B.5}$$

4. ***The log of the quotient of two numbers is the log of the numerator minus the log of the denominator.***

$$\log_{10} \frac{a}{b} = \log_{10} a - \log_{10} b \tag{B.6}$$

5. ***The log of a number taken to a power is equal to the product of the power and the log of the number.***

$$\log_{10} a^n = n \log_{10} a \tag{B.7}$$

Calculator Functions

On most calculators the log of a number is found by simply entering the number and pressing the [log] or [ln] key.

For example,

$$\log_{10} 80 = \boxed{8}\ \boxed{0}\ \boxed{\text{log}}$$

with a display of **1.903.**

For the reverse process, where N, or the antilogarithm, is desired, the function 10^x is used. On most calculators 10^x appears as a second function above the [log] key. For the case of

$$0.6 = \log_{10} N$$

the following keys are used:

[·] [6] [2ndF] [10^x]

with a display of **3.981.** Checking: $\log_{10} 3.981 = 0.6$.

EXAMPLE B.2

Evaluate each of the following logarithmic expressions.

a. $\log_{10} 0.004$

b. $\log_{10} 250\ 000$

c. $\log_{10}(0.08)(240)$

d. $\log_{10} \dfrac{1 \times 10^4}{1 \times 10^{-4}}$

e. $\log_{10}(10)^4$

Solutions:

a. $\mathbf{-2.398}$

b. $\mathbf{+5.398}$

c. $\log_{10}(0.08)(240) = \log_{10} 0.08 + \log_{10} 240 = -1.097 + 2.380$
$= \mathbf{1.283}$

d. $\log_{10} \dfrac{1 \times 10^4}{1 \times 10^{-4}} = \log_{10} 1 \times 10^4 - \log_{10} 1 \times 10^{-4} = 4 - (-4)$
$= \mathbf{8}$

e. $\log_{10} 10^4 = 4 \log_{10} 10 = 4(1) = \mathbf{4}$

Appendix C

The Greek Alphabet

Letter	Capital	Lowercase	Used to Designate
Alpha	A	α	Area, angles, coefficients
Beta	B	β	Angles, coefficients, flux density
Gamma	Γ	γ	Specific gravity, conductivity
Delta	Δ	δ	Density, variation
Epsilon	E	ϵ	Base of natural logarithms
Zeta	Z	ζ	Coefficients, coordinates, impedance
Eta	H	η	Efficiency, hysteresis coefficient
Theta	Θ	θ	Phase angle, temperature
Iota	I	ι	
Kappa	K	κ	Dielectric constant, susceptibility
Lambda	Λ	λ	Wavelength
Mu	M	μ	Amplification factor, micro, permeability
Nu	N	ν	Reluctivity
Xi	Ξ	ξ	
Omicron	O	o	
Pi	Π	π	3.1416
Rho	P	ρ	Resistivity
Sigma	Σ	σ	Summation
Tau	T	τ	Time constant
Upsilon	Υ	υ	
Phi	Φ	ϕ	Angles, magnetic flux
Chi	X	χ	
Psi	Ψ	ψ	Dielectric flux, phase difference
Omega	Ω	ω	Ohms, angular velocity

Appendix D
Magnetic Parameter Conversions

	SI (MKS)	CGS	English
Φ	webers (Wb) 1 Wb	maxwells $= 10^8$ maxwells	lines $= 10^8$ lines
B	Wb/m^2 1 Wb/m^2	gauss (maxwells/cm^2) $= 10^4$ gauss	lines/in.^{2} $= 6.452 \times 10^4$ lines/in.^{2}
A	1 m^2	$= 10^4$ cm^2	$= 1550$ in.^{2}
μ_o	$4\pi \times 10^{-7}$ Wb/Am	$= 1$ gauss/oersted	$= 3.20$ lines/Am
$\mathcal{F}$	NI (ampere-turns, At) 1 At	$0.4\pi NI$ (gilberts) $= 1.257$ gilberts	NI (At) 1 gilbert $= 0.7958$ At
H	NI/l (At/m) 1 At/m	$0.4\pi NI/l$ (oersteds) $= 1.26 \times 10^{-2}$ oersted	NI/l (At/in.) $= 2.54 \times 10^{-2}$ At/in.
H_g	$7.97 \times 10^5 B_g$ (At/m)	B_g (oersteds)	$0.313B_g$ (At/in.)

Appendix E

Maximum Power Transfer Conditions

Derivation of maximum power transfer conditions for the situation where the resistive component of the load is adjustable but the load reactance is set in magnitude.*

For the circuit of Fig. E.1, the power delivered to the load is determined by

$$P = \frac{V_{R_L}^2}{R_L}$$

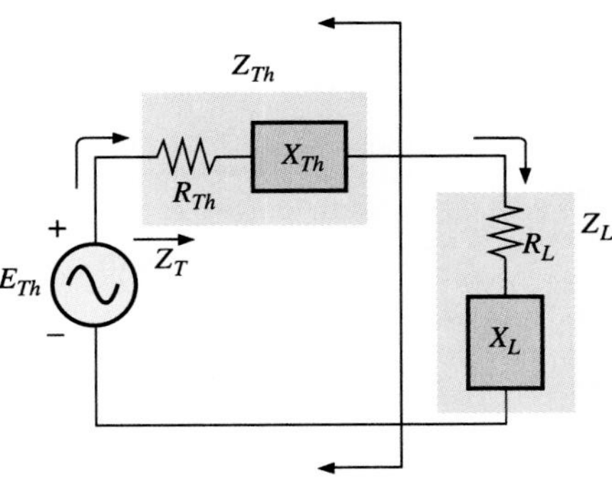

FIG. E.1

Applying the voltage divider rule:

$$\mathbf{V}_{R_L} = \frac{R_L \mathbf{E}_{Th}}{R_L + R_{Th} + X_{Th}\angle 90^\circ + X_L\angle 90^\circ}$$

The magnitude of $\mathbf{V}_{R_L}$ is determined by

$$V_{R_L} = \frac{R_L E_{Th}}{\sqrt{(R_L + R_{Th})^2 + (X_{Th} + X_L)^2}}$$

and

$$V_{R_L}^2 = \frac{R_L^2 E_{Th}^2}{(R_L + R_{Th})^2 + (X_{Th} + X_L)^2}$$

with

$$P = \frac{V_{R_L}^2}{R_L} = \frac{R_L E_{Th}^2}{(R_L + R_{Th})^2 + (X_{Th} + X_L)^2}$$

Using differentiation (calculus), maximum power will be transferred when $dP/dR_L = 0$. The result of the preceding operation is that

$$R_L = \sqrt{R_{Th}^2 + (X_{Th} + X_L)^2} \quad \text{[Eq. (18.21)]}$$

---

*With sincerest thanks for the input of Professor Harry J. Franz of the Beaver Campus of Pennsylvania State University.

The magnitude of the total impedance of the circuit is

$$Z_T = \sqrt{(R_{Th} + R_L)^2 + (X_{Th} + X_L)^2}$$

Substituting this equation for R_L and applying a few algebraic manipulations will result in

$$Z_T = 2R_L(R_L + R_{Th})$$

and the power to the load R_L will be

$$P = I^2R_L = \frac{E_{Th}^2}{Z_T^2}R_L = \frac{E_{Th}^2R_L}{2R_L(R_L + R_{Th})}$$

$$= \frac{E_{Th}^2}{4\left(\dfrac{R_L + R_{Th}}{2}\right)}$$

$$= \frac{E_{Th}^2}{4R_{av}}$$

with $$R_{av} = \frac{R_L + R_{Th}}{2}$$

Appendix F

Answers to Selected Problems

Chapter 1

7. **(a)** 10^4
(b) 10
(c) 10^9
(d) 10^{-2}
(e) 10
(f) 10^{31}
9. **(a)** 10^{-1}
(b) 10^{-4}
(c) 10^9
(d) 10^{-9}
(e) 1042
(f) 10^3
11. **(a)** 10^6
(b) 10^{-2}
(c) 10^{32}
(d) 10^{-63}
13. **(a)** 10^{-6}
(b) 10^{-3}
(c) 10^{-6}
(d) 10^9
(e) 10^{-16}
(f) 10^{-1}
15. **(a)** 0.006
(b) 400
(c) 5000, 5, 0.005
(d) 0.0003, 0.3, 300
17. **(a)** 90 s
(b) 144 s
(c) 50×10^3 μs
(d) 160 mm
(e) 120 ns
(f) 41.898 days
(g) 1.02 m
19. **(a)** 350 km
(b) 190 km/s
(c) 3.29 ps

Chapter 2

1. 230 N
3. **(a)** at $r = 1$ m: $F = 18$ mN
(b) *at* $r = 3$ m: $F = 2$ mN
(c) at $r = 10$ m: $F = 180$ μN
5. at $r = 0.5$ m, $F = 288$ N
at $r = 1$ m, $F = 72$ N
at $r = 5$ m, $F = 2.88$ N
at $r = 10$ m, $F = 0.72$ N
7. **(a)** 72 mN
(b) $Q_1 = 20$ μC, $Q_2 = 40$ μC
9. 3.1 A
11. 90 C
13. 0.5 A
15. 1.194 A > 1A (yes)
17. **(a)** 1.248 million
(b) 0.9363 million
Solution: (a) > (b)
19. 252 J
21. 4 C
23. 3.53 V
35. 600 C

Chapter 3

1. **(a)** 21.71 μΩ
(b) 35.59 μΩ
(c) increases
(d) decreases
3. 2.5 μcm
7. 2.409 Ω
9. 3.67 Ω
11. 46 mΩ
13. **(a)** 40.29°C
(b) −195.61°C
15. 2843.5°C
16. **(a)** $\alpha_{20} = 0.003929 \cong 0.00393$
(b) 83.61°C
18. 1.751 Ω
20. 142.86
22. −30°C : 10.2 kΩ
100°C : 10.15 kΩ
24. 6.5 kΩ
29. **(a)** 22 Ω = Red, Red, Black, Silver
(b) 4700 Ω = Yellow, Violet, Red, Silver
(c) 56 kΩ = Green, Blue, Orange, Silver
(d) 1 MΩ = Brown, Black, Green, Silver
31. no
33. **(a)** 156.6 mS
(b) 95.54 mS
(c) 21.95 mS
39. **(a)** −50°C : 0.3 MΩ-cm, 50°C : 3 kΩ-cm, 100°C : 0.4 kΩ-cm
(b) negative
(c) no—Log scales
(d) −321 Ω-cm/°C

Chapter 4

1. 15 Ω
3. 4 kΩ
5. 72 mΩ
7. 54.55 Ω
9. 28.571 Ω
11. 1.2 kΩ
14. **(a)** 12.632 Ω
(b) 4.1×10^6 J
18. 800 V
21. 1 W
23. **(a)** 57.6 kJ
(b) 16×10^{-3} kWh
25. 2 s
27. 196 μW
30. 4 A
32. 9.61 Ω
34. 0.833 A, 144.06 Ω
37. **(a)** 0.133 mA
(b) 66.5 mAh
39. **(c)** 70.7 mA
41. **(a)** 12 kW
(b) 10,130 < 12,000 W (Yes)
43. 16.34 A
46. **(a)** 238 W
(b) 17.36%
48. **(a)** 1657.78 W
(b) 15.07 A
(c) 19.38 A
49. 65.25%
53. 80%
55. **(a)** 17.9%
(b) 76.73%, 328.66% increase
57. **(a)** 1350 J
(b) W doubles, P the same
59. 6.67 h
62. **(a)** 50 kW
(b) 240.38 A
(c) 90 kWh
64. $2.19

Chapter 5

1. **(a)** 20 Ω, 3 A
(b) 1.63 MΩ, 6.135 μA
(c) 110 Ω, 318.2 mA
(d) 10 kΩ, 12 mA
3. **(a)** 16 V
(b) 4.2 V
5. **(a)** $I = 1$A, $V = 6$V
7. **(a)** 5 V
(b) 70 V
9. $V_1 = 45$V, $V_2 = -15$V
11. **(a)** 70.6 Ω, 85 mA (CCW)
$V_1 = 2.805$ V
$V_2 = 0.476$ V
$V_3 = 0.850$ V
$V_4 = 1.870$ V

(c) $P_1 = 238.4$ mW
$P_2 = 40.5$ mW
$P_3 = 72.3$ mW
$P_4 = 159$ mW
(d) All 1/2 W
13. (a) 225 Ω, 0.533 A
(b) 8 W
(c) 15 V
(d) All go out!
15. (a) $V_{ab} = 66.67$ V
(b) $V_{ab} = -8$ V
(c) $V_{ab} = 20$ V
(d) $V_{ab} = 0.18$ V
17. (a) 12 V
(b) 24 V
(c) 60 Ω
(d) 0.4 A
(e) 60 Ω
19. (a) 80 Ω
(b) 0.2 W < 1/4 W
21. $R_1 = 200\ \Omega$, $R_2 = 800\ \Omega$
23. (a) $R_1 = 0.4$ kΩ, $R_2 = 1.2$ kΩ,
$R_3 = 4.8$ kΩ
(b) $R_1 = 0.4$ MΩ, $R_2 = 1.2$ MΩ,
$R_3 = 4.8$ MΩ
25. (a) I(CW) = 6.667 A, $V = 20$ V
(b) I(CW) = 1 A, $V = 10$ V
27. (a) $V_a = 20$ V, $V_b = 26$ V,
$V_c = 35$ V, $V_d = -12$ V,
$V_e = 0$ V
(b) $V_{ab} = -6$ V, $V_{dc} = -47$ V,
$V_{cb} = 9$ V
(c) $V_{ac} = -15$ V, $V_{db} = -38$ V
29. $V_0 = 0$ V, $V_4 = +10$ V, $V_7 = 4$ V,
$V_{10} = 20$ V, $V_{23} = +6$ V
$V_{30} = -8$ V, $V_{67} = 0$ V,
$V_{56} = -6$ V, $I_{up} = 1.5$ A
31. 2 Ω
33. 100 Ω
35. 1.52%

Chapter 6

1. (a) 2, 3, 4 in parallel
(b) 2, 3 in parallel
(c) 2, 3 in series; 1, 4 in parallel
3. (a) 6 Ω, 0.1667 S
(b) 1 kΩ, 1 mS
(c) 2.076 kΩ, 0.4817 mS
(d) 1.333 Ω, 0.75 S
(e) 9.948 Ω, 100.525 mS
(f) 0.6889 Ω, 1.4516 S
5. (a) 18 Ω
(b) $R_1 = R_2 = 24\ \Omega$
7. 120 Ω
9. (a) 0.167 mS, 6 kΩ
(b) $I_s = 8$ mA
$I_1 = 6$ mA
$I_2 = 2$ mA
(d) $P_1 = 0.288$ W
$P_2 = 96$ mW
(e) both ½ W
11. (a) 0.8144 mS
(b) $I_s = 9.7728$ mA
$I_1 = 5.4545$ mA
$I_2 = 2.5532$ mA
$I_3 = 1.7647$ mA
(d) $P_1 = 65$ mW
$P_2 = 31$ mW
$P_3 = 21$ mW
(e) all ½ W
14. (a) $I_s = 7.5$ A, $I_1 = 1.5$ A
(b) $I_s = 9.6$ mA, $I_1 = 0.8$ mA
16. 1260 W
18. (a) 4 mA
(b) 24 V
(c) 18.4 mA
20. (a) $I_1 = 3$ mA, $I_2 = 1$ mA,
$I_3 = 1.5$ mA
(b) $I_2 = 4\ \mu$A, $I_3 = 1.5\ \mu$A,
$I_4 = 5.5\ \mu$A, $I_1 = 6\ \mu$A
22. (a) $R_1 = 5\ \Omega$, $R_2 = 10\ \Omega$
(b) $E = 12$ V, $I_2 = 1.333$ A,
$I_3 = 1$ A, $R_3 = 12\ \Omega$,
$I = 4.333$ A
(c) $I_1 = 64$ mA, $I_3 = 16$ mA,
$I_2 = 20$ mA, $R = 3.2$ kΩ,
$I = 36$ mA
(d) $E = 30$ V, $I_1 = 1$ A,
$I_3 = I_2 = 0.5$ A,
$R_2 = R_3 = 60\ \Omega$
$P_{R_2} = 15$ W
23. $I_1 = 14.98$ mA, $I_2 = 9.985$ A
25. (a) $I_1 = 4$ A, $I_2 = 8$ A
(b) $I_1 = 2$ A, $I_2 = 4$ A, $I_3 = 1$ A,
$I_4 = 1.333$ A
(c) $I_1 = 272.73$ mA,
$I_2 = 227.27$ mA,
$I_3 = 90.91$ mA,
$I_4 = 500$ mA
(d) $I_2 = 4.5$ A, $I_3 = 8.5$ A,
$I_4 = 8.5$ A
27. (a) $I = 4$ A, $I_2 = 4$ A,
$I_1 = 3$ A
(b) $I_3 = I = 6\ \mu$A, $I_2 = 2\ \mu$A,
$I_1 = 2\ \mu$A, $R = 9\ \Omega$
29. $R_1 = 6$ kΩ, $R_2 = 1.5$ kΩ,
$R_3 = 0.5$ kΩ
31. $I = 3$ A, $R = 2\ \Omega$
33. (a) 6.13 V
(b) 9 V
(c) 9 V
35. (a) 4 V
(b) 3.997 V
(c) 3.871 V
(d) 3 V
(e) R_m as large as possible
37. $I_5 = 5$ A, $I_4 = 8$ A

Chapter 7

1. (a) 5.764 Ω
(b) 3.609 Ω
(c) 7.22 Ω
3. (a) 11 Ω
(b) 6.25 Ω
5. (a) yes (KCL)
(b) 3 A
(c) yes (KCL)
(d) 4 V
(e) 2 Ω
(f) 5 A
(g) $P_{del} = 50$ W
$P_1 = 12$ W
$P_2 = 18$ W
7. (a) 4 Ω
(b) $I_s = 9$ A, $I_1 = 6$ A, $I_2 = 3$ A
(c) $V_a = 6$ V
9. $I_1 = 4$ A, $I_2 = 0.7175$ A
11. (a) $I_s = 5$ A, $I_1 = 1$ A, $I_3 = 4$ A,
$I_4 = 0.5$ A
(b) $V_a = 17$ V, $V_{bc} = 10$ V
13. (a) $I = 16$ mA, $I_6 = 2$ mA
(b) $V_1 = 28$ V, $V_5 = 7.2$ V
(c) $P = 8.64$ mW
15. (a) $V_{AB} = -26$ V
(b) $I = 0.1$ A (down)
17. (a) $I = 0.6$ A
(b) $V = 28$ V
19. (a) $I_2 = 1.667$ A, $I_6 = 1.111$ A,
$I_8 = 0$ A,
(b) $V_4 = 10$ V, $V_8 = 0$ V
21. (a) 1.882 Ω
(b) $V_1 = V_4 = E = 32$ V
(c) $I_3 = 8$ A ←
(d) $I_s = 17$ A, $R_T = 1.882\ \Omega$
23. (a) 6.75 A
(b) 32 V
25. 8.333 Ω
27. (a) 24 A
(b) $I_4 = 12$ A, $I_7 = 8$ A
(c) $V_3 = 48$ V, $V_5 = 24$ V,
$V_7 = 16$ V
(d) $P_{R7} = 128$ W, $P_E = 5760$ W
29. 4.44 W
31. $V_1 = 22$ V, $V_2 = 14.21$ V,
$V_3 = 0.66$ V
32. (a) 64 V
(b) $R_{L_2} = 4$ kΩ, $R_{L_3} = 3$ kΩ
(c) $R_1 = 0.5$ kΩ, $R_2 = 1.2$ kΩ,
$R_3 = 2$ kΩ
34. (a) yes
(b) $R_2 = 250\ \Omega$, $R_1 = 750\ \Omega$
(c) $R_2 = 255\ \Omega$, $R_1 = 745\ \Omega$
36. (a) 1 mA
(b) $R_{shunt} = 5$ mΩ
40. (a) $R_s = 300$ kΩ
(b) 20,000
42. 0.05 μA

Chapter 8

3. **(a)** $V = 23.982$ V
(b) $V = 24$ V
5. **(a)** $V_3 = 1.6$ V
(b) $I_2 = 0.1$ A
7. **(a)** $E = 4.5$ V, $R_S = 3\ \Omega$
(b) $E = 28.2$ V, $R_S = 4.7$ kΩ
9. **(a)** $E = 13.6$ V, $R_S = 6.8\ \Omega$
(b) $I_1 = 458.78$ mA
(c) $V_{ab} = 17.89$ V
11. **(a)** $I_1 = 3$ A, $R_1 = 3\ \Omega$, $I_2 = 10$ A, $R_2 = 2\ \Omega$
(b) $V_{ab} = 6.462$ V
(c) $I_\uparrow = 1.077$ A
13. **(a)** $I_{R_1} = -9/14$ A, $I_{R_2} = -12/7$ A, $I_{R_3} = -15/14$ A
(b) $I_{R_1} = 2.53$ A, $I_{R_2} = 0.24$ A, $I_{R_3} = 2.77$ A
15. **(a)** $I_2 = 8.548$ A, $V_{ab} = -22.75$ V
(b) $I_2 = 1.274$ A, $V_{ab} = -0.904$ V
17. **(a)** $I_B = 63.02$ mA, $I_C = 4.416$ mA, $I_E = 4.479$ mA
(b) $V_B = 2.985$ V, $V_E = 2.285$ A, $V_C = 10.285$ V
(c) $\beta \simeq 70.07$
19. **(a)** (CW): $I_{R_1} = -1/7$ A
(CW): $I_{R_2} = -5/7$ A
(down): $I_{R_3} = 4/7$ A
(b) (CW): $I_{R_1} = -3.0625$ A
(CW): $I_{R_3} = 0.1875$ A
(up): $I_{R_2} = -3.25$ A
21. **(a)** (CW): $I_1 = 1.8701$ A,
(CW): $I_2 = -8.5484$ A
$V_{ab} = -22.74$ V
(b) (CW): $I_2 = 1.274$ A,
(CW): $I_3 = 0.26$ A
$V_{ab} = -0.904$ V
23. **(a)** 72.16 mA, -4.433 V
(b) 1.953 A, -7.257 V
25. **(a)** all CW: $I_1 = 0.0321$ mA, $I_2 = -0.8838$ mA, $I_3 = -0.968$ mA, $I_4 = -0.639$ mA
(b) all CW: $I_1 = -3.8$ A, $I_2 = -4.2$ A, $I_3 = 0.2$ A,
27. **(a)** $I_1 = -9/14$ A
$I_2 = -12/7$ A
(b) $I_1 = -2.053$ A
$I_2 = +0.24$ A
29. (I): **(a)** CW
(b) $I_1 = 1.871$ A, $I_2 = -8.548$ A
(c) $I_{R_1} = 1.871$ A, $I_{R_2} = -8.548$ A, $I_{R_3} = 10.419$A
(II): **(a)** CW
(b) $I_2 = 1.274$ A, $I_3 = 0.26$ A
(c) $I_{R_2} = 1.274$ A, $I_{R_3} = 0.26$ A, $I_{R_4} = 1.014$ A, $I_{R_1} = 1.726$ A
31. CW: $I_{5\Omega} = 1.953$ A, $I_3 = -2.4186$ A, $V_a = -7.26$ V
33. **(a)** all CW: $I_1 = 0.0321$ mA, $I_2 = 0.8838$ mA, $I_3 = -0.968$ mA, $I_4 = -0.639$ mA
(b) all CW: $I_1 = 3.8$ A, $I_2 = -4.2$ A, $I_3 = 0.2$ A
35. $V_1 = 8.077$ V, $V_2 = 9.385$ V, symmetry is present
38. (I): $V_1 = 7.238$ V, $V_2 = -2.453$ V, $V_3 = 1.405$ V
(II): $V_1 = -6.642$ V, $V_2 = 1.293$ V, $V_3 = 10.664$ V
40. **(a)** $V_1 = 10.083$ V, $V_2 = 6.944$ V, $V_3 = -17.056$ V
(b) $V_1 = 48$ V, $V_2 = 64$ V
42. (I): **(b)** $V_1 = -14.86$ V, $V_2 = -12.57$ V
(c) $V_{R_2} = -12.57$ V
$V_{R_3} = 9.71$ V
(II): **(b)** $V_1 = -2.556$ V, $V_2 = 4.03$ V
(c) $V_{R_1} = -2.556$ V,
$V_{R_2} = V_{R_5}$ 4.03 V
$V_{R_3} = V_{R_4} = 6.586$ V
44. **(a)** $V_1 = -5.311$ V, $V_2 = -0.6219$ V, $V_3 = 3.751$ V, $V_{-5A} = -5.311$ V
(b) $V_1 = -6.917$ V, $V_2 = 12$ V, $V_3 = 2.3$ V, $V_{2A} = 9.7$ V, $V_{5A} = 18.917$ V
46. **(b)** $V_{R_5} = 0.1967$ V
(c) no
(d) no
48. **(b)** $I_{R_5} = 0$ A
(c) yes
49.

$$V_1\left[\frac{1}{1\text{ k}\Omega} + \frac{1}{100\text{ k}\Omega} + \frac{1}{200\text{ k}\Omega}\right] - \frac{1}{100\text{ k}\Omega}V_2 - \frac{1}{200\text{ k}\Omega}V_3 = 4\text{ mA}$$

$$V_2\left[\frac{1}{100\text{ k}\Omega} + \frac{1}{200\text{ k}\Omega} + \frac{1}{1\text{ k}\Omega}\right] - \frac{1}{100\text{ k}\Omega}V_1 - \frac{1}{1\text{ k}\Omega}V_3 = -9\text{ mA}$$

$$V_3\left[\frac{1}{200\text{ k}\Omega} + \frac{1}{100\text{ k}\Omega} + \frac{1}{1\text{ k}\Omega}\right] - \frac{1}{200\text{ k}\Omega}V_1 - \frac{1}{1\text{ k}\Omega}V_2 = 9\text{ mA}$$

50. **(a)** 3.33 mA
(b) 1.177A
51. $R_A = 137.5\ \Omega$, $R_B = 275\ \Omega$, $R_C = 91.67\ \Omega$
53. **(a)** 7.358 A
(b) 1.763 A
55. 2.143 A
57. **(b)** 5.714 mA

Chapter 9

1. **(a)** CW: $I_{R1} = 5/6$ A, $I_{R2} = 0$ A, $I_{R3} = 5/6$ A
(b) E_1: 5.333 W, E_2: 0.333 W
(c) 8.333 W
(d) no
3. **(a)** down: $I_1 = 4.4545$ mA
(b) down: $I_1 = 3.11$ A
5. **(a)** 8 Ω, 4 V
(b) 2 Ω, $I_1 = 0.4$ A
30 Ω, $I_2 = 0.105$ A
100 Ω, $I_3 = 0.037$ A
7. **(a)** 2 Ω, 84 V
(b) 1.579 kΩ, -1.149 V
9. **(a)** 45 Ω, -5 V
(b) 2.055 kΩ, 16.772 V
11. 4.041 kΩ, 9.733 V
13. **(a)** 14 Ω, 2.571 A
(b) 7.5 Ω, 1.333 A
15. **(a)** 9.756 Ω, 0.95 A
(b) 2 Ω, 30 A
17. **(a)** 10 Ω, 0.2 A
(b) 4.033 kΩ, 2.9758 mA
19. (I): **(a)** 14 Ω,
(b) 23.14 W
(II): **(a)** 7.5 Ω
(b) 3.33 W
21. **(a)** 9.756 Ω, 2.20 W
(b) 2 Ω, 450 W
23. 0 Ω
25. 500 Ω, $P_{max} = 1.44$ W
27. $I_L = 39.3\ \mu$A, $V_L = 220$ mV
29. $I_L = 2.25$ A, $V_L = 6.075$ V
35. **(a)** 0.357 mA
(b) 0.357 mA
(c) yes
37. 5 A
39. 5.24 kΩ
41. **(a)** 5.24 kΩ
(b) 0.00063 W

Chapter 10

1. 9×10^3 N/C
3. 0.05 μC
4. 70 μF
6. 50 V/m
9. 8×10^3 V/m
11. 937.5 pF
13. mica

15. **(a)** 10^6 V/m **(b)** 4.96 μC
(c) 0.0248 μF
17. 29,035 V
19. **(a)** 0.5 s **(b)** $20(1 - e^{-t/0.5})$
(c) 1τ: 12.64 V, 3τ: 19 V, 5τ: 19.87 V
(d) $i_C = 0.2 \times 10^{-3}e^{-t/0.5}$
$v_R = 20e^{-t/0.5}$
21. **(a)** 5.5 ms
(b) $100(1 - e^{-t/(5.5\times10^{-3})})$
(c) 1τ: 63.21 V, 3τ: 95.02 V, 5τ: 99.33 V
(d)
$i_C = 18.18 \times 10^{-3}e^{-t/(5.5\times10^{-3})}$
$v_R = 60e^{-t/(5.5\times10^{-3})}$
24. **(a)** 10 ms
(b) $50(1 - e^{-t/(10\times10^{-3})})$
(c) $10 \times 10^{-3}e^{-t/(10\times10^{-3})}$
(d) $v_C \cong 50$ V, $i_C = 0$ A
(e) $v_C = 50e^{-t/(4\times10^{-3})}$
$i_C = -25 \times 10^{-3}e^{-t/(4\times10^{-3})}$
27. **(a)** $80(1 - e^{-t/(1\times10^{-6})})$
(b) $0.8 \times 10^{-3}e^{-t/(1\times10^{-6})}$
(c) $v_C = 80e^{-t/(4.9\times10^{-6})}$
$i_C = 0.163 \times 10^{-3}e^{-t/(4.9\times10^{-6})}$
29. **(a)** 10 μs **(b)** 3 kA **(c)** yes
31. **(a)** $v_C = 52\text{ V} - 40\text{ V } e^{-t/123.8\text{ms}}$
$i_C = 2.198\text{ mA } e^{-t/123.8\text{ms}}$
33. 1.386 μs
35. $R = 54.567$ kΩ
37. **(a)** $v_C = 60(1 - e^{-t/0.2\text{s}})$,
0.5 s: 55.07 V, 1 s: 59.596 V
$i_C = 60 \times 10^{-3}\ e^{-t/0.2\text{s}}$
0.5 s: 4.93 mA,
1 s: 0.404 mA
$v_{R_1} = 60\ e^{-t/0.2\text{s}}$
0.5 s: 4.93 V, 1 s: 0.404 V
(b) $t = 0.405$ s, 1.387 s longer
39. **(a)** 19.634 V
(b) 2.31 s
(c) 1.155 s
41. **(a)** $v_C = 3.275(1 - e^{-t/52.68\text{ms}})$
$i_C = 1.216 \times 10^{-3}\ e^{-t/52.68\text{ms}}$
43. **(a)** $v_C = 27.2 - 25.2\ e^{-t/18.26\text{ms}}$
$i_C = 3.04\text{ mA } e^{-t/18.26\text{ms}}$
46. 0–4 ms: 0.3 mA,
4–6 ms: 0.9 mA,
6–7 ms: 3 mA,
7–10 ms: 0 mA,
10–13 ms: −3.2 mA,
13–15 ms: 1.8 mA
48. 0–4 ms: 0 V,
4–6 ms: −8 V,
6–16 ms: 20 V,
16–18 ms: 0 V,
18–20 ms: −12 V,
20–25 ms: 0 V
50. **(a)** $V_1 = 10$ V, $Q_1 = 60$ μC,
$V_2 = 6.67$ V, $Q_2 = 40$ μC,
$V_3 = 3.33$ V, $Q_3 = 40$ μC
53. **(a)** 56.54 V
(b) 42.405 V
(c) 14.135 V
(d) 43.46 V
(e) 433.44 ms
55. 8640 pJ
57. **(a)** 5 J
(b) 0.1 C
(c) 200 A
(d) 10 kW
(e) 10 s

Chapter 11

1. 0.625 m^2
3. **(a)** 0.04 T
5. 952.4×10^3 At/Wb
7. 1.35 N

Chapter 12

1. **(a)** 4.25 V
(b) 8.5 V
(c) 17 V
3. **(a)** 140
(b) 7
(c) 2.47
5. $1.421 * 10^{-5}$ H = 14.21 μH
7. **(a)** 2.5 V **(b)** 0.3 V
(c) 200 V
9. 0−3 ms: 0 V, 3−8 ms: 1.6 V,
8−13 ms: −1.6 V,
13−14 ms: 0 V,
14−15 ms: 8 V,
15−16 ms: –8 V,
16−17 ms: 0 V
11. 0−5 μs: 4 mA, 10 μs: −8 mA,
12 μs: 4 mA, 12−16 μs: 4 mA,
24 μs: 0 mA
13. **(a)** 2.27 μs
(b) $5.45 \times 10^{-3}(1 - e^{-t/2.27\mu s})$
(c) $v_L = 12e^{-t/2.27\mu s}$
$v_R = 12(1 - e^{-t/2.27\mu s})$
(d) i_L: $1\tau = 3.45$ mA,
$3\tau = 5.179$ mA,
$5\tau = 5.413$ mA,
v_L: $1\tau = 4.415$ V,
$3\tau = 0.598$ V,
$5\tau = 0.081$ V
15. **(a)** $i_L = 0.882 \times 10^{-3}\ (1 - e^{-t/0.735\mu s})$,
$v_L = 6e^{-t/0.735\mu s}$
(b) $i_L = 0.882 \times 10^{-3}e^{-t/0.333\mu s}$
$v_L = -13.23e^{-t/0.333\mu s}$
17. **(a)** $i_L = 1.765\text{ mA} - 4.765\text{ mA } e^{-t/588.24\mu s}$
$v_L = 16.2\text{ V } e^{-t/588.24\mu s}$
19. **(a)** $i_L = -0.692\text{ mA} - 2.308\text{ mA } e^{-t/19.23\mu s}$
$v_L = 24\text{ V } e^{-t/19.23\mu s}$
21. 25.68 μs
23. **(a)** $i_L = 3.638 \times 10^{-3}\ (1 - e^{-t/6.676\mu s})$,
$v_L = 5.45\ e^{-t/6.676\mu s}$
(b) 2.825 mA, 1.2186 V
(c) $i_L = 2.825 \times 10^{-3}e^{-t/2.128\mu s}$
$v_L = -13.27\ e^{-t/2.128\mu s}$
25. **(a)** 0.243 V
(b) 29.47 V
(c) 18.96 V
(d) 2.025 ms
27. **(a)** 20 V
(b) 12 μA
(c) 5.376 μs
(d) 0.366 V
29. $i_L = -3.478\text{ mA} - 7.432\text{ mA } e^{-t/173.9\mu s}$
$v_L = 51.28\text{ V } e^{-t/173.9\mu s}$
31. **(a)** 8 H
(b) 4 H
33. L: 4 H, 2 H
R: 5.7 kΩ, 9.1 kΩ
35. **(a)** $V_1 = 16$ V, $V_2 = 0$ V,
$I_1 = 4$ mA
(b) $I_1 = 3$ mA, $V_1 = 12$ V,
$V_2 = 0$ V
37. $V_1 = 10$ V
$I_1 = 2$ A
$I_2 = 1.33$ A
39. $W_C = 360$ μJ
$W_L = 12$ J

Chapter 13

1. **(a)** 10 ms **(b)** 2 **(c)** 100 Hz
(d) amplitude = 5 V,
$V_{p\text{-}p} = 6.67$ V
3. 10 ms, 100 Hz
5. **(a)** 60 Hz **(b)** 57.8 Hz
(c) 27 kHz **(d)** 79.37 MHz
7. **(a)** 0.04 s **(b)** 0.28 s
(c) 0.732 s **(d)** 25.14 s
9. $T = 50$ μs
11. **(a)** $\pi/4$ **(b)** $\pi/3$ **(c)** $\frac{2}{3}\pi$
(d) $\frac{3}{2}\pi$ **(e)** 0.989π **(f)** 1.228π
13. **(a)** 3.14 rad/s
(b) 20.94×10^3 rad/s
(c) 1.57×10^6 rad/s
(d) 157.1 rad/s
15. **(a)** 120 Hz, 8.33 ms
(b) 1.34 Hz, 746.27 ms
(c) 954.93 Hz, 1.05 ms
(d) 9.95×10^{-3} Hz, 100.5 s
17. **(a)** 104.7 rad/s
(b) 43.6 krad/sec
(c) $2.45*10^6$ rad/sec
(d) 0.75 rad/sec
23. **(a)** 0.4755 A
(b) 0.25 A

(c) -0.11 A
(d) -10.5 A
25. 11.537°, 168.463°
29. (a) v leads i by 10°
(b) i leads v by 70°
(c) i leads v by 80°
(d) i leads v by 150°
31. (a) $v = 25 \sin(\omega t + 30°)$
(b) $i = 3 \times 10^{-3} \sin(6.28 \times 10^3 t - 60°)$
33. $\frac{1}{3}$ ms
35. 0.388 ms
37. (a) 0.4 ms
(b) 2.5 kHz
(c) -25 mV
39. (a) 1.875 V (b) -4.778 mA
41. (a) 40 μs
(b) 25 kHz
(c) 17.13 mV
43. (a) $2 \sin 377t$
(b) $100 \sin 377t$
(c) $84.87 \times 10^{-3} \sin 377t$
(d) $33.95 \times 10^{-6} \sin 377t$
45. 2.16 V
47. 0 V
49. (a) $T = 40\ \mu s, f = 25$ kHz, $V_{av} = 20$ mV, $V_{eff} = 28.28$ mV
(b) $T = 100\ \mu s, f = 10$ kHz, $V_{av} = -0.3$ V, $V_{eff} = 0.212$ V

Chapter 14

3. (a) $3770 \cos 377t$
(b) $452.4 \cos (754t + 20°)$
(c) $4440.63 \cos (157t - 20°)$
(d) $200 \cos t$
5. (a) $210 \sin 754t$
(b) $14.8 \sin(400t - 120°)$
(c) $42 \times 10^{-3} \sin(\omega t + 88°)$
(d) $28 \sin(\omega t + 180°)$
7. (a) 1.592 H (b) 2.654 H
(c) 0.8414 H
9. (a) $100 \sin(\omega t + 90°)$
(b) $8 \sin(\omega t + 150°)$
(c) $120 \sin(\omega t - 120°)$
(d) $60 \sin(\omega t + 190°)$
11. (a) $1 \sin(\omega t - 90°)$
(b) $0.6 \sin(\omega t - 70°)$
(c) $0.8 \sin(\omega t + 10°)$
(d) $1.6 \sin(377t + 130°)$
13. (a) $\infty\ \Omega$ (b) 530.79 Ω
(c) 265.39 Ω (d) 17.693 Ω
(e) 1.327 Ω
15. (a) 9.31 Hz (b) 4.66 Hz
(c) 18.62 Hz (d) 1.59 Hz
17. (a) $6 \times 10^{-3} \sin(200t + 90°)$
(b) $33.96 \times 10^{-3} \sin(377t + 90°)$
(c) $44.94 \times 10^{-3} \sin(374t + 300°)$
(d) $56 \times 10^{-3} \sin(\omega t + 160°)$
19. (a) $1334 \sin(300t - 90°)$
(b) $37.17 \sin(377t - 90°)$
(c) $127.2 \sin 754t$
(d) $100 \sin(1600t - 170°)$
21. (a) C (b) $L = 254.78$ mH
(c) $R = 5\ \Omega$
25. (a) 318.47 mH
(b) 107.42 mH
(c) 79.58 μH
(d) 19.9 mH
27. (a) 5.067 nF
(b) 85.8 nF
(c) 36.6 mH
(d) 325.75 nF
29. (a) 0 W (b) 0 W
(c) 122.5 W
31. 192 W
33. $40 \sin(\omega t - 50°)$
35. (a) $2 \sin(157t - 60°)$
(b) 318.47 mH (c) 0 W
37. (a) $i_1 = 2.828 \sin(10^4 t + 150°)$, $i_2 = 11.312 \sin (10^4 t + 150°)$
(b) $i_s = 14.14 \sin(10^4 t + 150°)$
39. (a) $5 \angle 36.87°$
(b) $2.83 \angle 45°$
(c) $16.38 \angle 77.66°$
(d) $806.23 \angle 82.87°$
(e) $1077.03 \angle 21.80°$
(f) $0.00658 \angle 81.25°$
(g) $11.78 \angle -49.82°$
(h) $8.94 \angle 153.43°$
(i) $61.85 \angle -104.04°$
(j) $101.53 \angle -39.81°$
(k) $4326.66 \angle 123.69°$
(l) $25.495 \times 10^{-3} \angle -78.69°$
41. (a) $15.033 \angle 86.19°$
(b) $60.208 \angle 4.76°$
(c) $0.30 \angle 88.09°$
(d) $2002.5 \angle -87.14°$
(e) $86.182 \angle 93.73°$
(f) $38.694 \angle -94°$
43. (a) $11.8 + j\,7$
(b) $151.9 + j\,49.9$
(c) $4.72 \times 10^{-6} + j\,71$
(d) $5.2 + j\,1.6$
(e) $209.3 + j\,311$
(f) $-21.2 + j\,12$
(g) $7.03 + j\,9.93$
(h) $95.698 + j\,22.768$
45. (a) $6 \angle -50°$
(b) $0.2 \times 10^{-3} \angle 140°$
(c) $109 \angle -230°$
(d) $76.471 \angle -80°$
(e) $4 \angle 0°$
(f) $0.71 \angle -16.49°$
(g) $4.21 \times 10^{-3} \angle 161.1°$
(h) $18.191 \angle -50.91°$
47. (a) $x = 4, y = 3$
(b) $x = 4$
(c) $x = 3, y = 6$ or $x = 6, y = 3$
(d) 30°
49. (a) $56.569 \sin(377t + 20°)$
(b) $169.68 \sin 377t$
(c) $11.314 \times 10^{-3} \sin(377t + 120°)$
(d) $7.07 \sin(377t + 90°)$
(e) $1696.8 \sin(377t - 120°)$
(f) $6000 \sin(377t - 180°)$
51. $i_1 = 2.537 \times 10^{-5} \sin(\omega t + 96.79°)$
53. $i_s = 18 \times 10^{-3} \sin 377t$
55. $V_a = 3.07 \sin (377t + 126.29°)$
57. $i_1 = 17.8 \sin (377t - 15.87°)$

Chapter 15

1. (a) $6.8\ \Omega \angle 0°$
(b) $754\ \Omega \angle 90°$
(c) $15.7\ \Omega \angle 90°$
(d) $265.25\ \Omega \angle -90°$
(e) $318.47\ \Omega \angle -90°$
(f) $200\ \Omega \angle 0°$
3. (a) $88 \times 10^{-3} \sin \omega t$
(b) $9.045 \sin(377t + 150°)$
(c) $2547.02 \sin(157t - 50°)$
5. (a) $4.24\ \Omega \angle -45°$
(b) $3.04\ k\Omega \angle 80.54°$
(c) $1617.56\ \Omega \angle 88.33°$
7. (a) $10\ \Omega \angle 36.87°$
(c) $\mathbf{I} = 10\ A \angle -36.87°$, $\mathbf{V}_R = 80\ V \angle -36.87°$, $\mathbf{V}_L = 60\ V \angle 53.13°$
(f) 800 W (g) 0.8 lagging
9. (a) $1660.27\ \Omega \angle -73.56°$
(b) $8.517\ mA \angle 73.56°$
(c) $\mathbf{V}_R = 4.003\ V \angle 73.56°$, $\mathbf{V}_L = 13.562\ V \angle -16.44°$
(d) 34.09 mW, 0.283 leading
11. (a) $4.47\ \Omega \angle -63.43°$
(c) $\mathbf{L} = 16$ mH, $\mathbf{C} = 265\ \mu F$
(d) $\mathbf{I} = 11.19\ A \angle 63.43°$, $\mathbf{V}_R = 22.38\ V \angle 63.43°$, $\mathbf{V}_L = 67.14\ V \angle 153.43°$
(e) $\mathbf{P} = 205.43$ W
(h) 0.447 leading
(i) $i = 15.82 \sin (377t + 63.43°)$
$e = 70.7 \sin 377t$
$\mathbf{V}_R = 31.65 \sin (377t + 62.43°)$
$\mathbf{V}_L = 94.94 \sin (377t + 153.43°)$
$\mathbf{V}_C = 158.227 \sin (377t - 26.57°)$
17. (a) $\mathbf{V}_1 = 37.97\ V \angle -51.57°$, $\mathbf{V}_2 = 113.92\ V \angle 38.43°$
(b) $\mathbf{V}_1 = 55.80\ V \angle 26.55°$, $\mathbf{V}_2 = 12.56\ V \angle -63.45°$
19. (a) $\mathbf{I} = 39\ mA \angle 126.65°$, $\mathbf{V}_R = 1.17\ V \angle 126.65°$, $\mathbf{V}_C = 25.86\ V \angle 36.65°$

(b) 0.058 leading
(c) 45.63 mW
(g) $\mathbf{Z}_T = 30\ \Omega - j\ 512.2\ \Omega$

21. $\mathbf{Z}_T = 3.2\ \Omega + j\ 2.4\ \Omega$

27. (a) $\mathbf{Z}_T = 3\ \Omega + j\ 8\ \Omega$,
$\mathbf{Y}_T = 41.1$ mS $- j$ 109.5 mS
(b) $\mathbf{Z}_T = 60\ \Omega - j\ 70\ \Omega$,
$\mathbf{Y}_T = 7.1$ mS $+ j$ 8.3 mS
(c) $\mathbf{Z}_T = 200\ \Omega - j\ 100\ \Omega$,
$\mathbf{Y}_T = 4$ mS $+ j$ 2 mS

29. (a) $\mathbf{Y}_T = 538.52$ mS $\angle -21.8°$
(c) $\mathbf{E} = 3.71$ V $\angle 21.8°$,
$\mathbf{I}_R = 1.855$ A $\angle 21.8°$,
$\mathbf{I}_L = 0.742$ A $\angle -68.2°$
(f) 6.88 W
(g) 0.928 lagging
(h) $e = 5.25 \sin(377t + 21.8°)$,
$i_R = 2.62 \sin(377t + 21.8°)$,
$i_L = 1.049 \sin(377t - 68.2°)$,
$i_s = 2.828 \sin 377t$

31. (a) $\mathbf{Y}_T = 129.96$ mS $\angle -50.31°$
(c) $\mathbf{I}_s = 7.8$ A $\angle -50.31°$,
$\mathbf{I}_R = 5$ A $\angle 0°$
$\mathbf{I}_L = 6$ A $\angle -90°$
(f) 300 W
(g) 0.638 lagging
(h) $e = 84.84 \sin 377t$,
$i_R = 7.07 \sin 377t$,
$i_L = 8.484 \sin(377t - 90°)$,
$i_s = 11.03 \sin(377t - 50.31°)$

33. (a) $\mathbf{Y}_T = 0.416$ mS $\angle 36.897°$
(c) $L = 10.61$ H, $C = 1.326\ \mu$F
(d) $\mathbf{E} = 8.498$ V $\angle -56.897°$,
$\mathbf{I}_R = 2.833$ mA $\angle -56.897°$,
$\mathbf{I}_L = 2.125$ mA $\angle -146.897°$,
$\mathbf{I}_C = 4.249$ mA $\angle 33.103°$
(g) 24.078 mW
(h) 0.8 leading
(i) $e = 12.016 \sin(377t - 56.897°)$,
$i_R = 4 \sin(377t - 56.897°)$,
$i_L = 3 \sin(377t - 146.897°)$,
$i_C = 6 \sin(377t + 33.103°)$

35. (a) $\mathbf{I}_1 = 18.09$ A $\angle 65.241°$,
$\mathbf{I}_2 = 8.528$ A $\angle -24.759°$
(b) $\mathbf{I}_1 = 11.161$ A $\angle 0.255°$,
$\mathbf{I}_2 = 6.656$ A $\angle 153.690°$

37.

f	$\lvert Y_T \rvert$	θ_T
0 Hz	25.0 mS	0.0°
1 kHz	27.98 mS	26.67°
2 kHz	35.44 mS	45.14°
3 kHz	45.23 mS	56.44°
4 kHz	56.12 mS	63.55°
5 kHz	67.61 mS	68.30°
10 kHz	128.04 mS	78.75°
20 kHz	252.59 mS	89.86°

Chapter 16

1. (a) $R_p = 94.73\ \Omega$ (R),
$X_p = 52.1\ \Omega$ (C)
(b) $R_p = 4$k Ω (R),
$X_p = 4$k Ω (C)

3. (a) $E = 176.68$ V $\angle$ 36.44°
$I_R = 0.803$ A $\angle$ 36.44°
$I_L = 2.813$ A $\angle$ −53.56°
(b) $F_p = 0.804$ lagging
(c) $P = 141.86$ W
(f) $I_C = 1.11$ A $\angle$ 126.43° for both
(g) $Z_T = 142.15\ \Omega + j104.96\ \Omega$

5. $R = 4\ \Omega$ and, $X_L = 3.744\ \Omega$ in parallel

9. (a) 1.2 Ω ∠90°
(b) 10 A ∠−90°
(c) 10 A ∠−90°
(d) $\mathbf{I}_2 = 6$ A $\angle -90°$,
$\mathbf{I}_3 = 4$ A $\angle -90°$
(e) 60 V ∠0°

11. (a) $\mathbf{Z}_T = 3.87\ \Omega \angle -11.817°$,
$\mathbf{Y}_T = 0.258$ S $\angle 11.817°$
(b) 15.504 A ∠41.817°
(c) 3.985 A ∠82.826°
(d) 47.809 V ∠−7.174°
(e) 910.71 W

13. (a) 0.375 A ∠25.346°
(b) 70.711 V ∠−45°
(c) 33.9 W

15. (a) 1.423 A ∠18.259°
(b) 26.574 V ∠4.763°
(c) 54.074 W

17. (a) $\mathbf{Y}_T = 0.099$ S $\angle -9.709°$
(b) $\mathbf{V}_1 = 20.4$ V $\angle 30°$,
$\mathbf{V}_2 = 10.887$ V $\angle 58.124°$
(c) 1.933 A ∠11.109°

19. 33.201 A ∠38.89°

21. 139.71 mW

Chapter 17

3. (a) $\mathbf{Z} = 21.93\ \Omega \angle -46.85°$,
$\mathbf{E} = 10.97$ V $\angle 13.15°$
(b) $\mathbf{Z} = 5.15\ \Omega \angle 59.04°$,
$\mathbf{E} = 10.3$ V $\angle 179.04°$

5. (a) 5.15 A ∠−24.5°
(b) 0.442 A ∠143.48°

7. (a) 13.07 A ∠−33.71°
(b) 48.33 A ∠−77.57°

9. -3.165×10^{-3} **V** $\angle 137.29°$

11. $\mathbf{I}_{1k\Omega} = 10$ mA $\angle 0°$
$\mathbf{I}_{2k\Omega} = 1.667$ mA $\angle 0°$

13. $\mathbf{I}_L = 1.378$ mA $\angle -56.31°$

15. (a) $\mathbf{V}_1 = 19.86$ V $\angle 43.8°$,
$\mathbf{V}_2 = 8.94$ V $\angle 106.9°$
(b) $\mathbf{V}_1 = 19.78$ V $\angle 132.48°$,
$\mathbf{V}_2 = 13.37$ V $\angle 98.78°$

17. $\mathbf{V}_1 = 220$ V $\angle 0°$
$\mathbf{V}_2 = 96.664$ V $\angle -12.426°$
$\mathbf{V}_3 = 100$ V $\angle 90°$

19. (left) $\mathbf{V}_1 = 14.62$ V $\angle -5.86°$
(top) $\mathbf{V}_2 = 35.03$ V $\angle -37.69°$
(right) $\mathbf{V}_3 = 32.4$ V $\angle -73.34°$
(middle) $\mathbf{V}_4 = 5.677$ V $\angle 23.53°$

21. $\mathbf{V}_1 = 4.372$ V $\angle -128.66°$
$\mathbf{V}_2 = 2.253$ V $\angle 17.628°$

23. $\mathbf{V}_1 = -10.667$ V $\angle 0°$
$\mathbf{V}_2 = -6$ V $\angle 0°$

25. $-2451.92\mathbf{E}_i$

27. (a) No
(b) 1.76 mA ∠−71.54°
(c) 7.03 V ∠−18.46°

29. Balanced

31. $R_x = R_2R_3/R_1$
$L_x = R_2L_3/R_1$

33. (a) 11.57 A ∠−67.13°
(b) 36.9 A ∠23.87°

Chapter 18

1. (a) 6.095 A ∠−32.115°
(b) 3.77 A ∠−93.8°

3. $i = 0.5\text{A} + 1.581 \sin(\omega t - 26.565°)$ A

5. 6.261 mA ∠−63.43°

7. −22.09 V ∠6.34°

9. 19.62 V ∠53°

11. $\mathbf{V}_s = 10$ V $\angle 0°$

13. (a) $\mathbf{Z}_{Th} = 21.312\ \Omega \angle 32.196°$
$\mathbf{E}_{Th} = 2.131$ V $\angle 32.196°$
(b) $\mathbf{Z}_{Th} = 6.813\ \Omega \angle -54.228°$
$\mathbf{E}_{Th} = 57.954$ V $\angle 11.099°$

15. (a) $\mathbf{Z}_{Th} = 4\ \Omega \angle 90°$
$\mathbf{E}_{Th} = 4$ V $+ 10$ V $\angle 0°$
(b) $\mathbf{I} = 0.5\text{A} + 1.118$ A $\angle -26.565°$

17. (a) $\mathbf{Z}_{Th} = 4.472$ kΩ $\angle -26.565°$
$\mathbf{E}_{Th} = 31.31$ V $\angle -26.565°$
(b) $\mathbf{I} = 6.26$ mA $\angle -63.435°$

19. $\mathbf{Z}_{Th} = 4.44$ kΩ $\angle -0.031°$
$\mathbf{E}_{Th} = -444.45 \times 10^3 \mathbf{I} \angle 0.255°$

21. $\mathbf{Z}_{Th} = 5.099$ kΩ $\angle -11.31°$
$\mathbf{E}_{Th} = -50$ V $\angle 0°$

23. $\mathbf{Z}_{Th} = -39.215\ \Omega \angle 0°$
$\mathbf{E}_{Th} = 20$ V $\angle 53°$

25. $\mathbf{Z}_{Th} = 607.42\ \Omega \angle 0°$
$\mathbf{E}_{Th} = 1.62$ V $\angle 0°$

27. (a) $\mathbf{Z}_N = 21.312\ \Omega \angle 32.196°$,
$\mathbf{I}_N = 0.1$ A $\angle 0°$
(b) $\mathbf{Z}_N = 6.813\ \Omega \angle -54.228°$,
$\mathbf{I}_N = 8.506$ A $\angle 65.324°$

29. (a) $\mathbf{Z}_N = 9.66\ \Omega \angle 14.93°$,
$\mathbf{I}_N = 2.15$ A $\angle -42.87°$
(b) $\mathbf{Z}_N = 4.37\ \Omega \angle 55.67°$,
$\mathbf{I}_N = 22.83$ A $\angle -34.65°$

31. (a) $\mathbf{Z}_N = 9\ \Omega \angle 0°$,
$\mathbf{I}_N = 1.333$ A $+ 2.667$ A $\angle 0°$
(b) 12 V + 2.65 V ∠−83.66°

33. $\mathbf{Z}_N = 5.1$ kΩ $\angle -11.31°$,
$\mathbf{I}_N = -1.961 \times 10^{-3}$ **V** $\angle 11.31°$

35. $\mathbf{Z}_N = 5.1$ kΩ $\angle -11.31°$,
$\mathbf{I}_N = 9.81$ mA $\angle 11.31°$

37. $\mathbf{Z}_N = 6.63\text{ k}\Omega \angle 0°$
$\mathbf{I}_N = 0.792\text{ mA} \angle 0°$
39. (a) $\mathbf{Z}_L = 8.32\ \Omega \angle 3.18°$, 1198.2 W
(b) $\mathbf{Z}_L = 1.562\ \Omega \angle +14.47°$, 1.614 W
41. 40 kΩ, 25 W
43. (a) 9 Ω **(b)** 20 W
45. (a) 1.414 kΩ **(b)** 0.518 W
49. 25.77 mA ∠104.4°

Chapter 19

1. (a) 120 W
(b) $Q_T = 0$ VAR, $S_T = 120$ VA
(c) 0.5 A
(e) $I_1 = \frac{1}{6}$ A, $I_2 = \frac{1}{3}$ A
3. (a) 400 W, −400 VAR(C), 565.69 VA, 0.7071 leading
(c) 5.66 A ∠135°
5. (a) 500 W, −200 VAR(C), 538.52 VA
(b) 0.928 leading
(d) 10.776 A ∠81.875°
7. (a) R: 200 W, L,C: 0 W
(b) R: 0 VAR, C: 80 VAR, L: 100 VAR
(c) R: 200 VA, C: 80 VA, L: 100 VA
(d) 200 W, 20 VAR(L), 200.998 VA, 0.995 (lagging)
(f) 10.05 A ∠−5.73°
9. (a) R: 47.47 W, L: 0 W, C: 0 W
(b) R: 0 VAR, L: 135.91 VAR, C: 53.69 VAR
(c) R: 47.47 VA, L: 135.91 VA, C: 53.69 VA
(d) 47.47 W, 82.22 VAR(L), 94.94 VA, 0.5 (lagging)
(f) 0.79 J
(g) $W_L = 0.36$ J, $W_C = 0.14$ J
11. (a) $\mathbf{Z} = 2.30\ \Omega + j\ 1.73\ \Omega$
(b) 4000 W
13. (a) 900 W, 0 VAR, 900 VA, 1
(b) 9 A ∠0°
(d) $\mathbf{Z}_1$: $R = 0\ \Omega$, $X_C = 20\ \Omega$
$\mathbf{Z}_2$: $R = 2.83\ \Omega$, $X = 0\ \Omega$
$\mathbf{Z}_3$: $R = 5.66\ \Omega$, $X_L = 4.717\ \Omega$
15. (a) 1100 W, 2366.26 VAR, 2609.44 VA, 0.4215 (leading)
(b) 521.89 V ∠−65.07°
(c) $\mathbf{Z}_1$: $R = 1743.38\ \Omega$, $X_C = 1307.53\ \Omega$
$\mathbf{Z}_2$: $R = 43.59\ \Omega$, $X_C = 99.88\ \Omega$
17. (a) 7.81 kVA
(b) 0.640 (lagging)
(c) 65.08 A
(d) 1105 μF
(e) 41.67 A
19. (a) 128.14 W
(b) a–b: 42.69 W, b–c: 64.03 W, a–c: 106.72 W, a–d: 106.72 W, c–d: 0 W, d–e: 0 W, f–e: 21.34 W
21. (a) 5 Ω, 132.03 mH
(b) 10 Ω
(c) 15 Ω, 262.39 mH

Chapter 20

1. (a) $\omega_s = 250$ rad/s, $f_s = 39.79$ Hz
(b) $\omega_s = 3535.53$ rad/s, $f_s = 562.7$ Hz
(c) $\omega_s = 21{,}880$ rad/s, $f_s = 3482.31$ Hz
3. (a) $X_L = 40\ \Omega$
(b) $I = 10$ mA
(c) $V_R = 20$ mV, $V_L = 400$ mV, $V_C = 400$ mV
(d) $Q_s = 20$ (high)
(e) $L = 1.27$ mH, $C = 0.796\ \mu$F
(f) $BW = 250$ Hz
(g) $f_2 = 5.125$ kHz, $f_1 = 4.875$ kHz
5. (a) $BW = 400$ Hz
(b) $f_2 = 6200$ Hz, $f_1 = 5800$ Hz
(c) $X_L = X_C = 45\ \Omega$
(d) $P_{\text{HPF}} = 375$ mW
7. (a) $Q_s = 10$
(b) $X_L = 20\ \Omega$
(c) $L = 1.59$ mH, $C = 3.98\ \mu$F
(d) $f_2 = 2100$ Hz, $f_1 = 1900$ Hz
9. $L = 13.26$ mH, $C = 27.07$ nF
$f_2 = 8460$ Hz, $f_1 = 8340$ Hz
11. $R = 2.51\ \Omega$
$C = 202$ mF
$f_s = 12500$ Hz
$f_1 = 12250$ Hz
$f_2 = 12750$ Hz
13. (a) $f_p = 159.155$ kHz
(b) $V_C = 4$ V
(c) $I_L = I_C = 40$ mA
(d) $Q_p = 20$
15. (a) $f_s = 11{,}253.95$ Hz
(b) $Q_l = 1.77$ (no)
(c) $f_p = 9{,}280.24$ Hz, $f_m = 10{,}794.41$ Hz
(d) $X_L = 5.83\ \Omega$, $X_C = 8.57\ \Omega$
(e) $Z_{T_p} = 12.5\ \Omega$
(f) $V_C = 25$ mV
(g) $Q_p = 1.46$, $BW = 6.356$ kHz
(h) $I_C = 2.92$ mA, $I_L = 3.54$ mA
17. (a) $X_C = 30\ \Omega$
(b) $Z_{T_P} = 225\ \Omega$
(c) $\mathbf{I}_C = 0.6\text{ A} \angle 90°$, $\mathbf{I}_L \cong 0.6\text{ A} \angle -86.19°$
(d) $L = 0.239$ mH, $C = 265.26$ nF
(e) $Q_p = 7.5$, $BW = 2.67$ kHz
19. (a) $f_s = 7.118$ kHz, $f_p = 6.647$ kHz, $f_m = 7$ kHz
(b) $X_L = 20.88\ \Omega$, $X_C = 23.94\ \Omega$
(c) $Z_{T_P} = 55.56\ \Omega$
(d) $Q_p = 2.32$, $BW = 2.865$ kHz
(e) $I_L = 99.28$ mA, $I_C = 92.73$ mA
(f) $V_C = 2.22$ V
21. (a) $f_p = 3558.81$ Hz
(b) $V_C = 138.2$ V
(c) $P = 691$ mW
(d) $BW = 575.86$ Hz
23. (a) $X_L = 98.54\ \Omega$
(b) $Q_l = 8.21$
(c) $f_p = 8.05$ kHz
(d) $V_C = 4.83$ V
(e) $f_2 = 8.55$ kHz, $f_1 = 7.55$ kHz
25. $R_s = 3.244$ kΩ, $C = 31.66$ nF
27. (a) $f_p = 251.65$ kHz
(b) $Z_{T_p} = 4.444$ kΩ
(c) $Q_p = 14.05$
(d) $BW = 17.91$ kHz
(e) 20 nF: $f_p = 194.93$ kHz, $Z_{T_p} = 49.94\ \Omega$, $Q_p = 2.04$, $BW = 95.55$ kHz
(f) 1 nf: $f_p = 251.65$ kHz, $Z_{T_p} = 13.33$ kΩ, $Q_p = 21.08$, $BW = 11.94$ kHz
(g) Network; $L/C = 100 \times 10^3$
part (e): $L/C = 1 \times 10^3$
part (f): $L/C = 400 \times 10^3$
(h) yes, $L/C \uparrow$ BW$\downarrow$

Chapter 21

1. (a) 1.845
(b) 18.45
3. 13.01
5. 38.49
7. 24.08 dB_s
11. (a) $0.1f_c$: 0.995, $0.5f_c$: 0.894, f_c: 0.707, $2f_c$: 0.447, $10f_c$: 0.0995
(b) $0.1f_c$: −5.71°, $0.5f_c$: −26.57°, f_c: −45°, $2f_c$: −63.43°, $10f_c$: −84.29°
13. $C = 0.265\ \mu$F,
250 Hz: $A_v = 0.895$, $\theta = -26.54°$,
1000 Hz: $A_v = 0.4475$, $\theta = -63.41°$
15. $f_c = 3.617$ kHz,
f_c: $A_v = 0.707$, $\theta = 45°$,
$2f_c$: $A_v = 0.894$, $\theta = 26.57°$
$0.5f_c$: $A_v = 0.447$, $\theta = 63.43°$
$10f_c$: $A_v = 0.995$, $\theta = 5.71°$

$\frac{1}{10}f_c$: $A_v = 0.0995$,
$\theta = 84.29°$

17. $R = 795.77\ \Omega \rightarrow 797\ \Omega$,
f_c: $A_v = 0.707$, $\theta = 45°$
1 kHz: $A_v = 0.458$, $\theta = 63.4°$
4 kHz: $A_v \cong 0.9$, $\theta = 26.53°$

19. **(a)** $f_{c_1} = 795.77$ Hz,
$f_{c_2} = 1989.44$ Hz
f_{c_1}: $V_o = 0.656V_i$,
f_{c_2}: $V_o = 0.656V_i$
$f_{center} = 1392.60$ Hz: $V_o = 0.711V_i$
500 Hz: $V_o = 0.516V_i$,
4 kHz: $V_o = 0.437V_i$
(b) $BW \cong 2.9$ kHz,
$f_{center} = 1.94$ kHz

21. **(a)** $f_s = 100.658$ kHz
(b) $Q_s = 18.39$,
$BW = 5473.52$ Hz
(c) f_s: $A_v = 0.93$
$f_1 = 97{,}921.24$ Hz,
$f_2 = 103{,}394.76$ Hz,
$f = 95$ kHz: $A_v = 0.392$,
$f = 105$ kHz: $A_v = 0.5$
(d) $f = f_s$, $V_o = 0.93$ V,
$f = f_1$ or f_2, $V_o = 0.658$ V

23. **(a)** $Q_s = 12.195$
(b) $BW = 410$ Hz,
$f_2 = 5205$ Hz,
$f_1 = 4795$ Hz
(c) f_s: $V_o = 0.024V_i$
(d) f_s: V_o still $0.024V_i$

25. **(a)** $f_p = 726.44$ kHz (stop-band)
$f = 2.013$ MHz (pass-band)

27. **(a–b)** $f_c = 6772.55$ Hz
(c) f_c: -3 dB, $\frac{1}{2}f_c$: -6.7 dB,
$2f_c$: -0.969 dB,
$\frac{1}{10}f_c$: -20.04 dB,
$10f_c$: -0.043 dB
(d) f_c: 0.707, $\frac{1}{2}f_c$: 0.4472,
$2f_c$: 0.894
(e) f_c: 45°, $\frac{1}{2}f_c$: 63.43°, $2f_c$: 26.57°

29. **(a–b)** $f_c = 13.26$ kHz
(c) f_c: -3 dB, $\frac{1}{2}f_c$: -0.97 dB,
$2f_c$: -6.99 dB
$\frac{1}{10}f_c$: -0.043 dB,
$10f_c$: -20.04 dB
(d) f_c: 0.707, $\frac{1}{2}f_c$: 0.894,
$2f_c$: 0.447
(e) f_c: $-45°$, $\frac{1}{2}f_c$: $-26.57°$,
$2f_c$: $-63.43°$

Chapter 22

1. **(a)** 0.2 H
(b) $e_p = 1.6$ V, $e_s = 5.12$ V
(c) $e_p = 15$ V, $e_s = 24$ V

3. **(a)** 158.02 mH
(b) $e_p = 24$ V, $e_s = 1.8$ V
(c) $e_p = 15$ V, $e_s = 24$ V

5. 1.354 H

7. $\mathbf{I}_1 (R_1 + j X_{L_1}) + \mathbf{I}_2(j X_m) = \mathbf{E}_1$
$\mathbf{I}_1(j X_m) + \mathbf{I}_2(j X_{L_2} + R_L) = 0$

9. **(a)** 3.125 V **(b)** 391.02 μWb

11. 56.31 Hz

13. 400 Ω

15. $12{,}000t$

17. **(a)** 20 Ω
(b) 40 Ω
(d) 0.351 A $\angle -6.71°$
(e) 28.1 V $\angle -6.71°$
(g) 30 V

19. **(a)** $\mathbf{Z}_p = 280.71\ \Omega \angle -85.91°$
(b) $\mathbf{I}_p = 0.427$ A $\angle 85.91°$
(c) $\mathbf{V}_{R_e} = 8.54$ V $\angle 85.91°$
$\mathbf{V}_{X_e} = 17.08$ V $\angle 175.91°$
$\mathbf{V}_{X_C} = 136.64$ V $\angle -4.09°$

21. $\mathbf{Z}_i = 7980\ \Omega \angle 89.98°$

23. **(a)** 20 **(b)** 83.33 A
(c) 4.167 A
(d) $a = \frac{1}{20}$, $I_s = 4.167$ A,
$I_p = 83.33$ A

25. **(a)** 25 V $\angle 0°$, 5 A $\angle 0°$
(b) 80 Ω $\angle 0°$ **(c)** 20 Ω $\angle 0°$

27. **(a)** $\mathbf{E}_2 = 40$ V $\angle 60°$,
$\mathbf{I}_2 = 3.33$ A $\angle 60°$,
$\mathbf{E}_3 = 30$ V $\angle 60°$,
$\mathbf{I}_3 = 3$ A $\angle 60°$
(b) $R_1 = 64.52\ \Omega$

29. $[\mathbf{Z}_1 + \mathbf{X}_{L_1}]\mathbf{I}_1 - \mathbf{Z}_{M_{12}}\mathbf{I}_2 + \mathbf{Z}_{M_{13}}\mathbf{I}_3 = \mathbf{E}_1$,
$\mathbf{Z}_{M_{12}}\mathbf{I}_1 - [\mathbf{Z}_2 + \mathbf{Z}_3 + \mathbf{X}_{L_2}]\mathbf{I}_2 + \mathbf{Z}_2\mathbf{I}_3 = 0$,
$\mathbf{Z}_{M_{13}}\mathbf{I}_1 - \mathbf{Z}_2\mathbf{I}_2 + [\mathbf{Z}_2 + \mathbf{Z}_4 + \mathbf{X}_{L_3}]\mathbf{I}_3 = 0$

GLOSSARY

Absolute zero The temperature at which all molecular motion ceases; $-273.15°C$.

Active filter A filter that employs active devices such as transistors or operational amplifiers in combination with R, L, and C elements.

Admittance A measure of how easily a network will "admit" the passage of current through that system. It is measured in siemens, abbreviated S, and is represented by the capital letter Y.

Admittance diagram A vector display that clearly depicts the magnitude of the admittance of the conductance, capacitive susceptance, and inductive susceptance, and the magnitude and angle of the total admittance of the system.

Alternating waveform A waveform that oscillates above and below a defined reference level.

Ammeter An instrument designed to read the current through elements in series with the meter.

Amp-Clamp® A clamp-type instrument that will permit noninvasive current measurements and that can be used as a conventional voltmeter or ohmmeter.

Ampere (A) The SI unit of measurement applied to the flow of charge through a conductor.

Ampere-hour rating The rating applied to a source of energy that will reveal how long a particular level of current can be drawn from that source.

Angular velocity The velocity with which a radius vector projecting a sinusoidal function rotates about its center.

Apparent power The power delivered to a load without consideration of the effects of a power-factor angle of the load. It is determined solely by the product of the terminal voltage and current of the load.

Average (real) power The delivered power dissipated in the form of heat by a network or system.

Average value The level of a waveform defined by the condition that the area enclosed by the curve above this level is exactly equal to the area enclosed by the curve below this level.

Ayrton shunt for a multirange ammeter, composed of several resistors thay all carry current for every range.

Band (cutoff, half-power, corner) frequencies Frequencies that define the points on the resonance curve that are 0.707 of the peak current or voltage value. In addition, they define the frequencies at which the power transfer to the resonant circuit will be half the maximum power level.

Bandwidth The range of frequencies between the band, cutoff, or half-power frequencies.

Block impedance An equivalent impedance representing individual R, L, and C elements that may be in a series or parallel circuit, combined into a single impedance component. It is used to simplify calculations for ac circuits and networks.

Bode plot A plot of the frequency response of a system using straight-line segments called asymptotes.

Branch The portion of a circuit consisting of one or more elements in series.

Branch-current method A technique for determining the branch currents of a multiloop network.

Breakdown voltage Another term for dielectric strength, listed below.

Bridge network A network configuration having the appearance of a diamond in which no two branches are in series or parallel.

Capacitance A measure of a capacitor's ability to store charge; measured in farads (F).

Capacitance comparison bridge A bridge configuration having a galvanometer in the bridge arm that is used to determine an unknown capacitance and associated resistance.

Capacitive time constant The product of resistance and capacitance that establishes the required time for the charging and discharging phases of a capacitive transient.

Capacitive transient The waveforms for the voltage and current of a capacitor that result during the charging and discharging phases.

Capacitor A fundamental electrical element having two conducting surfaces separated by an insulating material and having the capacity to store charge on its plates.

Cell A fundamental source of electrical energy developed through the conversion of chemical or solar energy.

Choke A term often applied to an inductor, due to the ability of an inductor to resist a change in current through it.

Circuit A combination of a number of elements joined at terminal points providing at least one closed path through which charge can flow.

Closed loop Any continuous connection of branches that allows tracing of a path that leaves a point in one direction and returns to that same point from another direction without leaving the circuit.

Colour coding A technique employing bands of colour to indicate the resistance levels and tolerance of resistors.

Complex conjugate A complex number defined by simply changing the sign of an imaginary component of a complex number in the rectangular form.

Complex number A number that represents a point in a two-dimensional plane located with reference to two distinct axes. It defines a vector drawn from the origin to that point.

Conductance (*G*) An indication of the relative ease with which current can be established in a material. It is measured in siemens (S).

Conductors Materials that permit a generous flow of electrons with very little voltage applied.

Conventional current flow A defined direction for the flow of charge in an electrical system out of the positive terminal of a voltage source, and into the negative terminal.

Copper A material possessing physical properties that make it particularly useful as a conductor of electricity.

Coulomb (C) The fundamental SI unit of measure for charge. It is equal to the charge carried by 6.242×10^{18} electrons.

Coulomb's law An equation defining the force of attraction or repulsion between two charges.

Current divider rule A method by which the current through parallel elements can be determined without first finding the voltage across those parallel elements.

Current sources Sources that supply a fixed current to a network and have a terminal voltage dependent on the network to which they are applied.

Cycle A portion of a waveform contained in one period of time.

d'Arsonval movement An iron-core coil mounted on bearings between a permanent magnet. A pointer connected to the movable core indicates the strength of the current passing through the coil.

dc current source A source that will provide a fixed current level even though the load to which it is applied may cause its terminal voltage to change.

dc generator A source of dc voltage available through the turning of the shaft of the device by some external means.

Decibel A unit of measurement used to compare power levels.

Delta (Δ), pi (π) configuration A network structure that consists of three branches and has the appearance of the Greek letter delta (Δ) or pi (π).

Dependent (controlled) source A source whose magnitude and/or phase angle is determined (controlled) by a current or voltage of the system in which it appears.

Derivative The instantaneous rate of change of a function with respect to time or another variable.

Determinants method A mathematical technique for finding the unknown variables of two or more simultaneous linear equations.

Dielectric constant Another term for relative permittivity, listed below.

Dielectric strength An indication of the voltage required for unit length to establish conduction in a dielectric.

Dielectric The insulating material between the plates of a capacitor that can have a pronounced effect on the charge stored on the plates of a capacitor.

Diode A semiconductor device whose behavior is much like that of a simple switch; that is, it will pass current ideally in only one direction when operating within specified limits.

Direct current Current having a single direction (unidirectional) and a fixed magnitude over time.

Double-tuned filter A network having both a pass-band and a stop-band region.

Ductility The property of a material that allows it to be drawn into long thin wires.

Eddy currents Small, circular currents in a paramagnetic core causing an increase in the power losses and the effective resistance of the material.

Effective resistance The resistance value that includes the effects of radiation losses, skin effect, eddy currents, and hysteresis losses.

Effective value The equivalent dc value of any alternating voltage or current.

Efficiency (η) A ratio of output to input power that provides immediate information about the energy-converting characteristics of a system.

Electric field strength The force acting on a unit positive charge in the region of interest.

Electric flux lines Lines drawn to indicate the strength and direction of an electric field in a particular region.

Electrodynamometer meters Instruments that can measure both ac and dc quantities without a change in internal circuitry.

Electrolytes The contact element and the source of ions between the electrodes of the battery.

Electromagnetism Magnetic effects introduced by the flow of charge or current.

Electron The particle with negative polarity that orbits the nucleus of an atom.

Energy (*W*) A quantity whose change in state is determined by the product of the rate of conversion (P) and the period involved (t). It is measured in joules (J) or wattseconds (Ws).

Equivalent circuits For every series ac network there is a parallel ac network (and vice versa) that will be "equivalent" in the sense that the input current and impedance are the same.

Faraday's law A law relating the voltage induced across a coil to the number of turns in the coil and the rate at which the flux linking the coil is changing.

Ferromagnetic materials Materials having permeabilities hundreds and thousands of times greater than that of free space.

Filter Networks designed to either pass or reject the transfer of signals at certain frequencies to a load.

Flux density (B) A measure of the flux per unit area perpendicular to a magnetic flux path. It is measured in teslas (T) or webers per square meter (Wb/m2).

Free electron An electron unassociated with any particular atom, relatively free to move through a crystal lattice structure under the influence of external forces.

Frequency (f) The number of cycles of a periodic waveform that occur in one second.

Frequency counter An instrument that will provide a digital display of the frequency or period of a periodic time-varying signal.

Fringing An effect established by flux lines that do not pass directly from one conducting surface to another.

Hay bridge A bridge configuration used for measuring the resistance and inductance of coils in those cases where the resistance is a small fraction of the reactance of the coil.

High-pass filter A filter designed to pass high frequencies and reject low frequencies.

Horsepower (hp) Equivalent to 746 watts in the electrical system.

Hysteresis The lagging effect between the flux density of a material and the magnetizing force applied.

Hysteresis losses Losses in a magnetic material introduced by changes in the direction of the magnetic flux within the material.

Impedance diagram A vector display that clearly depicts the magnitude of the impedance of the resistive, reactive, and capacitive components of a network, and the magnitude and angle of the total impedance of the system.

Independent source A source whose magnitude is independent of the network to which it is applied. It displays its terminal characteristics even if completely isolated.

Inductor A fundamental element of electrical systems constructed of numerous turns of wire around a ferromagnetic or air core.

Inferred absolute temperature The temperature through which a straight-line approximation #for the actual resistance-versus-temperature curve will intersect the temperature axis.

Instantaneous value The magnitude of a waveform at any instant of time, denoted by lowercase letters.

Insulators Materials in which a very high voltage must be applied to produce any measurable current flow.

Integrated circuit (IC) A subminiature structure containing a vast number of electronic devices designed to perform a particular set of functions.

Internal resistance The inherent resistance found internal to any source of energy.

Joule (J) A unit of measurement for energy in the SI or MKS system.

Kelvin (K) A unit of measurement for temperature in the SI system. Equal to 273.15 + t°C.

Kilogram (kg) A unit of measure for mass in the SI system. Equal to 1000 grams in the CGS system.

Kilowatthour meter An instrument for measuring kilowatthours of energy supplied to a residential or commercial user of electricity.

Kirchhoff's current law The algebraic sum of the currents entering and leaving a node is zero.

Kirchhoff's voltage law The algebraic sum of the potential rises and drops around a closed loop (or path) is zero.

Ladder network A network that consists of a cascaded set of series-parallel combinations and has the appearance of a ladder.

Leading and lagging power factors An indication of whether a network is primarily capacitive or inductive in nature. Leading power factors are associated with capacitive networks, and lagging power factors with inductive networks.

Leakage current The current that will result in the total discharge of a capacitor if the capacitor is disconnected from the charging network for a sufficient length of time.

Lenz's law A law stating that an induced effect is always such as to oppose the cause that produced it.

Log-log paper Graph paper with vertical and horizontal log scales.

Low-pass filter A filter designed to pass low frequencies and reject high frequencies.

Magnetic flux lines Lines of a continuous nature that reveal the strength and direction of a magnetic field.

Magnetomotive force (^) The "pressure" required to establish magnetic flux in a ferromagnetic material. It is measured in ampere-turns (At).

Malleability The property of a material that allows it to be worked into many different shapes.

Maxwell bridge A bridge configuration used for inductance measurements when the resistance of the coil is large enough not to require a Hay bridge.

Megohmmeter An instrument for measuring very high resistance levels, such as in the megohm range.

Mesh analysis A method through which the loop (or mesh) currents of a network can be determined. The branch currents of the network can then be determined directly from the loop currents.

Mesh (loop) current A labeled current assigned to each distinct closed loop of a network that can, individually or in combination with other mesh currents, define all of the branch currents of a network.

Metre (m) A unit of measure for length in the SI and MKS system.

Microbar A unit of measurement for sound pressure levels that permits comparing audio levels on a dB scale.

Millman's theorem A method employing voltage-to-current source conversions that will permit the determination of unknown variables in a multiloop network.

Negative temperature coefficient of resistance The value revealing that the resistance of a material will decrease with an increase in temperature.

Neutron The particle having no electrical charge, found in the nucleus of the atom.

Newton (N) A unit of measurement for force in the SI system.

Nodal analysis A method through which the node voltages of a network can be determined. The voltage across each element can then be determined through application of Kirchhoff's voltage law.

Node A junction of two or more branches in a network.

Norton's theorem (ac) A theorem that permits the reduction of any two-terminal linear ac network to one having a single current source and parallel impedance. The resulting configuration can then be employed to determine a particular current or voltage in the original network or to examine the effects of a specific portion of the network on a particular variable.

Nucleus The structural center of an atom that contains both protons and neutrons.

Ohm (Ω) The unit of measurement applied to resistance.

Ohm's law An equation that establishes a relationship among the current, voltage, and resistance of an electrical system.

Ohm/volt rating A rating used to determine both the current sensitivity of the movement and the internal resistance of the meter.

Ohmmeter An instrument for measuring resistance levels.

Open circuit The absence of a direct connection between two points in a network.

Oscilloscope An instrument that will display, through the use of a cathode-ray tube, the characteristics of a time-varying signal.

Parallel ac circuits A connection of elements in an ac network in which all the elements have two points in common. The voltage is the same across each element.

Parallel circuit A circuit configuration in which the elements have two points in common.

Pass-band (band-pass) filter A network designed to pass signals within a particular frequency range.

Passive filter A filter constructed of series, parallel, or series-parallel R, L, and C elements.

Peak value The maximum value of a waveform, denoted by uppercase letters.

Peak-to-peak value The magnitude of the total swing of a signal from positive to negative peaks. The sum of the absolute values of the positive and negative peak values.

Period (T) The time interval between successive repetitions of a periodic waveform.

Periodic waveform A waveform that continually repeats itself after a defined time interval.

Permanent magnet A material such as steel or iron that will remain magnetized for long periods of time without the aid of external means.

Permeability (m) A measure of the ease with which magnetic flux can be established in a material. It is measured in Wb/Am.

Permittivity A measure of how well a dielectric will permit the establishment of flux lines within the dielectric.

Phase relationship An indication of which of two waveforms leads or lags the other, and by how many degrees or radians.

Phasor A radius vector that has a constant magnitude at a fixed angle from the positive real axis and that represents a sinusoidal voltage or current in the vector domain.

Phasor diagram A vector display that provides at a glance the magnitude and phase relationships among the various voltages and currents of a network.

Photoconductive cell A two-terminal semiconductor device whose terminal resistance is determined by the intensity of the incident light on its exposed surface.

Polar form A method of defining a point in a complex plane that includes a single magnitude to represent the distance from the origin, and an angle to reflect the counterclockwise distance from the positive real axis.

Positive ion An atom having a net positive charge due to the loss of one of its negatively charged electrons.

Positive temperature coefficient of resistance The value revealing that the resistance of a material will increase with an increase in temperature.

Potential difference The algebraic difference in potential (or voltage) between two points in an electrical system.

Potential energy The energy that a mass possesses by virtue of its position.

Potentiometer A three-terminal device through which potential levels can be varied in a linear or nonlinear manner.

Power An indication of how much work can be done in a specified amount of time; a rate of doing work. It is measured in joules/second (J/s) or watts (W).

Power factor (Fp) An indication of how reactive or resistive an electrical system is. The higher the power factor, the greater the resistive component.

Power-factor correction The addition of reactive components (typically capacitive) to establish a system power factor closer to unity.

PPM/°C Temperature sensitivity of a resistor in parts per million per degree Celsius.

Primary cell Sources of voltage that cannot be recharged.

Proton The particle of positive polarity found in the nucleus of the atom.

Quality factor (Q) A ratio that provides an immediate indication of the sharpness of the peak of a resonance curve. The higher the Q, the sharper the peak and the more quickly it drops off to the right and left of the resonant frequency.

Radian A unit of measure used to define a particular segment of a circle. One radian is approximately equal to 57.3°; 2p rad are equal to 360°.

Radiation losses The loss of energy in the form of electromagnetic waves during the transfer of energy from one element to another.

Reactance The opposition of an inductor or capacitor to the flow of charge that results in the continual exchange of energy between the circuit and magnetic field of an inductor or the electric field of a capacitor.

Reactive power The power associated with reactive elements that provides a measure of the energy associated with setting up the magnetic and electric fields of inductive and capacitive elements, respectively.

Reciprocal A format defined by 1 divided by the complex number.

Reciprocity theorem A theorem that states that for single-source networks, the current in any branch of a network, due to a single voltage source in the network, will equal the current through the branch in which the source was originally located if the source is placed in the branch in which the current was originally measured.

Rectangular form A method of defining a point in a complex plane that includes the magnitude of the real component and the magnitude of the imaginary component, the latter component being defined by an associated letter j.

Rectification The process by which an ac signal is converted to one that has an average dc level.

Rectifier-type ac meter An instrument calibrated to indicate the effective value of a current or voltage through the use of a rectifier network and d'Arsonval-type movement.

Relative permeability (mr) The ratio of the permeability of a material to that of free space.

Relative permittivity The permittivity of a material compared to that of air.

Reluctance (5) A quantity determined by the physical characteristics of a material that will provide an indication of the "reluctance" of that material to the setting up of magnetic flux lines in the material. It is measured in rels or At/Wb.

Resistance A measure of the opposition to the flow of charge through a material.

Resistivity (ρ) A constant of proportionality between the resistance of a material and its physical dimensions.

Resonance A condition established by the application of a particular frequency (the resonant frequency) to a series or parallel R-L-C network. The transfer of power to the system is a maximum, and, for frequencies above and below, the power transfer drops off to significantly lower levels.

Rheostat An element whose terminal resistance can be varied in a linear or nonlinear manner.

rms value The root-mean-square or effective value of a waveform.

Scientific notation A method for describing very large and very small numbers through the use of powers of 10, which requires that the multiplier be a number between 1 and 10.

Second (s) A unit of measurement for time in the SI system.

Secondary cell Sources of voltage that can be recharged.

Selectivity A characteristic of resonant networks directly related to the bandwidth of the resonant system. High selectivity is associated with small bandwidth (high Q's), and low selectivity with larger bandwidths (low Q's).

Self-inductance A measure of the ability of a coil to oppose any change in current through the coil and to store energy in the form of a magnetic field in the region surrounding the coil.

Semiconductor A material having a conductance value between that of an insulator and that of a conductor. Of significant importance in the manufacture of semiconductor electronic devices.

Semilog paper Graph paper with one log scale and one linear scale.

Series ac configuration A connection of elements in an ac network in which no two impedances have more than one terminal in common and the current is the same through each element.

Series circuit A circuit configuration in which the elements have only one point in common and each terminal is not connected to a third, current-carrying element.

Series ohmmeter A resistance-measuring instrument in which the movement is placed in series with the unknown resistance.

Series-parallel ac network A combination of series and parallel branches in the same network configuration. Each branch may contain any number of elements whose impedance is dependent on the applied frequency.

Sheet resistance Defined by ρ/d for thin-film and integrated circuit design.

Short circuit A direct connection of low resistive value that can significantly alter the behavior of an element or system.

SI system The system of units adopted by the IEEE in 1965 and the USASI in 1967 as the International System of Units (Système International d'Unités).

Sinusoidal ac waveform An alternating waveform of unique characteristics that oscillates with equal amplitude above and below a given axis.

Skin effect At high frequencies, a counter-induced voltage builds up at the center of a conductor, resulting in an increased flow near the surface (skin) of the conductor and a sharp reduction near the center. As a result, the effective area of conduction decreases and the resistance increases as defined by the basic equation for the geometric resistance of a conductor.

Solar cell Sources of voltage available through the conversion of light energy (photons) into electrical energy.

Source conversion The changing of a voltage source to a current source, or vice versa, which will result in the same terminal behavior of the source. In other words, the external network is unaware of the change in sources.

Specific gravity The ratio of the weight of a given volume of a substance to the weight of an equal volume of water at 4°C.

Static electricity Stationary charge in a state of equilibrium.

Stop-band filter A network designed to reject (block) signals within a particular frequency range.

Stray capacitance Capacitances that exist not through design but simply because two conducting surfaces are relatively close to each other.

Substitution theorem A theorem stating that if the voltage across and current through any branch of an ac bilateral network are known, the branch can be replaced by any combination of elements that will maintain the same voltage across and current through the chosen branch.

Superconductor Conductors of electric charge that have for all practical purposes zero ohms.

Superposition theorem A network theorem that permits considering the effects of each source independently. The resulting current and/or voltage is the algebraic sum of the currents and/or voltages developed by each source independently.

Surge voltage The maximum voltage that can be applied across the capacitor for very short periods of time.

Susceptance A measure of how "susceptible" an element is to the passage of current through it. It is measured in siemens, abbreviated S, and is represented by the capital letter B.

Thermistor A two-terminal semiconductor device whose resistance is temperature sensitive.

Thévenin's theorem A theorem that permits the reduction of any two-terminal linear ac network to one having a single voltage source and series impedance. The resulting configuration can then be employed to determine a particular current or voltage in the original network or to examine the effects of a specific portion of the network on a particular variable.

Transistor A three-terminal semiconductor electronic device that can be used for amplification and switching purposes.

Varistor A voltage-dependent, nonlinear resistor used to suppress high-voltage transients.

Volt (V) The unit of measurement applied to the difference in potential between two points. If one joule of energy is required to move one coulomb of charge between two points, the difference in potential is said to be one volt.

Voltage divider rule A method by which a voltage in a series circuit can be determined without first calculating the current in the circuit.

Voltage regulation (*VR*) A value, given as a percent, that provides an indication of the change in terminal voltage of a supply with a change in load demand.

Voltaic cell A storage device that converts chemical to electrical energy.

Voltmeter An instrument designed to read the voltage across an element or between any two points in a network.

VOM A multimeter with the capability to measure resistance and both ac and dc levels of current and voltage.

Wattmeter An instrument capable of measuring the power delivered to an element by sensing both the voltage across the element and the current through the element.

Waveform The path traced by a quantity, plotted as a function of some variable such as position, time, degrees, temperature, and so on.

Working voltage The voltage that can be applied across a capacitor for long periods of time without concern for dielectric breakdown.

Wye (Y), tee (T) configuration A network structure that consists of three branches and has the appearance of the capital letter Y or T.

INDEX

A

B

C

D

J

K

L

M

N

O

S

T

LICENSE AGREEMENT AND LIMITED WARRANTY

READ THE FOLLOWING TERMS AND CONDITIONS CAREFULLY BEFORE OPENING THIS DISK PACKAGE. THIS IS AN AGREEMENT BETWEEN YOU AND PRENTICE HALL CANADA INC. (THE "COMPANY"), BY OPENING THIS SEALED PACKAGE, YOU ARE AGREEING TO BE BOUND BY THESE TERMS AND CONDITIONS. IF YOU DO NOT AGREE, WITH THESE TERMS AND CONDITIONS, DO NOT OPEN THE DISK PACKAGE. PROMPTLY RETURN THE DISK PACKAGE AND ALL ACCOMPANYING ITEMS TO THE COMPANY.

1. GRANT OF LICENSE: In consideration of your adoption and/or other materials published by the Company, and your agreement to abide by the terms and conditions of the Agreement, the Company grants to you a nonexclusive right to use and display the copy of the enclosed software program (hereinafter "the SOFTWARE") so long as you comply with the terms of this Agreement. The Company reserves the rights not expressly granted to you under this Agreement. This license is not a sale of the original SOFTWARE or any copy to you.

2. USE RESTRICTIONS: You may not sell or license copies of the SOFTWARE or the Documentation to others. You may not transfer or distribute copies of the SOFTWARE or the Documentation, except to instructors and students in your school who are users of the adopted Company textbook that accompanies this SOFTWARE. You may not reverse engineer, disassemble, decompile, modify, adapt, translate or create derivative works based on the SOFTWARE or the Documentation without the prior written consent of the Company.

3. LIMITED WARRANTY AND DISCLAIMER OF WARRANTY: Because this SOFTWARE is being given to you without charge, the Company makes no warranties about the SOFTWARE, which is provided "AS-IS". The COMPANY DISCLAIMS ALL WARRANTIES, EXPRESS OR IMPLIED, INCLUDING WITHOUT LIMITATION, THE IMPLIED WARRANTIES OF MERCHANTABILITY AND FITNESS FOR A PARTICULAR PURPOSE. THE COMPANY DOES NOT WARRANT, GUARANTEE OR MAKE ANY REPRESENTATION REGARDING THE USE OR THE RESULTS OF THE USE OF THE SOFTWARE. IN NO EVENT, SHALL THE COMPANY OR ITS EMPLOYEES, AGENTS, SUPPLIERS OR CONTRACTORS BE LIABLE FOR ANY INCIDENTAL, INDIRECT, SPECIAL OR CONSEQUENTIAL DAMAGES ARISING OUT OF OR IN CONNECTION WITH THE LICENSE GRANTED UNDER THIS AGREEMENT INCLUDING, WITHOUT LIMITATION, LOSS OF USE, LOSS OF DATA, LOSS OF INCOME OR PROFIT, OR OTHER LOSSES SUSTAINED AS A RESULT OF INJURY TO ANY PERSON, OR LOSS OF OR DAMAGE TO PROPERTY, OR CLAIMS OF THIRD PARTIES, EVEN IF THE COMPANY OR AN AUTHORIZED REPRESENTATIVE OF THE COMPANY HAS BEEN ADVISED OF THE POSSIBILITY OF SUCH DAMAGES.

SOME JURISDICTIONS DO NOT ALLOW THE LIMITATION OF IMPLIED WARRANTIES OR LIABILITY FOR INCIDENTAL, INDIRECT, SPECIAL OR CONSEQUENTIAL DAMAGES. SO THE ABOVE LIMITATIONS MAY NOT ALWAYS APPLY. THE WARRANTIES IN THIS AGREEMENT GIVE YOU SPECIFIC LEGAL RIGHTS AND YOU MAY ALSO HAVE OTHER RIGHTS WHICH VARY IN ACCORDANCE WITH LOCAL LAW.